Handbook of Tissue Optical Clearing

Handbook of Tissue Optical Clearing

New Prospects in Optical Imaging

Edited by
Valery V. Tuchin, Dan Zhu, and Elina A. Genina

CRC Press
Taylor & Francis Group
Boca Raton London New York

CRC Press is an imprint of the
Taylor & Francis Group, an **informa** business

First edition published 2022
by CRC Press
6000 Broken Sound Parkway NW, Suite 300, Boca Raton, FL 33487-2742

and by CRC Press
2 Park Square, Milton Park, Abingdon, Oxon, OX14 4RN

Library of Congress Cataloging-in-Publication Data

Names: Tuchin, V. V. (Valeriĭ Viktorovich), editor. | Dan, Zhu, editor. |
Genina, Elina A., editor.
Title: Handbook of tissue optical clearing : new prospects in optical
imaging / edited by Valery Tuchin, Zhu Dan and Elina A. Genina.
Description: First edition. | Boca Raton : CRC Press, 2022. | Includes
bibliographical references and index.
Identifiers: LCCN 2021023380 | ISBN 9780367895099 (hardback) | ISBN
9781032118697 (paperback) | ISBN 9781003025252 (ebook)
Subjects: LCSH: Tissues--Imaging. | Tissues--Optical properties. | Imaging
systems in medicine.
Classification: LCC QP88 .H32 2022 | DDC 612.8/4--dc23
LC record available at https://lccn.loc.gov/2021023380

ISBN: 978-0-367-89509-9 (hbk)
ISBN: 978-1-032-11869-7 (pbk)
ISBN: 978-1-003-02525-2 (ebk)

DOI: 10.1201/9781003025252

Typeset in Times
by Deanta Global Publishing Services, Chennai, India

Contents

Part I Basic principles of tissue optical clearing

Part II Tissue optical clearing method for biology (3D imaging)

Part III   Towards *in vivo* tissue optical clearing

Part IV   Combination of tissue optical clearing and optical imaging/spectroscopy for diagnostics

Preface

As one of the fastest-growing fields in the life sciences, biomedical photonics connects research in physics, optics, and electrical engineering coupled with medical or biological applications, allowing structural and functional analysis of tissues and cells with resolution and contrast unattainable by any other method. The major challenges of many biophotonics techniques are associated with the need to enhance imaging resolution even further to the subcellular level as well as translate them for *in vivo* studies. In the last few decades, huge advances have been made in optical methods and optical molecular probes, paving the way towards real molecular imaging within cells, as well as deep-tissue imaging. However, the inherent opacity of most biological tissues, which contain various cellular and extracellular structures with different refractive indices as well as kinds of pigments, limits the penetration of light and causes blurring of the images. Both the imaging resolution and contrast decrease as light propagates deeper into the tissue and travels through different components. To perform deeper tissue imaging, great efforts have been made in terms of innovation in advanced optical imaging techniques and development in optical molecular. Alternatively, reducing the scattering and absorption of tissues probes can also significantly enhance the optical imaging of deep tissue. For example, Richard White and co-authors used careful breeding techniques to create a transparent adult zebrafish in 2008, which has been applied to the study of cancer pathology and development in real time, but this technique is unsuitable for studies in humans or other animals [1].

At the beginning of the 20th century, Werner Spalteholz used organic solvents to make large organs, such as the heart, transparent [2], pushing forward unprecedented advances in the field of anatomy. This was the first tissue clearing method of traceability. However, the severe damage it caused to outer tissue made it useful only for clearing very large samples. Since then, some improved or modified Spalteholz techniques have partially succeeded with the preservation of the histology [3].

Although tissue clearing technology dates back to a century ago, the formal concept of "tissue optical clearing" was proposed by Valery V. Tuchin and coauthors in 1996 [4]. It provides an innovative way to perform deep-tissue imaging by reducing the light attenuation and enhancing light penetration into the tissues. After almost 10 years of development, the book *Optical Clearing of Tissue and Blood* (SPIE Press) was first published in 2005 [5]; it described the optical clearing method based on reversible reduction of tissue scattering due to refractive index matching of scatterers and ground matter and overviewed the basic principles, recent results, advantages, limitations, and future of the optical immersion method applied to clearing of the naturally turbid biological tissues and blood.

In the last decades, tissue optical clearing techniques developed rapidly. Three special issues, two for the *Journal of Biomedical Optics* in 2008 [6] and 2016 [7], and one for the *Journal of Innovative Optical Health and Science* in 2010 [8] were organized. In two monographs, the problem of optical clearing is discussed in general with many practical examples [9] (see Chapter 9) as well as in terms of its specific aspects, with a focus on measuring and controlling the refractive index of individual tissue components [10]. In addition, more and more papers were published in journals, even top-level journals. Therefore, this book, *Handbook of Tissue Optical Clearing: New Prospects in Optical Imaging*, which differs from the published books, is conceived, discussed, and written with chapter contributions from the subject area specialists. With the development of advanced optical imaging technologies, numerous tissue optical clearing methods have sprung up in recent years, especially for *in vitro* tissue blocks, organs, and bodies, and even human tissue; and also *in vivo* skin, skull, and sclera. The combination of tissue optical clearing methods and advanced optical imaging techniques has been widely applied in different fields, such as neuroscience, anatomy, development, regeneration, and other physiological and pathological processes, boosting our understanding of many biological events and offering novel insights into fundamental questions in many discoveries, and it is also expected to play an essential role in clinical diagnosis applications.

This book is a self-contained introduction to tissue optical clearing, aiming to introduce the up-to-date developments in tissue optical clearing in both methodology and application to help readers understand the recent achievements of this technique systematically and comprehensively. This book covers four parts consisting of 37 chapters, including basic principles of tissue optical clearing (Part I), *in vitro* tissue optical clearing method for 3D imaging (Part II), *in vivo* tissue optical clearing methods (Part III), and the combination of tissue optical clearing and optical imaging/spectroscopy for diagnostics (Part IV).

This book starts from the basic principles of tissue optical clearing, aiming to elucidate optical clearing mechanisms from physical, chemical, molecular, and even mechanical aspects. Part I includes an overview of the fundamental physical principles of current tissue optical clearing via various chemical strategies, such as dissociation of collagen, dehydration, delipidation, decalcification, and hyperhydration, to reduce scattering as well as decolorization to reduce absorption. Some recent progress on Mueller matrix imaging applied in recording the optical clearing process is also reviewed. Then, the traditional and innovative optical clearing agents and the chemical enhancers for improving optical clearing efficacy are introduced, followed by a discussion of the impact of optical clearing agents on human skin autofluorescence. Then the molecular mechanism of the post-diffusion stage of tissue optical clearing is explained by presenting the modeling results of the interaction of optical clearing agents with collagen by classical molecular dynamics, molecular docking, and quantum chemistry. This part also presents methods for measuring the refractive index (RI) of biological tissues, blood, and their components in a wide range of wavelengths, including THz. The recent progress that has been made in the investigation of optical clearing influence on the water content in the skin using confocal Raman spectroscopy, refractometry, time-domain THz spectrometry, and other methods is also reviewed. Besides the soft tissues, the optical clearing of cartilage and its mechanical property changes are also introduced. In addition, compression optical clearing is also discussed by investigating the influence of external mechanical compression on the absorbing and scattering properties of skin tissue.

High-resolution mapping of three-dimensional (3D) structures in biological tissues is essential in various areas of biological research. With the revival of light-sheet microscopy, optical clearing for *in vitro* large-volume tissues has become a hot topic in recent years. Therefore, the second part of this book focuses on the *in vitro* tissue optical clearing method for 3D imaging and reconstruction. It first reviews the current *in vitro* tissue optical clearing methods from the end-users' perspective based on sample features in terms of size and age and presents comprehensive evaluation results of the performances for some typical clearing methods in various intact mouse organs. Classically, the taxonomy of current tissue optical clearing methods is oriented by the primary active chemical reagents used, in which two families of techniques can be identified: aqueous (or hydrophilic) solution-based clearing methods and solvent (or hydrophobic)-based clearing methods. The former is known for its good fluorescence preservation but long incubation time. Hence, in this part, some up-to-date ultrafast aqueous clearing protocols aiming at multiscale tissues are introduced, including hundreds-micron-thick tissue slices, several millimeter–thick tissue blocks, adult whole organs, and even whole bodies. Then, the detailed chemical structures and properties for hydrophilic tissue clearing compound, and the potential of modern hydrophilic tissue clearing methods in biomedical research and clinical practice are discussed. The latter techniques suffer from the loss of endogenous fluorescence; hence, some related work for better endogenous fluorescence by different strategies, involving the introduction of a resin formulation during archiving of cleared samples, screening the fluorescence-friendly compounds, and adjusting the conditions in the clearing and storage process, are also discussed. Then, two mainstream 3D fluorescence imaging techniques: the more traditional two-photon excitation microscopy (TPEM) imaging and the newly emerging light-sheet fluorescence microscopy (LSFM) are introduced, and their combination with tissue optical clearing for high-throughput imaging of entire organs and organisms is also demonstrated. The final chapter of Part II highlights the major challenges and proposed solutions for the application of tissue optical clearing from the biologists' perspective and covers the progress and application of this methodology in basic immunology and immuno-oncology.

So far, besides the rapid emergence of the *in vitro* tissue optical clearing methods that can render tissues, organs, and even bodies transparent, big advances towards *in vivo* tissue optical clearing techniques have also been achieved, facilitating the structural and functional monitoring of cells and vasculatures dynamically. The third part mainly introduces *in vivo* optical clearing methods for skin, skull, blood, and lymph flow. For *in vivo* skin optical clearing, three kinds of methods to enhance *in vivo* skin optical clearing efficacy in rodents are first introduced: physical enhancement methods, chemical enhancement methods, and the combination of both; this is followed by a chapter focusing on the features and possibilities of *in vivo* human skin optical clearing using biocompatible chemical and physical enhancers of epidermal permeability, and applications of the approaches to some optical diagnostic methods. For *in vivo* skull optical clearing, the first invention of this technique and several optimized optical clearing skull windows without craniotomy or with minimal craniotomy are presented, providing access to cortical neurovascular visualization within the brain. Then, the optical clearing of blood is described, followed by a review of current advances in the use of noninvasive optical imaging techniques for quantifying blood and lymph flow in combination with the optical clearing method.

As mentioned above, tissue optical clearing techniques can greatly reduce the attenuation of light in tissue and improve resolution and signal intensity, thereby increasing optical imaging/spectroscopy quality, including photoacoustic imaging, magnetic resonance imaging (MRI), optical coherence tomography (OCT), diffuse reflectance spectroscopy, and Raman spectroscopy, as well as the fluorescence imaging mentioned in Part II. Tissue optical clearing has been combined with these optical imaging/spectroscopy techniques to facilitate the observation and manipulation of physiological and pathological tissues.

The final Part IV of this book gives multiple examples to demonstrate the applications of the combination of tissue optical clearing and these techniques. It starts with a chapter summarizing the status of applying optical clearing in photoacoustic imaging *in vivo* and illustrating examples of enhancement in both optical resolution and acoustic resolution, followed by another chapter overviewing enhancement of contrast in photoacoustic-fluorescence tomography and cytometry based on the usage of contrast and optical clearing agents. Tissue optical clearing also shows its potential in the terahertz medical diagnosis; with a detailed discussion about this topic, future research in the demanding area of THz biophotonics with the aid of tissue optical clearing is anticipated. A following chapter focuses on the MRI study of diamagnetic and paramagnetic agents for optical clearing of tumor-specific fluorescent signal *in vivo*. Then, various optical clearing agents for tooth surfaces are introduced to enhance the imaging of caries lesions and internal structures in teeth at NIR and short wavelength IR (SWIR) wavelengths for transillumination and reflectance imaging with an emphasis on OCT. A review is provided to summarize diabetes-induced alterations in the optical and structural properties of tissues and the use of various optical methods for diagnosis and monitoring of the pathological effects of diabetes on tissues and the cardiovascular system. Then the use of skin and skull windows based on tissue optical clearing techniques for *in vivo* detection and imaging diabetes-induced changes in cells, vascular structure, and function are highlighted, and work on how the novel *in vivo* skull optical clearing technique facilitates cortical light operation is further introduced. Tissue optical clearing has been applied to enhance the contrast and imaging depth of OCT and the treatment depth in photodynamic therapy (PDT); the combination of tissue optical clearing and these techniques has facilitated the characterization, diagnosis, monitoring, and therapy of tumors, including melanoma. Recently, multimodal optical clearing methods, including optical clearing agent, compression, and computational optical clearing, have been applied to measure dermal beta-carotene using noninvasive diffuse reflectance spectroscopy. Besides the skin, skull, teeth, and tumor tissues, the optical clearing of other tissues, such as abdominal adipose tissue and human hard and soft oral tissues (gums and dentin), are also presented. Some *in vivo* applications based on the combination of tissue optical clearing and Raman spectroscopy are introduced in the last chapter.

This book is written by experts and scholars worldwide engaged in biomedical photonics and tissue optical clearing. We sincerely appreciate their contributions to this book. We believe that tissue optical clearing will make significant contributions in different fields, especially in light-based diagnostic and therapeutic techniques, as well as the life and health sciences.

As tissue optical clearing is a burgeoning technology, new theoretical methods, techniques, and application fields continue to emerge, and, coupled with the limited knowledge and time of the authors, there will inevitably be some ambiguities or deficiencies; readers are welcome to offer criticism and correction.

Dan Zhu
Huazhong University of Science and Technology, China

Elina A. Genina
Saratov State University, Russia
National Research Tomsk State University, Russia

Valery V. Tuchin
Saratov State University, Russia
National Research Tomsk State University, Russia
Institute of Precision Mechanics and Control of the
Russian Academy of Sciences
A.N. Bach Institute of Biochemistry of the Federal Research
Center of Biotechnology of the Russian Academy of Sciences

1. R.M. White, A. Sessa, C. Burke, et al., "Transparent adult zebrafish as a tool for in vivo transplantation analysis," *Cell Stem Cell* **2**(2), 183–189 (2008).

2. W. Spalteholz, *Uber das durchsichtigmachen von menschlichen und tierischen praparaten [About the transparency of human and animal preparations]*, Hierzal, Leipzig, 1914.

3. H. Steinke, W. Wolff, "A modified Spalteholz technique with preservation of the histology," *Ann. Anat.* **183**(1), 91–95 (2001).

4. V.V. Tuchin, I.L. Maksimova, D.A. Zimnyakov, I.L. Kon, A.Kh. Mavlutov, A.A. Mishin, "Light propagation in tissues with controlled optical properties," *Proc. SPIE* **2925**, 118–142, doi: 10.1117/12.260832 (1996).

5. V.V. Tuchin, *Optical Clearing of Tissues and Blood*, **PM 154**, SPIE Press, Bellingham, WA, 2005, 254 p. https://spie.org/Publications/Book/637760?SSO=1.

6. V.V. Tuchin, R.K. Wang, A.T. Yeh (Eds.), "Special section on optical clearing of tissues and cells," *J. Biomed. Opt.* **13**(March/April), 021101 (2008).

7. D. Zhu, B. Choi, E. Genina, V.V. Tuchin (Eds.), "Special section on tissue and blood optical clearing for biomedical applications," *J. Biomed. Opt.* **21**(8), 081201, doi: 10.1117/1.JBO.21.8.081201 (2016).

8. V.V. Tuchin, M. Leahy, D. Zhu (Eds.), Special Issue, "Optical clearing for biomedical imaging in the study of tissues and biological fluids," *J. Innov. Opt. Health Sci.* **3**(3), 147–219 (2010).

9. V.V. Tuchin, *Tissue Optics: Light Scattering Methods and Instruments for Medical Diagnostics*, 3rd ed., **PM 254**, SPIE Press, Bellingham, WA, 2015, 988 p. https://spie.org/Publications/Book/2175698.

10. L. Oliveira, V.V. Tuchin, "The optical clearing method: A new tool for clinical practice and biomedical engineering," Springer Nature Switzerland AG, Basel, 2019, 177 p. https://www.springer.com/gp/book/9783030330545.

Acknowledgments

First of all, we express our gratitude to our families for their indispensable support, understanding, and patience as we wrote and edited this book.

We are very grateful to all the authors of this book, who took the time and made significant efforts to write chapters containing the latest outstanding achievements in the field of optical clearing of biological tissues.

The editors would like to show their appreciation to the colleagues of the Wuhan National Laboratory of Optoelectronics (WNLO) of Huazhong University of Science and Technology (China) and the Department of Optics and Biophotonics of Saratov State University (Russia) for their collaboration and help.

Dan Zhu is grateful for support from grants from the NNSFC (Grant Nos. 61860206009, 81870934, 81701354, 31571002, 91232710, 81171376, 30770552), the NSFC-RFBR for International Cooperation (Grant Nos. 30911120074, 812111313) the National Key Research and Development Program of China (Grant No. 2017YFA0700501), the Foundation for Innovative Research Groups of the National Natural Science Foundation of China (Grant No. 61721092), and the Innovation Fund of WNLO.

Elina A. Genina is thankful for support from the grants RFBR 14-02-00526a, 17-02-00358, 18-52-16025 NTSNIL_a, 20-52-56005, and 20-32-90043; RF Presidential grant 14.Z57.16.7898-NSh; RF Governmental grants 14.Z50.31.0004, 14.Z50.31.0044, and 075-15-2021-615; RSF grants 14-15-00186 and 17-73-20172; and RF Ministry of Science and Higher Education 17.1223.2017/AP.

Valery V. Tuchin is thankful for support from the grants RFBR 14-02-00526a, 17-02-00358, KOMFI 17-00-00275 (17-00-00272), 18-52-16025 NTSNIL_a, 18-29-02060 MK; RF Presidential grant 14.Z57.16.7898-NSh; RF Governmental grants 14.Z50.31.0004, 14.Z50.31.0044, 14.W03.31.0023, 075-15-2019-1885, and 075-15-2021-615; RSF grants 14-15-00128 and 14-15-00186; RF Ministry of Science and Higher Education 17.1223.2017/AP; and 13.2251.21.0009 ARC Projects DP210103342, and grant of the Government of the Russian Federation for the Saratov State University "Priority-2030".

Editors

Dan Zhu is a distinguished professor of Huazhong University of Science and Technology (China), vice-director of Wuhan National Laboratory for Optoelectronics, and vice-director of Key Laboratory of Biomedical Photonics (HUST), Ministry of Education.

Her research interests include tissue optical imaging theory and methods, with a particular focus on tissue optical clearing imaging and applications. She has authored more than 150 peer-reviewed articles and reviews.

She was elected Fellow of SPIE, and secretary general and vice president of Biomedical Photonics Committee of the Chinese Optical Society. She has served as chair or co-chair at various international and domestic conferences, and as guest editor or editorial member of various journals, including *Biomedical Optics Express*, *Journal of Biomedical Optics*, *Scientific Reports*, *Journal of Innovative Optical Health Sciences*, and *Frontier of Optoelectronics*.

Valery V. Tuchin is a corresponding member of the Russian Academy of Sciences, professor and head of the Department of Optics and Biophotonics, and director of the Scientific Medical Center at Saratov State University. He is also the head of the laboratory for laser diagnostics of technical and living systems at the Institute of Precise Mechanics and Control of the RAS, and a supervisor of the Interdisciplinary Laboratory of Biophotonics at the National Research Tomsk State University and the Femtomedicine Laboratory of the ITMO University.

His research interests include biophotonics, biomedical optics, tissue optics, laser medicine, tissue optical clearing, and nanobiophotonics.

He is a member of SPIE, OSA, and IEEE, a visiting professor at Huazhong University of Science and Technology (China) and Tianjin Universities (China), and an adjunct professor at the University of Limerick (Ireland) and the National University of Ireland – Galway (Ireland).

Professor Tuchin was elected Fellow of SPIE and OSA and he was awarded many titles and awards, including Honored Scientist of the Russian Federation, Honored Professor of SSU, Honored Professor of Finland (FiDiPro), SPIE in the field of optical education, Chime Bell of Hubei province (China), Joseph Goodman (OSA / SPIE) for Outstanding Monograph (2015), Michael Feld (OSA) for Pioneering Research in Biophotonics (2019), the Medal of the D.S. Rozhdestvensky Optical Society (2018), and the Alexander Mikhailovich Prokhorov medal of the Academy of Engineering Sciences, named after A.M. Prokhorov (2021).

He is the author of over 700 articles (Web of Science) and 30 monographs and textbooks, he has over 60 patents, and his work has been cited over 30000 times.

Elina A. Genina is a professor in the Departments of Optics and Biophotonics at Saratov State University (Russia). Her research interests include biomedical optics, laser medicine, nanobiophotonics, and the development of methods for the control of tissue optical properties. She is a coauthor of more than 300 peer-reviewed publications, analytical reviews, book chapters, and patents on biomedical optics, and a guest editor of 18 special issues of journals and proceedings. She is a member of the editorial board of *J. Innovative Optical Health Sciences*, *Diagnostics*, *J. Biomedical Photonics & Engineering*, and *The Open Biomedical Engineering Journal*. She is a scientific secretary of the International Symposium on Optics and Biophotonics (Saratov Fall Meeting) and co-chair of a conference in the framework of the Symposium. She has more than 4000 citations and an h-index of 30.

Contributors

Arkady S. Abdurashitov
Laboratory of Remote Controlled Biomaterials, CNBR
Skolkovo Institute of Science and Technology
Moscow, Russia

Pavel D. Agrba
N. I. Lobachevsky State University of Nizhny Novgorod
Nizhny Novgorod, Russia

Yulia M. Alexandrovskaya
Institute of Photon Technologies, Federal Scientific Research
    Center "Crystallography and Photonics"
Russian Academy of Sciences
Moscow, Russia

Marine Amouroux
Université de Lorraine
CNRS, CRAN UMR 7039
Nancy, France

Mohammad Ali Ansari
Optical Bio Imaging Laboratory of Laser and Plasma
    Research Institute
Shahid Beheshti University
Tehran, Iran

Alexey N. Bashkatov
Science Medical Center and Research-Educational Institute
    of Optics and Biophotonics
Saratov State University
Saratov, Russia

and

Laboratories of Biophotonics and Laser Molecular Imaging
    and Machine Learning
National Research Tomsk State University
Tomsk, Russia

Olga I. Baum
Institute of Photon Technologies, Federal Scientific Research
    Center "Crystallography and Photonics"
Russian Academy of Sciences
Moscow, Russia

Kirill V. Berezin
Research-Educational Institute of Optics and Biophotonics
Saratov State University
Saratov, Russia

and

Department of General Physics
Astrakhan State University
Astrakhan, Russia

Walter Blondel
Université de Lorraine
CNRS, CRAN UMR 7039
Nancy, France

Alexei A. Bogdanov Jr.
Department of Radiology
University of Massachusetts Medical School
Worcester, MA

and

Laboratory of Molecular Imaging
Bach Institute of Biochemistry
Research Center of Biotechnology of the Russian Academy of
    Sciences
Moscow, Russia

Alla B. Bucharskaya
Saratov State Medical University named after
    V. I. Razumovsky
Saratov, Russia

and

"Smart Sleep" Laboratory of Science Medical Center
Saratov State University
Saratov, Russia

and

Laboratory of Laser Molecular Imaging and Machine
    Learning
National Research Tomsk State University
Tomsk, Russia

Maria L. Chernavina
Research-Educational Institute of Optics and Biophotonics
Saratov State University
Saratov, Russia

Victor Colas
Université de Lorraine
CNRS, CRAN UMR 7039
Nancy, France

Julijana Cvjetinovic
Center for Photonics and Quantum Materials
Skolkovo Institute of Science and Technology
Moscow, Russia

Maxim E. Darvin
Center of Experimental and Applied Cutaneous Physiology
Department of Dermatology, Venerology and Allergology
Charité – Universitätsmedizin Berlin, Corporate Member
of Freie Universität Berlin and Humboldt-Universität zu
Berlin
Berlin, Germany

and

Institute of Physics
Saratov State University
Saratov, Russia

Christian Daul
Université de Lorraine
CNRS, CRAN UMR 7039
Nancy, France

Valentin Demidov
Princess Margaret Cancer Center, University Health Network
and Department of Medical Biophysics
University of Toronto
Toronto, Canada

and

Geisel School of Medicine at Dartmouth
Hanover, NH

Irina N. Dolganova
Institute of Solid State Physics of the Russian
Academy of Sciences
Chernogolovka, Russia

and

Institute for Regenerative Medicine
Sechenov First Moscow State Medical University
(Sechenov University)
Moscow, Russia

Yixiang Duan
School of Mechanical Engineering
Research Center of Analytical Instrumentation
Sichuan University
Chengdu, China

Konstantin N. Dvoretskiy
Department of Medical and Biological Physics
Saratov Medical State University
Saratov, Russia

Polina A. Dyachenko (Timoshina)
Science Medical Center and Research-Educational Institute
of Optics and Biophotonics
Saratov State University
Saratov, Russia

and

Laboratories of Biophotonics and Laser Molecular Imaging
and Machine Learning
National Research Tomsk State University
Tomsk, Russia

Chunyu Fang
School of Optical and Electronic Information- Wuhan
National Laboratory for Optoelectronics
Huazhong University of Science and Technology
Wuhan, China

Peng Fei
School of Optical and Electronic Information- Wuhan
National Laboratory for Optoelectronics
Huazhong University of Science and Technology
Wuhan, China

Wei Feng
Britton Chance Center for Biomedical Photonics, Wuhan
National Laboratory for Optoelectronics
Huazhong University of Science and Technology
Wuhan, China

and

MoE Key Laboratory for Biomedical Photonics, School of
Engineering Sciences
Huazhong University of Science and Technology
Wuhan, China

and

Zhanjiang Institute of Clinical Medicine
Central People's Hospital of Zhanjiang
Zhanjiang, China

and

Zhanjiang Central Hospital
Guangdong Medical University
Zhanjiang, China

Daniel Fried
University of California
San Francisco, CA

Vadim D. Genin
Science Medical Center and Research-Educational Institute
of Optics and Biophotonics
Saratov State University
Saratov, Russia

and

Laboratories of Biophotonics and Laser Molecular Imaging
and Machine Learning
National Research Tomsk State University
Tomsk, Russia

Elina A. Genina
Science Medical Center and Research-Educational Institute
of Optics and Biophotonics
Saratov State University
Saratov, Russia

and

Laboratories of Biophotonics and Laser Molecular Imaging
and Machine Learning
National Research Tomsk State University
Tomsk, Russia

Jakub Gołąb
Department of Immunology
Medical University of Warsaw
Warsaw, Poland

Dmitry A. Gorin
Center for Photonics and Quantum Materials
Skolkovo Institute of Science and Technology
Moscow, Russia

Honghui He
Shenzhen Key Laboratory for Minimal Invasive Medical
    Technologies, Institute of Optical Imaging and Sensing,
    Graduate School at Shenzhen
Tsinghua University
Shenzhen, China

Yasunobu Iga
Olympus Corporation
Hachioji, Japan

Leszek Kaczmarek
Laboratory of Neurobiology, BRAINCITY
Nencki Institute of Experimental Biology of Polish Academy
    of Sciences
Warsaw, Poland

Natalia I. Kazachkina
Laboratory of Molecular Imaging
Bach Institute of Biochemistry
Research Center of Biotechnology of the Russian Academy of
    Sciences
Moscow, Russia

Mikhail Yu. Kirillin
Institute of Applied Physics
Russian Academy of Sciences
Nizhny Novgorod, Russia

Vyacheslav I. Kochubey
Research-Educational Institute of Optics and Biophotonics
Saratov State University
Saratov, Russia

Cristina Kurachi
Sao Carlos Institute of Physics
University of Sao Paulo
Sao Carlo, Brazil

Jürgen Lademann
Center of Experimental and Applied Cutaneous Physiology
Department of Dermatology, Venerology and Allergology
Charité – Universitätsmedizin Berlin, Corporate Member of
    Freie Universität Berlin and Humboldt-Universität zu Berlin
Berlin, Germany

Kirill V. Larin
Department of Biomedical Engineering
University of Houston
Houston, TX

and

Molecular Physiology and Biophysics
Baylor College of Medicine
Houston, TX

Ekaterina N. Lazareva
Science Medical Center and Research-Educational Institute
    of Optics and Biophotonics
Saratov State University
Saratov, Russia

and

Laboratories of Biophotonics and Laser Molecular Imaging
    and Machine Learning
National Research Tomsk State University
Tomsk, Russia

Xingde Li
Department of Biomedical Engineering
Johns Hopkins University
Baltimore, Maryland

Dongyu Li
Britton Chance Center for Biomedical Photonics, Wuhan
    National Laboratory for Optoelectronics
Huazhong University of Science and Technology
Wuhan, China

and

MoE Key Laboratory for Biomedical Photonics,
    School of Engineering Sciences
Huazhong University of Science and Technology
Wuhan, China

Yanmei Liang
Institute of Modern Optics
Nankai University
Tianjin, China

Anatoly M. Likhter
Department of General Physics
Astrakhan State University
Astrakhan, Russia

Qingyu Lin
School of Mechanical Engineering
Research Center of Analytical Instrumentation
Sichuan University
Chengdu, China

Hui Ma
Shenzhen Key Laboratory for Minimal Invasive Medical
    Technologies, Institute of Optical Imaging and Sensing,
    Graduate School at Shenzhen
Tsinghua University
Shenzhen, China

and

Department of Physics
Tsinghua University
Beijing, China

Galina N. Maslyakova
Saratov State Medical University named after V. I.
    Razumovsky
Saratov, Russia

and

"Smart Sleep" Laboratory of Science Medical Center
Saratov State University
Saratov, Russia

Paweł Matryba
Department of Immunology
Medical University of Warsaw
Warsaw, Poland

and

The Doctoral School of The Medical University of Warsaw
Warsaw, Poland

and

Laboratory of Neurobiology, BRAINCITY
Nencki Institute of Experimental Biology of
    Polish Academy of Sciences
Warsaw, Poland

Lev A. Matveev
Institute of Applied Physics
Russian Academy of Sciences
Nizhny Novgorod, Russia

Alexander L. Matveyev
Institute of Applied Physics
Russian Academy of Sciences
Nizhny Novgorod, Russia

Irina G. Meerovich
Laboratory of Molecular Imaging
Bach Institute of Biochemistry
Research Center of Biotechnology of the Russian Academy of
    Sciences
Moscow, Russia

Maksim Mokrousov
Center for Photonics and Quantum Materials
Skolkovo Institute of Science and Technology
Moscow, Russia

Guzel R. Musina
Prokhorov General Physics Institute of the
    Russian Academy of Sciences
Moscow, Russia

and

Bauman Moscow State Technical University
Moscow, Russia

Nikita A. Navolokin
Saratov State Medical University named after V. I.
    Razumovsky
Saratov, Russia

and

Pathological Department, State Healthcare Institution
    "Saratov City Clinical Hospital No. 1 named after Yu.Ya.
    Gordeev"
Saratov, Russia

and

"Smart Sleep" Laboratory of Science Medical Center
Saratov State University
Saratov, Russia

Maxim M. Nazarov
National Research Center "Kurchatov Institute"
Moscow, Russia

Alexander Novikov
Center for Photonics and Quantum Materials
Skolkovo Institute of Science and Technology
Moscow, Russia

Marina V. Novoselova
Center for Photonics and Quantum Materials
Skolkovo Institute of Science and Technology
Moscow, Russia

Daniil V. Nozdriukhin
Center for Photonics and Quantum Materials
Skolkovo Institute of Science and Technology
Moscow, Russia

Luís M. Oliveira
Physics Department – Polytechnic Institute of Porto, School
    of Engineering, Porto, Portugal and Centre of Innovation
    in Engineering and Industrial Technology (CIETI), School
    of Engineering
Polytechnic of Porto
Porto, Portugal

Layla Pires
Princess Margaret Cancer Center, University Health Network
    and Department of Medical Biophysics
University of Toronto
Toronto, Canada

Alexander B. Pravdin
Research-Educational Institute of Optics and Biophotonics
Saratov State University
Saratov, Russia

Yisong Qi
Britton Chance Center for Biomedical Photonics, Wuhan
National Laboratory for Optoelectronics
Huazhong University of Science and Technology
Wuhan, China

and

MoE Key Laboratory for Biomedical Photonics, School of
Engineering Sciences
Huazhong University of Science and Technology
Wuhan, China

and

Shenzhen Mindray Bio-Medical Electronic Co., Ltd
Shenzhen, China

Michelle Barreto Requena
Sao Carlos Institute of Physics
University of Sao Paulo
Sao Carlo, Brazil

Ana Gabriela Salvio
Amaral Carvalho Hospital
Jau, Brazil

Alexander P. Savitsky
Laboratory of Molecular Imaging
Bach Institute of Biochemistry
Research Center of Biotechnology of the Russian Academy of
Sciences
Moscow, Russia

Johannes Schleusener
Center of Experimental and Applied Cutaneous Physiology
Department of Dermatology, Venerology and Allergology
Charité – Universitätsmedizin Berlin, Corporate Member of
Freie Universität Berlin and Humboldt-Universität zu Berlin
Berlin, Germany

Anton Yu. Sdobnov
Optoelectronics and Measurement Techniques Unit
University of Oulu
Oulu, Finland

and

Science Medical Center and Research-Educational Institute
of Optics and Biophotonics
Saratov State University
Saratov, Russia

Alexey A. Selifonov
Research-Educational Institute of Optics and Biophotonics
Saratov State University
Saratov, Russia

and

Saratov State Medical University
Saratov, Russia

Oxana V. Semyachkina-Glushkovskaya
Deparment of Biology
Saratov State University
Saratov, Russia

and

"Smart Sleep" Laboratory of Science Medical Center
Saratov, Russia

Rui Shi
Britton Chance Center for Biomedical Photonics, Wuhan
National Laboratory for Optoelectronics
Huazhong University of Science and Technology
Wuhan, China

and

MoE Key Laboratory for Biomedical Photonics, School of
Engineering Sciences
Huazhong University of Science and Technology
Wuhan, China

and

Double Medical Technology Inc.
Xiamen, China

Alexander P. Shkurinov
Department of Physics
Lomonosov Moscow State University
Moscow, Russia

Yury P. Sinichkin
Research-Educational Institute of Optics
and Biophotonics
Saratov State University
Saratov, Russia

Olga A. Smolyanskaya
Institute of Photonics and Optical IT
ITMO University
Saint-Petersburg, Russia

Emil N. Sobol
Arcuo Medical Inc
Incline Village, NV

Ilya D. Solovyev
Laboratory of Molecular Imaging
Bach Institute of Biochemistry
Research Center of Biotechnology of the Russian Academy of
    Sciences
Moscow, Russia

Alexander A. Sovetsky
Institute of Applied Physics
Russian Academy of Sciences
Nizhny Novgorod, Russia

Etsuo A. Susaki
Department of Biochemistry and Systems Biomedicine
Graduate School of Medicine
Juntendo University
Bunkyo-ku, Tokyo, Japan

and

Laboratory for Synthetic Biology
RIKEN Center for Biosystems Dynamics Research
Suita, Japan

Shinichi Takimoto
Olympus Corporation
Hachioji, Japan

Yohei Tanikawa
Olympus Corporation
Hachioji, Japan

Georgy S. Terentyuk
Saratov State Medical University
Saratov, Russia

Valery V. Tuchin
Science Medical Center and Research-Educational Institute
    of Optics and Biophotonics
Saratov State University
Saratov, Russia

and

Laboratories of Biophotonics and Laser Molecular Imaging
    and Machine Learning
National Research Tomsk State University
Tomsk, Russia

and

Laboratory of Laser Diagnostics of Technical and Living
    Systems
Institute of Precision Mechanics and Control of the Russian
    Academy of Sciences
Saratov, Russia

and

Laboratory of Molecular Imaging
Bach Institute of Biochemistry
Research Center of Biotechnology of the Russian Academy of
    Sciences
Moscow, Russia

Daria K. Tuchina
Science Medical Center and Research-Educational Institute
    of Optics and Biophotonics
Saratov State University
Saratov, Russia

and

Laboratories of Biophotonics and Laser Molecular Imaging
    and Machine Learning
National Research Tomsk State University
Tomsk, Russia

Hiroki R. Ueda
Department of Systems Pharmacology, Graduate School of
    Medicine
The University of Tokyo
Bunkyo-ku, Japan

and

Laboratory for Synthetic Biology
RIKEN Center for Biosystems Dynamics Research
Suita, Japan

Nikita V. Chernomyrdin
Prokhorov General Physics Institute of the Russian Academy
    of Sciences
Moscow, Russia

I. Alex Vitkin
Princess Margaret Cancer Center, University Health Network
    and Department of Medical Biophysics
University of Toronto
Toronto, Canada

Lihong V. Wang
California Institute of Technology
Pasadena, CA

Brian C. Wilson
Princess Margaret Cancer Center, University Health Network
    and Department of Medical Biophysics
University of Toronto
Toronto, Canada

Jianyi Xu
Britton Chance Center for Biomedical Photonics, Wuhan
    National Laboratory for Optoelectronics
Huazhong University of Science and Technology
Wuhan, China

and

MoE Key Laboratory for Biomedical Photonics,
    School of Engineering Sciences
Huazhong University of Science and Technology
Wuhan, China

Irina Yu. Yanina
Research-Educational Institute of Optics and Biophotonics
Saratov State University
Saratov, Russia

and

Laboratories of Biophotonics and Laser Molecular Imaging
    and Machine Learning
National Research Tomsk State University
Tomsk, Russia

Tingting Yu
Britton Chance Center for Biomedical Photonics, Wuhan
    National Laboratory for Optoelectronics
Huazhong University of Science and Technology
Wuhan, China

and

MoE Key Laboratory for Biomedical Photonics,
    School of Engineering Sciences
Huazhong University of Science and Technology
Wuhan, China

Vladimir Yu. Zaitsev
Institute of Applied Physics
Russian Academy of Sciences
Nizhny Novgorod, Russia

Sergey M. Zaytsev
Université de Lorraine
CNRS, CRAN UMR 7039
Nancy, France

and

Research-Educational Institute of Optics and Biophotonics
Saratov State University
Saratov, Russia

Kirill I. Zaytsev
Prokhorov General Physics Institute of the Russian Academy
    of Sciences
Moscow, Russia

and

Bauman Moscow State Technical University
Moscow, Russia

Nan Zeng
Shenzhen Key Laboratory for Minimal Invasive Medical
    Technologies, Institute of Optical Imaging and Sensing,
    Graduate School at Shenzhen
Tsinghua University
Shenzhen, China

Chao Zhang
Britton Chance Center for Biomedical Photonics, Wuhan
National Laboratory for Optoelectronics
Huazhong University of Science and Technology
Wuhan, China

and

MoE Key Laboratory for Biomedical Photonics, School of
    Engineering Sciences
Huazhong University of Science and Technology
Wuhan, China

and

Zhanjiang Institute of Clinical Medicine
Central People's Hospital of Zhanjiang
Zhanjiang, China

and

Zhanjiang Central Hospital
Guangdong Medical University
Zhanjiang, China

Yanjie Zhao
Britton Chance Center for Biomedical Photonics, Wuhan
National Laboratory for Optoelectronics
Huazhong University of Science and Technology
Wuhan, China

and

MoE Key Laboratory for Biomedical Photonics, School of
    Engineering Sciences
Huazhong University of Science and Technology
Wuhan, China

and

Department of Radiology, Tongji Hospital, Tongji Medical
    College
Huazhong University of Science and Technology
Wuhan, China

Qingliang Zhao
State Key Laboratory of Molecular Vaccinology and
    Molecular Diagnostics & Center for Molecular Imaging
    and Translational Medicine, School of Public Health
Xiamen University
Xiamen, China

Vladimir P. Zharov
University of Arkansas for Medical Sciences
Little Rock, AR

Victoria V. Zherdeva
Laboratory of Molecular Imaging
Bach Institute of Biochemistry
Research Center of Biotechnology of the Russian Academy of
    Sciences
Moscow, Russia

Olga S. Zhernovaya
Research-Educational Institute of Optics and Biophotonics
Saratov State University
Saratov, Russia

Yong Zhou
Shenzhen Maidu Technology Co., Ltd
Shenzhen, China

Dan Zhu
Britton Chance Center for Biomedical Photonics, Wuhan
    National Laboratory for Optoelectronics
Huazhong University of Science and Technology
Wuhan, China

and

MoE Key Laboratory for Biomedical Photonics,
    School of Engineering Sciences
Huazhong University of Science and Technology
Wuhan, China

Jingtan Zhu
Britton Chance Center for Biomedical Photonics, Wuhan
    National Laboratory for Optoelectronics
Huazhong University of Science and Technology
Wuhan, China

and

MoE Key Laboratory for Biomedical Photonics
Huazhong University of Science and Technology
Wuhan, China

Olga A. Zyuryukina
Research-Educational Institute of Optics and Biophotonics
Saratov State University
Saratov, Russia

Part I

Basic principles of tissue optical clearing

1

Tissue optical clearing mechanisms

Tingting Yu, Dan Zhu, Luís Oliveira, Elina A. Genina, Alexey N. Bashkatov, and Valery V. Tuchin

CONTENTS

Introduction

Biomedical photonics is currently one of the fastest-growing fields in the life sciences, connecting research in physics, optics, and electrical engineering with medical or biological applications. It allows structural and functional analysis of tissues and cells with resolution and contrast unattainable by any other method. Advanced optical methods combined with various contrast agents pave the way towards real molecular imaging within living cells. However, major challenges are associated with the need to enhance imaging resolution even further to the subcellular level for larger tissue blocks or organs, as well as for *in vivo* studies.

Biological tissues are in general characterized by having strong light scattering, which is a significant obstacle to the application of optical methods in clinical practice [1–3]. Tissues have internal heterogeneous composition since protein fibers and globules, cells, lipid droplets, and phospholipid membranes are distributed in a background liquid, called *interstitial fluid* (ISF). Tissue cells also show similar heterogeneous composition, since the nucleus, mitochondria, and organelles are surrounded by the cytoplasm [4]. Both the ISF and cytoplasm are mainly composed of water (>90% in volume), wherein some salts, minerals, and proteins are dissolved [5–7]. Due to the vast water content in tissue and cell fluids, the refractive index (RI) is low when compared to the RI of tissue scatterers (or cell scatterers) [6, 8]. Such difference in the RI of tissue scatterers and fluids is significantly responsible for the strong light scattering tissues present. One way to quantify such scattering in a tissue is to evaluate the relative RI m, which is calculated by the ratio between the RI of tissue scatterers n_{scat}, and the RI of the ISF n_{ISF} [5, 6, 9]:

$$m = \frac{n_{scat}}{n_{ISF}} \qquad (1.1)$$

It is common to calculate m for 589.6 nm [5], but since both n_{scat} and n_{ISF} depend on wavelength (λ), m also depends on λ [6, 10]. Considering $\lambda = 589.6$ nm and tissues at a temperature of 20 °C, n_{ISF} will range between 1.35 and 1.37 [5]. These low values for the RI of the ISF occur because the ISF contains mainly water and a small quantity of dissolved organic compounds and salts [5, 7, 11]. Considering tissue scatterers, the values for n_{scat} will be significantly higher. Some examples are dry or hydrated muscle fibers that have been estimated to have a RI of 1.584 [7] or 1.530 [4, 11], respectively, and collagen fibers in scleral tissues, which at natural hydration have a RI of 1.474 [12]. For these examples, the corresponding m values range between 1.12 and 1.13 for the skeletal muscle, and between 1.08 and 1.09 for scleral collagen.

The examples above show that m has always values above unity; consequently, light scattering will occur in the tissues. The magnitude of m is a measure of that light scattering [6].

The high scattering of turbid tissues limits the penetration of visible and near-infrared light, and both the imaging resolution and contrast decrease as light propagates deeper into the tissue [1–3]. To perform deeper tissue imaging, various optical imaging techniques have been developed [13, 14]. Alternatively, reducing scattering and absorption of tissues could also significantly enhance optical imaging of deep-tissue structures

[5, 15, 16]. For example, White [17] created a transparent adult zebrafish, which has been used to study cancer pathology and development in real time. Still, this method is unsuitable for studies in other animals or humans.

Fortunately, the tissue optical clearing (TOC) technique, proposed by Tuchin, provides an innovative way to perform deep-tissue imaging by reducing the scattering of tissue and making tissue more transparent by using immersion of tissues into optical clearing agents (OCAs) [5, 15, 18]. The way to increase tissue transparency is to increase n_{ISF}, ideally, to match it to n_{scat}. In that case, m would reduce to 1, and the tissue would be completely transparent, or in other words, light scattering would be eliminated. An effective way to reduce m and light scattering in tissues is by replacing the interstitial (or cytoplasmic) water with an innocuous agent that presents a higher RI than water, better matched to the value of n_{scat}. Optical clearing studies with such kind of agents have been performed, where the m kinetics was calculated from thickness (d) and collimated transmittance (T_c) measurements made during optical clearing treatments [9, 10]. Figure 1.1 shows such kinetics for the human colorectal muscle under treatment with a 40% glycerol solution [10].

The data in Figure 1.1 was calculated assuming that during 30 min, glycerol only diffuses into the interstitial locations, meaning that the value of n_{scat} remains unchanged during treatment [9, 10]. To calculate the kinetics for n_{ISF} to use them in Equation (1.1), it was first necessary to retrieve the kinetics of the scattering coefficient ($\mu_s(\lambda)$) from T_c and thickness (d) measurements through the Bouguer–Beer–Lambert equation [1, 2]:

$$T_c = e^{-\mu_t d}, \quad (1.2)$$

where μ_t represents the attenuation coefficient, which is the sum of the absorption coefficient (μ_a) and the scattering coefficient μ_s [9, 10]. By using spectral T_c measurements and knowing the wavelength dependence for μ_a, the wavelength dependence of μ_s can be calculated through Equation (1.2).

Once $\mu_s(\lambda)$ was obtained, the kinetics for $n_{ISF}(\lambda)$ could be calculated with Equation (1.3) as reported in literature [5, 9, 10]:

$$n_{ISF}(\lambda, t) = \frac{n_{scat}(\lambda, t = 0)}{\left(\sqrt{\frac{\mu_s(\lambda, t) \times d(t)}{\mu_s(\lambda, t = 0) \times d(t = 0)}} \times \left(\frac{n_{scat}(\lambda, t = 0)}{n_{ISF}(\lambda, t = 0)} - 1 \right) + 1 \right)}$$

$$(1.3)$$

In Equation (1.3), $n_{scat}(\lambda, t=0)$ is the scatterers' dispersion, which remains unchanged during the treatment, and $n_{ISF}(\lambda, t=0)$ is the dispersion of the ISF in the untreated tissue; $\mu_s(\lambda, t=0)$ and $d(t=0)$ are the wavelength dependence for the scattering coefficient and sample thickness for the natural tissue. Correspondingly, $d(t)$ is the sample thickness at any time of treatment, which should be measured, and $\mu_s(\lambda, t)$ is the scattering coefficient for any wavelength and any time of treatment. Such values for $\mu_s(\lambda, t)$ were calculated from T_c and d kinetics measurements using Equation (1.2) [4].

According to this principle, the optical clearing agents (OCAs) with higher refractive indices should induce better optical clearing efficacy on biological tissues. After examining the clearing capabilities in the two tissue types – striated muscle and tendon – LaComb et al. [19] supported this mechanism, but the skin optical clearing efficacy did not correlate with the OCAs' refractive indices [20, 21]. Meanwhile, some results demonstrated that the optical clearing effects on the skin induced by alcohols were positively correlated to the number of hydroxyl groups [21], which could be explained by the dissociation of collagen caused by glycerol [22]. There is an additional clearing mechanism called protein dissociation, and in the interests of applying the optical clearing method in clinical practice, the reversibility of all these mechanisms has been studied and demonstrated [6].

Tuchin and collaborators [18, 23, 24] have proved that dehydration induced by OCAs is an important mechanism of tissue optical clearing. Tissues have various components of which water is the main one, comprising about 70%–80% of the volume. T_c and d measurements can be used in different calculation procedures to evaluate the tissue dehydration mechanism.

In addition, collagen and lipids with high RIs are the source of scattering in tissue. Therefore, the variety of optical clearing mechanisms under discussion should be due to complex interactions between agents and tissues, such as the penetrative ability of agents, dissociating collagen, removing lipids or calcium, dehydration, and so on, caused by interactions between OCAs and some components of tissues.

This chapter will introduce the mechanisms of tissue optical clearing from fundamental physical to chemical principles. The physical principle is RI matching, which results from the interactions between chemicals and biomolecules, including dissociation of collagen, dehydration, delipidation, decalcification, hyperhydration, etc. In addition, absorption is also an important factor in limiting optical imaging performance in deep tissue [25, 26], so decolorization will be briefly described.

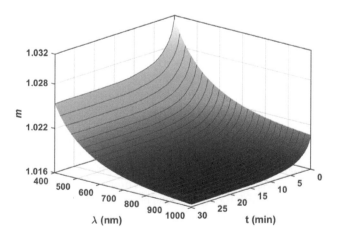

FIGURE 1.1 Relative index of refraction kinetics for muscle of the human colorectal wall under treatment with 40% glycerol. Adapted from Reference [10].

Refractive index matching: Phantom study

Tissues have heterogeneous structures, making the interactions between tissues and different OCAs very complex and preventing understanding of the optical clearing mechanisms. The tissue-simulating phantom with high homogeneity and stability should be an available and adequate model to block the biological effect of tissues and give a direct explanation for the physical mechanism of optical clearing. Among the different types of tissue-simulating phantom, Intralipid, one of the aqueous suspension phantoms, is the most suitable choice in the research. On the one hand, the scattering caused by the lipid bilayer vesicles of Intralipid is just close to that caused by the lipid bilayer membrane structures of cells, and the scattering property of Intralipid can be predicted based on the distribution of particle sizes. On the other hand, when this aqueous suspension phantom is applied to study the optical clearing effect, the diffuse process of agents can be neglected [27].

In order to explain the physical mechanism of optical clearing, Wen et al. [28] applied three methods – direct observation, Mie theory prediction, and spectral measurement – to investigate the change in the optical properties of Intralipid caused by different OCAs. Several commonly used OCAs, i.e., dimethyl sulfoxide (DMSO), glycerol, 1,4-butanediol, 1,2-propanediol, polyethylene glycol 200 (PEG-200), and polyethylene glycol 400 (PEG-400) at different concentrations were mixed with Intralipid, respectively. The results indicated that 5% Intralipid mixed with four kinds of OCAs, respectively, could keep the solution uniform. The optical clearing effect increased gradually in the order of 1,2-propanediol, 1,4-butanediol, glycerol, and DMSO. The RI of the above agents increased gradually. Their optical clearing effect increased in the same way, which suggested the RI matching dominated the optical clearing process. Meanwhile, they

also observed the mixtures of 5% Intralipid with PEG-200 or PEG-400 were no longer uniform but exhibited some aggregation of particles. This result suggests that the mechanism of RI matching is unsuitable for those interactions between the tissue phantom and OCA.

To make the tissue phantom accessible for theory prediction, the scattering particle size distribution of lipid droplets in Intralipid was measured by electron microscopy after the fixing and staining process, and they were counted by the imaging. Then Mie theory calculation [29] was used to quantitatively predict the scattering property of the Intralipid solution before and after mixing with OCAs according to the RI of the background medium. Meanwhile, the transmittance and reflectance of the solution in a fixed thickness sample cell were measured. Then an inverse adding-doubling method [30] was employed to get the scattering coefficients of the Intralipid solution before and after mixing with different OCAs with different concentrations. Figure 1.2 shows the comparison between the theoretical prediction and the experimentally measured scattering coefficient. The theoretical prediction is consistent with the experimentally measured scattering coefficient, which shows that the RI of the background medium directly affect the result of optical clearing. Moreover, the RI of the background medium can be directly derived from the RI of the OCA used and the mixing ratio with Intralipid.

In addition, Wen et al. [28] further investigated the relationship between the reduced scattering coefficient and the RI of the above agents with different concentrations. Figure 1.3 demonstrates the mean measurements of reduced scattering coefficients at 589 nm for the above samples. The solid line presents the prediction of Mie theory, and the dotted line is the fitting curve of the measurements. It can be found that the scattering of the mixture decreases with the increase of the background RI (n_b), and Mie theory predicted values match well with measurements. The correlation coefficient between

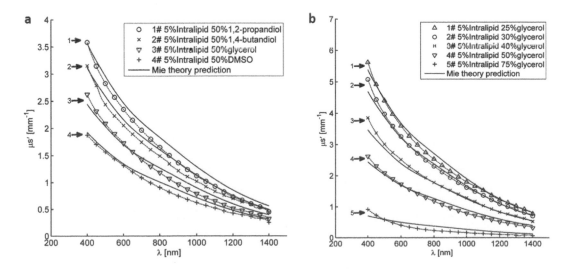

FIGURE 1.2 The comparison between the Mie theory predicted scattering coefficients and the experimental measured results in Wen's work [28]. (a) 5% Intralipid with 50% OCAs (DMSO, glycerol, 1,4-butanediol, 1,2-propanediol. (b) 5% Intralipid with 75%, 50%, 40%, 30%, 25% glycerol. Reprinted with permission from Reference [28].

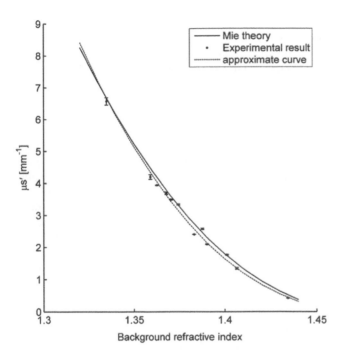

FIGURE 1.3 Reduced scattering coefficient of mixture versus the background RI at 589 nm. Reprinted with permission from Reference [28].

the reduced scattering coefficient and the background RI is about 0.97. By fitting the experiment data, we could obtain the following equation:

$$\mu'_s = k \left(\frac{1.475}{n_b} - 1 \right)^{2.09}, \quad (1.4)$$

where $k=739.8$. Therefore, we can use this formula to predict the change in scattering property caused by OCAs if OCAs keep the solution uniform. This reveals that increasing the background RI would make the RI matching of different components in the mixture. It is the reason that for tissue phantom, we do not consider the diffusion of agents, nor the interaction between Intralipid and OCAs. Therefore, in terms of physical principles, the mechanism of tissue optical clearing is to increase RI matching to reduce the light scattering of tissue.

Optical clearing mechanisms as a result of water and agent fluxes

Water content makes up 70%–80% of the volume of tissues and exhibits a much lower RI (~1.33) than that of proteins and lipids (>1.44). OCA-induced tissue dehydration was regarded as an important mechanism of tissue optical clearing. After comparing rat skin tissues after immersion in DMSO and glycerol with air-dried ones, it was found that transmittance in the air-dried rat skin tissues is also enhanced as it was for the OCA-immersed ones [31]. The phenomenon is quite similar between air-dried and OCA-immersed tissues, but air drying took a longer time to achieve an equally transparent effect. Their weight experiments show that water loss occured in both air-dried and OCA-immersed skin samples.

Transmission electron microscopy (TEM) can be employed to see the ultrastructure details. As shown in Figure 1.4, the ultrastructure after OCA immersion is similar to after air drying. Images reveal higher fibril packing density in the glycerol and air-immersed state versus the native state. The volume fraction of tendon fibrils increased from approximately 0.65 to approximately 0.90 due to glycerol- and air-immersion. The measured increase in scattering particle volume fraction corresponds to an approximately 60% decrease in reduced scattering coefficient, assuming no change in RI ratio between fibrils and surrounding fluid or change in scattering particle size or shape.

For the estimation of tissue degree of dehydration under drying by hot air and/or by the action of a hyperosmotic agent, the following equation can be used, which describes the flow of substance from a small volume with nonzero concentration into a reservoir with zero concentration [32, 33]:

$$H_D(t) = \frac{M(t=0) - M_0}{M(t=0)} \left(1 - \exp(-t/\tau_D) \right), \quad (1.5)$$

where M is a mass of the tissue (skin) sample (g), M_0 corresponds to the permanent mass of the sample (i.e., dry mass) (g), τ_D is a time constant which characterizes the rate of dehydration (s), and t is time of dehydration (s). The parameter M_0 includes mass of collagen, elastin, and other components of

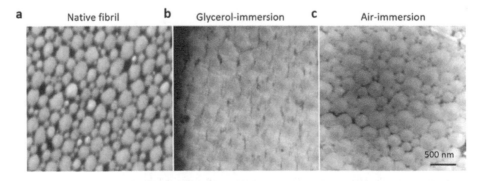

FIGURE 1.4 TEM images of tail tendon fibrils. (a) Native state. (b) 20 min glycerol immersion. (c) 2 h air immersion. Reprinted with permission from Reference [31].

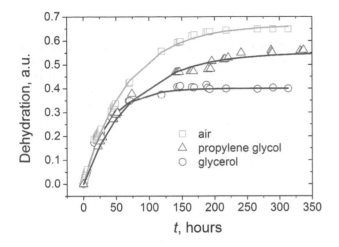

FIGURE 1.5 Kinetics of dehydration degree of human skin samples with damaged *stratum corneum*: squares correspond to dehydration in air, triangles are dehydration by propylene glycol, and rounds are dehydration by glycerol. Symbols and solid lines present experimental data and approximation using Equation (1.5), respectively. Adapted from Reference [34].

tissue after its total dehydration; bound water; and possibility residual free water. The difference between initial and permanent mass of the sample $\left(M(t=0) - M_0 \right)$ shows the amount of water that has escaped from tissue due to dehydration [23].

Skin dehydration kinetics induced by the following stimuli, evaporation and application of OCAs, is demonstrated in Figure 1.5. To enhance skin membrane permeability, the SC of the samples was perforated [23, 34].

The spectroscopy method can be employed to quantitatively analyze water loss in the optical clearing process because water has strong absorption bands in the near-infrared region. Xu et al. [35] used a dual-wavelength analysis method to estimate the water content of skin based on diffuse reflectance spectra over the wavelength range from 800 to 2200 nm and found that the water content changes show good consistency with the optical clearing effect.

The dual-wavelength method for water content prediction is usually applied in the food industry [36]. For biological tissue investigation, the results reflect the relative water content but not the absolute water content. Yu et al. [24] used a partial least-squares (PLS) regression model based on the reflectance spectra in the 1100–1700 nm region and indicated its good ability to calculate the water content in absolute values. They found the relative changes of reduced scattering coefficient and water content at 10, 30, and 60 min after the application of chemical agents, such as 1,2-propanediol, 1,4-butanediol, PEG-200, PEG-400, glycerol, 70% glycerol, and D-sorbitol (see Figure 1.6). As can be seen from the data, the water content reduced by 46%, 41%, 45%, 44%, 36%, 39%, and 32%, respectively; the optical clearing effect after a 60-min treatment with the OCAs decreased gradually on the order of glycerol, D-sorbitol, 70% glycerol, 1,2-propanediol, 1,4-butanediol, PEG-200, and PEG-400, with the reduced scattering coefficients decreasing by 62%, 58%, 54%, 52%, 43%, 42%, and 40%, respectively. Further statistical analysis demonstrates that there is an extremely significant difference ($P < 0.01$) between the relative changes in the reduced scattering coefficient and the water content, and there is a significant difference ($P < 0.05$) for PEG-400 after topical application of glycerol or D-sorbitol, whereas for 1,2-propanediol, 1,4-butanediol, or PEG-200, there is no statistical difference between the two data ($P > 0.05$). Hence, they concluded that dehydration is the main mechanism of skin optical clearing for 1,2-propanediol, 1,4-butanediol, or PEG-200 during the 60-min topical treatment, whereas for PEG-400, glycerol, or D-sorbitol, some other mechanisms might exist that lead to further clearing besides dehydration.

As mentioned above, replacing the interstitial water in a tissue with an OCA will lead to the occurrence of tissue dehydration and RI matching mechanisms. Such a replacement is initiated by the water flux out of and the OCA flux into the tissue. To understand how these fluxes occur and give rise to the OC mechanisms, we will consider an *ex vivo* slab-form tissue sample with a small thickness that is immersed in a solution containing a hyperosmotic agent, such as glycerol. The volume

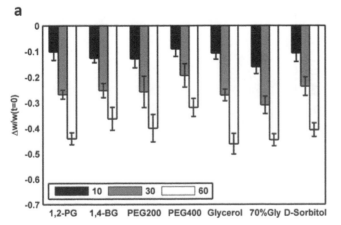

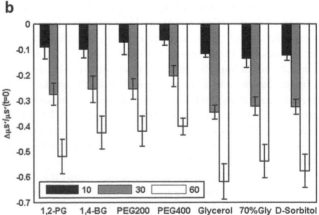

FIGURE 1.6 Relative reduction of (a) water content and (b) reduced scattering coefficient after 10-, 30-, and 60-min treatments with OCAs. Reprinted with permission from Reference [24].

of the treating solution must be significantly larger than the sample volume (about 10 × higher) to provide continuous flux both for water and OCA [6, 37].

As the tissue sample is immersed in the solution, an osmotic pressure is created over the sample by the OCA, which forces out the interstitial water that is either unbound or weakly bound to tissue scatterers. This is the water flux from the tissue to the outside. Such water loss induces a decrease in sample thickness and consequently leads to the approximation of tissue scatterers. As a result, a more compact and organized packing is created inside the tissue sample [1, 5, 37, 38]. These same changes in the internal organization of the tissue can also be obtained by mechanical compression or stretching, since mechanical forces also remove water from the compressed or stretched area of a tissue [34, 39]. By creating a denser scatterer distribution inside the tissue, we should expect an increase in the scattering and absorption coefficients, but due to the better ordering of scatterers inside and the decreased sample thickness that also occurs due to the water loss, the transparency of the samples increases [5, 38]. As a result of this initial transparency increase, the water flux out of the tissue is designated as the dehydration mechanism [38].

The OCA molecules in the treating solution also start to diffuse into the tissue at early stage of the treatment. At first, they interact only with the most superficial layers of the sample, but as the water molecules continue to flow out, the OCA molecules reach deeper inside and eventually fill all the interstitial areas of the tissue. The two fluxes (water going out and OCA going in) occur simultaneously, and it becomes necessary to discriminate and characterize them.

An increase in the mean RI of the ISF will occur with the diffusion of the OCA into the interstitial locations, bringing it closer to the RI of tissue scatterers. Such OCA diffusion into the ISF provides the RI matching mechanism [23, 37–39]. Both these mechanisms increase tissue transparency [2], allowing for the acquisition of better contrast images from deeper tissue layers [16, 40–42]. The tissue dehydration and RI matching mechanisms are reversible, since for *ex vivo* samples, the OCA can be washed out by immersing the tissue in physiological solution, and for the *in vivo* situation, the water in the adjacent tissues will flow into the treated tissue, washing the OCA out.

Several OCAs, such as ethylene glycol (EG) [37], glucose [2, 38, 43], dimethyl sulfoxide [44], glycerol [44–48], or Omnipaque™ [48], have been used to study the optical clearing treatments in various tissues with the objective of clarifying and characterizing the clearing mechanisms. The dehydration mechanism was also investigated under the application of mechanical forces and temperature variations to remove water from the tissue areas under study [2, 24, 31, 49–52]. It has been reported in these studies that tissue transparency increases due to the water loss as a result of the combination of smaller tissue thickness and better scatterer ordering inside.

As tissues lose water, a variation in the volume fractions of tissue scatterers (f_s) and ISF ($f_{ISF} = 1 - f_s$) will occur. Considering an *ex vivo* tissue slab again under treatment and also that only interstitial water flows out, the absolute scatterers' volume (V_s) remains unchanged, and the variations in sample volume (V) are driven by sample thickness (d) variations. At the beginning of the OC treatment, the dehydration mechanism is fast

and stronger than the RI matching mechanism, which results in a fast and strong decrease in d and ultimately in a strong V decrease ($V \propto d$). With the decrease in V, and since f_s is calculated as the ratio between V_s and V, f_s will increase and f_{ISF} decreases in the same proportion [4, 9, 10]. In reality, if all the interstitial water is lost, f_s reaches 1 and f_{ISF} reaches 0.

Considering a hypothetical tissue with a compact distribution of spherical scattering particles, it has been reported that during dehydration, the reduced scattering coefficient (μ'_s) is proportional to the product of the volume fractions of tissue components [31]:

$$\mu'_s = \frac{f_s(1-f_s)}{V_p}\sigma'_s, \tag{1.6}$$

with V_p representing the volume of a single scattering particle; and σ'_s, which depends on the wavelength of light in vacuum, on the scattering particle's radius and RI, and on the RI of the surrounding medium, representing the reduced scattering cross section (in cm²) for a single scattering particle. Considering any arbitrary values for V_p and σ'_s, Equation 1.6 shows that μ'_s has a parabolic dependence on f_s and that for both limiting situations, where $f_s = 0$ or $f_s = 1$, μ'_s will be zero. Figure 1.7 shows such dependence.

Due to the water loss, the scattering particle density and ordering inside the tissue change during the dehydration mechanism; Equation 1.4 shows this mechanism [31]. In opposition, during the RI matching mechanism, the tissue's scattering properties are expected to change, since the OCA may change the packing density and shape of scatterers, the scattering particle size, its RI, and the RI of the surrounding medium. These changes depend on the kinetics of water flux out and OCA flux in to the tissue and are to be accounted for in σ'_s [6].

For a system of scatterers having the shape of infinite cylinders, μ_s is expressed as [33, 53, 54]:

$$\mu_s(t) = \frac{f_s}{\pi a^2}\sigma_s(t)\frac{(1-f_s)^3}{1+f_s}, \tag{1.7}$$

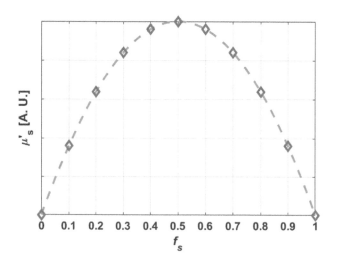

FIGURE 1.7 Relationship between μ'_s and f_s for a particular situation with unchanged light wavelength, unchanged scattering particle's radius and RI, and unchanged RI for the surrounding medium.

where σ_s is the cylinders scattering cross section. In the Rayleigh–Gans approximation for a system of infinite cylinders [29], it is given by the expression:

$$\sigma_s(t) = \frac{\pi^2 a x(t)^3}{8} \left(m(t)^2 - 1\right)^2 \left(1 + \frac{2}{\left(m(t)^2 + 1\right)^2}\right), \quad (1.8)$$

where $x(t) = 2\pi a n_{ISF}(t)/\lambda$ is the diffraction parameter, and $a \approx 50$ nm [33] is the radius of the scatterers.

Since we suppose that all variations of scattering kinetics during tissue dehydration are related only to the variations of the packing factor, the change of μ_s is determined by the behavior of the function $f_s(1 - f_s)^3 / (1 + f_s)$ [55], having a maximum at $f_s = 0.2$ (see Figure 1.8). In fibrous tissues, the value of f_s is commonly about 0.3 [33]. Therefore the increase of f_s leads to the decrease of μ_s, which, in turn, leads to the increase in the transmitted radiation intensity.

The RI matching mechanism has also been studied in various tissues, and valuable information was acquired. Different clearing potentials were observed in skin under treatment with some particular OCAs in different concentrations and osmolarities [20]. Different magnitudes of muscle transparency and different clearing kinetics were observed over 30 min for treatments with different concentrations of glucose and EG in aqueous solution [37, 56, 57]. Tissue permeability to water and OCA molecules determines the kinetics of the water and OCA fluxes. When comparing normal tissues and tissues with enhanced permeability, different kinetics for both OC mechanisms have been observed [23], and the increase in transparency is better when enhanced permeability is used. Increased tissue transmittance can also be obtained in treatments where alcohols are used as OCA diffusion enhancers, as demonstrated in other studies [7, 21]. The use of alcohol seems to be the less harmful option to enhance OCA diffusion into the tissues, since a small global variation in tissue thickness was observed between the untreated and treated samples [7, 21].

In general, dehydration can be realized by hydrophilic agents or the water-miscible polar solvents [25, 58]. For the former, some high concentration polyalcohols, such as glycerol [59, 60], sucrose [61, 62], fructose [63, 64], sorbitol [16, 65, 66], and contrast agents such as iohexol [65, 67–69] and iodixanol [70] could dehydrate the tissue samples by osmotic difference. Generally, these agents with high concentration produce a higher osmolarity external than internal tissues, thereby inducing efflux of water and influx of the outer molecules. As the main constituent, water forms 3D hydrogen-bonding networks with intracellular constituents, such as proteins [71]. The sugars, sugar-alcohols, and contrast agents have hydroxyls that can form hydrogen bonds with the proteins and show higher hydrogen-bonding ability than water [72]. Hence, the dehydrated samples with these agents demonstrated a certain degree of transparency.

The water-miscible polar solvents play an essential role in mediating the substitution of water with high-RI aromatic solvents inside tissues because the high-RI ($n > 1.5$) aromatic solvents are generally immiscible with water. Classically, biological tissues were dehydrated by alcohols, such as methanol [73, 74], ethanol [75–79], 1-propanediol [80], and tert-butanol [80–82]. Initially, the tissues were dehydrated by serially increasing the EtOH concentration up to 100% [75–77, 83]. Since EtOH has much less hydrogen bonding ability than water, after dehydration the tissues are shrunken and hardened with no sign of transparency like the hydrophilic agents. To achieve efficient clearing of the lipid-rich tissues, like the spinal cord and brainstem, the Dodt group selected a highly effective and GFP-friendly dehydrating reagent, THF, from thousands of chemicals for dehydration instead of ethanol [84, 85]. This polar ether-based protocol has achieved extremely rapid dehydration [84–86] due to its high permeable kinetics and has provided improved transparency of cleared large samples after RI matching due to its highly efficient lipid solubility.

In recent years, some other alcohols have been introduced as dehydration agents in solvent-based clearing methods for better fluorescence or structure preservation. For example, Schwarz et al. [80], using FluoClearBABB, proposed the use of 1-propanol or tert-butanol for dehydration; the Ertürk group developed uDISCO method [81] by using tert-butanol for dehydration to reveal better fluorescence; and Renier et al. [87] used methanol for dehydration to preserve the morphology of tissues.

There are some *ex vivo* methods, based on collimated transmittance (T_c) and thickness measurements performed during OC treatments that have been used to discriminate and characterize these mechanisms. The following two sections describe two of those methods.

Mechanism discrimination through the characterization of water and OCA fluxes

As indicated at the end of the previous section, the dehydration and the RI matching mechanisms are associated with the water flux out and with the OCA flux into the tissue, respectively. Due to these associations, if we can evaluate and characterize both these fluxes individually, it is possible to discriminate and

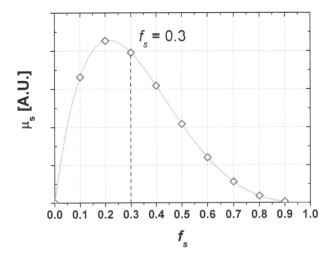

FIGURE 1.8 The scattering coefficient versus the volume fraction of scatterers. The radius of collagen fibers and the scattering cross section are assumed constant. Adapted from Reference [55].

characterize the two OC mechanisms. By performing thickness and T_c measurements from *ex vivo* tissue samples with slab form, immersed in solutions containing different concentrations of an OCA, such discrimination and characterization can be made, as reported in the literature [6, 88]. The diffusion time, τ, is estimated both for water and for OCA directly from the T_c measurements and further calculations that use thickness kinetics measurements allow the calculation of the diffusion coefficient, D, also both for water and OCA. Such estimations and calculations are made through Equations (1.9) and (1.10) [5, 6, 37, 38, 53, 56, 89]:

$$T_c\left(\lambda,t\right)=\frac{C_a\left(t\right)}{C_{a0}}\cong\left[1-\exp\left(-\frac{t}{\tau}\right)\right] \tag{1.9}$$

$$D_a=\frac{d^2}{\pi^2\times\tau} \tag{1.10}$$

In Equation (1.9), $C_a(t)$ represents the OCA concentration inside the tissue sample at a time t, while C_{a0} represents the OCA concentration in the immersion solution at the beginning of the OC treatment [6, 37].

Considering individual wavelengths within a range where the tissue presents no absorption bands, by fitting the T_c time dependencies for those wavelengths with Equation (1.9), the τ values for each wavelength can be estimated [56]. The mean τ value for each treatment can be calculated and represented as a function of the OCA concentration in the treating solution [88]. A spline connecting the estimated τ values will present a maximum for a treatment with a particular OCA concentration and a minimum for a treatment with a highly saturated solution. The maximum τ value obtained in such fitting indicates that the mobile water in the tissue is in equilibrium with the water in the treating solution, meaning that only the OCA flux in occurs. The minimum τ value obtained for a treatment with a highly saturated solution indicates that only the water flux out occurs as a result of the osmotic pressure of the OCA in the solution. This means that the maximum τ value in such representation is the diffusion time for the OCA and the minimum τ value obtained is the diffusion time for water [88]. Using these values in Equation (1.10), along with the corresponding thickness (d) obtained at those specific times of treatment and for the treatments with those particular OCA concentrations, allows for calculation of D for OCA and water [88]. If the water and OCA fluxes occur only through one slab surface, instead of Equation (1.10), Equation (1.11) should be used to calculate D [53, 90, 91]:

$$D_a=\frac{4d^2}{\pi^2\times\tau}. \tag{1.11}$$

Another solution of one-dimensional form of Fick's second law,

$$\frac{\partial C\left(x,t\right)}{\partial t}=D\frac{\partial^2 C\left(x,t\right)}{\partial x^2} \tag{1.12}$$

was presented in references [18, 53, 90–92]. The key approach in characterization of the transfer of a chemical agent is that a

set of boundary conditions defines the concentration profiles. Depending on the analytical solution used, the tissue type, and the experimental setup, several kinds of initial and boundary conditions are most commonly used to study agent transport in tissues.

The initial condition corresponds to the absence of an agent inside the tissue before the measurements, i.e.,

$$C\left(x,0\right)=0, \tag{1.13}$$

for all inner points of the tissue sample.

In case when a tissue is presented as a slab, the three kinds of boundary conditions are as follows:

    1) A tissue slab that is free of agent is immersed in solution with the agent concentration of C_0:

$$C\left(0,t\right)=C_0 \quad\text{and}\quad C\left(l,t\right)=C_0, \tag{1.14}$$

where l is the tissue sample thickness. The solution of Equation (1.12) with the initial [Equation (1.13)] and the boundary [Equation (1.14)] conditions has the form [18, 31, 53, 93]

$$C\left(x,t\right)=C_0\left(1-\sum_{i=0}^{\infty}\frac{4}{\pi\left(2i+1\right)}\sin\left(\frac{\left(2i+1\right)\pi x}{l}\right)\exp\left(-\frac{\left(2i+1\right)^2 D\pi^2 t}{l^2}\right)\right) \tag{1.15}$$

The integral of Equation (1.15) over x gives another physical quantity, average concentration (total solute entering the tissue), as:

$$C\left(t\right)=C_0\left(1-\frac{8}{\pi^2}\sum_{i=0}^{\infty}\frac{1}{\left(2i+1\right)^2}\exp\left(-\left(2i+1\right)^2 t\pi^2 D/l^2\right)\right), \tag{1.16}$$

where $C(t)$ is the volume-averaged concentration of an agent within tissue sample.

A more rigorous description of the diffusion process can be obtained by solution of diffusion equation in 3D geometry. In this geometry, the diffusion equation is

$$\frac{\partial C\left(x,y,z,t\right)}{\partial t}=D\left[\frac{\partial C^2\left(x,y,z,t\right)}{\partial x^2}+\frac{\partial C^2\left(x,y,z,t\right)}{\partial y^2}+\frac{\partial C^2\left(x,y,z,t\right)}{\partial z^2}\right] \tag{1.17}$$

$$t>0;-R_1<x<+R_1;-R_2<y<+R_2;-R_3<z<+R_3;$$

The initial condition corresponds to the absence of an agent inside the tissue before the measurements, i.e.,

$$C\left(x,y,z,0\right)=0, \tag{1.18}$$

for all inner points of the tissue sample.

In case when a tissue is presented as a parallelepiped, the boundary conditions are as follows:

$$C(\pm R_1, y, z, t) = C_0, \; C(x, \pm R_2, z, t) = C_0, \; C(x, y, \pm R_3, t) = C_0,$$
(1.19)

where $2R_1$ is the tissue sample width, $2R_2$ is the tissue sample length, and $2R_3$ is the tissue sample thickness.

The solution of Equation (1.17) with the initial [Equation (1.18)] and the boundary [Equation (1.19)] conditions has the form [94]

$$\frac{C_0 - C(x, y, z, t)}{C_0} = \sum_{n=1}^{\infty} \sum_{m=1}^{\infty} \sum_{k=1}^{\infty} \frac{A_n A_m A_k \cos \mu_n \frac{x}{R_1}}{\cos \mu_m \frac{y}{R_2} \cos \mu_k \frac{z}{R_3}}$$
$$\exp\left[-\left(\mu_n^2 K_1^2 + \mu_m^2 K_2^2 + \mu_k^2 K_3^2\right) F_0\right]$$

where

$$A_n = (-1)^{n+1} \frac{2}{\mu_n}, \; A_m = (-1)^{m+1} \frac{2}{\mu_m}, \; A_k = (-1)^{k+1} \frac{2}{\mu_k},$$

$$\mu_n = (2n-1)\frac{\pi}{2}, \; \mu_m = (2m-1)\frac{\pi}{2}, \; \mu_k = (2k-1)\frac{\pi}{2},$$

F_0 is the number of Fourier $\left(F_0 = \frac{Dt}{R^2}\right)$. R is the generalized size.

$$\frac{1}{R^2} = \frac{1}{R_1^2} + \frac{1}{R_2^2} + \frac{1}{R_3^2}, \qquad K_i = \frac{R}{R_i} \; (i = 1, 2, 3).$$
(1.20)

$$\frac{C_0 - C(x, y, z, t)}{C_0} = \sum_{n=1}^{\infty} \sum_{m=1}^{\infty} \sum_{k=1}^{\infty} \frac{A_n A_m A_k \cos \mu_n \frac{x}{R_1}}{\cos \mu_m \frac{y}{R_2} \cos \mu_k \frac{z}{R_3}}$$
$$\exp\left[-\left(\mu_n^2 K_1^2 + \mu_m^2 K_2^2 + \mu_k^2 K_3^2\right) F_0\right]$$

2) A tissue slab is free of agent, one side of the slab contacts a solution with the agent concentration of C_0, and the other side is isolated from agent penetration:

$$C(0, t) = C_0 \; \text{and} \; \frac{\partial C(l, t)}{\partial x} = 0.$$
(1.21)

The solution of Equation (1.12) with the initial [Equation (1.13)] and the boundary [Equation (1.21)] conditions has the form [53, 90, 91, 95]

$$C(x, t) = C_0 \left(1 - \sum_{i=0}^{\infty} \frac{\frac{4}{\pi(2i+1)} \sin\left(\frac{(2i+1)\pi x}{2l}\right)}{\exp\left(-\frac{(2i+1)^2 D\pi^2 t}{4l^2}\right)}\right).$$
(1.22)

The volume-averaged concentration in this case can be expressed as

$$C(t) = C_0 \left(1 - \frac{8}{\pi^2} \sum_{i=0}^{\infty} \frac{1}{(2i+1)^2} \exp\left(-(2i+1)^2 t \frac{\pi^2}{4} D \Big/ l^2\right)\right).$$
(1.23)

3) A tissue slab is free of agent, where one side contacts a solution with the agent concentration of C_0, and the other side is kept at zero concentration:

$$C(0, t) = C_0 \; \text{and} \; C(l, t) = 0.$$
(1.24)

The solution of Equation (1.12) with the initial [Equation (1.13)] and the boundary [Equation (1.24)] conditions has the form [53, 96–99]

$$C(x, t) - C_0 \left(1 - \frac{x}{l} - \sum_{i=1}^{\infty} \frac{2}{\pi i} \sin\left(\frac{i\pi x}{l}\right) \exp\left(-\frac{i^2 D\pi^2 t}{l^2}\right)\right).$$
(1.25)

The average concentration can be expressed as:

$$C(t) = \frac{C_0}{2} \left(1 - \frac{8}{\pi^2} \sum_{i=0}^{\infty} \frac{1}{(2i+1)^2} \exp\left(-(2i+1)^2 t \pi^2 D \Big/ l^2\right)\right).$$
(1.26)

4) When a penetrating agent is topically administered to tissue and the tissue is a semi-infinite medium, i.e. $x \in [0; \infty)$, the boundary conditions have the form

$$C(0, t) = C_0 \; \text{and} \; C(\infty, t) = 0.$$
(1.27)

The solution of Equation (1.12) with the initial [Equation (1.13)] and the boundary [Equation (1.27)] conditions, in this case, has the form [100]

$$C(x, t) = C_0 \left(1 - \operatorname{erf}\left(\frac{x}{2\sqrt{Dt}}\right)\right),$$
(1.28)

where $\operatorname{erf}(z) = \frac{2}{\sqrt{\pi}} \int_0^z \exp(-a^2) \, da$ is the error function.

5) The exact solution for the release kinetics of a solute from an infinite spherical reservoir with the burst effect initial condition into a finite external volume is developed. The governing diffusion equation for transport in the membrane can be solved by the time Laplace transform method [101]. In this study, the author has considered the diffusion release of a solute from a reservoir with spherical geometry into a finite external volume. A spherical reservoir consists of a core containing the drug (or OCA) and a coat, which is the rate-limiting element in the release process. The radius of the core is a, and the radius of the

coated sphere is b, giving the coat thickness of ($b - a$). The diffusion coefficient is assumed to be independent of concentration. This assumption is valid when (a) the solute has very low solubility in the membrane, or (b) the solute concentration gradient across the membrane is very low, or (c) the average value for the diffusion coefficient can be considered. The concentration in the membrane is a function of both time and the position variable r, and is determined by transient diffusion according to Fick's second law:

$$\frac{\partial C}{\partial t} = \frac{D}{r^2}\frac{\partial}{\partial r}\left(r^2 \frac{\partial C}{\partial r}\right), \qquad (1.29)$$

where D is the diffusion coefficient and C is the concentration of dissolved solute in the membrane. Assuming equilibrium between the surface and the external fluid at all times, without any mass transfer resistance in the fluid, the initial and boundary conditions are

$$C(r,0) = C_s \quad a < r < b, \qquad (1.30)$$

$$C(a,t) = C_s, \qquad (1.31)$$

$$C(b,t) = KC_b(t), \qquad (1.32)$$

where C_s is the solubility limit of the solute in the membrane, $C_b(t)$ is the external bulk concentration at time t, and K is the equilibrium distribution coefficient between the membrane and the external solution. Equation (1.30) states that the solute saturates the membrane at $t = 0$, which occurs when the device is either stored for some time before use or formed at elevated temperature for a certain period. Equation (1.31) describes that the solute concentration in the membrane at the reservoir side is equal to its solubility in the water, which is consistent with infinite reservoir, and Equation (1.32) gives the equilibrium relationship at the interface between the membrane and external solution.

The solution of Equation (1.29) with the initial and the boundary conditions has the form

$$\frac{M_t}{M_\infty} = 1 - \sum \frac{2\alpha\lambda_n^2}{\left(1 - \frac{a}{b}\right)\left[\lambda_n^2 + \left(1 + \alpha\lambda_n^2\right)^2\right] + \alpha\lambda_n^2 - 1}\exp\left[-\lambda_n^2 \tau\right],$$

$$(1.33)$$

where $\alpha = \dfrac{V}{4\pi b^3 K}$, V is the total external fluid volume, $\tau = \dfrac{Dt}{b^2}$, and λ_n are the positive roots of the characteristic equation:

$$\mathrm{Cotan}\left[\lambda_n\left(1 - \frac{a}{b}\right)\right] = \frac{1 + \alpha\lambda_n^2}{\lambda_n}.$$

Such an estimation/calculation method can be used for any *ex vivo* and slab-form tissue sample under treatment with aqueous solutions containing any OCA concentration. For a general treatment, the estimated τ and D values represent the effective diffusion time and diffusion coefficient for the combined flux of water going out and OCA going in [6]. Figure 1.9 presents the T_c kinetics for a few wavelengths obtained from human colorectal muscle under treatment with 40% glycerol and 60% glycerol [88].

As represented in Figure 1.9(a) for the treatment with 40% glycerol, the increase in T_c is smooth during the entire treatment for all wavelengths. This means that only glycerol is flowing into the tissue as a result of the water balance between the tissue and the treating solution. The representation in Figure 1.9(b) shows that due to the high glycerol concentration (60%) in the treating solution, the increase in T_c is fast and limited to the beginning of the treatment (first 2–3 min) and then stabilizes until approximately 10 min, before increasing again. The initial strong increase observed for this treatment indicates that only water is flowing out in this initial time interval, and as a result of a decreased sample thickness and better organization and packing for tissue scatterers, T_c increases strong as water flows out.

As previously indicated, since the maximum τ corresponds to a treatment where a balance is established between the

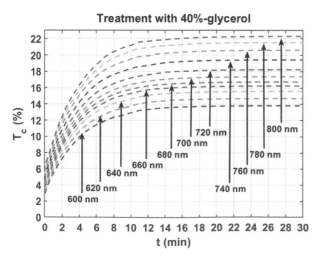

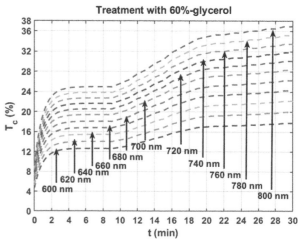

FIGURE 1.9 T_c kinetics for human colorectal muscle under treatment with 40% glycerol (a) and 60% glycerol (b). Adapted from Reference [88].

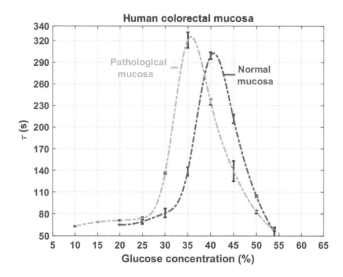

FIGURE 1.10 Mean τ as a function of glucose concentration in the treating solution for human normal and pathological colorectal mucosa tissues. Adapted from Reference [88, 89].

mobile water in the tissue and the water in the treating solution, this method has also allowed the evaluation of the mobile water content in tissues where it was unknown and discrimination between normal and pathological tissues [88]. By studying human normal and pathological colorectal mucosa under treatment with glucose solutions, it was found that the pathological tissues contain ~5% more mobile water than the normal tissues. As represented in Figure 1.10, the maximum τ occurs for different glucose concentrations in the treating solution for normal and pathological human colorectal mucosa [88, 89].

Various studies using this method have been made with different tissues under treatment with various OCAs to evaluate their diffusion properties and characterize both the water and OCA fluxes. Table 1.1 summarizes some of these results as reported in the literature.

Due to the nature of the measurements used in this method, evaluating the diffusion properties for water and OCAs can only be made for *ex vivo* tissues. To perform similar evaluations from *in vivo* tissues, a different procedure must be adopted. To evaluate τ, instead of measuring T_c, diffuse reflectance (R_d) or OCT measurements must be performed. To calculate D, the necessary thickness measurements can be made during treatment using OCT or confocal microscopy measurements [5, 57, 105].

Due to the great variety of human and animal tissues available and considering both normal and pathological versions (and also different pathologies), a large number of such studies is necessary to evaluate the diffusion properties of the various OCAs in as many tissues as possible. The above described method provides an easy and practical way to obtain such properties if the *ex vivo* estimations are acceptable.

Evaluation of the optical clearing mechanisms through the refractive index kinetics

According to the description presented in the previous section, which associates the OC mechanisms of tissue dehydration

and RI matching with the water and OCA fluxes, we know that these fluxes are the driving forces in any treatment and that they result in the partial exchange of tissue water by the OCA, leading to the elevation of the RI of the background material of the tissue – the interstitial fluid (ISF) [5, 38, 48, 88, 106–108].

The evaluation of the increase in the RI of the ISF can also be made from T_c and thickness measurements made during treatment [4, 10, 107], as already indicated at the beginning of this chapter. As referred to above, such measurements are sensitive to the water and OCA fluxes, meaning that they can be used to evaluate RI variations.

Considering also *ex vivo* tissues to perform both thickness and T_c measurements, we will now present the sequence of calculations that allow obtaining the RI kinetics of a tissue under treatment. To describe such a sequence of calculations, we present data obtained in a study with human normal colorectal mucosa tissues. The tissues used in the study were prepared from surgical resections obtained from patients under treatment at the Portuguese Oncology Institute of Porto, Portugal, who signed a written consent to the use of surgical specimens for research purposes. The colorectal mucosal layer is the inner layer of the colorectal wall, as represented in Figure 1.11.

A cryostat (Leica model CM 1850 UV) was used to prepare slab-form samples from the mucosa layer, with approximated circular form ($\phi = 1$ cm) and thickness of 0.5 mm [9]. Six samples were prepared to be used under treatment with an aqueous solution containing 40% (v/v) glucose. Three of these samples were used in the T_c kinetics measurements, and the other three were used in thickness kinetics measurements. Figure 1.12 presents the average results from these studies.

The data presented in Figure 1.12(a) were obtained for a wavelength range where the mucosa does not show any absorption bands.

To use these average results to calculate the RI kinetics, we considered a simple model for the mucosa tissues, which accounts for only two tissue components – the water in the ISF and tissue scatterers. Such a model is valid and useful for these calculations, since no matter how many tissue components the mucosa independently contains, it can always be represented as a simple two-component tissue with a completely dry matter (designated as scatterers) and water. In this model, the RI of the scatterers is the average RI that accounts for the individual RI of all scatterers. The model also accounts for the fact that only part of the tissue water will move out during treatment. Literature [48, 109] explains that biological tissues have various water states inside, according to their bounding strength: strongly bound, tightly bound, weakly bound, and free water. It is known that for short-term OC treatments, only weakly bound and free water are able to move out as a result of the osmotic pressure created by the OCA molecules [4, 6, 10]. These two states of tissue water are commonly called mobile water [6].

The first step in the calculations consists of determining the volume fractions (VFs) of the two components in the untreated mucosa. To obtain these VFs, we first considered the physical dimensions of the samples and calculated their volume with Equation (1.34):

$$V_{\text{sample}}(t = 0) = \left(\pi \times 0.5^2 \right) \times 0.05 \text{ cm}^3. \qquad (1.34)$$

TABLE 1.1

Diffusion properties for water and some OCAs in biological tissues and water.

Tissues	Water	Glucose	Fructose	Glycerol	Ethylene glycol (EG)	Polyethylene glycol (PEG-300/ 400)	Propylene glycol (PPG)
Skeletal muscle (rat – Wister Han)	τ=58.0 s; D=3.1×10⁻⁶ cm²/s [37]	τ=302.9 s; D=8.36×10⁻⁷ cm²/s [56]	τ=314.2 s; D=5.7×10⁻⁷ cm²/s [88]		τ=446 s; D=4.6×10⁻⁷cm²/s [37]		τ=411.4 s; D=5.1×10⁻⁷ cm²/s [88]
Colorectal muscle	τ=57.0 s; D=3.1×10⁻⁶ cm²/s [88]			τ=231.6 s; D=3.3×10⁻⁷ cm²/s [88]			
Normal colorectal mucosa (human)	τ=55.7 s; D=3.3×10⁻⁶ cm²/s [89]	τ=302.4 s; D=5.8×10⁻⁷ cm²/s [89]					
Pathological colorectal mucosa (human)	τ=58.4 s; D=2.4×10⁻⁶ cm²/s [91]	τ=325.1 s; D=4.4×10⁻⁷ cm²/s [91]					
Gastric mucosa (human)		τ=191.4 s; D=1.59×10⁻⁶ cm²/s [92]					
Liver (human)	τ=57.2 s; D=3.2×10⁻⁶ cm²/s [102]			τ=211.2 s; D=8.2×10⁻⁷ cm²/s [102]			
Skin (rat)		τ=~1800 s; D=1.1×10⁻⁶cm²/sec [103]		τ=~400 s; D=3.2×10⁻⁶cm²/s [95]		τ=~1800/ ~2100s; D=1.83×10⁻⁶/ 1.70×10⁻⁶ cm²/s [90]	τ=~2148 s; D=1.35×10⁻⁷cm²/s [104]

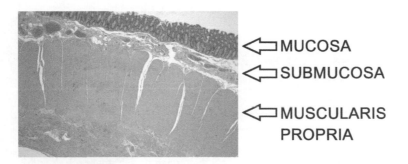

FIGURE 1.11 Histology photo of the various layers of the human colorectal wall – mucosa, submucosa, and muscularis propria.

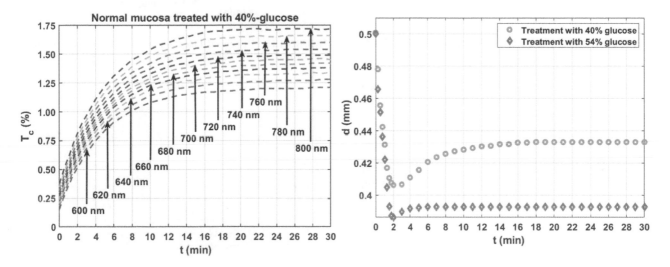

FIGURE 1.12 Average kinetics for Tc (a) and thickness (b) measurements for human colorectal mucosa under treatment with 40% glucose. Adapted from Reference [89].

We have determined that the mobile water in the mucosa is 60% (see Figure 1.10) [88, 110]. This result implies that the scatterers represent the remaining 40%. Using this value and the sample volume calculated by Equation (1.34), we could calculate the absolute volume for scatterers in the mucosa as [9]:

$$V_s = 0.4 \times V_{\text{sample}}(t = 0) \text{ cm}^3. \quad (1.35)$$

Since during short-time OC treatments no changes occur in scatterers, this volume remains unchanged. In opposition, the VFs of tissue components will change since the sample volume changes due to the partial exchange of interstitial water by the OCA. Due to the slab-form of the tissue samples, where the thickness is much smaller than the superficial area, we can calculate the kinetics for the VF of tissue scatterers from the sample thickness kinetics [6, 9, 10]:

$$f_s(t) = \frac{V_s}{\left(\pi \times 0.5^2\right) \times d(t)}. \quad (1.36)$$

Considering that for any time of treatment the sum of all VFs must be 1, the VF of the interstitial medium can be calculated from $f_s(t)$:

$$f_{\text{ISF}}(t) = 1 - f_s(t). \quad (1.37)$$

Figure 1.13 presents the kinetics of the VFs of the components in human colorectal mucosa, as calculated with Equations (1.36) and (1.37) [9].

Figure 1.13 shows that the strongest variations occur during the first 2 min of treatment, created by the dehydration mechanism. The later smooth and smaller magnitude variations seen in Figure 1.13 are created by the RI matching mechanism.

After obtaining the kinetics of the VFs for tissue components, we calculated the kinetics for the scattering coefficient (μ_s) of the mucosa. To perform this calculation, the average $T_c(t)$ and $d(t)$ data presented in Figure 1.12 and the estimated data for the absorption coefficient ($\mu_a(\lambda)$) of the untreated tissue [9] were used. This calculation was made using the Bouguer–Beer–Lambert equation – Equation (1.2), here reproduced in a different form [4, 9]:

$$\mu_s(\lambda, t) = -\frac{\ln\left[T_c(\lambda, t)\right]}{d(t)} - \mu_a(\lambda). \quad (1.38)$$

Due to the fact that for biological tissues, μ_s is much higher than μ_a [111–116], the calculation made with Equation (1.38), which does not account for μ_a variations, is reasonable, since those variations can be neglected compared to the variations of μ_s during treatment. Figure 1.14 presents the kinetics for the attenuation coefficient ($\mu_t = \mu_s + \mu_a$) and μ_s at 650 nm, as calculated by Equation (1.38).

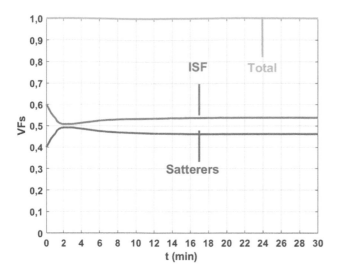

FIGURE 1.13 Time dependence of the VFs of tissue components in human normal colorectal mucosa during treatment with 40% glucose. Adapted from Reference [9].

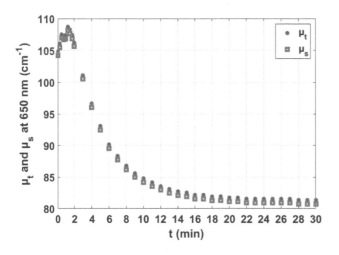

FIGURE 1.14 Kinetics of μ_t and μ_s at 650 nm for the human normal colorectal mucosa during treatment with 40% glucose.

Figure 1.14 presents high-magnitude variations during treatment for μ_t and μ_s [9]. For the native tissue ($t = 0$), the difference between these properties is small, indicating that μ_a is small when compared with μ_s, and that any variations in μ_a can be neglected in these calculations. Figure 1.14 also shows an increase for μ_t and μ_s at the beginning of the treatment (first 75 s), showing that due to water loss, scatterers approach each other (providing tissue shrinkage), which leads to an increase in both coefficients and a decrease in tissue sample thickness, as seen in Figure 1.12 (b).

After obtaining the μ_s kinetics, the following step is to calculate the variations in the RI of the ISF, which can be made at any wavelength using Equation (1.3) [4, 5, 10]. Considering the wavelength range between 400 and 1000 nm, this calculation was performed for the mucosa under treatment with 40% glucose, and Figure 1.15 presents the kinetics [9].

Figure 1.15 shows that the OC treatment has induced a smooth increase in the RI of the ISF of colorectal mucosa for

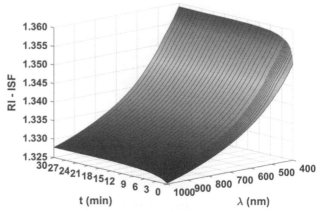

FIGURE 1.15 Dispersion kinetics for the human normal colorectal mucosa under treatment with 40% glucose. Adapted from Reference [9].

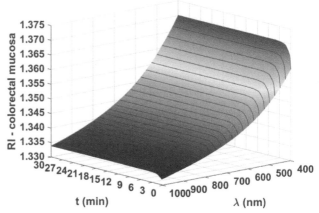

FIGURE 1.16 Kinetics of the RI of human normal colorectal mucosa under treatment with 40% glucose. Adapted from Reference [9].

all wavelengths, demonstrating the RI matching mechanism. The data in Figure 1.15 can be used in Gladstone and Dale law of mixtures [5, 115–119] to calculate the kinetics for the RI of the whole tissue ($n_{tissue}(\lambda,t)$) during treatment:

$$n_{tissue}\left(\lambda,t\right) = \bar{n}_0\left(\lambda,t\right)f_{ISF}\left(t\right) + n_s\left(\lambda\right)f_s\left(t\right). \quad (1.39)$$

The data calculated with Equation (1.39) is represented in Figure 1.16 [9].

The data in Figure 1.16 shows similar increasing behavior to the one presented in Figure 1.15 for the ISF, but with different magnitude in variations. The similar efficiency presented for all wavelengths in Figure 1.16 indicates that glucose has no absorption bands in the entire spectral range used in these calculations.

Three stages can be identified in Figure 1.16. Within the first minute, a strong increase in the RI of the tissue is created by the dehydration mechanism. Between 1 and 4 min, a transition between the dehydration and RI matching mechanisms occurs, and the final 26 min corresponds uniquely to the RI matching mechanism.

The study of the RI kinetics, both for the ISF and for the whole tissue, allows the detection of the tissue dehydration and the RI matching mechanisms of OC and also to identify the time intervals when they occur. By evaluating the magnitude of the variations in the kinetics presented in graphs of Figures 1.15 and 1.16, it is possible to quantify the transparency created by the treatment. A similar analysis can be made from the graph in Figure 1.14, where the reduction of μ_s was calculated only from T_c and thickness measurements were made during treatment.

In a number of studies, the reduction in the collimated transmittance of tissues after achieving the maximal values is described [32, 48, 55, 120]. The degree of the tissue swelling depends on the type of OCA and the concentration. The examples of the OCAs used are glycerol and glucose solutions (see Figure 1.17).

The tissue swelling affects the intensity of radiation transmitted through the tissue, which is caused by the increased sample volume and, therefore, the decreased volume fraction of scatterers. As shown above, the dependence of the scattering coefficient on the volume fraction of scatterers (collagen and elastin fibrils) is described by Equation (1.7). If, following References [31, 121], we assume that the tissue volume increases mainly because of larger separation between the fibrils while the fibril size does not change in the process of swelling during the observation time, then the parameters a and σ_s can be treated as constants (analogically with dehydration process). As follows from Figure 1.8, the decrease in the

volume fraction of the scatterers leads to the increase of the scattering coefficient, which, in turn, leads to the reduction of the transmitted radiation intensity.

Apparently, glycerol, glucose, and other hygroscopic OCAs bind water molecules in the interstitial space. For example, glycerol is a strongly hygroscopic compound, and, being placed in a humid environment, it adds water molecules until saturation level is achieved (55 vol %). It was shown that each glycerol molecule could add six water molecules [122]. Consequently, the tissue sample swells due to the added water. In this case, the RI of the OCA solution in the interstitial space decreases compared to its value at the earlier stage, which manifests itself as a decrease in the collimated transmittance of tissue.

The detection of the third mechanism of OC treatments, protein dissociation, is also possible from T_c measurements, and the analysis of such kinetics provides valuable information in the detection of certain pathologies, such as cancer. Further discussion about this matter will be made in Chapter 33.

Dissociation of collagen

Among the investigations on tissue optical clearing, various OCAs, alcohols [20, 21, 123], sugars [124, 125], organic acids [22, 126], and other organic solvents [86, 127] were used

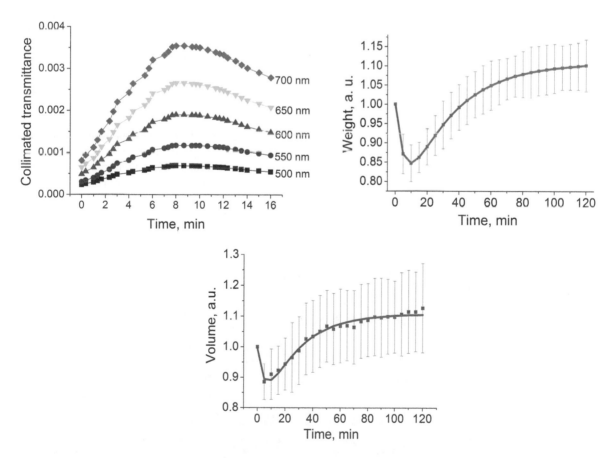

FIGURE 1.17 Typical temporal dependence of the collimated transmittance of the dura mater samples under the action of glucose aqueous solutions with the concentration 1.5 M (a) [55], the kinetics of changes in the weight (b) and volume (c) of myocardial samples under the action of 40% glucose solution. Reprinted with permission from Reference [120].

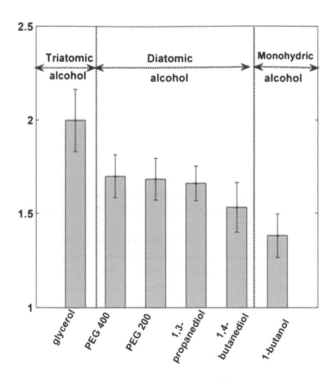

FIGURE 1.18 The maximal relative transmittance of skin samples treated by three categories of alcohol: triatomic, diatomic, and monohydric alcohol. Reprinted with permission from Reference [21].

widely. After comparing the dependence of skin optical clearing efficacy on RI and osmolality of three different groups of OCAs, i.e., hydroxy-terminated agents, organic solvents, and organic acids, Choi et al. [20] found that the optical clearing efficacy did not correlate with RI nor osmolality, but that the hydroxy-terminated organic compounds, called alcohols, demonstrated the highest optical clearing potential (OCP) from the three groups of OCAs.

Further, the optical clearing efficacy of porcine skin *in vitro* caused by six alcohol OCAs were investigated. Mao et al. [21] found that the alcohol OCAs with more hydroxyl groups have better optical clearing efficacy for porcine skin. According to the number of hydroxyl groups in the chemical structure, they divided the investigated alcohol OCAs into three categories:

monohydric alcohols (1-butanol), diatomic alcohols (PEG-400, PEG-200, 1,3-propanediol, and 1,4-butanediol), and triatomic alcohols (glycerol). As shown in Figure 1.18, glycerol, triatomic alcohol, has the greatest optical clearing effect with the highest maximal relative transmittance of 2.00; 1-butanol, monohydric alcohol, has the poorest effect with lowest maximal relative transmittance of 1.38; and the four diatomic alcohols are moderate.

The above experimental results demonstrate that skin optical clearing efficacy is correlated to the molecular structure of OCAs, which should be from some interaction between OCAs and the molecular composition of skin, i.e., collagen fibers. Collagen fibers have complex self-assembling structures, which make them strong scattering sources in biological tissues [128]. Collagen fiber is widely distributed in various biological tissues, such as the dermis of skin and cornea of eyes, its function being to provide structural support. In spite of nearly 30 years' research, there is still no exact explanation for its complex structure formation; synthetic collagen-like peptides have been used to demonstrate that the self-assembly of collagen I secondary and tertiary structures is driven by specific hydrophobic and electrostatic binding sites. The stability of these structures is enhanced by inductive (electronegativity) effects conferred by hydroxyproline residues. For higher-order structures, hydrogen bonding is the primary bonding force between collagen triple helices. OCAs with multiple hydroxyl groups may have strong electronegativity, which may screen the hydrogen bonds in collagen triple helices that may labilize the higher-order structure of collagen to its dissociation. As the screen of hydrogen bonds in collagen triple helices is a noncovalent effect, the OCAs' effect on the dissociation of collagen can be reversible. Yeh et al. [129] observed the dissociation of collagen fibers in *in vitro* immersion of tissues in OCAs by the second harmonics generation imaging (see Figure 1.19). Before the glycerol immersion, the fibrous structure of collagen can be clearly seen in the second harmonics generation image (Figure 1.19(a)). Following glycerol application, the dermis becomes optically transparent and the fibrous structure unravels into a matted morphology (Figure 1.19(b)). The removal of excess glycerol and the subsequent application of PBS lead to a recovery of the collagen's fibrous structure (Figure 1.19(c)).

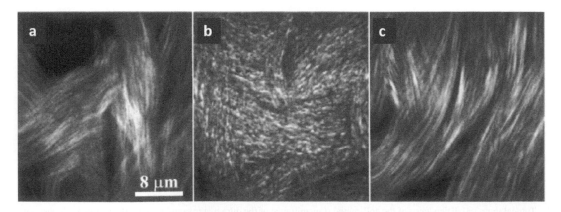

FIGURE 1.19 Reversible effect of glycerol on rodent dermis. Second harmonics generation images (a) prior to glycerol application, (b) after glycerol application, (c) after rehydration with PBS. Reprinted with permission from Reference [129].

Another study has reported that no collagen dissociation or fracturation was observed for *in vivo* dorsal rat skin under treatment with aqueous solutions containing different glycerol concentrations [130]. In this study, it was observed that skin thickness and collagen diameter decrease as a result of water loss. Other studies have been performed to study the clearing mechanisms. One of those studies used a time-lapse multiphoton microscopy technique to show that collagen in animal skin and tendon undergoes fast dehydration and shrinkage under treatment with high concentrated glycerol solutions [106]. This same study also demonstrated that glycerol penetration was slower than dehydration and accompanied by sample swelling.

Hirshburg et al. [72] used molecular dynamics (MD) simulation to explore the hydrogen bond formation between alcohols and collagen molecules. They divided the hydrogen bond bridges formatted between alcohols and collagens into different types. The index of the types is related to the number of carbon atoms of alcohols in a hydrogen bond bridge. Higher bridge types span further across the collagen surface and can potentially disrupt collagen–collagen and collagen–water interactions better than lower bridge types. Thus, alcohols with hydroxyl group pairs with longer distances on the carbon chain will have better optical clearing efficacy than alcohols that only have hydroxyl groups next to each other. The simulation results give a good explanation for the fact that 1,3-propanediol has twice the OCP than that of 1,2-propanediol even though they had identical molecular weights (76.10 Da), similar RI (1.44 versus 1.43), and osmolality (8.3 versus 8.7 Osm/kg). 1,2-propanediol can only form type I bridges, while 1,3-propanediol can form type II bridges. Since the MD simulation can reveal the details of the interaction between OCAs and tissue structure in the molecular scope, this will be a powerful tool in illustrating the mechanism of tissue optical clearing and the screening for new high efficacy optical clearing agents.

Noting that sugars have hydroxyl groups too, Zhu's group paid attention to monosaccharides and saccharides, respectively [124, 125]. The MD simulation was applied to evaluate the OCP of fructose, glucose, and ribose according to its

TABLE 1.2

Average number of hydrogen bonds and bridges formed between each sugar and collagen mimetic peptide per picosecond. Data from Reference [124].

Sugars	Fructose	Glucose	Ribose
Hydrogen bonds (#/ps)	1.25417	1.21333	1.03444
Hydrogen bond bridges (#/ps)	0.6517	0.5063	0.4067

propensity to form hydrogen bonds and bridges. Previous studies indicated that collagen tertiary structures and triple helices for higher-order assembly were stabilized and organized by water bridges and a hydration shell [131–134]. Further, the formation of hydrogen bonds as well as hydrogen bond bridges between chemical agents and collagen could disrupt the water bridges and hydration shell, destabilizing the higher order of collagen structures [22, 128]. Thus, the propensity of sugars to form hydrogen bonds and bridges can be considered a measurement of their optical clearing efficacy. The more hydrogen bonds and bridges a sugar forms, the more effectively the water bridges and hydration shell are disrupted. Table 1.2 shows the average numbers of hydrogen bonds and bridges formed between each sugar and the two peptides per picosecond. It can be found that fructose forms more hydrogen bonds and bridges than the other sugars, and ribose forms the fewest. *In vitro* skin was immersed in each sugar solution to evaluate the optical clearing efficacy, which showed a consistency with the results of the MD simulation [124].

Meanwhile, the authors found that an increase in OCA molecular size and the number of hydroxyl groups could increase the efficacy of TOC [135]. Considering disaccharides have a larger molecular size and more hydroxyl groups than the above three monosaccharides, Zhu's group further investigated the OCP of sucrose and maltose as the representatives of disaccharide by theoretical simulation, and compared them with fructose by *ex vivo* and *in vivo* experiments [125]. Figure 1.20 shows the hydrogen bonds formed between the three agents and collagen mimetic peptide. It is expected that the more hydrogen bonds and hydrogen bond bridges an

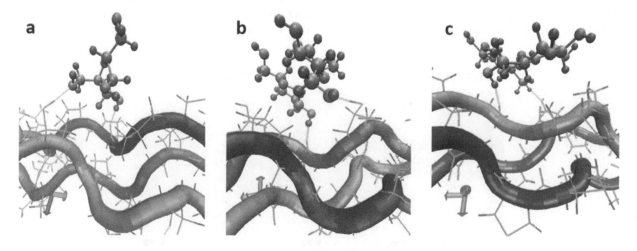

FIGURE 1.20 Schematic representation of hydrogen bond formation of fructose (a), sucrose (b) and maltose (c) with representative collagen mimetic peptides, respectively. Reprinted with permission from Reference [125].

agent forms, the more the hydration shell and water-mediated hydrogen bonds are disrupted. The average numbers of hydrogen bonds formed between each agent and two collagen mimetic peptides per ps are 1.254, 1.307, and 1.292 for fructose, sucrose, and maltose, respectively. That is, sucrose has the strongest ability to disrupt the hydration shell and water-mediated hydrogen bonds. Hence, it predicts that sucrose has the best skin OCP, maltose comes second, and fructose is the worst. The theoretical simulation would effectively avoid the time-consumption of blind experiments.

The correlations between the OCP and such intermolecular interaction parameters as the amount of time of the agent being in a hydrogen-bonded state with the collagen, the dependence of the volume occupied by the collagen in an aqueous solution on OCA concentration, and the probability of a double hydrogen bond formation between the agents and the collagen were carried out by the methods of classical MD [136]. Berezin et al. [136] showed that the effectiveness of interaction with the collagen depends both on the number of alcohol groups and on the distance between them. Moreover, the conformational mobility of the low molecular agents should be taken into consideration. The higher this mobility is, the better an agent can adapt to the molecular pocket. There are four groups available for classical hydrogen bonding formation in the site pocket of a collagen model under study: two carbonyl (one at a glycine residue, the other at a hydroxyproline residue of the same α-chain) and two alcohol ones at hydroxyproline residues of different α-chains [137].

Another informative parameter for characterizing the correlation between OCAs interacting with collagen and OCP is the protein volume change. Within the framework of molecular modeling, the relation between collagen peptides volume change and glycerol concentration was obtained. This relation is nonlinear; its maximum is reached at medium-range glycerol concentration (40%). Experimental data analysis of human skin OC by means of glycerol [95, 138] shows that the efficiency of OC is at its maximum at higher concentration (60%–70%). This difference can be explained in the following way: when this interaction is molecularly modulated, the glycerol concentration is set inside the intercellular space, while the experimental data indicate glycerol concentration prior to skin application. Since the tissue already has some amount of liquid inside, it can be expected that the glycerol concentration inside the tissue will be smaller than an applied one.

The influence of the concentration of the other OCAs on the collagen molecule volume is presented in Reference [138]. It reveals that the dependence of volume of the collagen peptide molecule is nonlinear. Almost all agents have a maximal impact on the volume of collagen fibers in the range of medium concentrations [137].

These studies have demonstrated that the clearing potential of an OCA is related to its protein solubility capability, indicating that protein solubility is also an OC mechanism. It was also demonstrated in these studies that OCA hyperosmolarity relative to a tissue forces sample dehydration. The disruption of the collagen hydration layer and consequent water replacement by the OCA to perform RI matching is obtained through the formation of hydrogen bond bridge connections [72].

Delipidation and decalcification

Except for the proteins such as collagen described above, lipids represent another cellular component with a high RI in most biological tissues. The lipid content reaches up to more than 70% in adipose tissue, and it is about 20% in white matter or tongue. Most soft tissues contain about 10% lipids. The lipid in tissues not only causes light scattering inside tissue due to its high RI, but also hinders the external medium from cells as the main component of membranes. Thus, tissue delipidation is important for the full clearing of tissues and permeation of external molecules.

Specific solvents (e.g., THF and DCM) or water-soluble molecules (e.g., detergents and aminoalcohols) in current tissue optical clearing techniques have mainly achieved the removal of lipid.

For solvent-based clearing methods, the delipidation was usually accompanied by dehydration [139]. Dodt's group firstly used ethanol as dehydration reagent and BABB as RI matching reagent to clear tissues [75–83]. However, it did not achieve sufficient transparency in adult tissues containing a high degree of lipids, such as the rodent spinal cord and brainstem. To overcome this limitation, they selected tetrahydrofuran (THF) to replace ethanol, and introduced dichloromethane (DCM) to improve the tissue transparency, due to their good lipid-solvating capability [84, 85]. As shown in Figure 1.21, THF-treated spinal cord was transparent while ethanol-treated

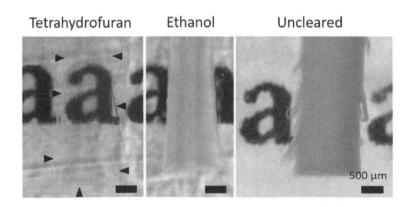

FIGURE 1.21 Photography of adult spinal cord cleared with tetrahydrofuran and ethanol. Reprinted with permission from Reference [85].

tissue remained opaque. Further, the combination of THF with dibenzyl ether (DBE) resulted in 3DISCO method, providing improved tissue transparency and fluorescence signal [86]. Generally, for delipidation, DCM was also used but only for short incubation times because of the trade-off between transparency and fluorescence preservation [81, 86]. Some other alcohols, like 1-propanediol and tert-butanol [80–82], were also used for delipidation in solvent-based clearing methods such as FluoClearBABB [80], uDISCO [81], and PEGASOS [82] method.

Among the water-soluble molecules, detergents and aminoalcohols play an important role in lipid removal.

The Sca*l*eA2 clearing agent proposed by Miyawaki's group was composed of urea, glycerol, and Triton X-100 [140]. Triton X-100, as a commonly used nonionic detergent in cell biology, can dissolve lipids to increase cell permeability by incorporating lipids into a micelle. Corresponding to the components of Sca*l*eA2, Ueda's group screened a lot of chemicals, including polyhydric alcohols, detergents, and hydrophilic small molecules [61]. To evaluate their ability to dissolve brain tissue, fixed brain homogenate was mixed with different solutions and measured the OD600, which was expected to correlate with the potential solubility to brain lipids [61]. As shown in Figure 1.22, all the detergents induced a decrease of OD600 of the

brain homogenate to a certain extent. Among the tested detergents, Triton X-100 and sodium deoxycholate (SDC) showed the highest level of solubilizing activity. Additionally, a series of aminoalcohols including Quadrol and triethanolamine were discovered to have considerable tissue solubilizing activity [61], indicating their potential clearing capability. It is speculated that the cationic amino group in aminoalcohols contributed to solvating anionic phospholipids. In other words, aminoalcohols presumably solubilize phospholipids by electrostatic interaction. Taking account of the clearing performance and fluorescence preservation, a cocktail of urea, Quadrol, and Triton X-100 was mixed as the first reagent in CUBIC protocol, i.e., Sca*l*eCUBIC-1 and the subsequent improved cocktail Sca*l*eCUBIC-1A (protocol available at http://cubic.riken.jp) [141], and triethanolamine was also included in Sca*l*eCUBIC-2 further to increase the tissue transmittance [26, 61]. Additionally, for whole-body profiling of cancer metastasis, the optimized cocktail of 10 w%/10 w% N-butyldiethanolamine/Triton X-100 (termed CUBIC-L) for highly effective delipidation was identified [142]. Recently, 1,3-bis(aminomethyl)cyclohexane (1,3-BAC), as an alphatic amine, also showed high lipid solubility and contributed to rapid delipidation in a CUBIC-HL cocktail [143].

Saponin, as a widely used mild nonionic detergent, was also utilized for delipidation to permeabilize cellular membranes in

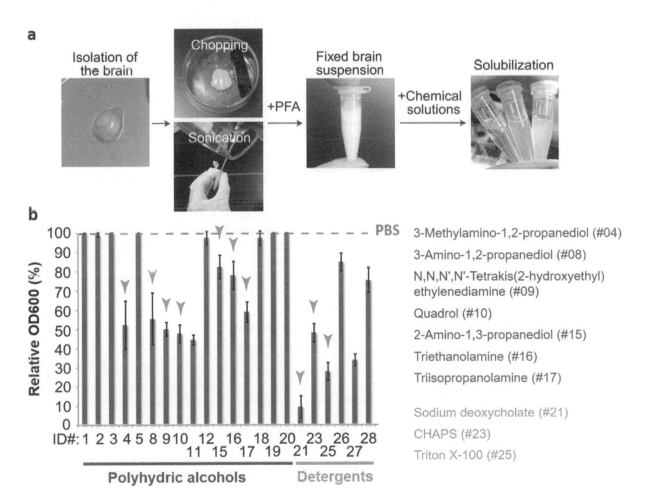

FIGURE 1.22 Screening chemicals for tissue-solubilizing activity using a fixed brain suspension. (a) Screening method. (b) The measured OD600 values for each chemical. Adapted with permission from Reference [61].

tissue clearing methods; for instance, it facilitated permeation of iohexol in SeeDB2 clearing [67] and diffusion of hydrogel monomer solution into the human or zebrafish tissue in CLARITY clearing [60].

Sodium dodecyl sulfate (SDS) is an ionic detergent and has strong lipid-dissolving capability. All CLARITY-related clearing methods are based on delipidation by SDS. Passive diffusion of the detergent micelles into tissues is very slow, presumably due to their large molecular size [25, 144]. To achieve sufficient delipidation in large tissues, acceleration of the diffusion of detergents inside tissues is necessary. The utilization of urea in Sca*l*e and CUBIC methods can promote the passive influx of detergent by increasing internal osmotic pressure [26, 61, 140, 145]. Some other strategies, including electric field potential, perfusion, and thermal energy, had also been utilized to promote the diffusion of detergent, such as SDS.

In 2013, Chung et al. [60] proposed a distinct clearing method, CLARITY, based on hydrogel embedding and electrophoresis. In this method, SDS was used to remove the lipids from hydrogel-embedded samples, which was accelerated by an electric field potential applied in the electrophoresis approach. However, its application had been limited to low electric fields because using high fields can damage tissue structures [146, 147]. Chung's group introduced stochastic electrotransport by using a rotational electric field [148], which can selectively transport highly electromobile molecules and therefore render tissue transparent rapidly without tissue damage. They further applied the stochastic electrotransport method to SHIELD [148] and eFLASH [149] for rapid delipidation or acceleration of probe penetration.

Perfusion is another strategy to reinforce detergent diffusion inside tissues. Yang et al. first introduced the perfusion-assisted agent release method PARS [65]. This method used the blood circulatory system or the cerebrospinal fluid route to deliver the detergents to the whole body or target organs for rapid delipidation, facilitating high-efficient whole-organ and whole-body clearing after RIMS treatment [65]. Moreover, Tainaka et al. [26] also utilized the transcardial perfusion of half-diluted Sca*l*eCUBIC-1 combined with external immersion for enhancing whole-organ and whole-body clearing.

Thermal activation of diffusion was also adopted to accelerate SDS permeation in several protocols due to the temperature dependency of solute diffusion kinetics. Yu et al. [150] demonstrated the elevated-temperature-induced acceleration of PACT and concluded 42–47°C as the alternative temperature range for PACT, and that higher temperatures would damage the sample by a strong surfactant solution. Chung's group introduced SWITCH-based glutaraldehyde fixation to secure tissue structures and the SWITCH-processed samples endured prolonged incubation in SDS at high temperatures, even at 80°C [70]. With SWITCH, thermal energy considerably increased the passive clearing speed without noticeable tissue damage. Notably, with the increase in temperature, the volume of SDS micelles decreases, and the number of micelles increases [151]. This would facilitate its diffusion into tissue, which might be another reason for the thermal reinforcement of SDS permeation except for the increased kinetics of molecules.

SDC, also an ionic detergent, has been used for delipidation in SeeNet introduced by Miyawaki et al. [152] Although the light transmittance of SDC-cleared tissues was slightly lower than that of SDS-cleared tissues, the SDC-cleared tissues exhibited a spatially uniform RI across the samples and had no opaque areas.

Recently, Zhao et al. [144] identified that the zwitterionic detergent CHAPS could form smaller micelles than Triton X-100 and SDS, which could penetrate more rapidly and deeply into tissue, allowing full permeabilization of aged human organs. Based on the CHAPS tissue permeabilization approach, SHANEL was proposed to clear and label stiff human organs.

Notably, for hard bone or calcified tissue, its mineral compositions make its refractive RI higher (1.55–1.65) than that of other tissue components [153]. Therefore decalcification is indispensable for high transparency of bone clearing or whole-body clearing. Generally, to clear bone tissues or the whole body, decalcification was carried out prior to the other steps for most clearing methods. Some mineral acids, such as nitric acid, hydrochloric acid, and weak organic acids such as formic acid, were used in histological decalcification [141, 154–156]. They can achieve fast decalcification, but would induce damage to the bone tissues [157–159].

The chelating reagents, such as ethylene diamine tetraacetic acid (EDTA), can also be used for decalcification. EDTA causes little damage to tissue and can preserve the activity of antigens and certain enzymes in tissues, which make it suitable for subsequent immunochemistry [157–159]. It has been utilized in the clearing methods such as PACT-deCal [160], Bone-CLARITY [161], PEGASOS [82], etc., for bone clearing or whole-body clearing. EDTA achieves very gentle decalcification mainly by capturing the calcium ions from the surface of the apatite crystal, and thus it weakens the microstructure of the bone [162]. By stimulated Raman scattering (SRS) imaging, Chen et al. [162] directly visualized microstructural variation of hydroxyapatite in the skull before and after treatment with EDTA. From the XY images at a depth of 40 μm in the same location of the skull, after 5–10 min of EDTA immersion, a large number of cavities formed around the bone lacunae, indicating hydroxyapatite corruption (Figure 1.23). The XZ projection images showed that the thickness of the hydroxyapatite in the skull was significantly reduced after being treated with EDTA.

Hyperhydration

Many tissue optical clearing methods benefited from the high hydration of the hyperhydration agents, like urea and urea-like chemicals, which can enhance the penetration of clearing reagents into the tissues and achieve good transparency.

Among the hyperhydration agents, urea was serendipitously discovered to have good clearing capability by Hama et al. [140]. As a small, uncharged molecule with two NH_2 groups, urea has strong hydration ability [163] because it can be both a proton donor and acceptor when forming hydrogen, thus disturbing hydrogen-bonding networks of proteins and nucleic acids as a relatively weak denaturant [164, 165]. Also, urea can

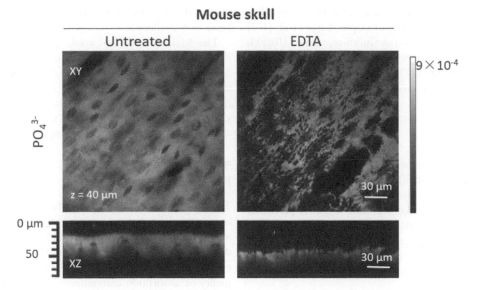

FIGURE 1.23 SRS images of the skull before and after treatment with EDTA. Adapted with permission from Reference [162].

increase membrane fluidity to enhance molecular flux [166, 167], contributing to enhanced permeation of reagents.

In hyperhydration-based clearing methods, the use of urea accounted for a large proportion. The ScaleA2 cocktail contains 4 M urea, 10% glycerol, and 0.1% nonionic detergent Triton X-100 [140]. It achieved good transparency on mouse brain and embryo but would cause obvious tissue swelling, presumably because urea relaxes protein scaffolds involving solid tissue frameworks such as collagen fibers, as described above. In addition, the interior osmotic pressure would be increased, thereby inducing an influx of small external molecules, including water and other ingredients. For instance, Hou et al. [64] utilized the cocktail of fructose and urea, the proposed FRUIT protocol, and achieved greater tissue transparency with a shorter clearing time than fructose alone in SeeDB. The Miyawaki group further proposed ScaleS by combination of urea, sorbitol, and DMSO, using a gradient concentration to enhance the penetration [168]. It not only accelerated the clearing process compared to the sorbitol/DMSO protocol proposed by Economo et al. [66], but also showed excellent preservation of fluorescence and ultrastructure. Both FRUIT and ScaleS demonstrated the importance of the gradient in solute concentration and the balance of ionic osmotic pressure. Notably, FRUIT and ScaleS were compatible with the lipophilic dye DiI, suggesting that urea would not disturb the lipid bilayer membrane.

Furthermore, Ueda's group developed CUBIC [61] based on Scale. In this method, aminoalcohol, Triton X-100, and urea were mixed to produce ScaleCUBIC-1 to remove lipids, loose tissue structure, and then ScaleCUBIC-2 with high RI was used to match the RI of tissue. The group also described some other CUBIC-derived protocols by utilizing the urea-like molecules, such as CUBIC-cancer using immersion of N-butyldiethanolamine and antipyrine/nicotinamide for whole-body profiling of cancer metastasis [142], and CUBIC-X based on imidazole and antipyrine reagents to expand and clear the samples [169], as well as a series of subsequent CUBIC protocols (I-IV) [143].

Kuwajima et al. [170] had developed a formamide-based clearing protocol termed *Clear^T*. Formamide is also a small, uncharged denaturant with both hydrogen donor and acceptor groups (Figure 1.24). The hydration energy of urea is expected to be much higher than that of formamide [163]. However, *Clear^T* demonstrated more efficient clearing performance than did ScaleA2, probably owing to the delipidation by the high concentration of formamide, which can also solubilize lipids as an aprotic polar solvent like N, N-dimethylformamide (DMF), and DMSO [171]. The 1,3-BAC in CUBIC-HL cocktail has two NH_2 groups similar to urea (Figure 1.24), suggesting high hydration ability, together with its high lipid solubility as described above, leading to the rapid clearing process.

Recently, the Zhu group firstly introduced m-xylylenediamine (MXDA) into tissue clearing and developed the MACS clearing method with ultrafast clearing speed, robust compatibility with lipophilic dyes, and fine morphology maintenance [16]. MXDA also has two NH_2 groups similar to urea (Figure 1.24), and it is expected to have great hyperhydration ability like urea [163]. Owing to the hyperhydration as well as a high RI, the MXDA contributed to the rapid clearing protocol MACS, showing much faster clearing than the Scale and CUBIC series methods.

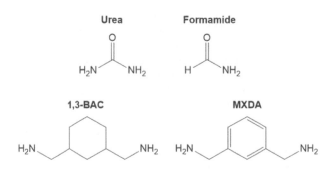

FIGURE 1.24 Chemical structures of urea, formamide, 1,3-bis (aminomethyl)cyclohexane (1,3-BAC), and m-xylylenediamine (MXDA).

Some other clearing methods also utilized the hyperhydration of urea or urea analogs. For example, Chen et al. [172] proposed a urea-based amino-sugar mixture, termed UbasM; Lai et al. [173] combined N-methylglucamine with 2,2'-thiodiethanol and iohexol for OPTIClear clearing; and Li et al. [69] reported C_e3D utilizing N-methylacetamide.

Decolorization

Except for the scattering, the penetration of light in tissue also suffers from the absorption of some endogenous pigments such as heme, melanin, and lipofuscin [174–176], which limits the imaging depth. As is known, simple buffer perfusion can effectively remove red blood cells in vessels. However, in certain tissues and organs, such as the spleen, there is still substantial residual blood after perfusion. Hence, decolorizing is important for the clearing of large tissues to achieve deep detection.

Generally, these pigments could be readily bleached by several oxidative treatments [177]. Hydrogen peroxide (H_2O_2) is one of the most widely used bleaching chemicals [178], which can directly degrade the pigment structure, but can also degrade the protein structure at the same time. Renier et al. [178] had used H_2O_2 to bleach the samples prior to immunostaining to reduce autofluorescence in the tissues in iDISCO protocols, and Zhao et al. [144] also utilized oxidative bleaching in SHANEL labeling of human tissues. In addition, the treatment of samples with acid acetone or strong alkalis (e.g., NaOH) can also dissociate heme, thereby achieving efficient pigment bleaching [179]. However, such harsh bleaching methods are not conducive to the preservation of fluorescent proteins and are not commonly used in fluorescence imaging. Thus, a milder decolorization strategy is needed for tissue clearing of whole organs or whole bodies.

In recent years, the Ueda group [26] had discovered that aminoalcohols, such as Quadrol, could achieve unexpected tissue decolorization by efficiently eluting the heme chromophore in blood. It was speculated that aminoalcohols had similar structures to that of the globulin chain and might tend to bind heme porphyrin instead of oxygen and histidine in hemoglobin, which might facilitate heme release and contribute to the decolorization effect. Recently, Zhu et al. [16] found that MXDA showed excellent decolorizing capability, and its decolorization principle was different from that of Quadrol (releasing hemin) as used in CUBIC-series methods but similar to that of NaOH (releasing Fe). In addition, the SDS used in CLARITY [60] and its derivatives [65] can also decolorize heme-rich tissues, such as kidney and spleen. This is probably because charged heme molecules could be eluted by denaturation of the heme-hemoglobin holoenzyme by SDS, which can also be aided by electrophoretic force with an electrophoretic tissue clearing (ETC) setup.

Except for heme, lipofuscin can be found in a wide range of cell types as the breakdown product of old red blood cells [176]. It may be quenched with either Copper(II) sulfate ($CuSO_4$) or Sudan Black [176]. The Gradinaru group demonstrated that Sudan Black B staining is successfully compatible with tissue clearing, fluorescent reporters, and immunostaining the

chemical strategies [65, 68]. However, decolorizing of melanin remains a challenge.

The Maillard reaction would induce browning and autofluorescence accumulation. In some clearing protocols, such as SeeDB [63] and FRUIT [64], utilizing fructose as the clearing agents, and SWITCH [70] based on glutaraldehyde fixation, the discoloration would reduce the light transmission. The thioglycerol and sodium sulfite [63, 64, 69, 70] have been used to partially suppress the Maillard reaction to improve tissue transparency.

Summary

The fundamental physical principle of tissue optical clearing is to enhance RI matching among various components in tissue, which can be realized by different chemical processes, such as dissociation of collagen, dehydration, delipidation, decalcification, hyperhydration, and so on. In addition, for tissues with plenty of absorption components, such as heme-rich tissues, decolorization is helpful for optical imaging in deep tissue.

Acknowledgments

T.Y. and D.Z. were supported by the National Key Research and Development Program of China grant number 2017YFA0700501, and NSFC grant number 81171736. E.A.G, A.N.B., and V.V.T were supported by RFBR grant number 18-52-16025. L. O. was supported by the Portuguese Science Foundation grant number FCT-UIDB/04730/2020. The authors are thankful to the staff of the Portuguese Oncology Institute of Porto (IPO-Porto), Portugal, for providing the human colorectal specimens used in the research presented in this chapter. The authors also appreciate the histological photograph of the human colorectal wall, which was supplied by Dr Sónia Carvalho from IPO-Porto.

REFERENCES

1. V.V. Tuchin, *Tissue Optics-Light Scattering Methods and Instruments for Medical Diagnostics*, 3rd ed., SPIE Press, Bellingham, WA (2015).
2. E.A. Genina, A.N. Bashkatov, and V.V. Tuchin, "Tissue optical immersion clearing," *Expert Rev. Med. Devices* **7**(6), 825–842 (2010).
3. A.Y. Sdobnov, M.E. Darvin, E.A. Genina, A.N. Bashkatov, J. Lademann, and V.V. Tuchin, "Recent progress in tissue optical clearing for spectroscopic application," *Spectrochim. Acta Part A: Mol. Biomol. Spectrosc.* **197**, 216–229 (2018).
4. L. Oliveira, M.I. Carvalho, E. Nogueira, and V.V. Tuchin, "Skeletal muscle dispersion (400–1000 nm) and kinetics at optical clearing," *J. Biophotonics* **11**(1), 1019 (2018).
5. V.V. Tuchin, *Optical Clearing of Tissues and Blood*, SPIE Press, Bellingham, WA (2006).
6. L. Oliveira, and V.V. Tuchin, *The Optical Clearing Method: A New Tool for Clinical Practice and Biomedical Engineering*, Springer, Cham-Switzerland (2019).
7. L. Oliveira, A. Lage, M. Pais Clemente, and V.V. Tuchin, "Rat muscle opacity decrease due to the osmosis of a simple mixture," *J. Biomed. Opt.* **15**(5), 055004 (2010).

8. V. Backman, R. Gurjar, K. Badizadegan, et al., "Polarized light scattering spectroscopy for quantitative measurement of epithelial cellular structures in situ," *IEEE J. Sel. Top. Quantum Electron.* **5**(4), 1019 (1999).

9. I. Carneiro, S. Carvalho, V. Silva, R. Henrique, L. Oliveira, and V.V. Tuchin, "Kinetics of optical properties of human colorectal tissues during optical clearing: A comparative study between normal and pathological tissues," *J. Biomed. Opt.* **23**(12), 121620 (2018).

10. I. Carneiro, S. Carvalho, R. Henrique, L. Oliveira, and V.V. Tuchin, "Kinetics of optical properties of colorectal muscle during optical clearing," *IEEE J. Sel. Top. Quantum Electron.* **25**(1), 7200608 (2019).

11. L. Oliveira, A. Lage, M. Pais Clemente, and V.V. Tuchin, "Optical characterization and composition of abdominal wall muscle from rat," *Opt. Lasers Eng.* **47**, 667–672 (2009).

12. A.N. Bashkatov, E.A. Genina, V.I. Kochubey, and V.V. Tuchin, "Estimation of wavelength dependence of refractive index of collagen fibers of scleral tissue," *Proc. SPIE* **4162**, 265–268 (2000).

13. E. Gratton, "Applied physics. Deeper tissue imaging with total detection," *Science* **331**(6020), 1016–1017 (2011).

14. T. Wang, D.G. Ouzounov, C. Wu, et al., "Three-photon imaging of mouse brain structure and function through the intact skull," *Nat. Methods* **15**(10), 789–792 (2018).

15. D. Zhu, K.V. Larin, Q. Luo, and V.V. Tuchin, "Recent progress in tissue optical clearing," *Laser Photonic Rev.* **7**(5), 732–757 (2013).

16. J. Zhu, T. Yu, Y. Li, et al., "MACS: Rapid aqueous clearing system for 3D mapping of intact organs," *Adv. Sci.* **7**(8), 1903185 (2020).

17. R.M. White, A. Sessa, C. Burke et al., "Transparent adult zebrafish as a tool for in vivo transplantation analysis," *Cell Stem Cell* **2**(2), 183–189 (2008).

18. V.V. Tuchin, I.L. Maksimova, D.A. Zimnyakov, I.L. Kon, A.H. Mavlyutov, and A.A. Mishin, "Light propagation in tissues with controlled optical properties," *J. Biomed. Opt.* **2**(4), 401–417 (1997).

19. R. LaComb, O. Nadiarnykh, S. Carey, and P.J. Campagnola, "Quantitative second harmonic generation imaging and modeling of the optical clearing mechanism in striated muscle and tendon," *J. Biomed. Opt.* **13**(2), 021109 (2008).

20. B. Choi, L. Tsu, E. Chen, et al., "Determination of chemical agent optical clearing potential using in vitro human skin," *Lasers Surg. Med.* **36**(2), 72–75 (2005).

21. Z. Mao, D. Zhu, Y. Hu, X. Wen, and Z. Han, "Influence of alcohols on the optical clearing effect of skin in vitro," *J. Biomed. Opt.* **13**(2), 021104 (2008).

22. J. Hirshburg, B. Choi, J.S. Nelson, and A.T. Yeh, "Correlation between collagen solubility and skin optical clearing using sugars," *Lasers Surg. Med.* **39**(2), 140–144 (2007).

23. E.A. Genina, A.N. Bashkatov, A.A. Korobko et al., "Optical clearing of human skin: Comparative study of permeability and dehydration of intact and photothermally perforated skin," *J. Biomed. Opt.* **13**(2), 021102 (2008).

24. T. Yu, X. Wen, V.V. Tuchin, Q. Luo, and D. Zhu, "Quantitative analysis of dehydration in porcine skin for assessing mechanism of optical clearing," *J. Biomed. Opt.* **16**(9), 095002 (2011).

25. K. Tainaka, A. Kuno, S.I. Kubota, T. Murakami, and H.R. Ueda, "Chemical principles in tissue clearing and staining protocols for whole-body cell profiling," *Annu. Rev. Cell Dev. Biol.* **32**, 713–741 (2016).

26. K. Tainaka, S.I. Kubota, T.Q. Suyama, et al., "Whole-body imaging with single-cell resolution by tissue decolorization," *Cell* **159**(4), 911–924 (2014).

27. V.V. Tuchin (Ed.), *Handbook of Optical Biomedical Diagnostics*, 2nd ed., PM262/263, SPIE Press, Bellingham, WA (2016).

28. X. Wen, V.V. Tuchin, Q. Luo, and D. Zhu, "Controling the scattering of intralipid by using optical clearing agents," *Phys. Med. Biol.* **54**(22), 6917–6930 (2009).

29. C.F. Bohren, and D.R. Huffman, *Absorption and Scattering of Light by Small Particles*, Willey, New York (1983).

30. S.A. Prahl, M.J.C. van Gemert, and A.J. Welch "Determining the optical properties of turbid media by using the adding-doubling method," *Appl. Opt.* **32**(4), 559–568 (1993).

31. C.G. Rylander, O.F. Stumpp, T.E. Milner, et al., "Dehydration mechanism of optical clearing in tissue," *J. Biomed. Opt.* **11**(4), 041117 (2006).

32. Y. Huang, and K.M. Meek, "Swelling studies on the cornea and sclera: The effect of pH and ionic strength," *Biophys. J.* **77**, 1655–1665 (1999).

33. A.N. Bashkatov, E.A. Genina, Yu.P. Sinichkin, V.I. Kochubey, N.A. Lakodina, and V.V. Tuchin, "Glucose and mannitol diffusion in human dura mater," *Biophys. J.* **85**(5), 3310–3318 (2003).

34. E.A. Genina, A.N. Bashkatov, K.V. Larin, and V.V. Tuchin, "Light–tissue interaction at optical clearing," Chapter 7 in: F.S. Pavone (ed.), *Laser Imaging and Manipulation in Cell Biology*, Wiley-VCH Verlag GmbH & Co. KGaA, Weinheim, Germany (2010).

35. X.Q. Xu, and R.K.K. Wang, "The role of water desorption on optical clearing of biotissue: Studied with near infrared reflectance spectroscopy," *Med. Phys.* **30**(6), 1246–1253 (2003).

36. B.G. Osborne, and T. Fearn, *Near-Infrared Spectroscopy in Food Analysis*, Longman Scientific and Technical, New York (1986).

37. L.M. Oliveira, M.I. Carvalho, E.M. Nogueira, and V.V. Tuchin, "Diffusion characteristics of ethylene glycol in skeletal muscle," *J. Biomed. Opt.* **20**(5), 051019 (2015).

38. L. Oliveira, M.I. Carvalho, E. Nogueira, and V.V. Tuchin, "Optical clearing mechanisms characterization in muscle," *J. Innov. Opt. Health Sci.* **9**(5), 1650035 (2016).

39. E.A. Genina, A.N. Bashkatov, Yu.P. Sinichkin, I. Yu. Yanina, and V.V. Tuchin, "Optical clearing of tissues: Benefits for biology, medical diagnostics and phototherapy," Chapter 10 in: V.V. Tuchin (ed.), *Handbook on Optical Biomedical Diagnostics*, Vol. 2: Methods, 2nd ed., SPIE Press, Bellingham, WA (2016).

40. R. Cicchi, D. Sampson, D. Massi, and F.S. Pavone, "Contrast and depth enhancement in two-photon microscopy of human skin ex vivo by use of optical clearing agents," *Opt. Express* **13**(7), 2337–2344 (2005).

41. S.V. Zaitsev, Y.I. Svenskaya, E.V. Lengert et al., "Optimized skin optical clearing for optical coherence tomography monitoring of encapsulated drug delivery through the hair follicles," *J. Biophotonics*, **13**(4), e201960020 (2020).

42. E.A. Genina, A.N. Bashkatov, O.V. Semyachkina-Glushkovskaya, and V.V. Tuchin, "Optical clearing of cranial bone by multicomponent immersion solutions and cerebral venous blood flow visualization," *Izv. Saratov Univ. (N.S.), Ser. Phys.*, **17**(2), 98–110 (2017).

43. H. Zheng, J. Wang, Q. Ye, et al., "Study on the refractive index matching effect of ultrasound on optical clearing of bio-tissues based on the derivative total reflection method," *Biomed. Opt. Express* **5**(10), 3482–3493 (2014).

44. J. Jiang, M. Boese, P. Turner, and R.K. Wang, "Penetration kinetics of dimethyl sulphoxide and glycerol in dynamic optical clearing of porcine skin tissue in vitro studied by Fourier transform infrared spectroscopic imaging," *J. Biomed. Opt.* **13**(2), 021105 (2008).

45. G. Vargas, E.K. Chan, J.K. Barton, H.G. Rylander, and A.J. Welch, "Use of an agent to reduce scattering in skin," *Laser Surg. Med.* **24**, 133–141 (1999).

46. G. Vargas, J.K. Barton, and A.J. Welch, "Use of hyperosmotic chemical agent to improve the laser treatment of cutaneous vascular lesions," *J. Biomed. Opt.* **13**(2), 021114 (2008).

47. E.A. Genina, A.N. Bashkatov, Yu.P. Sinichkin, and V.V. Tuchin, "Optical clearing of skin under action of glycerol: Ex vivo and in vivo investigations," *Opt. Spectrosc.* **109**(2), 225–231 (2010).

48. A.Y. Sdobnov, M.E. Darvin, J. Schleusener, J. Lademann, and V.V. Tuchin, "Hydrogen bound water profiles in the skin influenced by optical clearing molecular agents – quantitative analysis using confocal Raman microscopy," *J. Biophoton.* **12**(5), e201800283 (2019).

49. I.A. Nakhaeva, O.A. Zyuryukina, M.R. Mohammed, and Yu.P. Sinichkin, "The effect of external mechanical compression on in vivo water content in human skin," *Opt. Spectrosc.* **118**(5), 834–840 (2015).

50. C.W. Drew, C.G. Rylander, and ASME, "Mechanical compression for dehydration and optical clearing of skin," in: Proceedings of the ASME Summer Bioengineering Conference 2008, Pts a and B, 803–804 (2009).

51. A.A. Gurjarpadhye, W.C. Vogt, Y. Liu, and C.G. Rylander, "Effect of localized mechanical indentation on skin water content evaluated using OCT," *Int. J. Biomed. Imaging* **2011**, 817250 (2011).

52. Y. Tanaka, A. Kubota, M. Yamato, T. Okano, and K. Nishida, "Irreversible optical clearing of sclera by dehydration and cross-linking," *Biomaterials* **32**, 1080–1090 (2011).

53. A.N. Bashkatov, E.A. Genina, and V.V. Tuchin, "Measurement of glucose diffusion coefficients in human tissues," Chapter 19 in: V.V. Tuchin (ed.), *Handbook of Optical Sensing of Glucose in Biological Fluids and Tissues*, Taylor & Francis Group, LLC, CRC Press Boca Raton, FL (2009).

54. J.M. Schmitt, and G. Kumar, "Optical scattering properties of soft tissue: A discrete particle model," *Appl. Opt.* **37**(13), 2788–2797 (1998).

55. E.A. Genina, A.N. Bashkatov, and V.V. Tuchin, "Optical clearing of human dura mater under action of glucose solutions," *J. Biomed. Photon. Eng.* **3**(1), 010309 (2017).

56. L. Oliveira, M.I. Carvalho, E.M. Nogueira, and V.V. Tuchin, "The characteristic time of glucose diffusion measured for muscle tissue at optical clearing," *Laser Phys.* **23**, 075606 (2013).

57. E.A. Genina, A.N. Bashkatov, M.D. Kozintseva, and V.V. Tuchin, "OCT study of optical clearing of muscle tissue in vitro with 40% glucose solution," *Opt. Spectrosc.* **120**(1), 20–27 (2016).

58. H.R. Ueda, A. Erturk, K. Chung, et al., "Tissue clearing and its applications in neuroscience," *Nat. Rev. Neurosci.* **21**(2), 61–79 (2020).

59. T. Sekitani, T. Yokota, K. Kuribara, et al., "Ultraflexible organic amplifier with biocompatible gel electrodes," *Nat. Commun.* **7**, 11425 (2016).

60. K. Chung, J. Wallace, S.Y. Kim, et al., "Structural and molecular interrogation of intact biological systems," *Nature* **497**(7449), 332–337 (2013).

61. E.A. Susaki, K. Tainaka, D. Perrin, et al., "Whole-brain imaging with single-cell resolution using chemical cocktails and computational analysis," *Cell* **157**(3), 726–739 (2014).

62. P.S. Tsai, J.P. Kaufhold, P. Blinder, et al., "Correlations of neuronal and microvascular densities in murine cortex revealed by direct counting and colocalization of nuclei and vessels," *J. Neurosci.* **29**(46), 14553–14570 (2009).

63. M.T. Ke, S. Fujimoto, and T. Imai, "SeeDB: A simple and morphology-preserving optical clearing agent for neuronal circuit reconstruction," *Nat. Neurosci.* **16**(8), 1154–1161 (2013).

64. B. Hou, D. Zhang, S. Zhao, et al., "Scalable and DiI-compatible optical clearance of the mammalian brain," *Front. Neuroanat.* **9**, 19 (2015).

65. B. Yang, J.B. Treweek, R.P. Kulkarni, et al., "Single-cell phenotyping within transparent intact tissue through whole-body clearing," *Cell* **158**(4), 945–958 (2014).

66. M.N. Economo, N.G. Clack, L.D. Lavis, et al., "A platform for brain-wide imaging and reconstruction of individual neurons," *Elife* **5**, e10566 (2016).

67. M.T. Ke, Y. Nakai, S. Fujimoto, et al., "Super-resolution mapping of neuronal circuitry with an index-optimized clearing agent," *Cell Rep.* **14**(11), 2718–2732 (2016).

68. J.B. Treweek, K.Y. Chan, N.C. Flytzanis, et al., "Whole-body tissue stabilization and selective extractions via tissue–hydrogel hybrids for high-resolution intact circuit mapping and phenotyping," *Nat. Protoc.* **10**(11), 1860–1896 (2015).

69. W. Li, R.N. Germain, and M.Y. Gerner, "Multiplex, quantitative cellular analysis in large tissue volumes with clearing-enhanced 3D microscopy (Ce3D)," *Proc. Natl. Acad. Sci. U S A* **114**(35), E7321–E7330 (2017).

70. E. Murray, J.H. Cho, D. Goodwin, et al., "Simple, scalable proteomic imaging for high-dimensional profiling of intact systems," *Cell* **163**(6), 1500–1514 (2015).

71. S.K. Pal, J. Peon, and A.H. Zewail, "Biological water at the protein surface: Dynamical solvation probed directly with femtosecond resolution," *Proc. Natl. Acad. Sci. U S A* **99**(4), 1763–1768 (2002).

72. J.M. Hirshburg, K.M. Ravikumar, W. Hwang, and A.T. Yeh, "Molecular basis for optical clearing of collagenous tissues," *J. Biomed. Opt.* **15**(5), 055002 (2010).

73. G.D. Scott, E.D. Blum, A.D. Fryer, and D.B. Jacoby, "Tissue optical clearing, three-dimensional imaging, and computer morphometry in whole mouse lungs and human airways," *Am. J. Respir. Cell. Mol. Biol.* **51**(1), 43–55 (2014).

74. M.E. van Royen, E.I. Verhoef, C.F. Kweldam, et al., "Three-dimensional microscopic analysis of clinical prostate specimens," *Histopathology* **69**(6), 985–992 (2016).

75. H.-U. Dodt, U. Leischner, A. Schierloh, et al., "Ultramicroscopy: Three-dimensional visualization of neuronal networks in the whole mouse brain," *Nat. Methods* **4**(4), 331–336 (2007).

76. K. Becker, N. Jahrling, E.R. Kramer, F. Schnorrer, and H.-U. Dodt, "Ultramicroscopy: 3D reconstruction of large microscopical specimens," *J. Biophotonics* **1**(1), 36–42 (2008).

77. N. Jahrling, K. Becker, and H.-U. Dodt, "3D-reconstruction of blood vessels by ultramicroscopy," *Organogenesis* **5**(4), 227–230 (2009).

78. A. Klingberg, A. Hasenberg, I. Ludwig-Portugall, et al., "Fully automated evaluation of total glomerular number and capillary tuft size in nephritic kidneys using lightsheet microscopy," *J. Am. Soc. Nephrol.* **28**(2), 452–459 (2017).

79. W. Masselink, D. Reumann, P. Murawala, et al., "Broad applicability of a streamlined ethyl cinnamate-based clearing procedure," *Development* **146**(3), (2019).

80. M.K. Schwarz, A. Scherbarth, R. Sprengel, J. Engelhardt, P. Theer, and G. Giese, "Fluorescent-protein stabilization and high-resolution imaging of cleared, intact mouse brains," *PLoS One* **10**(5), e0124650 (2015).

81. C. Pan, R. Cai, F.P. Quacquarelli, et al., "Shrinkage-mediated imaging of entire organs and organisms using uDISCO," *Nat. Methods* **13**(10), 859–867 (2016).

82. D. Jing, S. Zhang, W. Luo, et al., "Tissue clearing of both hard and soft tissue organs with the PEGASOS method," *Cell Res.* **28**(8), 803–818 (2018).

83. N. Jährling, K. Becker, E.R. Kramer, and H.-U. Dodt, "3D-Visualization of nerve fiber bundles by ultramicroscopy," *Med. Laser Appl.* **23**(4), 209–215 (2008).

84. K. Becker, N. Jahrling, S. Saghafi, R. Weiler, and H.-U. Dodt, "Chemical clearing and dehydration of GFP expressing mouse brains," *PLoS One* **7**(3), e33916 (2012).

85. A. Erturk, C.P. Mauch, F. Hellal, et al., "Three-dimensional imaging of the unsectioned adult spinal cord to assess axon regeneration and glial responses after injury," *Nat. Med.* **18**(1), 166–171 (2012).

86. A. Erturk, K. Becker, N. Jahrling, et al., "Three-dimensional imaging of solvent-cleared organs using 3DISCO," *Nat. Protoc.* **7**(11), 1983–1995 (2012).

87. N. Renier, E.L. Adams, C. Kirst, et al., "Mapping of brain activity by automated volume analysis of immediate early genes," *Cell* **165**(7), 1789–1802 (2016).

88. I. Carneiro, S. Carvalho, R. Henrique, L. Oliveira, V.V. Tuchin, "A robust ex vivo method to evaluate the diffusion properties of agents in biological tissues," *J. Biophotonics* **12**(4), e201800333 (2019).

89. S. Carvalho, N. Gueiral, E. Nogueira, R. Henrique, L.M. Oliveira, and V.V. Tuchin, "Glucose diffusion in colorectal mucosa – a comparative study between normal and cancer tissues," *J. Biomed. Opt.* **22**(9), 091506 (2017).

90. D.K. Tuchina, V.D. Genin, A.N. Bashkatov, E.A. Genina, and V.V. Tuchin, "Optical clearing of skin tissue ex vivo with polyethylene glycol," *Opt. Spectrosc.* **120**(1), 28–37 (2016).

91. D.K. Tuchina, R. Shi, A.N. Bashkatov, et al., "Ex vivo optical measurements of glucose diffusion kinetics in native and diabetic mouse skin," *J. Biophotonics* **8**(4), 332–346 (2015).

92. V.D. Genin, E.A. Genina, S.V. Kapralov et al., "Optical clearing of the gastric mucosa using 40%-glucose solution," *J. Biomed. Photon. Eng.* **5**(3), 030302 (2019).

93. A.N. Bashkatov, E.A. Genina, Yu.P. Sinichkin, N.A. Lakodina, V.I. Kochubey, and V.V. Tuchin, "Estimation of the glucose diffusion coefficient in human eye sclera," *Biophysics*, **48**, 292–296 (2003).

94. A.V. Lykov, *Theory of Conduction of Heat*, High School, Moscow, RF (1967).

95. V.D. Genin, D.K. Tuchina, A.J. Sadeq, E.A. Genina, V.V. Tuchin, and A.N. Bashkatov "Ex vivo investigation of glycerol diffusion in skin tissue," *J. Biomed. Photon. Eng.* **2**(1), 010303 (2016).

96. F. Pirot, Y.N. Kalia, A.L. Stinchcomb, G. Keating, A. Bunge, and R.H. Guy, "Characterization of the permeability barrier of human skin in vivo," *Proc. Natl. Acad. Sci. U S A*, **94**, 1562–1567 (1997).

97. A.L. Stinchcomb, F. Pirot, G.D. Touraille, A.L. Bunge, and R.H. Guy, "Chemical uptake into human stratum corneum in vivo from volatile and non-volatile solvents," *Pharm. Res.* **16**, 1288–1293 (1999).

98. J.-C. Tsai, C.-Y. Lin, H.-M. Sheu, Y.-L. Lo, and Y.-H. Huang, "Noninvasive characterization of regional variation in drug transport into human stratum corneum in vivo," *Pharm. Res.* **20**, 632–638 (2003).

99. A.C. Watkinson, and K.R. Brain, "Basic mathematical principles in skin permeation," *J. Toxicol.-Cutan. Ocul. Toxicol.* **21**(4):371–402 (2002).

100. G.M. Prusakov, *Mathematical Models and Methods in PC Calculations*, Nauka, Moscow (1993).

101. M.J. Abdekhodaie, "Diffusion release of a solute from a spherical reservoir into a finite volume," *J. Pharmaceutical Sci.* **91**(8), 1803–1809 (2002).

102. I. Carneiro, S. Carvalho, R. Henrique, L. Oliveira, V.V. Tuchin, "Simple multimodal optical technique for evaluation of free/bound water and dispersion of human liver tissue," *J. Biomed. Opt.* **22**(12), 125002 (2017).

103. D.K. Tuchina, P.A. Timoshina, V.V. Tuchin, A.N. Bashkatov, and E.A. Genina, "Kinetics of rat skin optical clearing at topical application of 40% glucose: Ex vivo and in vivo studies," *IEEE J. Sel. Top. Quantum Electron.* **25**(1), 7200508 (2019).

104. V.D. Genin, A.N. Bashkatov, E.A. Genina, and V.V. Tuchin, "Measurement of diffusion coefficient of propylene glycol in skin tissue," *Proc. SPIE*, **9448**, 9448 0E (2015).

105. K. Schilling, V. Janve, Y. Gao, I. Stepniewska, B.A. Landman, and A.W. Anderson, "Comparison of 3D orientation distribution functions measured with confocal microscopy and diffusion MRI," *NeuroImage* **129**, 185–197 (2016).

106. V. Hovhannisyan, P.-S. Hu, S.-J. Chen, C.-S. Kim, C.-Y. Dong, Elucidation of the mechanisms of optical clearing in collagen tissue with multiphoton imaging. *J. Biomed. Opt.* **18**(4), 046004 (2013).

107. A. Kotyk, K. Janacek, *Membrane Transport: An interdisciplinary Approach*, Plenum Press, New York (1997).

108. L. Silvestri, I. Constantini, L. Scconi, F.S. Pavone, "Clearing of fixed tissue: A review from microscopist's perspective," *J. Biomed. Opt.* **21**(8), 081205 (2016).

109. C.-S. Choe, J. Lademann, M.E. Darvin, "Depth profiles of hydrogen bound water molecule types and their relation to lipid and protein interaction in the human stratum corneum in vivo," *Analyst* **141**(22), 6329–6337 (2016).

110. I. Carneiro, S. Carvalho, R. Henrique, L. Oliveira, and V.V. Tuchin, "Water content and scatterers dispersion evaluation in colorectal tissues," *J. Biomed. Phot. Eng.* **3**(4), 040301 (2017).

111. I. Carneiro, S. Carvalho, R. Henrique, L.M. Oliveira, V.V. Tuchin, "Optical properties of colorectal muscle in visible/NIR range," *Proc. SPIE* 10685, 106853D (2018).

112. A.N. Bashkatov, E.A. Genina, V.I. Kochubey, V.V. Tuchin, "Optical properties of human skin, subcutaneous and mucous tissues in the wavelength range from 400 to 2000 nm," *J. Phys. D Appl. Phys.* **38**(15), 2543–2555 (2005).

113. A.N. Bashkatov, E.A. Genina, V.I. Kochubey, V.S. Rubtsov, E.A. Kolesnikova, and V.V. Tuchin, "Optical properties of human colon tissues in the 350–2500 spectral range," *Quantum Electron.* **44**(8), 779–784 (2014).

114. S. Carvalho, N. Gueiral, E. Nogueira, R. Henrique, L. Oliveira, V.V. Tuchin, "Comparative study of the optical properties of colon mucosa and colon precancerous polyps between 400 and 1000 nm," *Proc. SPIE* **10063**, 100631L (2017).

115. A.N. Bashkatov, E.A. Genina, and V.V. Tuchin, "Tissue Optical Properties," Chapter 5 in D.A. Boas, C. Pitris, and N. Ramanujam (eds.), *Handbook of Biomedical Optics*, Taylor & Francis Group, LLC, CRC Press, Boca Raton, FL. (2011).

116. D.W. Leonard, and K.M. Meek, "Refractive indices of the collagen fibrils and extrafibrillar material of the corneal stroma," *Biophys. J.* **72**(3), 1382–1387 (1997).

117. K.M. Meek, S. Dennis, and S. Khan, "Changes in the refractive index of the stroma and its extrafibrillar matrix when the cornea swells," *Biophys. J.* **85**(4), 2205–2212 (2003).

118. K.M. Meek, D.W. Leonard, C.J. Connon, S. Dennis, and S. Khan, "Transparency, swelling and scarring in the corneal stroma," *Eye* **17**, 927–936 (2003).

119. O. Zernovaya, O. Sydoruk, V. Tuchin, and A. Douplik, "The refractive index of human hemoglobin in the visible range," *Phys. Med. Biol.* **56**, 4013–4021 (2011).

120. D.K. Tuchina, A.N. Bashkatov, E.A. Genina, and V.V. Tuchin, "Investigation of the impact of immersion agents on weight and geometric parameters of myocardial tissue *in vitro*," *Biophysics*, **63**(5), 791–797 (2018).

121. M.I. Ravich-Szherbo, and V.V. Novikov, *Physical and Colloid Chemistry*, Vysshaya Shkola, Moscow, RF (1975).

122. A.V. Rawlings and J.J. Leyden (eds.) *Skin Moisturization*, Taylor and Francis, London, (2009).

123. D. Zhu, J. Wang, Z. Zhi, X. Wen, and Q. Luo, "Imaging dermal blood flow through the intact rat skin with an optical clearing method," *J. Biomed. Opt.* **15**(2), 026008 (2010).

124. J. Wang, N. Ma, R. Shi, Y. Zhang, T.T. Yu, and D. Zhu, "Sugar-induced skin optical clearing: From molecular dynamics simulation to experimental demonstration," *IEEE J. Sel. Top. Quantum Electron.* **20**(2), 7101007 (2014).

125. W. Feng, R. Shi, N. Ma, D.K. Tuchina, V.V. Tuchin, and D. Zhu, "Skin optical clearing potential of disaccharides," *J. Biomed. Opt.* **21**(8), 081207 (2016).

126. S. Nagayama, S. Zeng, W. Xiong, et al., "In vivo simultaneous tracing and Ca(2+) imaging of local neuronal circuits," *Neuron* **53**(6), 789–803 (2007).

127. Y. Qi, T. Yu, J. Xu, et al., "FDISCO: Advanced solvent-based clearing method for imaging whole organs," *Sci. Adv.* **5**(1), eaau8355 (2019).

128. A.T. Yeh, and J. Hirshburg, "Molecular interactions of exogenous chemical agents with collagen – implications for tissue optical clearing," *J. Biomed. Opt.* **11**(1), (2006).

129. A.T. Yeh B. Choi, J.S. Nelson, and B.J. Tromberg, "Reversible dissociation of collagen in tissues," *J Invest Dermatol* **121**(6), 1332–1335 (2003).

130. X. Wen, Z. Mao, Z. Han, V.V. Tuchin, and D. Zhu, "In vivo skin optical clearing by glycerol solutions: Mechanism," *J. Biophotonics* **3**, 44–52 (2010).

131. J. Hirshburg, B. Choi, J.S. Nelson, and A.T. Yeh, "Collagen solubility correlates with skin optical clearing," *J. Biomed. Opt.* **11**(4), 040501 (2006).

132. N. Kuznetsova, S.L. Chi, and S. Leikin, "Sugars and polyols inhibit fibrillogenesis of type I collagen by disrupting hydrogen-bonded water bridges between the helices," *Biochemistry* **37**(34), 11888–11895 (1998).

133. K.M. Ravikumar, and W. Hwang, "Region-specific role of water in collagen unwinding and assembly," *Proteins* **72**(4), 1320–1332 (2008).

134. K.M. Ravikumar, J.D. Humphrey, and W. Hwang, "Spontaneous unwinding of a labile domain in a collagen triple helix," *J. Mech. Mater. Struct.* **2**(6), 999–1010 (2007).

135. J.M. Hirshburg, "Chemical agent induced reduction of skin light scattering," PhD diss., Texas A&M Univ. (2009).

136. K.V. Berezin, K.N. Dvoretski, M.L. Chernavina, et al., "Molecular modeling of immersion optical clearing of biological tissues," *J. Mol. Model.* **24**(2), 45 (2018).

137. A.N. Bashkatov, K.V. Berezin, K.N. Dvoretskiy, et al., "Measurement of tissue optical properties in the context of tissue optical clearing," *J. Biomed. Opt.* **23**(9), 091416 (2018).

138. E. Youn, T. Son, and B. Jung, "Determination of optimal glycerol concentration for optical tissue clearing," *Proc. SPIE* 8207, 82070J (2012).

139. D.S. Richardson, and J.W. Lichtman, "Clarifying tissue clearing," *Cell* **162**(2), 246–257 (2015).

140. H. Hama, H. Kurokawa, H. Kawano, et al., "Scale: A chemical approach for fluorescence imaging and reconstruction of transparent mouse brain," *Nat. Neurosci.* **14**(11), 1481–1488 (2011).

141. E.A. Susaki, and H.R. Ueda, "Whole-body and whole-organ clearing and imaging techniques with single-cell resolution: Toward organism-level systems biology in mammals," *Cell Chem. Biol.* **23**(1), 137–157 (2016).

142. S.I. Kubota, K. Takahashi, J. Nishida, et al., "Whole-body profiling of cancer metastasis with single-cell resolution," *Cell Rep.* **20**(1), 236–250 (2017).

143. K. Tainaka, T.C. Murakami, E.A. Susaki, et al., "Chemical landscape for tissue clearing based on hydrophilic reagents," *Cell Rep.* **24**(8), 2196–2210 e2199 (2018).

144. S. Zhao, M.I. Todorov, R. Cai, et al., "Cellular and molecular probing of intact human organs," *Cell* **180**(4), 796–812 e719 (2020).

145. E.A. Susaki, K. Tainaka, D. Perrin, H. Yukinaga, A. Kuno, and H.R. Ueda, "Advanced CUBIC protocols for whole-brain and whole-body clearing and imaging," *Nat. Protoc.* **10**(11), 1709–1727 (2015).

146. S.Y. Kim, J.H. Cho, E. Murray, et al., "Stochastic electrotransport selectively enhances the transport of highly electromobile molecules," *Proc. Natl. Acad. Sci. U S A* **112**(46), E6274–6283 (2015).

147. E. Lee, J. Choi, Y. Jo, et al., "ACT-PRESTO: Rapid and consistent tissue clearing and labeling method for 3-dimensional (3D) imaging," *Sci. Rep.* **6**, 18631 (2016).

148. Y.G. Park, C.H. Sohn, R. Chen, et al., "Protection of tissue physicochemical properties using polyfunctional crosslinkers," *Nat. Biotechnol.* **37**(1), 73–83 (2018).

149. D.H. Yun, Y.-G. Park, J.H. Cho, et al., "Ultrafast immunostaining of organ-scale tissues for scalable proteomic phenotyping," (2019). https://www.biorxiv.org/content/10.1101/660373v1

150. T. Yu, Y. Qi, J. Zhu, et al., "Elevated-temperature-induced acceleration of PACT clearing process of mouse brain tissue," *Sci. Rep.* **7**, 38848 (2017).

151. B. Hammouda, "Temperature effect on the nanostructure of SDS micelles in water," *J Res Natl. Inst. Stand. Technol.* **118**, 151–167 (2013).

152. T. Miyawaki, S. Morikawa, E.A. Susaki, et al., "Visualization and molecular characterization of whole-brain vascular networks with capillary resolution," *Nat. Commun.* **11**(1), 1104 (2020).

153. M. Ohmi, Y. Ohnishi, K. Yoden, and M. Haruna, "In vitro simultaneous measurement of refractive index and thickness of biological tissue by the low coherence interferometry," *IEEE Trans. Biomed. Eng.* **47**(9), 1266–1270 (2000).

154. G. Callis, and D. sterchi, "Decalcification of bone literature review and practical study of various decalcifying agents, methods, and their effects on bone histology," *J. Histotechnol.* **21**(1), 49–58 (1998).

155. S.A. Gomes, L.M. dos Reis, I.B. de Oliveira, I.L. Noronha, V. Jorgetti, and I.P. Heilberg, "Usefulness of a quick decalcification of bone sections embedded in methyl methacrylate[corrected]: An improved method for immunohistochemistry," *J. Bone. Miner. Metab.* **26**(1), 110–113 (2008).

156. F. Begum, W. Zhu, M.P. Namaka, and E.E. Frost, "A novel decalcification method for adult rodent bone for histological analysis of peripheral-central nervous system connections," *J. Neurosci. Methods* **187**(1), 59–66 (2010).

157. K. Sanjai, J. Kumarswamy, A. Patil, L. Papaiah, S. Jayaram, and L. Krishnan, "Evaluation and comparison of decalcification agents on the human teeth," *J. Oral Maxillofac. Pathol.* **16**(2), 222–227 (2012).

158. R. Sangeetha, K. Uma, and V. Chandavarkar, "Comparison of routine decalcification methods with microwave decalcification of bone and teeth," *J. Oral Maxillofac. Pathol.* **17**(3), 386–391 (2013).

159. S. Jimson, B.N.K.M.K. Masthan, and R. Elumalai, "A comparative study in bone decalcification using different decalcifying agents," *Int. J. Sci. Res.* **3**(8), 1226–1229 (2014).

160. J. Woo, M. Lee, J.M. Seo, H.S. Park, and Y.E. Cho, "Optimization of the optical transparency of rodent tissues by modified PACT-based passive clearing," *Exp. Mol. Med.* **48**(12), e274 (2016).

161. A. Greenbaum, K.Y. Chan, T. Dobreva, et al., "Bone CLARITY: Clearing, imaging, and computational analysis of osteoprogenitors within intact bone marrow," *Sci. Transl. Med.* **9**(387), (2017).

162. Y. Chen, S. Liu, H. Liu, et al., "Coherent raman scattering unravelling mechanisms underlying skull optical clearing for through-skull brain imaging," *Anal. Chem.* **91**(15), 9371–9375 (2019).

163. P. Jedlovszky, and A. Idrissi, "Hydration free energy difference of acetone, acetamide, and urea," *J. Chem. Phys.* **129**(16), 164501 (2008).

164. L. Hua, R. Zhou, D. Thirumalaic, and B.J. Berne, "Urea denaturation by stronger dispersion interactions with proteins than water implies a 2-stage unfolding," *Proc. Natl. Acad. Sci. U S A* **105**(44), 16928–16933 (2002).

165. U.D. Priyakumar, C. Hyeon, D. Thirumalai, and A.D. Mackerell, Jr., "Urea destabilizes RNA by forming stacking interactions and multiple hydrogen bonds with nucleic acid bases," *J. Am. Chem. Soc.* **131**(49), 17759–17761 (2009).

166. K.N. Barton, M.M. Buhr, and J.S. Ballantyne, "Effects of urea and trimethylamineN-oxide on fluidity of liposomes and membranes of an elasmobranch," *Am. J. Physiol.* **276**(2), R397–406 (1999).

167. Y. Feng, Z.-W. Yu, and P.J. Quinn, "Effect of urea, dimethylurea, and tetramethylurea on the phase behavior of dioleoylphosphatidylethanolamine," *Chem. Phys. Lipids* **114** 149–157 (2002).

168. H. Hama, H. Hioki, K. Namiki, et al., "ScaleS: An optical clearing palette for biological imaging," *Nat. Neurosci.* **18**(10), 1518–1529 (2015).

169. T.C. Murakami, T. Mano, S. Saikawa, et al., "A three-dimensional single-cell-resolution whole-brain atlas using CUBIC-X expansion microscopy and tissue clearing," *Nat. Neurosci.* **21**(4), 625–637 (2018).

170. T. Kuwajima, A.A. Sitko, P. Bhansali, C. Jurgens, W. Guido, and C. Mason, "ClearT: A detergent- and solvent-free clearing method for neuronal and non-neuronal tissue," *Development* **140**(6), 1364–1368 (2013).

171. C.H. Dapper, H.M. Valivullah, and T.W. Keenan, "Use of polar aprotic solvents to release membranes from milk lipid globules," *J. Dairy. Sci.* **70**(4), 760–765 (1987).

172. L. Chen, G. Li, Y. Li, et al., "UbasM: An effective balanced optical clearing method for intact biomedical imaging," *Sci. Rep.* **7**(1), 12218 (2017).

173. H.M. Lai, A.K.L. Liu, H.H.M. Ng, et al., "Next generation histology methods for three-dimensional imaging of fresh and archival human brain tissues," *Nat. Commun.* **9**(1), 1066 (2018).

174. R. Weissleder, "A clearer vision for in vivo imaging," *Nat. Biotechnol.* **19**, 316–317 (2001).

175. V.V. Tuchin, "Tissue optics and photonics: Light–tissue interaction," *J. Biomed. Photonics Eng.* **1**(2), 98–135 (2015).

176. S.A. Schnell, W.A. Staines, and M.W. Wessendorf, "Reduction of lipofuscin-like autofluorescence in fluorescently labeled tissue," *J. Histochem. Cytochem.* **47**(6), 719–730 (1999).

177. H. Lyon (Ed.), *Theory and Strategy in Histochemistry: A Guide to the Selection and Understanding of Techniques*, Springer-Verlag, Berlin (2011).

178. N. Renier, Z. Wu, D.J. Simon, J. Yang, P. Ariel, and M. Tessier-Lavigne, "iDISCO: A simple, rapid method to immunolabel large tissue samples for volume imaging," *Cell* **159**(4), 896–910 (2014).

179. H.G. Kristinsson, and H.O. Hultin, "Changes in trout hemoglobin conformations and solubility after exposure to acid and alkali pH," *J. Agric. Food. Chem.* **52**(11), 3633–3643 (2004).

2

Tissue optical clearing for Mueller matrix microscopy

Nan Zeng, Honghui He, Valery V. Tuchin, and Hui Ma

CONTENTS

Introduction

Optical imaging modalities in biomedicine offer insight into tissue structure (down to cellular or even subcellular level) and function (metabolic and compositional information, microvascular blood flow), such as absorption or scattering, which can be used to detect and differentiate tissue pathology [1, 2]. One of these optical characterization methods is polarimetry, which is the science of studying polarized light interaction with materials to infer information about their structure and composition.

Penetrating and propagating in a turbid medium, polarized light is multiply scattered and subsequently depolarized. The degree of depolarization significantly depends on the size and shape of the scatterers as well as the number of scattering events associated with the density and distribution of the scatterers in the medium. Numerous studies observing biological tissues using the polarization of backscattered light have been performed [3–5].

In the last few years, there have been significant developments in polarization technologies, computing, and widespread biomedical applications. In regard to the practical implication of the polarization method, its gating ability to select ballistic photons from diffuse ones gives rise to simplified schemes of optical medical tomography compared with time-resolved methods and provides enhanced image contrast and resolution, as well as additional information about tissue structure, absorption inclusions, and blood supply [6].

One particularly important tissue property, well suited for polarimetric measurements, is its microstructural anisotropy (asymmetry) [1]. Many diseases are associated with microstructural alterations, such as changes in collagen content and organization, muscular hypertrophy/atrophy, or perhaps cellular orientations. An emerging direction in recent years is the use of polarized light as a standalone modality for characterizing the structural properties of biological tissues.

Clinical interpretation of polarimetric images remains a challenge. Such interpretation would have to be based on a considerable investigation of the correlation of the polarimetric image contrast and its histological origins, which may involve physical interpretation of polarimetric data and modelling and simulation of polarized light propagation in complex-structured anisotropic tissues, assisted by dynamic observation of systematic control of tissue characteristics and statistical measures of clinical trials using polarimetric imaging [7].

Recent research works imply the potential of polarization features to characterize the substantial improvement of image quality and microphysical properties of tissues by elimination of multiple scattering effects at optical clearing (OC) [3, 6].

Mueller matrix imaging

Tissue scattering matrix

Polarization is a property that arises out of the transverse (and vector) nature of electromagnetic radiation and it describes the shape and the orientation of the locus of the electric field vector (E) extremity as a function of time at a given point of space. Note that in the corresponding quantum mechanical description, instead of the field vector, polarization is described for individual photons (energy quanta). It is assumed that individual photons are completely polarized, and accordingly their polarization is described by a state vector. The superposition of many such photon states yields the resulting polarization observed for the classical wave. Mathematical formalisms (in the classical approach) dealing with propagation of polarized light and its interaction with any optical system can be described by two formalisms: the Jones calculus, which is a field-based representation (assumes coherent addition of the amplitudes and phases of the waves); and the Stokes-Mueller calculus, which is an intensity-based representation (assumes an incoherent addition of wave intensities). A major drawback of the Jones formalism is that it deals with pure polarization states only and cannot handle partial polarizations and thus depolarizing interactions (which are common in biological tissues). Thus, the use of polarimetry in tissues can be better addressed by the Stokes-Mueller formalism [8].

In this formalism, the polarization state of the light beam is represented by four measurable quantities (intensities) grouped in a 4×1 vector, known as the Stokes vector. The four Stokes parameters are defined relative to the following six intensity measurements (I) performed with ideal polarizers: I_H, horizontal linear polarizer (0 deg); I_V, vertical linear polarizer (90 deg); I_P, 45 deg linear polarizer; I_M, 135 deg (−45 deg) linear polarizer; I_R, right circular polarizer, and I_L, left circular polarizer. The Stokes vector (S) is defined as

$$\mathbf{S} = \begin{bmatrix} I \\ Q \\ U \\ V \end{bmatrix} = \begin{bmatrix} I_H + I_V \\ I_H - I_V \\ I_P - I_M \\ I_R - I_L \end{bmatrix} \tag{2.1}$$

where I, Q, U, and V are Stokes vector elements. I is the total detected light intensity which corresponds to addition of the two orthogonal component intensities, Q is the difference in intensity between horizontal and vertical polarization states, U is the portion of the intensity that corresponds to the difference between intensities of linear +45 deg and −45 deg polarization states, and V is the difference between intensities of right circular and left circular polarization states.

Then we can define the degree of polarization by taking the ratio of the polarized components to the total intensity of light using the Stokes parameters as Equation 2.2. The degree of polarization ranges from 0 (unpolarized light) to 1 (pure polarized light). A value between 0 and 1 describes a partially polarized light.

$$P = \frac{\sqrt{Q^2 + U^2 + V^2}}{I} \tag{2.2}$$

When we describe the polarization states of both the incident and scattered light of a medium using Stokes vectors, the polarization properties of the medium can be represented by the 4×4 Mueller matrix as Equation 2.3. Since most biomedical tissue samples are depolarizing, the Stokes-Mueller formalism is a suitable tool for tissue polarimetry.

$$S_{out} = M \bullet S_{in}$$

$$\begin{pmatrix} I_{out} \\ Q_{out} \\ U_{out} \\ V_{out} \end{pmatrix} = \begin{pmatrix} m_{11} & m_{12} & m_{13} & m_{14} \\ m_{21} & m_{22} & m_{23} & m_{24} \\ m_{31} & m_{32} & m_{33} & m_{34} \\ m_{41} & m_{42} & m_{43} & m_{44} \end{pmatrix} \begin{pmatrix} I_{in} \\ Q_{in} \\ U_{in} \\ V_{in} \end{pmatrix} \tag{2.3}$$

The polarization of the scattered light from an object can be fully described by the Stokes vectors of the incident and scattered light and Mueller matrix. The matrix elements depend on the scattering angle θ, the wavelength, the shape, and the optical properties of the object. For scattering by a collection of randomly oriented particles, there are ten independent elements. The matrix for macroscopically isotropic and symmetric media has a well-known block-diagonal structure [9].

$$M(\theta) = \begin{bmatrix} M_{11}(\theta) & M_{12}(\theta) & 0 & 0 \\ M_{12}(\theta) & M_{22}(\theta) & 0 & 0 \\ 0 & 0 & M_{33}(\theta) & M_{34}(\theta) \\ 0 & 0 & -M_{34}(\theta) & M_{44}(\theta) \end{bmatrix} \tag{2.4}$$

The scattering matrix can be simplified for spherically symmetric particles, which are homogeneous or radially inhomogeneous, because in this case,

$$M_{11}(\theta) \equiv M_{22}(\theta), M_{33}(\theta) \equiv M_{44}(\theta) \tag{2.5}$$

Mie theory is an exact solution to Maxwell's electromagnetic field equations for a homogeneous sphere. Mie theory also can be extended to arbitrary coated spheres and to arbitrary cylinders.

Many imaging results show that the scattering depolarization characteristics of the samples are mainly manifested in the diagonal elements m_{11}, m_{22}, m_{33}, and m_{44}. The values of diagonal elements of the nondepolarizing medium are 1, and depolarization will reduce these elements' values. In addition, most tissues contain anisotropic structures such as collagen fiber and muscle fiber, which will cause changes in dichroism and birefringence. The anisotropy from the scattering structure of the sample is reflected in the 3×3 submatrix elements in the upper left corner.

If there is a positive or negative value in m_{12}, m_{13}, m_{21}, and m_{31}, it indicates that the medium has dichroism caused by scattering structure. The anisotropy from the birefringence of the sample is reflected in the 3×3 submatrix elements in the lower right corner. If the values of m_{24}, m_{34}, m_{42} and m_{43} are positive or negative, it indicates that there is retardation caused by birefringence in the medium.

Here is an example to roughly illustrate the capability of the Mueller matrixes to reflect tissue types. The backscattering Mueller matrix imaging results of four kinds of bulk tissues with typical structural characteristics (chicken heart muscle, bovine skeletal muscle, porcine liver, and porcine fat tissue) are shown in Figure 2.1. Mueller matrixes of chicken heart and bovine skeletal muscle are nondiagonal, and there are differences between m_{22} and m_{33}, which indicate that obvious anisotropy existed in the samples. In addition, from the distribution characteristics of m_{12}, m_{13}, m_{21}, and m_{31} values, it can be further seen that there are obvious differences in fiber arrangement patterns between chicken heart muscle tissue and bovine

skeletal muscle tissue, which lead to changes in dichroic distribution, and parameters can be quantitatively characterized by the Mueller matrix decomposition or transformation method mentioned later.

Compared with them, porcine liver and porcine fat tissue have approximately equal values of diagonal elements, indicating that the two tissues are isotropic, and further, the depolarization of porcine fat tissue is stronger. For the porcine liver tissue, there is a grid structure of m_{24}, m_{34}, m_{42}, and m_{43} elements, which indicates that the connective tissue around the isotropic liver lobule tissue has obvious birefringence.

Mueller imaging device and principle

Measurement of Mueller matrix

The simple optical scheme to monitor tissue changes with OC is based on two parallel or crossed polarizers. One linear polarizer is mounted on the light source and another is placed in front of the camera and provide either parallel or perpendicular orientation to the direction of polarization of the illuminating light. A calculation of the degree of polarization for each pixel of the polarization image gives a degree of polarization imaging (DPI). For thin sections, the above scheme can be developed to a polarized light microscopy, which can extract fiber orientation with μm-scale resolution over a broad field of view.

The most informative description of two-dimensional polarization patterns on light transmission and scattering in tissues is the Mueller matrix. For turbid tissue samples, when we record the polarization states of both the incident and scattered light, we can derive a 4×4 Mueller matrix to encode all the polarization related structural properties of the tissue samples. Mueller matrix polarimetry can be achieved by adding the polarization states generator (PSG) and analyzer (PSA) to the optical paths of existing nonpolarization techniques. It means that traditional optical devices such as microscopes and endoscopes can be upgraded to fulfill Mueller matrix imaging and measurement abilities to provide more structural information of the samples, particularly changes of subwavelength microstructures. As a label-free and noninvasive tool, Mueller matrix polarimetry certainly has broad application prospects in biomedical studies and clinical diagnosis [10].

A Mueller matrix is an information-rich tissue signature that contains the full polarization information reflecting its biophysical properties. There are several different Mueller matrix measurement schemes. For instance, for measuring the 4×4 Mueller matrix, six polarization states for the incident light can be achieved by rotating both the polarizers and retarders in PSG, i.e., horizontal linear (H), vertical linear (V), 45° linear (P), 135° linear (M), left circular (L), and right circular (R); then six polarization components corresponding to each incident state are measured by rotating PSA. After a total of 36 measurements are conducted, the Mueller matrix of a medium can be calculated as:

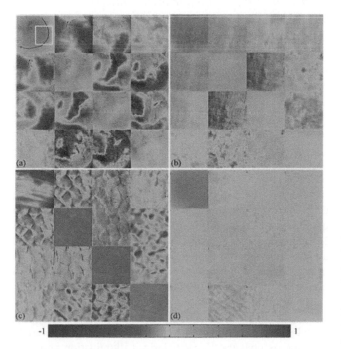

FIGURE 2.1 Backscattering Mueller matrices of biological samples [29].

$$M = \begin{pmatrix} m11 & m12 & m13 & m14 \\ m21 & m22 & m23 & m24 \\ m31 & m32 & m33 & m34 \\ m41 & m42 & m43 & m44 \end{pmatrix}$$

$$= \frac{1}{2} \begin{pmatrix} HH+HV+VH+VV & HH+HV-VH-VV & PH+PV-MP-MM & RH+RV-LH-LV \\ HH-HV+VH-VV & HH-HV-VH+VV & PH-PV-MH+MV & RH-RV-LH+LV \\ HP-HM+VP-VM & HP-HM-VP+VM & PP-PM-MP+MM & RP-RM-LP+LM \\ HR-LL+VR-RL & HR-VR+VL-HL & PR-MR+ML-PL & RR-RL-LR+LL \end{pmatrix}$$

(2.6)

In general, four independent polarization state measurements are necessary to obtain one element of the Mueller matrix. A Mueller matrix polarimeter based on polarization modulation of the incident and scattered light by mechanically rotating phase plates has been described [3]. The polarization modulation detection approach has been used for both Stokes vector and Mueller matrix measurements in complex tissue-like turbid media and in tissues. Among the various other modulation-based Mueller matrix polarimeters, the dual rotating retarder approach has been widely used in tissue polarimetry investigations.

In this scheme, polarization modulation of the incident state is generated by passing light first through a fixed linear polarizer and then through a rotating linear retarder (retardation δ_1) with angular speed ω_1. The analyzing optics contains another rotating retarder (retardation δ_2, synchronously rotating at angular speed ω_2) and a fixed linear polarizer. In the usual configuration, the retardation values of the two retarders are chosen to be the same, $\delta_1 = \delta_2 = \pi/2$; the axis of the polarizer and the analyzer are kept parallel; and the angular rotation speeds of the retarders are kept as $\omega_1 = \omega$ and $\omega_2 = 5\omega$, respectively.

The rotation of the retarders at these different rates results in a modulation of the detected intensity signal, as can be understood by sequentially writing the Mueller matrices corresponding to each optical element (polarizers, retarders, and the sample). Note that for this specific scheme, the modulation in the detected intensity arises due to harmonic variation of the orientation angle of the two retarders kept at the polarizing and analyzing end of the polarimeter (θ and 5θ, respectively). It has been shown that the five to one ratio of angular rotation speeds of the two retarders encodes all 16 Mueller matrix elements onto the amplitude and phases of 12 frequencies in the detected intensity signal. The detected signal is Fourier analyzed, and the Mueller matrix elements are constructed from the Fourier coefficients.

As the phase plates (QWPs) are rotated, the intensity recorded by a photodetector would depend on time. By performing the appropriate trigonometric transformations, one can find that the output intensity would be represented as a Fourier series, namely

$$I = \alpha_0 + \sum_{n=1}^{12} \left(\alpha_n \cos 2n\theta_1 + \beta_n \sin 2n\theta_1 \right) \qquad (2.7)$$

where α_n and β_n are the Fourier coefficients, and θ_1 is the rotation angle of the retarder 1 for the PSG. The coefficients of this series are defined by the values of the matrix M elements of the sample under study, and their measurement ensures a system of linear equations to determine the matrix M.

For the calibration of Mueller matrix polarimeter, some standard samples including air, a quarter-wave plate, and a polarizer can be used. For instance, for the transmission Mueller matrix microscope, we compensated the errors due to nonideal quarter-wave plates and polarizers. The experimental results testified that the maximum errors for the absolute values of the Mueller matrix elements are about 0.01. By simultaneously solving the above equations, we know that the Fourier amplitudes provide linear algebraic equations of the 16 unknown Mueller-matrix elements.

The rotation rates can also have other different values. More details on this Mueller matrix imaging method can be found in references [11, 12]. The optimal ratio between rotation rates of the phase plates is 1:5, allows one to get an optimally stipulated system of linear equations to find the full matrix M. Fast electro-optic modulation of the polarization state has been also used in Mueller matrix measurements.

Transmitted light Mueller matrix microscopes

An advantage of Mueller matrix polarimetry is that it can be upgraded from a nonpolarization optical technique by adding the PSG and PSA to the existing optical paths appropriately. Using modulus design, the PSG and PSA can be combined with commercial microscopes to obtain polarization images and quantitative Mueller matrix parameters of the samples.

Currently, the gold standard of abnormal tissues detection, including cancer diagnosis, is pathological observations of tissue slices using optical transmission microscopes. The process requires the specimen to be stained with certain dyes, and the observations of experienced pathologists. However, the staining process, especially immunohistochemical staining, can be time-consuming, and the pathological observations are often qualitative. As a label-free and quantitative method, in 1953 Inoue introduced a polarizer, a birefringent compensator, and an analyzer to an ordinary microscope to develop a prototype of a polarization microscope, which has enough sensitivity to image the spindle fibers of living cells in mitosis [13].

The application studies on biomedical and material specimens showed that the polarization microscope is a useful tool to reveal the optical anisotropy of samples. Traditional polarization microscopes use linear polarized illumination produced by a polarizer, and observe in cross- or parallel-polarized light mode using another analyzing polarizer qualitatively, which

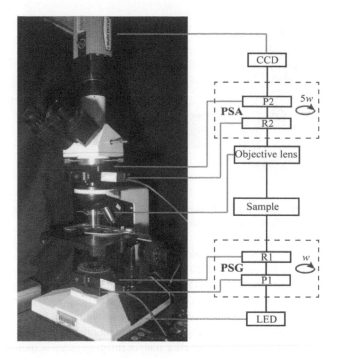

FIGURE 2.2 Photograph and schematic of the modulus designed Mueller matrix microscope [19].

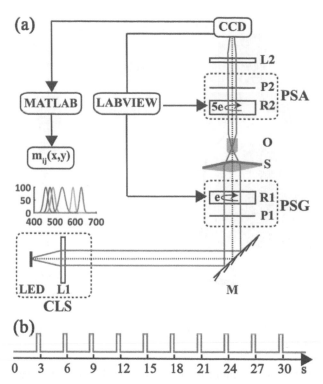

FIGURE 2.3 Schematic diagram of the multi-wavelength modulus designed Mueller matrix microscope [20].

can provide additional structural information of anisotropic components in cells [14]. However, when applied to specimens with depolarization such as tissues, the optical anisotropy information obtained can be quantitatively affected. It means that we should measure the Mueller matrix and get the accurate microstructural information of the sample.

Several Mueller matrix microscopes have been used in the past few years [15, 16]. In 2015, a modulus designed Mueller matrix microscope, shown in Figure 2.2, was reported by Wang et al.[16]. A commercial low-cost transmission light microscope was used as the main body. The light source module of the microscope was replaced by a new one to provide parallel LED beam. Compact PSG and PSA modules were designed for easy implementation and alignment with minimal disturbance to the existing optical path of the microscope. The modulus-designed Mueller matrix microscope was then applied to the detections of various pathological human tissue samples, including liver cirrhosis and cancer [17], cervical cancer [16], thyroid cancer [18], and breast ductal carcinoma tissues [19].

The preliminary imaging results demonstrated two merits of this microscope: first, as a label-free technique, the Mueller matrix microscope can provide different "polarization-staining" images of the specimen to highlight different structures without any interferences to the tissues; second, by simply adding the modules, the optical microscope can be upgraded to a quantitative tool to provide a lot more structural information of the sample. Recently, to obtain more structural information of tissue specimens, the monochrome red-light Mueller matrix microscope was upgraded to a multiwavelength version by redesigning the light source, using achromatic polarized optical elements, and taking extra cautions in calibrations, as shown in Figure 2.3 [20]. Comparisons between the Mueller matrix elements at different wavelengths may provide more detailed information helpful for biomedical diagnosis.

The polarization microscopes based on liquid crystal retarders or rotating wave plates for PSG and PSA are suitable for taking the Mueller matrix images of stationary samples such as pathological tissue slices. For dynamic samples or processes, the division of the focal plane (DoFP) polarimeters were introduced to the microscope [21]. For the DoFP polarimeter, an array consisting of numerous differently oriented pixel-size micropolarizers is fixed on the imaging sensor. Several kinds of DoFP polarimeter incorporated microscopes for simultaneous Stokes vector imaging have been reported [22, 23]. These DoFP polarimeters have the same instrument size as the conventional image sensor and can capture the Stokes vector images in a single shot.

A DoFP-based polarization microscope shown in Figure 2.4 was proposed by Chang et al., which can measure 3×4 Mueller matrix of the sample [21]. Considering that the m_{24} and m_{34} elements contain important structural information for the birefringence-dominant samples, right- or left-hand circularly polarized illumination light is used in this microscope, and the information can be obtained in a single shot. Some preliminary applications showed that the Mueller matrix microscope can provide information about structures smaller than the diffraction limit. Besides, circularly polarized light illumination in this equipment can avoid the orientation influence of the linearly or elliptically polarized light illuminations on the anisotropic fibrous microstructures existing in biological tissues [21]. Recently, a full polarization DoFP camera was developed to simultaneously detect linear polarized light of different orientations, or even circularly polarized light, and achieve real-time imaging. DoFP polarization camera can also be used in the backscattering setup, such as endoscope, to realize Mueller matrix *in vivo* imaging during surgery. With the

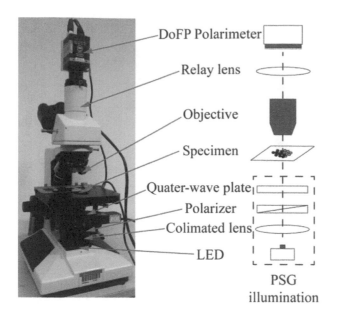

FIGURE 2.4 Photograph and configuration of the DoFP polarimeter based polarization microscope [21].

rapid improvement of hardware, the DoFP polarization camera is developing towards miniaturization.

Backscattering Mueller matrix imaging and microscopes

The transmission Mueller matrix microscopes can be potentially label-free tools for *ex vivo* imaging of tissue slices and cells. For bulk tissues, the backscattering Mueller matrix imaging devices are needed. In 1990s–2000s, some simple backscattering polarization imaging devices were developed for biomedical detection of superficial skin tissues. Anderson et al. applied the difference polarization method to the examination and photography of skin [24]; the collected backscattering light includes polarization component parallel with the illumination I_{co} and perpendicular to the illumination I_{cr}. The difference polarization DP and the normalized parameter by illuminating intensity, DOP, can be calculated as:

$$DP = I_{co} - I_{cr} = I(\theta_i, \theta_i) - I(\theta_i, \theta_i + \pi/2), \quad (2.8)$$

$$DOP = \frac{I_{co} - I_{cr}}{I_{co} + I_{cr}} = \frac{I(\theta_i, \theta_i) - I(\theta_i, \theta_i + \pi/2)}{I(\theta_i, \theta_i) + I(\theta_i, \theta_i + \pi/2)}. \quad (2.9)$$

Jacques et al. applied DOP to the detections of human skin basal cell carcinoma, squamous cell carcinoma tissues, and so on [25, 26]. The DOP imaging results showed good contrast between normal and abnormal areas to help decide the margins of cancerous tissues. However, the methods using linear polarized light partially reflect the structural properties of the sample and can be seriously affected by the orientation changes of fibrous structures in tissues, hindering their diagnostic applications.

Traditionally there are two different kinds of illumination for backscattering Mueller matrix imaging set-ups: point light source illumination using a pencil beam, shown in

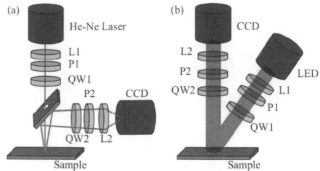

FIGURE 2.5 Setup for the backscattering Mueller matrix measurement with point light source and wide field illumination [10].

Figure 2.5(a); and wide field illumination using LED or lamps, shown in Figure 2.5(b) [10]. The set-up using point light illumination measures the 2D backscattering distribution patterns of Mueller matrix elements, which can provide the polarization and scattering properties of homogenous turbid media [27].

For most biomedical samples which are inhomogeneous, the wide-field illumination Mueller matrix imaging setup is more suitable.

The wide-field illumination Mueller matrix can be regarded as an accumulation of many point light source Mueller matrix patterns; therefore both the scattering and polarization information contained in 2D Mueller matrix patterns can also be retrieved from the wide-field Mueller matrix images. By analysis of these backscattering Mueller matrix patterns, combined with tissue models, we can better understand the relationships between Mueller matrix elements distributions and microstructures of turbid media. The backscattering Mueller matrix imaging equipment was applied to *ex vivo* detections of various abnormal tissues [5, 18, 28, 29].

It should be noted that for the backscattering Mueller matrix imaging, one of the ways to suppress strong light scattering and to get better signal-to-noise ratio in quantification of polarization properties of a tissue is the application of the immersion OC technique. A large amount of information about the microstructures of the tissues can be acquired from Mueller matrix elements using the OC technique [30].

The snapshot Mueller matrix polarimeter is another important development in modulation-based Mueller matrix polarimetry. The approach exploits wavelength polarization coding and decoding for high sensitivity and instantaneous measurement of all 16 Mueller matrix elements simultaneously [31]. In addition, polarimetric imaging systems based on liquid crystal variable retarders enable measurement of the Mueller matrix elements with higher sensitivity and precision [32].

Matrix feature extraction

Mueller matrix elements

As the Mueller matrix is a comprehensive description of polarization property and contains abundant structural information about the sample, it can be used as a potential tool for biomedical studies and medical practices. Information about tissue structure can be extracted from the registered depolarization degree of initially polarized light, the transformation of the polarization state, or the appearance of a polarized component in the scattered light [6].

The Mueller matrix imaging patterns for a point incident polarized light are constructed from polarization measurements for light emerging from different spatially separated points in the tissue sample at its illumination with a narrow laser beam. These patterns can be used to evaluate the scattering coefficient, the anisotropy factor, the mean particle size, and so on.

Sometimes by a few characteristic Mueller elements, we can get the interpretation of the measured polarization properties for tissue sections or dilute cell suspensions more or less. In general, as a comprehensive description of all the polarization related properties of the medium, the Mueller matrix can provide abundant microstructural information on both bulk tissues and thin slices of tissues. However, the dominant structural properties of bulk and thin tissue samples are different [10].

For thin tissue slices, their depolarization property is usually very limited. The illuminating polarized photons transmit the sample, meaning that a transmission Mueller matrix microscope is an appropriate choice. A lot of studies have shown that for thin tissue samples and cells, the birefringence induced by fibrous structures and internal force is the most prominent feature to bring the polarimetric imaging contrast. The appearance of the off-diagonal elements m_{24}, m_{34}, m_{42}, and m_{43} indicate the existence of birefringence whose values and orientations can be obtained using these elements.

Bulk tissues are usually highly depolarizing with complex microstructures, and backscattering Mueller matrix imaging equipment is often used. Also, most bulk tissues contain fibrous structures such as collagen, elastic, and muscle fibers, causing both anisotropic scattering and birefringence effects. The main characteristics of the Mueller matrix for a depolarization dominant medium can be summarized as follows [29, 33]: (1) The diagonal Mueller matrix elements indicate the depolarization ability of the sample. Depolarizing can reduce the values of m_{22}, m_{33}, and m_{44} elements. A comparison between linear and circular depolarization can be used to reveal the approximate size of the scatterers. (2) For an isotropic scattering medium, the Mueller matrix should be diagonal. The appearance of the off-diagonal elements m_{12}, m_{13}, m_{21}, and m_{31} indicate the existence of diattenuation, often by fibrous scatterers, and their values can be used to calculate the density and orientation distribution of fibers. (3) The birefringence induced anisotropy is mainly reflected in the lower right 3×3 elements. The appearance of the off-diagonal elements m_{24}, m_{34}, m_{42}, and m_{43} indicate the existence of birefringence, and they can be used to calculate the value and orientation of retardance.

However, for thick turbid tissues where polarization alteration can be due to multiple scattering effects, physical explanations of Mueller matrix measurements become complicated, and the individual Mueller matrix elements tell only how the polarization states vary during the light–tissue interaction, but cannot reveal explicit association with specific microstructures. Therefore, it is not easy to use images of the Mueller matrix elements directly, due to the lack of explicit relations to specific microstructural properties and Mueller matrix

elements seriously affected by the orientations of anisotropic structures [34, 35].

Characterization parameters derived from the Mueller matrix

Hence, for biomedical applications, some new parameters derived from Mueller matrix elements are needed to present clearer relationships to certain microstructures and to separate the microstructure-sensitive features from the azimuthal orientation dependence [10]. Recently, several approaches to transform Mueller matrix elements to quantitative parameters with explicit physics meanings have been proposed.

Mueller matrix decomposition

It is important to interpret Mueller matrices in terms of fundamental polarization properties with an intuitive physical explanation to 1) facilitate investigation and understanding of biophysical origin of the polarimetric signals, and 2) obtain azimuthal orientation angle independent parameters to maintain the reproducibility of results [7]. Several methods, including Lu–Chipman's polar decomposition [36], reverse polar decomposition [37], differential decomposition [38], and Mueller matrix transformation techniques [39] have been developed.

Tissue intrinsic polarization properties, including tissue depolarization, birefringence, and diattenuation, can convey morphological, microstructural and compositional information of tissue with great potential for label-free characterization of tissue pathological changes [7].

To extract tissue structural information in the presence of all these factors affecting the depolarization, retardance, and diattenuation is a difficult problem. Polar decomposition of the Mueller matrix [40] is an efficient way to quantify tissue polarization properties from the measured Mueller matrix, which is presented as a combination of a diattenuator $\mathbf{M}_D$, a retarder $\mathbf{M}_R$, and a depolarizer $\mathbf{M}_\Delta$ matrices. We can separate the effects of light scattering, birefringence, and dichroism on the tissue polarization properties by MMD method.

The decomposition of the Mueller matrix depends upon the order in which the diattenuator, depolarizer, and retarder matrices are multiplied. Based on the order of these matrices, six possible decompositions can be performed. Among these, the process in which the diattenuator matrix comes ahead of the retardance and the depolarization matrix $[\mathbf{M}=\mathbf{M}_\Delta \bullet \mathbf{M}_R \bullet \mathbf{M}_D]$ always leads to a physically realizable Mueller matrix [41]. The depolarizing matrix $\mathbf{M}_\Delta$ accounts for the depolarizing effects of the medium, the retarder matrix $\mathbf{M}_R$ describes the effects of linear birefringence and optical activity, and the diattenuator matrix $\mathbf{M}_D$ includes the effects of linear and circular dichroism. We can calculate directly diattenuation D from the Mueller matrix $\mathbf{M}$ as

$$D = \frac{\left(m_{01}{}^2 + m_{02}{}^2 + m_{03}{}^2 \right)^{1/2}}{m_{00}}. \qquad (2.10)$$

From the Mueller matrix $\mathbf{M}$, first we can construct a diattenuator Mueller matrix by taking diattenuation vector $\vec{D}$ as

$$\overline{D} = \frac{1}{m_{00}} \begin{pmatrix} m_{01} \\ m_{02} \\ m_{03} \end{pmatrix}. \tag{2.11}$$

Thus, the first row of $\mathbf{M}$ gives the diattenuation vector. Then from this diattenuation vector, the diattenuator Mueller matrix can be constructed as

$$\mathbf{M}_D = \begin{pmatrix} 1 & \overline{D}^T \\ \overline{D} & \mathbf{m}_D \end{pmatrix}, \tag{2.12}$$

$$\mathbf{m}_D = \sqrt{1-D^2}\,\mathbf{I} + \left(1 - \sqrt{1-D^2}\right)\hat{D}\hat{D}^T. \tag{2.13}$$

Further, a Mueller matrix $\mathbf{M}'$ is defined based on $\mathbf{M}$, as

$$\mathbf{M}' = \mathbf{M}\mathbf{M}_D^{-1}. \tag{2.14}$$

This $\mathbf{M}'$ contains only retardance and depolarization and no diattenuation. $\mathbf{M}'$ can be further decomposed as a retarder followed by a depolarizer

$$\mathbf{M}_\Delta \mathbf{M}_R = \begin{pmatrix} 1 & \vec{0} \\ \vec{P} & \mathbf{m}_\Delta \end{pmatrix} \begin{pmatrix} 1 & \vec{0} \\ \vec{0} & \mathbf{m}_R \end{pmatrix}$$

$$= \begin{pmatrix} 1 & \vec{0} \\ \vec{P}_\Delta & \mathbf{m}_\Delta \mathbf{m}_R \end{pmatrix} = \begin{pmatrix} 1 & \vec{0} \\ \vec{P}_\Delta & \mathbf{m}' \end{pmatrix}, \tag{2.15}$$

$$= \mathbf{M}'$$

$$\vec{P}_\Delta = \frac{\vec{P} - \mathbf{m}\vec{D}}{1-D^2}, \tag{2.16}$$

$$\mathbf{m}' = \mathbf{m}_\Delta \mathbf{m}_R, \tag{2.17}$$

where polarizance vector $\vec{P}$ can be expressed in terms of Mueller matrix elements as follows:

$$\vec{P} = \frac{1}{m_{00}} \begin{pmatrix} m_{10} \\ m_{20} \\ m_{30} \end{pmatrix}. \tag{2.18}$$

Let λ_1, λ_2, and λ_3 be the eigen values of $\mathbf{m}'(\mathbf{m}')^T$; $\mathbf{m}_\Delta$ has eigen values λ_1, λ_2, and λ_3; and $\mathbf{m}_\Delta$ can be obtained by

$$\mathbf{m}_\Delta = \pm \left[\mathbf{m}'\left(\mathbf{m}'\right)^T + \left(\lambda_1\lambda_2 + \lambda_2\lambda_3 + \lambda_3\lambda_1\right)^{1/2}\mathbf{I} \right]^{-1}$$

$$\times \left[\left\{ \left(\lambda_1\right)^{1/2} + \left(\lambda_2\right)^{1/2} + \left(\lambda_3\right)^{1/2} \right\} \mathbf{m}'\left(\mathbf{m}'\right)^T + \left(\lambda_1\lambda_2\lambda_3\right)^{1/2}\mathbf{I} \right]. \tag{2.19}$$

If determinant of $\mathbf{m}'$ is negative then minus sign is applied. Thus, $\mathbf{M}_\Delta$ can be determined. Once $\mathbf{M}_\Delta$ is determined, then we can evaluate depolarization power Δ as

$$\Delta = 1 - \frac{\left|\mathrm{tr}\left(\mathbf{M}_\Delta\right) - 1\right|}{3}. \tag{2.20}$$

Now $\mathbf{M}_R$ can be obtained by

$$\mathbf{M}_R = \mathbf{M}_\Delta^{-1}\mathbf{M}'. \tag{2.21}$$

From the retardance Mueller matrix, the retardance can be obtained by

$$R = \cos^{-1}\left[\frac{\mathrm{tr}\left(\mathbf{M}_R\right)}{2} - 1\right]. \tag{2.22}$$

In particular, the linear retardation δ and circular retardation Ψ can be separated from retardation matrix $\mathbf{M}_R$. The value and orientation of linear retardation, which are important indicators in biomedical studies, are given as [10]:

$$\delta = \cos^{-1}\left\{ \left[\left(\mathbf{M}_R(2,2) + \mathbf{M}_R(3,3)\right)^2 + \left(\mathbf{M}_R(3,2) - \mathbf{M}_R(2,3)\right)^2 \right]^{1/2} - 1 \right\} \tag{2.23}$$

$$\theta = 0.5\tan^{-1}(r_2/r_1). \tag{2.24}$$

$$r_i = (1/2\sin\delta) \times \sum_{j,k=1}^{3} \varepsilon_{ijk}\,\mathbf{m}_{LR}(j,k). \tag{2.25}$$

Mueller matrix transformation

A Mueller matrix transformation (MMT) method has been proposed to derive a set of new quantitative parameters [39] for describing the microstructural features of complex tissue more clearly by the measured Mueller matrix elements. By fitting the Mueller matrix elements into certain trigonometric functions, the MMT method can provide a set of parameters which is insensitive to the azimuth angle for the backscattering imaging of bulk tissues or transmission imaging of thin tissues with limited scattering, shown as:

$$\begin{cases} m22 = t_1\cos 4x + b \\ m33 = -t_1\cos 4x + b \\ m23 = m32 = t_1\sin 4x \\ m12 = m21 = 2t_2\cos 2x \\ m13 = m31 = 2t_2\sin 2x \end{cases}. \tag{2.26}$$

These derived new polarization parameters are functions of the Mueller matrix elements but are explicitly correlated to specific microstructures or the optical properties of the medium, such as the densities and sizes of subwavelength scatterers or the orientation and alignment of the fibers [39]. The MMT parameters b, x, t_1, t_2 include

$$t_1 = \frac{\sqrt{(m22 - m33)^2 + (m23 + m32)^2}}{2}$$

$$t_2 = \frac{\sqrt{(m21)^2 + (m31)^2}}{2}$$

$$b = \frac{m22 + m33}{2}$$

$$x = \frac{1}{2}\arctan\frac{m31}{m21}. \tag{2.27}$$

Experiments and simulations have shown that the MMT parameter x reveals the orientation of fibrous scatterers, t reveals the anisotropy degree, and b reveals the polarizance of the scattering samples. In particular, a normalized anisotropy parameter A can be obtained as [39]

$$A = \frac{2b \cdot t_1}{b^2 + t_1^2} \in [0,1]. \qquad (2.28)$$

For thin tissue slices however, the main polarization altering properties come from birefringence, the elements can be fitted as

$$\begin{cases} m42 = -m24 = 2t_3 \sin 2x_3 \\ m34 = -m43 = 2t_3 \cos 2x_3 \end{cases}. \qquad (2.29)$$

Then we can have the parameters x_3 and t_3 shown as (2.29) to present the orientation and value of linear retardation:

$$t_3 = \frac{\sqrt{(m42)^2 + (m43)^2}}{2}, \tan(2x_3) = \frac{m42}{-m43}. \qquad (2.30)$$

The MMT is a general concept. There can be many different ways to realize the transformation and obtain from the Mueller matrix new polarization parameters with more explicit connections to the physical properties. It has been shown that many MMPD or MMT parameters can separate these different contributions and are much less sensitive to the sample orientations [29].

By contrast, the MMT parameter t and MMPD (Mueller matrix polar decomposition) parameter δ are good indicators of the retardance of the media, while the MMT parameter x and MMPD parameter θ are related to the orientation directions of the aligned fibrous structures [17]. The MMPD parameters are slightly more sensitive to the fibers, whereas the MMT parameters can be calculated more easily and quickly [42]. In our recent studies, we have found that, for the thin tissue slices with anisotropic fibrous structures, the birefringence effect plays the dominant role for the polarization imaging contrast mechanism [42]. Therefore, for the cancerous liver tissue sections, the values of parameters δ and t and the distribution of parameters θ and x may be used to reflect the density and orientation of the fibrous structure.

Li et al. [43] confirmed that the above t and b parameters are invariant under azimuth rotation under the condition that the illumination and scattered lights are collinear. Detailed analysis based on the rotation and mirror symmetry properties of the Mueller matrix have revealed at least 15 polarization parameters which are independent of azimuth rotations in the sample, including the four corners m_{ij} (i, j = 1,4) and the four eigen vectors of the Mueller matrix, the modules of the four edges, and the trace, determinant, and Frobenius norm of the linear part of the Mueller matrix. Such azimuth invariant polarization parameters should help to identify polarization features in complex media. These conditions have potential applications in identifying contributions by multiple anisotropic components [43].

MMPD and MMT parameters can provide additional helpful quantitative information; through some more comprehensive transformation processes of the Mueller matrix elements, we may obtain more quantitative parameters for the extraction of the intrinsic microstructural characteristic features of pathological changes of tissues.

Polarization feature extraction using statistical analysis

Since the Mueller matrix contains abundant information on the tissue samples, it can be helpful to find a method to transform the 2D images of Mueller matrix elements into a group of quantitative or semiquantitative orientation-insensitive parameters, which are crucial for the extraction of the dominant microstructural information of samples. In recent studies, statistical analysis is used to transform the 2D Mueller matrix images to frequency distribution histogram (FDH) curves [33]. Experimental results obtained from different types of tissues have shown that the FDH curves can display the dominant microstructural features of tissues in a much clearer graphic form than 2D images of the Mueller matrix elements. By analyzing the peak positions, widths, and shapes of the FDHs of the Mueller matrix elements, one may learn the following information: (1) whether the sample is anisotropic or isotropic; (2) the depolarization power of the sample; (3) the orientation direction of the anisotropic structure; and (4) the origin of the anisotropy, e.g. if the anisotropic features are due to the fibrous scattering microstructures or the anisotropy in the interstitial medium.

To quantitatively evaluate the Mueller matrix elements, we can employ the central moment method for statistical analysis of frequency distributions [33]:

$$\mu = P1 = E(X),$$

$$\sigma^2 = P2 = \text{Var}(X),$$

$$\text{skewness} = P3 = \frac{E(X - \mu)^3}{\sigma^3} \qquad (2.31)$$

$$\text{kurtosis} = P4 = \frac{E(X - \mu)^4}{\sigma^4}.$$

Suppose we have a random variable X, whose central moments – expected value, variance, skewness, and kurtosis – are exactly represented by Equation 2.30. Here, the expected value $P1$ is the mean value of an FDH. The second central moment $P2$ is called the variance usually denoted by σ^2, where σ represents the standard deviation of the FDH. A small $P2$ indicates that the measured data tend to be distributed close to the expected value, while a large $P2$ indicates that the data points are spread out around the expected value and from each other. The third and fourth central moments $P3$ and $P4$ represent the skewness and kurtosis of the FDH, respectively. $P3$ (skewness) shows the asymmetry of the FDH. The skewness value can be positive or negative. A negative (or positive) skewness value indicates that the tail on the left side (or the right side) of the FDH is longer or fatter than the right side (or the left side). $P4$ (kurtosis) indicates the "peakedness" of the FDH. It is a descriptor of the shape of a probability distribution (Figure 2.6).

The horizontal axis of each FDH represents the value of the pixel from the corresponding Mueller matrix element, while

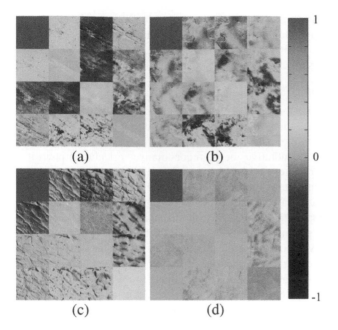

FIGURE 2.6 Two dimensional (2-D) images of backscattering Mueller matrices of biological samples [33].

the vertical axis represents the distributing probability. The influences of sample orientation on the Mueller matrix elements are serious. For the same sample placed along different orientations, the values of $P2$, $P3$, and $P4$ for the Mueller matrix elements almost do not change, while the value of $P1$ can be varied (for the anisotropic sample) or constant (for the isotropic sample)[33]. The central moment analysis of the FDHs provides us a tool to transform the complicated 2D

Mueller matrix images into a group of quantitative indicators of dominant structural properties of tissues. More importantly, using the $P2$, $P3$, and $P4$, we can obtain the main intrinsic properties of samples without influence from orientation variations.

We can see from Figure 2.7 that the FDHs of different tissues have very different distributions. First, the anisotropic and isotropic tissues can be distinguished by using the diagonal elements. The porcine liver (red lines) and fat (blue lines) tissues are predominantly isotropic; therefore, their m_{22} and m_{33} curves are almost the same. The anisotropic bovine skeletal muscle (black lines) and chicken heart (green lines) tissues, however, display differences between the m_{22} and m_{33}, which become more prominent as the anisotropy increases. This is because that the fibers in skeletal muscle samples are well aligned in almost the same direction, while in heart samples, the fibers are distributed in different orientations. Second, we also notice that the distribution widths of the FDHs (the values of $P2$) for bovine skeletal muscle, chicken heart, and porcine liver samples are larger than the fat sample, indicating more complicated microstructures for these metabolic exuberant tissues. The FDHs of the m24, m42, m34, and m43 elements for skeletal muscle, heart, and liver tissues show small positive or negative values, which are related to the birefringent structures in these tissues. The signs of the elements can be used to determine the aligned fibers directions. Finally, the different depolarization power of tissues can also be observed: the liver tissue sample has the largest $P1$ values of the diagonal elements, showing the smallest depolarization power, while the smallest $P1$ values of the diagonal elements indicate the most prominent depolarization property of the fat tissue.

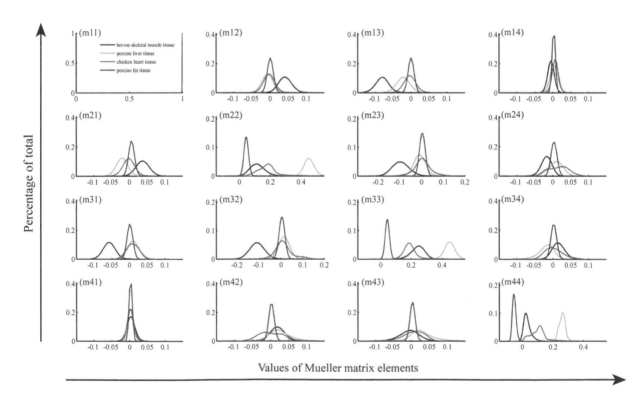

FIGURE 2.7 Frequency distribution histogram (FDH) of Mueller matrix elements of bovine skeletal muscle tissue along different orientation directions [33].

It has been shown that the parameter $P2$ should be sensitive to the complexity of a sample. A large value of $P2$ means that the measured polarization data are distributed in a wider range, indicating a complex structural feature of the tissue. The parameter $P3$ should be sensitive to the heterogeneity of a sample, and a large $P4$ shows that most measured polarization data are distributed very close to the mean value, meaning that the microstructural features are similar.

Polarization imaging and characterization applied in tissue OC

Partial polarization measurements for OC

In the past 10 years, polarization imaging techniques have increasingly become a powerful tool in biomedical diagnostics. Polarization measurements allow one to enhance the imaging contrast of superficial tissues and obtain new polarization-sensitive parameters for better descriptions of the micro- and macro structural and optical properties of complex tissues. It has been demonstrated that the simple linear degree of polarization (LDOP) imaging can mark the margins of cancerous tissues in both *ex vivo* and *in vivo* conditions [26]. Since the LDOP imaging of anisotropic tissues can be sensitive to the orientation of the samples [5], a rotating linear polarization imaging (RLPI) method has been developed to provide a set of orientation insensitive parameters, which can be used to differentiate different microstructural features in cancerous tissues [44, 45].

Noninvasive medical diagnostics using, e.g., polarization imaging is usually limited by poorer penetration depth due to the high turbidity of tissue. Tissues are low absorbing but highly scattering media. Scattering defines spectral and angular characteristics of light interacting with living objects. Optical clearing agents (OCAs) can reversibly change the light scattering properties of tissues and blood [46]. The immersion technique has a great potential for improving the quality of optical imaging in tissues. The challenge remains to understand OC mechanisms. For this purpose, imaging and monitoring various OC processes in fibrous and epithelial tissues (e.g., skin, sclera, cornea, dura mater, muscle, mucosa) as well as controlling of the tissue optical properties are extremely important for many biomedical applications.

With the help of OC, polarization techniques are capable of providing better imaging and more information on the cellular and subcellular details of superficial tissues. Owing to multiple scattering, the polarization properties are lost in turbid tissues, which can be improved through the reduction of scattering caused by OC [47]. On the other hand, pathological tissues show different tissue anisotropy because of the change in tissue structure. So, the kinetics of the polarization properties of the tissue sample at immersion can be used to observe and analyze tissue OC (TOC) process.

Figure 2.8 compares the kinetics of OC curves for the linearly polarized component of transmitted intensity $I_\parallel$, which is in parallel to the polarization of the incident beam, and the total transmitted intensity I_T of scleral tissue layer [6]. As a result, the kinetics of the average transmittance (I_T), the intensity of

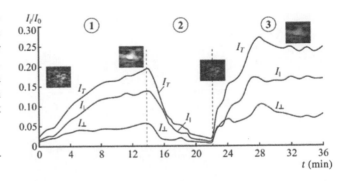

FIGURE 2.8 The time-dependent transmittance (I_t/I_0) of the human sclera specimen [6].

parallel polarization component $(I_\parallel)$ and the orthogonal polarization component $(I_\perp)$ are correlated, which can be explained as an inclusion of tissue birefringence that is revealed at low scattering conditions.

From Figure 2.8, it is well seen that the speckle patterns are transferred from small-size and homogeneously distributed speckles, characteristic for multiple scattering, to big-size inhomogeneously distributed speckles with a large portion of ballistic photons in the central part of the pattern. Reduction of the scattering at optical immersion makes it possible to detect the polarization anisotropy of tissues more precisely and to separate the effects of light scattering and intrinsic birefringence of the tissue polarization properties [6]. When immersion is strong, the RI of the tissue anisotropy structure will be close to the RI of the ground media, and the birefringence of form should be eliminated. A strong immersion condition is a way to evaluate a molecular intrinsic birefringence of collagen fibrils.

Figure 2.9 shows the reversible loss of turbidity and birefringence in rodent tail tendon observed at glycerol (13 M) application. Characteristic banding patterns observed in the tendon sample indicate ordered fibril organization. The distribution of pattern brightness corresponds to the distribution of a phase shift between orthogonal optical field components, and the background smooth brightness corresponds to light scattering. Loss of transmittance at the sample edges and the appearance of bright spots in the middle of the sample during glycerol action indicate RI matching of collagen fibers. The rehydration of the tissue sample in saline makes the banding structure fully visible in the crossed polarizers due to resumption of tissue birefringence and turbidity approximately to the initial states [6].

The effect of OC has been investigated on polarization properties of SHG emission from highly scattering muscle and tendon samples. The polarization dependences (laser polarization and SHG anisotropy) are significantly retained in both tendon and striated muscle through OC [48]. OC does not significantly disrupt the tissue structure in striated muscle and that in acellular collagenous tissues; the process is essentially reversible. It has also been reported that the depolarization of the laser beam leads to randomization of the SHG signal anisotropy [48].

Further, based on combination of polarization imaging and OC, we can develop functional quantitative assessment of Mueller-matrix elements and their physical interpretation,

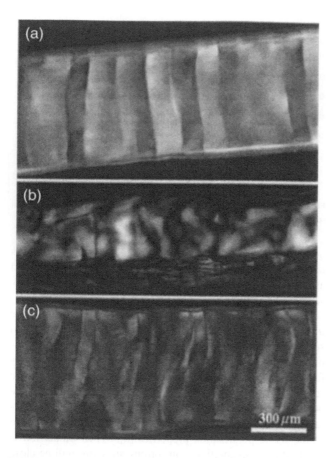

FIGURE 2.9 Reversible loss of turbidity and birefringence in rodent tail tendon following glycerol application [6].

where we modulate the optical properties of tissues by OC, which reduces scattering; and mechanical stretch, which induces birefringence. Reference [49] reports circular polarized light backscattered from a disperse random media influenced by OC. With the help of OC, how to probe a disperse random medium with backscattered circular polarized light is presented. For the first time, the polarization changes induced by OC can be clearly observed and quantitatively analyzed by tracking the polarization vectors on the Poincare sphere.

In reference [50], the isolated contributions of scattering and birefringence in the phase retardation of circularly polarized light propagated in biological tissues have been demonstrated with the help of tissue-mimicking phantoms and chicken skin *in vitro* with application of the OC and mechanical stretch. The decrease in tissue scattering due to OC enhances the degree of polarization up to 80%, making birefringence distinguishable on the background of the remaining scattering.

Full polarization imaging at OC

Variation characteristics of Mueller matrix with TOC

Mueller matrix imaging, including full polarization information, is a valuable method to quantify aspects of tissue structure, and may be a means to differentiate different types of microstructural changes and revealing the pathological microscopic origins [51, 52].

Although TOC has demonstrated its potential applications in diagnosis, its mechanism remains elusive [53]. Many groups have tried to explain how OCAs affect the microstructure of the tissues [54, 55]. Recently, due to their potential to extract much richer microstructural and optical information of tissues, full polarization optical techniques and methods have been introduced to characterize the monitoring and explanation of the TOC process [30, 56, 57].

Mueller matrix elements encode rich information about microstructures of samples. Previous studies have shown that, m_{22}, m_{33}, and m_{44} values are related with the depolarizing ability [29], m_{12}, m_{13}, m_{21}, and m_{31} values are connected to the anisotropy induced by aligned fibrous scatterers and m_{24}, m_{34}, m_{42}, and m_{43} values are closely connected to the anisotropy induced by the birefringence [18]. We can determine the axis direction of aligned fibers through the signs and intensities of m_{12} and m_{13} [18], and the direction of optical axis through those of m_{24} and m_{34}, similarly.

Reference [30] investigates the polarization features of nude mouse skin during immersion in an 80% glycerol solution for 0–30 min, and examines how the Mueller matrix elements vary with the immersion time. The backscattering Mueller matrices are measured using the dual rotating retarder configuration [12]. There is an oblique angle θ of 20 deg between the illumination light and the detection direction to avoid the sample surface reflection. Figure 2.10 is the photograph of skin samples before immersion (left) and after immersion for 20 min (right), respectively.

Two-dimensional images of backscattering Mueller matrices of the skin tissues at different immersion times are shown in Figure 2.11. The Mueller matrix images immediately show some characteristic features, e.g., their diagonal elements m_{22}, m_{33}, and m_{44} vary during the immersion, and the values of m_{34} and m_{43} are relatively large.

Reference [30] uses all the Mueller matrix elements, instead of the few parameters derived from the decomposition method. The frequency distribution histogram (FDH) of each Mueller matrix element shows the probabilities of the pixels of an image falling into a particular intensity range. The peak, width, and shape of FDH curves may all contain information on the tissues. In Figure 2.12, we choose an area of 100 × 100 pixels on the Mueller images to obtain the FDH curves. The following features are becoming evident as the immersion time increases: (1) The FDH curves of m_{22}, m_{33}, and m_{44}, and m_{34}, and m_{43} change significantly. (2) The values of the diagonal elements m_{22}, m_{33}, and m_{44} increase, indicating the

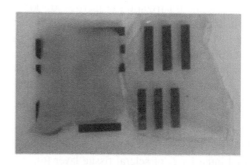

FIGURE 2.10 The photograph of nude mouse skin samples before and after immersion for 20 min in 80% glycerol solution respectively [30].

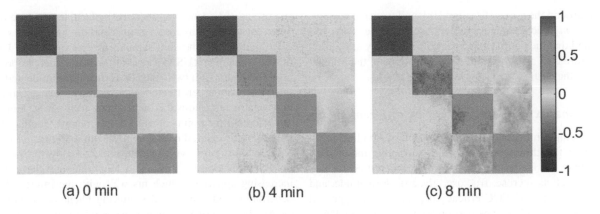

(a) 0 min   (b) 4 min   (c) 8 min

FIGURE 2.11   Two-dimensional backscattering Mueller matrix images of the skin tissues at different immersion times [30].

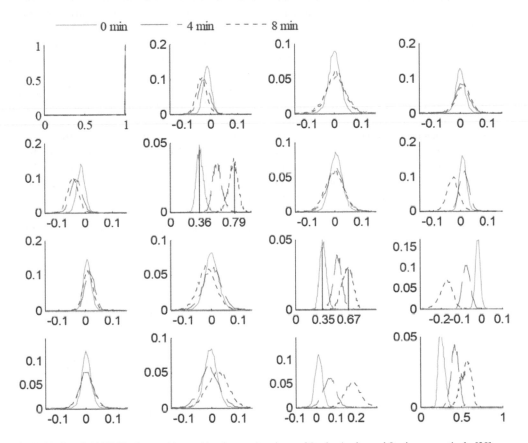

FIGURE 2.12   The FDHs for all 16 MMEs from m11 to m44 at immersion times of 0 min, 4 min, and 8 min, respectively [30].

decrease of depolarization by OC, and the values of m_{22} and m_{33} are greater than m_{44}, reflecting that the scattering characteristics of the tissue are dominated by Rayleigh scattering [58]. (3) An increase of the absolute values of m_{34} and m_{43} and m_{12} and m_{21} indicate an increase in the tissue anisotropy during TOC, which can also be confirmed by the increasing difference between m_{22} and m_{33} [59] (the difference between the central values of the FDH curves of m_{22} and m_{33}, which are marked by black solid lines). (4) The variations of m_{34} and m_{43} are much larger than those of m_{12} and m_{21}, meaning that the tissue anisotropy and its change during immersion are mainly due to the birefringence rather than scattering by well aligned fibrous structures [59]. Compared with the 2D images of the

MMEs, the semiquantitative FDH curves provide more characteristic information.

We further calculate the averages and standard deviations of the Mueller elements to examine quantitatively how the polarization properties of the skin vary with the immersion time. Figure 2.13 shows the kinetic changes in the FDHs as functions of the immersion time. The central value and the half width represent the average value and standard deviation (SD) of the FDH, respectively. The quantitative data confirm the qualitative conclusion obtained from the Mueller matrix images and FDH curves. In addition, from the negative averages of m_{12} and m_{21} and the averages of m_{13} and m_{31} being close to zero, it can be determined that the fibrous scatterers are aligned around the y-axis direction [18], and from the negative average of m_{34} and the much smaller absolute value of m_{24}, it is confirmed that the optical axis is also around the y-axis direction. Figure 2.13 also shows that the variations of the Mueller matrix of the cleared skin stay stable after 10 min.

Mueller matrix features for TOC by different agents

Usually, simple immersion is a passive OC method achieved by placing the thin tissue into high refractive index solutions. There are many OCAs which have been used for this purpose, such as sucrose, fructose, glycerol, formamide, and 2,2'-thiodiethanol. The OC efficacy depends on the type and concentration of OCAs as well as on the treatment time [56].

Further, for insight into the connection between OCAs and the microstructural and optical changes in tissues, a Mueller matrix microscope can be used to monitor the TOC processes, and various models are established to mimic the dynamic process of microphysical features. Consequently, by the feature exaction from Mueller matrixes, we can understand the major difference between cleared tissues due to different agents.

Polarization status is an effective way to study the microscopic changes in tissues and is especially sensitive to those optical anisotropic features, such as birefringence or fibrous microstructures. Collagen fibers are the main components of dermis, and show an apparent anisotropic scattering capability in our previous research work [56]. There are studies that show that the hydroxyl molecule in OCAs will interact with collagen and cause various possible microcosmic changes of fiber content and arrangement. We select formamide and saturated sucrose as our agents, because the molecular structure of the latter does contain hydroxyl groups and the former does not.

In Figure 2.14, the skin samples are put on a 1951 United States Air Force (USAF) resolution test target before and after treatment with two kinds of agents: formamide and saturated sucrose. When treated with these two agents, after 20 min treatments, mouse skin becomes transparent similarly. At different time points during OC, we record Mueller matrix images of the cleared mouse skin, shown in Figure 2.15.

From the figure, we can determine roughly the polarization changes due to TOC. The four diagonal elements are increasing apparent, which fits well with the improved light penetration and suppressed scattering phenomena shown by OCAs. Nondiagonal elements have relatively small values and changes; however, their trends and differences often provide key features of tissue microstructure and optical properties. When using formamide as agent, the values of m_{13}, m_{14}, m_{31}, m_{34}, m_{41}, and m_{41} are changed, but when using saturated sucrose as agent, the nondiagonal elements affected by OC are m_{23}, m_{24}, m_{32}, m_{34}, m_{42}, m_{43}. The most interesting difference in using these two types of OCAS is the opposite trends of m_{34} and m_{43}. According to our research works on Mueller matrix characterization, these two elements can be closed related with tissue anisotropy, especially from the intercellular birefringence effect. Except for m_{34} and m_{43}, the fluctuation of other nondiagonal elements is perhaps connected with the arrangement of fibrous structures, scatterer size, or optical rotation effect.

Figure 2.16 shows the tissue images and the values of depolarization and phase retardance with OC time. When using

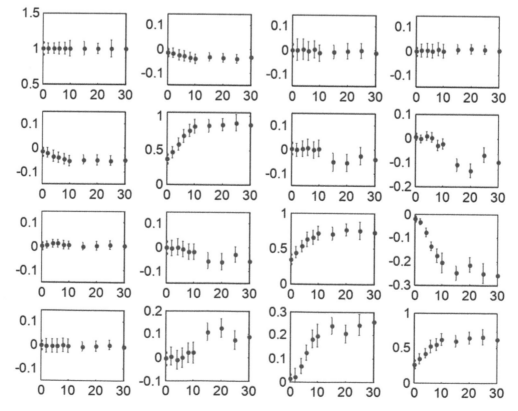

FIGURE 2.13   The average values and SDs of all 16 MMEs change with the immersion time (min) [30].

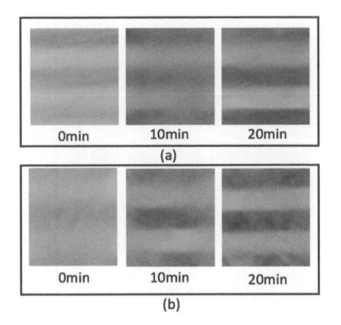

FIGURE 2.14 White-light images of in vitro rat skin before and after clearing by formamide (a), saturated sucrose (b) [56].

formamide as an OCA, both the depolarization parameter and the retardation parameter decrease dramatically. However, when using saturated sucrose as agent, by the same experiment setup, the depolarization parameter also apparently decreases, but the phase retardance increases with the oscillatory behavior. It looks like it is a key factor to explain the opposite tendency of retardance with OC due to different agents.

The above studies investigate the variations of polarization features of tissue samples with OC process by measuring the backscattering or transmission microscopic Mueller matrices.

The experimental results show the change regularity of Mueller matrix elements during immersion, and provide a qualitative connection with the depolarizing ability, the interstitial birefringence and the aligned fiber content of tissue samples. However, to quantitatively analyze and explain the polarization features corresponding to microstructural changes during the TOC process, we need some efficient tissue modeling and polarization scattering simulations combined with experimental results. Monte Carlo simulations using properly designed tissue models provide insight into the relations between the polarization features and the microstructure of tissues during the immersion. Comparisons between the experimental and the simulated Mueller matrices provide a way to understand the OC process according to the connection between polarization changes and microphysical features of tissues, and verify that Mueller matrix imaging is potentially a powerful method applied in TOC.

Polarized light interactions with tissues

Monte Carlo simulations of tissue scattering

Introduction

Many groups have studied the polarization characteristics of light propagation in turbid media, and reported different ways of tracking the polarization status of scattering photons in isotropic media [60, 61]. This helps in gaining physical insight, designing, and optimizing experiments, and interpreting the measured data. The use of electromagnetic theory with Maxwell's equations is the most rigorous and best suited method for polarimetry analysis in a transparent medium with well-defined optical interfaces. However, tissue is a turbid medium possessing microscopically inhomogeneous complex

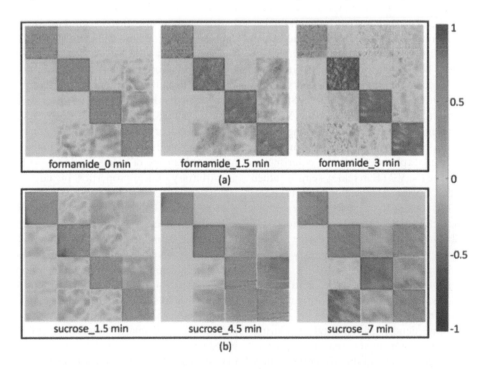

FIGURE 2.15 Pesudo-color images of the Mueller matrix elements at three time points during clearing process using formamide (a) and saturated sucrose (b) as OCAs [56].

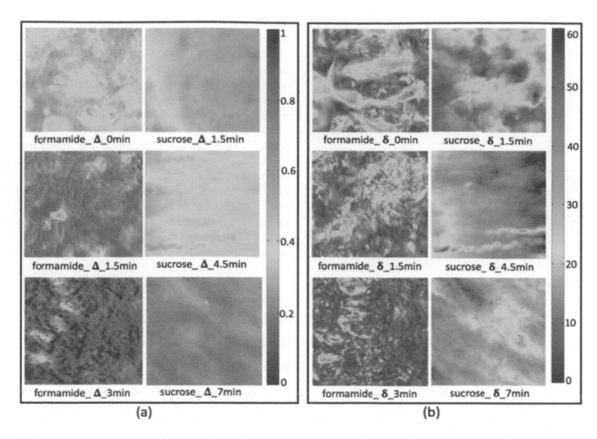

FIGURE 2.16   Pseudo-color images of the MMPD parameter: (a) depolarization parameter and (b) retardation parameter for the unstained mouse skin tissue slice [56].

dielectric structures (macromolecules, cell organelles, organized cell structures, blood and lymphatic networks, extracellular matrix, interstitial layers, etc.). Due to the resulting complexity, the Maxwell's equations approach for polarized light propagation in such a complex turbid medium is impractical and is not presently feasible. Although the scalar radiative transport theory and its simplified approximation, the diffusion equation, has been successfully used to model light transport in tissue (specifically light intensity distribution in tissue volume, diffuse reflectance, etc.), both are intensity-based techniques, and hence typically neglect polarization [8].

Alternatively, the vector radiative transfer equation (VRTE), which includes polarization information by describing transport of the Stokes vectors of light (photon packet) through a random medium, has been explored for tissue polarimetry modeling. However, solving the VRTE in real systems is rather complex. Some analytical and numerical techniques have been developed to solve VRTE, such as the transfer matrix and the singular eigenfunction. Unfortunately, these are often too slow and insufficiently flexible to incorporate the necessary boundary conditions for arbitrary geometries and arbitrary optical properties as desirable in case of tissue [8].

Some approximate analytical approaches, like approximate analytical/heuristic approaches modeling depolarization of multiply scattered light in a turbid medium, are mainly aimed at understanding the overall depolarization trends, exploring the dependence of depolarization on the scattering properties of the media, and designing general polarization schemes to discriminate against multiply scattered photons for tissue

imaging in "simple" geometries. However, these approaches typically neglect other simultaneously occurring complex tissue polarimetry events (such as linear birefringence, optical activity, etc.) [8]. A more encompassing, accurate method is polarization-sensitive Monte Carlo (PSMC) techniques.

The Monte Carlo (MC) technique is a general and robust approach for modeling light transport in a random medium, and it has been an effective method for the investigation of light propagation in biological tissues [8]. In this statistical approach to radiative transfer, the multiple scattering trajectories of individual photons are determined using a random number generator to predict the probability of each scattering event. The superposition of many photon paths approaches the actual photon distribution in time and space. This approach has the advantage of being applicable to arbitrary geometries and arbitrary optical properties, including the ability to simulate heterogeneous media.

Most MC models were developed for intensity calculations only and neglected polarization information, the most commonly used being the code of Wang et al. [62] . More recently, a few implementations have incorporated polarization into the MC approach [60, 61, 63, 64]. To better understand the polarimetric images of tissues, MC simulations for multicomponent or multilayered scattering media are often used as a theoretical prediction basis. The changes in photon position, direction of propagation, and polarization change of light can be calculated based on various tissue models and programs [64]. These simulations usually employ the following physical assumptions: 1) All scattering particles are in the far-field zones of other particles. The observation point is also located in the far-field zones of all scattering particles. 2) The positions of all particles are independent of each other, particles are randomly and uniformly distributed over the scattering medium volume, and there is no coherent scattering.

MC modeling can help the interpretation of experimental data. A three-dimensional Monte Carlo has been developed to study optically active molecules in turbid media [63]. These implementations describe polarization in terms of the Stokes vector. As shown in reference [58], the MC simulations of the backscattering Mueller matrix images of three different tissue phantoms showed that not only the size of scatterers but also the optical index contrast affect the ratio of linear to circular polarization of the backscattered light and, consequently, the contrast of polarimetric images. The Rayleigh-like optical response of cancerous tissues can be attributed to the light scattering on both small and large scatterers, as the optical index contrast in biological tissues is quite small.

The PSMC modeling has the following major advantages: the possibility of employing any scattering matrix; the fact that one is not able to use strongly forward-directed phase functions or experimental single-scattering matrices; and the possibility of modeling media with complex geometry. The main challenges of MC simulation are the accuracy and the computation time [6].

To illustrate a MC simulation technique, we will briefly introduce the framework of program described in reference [64]. To simulate anisotropic behavior of the media, the new program introduces infinitely long cylinders into the scattering media. The number of different cylinders, as well as their sizes, concentrations, and orientation functions, are preset parameters in the input data file. New subroutines are added for calculations of cylindrical scatters, such as scattering probability, phase function, rotation of the Stokes vector reference plane, etc. For PSMC, the polarization states of photons are tracked step by step in addition to their spatial positions and direction of motions as follows (Figure 2.17):

In Monte Carlo simulations of anisotropic tissues [65], by solving the scalar wave equation analytically, we can calculate the Mueller matrix of an infinitely long cylinder. The program starts by launching the normal incident photon whose polarization state is represented by a Stokes vector. At each scattering event, we make a statistical choice as to what type of scatterers the photon hits. Then the program rotates the Stokes vector reference frame and computes the phase function using the precalculated scattering matrices. The scattering direction is determined according to the phase function together with a random number, and then the Stokes vector is updated. The simulation process continued until a photon is completely absorbed or move out of the sample. Locations, polarization states, and other information on the emitted photons are stored.

The Monte Carlo simulation for anisotropic tissue model tracks the trajectory and polarization state of photons. After each scattering event, the photon moves along a new direction determined by the Mie theory and may lose part of its energy because of the absorption. A statistical method was designed to determine whether the photon is scattered by the cylinders or the spherical particles. For anisotropic tissues, as the polarized photons transmit in the anisotropic medium, they alternately experience the transmission process in the birefringent

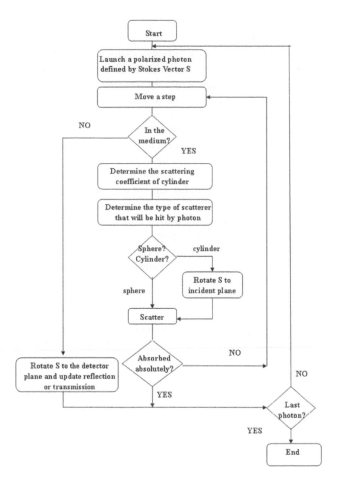

FIGURE 2.17 Flow chart of Monte Carlo program of polarized light scattering in medium [64].

medium and the scattering process by the spherical or infinitely long cylindrical scatterers. With the help of PSMC simulation, we try to reveal the details of interaction between OCAs and tissue structure on the microscopic (molecular) scale, and this can be a powerful tool in exploring the mechanisms of TOC.

For a better understanding of the relationships between the Mueller matrix parameters and the microstructural changes in the skeletal muscle tissues, we can use the MC simulation program based on the SCDM to track the trajectories and polarization states of backscattered photons as they propagate in the tissues. For example, by comparing the MC simulated results with the experiment, Reference [59] proposes the possible structural variations in the skeletal muscle tissues during the rigor mortis and proteolysis processes. Some research works demonstrate that by using Mueller matrix images, MMD or MMT parameters, and FDH curves, combined with MC simulations, abundant microstructural information can be extracted from the tissue samples.

Anisotropic scatterers and optical effects

Optical applications for therapeutic and diagnostic purposes motivate efforts in improving numerical simulations for real biological samples, which often contain complex tissue microstructure.

PSMC models deal with various intrinsic tissue polarimetry characteristics in polarized light–tissue interactions.

Depolarization caused by multiple scattering is the most prominent polarimetry effect in biological tissues, which is caused by the high density of tissue scattering centers, originating from the random fluctuations of the local refractive index in the tissue microstructure (inside the cell and in the extracellular matrix). The tissue scattering centers vary in size (and shape) from micrometer scale and below (subcellular structures such as mitochondria, ribosomes, lysosomes, Golgi apparatus, etc.) to several tens of micrometers (whole cells, collagen fibers, etc.). The typical refractive index of these scattering structures is about 1.4–1.5 (the average background refractive index of cytoplasm and interstitial fluid is about 1.34) [66]. Light scattering from all of these microscopic scattering structures contributes in a complex fashion to the observed depolarization of light in tissue. Note that the underlying mechanism of depolarization due to multiple scattering is the scrambling of the photon's reference frame (scattering plane) as a consequence of the random sequence of scattering events in a variety of scattering directions.

Linear retardance is the other important tissue polarimetry characteristic, and it is usually closely related to birefringence, which is defined as the difference in refractive indices, $\Delta n = n_e - n_o$, where n_e and n_o are extraordinary and ordinary refractive indices. Although not as pervasive as multiple scattering, the anisotropic organized nature of many tissues exhibits phase retardation, that is, the retardance $\delta = (2\pi/\lambda) \times \Delta n \times L$, L is the optical pathlength between two orthogonal linear polarization states. Various types of tissues, such as muscle, skin, myocardium, bone, teeth, cornea, tendon, cartilage, eye sclera, dura mater, nerve, retina, and myelin, possess linear retardance. The typical values of linear birefringence of these biological fibers in the visible wavelength range are in the range $\Delta n = 10^{-3}$ to 10^{-2} [67]. Interestingly, even though uniform uniaxial birefringence may not be a direct contributor to depolarization, randomly oriented spatial domains of uniform uniaxial birefringent properties may cause polarization loss [3].

Similarly, circular birefringence (retardance, also called *optical rotation* in this context) in tissue arises due to the presence of asymmetric optically active chiral molecules like glucose, proteins, and lipids [68]. Finally, many biological molecules (such as amino acids, proteins, and nucleic acids) also exhibit dichroism or diattenuation effects. The magnitude of diattenuation effects in tissue is, however, much lower compared to the other polarization phenomena described above.

Many biological tissues are structurally anisotropic. Such tissues can be locally considered as a short-range spatially ordered fibrillar system, and aging and pathology may change such micro-ordering. Polarized photon scattering of fibrous microstructures can be calculated by the scattering theory of an infinitely long cylinder. The scattered fields in matrix form and relevant parameters are given as [64]:

$$\begin{pmatrix} E_s \\ E_{\perp s} \end{pmatrix} = e^{i3\pi/4}\sqrt{\frac{2}{\pi kr\sin\zeta}}e^{ik(r\sin\zeta - z\cos\zeta)}\begin{pmatrix} T_1 & T_4 \\ T_3 & T_2 \end{pmatrix}\begin{pmatrix} E_i \\ E_{\perp i} \end{pmatrix} \quad (2.32)$$

$$\begin{cases} T_1 = \sum_{-\infty}^{\infty} b_{nI}e^{-in\Theta} = b_{0I} + 2\sum_{n=1}^{\infty} b_{nI}\cos(n\Theta) \\[2mm] T_2 = \sum_{-\infty}^{\infty} a_{nII}e^{-in\Theta} = a_{0II} + 2\sum_{n=1}^{\infty} a_{nII}\cos(n\Theta) \\[2mm] T_3 = \sum_{-\infty}^{\infty} a_{nI}e^{-in\Theta} = -2i\sum_{n=1}^{\infty} a_{nI}\sin(n\Theta) \\[2mm] T_4 = \sum_{-\infty}^{\infty} b_{nII}e^{-in\Theta} = -2i\sum_{n=1}^{\infty} b_{nII}\sin(n\Theta) \end{cases} \quad (2.33)$$

$$\begin{cases} a_{nI} = \dfrac{C_nV_n - B_nD_n}{W_nV_n + iD_n^2}, \quad a_{nII} = -\dfrac{A_nV_n - iC_nD_n}{W_nV_n + iD_n^2} \\[4mm] b_{nI} = \dfrac{W_nB_n + iD_nC_n}{W_nV_n + iD_n^2}, \quad b_{nII} = -i\dfrac{C_nW_n + A_nD_n}{W_nV_n + iD_n^2} \end{cases} \quad (2.34)$$

$$\begin{cases} A_n = i\xi\left[\xi J_n'(\eta)J_n(\xi) - \eta J_n(\eta)J_n'(\xi)\right] \\[2mm] B_n = \xi\left[m^2\xi J_n'(\eta)J_n(\xi) - \eta J_n(\eta)J_n'(\xi)\right] \\[2mm] C_n = n\cos\zeta\eta J_n(\eta)J_n(\xi)\left(\dfrac{\xi^2}{\eta^2} - 1\right) \\[2mm] D_n = n\cos\zeta\eta J_n(\eta)H_n^{(1)}(\xi)\left(\dfrac{\xi^2}{\eta^2} - 1\right) \\[2mm] W_n = i\xi\left[\eta J_n(\eta)H_n^{(1)'}(\xi) - \xi J_n'(\eta)H_n^{(1)}(\xi)\right] \\[2mm] V_n = \xi\left[m^2\xi J_n'(\eta)H_n^{(1)}(\xi) - \eta J_n(\eta)H_n^{(1)'}(\xi)\right] \end{cases} \quad (2.35)$$

$$\begin{cases} \Theta = \pi - \theta \\ \xi = x\sin\zeta \\ \eta = x\sqrt{m^2 - \cos^2\zeta} \\ x = ka \end{cases} \quad (2.36)$$

Where $E_{\|s}$ and $E_{\perp s}$ represent the incident electrical field, r and z are cylindrical polar coordinates, a is the radius of the cylinder, m is the refractive index of the cylinder relative to that of the surrounding medium, k is the wave number, and ζ and θ are the incident angle and the angle between the incident and scattering planes, as shown in Figure 2.18. The Mueller matrix can be derived according to the definition of Stokes vector:

$$M(\zeta,\theta) = \frac{2}{\pi kr\sin\zeta}\begin{pmatrix} m_{11} & m_{12} & m_{13} & m_{14} \\ m_{21} & m_{22} & m_{23} & m_{24} \\ m_{31} & m_{32} & m_{33} & m_{34} \\ m_{41} & m_{42} & m_{43} & m_{44} \end{pmatrix}. \quad (2.37)$$

The spatial distribution of scattered photons for oblique incidence at an infinitely long cylinder follows a conical shape around the cylinder with the half angle ζ, which is the same as the incident angle. The scattering probability around the cone depends on the scattering direction defined by ζ. The scattering

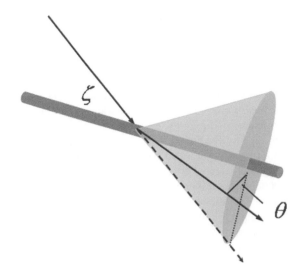

FIGURE 2.18 Scheme of light scattering characteristics by an infinitely long cylinder [64].

coefficient of the microspheres $\mu_{s,sph}$ is a constant once their size and refractive index are set. However, the scattering coefficient of cylinders, $\mu_{s,cyl}$, varies with the angle between the direction of incident photon and the cylinder.

Tissue anisotropy plays an important role in depolarization and birefringence, which can be reflected by polarization characterization. Tissue birefringence results primarily from the linear anisotropy of fibrous structures. A tissue system composed of parallel cylinders is to some extent closer to a uniaxial birefringent medium with the optic axis parallel to the cylinder axes. This type of birefringence is called *birefringence of form* (Figure 2.19(a)). A large variety of tissues are involved, such as eye cornea, tendon, cartilage, eye sclera, muscle, artery wall, nerve, retina, and teeth. Besides anisotropy of form, collagen fibrils as complex molecular structures have intrinsic linear birefringence arising due to anisotropy at the molecular scale and are fundamentally determined by the anisotropic distribution of electrical charge. Diattenuation (linear dichroism) is the difference in attenuation of two waves with orthogonal polarizations traveling in an anisotropic medium, which is described by the difference between the imaginary parts of the effective indices of refraction for two orthogonal directions.

In addition to linear birefringence and diattenuation, many tissue components show optical activity (circular birefringence) and circular diattenuation. In complex tissue structures, chiral aggregates of particles may be responsible for tissue optical activity (Figure 2.19(c)), and the molecule's chirality, which stems from its asymmetric molecular structure (Figure 2.19(d)), also results in optical activity.

Some groups have investigated light scattering from nonspherical particles, such as disc [69] and ellipsoid [70]. Many tissues contain a microstructure of ordered elongated subunits, such as the myofibrils in muscles or the collagen fibers in skin, tendons, or ligaments. They are anisotropic and cannot be described by an isotropic scattering model, such as randomly distributed spherical or nonspherical scatterers. Recently, MC simulations of unpolarized photon propagation in anisotropic media, such as dentin [71], have been reported. The anisotropic media were approximated to a mixture of aligned cylinders of infinite length and solid microspheres [72]. More sophisticated anisotropic tissue models can also be found. For example, the eye cornea can be represented as a system of plane anisotropic layers (plates, i.e., lamellas), each of which is composed of densely packed long cylinders (fibrils).

Tissue models and phantoms

A major goal of biomedical optics studies is to understand the interactions between photons and complicated biological tissues, which usually have to be characterized by simplified models. Biological tissue is an inhomogeneous medium with different levels of organization that include cells, cell organelles, inclusions, and different fiber and tubular/lamellar structures. In view of the great diversity and structural complexity of tissues, an adequate optical model to describe a specific tissue type should include reasonable consideration of light scattering and absorption. Many tissues are composed of structures with a wide range of sizes and can be represented as a system of discrete scattering particles.

Since many cells, cell organelles, and biological macromolecules are close in shape to spheres or ellipsoids, it is common to model tissues as ensembles of homogeneous spherical particles. The Rayleigh and Mie theories can be used to calculate the light–matter interaction in tissues. A tissue model composed of several proportions of large and small spheres can provide a simulated situation of pathological tissue, e.g., a cataract eye lens or enlargement of a cell nucleus in cancer.

For fibrous microstructures, such as tendon, skin dermis, and muscular tissue, a system of long cylinders is the most appropriate model to describe light scattering. Typical scatterers

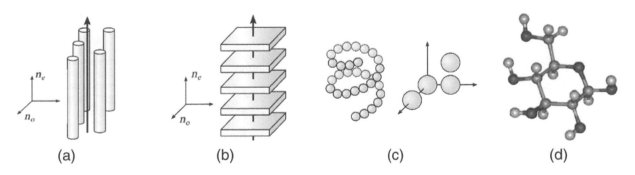

FIGURE 2.19 Examples of structurally anisotropic models of tissues and tissue components [6].

in an anisotropic tissue model include spherical particles or cylinders with a refractive index, and they are randomly or regularly distributed in the isotropic base matter with a smaller refractive index than scatterers.

In general, if considering parameter setting of tissue optical model, the scattering coefficient of a tissue, μ_s, is a basic and key factor in tissue modeling, depending on the refractive index mismatch between cellular tissue components: cell membrane, cytoplasm, cell nucleus, cell organelles, melanin granules, and extracellular fluid (ECF). The other common optical transport parameters in modeling tissues include the absorption coefficient (μ_a) and scattering anisotropy (g).

To mimic the anisotropic media, several phantoms have been reported. A spherical birefringence model (SBM) can be used to model spherical scatterers embedded in a linearly birefringence medium. The SBM phantom consists of polystyrene microspheres immersed in a polymer hydrogel called polyacrylamide [73]. Another model closer to real tissues is a sphere-cylinder birefringence model (SCBM), where we approximate the fibrous tissues to a mixture of polydispersed microspheres and cylindrical scatterers of infinite length. A sample with polystyrene microspheres and well-aligned silk fibers immersed in water can be used as the phantom for a sphere-cylinder scattering model (SCSM). The phantom consists of a slab of well aligned silk fibers submerged in microsphere solution as shown in Figure 2.20.

In SCSM, the sizes and concentrations of the spheres and cylinders can be varied independently. The cylinders are aligned following a Gaussian distribution. Variables of SCSM include parameters of the scatterers and the surrounding medium. Those for the scatterers are the numbers of spherical and cylindrical components, their number densities and sizes, and the mean value and standard deviation of the direction distribution function for the cylinders. Those for the surrounding medium are birefringence, dichroism, refractive index, and absorption coefficient, which may or may not be polarization dependent. The orientation and standard deviation $\Delta\eta$ of the distribution function are also variable. Based on such a simple sphere-cylinder model, more complicated structures can be simulated.

By comparing experiments and MC simulations, a SCSM can reproduce all the characteristic features in spatially resolved unpolarized, polarized reflectance, and Mueller matrix elements of typical fibrous tissues like skeletal

muscle [65]. The evidence indicates that SCSM may be used to characterize both the anisotropic optical properties (i.e. parameters of the scatterers and the ambient medium), and the anisotropic structure (i.e. alignment of the fibers) of skeletal muscle.

In the above phantom, the first and third layers are solutions of polystyrene microspheres in water as an equivalent approximation of the cellular organelles. The second layer contains only well aligned silk fibers. During experiments, the depth of submersion can be adjusted to vary the ratio of scattering coefficients between spheres and cylinders. As the depth of submersion increases, the thickness of the first layer increases. Photons interact with more spheres before reaching the well aligned silk fibers, corresponding to an increase in the ratio of scattering coefficients between spheres and cylinders. Although such a separately spaced sphere-cylinder sample is different from a true SCSM in which spheres and cylinders are mixed together, such a layered microsphere-silk sample generates very similar results as the homogeneous sphere-cylinder scattering medium [65].

Figure 2.21. shows the polarized reflectance images of the microsphere-silk sample (Figure 2.21(b)) containing the same characteristic features of skeletal muscle (Figure 2.21(a)). The differences in "VV" and "HH" images can be explained using Mie scattering theory for spheres and infinitely long cylinders. Polarization-maintaining photons tend to be scattered to the perpendicular direction of cylindrical scatterers, i.e. the x-axis here. Hence, compared to a sphere-only medium, all nine reflectance images of the microsphere-silk sample are elongated along the x-axis due to the scattering of cylinders. The polarized reflectance images of skeletal muscle can be considered as a combination of the contributions of both spherical and cylindrical scatterers.

In SCSM, the polarization states of the photons are altered by scattering only. If the medium around the scatterers is birefringent, a scattered photon undergoes retardation between successive scatterings. The new module, SCBM, approximates the anisotropic turbid medium to a mixture of solid spherical and infinitely long cylindrical scatterers embedded in a linearly birefringent medium. In SCBM, the changes of Stokes vector are calculated due to the birefringent medium:

$$S' = R(-\beta)M(\delta)R(\beta)S, \qquad (2.38)$$

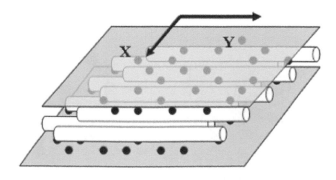
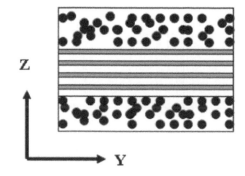

FIGURE 2.20 Schematics of the three-layer microsphere-silk sample [65].

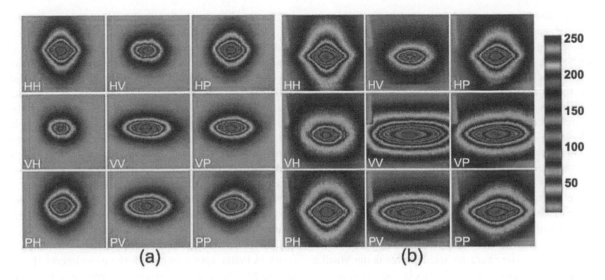

FIGURE 2.21 Polarized reflectance images of (a) fresh skeletal muscle and (b) microsphere-silk sample [65].

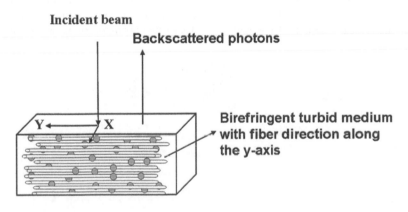

FIGURE 2.22 Schematics of the sphere-cylinder birefrngence model (SCBM) [74].

where $M(\delta)$ is the Mueller matrix of the standard retarder and $R(\beta)$ is the rotational matrix. $M(\delta)$ can be expressed by

$$M(\delta) = \begin{bmatrix} 1 & 0 & 0 & 0 \\ 0 & 1 & 0 & 0 \\ 0 & 0 & \cos\delta & \sin\delta \\ 0 & 0 & -\sin\delta & \cos\delta \end{bmatrix}, \quad (2.39)$$

where δ is the retardation, which can be obtained by the transport path length between two successive scattering events, the average refractive index, the wavelength of light, and $\Delta n'$, the difference in refractive indices, expressed by

$$\Delta n' = n_e'(\theta) - n_o = \frac{n_o n_e}{\left(n_o^2 \sin^2\theta + n_e^2 \cos^2\theta\right)^{1/2}} - n_o, \quad (2.40)$$

where θ is the angle between the propagation direction of the photon and the extraordinary axis. The birefringence value is defined by $\Delta n = n_e - n_o$, where n_e and n_o are the refractive indices along the extraordinary and ordinary axes, respectively. $R(\beta)$ can be expressed by

$$R(\beta) = \begin{bmatrix} 1 & 0 & 0 & 0 \\ 0 & \cos2\beta & \sin2\beta & 0 \\ 0 & -\sin2\beta & \cos2\beta & 0 \\ 0 & 0 & 0 & 1 \end{bmatrix}, \quad (2.41)$$

where β is the rotational angle.

In SCBM, parameters for the scatterers are the same as those for the SCSM. Additional parameters for the surrounding medium include the value and axis direction of birefringence. Both the cylindrical orientation and the axis direction of birefringence can be adjusted in the three-dimensional (3D) space. In SCBM, we assumed that the birefringence effect is the property of the surrounding medium and does not affect the scattering phase function, but the birefringence does alter the polarization states of the photons as they propagate between two successive scattering events (Figure 2.22).

During the experiments, the polyacrylamide sample was strained via extension to create birefringence along the direction of strain. In SCBM, scattering is produced through the addition of polystyrene microspheres and silk fibers before the polymerization of the polyacrylamide, and birefringence is produced through the straining of the polyacrylamide. In addition, the direction of both the strain (the extraordinary axis of birefringence) and direction of silk fibers, the scattering coefficient of spherical scatterers, and the birefringence value can be adjusted.

To discriminate the anisotropy due to the birefringence effect of the surrounding medium and the scattering by aligned cylinders, we aligned the silk fibers along the y axis and the direction of strain (the extraordinary axis of birefringence) along the 45-degree angle axis on the *x–y* plane. The polyacrylamide sample was strained with the extension of 5 mm (the maximum extension is 6 mm), and the difference in refractive indices was about 1×10^{-5}.

The Mueller matrix elements of the skeletal muscle were compared with the simulations using SCBM, SCSM, and SBM as shown in Figure 2.23. Simulations of the three anisotropic models (SCBM, SCSM, and SBM) had respective similarities and differences. The characteristic features in the Mueller matrix patterns of anisotropic turbid media showed contributions by both optical anisotropy due to birefringence and scattering anisotropy due to cylindrical scatterers.

As shown in Figure 2.23, for the Mueller matrix elements of a muscle sample, m_{11} has the typical rhombus shape, and m_{22}

has a cross-like pattern with the dominant distribution along the *x* axis. The simulations using SCBM and SCSM regenerated similar features in these two elements. However, simulations using SBM resulted in totally different shapes. Thus, one has to include cylindrical scatterers in the model of anisotropic tissues. These features of m_{11}, m_{12}, m_{21}, and m_{22} are due to the anisotropy of the cylinders [74]. On the other hand, the third and fourth rows and columns of the experimentally obtained Mueller matrix for the muscle samples show only weak patterns. Simulations using both SCBM and SBM regenerated such features, but simulations using SCSM resulted in much clearer patterns. More detailed simulations using SCBM showed that the birefringence effect is responsible for smoothing out the distinctive features in the third and fourth rows and columns of the Mueller matrix elements.

The above analysis proves that simulations using SCBM result in better agreements with the experiments than SCSM and SBM. One has to consider contributions from both the cylindrical scatterers and the birefringence effect to explain the characteristic features of polarized photon scattering in complicated anisotropic turbid media such as skeletal muscles. These two terms on tissue anisotropy perhaps correspond to

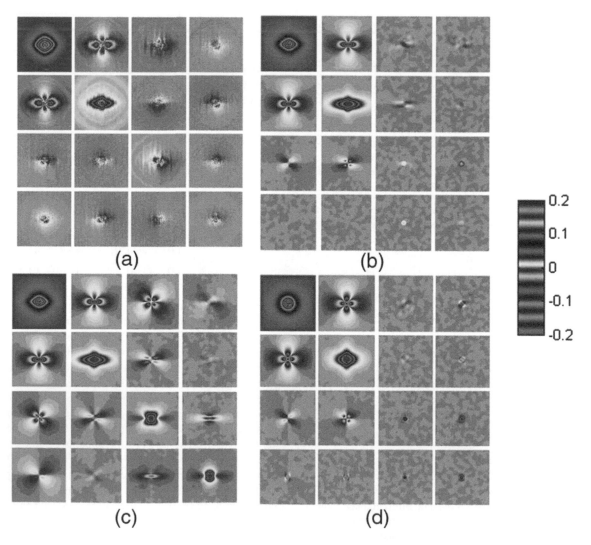

FIGURE 2.23 The backscattering Mueller matrices of fresh skeletal muscle (a), the SCBM simulation (b), the SCSM simulation (c), and the SBM simulation (d) [74].

the anisotropy of form defined by the tissue microstructure and the anisotropy of material controlled by the intrinsic anisotropic character of metabolic molecules.

Optical simulations combined with specific tissue model and phantoms have shown great potential in Mueller matrix polarimetry [75]. MC simulations based on the SCBM can provide a better understanding of the relationship between the polarization parameters and the structural parameters of the scattering models, corresponding to morphological and pathological changes in cancerous tissues at the cellular level.

Simulations of OC process

Mechanisms and models

To understand how to improve OC efficacy, the mechanisms of TOC have been carefully investigated from the macroscopic (e.g. tissues, tissue phantoms) to microscopic (molecular) level and at both *in vitro* and *in vivo* studies [53].

There are a few main mechanisms of light scattering reduction induced by an OCA: dehydration of tissue constituents; partial replacement of the interstitial ECF by the immersion substance; and structural modification or dissociation of collagen [6, 46]. The refractive index matching was regarded as the major mechanism of TOC. Both the first and the second processes mostly cause matching of the refractive indices of the tissue scatterers (cell compartments, collagen, and elastin fibers) and the cytoplasm and ECF.

The diffusion of OCAs with higher refractive indices and higher osmolality into tissues will match the refractive indices of tissue components with extracellular fluid, thus reducing the scattering of tissue. This is regarded as the major mechanism of tissue OC [53]. The refractive index matching is manifested in the reduction of the scattering coefficient. Structural modification can lead to tissue shrinkage; as a result, the increased constructive interference of the elementary scattered fields may significantly increase tissue transmittance. Tissue dehydration can be an important mechanism of TOC [53]. Application of a hyperosmotic OCA to the tissue surface induces water flux from the interstitial space to the tissue surface and even out of the tissue, and, consequently, draws water from cells and/ or collagen fibers. Due to the reduction of water content in the interstitial space, the overall thickness of tissue is reduced, making it denser (more ordered). For tissue initial structure, in particular, the ability to order its scattering components at dehydration may have a considerable inclusion in the OC effect. Addition mechanisms include the dissociation of collagen fibers. Collagen fibers have complex self-assembled structures and are the major scattering centers in tissues [53]. If an OCA, like sugars, is applied to living tissue for a limited time, it could reversibly destabilize the collagen structure by interactions of hydrogen bonds between collagen and OCA molecules, i.e. it could lead to reversible collagen solubility that correlates with TOC potential. A lesser hydrodynamic radius of tissue scatterers (collagen fibers) in that case could be the main reason for a reduction in tissue light scattering [6].

All these mechanisms usually coexist and lead to a significant decrease in the reduced scattering coefficient of fibrous connective tissues, such as skin dermis, eye sclera, tendon, and skeletal muscle. Therefore, it will be important and helpful to understand the microregulation mechanism and establish an optimized OC process by monitoring the dynamics of tissue microstructures at OCA immersion using some optical scheme.

There are some recent research works focusing on the polarization features during the TOC process. Based on the above anisotropic tissue model and polarization sensitive Monte Carlo simulation program, the influence of refractive index matching on polarized photons behaviors and how polarization parameters change with the OC process is simulated.

MC simulations based on anisotropic sclera-mimicking models is conducted to examine the polarization features in Mueller matrix polar decomposition (MMPD) parameters during the refractive index matching process, which is one of the major mechanisms of OC. By changing the parameters of the models, wavelengths, and detection geometries, simulations demonstrate how the depolarization coefficient and retardance vary during the refractive index matching process and explain the polarization features using the average value and standard deviation of scattering numbers of the detected photons. The depth-resolved polarization features are also simulated during the gradual progression of the refractive index matching process. These simulations indicate that the refractive index matching process increases the depth of polarization measurements and may lead to higher contrast between tissues of different anisotropies in deeper layers. MMPD-derived polarization parameters can characterize the refractive index matching process qualitatively [57].

To model the scattering and polarization properties of cleared tissue, Chen et al. proposed an SCBM and a corresponding MC simulation program, as mentioned here. The SCBM consists of three key components: spherical scatterers, infinitely long cylindrical scatterers, and a birefringent interstitial medium. Different models can be developed by various combinations of the two types of scatterers and a birefringent interstitial medium according to the characteristics of the tissues under study.

Reference [57] simulates the sclera using an SCSM. The sclera mainly consists of long collagen fibrils with variable diameters, which form lamellar bundles There are occasionally large empty spaces between the bundles. Elastic fibers and microfibrils are occasionally found between the collagen bundles. There also exist proteoglycan filaments in association with the collagen D-band throughout all levels of the sclera, and the filaments appear in three orientations. In the model, cylindrical scatterers represent long collagen fibrils, and spherical scatterers account for the inclusion of short collagen fibrils randomly distributed among basic aligned, much longer fibrils, the proteoglycans, distributed between the collagen bundles and cell organelles. The ratio between scattering coefficients induced by spherical scatterers and cylindrical scatterers (represented by S/C) is varied to create systems of different anisotropy degrees.

According to the forward and backward simulated detection configuration, the average value and standard deviation of scattering numbers get smaller with the refractive index matching process as shown in Figure 2.24 [57]. By contrast, the MMPD parameters for 800 nm are smaller, and is more

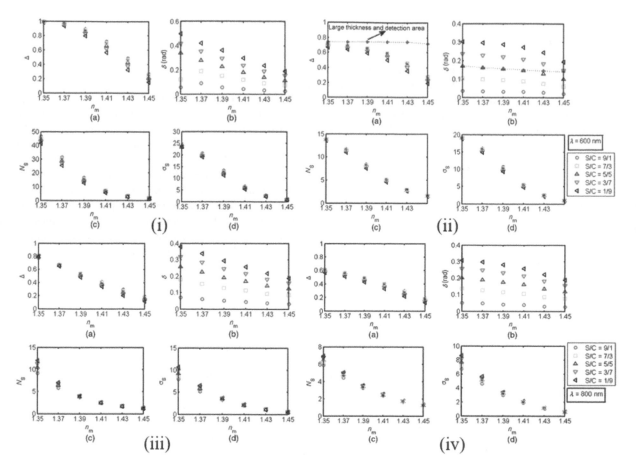

FIGURE 2.24 The single-layer sclera model for the forward configuration and backward configuration with incident wavelength of 600nm and 800nm respectively [30].

obvious in the initial stage of the refractive index matching, because the depolarization is induced by scattering of both types of scatterers and retardance by scattering of the cylindrical scatterers. Further, to study the gradual refractive index matching process with depth, we use a three-layer sclera scattering model. The process of the refractive index matching is divided into four stages with the refractive indexes of the three layers. MMPD parameters from photon backscattered from different layers at different OC stages is shown in Figure 2.25 and Figure 2.26.

According to these simulated figures, the refractive index matching changes the polarization parameters from the deeper layer to a greater extent, probably due to longer optical path of photons from the deeper layer. Meanwhile, the discrepancy of polarization parameters among different layers decreases with the refractive index matching as the whole sample gets more and more transparent. It can be observed that the trend of polarization parameters with depth gets more and more linear with OC stages, indicating that the retardance effect of cylindrical scattering approaches the effect of birefringence during the refractive index matching process. By Figure 2.25, it is observed that there is increasing contrast of depolarization coefficients for different anisotropic samples in the deep layer, making it possible to differentiate different tissue anisotropy by polarization methods. It also indicates the improvement of polarization imaging depth by refractive index matching.

With the help of Monte Carlo simulations on OC, the refractive index matching process suppresses scattering in both forward and backward detection configurations, reducing the depolarization by scattering of both types of scatterers, and the retardance induced by cylindrical scatterers. It also indicates that OC can enhance the imaging depths of polarization measurements and improve the contrast of specific polarization parameters in deeper layers.

In addition, by comparing experiments and simulations, we can further examine how the polarization contrast changes with the OC and what characteristics can be suitable for the polarization imaging applied in TOC. Experimental results show the decreased depolarization and the increased retardance of an anisotropic tissue sample immersed in glycerol solution. Here, the retardance is due to molecular birefringence and fibrous microstructures, so we can investigate the polarization image contrast based on these two anisotropic factors in the following simulations [76].

As shown in Figure 2.27, the image contrast of the retardance parameter due to interstitial birefringence will be improved with the immersion process, and meantime the difference of depolarization parameter with OC is also slightly increased, which implies the possible improvement of polarization detection based on birefringence-induced anisotropy by OC. As shown in Figure 2.28, the image contrast of retardance parameter due to cylindrical scatterers will be improved in the early stages of OC, and the difference between depolarization will

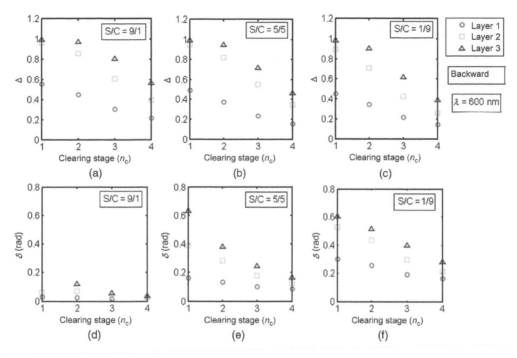

FIGURE 2.25 The three-layer sclera model for the backward configuration with the incident wavelength of 600 nm [30].

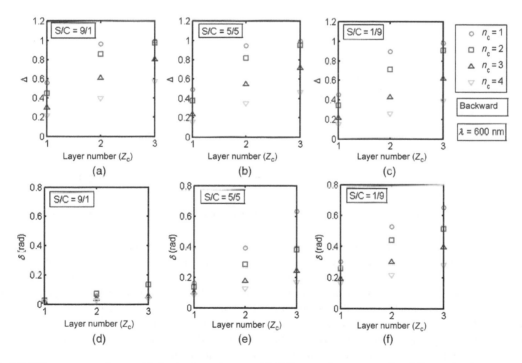

FIGURE 2.26 MMPD parameters change with the depth number zc, with different makers representing different clearing stages nc, the ratios of scattering coefficients induced by S/C are 9/1, 5/5, and 1/9, respectively [30].

be apparent in the later time of OC, which implies the possible improvement of polarization detection based on cylinder induced anisotropy by OC.

Comparison between simulations and experiments

Due to the complexity of the structures and properties of biological tissues, it is hard to solve the Mueller matrix of light transport in turbid biological tissues with electromagnetic theory. MC simulations offer a flexible yet rigorous approach to examining in detail the polarization behaviors of photons as they transport in turbid tissues, unaffected by the morphology and boundary conditions of the sample [30].

In previous studies, the SCBM and corresponding PSMC simulation program have been developed for anisotropic tissues to investigate the polarization features and explore the microstructural origin of pathological abnormality in several tissues [29, 42, 64].

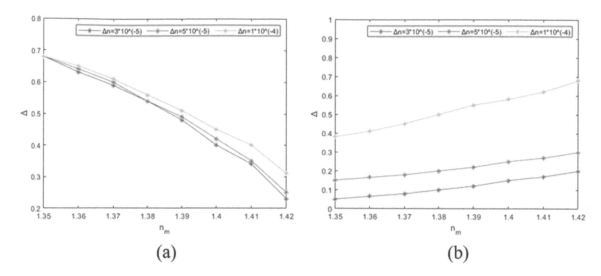

FIGURE 2.27 The effect of birefringence on depolarization during refractive index matching.

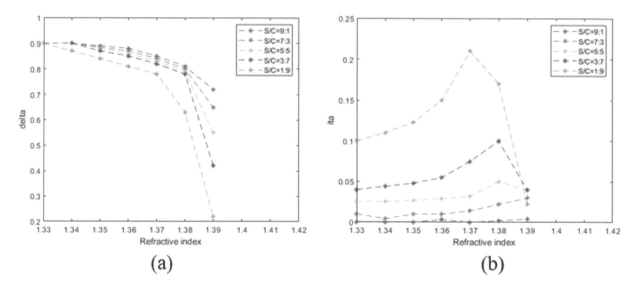

FIGURE 2.28 The influence of the change of the ratio of sphere and cylinder on depolarization in the process of refractive index matching.

Aiming for a deeper insight into how the microstructures of the skin vary during the TOC, we set up an SCBM of the skin and carried out simulations corresponding to different TOC mechanisms. We can trace the spatial and polarization features of the photons and systematically study the relationship between the polarization and structural changes during TOC.

It has been demonstrated by simulations that apart from the birefringence of the interstitial media, the size, density [77], and birefringence [78] of well aligned cylindrical scatterers may also contribute to the anisotropic optical properties of the sample. MC simulations allow us to separate these different contributions and help us to understand how the skin's microstructure changes during OCA immersion.

According to the literature [79], the average scattering properties of the skin are defined by the scattering properties of the dermis, which mainly consists of fibrous structures where collagen fibrils are packed in collagen bundles and form a lamellae structure. To understand how immersion in the glycerol solution changes the microstructure and hence the polarization

characteristics of the skin sample, we simulate the backscattering Mueller matrices using an SCBM model which approximates the skin to a mixture of spherical and infinitely long cylindrical scatterers embedded in a birefringent interstitial medium. A GPU program is used, and the photon scattering calculations of 10^7 photons only take about 38.8 s [80].

The SCBM model parameters are set as follows. The diameters of the scatterers are 0.2 μm for the spheres and 1.5 μm for the cylinders (bundles of collagen fibrils). The scattering coefficients for spherical and cylindrical scatterers are 20 cm^{-1} and 180 cm^{-1}, respectively. The cylindrical scatterers are aligned along the y-axis direction. Considering that the collagen fibers are packed in bundles and arranged in a lamellae structure, the directions of the cylinders are allowed to fluctuate by 40 degrees within the lamellae but by 5 degrees perpendicular to the lamellae. The refractive index (RI) is 1.43 for both types of scatterers. For the interstitial medium, RI is 1.35, birefringence is 3.0×10^{-5}, and the optical axis is in the y-axis direction. The thickness of the sample is 0.5 mm.

Refractive index matching (RIM) is one of the main mechanisms of TOC. When a nude mouse skin sample is immersed in a glycerol solution, glycerol molecules will migrate into the interstitial fluid and increase its RI, resulting in reduced tissue scattering. We simulate the RIM process by increasing the RI of the medium from 1.35 to 1.40 at 0.01 intervals, corresponding to the RIM stage progress from the first to the sixth stage, as shown in Figure 2.29. The behaviors revealed in the simulations can be compared with the experimental results, and we can see that the simulations on RIM can better reflect the change of m_{22} and m_{33} and m_{44} and m_{34}, and m_{43}. However, in terms of quantitative change, the simple RIM model cannot explain quantitatively the variations of m_{34} and m_{43} and m_{12} and m_{21}, and the difference between m_{22} and m_{33} well.

Since m_{34} and m_{43} and m_{12} and m_{21} are related mainly to the anisotropic properties induced by birefringence and scattering of the aligned cylinders, respectively [59], we need to consider effects due to the variations of the interstitial birefringence and cylindrical scatterers during the immersion. It has been reported [54] that the immersion in OCA will make the fibrils in the skin become more densely packed, which should improve the alignment of cylindrical scatterers and increase the birefringence value. Therefore, we should consider these effects in the simulations.

In Figure 2.29 (b) and (c), we simulate another two cases of single mechanism OC model: the increasing extracellular interstitial birefringence and the decreasing fluctuation of fiber orientation with the immersion time. These two simulation results can better show the difference of principal diagonal elements and the change trend and amplitude of off-diagonal elements using the experimental results as a reference. From the above results, we can infer that the polarization changes with OC not only come from the refractive index matching, but may also be due to the interstitial birefringence and the enhanced order degree of fibrous microstructures. Therefore, the comprehensive simulation of three physical processes at the same time can more reasonably explain the skin soaking in glycerin solution. Figure 2.30 shows the simulated Mueller

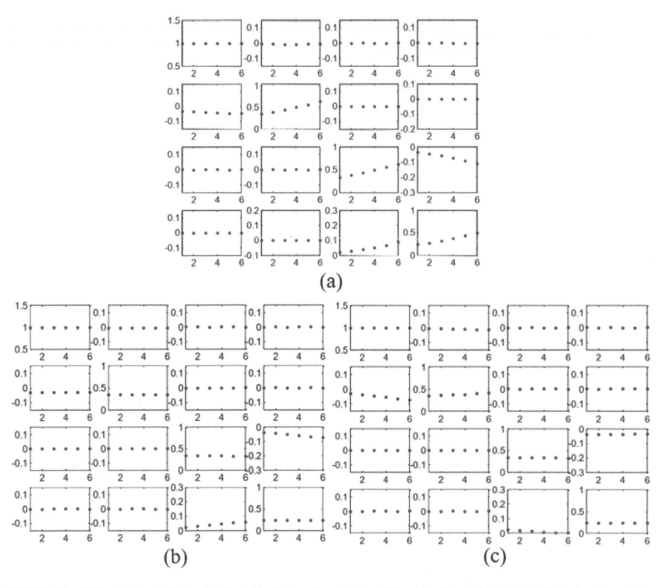

FIGURE 2.29 (a) The variation of MMEs with RI matching between the scatterers and the surrounding medium by Monte Carlo simulations, (b) the increasing extracellular interstitial birefringence and (c) the decreasing fluctuation of fiber orientation [30].

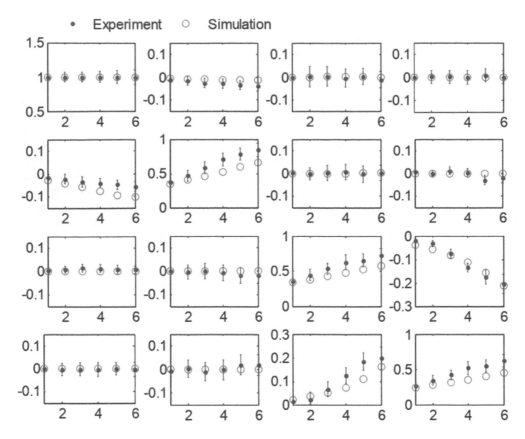

FIGURE 2.30 The MMEs by the TOC experiment and simulation, the map in i^{th} row and j^{th} column corresponds to m_{ij} [30].

matrices with all the three possible TOC mechanisms, i.e. increasing RI of the medium from 1.35 to 1.40 at 0.01 intervals, increasing the interstitial birefringence from 3.0×10^{-5} to 6.0×10^{-5} at 0.5×10^{-5} intervals and decreasing the orientation fluctuation of the cylindrical scatterers in the y-axis direction from 40 degrees to 15 degrees at 5-degree intervals. The experimental results are also plotted in the same graph for direct comparison.

The results show that the simulations accurately reproduce some of the experimental features, such as the varying trends and amplitudes of the diagonal elements m_{22} and m_{33} and m_{44}, and off-diagonal elements m_{34} and m_{43}. The relative values of m_{12} and m_{21} and m_{34} and m_{43} and m_{22} and m_{33} and m_{44} by both simulations and experiments also agree qualitatively with each other.

During the TOC process, the increasing degree of RIM between the scatterers and the interstitial medium reduces the tissue scattering coefficient, which results in a reduction in depolarization as well as an increase in the mean free path of the photons. In the meantime, the interstitial birefringence due to the increase of mean free path contributes more anisotropic features in tissues, while the improving alignment of fibrous scatterers makes a minor contribution. Some other possible microstructural and optical changes during OC are also investigated and show rather small impacts, such as variations in the sample thickness, the diameter of cylindrical scatterers, and the ratio between the spherical and cylindrical scatterers [30]. All these simulations verify that the feasible polarization scattering explanation is the co-action of RIM, the increase of

the interstitial birefringence and the alignment of the cylindrical scatterers.

Further, MC simulation also can help us to estimate the microstructural changes with the OC process under different settings of skin model parameters. Figure 2.31 shows the various influences of several model parameters on the Mueller matrix, including the ratio of spherical to cylindrical scatterers, diameter of cylinders, total scattering coefficient, birefringence, and thickness of tissue. Based on these simulation results, we can infer that these typical microphysical parameters in skin samples have little effect on the general characteristics of the Mueller matrix.

To summarize, based on the compound action of several possible mechanisms, MC simulations predict the corresponding Mueller matrices. Comparisons between the experiments and the simulations help us to analyze which mechanism or combination of mechanisms can describe the polarization features of TOC very well. Therefore, it is difficult to use an individual possible mechanism to explain the polarization phenomena due to OC, so in Reference [56], when we consider different agents, the comparison between the experimental results with simulations is necessary based on a single mechanism model and a combined model, respectively.

To mimic TOC process by different agents, various dynamic models corresponding to several possible OC mechanisms are considered, including RIM, tissue shrinkage by dehydration, the fluctuation of the birefringence effect in an intercellular substance, and ordering of fiber arrangement. To speculate how the microstructural behavior of tissues has been changed

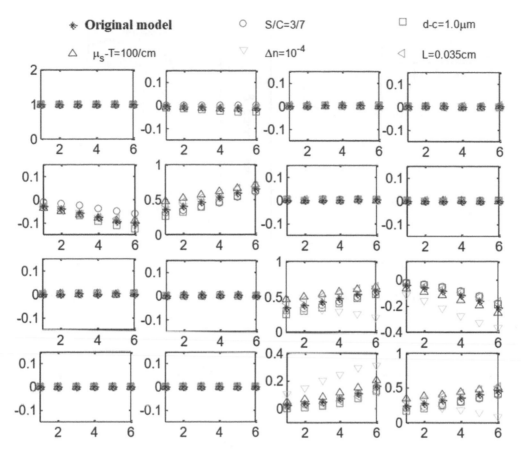

FIGURE 2.31 The influence of different skin model parameters on Mueller matrix element changes with the permeability process.

by OCAs and then understand the different OC mechanisms of two OCAs, Reference [56] establishes dynamic tissue models corresponding to various possible TOC mechanisms, and then uses MC simulations to verify the influence of different agents on the microstructural properties of cleared tissues.

The simulations of single mechanism, including RIM, collagen shrinkage, and more orderly fibers, are compared with experiments [56]. In the simulation model mimicking RIM, we set a dynamic process of increasing refractive index from 1.35 to 1.40 with an interval of 0.01. In the simulation model mimicking more orderly fibers, the FWHM of the orientation distribution of cylindrical scatterers is reduced from 18 degrees to 8 degrees with an interval of 2 degrees. In the case of the collagen shrinkage OC model, we reduce the cylinder diameter step by step from 1.5 μm to 0.5 μm. Correspondingly, in simulations, the variation in scattering coefficient due to the change in scatterer size and refractive index of the intercellular substance has been considered.

According to the above experiments, the phase retardance of tissue samples has an opposite trend with OC of two kinds of OCAs. By contrast, the simulated retardance changes just a little, inconsistent with the obvious fluctuation in experiments (solid line in Figure 2.32). Therefore, those simulations that only consider one single mechanism cannot cause an apparent polarization change as experimental phenomena. Figure 2.32 implies that the polarization changes by different agents during TOC of OCAs should be understood by a comprehensive effect of multiple possible mechanisms. In the next simulations, two

combined models are established to mimic the dynamic influence of formamide: one is RIM plus decreased birefringence; the other is collagen shrinkage plus birefringence reduction. In the case of using saturated sucrose as the agent, we establish three models: RIM plus increased birefringence model, collagen shrinkage plus increased birefringence model, and fiber ordering plus increased birefringence model.

The simulated MMPD with OC has been shown in Figure 2.33 and can be compared with experimental results marked by solid lines. Compared with single mechanism models, we introduce the variation of birefringence due to the immersion of OCAs into the intercellular substance. The birefringence setting is decreased or increased gradually with an interval of $1e^{-5}$.

As shown in Figure 2.33 (a) and (b), both OC models using formamide can describe the trends of experimental phenomena qualitatively. Relatively speaking, the model including collagen shrinkage seems more consistent with the experiments. From Figure 2.33 (c) and (d), the major polarization variation is the apparent decline in depolarization parameter, Δ, with OC time. Among three models using saturated sucrose, the one combining RIM and increased birefringence agrees well with experiments.

According to the above simulations, it has been confirmed again that a combined model involving multiple mechanisms can better explain the trend of phase retardance with OC than any single mechanism model. Considering the difference between simulation results of various combined models is depolarization.

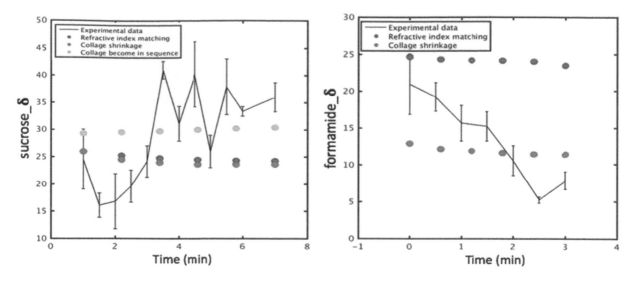

FIGURE 2.32 Comparison of phase retardance between experimental results and MC simulations based on one single mechanism models: (a) OC using sucrose; (b) OC using formamide [56].

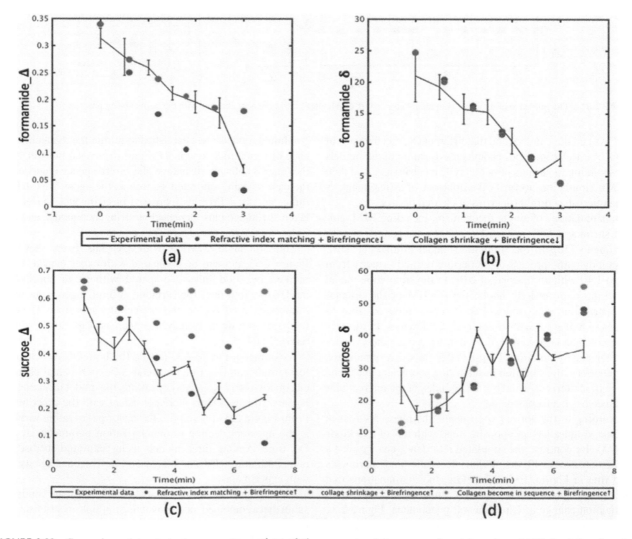

FIGURE 2.33 Comparison of depolarization parameters and retardation parameters between experimental results and MC simulations based on multiple mechanism models: (a), (b) with OC using formamide; (c), (d) with OC using sucrose [56].

the change of depolarization parameter. Next, we can investigate three diagonal elements of the Mueller matrix, which are closely related to the depolarization phenomena of measured tissues. From Figure 2.34, by comparing experimental data with simulation results, we can deduce similar possible OC models. Specifically, the experimental depolarization parameter, Δ, cannot be close to 1 after formamide OC (shown in Figure 2.34(a)), which supports the model involving collagen shrinkage plus birefringence reduction again. From Figure 2.34(b), only the combined model involving RIM and increased birefringence can better mimic the increase of Δ using saturated sucrose as an agent than the other two models.

The above experimental results (shown in Figure 2.33(c)) have demonstrated rather regular oscillations of depolarization Δ of the skin sample with OC. Similar phenomena can be observed on M_{44} image of two kinds of animal skeletal muscle samples in Figure 2.35. These oscillations occurred with a time period of about 40 s. Such quasiperiodic oscillations were described for the first time in Reference [81]. It was also hypothesized that the oscillations are the result of temporal-spatially irregular OCA diffusion driven by a local multistep dehydration of collagen and dilution of the interstitial fluid.

To find out some possible explanations for such regular oscillations of polarization features, some dynamic parameter modulations can be introduced to the original MC simulation to describe the TOC. These further simulations consider four cases: 1) the fluctuation of optical birefringence of intercellular substance; 2) alternating shrinkage and expansion of fibrous and cellular level structures; 3) alternation of fiber content; and 4) alternation of fiber orientation. The corresponding simulation results can be seen in Figure 2.36 (a)~(d).

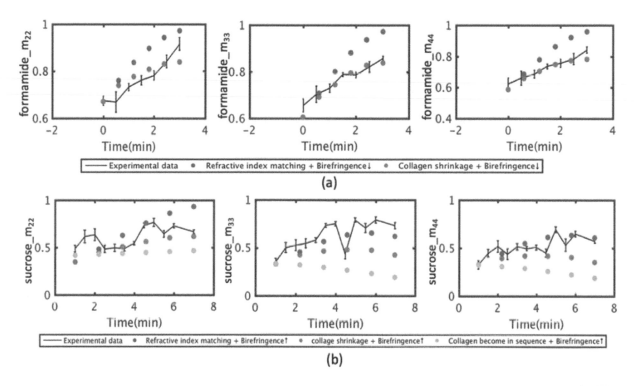

FIGURE 2.34 MC simulation results of the three Mueller matrix elements: m22, m33 and m34 using a sphere-cylinder birefringence model (SCBM) [56].

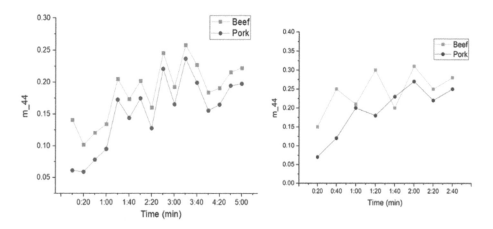

FIGURE 2.35 The oscillating M_{44} of experimental results of animal skeletal muscle.

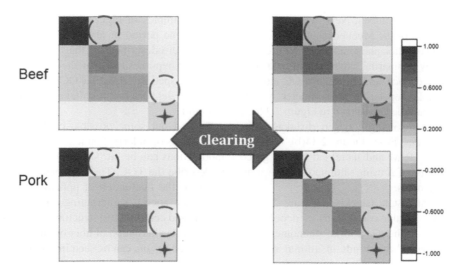

FIGURE 2.36 Notable experimental Mueller matrix elements as a reference for the simulation results.

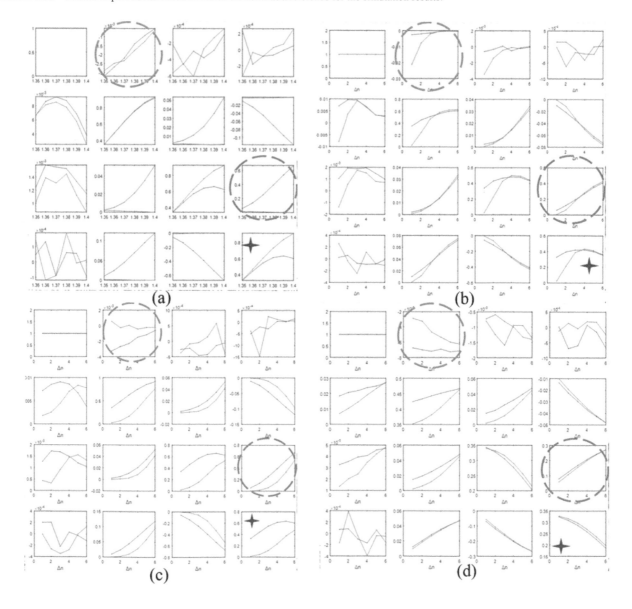

FIGURE 2.37 MC simulations based on TOC models with different oscillating parameter setting.

On the oscillation of polarization parameters, we need to compare various simulated Mueller matrices based on the above assumptions with the key changes of experimental Mueller matrix, including the oscillating M_{44}, the increasing absolute values of M_{34} related to optical birefringence, and M_{21} related to fibrous scatterers, as shown in Figure 2.36. By comparison, for the former three cases, the simulated M_{21} or M_{44} cannot agree with the experimental phenomena. Only when we introduce the fluctuation of fiber orientation in the dynamic model of the TOC process can both the variation of off-diagonal elements and the oscillating M_{44} be simulated (Figure 2.37).

Summary

Recent research progress made through a combination of experiments and simulations has demonstrated that the Mueller matrix microscope is a good tool to observe the differences of polarization features due to the respective OC mechanisms of different agents. It has been also indicated that different agents have respective influence on the cleared tissue, such as shrinkage due to dehydration, changes in fiber orientation, and birefringence variation in intercellular substance. Based on several possible OC mechanisms, the corresponding single factor models and combined models are designed and compared with experimental results, respectively. These investigations, including experiments and simulations, provide a way to understand the OC process according to the connection between polarization changes and the microphysical features of tissues.

Acknowledgments

The work has been supported in part by the National Natural Science Foundation of China(NSFC) under Grants 61405102, 61527826, 41475125,11174178,11374179, 10974114, 60778044, 60578003, 61205199, and 41106034; in part by the National Key Research and Development Program of China under Grants 2016YFF0103000 and 2016YFC0208600; and in part by the Science and Technology Project of Shenzhen under Grants JCYJ20170412170814624 and GJHZ20150316160614844. Valery V. Tuchin is grateful for support from RFBR grant # 18-52-16025 and from the Government of the Russian Federation (grant # 14.W03.31.0023).

REFERENCES

1. S. Alali and A. Vitkin, "Polarized light imaging in biomedicine: Emerging Mueller matrix methodologies for bulk tissue assessment." *J. Biomed. Opt.* **20**(6), 061104 (2015).
2. N. Nishizawa, A. Hamada, K. Takahashi, T. Kuchimaru, and H. Munekata, "Monte Carlo simulation of scattered circularly polarized light in biological tissues for detection technique of abnormal tissues using spin-polarized light emitting diodes." *Jpn. J. Appl. Phys.* **59**(SE), SEEG03 (2020).
3. V.V. Tuchin, L. Wang, and D. Zimnyakov, *Optical Polarization in Biomedical Applications*, Springer, Berlin, Germany (2006).
4. D. Pallavi, A. Karim, S.J. Cecil, and H.G. Robert, "Detection of colon cancer by continuous-wave terahertz polarization imaging technique." *J. Biomed. Opt.* **18**(9), 090504 (2013).
5. A. Pierangelo, A. Nazac, A. Benali, et al., "Polarimetric imaging of uterine cervix: A case study." *Opt. Express.* **21**(12), 14120–14130 (2013).
6. V.V. Tuchin, "Polarized light interaction with tissues." *J. Biomed. Opt.* **21**(7), 071114 (2016).
7. J. Qi and D.S. Elson, "Mueller polarimetric imaging for surgical and diagnostic applications: A review." *J Biophotonics.* **10**(8), 950–982 (2017).
8. N. Ghosh and I.A. Vitkin, "Tissue polarimetry: Concepts, challenges, applications, and outlook." *J. Biomed. Opt.* **16**(11), 110801 (2011).
9. M.I.M.J.W. Hovenier, "Light scattering by nonspherical particles: Theory, measurements, and applications." *Meas. Sci. Technol.* **11**(12), 1827–1827 (2000).
10. H. He, R. Liao, N. Zeng, et al., "Mueller matrix polarimetry: An emerging new tool for characterizing the microstructural feature of complex biological specimen." *J. Lightwave Technol.* **37**(11), 2534–2548 (2019).
11. D.H. Goldstein and R.A. Chipman, "Error analysis of a Mueller matrix polarimeter." *J Opt Soc Am A.* **7**(4), 693–700 (1990).
12. D.H. Goldstein, "Mueller matrix dual-rotating retarder polarimeter." *Appl. Opt.* **31**(31), 6676–6683 (1992).
13. S. Inoué, "Polarization optical studies of the mitotic spindle." *Chromosoma.* **5**(5), 487–500 (1953).
14. L. Liu, R. Oldenbourg, J.R. Trimarchi, and D.L. Keefe, "A reliable, noninvasive technique for spindle imaging and enucleation of mammalian oocytes." *Nat. Biotechnol.* **18**(2), 223–225 (2000).
15. O. Arteaga, M. Baldris, J. Anto, et al., "Mueller matrix microscope with a dual continuous rotating compensator setup and digital demodulation." *Appl. Opt.* **53**(10), 2236–2245 (2014).
16. Y. Wang, H. He, J. Chang, et al., "Differentiating characteristic microstructural features of cancerous tissues using Mueller matrix microscope." *Micron.* **79**, 8–15 (2015).
17. Y. Wang, H. He, J. Chang, et al., "Mueller matrix microscope: A quantitative tool to facilitate detections and fibrosis scorings of liver cirrhosis and cancer tissues." *J. Biomed. Opt.* **21**(7), 071112 (2016).
18. H. He, M. Sun, N. Zeng, et al., "Mapping local orientation of aligned fibrous scatterers for cancerous tissues using backscattering Mueller matrix imaging." *J. Biomed. Opt.* **19**(10), 106007 (2014).
19. Y. Dong, J. Qi, H. He, et al., "Quantitatively characterizing the microstructural features of breast ductal carcinoma tissues in different progression stages by Mueller matrix microscope." *Biomed Opt Express.* **8**(8), 3643–3655 (2017).
20. J. Zhou, H. He, Z. Chen, Y. Wang, and H. Ma, "Modulus design multiwavelength polarization microscope for transmission Mueller matrix imaging." *J. Biomed. Opt.* **23**(1), 016007 (2018).
21. J. Chang, H. He, Y. Wang, et al., "Division of focal plane polarimeter-based 3×4 Mueller matrix microscope: A potential tool for quick diagnosis of human carcinoma tissues." *J. Biomed. Opt.* **21**(5), 056002 (2016).

22. Y. Liu, T. York, W. Akers, et al., "Complementary fluorescence-polarization microscopy using division-of-focal-plane polarization imaging sensor." *J. Biomed. Opt.* **17**(11), 116001 (2012).

23. W.L. Hsu, J. Davis, K. Balakrishnan, et al., "Polarization microscope using a near infrared full-Stokes imaging polarimeter." *Opt. Express.* **23**(4), 4357–68 (2015).

24. R.R. Anderson, "Polarized light examination and photography of the skin." *Arch. Dermatol.* **127**(7), 1000–1005 (1991).

25. S.L. Jacques, J.R. Roman, and K. Lee, "Imaging superficial tissues with polarized light." *Lasers Surg. Med.* **26**(2), 119–129 (2000).

26. S.L. Jacques, J.C. Ramella-Roman, and K. Lee, "Imaging skin pathology with polarized light." *J. Biomed. Opt.* **7**(3), 329–340 (2002).

27. H. He, N. Zeng, W. Li, et al., "Two-dimensional backscattering Mueller matrix of sphere-cylinder scattering medium." *Opt. Lett.* **35**(14), 2323–2325 (2010).

28. A. Pierangelo, S. Manhas, A. Benali, et al., "Multispectral Mueller polarimetric imaging detecting residual cancer and cancer regression after neoadjuvant treatment for colorectal carcinomas." *J. Biomed. Opt.* **18**(4), 046014 (2013).

29. M. Sun, H. He, N. Zeng, et al., "Characterizing the microstructures of biological tissues using Mueller matrix and transformed polarization parameters." *Biomed. Opt. Express.* **5**(12), 4223–4234 (2014).

30. D. Chen, N. Zeng, Q. Xie, et al., "Mueller matrix polarimetry for characterizing microstructural variation of nude mouse skin during tissue optical clearing." *Biomed. Opt. Express.* **8**(8), 3559–3570 (2017).

31. M. Dubreuil, S. Rivet, B. Le Jeune, and J. Cariou, "Snapshot Mueller matrix polarimeter by wavelength polarization coding." *Opt. Express.* **15**(21), 13660–13668 (2007).

32. B. Laude-Boulesteix, A. De Martino, B. Drevillon, and L. Schwartz, "Mueller polarimetric imaging system with liquid crystals." *Appl. Opt.* **43**(14), 2824–2832 (2004).

33. C. He, H. He, X. Li, et al., "Quantitatively differentiating microstructures of tissues by frequency distributions of Mueller matrix images." *J. Biomed. Opt.* **20**(10), 105009 (2015).

34. M. Sun, H. He, N. Zeng, et al., "Probing microstructural information of anisotropic scattering media using rotation-independent polarization parameters." *Appl. Opt.* **53**(14), 2949–2955 (2014).

35. P. Ghassemi, L.T. Moffatt, J.W. Shupp, and J.C. Ramella-Roman, "A new approach for optical assessment of directional anisotropy in turbid media." *J Biophotonics.* **9**(1–2), 100–108 (2016).

36. S.Y. Lu and R.A. Chipman, "Interpretation of Mueller matrices based on polar decomposition." *J. Opt. Soc. Am. A.* **13**(5), 1106–1113 (1996).

37. R. Ossikovski, A. De Martino, and S. Guyot, "Forward and reverse product decompositions of depolarizing Mueller matrices." *Opt. Lett.* **32**(6), 689–691 (2007).

38. N. Agarwal, J. Yoon, E. Garcia-Caurel, et al., "Spatial evolution of depolarization in homogeneous turbid media within the differential Mueller matrix formalism." *Opt. Lett.* **40**(23), 5634–5637 (2015).

39. H. He, N. Zeng, E. Du, et al., "A possible quantitative Mueller matrix transformation technique for anisotropic scattering media." *Photonics Lasers Med.* **2**(2), 129–137 (2013).

40. A. Vitkin, N. Ghosh, and A.d. Martino, "Tissue polarimetry." *Photonics.* 239–321 (2015).

41. P. Shukla and A. Pradhan, "Mueller decomposition images for cervical tissue: Potential for discriminating normal and dysplastic states." *Opt. Express.* **17**(3), 1600–1609 (2009).

42. C. He, H. He, J. Chang, et al., "Characterizing microstructures of cancerous tissues using multispectral transformed Mueller matrix polarization parameters." *Biomed. Opt. Express.* **6**(8), 2934–2945 (2015).

43. P. Li, D. Lv, H. He, and H. Ma, "Separating azimuthal orientation dependence in polarization measurements of anisotropic media." *Opt. Express.* **26**(4), 3791–3800 (2018).

44. N. Zeng, X. Jiang, Q. Gao, Y. He, and H. Ma, "Linear polarization difference imaging and its potential applications." *Appl. Opt.* **48**(35), 6734–6739 (2009).

45. R. Liao, N. Zeng, X. Jiang, et al., "Rotating linear polarization imaging technique for anisotropic tissues." *J. Biomed. Opt.* **15**(3), 036014 (2010).

46. K.V. Larin, M.G. Ghosn, A.N. Bashkatov, et al., "Optical clearing for OCT image enhancement and in-depth monitoring of molecular diffusion." *IEEE J Sel Top Quantum Electron.* **18**(3), 1244–1259 (2012).

47. E.A. Genina, A.N. Bashkatov, and V.V. Tuchin, "Tissue optical immersion clearing." *Expert Rev. Med. Devices.* **7**(6), 825–842 (2010).

48. O. Nadiarnykh and P.J. Campagnola, "Retention of polarization signatures in SHG microscopy of scattering tissues through optical clearing." *Opt. Express.* **17**(7), 5794–5806 (2009).

49. C. Macdonald and I. Meglinski, "Backscattering of circular polarized light from a disperse random medium influenced by optical clearing." *Laser Phys. Lett.* **8**(4), 324–328 (2011).

50. M. Borovkova, A. Bykov, A. Popov, and I. Meglinski, "Role of scattering and birefringence in phase retardation revealed by locus of Stokes vector on Poincare sphere." *J. Biomed. Opt.* **25**(5), 057001 (2020).

51. M. Dubreuil, P. Babilotte, L. Martin, et al., "Mueller matrix polarimetry for improved liver fibrosis diagnosis." *Opt. Lett.* **37**(6), 1061–1063 (2012).

52. T. Novikova, A. Pierangelo, S. Manhas, et al., "The origins of polarimetric image contrast between healthy and cancerous human colon tissue." *Appl. Phys. Lett.* **102**(24), 241103 (2013).

53. D. Zhu, K.V. Larin, Q. Luo, and V.V. Tuchin, "Recent progress in tissue optical clearing." *Laser Photonics Rev..* **7**(5), 732–757 (2013).

54. C.G. Rylander, O.F. Stumpp, T.E. Milner, et al., "Dehydration mechanism of optical clearing in tissue." *J. Biomed. Opt.* **11**(4), 041117 (2006).

55. V.V. Tuchin, "Optical immersion as a new tool for controlling the optical properties of tissues and blood." *Laser Phys.* **15**(8), 1109–1136 (2005).

56. Q. Xie, N. Zeng, Y. Huang, V.V. Tuchin, and H. Ma, "Study on the tissue clearing process using different agents by Mueller matrix microscope." *Biomed. Opt. Express.* **10**(7), 3269–3280 (2019).

57. D. Chen, N. Zeng, Y. Wang, et al., "Study of optical clearing in polarization measurements by Monte Carlo simulations with anisotropic tissue-mimicking models." *J. Biomed. Opt.* **21**(8), 081209 (2016).

58. M.-R. Antonelli, A. Pierangelo, T. Novikova, et al., "Mueller matrix imaging of human colon tissue for cancer diagnostics: How Monte Carlo modeling can help in the interpretation of experimental data." *Opt. Express.* **18**(10), 10200–10208 (2010).

59. H. He, C. He, J. Chang, et al., "Monitoring microstructural variations of fresh skeletal muscle tissues by Mueller matrix imaging." *J Biophotonics.* **10**(5), 664–673 (2017).

60. J.C. Ramella-Roman, S.A. Prahl, and S.L. Jacques, "Three Monte Carlo programs of polarized light transport into scattering media: Part II." *Opt. Express.* **13**(25), 10392–10405 (2005).

61. J.C. Ramella-Roman, S.A. Prahl, and S.L. Jacques, "Three Monte Carlo programs of polarized light transport into scattering media: Part I." *Opt. Express.* **13**(12), 4420–4438 (2005).

62. L.H. Wang, S.L. Jacques, and L.Q. Zheng, "MCML: Monte Carlo modeling of light transport in multilayered tissues." *Comput. Methods Programs Biomed.* **47**(2), 131–146 (1995).

63. D. Cote and I.A. Vitkin, "Robust concentration determination of optically active molecules in turbid media with validated three-dimensional polarization sensitive Monte Carlo calculations." *Opt. Express.* **13**(1), 148–163 (2005).

64. T. Yun, N. Zeng, W. Li, et al., "Monte Carlo simulation of polarized photon scattering in anisotropic media." *Opt. Express.* **17**(19), 16590–16602 (2009).

65. H. He, N. Zeng, R. Liao, et al., "Application of sphere-cylinder scattering model to skeletal muscle." *Opt. Express.* **18**(14), 15104–15112 (2010).

66. A.J. Welch and M.J.C. van Gemert, *Optical-Thermal Response of Laser-Irradiated Tissue.* Springer, Netherlands (2011).

67. L.H.V. Wang, G.L. Cote, and S.L. Jacques, "Special section guest editorial: tissue polarimetry." *J. Biomed. Opt.* **7**(3), 278 (2002).

68. R.J. McNichols and G.L. Cote, "Optical glucose sensing in biological fluids: An overview." *J. Biomed. Opt.* **5**(1), 5–16 (2000).

69. J.P. He, A. Karlsson, J. Swartling, and S. Andersson-Engels, "Light scattering by multiple red blood cells." *J. Opt. Soc. Am. A.* **21**(10), 1953–1961 (2004).

70. J.D. Keener, K.J. Chalut, J.W. Pyhtila, and A. Wax, "Application of Mie theory to determine the structure of spheroidal scatterers in biological materials." *Opt. Lett.* **32**(10), 1326–1328 (2007).

71. A. Kienle and R. Hibst, "Light guiding in biological tissue due to scattering." *Phys. Rev. Lett.* **97**(1), 018104 (2006).

72. A. Kienle, F.K. Forster, and R. Hibst, "Anisotropy of light propagation in biological tissue." *Opt. Lett.* **29**(22), 2617–2619 (2004).

73. M.F. Wood, X. Guo, and I.A. Vitkin, "Polarized light propagation in multiply scattering media exhibiting both linear birefringence and optical activity: Monte Carlo model and experimental methodology." *J. Biomed. Opt.* **12**(1), 014029 (2007).

74. E. Du, H. He, N. Zeng, et al., "Two-dimensional backscattering Mueller matrix of sphere-cylinder birefringence media." *J. Biomed. Opt.* **17**(12), 126016 (2012).

75. E. Du, H. He, N. Zeng, et al., "Mueller matrix polarimetry for differentiating characteristic features of cancerous tissues." *J. Biomed. Opt.* **19**(7), 076013 (2014).

76. Y. Huang, N. Zeng, D. Chen, et al. "Study on the influence of optical clearing on polarization imaging contrast." In International Conference on Photonics and Imaging in Biology and Medicine. W3A.96 suzhou China (2017).

77. Y. Guo, N. Zeng, H. He, et al., "A study on forward scattering Mueller matrix decomposition in anisotropic medium." *Opt. Express.* **21**(15), 18361–18370 (2013).

78. Y. Guo, C. Liu, N. Zeng, et al., "Study on retardance due to well-ordered birefringent cylinders in anisotropic scattering media." *J. Biomed. Opt.* **19**(6), 065001 (2014).

79. A.N. Bashkatov, E.A. Genina, V.I. Kochubey, and V.V. Tuchin, "Optical properties of human skin, subcutaneous and mucous tissues in the wavelength range from 400 to 2000 nm." *J. Phys. D.* **38**(15), 2543–2555 (2005).

80. P. Li, C. Liu, X. Li, H. He, and H. Ma, "GPU acceleration of Monte Carlo simulations for polarized photon scattering in anisotropic turbid media." *Appl. Opt.* **55**(27), 7468–7476 (2016).

81. V.V. Tuchin, I.L. Maksimova, D.A. Zimnyakov, et al., "Light propagation in tissues with controlled optical properties." *J. Biomed. Opt.* **2**(4), 401–417 (1997).

3

Traditional and innovative optical clearing agents

Elina A. Genina, Vadim D. Genin, Jingtan Zhu, Alexey N. Bashkatov, Dan Zhu, and Valery V. Tuchin

CONTENTS

Introduction

Currently, tissue optical clearing (TOC) is a confirmed and approved method for controlling the scattering properties of biological tissues [1–7]. The problem of increasing the accuracy of the optical diagnostic methods that scientists meet is successfully solved through the use of a well-developed optical clearing technique using optical clearing agents (OCAs) and methods for their incorporation into biological tissues.

The ultimate goal of TOC techniques is to match the refractive index throughout the tissue to reduce optical inhomogeneities causing light scattering. However, the approaches to this problem can differ significantly depending on the object of study and the OCA used. Nowadays the term *OCA* includes not only the substance (or substances) but also the technology used (TOC protocol).

TOC is developing continuously in multiple *in vitro* studies to improve the quality of optical images of fixed thick tissues and animal organs [8–14]. Optical clearing with different microscopic modalities has been demonstrated for imaging brain [9], spinal cord [8], nervous system [12, 14], bone [11], skull [14] and other organs or even the whole body of mice, insects, and embryos using specially optimized technologies of specimen preparation and multicomponent OCAs. These technologies, developed in less than a decade, have allowed us to almost completely overcome light scattering by turning the studied excised thick tissue samples, organs, or small animals into optically transparent materials. The treatment of the samples, in advance of the refractive index matching agent application step, can include a few preliminary steps with tissue fixation, dehydration or hyperhydration, delipidation, and decolorization by organic solvents, detergents, and special additives [3, 4, 6, 7].

In *in vivo* (*ex vivo*) studies, there is a need to significantly reduce the time of the treatment (not more than dozens of minutes) and use only biologically compatible OCAs [15–23] with safe penetration enhancers [15, 18, 19, 24–28] and such noninvasive or minimally invasive treatments as compression [29], *stratum corneum* stripping [30], microdermabrasion [31], microperforation [32], sonophoresis [33], laser irradiation [34], and combinations of these treatments [35–40] for enhancement of tissue permeability.

This chapter focuses on features of traditional and innovative OCAs and the development and specialization of *in vitro* and *in vivo* (*ex vivo*) optical clearing approaches and diffusion coefficients of OCAs in tissues.

Classification of OCAs

Nowadays, optical clearing techniques include some approaches dependent on certain scientific tasks: i) simple immersion with the use of hyperosmotic or lipophilic OCAs as one-component or multicomponent solutions with different chemicals and/or physical enhancers; ii) immersion with previous dehydration and delipidation with the use of organic solvents; iii) hyperhydration with the use of delipidation and partial denaturation of tissue proteins; and iv) hydrogel embedding with the use of fixation and crosslinking [3]. Some TOC protocols allow for simultaneous tissue decolorization [41, 42]. The scheme of these directions in tissue optical clearing is presented in Figure 3.1. The first approach is useful for *in vitro*, *ex vivo*, and *in vivo* TOC, while other approaches are applied for the study of isolated tissue samples and small animals *in vitro* only.

DOI: 10.1201/9781003025252-4

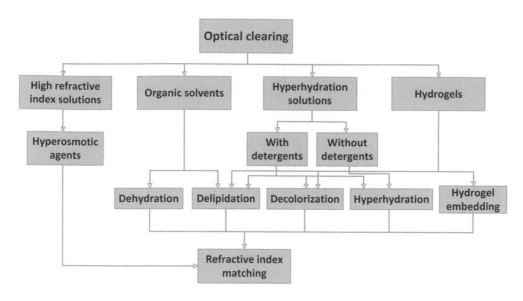

FIGURE 3.1 Scheme of approaches to tissue optical clearing.

Most simple immersion techniques do not remove lipids and simply try to match the average refractive index of a tissue (>1.45) by replacing the liquid within tissue with a high refractive index solution. For simple immersion, the tissue sample to be cleared is placed in clearing solution for a period of from minutes to months in dependence on the type of the tissue and tasks. The immersion OCAs used can be roughly divided into polyatomic alcohols (glycerol, ethylene glycol, polyethylene glycols (PEGs), propylene glycol, polypropylene glycol, combined mixtures on the base of polypropylene glycols and polyethylene glycols, mannitol, sorbitol, xylitol, etc.) [17, 40, 43–55], solutions of sugars (glucose, fructose, ribose, sucrose, etc.) [22, 23, 46, 56–69], organic acids (oleic, linoleic, salicylic acids) [18, 26, 39, 48, 49, 70], oils, [20], organic solvents (dimethyl sulphoxide (DMSO), [48, 70–74], and x-ray contrast agents (Verografin™, Trasograf™, Hypaque™, Omnipaque™, etc.) [1, 5, 75–77]. All these OCAs are used separately or in different combinations for enhancement of penetration. In the case of *in vivo* and *ex vivo* study, biological safe solvents as ethanol, azone, thiazone, DMSO, and propylene glycol in small concentration are also included in the OCAs [24–26, 28, 35, 37, 70, 72, 73, 78, 79]. It is important that this approach provides tissue viability, and after removing of OCAs by saline diffusion (*ex vivo*) or during natural water diffusion from surrounding tissues (*in vivo*), the tissue structure, functions, and optical properties are totally restored [7, 17–40, 51, 54, 56–64, 68–73, 80, 81].

TOC *in vitro* are realized using complex solutions. TOC by simple immersion mainly uses hyperosmotic agents with high refractive index including sugars [82–85], diatrizoic acid [86], formamide and formamide/PEG [87], 2,2'-thiodiethanol [88–90], etc. Optical clearing is executed by passive diffusion and finishes after the total replacement of interstitial fluid by OCA; therefore the time to clearing varies from hours to several days. Examples of the developed OCAs for the simple immersion are FocusClear (the key component being diatrizoic acid) [86], SeeDB (fructose/thioglycerol) [83], TDE (2,2'-thiodiethanol) [88–90], sucrose [82] etc. The FRUIT OCA mixes urea with fructose to lower the overall viscosity

of the SeeDB and improve tissue penetration and clearing [84] (see Table 3.1).

Solvent-based clearing techniques are most commonly comprised of two steps: i) dehydration with partial lipid solvation and ii) additional lipid solvation and clearing by refractive index matching to the remaining dehydrated tissue's index [3]. Most commonly, dehydration is performed using methanol, ethanol, tetrahydrofurane (THF), and tert-Butanol [8, 91–93]. While these agents remove water, they also solvate and remove some of the lipids. Removal of water and lipids results in a fairly homogenous, primarily proteinaceous, dense (i.e., high index of refraction) sample. Dehydrated protein has a refractive index of >1.5 (higher than water or lipid). Therefore, the dehydration step must be followed by a second set of agents that solvate additional lipids and intercalate homogeneously throughout the sample to clear it by matching the higher refractive index of the defatted and dehydrated tissue. To date, dichloromethane has often been used for additional delipidation after the use of dehydration, and methyl salicilate, benzyl alcohol, benzyl benzoate, and dibenzyl ether as final clearing solutions [8, 91–95] (see Table 3.1). For this approach, a number of optical clearing protocols have been developed, such as Adipo-Clear [91], BABB [94], DISCO family [8, 92, 93], etc. The solvent-based clearing techniques are robust and work on a number of different tissue types. However, the toxic nature of many solvents, their capacity to dissolve glues used in the construction of objective lenses, the substantial shrinkage of tissue during dehydration (up to 50% [95]), and the quenching of fluorescent protein emission reduce their utility [3].

Hyperhydration-based clearing techniques use urea and urea-like molecules to dissolve dense fibers and relax protein scaffolds, thereby reducing the refractive index of tissue samples during the clearing process [3]. Some clearing methods also remove lipids by extensive incubation lasting from days to months with detergent and multiple solution changes. At the same time, urea is used to penetrate, partially denature, and thus hydrate even the hydrophobic regions of high refractive index proteins [96]. Hyperhydration reduces the overall refractive index to ~1.38 [9]. As a result of hyperhydration, swelling

TABLE 3.1

Main characteristics of OCAs as clearing techniques

OCA	Chemical composition	Refractive index (RI)	Tissue tested	Tissue preparation	Reference
Simple immersion					
1,2-propanediol	Pure	1.433[a]	Porcine skin	Ex vivo	[103]
1,3-butanediol	Pure	1.44[a]	Human skin	In vitro	[49]
1,4-butanediol	Pure	1.446[a]	Porcine and human skin	Ex vivo, in vitro	[49, 103]
ClearT	Formamide	1.44[a]	Mouse brain and embryos	In vitro	[87]
ClearT2	Formamide and PEG	1.44[a]	Liver	In vitro	[87]
FocusClear™	Diatrizoic acid, Tween 20	1.47[a]	Insect, mouse brain, transgenic mouse skin, lymph node cortex	In vitro	[86, 104–106]
Fructose	1–6 mol/L aqueous solutions	1.3601–1.4601[a]	Rodent skin	In vivo, ex vivo	[22, 65, 67]
	78.9% (wt/wt) aqueous solution		Mouse brain	In vitro	[107]
Fructose/Ethanol	25% water, 25% ethanol, 50% fructose solution	1.3850[b]	Pig skin	Ex vivo	[108]
FRUIT	Fructose/urea/thioglycerol in different proportions	1.48[a]	Rabbit brain	In vitro	[84, 109]
FTP	78.9% fructose solution, tiazone, PEG-400: 11/1/8	1.4801[a]	Rat skin	In vivo	[67]
DMSO	100%	1.47[a]	Human and porcine skin	Ex vivo, in vitro	[49, 70, 108]
	50% aqueous solution (vol/vol)	1.4699[b]			
		1.4[a]			
DMSO/ethanol	50%/50% (vol/vol)	–	Mouse ear	Ex vivo	[74]
D-sorbitol	70% (wt/wt) aqueous solution	1.473[a]	Porcine skin	Ex vivo	[103]
Ethylene glycol	Pure	1.43[a]	Human skin, rat abdominal wall muscle	Ex vivo, in vitro	[49, 52]
	20–60% aqueous solutions	1.3525–1.3925[a]			
Glucose	3.5 M mol/L aqueous solution	1.4065[a]	Rat skin	In vitro	[67]
	40% (wt/wt), 30%, 43%, 56% (wt/vol) aqueous solutions	1.391, 1.379, 1.398, 1.418[a]	Rat, monkey and human skin, native and diabetic mouse skin, human dura mater, arterial wall, rabbit cornea, rabbit and human sclera, lung, human normal and cancerous esophagus tissues, human gastric mucosa	In vivo, ex vivo	[23, 46, 56–65, 67–69, 110]
Glycerol	30–100% (vol/vol) aqueous solution	1.378–1.474[a]	Human normal and cancer breast tissues, native and diabetic rat skin and myocardium, human and porcine skin, human and porcine cranial bone, mouse brain	In vivo, ex vivo, in vitro	[17, 30, 44, 47, 49, 51, 55, 73, 103, 107, 111, 112]
Linoleic acid	Pure	1.47[a]	Human skin	In vitro	[49]
Maltose	1, 1.5, 2 mol/L aqueous solutions	1.3830, 1.4108, 1.4471[a]	Rat skin	In vivo, ex vivo	[22]
D-Mannitol	0.16 g/mL aqueous solution	1.357[a]	Human dura mater	In vitro	[46]

(Continued)

TABLE 3.1 (CONTINUED)

Main characteristics of OCAs as clearing techniques

OCA	Chemical composition	Refractive index (RI)	Tissue tested	Tissue preparation	Reference
Mineral oil	Pure	1.4530[c]	Rat skin	Ex vivo	[113]
MO/DMSO	Mineral oil, DMSO: 9/1, 8/2	1.4539, 1.4548[c]	Rat skin	Ex vivo	[113]
MOUSE (Microdermabrasion, Oleic acid, UltraSound Effect)	Oleic acid	1.4500[c]	Light and dark human skin	In vivo, ex vivo	[39]
Oleic acid (OA)	Pure	1.46[a] 1.4500[c]	Human and rat skin	Ex vivo, in vitro	[49, 113]
Omnipaque	Pure	1.4327[a]	Pig skin	Ex vivo	[108]
Omnipaque/DMSO	90% Omnipaque, 10% DMSO	1.4364[b]	Pig skin	Ex vivo	[108]
Polyethylene glycol 200 (PEG-200)	Pure	1.461[a]	Rat and porcine skin	Ex vivo	[103]
Polyethylene glycol 300 (PEG-300)	Pure	1.4631[a], 1.456[c]	Rat skin	In vivo, ex vivo	[40, 53, 54]
Polyethylene glycol 400 (PEG-400)	Pure	1.4649[a], 1.4487[c]	Rat skin	Ex vivo	[53, 103, 113]
PEG/DMSO	PEG-400, DMSO: 9/1, 8/2	1.4500, 1.4514[c]	Rat skin	Ex vivo	[113]
PEG/OA	PEG-400, OA: 8/2	1.4490[c]	Rat skin	Ex vivo	[113]
PEG/glycerol	50% PEG-400, 50% glycerol	1.421[c]	Rat skin	In vivo	[37]
PEG/glycerol/DMSO	Mixture of 50% PEG-400, 50% glycerol with 9% DMSO	1.423[c]	Rat skin	In vivo	[37]
PEG/Th	PEG-400, tiazone: 9/1, 8/2	1.4694[a], 1.4563[c]	Human and rat skin	In vivo, ex vivo	[35, 113]
Propylene glycol (PG)	Pure 80%PG aqueous solution	1.4312[a], 1.411[a]	Rat, porcine and human skin	Ex vivo	[70, 114, 115]
PG/DMSO	80%PG and 50%DMSO mixture	–	Porcine skin	Ex vivo	[70]
PG/OA	0.1 M OA in 40% PG	–	Porcine skin	Ex vivo	[70]
Ribose	3.5 M mol/L aqueous solution	1.3985[a]	Rat skin	In vitro	[67]
SeeDB (See Deep Brain)	80.2% fructose with a 0.5% α-thioglycerol in water	1.50[a]	Brain, mammary gland	In vitro	[10, 83]
SeeDB2	Iohexol, saponin and Tris-EDTA	1.52[a]	Mouse brain	In vitro	[10]
SOCS (Skull Optical Clearing Solution)	0-60% laurinol, weak alkaline substances, EDTA, DMSO, sorbitol, alcohol, and glucose	–	Mouse skull	Ex vivo, in vivo	[15, 118]
Sorbitol	70.1% (wt/wt) aqueous solution	–	Rodent skin, mouse brain	Ex vivo, in vitro	[65, 107]
Sucrose	1-3 mol/L aqueous solutions	1.3875–1.4773[a]	Rat skin	In vivo, ex vivo	[22, 65]
	67.1% (wt/wt) aqueous solution	–	Mouse brain	In vitro	[107]
	Sucrose/Triton (2%)	1.44[a]	Mouse cortex	In vitro	[82]
TDE	97% 2,2'- thiodiethanol in water	1.42[a]	Brain	In vitro	[88–90]
Solvent based					
Adipo-Clear	Methanol, dichloromethane (DCM), dibenzyl ether (DBE), TritonX	–	Adipose tissue	In vitro	[91]

(Continued)

TABLE 3.1 (CONTINUED)

Main characteristics of OCAs as clearing techniques

OCA	Chemical composition	Refractive index (RI)	Tissue tested	Tissue preparation	Reference
BABB	33% benzyl alcohol (BA), 67% benzyl benzoate (BB)	1.55[a]	Mouse embryo, brain, tumor, kidney, human aorta, lung	In vitro	[94]
DBE	Pure	1.562[a]	Mouse brain	In vitro	[95]
2ECi	1-propanol, ethyl cinnamate	–	Human cerebral organoids, whole axolotl, Xenopus and Drosophila	In vitro	[119]
Ethanol-ECi	Ethyl-3-phenylprop-2-enoate	1.56[a]	Kidney, heart	In vitro	[120]
EyeCi	ECi clearing+ iDISCO immunolabeling	–	Mouse eye	In vitro	[121]
FDISCO	4°C/pH 9.0 tetrahydrofuran (THF) and 4°C DBE	–	Brain, kidney, muscle	In vitro	[122]
3DISCO (3D Imaging of Solvent-Cleared Organs)	THF, DCM, DBE	1.56[a]	Most mouse organs	In vitro	[8, 92]
iDISCO (Immunolabeling-enabled DISCO)	Methanol, hydrogen peroxide/methanol (ratio of 1:5), PBS with 0.2% Triton X100, DCM, DBE	1.56[a]	Mouse organs and brain tumors	In vitro	[93, 123]
iDISCO+	Methanol, DCM, DBE	1.56[a]	Mouse and rat brains	In vitro	[124]
sDISCO	Hydroxytoluene, THF, BABB, DBE, propyl gallate	–	Mouse brain	In vitro	[125]
uDISCO (Ultimate DISCO)	Diphenyl ether, BABB, Vitamin E, tert-butanol (tB)	1.56[a]	Brain, liver, lung, bone, whole rat	In vitro	[126]
uDISCO modified	uDISCO, DCM and triethylamine (pH adjustment to 9.0-9.5)).	–	Mouse brain, muscle, stomach, and lung. Not recommended for heme-rich tissues such as heart, spleen, liver, and kidney	In vitro	[127]
vDISCO	tB 30%-100%, DCM	–	Mouse brain, head and meninges	In vitro	[128]
FluoClearBABB	1-propanol, tB	1.56[a]	Mouse brain	In vitro	[129]
MASH (Multiscale Architectonic Staining of Human cortex)	DISCO with RI matching solution substituted by trans-cinnamaldehyde, TDE, and wintergreen oil	1.56[a]	Human cortex	In vitro	[130]
PEGASOS(PolyEthylene Glycol Associated SOlvent System)	Polyethylene glyco., 20% EDTA, 25% N, N,N´, N´-Tetrakis (2-Hydroxypropyl) ethylenediamine, ammonium, 75% tB + 22% PEG methacrylate + 3% Quadrol, 75% BB	1.543[a]	Whole adult mouse body including bones and teeth but excluding pigmented epithelium	In vitro	[131]
Hyperhydration					
DEEP-Clear	FlyClear solution 1 1 (N,N,N´,N´-Tetrakis(2-hydroxyethyl) ethylenediamine (THEED), Triton X-100, acetone, hydrogen peroxide and solution 2 (meglumine diatrizoate)	–	Bristle worm, squid, zebrafish, axolotl	In vitro	[132]
C₃3D (Clearing-Enhanced 3D microscopy)	N-methyl acetamide, Histodenz, 1-thioglycerol, Triton X-100	1.48[a]	Lymph nodes, liver, muscle, thymus, brain	In vitro	[133]

(Continued)

TABLE 3.1 (CONTINUED)

Main characteristics of OCAs as clearing techniques

OCA	Chemical composition	Refractive index (RI)	Tissue tested	Tissue preparation	Reference
CUBIC (Clear, Unobstructed Brain/ Body Imaging Cocktails and Computational Analysis)	ScaleCUBIC-1 (25 wt% urea, 25 wt% N,N,N',N'-tetra kis(2-hydroxypropyl)ethylenediamine, 15 wt% Triton X-100) ScaleCUBIC-2 50 wt% sucrose, 25 wt% urea, 10 wt% 2,2,2'-nitrilotriethanol, 0.1% (v/v) Triton X-100	1.48[a]	Whole body, mouse and chicken embryo, mouse brain, lung, human intestine, adult mouse heart, mammary gland, liver, lung carcinoma	In vitro	[85, 134]
CUBIC-Perfusion	ScaleCUBIC-1, ScaleCUBIC-2	1.48[a]	Whole mouse body, liver, intestine, pancreas, lung, brain	In vitro	[41]
Upgraded-CUBIC	CUBIC-L (10 wt% Nbutyldiethanolamine, 10 wt% Triton X-100), CUBIC-P (5 wt% 1-methylimidazole, 10 wt% Nbutyldiethanolamine10 wt% Triton X-100), CUBIC-B (10 wt% EDTA, 15 wt% imidazole),CUBIC-HL (10 wt% 1,3-bis(aminomethyl)cyclohexane, 10 wt% sodium dodecylbenzene sulfonate, pH 12.0 adjusted by ptoluene sulfonic acid)CUBIC-R (5 wt% antipyrine, 30 wt% nicotinamide, optionally pH 8-9 adjusted by Nbutyldiethanolamine), or CUBIC-RA 5 wt% antipyrine, 30 wt% Nmethylnicotinamide, optionally pH 8-9 adjusted by Nbutyldiethanolamine).	1.52[a]	Mouse brain and kidney	In vitro	[135]
CUBIC-cancer	CUBIC-L, modified CUBIC-R (45 w% antipyrine and 30 w% nicotinamide)	1.44–1.52[a]	Mouse liver, intestine, pancreas, lung, brain	In vitro	[136]
CUBICHistoVIsion	CUBIC clearing, 500 mM NaCl, HEPES buffer, 0.5% casein, Quadrol, urea, collagenase P	–	Marmoset whole body, mouse brain, human cerebellum	In vitro	[137]
CUBIC-X	CUBIC-R1 (polyalcohols, Triton, urea), CUBIC-X1 (20% imidazole), CUBIC-X2 (5% imidazole, 55% antipyrine cocktail)	–	Mouse brain	In vitro	[138]
IsoScaleSQ	9.1 M urea, 22.5% (vol/wt) d-sorbitol, 200 mM sodium chloride, 2% (vol/wt) Triton X-100	–	Mouse brain	In vitro	[139]
MACS	MACS-R0 (20% (vol/vol) MXDA, 15% (wt/vol) sorbitol), MACS-R1 (40% (vol/vol) MXDA, 30% (wt/vol) sorbitol) and MACS-R2(40% (vol/vol) MXDA, 50% (wt/vol) sorbitol)	1.51[a]	Mouse brain, embryo, kidney, spleen, lung and whole body	In vitro	[97]
Scale	4M urea, 10% glycerol, Triton-X	1.38[a]	Mouse embryo, heart	In vitro	[9]
Scale A2	4M urea, 10% glycerol, 0.1%Triton X-100	1.38[a]	Mouse brain	In vitro	[9, 140]
ScaleS	Sorbitol based Scale	1.44[a]	Mouse brain	In vitro	[141, 142]
ScaleSQ	9.1 M urea, 22.5% (wt/wt) sorbitol	–	Mouse brain	In vitro	[142, 143]
Scale U2	4M urea, 30% glycerol,0.1%Triton X-100	1.38[a]	Mouse brain	In vitro	[9]
UbasM	Ub-1 (1,3-dimethyl-2-imidazolidinone, meglumine, Triton-X, urea) and Ub-2 (urea, sucrose and 1,3-dimethyl-2-imidazolidinone)	–	Mouse intestine, spleen, pancreas, heart, kidney, lung, liver, and brain	In vitro	[144]

(Continued)

TABLE 3.1 (CONTINUED)

Main characteristics of OCAs as clearing techniques

OCA	Chemical composition	Refractive index (RI)	Tissue tested	Tissue preparation	Reference
Hydrogel embedding					
ACT-PRESTO (Active Clarity Technique–Pressure Related Efficient and Stable Transfer of macromolecules into Organs)	Hydrogel Histodenz (acrylamide+ 8%SDS)	1.43–1.48[a]	Brain,	In vitro	[101]
Bone CLARITY	10% EDTA, 4% acrylamide, SDS, refractive index matching solution (RIMS)	1.45[a]	Mouse femur, tibia and vertebral column	In vitro	[11]
CLARITY (Clear Lipid-exchanged Acrylamide-hybridized Rigid Imaging/Immunostaining/In situ hybridization-compatible Tissue-hYdrogel)	FocusClear, 80% glycerol, 8%SLS	1.45[a]	Whole body, mouse embryo, mouse heart, brain, lung,	In vitro	[99, 123]
EDC-CLARITY	1-ethyl-3-3-dimethyl-aminopropyl carbodiimide-CLARITY	1.45[a]	Human, mouse	In vitro	[145]
ePACT (expansion-PACT)	Hydrogel + 10% SDS-PBS (pH 8)	–	Mouse brain	In vitro	[146]
FACT (Free-of-Acrylamide Clearing Tissue)	8%-sodium dodecyl sulfate(SDS), Focus Clear	–	Mouse brain	In vitro	[147]
FASTClear	8%-sodium dodecyl sulfate		Human myocardium	In vitro	[13]
MYOCLEAR	Hydrogel (VA-044, A4P0), PTwH,	–	Mice diaphragm, muscle.	In vitro	[148]
PACT (Passive CLARITY Technique)	Histodenz, RIMS can be added	1.38–1.48[a]	Lung mouse embryo	In vitro	[100]
PACT-deCAL	0.1 M EDTA in 1X PBS, 10% SLS, PBS, hydrogel	–	Tibia bone, vertebral column	In vitro	[146]
PARS (Perfusion-assisted Agent Release in Situ)	Histodenz	1.38–1.48[a]	Whole mouse, human tumor tissue	In vitro	[100]
SCM (Simplified CLARITY method)	CLARITY with a larger concentration of VA-044	1.46[a]	Mouse heart	In vitro	[149]
SUT (Scheme Update on tissue Transparency)	Urea, SDS, Triton-X, acrylamide, and PBS	–	Mouse, rat, and pigs' heart	In vitro	[150]
SWITCH (System-Wide control of Interaction Time and kinetics of CHemicals)	Glutaraldehyde, sodium sulphite, 1-thioglycerol, SWITCH-off (PBS+0.5 mM SDS), SWITCH-on (PBST)	1.47[a]	Human and rat brains. Murine lung, heart, kidney, liver and spinal cord.	In vitro	[151]
Decolorization					
CUBIC/CUBIC-Perfusion	ScaleCUBIC-1		Whole mouse body, liver, intestine, pancreas, lung, brain	In vitro	[41, 85]
Upgraded-CUBIC/CUBIC-Cancer	CUBIC-L (10 wt% Nbutyldiethanolamine, 10 wt% Triton X-100)		Mouse brain and kidney	In vitro	[135, 136]
CLARITY	SDS (8%)		Whole body, mouse embryo, mouse heart, brain, lung,	In vitro	[99]

(Continued)

TABLE 3.1 (CONTINUED)

Main characteristics of OCAs as clearing techniques

OCA	Chemical composition	Refractive index (RI)	Tissue tested	Tissue preparation	Reference
DEEP-Clear	FlyClear solution 1 1 (N,N,N′,N′-Tetrakis(2-hydroxyethyl) ethylenediamine (THEED), Triton X-100, urea), acetone, hydrogen peroxide and solution 2 (meglumine diatrizoate)	–	Bristle worm, squid, *Zebrafish*, axolotl	In vitro	[132]
iDISCO	Hydrogen peroxide/methanol (ratio of 1:5),		Mouse organs and brain tumors	In vitro	[93]
PEGASOS	25% N, N,N′, N′-Tetrakis (2-Hydroxypropyl) ethylenediamine, ammonium,		Whole adult mouse body including bones and teeth but excluding pigmented epithelium	In vitro	[131]

[a] RI was measured at the wavelength 589 nm

[b] RI was measured at the wavelength 785 nm

[c] RI was measured at the wavelength 930 nm

of these samples is observed. This expansion of the sample volume can be controlled by subsequent refractive index matching using solutions with high osmotic pressure. In OCAs, urea or urea-like molecules are used, for example, Sca*l*e family (4M urea, glycerol) [9], CUBIC family (4M urea/sucrose) [41, 85], MACS [97], etc. (see Table 3.1).

These techniques, utilizing harsh solvents or a high concentration of denaturants and detergents, risk the removal of large percentages of the protein content of the tissue. The CLARITY and PACT/PARS methods attempt to address these issues by first embedding the tissue in hydrogel [41, 98–101]. A hydrogel–tissue hybridization method secures proteins and nucleic acids at their physiological locations by covalently linking the molecules to an acryl-based hydrogel (Histodenz) [7]. After hydrogel embedding is complete, lipids are removed passively by a week(s)-long incubation in detergent or rapidly (in days) via electrophoresis. A final step involving immersion in either FocusClear, the Histodenz-based clearing solution, or 80% glycerol produces a sample that is well cleared [3, 98, 100] (see Table 3.1).

Decolorization can be an adjuvant in tissue optical clearing for preventing light absorption by endogenous pigments such as heme, melanin, and lipofuscin [93, 101, 102]. The basis of decolorization is oxidation [42, 93], dissociation [101], or eluting [41, 98, 99] of tissue chromophores using special bleaching agents.

Table 3.1 summarizes the main characteristics of representatives of traditional [17, 22, 23, 30, 35, 37, 44, 46, 47, 49, 51–65, 67–69, 103–115] and innovative [3, 4, 10, 15, 39, 116–151] OCAs.

History of tissue optical clearing technique development

The history of TOC starts at the beginning of the 20th century with a work by Spalteholz [7, 152] that used an organic solvent-based technique to clear large tissue samples. The method was intensive, requiring various dehydration, tissue-bleaching, and clearing steps. However, it produced samples that were unprecedented at the time and helped to push forward the field of anatomy. Unfortunately, this approach damaged the superficial few centimeters of tissue and therefore was useful only for clearing the largest samples [153]. Nevertheless, this method provided a new methodology for the study of anatomy [3, 7].

Following the development of immersion refractometry applied to cells, in 1955 Barer et al. [154] first proposed the optical clearing of cell suspension by means of the protein solution having the same refractive index as the cell cytoplasm. TOC was not developed until the late 1980s, when the method was first applied to the eye sclera and cornea [155]. Then some studied were published testing a series of chemicals to increase the transparency of different biological materials [75, 76, 156, 157].

Subsequently, some research groups started intense studies of the specific features and mechanisms of the *ex vivo* and *in vivo* OC phenomenon and demonstrated the capabilities of the method to increase the probing depth or image contrast of optical inhomogeneities inside a scattering medium [17, 36, 43–52, 55, 56, 58–62, 64–67, 75, 76, 80, 158–176].

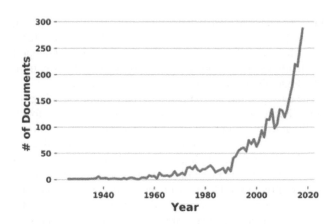

FIGURE 3.2 The number of published papers on tissue optical clearing obtained from Scopus (www.scopus.com). Reprinted from Reference [7] by permission of © The Optical Society.

Development of light sheet microscopy that, together with two-photon fluorescence microscopy, provided the possibility of performing large volume sample imaging and of obtaining the three-dimensional reconstruction of the tissue anatomy created a new direction in TOC evolution [177–179]. In 2013, a tissue transformation technique [98] opened the way to a new approach: clearing was obtained with a manipulation of the sample chemical structure by means of hydrogel embedding and lipid removal [7].

Figure 3.2 shows the trend of published papers on tissue optical clearing. The development of immersion OCAs on the basis of biocompatible components for *ex vivo* and *in vivo* optical clearing, as well as the development of the tissue transformation technique, provoked the widening of the field, identified by the two abrupt variations in slope occurring in the mid-1990s and 2013, respectively [7].

Now, one-component OCAs, multi-component OCAs, and innovative optical clearing protocols are being developed in parallel. Below we will briefly focus on these directions using examples of some typical OCAs.

One-component OCAs

Penetration of one-component OCAs as glycerol and its aqueous solutions [17, 30, 44, 47, 51, 55, 73, 103, 107, 111, 112], aqueous solutions of glucose [23, 46, 56–65, 67–69, 110] and other sugars [21, 22, 65, 67, 107, 108], ethylene and polyethylene glycols [40, 49, 52–54, 103, 113], and propylene glycol [70, 114, 115] were investigated in detail in different tissues. They helped the study of such mechanisms of TOC as dehydration, refractive index matching, collagen dissociation, and swelling and effects on blood vessels [1, 2, 5, 7, 21, 23, 25, 27, 30, 37, 43, 45, 46, 49–53, 65–69, 156–176, 180–182]. The study of the kinetics of tissue optical clearing allowed one to evaluate the OCA diffusion and permeability coefficients [16, 46, 47, 52, 53, 55, 57–64, 69,113, 116, 181, 182]. Using these techniques, based mostly on collimated light transmittance measurements for no-fixed *in vitro* and *ex vivo* studies and OCT for *in vivo*, the diffusion rates of the OCAs were determined in eye tissues [59–62], muscle [52, 180, 181], myocardium [55], skin [16, 53, 55, 61], *dura mater* [46], arterial tissues [57], lung tissues [63], mucosa [69, 182], and other tissues (see Table 3.2). The monitoring of OCA diffusion with high temporal and depth resolution allowed one to differentiate healthy from pathologically modified tissues [47, 48, 55, 64, 182].

The most popular biocompatible hyperosmotic OCA is glycerol. Glycerol ($C_3H_8O_3$) is the simplest triol. It can be found in all natural fats and oils as fatty esters and is an important intermediate in the metabolism of living organisms. Glycerol is a viscous hygroscopic liquid with molecular weight (MW) 92.1 Da [183]. Using glycerol as an OCA opens the way for the therapy of tissues hidden under bones, cartilages, tendons, etc. [2, 5]. The possibility of observation of subepidermal cavity, malignant melanoma diagnostics, and control of human skin scattering properties under the application of glycerol to its surface was shown by the example of OCT imaging *in vitro* and *in vivo*. Cicchi et al. [114] presented the first studies of two-photon in-depth enhancement under treatment with OCAs. Human dermis was treated with hyperosmotic agents (glucose, glycerol, and polypropylene glycol) for a few minutes, and glycerol showed the best but slowest clearing result (with relative contrast 16.3 at 20 μm). Huang et al. [184] investigated the OC effect of glycerol on the porcine skin tissue *in vitro* by Raman microspectroscopy at time intervals from 0 to 75 min. It was found that application of glycerol significantly improved the depth of Raman measurements. Glycerol application also contributed to a better recovery of skin tissue Raman spectra that were not overlapped with the glycerol Raman spectra over time. Since it was shown that glycerol solutions with higher concentrations provided a better clearing effect, the intensity of the band at 1003 cm^{-1} could be increased for improved cancer detection [185].

Schulmerich et al. [186] first demonstrated Raman spectroscopic diffuse tomographic imaging at optical clearing and obtained *in vivo* Raman images for canine bone under glycerol treatment. The glycerol OC effect on the bone tissue was also studied [187]. It was shown that glycerol reduced the noise in the raw spectra and significantly improved the crosscorrelation between the recovered bone factor and the exposed bone measurement in a low signal-to-noise region of the bone spectra.

In Reference [44], the optical immersion clearing of the cranial bone under action of anhydrous glycerol was studied. It was shown that the reduced scattering coefficient of superficial tissue layers of a cranial bone sample decreased by approximately 25% in the wavelength range of 1400–2000 nm after 1 h of exposure. Laser speckle contrast imaging (LSCI) was used for imaging of cerebral blood flow in newborn mice brain during topical OCA application in the area of the fontanelle. These results demonstrated the effectiveness of aqueous 60% glycerol solution for the investigation of cerebral blood flow in newborn mice without scalp removal and skull thinning [188].

Song et al. [106] presented their investigation of glycerol on mouse skin also by confocal microscopy. For a 3-h pinna treatment with glycerol, no significant enhancement in the imaging depth and contrast was observed. However, a noticeable optical clearing with enhanced factor of two was obtained on dorsal skin.

The dependence of TOC's effect on glycerol concentration was investigated in Refs. [112, 189–191]. Carneiro et al. evaluated free/bound water in tissue using 20%–60% glycerol solutions [189] and evaluated the kinetics of the optical properties for colorectal muscle under 40% glycerol action to characterize the dehydration and refractive index matching mechanisms [192]. Genin et al. [112] studied the change in optical parameters of rat skin *ex vivo* under the action of aqueous 30%, 50%, 60%, 70%, 85%, and 100% glycerol solutions. The most efficient optical clearing within the spectral range 50–900 nm was demonstrated by the 60% and 100% solutions. Sufficiently high diffusion rate in combination with high efficiency of optical clearing was demonstrated by the 85% solution of glycerol. Accordingly to reference [190], the optimal concentration of glycerol to maximize TOC was found to be 70%. The efficacy of glycerol was quantitatively evaluated for concentrations from 50% to 90% using OCT to evaluate light scattering and ultrasound imaging modality to evaluate collagen dissociation. A 70% glycerol solution was determined to be the optimal concentration for the combination method, which utilized both microneedling and sonophoresis to further enhance the transdermal delivery of the OCA in Reference [191].

Glycerol diffusion rate in tissues can be a biomarker for differentiation between normal and pathological tissues as demonstrated in Reference [55], which reported a comparative study of 70% glycerol diffusion in skin and myocardium of rats *ex vivo* in the normal condition and the conditions of alloxan-induced diabetes.

Although glycerol is generally a nontoxic agent, prolonged treatment with a highly concentrated solution of it can cause negative effects on tissue such as local hemostasis, shrinkage, and even tissue necrosis. Topical application or injection of glycerol into skin also influences the state of blood microcirculation in the dermis. The OCA diffuses to the vessel net area, partly penetrates vessel walls, interacts with endothelial and blood cells, and leads to local osmotic stress and follow-up dehydration of tissue and cells [166]. It was found that glycerol causes an anhydrous effect on cutaneous vasculature, but that the effect on vessels is reversible with hydration [162]. In addition, the transition from an oxygenated form of Hb to a deoxygenated form for rat skin related to local hemostasis during 84.4% glycerol treatment *in vivo* has been investigated [51]. Moreover, it was found that the topical application 75% glycerol solution on the mesenteric microvessels of a rat in vivo within 1–3 s led to a reduction in blood flow velocity in all microvessels and to stasis appearing after 20 s. The results have shown that after glycerol treatment, the flow velocity in blood vessels was recovered to different extents, and new blood vessels developed after 2 days [166]. Zhu et al. [168] used the LSCI technique to investigate the long-term and short-term effects of glycerol and showed that direct topical application of the OCAs to blood vessels in the chick chorioallantoic membrane can decrease blood-flow velocity. Blocking of blood vessels has also been noticed. Long-term observations indicate that glycerol slowed down the development of blood vessels. The blood flow can be recovered to different extents if the blood vessel is not blocked completely. Mao et al. [193] investigated the influence of a 30% glycerol solution on the dermal blood vessels of flap window of rat skin also using LSCI. It was found that the blood flow velocity decreased initially and began to recover after 16 min.

TABLE 3.2

Properties of the optical clearing agents (OCAs) (molecular weight (MW), dynamic viscosity (V), their efficiency

$(E = \frac{\mu_{s0} - \mu_{smin}}{\mu_{s0}} \times 100\%)$, penetration (P) and diffusion (D) coefficient in tissues. Glucose (Gl), fructose (F), sucrose (S), mannitol (M), glycerol (G), ethylene glycol (EG), polyethylene glycol (PEG), propylene glycol (PG), polypropylene glycol (PPG), human hemoglobin (HG), temperature (T)

OCA	MW (Da)	V (mPa×s)	E (%)	P×10⁻⁵ (cm/s)	D×10⁻⁶ (cm²/s)	Reference
Human eye sclera in vitro						
Gl40% (wt/vol)	180.16	5.5[a]	n/a	n/a	1.52 ± 0.05	[62]
Gl30% (wt/vol)		2.9[a]	n/a	n/a	1.47 ± 0.36	
Gl20% (wt/vol)		1.9[a]	n/a	n/a	0.57 ± 0.09	
Rabbit cornea in vitro						
M20% (wt/vol)	182.17	n/a	n/a	0.899 ± 0.143	n/a	[59]
Rabbit sclera in vitro						
M20% (wt/vol)	182.17	n/a	n/a	0.618 ± 0.108	n/a	[59]
Gl20% (wt/vol)	180.16	n/a	n/a	0.864 ± 0.112	n/a	
Human dura mater in vitro						
M16% (wt/vol)	182.17	≥1	30.0±1.5	n/a	1.31±0.41	[46]
Gl20% (wt/vol)	180.16	1.9	22.0±1.1	n/a	1.63+0.29	
Rat skin ex vivo						
G	92.1	1410[b]	37.7±12.9	n/a	3.23±2.21	[55, 112]
G85% (vol/vol)		109[b]	24.6±5.2	n/a	1.81±1.13	
G70% (vol/vol)		22.5[b]	29.0±10.0	1.68±0.88	0.86±0.73	
G60% (vol/vol)		10.8[b]	33.5±11.6	n/a	1.098±0.62	
G50% (vol/vol)		6[b]	22.0±7.0	n/a	1.04±0.52	
G30% (vol/vol)		2.5[b]	8.0±5.0	n/a	2.79±1.12	
PEG-300	285-315	~110[c]	56.7±7.9	n/a	1.83±2.22	[53]
PEG-400	380-420	~135[c]	69.1±4.5, 21.0±6.0	n/a	1.70±1.47	[53, 49]
PG	76.09	42[d]	50.4±7.8	n/a	0.135±0.095	[115]
Gl40% (wt/wt)	180.16	n/a	27±12	n/a	1.11±0.78	[23]
Rat skin in vivo						
Gl40% (wt/wt)	180.16	n/a	n/a	n/a	1.54±0.28	[23]
Human skin in vivo						
Gl40% (wt/wt)	180.16	n/a	n/a	n/a	2.56±0.13	[158]
Murine skin ex vivo						
Gl56% (wt/vol)	180.16	23.6[a]	62.0±4.0	8.24 ± 5.64	1.40 ± 0.96	[110]
Gl43% (wt/vol)		6.4[a]	52.0±9.0	10.29 ± 1.06	2.70 ± 2.22	
Gl30% (wt/vol)		2.9	34.0±7.0	10.15 ± 0.61	2.87 ± 1.53	
Diabetic rat skin ex vivo						
G70% (vol/vol)	92.1	22.5	63.0±1.0	1.20±0.33	0.68±0.21	[55]
Diabetic murine skin ex vivo						
Gl56% (wt/vol)	180.16	23.6	53.0±12.0	6.58±2.84	1.02±0.44	[110]
Gl43% (wt/vol)		6.4	51.0±5.0	5.90±3.23	1.15±0.63	
Gl30% (wt/vol)		2.9	42.0±6.0	5.43±2.82	1.06±0.55	
Monkey skin in vivo						
Gl20% (wt/vol)	180.16	1.9	n/a	0.441±0.028	n/a	[16]
Rat myocardium ex vivo						
G70% (vol/vol)	92.1	22.5	57.0±3.0	11.8±6.1	0.79±0.36	[55]
Diabetic rat myocardium ex vivo						
G70% (vol/vol)	92.1	22.5	61.0±5.0	8.60±3.21	0.51±0.21	[55]

(Continued)

TABLE 3.2 (CONTINUED)

Properties of the optical clearing agents (OCAs) (molecular weight (MW), dynamic viscosity (V), their efficiency

$(E = \frac{\mu_{s0} - \mu_{smin}}{\mu_{s0}} \times 100\%)$, penetration (P) and diffusion (D) coefficient in tissues. Glucose (Gl), fructose (F), sucrose (S), mannitol (M), glycerol (G), ethylene glycol (EG), polyethylene glycol (PEG), propylene glycol (PG), polypropylene glycol (PPG), human hemoglobin (HG), temperature (T)

OCA	MW (Da)	V (mPa×s)	E (%)	P×10⁻⁵ (cm/s)	D×10⁻⁶ (cm²/s)	Reference
Porcine myocardium in vitro						
Gl40% (wt/vol)	180.16	5.5	39.5±3.1	87.8±72.2	0.48±0.34	[209]
G58% (vol/vol)	92.1	8.5ᵇ	37.5±0.6	1.59±0.99	0.77±0.46	
Rat skeletal muscle ex vivo						
F	180.16	-	n/a	n/a	0.57	[181]
PPG	92 – 425	n/a	n/a	n/a	0.51	
EG	62	n/a	n/a	n/a	0.46	[52]
Human colorectal muscle ex vivo						
G	92.1	n/a	n/a	n/a	0.33	[181]
Bovine muscle tissue ex vivo						
Gl40% (wt/wt)	180.16	n/a	n/a	n/a	2.98±0.94	[199]
Human breast tissue in vitro						
G60% (vol/vol)	92.1	10.8	n/a	0.89±0.02	n/a	[47]
Gl20% (wt/vol)	180.16	1.9	n/a	0.845±0.16	n/a	[48]
Gl40% (wt/vol)	180.16	5.5	n/a	0.576±0.089	n/a	
M20% (wt/vol)	182.17	n/a	n/a	0.715±0.112	n/a	
Cancerous human breast tissue in vitro						
G60% (vol/vol)	92.1	10.8	n/a	3.14±0.07	n/a	[47]
Gl20% (wt/vol)	180.16	1.9	n/a	1.92±0.15	n/a	[48]
Gl40% (wt/vol)	180.16	5.5	n/a	1.09±0.11	n/a	
M20% (wt/vol)	182.17	n/a	n/a	1.39±0.22	n/a	
Pig aorta tissues in vitro						
Gl20% (wt/vol)	180.16	1.9	n/a	1.43±0.24	n/a	[57]
Human lung tissuesin vitro						
Gl30% (wt/vol)	180.16	2.9	n/a	1.35±0.13	n/a	[63]
Cancerous human lung tissues in vitro						
Gl30% (wt/vol)	180.16	2.9	n/a	1.78±0.21 (benign granulomatosis) 2.88±0.19 (adenocarcinoma) 3.53±0.25 (squamous cell carcinoma)	n/a	[63]
Human esophagus ex vivo						
Gl40% (wt/vol)	180.16	5.5	n/a	1.74±0.04	n/a	[64]
Cancerous human esophagus ex vivo						
Gl40% (wt/vol)	180.16	5.5	n/a	2.45±0.06 (squamous cell carcinoma)	n/a	[64]
Human gastric mucosa ex vivo						
Gl40% (wt/wt)	180.16	n/a	n/a	2.81±0.9	1.59±0.96	[69]
Human colorectal mucosa ex vivo						
Gl40% (wt/wt)	180.16	n/a	n/a	n/a	0.58	[182]
Cancerous human colorectal mucosa ex vivo						
Gl40% (wt/wt)	180.16	n/a	n/a	n/a	0.44	[182]

(Continued)

TABLE 3.2 (CONTINUED)

Properties of the optical clearing agents (OCAs) (molecular weight (MW), dynamic viscosity (V), their efficiency

($E = \frac{\mu_{s0} - \mu_{smin}}{\mu_{s0}} \times 100\%$), penetration (P) and diffusion (D) coefficient in tissues. Glucose (Gl), fructose (F), sucrose (S), mannitol (M), glycerol (G), ethylene glycol (EG), polyethylene glycol (PEG), propylene glycol (PG), polypropylene glycol (PPG), human hemoglobin (HG), temperature (T)

OCA	MW (Da)	V (mPa×s)	E (%)	P×10⁻⁵ (cm/s)	D×10⁻⁶ (cm²/s)	Reference
Water						
PEG-200 (T = 303 °K)	180–220	n/a	-	n/a	6.53±0.08	[210]
PEG-200 (T = 308 °K)	180–220	n/a	-	n/a	7.34±0.09	
PEG-300 (T = 303 °K)	285–315	n/a	-	n/a	5.44±0.07	
PEG-300 (T = 308 °K)	285–315	n/a	-	n/a	6.15±0.07	
PEG-400 (T = 303 °K)	380–420	n/a	-	n/a	5.13±0.08	
PEG-400 (T = 308 °K)	380–420	n/a	-	n/a	5.63±0.08	
PEG-600 (T = 303 °K)	550–650	n/a	-	n/a	4.75±0.07	
PEG-600 (T = 308 °K)	550–650	n/a	-	n/a	6.11±0.07	
PPG-400 (T = 303 °K)	380–420	n/a	-	n/a	4.4±0.07	
PPG-400 (T = 308 °K)	380–420	n/a	-	n/a	4.87±0.07	
Gl (T = 298 °K)	180.16	n/a	-	n/a	6.9, 6.728 (0.78 wt%)	[211, 212]
Fructose (T = 298 °K)	180.16	n/a	-	n/a	6.9	[211]
Sucrose (T = 298 °K)	342	n/a	-	n/a	5.6, 5.209 (0.769 wt%)	[211, 212]
HG9.7% (wt/vol)	64458	n/a	-	n/a	0.453±0.008	[213]
HG21.9% (wt/vol)	64458	n/a	-	n/a	0.223±0.002	
HG31.5% (wt/vol)	64458	n/a	-	n/a	0.0953±0.002	
HG36.5% (wt/vol)	64458	n/a	-	n/a	0.045±0.001	
HG43.3% (wt/vol)	64458	n/a	-	n/a	0.0138±0.0004	
G10% (vol/vol) (T = 298 °K)	92.1	n/a	-	n/a	10.2	[214]
G20% (vol/vol) (T = 298 °K)	92.1	n/a	-	n/a	1.02	
G30% (vol/vol) (T = 298 °K)	92.1	n/a	-	n/a	9.89	
G40% (vol/vol) (T = 298 °K)	92.1	n/a	-	n/a	9.39	
G50% (vol/vol) (T = 298 °K)	92.1	n/a	-	n/a	9.19	
G60% (vol/vol) (T = 298 °K)	92.1	n/a	-	n/a	9.44	
G70% (vol/vol) (T = 298 °K)	92.1	n/a	-	n/a	10.6	
G80% (vol/vol) (T = 298 °K)	92.1	n/a	-	n/a	15.8	
G90% (vol/vol) (T = 298 °K)	92.1	n/a	-	n/a	30.9	
G100% (T = 293.2 °K)	92.1	n/a	-	n/a	0.0137	[215]

[a] "The functional properties of sugar," Nordic Sugar, http://www.nordicsugar.dk/fileadmin/Nordic_Sugar/Brochures_factsheet_policies_news/Download_center/Functional_properties_of_sugar_on_a_technical_level/Functional_prop_on_tech_level_uk.pdf.

[b] Segur, J. B. and H. E. Oberstar, 1951, Viscosity of glycerol and its aqueous solutions. *Ind Eng Chem* 43:2117–20.

[c] Wen, X., S. L. Jacques, V. V. Tuchin, and D. Zhu. 2012. Enhanced optical clearing of skin in vivo and optical coherence tomography in-depth imaging. *J Biomed Opt* 17:066022.

[d] Dynamic Viscosity of Common Liquids. https://www.engineeringtoolbox.com/absolute-viscosity-liquids-d_1259.html.

It was found that glycerol can cause alterations in skin morphology due to a dissociation of the collagen fibers [66, 80]. The molecular mechanism of glycerol action leads to a loss of order in fibril organization that was first investigated by Yeh et al. [80] using multiphoton microscopy. The change in collagen organization and size can lead to a significant reduction in tissue light scattering. Also, it was shown that a 75% glycerol solution does not induce any loss of collagen organization [17]. Pure glycerol influence on dorsal mouse skin has been investigated by confocal scanning laser microscopy [194]. Reflectance images measured at a wavelength of 488 nm showed that glycerol significantly increased the anisotropy of dermis scattering (from 0.7 to 0.99) with little change in the scattering coefficient. It was suggested that the glycerol-related clearing effect starts when reducing the angular deviation of scattering. Moreover, an increase in anisotropy of scattering with a minor change in the scattering coefficient should cause an increase in the size of the scattering particles, meaning swelling of the collagen fibers in the dermis. In Reference [188], the authors reported that the application of an 85 vol% aqueous glycerol solution to rat tail tendon fascicle significantly changes not only the scattering properties of the tissue, but also its birefringence and diattenuation. These changes in the collagen structure can be irreversible [80].

Glucose solutions with different concentrations also play important role as OCAs due to their biocompatibility and low cost. Glucose ($C_6H_{12}O_6$) is a simple sugar or monosaccharide, a subcategory of carbohydrates. The molecular weight of glucose is 180.16 Da. Most often, glucose solutions are used for optical clearing of connective tissues, such as the derma, sclera, dura mater, etc. In addition, the optical clearing of blood in the presence of glucose contributes to development of methods for blood glucose monitoring [56, 195–198].

Genina et al. [68] demonstrated OC of dura mater using glucose solutions. The OC efficiency of 3.0 M-glucose solution in the 500–600 nm spectral range was 60.9% ± 0.7%. Tuchina et al. [23] studied OC effect of 40% glucose solution on rat skin. Increase of skin collimated transmittance *ex vivo*, transverse, and along skin shrinkage and weight loss was observed for the first 20–60 min of immersion; in the case of the longer time, tissue swelling was found. The investigation of the influence of a 40% glucose solution on bovine skeletal muscle in vitro showed a four-fold increase in the image contrast for OCT measurements at the depth of 360 μm [199].

Significant increase in image contrast (2.4-fold) has been obtained at the depths up to 800 μm. Liu et al. [200] demonstrated that increments of glucose concentration cause a decrease in stiffness of cartilage by 44%, 55%, and 76% for 30%, 40%, and 70% glucose solutions, respectively. It was suggested that stiffness decreased due to the OCA replacing water in the extracellular matrix and partly dehydrating the cartilage.

In Refs. [69, 182], the kinetics of collimated transmittance of the gastric and colorectal mucosa under the action of aqueous glucose solutions was studied. It was shown that the introduction of the 40% glucose solution into the healthy mucosa reduced the light scattering coefficient by approximately 5%–10%. The increase in the depth of light penetration was from 5% to 15%, depending on the selected spectral range [69]. It was found that the mean diffusion time as a function of the glucose concentration in solution differed for healthy and pathologically changed tissue [182] and allowed differentiation of the healthy mucosa and tumor. Zhao et al. [201] found that the permeability of glucose in cancerous tissue was 1.95-fold greater than in normal tissue within the same OCA treatment time. The efficiency of optical clearing of healthy and diabetic mouse skin expressed as relative change of averaged collimated transmittance T_c was presented in Reference [110]. Considerable differences in the optical and kinetic properties of diabetic and nondiabetic skin were found: clearing efficiency was 1.5-fold better for diabetic skin (see Figure 3.3).

Similar to the case with glycerol, topical application or injection of glucose into skin influence the state of blood microcirculation in the dermis. Galanzha et al. [166] first showed that topical application of glucose on mesenteric microvessels caused a decrease in blood flow. After 5 s interaction with OCA, the blood flow rate in venule with 11 μm diameter decreased from 1075 μms^{-1} to 202 μms^{-1}. Moreover, after 20 s of interaction, stasis appeared in venule. Zhu et al. [168], using the LSCI technique, showed that the long-term application of glucose had stronger negative effects in terms of decreasing blood flow velocity than short-term application. Moreover, the investigation of glucose's long-term impact on blood perfused tissues showed that no new blood vessels had developed after 2 days, unlike in the case of glycerol. The results obtained in Reference [23] allowed one to evaluate glucose's impact on blood microcirculation in skin. The decrease in the average rate of microcirculation by 2.2-fold was observed in rat skin *in vivo*.

Polyethylene glycol (PEG, chemical formula $C_{2n}H_{4n+2}O_{n+1}$) is a polymer of ethylene glycol ($C_2H_6O_2$) belonging to the class of diols. Depending on the molecular weight, polyethylene can be a viscous liquid, gel, or solid. The most applicable PEGs are PEG-300 and PEG-400, which are transparent, viscous, colorless liquids with a molecular weight of between 300 and 400 Da and exhibit strong hygroscopic properties that decrease with increasing molecular weight. PEG is actively used in medicine and cosmetology as a basis for ointments and was registered as a food additive E1521. It is also used as a solvent, extractant, preservative, and strong osmotic agent [202].

Mechanisms of ethylene and polyethylene glycol interaction with different tissues are presented in Refs. [40, 52–54, 103, 113, 203]. The results shown by Tuchina et al. [53] demonstrated the comparative efficacy of PEG-300 and PEG-400 as clearing agents to control the scattering characteristics of the rat skin *ex vivo*. In particular, an increase in the collimated light transmittance through the skin in the spectral range of 500–900 nm, a decrease in the diffusion rate of PEG, an increase in the efficiency of optical clearing of the skin, and an increase in the enhancement of the transversal compression of the skin were observed with the increasing molecular weight of PEG. The resulting value of the diffusion coefficients of PEG-300 and PEG-400 is well matched to results obtained by Jakasa et al. [203], who studied the diffusion coefficients of PEG-300 and PEG-400 in the corneal layer of the skin epidermis.

Propylene glycol (PG) is widely used as an OCA due to its efficiency, availability, and biocompatibility [70, 114, 115, 204–207]. It is a viscous colorless water-soluble hygroscopic liquid from the class of diols (chemical formula: $CH_3CH(OH)CH_2OH$) having a molecular weight 76.09 [208]. PG is included in the composition of a large number of pharmaceutical and cosmetic preparations as a solvent, extractant, and preservative [208].

Cicchi et al. [114] demonstrated that the application of PG to human dermis approved to be similarly efficient (relative contrast is 12.6 at 20 μm) to glycerol. Genin et al. [115] studied optical clearing of the rat skin *ex vivo* under the action of PG. It was found that collimated transmittance of skin samples increased, whereas weight and thickness of the samples decreased during PG penetration in skin tissue.

The use of PG as an enhancer for skin transport of other OCAs was proposed in Reference [206]. The results indicated that the decrease in reduced scattering coefficient caused by OCA/PG was significantly larger than that by pure OCA.

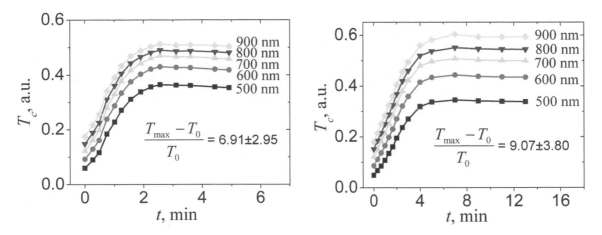

FIGURE 3.3 Collimated transmittance kinetics of nondiabetic (A) and diabetic (B) mouse skin samples during treatment by 56%-aqueous glucose solutions measured for different wavelengths. T_{max} is the maximal averaged collimated transmittance, T_0 is the initial averaged collimated transmittance.

The knowledge of the diffusion rate of OCAs is very important for the development of mathematical models describing the interaction of OCAs with tissues and creating models for effective TOC. The optical methods of evaluation of OCA diffusion and penetration coefficient in a tissue are based on the measurements of temporal changes of tissue scattering properties [61]. Some characteristics of one-component OCAs and the values of both diffusion and penetration coefficients in different tissues and water are presented in Table 3.2.

Multicomponent OCAs

Development of multicomponent OCAs was provoked by the need to enhance tissue permeability for the immersion component. For example, skin is a very important object for optical clearing in many diagnostical and therapeutical applications. However, in the case of noninvasive delivery of OCAs into the dermis, transepidermal penetration is inhibited. Topically applied OCAs exhibit slow uptake of immersion agent into the dermis due to the passive delivery induced by a concentration gradient, in which the diffusion properties are a function of the skin characteristics and the solution molecules [216, 217]. A living epidermis of a thickness 75–150 μm and its upper-layer *stratum corneum* (SC) with a thickness in the range of 10–20 μm, which consists of approximately 15 layers of denuclearized corneocytes embedded in a lipid-enriched extracellular matrix [218], represents a major barrier. The composition and organization of the lipid matrix provide an effective protection function as a skin barrier to body water loss and influx of exogenous substances [219, 220]; therefore, the effective delivery of OCAs into dermis remains a challenge. Results obtained in Reference [221] showed about a 5-fold decrease in glycerol concentration at a depth 10 μm from the skin surface after 1-hour topical exposition on the human skin *in vivo* measured using two-photon excited autofluorescence. The amount of glycerol penetrated through the skin sample was estimated in Reference [43] using a Franz cuvette. The volume of glycerol penetrated through an intact skin sample with a diameter of 25 mm and thickness of about 1.5 mm from reservoir with 20 ml

of 88% aqueous solution of glycerol into reservoir with water was about 60 μl over 6 h. In Reference [40], it was shown that PEG-300 decreased the attenuation coefficient of the upper dermis *in vivo* (~50 – 200 μm) by 23 ± 6% during about 30 min and the middle dermis (~200–400 μm) by 17 ± 4% during about 80 min, apparently due to dehydration.

Thus, additional components are needed to include to hydrophilic OCA formulation for increasing their penetration speed. In particular, it is shown that azone, oleic acid, DMSO, ethanol, propylene glycol, and thiazone considerably enhance SC permeability for OCAs and increase the efficiency of *ex vivo* and *in vivo* reversible skin optical clearing [24–26, 28, 35, 37, 70, 72–74, 113, 118, 195, 196, 206, 222–226]. For example, Zhao et al. [26] found that salicylic acid with azone exhibited a synergistic effect on enhancing light penetration and OCT imaging depth; a 40% increase in depth assessment with OCT was achieved after skin optical clearing *in vivo* over 80 min. Zhu et al. developed a number of TOC methods for improvement of LSCI for blood-flow monitoring [222–225] In Reference [222] they presented an optical clearing method to make the skin transparent and imaged the dermal blood flow through the skin during 40 min of treatment using the mixture of PEG 400 and thiazone. It was demonstrated that the speckle contrast improves after topical application of the mixture, and more details of the dermal blood vessels or blood flow can be seen [223]. An ear skin optical clearing agent (ESOCA) including PEG-400, fructose, and thiazone was developed that could make the ear skin of mice transparent, and the transmittance of the mouse ear was enhanced by approximately 111% at 633 nm [224].

Optical clearing of hard tissues, such as bone, is of great practical interest. The reduction of scattering in this tissue offers the possibility of developing minimally invasive methods of laser diagnostics and therapy of the brain. The possibility of optical monitoring of cerebral blood flow through an intact mouse skull at optical clearing was demonstrated by Wang et al. [225]. An innovative skull optical clearing solution (SOCS), whose main components included laurinol, weak alkaline substances, ethylenediaminetetraacetic acid (EDTA), DMSO, sorbitol, alcohol, and glucose, was invented that made the mouse skull transparent within 25 min [118] (see Figure 3.4). Experiments with laboratory mice *in vivo* have shown that the minimal diameter of microvessels that could be resolved at optical imaging through the optically cleared skull amounts to 14.4 ± 0.8 μm. Optical clearing of rat skull bone *in vivo* using multicomponent OCA (45% ethanol, 25% dehydrated glycerol, 10% PEG-300, and 20% water) was studied in Reference [226]. The use of the solution contributed to significant improvement in visualization of *vena cerebri magna* using Doppler OCT without removing the cranial bone and allowed the determination of the velocity of blood flow in the vein in the normal state (see Figure 3.4) as well as under the action of adrenaline.

OCAs developed for advanced microscopic study include complex combinations of organic solvents (urea, methanol, etc.) and detergents (e.g. Triton, Tween 20, SDS) providing total optical clearing (see Table 3.1).

Innovative technologies for optical clearing

Optical clearing protocols were developed in two different directions: i) the use of complex two- or three-step processes including dehydration, delipidation (with or without electrophoresis), decolorization, and refractive index matching; and ii) the use of multimodal chemical and physical treatments.

Various optical clearing methods based on tissue dehydration and solvent-based clearing or innovative techniques were developed to achieve complete transparency of the sample. Specially optimized multicomponent solutions with optical clearing protocols as DISCO family [8, 92, 93, 123–128], MASH [130], PEGASOS [131], CUBIC family [41, 85, 117, 134–138], Sca*l*e family [9, 139–143], CLARITY [11, 99, 123, 145, 148], PACT [100, 146], PARS [100], ACT-PRESTO [101], SWITCH [151], MACS [97], and many others were suggested

for application *in vitro* in studies with modern microscopic techniques.

The choice of the optical clearing protocol depends on the size and type of tissue, type of fluorescence, importance of tissue shrinkage, and clearing time [117]. Aqueous reagent–based clearing methods, including CLARITY, PACT, CUBIC, and MACS, efficiently cleared soft tissue [41, 100, 136]. Ultimate DISCO (uDISCO) was supposed to clear the whole mouse body [126], including calcified bones, with no need for a previous decalcification treatment. However, it was not efficient on clearing highly colorized organs including liver and spleen, and achieved only partial success on clearing hard tissue [227]. Development of a whole-body immunolabeling method to boost the signal of fluorescent proteins (named vDISCO) allowed the visualization of whole-body neuronal connectivity, meningeal lymphatic vessels, and immune cells through the intact skull and vertebra in naive animals and trauma models [227]. An example of an "invisible mouse" generated by vDISCO is presented in Reference [227].

Bone CLARITY was recently developed for clearing bones, but the clearing effect of this method on soft tissue organs has not been demonstrated [11].

The PEGASOS method consists of multiple steps including fixation, decalcification (hard tissue only), decolorization, delipidation, dehydration, and clearing [131]. The authors demonstrated that the PEGASOS method renders nearly all types of tissues transparent, except for pigmented epithelium.

In reference [140], tissue clearing protocols including ScaleA2, ClearT2, and 3DISCO for visualizing native and tissue engineered muscle by confocal microscopy were compared. It was found that 3DISCO had better clearing properties.

The CUBIC protocol was shown to be effective for the improvement of confocal imaging of mouse heart sections [228] and 3D vasculature image of whole mouse heart [229]. A combination of CUBIC and CLARITY, named SUT (Scheme Update

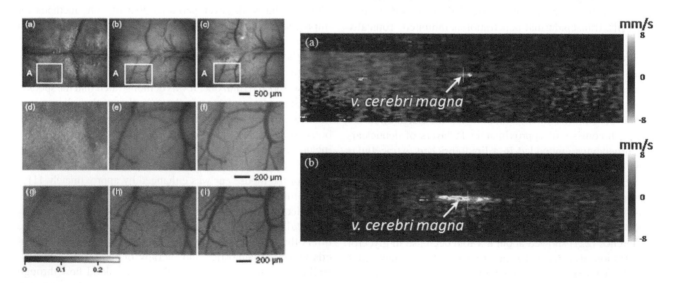

FIGURE 3.4 Left – White-light images obtained from intact skull (a), the transparent skull after SOCS treatment for 25 min without (b) and with exposed cortex of rectangle area A (c). (d)–(f) are the corresponding magnified images of the rectangle area A shown in (a), (b), and (c); (g)–(i) demonstrate the speckle contrast images corresponding to the area in (d)–(f), respectively (© Astro Ltd. Reproduced from Reference [118] by permission of IOP Publishing. All rights reserved). Right – Doppler OCT images of rat cranial bone *in vivo* without optical clearing (j) and after OCA exposure during an hour (k). Bars correspond to 500 μm. Adapted from Reference [226] by permission of SSU Publishing.

on tissue Transparency), was used to evaluate spatial distribution and phenotype of fibroblasts in mice left ventricles in a post-myocardial infarction [150]. Additionally, a modified CLARITY method named SCM (Simplified CLARITY method), was applied to mouse heart to study protein distribution [149].

An approach combining CLARITY with double photon microscopy to enable visualization of gray matter in the intact brain was proposed in reference [230]. Mouse brains were rendered transparent with an upgraded CUBIC protocol that used CUBIC-L and CUBIC-P for the delipidation and decolorization step [231]. A high resolution whole mouse brain atlas was achieved by using a newly designed expansion CUBIC pipeline, named CUBIC-X [138].

ACT-PRESTO (active CLARITY technique-pressure related efficient and stable transfer of macromolecules into organs) was also used for brain and whole-body clearing with shorter protocols [101]. It used a highly porous hydrogel that allowed better removal of lipids and a more rapid diffusion of macromolecules compared to CLARITY. In order to better preserve tissue architecture and fluorescence, an alternative hydrogel-based method named SHIELD (stabilization under harsh conditions via intramolecular epoxide linkages to prevent degradation) was used to evaluate synaptic architecture of virally labeled murine neurons at single cell resolution [232].

The MACS protocol was recently designed based on a new hyperhydration reagent – M-xylylenediamine (MXDA), which possesses superior hyperhydration ability and a super high refractive index up to 1.57 [97]. MACS enables efficient clearing for many kinds of tissues and is compatible with various fluorescent probes. Notably, MACS shows good compatibility with the commonly used lipophilic dye DiI, which cannot be preserved by other clearing methods with delipidation.

Recently, a new hydrogel-based tissue clearing method, MYOCLEAR, was described for the labeling and detection of all neuromuscular junctions and diaphragm muscles in mice [148].

Lloyd-Lewis et al. compared PACT-RC, PACT-RIMS, 3DISCO, CUBIC, and SeeDB and concluded that SeeDB and CUBIC rendered better results in terms of transparency of a mouse mammary gland [233]. These methods were used for the identification of the contribution of different cell clones in the development of this tissue. They also allowed healthy and tumor tissue in the gland to be compared and mammary cells to be phenotyped within the tissue.

The second approach was developed for *ex vivo* and *in vivo* applications. The main object of the treatment was skin. Besides chemically based approaches, different physically based ones were proposed to improve skin optical clearing *in vivo*. For example, low-frequency sonophoresis has demonstrated the effectiveness of transepidermal delivery of drugs and OCAs [33, 37, 205]. A cavitation is the basis for the delivery enhancement of skin sonication with frequency of around 1 MHz due to induced disruption of SC integrity superficially and ultrasound massage of in-depth tissue [234] that provided OCA delivery through the epidermis. Ultrasound (US) exhibited an enhancing skin clearing effect when topically applied with glycerol, polyethylene glycol, propylene glycol, and other agents [111, 205]. A significant US-induced enhancement in OCT imaging depth and contrast of *in vitro* porcine skin and *in vivo* human skin was found [33]. In Reference [39], the authors demonstrated a rapid ultrasound-based optical clearing of human skin without chemical agents that provides dramatic (up to 10-fold) reducing processing time down to 2–5 min.

The comparison of the total attenuation coefficients measured for rat skin *in vivo* at application of physical and chemical penetration enhancers was carried out in Reference [37]. Maximal efficacy of optical clearing was achieved after 20 min application of a complex OCA (a mixture of equal amounts of glycerol and PEG-400 with 9% DMSO) and 4-min sonophoresis (about 35%).

Removal of upper layers of SC using type stripping [235] or microdermabrasion [31] was also used for the enhancement of epidermal permeability. The combined use of surface epidermis stripping with chemical enhancers of diffusion, such as thiazone, azone, and propylene glycol, was studied for rat skin *in vivo* [236].

Optical clearing protocol based on sequent application of microdermabrasion, oleic acid, and US effect (MOUSE) [39] allowed increasing OCT signal amplitude up to 3.3-fold with more than twice the improved depth penetration for human skin *in vivo* that was achieved with other approaches. The use of lipophylic OCA did not cause dehydration and shrinkage of the skin.

US processing was also used in combination with thiazone to improve the penetration of PEG-400 into the skin [35]. It was shown that after complex processing with thiazone, PEG-400, and US, the diffuse reflection coefficient decreased by 33.7% as compared to the control measurements (without TOC).

Different radiation sources (e.g., CO_2 and Nd:YAG lasers, operating at the wavelengths 532 and 1064 nm, respectively, and broadband sources of intense pulsed light, operating in the ranges 650–1200, 525–1200, and 470–1400 nm) were used to irradiate the skin *in vivo* before the application of OCA with different doses and exposures. The measurements of reflection spectra before and after the exposure showed that the radiation of Nd:YAG laser in the Q-switch and long pulse modes can improve transepidermal OCA penetration by from eight to nine fold as compared to intact skin [34]. In another study by Stumpp et al. [237], a 980 nm-diode laser was used for rodent skin irradiation using absorbing substrates on the skin surface, providing heating of the *stratum corneum* and causing a failure of the protective barrier function. After the laser action and removal of absorbing substrates, the skin surface was subjected to glycerol application. The OCT study of skin in the exposed regions has shown an increase of light penetration depth of up to 42%.

Laser ablation of epidermis to a depth of ~50 μm was also used to enhance OCA diffusion into the dermis [54]. Fractional laser microablation (FLMA) is a method allowing direct delivery of drugs and micro- and nanoparticles into the dermis [238, 239]. It is characterized by efficacy and safety for humans, and provides controlled microscopic channels in skin with different depths and diameters [238–240]. In this method, the damaged area of the skin surface is smaller than that after laser ablation by a wide beam with the same parameters. FLMA was suggested as an approach with minimal side effects and quicker skin healing in comparison with a conventional totally ablative one [40].

One alternative to FLMA is the use of microneedles, which are effective for breaking through SC. Son *et al.* [32, 241] used a physicochemical tissue optical clearing (PCTOC) method, which combined microneedle rolling and topical glycerol application on mice skin *in vivo* for blood flow imaging using the LSCI technique. Other injection alternatives developed over the last decades, such as high-pressure injections and needle-assisted jet injectors, have shown the advantages of needle-free injections, in which a pressurized liquid jet penetrates the skin [217]. Stumpp et al. proposed using a pistol for the needle-less injection of glycerol with concentrations of 100%, 50%, and 25% into the porcine skin *ex vivo* [242].

Conclusion

We have briefly reviewed the specific features and approaches to optical clearing using OCAs. The impact of an OCA on a tissue allows for efficient control of the optical properties, particularly the reduction of the tissue scattering coefficient. The field of optical clearing of tissues is extremely active. In the last few years, OCAs have been largely and successfully used for deep visualization of both fixed animal tissue samples and live animal and human tissues. Along with the development of new multimodal approaches and combinations of multigoal agents, the actual direction is still the use of one-component OCAs for diffusion coefficient estimation and differentiation of healthy and pathologically changed tissues.

Acknowledgments

E.A.G, A.N.B., and V.V.T were supported by RFBR grant number 18-52-16025. V.D.G. was supported by RFBR grant number 19-32-90224. D.Z and J.T.Z were supported by the National Natural Science Foundation of China (Grant Nos. 61860206009, 81870934, 81701354), the National Key Research and Development Program of China (Grant No. 2017YFA0700501), the Foundation for Innovative Research Groups of the National Natural Science Foundation of China (Grant No. 61721092), and the Innovation Fund of WNLO.

REFERENCES

1. Zhu, D., K.V. Larin, Q. Luo, and V.V. Tuchin. 2013. Recent progress in tissue optical clearing. *Laser Photonics Rev* 7:732–57.
2. Genina, E.A., A.N. Bashkatov, Yu.P. Sinichkin, I. Yu. Yanina, and V.V.Tuchin. 2015. Optical clearing of biological tissues: Prospects of application in medical diagnostics and phototherapy. *J Biomed Photonics Eng* 1:22–58.
3. Richardson, D.S., and J.W. Lichtman. 2015. Clarifying tissue clearing. *Cell* 162:246–57.
4. Yu, T., Y. Qi, H. Gong, Q. Luo, and D. Zhu. 2018. Optical clearing for multiscale biological tissues. *J Biophotonics* 11:e201700187.
5. Sdobnov, A.Yu., M.E. Darvin, E.A. Genina, A.N. Bashkatov, J. Lademann, and V.V. Tuchin. 2018. Recent progress in tissue optical clearing for spectroscopic application. *Spectrochim Acta A Mol Biomol Spectrosc* 197:216–29.
6. Matryba, P., L. Kaczmarek, and J. Gołąb. 2019. Advances in ex situ tissue optical clearing. *Laser Photonics Rev* 13:1800292.
7. Costantini, I., R. Cicchi, L. Silvestri, F. Vanzi, and F.S. Pavone. 2019. In-vivo and ex-vivo optical clearing methods for biological tissues: Review. *Biomed Opt Express* 10:5251–67.
8. Ertürk, A., C.P. Mauch, F. Hellal, et al. 2012. Three-dimensional imaging of the unsectioned adult spinal cord to assess axon regeneration and glial responses after injury. *Nat Med* 18:166–71.
9. Hama, H., H. Kurokawa, H. Kawano, et al. 2011. Scale: A chemical approach for fluorescence imaging and reconstruction of transparent mouse brain. *Nat Neurosci* 14:1–10.
10. Ke, M.-T., Y. Nakai, S. Fujimoto, et al. 2016. Super-resolution mapping of neuronal circuitry with index-optimized clearing agent. *Cell Rep* 14:2718–32.
11. Greenbaum, A., K.Y. Chan, T. Dobreva, et al. 2017. Bone CLARITY: Clearing, imaging, and computational analysis of osteoprogenitors within intact bone marrow. *Sci Transl Med* 9:eaah6518.
12. Pende, M., K. Becker, M. Wanis, et al. 2018. High-resolution ultramicroscopy of the developing and adult nervous system in optically cleared Drosophila melanogaster. *Nat Commun* 9:4731.
13. Perbellini, F., A.K.L. Liu, S.A. Watson, I. Bardi, S.M. Rothery, and C.M. Terracciano. 2017. Free-of-Acrylamide SDS-based Tissue Clearing (FASTClear) for three dimensional visualization of myocardial tissue. *Sci Rep* 7:5188.
14. Kim, J.H., M.J. Jang, J. Choi, et al. 2018. Optimizing tissue-clearing conditions based on analysis of the critical factors affecting tissue clearing procedures. *Sci Rep* 8:12815.
15. Yang, X., Y. Zhang, K. Zhao, et al. 2016. Skull optical clearing solution for enhancing ultrasonic and photoacoustic imaging. *IEEE Trans Med Imaging* 35:1903–6.
16. Ghosn, M.G., N. Sudheendran, M. Wendt, A. Glasser, V.V. Tuchin, and K.V. Larin. 2010. Monitoring of glucose permeability in monkey skin in vivo using optical coherence tomography. *J Biophotonics* 3:25–33.
17. Wen, X., Z. Mao, Z. Han, V.V. Tuchin, and D. Zhu. 2010. In vivo skin optical clearing by glycerol solutions: Mechanism. *J Biophotonics* 3:44–52.
18. Ding, Y., J. Wang, Z. Fan, et al. 2013. Signal and depth enhancement for in vivo flow cytometer measurement of ear skin by optical clearing agents. *Biomed Opt Express* 4:2518–26.
19. Deng, Z., L. Jing, N. Wu, et al. 2014. Viscous optical clearing agent for in vivo optical imaging. *J Biomed Opt* 19:076019.
20. Choe, C., J. Lademann, and M E. Darvin. 2015. Confocal Raman microscopy for investigating the penetration of various oils into the human skin in vivo. *J Dermatol Sci* 79:171–8.
21. Enfield, J., J. McGrath, S.M. Daly, and M. Leahy. 2016. Enhanced in vivo visualization of the microcirculation by topical application of fructose solution confirmed with correlation mapping optical coherence tomography. *J Biomed Opt* 21:081212.
22. Feng, W., R. Shi, N. Ma, D.K. Tuchina, V.V. Tuchin, and D. Zhu. 2016. Skin optical clearing potential of disaccharides. *J Biomed Opt* 21:081207.

23. Tuchina, D.K., P.A. Timoshina, V.V. Tuchin, A.N. Bashkatov, and E.A. Genina. 2019. Kinetics of rat skin optical clearing at topical application of 40% glucose: Ex vivo and in vivo studies. *IEEE J Sel Top Quantum Electron* 25:7200508.

24. Xu, X., and Q. Zhu. Evaluation of skin optical clearing enhancement with Azone as a penetration enhancer. 2007. *Opt Commun* 279:223–8.

25. Wen, X., S.L. Jacques, V.V. Tuchin, and D. Zhu. 2012. Enhanced optical clearing of skin in vivo and optical coherence tomography in-depth imaging. *J Biomed Opt* 17:066022.

26. Zhao, Q., C. Dai, S. Fan, J. Lv, and L. Nie. 2016. Synergistic efficacy of salicylic acid with a penetration enhancer on human skin monitored by OCT and diffuse reflectance spectroscopy. *Sci Rep* 6:34954.

27. Feng, W., R. Shi, C. Zhang, S. Liu, T. Yu, and D. Zhu. 2019. Vizualization of skin microvascular dysfunction of type I diabetic mice using in vivo skin optical clearing method. *J Biomed Opt* 24:031003.

28. Feng, W., C. Zhang, T. Yu, O. Semyachkina-Glushkovskaya, and D. Zhu. 2019. In vivo monitoring blood–barrier permeability using spectral imaging through optical clearing skull window. *J Biophotonics* 12:e201800330.

29. Agrba, P.D., and M.Yu. Kirillin. 2016. Effect of temperature regime and compression in OCT imaging of skin in vivo. *Photonics Lasers Med* 5:161–8.

30. Fox, M.A., D.G. Diven, K. Sra, et al. 2009. Dermal scatter reduction in human skin: A method using controlled application of glycerol. *Lasers Surg Med* 41:251–5.

31. Lee, W.-R., R.-Y. Tsai, C.-L. Fang, C.-J. Liu, C.-H. Hu, and J.-Y. Fang. 2006. Microdermabrasion as a novel tool to enhance drug delivery via the skin: An animal study. *Dermatol Surg* 32:1013–22.

32. Yoon, J., T. Son, E. Choi, B. Choi, J.S. Nelson, and B. Jung. 2008. Enhancement of optical clearing efficacy using a microneedle roller. *J Biomed Opt* 13:021103.

33. Xu, X., and Q. Zhu. 2008. Sonophoretic delivery for contrast and depth improvement in skin optical coherence tomography. *IEEE J Sel Top Quantum Electron* 14:56–61.

34. Liu, C., Z. Zhi, V.V. Tuchin, Q. Luo, and D. Zhu. 2010. Enhancement of skin optical clearing efficacy using photo-irradiation. *Lasers Surg Med* 42:132–40.

35. Zhong, H., Z. Guo, H. Wei, et al. 2010. Synergistic effect of ultrasound and thiazone–PEG 400 on human skin optical clearing in vivo. *Photochem Photobiol* 86:732–7.

36. Menyaev, Yu.A., D.A. Nedosekin, M. Sarimollaoglu, et al. 2013. Optical clearing in photoacoustic flow cytometry. *Biomed Opt Express* 4:3030–41.

37. Genina, E.A., A.N. Bashkatov, E.A. Kolesnikova, M.V. Basko, G.S. Terentyuk, and V.V. Tuchin. 2014. Optical coherence tomography monitoring of enhanced skin optical clearing in rats in vivo. *J Biomed Opt* 19:021109.

38. Damestani, Y., B. Melakeberhan, M.P. Rao, and G. Aguilar. 2014. Optical clearing agent perfusion enhancement via combination of microneedle poration, heating and pneumatic pressure. *Lasers Surg Med* 46:488–98.

39. Genina, E.A., Yu.I. Surkov, I.A. Serebryakova, A.N. Bashkatov, V.V. Tuchin, and V.P. Zharov. 2020. Rapid ultrasound optical clearing of human light and dark skin. *IEEE Trans Med Imaging* 39:3198–206.

40. Genina, E.A., A.N. Bashkatov, G.S. Terentyuk, and V.V. Tuchin. 2020. Integrated effects of fractional laser microablation and sonophoresis on skin immersion optical clearing in vivo. *J Biophotonics* 13:e202000101.

41. Tainaka, K., S.I. Kubota, T.Q. Suyama, et al. 2014. Whole-body imaging with single-cell resolution by tissue decolorization. *Cell* 159:911–24.

42. Zhao, S., M.I. Todorov, R. Cai, et al. 2020. Cellular and molecular probing of intact human organs. *Cell* 180:796-812e19.

43. Genina, E.A., A.N. Bashkatov, A.A. Korobko, et al. 2008. Optical clearing of human skin: Comparative study of permeability and dehydration of intact and photothermally perforated skin. *J Biomed Opt* 13:021102.

44. Genina, E.A., A.N. Bashkatov, and V.V. Tuchin. 2008. Optical clearing of cranial bone. *Adv Opt Technol* 2008:267867.

45. Hovhannisyan, V., P.-S. Hu, S.-J. Chen, C.-S. Kim, and C.-Y. Dong. 2013. Elucidation of the mechanisms of optical clearing in collagen tissue with multiphoton imaging. *J Biomed Opt* 18:046004.

46. Bashkatov, A.N., E.A. Genina, Yu.P. Sinichkin, V.I. Kochubey, N.A. Lakodina, and V.V. Tuchin. 2003. Glucose and mannitol diffusion in human dura mater. *Biophys J* 85:3310–8.

47. Zhong, H.Q., Z.Y. Guo, H.J. Wei, et al. 2010. Quantification of glycerol diffusion in human normal and cancer breast tissues in vitro with optical coherence tomography. *Laser Phys Lett* 7:315–20.

48. Zhu, Z., G. Wu, H. Wei, et al. 2012. Investigation of the permeability and optical clearing ability of different analytes in human normal and cancerous breast tissues by spectral domain OCT. *J Biophotonics* 5:1–8.

49. Choi, B., L. Tsu, E. Chen, et al. 2005. Determination of chemical agent optical clearing potential using in vitro human skin. *Lasers Surg Med* 36:72–5.

50. Mao, Z., D. Zhu, Y. Hu, X. Wen, and Z. Han. 2008. Influence of alcohols on the optical clearing effect of skin in vitro. *J Biomed Opt* 13:021104.

51. Genina, E.A., A.N. Bashkatov, Yu.P. Sinichkin, and V.V. Tuchin. 2010. Optical clearing of skin under action of glycerol: Ex vivo and in vivo investigations. *Opt Spectrosc* 109:225–31.

52. Oliveira, L.M., M.I. Carvalho, E.M. Nogueira, and V.V. Tuchin. 2015. Diffusion characteristics of ethylene glycol in skeletal muscle. *J Biomed Opt* 20:051019.

53. Tuchina, D.K., V.D. Genin, A.N. Bashkatov, E.A. Genina, and V.V. Tuchin. 2016. Optical clearing of skin tissue ex vivo with polyethylene glycol. *Opt Spectrosc* 120:28–37.

54. Genina, E.A., N.S. Ksenofontova, A.N. Bashkatov, G.S. Terentyuk, and V.V. Tuchin. 2017. Study of the epidermis ablation effect on the efficiency of optical clearing of skin in vivo. *Quantum Electron* 47:561–6.

55. Tuchina, D.K., A.N. Bashkatov, A.B. Bucharskaya, E.A. Genina, and V.V. Tuchin. 2017. Study of glycerol diffusion in skin and myocardium ex vivo under the conditions of developing alloxan-induced diabetes. *J Biomed Photon Eng* 3:020302.

56. Kinnunen, M., R. Myllylä, and S. Vainio. 2008. Detecting glucose-induced changes in in vitro and in vivo experiments with optical coherence tomography. *J Biomed Opt* 13:021111.

57. Larin, K.V., M.G. Ghosn, S.N. Ivers, A. Tellez, and J.F. Granada. 2007. Quantification of glucose diffusion in arterial tissues by using optical coherence tomography. *Laser Phys Lett* 4:312–7.

58. Ghosn, M.G., V.V. Tuchin, and K.V. Larin. 2006. Depth-resolved monitoring of glucose diffusion in tissues by using optical coherence tomography. *Opt Lett* 31:2314–6.

59. Ghosn, M.G., V.V. Tuchin, and K.V. Larin. 2007. Nondestructive quantification of analyte diffusion in cornea and sclera using optical coherence tomography. *Invest Ophthalmol Vis Sci* 48:2726–33.

60. Ghosn, M.G., E.F. Carbajal, N. Befrui, V.V. Tuchin, and K.V. Larin. 2008. Differential permeability rate and percent clearing of glucose in different regions in rabbit sclera. *J Biomed Opt* 13:021110.

61. Bashkatov, A.N., E.A. Genina, and V.V. Tuchin. 2009. Measurement of glucose diffusion coefficients in human tissues. In *Handbook of Optical Sensing of Glucose in Biological Fluids and Tissues*, ed. V.V. Tuchin, 587–621. Taylor & Francis Group LLC/CRC Press.

62. Bashkatov, A.N., E.A. Genina, Yu.P. Sinichkin, V.I. Kochubei, N.A. Lakodina, and V.V. Tuchin. 2003. Estimation of the glucose diffusion coefficient in human eye sclera. *Biophysics* 48:292–6.

63. Guo, X., G. Wu, H. Wei, et al. 2012. Quantification of glucose diffusion in human lung tissues by using Fourier domain optical coherence tomography. *Photochem Photobiol* 88:311–6.

64. Zhao, Q.L., J.L. Si, Z.Y. Guo, et al. 2011. Quantifying glucose permeability and enhanced light penetration in ex vivo human normal and cancerous oesophagus tissues with optical coherence tomography. *Laser Phys Lett* 8:71–7.

65. Hirshburg, J., B. Choi, J.S. Nelson, and A.T. Yeh. 2007. Correlation between collagen solubility and skin optical clearing using sugars. *Lasers Surg Med* 39:140–4.

66. Hirshburg, J.M., K.M. Ravikumar, W. Hwang, and A.T. Yeh. 2010. Molecular basis for optical clearing of collagenous tissues. *J Biomed Opt* 15:055002.

67. Wang, J., N. Ma, R. Shi, Y. Zhang, T. Yu, and D. Zhu. 2014. Sugar-induced skin optical clearing: From molecular dynamics simulation to experimental demonstration. *IEEE J Sel Top Quantum Electron* 20:7101007.

68. Genina, E.A., A.N. Bashkatov, and V.V. Tuchin. 2017. Optical clearing of human dura mater under action of glucose solutions. *J Biomed Photon Eng* 3:010309.

69. Genin, V.D., E.A. Genina, S.V. Kapralov, et al. 2019. Optical clearing of the gastric mucosa using 40%-glucose solution. *J Biomed Photon Eng* 5:030302.

70. Jiang, J., and R.K. Wang. 2004. Comparing the synergistic effects of oleic acid and dimethyl sulfoxide as vehicles for optical clearing of skin tissue in vitro. *Phys Med Biol* 49:5283–94.

71. Liu, Y., X. Yang, D. Zhu, and Q. Luo. 2013. Optical clearing agents improve photoacoustic imaging in the optical diffusive regime. *Opt Lett* 38:4236–9.

72. Xu, X., and R.K. Wang. 2004. Synergistic effect of hyperosmotic agents of dimethyl sulfoxide and glycerol on optical clearing of gastric tissue studied with near infrared spectroscopy. *Phys Med Biol* 49:457–68.

73. Jiang, J., M. Boese, P. Turner, and R.K. Wang. 2008. Penetration kinetics of dimethyl sulphoxide and glycerol in dynamic optical clearing of porcine skin tissue in vitro studied by Fourier transform infrared spectroscopic imaging. *J Biomed Opt* 13:021105.

74. Masoumi, Sh., M.A. Ansari, E. Mohajerani, E.A. Genina, and V.V. Tuchin. 2018. Combination of analytical and experimental optical clearing of rodent specimen for detecting beta-carotene: Phantom study. *J Biomed Opt* 23:095002.

75. Tuchin, V.V., I.L. Maksimova, D.A. Zimnyakov, I.L. Kon, A.H. Mavlutov, and A.A. Mishin. 1997. Light propagation in tissues with controlled optical properties. *J Biomed Opt* 2:401–17.

76. Bashkatov, A.N., E.A. Genina, V.I. Kochubey, V.V. Tuchin, and Yu.P. Sinichkin. 1998. The influence of osmotically active chemical agents on the transport of light in the scleral tissue. *Proc SPIE* 3726:403–9.

77. Sdobnov, A.Y., M.E. Darvin, J. Schleusener, J. Lademan, and V.V. Tuchin. 2019. Hydrogen bound water profiles in the skin influenced by optical clearing molecular agents: Quantitative analysis using confocal Raman microscopy. *J Biophotonics* 12:e201800283.

78. Funke, A.P., R. Schiller, H.W. Motzkus, C. Gunther, R.H. Muller, and R. Lipp. 2002. Transdermal delivery of highly lipophilic drugs: in vitro fluxes of antiestrogens, permeation enhancers, and solvents from liquids formulations. *Pharm Res* 19:661–8.

79. Lane, M.E. 2013. Skin penetration enhancers. *Int J Pharm* 447:12–21.

80. Yeh, A.T., B. Choi, J.S. Nelson, and B.J. Tromberg. 2003. Reversible dissociation of collagen in tissues. *J Invest Dermatol* 121:1332–5.

81. Shi, R., W. Feng, C. Zhang, Z. Zhang, and D. Zhu. 2017. FSOCA-induced switchable footpad skin optical clearing window for blood flow and cell imaging in vivo. *J Biophotonics* 10:1647–56.

82. Tsai, P.S., J.P. Kaufhold, P. Blinder, et al. 2009. Correlations of neuronal and microvascular densities in murine cortex revealed by direct counting and colocalization of nuclei and vessels. *J Neurosci* 29:14553–70.

83. Ke, M.T., S. Fujimoto, and T. Imai. 2013. SeeDB: A simple and morphology preserving optical clearing agent for neuronal circuit reconstruction. *Nat Neurosci* 16:1154–61.

84. Hou, B., D. Zhang, S. Zhao, et al. 2015. Scalable and DiI-compatible optical clearance of the mammalian brain. *Front Neuroanat* 9:19.

85. Susaki, E.A., K. Tainaka, D. Perrin, et al. 2014. Whole-brain imaging with single-cell resolution using chemical cocktails and computational analysis. *Cell* 157:726–39.

86. Chiang, A.S., W.Y. Lin, H.P. Liu, et al. 2002. Insect NMDA receptors mediate juvenile hormone biosynthesis. *Proc Natl Acad Sci USA* 99:37–42.

87. Kuwajima, T., A.A. Sitko, P. Bhansali, C. Jurgens, W. Guido, and C. Mason. 2013. ClearT: A detergent- and solvent-free clearing method for neuronal and non-neuronal tissue. *Development* 140:1364–8.

88. Aoyagi, Y., R. Kawakami, H. Osanai, T. Hibi, and T. Nemoto. 2015. A rapid optical clearing protocol using 2,2′-thiodiethanol for microscopic observation of fixed mouse brain. *PLoS One* 10:e0116280.

89. Costantini, I., J.P. Ghobril, A.P. Di Giovanna, et al. 2015. A versatile clearing agent for multi-modal brain imaging. *Sci Rep* 5:9808.

90. Staudt, T., M.C. Lang, R. Medda, J. Engelhardt, and S.W. Hell. 2007. 2,2′-thiodiethanol: A new water soluble mounting medium for high resolution optical microscopy. *Microsc Res Tech* 70:1–9.

91. Chi, J., Z. Wu, C.H.J. Choi, et al. 2018. Three-dimensional adipose tissue imaging reveals regional variation in beige fat biogenesis and PRDM16-dependent sympathetic neurite density. *Cell Metab* 27:226-36.e223.

92. Ertürk, A., K. Becker, N. Jährling, et al. 2012. Three-dimensional imaging of solvent-cleared organs using 3DISCO. *Nat Protoc* 7:1983–95.

93. Renier, N., Z. Wu, D.J. Simon, J. Yang, P. Ariel, and M. Tessier-Lavigne. 2014. Resource iDISCO: A simple, rapid method to immunolabel large tissue samples for volume imaging. *Cell* 159:896–910.

94. Dodt, H.U., U. Leischner, A. Schierloh, et al. 2007.Ultramicroscopy: three-dimensional visualization of neuronal networks in the whole mouse brain. *Nat Methods*4:331–6.

95. Becker, K., N. Jährling, S. Saghafi, R. Weiler, and H.U. Dodt. 2012. Chemical clearing and dehydration of GFP expressing mouse brains. *PLoS One* 7:e33916.

96. Hua, L., R. Zhou, D. Thirumalai, and B.J. Berne. 2008. Urea denaturation by stronger dispersion interactions with proteins than water implies a 2-stage unfolding. *Proc Natl Acad Sci U S A* 105:16928–33.

97. Zhu, J.,T. Yu, Yu. Li, et al. 2020.MACS: Rapid aqueous clearing system for 3D mapping of intact organs. *Adv Sci*7:1903185.

98. Chung, K., J. Wallace, S.Y. Kim, et al. 2013. Structural and molecular interrogation of intact biological systems. *Nature* 497:332–7.

99. Tomer, R., L. Ye, B. Hsueh, and K. Deisseroth. 2014. Advanced CLARITY for rapid and high-resolution imaging of intact tissues. *Nat Protoc* 9:1682–97.

100. Yang, B., J.B. Treweek, R.P. Kulkarni, et al. 2014. Single-cell phenotyping within transparent intact tissue through whole-body clearing. *Cell* 158:945–58.

101. Lee, E., J. Choi, Y. Jo, et al. 2016. ACT-PRESTO: Rapid and consistent tissue clearing and labeling method for 3-dimensional (3D) imaging. *Sci Rep* 6:18631.

102. Kristinsson, H.G., and H.O. Hultin. 2004. Changes in trout hemoglobin conformations and solubility after exposure to acid and alkali pH. *J Agric Food Chem* 52:3633–43.

103. Yu, T., X. Wen, V.V. Tuchin, Q. Luo, D. Zhu. 2011. Quantitative analysis of dehydration in porcine skin for assessing mechanism of optical clearing. *J Biomed Opt* 16:095002.

104. Moy, A.J., B.V. Capulong, R.B. Saager, et al. 2015. Optical properties of mouse brain tissue after optical clearing with FocusClear™. *J Biomed Opt* 20:095010.

105. Song, E., Y. Ahn, J. Ahn, et al. 2015. Optical clearing assisted confocal microscopy of ex vivo transgenic mouse skin. *Opt Laser Technol* 73:63–76.

106. Song, E., H. Seo, K. Choe, et al. 2015. Optical clearing based cellular-level 3D visualization of intact lymph node cortex. *Biomed Opt Express* 6:4154–64.

107. Yu, T., Y. Qi, J. Wang, et al. 2016. Rapid and prodium iodide-compatible optical clearing method for brain tissue based on sugar/sugar-alcohol. *J Biomed Opt* 21:081203.

108. Yanina, I. Yu., J. Schleusener, J. Lademann, V.V. Tuchin, and M.E. Darvin. 2020. Confocal Raman microspectroscopy for evaluation of optical clearing efficiency of the skin ex vivo. *Proc SPIE* 11239:112390W.

109. Xu, J., Y. Ma, T. Yu, and D. Zhu. 2019. Quantitative assessment of optical clearing clearing methods in various intact mouse organs. *J Biophotonics* 12:e201800134.

110. Tuchina, D.K., R. Shi, A.N. Bashkatov, et al. 2015. Ex vivo diffusion kinetics of glucose in native and in vivo glycated mouse skin. *J Biophotonics* 8:332–46.

111. Zhong, H., Z.Y. Guo, H. Wei, et al. 2010. In vitro study of ultrasound and different-concentration glycerol-induced changes in human skin optical attenuation assessed with optical coherence tomography. *J Biomed Opt* 15:036012.

112. Genin, V.D., D.K. Tuchina, A.J. Sadeq, E.A. Genina, V.V. Tuchin, and A.N. Bashkatov. 2016. Ex vivo investigation of glycerol diffusion in skin tissue. *J Biomed Photon Eng* 2:010303.

113. Zaitsev, S.V., Y.I. Svenskaya, E.V. Lengert, et al. 2020. Optimized skin optical clearing for optical coherence tomography monitoring of encapsulated drug delivery through the hair follicles. *J Biophotonics* 13:e201960020.

114. Cicchi, R., F.S. Pavone, D. Massi, and D.D. Sampson. 2005. Contrast and depth enhancement in two-photon microscopy of human skin ex vivo by use of optical clearing agents. *Opt Express* 13:2337–44.

115. Genin, V.D., A.N. Bashkatov, E.A. Genina, and V.V. Tuchin. 2015. Measurement of diffusion coefficient of propylene glycol in skin tissue. *Proc SPIE* 9448:94480E.

116. Schnell, A., W.A. Staines, and M.W. Wessendorf. 1999. Reduction of lipofuscin-like autofluorescence in fluorescently labeled tissue. *J Histochem Cytochem* 47:719–30.

117. Gómez-Gaviro, M.V. D. Sanderson, J. Ripoll, and M. Desco. 2020. Biomedical applications of tissue clearing and three-dimensional imaging in health and disease. *iScience* 23:101432.

118. Wang, J., Y. Zhang, T. Xu, Q. Luo, and D. Zhu. 2012. An innovative transparent cranial window based on skull optical clearing. *Laser Phys Lett* 9:469–73.

119. Masselink, W., D. Reumann, P. Murawala, et al. 2019. Broad applicability of a streamlined ethyl cinnamate-based clearing procedure. *Development* 146:dev166884.

120. Klingberg, A., A. Hasenberg, I. Ludwig-Portugall, et al. 2017. Fully automated evaluation of total glomerular number and capillary tuft size in nephritic kidneys using light-sheet microscopy. *J Am Soc Nephrol* 28:452–9.

121. Henning, Y., C. Osadnik, and E.P. Malkemper. 2019. EyeCi: optical clearing and imaging of immunolabeled mouse eyes using light-sheet fluorescence microscopy. *Exp Eye Res* 180:137–45.

122. Qi, Y., T. Yu, J. Xu, et al. 2019. FDISCO: Advanced solvent-based clearing method for imaging whole organs. *Sci Adv* 5:eaau8355.

123. Lagerweij, T., S.A. Dusoswa, A. Negrean, et al. 2017. Optical clearing and fluorescence deep-tissue imaging for 3D quantitative analysis of the brain tumor microenvironment. *Angiogenesis* 20:533–46.

124. Perin, P., F.F. Voigt, P. Bethge, F. Helmchen, and R. Pizzala. 2019. iDISCO+ for the Study of Neuroimmune Architecture of the Rat Auditory Brainstem. *Front Neuroanat* 13:15.

125. Hahn, C., K. Becker, S. Saghafi, et al. 2019. High-resolution imaging of fluorescent whole mouse brains using stabilised organic media (sDISCO). *J Biophotonics* 12:e201800368.

126. Pan, C., R. Cai, F.P. Quacquarelli, et al. 2016. Shrinkage-mediated imaging of entire organs and organisms using uDISCO. *Nat Methods* 13:859–67.

127. Li, Y., J. Xu, P. Wan, T. Yu, and D. Zhu. 2018. Optimization of GFP Fluorescence Preservation by a Modified uDISCO Clearing Protocol. *Front Neuroanat* 12:67.

128. Cai, R., C. Pan, A. Ghasemigharagoz, et al. 2019. Panoptic imaging of transparent mice reveals whole-body neuronal projections and skull-meninges connections. *Nat Neurosci* 22:317–327.

129. Schwarz, M.K., A. Scherbarth, R. Sprengel, J. Engelhardt, P. Theer, and G. Giese. 2015. Fluorescent-protein stabilization and high-resolution imaging of cleared, intact mouse brains. *PLoS One* 10:e0124650.

130. Hildebrand, S., A. Schueth, A. Herrler, R. Galuske, and A. Roebroeck. 2019. Scalable labeling for cytoarchitectonic characterization of large optically cleared human neocortex samples. *Sci Rep* 9:10880.

131. Jing, D., S. Zhang, W. Luo, et al. 2018. Tissue clearing of both hard and soft tissue organs with the PEGASOS method. *Cell Res* 28:803–18.

132. Pende, M., K. Vadiwala, H. Schmidbaur, et al. 2020. A versatile depigmentation, clearing, and labeling method for exploring nervous system diversity. *Sci Adv* 6:eaba0365.

133. Li, W., R.N. Germain, and M.Y. Gerner. 2017. Multiplex, quantitative cellular analysis in large tissue volumes with clearing-enhanced 3D microscopy (C$_e$3D). *Proc Natl Acad Sci USA* 114:E7321–30.

134. Susaki, E.A., and H.R. Ueda. 2016. Whole-body and whole-organ clearing and imaging techniques with single-cell resolution: Toward organism-level systems biology in mammals. *Cell Chem Biol* 23:137–57.

135. Tainaka, K., T.C. Murakami, E.A. Susaki, et al. 2018. Chemical landscape for tissue clearing based on hydrophilic reagents. *Cell Rep* 24:2196-210.e2199.

136. Kubota, S.I., K. Takahashi, J. Nishida, et al. 2017. Whole-body profiling of cancer metastasis with single-cell resolution. *Cell Rep* 20:236–50.

137. Susaki, E.A., C. Shimizu, A. Kuno, et al. 2020. Versatile whole-organ/body staining and imaging based on electrolyte-gel properties of biological tissues. *Nat Commun* 11:1982.

138. Murakami, T.C., T. Mano, S. Saikawa, et al. 2018. A three-dimensional single-cell-resolution whole-brain atlas using CUBIC-X expansion microscopy and tissue clearing. *Nat Neurosci* 21:625–37.

139. Sato, Y., T. Miyawaki, A. Ouchi, A. Noguchi, S. Yamaguchi, and Y. Ikegaya. 2019. Quick visualization of neurons in brain tissues using an optical clearing technique. *Anat Sci Int* 94:199–208.

140. Decroix, L., V. Van Muylder, L. Desender, M. Sampaolesi, and L. Thorrez. 2015. Tissue clearing for confocal imaging of native and bio-artificial skeletal muscle. *Biotech Histochem* 90:424–31.

141. Hama, H., H. Hioki, K. Namiki, et al. 2015. ScaleS: An optical clearing palette for biological imaging. *Nat Neurosci* 18:1518–29.

142. Wan, P., J. Zhu, J. Xu, Y. Li, T. Yu, and D. Zhu. 2018. Evaluation of seven optical clearing methods in mouse brain. *Neurophotonics* 5:035007.

143. Costa, E.C., D.N. Silva, A.F. Moreira, and I.J. Correia. 2019. Optical clearing methods: An overview of the techniques used for the imaging of 3D spheroids. *Biotechnol Bioeng* 116:2742–63.

144. Chen, L., G. Li, Y. Li, et al. 2017. UbasM: An effective balanced optical clearing method for intact biomedical imaging. *Sci Rep* 7:12218.

145. Sylwestrak, E.L., P. Rajasethupathy, M.A. Wright, A. Jaffe, and K. Deisseroth. 2016. Multiplexed intact-tissue transcriptional analysis at cellular resolution. *Cell* 164:792–804.

146. Treweek, J.B., K.Y. Chan, N.C. Flytzanis, et al. 2015. Whole-body tissue stabilization and selective extractions via tissue-hydrogel hybrids for high-resolution intact circuit mapping and phenotyping. *Nat Protoc* 10:1860–96.

147. Xu, N., A. Tamadon, Y. Liu, et al. 2017. Fast free-of-acrylamide clearing tissue (FACT) – an optimized new protocol for rapid, high-resolution imaging of three-dimensional brain tissue. *Sci Rep* 7:9895.

148. Williams, M.P.I., M. Rigon, T. Straka, et al. 2019. A novel optical tissue clearing protocol for mouse skeletal muscle to visualize endplates in their tissue context. *Front Cell Neurosci* 13:49.

149. Sung, K., Y. Ding, J. Ma, et al. 2016. Simplified three-dimensional tissue clearing and incorporation of colorimetric phenotyping. *Sci Rep* 6:30736.

150. Wang, Z., J. Zhang, G. Fan, et al. 2018. Imaging transparent intact cardiac tissue with single-cell resolution. *Biomed Opt Express* 9:423–36.

151. Murray, E., J.H. Cho, D. Goodwin, et al. 2015. Simple, scalable proteomic imaging for high-dimensional profiling of intact systems. *Cell* 163:1500–14.

152. Spalteholz, W. 1914. *Über das durchsichtigmachen von menschlichen und tierischen präparaten und seine theoretischen bedingungen*. Leipzig: S. Hirzel.

153. Steinke, H., and W. Wolff. 2001. A modified Spalteholz technique with preservation of the histology. *Ann Anat* 183:91–5.

154. Barer, R. 1955. Spectrophotometry of clarified cell suspensions. *Science* 121:709–15.

155. Bakutkin, V.V., I.L. Maksimova, P.I. Saprykin, V.V. Tuchin, and L.P. Shubochkin. 1987. Light scattering by the human eye sclera. *J Appl Spectrosc* 46:104–7.

156. Chance, B., H. Liu, T. Kitai, and Y. Zhang. 1995. Effects of solutes on optical properties of biological materials: Models, cells, and tissues. *Anal Biochem* 227:351–62.

157. Tuchin, V.V., I.L. Maksimova, D.A. Zimnyakov, I.L. Kon, A.H. Mavlutov, A. and A. Mishin. 1996. Light propagation in tissues with controlled optical properties. *Proc SPIE* 2925:118–42.

158. Tuchin, V.V., A.N. Bashkatov, E.A. Genina, Yu.P. Sinichkin, and N.A. Lakodina. 2001. In vivo investigation of the immersion-liquid-induced human skin clearing dynamics. *Tech Phys Lett* 27:489–90.

159. Meglinski, I.V., A.N. Bashkatov, E.A. Genina, D.Y. Churmakov, and V.V. Tuchin. 2003. The enhancement of confocal images of tissues at bulk optical immersion. *Laser Phys* 13:65–9.

160. Vargas, G., E.K. Chan, J.K. Barton, H.G. Rylander, and A.J. Welch. 1999. Use of an agent to reduce scattering in skin. *Lasers Surg Med* 24:133–41.

161. Vargas, G., K.F. Chan, S.L. Thomsen, and A.J. Welch. 2001. Use of osmotically active agents to alter optical properties of tissue: Effects on the detected fluorescence signal measured through skin. *Lasers Surg Med* 29:213–20.

162. Vargas, G., A. Readinger, S.S. Dosier, A.J. Welch. 2003. Morphological changes in blood vessels produced by hyperosmotic agents and measured by optical coherence tomography. *Photochem Photobiol* 77:541–9.

163. Rylander, C.G., O.F. Stumpp, T.E. Milner, et al. 2006. Dehydration mechanism of optical clearing in tissue. *J Biomed Opt* 11:041117.

164. Wang, R.K., X. Xu, V.V. Tuchin, and J.B. Elder. 2001. Concurrent enhancement of imaging depth and contrast for optical coherence tomography by hyperosmotic agents. *J Opt Soc Am B* 18:948–53.

165. Xu, X., and R.K. Wang. 2003. The role of water desorption on optical clearing of biotissue: Studied with near infrared reflectance spectroscopy. *Med Phys* 30:1246–53.

166. Galanzha E.I., V.V. Tuchin, A.V. Solovieva, T.V. Stepanova, Q. Luo, and H. Cheng. 2003. Skin backreflectance and microvascular system functioning at the action of osmotic agents. *J Phys D Appl Phys* 36:1739–46.

167. Zhu, D., Q. Luo, and J. Cen. 2003. Effects of dehydration on the optical properties of in vitro porcine liver. *Lasers Surg Med* 33:226–31.

168. Zhu, D., J. Zhang, H. Cui, Z. Mao, P. Li, and Q. Luo. 2008. Short-term and long-term effects of optical clearing agents on blood vessels in chick chorioallantoic membrane. *J Biomed Opt* 13:021106.

169. Wang, J., D. Zhu, M. Chen, and X. Liu. 2010. Assessment of optical clearing induced improvement of laser speckle contrast imaging. *J Innov Opt Health Sci* 3:159–67.

170. Khan, M.H., B. Choi, S. Chess, K.M. Kelly, J. McCullough, and J.S. Nelson. 2004. Optical clearing of in vivo human skin: Implications for light-based diagnostic imaging and therapeutics. *Lasers Surg Med* 34:83–5.

171. Yeh, A.T., J. Hirshburg. 2006. Molecular interactions of exogenous chemical agents with collagen – implications for tissue optical clearing. *J Biomed Opt* 11:014003.

172. Plotnikov, S., V. Juneja, A.B. Isaacson, W.A. Mohler, and P.J. Campagnola. 2006. Optical clearing for improved contrast in second harmonic generation imaging of skeletal muscle. *Biophys J* 90:328–39.

173. Nadiarnykh O., P.J. Campagnola. 2009. Retention of polarization signatures in SHG microscopy of scattering tissues through optical clearing. *Opt Express* 17:5794–806.

174. Proskurin, S.G., and I.V. Meglinski. 2007. Optical coherence tomography imaging depth enhancement by superficial skin optical clearing. *Laser Phys Lett* 4:824–6.

175. Bonesi, M., S.G. Proskurin, and I.V. Meglinski. 2010. Imaging of subcutaneous blood vessels and flow velocity profiles by optical coherence tomography. *Laser Phys* 20:891–9.

176. Larina, I.V., E.F. Carbajal, V.V. Tuchin, M.E. Dickinson, and K.V. Larin. 2008. Enhanced OCT imaging of embryonic tissue with optical clearing. *Laser Phys Lett* 5:476–9.

177. Keller, P.J., and H.U. Dodt. 2012. Light sheet microscopy of living or cleared specimens. *Curr Opin Neurobiol* 22:138–43.

178. Dunn, A.K., V.P. Wallace, M. Coleno, M.W. Berns, and B.J. Tromberg. 2000. Influence of optical properties on two-photon fluorescence imaging in turbid samples. *Appl Opt* 39:1194–201.

179. Gu, M., X.S. Gan, A. Kisteman, and M.G. Xu. 2000. Comparison of penetration depth between two-photon excitation and single-photon excitation in imaging through turbid tissue media. *Appl Phys Lett* 77:1551–3.

180. Oliveira, L., M.I. Carvalho, E. Nogueira, and V.V. Tuchin. 2016. Optical clearing mechanisms characterization in muscle. *J Innov Opt Health Sci* 9:1650035.

181. Carneiro, I., S. Carvalho, R. Henrique, L. Oliveira, and V.V. Tuchin. 2019. A robust ex vivo method to evaluate the diffusion properties of agents in biological tissues. *J Biophotonics* 12:e201800333.

182. Carvalho, S., N. Gueiral, E. Nogueira, R. Henrique, L.M. Oliveira, and V.V. Tuchin. 2017. Glucose diffusion in colorectal mucosa: A comparative study between normal and cancer tissues. *J Biomed Opt* 22:091506.

183. Christoph, R., B. Schmidt, U. Steinberner, W. Dilla, and R. Karinen. 2012. Glycerol. In *Ullmann's Encyclopedia of Industrial Chemistry*, 67–81. Weinheim: Wiley-VCH Verlag GmbH & Co. KGaA.

184. Huang, D., W. Zhang, H. Zhong, H. Xiong, X. Guo, and Z. Guo. 2012. Optical clearing of porcine skin tissue in vitro studied by Raman microspectroscopy. *J Biomed Opt* 17:015004.

185. Nijssen, A., T.C.B. Schut, F. Heule, et al. 2002. Discriminating basal cell carcinoma from its surrounding tissue by Raman spectroscopy. *J Invest Dermatol* 119:64–9.

186. Schulmerich, M.V., K.A. Dooley, T.M. Vanasse, S.A. Goldstein, and M.D. Morris. 2007. Subsurface and transcutaneous Raman spectroscopy and mapping using concentric illumination rings and collection with a circular fiber optic array. *Appl Spectrosc* 61:671–8.

187. Schulmerich, M.V., J.H. Cole, K.A. Dooley, M.D. Morris, J.M. Kreider, and S.A. Goldstein. 2008. Optical clearing in transcutaneous Raman spectroscopy of murine cortical bone tissue. *J Biomed Opt* 13:021108.

188. Bashkatov, A.N., K.V. Berezin, K.N. Dvoretskiy, et al. 2018. Measurement of tissue optical properties in the context of tissue optical clearing. *J Biomed Opt* 23:091416.

189. Carneiro, I., S. Carvalho, R. Henrique, R. Oliveira, and V.V. Tuchin. 2017. Simple multimodal optical technique for evaluation of free/bound water and dispersion of human liver tissue. *J Biomed Opt* 22:125002.

190. Son, T., and B. Jung. 2015. Cross-evaluation of optimal glycerol concentration to enhance optical clearing efficacy. *Skin Res Technol* 21:327–32.

191. Yoon, J., D. Park, T. Son, J. Seo, J.S. Nelson, and B. Jung. 2010. A physical method to enhance transdermal delivery of a tissue optical clearing agent: Combination of microneedling and sonophoresis. *Lasers Surg Med* 42:412–7.

192. Carneiro, I., S. Carvalho, R. Henrique, R. Oliveira, and V.V. Tuchin. 2019. Kinetics of optical properties of colorectal muscle during optical clearing. *IEEE J Sel Top Quantum Electron* 25:7200608.

193. Mao, Z., X. Wen, J. Wang, and D. Zhu. 2009. The biocompatibility of the dermal injection of glycerol in vivo to achieve optical clearing. *Proc SPIE* 7519:75191.

194. Samatham, R., K.G. Phillips, S.L. Jacques. 2010. Assessment of optical clearing agents using reflectance-mode confocal scanning laser microscopy. *J Innov Opt Health Sci* 3:183–8.

195. Genina, E.A., A.N. Bashkatov, and V.V. Tuchin. 2010. Tissue optical immersion clearing. *Expert Rev Med Devices* 7:825–42.

196. Genina, E.A., A.N. Bashkatov, Yu.P. Sinichkin, I. Yu. Yanina, and V.V. Tuchin. 2016. Optical clearing of tissues: Benefits for biology, medical diagnostics, and phototherapy. In *Handbook of Optical Biomedical Diagnostics, 2nd Edition*, ed. V.V. Tuchin, 565–637. Bellingham: SPIE Press.

197. Ullah, H., F. Hussain, and M. Ikram. 2015. Optical coherence tomography for glucose monitoring in blood. *Appl Phys B* 120:355–66.

198. Kuranov, R.V., V.V. Sapozhnikova, D.S. Prough, I. Cicenaite, and R.O. Esenaliev. 2006. In vivo study of glucose-induced changes in skin properties assessed with optical coherence tomography. *Phys Med Biol* 51:3885–900.

199. Genina, E.A., A.N. Bashkatov, M.D. Kozintseva, and V.V. Tuchin. 2016. OCT study of optical clearing of muscle tissue in vitro with 40% glucose solution. *Opt Spectrosc* 120:20–7.

200. Liu, C.H., M. Singh, J. Li, et al. 2015. Quantitative assessment of hyaline cartilage elasticity during optical clearing using optical coherence elastography. *Sovrem Tekhnologii Med* 7:44–51.

201. Zhao, Q., H. Wei, Y. He, Q. Ren, and C. Zhou. 2014. Evaluation of ultrasound and glucose synergy effect on the optical clearing and light penetration for human colon tissue using SD-OCT. *J Biophotonics* 7:938–47.

202. Gao, J.K. 1993. *Polyethylene Glycol as an Embedment for Microscopy and Histochemistry*. Boca Raton, FL: CRC Press.

203. Jakasa, I., M.M. Verberk, M. Esposito, J.D. Bos, and S. Kezic. 2007. Altered penetration of polyethylene glycols into uninvolved skin of atopic dermatitis patients. *J Invest Dermatol* 127:129.

204. Tuchin, V.V., G.B. Altshuler, A.A. Gavrilova, et al. 2006. Optical clearing of skin using flashlamp-induced enhancement of epidermal permeability. *Lasers Surg Med* 38:824–36.

205. Xu, X., and Q. Zhu. 2008. Feasibility of sonophoretic delivery for effective skin optical clearing. *IEEE Trans Biomed Eng* 55:1432–7.

206. Zhi, Z., Z. Han, Q. Luo, and D. Zhu. 2009. Improve optical clearing of skin in vitro with propylene glycol as a penetration enhancer. *J Innov Opt Health Sci* 2:269–78.

207. Xu, X., Q. Zhu, and C. Sun. 2009. Assessment of the effects of ultrasound-mediated alcohols on skin optical clearing. *J Biomed Opt* 14:034042.

208. Rowe, R.C., P.J. Sheskey, and M.E. Quinn. 2009. *Handbook of Pharmaceutical Excipients*. Midland: Pharmaceutical Press and American Pharmacists Association.

209. Tuchina, D.K., A.N. Bashkatov, E.A. Genina, and V.V. Tuchin. 2015. Quantification of glucose and glycerol diffusion in myocardium. *J Innov Opt Health Sci* 8:1541006.

210. Chin, K.P., S.F.Y. Li, Y.J. Yao, and L.S. Yue. 1991. Infinite dilution diffusion coefficients of poly(ethylene glycol) and poly(propylene glycol) in water in the temperature range 303–318 K. *J Chem Eng Data* 36:329–31.

211. Gekas V. and N. Mavroudis. 1998. Mass transfer properties of osmotic solutions. II. Diffusivities. *Int J Food Prop* 1:181–95.

212. Longsworth L.G. 1953 Diffusion measurement, at 25°, of aqueous solutions of amino acids, peptides and sugars. *J Am Chem Society* 75:5705–09.

213. Gros G. 1978 Concentration dependence of the self-diffusion of human and *Lumbricus terrestris* hemoglobin. *Biophys J* 22:453–68.

214. Nishijima Y. and G. Oster 1960 Diffusion of glycerol–water mixture. *Bulletin Chem Soc Jpn* 33:1649–51.

215. Tomlinson D.J. 1972 Temperature dependent self-diffusion coefficient measurements of glycerol by the pulsed N.M.R. technique. *Mol Phys* 25: 735–8.

216. Prausnitz, M.R., and R. Langer. 2008. Transdermal drug delivery. *Nat Biotechnol* 26:1261–8.

217. Cu, K., R. Bansal, S. Mitragotri, and D.F. Rivas. 2020. Delivery strategies for skin: Comparison of nanoliter jets, needles and topical solutions. *Ann Biomed Eng* 48:2028–39.

218. Schaefer, H., and T.E. Redelmeie. 1996. *Skin Barrier*. Basel: Karger.

219. van Smeden, J., M. Janssens, G.S. Gooris, and J.A. Bouwstra. 2014. The important role of stratum corneum lipids for the cutaneous barrier function. *Biochim Biophys Acta Mol Cell Biol Lipids* 1841:295–313.

220. Barba, C., C. Alonso, M. Marti, A. Manich, and L. Coderch. 2016. Skin barrier modification with organic solvents. *Biochim Biophys Acta* 1858:1935–1943.

221. Sarri, B., X. Chen, R. Canonge, et al. 2019. In vivo quantitative molecular absorption of glycerol in human skin using coherent anti-Stokes Raman scattering (CARS) and two-photon autofluorescence. *J Controlled Release* 308:190–6.

222. Zhu, D., J. Wang, Z. Zhi, X. Wen, and Q. Luo. 2010. Imaging dermal blood flow through the intact rat skin with an optical clearing method. *J Biomed Opt* 15:026008.

223. Wang, J., R. Shi, and D. Zhu. 2013. Optical clearing method induced switchable skin window for dermal blood flow imaging. *J Biomed Opt* 18:061209.

224. Wang, J., R. Shi, Y. Zhang, and D. Zhu. 2013. Ear skin optical clearing for improving blood flow imaging. *Photon Lasers Med* 2:37–44.

225. Wang, J., Y. Zhang, P. Li, Q. Luo, and D. Zhu. 2014. Review: Tissue optical clearing window for blood flow monitoring. *IEEE J Sel Top Quantum Electron* 20:6801112.

226. Genina, E.A., A.N. Bashkatov, O.V. Semyachkina-Glushkovskaya, and V.V. Tuchin. 2017. Optical clearing of cranial bone by multicomponent immersion solutions and cerebral venous blood flow visualization. *Izv. Saratov Univ (N S) Ser Phys* 17:98–110. (in Russian)

227. Cai, R., C. Pan, A. Ghasemigharagoz, et al. 2018. Panoptic vDISCO imaging reveals neuronal connectivity, remote trauma effects and meningeal vessels in intact transparent mice. bioRxiv374785.

228. Nehrhoff, I., D. Bocancea, J. Vaquero, et al. 2016. 3D imaging in CUBIC-cleared mouse heart tissue: Going deeper. *Biomed Opt Express* 7:3716–20.

229. Nehrhoff, I., J. Ripoll, R. Samaniego, M. Desco, M.V. Gómez-Gaviro. 2017. Looking inside the heart: a see-through view of the vascular tree. *Biomed Opt Express* 8:3110–8.

230. Chang, E.H., M. Argyelan, M. Aggarwal, et al. 2017. The role of myelination in measures of white matter integrity: Combination of diffusion tensor imaging and two-photon microscopy of CLARITY intact brains. *NeuroImage* 147:253–61.

231. Ueda, H.R., H.-U. Dodt, P. Osten, M.N. Economo, J. Chandrashekar, and P.J. Keller. 2020. Whole-brain profiling of cells and circuits in mammals by tissue clearing and light-sheet microscopy. *Neuron* 106:369–87.

232. Park, Y.-G., C.H. Sohn, R. Chen, et al. 2019. Protection of tissue physicochemical properties using polyfunctional crosslinkers. *Nat Biotechnol* 37:73–83.

233. Lloyd-Lewis, B., F.M. Davis, O.B. Harris, et al. 2016. Imaging the mammary gland and mammary tumours in 3D: Optical tissue clearing and immunofluorescence methods. *Breast Cancer Res* 18:127.

234. Polat, B.E., D. Hart, R. Langer, and D. Blankschtein. 2011. Ultrasound-mediated transdermal drug delivery: Mechanisms, scope, and emerging trends. *J Controlled Release* 152:330–48.

235. Alkilani, A.Z., M.T.C. McCrudden, and R.F. Donnelly. 2015. Transdermal drug delivery: innovative pharmaceutical developments based on disruption of the barrier properties of the stratum corneum. *Pharmaceutics* 7:438–70.

236. Wang, J., X. Zhou, S. Duan, Z. Chen, and D. Zhu. 2011. Improvement of in vivo rat skin optical clearing with chemical penetration enhancers. *Proc SPIE* 7883:78830Y.

237. Stumpp, O., A.J. Welch, and J. Neev. 2005. Enhancement of transdermal skin clearing agent delivery using a 980 nm diode laser. *Lasers Surg Med* 37:278–85.

238. Lin, C.-H., I.A. Aljuffali, and J.-Y. Fang. 2014. Lasers as an approach for promoting drug delivery via skin. *Expert Opin Drug Deliv* 11:599–614.

239. Genina, E.A., L.E. Dolotov, A.N. Bashkatov, and V.V. Tuchin. 2016. Fractional laser microablation of skin: Increasing the efficiency of transcutaneous delivery of particles. *Quantum Electron* 46:502–9.

240. Trelles, M.A., S. Mordon, M. Velez, F. Urdiales, and J.L. Levy. 2009. Results of fractional ablative facial skin resurfacing with the erbium:yttrium-aluminium-garnet laser 1 week and 2 months after one single treatment in 30 patients. *Lasers Med Sci* 24:186–94.

241. Son, T., J. Lee, and B. Jung. 2013. Contrast enhancement of laser speckle contrast image in deep vasculature by reduction of tissue scattering. *J Opt Soc Korea* 17:86–90.

242. Stumpp, O., and A.J. Welch. 2003. Injection of glycerol into porcine skin for optical skin clearing with needle-free injection gun and determination of agent distribution using OCT and fluorescence microscopy. *Proc SPIE* 4949:44–50.

4

Chemical enhancers for improving tissue optical clearing efficacy

Dan Zhu, Yanmei Liang, Xingde Li, and Valery V. Tuchin

CONTENTS

Introduction

Optical diagnostic and therapeutic techniques have been a hotspot in the area of life sciences and biomedical technologies. However, the strong scattering of biological tissue limits the penetration depth of visible and near-infrared light in tissue, which greatly restricts the clinical application of optical techniques. The tissue optical clearing (TOC) technique [1, 2] can reduce the scattering and improve light penetration depth in tissue by introducing a hyperosmotic, high refractive index chemical agent termed an optical clearing agent (OCA) into tissue. It brings a new opportunity to the development of biomedical optical diagnosis and therapy.

Various researches have been conducted on the optical clearing of different tissues, among which skin has attracted more attention [3–6]. However, the outermost layer of the skin, the stratum corneum (SC), presents a significant barrier for the topical application of most OCAs and is hence responsible for the poor optical clearing effect. To reduce the barrier function of the SC, numerous methods, such as physical methods to remove the SC (photothermal [7], ultrasound [8], sandpaper [9], and microneedle arrays [10]), have been studied.

Compared to the physical methods, a nonphysical way is to incorporate a chemical penetration enhancer (PE) in the transdermal formulation to reduce the SC barrier and enhance the permeability of the skin to drugs [11]. In fact, chemical PEs have been commonly used for the enhancement of drug penetration into tissue in clinical medicine. Recently, PEs were introduced into the research area of skin optical clearing. Jiang *et al.* combined DMSO or oleic acid with propylene glycol (PG) [12] and Xu *et al.* mixed Azone with glycerol or PG, respectively [13], and then they applied the mixtures topically on porcine skin *in vitro* and found that all the mixtures were able to improve optical clearing of skin effectively compared with the pure agent. Nevertheless, the commonly used penetration enhancer DMSO has potential toxicity, and the penetration enhancing effect of Azone is not ideal [14]. Although PG was always used as an OCA in tissue optical clearing research, its effect is not as ideal as glycerol or PEG400, etc., when applied at the same concentration. In fact, PG was also considered to be a useful penetration enhancer in clinical application, especially for alcohol-soluble drugs [11, 15, 16]. Zhu's group considered PG and Thiazone as chemical penetration enhancers to improve the skin optical clearing efficacy from *in vitro* to *in vivo* [17, 18].

Liquid paraffin is a kind of highly refined mineral oil that has been used in cosmetics and pharmaceutical excipients to retain water in the skin for more than 100 years, and it has been approved by Food and Drug Administration (FDA) as a safe pharmaceutical vehicle and was listed in many different pharmacopeia [19]. It can be used to treat constipation and encopresis, and as an eye ointment to relieve dryness and irritation. As a solvent for dermatology medicine and a

DOI: 10.1201/9781003025252-5

humectant, it can plasticize the stratum corneum and improve the surface smoothness of skin [20]. It has been proven that lipophilic agents such as liquid paraffin are capable of penetrating through epithelium tissue and can serve as a carrier for drug delivery [19, 21–23]. The optical clearing results of liquid paraffin as the enhancer of glycerol on various biotissues were evaluated with different optical technologies [24–28].

PG/Azone/Thiazone for improving *in vitro* skin optical clearing efficacy

Chemical agents

Three penetration enhancers, PG, Azone, and Thiazone, have been chosen to evaluate the improvement of skin optical clearing efficacy. Among them, Azone is commonly used and widely recognized as chemical skin penetration enhancer [13, 29], whereas Thiazone (1,2- benzisothiazole-3 (2H)-2- butyl-1,1-dioxide is the chemical name), a derivative of Azone, is a newly designed chemical PE developed by the Applied Chemistry Institute of Beijing Normal University in China, which has a penetration enhancing effect three times higher than Azone [30]. PG is what we mainly want to study as a PE.

The previous study shows that alcohols can induce good optical clearing effect on the skin, and the more hydroxyl groups alcohol contains, the better the optical clearing effect on the skin is [14]. Meanwhile, the solubility of OCA and PEs was also considered. Hence, three multihydric alcohols were chosen as OCAs, i.e., D-sorbitol (hexahydric alcohol), glycerol (trihydric alcohol), and polyethylene glycol (PEG400, dihydric alcohol) (Qiangsheng Chinese Chemicals, Limited, China), which are all mixable with the PEs. In order to compare how the chemical penetration enhancer improved different OCAs' optical clearing effect of skin *in vitro*, the OCAs, the OCAs with water or PEs at a ratio of 19:1 were designed and listed in Table 4.1. In the table, the numbers represent the percentages of chemical substances and water. Among them, pure D-sorbitol is solid, and its saturated

concentration is only 70%. Hence, 70% D-sorbitol was used as pure OCA to mix with water or PG in the volume ratio of 19:1.

Different OCAs for optical clearing of skin

In order to directly observe the optical skin-clearing efficacy of OCAs, photos of background trough skin with different OCA treatments were taken at time intervals of 0 (native state), 15, 30, 60, and 120 min. The typical results were shown in Figure 4.1, where (a), (b), (c), (d), (e) represent the group of glycerol, D-sorbitol, PEG400, PG, and saline, respectively and each group includes OCA, OCA/water, and OCA/PE with the same volume ratio of 19:1, except for the PG and saline group.

Among each group, obvious discrepancies can be found between the samples treated with pure OCA, with water, or with PE. As shown in Figure 4.1(a), for the sample treated by glycerol/PG, the background could be identified at 30 min and turned more and more clear; the sample treated by glycerol turned clear more or less at 30 min; whereas the sample treated by glycerol/water was still turbid and the background could not be identified at 30 min.

In Figure 4.1(b), the saturated D-sorbitol (70%) was mixed with water and PG (V/V, 19:1). The agent with PG induced the best clearing effect, which enabled us to observe the pattern underneath the skin at 30 min more or less; the effect of 70% D-sorbitol/water was the worst.

PEG 400 can be mixed with different PEs, and Figure 4.1(c) shows the clearing effects of PEG 400 group. The results indicated that treatment with PEG400/water did not induce a good clearing effect even at 120 min; pure PEG or PEG/Azone could induce optical clearing at 120 min. In contrast, PEG/PG or PEG/Thiazone treatment enabled us to see the pattern behind the skin sample at 60 min and more clearly at 120 min, but the difference between these two agents was not obvious.

By comparing the effect of the first three groups, it can be found that, for pure OCAs, glycerol induced the optimal clearing effect, followed by 70% D-sorbitol, while PEG400 was the

TABLE 4.1

Different agents and their refractive indexes. Data from reference [18].

Solutions	G	D-sorbitol	PEG	Azone	Thiazone	PG	H$_2$O	RI*
Glycerol(100%)	100							1.471
Glycerol/water(19:1)	95						5	1.458
Glycerol/PG(19:1)	95					5		1.468
D-sorbitol(70%)		70					30	1.466
70%D-sorbitol/water(19:1)		65					35	1.458
70%D-sorbitol/PG(19:1)		66.5				5	28.5	1.462
PEG400(100%)			100					1.469
PEG400/water(19:1)			95				5	1.460
PEG/Azone(19:1)			95	5				1.461
PEG/Thiazone(19:1)			95		5			1.462
PEG/PG(19:1)			95			5		1.470

*RI represents Refractive Indices of agents measured by Abbe Digital Refractometer.

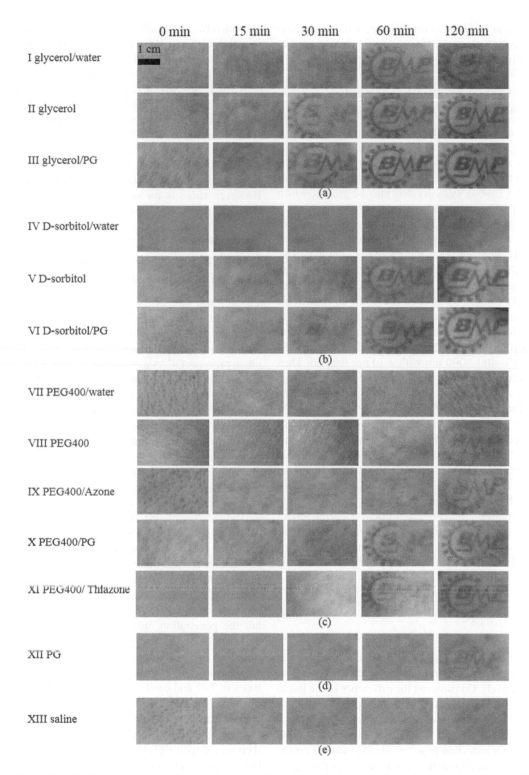

FIGURE 4.1 Changes in optical transparency of porcine skin samples after treatments with different agents after 15, 30, 60, and 120 min. Among them (a) the group of glycerol, (b) the group of 70% D-sorbitol, (c) the group of PEG400, (d) pure PG, (e) saline. Reprinted with permission from reference [18].

worst. If water was added into the OCAs, the optical clearing effect turned out to be worse; however, when PG was added into the OCAs, the clearing effect turned out to be better, and the order did not change, i.e., glycerol/PG induced the best clearing effect, 70% D-sorbitol/ PG the second, and PEG400/ PG the worst.

The PG and saline groups were set as control groups, as shown in Figure 4.1(d) and (e); saline hardly induced any change on the clearing effect on the skin, and PG induced a very limited effect. As can be seen from each row, with the continuous treatment time, the skin samples become clearer and clearer, which finally enables us to see the background partly or totally, except saline. Furthermore, at the same treatment time, the optical clearing effect varies for different agents as well as for different groups.

Quantitative evaluation of enhancement of skin optical clearing efficacy

In order to quantitatively evaluate the enhancement of skin optical clearing efficacy by different PEs, fresh porcine skin samples, 3 cm × 3 cm each, were sealed to prevent natural dehydration and stored at 4°C for no longer than 4 h before use. All samples (n=64) were divided into 12 groups, which included 11 experimental groups (agents listed in Table 4.1) and a control group (saline). Between five and six samples for each group were used to experiment with the same agent. A commercially available spectrophotometer (Lambda 950, PerkinElmer, USA) with an integrating sphere was applied to measure the transmittance and reflectance spectra of the sample after topical treatment using different OCAs listed in Table 4.1 at the time intervals of 0, 15, 30, 60, and 120 min, respectively. In the meanwhile, the thickness of samples was measured by a micrometer. The scanning wavelength range of the measurement was 400–1000 nm with a 10 nm interval. Then the reduced scattering coefficient (μ_s') of the skin samples at different times was calculated with the IAD program (2007), developed by Dr Prahl [22–23].

Figure 4.2 shows that the reduced scattering spectra of skin (μ_s') (1/mm) decreased after different treatment times for all test agents. It can be found that the μ_s' significantly decreased in the whole wavelength band with the increasing treatment time of OCAs, but the changes in reduced scattering coefficient varied with different agents.

In order to better compare the optical clearing effect of different agents quantitatively, the normalized relative changes in μ_s' at 630 nm after the application of agents during 60 min and 120 min were analyzed using the following formula:

$$\Delta\mu_s' = \frac{\mu_{s_{treated}}' - \mu_{s_{native}}'}{\mu_{s_{native}}'} \tag{4.1}$$

Here the subscripts *native* and *treated* refer to the samples before and after the application of agents at the different time intervals, respectively. Considering the difference in the initial value for different skin samples, the statistical analysis was performed to better evaluate the effect of different agents.

It can be found in Figure 4.3(a) that saline application induced only minor reduction of μ_s', while PG induced a more pronounced reduction, although less than the other three pure OCAs. This means that PG is not a good OCA compared with the other three alcohols. In Figure 4.3(b), application of glycerol/PG resulted in a more pronounced reduction of μ_s' compared to glycerol and glycerol/water; the reduction was enhanced by 7% and 16%, respectively. The differences are significant between glycerol/PG and glycerol (p<0.05) and between glycerol/PG and glycerol/water (p<0.05). In Figure 4.3(c), application of 70% D-sorbitol/PG also resulted in a more pronounced reduction of μ_s' and compared to 70% D-sorbitol and 70% D-sorbitol/water; the reduction was enhanced by 9% and 22%,

respectively. The differences are significant between 70% D-sorbitol/PG and 70% D-sorbitol (p<0.05), and between 70% D-sorbitol/PG and 70% D-sorbitol/water (p<0.01).

Figure 4.3(d) demonstrates the differences that exist not only for agents with PE and water, but also for agents with different PEs. Compared to pure PEG400, application of PEG/Azone, PEG/PG, and PEG/Thiazone induced μ_s' are further reduced by 2%, 8%, and 17%, respectively. Statistical analysis shows that there is a significant difference between PEG/PG and PEG400 (p<0.05) and an extremely significant difference between PEG/Thiazone and PEG400 (p<0.01). However, there is no significant difference between PEG/Azone and PEG400 (p>0.05).

Chemical penetration enhancers for improving *in vivo* skin optical clearing

The ultimate goal of the optical clearing technique is for use in clinical applications. Still, there exists a certain degree of difference in the physiology and metabolism state between *in vitro* and *in vivo* skin. For *in vivo* skin optical clearing, metabolism prevents OCAs assembling in local tissue, which makes the optical clearing of skin *in vivo* more difficult to achieve than that of skin *in vitro*. Hence, hypodermic injection became a common method to improve optical clearing of skin *in vivo*, but it has always been associated with some degree of scathe to skin or even the blockage of blood vessels [31]. Chemical penetration enhancers used in medicine and cosmetics were also applied in TOC. Azone [13], oleic acid [32, 33], DMSO [34], PG [18], and Thiazone [14, 18] were shown to accelerate the permeability of OCAs into the skin, improving the optical clearing of skin. The above *in vitro* experiments demonstrated that PG and Thiazone has an excellent penetration effect for OCAs. In contrast, Thiazone is new skin penetration enhancer with nonirritant properties and does not cause immune responses in the skin [14]. It is a white crystalline substance at room temperature, and can be melted at 40°C. Its mechanism of penetration enhancement is similar to Azone, but the penetration enhancing effect is almost three times higher than Azone, and it is considered to be an ideal substitute for Azone. PEG400, as a kind of OCA, is widely used in the study of skin optical clearing [35]. PEG400 has a refractive index of 1.47 and can be miscible with Thiazone [14]. To optimize the enhancing effect, Thiazone was usually mixed with chemical agents as a volume percent of 0.5%–10% [36]. Before the experiment, Thiazone was heated to melt at 40–45°C and then mixed with PEG400 as a volume ratio of 1:9; after that, the mixture solution was kept at room temperature (20–25°C). The mixture was able to maintain a liquid state when the temperature was higher than 15°C.

Accessing rat cutaneous vessels based on treatment with PEG400 and Thiazone

Figure 4.4 shows typical morphological images of rat skin *in vivo* at different times after the application of the mixture of PEG400 and Thiazone (PT). In the beginning, the dermal blood vessel is invisible to the naked eye. After having

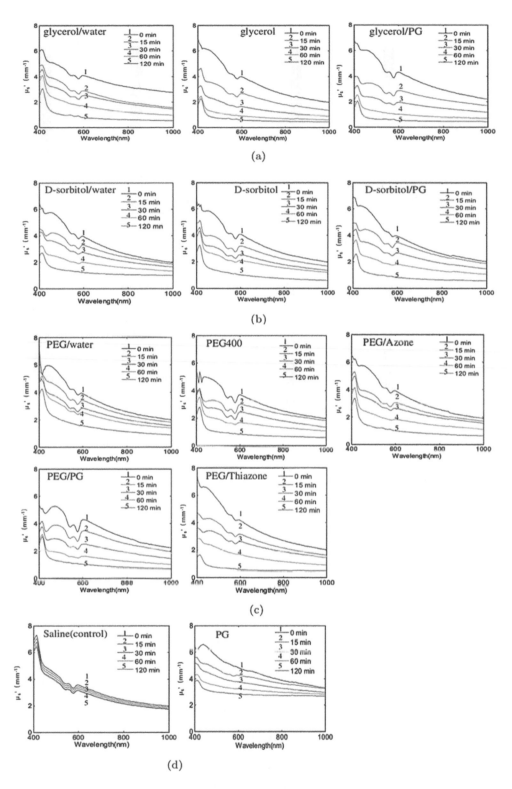

FIGURE 4.2 Typical changes of calculated μ_s' (1/mm) spectra: (a) Glycerol group, (b) D-sorbitol group, (c) PEG400 group, and (d) control. The curves were obtained at time intervals of 0, 15, 30, 60, and 120 min from top to bottom. Reprinted with permission from Ref. [18].

topically applied the PT mixture on the rat skin for 4 min, some large vessels show through the intact skin. At up to 12 min, even the small branch vessels can be distinguished clearly, just as the region in the rectangle shows. Moreover, this situation will be allowed to continue. After treatment with

the mixture for 40 min, the saline was applied onto the interested area. Immediately, the skin recovers to the turbid state and the blood vessels are out of view [36].

The laser speckle contrast imaging (LSCI) technique was used to monitor the same area in Figure 4.4. The speckle

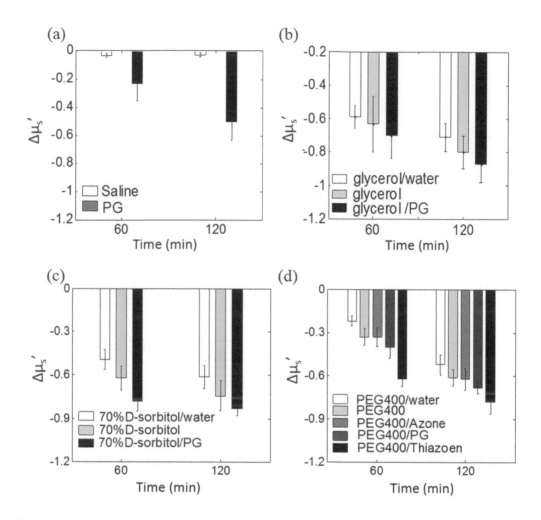

FIGURE 4.3 The relative changes of μ_s' at 630 nm after treatment with different agents. (a) Control group including saline and PG, (b) glycerol group, (c) D-sorbitol group, and (d) PEG group. Reprinted with permission from reference [18].

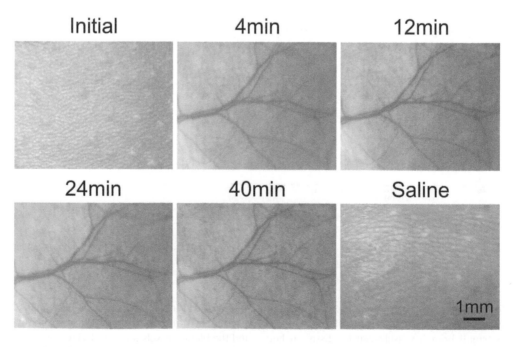

FIGURE 4.4 Visible optical clearing process of *in vivo* rat skin before and after treatment with mixture of PEG400 and Thiazone on skin surface. Reprinted with permission from Ref. [37].

temporal contrast image was constructed by calculating the speckle temporal contrast of each pixel in the time sequence. Figure 4.5 shows the corresponding speckle temporal contrast maps and profiles of speckle contrast values along the horizontal line before and after topical application of the mixture of PT to rat dorsal skin at initial state, at 4, 12, 24, and 40 min, and after treatment with saline. The temporal contrast maps demonstrate that the topical application of the PT mixture can decrease the temporal contrast of the vessels, which provides better visualization of the blood flow. After 40 min of PT treatment, a saline solution was applied to the area of interest; the skin immediately recovered to the initial state, and the information on both the blood vessels and blood flow was again concealed by the turbid skin.

A comparison of the speckle contrast values before and 12 min after PT treatment reveals that the laser speckle contrast values decreased by $5.0 \pm 0.4\%$, $40.0\% \pm 3.2\%$, and $59.0\% \pm 4.1\%$ in the nonvascular, small vessel (30–80 µm) and large vessel (80–150 µm) regions, respectively. Rat skin is well

known to be much thicker than mouse skin; however, the enhancement of the image contrast in the vascular regions caused by PT treatment of rat skin is still larger than that caused by a combination of microneedle rolling and topical glycerol application on mouse skin.

Comparison of PG and Thiazone for enhancing *in vivo* skin optical clearing

The above experimental results demonstrate that the PT mixture can induce better *in vitro* skin optical clearing efficacy than the mixture of PEG and PG (PPG) by enhancing the penetration of agents in the skin and decreasing the scattering of skin [18]. PT can also evidently enhance the image contrast of LSCI for cutaneous blood flow. The following will introduce another evaluating method to enhance the imaging depth of Angio-OCT by PPG and PT and a mixture with sucrose (Suc) or fructose (Fruc), respectively [17]. Table 4.2 shows the six kinds of skin optical clearing agents

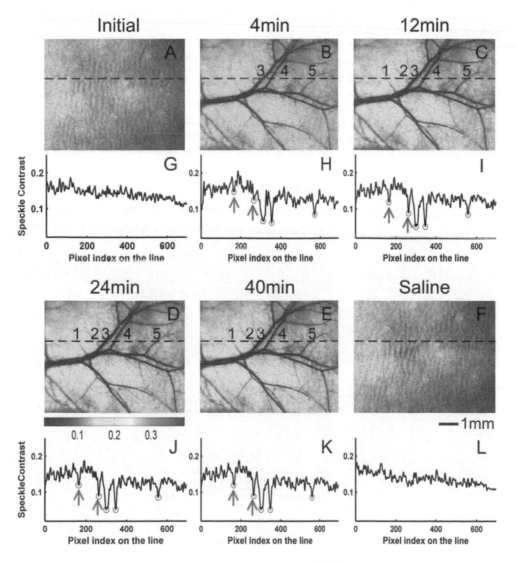

FIGURE 4.5 Laser speckle temporal contrast maps of the same areas in Figure 4.4 at initial state, 4, 12, 24, and 40 min after treatment of different OCAs, and 2 min after treatment of saline, and corresponding profiles of speckle temporal contrast values along horizontal line. Numbers 1–5 indicate cutaneous blood vessels of interest. Reprinted with permission from reference [37].

TABLE 4.2

The schematic design for preparing different SOCAs. Data from reference [17].

PEG			
(+) PG: PEG+PG (PPG)		(+) Thiaz: PEG+Thiaz (PT)	
(+) Suc: PPG+Suc	(+) Fruc: PPG+Fruc	(+)Suc: PT+Suc	(+) Fruc: PT+Fruc

(SOCAs), namely PPG, PT, PPG+Fruc, PT+Fruc, PPG+Suc, and PT+Suc. It is worth noting that Suc and Fruc were prepared to form a saturated aqueous solution (67.1% for Suc and 78.9% for Fruc (wt/wt), respectively) prior to mixing with others.

As for the Angio-OCT system, the light source is a broadband superluminescent diode (SLD; Superlum, Carrigtwohill, Ireland, Broadlighters D855-HP2) with a central wavelength of 850 nm, a bandwidth of 100 nm, and an incident optical power of 2 mW upon the sample surface, theoretically offering an axial resolution of ~3.2 μm in air. An objective lens with a focal length of 40 mm was used to focus the probing light beam, yielding a lateral resolution of 15 μm. The OCT detection unit was a highspeed spectrometer, equipped with a fast line-scan CMOS camera (Basler, Ahrensburg, Germany, Sprint spL4096-140k), providing an imaging range of ~2.9 mm in air.

Figure 4.6(a) exhibits the cross-section skin structural information obtained using Angio-OCT before or after treatment with different SOCAs, respectively. It demonstrates that after

SOCA treatment, some changes in skin structural texture occur to various extents, e.g., in terms of structural tortuosity, an interface formed among some layered cutaneous components, some hollow structures, etc. In particular, the interface formed among some layered cutaneous components derives from the different extents of dehydration and RI matching caused by the hyperosmotic SOCAs. What is interesting is that the decreased RI mismatching extent enhances the imaging depth and signal intensity in the deeper tissue. In contrast, the signal intensity in the superficial tissue instead decreases because of skin transparency for all the SOCAs.

Angio-OCT can not only provide the cross-section skin structural information with superior resolution and contrast but also can represent the blood flow distribution information. Figure 4.6(b) shows the blood flow distribution information before or after treatment with different SOCAs . It indicates that the vasculature network is invisible to the naked eye and the blood flow distribution information cannot yet be acquired by Angio-OCT clearly through the turbid skin, whereas this situation changes dramatically after treatment with SOCAs for all the SOCAs. In particular, PT+Suc and PPG induce the strongest and weakest optimization in Angio-OCT imaging performances for blood flow imaging, respectively. In addition, PPG+Suc has a roughly similar capacity to optimize the Angio-OCT imaging performance as PT+Fruc, as do PPG+Fruc and PT, but both are weaker than PPG+Suc and PT+Fruc for optimizing blood flow imaging.

Further, the depth at the 1/e of normalized Angio-OCT signal profile was defined as the imaging depth, as shown in Figure 4.7(a). To quantify the imaging performances induced

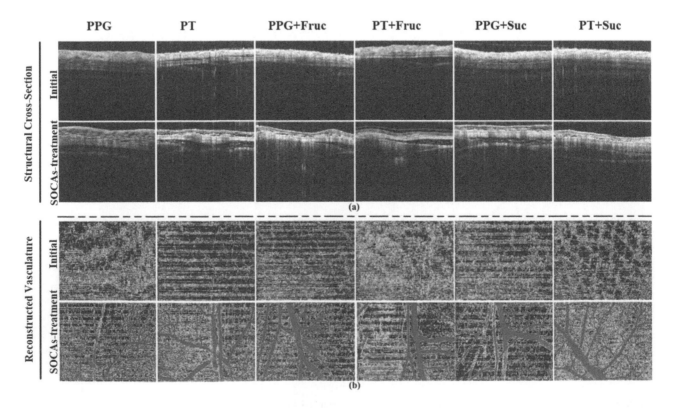

FIGURE 4.6 The cross-section skin structure (B-scan) (a), and reconstructed blood flow distribution information (en-face scan) (b), before or after treatment with different SOCAs. Reprinted with permission from reference [17].

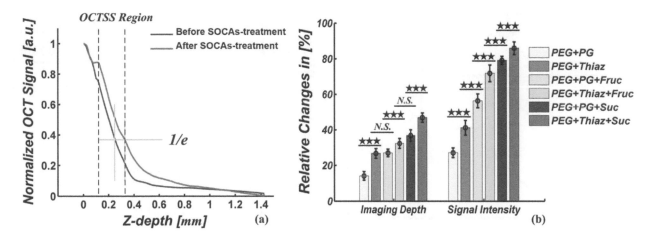

FIGURE 4.7 Normalized Angio-OCT signal profiles (a), relative changes in imaging depth and signal intensity at that depth calculated according to the OCT profile (1/e of normalized signal profile) induced by treatment with different SOCAs (b); ★★★ refers to (P < 0.001), and N.S. refers to No Significance. OCTSS means OCT signal slope, and OCTSS region indicates the actual effective imaging depth range labeled with two dotted lines. Reprinted with permission from reference [17].

by different SOCAs, the enhancements in imaging depth and signal intensity at that depth were calculated according to the OCT depth profile (see Figure 4.7(b)). Figure 4.7 indicates that the increase in signal intensity is largest caused by PT+Suc, followed by PPG+Suc, PT+Fruc, PPG+Fruc, PT, and PPG, consecutively. However, this descending order is not rigidly adapted for the enhancements in the imaging depth, which shows the biggest and smallest improvements for PT+Suc and PPG, respectively. What is interesting is that it demonstrates that PPG+Suc has no significant difference in improving the imaging depth than PT+Fruc, and the same for PPG+Fruc and PT.

In addition, quantitative comparisons were performed using Angio-OCT in terms of the changes in skin optical properties and RI mismatching extent, as well as permeability rate [17]. Experimental results with PT+Suc show optimal capacity in enhancing the imaging performances, decreasing the scattering and the RI mismatching extent, which is due to the fact that Thiazone is superior to PG and Suc is superior to Fruc for optical clearing efficacy. Nonetheless, both chemical penetration enhancers can improve *in vivo* skin structural and vascular functional imaging.

Liquid paraffin as a penetration enhancer for glycerol

An ideal OCA should be safe and have a short effective time for *in vivo* applications. Glycerol is one of the most used biocompatible agents for optical clearing. However, it is difficult for hydrophilic agents such as glycerol to penetrate the dense stratum corneum (SC) of skin or the dense stratified squamous epithelium of internal organs into deeper layers within a short time.

Liquid paraffin, as a lipophilic agent, has various medicinal values, and its refractive index is 1.4691 ± 0.0003 (measured with Abbe refractometer under visible light) and is close to 1.4689 ± 0.0003 of glycerol. Mixing them in different ratios will not change the refractive index of the mixture and will

not alter the refractive match effect of glycerol in bio-tissue. Optical technologies such as OCT [24–26], near infrared spectrometry [27], and multiphoton microscopy [28] have been used to evaluate the effect of liquid paraffin as the penetration enhancer of glycerol.

Effect of different concentrations based on skin by OCT imaging

First, a fiber-based OCT system with 1550 nm central wavelength was used, which has a lateral resolution of ~12 μm and an axial resolution of ~12 μm in bio-tissue. Different concentrations of liquid paraffin with its volume ratio in mixture increasing from 0% to 70% by increments of 10% were used. The volume ratio was calculated as $V_{lp} / (V_{lp} + V_{gl})$, where V_{lp} is the volume of liquid paraffin and V_{gl} is the volume of glycerol. A concentration of 0% denotes denotes anhydrous glycerol. A concentration of 60% denotes that the ratio of the volume between liquid paraffin and glycerol is 6:4.

Hair-depilated rat skin was scanned to observe the variation of imaging contrast and the penetration depth of the combined liquid paraffin with glycerol mixture. The size of sample in each test was about 8 mm × 10 mm. The skin was kept cooled in 0.9% NaCl solution to avoid dehydration in air.

The optical clearing effect of OCAs can be revealed by the contrast variation of the internal tissue in OCT images. According to the characteristics of OCT images, the intensity ratio of regions (RIR) was proposed to evaluate the contrast of OCT images [24–26]. It is expressed as

$$RIR = \frac{\dfrac{1}{N_1} \displaystyle\sum_{x,y \in internal} g(x,y)}{\dfrac{1}{N_2} \displaystyle\sum_{x,y \in surface} g(x,y)}. \quad (4.2)$$

Where N_1 and N_2 are the number of pixels in the internal and surface regions, respectively, $g(x,y)$ is the gray level in the OCT image.

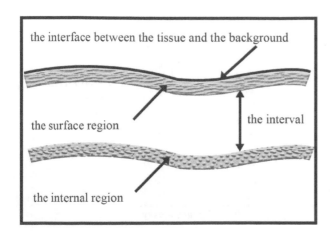

FIGURE 4.8   A sketch map of the two regions in calculating RIR. The thick real curve is the interface between the tissue and the background, and the layered and the dotted structure regions are the surface and the internal regions, respectively. Adapted from reference [25].

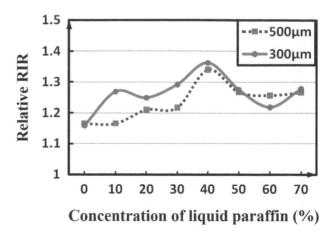

FIGURE 4.9   Relative RIRs by applying different concentrations of liquid paraffin. Reprinted with permission from reference [24].

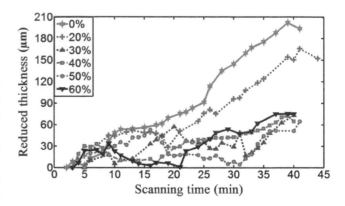

FIGURE 4.10   The thickness variation of rat skin with time elapsing by applying mixtures of liquid paraffin and glycerol at different concentrations. Reprinted with permission from reference [24].

By finding the interface between the tissue and the background using a simple method, the surface and internal regions can be selected. A sketch map of the regions is given in Figure 4.8 [25]. Pixels whose gray levels are larger than a certain threshold can be considered as the interface pixels between the tissue and the background. A region with a given thickness below the interface pixels was selected as the surface region. Under the surface region, a region with the same thickness was selected as the internal region.

The time-dependent optical clearing effect can be evaluated by calculating the ratio of RIR with and without the agent to obtain the relative RIRs [24]. Generally, a larger relative RIR value means a better optical clearing effect.

For comparison, three OCT images of each sample were firstly collected without applying any agent, and their RIRs were averaged as the benchmark. Then, the mixture of liquid paraffin and glycerol was applied on the same scanning area, and more OCT images with time elapsing were thereafter obtained in succession, and their RIRs were calculated. The maximum RIR of these images was divided by the benchmark, which was regarded as the relative RIR of this sample. Five samples for each concentration were tested, whose average of relative RIR was used to evaluate its improvement in contrast.

The average relative RIRs versus concentration are shown in Figure 4.9. The thickness of the regions is 50 μm and their width is 1 mm in calculating the RIRs. Two intervals between the internal and surface region, 300 μm and 500 μm, were used. The red solid curve and the blue dotted curve represent the average relative RIR at intervals of 300 μm and 500 μm, respectively.

As shown in Figure 4.9, it can be seen that all relative RIRs are larger than 1, which means that all agents improve the contrast. Furthermore, the optical clearing effects of the mixture are all better than glycerol, and the 40% concentration exhibits the best optical clearing effect at both 300 μm and 500 μm intervals.

The variation of tissue thickness, along with the immersion of the agent, can be used to qualitatively evaluate the degree of tissue structure deformation. By keeping the distance of OCT probe and sample platform unchanged after applying the agent and recording the height variation of the interface between the tissue and the background in OCT images, the variation of the tissue thickness was measured indirectly. In order to simulate the environment of the living organism and avoid dehydration in air, the sample was soaked in a physiological saline solution as in Refs. [38, 39].

A width of about 1 mm was selected from the same lateral position in all images of each group. Axial positions of the surface point in this region were averaged. The variation of the average value in different images represents the change of the surface position, i.e., the change of the tissue thickness. The typical results are shown in Figure 4.10. The starting time 0 is the time of applying the agent, and the times of the first scanning image for different concentrations are slightly different. The time interval between two scans was about 1 min.

The slopes of curves shown in Figure 4.10 reflect the speed of water loss. Due to the dehydrating effect of glycerol, it can be seen that tissue thickness has obvious variation during the whole scanning time for the agents whose concentrations are 20% and anhydrous glycerol. With the increase of fractions of liquid paraffin in the mixture, the thickness has only slight variation within 35 min.

From Figures 4.9 and 4.10, it can be concluded that adding liquid paraffin in glycerol can not only improve the optical clearing effect but also reduce water loss with time elapsing. The mixture within 30%–50% concentration shows the best comprehensive effect.

The evaluation experiments of effective time and the optimal duration of optical clearing based on *ex vivo* rat skin and *in vivo* human skin demonstrated that the optical clearing effect took place very rapidly during the first 10 min and can last more than 30 min [24].

Evaluation of optical clearing based on diseased skin by OCT imaging

This mixture was further applied to some of the most familiar and typical skin diseases, such as a fibroma, pigmented nevus, seborrheic keratosis, sebaceous cyst, hemangioma, etc. Figure 4.11 is a group of OCT images of an epidermoid cyst [25]. Figure 4.11(a) is the original image without the mixture. Figures 4.11(b)–4.11(f) are the images after application of the mixture of 40% liquid paraffin on the same area after 5, 10, 15, 20, and 25 min, respectively. Before applying the mixture, it was hard to see anything under the epidermis. However, after applying the mixture from 5–10 min, a structure clearly emerged in the dermis, which was similar to a cyst as indicated with a white arrow in each image in Figure 4.11. As shown in Figure 4.11, this kind of cystiform structure moves away from the surface gradually as time elapses. It is considered that the contents of the cyst are dehydrated with glycerol permeating, leading to the shrinkage of the cyst. It is concluded that the mixture of liquid paraffin and glycerol enhanced the penetration of light for this kind of diseased skin. In addition, the skin surfaces are at the nearly same longitudinal position in Figures 4.11(a)–4.11(f), and it is concluded that liquid paraffin reduces the loss of water.

The statistical results of the optical clearing evaluation of fibroma, pigmented nevus, and seborrheic keratosis are shown in Figure 4.12 [25], in which the relative RIRs with time elapsing after applying the mixture on the surface of the specimen are given. The two intervals between the internal and surface regions in Figure 4.12 are 500 μm and 700 μm, respectively. The thickness and width of these two regions are 50 μm and 1 mm, respectively. The values of the starting time 0 represent the normalized benchmark. For each sample, images were taken every 5 min. The bars on the curve show the maximum and minimum relative RIRs, which correspond to the best and worst optical clearing effects for a given skin disease.

As shown in Figure 4.12, it can be seen that after applying the mixture, the relative RIRs of fibroma are all larger than 1, but those of pigmented nevus and seborrheic keratosis are around 1, which means that the mixture of liquid paraffin and glycerol has optical clearing effect for fibroma but little effect on the other two diseases. By introducing biocompatible chemical reagents to reduce multiple scattering and increase the optical penetration intensity in bio-tissue, typical hypotheses of the mechanism of optical clearing technique include dehydration, index matching, etc. Different bio-tissues have different structures, compositions, and refractive indices. Therefore, the mechanism and effect of optical clearing will be different. Therefore, it is supposed that optical clearing technique has selectivity and that one method cannot be suitable for all skin diseases.

Evaluation based on luminal organs by ultra-high resolution OCT imaging

Optical imaging depth in high scattering biological tissues is limited, especially at shorter wavelength regions. The optical clearing effect of a mixture of liquid paraffin and glycerol on *ex vivo* luminal organs was explored with an ultrahigh-resolution spectral domain OCT (SD-OCT) at 800 nm [26]. This SD-OCT system [40, 41], with a home-made Ti: Sapphire laser, has an axial resolution of ~2.8 μm (in air) at the central wavelength of ~825 nm and a lateral resolution of ~16 μm. Sample imaging is 70k A-scans/sec, and imaging length is ~1.2 mm.

Figure 4.13 is a series of representative B-scan OCT images of the guinea pig esophagus, which were obtained after the application of the liquid paraffin (at 30% volume concentration) and glycerol mixture. As shown in Figure 4.13(b), the optical clearing effect of the mixture on the esophagus tissue

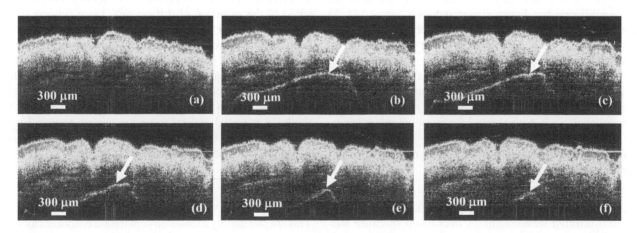

FIGURE 4.11 A group of OCT images of epidermoid cyst: (a) OCT image without the mixture, (b) 5 min, (c) 10 min, (d) 15 min, (e) 20 min, and (f) 25 min OCT images after application of the mixture on the same area, respectively. The cyst is pointed out with a white arrow in each figure. Reprinted with permission from reference [25].

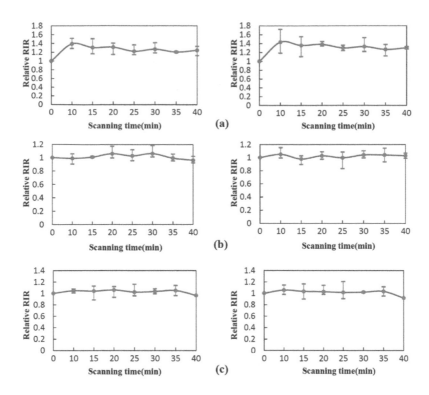

FIGURE 4.12 Relative RIRs of different diseased skins with time elapsing: (a) fibroma, (b) pigmented nevus, (c) seborrheic keratosis. Left and right curves in (a)–(c) are the results at 500 μm and 700 μm intervals, respectively. Bars on the curves are the maximum and minimum values of relative RIRs. Reprinted with permission from reference [25].

is evident. The increased transparency of upper layers, such as epithelium and lamina propria, led to a pronounced muscle layer at even only 10 min after the application of the mixture.

The relative RIRs of 30%–50% concentration of *ex vivo* pig esophagus tissues are shown in Figure 4.14. Since it was difficult to ensure the same B-scan position before and after the application of the clearing agent, the B-scan image used as benchmark was acquired at the first minute after the application of the mixture. Three regions with 400 × 100 pixels starts from pixels 1, 600, and 1000 below the tissue surface along the imaging depth were selected, which correspond to 0 μm, ~360 μm, and ~600 μm below the tissue surface, respectively, as shown in Figure 4.13. The standard deviations were calculated with three pig esophagus tissues and displayed as error bars in Figure 4.14. As shown in Figure 4.14, optical clearing occurred after ~10 min with the relative RIRs > 1 for all of these mixtures, and the effect lasted for more than 30 min.

The above results verified that liquid paraffin could facilitate the penetration of glycerol through the dense stratified squamous epithelium of esophagus tissue into deeper layers and work collaboratively with glycerol for tissue optical clearing.

Evaluation of optical clearing based on skin reflection by spectroscopy

A liquid paraffin/glycerol mixture was further studied by spectroscopy to evaluate its synergistic effect within visible and near-infrared wavelengths. A fiber-based spectrum analyzer system [27] was set up to measure the surface reflection of *in vivo* human skin before and after the application of the mixture.

An example of the diffuse reflection spectra over a range from 600 nm to 1400 nm is shown in Figure 4.15 [27]. Figures 4.15(a)–(d) correspond to the results of anhydrous glycerol at 30%, 40%, and 50% concentration, respectively. The curves in each figure were obtained at different time intervals. It can be seen that the diffuse reflection decreases gradually with time elapsing, and these curves have similar trends qualitatively over the whole wavelength range investigated. Liquid paraffin/glycerol mixtures have a much better effect than that of anhydrous glycerol alone, and 50% is the best one. It means the more liquid paraffin is added, the larger decrease of diffuse reflectance is caused, which is similar to the effect of DMSO, as shown in Ref. [42]. Furthermore, the mixtures improve the speed of the reduction of diffuse reflectance because of the addition of liquid paraffin. The experiment further proved that combining hydrophilic agents with lipophilic agents can improve the speed of percutaneous penetration over the whole spectrum investigated. As shown in Figure 4.15, there is a better effect at short wavelengths than that at long wavelengths, which is similar to the results of many agents [36, 42, 43]. We think the potential underpinning mechanism responsible for the decrease in diffuse reflection intensity is a reduced refractive index mismatch and thus reduced scattering.

Summary and prospect

This chapter mainly focuses on penetration enhancers for optical clearing of skin tissues, from *in vitro* to *in vivo* applications. The effective chemical penetration enhancers include PG, Thiazone, and liquid paraffin.

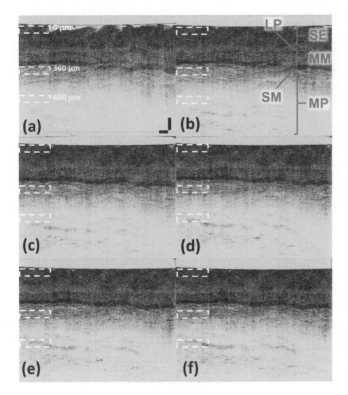

FIGURE 4.13 Representative B-scan images of guinea pig esophagus at (a) 1 min, (b) 10 min, (c) 20 min, (d) 30 min, (e) 40 min, and (f) 50 min after the application of 30% paraffin/glycerol mixture. Yellow rectangle boxed regions correspond to 0 μm, ~360 μm, and ~600 μm below the tissue surface, respectively, and are selected to calculate RIRs. SE: stratified epithelium; LP: lamina propria; MM: muscularis mucosae; SM: submucosa; MP: muscularis propria. Scale bars in (a): 100 μm. Reprinted with permission from reference [26].

with water for the hydrogen bond binding sites and the intercalation in the polar head groups of the lipid bilayers by propylene glycol are postulated as mechanisms for the penetration enhancing effects of propylene glycol [15, 18]. For Thiazone and Azone, a possible mechanism would be the increase of fluidity of the hydrophobic stratum corneum regions and reduction of its permeation resistance against drug substances [18, 32]. Although the penetration enhancing effect of PG is suboptimal to Thiazone, PG can be used more widely as it is a cosolvent and can be mixed with various OCAs.

The *in vivo* skin optical clearing experiments also demonstrated similar results. Thiazone is superior to PG for obtaining cutaneous vascular structure with higher contrast for same OCA by using an optical coherence tomography angiography. The mixture of Thiazone and PEG400 induces the cutaneous vessels visible clearly and permits LSCI to get blood flow distribution with high contrast and resolution. Of course, the optimal OCA could induce a better *in vivo* skin optical clearing effect for the same chemical penetration enhancers, i.e., sucrose is superior to fructose.

Liquid paraffin also has a good synergistic optical clearing effect with glycerol on skin and esophagus tissues within a short effective time of ~10 min. The mixtures of 30%–50% of liquid paraffin have the optimal comprehensive effect. The fine balance between the dehydration effect of glycerol and the water retention capability of the liquid paraffin can avoid tissue deformation. Based on ultrahigh-resolution OCT imaging, the phenomena of a decrease in scattering in upper layers and an increase in scattering in lower layer with time were observed [26]. The attenuation coefficients experienced the most dynamic change in the first 10 min after the application of the mixture, which means glycerol rapidly penetrates into inner tissue with the help of liquid paraffin. After 20 min, the change slowed down with time before becoming homogeneous. This change can be attributed to the index matching effect of the liquid paraffin/glycerol mixture, which leads to a more homogeneous tissue refractive index that in turn reduces tissue attenuation (mainly scattering) and consequently enhances light penetration in the tissue. In addition, the little effect of liquid paraffin/glycerol on some tissues indicates that optical clearing technique has selectivity for biological tissues, and more detailed studies on theoretical and experimental

Compared to pure OCAs, mixtures with PG could induce a better skin optical clearing effect owing to the penetration enhancer. The penetration enhancing effect of PG was also compared to that of Thiazone and Azone, with descends in the order of Thiazone, PG, and Azone. This result indicates that differences exist in the enhancement of OCAs' permeability by different PEs. The reason may be due to different mechanisms. The solvation of keratin within the SC by competition

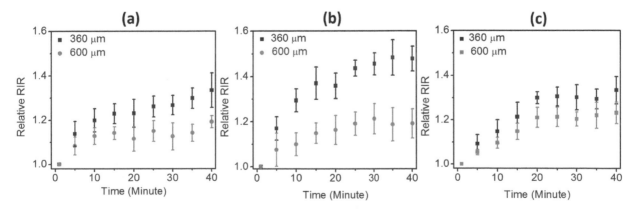

FIGURE 4.14 Relative RIR variation versus time after applying the mixture of liquid paraffin and glycerol on pig esophagus tissues with different volume ratios of (a) 30%, (b) 40%, and (c) 50%. Reprinted with permission from reference [26].

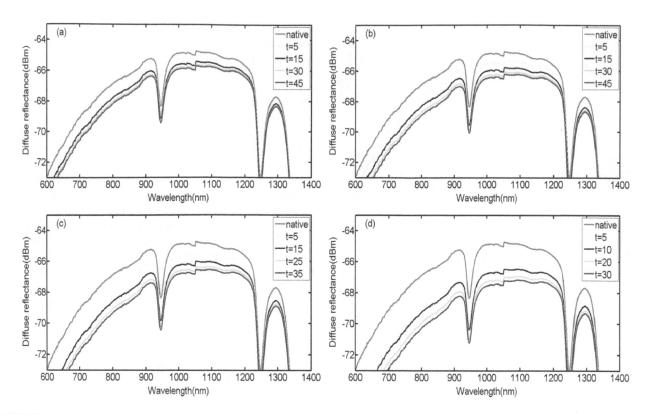

FIGURE 4.15 Changes of diffuse reflectance for human finger skin before and after application of different mixtures over the range of 600 nm–1400 nm. (a)–(d) are anhydrous glycerol, 30%, 40%, and 50% liquid paraffin/glycerol mixtures, respectively. Reprinted with permission from reference [27].

conditions should be done until the optical clearing technique can be used in clinic.

Despite the advantages of all the kinds of PE, it still has potential shortcomings. Since most of the PE aim to help OCAs penetrate corneum, their use may result in the stripping of the stratum corneum and dehydration, which will quickly make the epidermis become thin. In this case, the repeated performance of skin optical clearing might cause epidermal cell proliferation and dermal cell dehydration. Future efforts should be performed on developing PE with high penetration enhancement efficacy and negligible influence on corneum, as well as better medicine for repairing the skin, for repeatedly imaging dermal cells and vasculature.

In a broad sense, penetration enhancement plays an indispensable role in both *in vivo* and *in vitro* tissue optical clearing. For *in vitro* clearing, especially for large-volume tissue clearing, the permeation of the clearing agents into tissue is necessary for achieving tissue transparency throughout the entire volume. To improve the clearing efficacy, the main chemical strategies used in current *in vitro* clearing protocols include three aspects [44]. First, some water-miscible polar solvents, such as alcohols (e.g., ethanol [45]) and tetrahydrofuran [46], were used to mediate the replacement of water with high-RI aromatic solvents, partially accompanied by effective delipidation. Second, hyperhydration chemicals such as urea [47, 48] and formamide [49] could increase membrane fluidity and the internal osmotic pressure, which could facilitate the influx of the outer agents. Third, some detergents, including nonionic detergent (e.g., Triton X-100 [47, 48]) and ionic detergent

(e.g., SDS [50, 51]), were adopted to solubilize the lipid component, hence removing the barriers to clearing agents. Overall, all these penetration enhancement strategies significantly contributed to the highly effective clearing of large tissues, and the diverse clearing protocols that were produced provided various alternatives for wide applications.

Acknowledgments

D.Z. was supported by the National Key Research and Development Program of China (grant No. 2017YFA0700501), and the National Natural Science Foundation of China (grant Nos. 61860206009). Y.L. was supported by the National Natural Science Foundation of China (grant No. 61875092), and the Beijing-Tianjin-Hebei Basic Research Cooperation Special Program (grant No. 19JCZDJC65300). V.V.T. was supported by the Government of the Russian Federation (grant No. 14.W03.31.0023) and RFBR grant # 18-52-16025.

REFERENCES

1. V.V. Tuchin, I.L. Maksimova, D.A. Zimnyakov, I.L. Kon, A.H. Mavlyutov, and A.A. Mishin, "Light propagation in tissues with controlled optical properties," *J. Biomed. Opt.* **2**(4), 401–417 (1997).
2. V.V. Tuchin, "Optical immersion as a new tool for controlling optical properties of tissues and blood," *Laser Phys.* **15**(8), 1109–1136 (2005).

3. G. Vargas, J.K. Barton, and A.J. Welch, "Use of hyperosmotic chemical agent to improve the laser treatment of cutaneous vascular lesions," *J. Biomed. Opt.* **13**(2), 021114 (2008).

4. B. Choi, L. Tsu, E. Chen, et al., "Determination of chemical agent optical clearing potential using *in vitro* human skin," *Lasers Surg. Med.* **36**(2), 72–75 (2005).

5. G. Vargas, K.F. Chan, S.L. Thomsen, and A.J. Welch, "Use of osmotically active agents to alter optical properties of tissue: Effects on the detected fluorescence signal measured through skin," *Lasers Surg. Med.* **29**(3), 213–220 (2001).

6. M.H. Khan, B. Choi, S. Chess, K.M. Kelly, J. McCullough, and J.S. Nelson, "Optical clearing of *in vivo* human skin: Implications for light-based diagnostic imaging and therapeutics," *Lasers Surg. Med.* **34**(2), 83–85 (2004).

7. V.V. Tuchin, G.B. Altshuler, A.A. Gavrilova, et al., "Optical clearing of skin using flashlamp-induced enhancement of epidermal permeability," *Lasers Surg. Med.* **38**(9), 824–836 (2006).

8. X. Xu, Q. Zhu, and C. Sun, "Combined effect of ultrasound-SLS on skin optical clearing," *IEEE Photonics Technol. Lett.* **20**(24), 2117–2119 (2008).

9. O. Stumpp, B. Chen, and A.J. Welch, "Using sandpaper for noninvasive transepidermal optical skin clearing agent delivery," *J. Biomed. Opt.* **11**(4), 041118 (2006).

10. J. Yoon, T. Son, E. Choi, B. Choi, J.S. Nelson, and B. Jung, "Enhancement of optical skin clearing efficacy using a microneedle roller," *J. Biomed. Opt.* **13**(2), 021103 (2008).

11. H. Trommer and R.H.H. Neubert, "Overcoming the stratum corneum: the modulation of skin penetration," *Skin Pharmacol. Physiol.* **19**(2), 106–121 (2006).

12. J. Jiang, and R.K. Wang, "Comparing the synergistic effects of oleic acid and dimethyl sulfoxide as vehicles for optical clearing of skin tissue in vitro," *Phys. Med. Biol.* **49**(23), 5283–5294 (2004).

13. X. Xu, and Q. Zhu, "Evaluation of skin optical clearing enhancement with Azone as a penetration enhancer," *Opt. Commun.* **279**(1), 223–228 (2007).

14. Z. Mao, Y. Hu, Y. Zheng, W. Lu, Q. Luo, and D. Zhu, "Experimental study on influence of Thiazone on optical clearing of piglet skin *in vitro*," *Acta Optica Sinica*, **27**(3), 526–530 (2007).

15. A.C. Williams and B.W. Barry, "Penetration enhancers," *Adv. Drug Deliv. Rev.* **56**(5), 603–618 (2004).

16. A.P. Funke, R. Schiller, H.W. Motzkus, C. Günther, R.H. Müller, and R. Lipp, "Transdermal delivery of highly lipophilic drugs: In vitro fluxes of antiestrogens, permeation enhancers, and solvents from liquid formulations," *Pharm. Res.* **19**(5), 661–668 (2002).

17. R. Shi, L. Guo, C. Zhang, et al., "A useful way to develop effective in vivo skin optical clearing agents," *J. Biophotonics*, **10**(6–7), 887–895 (2017).

18. Z. Zhi, Z. Han, Q. Luo, and D. Zhu, "Improve optical clearing of skin in vitro with propylene glycol as a penetration enhancer," *J. Innov. Opt. Heal. Sci.* **2**(3), 269–278 (2009).

19. A.V. Rawlings, and K.J. Lombard, "A review on the extensive skin benefits of mineral oil," *Int. J. Cosmetic Sci.* **34**(6), 511–518 (2012).

20. F. Sharif, E. Crushell, K. O'Driscoll, and B. Bourke, "Liquid paraffin: A reappraisal of its role in the treatment of constipation," *Arch. Dis. Child.* **85**(2), 121–124 (2001).

21. N. Urganci, B. Akyildiz, and T.B. Polat, "A comparative study: the efficacy of liquid paraffin and lactulose in management of chronic functional constipation," *Pediatr. Int.* **47**(1), 15–19 (2005).

22. N. Concin, G. Hofstetter, B. Plattner, et al., "Mineral oil paraffins in human body fat and milk," *Food Chem. Toxicol.* **46**(2), 544–552 (2008).

23. A. Patzelt, J. Lademann, H. Richter, et al., "*In vivo* investigations on the penetration of various oils and their influence on the skin barrier," *Skin Res. Technol.* **18**(3), 364–369 (2011).

24. J. Wang, Y. Liang, S. Zhang, Y. Zhou, H. Ni, and Y. Li, "Evaluation of optical clearing with the combined liquid paraffin and glycerol mixture," *Biomed. Opt. Express* **2**(8), 2329–2338 (2011).

25. H. Shan, Y. Liang, J. Wang, and Y. Li, "Study on application of optical clearing technique in skin diseases," *J. Biomed. Opt.* **17**(11), 115003 (2012).

26. Y. Liang, W. Yuan, J. Mavadia-Shukla, and X. Li, "Optical clearing for luminal organ imaging with ultrahigh-resolution optical coherence tomography," *J. Biomed. Opt.* **21**(8), 081211 (2016).

27. K. Chen, Y. Liang, and Y. Zhang, "Study on reflection of human skin with liquid paraffin as the penetration enhancer by spectroscopy," *J. Biomed. Opt.* **18**(10), 105001 (2013).

28. J.W. Wilson, S. Degan, W.S. Warren, and M.C. Fischer, "Optical clearing of archive-compatible paraffin embedded tissue for multiphoton microscopy," *Biomed. Opt. Express* **3**(11), 2752–2760 (2012).

29. J.W. Wiechers, and R.A. de Zeeuw, "Transdermal drug delivery: Efficacy and potential applications of the penetration enhancer Azone," *Drug Des. Deliv.* **6**(2), 87–100 (1990).

30. H. Li, and Z. Ye, "Developments on synthesis of a new skin penetration enhancer: Thiazone," *Jiangsu Chem. Ind.* **28**, 15–18 (2006).

31. G. Vargas, A. Readinger, S.S. Dozier, and A.J. Welch, "Morphological changes in blood vessels produced by hyperosmotic agents and measured by optical coherence tomography," *Photochem. Photobiol.* **77**(5), 541–549 (2003).

32. J. Kalbitz, R. Neubert, and W. Wohlrab, "Modulation of drug penetration in the skin," *Pharmazie* **51**(9), 619–637 (1996).

33. J. Jiang, and R.K. Wang, "How different molarities of oleic acid as enhancer exert its effect on optical clearing of skin tissue in vitro," *J. X-Ray Sci. Technol.* **13**(3), 149–159 (2005).

34. S. Karma, J. Homan, C. Stoianovici, and B. Choi, "Enhanced fluorescence imaging with DMSO-mediated optical clearing," *J. Innov. Opt. Heal. Sci.* **3**(3), 153–158 (2010).

35. Z. Mao, D. Zhu, Y. Hu, X. Wen, and Z. Han, "Influence of alcohols on the optical clearing effect of skin in vitro," *J. Biomed. Opt.* **13**(2), 021104 (2008).

36. D. Zhu, J. Wang, Z. Zhi, X. Wen, and Q. Luo, "Imaging dermal blood flow through the intact rat skin with an optical clearing method," *J. Biomed. Opt.* **15**(2), 026008 (2010).

37. J. Wang, Y. Zhang, P. Li, Q. Luo, and D. Zhu, "Review: Tissue optical clearing window for blood flow monitoring," *IEEE J. Sel. Top. Quantum Electron.* **20**(2), 6801112 (2014).

38. G. Vargas, E.K. Chan, J.K. Barton, H.G. Rylander 3rd, and A.J. Welch, "Use of an agent to reduce scattering in skin," *Lasers. Surg. Med.* **24**(2), 133–141 (1999).

39. M.G. Ghosn, V.V. Tuchin, and K.V. Larin, "Depth-resolved monitoring of glucose diffusion in tissues by using optical coherence tomography," *Opt. Lett.* **31**(15), 2314–2316 (2006).

40. J. Xi, A. Zhang, Z. Liu, W. Liang, L.Y. Lin, S. Yu, and X.D. Li, "Diffractive catheter for ultrahigh-resolution spectral-domain volumetric OCT imaging," *Opt. Lett.* **39**(7), 2016–2019 (2014).

41. W. Yuan, J. Mavadia-Shukla, J. Xi, W. Liang, X. Yu, S. Yu, and X.D. Li, "Optimal operational conditions for supercontinuum-based ultrahigh-resolution endoscopic OCT imaging," *Opt. Lett.* **41**(2), 250–253 (2016).

42. X. Xu and R.K. Wang, "Synergistic effect of hyperosmotic agents of dimethyl sulfoxide and glycerol on optical clearing of gastric tissue studied with near infrared spectroscopy," *Phys. Med. Biol.* **49**, 457–468 (2004).

43. X. Xu, and Q. Zhu, "Feasibility of sonophoretic delivery for effective skin optical clearing," *IEEE Trans. Biomed. Eng.* **55**(4), 1432–1437(2008).

44. K. Tainaka, A. Kuno, S.I. Kubota, T. Murakami, and H.R. Ueda, "Chemical principles in tissue clearing and staining protocols for whole-body cell profiling," *Annu. Rev. Cell Dev. Biol.* **32**, 713–741 (2016).

45. A. Erturk, C.P. Mauch, F. Hellal, et al., "Three-dimensional imaging of the unsectioned adult spinal cord to assess axon regeneration and glial responses after injury," *Nat. Med.* **18**(1), 166–171 (2012).

46. A. Erturk, K. Becker, N. Jahrling, et al., "Three-dimensional imaging of solvent-cleared organs using 3DISCO," *Nat. Protoc.* **7**(11), 1983–1995 (2012).

47. E.A. Susaki, K. Tainaka, D. Perrin, et al., "Whole-brain imaging with single-cell resolution using chemical cocktails and computational analysis," *Cell* **157**(3), 726–739 (2014).

48. E.A. Susaki, K. Tainaka, D. Perrin, H. Yukinaga, A. Kuno, and H.R. Ueda, "Advanced CUBIC protocols for whole-brain and whole-body clearing and imaging," *Nat. Protoc.* **10**(11), 1709–1727 (2015).

49. T. Kuwajima, A.A. Sitko, P. Bhansali, C. Jurgens, W. Guido, and C. Mason, "ClearT: A detergent- and solvent-free clearing method for neuronal and non-neuronal tissue," *Development* **140**(6), 1364–1368 (2013).

50. K. Chung, J. Wallace, S.Y. Kim, et al., "Structural and molecular interrogation of intact biological systems," *Nature* **497**(7449), 332–337 (2013).

51. J.B. Treweek, K.Y. Chan, N.C. Flytzanis, et al., "Whole-body tissue stabilization and selective extractions via tissue-hydrogel hybrids for high-resolution intact circuit mapping and phenotyping," *Nat. Protoc.* **10**(11), 1860–1896 (2015).

5

Human skin autofluorescence and optical clearing

Walter Blondel, Marine Amouroux, Sergey M. Zaytsev, Elina A. Genina, Victor Colas,
Christian Daul, Alexander B. Pravdin, and Valery V. Tuchin

CONTENTS

Introduction

Preamble

Autofluorescence (AF) is increasingly considered as a powerful diagnostic intrinsic biomarker because of its direct dependence on the nature, amount, and microenvironment of the fluorophores naturally present in the tissues, and hence the related metabolic processes and structural organization of cells and tissues. AF intensity spectra carry multiple pieces of information that can be decrypted by a proper analysis of the overall emission signal, allowing for monitoring tissue and cell metabolism *in situ*, in real time, and in the absence of perturbation from exogenous markers [1]. Most optical clearing agents (OCA) have been developed for *ex vivo* tissue bioimaging and especially fluorescent stains–based imaging modalities exploiting the signals emitted by artificially added fluorescent markers that can be fluorescent protein-based reporters (e.g. green fluorescent protein, GFP) or exogenous fluorescent dyes. Indeed, the preservation during the OC process of such labelling-based "endogenous" and "exogenous" fluorescence has long been a major concern while developing new OCA. For the sake of clarity in this chapter, the terms *autofluorescence*, *endogenous-tagged fluorescence*, and *exogenous-tagged*

fluorescence will refer to fluorescence emitted by cells' and tissues' intrinsic constituents ("label-free"), proteinaceous label dyes' and exogenous probes, respectively. Concerning the issue of investigating skin AF modifications during optical clarification (OC), only a limited number of scientific contributions have been published, from the first study in 2010 by [2] on *ex vivo* rat skin to the latest works of [3] and [4, 5] analyzing AF intensity spectra changes under the action of different OCA applied either on *ex vivo* immersed porcine ear samples or topically on *ex vivo* human skin samples, respectively.

Motivations: Issues and aims

Optical spectro-imaging methods have been intensively developed for human skin disease investigations in *ex vivo* and *in vivo* (including clinical) conditions [6] in order to (i) analyze the structure and the metabolism of the cutaneous tissues and characterize the optical properties of their constituents; (ii) extract, from the measured data sets, new "optical biomarkers" of diagnostic interest in correlation with targeted anatomo- and physio-pathological features and skin tissue states; (iii) provide noninvasive, atraumatic; and real-time clinical tools to help surgical guiding (margin delineation, lesion localization, and identification); or to improve screening methods, tumor relapse follow-up, and biopsy sampling efficiency (early detection of diseases) [7, 8].

The interest in exploiting UV-visible light-induced AF emitted by the cutaneous tissue and cell intrinsic fluorescing compounds is that it gives noninvasive and nondestructive access to information of diagnostic interest about the structural organization and the metabolism of cells and tissues *in vivo*.

The skin is the largest organ in the human body, and skin cancer is by far the most common of all cancers. Among skin cancers, nonmelanoma skin cancer (NMSC), including basal cell carcinoma (BCC) and squamous cell carcinoma (SCC), is the most commonly diagnosed malignancy in humans, while melanoma skin cancer (MSC), which accounts for about 1% of all skin cancers diagnosed, causes most deaths attributed to skin cancers. Furthermore, the worldwide incidence of both types of cancers has steadily increased over the past decades and continues to rise annually as rapidly as 3%–8% and 4%–6% for NMSC and MSC respectively, with considerable public health impacts [7–10].

The main skin intrinsic fluorophores targeted in the early detection of biochemical and morphological properties modifications related to premalignant and malignant skin tissues are extra-cellular matrix (ECM) proteins such as collagen and elastin, and biological molecules found within cells such as keratin, tryptophan (Trp), a reduced form of nicotinamide adenine dinucleotide (NADH), flavin adenine dinucleotide (FAD), and protoporphyrin IX (PpIX) [11]. When analyzing AF intensity spectral data collected from healthy skin and skin with different pathological states, discriminant spectral features can be extracted which improve the classification efficiency towards noninvasive diagnostics [1, 8–9, 11–22]. Besides cutaneous cancer applications, skin AF ratio of the average intensity measured in the 420–600 nm-spectral range to the 300–420 nm-spectral range was also recently identified as a novel biomarker of cardiovascular disease (CVD) related to the accumulation of advanced glycation end products (AGEs), which are known to progress under hyperglycemic, inflammatory, and oxidative stress conditions [23].

Among the wide spectrum of *in vivo* optical spectroscopy and imaging methods available for skin examination [6], those relying on the exploitation of AF-based information are mainly confocal and multiphoton excitation fluorescence imaging [9, 24–29], fluorescence lifetime imaging [30, 31], light sheet fluorescence microscopy [26, 32, 33], or light-induced fluorescence spectroscopy [7, 8, 11, 14–17, 34].

However, the intrinsic optical heterogeneity of the cutaneous tissues and the very high values of the scattering coefficients characterizing the various skin layers within the UV-visible–NIR wavelength range, strongly restrict the penetration depth of the incident fluorescence excitation light source into the skin and the depth sensitivity of the spectro-imaging detection systems [33, 35]. It also leads to the acquisition of distorted AF intensity spectra (compared to undistorted intrinsic fluorophore emission spectra) [6].

The application of OCA in *ex vivo* and *in vivo* skin studies using optical spectro-imaging methods provides evident enhancement of their probing depth capabilities and of the collected data quality, *e.g.* image contrast and resolution and

spectral signal-to-noise ratio [6, 25, 33, 35, 36]. The development of biocompatible (safe, without any side effects) and optimized efficiency OCAs (in terms of clarification kinetics, depth penetration, reversibility, AF preservation/enhancement) is at stake for clinical applications [24, 33, 37, 38] and for the understanding of the skin OC's underlying mechanisms during the various time phases of this process [6]. Indeed, the interest in controlling the OC within the cutaneous tissue upper layers specifically, that is to say the *stratum corneum* (SC) and the living epidermis (LE), is to overcome their related "optical shield effect" and to access *in vivo* information of diagnostic interest within both the epidermis and the upper dermis volumes of the human skin, *e.g.* elastosis, collagenase, epidermis thickening, *etc.*

This is why it is important to study the way cutaneous tissues' AF properties may be modified during the OC process as well as the capacity of OCAs to preserve skin intrinsic fluorescence related to tissue structures and metabolism. These investigations should strengthen the clinical transferability potential of AF-based diagnostic spectro-imaging methods combined with skin OC.

This chapter provides a review of the research works studying the impact of OCA on skin AF emission features and describes some of the recent progress made in the analysis of OC-induced skin AF modifications using *ex vivo* and *in vivo* spectro-imaging modalities as well as related optical property modelling.

Skin intrinsic fluorophores, optical properties, and optical clearing

Skin intrinsic fluorophores and in vivo tissue state optical biomarkers

The skin is a complex opaque organ characterized by an anisotropic structure of multiple layers within three main layers: epidermis, dermis, and hypodermis. The optical properties of the different skin layers, *i.e.* absorption, scattering and fluorescence, depend on the density, distribution, orientation, physiological processes, and metabolic activity of the chromophores (melanin, hemoglobin, water) and fluorophores therein [39]. Skin AF emitted under excitations in the near-ultraviolet (UV)–blue wavelength range arise from different fluorescing proteins and cellular compounds located at various depths within the cutaneous layers. The main skin fluorophores, also considered as intrinsic photophysical biomarkers of skin pathologies, are related to the metabolic processes (Trp, NADH, FAD, prophyrins) and structural arrangement of cells and tissues (keratin, collagen, elastin) [11].

In the superficial layer of the skin, the epidermis has varying thicknesses depending on the anatomical site and ranging from dozens to hundreds of micrometers. Of the epidermal cells, 90% are keratinocytes, which, along to their cellular maturation, migrate throughout the epidermis from their originating *stratum basal* layer up to the SC, the typical thickness of which is ~10 µm but which may reach 100 µm for the thickest skin sites. The latter outermost layer of the epidermis is a fully keratinized layer made of dead cells (corneocytes) embedded in a lipid-rich intercellular matrix and organized in an orthorhombic lateral packing order that determines the skin barrier function [40]. The high impermeability of the human SC to lipophilic and hydrophilic substances is also an issue to be resolved in the development and clinical application of efficient *in vivo* skin OC methods. During the cellular differentiation process of keratinization, keratinocytes accumulate a fibrous protein in their cytoplasm, the keratin, which emits strong fluorescence in the 360–700 nm-spectral range when excited with a monochromatic light ranging from 250 to 460 nm (see Table 5.1) [46]. Indeed, unmixing this collagen-like fluorescence contribution from bulk AF spectra measured without any complementary depth resolution information is a serious challenge [19]. Within the basal layer of the epidermis are also found melanocytes, which strongly absorb UV light but without emitting fluorescence; they synthetize melanin, which is embedded into melanosomes and then transferred to keratinocytes for the nucleic acids' protection (ribo- and deoxyribose-nucleic acids, RNA and DNA) against mutagenic UV-B radiations.

Nonmelanoma skin cancers (BCC and SCC) and precancers (actinic keratosis AK) arise from keratinocytes wherein the DNA has been damaged due to sun overexposure. Although rarely lethal, these are recurrent lesions spreading over large sun-exposed areas of the body as well as in nerves and bones [9]. MSC come from melanocytes related to risk factors such as genetic history and early childhood sunburn. They are much less frequent than NMSC but have high metastatic potential and death rates.

The dermis (1.5–4 mm thick) is mainly composed of structural protein molecules and blood capillaries which give the skin its biomechanical properties and supply the epidermis with nutrient-saturated blood, respectively. In the dermis, the ECM fibrous proteins elastin and collagen (filled with hyaluronic acid and glycosaminoglycans) and their crosslinks exhibit fluorescence emission between 420 and 510 nm under excitations (λ^{exc}) in the wavelength range 350–420 nm. Connective tissue diseases and CVD complications involve

changes on the molecular level in the papillary dermis (*e.g.* alteration of collagen I and III content) and in the capillary structure [30]. During the carcinogenesis process from healthy skin to dysplasia or BCC, a reduced amplitude of the collagen-emitted AF was reported due to an increased production of collagenase enzymes, which break the collagen cross-links [34]. The same authors observed the contrary, *i.e.* an increased level of collagen AF during the carcinogenesis from normal to AK and SCC. Until now, however, no study has been published on the optical characterization of solar elastosis underlying skin carcinogenesis. Indeed, dermal elastic fibers in photoaged skin are replaced with thickened, tangled, and ultimately granular amorphous elastotic materials called *solar elastosis*, which are the most prominent histologic features of photoaged skin [47].

At the cellular level, NADH and FAD ratio modifications are indicators of cellular metabolic alterations associated with large-scale cell proliferation or carcinogenesis process, leading to significant modifications in their AF emissions [34]. NAD is a coenzyme involved in the cell metabolism through redox reactions where NAD^+ (oxidized form) and NADH (reduced form) act as oxidation and reduction agents, respectively. NADH is highly fluorescent with emissions peaking in the 440–460 nm range when excited between 290 and 350 nm. FAD features wide absorption and fluorescence emission spectra between 350 and 500 nm and 450 and 600 nm respectively, and with characteristic maximum absorption peaks at 380 and 450 nm and emission peak at 525 nm (see Table 5.1) [1, 14, 41].

Additional skin intrinsic fluorophores such as Trp and PpIX can be exploited as photophysical biomarkers of aging processes, metabolic disorders, or cell transformation processes. Trp is an aromatic amino acid which fluoresces under deep UV excitation (λ^{exc} = 295 nm) with an emission spectrum (λ^{em}) ranging from 300 to 350 nm [42]. The fluorescence intensity reduction of the band assigned to tryptophan moieties reflects the slowing down of the epidermal cell turnover rates [48]. Endogenous PpIX is used as a diagnostic biomarker of cancer, *e.g.* in skin or bladder, but only when its production is stimulated by exogenous 5-ALA.

TABLE 5.1

Skin primary intrinsic fluorophores including peaks/range of excitation and emission wavelengths (λ^{exc} and λ^{em}), location in skin layers. Data from references [8, 15, 17, 41–45]

Category	Intrinsic fluorophores	λ^{exc} and λ^{em} peaks/range (nm)	Skin layer location
Structural protein	Collagen	325 and 390–405	Dermis – ECM*
	Collagen cross-links	335, 370, and 460–490	
	Elastin	290, 325, and 340, 400–420, 460	
	Elastin cross-links	420–460 and 500–540	
	Keratin	405 and 550–600	Epidermis – cells
		380–400 and 500–550	
		370 and 460–500	
		280 and 380	
Electron carrier co-enzymes	FAD	405, 450 and 500–520, 535	
	NADH	290, 350 and 440–470	
Cofactor for oxygen transfer	PpIX	400–405 and 630, 700	
Amino acid	Tryptophan	280–295 and 345–350	Epidermis – cells
			Dermis collagen
Lipids	Phospholipids	340–440 and 430–460, 540–560	Dermis fibroblasts and epidermis keratinocytes

ECM: Extra-cellular matrix

TABLE 5.2

Modifications observed in *in vivo* human skin cancer AF intensity spectra amplitudes compared to healthy skin ones, data from references given in the table and from [8, 43]

λ_{exc} (nm)	Modifications in AF spectra	Skin lesion type, phototypes and ref.
295	↑↑	BCC and SCC, [21]
337	↓↓	MSC, II–III, [18]
	↓	BCC, I–II, [34]
	↑	AK, SCC, I–II, [34]
350	↓	BCC and SCC, [21]
355, 370, 375, 395	↓↓	BCC, I–III, [13, 49]
385	↓↓	BCC, II–III, [17]
365, 405	↓	BCC, II–III, [17]
410	↓	BCC, SCC, I–III [20]
442	↓	BCC, [50]

In a clinical study on NMSC using LIFS with excitation wavelengths at (i) 295 and (ii) 350 nm and fluorescence microscopic imaging, [21] observed that the collected AF intensities due to (i) tryptophan and (ii) collagen crosslinks were (i) between two- and threefold higher and (ii) between 1.25 and 1.4-fold lower in BCC and SCC than in healthy skin tissues, because of (i) epidermal thickening related to pathological cell hyperplasia and (ii) connective tissue enzymatic degradation due to tumor activity in the dermis, respectively. In the work done by [49] on BCC and normal skin, the authors measured 420–700 nm AF intensity spectra with main emission peak around 455 nm for both tissue states but with PpIX fluorescence peaks at 630 and 700 nm only visible for BCC. These results were further confirmed by [17], who observed increased PpIX fluorescence peaks with advanced stages of BCC growth.

From AF studies conducted *in vivo* on BCC [13, 17, 20–21, 49, 50], SCC [20, 21], and MSC [18] *vs.* normal human skin (phototypes I–III), it can be noticed (see Table 5.2) that excitations above 337 nm and up to 442 nm allow consistent results to be obtained of decreased AF levels in pathological cutaneous tissues compared to healthy ones. The strongest differences between BCC and healthy skin AF intensity spectra were obtained for excitations wavelengths in the range 370–395 nm *i.e.* in the highest part of the near-UV spectral region (UV-A), which favors clinical settings compliance. [8] reported results from several other studies showing that BCC lesions excited at 370 or 442 nm emit decreased AF intensity compared to healthy skin. Higher AF intensities observed in BCC compared to normal skin were related to decreased fluorescence contributions of collagen and elastin, but also of NADH due to an increased NAD$^+$/NADH ratio in tumor tissues.

It is worth mentioning the indirect but enhanced effect of hemoglobin absorption, related to UV-induced erythema, tumor neovascularization, and blood supply increase, on reducing the penetration depth of the excitation radiations and the propagation of the emitted fluorescence to outside the tissue, leading to reduced amplitudes of the AF intensity spectra collected for BCC compared to healthy skin tissues [20]. As reported by [17, 41, 51], AF spectral curves collected on human BCC *in vivo* highlight (i) enhanced wavelengths troughs at 420, 543, and 575 nm corresponding to characteristic hemoglobin pigment absorption peaks, but also (ii) strong decrease of the short- *vs.* long-wavelength intensities due to melanin.

The changes observed in skin AF emission signals, as summarized in Table 5.2, under excitations within the 295–442 nm wavelength range and related to the biochemical activity and metabolic state modifications between NMSC or MSC and normal tissues, can be exploited to extract optical biomarkers of diagnostic interest [8].

Skin is a complex, multilayered, inhomogeneous, and optically turbid medium, featuring spatially varying absorption and scattering optical properties that make skin *in vivo* bulk AF emission difficult to analyze quantitatively. Skin AF intensity spectra collected under excitations in the near-UV-visible spectral range carry the contributions of several major fluorophores (keratin, collagen, elastin, proteins' cross-links, NADH, FAD, and PpIX), which need to be unmixed for diagnostic purposes. In this regard, elastosis and fibrosis processes within elastin and collagen networks located in the upper dermis need to be further investigated *in vivo*.

Therefore, the present limitations related to (i) nontraumatically breaking the SC optical barrier and discarding the optical absorption screen effect of epidermis melanin and dermal blood and (ii) *in vivo* accessing the intrinsic spectra of dermal structural proteins need to be overcome to improve the diagnostic efficiency and clinical application potential of AF-based spectro-imaging techniques.

In this context, *in vivo* skin OC appear to be an interesting and very promising way to counter the protective function of the skin barrier to body water outflow and exogenous substance inflow [52].

Optical clearing mechanisms and skin optical properties

Several recently published exhaustive reviews [6, 24, 25, 33, 53, 54] highlighted the importance of skin OC methods use nowadays in the scientific community and the current research into more and more efficient, fast, and clinically safe (non-toxic) solutions.

Skin optical clearing principle and agents

The majority of the commonly developed OCA are for immersion OC of fixed or *ex vivo* tissue samples and whole small organs, but are toxic or chemically very aggressive [33]. The growing need for applying OC methods to *in vivo* skin tissues in the frame of preclinical and clinical studies requires the development of (i) biocompatible and reversible alternative OCAs applicable onto the tissue surface, *i.e.* topically, and (ii) efficient methods of delivering the OCA into the epidermis and dermis [52].

The tissue dehydration and lipid solvation–induced OC process, related to the penetration of OCA within skin and resulting in a decrease in the scattering of skin tissues, involves three complementary mechanisms [6, 24]. First, most OCA are hyperosmotic agents able to cause tissue dehydration [55]. The latter is due to osmotic diffusion of bound water out of cells and ECM, which induces a reduction of the highest refractive indices (RI) gradients existing between the tissue constituents, that is to say the mismatching between the low RI values of water or interstitial fluids (RI = 1.33–1.34 for water in the NUV-visible spectral range) and the higher RI values of the other tissue compounds, typically from 1.4 up to 1.57, including ECM protein fibers (elastin and collagen RI = 1.47–1.55) and high density biochemical constituents in cells such as mitochondria (RI = 1.4–1.42), nucleolus (RI > 1.5), or lipid bilayer membranes (RI = 1.45–1.48) [33, 56]. It is to be noticed that this dehydration process also leads to (i) skin tissue shrinkage, *i.e.* structural thinning and reordering, which further increases tissue transparency; and (ii) fluorescence quenching due to water molecule outflow from the sample (water maintains endogenous fluorescent protein folding). Second, the penetration of the OCA into the tissue "water-freed" spaces reinforces the overall RI matching considering the RI values of most OCAs are between 1.43 and 1.53. Third, the interaction of the OCA within the dermal collagen structure through a reversible process of collagen fiber dissociation leads to a reduction of the size of the scattering elements within this collagen structure and hence decreased light scattering.

A large number of OC methods have been developed for fixed tissue imaging and immunolabelling applications, including aqueous- and solvent-based clearing agents or protocols. Primary hydrophilic water-based OCA have the advantage of preserving tissue structure and fluorescence emission of labelling fluorescent dyes, while organic solvents provide favorable clearing speed. The former approach exhibits inferior tissue OC performance as well as massive swelling and alteration, while the latter provides strong shrinkage and appears to be too aggressive for *in vivo* applications considering skin lipid matrix dissolution and SC integrity damaging actions [52, 53].

In [57], CUBIC* aqueous-based hyperhydration OC (RI = 1.48) of mice skin *ex vivo* samples allowed confocal microscopic imaging observations down to 300 µm deep. Although the OC procedure last for 12 days and the CUBIC OCA was nontoxic, the authors mentioned that it was not possible to clarify melanin-containing tissues, a problem which should be addressed in further studies. The efficiency of

Scale (aqueous-based OCA) and ethanol-BABB[†] (RI = 1.56), ethanol-DBE[‡] (solvent-based OCA) and 3DISCO[§] protocol, including immersion in DBE (RI=1.56), for clearing human skin biopsy samples *ex vivo* was evaluated by [26] through enhanced in-depth optical sectioning of SC, dermis, and epidermis appendages using LSFM imaging. The worst performance was obtained for Scale (no clearing) and the best for ethanol-BABB, thus demonstrating the usefulness of the method for better quantitative analysis of histological anomalies such as epidermal hyperplasia.

Foster *et al.* in [24] recently proposed a pH-balanced modified BABB clearing methodology validated on whole-mount and sectioned uninjured and wound healing skin tissues from transgenic mouse models using fluorescence confocal microscopic imaging. This protocol caused rapid and efficient clearing with minimal tissue shrinkage allowing for (i) detailed visualization of epidermal and dermal biology at homeostasis and after injury and (ii) preservation of the endogenous-tagged fluorescence expression from mice skin *ex vivo* samples.

Considering (i) the variability of the skin physiological and functional conditions (permeability to OCA, homeostasis, blood and lymph circulations, *etc.*); (ii) their changes with pathology, both affecting the OCA-induced skin transparency; and (iii) the selection of OCA application optimal conditions including concentration and exposure time, the proper quantification and control of skin OC *in vivo* is still a challenging problem.

Most popular aqueous biocompatible agents of particular interest for skin *in vivo* reversible OC are (i) polyatomic alcohols such as glycerol (RI = 1.41/1.47 for 60%/100% glycerol aqueous solution), poly-/propylene Glycol (PPG/PG, RI = 1.44), and polyethylene glycol with various molecular weights (PEG-300/PEG-400, RI = 1.47); (ii) sugars, *e.g.* sucrose (RI = 1.39/1.47 for 1M/3M concentrations in water); (iii) organic acids such as oleic (RI = 1.46) and acetic acid (RI = 1.36–1.41); (iv) organic solvents like DMSO[¶] (RI = 1.48); and (v) radiocontrast agents, namely Omnipaque™ (RI = 1.39/1.43 for 60%/100% Omnipaque™300) [6, 28, 33, 53, 58–65].

Among sugars investigated for skin OC, disaccharides (sucrose, maltose) perform better than monosaccharides (glucose, fructose, ribose) because their larger molecular size, *i.e.* collagen surface wider coverage, allows for more water–collagen hydrogen bond breakings, thus increasing collagen disruption. For instance, a saturated sucrose–based OCA mixture applied on rat skin *ex vivo* and *in vivo* provides more effective clearing results compared to maltose and fructose at similar concentrations. The neutral pH of sucrose (pH = 7 at 3M) also favors highest AF preservation potential compared to the increasing acidity of fructose with concentration [58]. [6] reported the successful results obtained by several groups using Omnipaque™ –based OC of porcine skin samples. The latter produced a rapid decrease in the fraction of strongly

---

* CUBIC = Clear, unobstructed brain imaging cocktails and computational analysis

[†] ethanol-BABB = ethanol dehydration followed by incubation in Benzoic Acid Benzyl Benzoate

[‡] ethanol-DBE = ethanol dehydration followed by incubation in DiBenzyl Ether

[§] 3DISCO = 3D Imaging of Solvent-Cleared Organs

[¶] DMSO = dimethyl sulphoxide

bound water deeper into the epidermis compared to glycerol and without skin layer deformation (due to its low osmolarity) or strong dehydration.

Chemical and physical enhancers-based Skin OC approaches

In order to further accelerate the OCA penetration through the SC and enhance their transcutaneous delivery, chemical and physical enhancing strategies have been developed for reducing the natural SC barrier function and promoting water outflow and hyperosmotic agent inflow within the epidermis. The use of chemical enhancers, such as ethanol (thiaz, RI = 1.47), and hyaluronic acid (HA) or even DMSO, PG, and oleic acids (which are already OCA by themselves) allows the modification of the properties of SC through molecular binding with SC lipid molecules and the acceleration of the OC process thanks to skin surface oil removal and direct epidermal penetration of OCA via skin appendages. The implementation of physical treatments before or during the OCA application consist of mechanical peeling of the SC layer using tape-stripping and/ or microdamage of epidermis, microdermabrasion, photothermal and mechanical perforation, mechanical compression and strain, fractional laser microablation (FLMA), electrophoresis, ultrasounds, pressure, etc. [6, 66].

As reported in [67], fast OC of *in vivo* rat skin tissues was obtained by topical application of (i) PEG-400 + thiaz mixture including SC preliminary tape-stripping and physical massage [68] or (ii) PEG-400 + PG [69] and assessed by optical coherence tomography (OCT) and laser speckle contrast imaging (LSCI) modalities, respectively. The authors compared the clearing efficacy of six different combinations of OCAs (sucrose, fructose, PEG-400) and enhancers (PG, thiaz) applied on *in vivo* mouse depilated skin sites and monitored using angio-OCT. They showed that the mixture PEG + thiaz + sucrose provided superior performance in comparison to others, namely because of the higher enhancement and OC efficiency of thiaz and sucrose *vs.* PG and fructose, respectively.

In [70], HA was topically applied to pig skin samples *in vitro* prior to the application of PPG or PEG-400 OCAs, which resulted in significantly increased red-NIR light transmission and enhanced optoacoustic image contrast. This combination

of lipo- and hydro-philic PPG and PEG polymers supported by amphiphilic HA demonstrates an optimal base for coupling OC gel.

Recently, several studies on *in vivo* rat and human skin OC further emphasized the importance and interest of low-frequency (1 MHz) sonophoresis-based physical enhancement in combination with FLMA and PEG-300 [52], or microdermabrasion and oleic acid, respectively.

As mentioned by several authors, it is also important to note that skin compression or flattening also act as physical methods of slow OC, without OCA but similar to the OCA-induced dehydration process [28, 61].

Skin AF and OC studies using spectro-imaging techniques and modelling tools

This section presents the main progress made over the past 10 years, since the pioneer work of [2], in the analysis of skin AF modifications following and/or during OC, using different cellular and tissue spectroscopic and imaging modalities (Table 5.3).

Fluorescence confocal microscopy

Confocal microscopy or confocal laser scanning microscopy (CLSM) provides highly detailed optical sectioning of sliced or whole-mounting tissue samples *ex vivo* but for penetration depths limited to 4–6 μm [24]. Indeed, the main limitation of using CLSM for skin studies is high scattering, which reduces the quality of CLSM images [25]. Based on fluorescence CLSM imaging, [24] demonstrated the efficiency of modified BABB to clarify whole-mount and sectioned skin tissues from various transgenic mouse models and assessed endogenous-tagged fluorescence preservation allowing detailed visualization of epidermal and dermal biology at homeostasis and after injury.

Similarly, [74] used optical clearing–assisted confocal microscopy to successfully enhance in-depth imaging of *ex vivo* human skin endogenous-tagged fluorescence, thus providing high-resolution images of the whole internal epidermis architecture.

TABLE 5.3

Skin AF and OC-based studies using spectro-imaging modalities

Spectro-imaging modality	Excitation wavelength (nm)	Optical clearing agents	Skin conditions	Ref.
2-Photon microscopy and second harmonic generation	750	Glycerol, PG, and glucose	Immersion of *ex vivo* human skin	[71]
	790	Glycerol	*Ex vivo* and *in vitro* rat skin after *in vivo* dermal injection or immersion	[28, 61]
	760	Glycerol and Omnipaque™300	Topical application on *ex vivo* porcine skin	[3]
Light sheet fluorescence microscopy	405, 488, 561 and 642	ethanol-BABB and DBE	*ex vivo* human skin	[26]
Light induced fluorescence Spectroscopy	337	Glycerol, Sucrose and PG	Topical (dermal side) application on *ex vivo* rat skin	[2]
Spatially resolved light induced fluorescence spectroscopy	365, 385, 395, 405, and 415	Sucrose, PG, PEG-400 and DMSO	Topical application on *ex vivo* human skin	[4, 54, 72, 73]

In vivo CLSM, commonly called reflectance confocal microscopy (RCM), was early and recently reviewed by [29] and [9] respectively, in the clinical application frame of skin lesion examination. In the latest, the authors identified the recent updates on RCM techniques for skin cancer diagnosis, among which none were exploiting skin AF signals. Finally, among the existing literature, we did not identify any contribution on skin AF and OC-assisted CLSM or RCM.

Multiphoton excitation autofluorescence microscopy

As stated by [26], two-photon microscopy (2PM) has emerged as a useful noninvasive skin imaging modality, providing subcellular resolved 3D images of the epidermis and superficial dermis structure under two complementary modes: two-photon excitation fluorescence microscopy (2PEFM), enabling the visualization of epidermal and dermal fluorophores; and second harmonic generation (SHG), enabling the visualization of intrinsic sources of contrast, such as collagen type I in the dermis. Thanks to higher tissue transparency at red-NIR excitation wavelengths, the imaging depth of 2PM may reach ~200 μm in normal skin and ~400–500 μm in cleared skin [6, 25]. 2PEFM on NADH and FAD and both 2PEFM and SHG of collagen have been the most investigated techniques in the study of intrinsic fluorophores in cancer research [27, 75].

Pioneered works of [71] on *ex vivo* human skin OC-assisted 2PEFM imaging (using 2P-excitation wavelength 750 nm) demonstrated significant contrast enhancement as well as increased skin AF signal intensity and imaging depth (from 40 to 80 μm). Comparing glycerol, PG, and glucose, they identified glucose as the least efficient but fastest OCA, while it was the reverse case for glycerol. The latter additionally exhibited a partially reversible OC effect, supporting collagen structure alteration to significantly contribute to the clearing effect.

This hypothesis was further investigated by [28] using SHG (2P excitation wavelength 790 nm) to image mouse skin *ex vivo* and *in vitro* samples under glycerol OC. Collagen dissociation was observed for the *in vitro* condition only, thus confirming, as for other studies [61, 76], that one of the most important mechanisms of *ex vivo* skin glycerol-induced OC was associated with (i) more dense and regular packing of collagen fibers, *i.e.* dermis thickness compression; and (ii) dehydration of the interfibrillar matter and collagen fibers themselves, *i.e.* collagen fiber diameter decrease [61].

In the work of [3], two-photon excitation AF microscopy (2PEAFM, with 760 nm excitation wavelength) and SHG were used to demonstrate the efficacy of glycerol and Omnipaque™300 topically applied onto *ex vivo* porcine skin samples. Both OCAs improved the imaging probing depth and contrast of skin AF and SHG signals. Omnipaque™300 was the most efficient in reducing skin-bound water content with less skin dehydration, thus allowing increased 2PEAFM signal at skin depth down to 300 μm; the authors therefore suggested it as a safe and fast acting OCA for clinical tests.

As reported in [6], OC-induced reduction of skin tissue scattering significantly increases the fluorescence yield until the clearing window reaches the fluorophores. With scattering loss, the probability that a fluorophore will be excited is decreased while the probability of the detection of excited fluorophore emitted photons is increased; thus strong OC would result in a loss of signal.

Light sheet microscopy

Light sheet fluorescence microscopy (LSFM) performs rapid and whole-organ volumetric imaging with sectioning capability, thus drastically reducing acquisition time and photobleaching compared to CLSM and 2PM [32, 33]. Although these very attractive assets make this technique of major interest for exploiting *ex vivo* clearing of fixed bulk specimens, only a very limited number of scientific publications have been found on tissue OC studies using LSFM and only one on skin OC [26]. The OC effect of ethanol-BABB and DBE on 5 mm-thick human skin biopsies was investigated using LSFM. BAAB-cleared skin AF 3D images were collected at four different excitation wavelengths. At $\lambda^{exc} = 405$ nm, AF emitted from SC and dermis layers was visualized with a low and rapidly decreasing resolution in depth related to strong skin absorption and reduced light diffusion in skin. Increased imaging depth was observed for both 488 and 561 nm excitations, with stronger ECM AF emission in dermis and stronger keratin AF intensity in SC, respectively. At $\lambda^{exc} = 642$ nm, only SC produced a strong AF signal owing to the high concentration of keratin. Although the authors noticed the unexpected absence of AF signal in the epidermis whatever the excitation wavelengths, it was demonstrated that the method was useful to visualize and quantify histological features of the various skin layers and anomalies such as epidermal hyperplasia.

Light-induced AF spectroscopy

OC-induced skin AF spectra modifications

Light-induced autofluorescence spectroscopy (LIFS or AFS) is a UV-visible point optical biopsy technique advantageously used for real-time *in vivo* skin investigations, including in clinics, in a wide area of applications such as the estimation of skin photodegradation, determination of areas with melanin content, estimation of skin erythema extent, discrimination between skin disorders (*e.g.* BCC, SCC, MSC, AK, psoriasis, Bowen's disease, sun-exposed normal skin), and skin cancer screening, diagnosis, staging, and surgical delineation. [7–9, 11, 14–17, 34]. For truly quantitative analysis, the distorting effects of skin absorption and scattering need to be considered for accessing the accurate measurement of intrinsic fluorophore concentrations, as investigated in several works [77–80].

However, only a few contributions have been published until now that study the modifications in skin AF intensity spectra during the OC process considering multiple excitation wavelengths [5]. Early works of [2] highlighted a decreasing AF spectrum intensity, using 337 nm-excited LIFS, with increasing immersion time of *ex vivo* rat skin in different OCAs (glycerol, sucrose, and PG), applied from the dermal side of the skin samples.

Recent works by [4, 54, 72] used spatially-resolved multiply excited AF spectroscopy (SR-mAFS) to investigate the time kinetics of *ex vivo* human skin (phototype 2 female abdominal skin) AF spectra following topical application of an OCA combining sucrose, PG, and PEG-400 (solution S1) or PEG-400 and DMSO (solution S2). The latter SR-mAFS device is described in [81]. Briefly, it features (i) five different excitation wavelengths at $\lambda_{j=1,2,3,4,5}^{exc} = 365, 385, 395, 405, 415$ nm, allowing for scanning the relative contributions of various fluorophores according to their respective excitation and emission spectra; and (ii) four source-to-detector separations (SDS) $D_{i=1,2,3,4} = 400, 600, 800, 1000\,\mu m$, allowing for collecting fluorescence signals originating from various skin depths. In order to have complementary information on the OC-induced transparency of the skin sample [82], the latter was placed on top of a fluorescing 5 mm-thick agarose gel substrate composed of Chinese ink, intralipids-20%, and chlorin-e6 (Ce6). A multifiber optics probe was positioned with pressure-controlled contact onto the skin's outer surface with topically lying OC solution (Figure 5.1). The data set collected consisted in fluorescence intensity spectra $x\left(\lambda; D_i, \lambda_j^{exc}, t_k\right)$ measured as a function of the four SDS D_i ($i \in \{1, ..., N_D = 4\}$) and of the five excitation wavelengths λ_j^{exc} ($j \in \{1, ..., N_{exc} = 5\}$), and at eight time points t_k ($k \in \{1, ..., N_t = 8\}$) corresponding to T_0 (before OCA application), T_0+7, T_0+13, T_0+20, T_0+25, T_0+32, and T_0+37 minutes after OCA application, respectively.

From the collected fluorescence intensity spectra, skin AF and Gel Ce6 fluorescence contributions were separated using area under the curve (AUC) *i.e.* integrated values calculated in their respective 425–525 nm (AF–AUC) and 650–700 nm (Ce6–AUC) spectral ranges, as represented in Figure 5.2. AUC values were all normalized to their initial values (at T_0) before plotting their time kinetics during the OC process, which are represented in Figure 5.3 for $\lambda_{2,5}^{exc}$ and for $D_{1,2}$.

Overall results of T_0-normalized AUC kinetics during OC, partially shown in Figure 5.3a, highlighted enhanced light penetration and depth probing through increasing gel Ce6 fluorescence up to 63% *vs.* decreasing skin AF down to 35%, both at D_1, for all λ_j^{exc}, and for S1 OCA. For S2 OCA, the same increase and decrease were observed for Ce6 fluorescence (94%) and skin AF (21%), respectively. It is to be noticed first that saline solution (used as control against OCA and dry

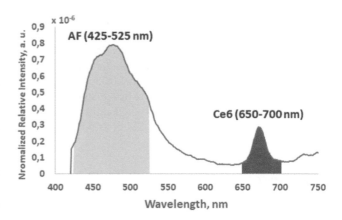

FIGURE 5.2 Example of fluorescence emission spectrum (red curve) collected at $D_2 = 600\,\mu m$ under $\lambda_2^{exc} = 385\,nm$, and showing the spectral ranges defined for the calculation of the areas under the curve (AUC) of skin AF (425–525 nm, yellow) and Gel Ce6 fluorescence (650–700 nm, green) intensities, from [4].

conditions) induced slight or limited decrease of the skin AF (~6%) and gel Ce6 fluorescence (~14%) collected during time at D_1 and for all λ_j^{exc} (Figure 5.3b). Second, tests performed in dry conditions, *i.e.* neither OCA nor saline solution, revealed a ~12% overall decline in skin AF for all λ_j^{exc} at D_1 but a strong increase of up to 59% of the Ce6 fluorescence coming from the underlying gel layer, which confirmed the similar clearing effect of more densely and compacted skin compound structures resulting from both (i) probe pressure–induced strains and (i) dry-induced constriction of the *ex vivo* sample [86].

Numerical modelling: Skin AF spectra unmixing

In order to further analyze the diversity of skin AF spectra modifications during the (topically applied) OCA in-depth penetration and propagation process, the individual but spectrally overlapping contributions of the various skin fluorophores that form and shape the LIFS collected "bulk" AF spectra need to be disentangled. To that end, a nonnegative matrix factorization (NMF) [83, 84] blind source separation unmixing method was recently applied on *ex vivo* human skin AF spectra measured by SR-mAFS [5]. The multidimensionality of the

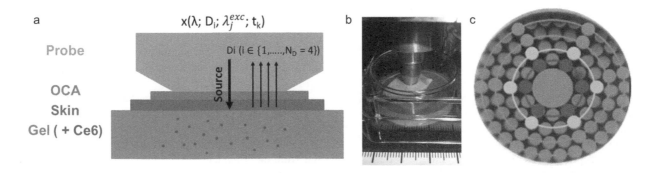

FIGURE 5.1 Schematic representation (left) and picture (middle) of the experimental configuration showing the SR-mAFS multifiber optics probe tip on top of the skin-gel model, with $x\left(\lambda; D_i, \lambda_j^{exc}, t_k\right)$ the fluorescence spectra collected at the 4 source-to-detector separations (SDS) D_i ($i \in \{1, ..., N_D = 4\}$), under 5 excitations λ_j^{exc} ($j \in \{1, ..., N_{exc} = 5\}$), and at 8 time points t_k ($k \in \{1, ..., N_t = 8\}$). The probe tip front view (on the right) shows the four fiber rings (red, blue, yellow, and green) corresponding to the four SDS respectively, adapted from [4, 5].

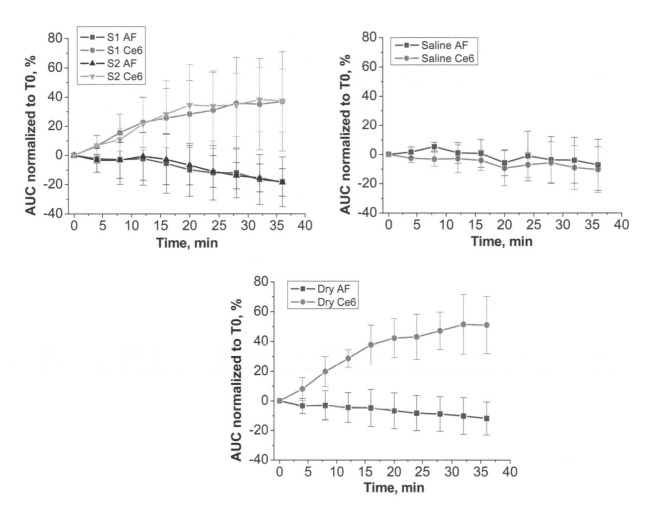

FIGURE 5.3 Time curves of normalized AUC of AF and Ce6 spectra measured (a) at $D_2 = 600\ \mu m$ under $\lambda_2^{exc} = 385\,nm$, for two OCA (S1 and S2); (b) at $D_1 = 400\ \mu m$ under $\lambda_5^{exc} = 415\,nm$, for saline application; and (c) at $D_2 = 600\ \mu m$ under $\lambda_2^{exc} = 385\,nm$, for "dry" condition, adapted from [4].

spectral data set was exploited to reconstruct the skin intrinsic fluorophores "source" signals $s_m(\lambda)$ from the observed (measured) fluorescence spectra $x(\lambda)$, modelled as linear mixing of the latter such as

$$x(\lambda) = \sum_{m=1}^{N_s} a_m s_m(\lambda) \qquad (5.1)$$

with $N_s = 4$ the total number of sources ($m \in \{1,\ldots,N_s\}$) and a_m the abundance, *i.e.* the weighting coefficient corresponding to the source $s_m(\lambda)$.

The NMF-based source separation applied to the SR-mAFS data set was formulated as the following matrix factorization problem:

$$\mathbf{X} \cong \mathbf{AS} \therefore \begin{pmatrix} x_1(\lambda) \\ \vdots \\ x_l(\lambda) \\ \vdots \\ x_{N_{sp}}(\lambda) \end{pmatrix} \cong$$

(5.2), adapted from [5]

where $\mathbf{X} \in R_+^{N_{sp} \times N_\lambda}$ is the matrix of the N_{sp} observed fluorescence intensity spectra $x_l(\lambda)$ of size N_λ; the vertical size of $\mathbf{X}$ ($l \in \{1,\ldots,N_{sp}\}$) corresponding to the total number of spectra acquired for N_D SDS, N_{exc} fluorescence excitation wavelengths, and N_t time steps, that is, $N_{sp} = N_D \times N_{exc} \times N_t$. Matrices $\mathbf{A}$ and $\mathbf{S}$ are unknowns to be estimated, with $S \in R_+^{N_s \times N_\lambda}$, the source matrix comprising the N_s source spectra $x_m(\lambda)$ and $A \in R_+^{N_{sp} \times N_s}$, the mixing matrix of the $N_{sp} \times N_s$ abundances $a_{l,m}$. Vector $a_{[l,.]}$, refers to the lth observation (l^{th} line in $\mathbf{X}$) corresponding to a mixture of source signals $s_1(\lambda),\ldots,s_{N_s}(\lambda)$.

To solve Equation 5.2, the optimal values $(\tilde{\mathbf{A}},\tilde{\mathbf{S}})$ of abundances and sources in matrices $\mathbf{A}$ and $\mathbf{S}$ were estimated through a minimization procedure of the following cost-function $f(\mathbf{A},\mathbf{S})$:

$$f(\mathbf{A},\mathbf{S}) = \|\mathbf{X} - \mathbf{AS}\|_F^2 = \sum_{l=1}^{N_{sp}} \left\| x_l(\lambda) - \sum_{m=1}^{N_s} a_{l,m} s_m(\lambda) \right\|_2^2$$

$$= \sum_{l=1}^{N_{sp}} \left(x_l(\lambda) - \sum_{m=1}^{N_s} a_{l,m} s_m(\lambda) \right)^2 \qquad (5.3),$$

adapted from [5]

with $\left\|\cdot\right\|_F^2$ and $\left\|\cdot\right\|_2^2$ denoting the squared Frobenius and the squared Euclidean norms, respectively.

Because of the nonconvexity of $f(\mathbf{A},\mathbf{S})$ against joint matrices $(\mathbf{A},\mathbf{S})$, *i.e.* a nonunique solution, but the convexity of quadratic functions $\mathbf{A} \mapsto f(\mathbf{A},\mathbf{S})$ and $\mathbf{S} \mapsto f(\mathbf{A},\mathbf{S})$ for $\mathbf{S}$ and $\mathbf{A}$ fixed respectively, an alternating least squares (ALS) strategy with multiplicative update method was implemented for minimizing $f(\mathbf{A},\mathbf{S})$ which consists in alternating between the minimization of $f(\mathbf{A},\mathbf{S})$ against $\mathbf{A}$ for $\mathbf{S}$ fixed and against $\mathbf{S}$ for $\mathbf{A}$ fixed [5].

On the experimental ($\mathbf{X}$) and estimated ($\tilde{\mathbf{A}}\tilde{\mathbf{S}}$) fluorescence intensity spectral curves plotted in Figure 5.4, the presence of both skin AF and gel-Ce6 fluorescence can be observed in the wavelength ranges below 600 nm and above 650 nm, respectively. This representation allows for observing qualitatively the relative amplitude variations of the latter contributions as a function of the five excitation wavelengths $\lambda_{1\ldots5}^{exc}$ and four time

points during the OC process T_0, +13, +20, and +25 minutes. As represented in Figure 5.5, the four spectral sources estimated, $s_1(\lambda)$, $s_2(\lambda)$, $s_3(\lambda)$, and $s_4(\lambda)$, were identified to correspond to skin dermis collagen and elastin cross-links AF, to skin epidermis FAD AF, to gel Ce6 fluorescence, and to a model residual noise component, respectively.

Finally, the quantitative analysis of the contribution of each of these sources to the bulk fluorescence spectra during the OC process was made possible thanks to the $D_{1\ldots4}$-, $\lambda_{1\ldots5}^{exc}$-, and time-dependent curves of the weight coefficients (normalized to their initial values at the OC process start time T_0) shown in Figure 5.6. This approach allowed it to be identified that the 450 towards 500 nm red-shift of the AF peak with increasing excitation wavelengths was related to the change of the relative contributions of sources 1 (collagen/elastin) and 2 (FAD) within the spectral mixing.

It was revealed that the skin AF contributions (sources 1 and 2) strongly increased between T_0 and T_0+13 minutes (at least

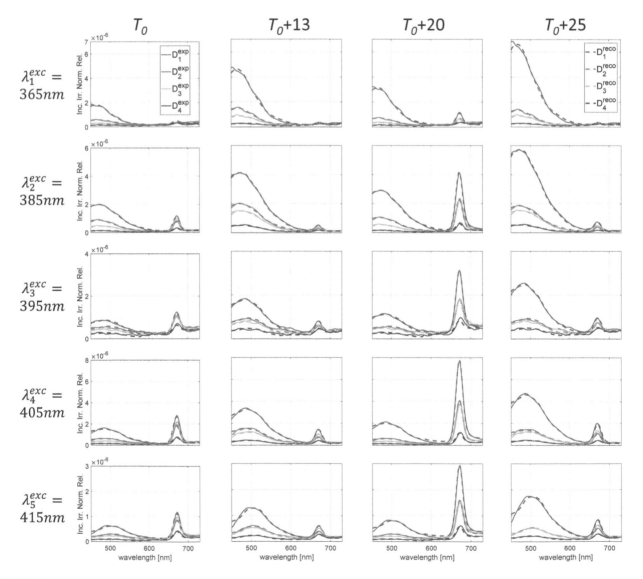

FIGURE 5.4 Measured (continuous lines) and NMF-based recovered (dashed lines) fluorescence spectra with $N_s = 4$ sources. Results are plotted for the four SDS $D_{1\ldots4}$ in blue, red, yellow and purple colors, the five excitations $\lambda_{1\ldots5}^{exc}$ (rows) and four time points during the OC process at T_0, +13, +20, and +25 minutes (columns), reprinted/adapted with permission from [5] © The Optical Society.

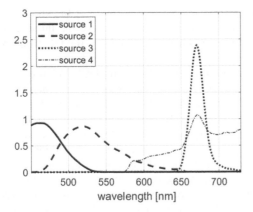

FIGURE 5.5 NMF-based estimated spectral sources $s_{1...4}(\lambda)$ ($Ns = 4$) obtained according to all SDS $D_{1...4}$, all excitations $\lambda_{1...5}^{exc}$, and all time points $t_{1...8}$ during the OC process, reprinted/adapted with permission from [5] © The Optical Society.

two-fold up to five-fold for shortest $\lambda_{1,2}^{exc}$), then decreased between T_0+13 and at T_0+20 minutes by 35% to 40% down with rising SDS, then raised up again at equivalent and higher magnitudes for longest SDS $D_{2,3,4}$ and shortest one $D1$, respectively. During the same periods of time, the fluorescence contribution of Ce6 showed an opposite behavior with (i) a distinct decrease (down to almost zero value) in related skin optical property modifications (higher AF), leading to limited collection of Ce6 fluorescence; (ii) a sharp increase at $T_0 + 20$ minutes, correlated with maximum transparency and lowered skin AF at this moment; and (iii) a final decrease and return to unity value (starting level). The behavior of these three phases was related to the progressive diffusion of the topically applied OCA towards the deepest layers of the skin sample, including the skin initial dehydration and subsequent rehydration steps occurring during the OC process, and the in-depth progressive RI matching and loss of scattering between tissue components and interstitial fluid related to the first 20 minutes of OC action

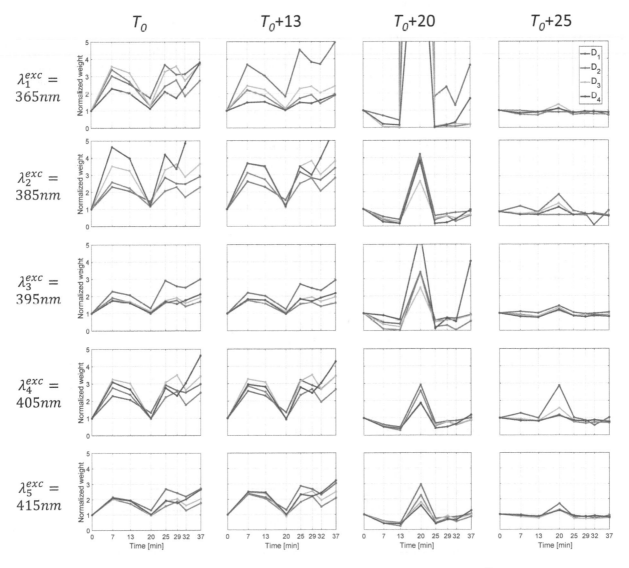

FIGURE 5.6 Time evolution of the T_0-normalized weight coefficients $a_{l,m}$ in the NMF-based abundance matrix $\tilde{A}$ estimated with N_s = four sources, plotted for the four SDS $D_{1...4}$ in blue, red, yellow, and purple colors, the five excitations $\lambda_{1...5}^{exc}$ (rows) and the four sources $s_{1...4}(\lambda)$ (columns), reprinted/adapted with permission from [5] © The Optical Society.

with maximum skin transparency achieved at T_0+20 minutes, and finally a return to equilibrium.

Experimental modeling: Three-layer fluorescent phantoms

In parallel, the OC-induced changes in detected fluorescence as a function of SDS were studied using scattering multilayer tissue-simulating phantoms incorporating a fluorescent layer located at different depths [72, 85]. The phantom consisted of three stacked 1mm-thick gelatinous (10% gelatin) layers placed on a strongly scattering optically semi-infinite gelatinous substrate. To simulate light scattering within tissue, neutral (nonfluorescent and nonabsorbing in the visible spectral range) 410-nm TiO2 scattering particles were added into those layers (and substrate). As represented in Figure 5.7 (top), four phantom models (0–3) were designed to mimic the depth gradual (layer by layer) tissue OC upon OCA topical application corresponding to successive decreasing reduced scattering coefficient values within layers from 0.49 (no clearing) to 0.39, 0.29, and 0.19 cm⁻¹ (maximum clearing), *i.e.* lowering TiO2 concentration ([TiO2]) in gelatin layers from 0.4 to 0.35, 0.3, and 0.2 mg/ml, respectively.

In the phantoms, one of the 1-mm layers was additionally made fluorescent with rhodamine 6G (R6G). Thus, three series of phantoms were prepared corresponding to the three depth locations of the fluorescent layer (see Figure 5.7 bottom), with each series comprising all four OC models 0–3. The fluorescence spectroscopic measurements were performed under 532 nm laser excitation and at 13 different SDS, from 2 to 8 mm, between the excitation (illuminating) 600-μm fiber and collecting (receiving) fiber (actually a bundle of six 200-μm fibers), see Figure 5.8.

The measured emitted fluorescence intensity curves $I_{flu} = f(D/L)$, defined as a function of the reduced source-detector distance D/L (where $D = SDS$ and $L = 1$-mm layer thickness), are plotted in Figure 5.9 for different depths of the fluorescent layer and for different penetration depths of the OC front (Models 0-3).

It was observed that "without OC" (Figure 5.9a), the recorded fluorescence intensity was highest when the second layer was fluorescent and lowest when the third layer was. In the case of the initial OC step, when only the first (upper) layer was "cleared" (Figure 5.9b), the detected fluorescence intensity was the highest when this very layer was fluorescent. In the next configuration where the two upper layers were "cleared" (Figure 5.9c), the fluorescence intensity was the highest (i) at small D/L for the fluorescent second layer and (ii) at large D/L (>5) for the fluorescent upper layer. In the case of all layers "cleared" (Figure 5.9d), the highest and lowest fluorescence intensities were obtained for the fluorescent third and first layers, respectively.

These results highlighted that after topical application of OCA, fluorescence of upper layers should be predominant at the initial stage of the OC process and the fraction of deeper layer fluorescence in the total recorded intensity should increase with clearing depth.

Such results obtained on phantoms indicated the feasibility of employing (i) OC to estimate the depth of fluorophore localization in the subsurface layers of tissue and (ii) SR-LIFS to evaluate the dynamics of tissue clearing in the case of known localization of fluorophores.

Conclusion and perspectives

The main skin intrinsic fluorophores, related to metabolic processes and structural arrangement of cells and tissues, are

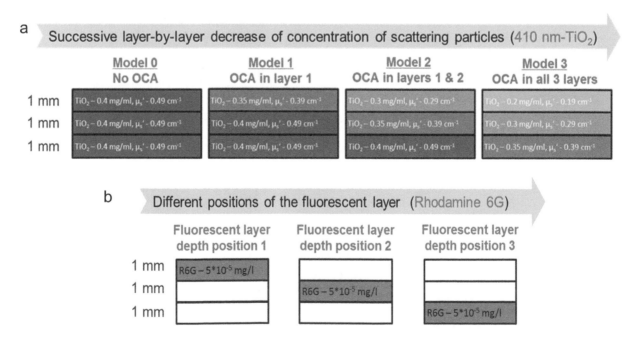

FIGURE 5.7 Three-layer gel phantoms mimicking OC of scattering layered tissues with a fluorescent dye in one of the layers: (top) models mimicking temporal-spatial propagation of the OC front and (bottom) models mimicking fluorophore location at different tissue depths (with scatterers' concentrations within the layers given by the models 0–3), adapted from [72].

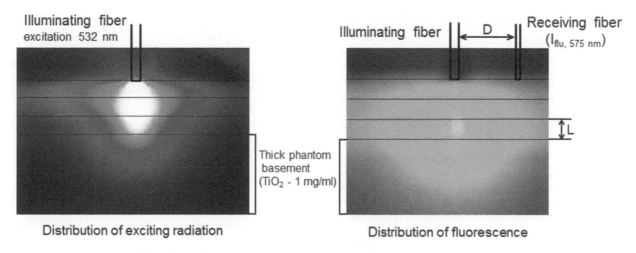

FIGURE 5.8 The spatial distribution of (a) excitation light (blue-green band-pass filter used) and (b) fluorescence emission (red band-pass filter used) observed in the phantom corresponding to OC Model 3 (OCA in all three layers and third layer fluorescent, see Figure 5.7), adapted from oral presentation in [72].

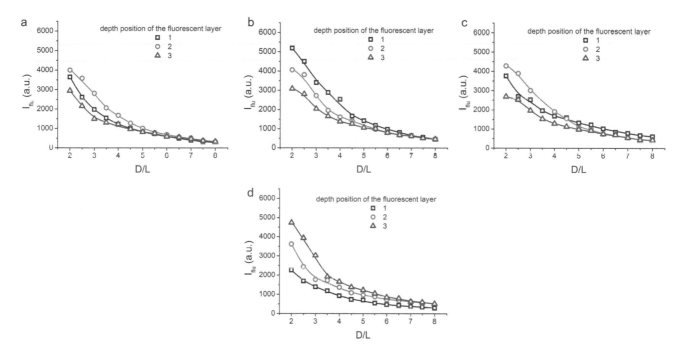

FIGURE 5.9 Fluorescence intensity I_{flu} measured on the phantoms as a function of the reduced SDS for the three in-depth positions of the fluorescent layer (first–red, second–black, third–blue) within the three-layer structures and the four OC Models (see Figure 5.7) with (a) no OCA, (b) OCA in first (upper) layer, (c) OCA in two layers, and (d) OCA in all layers, adapted from [72].

valuable and noninvasively accessible photophysical biomarkers of skin pathology states and progress. Skin AF emitted under NUV–blue wavelength range excitations arises from different fluorescing proteins and cellular compounds, located at various depths within the cutaneous layers.

The access to these intrinsic, *i.e.* "undistorted," fluorescence in-depth information can be facilitated by OC methods allowing for reversible modifications of the skin optical properties, particularly scattering strong decrease. In this regard, further works need to be conducted for developing safe, reversible, and biocompatible OCA for *in vivo* applications at the research and clinical dermatology transfer levels. The latter should be supported by experimental characterizations of OC-related

modifications in *ex vivo* skin samples (layer thickening/thinning, dehydration) and in *in vivo* skin tissues (OCA formulation and operative protocol).

Among the few optical spectro-imaging techniques currently available for the *in vivo* measurement of skin tissue AF, fluorescence confocal, multiphoton and light sheet microscopies, and light-induced AF spectroscopy have been the main modalities implemented for studying the impact of the OC process on the skin intrinsic fluorophores, depending on their excitation-emission features and locations in cells and tissues. Consequently, further developments are needed to validate and transfer into clinical settings noninvasive, cheap, and easy to use AF spectro-imaging tools for allowing *in vivo* real-time preoperative tissue diagnosis and/or helping surgical guiding.

Successful application and clinical transfer of skin OC and *in vivo* tissue characterization spectro-imaging modalities require a compromise to be found between the clearing efficiency throughout the cutaneous layers, *i.e.* optimized composition and concentration of topically applied OCA, and the collecting efficiency (considering spatial, spectral, and temporal resolutions) of the AF signal originating from epidermis and dermis intrinsic fluorophores.

Consequently, combining skin OC and AF-based spectro-imaging techniques is of interest (i) to improve the understanding of the skin OC mechanisms and help the development of *in vivo* biocompatible agents, and (ii) to investigate the OC-related modifications of skin AF optical properties and increase the *in vivo* diagnostic potential of AF-based methods.

Finally, the development of (i) multidimensional data processing, *e.g.* spectral signal unmixing; (ii) numerical simulation-based modelling and optical property estimation considering realistic skin model complexity and precision (including absorption, scattering, and autofluorescence); and (iii) optical phantom-based experimental modelling approaches are also at stake if deeper insight is to be gained into the interpretation of the AF signal collected and thus the understanding of the underlying OC-related photophysical phenomenon.

Acknowledgments

- French Région Grand-Est, CNRS and Metz-Thionville Regional Hospital (2016 clinical research Award) in the frame of the SpectroLive project
- Contrat de Plan Etat – Région Grand Est 2015-2020 (CPER IT2MP, plateforme IMTI) including the European Regional Development Fund FEDER
- Ligue Contre le Cancer
- Lorraine Université d'Excellence (LUE) Doctorate 2019 under the PhD grant R01PJYRX-PHD19-COLA-CRAN (Victor Colas, 2019-2022) and LUE Graduate international mobility program 2018 under the DREAM grant R01PJJFX-DREAM-ZAYT-CRAN (Sergey Zaytsev, 2020-21
- French Embassy in Russia under the Vernadski international joint PhD program (Sergey Zaytsev, 2018–2021)
- Russian Foundation for Basic Research (18-52-16025 and 20-32-90043)
- Government of the Russian Federation (14. W03.31.0023)

REFERENCES

1. Croce and Bottiroli, "Autofluorescence spectroscopy for monitoring metabolism in animal cells and tissues " *Methods Mol. Biol.* **1560**, 15–43 (2017).
2. E. Migacheva, A. Pravdin and V. Tuchin, "Alterations in autofluorescence signal from rat skin *ex vivo* under optical immersion clearing," *J. Innov. Opt. Health Sci.* **3**(3), 147–152 (2010).
3. A. Sdobnov, M. Darvin, J. Lademanfn and V. Tuchin, "A comparative study of *ex vivo* skin optical clearing using two-photon microscopy," *J. Biophoton.* **10**, 1115–1123 (2017).
4. S. Zaytsev, W. Blondel, M. Amouroux et al., "Optical spectroscopy as an effective tool for skin cancer features analysis: Applicability investigation," *Proc. SPIE* **11457**, 1145706 (2020).
5. P. Rakotomanga, C. Soussen, G. Khairallah et al., "Source separation approach for the analysis of spatially resolved multiply excited autofluorescence spectra during optical clearing of *ex vivo* skin," *Biomed. Opt. Express* **10**(7), 364338 (2019).
6. A. Sdobnov, J. Lademann, M. Darvin, and V. Tuchin, "Methods for optical skin clearing in molecular optical imaging in dermatology," *Biochemistry*, **84**(1), S144–S158 (2019).
7. D. Carpenter, M. Sajisevi, N Chapurin et al., "Noninvasive optical spectroscopy for identification of non-melanoma skin cancer: Pilot study," *Lasers Surg. Med.* **50**(3), 246–252 (2018).
8. E. Drakaki, C. Dessinioti, A. Stratigos, C. Salavastru and C. Antonioua, "Laser-induced fluorescence made simple: Implications for the diagnosis and follow-up monitoring of basal cell carcinoma," *J. Biomed. Opt.* **19**(3), 030901 (2014).
9. V. Narayanamurthy, P. Padmapriya, A. Noorasafrin et al., "Skin cancer detection using non-invasive techniques," *RSC Adv.* **8**, 28095 (2018).
10. N. Matthews, W. Li, A. Qureshi et al., "Epidemiology of melanoma," in Chapter 1, *Cutaneous Melanoma: Etiology and Therapy* [Internet], W. Ward and J. Farma, Eds., Brisbane, AU: Codon Publications; Available from: https://www.ncbi.nlm.nih.gov/books/NBK481862/ (2017).
11. V. Maciel, W. Correr, C. Kurachi, V. Bagnato and C. da Silva Souza, "Fluorescence spectroscopy as a tool to *in vivo* discrimination of distinctive skin disorders," *Photodiagn. Photodyn. Ther.* **19**, 45–50 (2017).
12. H. Liu, H. Gisquet, F. Guillemin and W. Blondel, "Bimodal spectroscopy for *in vivo* characterization of hypertrophic skin tissue: pre-clinical experimentation, data selection and classification," *Biomed. Opt. Express* **3**(12), 3278–3290 (2012).
13. J. Thompson, S. Coda, M. Sørensen et al., "*In vivo* measurements of diffuse reflectance and time-resolved autofluorescence emission spectra of basal cell carcinomas," *J. Biophoton.* **5**(3), 240–254 (2012).
14. E. Borisova, P. Pavlova, E. Pavlova, P. Troyanova and L. Avramov, "Optical biopsy of human skin: A tool for cutaneous tumours' diagnosis," *Int. J. Bioautom.* **16**(1), 53–72 (2012).
15. Q. Liu, "Role of optical spectroscopy using endogenous contrasts in clinical cancer diagnosis," *World J. Clin. Oncol.* **2**(1), 50–63 (2011).
16. M. Amouroux, G. Diaz-Ayil, W. Blondel, G. Bourg-Heckly, A. Leroux and F. Guillemin, "Classification of ultra-violet irradiated mouse skin histological stages by bimodal spectroscopy (multiple excitation autofluorescence and diffuse reflectance)," *J. Biomed. Opt.* **14**(1):014011 (2009).
17. E. Borisova, E. Carstea, L. Cristescu et al., "Light induced fluorescence spectroscopy and optical coherence tomography of basal cell carcinoma," *J. Innov. Opt. Health Sci.* **2**(3), 261–268 (2009).

18. E. Borisova, P. Troyanova, P. Pavlova and L. Avramov, "Diagnostics of pigmented skin tumors based on laser-induced autofluorescence and diffuse reflectance spectroscopy," *Quantum Electron.* **38**(6), 597–605 (2008).

19. Y. Wu and J. Qu, "Autofluorescence spectroscopy of epithelial tissues," *J. Biomed Opt.* **11**(5), 054023 (2006).

20. M. Panjehpour, C. Julius, M. Phan, T. Vo-Dinh and S. Overholt, "Laser-induced fluorescence spectroscopy for *in vivo* diagnosis of non-melanoma skin cancers," *Lasers Surg. Med.* **31**(5), 367–373 (2002).

21. L. Brancaleon, A. Durkin, J. Tu et al., "*In vivo* fluorescence spectroscopy of non-melanoma skin cancer," *Photochem. Photobiol.* **73**(2), 178–183 (2001).

22. R. Gillies, G. Zonios, R. Anderson, N. Kollias, "Fluorescence excitation spectroscopy provides information about human skin *in vivo*," *J. Invest. Dermatol.* **115**, 704–707 (2000).

23. Y. Fujino, G. Attizzani, S. Tahara et al., "Association of skin autofluorescence with plaque vulnerability evaluated by optical coherence tomography in patients with cardiovascular disease," *Atherosclerosis* **274**, 47e53 (2018).

24. D. Foster, A. Nguyen, M. Chinta et al., "A clearing technique to enhance endogenous fluorophores in skin and soft tissue" *Sci. Rep.* **9**,15791 (2019).

25. A. Sdobnov, M. Darvin, E. Genina, A. Bashkatov, J. Lademann and V. Tuchin, "Recent progress in tissue optical clearing for spectroscopic application," *Spectrochim. Acta A* **197**, 216–229 (2018).

26. S. Abadie, C. Jardet, J. Colombelli et al., "3D imaging of cleared human skin biopsies using light-sheet microscopy: A new way to visualize in-depth skin structure," *Skin Res. Technol.* **24**:294–303 (2018).

27. R. Cicchi, D. Kapsokalyvas and F. Pavone, "Clinical non-linear laser imaging of human skin: A review," *BioMed. Res. Int.* **2014**, 903589 (2014).

28. D. Zhu, K.V. Larin, Q. Luo, and V. Tuchin, "Recent progress in tissue optical clearing," *Laser Photonics Rev.* **7**(5), 732–757 (2013).

29. P. Calzavara-Pinton, C. Longo, M. Venturini, R. Sala and G. Pellacani, "Reflectance Confocal Microscopy for *in vivo* skin imaging," *Photochem. Photobiol.* **84**, 1421–1430 (2008).

30. E. Shirshin, Y. Gurfinkel, A. Priezzhev, V. Fadeev, J. Lademann and M. Darvin, "Two-photon autofluorescence lifetime imaging of human skin papillary dermis *in vivo*: assessment of blood capillaries and structural proteins localization," *Sci. Rep.* **7**, 1171 (2017).

31. I. Ferulova, A. Lihachev and J. Spigulis, "Photobleaching effects on *in vivo* skin autofluorescence lifetime," *J. Biomed. Opt.* **20**, 051031 (2015).

32. B. Lloyd-Lewis, "Multidimensional imaging of mammary gland development: A window into breast form and function," *Frontiers Cell Dev. Biol.* **8**, 203 (2020).

33. I. Costantini, R. Cicchi, L. Silvestri, F. Vanzi and F. Pavone, "*In-vivo* and *ex-vivo* optical clearing methods for biological tissues: Review," *Biomed. Opt. Express* **10**(10), 5251 (2019).

34. N. Rajaram, J. Reichenberg, M. Migden, T. Nguyen and J. Tunnell, "Pilot clinical study for quantitative spectral diagnosis of nonmelanoma skin cancer," *Lasers Surg. Med.* **42**(10), 716–727 (2010).

35. R. Shi, L. Guo, C. Zhang et al., "A useful way to develop effective *in vivo* skin optical clearing agents," *J. Biophoton.* **10**(6–7), 887–895 (2016).

36. D. Zhu, B. Choi, E. Genina and V. Tuchin, "Tissue and blood optical clearing for biomedical applications," *J. Biomed. Opt.* **21**(8), 081201 (2016).

37. T. Yu, Y. Qi, J. Zhu et al., "Elevated-temperature-induced acceleration of PACT clearing process of mouse brain tissue," *Sci. Rep.* **7**, 38848 (2017).

38. E. Genina, A. Bashkatov, Y. Sinichkin, I. Yanina and V. Tuchin, "Optical clearing of biological tissues: Prospects of application in medical diagnostics and phototherapy," *J. Biomed. Photon. Eng.* **1**(1), 22–58 (2015).

39. N. Bashkatov, E. Genina, V. Kochubey and V. Tuchin, "Optical properties of human skin, subcutaneous and mucous tissues in the wavelength range from 400 to 2000 nm," *J. Phys. D Appl. Phys.* **38**(15), 2543–2555 (2005).

40. C. Choe, J. Schleusener, J. Lademann and M. Darvin, "Human skin *in vivo* has a higher skin barrier function than porcine skin *ex vivo*: Comprehensive Raman microscopic study of the stratum corneum," *J. Biophoton.* **11**, e201700355 (2018).

41. Y. Sinichkin, N. Kollias, G. Zonios, S. Utz, and V. Tuchin, "Reflectance and fluorescence spectroscopy of the human skin *in vivo*," in *Handbook on Optical Biomedical Diagnostics. Methods*, vol.2 (PM263), pp.99–190, 2nd ed., V.V. Tuchin, Ed., Bellingham, SPIE Press (2016).

42. B. Brancaleon, G. Lin, N. Kollias, "The *in vivo* fluorescence of tryptophan moieties in human skin increases with UV exposure and is a marker for epidermal proliferation," *J. Invest. Dermatol.* **113**, 977–982 (1999).

43. E. Drakaki, T. Vergou, C. Dessinioti, A. Stratigos, C. Salavastru and C. Antoniou, "Spectroscopic methods for the photodiagnosis of nonmelanoma skin cancer," *J. Biomed. Opt.* **18**(6), 061221 (2013).

44. I. Bliznakova, E. Borisova and L. Avramov, "Laser- and light-induced autofluorescence spectroscopy of human skin in dependence on excitation wavelengths," *Acta Phys. Pol. A* **112**(5), 1131–1136 (2007).

45. N. Kollias, G. Zonios and G. Stamatas, "Fluorescence spectroscopy of skin," *Vib. Spectrosc.* **28**(1), 17–23 (2002).

46. A. Pena, M. Strupler, T. Boulesteix and M. Schanne-Klein, "Spectroscopic analysis of keratin endogenous signal for skin multiphoton microscopy," *Opt. Express* **13**(16), 6268–(2005).

47. K. Kawabata, M. Kobayashi, A. Kusaka-Kikushima et al., "A new objective histological scale for studying human photoaged skin," *Skin Res. Technol.* **20**(2), 155–163 (2014).

48. P. Bargo, A. Doukas, S. González et al., "The Kollias legacy: Skin autofluorescence and beyond," *Exp. Dermatol.* **26**(10), 858–860 (2017).

49. R. Na, I. Stender, and H. Wulf, "Can autofluorescence demarcate basal cell carcinoma from normal skin? A comparison with protoporphyrin IX fluorescence," *Acta Derm. Venerol.* **81**(4), 246–249 (2001).

50. H. Zeng et al., "Autofluorescence of basal cell carcinoma," *Proc. SPIE* **3245**, 314–317 (1998).

51. Y. Sinichkin, S. Uts, I. Meglinskii and E. Pilipenko, "Spectroscopy of human skin *in vivo*: II. Fluorescence spectra," *Opt. Spectrosc.* **80**(3), 383–389 (1996).

52. E. Genina, A. Bashkatov, G. Terentyuk and V. Tuchin, "Integrated effects of fractional laser microablation and sonophoresis on skin immersion optical clearing *in vivo*," *J. Biophoton.* **13**, e2020000101 (2020).

53. P. Matryba, L. Kaczmarek and J. Golab, "Advances in *ex situ* tissue optical clearing," *Laser Photonics. Rev.* **13**, 1800292 (2019).

54. V. Genin, P. Rakotomanga, S. Zaytsev et al., "Research and development of effective optical technologies for diagnostics in dermatology," *Proc. SPIE* **11065**, 1106505 (2019).

55. E. Lazareva, P. Dyachenko, A. Bucharskaya, N. Navolokin and V. Tuchin, "Estimation of dehydration of skin by refractometric method using optical clearing agents," *J. Biomed. Photon. Eng.* **5**(2), 020305 (2019).

56. Q. Zhang, L. Zhong, P. Tang et al., "Quantitative refractive index distribution of single cell by combining phase-shifting interferometry and AFM imaging," *Sci. Rep.* **7**, 2532 (2017).

57. H. Liang, B. Akladios, C. Canales, R. Francis, E. Hardeman, and A. Beverdam, "CUBIC protocol visualizes protein expression at single cell resolution in whole mount skin preparations," *J. Vis. Exp.* **114**, e54401 (2016).

58. W. Feng, R. Shi, N. Ma, D. Tuchina, V. Tuchin and D. Zhu, "Skin optical clearing potential of disaccharides," *J. Biomed. Opt.* **21**(8), 081207 (2016).

59. Z. Deng, L. Jing, N. Wu et al., "Viscous optical clearing agent for *in vivo* optical imaging," *J. Biomed. Opt.* **19**(7), 76019 (2014).

60. D. Zhu, J. Wang, Z. Zhi, X. Wen, and Q. Luo, "Imaging dermal blood flow through the intact rat skin with an optical clearing method," *J. Biomed. Opt.* **15**(2), 026008 (2010).

61. X. Wen, Z. Mao, Z. Han, V. Tuchin, and D. Zhu, "*In vivo* skin optical clearing by glycerol solutions: Mechanism," *J. Biophoton.* **3**(1–2), 44–52 (2010).

62. S. Karma, J. Homan, C. Stoianovici and B. Choi, "Enhanced fluorescence imaging with DMSO-mediated optical clearing," *J. Innov. Opt. Health Sci.* **3**(3), 153–158 (2010).

63. S. Millon, K. Roldan-Perez, K. Riching, G. Palmer, and N. Ramanujam, "Effect of optical clearing agents on the *in vivo* optical properties of squamous epithelial tissue," *Lasers Surg. Med.* **38**(10), 920–927 (2006).

64. X. Xu, Y. He, S. Proskurin, R. Wang, and J. Elder, "Optical clearing of *in vivo* human skin with hyperosmotic chemicals investigated by optical coherence tomography and near infrared reflectance spectroscopy," *Proc. SPIE* **5486**, 129–135 (2003).

65. V. Tuchin, A. Bashkatov, E. Genina, Y. Sinichkin, and N. Lakodina, "*In vivo* investigation of the immersion-liquid-induced human skin clearing dynamics," *Tech. Phys. Lett.* **27**(6), 489–490 (2001).

66. X. Liu and B. Chen, "*In vivo* experimental study on the enhancement of optical clearing effect by laser irradiation in conjunction with a chemical penetration enhancer," *Appl. Sci.* **9**(3), 542 (2019).

67. R. Shi, L. Guo, C. Zhang et al., "A useful way to develop effective *in vivo* skin optical clearing agents," *Biophotonics* **10**, 887–895 (2017).

68. X. Wen, S. Jacques, V. Tuchin and D. Zhu, "Enhanced optical clearing of skin *in vivo* and optical coherence tomography in-depth imaging," *J. Biomed. Opt.* **17**(6), 066022 (2012).

69. J. Wang, R. Shi and D. Zhu, "Switchable skin window induced by optical clearing method for dermal blood flow imaging," *J. Biomed. Opt.* **18**(6), 061209 (2013).

70. A. Liopo, R. Su, D. Tsyboulski and A. Oraevsky, "Optical clearing of skin enhanced with hyaluronic acid for increased contrast of optoacoustic imaging," *J. Biomed. Opt.* **21**(8), 081208 (2016).

71. R. Cicchi, F. Pavone, D. Massi and D. Sampson, "Contrast and depth enhancement in two-photon microscopy of human skin *ex vivo* by use of optical clearing agents," *Opt. Express* **13**(7), 2337–2344 (2005).

72. P. Rakotomanga, G. Khairallah, M. Shvachkina et al. "Investigation of human skin autofluorescence modifications due to optical clearing using spatially resolved spectroscopy and multiple excitations," In OSA/IEEE: Asia Communications and Photonics Conference (ACP) 2018, Hangzhou, China, October 26–29 (2018).

73. P. Rakotomanga, "Model inversion and separation of optical spectroscopic signals for the *in vivo* characterization of skin tissues," PhD thesis [in French], Ecole Doctorale IAEM, Doctorat de l'Université de Lorraine, December 18th 2019.

74. E. Fernandez and S. Marull-Tufeu, "3D imaging of human epidermis micromorphology by combining fluorescent dye, optical clearing and confocal microscopy," *Skin Res. Technol.* **25**, 735–742 (2019).

75. S. Perry, R. Burke and E. Brown, "Two-photon and second harmonic microscopy in clinical and translational cancer research," *Ann. Biomed. Eng.* **40**(2), 277–291 (2012).

76. V. Hovhannisyan, P.-S. Hu, S.-J. Chen, C.-S. Kim and C.-Y. Dong, "Elucidation of the mechanisms of optical clearing in collagen tissue with multiphoton imaging," *J. Biomed. Opt.* **18**(4), 046004 (2013).

77. H. Lu, F. Floris, M. Rensing and S. Andersson-Engels, "Fluorescence spectroscopy study of Protoporphyrin IX in optical tissue simulating liquid phantoms," *Materials* **13**, 2105 (2020).

78. S. Brooks, C. Hoy, A. Amelink, D. Robinson and T. Nijsten, "Sources of variability in the quantification of tissue optical properties by multidiameter single-fiber reflectance and fluorescence spectroscopy," *J. Biomed. Opt.* **20**, 57002 (2015).

79. E. Pery, W. Blondel, S. Tindel, M. Ghribi, A. Leroux and F. Guillemin, "Spectral features selection and classification for bimodal optical spectroscopy applied to bladder cancer *in vivo* diagnosis," *IEEE Trans. Biomed. Eng.* **61**(1), 207–216 (2014).

80. A. Kim, M. Khurana, Y. Moriyama and B. Wilson, "Quantification of *in vivo* fluorescence decoupled from the effects of tissue optical properties using fiber-optic spectroscopy measurements," *J. Biomed. Opt.* **15**(6), 067006 (2010).

81. M. Amouroux, W. Blondel and A. Delconte, "Medical device for fibered bimodal optical spectroscopy," WO2017093316 (A1), filed Nov. 30th 2015, and issued June 8th 2017.

82. G. Vargas, K. Chan, S. Thomsen and A Welch, "Use of osmotically active agents to alter optical properties of tissue: Effects on the detected fluorescence signal measured through skin" *Lasers Surg. Med.* **29**,213–220 (2001).

83. A. Montcuquet, L. Hervé, F. Navarro, J. Dinten, and J. Mars, "*In vivo* fluorescence spectra unmixing and autofluorescence removal by sparse nonnegative matrix factorization," *IEEE Trans. Biomed. Eng.* **58**(9), 2554–2565 (2011).

84. D. Lee and H. Seung, "Learning the parts of objects by non-negative matrix factorization," *Nature* **401**, 788–791 (1999).

85. A. Pravdin, G. Filippidis, G. Zacharakis, T. Papazoglou, V. Tuchin. "Tissue phantoms," Chapter 5 in *Handbook of Optical Biomedical Diagnostics. Light–Tissue Interaction*, vol.1 **(PM262)**, pp. 335–395, 2nd ed., Bellingham, SPIE Press, 2016.

86. S. Zaytsev, M. Amouroux, G. Khairallah, A. Bashkatov, V. Tuchin, W. Blondel and E. Genina, "Impact of optical clearing on ex vivo human skin optical properties characterized by spatially resolved multimodal spectroscopy," *Journal of Biophotonics*, e202100202 (2021).

6

Molecular modeling of post-diffusion phase of optical clearing of biological tissues

Kirill V. Berezin, Konstantin N. Dvoretskiy, Maria L. Chernavina, Anatoly M. Likhter, and Valery V. Tuchin

CONTENTS

Introduction

It is known that immersion optical clearing (OC) of biological tissues is a complex and multiple-stage process. The first stage is associated with the hyperosmotic properties of immersion agents. The contact of agents with the biological tissue surface causes its dehydration. At the second stage, the immersion agent diffuses into the biological tissue. At the third stage, the immersion agents interact with the collagen protein.

It was shown that correlation between an experimentally observed degree of OC and such parameters as osmolarity and the refraction index of an immersion agent does not exist. At the same time, a correlation with such parameters as the time of immersion agent being in a water-bonded state with collagen was identified [1–3]. Results of this modeling indicate the necessity for more detailed consideration of a molecular mechanism of interaction with the use of quantum chemistry means. In the present chapter, the authors build molecular models that allow one to explain a process taking place, the third stage of which, in its turn, makes it possible to predict – at a theoretical level – a degree of OC for different immersion agents [4–6].

Stages of molecular modeling

A mimetic peptide of collagen $((GPH)_9)_3$ [7], forming the basis of a great part of regular domains of human collagen, was used as a molecular model of collagen. Such relatively small synthetic peptides are often used for collagen molecular modeling. The molecular modeling of interaction of OCA_S with collagen was carried out in several stages.

At the first stage, a peptide 3D model was built according to the Protein Data Bank (PDB) data with further addition of hydrogen atoms and optimization of the structure by the molecular mechanics method [8]. Then the methods DFT/B3LYP/6-311 + G(d, p) [9, 10] and Gaussian program [11] were used to identify and calculate all the lowest energy conformers of the considered monosaccharides in their isolated state. The calculated geometric parameters and values of the Mulliken atomic charges were further used in modeling these systems within the framework of classical molecular dynamics. Vibrational transition wavenumbers were also calculated and appeared to be positive, which gives another evidence of molecular systems being at the local minimum.

At the second stage of the modeling, methods of classical molecular dynamics were used to study the probability of hydrogen bonding between collagen peptide $((GPH)_9)_3$ and the chosen molecular agents. We also measured the time of their being in a water-bonded state with collagen and their effect on its structure. Molecular modeling of interaction of these agents with collagen was carried out with GROMACS package of classical molecular dynamics [12] with AMBER-03 force field [13]. A modeling scene was a 3D cell having the form of a rectangular parallelepiped with the side lengths of 3nm × 3nm × 9nm. The boundaries of the cell were chosen periodically (when colliding with a boundary, a molecule passes through it, appearing on the side of the opposite boundary). Before the start of each modeling, 20 molecules of the agent are distributed randomly within the cell. Initial velocities of atoms were set with the use of random-number generators of the GROMACS package and had the Maxwellian distribution, corresponding to the chosen temperature. To model the system, the Berendsen thermostat and barostat were used [14], ensuring convergence of temperature and pressure of the system to the set values $T_0 = 300K$ and $p_0 = 1$ bar. The time step of modeling was equal to 0.0001 ps, while the total time

DOI: 10.1201/9781003025252-7

of modeling was 100 ps. The system state was recorded each 0.1 ps. Recorded tracks of molecular motion were processed by GROMACS tools and with the use of the VMD (visual molecular dynamics) program [15]. Each system under study was modeled 30 times, and the obtained results were averaged. A standard error of the obtained arithmetical means was also calculated.

Then, to model a molecular state closer to native, the modeling scene was filled with a water solution of the chosen OCAs. Their concentration in the water solution changed from 0% (pure solvent of water – model SPC/E [16]) to 60%–70% depending on the type of the agent. All the starting parameters for molecular modeling were similar to the previous stage, except for the modeling time, which was 50 ps. At this stage, the effect of the OCA, added to the water environment of collagen, on the geometric parameters of peptide α-chains was assessed.

At the third stage, a minimum fragment of the mimetic peptide, preserving the regular structure $((GPH)_2)_3$, consisting of 231 atoms, and having a structure that was preliminary optimized within the semiempirical method PM6, was used to evaluate the energy of the intermolecular interaction of the chosen OCA with collagen [17]. This optimized structure of the collagen model was used to carry out molecular docking with OCAs within the AutoDockVina program [18]. When the molecular docking was carried out, the ten most suitable configurations were selected for each interacting system, and then they were optimized with the semiempirical method PM6. Then the total electronic energy of the complexes was calculated with the method DFT/B3LYP/6-31G(d)

and through the single SCF procedure. A similar procedure was used to obtain values of the total electronic energy of the OCAs and the peptide fragment. The intermolecular interaction energy was calculated as the difference between the complex total energies and the sum of energies of its specific components. The largest values of intermolecular interaction energies, corresponding to the most probable complex structures, were chosen to build a correlation with the OC potential.

Immersion agents (alcohols)

Three diatomic alcohols (ethylene glycol 1.2- and 1.3-propanediol) and three polyatomic alcohols (glycerol, xylitol, and sorbitol) were examined as immersion agents, for which there are experimental data on OC potential, contained in paper [1]. Figure 6.1 shows the spatial configurations of the lowest energy conformers of the OCAs under study. It should be noted that the values of relative populations were measured for all the calculated conformers, and if the mixture of conformers had, for example, two or more significant (more than 10%) conformers, all of them participated in the molecular modeling, carried out with the molecular dynamics method, in respective proportions. At the final stage of the modeling, only the lowest energy conformers of the OCAs participated in the calculation of energies of intermolecular modeling with quantum chemistry methods.

As is seen from Figure 6.1, the most favorable conformations in an isolated state for all the given compounds are those

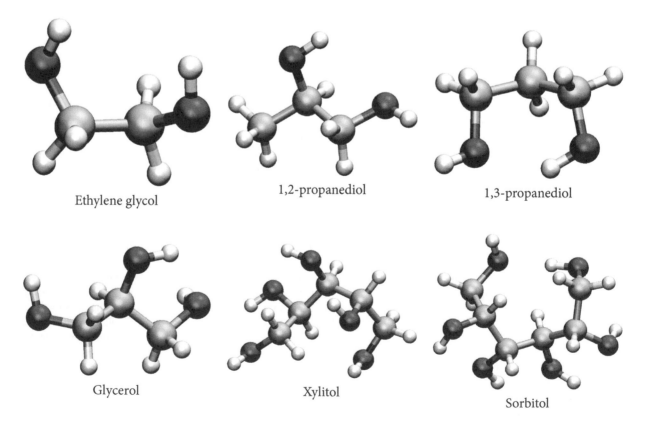

Ethylene glycol   1,2-propanediol   1,3-propanediol

Glycerol   Xylitol   Sorbitol

FIGURE 6.1 Spatial configurations of the lowest energy conformers of the clearing agents, calculated with the method DFT/B3LYP/6-311 + G (d, p).

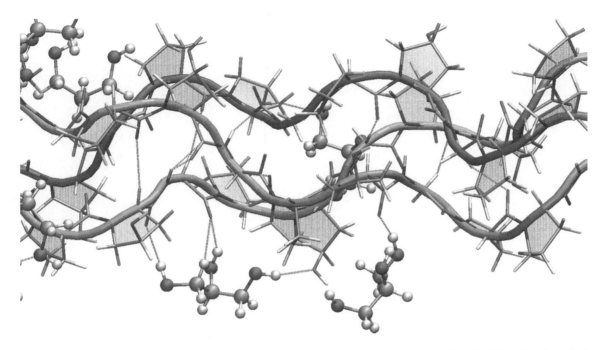

FIGURE 6.2 Example of interaction of collagen peptide ((GPH)9)3 with sorbitol molecules without regard to the water environment in the method of classical molecular dynamics.

in which hydroxyl groups are as close to each other as possible, which leads to the formation of intramolecular hydrogen bonds.

The second stage of the modeling included measurement of the average time the molecules of the OCAs were in a state of hydrogen bond with a peptide molecule. The time in a bonded state means the sum of times of life of all the hydrogen bonds that were formed between low-molecular agents and collagen in the time interval of 100 ps. It was assumed that a hydrogen bond is formed between atoms if the following geometric criteria are observed: $R \leq 3.5$ Å [19] and $\varphi \leq 30°$, where R is the distance between the "donor" atom A, covalently bonded with hydrogen atom H and "acceptor" atom B of another molecule (or a functional group of the same molecule); and φ is the angle formed by bonds AH and AB. Figure 6.2 presents an example of such an interaction.

Figure 6.3 shows the dependence between the maximum change in the collagen molecule volume and time of agents being in a water-bonded state for different OCAs.

As seen in Figure 6.3, an increase in the number of hydroxyl groups in the OCAs under study leads to an increase in the time of their being in a hydrogen bond state, which results in respective changes in collagen structure, which is reflected in changes in its volume. The correlation coefficient between these parameters equals 0.9. It should be noted that the data we obtained at the time of the agents bonding with collagen correlate well with similar data, obtained earlier in paper [3] with the CHARMM program.

The next subject of study was the ability of the chosen alcohols to produce an effect on the spatial structure of a model collagen in a water medium, *i.e.* in a state close to native. Figure 6.4 shows an effect of OCA concentration on collagen molecule volume, assessed with GROMACS package tools [20].

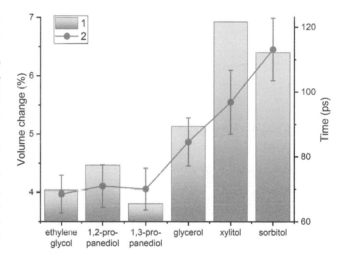

FIGURE 6.3 Dependence between the maximum change in the collagen molecule volume (1) and the time of the agents being in a water-bonded state (2) (right scale) for different types of OCAs. Vertical bars are the level of the standard error of the obtained arithmetical means.

A calculated change in collagen peptide volume is possible only with changes in the lengths of covalent bonds of its α-chains, which may serve one of the parameters describing a degree of interaction of OCAs on its structure. As is seen from Figure 6.4, the dependence of collagen molecule volume on OCA concentration is nonlinear. The maximum effect on the geometric parameters of collagen chains falls in the area of average concentrations almost for all the types of agent.

This stage of modeling also included analysis of relative probability of formation of double hydrogen bonds between alcohol molecules and peptide. For this purpose, radial distribution functions of atoms g(r), participating in hydrogen bonding, were calculated with GROMACS package tools.

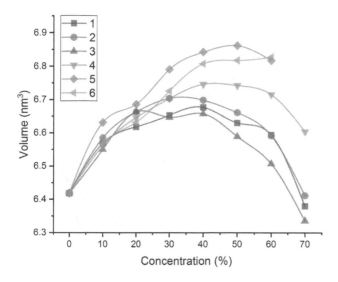

FIGURE 6.4 Dependence of the volume of a collagen peptide molecule, being in the water solution, on the concentration of different OCAs: ethylene glycol (1), 1,2-propanediol (2), 1,3-propanediol (3), glycerol (4), xylitol (5) and sorbitol(6).

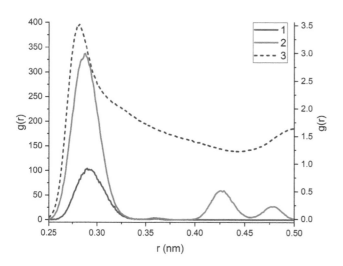

FIGURE 6.5 Radial distribution functions of oxygen atoms, participating in formation of hydrogen bonds for molecules of ethylene glycol (1), glycerol (2) and collagen peptide (3) (right scale).

Figure 6.5 shows examples of the obtained radial distribution functions for molecules of ethylene glycol and glycerol against a similar radial distribution function for collagen.

As these functions are measures of probability of detection of one atom at the distance r from the given atom, the obtained results (Figure 6.5) show that the possibilities that a glycerol molecule will form double hydrogen bonds (three hydroxyl groups) are significantly higher than for a molecule of ethylene glycol which has only two such groups. The product of the radial distribution function of an OCA on a similar function for a collagen peptide molecule will be a measure of the appearance of two pairs of atoms at a given distance from each other that are capable of forming a hydrogen bond, *i.e.* a measure of the probability of formation of two hydrogen bonds at the same time. The obtained results are presented in Figure 6.6.

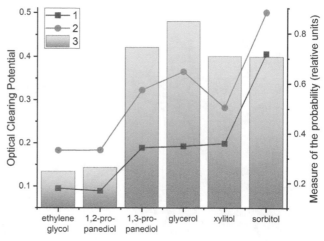

FIGURE 6.6 Dependence of the value of the OC potential of rat skin (1) and human skin (2) [1], as well as relative probability of formation of intermolecular double hydrogen bonds of a collagen peptide molecule with a molecule of an OCA (3), on the agent type. The potential of biological tissue OC was introduced in [1–3] as a rate of change of the reduced scattering coefficient after 45 min of immersion agent action to the agent concentration value, expressed in moles.

As seen in Figure 6.6, the probability values correlate well with the values of the OC potential. It is explained by the fact that for these alcohols, the prevalent type of their interaction with collagen is when a significant contribution to interaction energy is made by relatively strong double hydrogen bonds. An exception here is a sorbitol molecule, which, as will be shown later, forms a lot of relatively strong hydrogen bonds with collagen, and for this reason it is not enough to take into account formation of only paired hydrogen bonds, as it does not reflect the correct mode of interaction.

The third stage of the modeling included calculation of energies of intermolecular interactions of the OCAs with a fragment of the mimetic peptide of collagen ((GPH)$_2$)$_3$, the spatial structure of which is shown in Figure 6.7.

As is seen from Figure 6.7, an entry molecular pocket is a peptide region with an approximate size of 10 × 12 Å that has four functional groups available for intermolecular bonding: two carbonyl groups (one at the glycine residue-2 and the other at the hydroxyproline residue-3 of the same α-chain) and two alcohol groups, -1 and -4 at the hydroxyproline residues of different α-chains. When collagen interacts with low-molecular agents, a certain spatial adjustment of the molecular pocket takes place for formation of the maximum number of possible hydrogen bonds. For example, during the formation of a complex with 1,2-propanediol, the distance between atoms of oxygen from groups one and four is reduced to 10.8 Å, while with sorbitol it is reduced to 9.3 Å. The distance between oxygen atoms from carbonyl groups two and three changes as well – from 3.8 Å in an isolated model to 3.0 Å in complex with sorbitol.

Figure 6.8 shows the spatial structure of hydrogen-bonded complexes formed by the collagen fragment ((GPH)$_2$)$_3$ and the OCAs, which was obtained with PM6 method.

As seen in Figure 6.8a, a molecule of 1,2-propanediol has the least distance between alcohol groups of all the diatomic

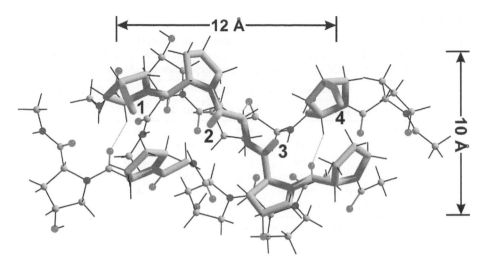

FIGURE 6.7 Spatial structure of a fragment of the mimetic peptide ((GPH)2)3 optimized within the semi-empirical method PM6. Figures stand for molecular groups participating in formation of hydrogen bonds with the OCAs. Hydrogen bonds between different α-chains are shown with dash lines.

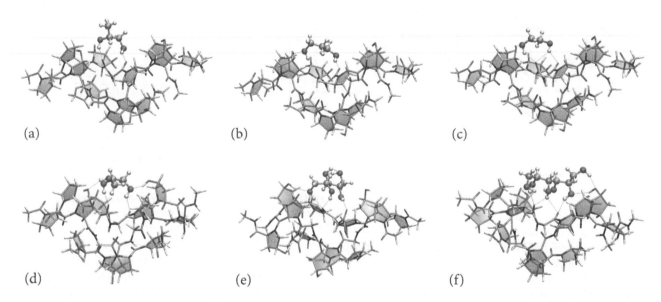

FIGURE 6.8 Structure of hydrogen-bonded complexes formed by the collagen fragment ((GPH)2)3 and immersion OCAs, which was obtained with PM6 method: 1,2-propanediol (a), 1,3-propanediol (b), ethylene glycol (c), glycerol (d), xylitol (e) and sorbitol (f).Classical intermolecular hydrogen bonds are shown with dash lines.

alcohols under study, which is why it forms two hydrogen bonds only with oxygen atoms of carbonyl groups. The calculated value of the energy of such intermolecular interaction is equal to $\Delta E = -23.4$ kJ/mol. To make the discussion of quantitative parameters of the intermolecular interactions more convenient, Table 6.1 presents the values of lengths of classical hydrogen bonds, formed, according to calculations, between active groups of the collagen molecular pocket and hydroxyl groups of the OCAs, as well as the calculated values of the energies of intermolecular interactions.

The distance between the alcohol groups of molecules of 1,3-propanediol and those of ethylene glycol is already enough to form three and four hydrogen bonds with collagen respectively; the third bond is formed between a proton of the alcohol group of the hydroxyproline residue and oxygen of the agent alcohol group, and the fourth bond is formed in ethylene glycol

between a proton of another alcohol group and oxygen of the carbonyl group of the hydroxyproline residue. Trialcohol glycerol forms four relatively strong hydrogen bonds and one weaker bond with all active groups of the entry pocket; however, the molecule length is insufficient to make all the hydrogen bonds effective, which is why the energy of bonding grows insufficiently (see Table 6.1). Pentatomic alcohol xylitol forms four hydrogen bonds with collagen, though only three active groups of the pocket are involved, and two alcohol groups of xylitol form an intramolecular hydrogen bond. The energy of such an interaction appears to be a shade more than in glycerol, which correlates well with the data on OC potential [1].

It should be noted that changing the spatial conformation of xylitol, so that all active groups of the pocket are involved (see Figure 6.9) but an intramolecular hydrogen bond is not formed, leads to an increase in the total energy of the complex in this

TABLE 6.1

Lengths of hydrogen bonds (in angstroms), energies of intermolecular interactions (in kJ/mol) between the fragments of collagen $((GPH)_2)_3$ and different OCAs, calculated with the use of the method PM6/B3LYP/6-31G(d), as well as experimental values of OC potential

| OCA | Molecular groups of collagen | | | | ΔE | OC potential [1] | |
	(OH) 1*	(CO) 2	(CO) 3	(OH) 4		Rat	Human
1,2-propanediol	-	1.86	1.79	-	–23.4	0.0892	0.1831
Ethylene glycol	-	2.15	1.85	1.89	–37.7	0.0949	0.1826
1,3-propanediol	-	1.89	1.75	1.97	–40.6	0.1887	0.3221
Glycerol	1.92	1.91	1.74; 2.44	1.93	–42.8	0.1924	0.3649
Xylitol	2.16	1.80; 1.95	1.82	-	–48.8	0.1987	0.2818
Sorbitol	1.85	1.9; 2.44	1.91; 1.75	1.86; 2.25	–80.1	0.4060	0.5001

*Note: Numbering of molecular groups corresponds to the one in Figure 6.7.

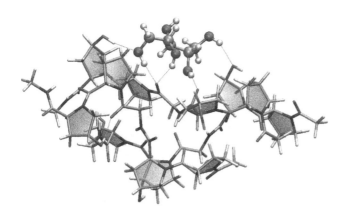

FIGURE 6.9 Spatial configuration of a high-energy intermolecular complex of collagen and xylitol, calculated with PM6 method.

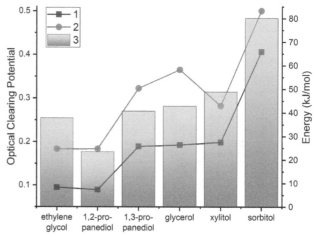

FIGURE 6.10 Dependence of the OC potential of rat skin (1) and human skin (2) [4], as well the energies of interaction between a collagen peptide molecule and a molecule of OCAs (3), on the agent type.

case and, consequently, to a decrease in the interaction energy. However, the length of hexatomic sorbitol is already enough to form seven hydrogen bonds, five of which are relatively strong, which has a positive effect on the interaction energy. This circumstance allows one to explain almost a double increase in the value of OC potential of sorbitol in comparison with xylitol, given a slight difference in the molecular structure of these compounds.

Figure 6.10 shows dependences of the OC potential of rat and human skin [1], as well as of the interaction energy calculated with PM6/B3LYP/6-31G method, on the OCA type. It is clearly seen that all the dependences correlate well with each other. The calculated correlation coefficients were 0.89 and 0.97 for human skin and rat skin, respectively. Therefore, high values of the correlation coefficients make it possible to speak about fundamental importance of a postdiffusion stage, where collagen interacts with OCAs and influences the OC of biological tissues. The results of the study allow one to suggest that during the process of this interaction, the partial substitution of collagen-related water occurs. It leads to disturbance in a binding net of hydrogen bonds and, consequently, to a reversible process of collagen fibril dissociation, which, in its turn, decreases their index of refraction and equalizes it to the surrounding interstitial medium. The higher collagen affinity to an OCA, the more effective the process is. We calculated the coefficient of numerical correspondence between the energy

of intermolecular interaction of an OCA with collagen and the potential of OC for human and rat skin, which were equal to 0.0069 and 0.0041, respectively. In order to check the predicted possibilities of the determined correlation, we calculated the energy of intermolecular interaction of a dextrose molecule with the chosen model of collagen, which was –94.5 kJ/mol. The predicted value of the dextrose OC potential was 0.388 for rat skin, which conforms well with the experimental data of 0.387 [2].

Immersion agents (sugars)

Along with alcohols, various mono- and disaccharides are used as immersion agents for optical illumination of biotissues [2, 21]. In order to assess the efficiency of OC with the use of several monosaccharides (ribose, glucose, and fructose), experimental data on *in vivo* OC of human skin were obtained [5]. Aqueous solutions (60%) of the sugars were used for the measurements, as well as glycerol – a triatomic alcohol – for comparison.

Thorlabs OCP930SR (Thorlabs, USA), an optical coherence tomography (OCT) system, was used to determine OC properties of the OCAs on skin. The measurements were carried out on a skin area of the volar side of the forearm. OCT scans were recorded before the exposure with the OCA, then at 1-min intervals during 40 min of exposure. The measurements involved four volunteers, and five measurements were carried out in total for each OCA.

The efficiency of OC was determined as the scattering coefficient magnitude, obtained with the use of the averaged A-scan in the dermal region with the depth of 350–700 μm. Figure 6.11 presents temporal dependences of the obtained scattering coefficients under skin exposure to OCAs. As is seen in Figure 6.11, the temporal dependence of OC is nonlinear on the long time interval and is described well with an exponential regression model (the determination coefficient R^2 is within the range from 60%–90%). However, at a short time segment, this dependence also can be described well within a simpler linear regression model. The interval from 7 to 24 min appeared to be the best time segment for the dermal region, chosen within the framework of Ref. [5]. The determined coefficient R^2 for the linear regression model is within the range from 50% to 80%, which makes this model acceptable to describe the obtained dependence.

We used the values of the modulus of the average rate of change of the scattering coefficient over the selected time interval as a numerical expression of the skin OC efficiency. These rates are represented in the graphs as a coefficient for variable x (slope) in the regression equations.

Spatial configurations of monosaccharides, shown in Figure 6.12, were used in the modeling of intermolecular interaction with the collagen fragment.

The next stage of the modeling included calculation of the energy of intermolecular interaction of monosaccharides with the fragment of the collagen mimetic peptide $((GPH)_2)_3$, the spatial structure of which is shown in Figure 6.13.

To make discussion of the obtained results more convenient, the qualitative parameters of intermolecular interactions (the values of length of classical hydrogen bonds, formed according to calculations between active groups of the collagen molecular pocket and hydroxyl groups of OCAs, and calculated values of intermolecular interaction energies), as well as the module of average slope of scattering coefficient temporal dependence, experimentally obtained using OCT, are presented in Table 6.2. Experimental data on the efficiency of OC (slope of regression line) and the energy of triatomic glycerol obtained [5], are also presented in Table 6.2 for comparison.

As can be seen from Table 6.2, a molecule of trialcohol glycerol forms four relatively strong hydrogen bonds and one weaker bond with all active groups of the entry pocket; however, the molecule length is insufficient to make all the hydrogen bonds maximally efficient. The transition from trialcohol glycerol to glucose monosaccharide demonstrates a significant increase in the skin OC efficiency. It can be explained by the fact that, according to the calculation, glucose forms stronger hydrogen bonds with collagen than glycerol does. Table 6.1

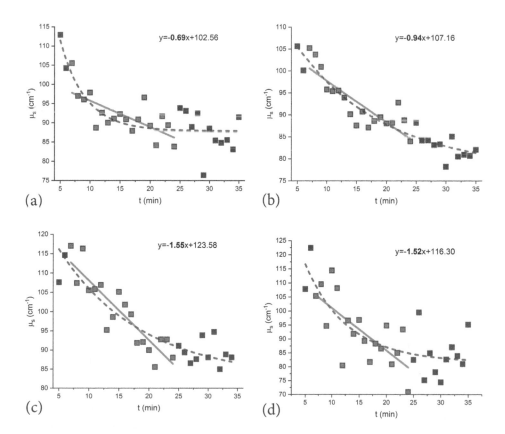

FIGURE 6.11 Temporal dependences of the scattering coefficient μs estimated in the skin dermal region of volunteers within the depth of 350 to 700 μm, based on the analysis of the depth distribution of the averaged OCT signal with the use of the single scattering model, on the OCA action: (a) – glycerol; (b) – ribose; (c) – glucose and (d) – fructose. Exponential approximation of this dependence is marked with a dash line. The linear approximation at the time segment from 7 to 24 min is shown with a solid line (the data are marked in red squares).

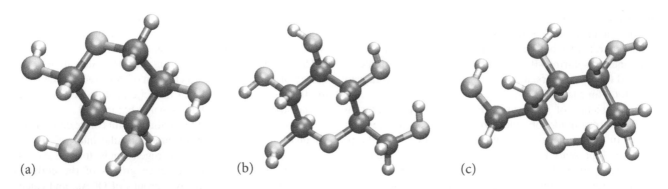

FIGURE 6.12   Spatial configurations of the lowest energy conformers of the monosaccharides, calculated with the method DFT/B3LYP/6-311 + G(d, p).

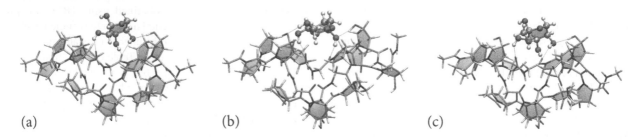

FIGURE 6.13   Structure of hydrogen-bonded complexes, formed by the collagen fragment ((GPH)2)3 and OCAs–ribose (a), glucose (b), and fructose (c), which were calculated with PM6method. Dash lines mark classical intermolecular hydrogen bonds.

TABLE 6.2

Lengths of hydrogen bonds (in angstroms), energies of intermolecular interactions ΔE (in kJ/mol) between the fragments of collagen ((GPH)$_2$)$_3$ and different OCAs, calculated with the use of the PM6/B3LYP/6-31G(d) method, as well as experimental data for OC efficiency of human skin *in vivo* and their standard errors [5]

OCA	Hydrogen bond lengths, Å	ΔE, kJ/mol	Efficiency of OC
Ribose	1.84; 1.90; 1.91; 1.95	−80.9	0.94 ± 0.15
Glucose	1.68; 1.71; 1.84; 1.94	−94.5	1.55 ± 0.20
Fructose	1.82; 1.84; 1.90; 1.96	−89.2	1.52 ± 0.37
Glycerol	1.74; 1.91; 1.92; 1.93; 2.44	−42.8	0.69 ± 0.17

shows that, despite fewer hydrogen bonds, their length is notably shorter, which is a determining factor for bonding energy. As is seen from Figure 6.3e, stronger hydrogen bonds with carbonyl groups are formed due both to a compact ring structure of glucose that enables its lowering into a collagen molecular pocket and to good mutual disposition of interacting groups.

Various parameters are used by different authors to assess the degree (efficiency) of OC with immersion OCAs. For example, in [1–3] they use the OC potential, introduced as a slope of the dependence of reduced scattering coefficient after 45 min of immersion agent effect on its concentration, expressed in moles, *i.e.* for its determination, it is necessary to study the effect of OCAs with different initial concentrations.

In this paper [5], the module of average slope of scattering coefficient change under effect of the water solution of an OCA with a moderate initial concentration (60%) was used for numerical expression of the skin OC efficiency.

As seen in Figure 6.14, these two parameters correlate well with each other. It allows one to use the module of average slope of scattering coefficient temporal dependence, experimentally

obtained using OCT, in further research as a means to evaluate the efficiency of OC with immersion OCAs.

Figure 6.14 also shows that the energies of interaction between a collagen peptide molecule and molecules of different OCAs, calculated with the use of the method PM6/B3LYP/6-31G (the values for alcohols are taken from [4]), correlate well both with the OC potential of rat and human skin [1, 2] and with the experimentally obtained average slope of scattering coefficient temporal dependence. The values of linear correlation coefficients are 0.94 and 0.88 respectively. It is therefore possible to speak of importance of a postdiffusion stage, where collagen interacts with OCAs providing tissue OC.

Assessment of OCA Concentration on spatial configuration of the collagen microfibril fragment

At this stage of the modeling, a collagen mimetic peptide ((GPH)$_9$)$_3$ was used as a molecular model of collagen, as well a

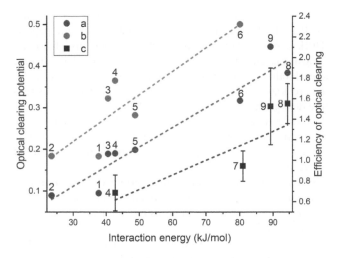

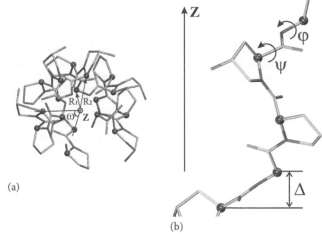

(a)

(b)

FIGURE 6.14 Dependence of the OC potential (left axis) of rat (a) and human (b) skin [1–2,4], as well as the efficiency of OC (right axis) of the human skin [5](c) on the energy of interaction between a collagen peptide molecule and molecules of OCAs. Immersion agents are indicated with numbers: 1 – ethylene glycol; 2 – 1,2-propanediol; 3 – 1,3-propanediol; 4 – glycerol; 5 – xylitol; 6 – sorbitol; 7 –ribose; 8 – glucose, and 9 – fructose. Linear approximations of these dependences are marked with dashlines.

FIGURE 6.16 Geometric parameters of α-chains of collagen peptide. A single α-chain is shown in frontal (a) and side (b) views.

fragment of collagen microfibril (Figure 6.15), an ensemble of five mimetic peptides ((GPH)$_{12}$)$_3$ [22]. Trialcohol glycerol was considered as an OCA for which experimental data on OC of human skin depending on glycerol concentration in water are available elsewhere [23, 24].

Methods of classical molecular dynamics were used to study the effect of glycerol water solution on geometric parameters of α-chains of collagen peptide (Figure 6.16). A modeling scene was a 3D cell having the form of a rectangular parallelepiped with the side lengths of 3 × 9 ×3 nm for peptide ((GPH)$_9$)$_3$ and 5 × 13 × 5 nm for peptide 5((GPH)$_{12}$)$_3$ with a respective collagen peptide put in the center. The rest of the space was filled with glycerol water solution, the concentration of which changed from 0% to 60% in increments of 10%.

At the second stage of the modeling, the following geometric parameters of peptide α-chains were chosen for analysis of the OCA effect on tertiary and quaternary structures of collagen (Figure 6.16): spiral length L and its radius R (calculated as an average distance from axis Z to C_α–chain atoms); shift

Δ and spiral rotation ω (the distance in the projection of the spiral axis from one residue to another and the angle between them); angles φ and ψ (characterizing the rotation about ordinary bonds C_α–N and C_α–C, respectively); value of dihedral angle θ (formed by two semiplanes $C_{\alpha 1}$–N–C and N–C–$C_{\alpha 2}$). Changes in the distance between different chains of collagen peptide (by changes in the sum of distances between C_α atoms of identical residues of different α-chains) were evaluated as well. The tertiary structure of collagen peptide is stabilized by intrapeptide hydrogen bonds, and its quaternary structure, in addition, by hydrogen bonds formed between the tertiary structures. The average time of amino-acid residues being in a hydrogen-bonded state may be used as an additional parameter to describe the system stability. Analysis of changes in all the above-mentioned parameters allows one to have a better understanding of the immersion agent effect on collagen structure. However, such a parameter as molecule volume will be the best to connect changes of collagen peptide structure under the OCA effect and tissue OC, because these are changes in physical dimensions of scatterers, along with changes in their shape and refraction index, that have the maximum effect on tissue optical properties.

All the above-mentioned parameters were evaluated with the use of standard procedures, provided in GROMACS package:

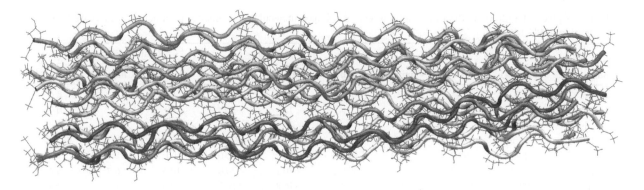

FIGURE 6.15 Spatial configuration of collagen peptide 5((GPH)12)3.

geometric parameters of α-chains, a distance between them and collagen molecule volume were evaluated with the procedures gmx_helix, gmx_distance, and gmx_freevolume [20], respectively. Hydrogen bonds were analyzed with the procedure hydrogen_bonds of VMD package. It was assumed that a hydrogen bond is formed between atoms if the following geometric criteria are observed: $R \leq 3.5$ Å [19] and $\varphi \leq 30°$, where R is the distance between the "donor" atom A, covalently bonded with hydrogen atom H, and the "acceptor" atom B of another molecule (or a functional group of the same molecule), and γ is the angle formed by bonds AH and AB.

The geometric parameters of three α-chains of peptide $(GPH_9)_3$, obtained with molecular modeling, are shown in Table 6.3. As seen in Table 6.3, the addition of glycerol to the water medium has the strongest effect on the geometric parameters of α-chains, characterizing the rotation of amino-acid residues relative to each other – angles φ and ψ. By contrast, the geometric parameters of α-chains, characterizing the displacement of amino-acid residues in the chain relative to each other, change to a much lesser extent – L and Δ. This is explained by the fact that when hydrogen bonds are formed during the interaction of collagen peptide with glycerol molecules, it is more favorable (in terms of energy) to turn a group of atoms, directly participating in the interaction, than to displace it. As all the three α-chains of collagen are spiraled in a single helix, such parameters as R and ω reflect changes not only in a certain chain but also in the whole tertiary structure of the collagen molecule in general. All this makes them more sensitive to all changes in the spatial structure of a collagen molecule.

Analysis of the distances between different chains using the example of collagen peptide $(GPH_9)_3$ showed that addition of an immersion agent to the system does not lead to significant changes in them and, consequently, to weakening of intrapeptide hydrogen bonds. Exceptions are the least stable regions of the molecule, located on their terminals.

Similar results, obtained for three α-chains of peptide $(GPH_{12})_3$ from the collagen microfibril fragment, are given in Table 6.4.

Comparative analysis of the results, presented in Table 6.3, with the similar data from Table 6.4 shows that this structure is more resistant to external effects. Geometric parameters of α-chains, characterizing rotation of amino-acid residues in the

chain relative to each other, are between three and four times smaller (in absolute terms) than similar values for peptide $(GPH_9)_3$. While geometric parameters of α-chains, characterizing displacement of amino-acid residues in the chain relative to each other, do not change significantly, as was the case for the previous results of the modeling. As this molecule structure demonstrates spatial structure elements of a higher order (five triple helices $(GPH_{12})_3$ are also spiraled in a single helix), such parameters as R and ω are exposed to all changes not only in the tertiary structure but also in the quaternary structure of collagen.

Besides, in molecular modeling, periods of time of amino-acid residues being in a hydrogen-bonded state with each other were used for comparative analysis of stability of the collagen peptides (Table 6.5).

As seen in Table 6.5, the average time of an ensemble of five collagen peptides being in a hydrogen-bonded state is more than twice as much as in a single peptide. It is explained by the fact that hydrogen bonds of peptide $5(GPH_{12})_3$ are formed both between α-chains inside one peptide and between side chains of different peptides. The table also shows that no unique dependence between the time of peptides being in a hydrogen-bonded state and glycerol concentration is traced. In addition, this parameter does not take into account the hydrogen-bond force.

Changes in protein volume are the most informative parameter for determining correlation between interaction of immersion agents with collagen protein and OC efficiency. At present, a large amount of experimental data has been accumulated on the assessment of the efficiency of tissue OC with glycerol [23, 24]. It also should be noted that different authors still use different parameters for the assessment of OC efficiency. For example, in Ref. [23], optical clearing efficiency factor is used as such a parameter and was calculated using the formula:

$$OC_{eff} = \frac{\mu_t(t=0) - \mu_t(\min)}{\mu_t(t=0)}, \quad (6.1)$$

where μ_t is the attenuation coefficient calculated from the measured time dependence of the collimated transmission using the Bouguer-Lambert law; and $\mu_t(min)$ is the minimal value of the attenuation coefficient. While in reference [24], authors

TABLE 6.3

Effect of glycerol addition to the water environment of collagen molecule on geometric parameters of three α-chains of peptide $(GPH_9)_3$

α-chain parameter	0% glycerol			40% glycerol			Relative difference, %		
	Chain #1	Chain #2	Chain #3	Chain #1	Chain #2	Chain #3	Chain #1	Chain #2	Chain #3
R, nm	0.3	0.4	0.4	0.3	0.3	0.4	7.0	−8.1	8.6
L, nm	7.3	7.5	7.2	7.4	7.1	7.2	1.0	−4.5	0.7
Δ, nm	0.3	0.3	0.3	0.3	0.3	0.3	0.8	−0.9	0.8
ω, deg.	25.1	19.3	22.4	24.5	26.4	17.4	−2.3	36.5	−22.4
θ, deg.	−89.3	−90.2	−88.9	−90.2	−92.1	−90.0	1.1	2.1	1.3
φ, deg.	−63.5	−65.7	−65.7	−70.2	−70.1	−70.0	10.5	6.7	6.4
ψ, deg.	97.9	114.8	99.5	119.8	116.8	118.2	22.3	1.8	18.8

TABLE 6.4

Effect of glycerol addition to the water environment of collagen molecule on geometric parameters of three α-chains of the microfibril fragment – peptide $5(GPH_{12})_3$

α-chain parameter	0% glycerol			40% glycerol			Relative difference, %		
	Chain #1	Chain #2	Chain #3	Chain #1	Chain #2	Chain #3	Chain #1	Chain #2	Chain #3
R, nm	0.3	0.4	0.5	0.3	0.5	0.4	2.0	28.9	−7.2
L, nm	10.5	10.3	10.4	10.5	10.4	10.4	0.0	0.9	0.3
Δ, nm	0.3	0.3	0.3	0.3	0.3	0.3	−0.1	1.0	0.5
ω, deg.	21.0	16.4	13.3	17.1	11.5	11.4	−18.8	−29.8	−14.1
θ, deg.	−95.9	−93.8	−95.2	−94.5	−95.2	−94.2	−1.5	1.5	−1.0
φ, deg.	−72.4	−67.7	−68.9	−71.1	−65.3	−69.6	−1.8	−3.7	0.9
ψ, deg.	125.0	133.4	133.8	134.1	130.3	130.6	7.3	−2.3	−2.4

TABLE 6.5

Average time (in % of the total simulation time) of peptide α-chains of collagen being in a hydrogen-bonded state for one amino-acid residue

Glycerol concentration, %	Average time, %	
	Peptide $(GPH_9)_3$	Peptide $5(GPH_{12})_3$
0	9.9	22.3
10	8.8	21.4
20	11.0	19.8
30	10.6	19.8
40	8.2	22.6
50	8.5	20.5
60	8.9	24.2

assessed the efficiency of OC by a fraction of light passing through tissue, and the increase of penetration depth in OCT measurements. Within the framework of the molecular modeling, we studied the dependence of changes in collagen peptide volume on glycerol concentration (Figure 6.17).

As seen in Figure 6.17, the dependence of the change in collagen peptide volume on glycerol concentration is nonlinear: it has a typical maximum in the area of average concentrations (40%). Analysis of the experimental data on rat [23] and porcine [24] skin *ex vivo* OC with glycerol shows that the OC efficiency reaches its maximum under higher concentrations of the OCA (60%–70%). This difference can be explained by the fact that during the molecular modeling of this interaction, glycerol concentration is already determined inside the intramolecular space, while experimental data contain glycerol concentration before application to the skin. As there is a certain amount of interstitial fluid inside the tissue, one can expect that glycerol concentration inside the tissue will be less than the applied one.

The presence of the maximum collagen molecule volume change for 40% glycerol can be explained as follows. Firstly, the collagen surface has a finite number of seats – "molecular pockets," suitable for effective addition of glycerol molecules. Secondly, as the molecular modeling showed, an increase in glycerol concentration also increases the probability of formation of hydrogen-bonded self-assemblies (Figure 6.18). After a while, these self-assemblies form cluster structures, which

leads to an uneven effect on different regions of collagen molecules. In its turn, it reduces the efficiency of hydration shell destruction around collagen and thus decreases the degree of glycerol influence on collagen.

Based on the conducted modeling, the mechanism of collagen fibrils swelling under the glycerol effect can be presented as follows: glycerol molecules, having high affinity to collagen, partially displace the bonded water during their addition. It disturbs a hydrogen-bond net between collagen fibrils and leads to fibrillary protein swelling. Disturbance in a hydrogen-bond net happens because glycerol molecules, bonded with collagen through alcohol groups, stick out with their hydrophobic part (CH_2-groups) and prevent the formation of new hydrogen bonds.

Summary

The mechanism of the molecular interaction of collagen with various immersion OCAs was investigated, and as a result of the study, a numerical correlation was established between the energy parameters of interaction and the experimentally established parameters of the efficiency of OC of tissues. The experimentally observed dependence of the efficiency of OC on the concentration of OCAs is theoretically explained. The results obtained allow, without carrying out a number of costly *in vivo* and *in vitro* studies, an improvement in the reliability of the theoretical prediction of the parameters of tissue OC using

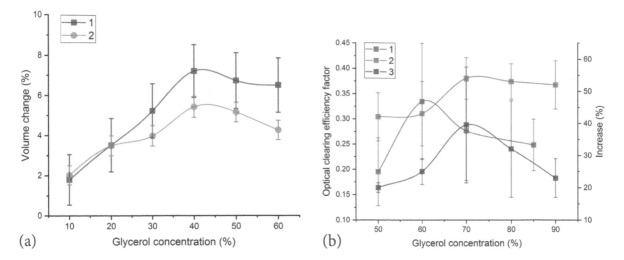

FIGURE 6.17 Dependence of collagen peptide volume change on glycerol concentration (a): 1 – peptide (GPH9)3, 2 – peptide 5(GPH12)3. Experimental data on the dependence of the OC efficiency on glycerol concentration (b): 1 – optical clearing efficiency factor (left axis) in the region 600–700 nm for rat skin *ex vivo* [23], 2 and 3 – percent increase (right axis) of light transmittance (615 nm) and light penetration depth in OCT measurements for porcine skin ex vivo [24].

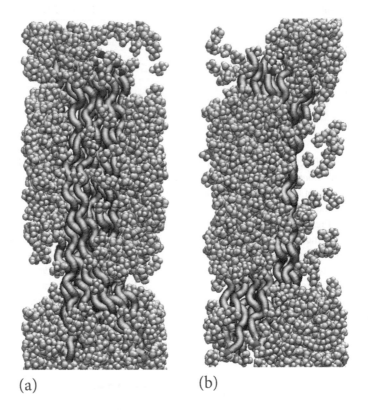

FIGURE 6.18 Spatial configuration of glycerol molecules around collagen peptide 5(GPH12)3 at the 5th ns of the observation in two different side views.

various immersion agents. This opens the way for the selection and synthesis of new agents with a predicted efficiency of OC.

Acknowledgments

V. V. Tuchin was supported by the Government of the Russian Federation (grant no. 075-15-2021-615).

REFERENCES

1. J. Hirshburg, B. Choi, J.S. Nelson and A.T. Yeh, "Collagen solubility correlates with skin optical clearing," *J. Biomed. Opt.* **11**(4), 040501 (2006). Doi: 10.1117/1.2220527
2. J. Hirshburg, B. Choi, J.S. Nelson and A.T. Yeh, "Correlation between collagen solubility and skin optical clearing using sugars," *Lasers Surg Med.* **39**(2), 140–144 (2007). Doi: 10.1002/lsm.20417

3. J.M. Hirshburg, K.M. Ravikumar, W. Hwang and A.T. Yeh, "Molecular basis for optical clearing of collagenous tissues," *J. Biomed. Opt.* **15**(5), 055002 (2010). Doi: 10.1117/1.3484748

4. K.V. Berezin, K.N. Dvoretski, M.L. Chernavina et al., "Molecular modeling of immersion optical clearing of biological tissues," *J. Mol. Model.* **24**(2), 45 (2018). Doi: 10.1007/s00894-018-3584-0

5. K.V. Berezin, K.N. Dvoretski, M.L. Chernavina et al., "Optical clearing of human skin using some monosaccharides in vivo," *Opt. Spectrosc.* **127**(2), 352–358 (2019). Doi: 10.1134/S0030400X19080071

6. K.N. Dvoretski, K.V. Berezin, M.L. Chernavina et al., "Molecular modeling of the post-diffusion stage of surface biotissue layers immersion optical clearing," *J. Synch. Investig.* **12**(5), 961–967 (2018). Doi: 10.1134/S1027451018050233

7. K. Okuyama, K. Miyama, K. Mizuno and H.P. Bachinger, "Crystal structure of (Gly-pro-Hyp)9: implications for the collagen molecular model," *Biopolymers* **97**(8), 607–616 (2012). Doi: 10.1002/bip.22048

8. W.D. Cornell, P. Cieplak, C.I. Bayly, et al., "A second generation force field for the simulation of proteins, nucleic acids, and organic molecules," *J. Am. Chem. Soc.* **117**(19), 5179–5197 (1995). Doi: 10.1021/ja00124a002

9. A.D. Becke, "Density-functional thermochemistry. III. The role of exact exchange," *J. Chem. Phys.* **98**(7), 5648–5652 (1993). Doi: 10.1063/1.464913

10. C. Lee, W. Yang and R.G. Parr, "Development of the Colle-Salvetti correlation-energy formula into a functional of the electron density," *Phys. Rev.* **37B**(2), 785–789 (1988). 10.1103/PhysRevB.37.785

11. M.J. Frisch, G.W. Trucks, H.B. Schlegel et al., *Gaussian03, Revision B.03*, Gaussian Inc., Pittsburgh PA (2003).

12. D. van der Spoel, E. Lindahl , B. Hess, G. Groenhof, A.E. Mark and H.J. Berendsen, "GROMACS: fast, flexible, and free," *J. Comput. Chem.* **26**(16), 1701–1718 (2005). Doi: 10.1002/jcc.20291

13. Y. Duan, C. Wu, S. Chowdhury et al., "A point-charge force field for molecular mechanics simulations of proteins based on condensed-phase quantum mechanical calculations," *J. Comput. Chem.* **24**(16), 1999–2012 (2003). Doi: 10.1002/jcc.10349

14. H.J.C. Berendsen, J.P.M. Postma, W.F. van Gunsteren, A. DiNola and J.R. Haak, "Molecular dynamics with coupling to an external bath," *J. Chem. Phys.* **81**(8), 3684–3690 (1984). Doi: 10.1063/1.448118

15. W. Humphrey, A. Dalke and K. Schulten, "VMD: visual molecular dynamics," *J. Mol. Graphics* **14**(1), 33–38 (1996). Doi: 10.1016/0263-7855(96)00018-5

16. H.J.C. Berendsen, J.R. Grigera and T.P. Straatsma, "Themissing term in effective pair potentials," *J. Phys. Chem.* **91**(24), 6269–6271 (1987). Doi: 10.1021/j100308a038

17. J.J.P. Stewart, "Optimization of parameters for semiempirical methods V: modification of NDDO approximations and application to 70 elements," *J. Mol. Model.* **13**(12), 1173–1213 (2007). Doi: 10.1007/s00894-007-0233-4

18. O. Trott and A.J. Olson, "AutoDock Vina: Improving the speed and accuracy of docking with a new scoring function, efficient optimization, and multithreading," *J. Comput. Chem.* **31**(2), 455–461 (2010). Doi: 10.1002/jcc.21334

19. H.D. Loof, L. Nilsson and R. Rigler, "Molecular dynamics simulation of galanin in aqueous and nonaqueous solution," *J. Am. Chem. Soc.* **114**(11), 4028–4035 (1992). Doi: 10.1021/ja00037a002

20. A. Bondi, "van der Waals volumes and radii," *J. Phys. Chem.* **68**(3), 441–451 (1964). Doi: 10.1021/j100785a001

21. W. Feng, R. Shi, N. Ma, D.K. Tuchina, V.V. Tuchin and D. Zhu, "Skin optical clearing potential of disaccharides," *J. Biomed. Opt.* **21**(8), 081207 (2016). Doi: 10.1117/1.JBO.21.8.081207

22. J.M. Chen, C.E. Kung, S.H. Feairheller and E.M. Brown, "An energetic evaluation of a 'Smith' collagen microfibril model," *J. Protein Chem.* **10**(5), 535–552 (1991). Doi: 10.1007/BF01025482

23. V.D. Genin, D.K. Tuchina, A.J. Sadeq, E.A. Genina, V.V. Tuchin and A.N. Bashkatov, "Ex vivo investigation of glycerol diffusion in skin tissue," *J. Biomed. Photon. Eng.* **2**(1), 010303 (2016). Doi: 10.18287/JBPE16.02.010303

24. E. Youn, T. Son, H.-S. Kim and B. Jung, "Determination of optimal glycerol concentration for optical tissue clearing," *Proc. SPIE* **8207**, 82070J (2012). Doi: 10.1117/12.909790

7

Refractive index measurements of tissue and blood components and OCAs in a wide spectral range

Ekaterina N. Lazareva, Luís Oliveira, Irina Yu. Yanina, Nikita V. Chernomyrdin, Guzel R. Musina,
Daria K. Tuchina, Alexey N. Bashkatov, Kirill I. Zaytsev, and Valery V. Tuchin

CONTENTS

Introduction

The evaluation of the refractive index (RI) of tissues and their components is highly important for various applications, especially for the study of optical clearing (OC) treatments. Since the 19th century, the most common equipment used to measure the RI of materials is the Abbe refractometer. This piece of equipment allows one to obtain the RI of solid and liquid samples at 589 nm [1].

With the recent development of biophotonics methods, the measurement of the RI of biological tissues and their components at a single wavelength is not sufficient, since it depends strongly on the wavelength [1]. New methods and equipment to quantify RI in a wide wavelength range need to be developed. Considering the technique of OC of biological tissues and fluids through the immersion method, where one of the main mechanisms is the RI matching (see Chap. 1), the evaluation of the RI of the optical clearing agents (OCAs) is also necessary [2, 3].

At present, the evaluation of the RI at various wavelengths is possible through different methods, which allows for interpolation these discrete experimental data in a wide spectral range [1, 4–6, 7]. An alternative indirect method based on the Kramers-Kronig relations [8] allows one to calculate the wavelength dependence for biological materials and OCAs directly from the wavelength dependence of the absorption coefficient ($\mu_a(\lambda)$) reconstructed from the spectral intensity measurements [9]. During OC, it is also of interest to evaluate the efficiency of the RI matching mechanism, meaning that if RI kinetics of

the interstitial fluid (ISF) and of the tissue sample as a whole can be estimated, then evaluation of the efficiency is possible [10–11].

The following sections describe such techniques and present the most complete data for RI wavelength dependences of OCAs, biological tissues, and their components.

Optical coherence tomography

Optical coherence tomography (OCT) is a noncontact imaging technique which provides 3D imaging within optical scattering media (e.g., biological tissues) with a high microscopic resolution [12–14]. Typically, OCT systems have a resolution of $5 \div 20$ μm [14]. The introduction of spectral domain OCT (SD-OCT) was able to overcome the limitations of time-domain (TD-OCT). Image quality and imaging speed were significantly improved by SD-OCT, which is able to capture the whole depth information simultaneously.

As OCT measures optical delays, all axial distances are optical distances. To achieve scaling in geometrical distances to allow for instance thickness measurements, the RI n of the medium needs to be known, and axial distances measured in OCT scans are divided by the RI n [15].

For comparison of RI measurements performed by OCT and Abbe refractometry, it should be considered that due to a broadband light source used in OCT, it measures the group RI of a material, as a single wavelength measurements of Abbe refractometry give the phase RI. The group and phase RIs are related; however, in dispersive media they are different. Simultaneous measurements of both phase and group RI and sample thickness are possible using combination of OCT and confocal microscopy [16, 17] or low-coherence interferometry at multiple angles of incidence enabling bulk RI measurement of scattering and soft samples [18].

Using OCT technology, the RIs of various biological tissues were measured. For example, fresh rat cornea *ex vivo* [17, 19] and human cornea [20], fish epithelium and stroma I and II of the cornea (see Table 7.1) [17], acute rat brain tissue [21], rat dermis (1.25 ± 0.03), dermoepidermal junction (1.13 ± 0.05) and deeper layers (1.19 ± 0.07) [22], human teeth [23,24], multicellular tumor spheroids [25], skin [26–30], adipose tissue [31, 32], articular cartilage [33, 34], muscle tissue [35], jawbone of pig [36], blood [37,38], lymph [39, 40], nerve fibers [41, 42], *etc.*

Total internal reflection method and dispersion calculation

The measurement of the RI of tissues and fluids at discrete wavelengths can be made using a total internal reflection setup [1, 4–6]. In such a setup, which is represented in Figure 7.1, different lasers with various wavelengths in a wide spectral range can be used to obtain the sample's RI at those wavelengths.

In this setup, an equilateral prism with the tissue sample attached to it is placed on a rotating stage. For fluids, such as OCAs, a plastic cuvette with one surface missing to allow the fluid to be in direct contact with the prism surface may be glued to the prism. A laser is pointed to one of the prism surfaces, and after being reflected at the prism/sample interface, the beam will exit by the third prism surface, where a detector connected to an electrical multimeter is placed to collect the signal. Several studies have been performed with this setup to measure the RI of tissues at different wavelengths [1, 4, 5, 43, 44].

To explain the measuring procedure in detail, we will consider a study made with human colorectal muscle samples [1]. In this study, three muscle samples were prepared with approximated rectangular shape (~4 × 3 cm²). One of the surfaces of the samples was flattened with a cryostat, so that it perfectly adheres to the prism surface. The dispersion prism used in this study is made of SCHOTT N-SF11 glass, and the wavelength dependence for its RI is represented in Figure 7.2 [45].

The curve presented in Figure 7.2 is described by the following Sellmeyer equation [45].

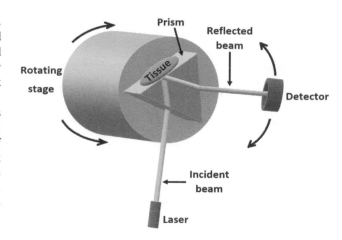

FIGURE 7.1   Total internal reflection setup for the RI measurements.

$$n^2 - 1 = \frac{K_1 \lambda^2}{\lambda^2 - L_1} + \frac{K_2 \lambda^2}{\lambda^2 - L_2} + \frac{K_3 \lambda^2}{\lambda^2 - L_3}. \quad (7.1)$$

Ref. [45] contains the Sellmeyer coefficients for this glass, which are necessary in Equation (7.1). Those values are K_1=1.7376, K_2=0.3137, K_3=1.8988, L_1=0.0132, L_2=0.0623, and L_3=155.2363. Considering Figure 7.1, and since the incident and reflected angles for the beam can only be measured outside the prism at the prism/air interfaces, Snell-Descartes equation was used to convert the angle of the incident (or reflected) beam as measured outside the prism (α) to the incident (or reflected) angle at the prism/sample interface (θ) [46]:

$$\theta = \beta - \arcsin\left[\frac{1}{n_1} \times \sin(\alpha)\right], \quad (7.2)$$

with β representing the internal angle of the prism (60 deg in this case) and n_1 representing the RI of the prism at the wavelength of the laser in use, as retrieved from Figure 7.2, or calculated with Equation (7.1).

For each set of measurements, performed with a particular laser, a reflectance curve was calculated for the prism/sample interface as [43]:

$$R(\theta) = \frac{V(\theta) - V_{\text{noise}}}{V_{\text{laser}} - V_{\text{noise}}}, \quad (7.3)$$

with the potential measured at angle θ represented by $V(\theta)$, the potential measured with background light represented by V_{noise}, and the potential measured directly from the laser represented by V_{laser}. A representation of $R(\theta)$ is made in Figure 7.3

TABLE 7.1

RIs and thickness of fish cornea at wavelength 800 nm

	Specimen 1		Specimen 2	
Layer	**Refractive index**	**Average thickness, μm**	**Refractive index**	**Average thickness, μm**
Epithelium (L1)	1.448 (0.015)	43.8	1.446 (0.011)	33.5
Stroma I (L2)	1.345 (0.002)	252.9	1.372 (0.005)	219.0
Stroma II (L3)	1.436 (0.009)	41.2	1.392 (0.012)	39.9
Overall	1.370 (0.004)	337.9	1.386 (0.005)	292.4

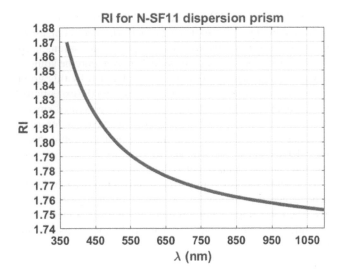

FIGURE 7.2 Wavelength dependence for the RI of the N-SF11 prism.

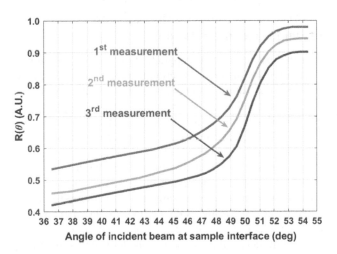

FIGURE 7.3 Reflectance curves obtained with the 668.1-nm laser from the colorectal muscle.

for the three measurements (3 samples) made with the 668.1-nm laser.

Similar curves to the ones presented in Figure 7.3 were obtained from measurements with lasers at wavelengths of 401.4, 532.5, 782.1, 820.8, and 850.5 nm [43]. To identify the critical angle of reflectance at the prism/sample interface, the first derivative of the curves presented in Figure 7.3 needed to be calculated. Such a calculation was made according to [43]:

$$deriv(\theta) = \frac{\text{Ref}(\theta_i) - \text{Ref}(\theta_{i-1})}{\theta_i - \theta_{i-1}}, \qquad (7.4)$$

where θ_i and θ_{i-1} are the consecutive angles of measurement and $\text{Ref}(\theta_i)$ and $\text{Ref}(\theta_{i-1})$ are the corresponding reflectance at those angles.

Such derivatives, which are presented in Figure 7.4, give a strong peak, whose central angle identifies the critical angle for each measurement.

The critical reflectance angles for each curve in Figure 7.4 were obtained as $\theta_{c1} = 50.07°$, $\theta_{c2} = 50.14°$, and $\theta_{c3} = 50.22°$. The

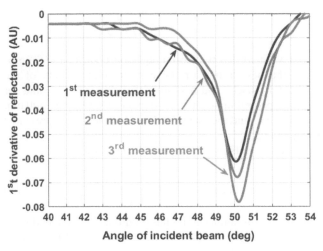

FIGURE 7.4 First derivative curves of the reflectance measurements presented in Figure 7.3.

corresponding RI values of the muscle tissue could be calculated from these values using [43]:

$$n_{\text{muscle}}(\lambda) = n_{\text{prism}}(\lambda) \times \sin(\theta_c), \qquad (7.5)$$

where n_{prism} represents the RI of the dispersion prism at the laser wavelength (1.775 at $\lambda = 668.1$ nm – see Figure 7.2). The RI values of the muscle at this wavelength, as calculated from Equation (7.5), were 1.3611, 1.3625, and 1.3641. The mean and standard deviation of these values were calculated. A similar procedure was performed with measurements made with the other lasers, and the final results are presented in Table 7.2.

The wavelength dependence for the RI of colorectal muscle was obtained by fitting the mean data in Table 7.2 with an appropriate curve using the MATLAB™ curve fitting tool. According to the literature [4, 5], the Cauchy, the Conrady, and the Cornu equations (Equations (7.6), (7.7), and (7.8)) are the most appropriate to describe such wavelength dependencies for biological tissues:

$$n_{\text{tissue}}(\lambda) = A + \frac{B}{\lambda^2} + \frac{C}{\lambda^4}, \qquad (7.6)$$

$$n_{\text{tissue}}(\lambda) = A + \frac{B}{\lambda} + \frac{C}{\lambda^{3.5}}, \qquad (7.7)$$

TABLE 7.2

Experimental RI data of human colorectal muscle tissue at different wavelengths [44]

λ, nm	Mean RI	SD
401.4	1.3801	0.0012
532.5	1.3693	0.0024
668.1	1.3626	0.0014
782.1	1.3590	0.0018
820.8	1.3575	0.0014
850.7	1.3566	0.0014

$$n_{\text{tissue}}(\lambda) = A + \frac{B}{(\lambda - C)}. \tag{7.8}$$

It was verified that the Cornu equation (Equation (7.8)) provides the best fitting for the experimental RI data of the colorectal muscle, and assuming that this curve is valid for a wider spectral range, it was extended from 850 to 1000 nm. Such a curve and the mean RI data in Table 7.2 are presented in Figure 7.5.

The curve in Figure 7.5 is described by [1]:

$$n_{\text{muscle}}(\lambda) = 1.3346 + \frac{19.48}{(\lambda - 27.17)}. \tag{7.9}$$

This method is simple and relies on direct experimental measurements. Similar studies have been performed to estimate the wavelength dependence for the RI of other tissues, such as human normal and pathological colorectal mucosa ($n_{\text{n-mucosa}}$ and $n_{\text{p-mucosa}}$) [1, 10, 117] or human normal and pathological liver ($n_{\text{n-liver}}$ and $n_{\text{p-liver}}$) [44]. The resulting dispersion curves for these tissues were obtained for the spectral range between 400 and 1000 nm as

$$n_{\text{n-mucosa}}(\lambda) = 1.315 + \frac{16.73}{(\lambda - 38.84)}, \tag{7.10}$$

$$n_{\text{p-mucosa}}(\lambda) = 1.315 + \frac{19.06}{(\lambda - 49.42)}, \tag{7.11}$$

$$n_{\text{n-liver}}(\lambda) = 1.354 + \frac{13.56}{(\lambda - 37.24)}, \tag{7.12}$$

$$n_{\text{p-liver}}(\lambda) = 1.343 + \frac{10.63}{(\lambda - 103.7)}. \tag{7.13}$$

Although this method is reliable, it does not provide continuous information for the wavelength dependence of tissue RI. To obtain continuous wavelength dependence for the RI of

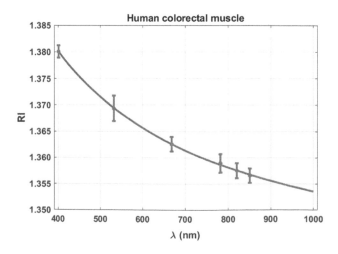

FIGURE 7.5 Experimental RI data and calculated dispersion curve for the human colorectal muscle tissue.

tissues, Deng *et al.* [47] have upgraded the experimental setup presented in Figure 7.1, so that it can perform spectral measurements. Using this lab-built apparatus, spectral measurements of the RI of various tissues, such as chicken liver, porcine adipose, bovine muscle, and hemoglobin, were obtained by this group between 400 and 750 nm [47]. In all cases, with the exception of hemoglobin, the obtained dispersion curves presented smooth decreasing behavior, which is well fitted with curves described by Equations (7.6) to (7.8). For the case of hemoglobin, peaks were detected near 418, 540, and 570 nm, which correlate with the Soret and *Q*-bands of hemoglobin [47].

This study shows the importance of continuous wavelength dependence measurements for the RI of biological materials. The following section presents an alternative method for obtaining such continuous dispersion of biological tissues and fluids.

Tissue dispersion calculation from spectral measurements through the Kramers–Kronig relations

With the exception of the upgrade described by the authors of reference 47, the total reflection method to measure the RI of tissues and fluids presents one serious limitation: it does not provide continuous but smooth-fitted wavelength dependence. Consequently, it may lack spectral information, such as absorption signatures of tissue chromophores as reported in the study of hemoglobin by Deng *et al.* [47].

An alternative method that focuses on the calculation of the RI as a sum of absorption contributions of water and other tissue components (designated as scatterers) is based on the Kramers–Kronig (KK) relations [8, 48]. This method has been widely used on various solid, liquid, and gaseous materials to derive the real part of the RI from its imaginary part [49]. The KK relations are mathematical transformations that relate the real ($n(\lambda)$) and imaginary ($\kappa(\lambda)$) parts of the RI ($n(\lambda)$) [8]:

$$n(\lambda) = n(\lambda) + i\kappa(\lambda). \tag{7.14}$$

Such transformation relations can be expressed as [8]

$$n(\lambda) - 1 = +\frac{2}{\pi}\int_0^\infty \frac{\lambda}{\Lambda}\cdot\frac{\lambda}{\Lambda^2 - \lambda^2}\kappa(\lambda)d\Lambda, \tag{7.15}$$

$$\kappa(\lambda) = +\frac{2}{\pi}\int_0^\infty \frac{\lambda}{\Lambda^2 - \lambda^2}n(\lambda)d\Lambda, \tag{7.16}$$

where Λ represents the integrating variable over a wavelength spectral range under consideration and λ is a fixed wavelength in that range that can be tuned for better adjustment of the calculated dispersion. From Equations (7.15) and (7.16), we see that $n(\lambda)$ is a transformation of $\kappa(\lambda)$, and vice-versa. Although Equations (7.15) and (7.16) allow the calculation of $n(\lambda)$ and $\kappa(\lambda)$ one from the other, alternative determination of $n(\lambda)$ for a tissue can be made if in Equation (7.15), $\kappa(\lambda)$ is replaced by the following expression [8]:

$$\kappa(\lambda) = \frac{\mu_a(\lambda)\lambda}{4\pi}, \tag{7.17}$$

where $\mu_a(\lambda)$ is the wavelength dependence for the absorption coefficient of the tissue, which can be calculated by various methods. One of the simplest methods to obtain $\mu_a(\lambda)$ consists of direct calculation from the total transmittance (T_t) and total reflectance (R_t) spectral measurements, obtained from a sample with thickness d [9]:

$$\mu_a(\lambda) = \left[1 - \left(T_t(\lambda) + R_t(\lambda)\right)\right]/d. \tag{7.18}$$

In most tissues, the water volume content is very high (>70%), but for the wavelength range between 200 and 1000 nm, water absorption is significantly low when compared to the absorption of other tissue components [8, 48]. Such fact suggests that we can write another expression for $\kappa(\lambda)$, where the water and scatterers' contributions for the imaginary part of the RI are added [48]:

$$\kappa(\lambda) = \kappa_{scat}(\lambda) + \kappa_{H_2O}(\lambda). \tag{7.19}$$

By using this relation in Equation (7.15), we can express the real part of the RI of the tissue as [8, 48]

$$n(\lambda) = 1 + \frac{2}{\pi}\int_0^\infty \frac{\lambda}{\Lambda} \cdot \frac{\lambda}{\Lambda^2 - \lambda^2}\kappa_{H_2O}(\lambda)d\Lambda + \frac{2}{\pi}\int_0^\infty \frac{\lambda}{\Lambda}\frac{\lambda}{\Lambda^2 - \lambda^2}\kappa_{scat}(\lambda)d\Lambda, \tag{7.20}$$

where the first two terms represent the RI of water, and the last term represents the contribution of tissue scatterers, which will provide information of the major absorption bands in the tissue. Consequently, Equation (7.20) can be rewritten as [48]

$$n(\lambda) = n_{H_2O} + \frac{2}{\pi}\int_0^\infty \frac{\lambda}{\Lambda}\frac{\lambda}{\Lambda^2 - \lambda^2}\kappa_{scat}(\lambda)d\Lambda. \tag{7.21}$$

Since in the spectral range between 200 and 1000 nm, the major absorption bands of tissue chromophores are identified

in $\mu_a(\lambda)$, we can replace $\kappa_{scat}(\lambda)$ in Equation (7.21) by the expression in Equation (7.17) to calculate the dispersion of a tissue.

Using this method, the real part of the RI of human normal and pathological (metastatic carcinoma) liver tissues was calculated and compared with the Cornu-type equations (see Equations (7.12) and (7.13)) previously calculated to fit experimental RI values [9]. These results are presented in Figure 7.6.

From graphs in Figure 7.6, we see a good agreement between the dispersion curves that were calculated through the KK relations and the Cornu-type curves that were obtained from experimental discrete RI data. For the normal liver, such good agreement is seen between 300 and 1000 nm, and for the pathological liver it is seen in the entire spectral range. The curve calculated by KK relations provides information about the absorption bands of hemoglobin, since as in the study by Deng *et al.* [47], we see the Soret band at 418 nm and a single band at 550 nm that corresponds to deoxygenated hemoglobin. Comparison between the magnitudes of these bands in graphs (a) and (b) of Figure 7.6 shows higher blood content in the pathological liver tissue.

These results show that the dispersion calculation through KK relations is reliable and provides continuous information on the wavelength dependence of tissue RI from the deep-ultraviolet to the near infrared. To provide such good estimation, it is necessary that some experimental RI data is available for some discrete wavelengths within the desired spectral range, since during the KK calculation process, λ in Equation (7.21) must be adjusted so that the calculated dispersion matches experimental data. Yet, such measurements are always possible through the method described, which allows one to optimize and validate the dispersion calculation through KK relations. This method can also be used to calculate the dispersion of OCAs and other fluids such as blood or ocular fluids.

After describing the methods for obtaining the dispersion of biological tissues, it is also of interest to evaluate tissue dispersion kinetics during OC treatments. The following section describes a simple method to obtain such kinetics, based only on thickness and collimated transmittance (T_c) measurements performed during treatment. RI at different wavelengths for different biological tissues are shown in Table 7.3 [5, 45, 50–65].

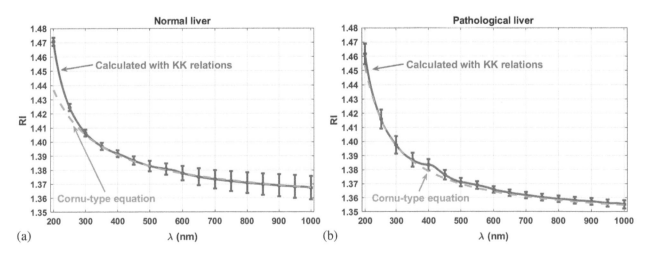

FIGURE 7.6 Wavelength dependence for the RI, using Cornu-type equation and estimated through KK relations for human normal (a) and pathological liver (b).

TABLE 7.3

RI of biological tissues

Tissue	λ, nm	Refractive index	Notes (sample, method)	Ref.
Human skin	980	1.5(0.02)	*In vitro*, human stratum corneum of the dorsal surface of the thumb	[50]
	300–780	1.419(0.002)	*In vivo*, the human stratum corneum, ellipsometry	[51]
	325	1.492	*In vitro*, the human skin epidermis, method of coherent reflectance curve measurement $$n(\lambda) = 1.41188 + \frac{9091.11}{\lambda^2} - \frac{7.64945 \times 10^7}{\lambda^4}, \lambda \text{ in nm}$$	[4]
	442	1.451		
	532	1.449		
	633	1.435		
	850	1.418		
	1064	1.432		
	1310	1.424		
	1557	1.403		
	325	1.403	*In vitro*, the human skin dermis method of coherent reflectance curve measurement $$n(\lambda) = 1.36578 + \frac{9463.72}{\lambda^2} - \frac{5.93644 \times 10^8}{\lambda^4}, \lambda \text{ in nm}$$	[4]
	442	1.399		
	532	1.380		
	633	1.395		
	850	1.387		
	1064	1.380		
	1310	1.362		
	1557	1.365		
	1300	1.41(0.03)	*In vitro*, human dermis, OCT	[52]
	1300	1.51(0.02)	*In vivo*, human stratum corneum, OCT	[52]
	1300	1.34(0.01)	*In vivo*, human epidermis, OCT	[52]
Porcine skin	325	1.413	*In vitro*, the porcine skin epidermis (s-polarization), the method of measuring the critical angle of total reflection	[53]
	442	1.379		
	532	1.407		
	633	1.389		
	850	1.402		
	1064	1.419		
	1310	1.386		
	1557	1.372		
	325	1.392	*In vitro*, the porcine skin epidermis (p-polarization), the method of measuring the critical angle of total reflection	[53]
	442	1.355		
	532	1.393		
	633	1.383		
	850	1.401		
	1064	1.420		
	1310	1.384		
	1557	1.364		
	325	1.392	*In vitro*, the porcine skin epidermis (p-polarization), the method of measuring the critical angle of total reflection	[53]
	442	1.355		
	532	1.393		
	633	1.383		
	850	1.401		
	1064	1.420		
	1310	1.384		
	1557	1.364		
	325	1.393	*In vitro*, the porcine skin dermis (s-polarization), the method of measuring the critical angle of total reflection	[53]
	442	1.374		
	532	1.359		
	633	1.355		
	850	1.364		
	1064	1.360		
	1310	1.360		
	1557	1.362		

(Continued)

TABLE 7.3 (CONTINUED)

RI of biological tissues

Tissue	λ, nm	Refractive index	Notes (sample, method)	Ref.
	325	1.393	*In vitro*, the porcine skin dermis (p-polarization), the method	[53]
	442	1.378	of measuring the critical angle of total reflection	
	532	1.360		
	633	1.354		
	850	1.364		
	1064	1.360		
	1310	1.359		
	1557	1.361		
Chicken skin	1310	1.424(0.022)	*In vitro*, the chicken skin, OCT	[54]
Porcine adipose	650	1.382(0.006)	*In vitro*, measuring the critical angle of total reflection	[55]
	488	1.510(0.002)	*In vitro*, measuring the critical angle of total reflection	[56]
	632.8	1.492(0.003)	In the spectral range:	
	1079.5	1.482(0.002)		
	1341.4	1.478(0.004)	$n(\lambda) = 1.4753 + \dfrac{4.3902 \times 10^{-3}}{\lambda^2} + \dfrac{9.2385 \times 10^{-4}}{\lambda^4}$, [λ] in μm	
	432		*In vitro*, measuring the critical angle of total reflection	[57]
	532.1		In the spectral range (Cauchy equation):	
	632.9			
	732.2		$n(\lambda) = 1.45978 + \dfrac{2.83095 \times 10^3}{\lambda^2} + \dfrac{1.53151 \times 10^8}{\lambda^4}$ or (Cornu	

equation):

$$n(\lambda) = 1.45619 + \frac{4.71521}{\lambda - 228.085}$$

or (Condrary equation):

$$n(\lambda) = 1.4564 + \frac{5.54824}{\lambda} + \frac{1.69195 \times 10^7}{\lambda^{3.5}},\ [\lambda]\ \text{in nm}$$

Tissue	λ, nm	Refractive index	Notes (sample, method)	Ref.
	632.8	1.4699(0.0003)	*In vitro*, measuring the critical angle of total reflection	[45]
	632.8	1.464	*In vitro*, measuring the critical angle of total reflection	[58]
	632.8	1.4663(0.003) (s-polarized) 1.4676(0.003) (p-polarized)	*In vitro*, extended differential total reflection method	[59]
	480	1.478(0.0035)	*In vitro*, melted, Abbe refractometer	[60]
	589	1.4707(0.0036)	In the spectral range (Sellmeier equation):	
	680	1.4667(0.0036)		
	930	1.4635(0.0034)	$n^2(\lambda) = 1 + \dfrac{1.1313 \times \lambda^2}{\lambda^2 - 10348.0147} + \dfrac{0.0357 \times \lambda^2}{\lambda^2 - 0.6068}$, [λ] in nm	
	1100	1.4612(0.0034)		
	1300	1.4564(0.0033)		
	1550	1.4538(0.0025)		
	480	1.4825(0.004)	*In vitro*, slice, Abbe refractometer	[60]
	589	1.4726(0.0052)	In the spectral range (Sellmeier equation):	
	680	1.4695(0.0052)		
	930	1.4662(0.0047)	$n^2(\lambda) = 1 + \dfrac{1.1236 \times \lambda^2}{\lambda^2 - 10556.6963} + \dfrac{0.2725 \times \lambda^2}{\lambda^2 - 1.8867}$	
	1100	1.4628(0.0055)		
	1300	1.4589(0.0053)		
	1550	1.4576(0.0025)		
Human adipose	480	1.4746(0.0008)	*In vitro*, slice, human abdominal adipose, Abbe refractometer	[60]
	589	1.4677(0.0004)	In the spectral range (Sellmeier equation):	
	680	1.4632(0.0010)		
	930	1.4581(0.0011)	$n^2(\lambda) = 1 + \dfrac{1.1236 \times \lambda^2}{\lambda^2 - 10556.6963} + \dfrac{0.2725 \times \lambda^2}{\lambda^2 - 1.8867}$, [λ] in nm	
	1100	1.4545(0.0013)		
	1300	1.4499(0.0003)		
	1550	1.4454(0.0011)		
	1300	1.467(0.008)	*In vitro*, human mesenteric adipose, OCT	[52]
Rat adipose	800	1.467 (0.026)	*In vitro*, OCT	[5]
Chicken adipose	1310	1.45(0.014)	*In vitro*, low-coherence interferometry	[54]
Bovine adipose	589	1.4536 (intramuscular adipose) 1.4522 (inner subcutaneous adipose) 1.4523 (outer subcutaneous adipose)	*In vitro*, Abbe refractometer	[61]

(Continued)

TABLE 7.3 (CONTINUED)

RI of biological tissues

Tissue	λ, nm	Refractive index	Notes (sample, method)	Ref.
Natural eumelanin	350	1.791(0.001)	Natural eumelanin isolated and purified from the ink of the cuttlefish *Sepia officinalis* (Sigma, M-2649) $$n(\lambda) = 1.68395 - \frac{1.87232 \cdot 10^4}{\lambda^2} + \frac{1.09644 \cdot 10^{10}}{\lambda^4} - \frac{8.64842 \cdot 10^{14}}{\lambda^6}$$	[62]
	400	1.784(0.004)		
	450	1.753(0.003)		
	500	1.733(0.007)		
	550	1.710(0.002)		
	600	1.697(0.001)		
	650	1.686(0.006)		
	700	1.682(0.004)		
	750	1.682(0.003)		
	800	1.680(0.003)		
Natural eumelanin	600	1.7	Natural eumelanin isolated and purified from the ink of the cuttlefish *Sepia officinalis*	[63]
Melanosomes	300	1.605	Melanosomes of retinal pigmented epithelium	[64]
	400	1.644		
	500	1.650		
	600	1.651		
	700	1.651		
	800	1.651		
	900	1.651		
	1000	1.65		
	1100	1.65		
	1200	1.65		
	1300	1.65		
	1400	1.65		
	1500	1.65		
	1600	1.65		
	1700	1.65		
	1800	1.65		
	1900	1.65		
	2000	1.65		
Myoglobin	589	1.335	Myoglobin 460 mg/ml, refractometer Atago R-5000	[65]
Porcine muscle	488	1.402(0.002)	*In vitro*, measuring the critical angle of total reflection $$n(\lambda) = 1.3694 + \frac{7.3223 \times 10^{-5}}{\lambda^2} + \frac{1.8317 \times 10^{-3}}{\lambda^4}$$	[56]
	632.8	1.381(0.002)		
	1079.5	1.372(0.003)		
	1341.4	1.370(0.003)		
	488	1.399(0.002)	*In vitro*, measuring the critical angle of total reflection $$n(\lambda) = 1.3657 + \frac{1.5123 \times 10^{-3}}{\lambda^2} + \frac{1.5291 \times 10^{-3}}{\lambda^4}$$	[56]
	632.8	1.379(0.002)		
	1079.5	1.370(0.002)		
	1341.4	1.367(0.003)		
Ovine muscle	488	1.404(0.003)	*In vitro*, measuring the critical angle of total reflection $$n(\lambda) = 1.3716 + \frac{5.8677 \times 10^{-3}}{\lambda^2} + \frac{4.3999 \times 10^{-4}}{\lambda^4}$$	[56]
	632.8	1.389(0.002)		
	1079.5	1.378(0.004)		
	1341.4	1.375(0.003)		
	488	1.402(0.002)	*In vitro*, measuring the critical angle of total reflection $$n(\lambda) = 1.3682 + \frac{8.7456 \times 10^{-3}}{\lambda^2} - \frac{1.6532 \times 10^{-4}}{\lambda^4}$$	[56]
	632.8	1.389(0.002)		
	1079.5	1.375(0.003)		
	1341.4	1.373(0.003)		
Chicken muscle	1310	1.399(0.013)	*In vitro*, low-coherence interferometry	[54]
Chicken liver	1310	1.410(0.014)	*In vitro*, low-coherence interferometry	[54]

Refractive properties of blood and its components

Optical properties of blood can be considered in a microscopic or macroscopic way. As a microscopic object, blood is a strongly scattering medium. It is a heterogeneous medium, consisting of plasma and blood corpuscles. The blood plasma consists of 90% water and 10% proteins. The blood corpuscles are mainly red blood cells (RBCs) (about 45% of the blood volume), leucocytes (about 1%), and platelets (less than 1%).

Plasma makes up the rest of the blood. The RBCs possess the largest geometric size, as a rule being 6.2–8.2 μm, and mainly determine the optical properties of blood [66–68]. The quantitative and qualitative information on the blood optical properties, in particular, on the RI, is of great interest for many fields of biomedical studies and practical medicine, since the noninvasive or less-invasive optical technologies become more and more widely used in diagnostics and therapy [66–71]. It is well known that the optical properties of blood are determined by such physiologic and biologic parameters as hematocrit, temperature, osmolarity, saturation with oxygen or other gases, and membrane rigidity of RBCs, and depend on the wavelength in a complex way [66, 69–77]. The visible and NIR spectral regions are often referred to as "therapeutic/diagnostic window," since just in this wavelength range, water, which is a major component of many biological tissues, absorbs weakly. At present there is a revival of interest to the problem of measuring the RI of different biological tissues and blood in a wide range of wavelengths, since just the RI is proposed for using as an endogenous diagnostic marker of different diseases [78–82]. For example, in Ref. [82], it is shown that the RI of blood serum can be used as additional criterion to assess the dynamics of the variation of blood serum properties under the antitumor therapy. The knowledge of the RI of blood in a wide wavelength range (optical dispersion) is necessary to describe the optical properties of different layers of blood-saturated tissues, *e.g.*, using the Monte Carlo method of statistical modeling [66, 83–86]. The knowledge of tissue and blood optical properties allows for determination of the optimal wavelength providing the maximal penetration depth of laser radiation. This is important for modeling the interaction between the tissue and laser radiation, *e.g.*, in planning such clinical procedures as laser interstitial thermotherapy or photodynamic therapy, as well as in choosing the operating wavelengths of pulse oximeters widely used in different fields of medicine for monitoring blood oxygen saturation [66–71].

The RI of a multicomponent biological medium can be calculated using the Gladstone–Dale law, according to which, in the absence of chemical interaction between the components of the medium, the resulting RI is the average of RIs of the components with their volume fractions as weighting factors, *i.e.* [87–90]

$$n = \sum_{i=1}^{N} n_i f_i, \tag{7.22}$$

where n_i and f_i are the RI and the volume fractions of individual components, respectively; and N is the number of components.

In a simplified model, the blood can be presented as a two-component medium consisting of RBCs suspended in plasma. Then Equation (7.22) can be applied to the blood in the form

$$n_{\text{blood}} = n_{\text{RBC}} f_{\text{RBC}} + n_{\text{Pl}} f_{\text{Pl}}, \tag{7.23}$$

where n_{blood}, n_{RBC}, and n_{Pl} are the RIs of blood, RBCs, and plasma, respectively; and f_{RBC} and f_{Pl} are the volume fractions of RBCs and plasma, respectively.

Since the plasma consists of 90% water and 10% proteins, the main one being albumin, its RI can be calculated using the Gladstone–Dale equation written in the form

$$n_{\text{Pl}} = 0.9 n_{\text{Water}} + 0.1 n_{\text{Alb}}, \tag{7.24}$$

where n_{Pl}, n_{Water}, and n_{Alb} are the RIs of plasma, water, and albumin, respectively.

The RI of RBCs can be also calculated using Equation (7.23) keeping in mind that hemoglobin, the main protein in the RBC, occupies 25% of its volume, and the rest volume is filled with water [66–68]:

$$n_{\text{RBC}} = 0.75 n_{\text{Water}} + 0.25 n_{\text{Hb}}, \tag{7.25}$$

where n_{RBC}, n_{Water}, and n_{Hb} are the RIs of the RBC, water, and hemoglobin, respectively.

The values of RI for the albumin and hemoglobin solutions are presented in Table 7.4 [56, 91–94] and Table 7.5 [47, 48, 72–74, 76, 95–98].

As already mentioned, the measurement of blood RI is a complex problem solved using complementary direct and

TABLE 7.4

RIs of albumin solutions and plasma

λ, nm	Refractive index	Notes (sample, method)	Ref.
480	1.3480(0.0002)	Albumin solution (55 g/l) prepared from dry	[91,92]
486	1.3478(0.0002)	human serum albumin (Sigma Aldrich, USA),	
546	1.3449(0.0002)	multi-wavelength Abbe refractometer	
589	1.3434(0.0002)	DR-M2/1550 (Atago, Japan)	
644	1.3416(0.0002)		
656	1.3414(0.0002)		
680	1.3405(0.0002)		
930	1.3380(0.0004)		
1100	1.3361(0.0003)		
1300	1.3325(0.0004)		
1550	1.3285(0.0004)		
480	1.3444(0.0004)	BSA solution (40 g/l), multi-wavelength Abbe	[93]
486	1.3436(0.0003)	refractometer DR-M2/1550 (Atago, Japan)	
546	1.3406(0.0002)		
589	1.3392(0.0003)		
644	1.3375(0.0001)		
656	1.3373(0.0002)		
680	1.3365(0.0001)		
800	1.3343(0.0003)		
600	1.347	plasma	[94]
488	1.350(0.002)	Measuring the critical angle of total reflection	[56]
632.8	1.345(0.002)	$n(\lambda) = 1.3194 + \dfrac{1.4578 \times 10^{-2}}{\lambda^2} - \dfrac{1.7383 \times 10^{-3}}{\lambda^4}$	
1079.5	1.332(0.003)		
1341.4	1.327(0.004)		

TABLE 7.5

The RI of the hemoglobin solutions

λ, nm	g/l	Refractive index	Notes	Ref.
250	46	1.398	Human hemoglobin from fresh RBC suspensions of donors; VIS-NIR-spectrometer	[95], [96]
	104	1.406		
	165	1.435		
	287	1.470		
300	46	1.373		
	104	1.389		
	165	1.405		
	287	1.441		
400	46	1.354		
	104	1.367		
	165	1.383		
	287	1.409		
400	20	1.35223	Bovine hemoglobin (dry); Hb; pH 7.4; Room temperature; Continuous refractive index dispersion (CRID)	[72]
	40	1.35495		
	60	1.35806		
	80	1.36078		
	120	1.36369		
	140	1.36600		
	280	1.37010		
	320	1.38621		
400	20	1.35107	Bovine hemoglobin (dry); HbO_2; pH 7.4; Room temperature; CRID	
	40	1.35417		
	60	1.35767		
	80	1.36039		
	120	1.36369		
	140	1.36602		
	280	1.36951		
	320	1.38660		
400	320	1.3822	Bovine hemoglobin (lyophilized powder); 0.5% HbO_2; T=20°C; Fiber spectrometer	[47]
400	320	1.3775	Bovine hemoglobin (lyophilized powder); Hb; T=20°C; Fiber spectrometer	
401	140	1.365	Human hemoglobin (lyophilized powder); Hb; T=20°C; pH 7.4; TIR (total internal reflection)	[48], [73]
435.8		1.367		
401	140	1.369	Human hemoglobin (lyophilized powder); HbO_2; T=20°C; pH 7.4;TIR	
435.8		1.366		
436	150	1.36481	Human hemoglobin (dry); T=20°C; pH 7.4; Abbe Refractometer	[74]
438	140	1.374	Bovine hemoglobin(dry); Hb; HbO_2; Room temperature; pH 7.4	[72]
440	50	1.3562	Human hemoglobin (lyophilized powder); Spectroscopic phase microscopy	[97]
	150	1.3780		
	300	1.4187		
450	320	1.3888	Bovine hemoglobin (lyophilized powder); 0.5% HbO_2;T=20°C; Fiber spectrometer	[47]
450	320	1.3933	Bovine hemoglobin (lyophilized powder); Hb; T=20°C; Fiber spectrometer	
480	65	1.3476(0.0003)	Human hemoglobin from whole blood; HbO_2; T=23°C; Multiwavelength Abbe Refractometer	[98]
	87	1.3571(0.0003)		
	173	1.3728(0.0003)		
	260	1.3879(0.0002)		
486	65	1.3478(0.0002)	Human hemoglobin from whole blood; HbO_2; T=23°C; Multiwavelength Abbe Refractometer	[98]
	87	1.3563(0.0002)		
	173	1.3721(0.0002)		
	260	1.3871(0.0004)		
486.1	140	1.361	Human hemoglobin (lyophilized powder); Hb; T=20°C; pH 7.4; TIR	[48], [73]
486.1	140	1.361	Human hemoglobin (lyophilized powder); HbO_2; T=20°C; pH 7.4; TIR	
500	287	1.413	Human hemoglobin from fresh RBC suspensions of donors; VIS-NIR-spectrometer	[95], [96]
	165	1.383		
	104	1.363		
	46	1.348		

(Continued)

TABLE 7.5 (CONTINUED)

The RI of the hemoglobin solutions

λ, nm	g/l	Refractive index	Notes	Ref.
500	20	1.34583	Bovine hemoglobin (dry);Hb;	[72]
	40	1.34913	pH 7.4;	
	60	1.35223	Room temperature;	
	80	1.35592	CRID	
	120	1.35922		
	140	1.36175		
	280	1.36544		
	320	1.38408		
500	20	1.34505	Bovine hemoglobin (dry); HbO_2;	
	40	1.34854	pH 7.4; Room temperature;	
	60	1.35262	CRID	
	80	1.35573		
	120	1.35845		
	140	1.36214		
	280	1.36544		
	320	1.38505		
513.9	150	1.36053	Human hemoglobin (dry); T=20°C;	[74]
			pH 7.4; Abbe Refractometer	
532	1.7	1.3400	Human hemoglobin (fresh human blood); T=25°C; TIR	[76]
	2.5	1.3431		
	4	1.3485		
	7	1.3604		
	12.97	1.3871		
546	65	1.3448(0.0002)	Human hemoglobin from whole blood; HbO_2; T=23°C;	[98]
	87	1.3533(0.0002)	Multiwavelength Abbe Refractometer	
	173	1.3681(0.0007)		
	260	1.3836(0.0002)		
546	50	1.3472	Human hemoglobin (lyophilized powder); Spectroscopic phase microscopy	[97]
	150	1.3700		
	300	1.4051		
546.1	140	1.357	Human hemoglobin (lyophilized powder); Hb; T=20°C;	[48], [73]
			pH 7.4; TIR	
546.1	140	1.357	Human hemoglobin (lyophilized powder); HbO_2; T=20°C;	
			pH 7.4; TIR	
550	320	1.3724	Bovine hemoglobin (lyophilized powder); 0.5% HbO_2; T=20°C; Fiber spectrometer	[47]
550	320	1.3738	Bovine hemoglobin (lyophilized powder); Hb; T=20°C; Fiber spectrometer	
560	50	1.3466	Human hemoglobin (lyophilized powder); Spectroscopic phase microscopy	[97]
	150	1.3687		
	300	1.4033		
580	50	1.3451	Human hemoglobin (lyophilized powder); Spectroscopic phase microscopy	[97]
	150	1.3668		
	300	1.4025		
587.6	140	1.356	Human hemoglobin (lyophilized powder); Hb; T=20°C;	[74], [97]
			pH 7.4; TIR	
587.6	140	1.357	Human hemoglobin (lyophilized powder); HbO_2; T=20°C;	
			pH 7.4; TIR	
589	65	1.3438(0.0002)	Human hemoglobin from whole blood; HbO_2; T=23°C;	[98]
	87	1.3519(0.0003)	Multiwavelength Abbe Refractometer	
	173	1.3667(0.0004)		
	260	1.3821(0.0004)		
589	46	1.343	Human hemoglobin from fresh RBC suspensions of donors;	[95], [96]
	104	1.357	VIS-NIR-spectrometer	
	165	1.375		
	287	1.406		
589.2	150	1.35724	Human hemoglobin (dry);T=20°C; pH 7.4; Abbe Refractometer	[74]

(Continued)

TABLE 7.5 (CONTINUED)

The RI of the hemoglobin solutions

λ, nm	g/l	Refractive index	Notes	Ref.
589.3	140	1.356	Human hemoglobin (lyophilized powder); Hb; T=20°C; pH 7.4;TIR	[48], [73]
589.3	140	1.357	Human hemoglobin (lyophilized powder); HbO$_2$; T=20°C; pH 7.4;TIR	
600	50	1.3443	Human hemoglobin (lyophilized powder); Spectroscopic phase microscopy	[97]
	150	1.3666		
	300	1.4014		
600	20	1.34233	Bovine hemoglobin (dry); Hb;	[72]
	40	1.34485	pH 7.4; Room temperature;	
	60	1.34874	CRID	
	80	1.34835		
	120	1.3520		
	140	1.35495		
	280	1.36155		
	320	1.38233		
600	20	1.34136	Bovine hemoglobin (dry); HbO$_2$;	
	40	1.34447	pH 7.4; Room temperature;	
	60	1.34874	CRID	
	80	1.35068		
	120	1.35456		
	140	1.35767		
	280	1.36155		
	320	1.38058		
600	320	1.3684	Bovine hemoglobin (lyophilized powder); 0.5% HbO$_2$; T=20°C; Fiber spectrometer	[47]
600	320	1.3702	Bovine hemoglobin (lyophilized powder); Hb; T=20°C; Fiber spectrometer	
632	1.7	1.3626	Human hemoglobin (fresh human blood); T=25°C; TIR	[76]
	2.5	1.3360		
	4	1.3425		
	7	1.3538		
	12.97	1.3800		
632.8	140	1.354	Human hemoglobin (lyophilized powder); Hb; T=20°C;	[48], [73]
656.3		1.354	pH 7.4; TIR	
632.8	140	1.355	Human hemoglobin (lyophilized powder); HbO$_2$; T=20°C;	
656.3		1.354	pH 7.4; TIR	
633.2	150	1.35601	Human hemoglobin (dry); T=20°C; pH 7.4; Abbe Refractometer	[74]
657.2		1.35587		
633	104	1.3600	Human hemoglobin (dry); T=20°C; pH 7.4; Abbe Refractometer	[74]
	165	1.3750		
644	65	1.3419(0.0002)	Human hemoglobin from whole blood; HbO$_2$; T=23°C;	[98]
	87	1.3497(0.0002)	Multiwavelength Abbe Refractometer	
	173	1.3640(0.0003)		
	260	1.3801(0.0003)		
650	320	1.3652	Bovine hemoglobin (lyophilized powder); 0.5% HbO$_2$; T=20°C; Fiber spectrometer	[47]
650	320	1.3668	Bovine hemoglobin (lyophilized powder); Hb; T=20°C; Fiber spectrometer	
655	50	1.3408	Human hemoglobin (lyophilized powder); Spectroscopic phase microscopy	[97]
	150	1.3642		
	300	1.3969		
656	65	1.3414(0.0002)	Human hemoglobin from whole blood; HbO$_2$;T=23°C;	[98]
	87	1.3493(0.0002)	Multiwavelength Abbe Refractometer	
	173	1.3647(0.0003)		
	260	1.3792(0.0009)		
680	65	1.3403(0.0003)	Human hemoglobin from whole blood; HbO$_2$; T=23°C;	[98]
	87	1.3482(0.0003)	Multiwavelength Abbe Refractometer	
	173	1.3633(0.0003)		
	260	1.3771(0.0002)		
700	50	1.3405	Human hemoglobin (lyophilized powder); Spectroscopic phase microscopy	[97]
	150	1.3634		
	300	1.3971		

(Continued)

TABLE 7.5 (CONTINUED)

The RI of the hemoglobin solutions

λ, nm	g/l	Refractive index	Notes	Ref.
700	20	1.33961	Bovine hemoglobin (dry); Hb; pH 7.4; Room temperature; CRID	[72]
	40	1.34252		
	60	1.34602		
	80	1.34874		
	120	1.35184		
	140	1.35456		
	280	1.35806		
	320	1.37709		
700	20	1.33883	Bovine hemoglobin (dry); HbO_2; pH 7.4; Room temperature; CRID	
	40	1.34175		
	60	1.34583		
	80	1.34835		
	120	1.35107		
	140	1.35476		
	280	1.35748		
	320	1.3767		
700	320	1.3612	Bovine hemoglobin (lyophilized powder); 0.5% HbO_2; T=20°C; Fiber spectrometer	[47]
700	320	1.3637	Bovine hemoglobin (lyophilized powder); Hb; T=20°C; Fiber spectrometer	
700	46	1.341	Human hemoglobin from fresh RBC suspensions of donors;	[95], [96]
	104	1.356	VIS-NIR-spectrometer	
	165	1.374		
	287	1.404		
706.5	140	1.352	Human hemoglobin (lyophilized powder); Hb; T=20°C; pH 7.4; TIR	[48], [73]
706.5	140	1.352	Human hemoglobin (lyophilized powder); HbO_2; T=20°C; pH 7.4; TIR	
750	320	1.3589	Bovine hemoglobin (lyophilized powder); 0.5% HbO_2; T=20°C; Fiber spectrometer	[47]
750	320	1.3599	Bovine hemoglobin (lyophilized powder); Hb; T=20°C; Fiber spectrometer	
800	46	1.338	Human hemoglobin from fresh RBC suspensions of donors;	[95], [96]
	104	1.353	VIS-NIR-spectrometer	
	165	1.370		
	287	1.400		
900	46	1.338	Human hemoglobin from fresh RBC suspensions of donors;	[95], [96]
	104	1.352	VIS-NIR-spectrometer	
	165	1.369		
	287	1.401		
930	65	1.3360(0.0002)	Human hemoglobin from whole blood; HbO_2; T=23°C;	[98]
	87	1.3440(0.0002)	Multiwavelength Abbe Refractometer	
	173	1.3572(0.0003)		
	260	1.3735(0.0007)		
1000	46	1.338	Human hemoglobin from fresh RBC suspensions of donors;	[95], [96]
	104	1.353	VIS-NIR-spectrometer	
	165	1.370		
	287	1.401		
1100	46	1.337	Human hemoglobin from fresh RBC suspensions of donors;	[95], [96]
	104	1.352	VIS-NIR-spectrometer	
	165	1.369		
	287	1.400		
1100	65	1.3329(0.0002)	Human hemoglobin from whole blood; HbO_2; T=23°C;	[98]
	87	1.3411(0.0002)	Multiwavelength Abbe Refractometer	
	173	1.3542(0.0002)		
	260	1.3690(0.0006)		
1300	65	1.3280(0.0005)	Human hemoglobin from whole blood; HbO_2; T=23°C;	[98]
	87	1.3364(0.0002)	Multiwavelength Abbe Refractometer	
	173	1.3503(0.0002)		
	260	1.3642(0.0004)		
1550	65	1.3244(0.0004)	Human hemoglobin from whole blood; HbO_2; T=23°C;	[98]
	87	1.3314(0.0003)	Multiwavelength Abbe Refractometer	
	173	1.3458(0.0002)		
	260	1.3598(0.0004)		

indirect methods. The advantage of direct methods is the measurement of the RI in the presence of all blood components. However, the inhomogeneity of the cell suspension and the strong absorption and scattering do not allow for the measurement of whole blood using a direct method with a high accuracy; RI values can vary significantly depending on the sample preparation and measurement technique used.

Since blood is a turbid biological medium with a high scattering anisotropy ($g = 0.9996$) [68], possessing strong scattering and absorption in the visible region, different models of light scattering by equivalent particles and other indirect methods are used to determine the RI and its dispersion. They include empirical methods based on the calculation of blood RI from the experimental values of the RIs of the components [74, 91] or theoretical methods that allow for calculation

of the real part of the RI from the measured spectra of its imaginary part (absorption spectra) using the KK relations [75–77]. Model function to calculate the refractive index of native hemoglobin in the wavelength range of 250 to 1100 nm dependent on concentration and the data for complex refractive index of highly concentrated hemoglobin solutions determined using transmittance and reflectance measurements can be found in Refs. [95, 96].

Among the direct methods, the most frequently used ones are based on different modifications of OCT [53, 66, 78, 97, 99–102], phase microscopy [80, 103, 104], laser refractometry with a hollow prism [85, 86], and the total internal reflection phenomenon [73, 76, 87]. These methods possess a number of advantages, but are not free of drawbacks. The OCT-based methods allow for RI measurements in scattering media, but

TABLE 7.6

The RI of blood

Wavelength, nm	Refractive index	Notes	Ref.
488	1.395(0.003)	Whole human blood,	[56]
632.8	1.373(0.004)	measuring the critical angle of total reflection	
1079.5	1.363(0.004)	$n(\lambda) = 1.3587 + \dfrac{1.4744 \times 10^{-3}}{\lambda^2} + \dfrac{1.7103 \times 10^{-4}}{\lambda^4}$	
1341.4	1.360(0.005)		
632.8	1.4	Whole human blood, fiberoptic laser refractometer	[84]
632,8	1.37	60% blood solution, laser refractometer with a hollow prism	[85]
370	1.44795	Refractive dispersions of whole blood with type A, B and 0, $T_{sample} =$	[86]
400	1.42981	27-28°C, total internal reflection with CCD camera	
420	1.42054	$n = 1.357 + \dfrac{6.9*10^3}{\lambda^2} + \dfrac{7.6*10^8}{\lambda^4}$	
450	1.40961		
480	1.40126		
500	1.39676		
530	1.3912		
560	1.38673		
590	1.38309		
620	1.38009		
650	1.37759		
680	1.37548		
710	1.37368		
740	1.37213		
770	1.3708		
800	1.36964		
830	1.36862		
850	1.36801		
820	1.475	Whole human blood, OCT	[104]
436.1	1.36841	Modeling method	[74]
513.9	1.36053		
589.1	1.35724		
633.2	1.35601		
657.2	1.35587		
480	1.3690(0.0002)	Modeling method based on the Gladstone-Dale formula	[91]
486	1.3686(0.0002)	$n^2(\lambda) = 1 + \dfrac{0,83423*\lambda^2}{\lambda^2 - 10775,44775} + \dfrac{0,04296*\lambda^2}{\lambda^2 - 6,13587*10^6}$	
546	1.3645(0.0002)		
589	1.3635(0.0002)		
644	1.3613(0.0002)		
656	1.3610(0.0002)		
680	1.3600(0.0002)		
930	1.3561(0.0003)		
1100	1.3537(0.0003)		
1300	1.3497(0.0003)		
1550	1.3456(0.0003)		

with the accuracy restricted to 0.01–0.001, which is not always sufficient in analytical applications. However, in the *in vivo* measurements the OCT offers wide possibilities for medical applications, where the above accuracy of the RI measurement is sufficient [53, 66, 78, 97, 99–102]. Various modifications of phase microscopy are most frequently used to study the refraction properties of individual cells, including blood cells, e.g., RBCs [80, 103, 104]. The application of the laser refractometry with a hollow prism allows for RI measurement in fluids having large coefficients of scattering and absorption to which the blood belongs. The method provides *in vitro* measurements with the accuracy of 0.01–0.0001. The drawbacks of the method include the need to use laser sources of radiation with relatively high power [85, 86]. The methods based on total internal reflection allow for *in vitro* measurement of blood RI and its individual components with the accuracy of 0.0001–0.00001. The simplicity and the small amount of the required sample material make them available for online monitoring of RI [73, 74, 87].

In the visible and NIR regions, blood RI dispersion demonstrates nonlinear behavior, with RI decreasing with the wavelength increase. The RI values of the blood are presented in Table 7.6 [56, 74, 84–86].

The results of direct measurement of blood RI are also affected by two additional important phenomena: aggregation and sedimentation. The necessary usage of anticoagulant scan leads to the deformation of some blood corpuscles, which affects the optical properties. For example, it is known that the use of heparin can change the size and shape of platelets and leucocytes, and the excess amount of K2EDTA can be hypertonic and lead to osmotic stress. The sedimentation is related to the fact that the density of blood cells is higher than the density of plasma and saline. Although the sedimentation rate in normal blood is not high (up to 30 mm/h), it can be essential in the long-term measurements of the sample optical properties [68, 105, 106]. Bolin *et al.* obtained blood RI equal to 1.400 using a fiberoptic laser refractometer at the wavelength 632.8 nm [84]. Sardar *et al.* measured the RI of a

few concentrations of the diluted solutions of whole blood by means of the laser refractometer with a hollow prism. Thus, for the 60% blood solution, they found the RI = 1.37 at the wavelength 632.8 nm [85]. Li *et al.* performed the RI measurement for the blood samples of different groups (O(I), A(II), and B(III)) using the total internal reflection technique; they presented a dispersion equation for the calculation the mean RI of blood in the visible/NIR spectral range. According to this equation, RI = 1.4480–1.3680 for the wavelength range 370–850 nm [86]. Cheng *et al.* measured the RI of whole blood using the total internal reflection technique with several lasers. They obtained the following values of the RI: 1.395 for 488 nm, 1.373 for 632.8 nm, 1.363 for 1079.5 nm, and 1.360 for 1341.4 nm [56].

Liu *et al.* measured the continuous complex refractive index dispersion (CRID) of whole blood and blood component solutions in the spectral region of 400–750 nm, and spectral resolution is about 0.263 nm [107] (see Figure 7.7).

Measurement of RIs of biological tissues prior OC

To select the most suitable immersion OCA for providing OC of tissue in a certain wavelength range, rapid measurements of the RI of tissue samples in a wide wavelength range are required. These measurements can be made quickly and reliably with a commercial multiwavelength refractometer, the Abbe DR-M2/1550 (Atago, Japan) (Figure 7.8), working at 12 wavelengths in the range from 450 to 1550 nm with the accuracy of ±0.0002. The measurement results and their interpolation are presented in Table 7.7 and Figure 7.9 for a few examples of *ex vivo* samples of brain, skin, and muscle of Wistar laboratory rats; and pig muscle. Skin samples of approximately 10×20 mm^2 were obtained using surgical scissors. The subcutaneous fat layer was removed from the samples. Thin muscle tissue samples were obtained using a scalpel.

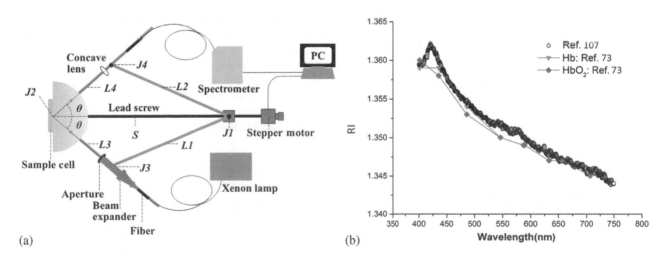

(a) (b)

FIGURE 7.7 The schematic of the experimental setup for RI measurements continuously in a wide spectral range (a); the continuous complex refractive index dispersion of hemoglobin solution, compared with data from reference 73 (b) [107].

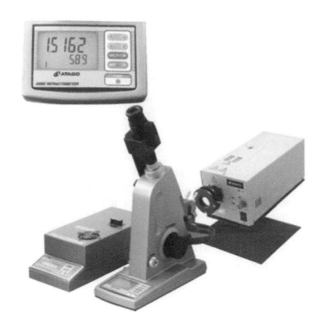

FIGURE 7.8 Multiwavelength refractometer Abbe DR-M2/1550 (Atago, Japan).

RI kinetics during optical clearing treatments

During OC treatments, the partial exchange of interstitial water by the OCA will induce an elevation of the RI of the ISF, approaching it to the RI of tissue scatterers [10, 11]. If the dispersion of the tissue is known to have a wide spectral range, and sample thickness measurements and T_c spectra measured for the same range during the treatment, it is possible to calculate the dispersion kinetics both for the ISF and for the tissue.

Considering human normal colorectal mucosa as an example, Carneiro *et al.* [10] have reported how to make such calculations for a treatment with an aqueous solution containing 40% glucose. To initiate those calculations, it is necessary to reconstruct the dispersion of the scatterers and ISF in the untreated colorectal mucosa. By measuring the total water content in the mucosa as 73.96% $\pm$ 0.02%, the dispersion for the completely dry mucosa could be calculated according to [10]

$$n_{\text{dry}}\left(\lambda\right) = \frac{n_{\text{tissue}}\left(\lambda\right) - 0.7396 \times n_{\text{H}_2\text{O}}\left(\lambda\right)}{1 - 0.7396}, \quad (7.26)$$

where $n_{\text{tissue}}(\lambda)$ is the dispersion of the mucosa tissue, as described by Equation (7.10), and $n_{\text{H}_2\text{O}}\left(\lambda\right)$ is the dispersion of water at 20°C, the temperature at which measurements were performed. Such water dispersion was retrieved from literature [108]. By combining the dispersion of dry mucosa, as calculated with Equation (7.26) with $n_{\text{H}_2\text{O}}\left(\lambda\right)$ in the Gladstone and Dale equation [2, 73, 109–111], it was possible to obtain the scatterers' dispersion:

$$n_s\left(\lambda\right) = n_{\text{H}_2\text{O}}\left(\lambda\right)f_{\text{H}_2\text{O}} + n_{\text{dry}}\left(\lambda\right)f_{\text{dry}}$$

$$f_{\text{H}_2\text{O}} + f_{\text{dry}} = 1. \quad (7.27)$$

Some arbitrary values for the volume fractions (typically, f_{dry}=0.9 and $f_{\text{H}_2\text{O}}$=0.1) were used to perform this initial estimation for $n_s(\lambda)$.

The following step was to use the Gladstone and Dale equation again to calculate the dispersion of the ISF ($n_{\text{ISF}}(\lambda)$) from the dispersions of the mucosa tissue (calculated with Equation (7.10)) and scatterers (as calculated with Equation (7.27)). In this calculation, the mobile water content (59.4%), which was previously estimated [112], was used as the volume fraction (VF) for the ISF [10]:

$$n_{\text{ISF}}\left(\lambda\right) = \frac{n_{\text{tissue}}\left(\lambda\right) - (1 - 0.594) \times n_s\left(\lambda\right)}{0.594}. \quad (7.28)$$

This means that the VF for the scatterers in the mucosa is 1 −0.594 = 0.406. Considering this two-component model for the mucosa, one way to check and correct the VFs in Equation (7.27) was to use both $n_{\text{ISF}}(\lambda)$ and $n_s(\lambda)$ in Equation (7.29) [11, 113, 114] to reconstruct the wavelength dependence of the reduced scattering coefficient ($\mu'_s(\lambda)$) that was previously estimated using inverse adding-doubling simulations [115]:

$$\mu'_s\left(\lambda\right) = \frac{3f_s(1-f_s)}{4\pi a^3} \times 3.28\pi a^2 \left(\frac{2\pi \bar{n}_0 a}{\lambda}\right)^{0.37} \left(\frac{n_s\left(\lambda\right)}{n_{\text{ISF}}\left(\lambda\right)} - 1\right)^{2.09}, \quad (7.29)$$

where f_s is the VF of scatterers (=0.406) and a is the mean scatterer size in the mucosa [113, 114], which was estimated in that study as 46 nm [43]. Changes were made in the VFs of Equation (7.27), so that both $n_{\text{ISF}}(\lambda)$ and $n_s(\lambda)$ were changed to allow a good reconstruction of spectral μ'_s data with Equation (7.29). After obtaining the good reconstruction of $\mu'_s(\lambda)$ with Equation (7.29), the dispersions of scatterers and ISF, as calculated with Equations (7.27) and (7.28), were optimized for the colorectal

TABLE 7.7

Examination of RIs of *ex vivo* biological tissues prior OC

λ, nm	450	480	486	546	589	644	656	680	930	1100	1300	1550
Rat brain	-	1.3658	1.3645	1.3620	1.3584	1.3576	1.3570	1.3564	1.3523	1.3479	1.3458	1.3407
		(0.0033)	(0.0030)	(0.0032)	(0.0037)	(0.0036)	(0.0044)	(0.0039)	(0.0047)	(0.0041)	(0.0037)	(0.0038)
Rat skin	-	1.4730	1.4715	1.4685	1.4660	1.4637	1.4623	1.4599	1.4569	-	-	-
Rat muscle	-	1.3979	1.3959	1.3939	1.3926	1.3926	1.3934	1.393	1.3889	1.3882	-	-
Pig muscle	1.5111	1.4910	1.4824	1.4701	1.4660	1.4566	1.4692	1.4472	1.3952	1.3788	-	-

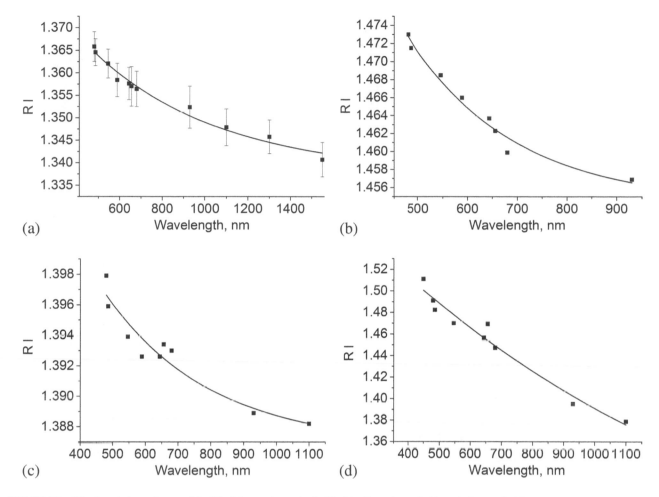

FIGURE 7.9 Wavelength dependences of the RI of the *ex vivo* rat brain (a), skin (b), and muscle (c); and pig muscle (d).

mucosa. Figure 7.10 presents those dispersions with $n_{tissue}(\lambda)$ and $n_{H_2O}(\lambda)$ for comparison.

Using a Cornu-type equation, $n_s(\lambda)$ and $n_{ISF}(\lambda)$ can be described as [10]

$$n_s(\lambda) = 1.325 + \frac{17.32}{(\lambda - 68.94)}, \tag{7.30}$$

$$n_{ISF}(\lambda) = 1.309 + \frac{16.58}{(\lambda - 9.636)}. \tag{7.31}$$

At this stage, and after calculating $n_s(\lambda)$ and $n_{ISF}(\lambda)$, time dependence T_c spectral data ($T_c(\lambda,t)$) and thickness kinetics ($d(t)$) measurements were used to obtain the kinetics of the RI for the ISF and for the whole tissue during the treatment with glucose 40%. For short-term treatments, where it may be considered that the OCA only diffuses into the interstitial locations, it is assumed that scatterers hydration remains unchanged, and consequently no changes are expected for $\mu_a(\lambda)$ and for $n_s(\lambda)$. This way, and using the T_c and thickness measurements made during treatment, the kinetics for the scattering coefficient could be calculated as [10, 11, 116]:

$$\mu_s(\lambda,t) = -\frac{\ln[T_c(\lambda,t)]}{d(t)} - \mu_a(\lambda). \tag{7.32}$$

This data can be combined with $d(t)$ to calculate the kinetics of the RI of the ISF ($n_{ISF}(\lambda,t)$) as [10, 11, 116]

$$n_{ISF}(\lambda,t) = \frac{n_s(\lambda)}{\left[\sqrt{\frac{\mu_s(\lambda,t) \times d(t)}{\mu_s(\lambda,t-0) \times d(t=0)}} \times \left(\frac{n_s(\lambda)}{n_{ISF}(\lambda,t=0)} - 1\right) + 1\right]}, \tag{7.33}$$

where $\mu_s(\lambda, t=0)$ and $d(t=0)$ are the wavelength dependence for scattering coefficient and sample thickness for the untreated mucosa and $n_{ISF}(\lambda,t=0)$ is the RI of the ISF before treatment as described by Equation (7.31).

Since for the colorectal mucosa, the wavelength dependence for the optical properties between 400 and 1000 nm were previously estimated [117], these calculations were possible. Figure 7.11 presents the RI kinetics for the ISF of the mucosa.

The data in Figure 7.11 shows that due to the partial replacement of water by glucose leads to an increase in the RI of the ISF. Such variations reduce the RI mismatch between the ISF and scatterers in the colorectal mucosa. During treatment, not only n_{ISF} changes but also f_s and f_{ISF} are altered as a consequence of sample thickness variations. To calculate the kinetics for the

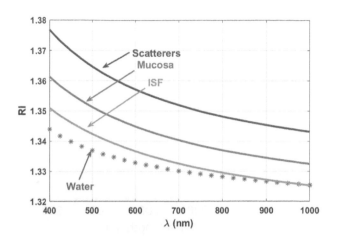

FIGURE 7.10 Dispersions for water [108], colorectal mucosa, its scatterers and ISF at 20°C.

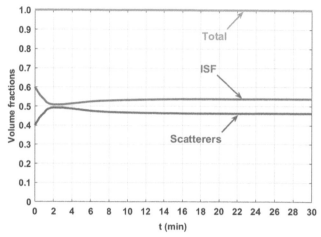

FIGURE 7.12 Time dependence for the VFs of colorectal mucosa during treatment with 40% glucose.

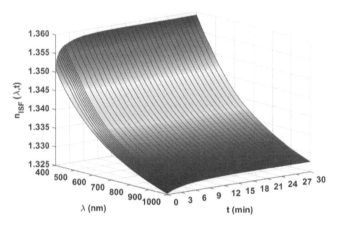

FIGURE 7.11 Kinetics of $n_{\mathrm{ISF}}(\lambda)$ for the human colorectal mucosa under treatment with glucose –40%.

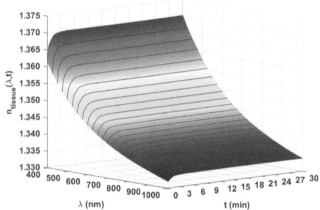

FIGURE 7.13 Kinetics of $n_{\mathrm{tissue}}(\lambda)$ for the human colorectal mucosa under treatment with glucose –40%.

RI of the whole tissue, one needs to account for the variations in f_s and f_{ISF}. Those changes in the VFs can be calculated from the thickness kinetics measured during the treatment.

Since the untreated tissue samples were prepared with in approximately circular slab form ($\phi \approx 1$ cm, d=0.5 mm), the volume of the sample could be calculated as [10]

$$V_{\mathrm{sample}}\left(t = 0\right) = \left(\pi \times 0.5^2\right) \times 0.05 \ \mathrm{cm}^3, \qquad (7.34)$$

which allows one to obtain the absolute volume of scatters according to [10]

$$V_{\mathrm{scat}} = V_{\mathrm{sample}}\left(t = 0\right) \times f_s\left(t = 0\right) \mathrm{cm}^3, \qquad (7.35)$$

where $f_s(t=0)$ equals 0.406, as previously indicated. This volume for scatterers remains unchanged during treatment, whereas f_s changes according to [10]

$$f_s\left(t\right) = \frac{V_{\mathrm{scat}}}{\left(\pi \times 0.5^2\right) \times d\left(t\right)}, \qquad (7.36)$$

and considering $d(t)$ as the time dependence of sample thickness during treatment. Since, according to the Gladstone and Dale law [2, 73, 109–111], $f_{\mathrm{ISF}}(t)+f_s(t)=1$, the time dependence

for the VF of the ISF could be obtained from $f_s(t)$. Figure 7.12 presents the kinetics for the two VFs during the treatment [10].

Using the data in Figures 7.11 and 7.12 with the unchanged $n_s(\lambda)$ in the Gladstone and Dale equation [2, 73, 109–111], the kinetics for $n_{\mathrm{tissue}}(\lambda)$ could be calculated as [10]

$$n_{\mathrm{tissue}}\left(\lambda, t\right) = n_{\mathrm{ISF}}\left(\lambda, t\right) f_{\mathrm{ISF}}\left(t\right) + n_s\left(\lambda, t\right) f_s\left(t\right)$$

$$f_{\mathrm{ISF}}\left(t\right) + f_s\left(t\right) = 1. \qquad (7.37)$$

The result of this calculation is presented in Figure 7.13.

As a result of the treatment, the RI of the whole mucosa tissue increases uniformly in the entire spectral range between 400 and 1000 nm. According to the data in Figure 7.13, this increase is higher for the lowest wavelengths. Similar studies were reported in Ref. [10] for the pathological colorectal mucosa under treatment with a glucose 35% solution and in Ref. [11] for the human colorectal muscle under treatment with a solution containing 40% glycerol. These studies reported similar kinetics to the data presented in Figure 7.11 and Figure 7.12. This methodology can be applied to other tissues and treatments with other OCAs to quantify the variations in the RI of the tissues.

THz pulsed spectroscopy of OCAs

Terahertz (THz) pulsed spectroscopy optical properties can be used to study optical properties – RI *n* and amplitude absorption coefficient α – of common OCAs at THz range [7]. Among the analyzed OCAs: glycerol ("SpektrChem," Russia); propylene glycol (PG) ("Chemical Line Co. Ltd," Russia); dimethyl sulfoxide (DMSO) ("SpektrChem," Russia); polyethylene glycol (PEG) 200 ("Nizhnekamskneftekhim," Russia); PEG 400 ("Nizhnekamskneftekhim," Russia); PEG 600 ("Norchem," Russia); fructose ("PanReac," Spain), sucrose ("PanReac," Spain); glucose ("Hungrana KFT," Hungary); Dextran 40 ("AppliChem," Germany); Dextran 70 ("AppliChem," Germany). Along with the pure agents, their aqueous solutions with different concentrations were evaluated. Aqueous solutions of agents were prepared in a laboratory. For this aim, the volume/volume method (for liquid agents) and the weight/volume method (for powders) were used, along with a deionized water and an analytical laboratory weight AND HR-250AZG, featuring a weight limit of 210 mg and a resolution of 0.1 mg.

The THz pulsed spectrometer, which is shown in Figure 7.14, with LT-GaAs photoconductive antennas and vacuum-capable sample chamber, works in transmission mode in the frequency range of 0.1–4.0 THz (when the THz beam path is empty), and its maximal spectral resolution is about 0.015 THz [118]. It is used for studying hyperosmotic agents and biological tissues; the spectral operation range is of 0.25–2.5 THz, applying a cuvette for measurements of liquids. Such range is conditioned by the THz beam diffraction at the aperture of the sample cuvette and Fresnel losses at interfaces, as well as THz beam absorption in the bulk of samples. For focusing of the THz beam on the sample cuvette and its collimation after interaction with the sample, a pair of off-axis parabolic

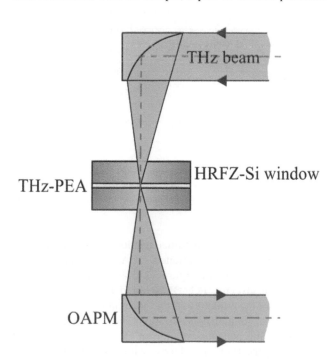

FIGURE 7.14 A scheme of THz beam path and the cuvette for THz pulsed spectroscopy [120].

mirrors (OAPM) is used. To reduce an impact of water vapor absorption on measured data, THz measurements are usually performed in a vacuum. There are two windows from high-resistivity float-zone silicon (HRFZ-Si) in a sample cuvette. They are needed for placing the OCA layer and a spacer in order that the thickness of OCA layer might be controlled [7, 119–121].

The algorithm of dielectric parameter reconstruction is based on the reference and sample time-domain spectral (TDS) waveforms [7, 122]. A waveform propagated through both HRFZ-Si windows without any free space between them is used as reference one $\tilde{E}_r(t)$. At the same time, a signal propagated through the assembled cuvette filled with OCA is used as a sample waveform $\tilde{E}_s(t)$. The higher RI and geometrical thickness the material has, the higher the optical thickness. Thus, HRFZ-Si windows have considerably higher optical thickness than the OCA layer. This feature allows the interference pattern in HRFZ-Si in the data processing to be ignored. The excess satellite pulses from windows are filtered out using the Tukey apodization in the time-domain.

The aforementioned algorithm allows the reconstruction of a complex RI of a liquid sample

$$\tilde{n} = n - i\frac{c_0}{2\pi\nu}\alpha, \tag{7.38}$$

where *n* is the RI; α is the THz-wave absorption coefficient (by amplitude), ν is the frequency, and $c_0 = 3 \times 10^8$ m/s is the speed of light in free space. The reconstruction procedure is based on a minimization of vector error functional:

$$\tilde{n}(\nu) = argmin_{\tilde{n}}\{\Phi\}, \Phi(\nu,\tilde{n}(\nu)) = \begin{pmatrix} \left|\tilde{H}_{exp}\right| - \left|\tilde{H}_{th}(\tilde{n})\right| \\ \varphi\left[\tilde{H}_{exp}\right] - \varphi\left[\tilde{H}_{th}(\tilde{n})\right] \end{pmatrix}, \tag{7.39}$$

where $\tilde{H}_{exp}$ and $\tilde{H}_{th}$ are the theoretical and experimental frequency-dependent transfer functions, and $|...|$ and $\varphi[...]$ are the modulus and the phase operators. The experimental transfer function is the ratio of the sample signal and the reference signal:

$$\tilde{H}_{exp} = \frac{\tilde{E}_s(\nu)}{\tilde{E}_r(\nu)}. \tag{7.40}$$

The theoretical transfer function, taking into account the obtained mathematical signal models, is

$$\tilde{H}_{th} = \tilde{T}_{1,2}\tilde{T}_{2,1}\frac{\tilde{P}_2(l_2)}{\tilde{P}_0(l_2)}\sum_{j=0}^{N}\left(\tilde{P}_2(l_2)\tilde{R}_{2,1}\right)^{2j}. \tag{7.41}$$

The interaction of THz radiation with the cuvette interfaces is described by Fresnel formulas for normal incidence; they determine the transmission and reflection of the THz beam at the interface between the *m* and *k* media, respectively:

$$\tilde{T}_{m,k} = \frac{2\tilde{n}_m}{\tilde{n}_m + \tilde{n}_k}, \tilde{R}_{m,k} = \frac{\tilde{n}_m - \tilde{n}_k}{\tilde{n}_m + \tilde{n}_k}, \tag{7.42}$$

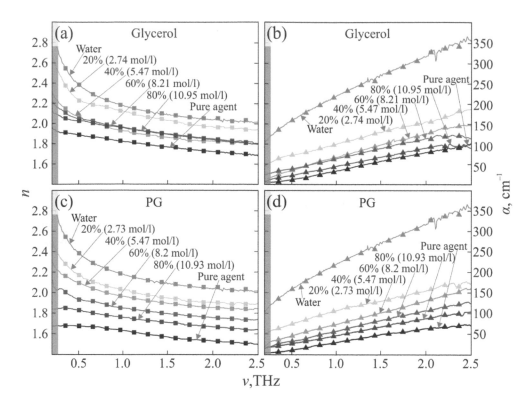

FIGURE 7.15 THz optical properties of glycerol and PG, and their aqueous solutions: (a),(b) refractive index n and amplitude absorption coefficient α of Glycerol and its aqueous solutions; (c),(d) equal data set for PG. Here, the concentration presented as percentages (20%-80%) and molar concentrations in brackets defines an amount of hyperosmotic agent (by volume) solute in deionized water, while a yellow-colored area at low frequencies defines the spectral range where distortions of experimental data due to the THz-beam diffraction at the cuvette aperture might be expected.

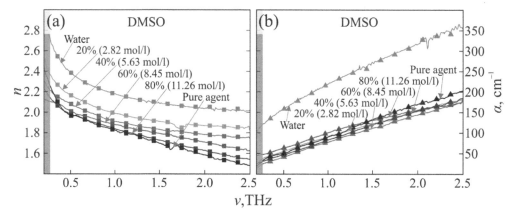

FIGURE 7.16 THz optical properties of DMSO and its aqueous solutions: (a) refractive index n, (b) amplitude absorption coefficient α. Here, the concentration presented as percentages (20%–80%) and molar concentrations in brackets defines an amount of hyperosmotic agent (by volume) solute in deionized water, while a yellow-colored area at low frequencies defines the spectral range where distortions of experimental data due to the THz-beam diffraction at the cuvette aperture might be expected.

where $\tilde{n}_{\mathrm{m}}$ and $\tilde{n}_{\mathrm{k}}$ are the complex RIs of two media. In turn, the modified Bouguer–Lambert–Beer law describes electromagnetic-wave attenuation and phase delay affected by wave propagation through a bulk medium with a complex RI $\tilde{n}$ and a length l:

$$\tilde{P}(\tilde{n}, l, \nu) = \exp\left(-i\frac{2\pi\nu}{c_0}\tilde{n}l\right). \qquad (7.43)$$

In Figures 7.15–7.19, the obtained results of THz pulsed spectroscopy of the abovementioned OCAs and their aqueous solutions are shown. In Figures 7.15–7.17, the THz optical properties of glycerol, PG, DMSO, and PEG 200, 400, 600 are shown for the total 20%–80% range (molar concentrations in brackets) of solution concentrations and pure agent. In turn, since obtaining aqueous solution of sugars Dextran 40 and 70, sucrose, fructose, and glucose with concentrations above 50% is a daunting task, in Figures 7.18 and 7.19, results are presented for a limited range of concentrations. It should be

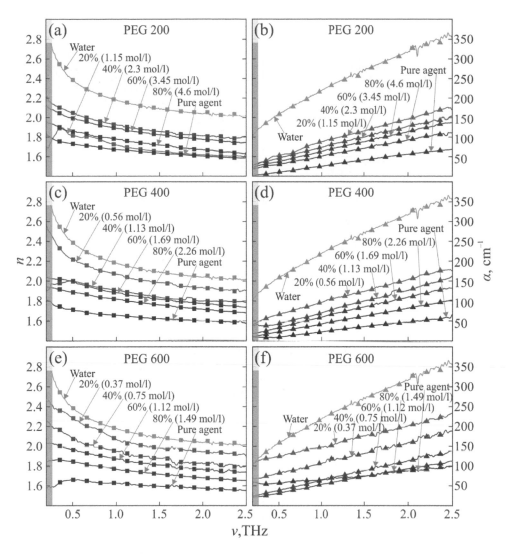

FIGURE 7.17 THz optical properties of PEG 200, 400, and 600, and their aqueous solutions: (a),(b) refractive index *n* and amplitude absorption coefficient *α* of PEG 200 and its aqueous solutions; (c)–(f) equal data sets for PEG 400 and 600. Here, the concentration presented as percentages (20–80%) and molar concentrations in brackets defines an amount of hyperosmotic agent (by volume) solute in deionized water, while a yellow colored area at low frequencies defines the spectral range where distortions of experimental data due to the THz-beam diffraction at the cuvette aperture might be expected.

noticed that THz optical properties of agents and their solutions are compared with a deionized water, as shown in gray.

From Figures 7.15–7.19, one notices that RI *n* and the amplitude absorption coefficient *α* of all agents is less than that of a liquid water, which makes them potentially attractive for immersion OC of tissues in the THz range. Such a reduction in THz optical properties of aqueous solutions is due to the partial substitution of very polar water molecules, with a high RI and amplitude absorption coefficient in a liquid state, by less polar molecules of an agent, featuring much lower RI and absorption coefficient in a liquid state. Furthermore, part of the water molecules in a solution can be bounded by the hydrophilic parts of an agent molecule, thus leading to an additional reduction in its THz optical properties [123, 124]. Nevertheless, the detailed analysis of the THz dielectric response of aqueous solution of hyperosmotic OCAs, including hydration of agent

molecules, formation of the macromolecular complex, and their evolution in time, as well as the development of a related physical and mathematical model of a complex dielectric permittivity, deserve further comprehensive studies.

From Figures 7.15–7.19, one notices that, among the considered OCAs, glycerol, PG, PEG 200, 300, 400 possess the lowest THz-wave amplitude absorption coefficient in the considered frequency range of 0.2–0.25 THz. Their THz-wave absorption (~60–70 cm^{-1} @ 1.0 THz) is a few times smaller than that of a liquid water (~210 cm^{-1}). This yields a raw estimation of the maximal achievable enhancement of the THz-wave penetration depth in hydrated tissues when these agents are used – namely, the penetration enhancement can reach ~3.0 times. In turn, other hyperosmotic agents and their water solutions possess much higher THz-wave absorption, which makes them suboptimal for tissue OC in THz biophotonics.

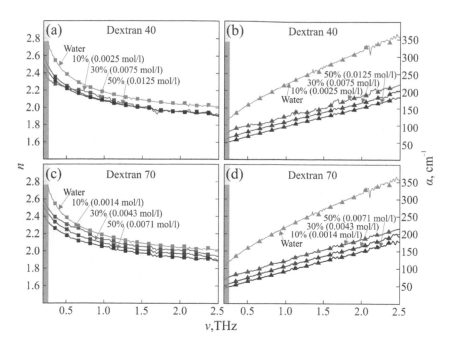

FIGURE 7.18 THz optical properties of aqueous solutions of Dextran 40 and 70: (a),(b) refractive index *n* and amplitude absorption coefficient *α* of Dextran 40 and its aqueous solutions; (c),(d) equal data set for Dextran 70. Here, the concentration presented as percentages (10%–50%) and molar concentrations in brackets defines an amount of hyperosmotic agent (by mass) solute in deionized water, while a yellow-colored area at low frequencies defines the spectral range where distortions of experimental data due to the THz-beam diffraction at the cuvette aperture might be expected.

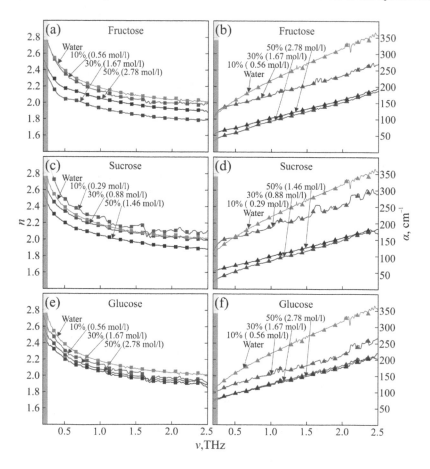

FIGURE 7.19 THz optical properties of aqueous solutions of fructose, sucrose, and glucose: (a),(b) refractive index *n* and amplitude absorption coefficient *α* of fructose aqueous solutions; (c)–(f) equal data sets sucrose and glucose. Here, the concentration presented as percentages (10%–50%) and molar concentrations in brackets defines an amount of hyperosmotic agent (by mass) solute in deionized water, while a yellow-colored area at low frequencies defines the spectral range where distortions of experimental data due to the THz-beam diffraction at the cuvette aperture might be expected.

Conclusion

The active development of photonics and intensive research and development in the field of biomedical diagnostics and therapy have contributed to the growth of interest in the quantitative assessment of the optical properties of tissues. This chapter provides an overview of methods for measuring one of the most important optical parameters of biological media – the refractive index. As has been demonstrated, refractive index quantification is equally important for the development of tissue OC technology, which can significantly improve the efficiency of optical imaging methods used in diagnostics, laser surgery, and phototherapy.

Acknowledgments

E.N.L. was supported by RFBR grant 20-32-90058-postgraduate students; V.V.T was supported by grant of the Government of the Russian Federation 075-15-2021-615; I.Y.Y. and A.N.B. were supported by RFBR grant 18-52-16025 NTSNIL_a, and D.K.T. was supported by the RF Presidential grant for young scientists SP-3507.2018.4.

L.O. was supported by the Portuguese Science Foundation, grant # FCT-UIDB/04730/2020.

The work of K.I.Z. on Sect. 7.8 was supported by the Russian Science Foundation (RSF) project #17-79-20346.

REFERENCES

1. I. Carneiro, S. Carvalho, R. Henrique, L. Oliveira, V.V. Tuchin, "Water content and scatterers dispersion evaluation in colorectal tissues," *J. Biomed. Phot. Eng.* **3**(4), 040301 (2017).
2. V.V. Tuchin, *Optical Clearing of Tissues and Blood*, SPIE Press, Bellingham, 2006.
3. L. Oliveira, V.V. Tuchin, *The Optical Clearing Method: A New Tool for Clinical Practice and Biomedical Engineering*, Springer, Cham-Switzerland, 2019.
4. H. Ding, J.Q. Lu, W.A. Wooden, P.J. Kragel, and X.-H. Hu, "Refractive indices of human skin tissues at eight wavelengths and estimated dispersion relations between 300 and 1600 nm," *Phys. Med. Biol.* **51**(6), 1479–1489 (2006).
5. Q. Ye, J. Wang, Z.-C. Deng, W.-Y. Zhou, C.-P. Zhang, and J.-G. Tian, "Measurement of the complex refractive index of tissue-mimicking phantoms and biotissue by extended differential total reflection method," *J. Biomed. Opt.* **16**(9), 097001 (2011).
6. P. Giannios, S. Koutsoumpos, K.G. Toutouzas, M. Matiatou, G.C. Zografos, and K. Moutzouris, "Complex refractive index of normal and malignant human colorectal tissue in the visible and near-infrared," *J. Biophoton.* **10**(2), 303–310 (2017).
7. G.R. Musina, I.N. Dolganova, N.V. Chernomyrdin, A.A. Gavdush, V.E. Ulitko, O.P. Cherkasova, D.K. Tuchina, P.V. Nikitin, A.I. Alekseeva, N.V. Bal, G.A. Komandin, V.N. Kurlov, V.V. Tuchin, K.I. Zaytsev, "Optimal hyperosmotic agents for tissue immersion optical clearing in terahertz biophotonics," *J. Biophoton.* **13**(12), e202000297 (2020).

8. J. Gienger, H. Groβ, J. Neukammer, and M. Bär, "Determining the refractive index of human hemoglobin solutions by Kramers-Kronig relations with an improved absorption model," *Appl. Opt.* **55**(31), 8951–8961 (2016).
9. I. Carneiro, S. Carvalho, R. Henrique, L. Oliveira, and V.V. Tuchin, "Measurement of optical properties of normal and pathological human liver tissue from deep-UV to NIR, in tissue optics and photonics," *Proc. SPIE* **11363**, 113630G (2020).
10. I. Carneiro, S. Carvalho, V. Silva, R. Henrique, L. Oliveira, V.V. Tuchin, "Kinetics of optical properties of human colorectal tissues during optical clearing: a comparative study between normal and pathological tissues," *J. Biomed. Opt.* **23**(12), 121620 (2018).
11. I. Carneiro, S. Carvalho, R. Henrique, L. Oliveira, V.V. Tuchin, "Kinetics of optical properties of colorectal muscle during optical clearing," *IEEE J. Sel. Top. Quantum Electron.* **25**(1), 7200608 (2019).
12. D. Huang, E.A. Swanson, C.P. Lin, *et al.*, "Optical coherence tomography," *Science* **254** (5035), 1178–1181 (1991).
13. J.G. Fujimoto, C. Pitris, S.A. Boppart, M.E. Brezinski, "Optical coherence tomography: An emerging technology for biomedical imaging and optical biopsy," *Neoplasia* **2** (1–2), 9–25 (2000).
14. J.F. Bille (ed.) *High Resolution Imaging in Microscopy and Ophthalmology*. Springer, Cham (2019).
15. J. Stritzel, M. Rahlves, and B. Roth, "Refractive-index measurement and inverse correction using optical coherence tomography," *Opt. Lett.* **40**, 5558–5561 (2015).
16. S. Kim, J. Na, M.J. Kim, and B.H. Lee, "Simultaneous measurement of refractive index and thickness by combining low-coherence interferometry and confocal optics," *Opt. Express* **16**, 5516–5526 (2008).
17. Y. Zhou, K.K.H. Chan, T. Lai, and S. Tang, "Characterizing refractive index and thickness of biological tissues using combined multiphoton microscopy and optical coherence tomography," *Biomed. Opt. Express* **4**, 38–50 (2013).
18. P.H. Tomlins, P. Woolliams, C. Hart, A. Beaumont, and M. Tedaldi, "Optical coherence refractometry," *Opt. Lett.* **33**, 2272–2274 (2008).
19. H. Tang, X. Liu, S. Chen, X. Yu, Y. Luo, J. Wu, X. Wang, L. Liu, "Estimation of refractive index for biological tissue using micro-optical coherence tomography," *IEEE Trans. Biomed. Eng.* **66**(6), 1803–1809 (2019).
20. R.C. Lin, M.A. Shure, A.M. Rollins, J.A. Izatt, D. Huang, "Group index of the human cornea at 1. 3-microm wavelength obtained in vitro by optical coherence domain reflectometry," *Opt. Lett.* **29**, 83–85 (2004).
21. J. Sun, S.J. Lee, L. Wu, M. Sarntinoranont, and H. Xie, "Refractive index measurement of acute rat brain tissue slices using optical coherence tomography," *Opt. Express* **20**(2), 1084–1095 (2012).
22. Z. Turani, E. Fatemizadeh, Q. Xu, S. Daveluy, D. Mehregan, and M.R.N. Avanaki, "Refractive index correction in optical coherence tomography images of multilayer tissues," *J. Biomed. Opt.* **23**(7), 070501 (2018).
23. Z. Meng, X.S. Yao, H. Yao, Y. Liang, T. Liu, Y. Li, G. Wang, S. Lan, "Measurement of the refractive index of human teeth by optical coherence tomography," *J. Biomed. Opt.* **14**, 034010 (2009).

24. I. Hariri, A. Sadr, Y. Shimada, J. Tagami, Y. Sumi, "Effects of structural orientation of enamel and dentine on light attenuation and local refractive index: An optical coherence tomography study," *J. Dent.* **40**(5), 387–396 (2012).

25. N. Hari, P. Patel, J. Ross, K. Hicks and F. Vanholsbeeck, "Optical coherence tomography complements confocal microscopy for investigation of multicellular tumour spheroids," *Sci. Rep.* **9**, 10601 (2019).

26. T. Gambichler, G. Moussa, M. Sand, D. Sand, P. Altmeyer, K. Hoffmann, "Applications of optical coherence tomography in dermatology," *J. Dermatol. Sci.* **40**(2), 85–94 (2005).

27. J. Welzel, "Optical coherence tomography in dermatology: A review," *Skin Res. Technol.* **7**, 1–9 (2001).

28. T. Gambichler, V. Jaedicke, S. Terras, "Optical coherence tomography in dermatology: Technical and clinical aspects," *Arch Dermatol Res* **303**, 457–473(2011).

29. E.C. Sattler, R. Kästle, and J. Welzel, "Optical coherence tomography in dermatology," *J. Biomed. Opt.* **18**(6), 061224 (2013).

30. J. Olsen, J. Holmes, and G.B.E. Jemec, "Advances in optical coherence tomography in dermatology: A review," *J. Biomed. Opt.* **23**(4), 040901 (2018).

31. I. Yu. Yanina, N.A. Trunina, V.V. Tuchin, "Time variation of adipose tissue refractive index under photodynamic treatment: in vitro study using OCT," *Proc. SPIE* **8222**, 82221G-1-6 (2012).

32. I.Y. Yanina, A.P. Popov, A.V. Bykov, I.V. Meglinski, V.V. Tuchin, "Monitoring of temperature-mediated phase transitions of adipose tissue by combined optical coherence tomography and Abbe refractometry," *J. Biomed. Opt.* **23**(1), 016003 (2018).

33. S. Saarakkala, S.-Z. Wang, Y.-P. Huang and Y.-P. Zheng, "Quantification of the optical surface reflection and surface roughness of articular cartilage using optical coherence tomography," *Phys. Med. Biol.* **54**, 6837(2009).

34. S.Z. Wang, Y.P. Huang, Q. Wang, Y.P. Zheng, Y.H. He, "Assessment of depth and degeneration dependences of articular cartilage refractive index using optical coherence tomography in vitro," *Connect Tissue Res.* **51**(1), 36–47 (2010).

35. B.R. Klyen, L. Scolaro, T. Shavlakadze, M.D. Grounds, and D.D. Sampson, "Optical coherence tomography can assess skeletal muscle tissue from mouse models of muscular dystrophy by parametric imaging of the attenuation coefficient," *Biomed. Opt. Express* **5**, 1217–1232 (2014).

36. N. Tachikawa, R. Yoshimura, and K. Ohbayashi, "Cross-sectional imaging of extracted jawbone of a pig by optical coherence tomography," *Proc. SPIE* **7884**, 78840F (2011).

37. O.S. Zhernovaya, V.V. Tuchin, M.J. Leahy, "Blood optical clearing studied by optical coherence tomography," *J. Biomed. Opt.* **18**(2) 026014 (2013).

38. M. Brezinski, K. Saunders, C. Jesser, X. Li, J. Fujimoto, "Index matching to improve optical coherence tomography imaging through blood," *Circulation* **103**(15), 1999–2003 (2001).

39. L. Scolaro, R.A. McLaughlin, B.R. Klyen *et al.*, "Parametric imaging of the local attenuation coefficient in human axillary lymph nodes assessed using optical coherence tomography," *Biomed. Opt. Express* **3**, 366–379 (2012).

40. F.T. Nguyen, A.M. Zysk, E.J. Chaney *et al.*, "Optical coherence tomography: The intraoperative assessment of lymph nodes in breast cancer," *IEEE Eng. Med. Biol. Mag.* **29**(2), 63–70 (2010).

41. X. Ma, Y. Chen, X. Liu, H. Ning, "Effect of refractive correction error on retinal nerve fiber layer thickness: A spectralis optical coherence tomography study," *Med. Sci. Monit.* **22**, 5181–5189 (2016).

42. F. Troiani, K. Nikolic, T.G. Constandinou, "Simulating optical coherence tomography for observing nerve activity: A finite difference time domain bi-dimensional model," *PLOS ONE* **13**(7), e0200392 (2018).

43. I. Carneiro, S. Carvalho, R. Henrique, L. Oliveira, V.V. Tuchin, "Simple multimodal optical technique for evaluation of free/bound water and dispersion of human liver tissue," *J. Biomed. Opt.* **22**(12), 125002 (2017).

44. I. Carneiro, S. Carvalho, R. Henrique, L. Oliveira, and V.V. Tuchin, "Measuring optical properties of human liver between 400 and 1000 nm," *Quant. Elect.* **49**(1), 13–19 (2019).

45. S. Glass, "Schott optical glass data sheets," http://refractiveindex.info, accessed 7 April 2020.

46. Q.W. Song, C.-Y. Ku, C. Zhang, R.B. Gross, R.R. Birge, and R. Michalak, "Modified critical angle method for measuring the refractive index of bio-optical materials and its application to bacteriorhodopsin," *J. Opt. Soc. Am. B* **12**(5), 797–803 (1995).

47. Z. Deng, J. Wang, Q. Ye *et al.*, "Determination of continuous complex refractive index dispersion of biotissue based on internal reflection," *J. Biomed. Opt.* **21**(1), 015003 (2016).

48. O. Sydoruk, O. Zhernovaya, V. Tuchin, and A. Douplik, "Refractive index od solutions of human hemoglobin from the near-infrared to the ultraviolet range: Kramers-Kronig analysis," *J. Biomed. Opt.* **17**(11), 115002 (2012).

49. V. Lucarini, K.-E. Peiponen, J.J. Saarinen, and E.M. Vartainen, *Kramers-Kronig Relations in Optical Materials Research*, Springer, Berlin, 2005.

50. S.A. Alexandrov, A.V. Zvyagin, K.K.M.B.D. Silva, D.D. Sampson, "Bifocal optical coherence refractometry of turbid media," *Opt. Lett.* **28**(2), 117–119 (2003).

51. H. Ding, J.Q. Lu, K.M. Jacobs, X.-H. Hu, "Determination of refractive indices of porcine skin tissues and Intralipid at eight wavelengths between 325 and 1557 nm," *J. Opt. Soc. Am. A* **22**(6), 1151–1157 (2005).

52. G.J. Tearney, M.E. Brezinski, J.F. Southern, B.E. Bouma, M.R. Hee, J.G. Fujimoto, "Determination of the refractive index of highly scattering human tissue by optical coherence tomography," *Opt. Lett.* **20**(21), 2258–2260 (1995).

53. D. Chan, B. Schulz, K. Gloystein, H.H. Muller, M. Rubhausen, "In vivo spectroscopic ellipsometry measurements on human skin," *J. Biomed. Opt.* **12**(1), 014023 (2007).

54. A.M. Zysk, S.G. Adie, J.J. Armstrong *et al.*, "Needle-based refractive index measurement using low-coherence interferometry," *Opt. Lett.* **32**(4), 385–387 (2007).

55. P. Sun, Y. Wang, "Measurements of optical parameters of phantom solution and bulk animal tissues *in vitro* at 650 nm," *Opt. Laser Technol.* **42**, 1–7 (2010).

56. S. Cheng, H.Y. Shen, G. Zhang, C.H. Huang, X.J. Huang, "Measurement of the refractive index of biotissue at four laser wavelengths," *Proc. SPIE* **4916**, 172–176 (2002).

57. J. Lai, Z. Li, C. Wang, A. He, "Experimental measurement of the refractive index of biological tissues by total internal reflection," *Appl. Opt.* **44**(10), 1845–1849 (2005).

58. L. Lin, H. Li, S. Xie, "Linear method of determining the refractive index of biotissue," *Proc. SPIE* **3863**, 177–181 (1999).

59. A.M. Zysk, E.J. Chaney, S.A. Boppart, "Refractive index of carcinogen-induced rat mammary tumours," *Phys. Med. Biol.* **51**, 2165–2177 (2006).

60. I. Yu. Yanina, E.N. Lazareva, V.V. Tuchin, "Refractive index of adipose tissue and lipid droplet measured in wide spectral and temperature range," *Appl. Opt.* **57**(17), 4839–4848 (2018).

61. C.F. Cook, R.W. Bray, K.G. Weckel, "Variations in the chemical and physical properties of three bovine lipid depots," *J. Animal Sci.* **24**(4), 1192–1194 (1965).

62. A.N. Bashkatov, E.A. Genina, V.I. Kochubey *et al.*, "Optical properties of melanin in the skin and skin-like phantoms," *Proc. SPIE* **4162**, 219–226 (2000).

63. M. Rajadhyaksha, M. Grossman, D. Esterrowitz, R.H. Webb, R.R. Anderson, "*In vivo* confocal scanning laser microscopy of human skin: Melanin provides strong contrast," *J. Invest. Dermatol.* **104**(6), 946–952 (1995).

64. W. Song, L. Zhang, S. Ness, J. Yi, "Wavelength-dependent optical properties of melanosomes in retinal pigmented epithelium and their changes with melanin bleaching: A numerical study," *Biomed. Opt. Express* **8**(9), 3966–3980 (2017).

65. B.S. Goldschmidt, S. Mehta, J. Mosley *et al.*, "Photoacoustic measurement of refractive index of dye solutions and myoglobin for biosensing applications," *Biomed. Opt. Express* **4**, 2463–2476 (2013).

66. V.V. Tuchin, *Tissue Optics: Light Scattering Methods and Instruments for Medical Diagnostics*, 3rd ed., SPIE Press, Bellingham, WA, 2015.

67. J. Heijmans, L. Cheng, and F. Wieringa, "Optical fiber sensors for medical applications: Practical engineering considerations," *IFMBE Proc.* **22**, 2330–2334 (2009).

68. V.V. Tuchin (Ed.), *Handbook of Optical Biomedical Diagnostics. Light–Tissue Interaction*, vol. 1, 2nd ed., SPIE Press, Bellingham, WA, 2016.

69. A.N. Yaroslavsky, I.V. Yaroslavsky, T. Goldbach, and H.-J. Schwarzmaier, "Optical properties of blood in the near infrared spectral range," *Proc. of SPIE* **2678**, 314–324 (1996).

70. S.L. Jacques, "Corrigendum: Optical properties of biological tissues: A review," *Phys. Med. Biol.* **58**(13), 5007–5008 (2013).

71. N. Bosschaart, G.J. Edelman, M.C.G. Aalders, T.G. van Leeuwen, and D.J. Faber, "A literature review and novel theoretical approach on the optical properties of whole blood," *Lasers Med. Sci.* **29**(2), 453–479 (2014).

72. J. Wang, Zh. Deng, X. Wang *et al.*, "Measurement of the refractive index of hemoglobin solutions for a continuous spectral region," *Biomed. Opt. Express* **6**(7), 2536–2541 (2015).

73. O. Zhernovaya, O. Sydoruk, V. Tuchin, and A. Douplik, "The refractive index of human hemoglobin in the visible range," *Phys. Med. Biol.* **56**(13), 4013–4021 (2011).

74. M. Yahya, and M.Z. Saghir, "Empirical modelling to predict the refractive index of human blood," *Phys. Med. Biol.* **61**(4), 1405–1415 (2016).

75. J.D. Rowe, D. Smith, and J.S. Wilkinson, "Complex refractive index spectra of whole blood and aqueous solutions of anticoagulants, analgesics and buffers in the mid-infrared," *Sci. Rep.* **7**, 7356 (2017).

76. Y.L. Jin, J.Y. Chen, L. Xu, and P.N. Wang, "Refractive index measurement for biomaterial samples by total internal reflection," *Phys. Med. Biol.* **51**(20), 371–379 (2006).

77. D.J. Faber, M.C.G. Aalders, E.G. Mik, B.A. Hooper, M.J.C. van Gemert, and T.G. van Leeuwen, "Oxygen saturation-dependent absorption and scattering of blood," *Phys. Rev. Lett.* **93**, 028102 (2004).

78. M. Jedrzejewska-Szczerska, "Measurement of complex refractive index of human blood by low-coherence interferometry," *Eur. Phys. J. Spec. Top.* **222**(9), 2367–2372 (2013).

79. Zh. Wang, K. Tangella, A. Balla, and G. Popescu, "Tissue refractive index as marker of disease," *J. Biomed. Opt.* **16**(11), 116017 (2011).

80. H. Majeed, Sh. Sridharan, M. Mir *et al.*, "Quantitative phase imaging for medical diagnosis," *J. Biophot.* **10**(2), 177–205 (2017).

81. G. Mazarevica, T. Freivalds, and A. Jurka, "Properties of erythrocyte light refraction in diabetic patients," *J. Biomed. Opt.* **7**(2), 244–247 (2002).

82. L.V. Plotnikova, A.M. Polyanichko, M.O. Kobeleva *et al.*, "Analysis of blood serum by the method of refractometry in antitumor therapy in patients with multiple myeloma," *Opt. Spectrosc.* **124**(1), 140–142 (2018).

83. A.N. Bashkatov, E.A. Genina, and V.V. Tuchin, "Optical properties of skin, subcutaneous, and muscle tissues: A review," *J. Innov. Opt. Health Sci.* **4**(1), 9–38 (2011).

84. F.P. Bolin, L.E. Preuss, R.C. Taylor, and R.J. Ference, "Refractive index of some mammalian tissues using a fiber optic cladding method," *Appl. Opt.* **28**(12), 2297–2303 (1989).

85. D.K. Sardar, and L.B. Levy, "Optical properties of whole blood," *Lasers Med. Sci.* **13**(2), 106–111 (1998).

86. H. Li, L. Lin, and S. Xie, "Refractive index of human whole blood with different types in the visible and near-infrared ranges," *Proc. SPIE* **3914**, 517–521 (2000).

87. R. Barer, "Refractometry and interferometry of living cells," *J. Opt. Soc. Am.* **47**(6), 545–556 (1957).

88. M.V. Volkenshtein, *Molecular Optics*, Gosteskhizdat, Moscow, 1951 [in Russian].

89. D. Segelstein, "The complex refractive index of water," M.S. thesis, Department of Physics, University of Missouri, Kansas City (1981).

90. W. Heller, "Remarks on refractive index mixture rules," *J. Phys. Chem.* **69**(4), 1123–1129 (1966).

91. E.N. Lazareva, V.V. Tuchin, "Blood refractive index modelling in the visible and near infrared spectral regions," *J. Biomed. Photon. Eng.* **4**(1), 010503 (2018).

92. A.N. Bashkatov, K.V. Berezin, K.N. Dvoretskiy *et al.*, "Measurement of tissue optical properties in the context of tissue optical clearing," *J. Biomed. Opt.* **23**(9), 091416 (2018).

93. T. Ermatov, R.E. Noskov, A.A. Machnev *et al.*, "Multispectral sensing of biological liquids with hollow-core microstructured optical fibres," *Light Sci. Appl.* **9**, 173 (2020).

94. L.Y. Mattley, G. Leparc, R. Potter, and L. Garcia-Rubio, "Light scattering and absorption model for the qualitative interpretation of human blood platelets spectral data," *Photochem. Photobiol.* **715**, 610–619 (2000).

95. M. Friebel and M. Meinke, "Model function to calculate the refractive index of native hemoglobin in the wavelength range of 250 to 1100 nm dependent on concentration," *Appl. Opt.* **45**(12), 2838–2842 (2006).

96. M. Friebel, "Determination of the complex refractive index of highly concentrated hemoglobin solutions using transmittance and reflectance measurements," *J. Biomed. Opt.* **10**(6), 064019 (2006).

97. Y.K. Park, T. Yamauchi, W. Choi, R. Dasari, and M.S. Feld, "Spectroscopic phase microscopy for quantifying hemoglobin concentrations in intact red blood cells," *Opt. Lett.* **34**(23), 3668–3670 (2009).

98. E.N. Lazareva, V.V. Tuchin, "Measurement of refractive index of hemoglobin in the visible/NIR spectral range," *J. Biomed. Opt.* **23**(3), 035004 (2018).

99. H.-C. Cheng and Y.-C. Liu, "Simultaneous measurement of group refractive index and thickness of optical samples using optical coherence tomography," *Appl. Opt.* **49**, 790–797 (2010).

100. I. Yu. Yanina, N.A. Trunina, and V.V. Tuchin, "Photoinduced cell morphology alterations quantified within adipose tissues by spectral optical coherence tomography," *J. Biomed. Opt.* **18**(11), 111407 (2013).

101. J.H. Jung, K. Kim, J. Yoon, and Y.K. Park, "Hyperspectral optical diffraction tomography," *Opt. Express* **24**(3), 2006–2012 (2016).

102. F.E. Robles, C. Wilson, G. Grant, and A. Wax, "Molecular imaging true-colour spectroscopic optical coherence tomography," *Nat. Photonocs* **5**(12), 744–747 (2011).

103. F.E. Robles, L.L. Satterwhite, and A. Wax, "Non-linear phase dispersion spectroscopy," *Opt. Lett.* **36**(23), 4665–4667 (2011).

104. J. Singh, *Optical Properties of Condensed Matter and Applications*, Wiley, Chichester, 2006.

105. X. Xu, R.K. Wang, J.B. Elder and V.V. Tuchin, "Effect of dextran-induced changes in refractive index and aggregation on optical properties of whole blood," *Phys. Med. Biol.* **48**, 1205–1221 (2003).

106. J.A. Lamasso, "Error in hematocrit value produced by excessive ethylenediaminetetraacetate," *Am. J. Clin. Pathol.* **44**, 109–110 (1965).

107. S. Liu, Z. Deng, J. Li *et al.*, "Measurement of the refractive index of whole blood and its components for a continuous spectral region," *J. Biomed. Opt.* **24**(3), 035003 (2019).

108. M. Daimon, and A. Masumura, "Measurement of the refractive index of distilled water from the near-infrared region to the ultraviolet region," *Appl. Opt.* **46**(18), 3811–3820 (2007).

109. D.W. Leonard, and K.M. Meek, "Refractive indices of the collagen fibrils and extrafibrillar material of the corneal stroma," *Biophys. J.* **72**(3), 1382–1387 (1997).

110. K.M. Meek, S. Dennis, and S. Khan, "Changes in the refractive index of the stroma and its extrafibrillar matrix when the cornea swells," *Biophys. J.* **85**(4), 2205–2212 (2003).

111. K.M. Meek, D.W. Leonard, C.J. Connon, S. Dennis, and S. Khan, "Transparency, swelling and scarring in the corneal stroma," *Eye* **17**, 927–936 (2003).

112. S. Carvalho, N. Gueiral, E. Nogueira, R. Henrique, L. Oliveira, and V.V. Tuchin, "Glucose diffusion in colorectal mucosa – a comparative study between normal and cancer tissues," *J. Biomed. Opt.* **22**(9), 091506 (2017).

113. H. Liu, B. Beauvoit, M. Kimura, and B. Chance, "Dependence of tissue optical properties on solute-induced changes in refractive index and osmolarity," *J. Biomed. Opt.* **1**(2), 200–211 (1996).

114. R. Graaff, J.G. Aarnoudse, J.R. Zijp *et al.*, "Reduced light-scattering properties for mixtures of spherical particles: A simple approximation derived from Mie calculations," *Appl. Opt.* **31**(10), 1370–1376 (1992).

115. S.A. Prahl, M.J. van Gemert, and A.J. Welch, "Determining the optical properties of turbid media by using the adding-doubling method," *Appl. Opt.* **32**(4), 559–568 (1993).

116. L. Oliveira, M.I. Carvalho, E. Nogueira, and V.V. Tuchin, "Skeletal muscle dispersion (400–1000 nm) and kinetics at optical clearing," *J. Biophoton.* **11**(1), 1019 (2018).

117. S. Carvalho, N. Gueiral, E. Nogueira, R. Henrique, L. Oliveira, and V.V. Tuchin, "Comparative study of the optical properties of colon mucosa and colon precancerous polyps between 400 and 1000 nm," *Proc. SPIE* **10063**, 100631L (2017).

118. G.A. Komandin, V.B. Anzin, V.E. Ulitko *et al.*, "Optical cryostat with sample rotating unit for polarization-sensitive terahertz and infrared spectroscopy," *Opt. Eng.* **59**(6), 61603 (2019).

119. G.R. Musina, I.N. Dolganova, K.M. Malakhov *et al.*, "Terahertz spectroscopy of immersion optical clearing agents: DMSO, PG, EG, PEG," *Proc. SPIE* **10800**, 108000F (2018).

120. G.R. Musina, A.A. Gavdush, D.K. Tuchina *et al.*, "A comparison of terahertz optical constants and diffusion coefficients of tissue immersion optical clearing agents," *Proc. SPIE* **11065**, 110651Z (2019).

121. G.R. Musina, A.A. Gavdush, N.V. Chernomyrdin *et al.*, "Optical properties of hyperosmotic agents for immersion clearing of tissues in the terahertz range," *Opt. Spectrosc.* **128**(7), 1020 (2020).

122. A.A. Gavdush, V.E. Ulitko, G.R. Musina *et al.*, "A method for reconstruction of terahertz dielectric response of thin liquid samples," *Proc. SPIE* **11060**, 110601G (2019).

123. O. Smolyanskaya, N. Chernomyrdin, A. Konovko *et al.*, "Terahertz biophotonics as a tool for studies of dielectric and spectral properties of biological tissues and liquids," *Prog. QuantUM ElectrON.* **62**, 1–77 (2018).

124. O.P. Cherkasova, M.M. Nazarov, M. Konnikova, and A.P. Shkurinov, "THz spectroscopy of bound water in glucose: direct measurements from crystalline to dissolved state," *J. Infrared Millimeter Terahertz Waves* **41**, 1057–1068 (2020).

8

Water migration at skin optical clearing

Anton Yu. Sdobnov, Johannes Schleusener, Jürgen Lademann, Valery V. Tuchin, and Maxim E. Darvin

CONTENTS

Introduction

The skin, the outer organ of vertebrates, plays an essential role in providing a barrier between the environment and the living organism. The skin protects the internal organs from external mechanical damage and prevents an organism from penetration by pathogens [1]. For humans, the skin is the largest organ, covering an area of 1.5–2 m² of the body. Skin is a multifunctional organ that also performs thermoregulatory, receptor, endocrine, metabolic, secretive, respiratory, and immune functions. It also helps in regulation of water loss and metabolism [2]. Changes in skin morphology and disturbances of the skin's barrier function can be an evidence of skin disease progression [3–5]. Therefore, an early diagnosis of skin diseases, monitoring of disease development, and treatment efficiency, as well as age-related changes in the skin structure, are urgent problems in dermatology and cosmetology. However, for precise diagnosis of skin conditions, a biopsy is often necessary. Biopsy is an invasive surgical procedure and can be painful. Thus, the development of noninvasive methods for monitoring of skin properties and conditions is a very promising and relevant field of research. In the last decades, a large amount of optical techniques, such as confocal Raman microscopy (CRM) [6, 7], multiphoton tomography (MPT) [8–10], Raman and coherent anti-Stokes Raman spectroscopy [11–13], laser speckle contrast imaging [14, 15], optical coherence tomography (OCT) [16, 17], laser scanning microscopy [18, 19], fluorescence lifetime imaging (FLIM) [20–22] have been successfully implemented for noninvasive and partially noninvasive skin imaging/assessment in dermatology and skin physiology. However, the main problem arising during optical skin measurement is the limited probing depth of excitation and informative light due to high absorption and multiple scattering of the light [23, 24].

To overcome this limitation and increase the probing depth in the skin, an optical clearing (OC) technique was introduced [23, 24], which used a variety of optical clearing agents (OCAs) topically applied on the skin. Among them are glycerol, glucose, polyethylene glycol (PEG), fructose, iohexol, and other substances. OC allows one to significantly increase the probing depths in skin, as well providing better resolution and image quality [25, 26]. Most research suggests that three main mechanisms contribute to OC efficiency: tissue dehydration, refractive index (RI) matching, and collagen dissociation [25, 26]. All three mechanisms proceed simultaneously with different relative contributions depending on the OCA and its concentration, tissue type, OCA delivery method, etc. In general, most OCAs are considered nontoxic for usage on the skin. However, prolonged treatment with some OCAs can cause negative effects such as skin compression, local hemostasis, and even necrosis in case of intradermally injected OCA. For instance, it was demonstrated that glycerol could result in dissociation of collagen fibrils, which can cause morphological changes in the dermis [27]. Also, glycerol can affect perfusion of skin blood in particular, causing venous and arterial stasis [28]. In fact, all three mechanisms are associated with the water migration in the skin. As one of the main OC mechanisms is dehydration, the application of OCA affects the water content in the skin. The application of OCA leads to an increase of the water flux from the skin interstitial fluid (ISF) towards the opposite direction of OCA presence. This is due to high osmotic pressure caused by the interaction with the OCA [29–31]. At the same time, it was suggested that water leaked from the skin, was replaced by the OCA and diffused into the space between collagen fibrils, thus, decreasing the tissue RI mismatch [32, 33]. For example, the RI of human epidermis at 589 nm wavelength is 1.44, of dermis is 1.39, while the RI of water at the same wavelength is 1.33. At the same time, the RI of 100% Omnipaque and 70% glycerol is 1.43, which is much closer to the RI of the described skin layers [9, 34, 35]. Also, the collagen dissociation is connected with changes in collagen hydration [36]. This way, the OCA application can lead to water fluxes both outside and inside the skin, which occurs simultaneously and affects the skin water content. As well as the OC efficiency, the influence of the water content of skin depends on the OCA parameters, treatment time, and skin properties.

Skin is always hydrated, and therefore the water concentration and its binding strength is one of the most important parameters of a physiologically healthy state [37]. The skin hydration level is critical for maintaining healthy skin conditions. Therefore, the investigation of the influence of OCA on skin water content changes and its reversibility is a critical goal for modern dermatology and represents one of the current tasks of investigation.

This chapter provides a brief summary and description of the optical properties of skin and its components, various methods of skin water content measurement, and recent investigations on the influence of different OCAs on skin water content.

Structure and optical properties of skin

The skin is a heterogeneous organ containing complex cell-fibrous structures [38] consisting of three main layers: the epidermis, the dermis, and subcutaneous fat.

The epidermis is the outer skin layer, which acts as protective barrier between the environment and the living organism and is mainly responsible for water balance regulation in the body. The epidermis can be divided into five sublayers [39]. The deepest of them is the *stratum basale* (also known as basal layer or *stratum germinativum*), which separates the epidermis from the dermis. The *stratum basale* is a single layer of columnar or cuboidal epithelium containing continuously proliferating keratinocytes. Above this layer, the *stratum spinosum* (also known as spinous layer) is located, which consists of between three and eight rows of keratinocytes with cytoplasmic outgrowths. This layer is the bulk mass of the epidermis and has a thickness of 20–150 μm depending on the area [39]. The next layer is the *stratum granulosum* (granular layer), which consist of between one and four cell rows. The cells of the *stratum granulosum* layer contain keratogyalin and lipid granules, whose main role is to synthetize keratin filaments and lipids [39]. Above this, the *stratum lucidum* is located, which consists of between two and four rows of dead cells without nuclei. The *stratum lucidum* is not always observed, and therefore some authors ignore it. The outermost layer of the epidermis is the *stratum corneum* (SC). This layer is composed of flattened dead keratinocytes (corneocytes) with a thickness of 70–100 nm and a typical length of 15–20 μm [40]. The main component of the SC is keratin, which exceeds 50% of its mass. Intercellular lipids located between the corneocytes in form of low-permeable lamellas [41] whose lateral organization determined the skin barrier function [42, 43]. The thickness of the SC can significantly vary for different body areas. For example, the typical SC thickness on the human forearm does not exceed 20 μm, while on the foot and palm, it can be up to 200 μm.

The epidermis mostly protects the living organism from external threats. Besides, it is also involved in respiration, regulation of water evaporation, thermoregulation, metabolic processes, and protection of the organism from solar radiation [1, 2, 44].

The dermis is located beneath the epidermis. The dermis is a connective tissue mainly consisting of collagen and elastin fibers and intracellular fluids [39, 45–47]. The thickness of the dermis can vary for different body parts. The average thickness ranges from 1.0 mm to 1.5 mm and is a maximum of 3 mm [47]. The dermis can be divided into two sublayers. The uppermost one is the papillary dermis, intertwining with rete ridges of the epidermis. The lower layer is the reticular dermis, which contains most of the dermal elastic fibers [45, 46].

Beneath the dermis is subcutaneous fat, which mainly consists of areolar tissue, adipocytes, glycosaminoglycans, collagen, and elastin. The main function of this layer is to connect skin with deep fascia through connective fibers [48, 49]. The subcutaneous fat contains nerve endings and hair follicles, as well as blood and lymph vessels. This layer is involved in homeostasis, protecting the body from heat loss.

One of the main components responsible for the structure and mechanical properties of skin is water [50]. The study of skin hydration and changes in water content is one of the major topics in modern dermatology and cosmetology [51, 52]. In a series of works, it was shown that the SC serves as a barrier protecting skin from penetration of xenobiotics [53–55]. Also, it controls the penetration of water into the skin and its evaporation [56–59]. The skin water content affects the diffusion velocity of drugs, OCAs, and cosmetic products applied on the skin [60, 61]. Also, the water content and its binding properties are affected by aging [62]. Thus, an investigation of different types of water molecules, depending on the strength of their bonding, is up to date [40, 57, 63]. Particularly, the water molecules in the SC are mainly bound to the amino acids and proteins and partly to intercellular lipids due to varying strengths of hydrogen bonds, specified as tightly, strongly, and weakly bound water [64]. It was shown that changes in bound water affect skin elasticity [65] and its vital functions [66]. Also, a small amount of the water molecules in the skin are free and in a very weakly bound state, specified as unbound water [64]. It was shown that the content of unbound water could depend on environmental humidity [67–69], as well as on the general physiological and pathological condition of the skin.

The main skin parameters responsible for its optical properties and possible probing depth using optical imaging methods are scattering, absorption, and the refractive index RI [70–72].

Absorption can be characterized as the loss of the light energy after the interaction of light with tissue. It is generally accepted that the skin chromophores hemoglobin, bilirubin, carotenoids, and melanin mainly determine the absorption properties of the skin in the visible range [73, 74].

The dermis is well supplied with blood, which contains hemoglobin. Hemoglobin is the pigment responsible for oxygen delivery and serves one of the main visible light absorber in the dermis [39]. For a healthy adult hemoglobin (Hb A) is a protein consisting of four polypeptide chains connected in heme [73]. In turn, heme in Hb A is the main absorber of visible light in the blood. The absorption spectra of Hb A contain three unique peaks. The first one is the Soret band, located in the blue region of the spectrum. Two other peaks are located in the green-yellow region between 500 and 600 nm. Thus, the high absorption of these peaks results in the visual appearance of Hb A as red.

The next light absorber in skin is melanin, which is present in the epidermis. The absorption spectra of melanin gradually decrease from the ultraviolet to the infrared range. Despite the large variety of research works, the melanin absorption spectra are still poorly investigated [75, 76]. Currently, the most widespread theory of high melanin absorption is that it consists of a set of randomly arranged oligomers and polymers with different shapes. Such a structure results in the appearance of multiple absorption peaks, leading to the effect of broadband absorption [76, 77].

In addition to melanin and hemoglobin, chromophores (bilirubin and carotenoids [78, 79], fats [78], cell nuclei, and filamentous proteins [80] also contribute to strong skin absorption in the visible range. Despite the fact that chromophores' contribution to skin absorption can be considered as a separate factor [81]. Its contribution is usually combined in one group with other minor factors during modeling of light interaction with skin [82, 83]. While bilirubin is a blood chromophore, i.e. can be found in the dermis [84], carotenoids are mainly concentrated in the SC [85]. Both of these pigments make the skin color appear yellowish.

Despite the presence of a high water content in the skin, water strongly absorbs light only in the infrared range and has no significant absorption in the visible range. This way, the absorption of water is mainly responsible for the skin absorption properties in the near infrared range of the spectrum [29, 86, 87].

Scattering is a term used to describe changes in the direction of propagation, polarization and phase of a light wave when it interacts with locally inhomogeneous media. Usually, scattering is considered an effect arising on the surface of the tissue (as well as refraction and reflection) or as light interaction with small areas which have optical properties different from their surroundings. It was shown that only 4%–7% of visible light reflects from the skin's surface, while the rest of the light scatters inside the skin [88, 89]. For skin, scattering can be described as a relative contribution of Rayleigh and Mie scattering.

The main skin components responsible for scattering are filamentous proteins. For the epidermis, filamentous proteins are represented by keratin, while for the dermis they are mostly collagen, which makes up 70% of the dermis [39]. Also, melanosomes, cell nuclei, and cell membranes partly contribute to skin scattering [90–92]. In general, the epidermis is close to the dermal scattering. At the same time, due to the thinness of the epidermis, the skin scattering can be described just by the dermal scattering properties [93].

The scattering coefficient μ_s and scattering anisotropy factor g can be determined by the differences between RI (n) of tissue cell components (e.g. organelles, membranes, nuclei, etc.) and interstitial fluid and/or scleroprotein fibrous structures

of the dermis, muscle, and cartilage [94]. Currently, refractive indices for various biological tissues and their components (collagen [95], cells organelles [96], and non-oxygenated hemoglobin [97, 98]) have been experimentally measured in the visible and infrared ranges. Tables 8.1 and 8.2 show data on the measurements of scattering coefficients, absorption coefficients μ_a, and refractive indices for different layers of human and porcine skin measured *in vitro* and *ex vivo*.

Methods of water content estimation

A variety of biophysical techniques are widely used for assessment of the water content in the skin. One of the pioneering works presented the measurements of bound water in the SC using differential scanning calorimetry [116]. The results showed that bound water content is 0.34 wt% of dry SC. Also, the investigation of SC hydration using the same method has been presented in a series of works [63, 117], where the bound water content was measured to be around 25 wt%. It was shown [118] that bound water in SC was 34% of its dry weight, meaning that bound water makes up 80% of total water in the SC.

Further, the noninvasive high impedance method for skin surface hydration assessment *in vivo* has been introduced [119]. This method is based on the measurement of changes in capacitive reactance and output voltage of the resistance detector caused by changes in skin hydration. Despite the high sensitivity, this method is able to measure only indirect changes in hydration but not the quantitative amount of the water content. Currently, several commercial devices are available for skin hydration assessment based on the electrical properties of the skin [120–123].

Nuclear magnetic resonance (NMR) spectroscopy has been used for the determination of the bound water content in the skin [124]. Further, the results of bound water content measurements of hairless rat SC using proton nuclear magnetic resonance (^{1}H-NMR) spectroscopy has been presented for different temperature conditions [125]. However, NMR spectroscopy also provides only indirect information on the water content by assessing relaxation time in nuclear magnetic resonance. The same disadvantage may be attributed to the differential thermal analysis method, where bound water content is estimated as the fraction of nonfreezable water [63, 116–118, 126] and to optical coherence tomography (OCT), where changes in skin hydration are estimated as changes in refractive index [113, 127].

Several works presented results on the investigation of skin hydration, skin permeability, and changes in bound water [128–133] using infrared and near-infrared spectroscopy. Hansen and Yelin [134] showed that 30 wt% bound water in the SC can be separated into primary and secondary bound water. Also, it was shown that even water which is weakly bound to proteins can freeze at between −50°C and −30°C [134, 135]. Further, the modified near-infrared multispectral imaging method has been introduced for measurement of skin hydration [136]. Comparing the methods based on conductance and capacitance, the proposed method showed much better sensitivity to hydration changes. However, due to high absorption

TABLE 8.1

Scattering and absorption coefficients for different layers of human and porcine skin measured *in vitro* and *ex vivo*

Tissue	Wavelength, nm	μ_a, cm^{-1}	μ_s, cm^{-1}	g	Reference
Human *SC*	400	17.28	500	0.903	[93, 99–102]
	450	11.63	500	0.910	
	500	10.47	500	0.916	
	550	9.83	500	0.923	
	600	8.67	500	0.930	
	650	8.21	500	0.936	
	700	8.15	500	0.942	
Human epidermis	400	6.77	156.3	0.712	[93, 99–102]
	450	4.41	121.6	0.728	
	500	2.58	93.01	0.745	
	550	1.63	74.70	0.759	
	600	1.47	63.76	0.774	
	650	1.2	55.48	0.787	
	700	1.06	54.66	0.804	
Human bloodless epidermis	450	5.13	134.9	0.054	[103]
	500	3.45	119.9	0.120	
	550	2.28	108.1	0.288	
	600	1.81	97.38	0.410	
	650	1.44	87.89	0.461	
	700	1.16	78.48	0.500	
	750	1.03	72.29	0.519	
	800	0.88	65.89	0.531	
Human dermis	400	13.82	159.9	0.715	[93, 99–102]
	450	9.31	124.1	0.715	
	500	8.37	92.24	0.715	
	550	7.86	77.22	0.715	
	600	6.94	63.09	0.715	
	650	6.57	55.98	0.715	
	700	6.52	53.62	0.715	
Human bloodless dermis	633	2.7	187	0.820	[101]
Human bloodless dermis	350	23.2	147.2	0.140	[104]
	400	9.5	136.1	0.220	
	450	6.3	130.8	0.380	
	633	2.7	90.3	0.620	
Piglet dermis	632.8	0.89	289	0.926	[105]
	790	1.8	254	0.945	
	850	0.33	285	0.968	
Piglet skin (epidermis + dermis)	632.8	1.0	492	0.953	[105]
	790	2.4	409	0.952	
	850	1.6	403	0.962	
Porcine dermis	900	0.06	282.6	0.904	[106]
	1000	0.12	270.4	0.904	
	1100	0.17	267.2	0.904	
	1200	1.74	263.9	0.903	
	1300	1.04	262.8	0.903	
	1400	9.11	246.2	0.872	
	1500	7.32	259.6	0.873	
Porcine dermis	442	1.9	59	0.360	[107]
	532	1.4	69	0.640	
	633	0.7	58	0.720	
	850	1.6	90	0.880	
	1064	3.1	26	0.860	
	1310	6.2	40	0.870	
	1557	10.4	21	0.820	

TABLE 8.2

Refractive indices for different layers of human and porcine skin

Tissue	Wavelength, nm	n	Reference
Human SC	400–700	1.55	[108]
Human SC	1300	1.51	[109]
Human SC	1300	1.47	[110]
Human SC	980	1.50	[111]
Human epidermis	980	1.34	[112]
Human epidermis	1300	1.43	[110]
Human epidermis	1300	1.34	[109]
Human dermis	1300	1.41	[109]
Human dermis	1300	1.41	[109]
Human epidermis	442	1.45	[34]
	532	1.45	
	633	1.453	
	850	1.42	
	1064	1.43	
	1310	1.42	
	1557	1.40	
Human epidermis	1300	1.39	[113]
Human dermis	442	1.44	[34]
	532	1.38	
	633	1.39	
	850	1.38	
	1064	1.37	
	1310	1.36	
	1557	1.36	
Porcine skin	1300	1.415	[114]
Porcine dermis	442	1.376	[115]
	532	1.359	
	633	1.354	
	850	1.364	
	1064	1.360	
	1310	1.357	
	1557	1.361	

of water in the infrared region, the possible penetration depth is limited to a few micrometers by using mid- and far-infrared spectroscopy. This way, investigation of the water content by this method is only possible in the SC.

The X-ray microanalysis (XRMA) technique allowed the local water content in different skin layers from the SC to the papillary dermis to be quantitatively determined for the first time [137]. It was shown that the water content increases from the SC towards the *stratum granulosum*. A similar result has been shown in [138] using a scanning transmission electron microscope (STEM). At the same time, the water content at the deeper layers of skin remains constant [139]. The water regulation mechanisms in the SC have been studied using the X-ray diffraction technique [140]. The authors showed that the water content in human SC varies between 20 and 30 wt%, depending on the skin state. Also, they suggested that stabilization of the short lamellar structure of the intercellular lipids of SC plays a key role in the regulation of the water content.

Kelman *et al.* [141] suggested that analysis of back-reflected speckle patterns registered from skin allows one to obtain information about the hydration level based on the behavior of

acoustic waves in skin. A verification of the proposed method on human skin showed high accuracy compared to commercially available devices for skin moisture assessment.

Another approach for the analysis of skin water content is the use of diffuse reflectance spectroscopy (DRS). Water absorption bands with a maximum at $\lambda=1450$ nm and $\lambda=1920$ nm are mostly used for estimation of skin moisturization [142]. In work [143], the bands with maxima at $\lambda=1190$ nm and $\lambda=1450$ nm were used for monitoring of porcine skin dehydration caused by OCA application. However, strong absorption of water in the region of 1400–2000 nm ($\mu_a = 30$–50 cm^{-1}) limits the probing depth of this method to ≈100 μm. Yakimov *et al.* [144] investigated algorithms for human skin water content assessment using water absorption bands at $\lambda=980$, 1190, 1450, and 1920 nm. It was shown that water content estimation using $\lambda=980$ nm weakly correlates with algorithms using other wavelengths due to the difference in penetration depths. Also, it was demonstrated that results of water estimation using $\lambda=980$ nm are affected by oxyhemoglobin concentration. However, the authors suggested that this algorithm can still be used to determine the water content in deep skin layers.

Along all the methods for water content assessment in the skin, Raman spectroscopy takes a special place. The effect of Raman scattering was discovered in 1928 [145]. The simplest way to describe this effect is to use the quantum theory. Thus, light can be considered as photons, which are scattered after interaction with tissue molecules. The amount of scattering molecules is proportional to the amount of bonds in molecules [146]. This way, Raman spectroscopy is based on the detection of inelastic scattered light during the exchange of vibrational and rotational energy between probing radiation and tissue molecules. Figure 8.1 shows a diagram representing different types of scattering.

As can be seen from Figure 8.1, light interaction with molecules causes excitation of photons and its following transition to the virtual state. Further, three potential events can occur:

- The molecule can return from the virtual to the ground state, at the same time emitting a photon with energy $h\nu_0$, which is equal to the energy of the exciting photon. This process is called Rayleigh scattering (elastic scattering without changes in the wavelength of scattered light).

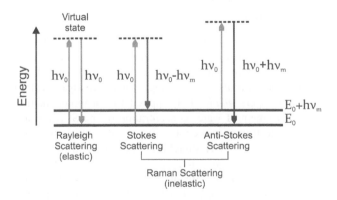

FIGURE 8.1 Schematic representation of Rayleigh and Raman scattering.

- The molecule can transit to the first excited vibrational state, at the same time emitting a photon with energy $h\nu_0 - h\nu_m$, which is less than the energy of the exciting photon. This process is called Stokes scattering (inelastic scattering with an increasing wavelength of scattered light).

- The molecule is in the first excited vibrational state and transits to the virtual state. After that, it returns to the ground state, at the same time emitting a photon with the energy $h\nu_0 + h\nu_m$, which is greater than the energy of the exciting photon. This process is called anti-Stokes scattering (inelastic scattering with decreasing wavelength of scattered light).

At physiological temperatures, most molecules of the skin are in the ground state, whose energy-related distribution is described by the Boltzmann rule. Therefore, the probability of anti-Stokes scattering is extremely low, and hence most Raman spectrometers only measure the Stokes scattering lines. Also, the measurements are based on detection of the registered wavelength changes. Thus, a monochromatic light source with a highly stable working frequency is necessary for the optical setup. Additionally, a narrowband filter is used to maintain a narrow exciting radiation band. For measurement in the skin the light sources in the infrared range (typically 785 nm) are usually used to avoid the effects of fluorescence, which is extremely high for shorter excitation wavelengths and achieve deep sampling depths. As the Raman signal from tissue can contain lines located close to each other, spectrometers with high spatial resolution are required in order to reduce superposition.

The Raman spectrum of skin is a molecular fingerprint, which contains information about its chemical composition. This way, even the slightest changes in cells and other compartments can cause changes in Raman spectra – the position and intensity of Raman lines. This feature allows one to use Raman spectroscopy as an effective tool for the measurement of the chemical composition of different tissues [147–149]. In most cases, Raman peaks correspond to specific types of molecules and their molecular vibrations, which allows one to obtain information about biochemical properties of the skin [150].

Raman spectroscopy for skin water content assessment

Huizinga *et al.* [151] firstly demonstrated that the water-to-protein ratio in eye lenses can be estimated as the ratio between the Raman peak intensities at 3390 cm^{-1} and 2935 cm^{-1}. Later, this approach was also used during cornea investigation [152]. Further, Caspers *et al.* [153] applied the proposed method for measurement of water concentration in the SC as a ratio between the Raman peak intensity of water integrated from 3350 to 3550 cm^{-1} (OH-stretching vibration) and the Raman peak intensity of protein integrated from 2910 to 2965 cm^{-1}. These spectral ranges were suggested to increase the signal-to-noise ratio as well as to avoid superposition between water and N–H vibration peak of protein at ≈3330 cm^{-1}. The water content was calculated using the following equation:

$$water\ content\,(\%) = \frac{m_w}{m_w + m_p} = \frac{W/P}{\dfrac{W}{P} + R}\,100\%, \qquad (8.1)$$

where m_w and m_p are water and protein masses in the sampling volume, W is the integrated water signal, P is the integrated protein signal, and R is a proportionality constant, which was experimentally determined and is equal to two. For the first time, authors showed depth-dependent distribution of water content in the SC using confocal Raman microspectroscopy. Currently, the proposed method is widely used for the determination of water concentration profiles in the skin [57, 154–156]. However, a recent study shows that keratin, which is always used for the depth normalization of Raman spectra, is nonhomogeneously distributed in the SC [157], which gives rise to the overestimation of water concentration values in the SC, especially in the bottom regions of SC [158].

Water is bound with surrounding molecules with hydrogen bonds, whose energy could vary substantially. Sun *et al.* [159] analyzed water at 290 K temperature and 0.1 MPa pressure to show the distribution of water types depending on local hydrogen-bound networks. Using decomposition of the Raman spectrum (2900–3750 cm^{-1}), they found that the strongest hydrogen bonds, where water molecules are bound with single donors and double acceptors (DAA), correspond to the Raman band at 3014 cm^{-1}. Consequently, with increasing wavenumbers, hydrogen bonds of water lose their strength as follows: the Raman bands at 3226, 3432, 3572, and 3636 cm^{-1} corresponds to double donor–double acceptor (DDAA, strongly bound water), single donor–single acceptor (DA, weakly bound water), double donor–single acceptor (DDA, very weakly bound water), hydrogen bonding, and free water molecules, respectively. Further, Sun *et al.* [160] analyzed different types of water at different temperatures. It was also shown that a decrease in water temperature leads to a structural transition from DA to DDAA water type.

Vyumvuhore *et al.* [57] suggested assessing features associated with the different types of water depending on their bonding properties in skin biopsies using decomposition of Raman spectra in the 3100–3700 cm^{-1} region with four Gaussian functions, centered at around 3210, 3280, 3345, and 3470 cm^{-1}. Based on their classification, the 3210 cm^{-1} subband represented totally or primary bound water, where water molecules are characterized by DDAA bonding (four hydrogen bonds). The 3280 cm^{-1} and 3345 cm^{-1} subbands represented partially bound water, where water molecules are involved in interaction with neighboring molecules using two or three hydrogen bonds instead of four. The 3470 cm^{-1} subband represented unbound water, where water molecules have no hydrogen bonds with amino acids, proteins, or lipids. The common content for introduced water types was calculated as the sum of areas under the curve (AUC) for corresponding subbands.

Later, Boireau-Adamzyk *et al.* [156] adapted this idea for analysis of human skin Raman spectra *in vivo* without using complicated decomposition and also separated water molecules in three categories. The bound water was determined as the mean of the AUC for the 3100–3250 cm^{-1} region. The intermediately mobile water was determined as the mean of the 3250–3420 cm^{-1} AUC. The most mobile water was determined as the mean of the 3420–3620 cm⁻¹AUC. Further, the authors analyzed different types of water in human skin from the cheek, dorsal forearm, and upper inner arm *in vivo* for two age groups. It was shown that the concentration profiles for each water type behave similarly with the total water content. Particularly, it has low values close to the skin surface and increases to a plateau at the base of the SC. However, no significant differences have been found for determined water types between two age groups.

Further, based on results of Sun [159, 160] on water investigation at different temperatures, Choe *et al.* [64] modified the method described by Vyumvuhore *et al.* [57] and presented the depth-dependent profiles of four different types of water depending on its bonding strength in the SC *in vivo*, i.e. tightly, strongly, and weakly bound and unbound (free) water. The authors applied the decomposition procedure of Raman spectrum in 2770–3900 cm⁻¹ region using ten Gaussian functions, which take the contribution of keratin-related bands in the "water region" (3000–3700 cm⁻¹) into consideration (see Figure 8.2).

The 3000–3700 cm⁻¹ region represents OH vibrations of water with low contribution of proteins. For the decomposition procedure, four water-related Gaussian functions were centered around 3005, 3277, 3458, and 3604 cm⁻¹. These functions correspond to the tightly (DAA-OH), strongly (DDAA-OH), and weakly hydrogen bound water molecules (DA-OH) and to the free water molecules, respectively. The free water represented a superposition of DDA-OH and unbound water molecules due to the impossibility of resolving these close located contributions [62]. Two Gaussian functions centered around 3060 and 3330 cm⁻¹ corresponded to the unsaturated methylene stretching band and the NH vibration band of proteins, respectively. Also, four Gaussian functions were centered around 2850, 2880, 2930, and 2980 cm⁻¹, which correspond to lipids and keratin. The ratio between DA-OH and DDAA-OH is an important parameter related to the hydrogen bonding states of water [64, 139, 161]. Maeda *et al.* [161] showed that this value corresponds to the water structures in polymer systems. In particular, phase transformation of the coil-globule leads to significant increase in DA-OH/DDAA-OH ratio values, due to the fact that water molecules were released from a polymer matrix and thus became unbound. At the same time, the water molecules remaining inside the globule were surrounded by hydrophobic molecular groups with less hydrogen bonds. This way, lower values of the DA-OH/DDAA-OH ratio correspond to increased amount of hydrogen bonds between water and the surroundings (intercellular lipids, natural moisturizing factor, keratin).

The described approach using the decomposition procedure of Raman spectra was successfully used for the measurement of different water types in the skin. In particular, the depth profiles of water type concentrations depending on the strength of hydrogen bonds and the hydrogen bonding state of water molecules were determined and compared with depth profiles of the concentration of natural moisturizing factor, keratin folding, and lateral organization of intercellular lipids for *ex vivo* porcine and *in vivo* human SC [40]. It was shown that in human SC *in vivo*, water is in a more bound state at 10%–30% of the SC depth compared to porcine SC *ex vivo*. Also, the concentration of natural moisturizing factor is higher at all SC depths in human skin *in vivo*. Further, the depth-dependent profiles of water type concentrations depending on the strength of hydrogen bonds, the hydrogen bonding state of water molecules, concentration of natural moisturizing factor, and lateral organization of intercellular lipids were investigated *in vivo* for the SC of humans of different age groups [62]. It was shown that water in the SC of the older group is stronger bound at 10%–30% SC depth than in the younger group. As the concentration of natural moisturizing factor was higher for the older group at 20%–40% SC depth, its role in binding water molecules was proposed. In the intermediate SC region (30%–70% SC depths), keratin is mainly responsible for binding water molecules [162]. Further, a normalization procedure considering the nonhomogeneous distribution of keratin in human SC was introduced for the correction of Raman spectral depth profiles and, thus, for more precise measurement of water concentration [158]. Also, a modified method for the calculation of the water mass percentage for oil-treated skin has been introduced [163]. The proposed method is based on the calculation of the ratio between band intensities of water in the 3350–3550 cm⁻¹ region and the keratin at 1650 cm⁻¹, where oils have no or minimal influence. This method was criticized by Puppels *et al.* [164]. The authors claimed that the method suggested by Choe *et al.* [163] is incorrect and does not calculate the water mass percentage in the SC as far as its ignores the presence of the oil in SC. Also, they suggested that normalization of SC thickness is not correct. Further, Darvin *et al.* [165] answered critics and showed that method proposed by his group is relevant and correct.

Measurement of collagen hydration affected by OC using Raman spectroscopy

Besides the direct measurement of water concentration and bonding state depth profiles, confocal Raman microspectroscopy allows for the measurement of collagen hydration. Zhang *et al.* [166] showed that an increase in the relative humidity of human collagen type I and porcine dermis results in shift of the Amide I band from 1672 to 1665 cm⁻¹ and a shift of the Amide

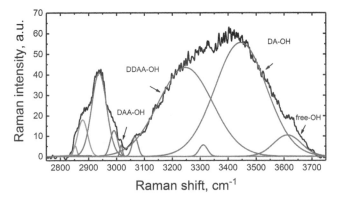

FIGURE 8.2 Gaussian function–based decomposition procedure of the Raman spectrum of porcine SC (black – skin; green – lipids; red – proteins; blue – DAA-OH, DDAA-OH, DA-OH, and free-OH water types).

III band at 1245 cm^{-1} towards higher wavenumbers. Also, the Raman band intensity at 938 cm^{-1} corresponding to v(C–C) vibrations of the protein backbone significantly increased with an increase in relative humidity. The authors suggested that the ratio between integrated areas of 988–898 cm^{-1} and 922–898 cm^{-1} can be sensitive to a hydration state of both skin collagen type I and porcine dermis. They showed that their hydration increased with relative humidity. Nguyen *et al.* [167] also showed that intensity of 938 cm^{-1} peak increased with increase of relative humidity. The authors suggested that this peak can be considered as a marker for collagen hydration. They showed that the 938/922 cm^{-1} ratio increases almost linearly with an increase in relative humidity.

In the work [36], the OCA influence on porcine skin collagen hydration was investigated. The same OCAs (70% glycerol and 100% Omnipaque) were used in the experiment. To investigate the changes in collagen hydration, the ratio between the 938 cm^{-1} and 922 cm^{-1} Raman peak intensities was measured for different depths in the skin before and after the application of OCAs. The obtained values are summarized in Table 8.3. It can be clearly seen that glycerol application leads to a strong dehydration of collagen down to the depth of 120 μm compared to Omnipaque at both 30 and 60 min treatment times. The intensity ratio between the peaks at 938 and 922 cm^{-1} (I_{938}/I_{922}) is significantly different between glycerol and Omnipaque from 0 to 80 μm depths. In general, this result is in good agreement with the result of the hydrogen bonding water measurements. In spite of the fact that Omnipaque has a lower OC efficiency compared to glycerol, it is a promising OCA for skin application due to its lower aggressive dehydration effect.

Assessment of OC influence on skin water content dependent on the bonding strength using Raman spectroscopy

In work [139], the influence of the OC on the different types of bound water in the skin *ex vivo* has been shown for the first time. Porcine ear skin was chosen for the experiment due to its high morphological [168] and histological [19] similarity to human skin. 70% glycerol and 100% Omnipaque™ (300) solutions were chosen as OCAs. For investigation of the water migration in skin after OCA application, the method suggested by Choe *et al.* [64] was implemented. The AUC values for DAA-OH, DDAA-OH, DA-OH, and free-OH were calculated for the different depths in skin samples and further normalized

to the intensity of the protein band at 2930 cm^{-1}. The obtained normalized values are summarized in Table 8.4. The sum of the values for all four water molecule types represents the full water content of the skin. For a better understanding of the changes in each water type of skin caused by OCA application, the relative changes in water percentage were calculated. Figure 8.3 shows relative percentages for all water types before and after OCA treatment within 30 and 60 min.

It can be clearly seen that for skin after treatment with both OCAs, the sum of all water type percentages is less than 100%. This means that water has been displaced from the skin due to the dehydration caused by OCA application. The application of both OCAs caused reduction in percentages of tightly bound water (DAA-OH). However, in comparison to the changes in the other water types, DAA-OH is least affected by the OCA application. Also, it is noticeable that for untreated skin, the DAA-OH percentage decreased for higher depths, while DDAA-OH and DA-OH showed no significant changes for all depths. Omnipaque caused significant changes in DDAA-OH and DA-OH percentages only at the 80–200 μm depths, while glycerol caused an even greater effect at 40–200 μm depths. The curves for both DDAA-OH and DA-OH concentration show a similar tendency. It is also noticeable that for both water types, the percentage at the 40 μm depth for OCA-treated samples is higher than for the control samples. This effect can be explained by several factors. The first is the fact that Omnipaque penetrates into the skin up to 40 μm within 60 min of treatment [36]. Therefore, the increase in the DDAA-OH and DA-OH percentages may be due to the superposition of the Omnipaque and skin-based water content. The second reason could be the not-fully-correct normalization of the AUC values on the protein peak at the 40 μm depth. The values for all water types at the skin surface are in excellent agreement with the values presented in [40] for porcine SC. In general, it can be seen that glycerol has a much greater dehydration effect on skin than Omnipaque.

The most prominent water types at all investigated depths for untreated skin are DDAA-OH and DA-OH water molecules, which represent more than 90% of the skin water content. The free-OH and DAA-OH water represent the remaining <10%. This is in a good agreement with the results of Choe *et al.* [40, 64]. It can clearly be seen that both OCAs mostly affect weakly bound and strongly bound water. Thus, these water types are mostly responsible for OC efficiency and are preferentially involved in the water flux from the skin caused by OCA application.

TABLE 8.3

Depth-dependent AUC values of DAA–OH, DDAA–OH, DA–OH, and free–OH water types and full water content normalized to the protein band. Mean values ±SD. Adapted from Ref. [36]

Agent/Depth	0 μm	40 μm	80 μm	120 μm	160 μm	200 μm	240 μm
Untreated	1.19±0.1	1.27±0.06	1.65±0.04	1.67±0.04	1.70±0.13	1.62±0.14	1.54±0.25
Omnipaque, 30 min	1.23±0.03	1.34±0.04	1.46±0.03	1.21±0.05	1.37±0.03	1.48±0.05	1.25±0.20
Omnipaque, 60 min	1.10±0.03	1.34±0.03	1.59±0.04	1.43±0.06	1.50±0.05	1.37±0.08	1.52±0.15
Glycerol, 30 min	0.90±0.03	1.11±0.04	1.02±0.03	1.27±0.05	1.34±0.04	1.32±0.01	1.29±0.10
Glycerol, 60 min	0.78±0.04	0.88±0.05	0.99±0.05	1.34±0.04	1.45±0.05	1.42±0.03	1.28±0.07

TABLE 8.4

Depth-dependent AUC values of DAA–OH, DDAA–OH, DA–OH, and free–OH water types and full water content normalized to the protein peak. Mean values ± SD. Adapted from reference [139]

Depth, μm	Untreated	Omnipaque™, 30 min	Omnipaque™, 60 min	Glycerol, 30 min	Glycerol, 60 min
DAA–OH water band to protein band ratio (tightly hydrogen bound water)					
0	0.14±0.02	0.11±0.02	0.13±0.01	0.08±0.01	0.07±0.01
40	0.17±0.01	0.15±0.02	0.13±0.01	0.09±0.01	0.08±0.01
80	0.17±0.01	0.14±0.01	0.11±0.01	0.10±0.01	0.06±0.01
120	0.16±0.01	0.14±0.01	0.11±0.02	0.11±0.01	0.05±0.02
160	0.16±0.01	0.13±0.01	0.12±0.01	0.09±0.01	0.06±0.01
200	0.16±0.02	0.14±0.02	0.12±0.01	0.08±0.01	0.05±0.01
DDAA–OH water band to protein band ratio (strongly hydrogen bound water)					
0	2.93±0.24	3.12±0.21	3.20±0.24	3.44±0.25	2.91±0.24
40	4.79±0.19	5.08±0.14	5.26±0.18	3.49±0.17	2.71±0.18
80	4.86±0.14	4.51±0.18	4.25±0.18	3.23±0.16	2.54±0.15
120	4.95±0.2	4.63±0.14	4.53±0.17	3.13±0.19	2.19±0.16
160	4.88±0.18	4.75±0.17	4.27±0.17	3.40±0.16	2.37±0.16
200	4.96±0.29	4.62±0.25	4.16±0.24	3.12±0.27	2.29±0.25
DA–OH water band to protein band ratio (weakly hydrogen bound water)					
0	3.37±0.24	3.55±0.24	3.66±0.24	3.65±0.27	3.33±0.27
40	5.40±0.23	5.70±0.25	6.20±0.19	4.02±0.18	2.99±0.17
80	5.59±0.21	5.23±0.17	4.95±0.14	3.76±0.17	2.89±0.16
120	5.69±0.2	5.0±0.2	5.25±0.17	3.65±0.17	2.52±0.2
160	5.91±0.17	5.61±0.2	5.09±0.21	3.91±0.20	2.93±0.26
200	6.12±0.38	5.1±0.27	5.00±0.25	3.81±0.30	2.79±0.25
Free–OH water band to protein band ratio (unbound water)					
0	0.32±0.04	0.42±0.04	0.38±0.05	0.44±0.05	0.37±0.04
40	0.59±0.05	0.57±0.04	0.62±0.05	0.46±0.04	0.32±0.04
80	0.63±0.05	0.60±0.04	0.55±0.04	0.51±0.04	0.25±0.04
120	0.59±0.05	0.57±0.04	0.55±0.06	0.48±0.05	0.27±0.05
160	0.60±0.05	0.57±0.07	0.50±0.05	0.40±0.05	0.26±0.04
200	0.62±0.05	0.55±0.05	0.49±0.05	0.40±0.04	0.25±0.04
Full water content to protein band ration					
0	6.76±0.59	7.22±0.5	7.37±0.58	7.62±0.6	6.68±0.62
40	10.97±0.46	11.52±0.49	12.22±0.43	8.06±0.41	6.10±0.36
80	11.26±0.45	10.49±0.45	9.87±0.36	7.50±0.41	5.74±0.41
120	11.40±0.46	10.75±0.42	10.45±0.46	7.37±0.42	5.04±0.44
160	11.55±0.43	11.06±0.47	10.00±0.44	7.82±0.41	5.53±0.5
200	11.87±0.74	10.72±0.57	9.78±0.58	7.42±0.6	5.38±0.58

Figure 8.4 shows the influence of OCA application on the total water content in skin represented as a depth-dependent ratio between total water (sum of AUC values for each water type) and protein. It can be seen that the water content for untreated skin samples significantly increases from 0 to 40 μm depth and reaches a maximum at 200 μm depth. The 60-min application of Omnipaque caused a significant reduction in the water content from 80 to 200 μm depths. Glycerol caused a much stronger reduction in the water content even after 30-min application from 40 to 200 μm depths due to dehydration caused by the OCA. In general, 30 and 60 min treatment with Omnipaque caused a reduction of the total water content in the skin to 94% and 87% of the initial concentration, respectively. Treatment times of 30 and 60 min with glycerol solution caused a reduction of the total water content in skin to 65 and 47% of the initial concentration, respectively. Therefore, it can be suggested that the dehydration efficiency is directly related to the OC efficiency.

The osmotic pressure caused by glycerol treatment leads to conversion of bound into unbound water [169, 170]. Also, bound water can be converted into unbound water after mechanical pressure on the skin [127] and by occlusion with topically applied formulations [171]. The penetration of glycerol solution into the skin results in bonding of tissue water molecules with glycerol molecules. As a result, the flux between skin water and OCA-bound water occurs. Carniero *et al.* [172] showed the dependence between the OCA's water content and water flux from liver tissue during OC. It was shown that the water flux

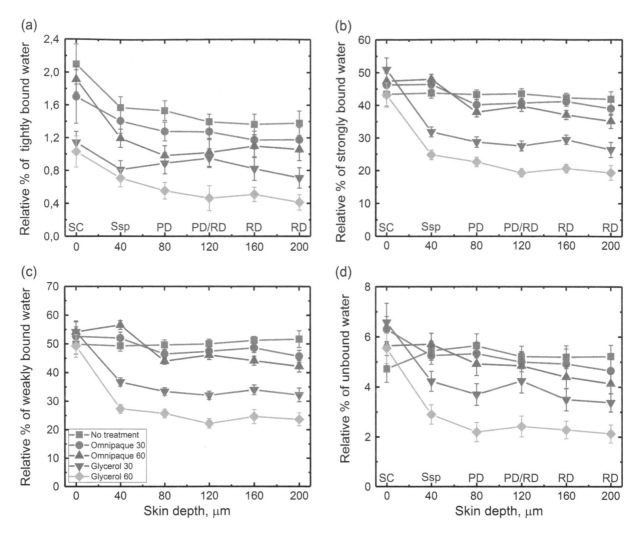

FIGURE 8.3 The depth dependent distribution of the relative percentage of DAA–OH (A), DDAA–OH (B), DA–OH (C), and free–OH (D) water types in the untreated skin (turquoise) and in the skin treated with Omnipaque for 30 min/60 min (orange/lilac) and in the skin treated with glycerol for 30 min/60 min (purple/yellow). SC, stratum corneum; SSp, stratum spinosum; PD, papillary dermis; RD, reticular dermis. Adapted from reference [139].

between the OCA and tissue is minimal when the concentration of mobile water in liver tissue and water in the OCA are equal. In case of porcine ear skin, weakly bound water and unbound water ($\approx$55% of total water content) are mostly involved in the water flux during OC of the skin. Strongly bound water ($\approx$44% of total water content) can also be involved in the water flux, but with lower mobility. Also, the described results are in a good agreement with results presented by Yanina *et al.* [173]. Authors investigated the influence of 40%, 60%, and 100% glycerol in water solutions, as well as 100% DMSO, 80% glycerol +20% DMSO, and 90% glycerol +10% DMSO solutions on porcine skin Raman spectra.

It is important to note that treatment with both OCAs leads to a significant decrease in fluorescence background at the first 50 μm depths due to the penetration of the OCA into the upper skin layers (see Figure 8.5). For deeper skin areas, the fluorescence background in the skin after Omnipaque application was comparable to the fluorescence of untreated skin, while for skin after glycerol application, the fluorescence background increased. This effect caused by glycerol application can be associated with the finding that decreased scattering after OCA application allows more excitation light to reach

the deeper located fluorophores. Thus, the fluorescence intensity is additionally enhanced due to decreased scattering. The fluorescence intensity was calculated without considering the effect of fluorescence photobleaching, which is pronounced in all skin layers [174].

Other methods for assessment of OC influence on skin water content

Few other methods such as refractometry and measurement of transmittance spectra were also used for the estimation of the skin dehydration degree after OCA treatment. Particularly, in work [175] the modified refractometric method for the assessment of the dehydration degree has been implemented *in vivo*. The influence of 70% glycerol and 40% glucose solutions on rat skin *ex vivo* were investigated. The measurement of skin dehydration during OC was performed using the Gladstone-Dale relation for multicomponent solutions [176], as described by the equation

$$n(\lambda, t) = \sum_i n_i(\lambda) f_i(t), \tag{8.2}$$

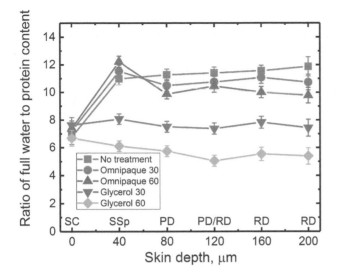

FIGURE 8.4 Total water content (determined by the ratio of full water to proteins content). Turquoise – untreated skin; orange/lilac – skin treated with Omnipaque for 30 min/60 min; purple/yellow – skin treated with glycerol for 30 min/60 min. SC, stratum corneum; SSp, stratum spinosum; PD, papillary dermis; RD, reticular dermis. Adapted from reference [139].

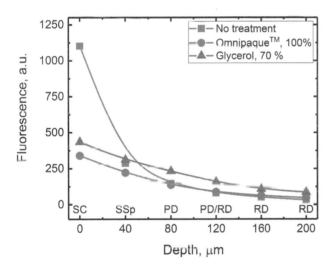

FIGURE 8.5 Background fluorescence of the Raman spectra (excitation wavelength, 785 nm) for the control skin samples (turquoise) and samples treated with 70% glycerol (lilac) and 100% Omnipaque (orange). SC, stratum corneum; SSp, stratum spinosum; PD, papillary dermis; RD, reticular dermis. Adapted from reference [26].

where n is the refractive index of the solution, i is the number of solution components, n_i is the refractive index of the i-th component, and f_i is the volume fraction of the i-th component.

This relation can be rewritten in the case of OCA application, and the volume fraction of the extracted fluid is described as

$$f_{ext\,fl}(t) = \frac{n_{exp}(\lambda,t) - n_{OCA}(\lambda) + \left(n_{OCA}(\lambda) - n_{NaCl}(\lambda)\right)f_{NaCl}(t)}{\left(n_{ext\,fl}(\lambda) - n_{OCA}(\lambda)\right)},$$

(8.3)

where $f_{ext\,fl}$ is the volume fraction of the extracted fluid, n_{exp} is the refractive index of the experimental solution, n_{OCA} is the refractive index of the OCA, n_{NaCl} is the refractive index of the saline, f_{OCA} is the volume fraction of the OCA, and f_{NaCl} is the volume fraction of the saline.

The measured refractive index of the extracted fluid is described by

$$n_{ext\,fl}(\lambda) = \frac{n_{exp}(\lambda,t) - n_{OCA}(\lambda) + n_{OCA}(\lambda)f_{ext\,fl}(t)}{f_{ext\,fl}(\lambda)}$$

$$+ \frac{\left(n_{OCA}(\lambda) - n_{NaCl}(\lambda)\right)f_{NaCl}(t)}{f_{ext\,fl}(\lambda)}$$

(8.4)

Experimental results showed that the refractive index of glycerol solution decreased strongly compared to the glucose solution within 60 min treatment. Thus, glycerol acts as a hyperosmotic agent causing intense flux of interstitial free and weakly bound water.

The results of refractive index and volume fraction of water calculation in the skin before and after OCA treatment are summarized in Table 8.5 for five excitation wavelengths. It can be seen that glycerol solution caused greater dehydration of skin compared to glucose.

Further, it was shown that the volume fractions of extracted water for healthy and cancerous skin areas after glycerol application are different. The dehydration of healthy and cancerous skin was 3.0 ± 0.3 and $1.1 \pm 0.2\%$, respectively. This is in a good agreement with results presented by Chung *et al.* [177], who investigated the water content in breast cancer tissue. Also, Barroso *et al.* [178] showed that the water content in cancerous tissue is significantly higher than in healthy tissue.

Genin *et al.* [179] presented diffusion results by investigating 30%, 50%, 60%, 70%, 85%, and 100% glycerol solutions in rat skin. The method of diffusion coefficient assessment was based on the measurement of time dependence of collimated transmittance in skin samples. The authors suggested that only the refractive index of the interstitial fluids of the skin was changed during OCA treatment due to OCA diffusion into the skin and its simultaneous dehydration. The matching of refractive indices between OCA and skin occurred, leading to a reduction in the skin scattering coefficient. This way, the measurement of diffusion coefficient can be a measure of tissue dehydration as the mean rate of the total exchange of liquid penetrating into the skin and the water leaking from the tissue.

The authors [179] showed that the diffusion coefficient depends on the OCA concentration. In particular, the diffusion coefficient decreases in glycerol solutions of 30%–70%. This effect can be explained by specific features of the hydrodynamic radius of diffusing molecules as well as an increase in solution viscosity. For 85% and 100% concentrations, the diffusion coefficient increased again. However, the authors showed that higher values of OCA diffusion coefficient did not guarantee higher OC efficiency.

Kolesnikov *et al.* [180] showed the results of *in vivo* investigation of human healthy and tumorous skin dehydration after treatment with 30% and 40% glycerol solutions, polyethylene

TABLE 8.5

Data on refractive index and volume fraction of extracted water for skin samples treated with 70% glycerol and 40% glucose solutions. Adapted from Ref. [175]

	Sample weight (mg)	Sample width (mm)	Refractive index	λ (nm)	The volume fraction of the water and extracted fluid in OCA, %	The average volume fraction of the water and extracted fluid in OCA, %
40% glucose solution						
No treatment	756±35	10.05±0.20	1.3747±0.0002	486	60	60
			1.3898±0.0002	589	60	
			1.3865±0.0002	680	60	
			1.3826±0.0003	930	60	
			1.3740±0.0004	1300	60	
60 min treatment	640±25	7.00±0.18	1.3913±0.0002	486	62.39	62.02±0.27
			1.3870±0.0002	589	62.01	
			1.3840±0.0003	680	61.80	
			1.3796±0.0003	930	62.15	
			1.3718±0.0004	1300	61.74	
70% glycerol solution						
No treatment	805±41	10.50±0.23	1.4373±0.0002	486	30	30
			1.4325±0.0002	589	30	
			1.4290±0.0002	680	30	
			1.4250±0.0003	930	30	
			1.4166±0.0003	1300	30	
60 min treatment	530±34	8.10±0.21	1.4260±0.0002	486	37.96	38.11±0.17
			1.4210±0.0003	589	38.16	
			1.4179±0.0002	680	37.91	
			1.4134±0.0003	930	38.27	
			1.4056±0.0004	1300	38.26	

glycol (PEG), 100% glycerol, and propylene glycol (PG), using a time domain THz spectrometer. The authors showed that skin transmittance increased after the application of each OCA. It is important to notice that for tumorous skin, the transmittance increased more. Also, the authors showed that the degree of human skin hydration significantly decreased 13–17 min after OCA application. Then, the hydration level sharply increased 20 min after OCA application, possibly due to rehydration of the upper skin layers caused by a water flow from water-saturated deep layers of the skin. Similar results were presented in work [181]. The influence of 30%, 40%, and 50% glucose solutions as well as 50% fructose solution on the degree of skin hydration using the skin assessment system Soft Plus (Callegari, Italy) was presented. It was also shown that the hydration level decreased within the first 20 min of OCA application, while after this time, the skin became even more hydrated compared to the beginning of the experiment. A possible reason for this effect could be the flux of free water from deeper skin layers towards the surface.

In general, all results on the investigation of skin water content changes during OCA application are more or less consistent with similar investigations on tissues such as mucosa and skeletal muscle [169, 182, 183]. The difference can be explained by the different skin water bonding properties compared to other tissues as well as skin heterogeneity.

investigations of OCA influence on skin hydration. Insofar as the OC technique allows one to increase the light penetration depth in skin and the quality of the obtained information for optical imaging systems, the investigation of OCA influence on skin water content is the key goal for finding a safe and effective protocol for *in vivo* OC of human skin. However, currently only a few works have presented the results of the investigation of OCA application influence on skin hydration. In general, only confocal Raman microspectroscopy allows one to obtain the most precise results on different water types depending on hydrogen bonding strength changes in skin on the molecular level during OCA application as well as to track OCA penetration depth in the skin, while the other methods allow one to obtain only indirect information on changes in the degree of hydration. We believe that this chapter will help in further investigations of OCA interaction with the skin.

Acknowledgments

V.V.T. was supported by project No. 13.2251.21.0009 of the Ministry of Science and Higher Education of the Russian Federation. The work of J.S., J.L., and M.E.D. was partly supported by the Foundation for Skin Physiology of the Donor Association for German Science and Humanities.

REFERENCES

1. M. Yokota and Y. Tokudome, "The effect of glycation on epidermal lipid content, its metabolism and change in barrier function," *Skin Pharmacol. Physiol.* **29**(5), 231–242, Karger Publishers (2016).

Summary

Here, we presented a brief review of optical properties of the skin, methods of skin water content measurement, and the results of

2. E. Proksch, J.M. Brandner, and J. Jensen, "The skin: An indispensable barrier," *Exp. Dermatol.* **17**(12), 1063–1072, Wiley Online Library (2008).

3. M. Janssens, J. van Smeden, G.J. Puppels, et al., "Lipid to protein ratio plays an important role in the skin barrier function in patients with atopic eczema," *Br. J. Dermatol.* **170**(6), 1248–1255, Wiley Online Library (2014).

4. M. Weinigel, H.G. Breunig, M. Kellner-Höfer, et al., "In vivo histology: Optical biopsies with chemical contrast using clinical multiphoton/coherent anti-Stokes Raman scattering tomography," *Laser Phys. Lett.* **11**(5), 55601, IOP Publishing (2014).

5. F.F. Sahle, T. Gebre-Mariam, B. Dobner, J. Wohlrab, and R.H.H. Neubert, "Skin diseases associated with the depletion of stratum corneum lipids and stratum corneum lipid substitution therapy," *Skin Pharmacol. Physiol.* **28**(1), 42–55, Karger Publishers (2015).

6. R.J.H. Richters, D. Falcone, N.E. Uzunbajakava, et al., "Sensitive skin: Assessment of the skin barrier using confocal Raman microspectroscopy," *Skin Pharmacol. Physiol.* **30**(1), 1–12, Karger Publishers (2017).

7. C. Choe, J. Lademann, and M.E. Darvin, "Analysis of human and porcine skin in vivo/ex vivo for penetration of selected oils by confocal Raman microscopy," *Skin Pharmacol. Physiol.* **28**(6), 318–330, Karger Publishers (2015).

8. M. Balu, H. Mikami, J. Hou, E.O. Potma, and B.J. Tromberg, "Rapid mesoscale multiphoton microscopy of human skin," *Biomed. Opt. Express* **7**(11), 4375–4387, Optical Society of America (2016).

9. A. Sdobnov, M.E. Darvin, J. Lademann, and V. Tuchin, "A comparative study of ex vivo skin optical clearing using two-photon microscopy," *J. Biophoton.* **10**(9), 1115–1123, Wiley Online Library (2017).

10. M. Klemp, M.C. Meinke, M. Weinigel, et al., "Comparison of morphologic criteria for actinic keratosis and squamous cell carcinoma using in vivo multiphoton tomography," *Exp. Dermatol.* **25**(3), 218–222, Wiley Online Library (2016).

11. R. Vyumvuhore, A.M. Tfayli, O. Piot, et al., "Raman spectroscopy: In vivo quick response code of skin physiological status," *J. Biomed. Opt.* **19**(11), 111603, International Society for Optics and Photonics (2014).

12. C. Krafft and J. Popp, "The many facets of Raman spectroscopy for biomedical analysis," *Anal. Bioanal. Chem.* **407**(3), 699–717, Springer (2015).

13. H.G. Breunig, M. Weinigel, R. Bückle, et al., "Clinical coherent anti-Stokes Raman scattering and multiphoton tomography of human skin with a femtosecond laser and photonic crystal fiber," *Laser Phys. Lett.* **10**(2), 25604, IOP Publishing (2013).

14. M. Roustit, C. Millet, S. Blaise, B. Dufournet, and J.L. Cracowski, "Excellent reproducibility of laser speckle contrast imaging to assess skin microvascular reactivity," *Microvasc. Res.* **80**(3), 505–511, Elsevier (2010).

15. V. Kalchenko, I. Meglinski, A. Sdobnov, Y. Kuznetsov, and A. Harmelin, "Combined laser speckle imaging and fluorescent intravital microscopy for monitoring acute vascular permeability reaction," *J. Biomed. Opt.* **24**(6), 1 (2019) [doi:10.1117/1.jbo.24.6.060501].

16. A.F. Fercher, "Optical coherence tomography," *J. Biomed. Opt.* **1**(2), 157–174, International Society for Optics and Photonics (1996).

17. T. Gambichler, A. Pljakic, and L. Schmitz, "Recent advances in clinical application of optical coherence tomography of human skin," *Clin. Cosmet. Investig. Dermatol.* **8**, 345, Dove Press (2015).

18. M. Ulrich, M. Klemp, M.E. Darvin, et al., "In vivo detection of basal cell carcinoma: Comparison of a reflectance confocal microscope and a multiphoton tomograph," *J. Biomed. Opt.* **18**(6), 61229, International Society for Optics and Photonics (2013).

19. M.E. Darvin, H. Richter, Y.J. Zhu, et al., "Comparison of in vivo and ex vivo laser scanning microscopy and multiphoton tomography application for human and porcine skin imaging," *Quantum Electron.* **44**(7), 646, IOP Publishing (2014).

20. K. König, "Clinical multiphoton tomography," *J. Biophoton.* **1**(1), 13–23, Wiley Online Library (2008).

21. E.A. Shirshin, Y.I. Gurfinkel, A.V. Priezzhev, et al., "Two-photon autofluorescence lifetime imaging of human skin papillary dermis in vivo: Assessment of blood capillaries and structural proteins localization," *Sci. Rep.* **7**(1), 1–10, Nature Publishing Group (2017).

22. E.A. Shirshin, B.P. Yakimov, M.E. Darvin, et al., "Label-free multiphoton microscopy: The origin of fluorophores and capabilities for analyzing biochemical processes," *Biochem.* **84**(1), 69–88, Springer (2019).

23. V.V. Tuchin, "Optical clearing of tissues and blood using the immersion method," *J. Phys. D. Appl. Phys.* **38**(15), 2497, IOP Publishing (2005).

24. V.V Tuchin, I.L. Maksimova, D.A. Zimnyakov, et al., "Light propagation in tissues with controlled optical properties," *J. Biomed. Opt.* **2**(4), 401–418, International Society for Optics and Photonics (1997).

25. A.Y. Sdobnov, M.E. Darvin, E.A. Genina, et al., "Recent progress in tissue optical clearing for spectroscopic application," *Spectrochim. Acta Part A* **197**, 216–229, Elsevier (2018).

26. A.Y. Sdobnov, J. Lademann, M.E. Darvin, and V.V. Tuchin, "Methods for optical skin clearing in molecular optical imaging in dermatology," *Biochemistry* **84**(1), 144–158, Springer (2019).

27. J.M. Hirshburg, K.M. Ravikumar, W. Hwang, and A.T. Yeh, "Molecular basis for optical clearing of collagenous tissues," *J. Biomed. Opt.* **15**(5), 55002, International Society for Optics and Photonics (2010).

28. G. Vargas, A. Readinger, S.S. Dozier, and A.J. Welch, "Morphological changes in blood vessels produced by hyperosmotic agents and measured by optical coherence tomography," *Photochem. Photobiol.* **77**(5), 541–549, Wiley Online Library (2003).

29. V.V. Tuchin, *Optical Clearing of Tissues and Blood*, Society of Photo Optical (2005).

30. L. Oliveira, M.I. Carvalho, E. Nogueira, and V.V. Tuchin, "Optical clearing mechanisms characterization in muscle," *J. Innov. Opt. Health Sci.* **9**(05), 1650035, World Scientific (2016).

31. L.M. Oliveira, M.I. Carvalho, E.M. Nogueira, and V.V. Tuchin, "Diffusion characteristics of ethylene glycol in skeletal muscle," *J. Biomed. Opt.* **20**(5), 51019, International Society for Optics and Photonics (2014).

32. A. Bykov, T. Hautala, M. Kinnunen, et al., "Imaging of subchondral bone by optical coherence tomography upon optical clearing of articular cartilage," *J. Biophoton.* **9**(3), 270–275, Wiley Online Library (2016).

33. E.K. Chan, B. Sorg, D. Protsenko, et al., "Effects of compression on soft tissue optical properties," *IEEE J. Sel. Top. Quantum Electron.* **2**(4), 943–950, IEEE (1996).

34. H. Ding, J.Q. Lu, W.A. Wooden, P.J. Kragel, and X.-H. Hu, "Refractive indices of human skin tissues at eight wavelengths and estimated dispersion relations between 300 and 1600 nm," *Phys. Med. Biol.* **51**(6), 1479, IOP Publishing (2006).

35. L.F. Hoyt, "New table of the refractive index of pure glycerol at 20 C," *Ind. Eng. Chem.* **26**(3), 329–332, ACS Publications (1934).

36. A.Y. Sdobnov, V.V. Tuchin, J. Lademann, and M.E. Darvin, "Confocal Raman microscopy supported by optical clearing treatment of the skin: Influence on collagen hydration," *J. Phys. D. Appl. Phys.* **50**(28), aa77c9, IOP Publishing (2017) [doi:10.1088/1361-6463/aa77c9].

37. I.H. Blank, "Factors which influence the water content of the stratum corneum," *J. Invest. Dermatol.* **18**(6), 433–440, Elsevier (1952).

38. L.A. Goldsmith, *Physiology, Biochemistry, and Molecular Biology of the Skin*, Oxford University Press (1991).

39. T. Igarashi, K. Nishino, and S.K. Nayar, "The appearance of human skin: A survey," *Comput. Graph. Vis.* **3**(1), 1–95, Now Foundations and Trends Publishers, Inc. (2007).

40. C. Choe, J. Schleusener, J. Lademann, and M.E. Darvin, "Human skin in vivo has a higher skin barrier function than porcine skin ex vivo: Comprehensive Raman microscopic study of the stratum corneum," *J. Biophoton.* **11**(6), e201700355, Wiley Online Library (2018).

41. K.C. Madison, D.C. Swartzendruber, P.W. Wertz, and D.T. Downing, "Presence of intact intercellular lipid lamellae in the upper layers of the stratum corneum," *J. Invest. Dermatol.* **88**(6), 714–718 (1987).

42. J. Van Smeden, M. Janssens, G.S. Gooris, and J.A. Bouwstra, "The important role of stratum corneum lipids for the cutaneous barrier function," *Biochim. Biophys. Acta Mol. Cell Biol. Lipids* **1841**(3), 295–313, Elsevier (2014).

43. C. Choe, J. Lademann, and M.E. Darvin, "A depth-dependent profile of the lipid conformation and lateral packing order of the stratum corneum in vivo measured using Raman microscopy," *Analyst* **141**(6), 1981–1987, Royal Society of Chemistry (2016).

44. J. van Smeden, M. Janssens, E.C.J. Kaye, et al., "The importance of free fatty acid chain length for the skin barrier function in atopic eczema patients," *Exp. Dermatol.* **23**(1), 45–52, Wiley Online Library (2014).

45. F. Quondamatteo, "Skin and diabetes mellitus: What do we know?," *Cell Tissue Res.* **355**(1), 1–21, Springer (2014).

46. C.L. Simpson, D.M. Patel, and K.J. Green, "Deconstructing the skin: Cytoarchitectural determinants of epidermal morphogenesis," *Nat. Rev. Mol. Cell Biol.* **12**(9), 565–580, Nature Publishing Group (2011).

47. M.A. Farage, K.W. Miller, P. Elsner, and H.I. Maibach, "Structural characteristics of the aging skin: A review," *Cutaneous Ocul. Toxicol.* **26**(4), 343–357, Taylor & Francis (2007).

48. W.D. Glanze, K. Anderson, and L.E. Anderson, *Mosby's Medical, Nursing, and Allied Health Dictionary*, Mosby (1990).

49. R. O'Rahilly and F. Müller, *Basic Human Anatomy: A Regional Study of Human Structure*, WB Saunders Company (1983).

50. M. Gniadecka, S. Wessel, M. Heidenheim, et al., "Water and protein structure in photoaged and chronically aged skin," *J. Invest. Dermatol.* **111**(6), 1129–1133, Elsevier (1998).

51. M.O. Visscher, G.T. Tolia, R.R. Wickett, and S.B. Hoath, "Effect of soaking and natural moisturizing factor on stratum corneum water-handling properties," *J. Cosmet. Sci.* **54**(3), 289–300, Society of Cosmetic Chemists (2003).

52. M. Egawa, M. Yanai, N. Maruyama, Y. Fukaya, and T. Hirao, "Visualization of water distribution in the facial epidermal layers of skin using high-sensitivity near-infrared (nir) imaging," *Appl. Spectrosc.* **69**(4), 481–487, SAGE Publications (2015).

53. J. Lademann, T. Vergou, M.E. Darvin, et al., "Influence of topical, systemic and combined application of antioxidants on the barrier properties of the human skin," *Skin Pharmacol. Physiol.* **29**(1), 41–46, Karger Publishers (2016).

54. S.M. Ascencio, C. Choe, M.C. Meinke, et al., "Confocal Raman microscopy and multivariate statistical analysis for determination of different penetration abilities of caffeine and propylene glycol applied simultaneously in a mixture on porcine skin ex vivo," *Eur. J. Pharm. Biopharm.* **104**, 51–58, Elsevier (2016).

55. J. Lademann, M.C. Meinke, S. Schanzer, et al., "In vivo methods for the analysis of the penetration of topically applied substances in and through the skin barrier," *Int. J. Cosmet. Sci.* **34**(6), 551–559, Wiley Online Library (2012).

56. J.A. Bouwstra, A. de Graaff, G.S. Gooris, et al., "Water distribution and related morphology in human stratum corneum at different hydration levels," *J. Invest. Dermatol.* **120**(5), 750–758, Elsevier (2003).

57. R. Vyumvuhore, A. Tfayli, H. Duplan, et al., "Effects of atmospheric relative humidity on stratum corneum structure at the molecular level: Ex vivo Raman spectroscopy analysis," *Analyst* **138**(14), 4103–4111, Royal Society of Chemistry (2013).

58. L. Norlen, "Current understanding of skin barrier morphology," *Skin Pharmacol. Physiol.* **26**(4–6), 213–216, Karger Publishers (2013).

59. M.D.A. van Logtestijn, E. Domínguez-Hüttinger, G.N. Stamatas, and R.J. Tanaka, "Resistance to water diffusion in the stratum corneum is depth-dependent," *PLoS One* **10**(2), Public Library of Science (2015).

60. L.M. Oliveira, M.I. Carvalho, E.M. Nogueira, and V.V. Tuchin, "The characteristic time of glucose diffusion measured for muscle tissue at optical clearing," *Laser Phys.* **23**(7), 75606, IOP Publishing (2013).

61. L. Oliveira, M.I. Carvalho, E. Nogueira, and V.V. Tuchin, "Optical measurements of rat muscle samples under treatment with ethylene glycol and glucose," *J. Innov. Opt. Health Sci.* **6**(02), 1350012, World Scientific (2013).

62. C. Choe, J. Schleusener, J. Lademann, and M.E. Darvin, "Age related depth profiles of human stratum corneum barrier-related molecular parameters by confocal Raman microscopy in vivo," *Mech. Ageing Dev.* **172**, 6–12, Elsevier (2018).

63. G. Imokawa, H. Kuno, and M. Kawai, "Stratum corneum lipids serve as a bound-water modulator," *J. Invest. Dermatol.* **96**(6) 845–851 (1991).

64. C. Choe, J. Lademann, and M.E. Darvin, "Depth profiles of hydrogen bound water molecule types and their relation to lipid and protein interaction in the human stratum corneum in vivo," *Analyst* **141**(22), 6329–6337, Royal Society of Chemistry (2016).

65. A. Alonso, N.C. Meirelles, and M. Tabak, "Effect of hydration upon the fluidity of intercellular membranes of stratum corneum: An EPR study," *Biochim. Biophys. Acta Biomembr.* **1237**(1), 6–15, Elsevier (1995).

66. R. Vyumvuhore, A. Tfayli, K. Biniek, et al., "The relationship between water loss, mechanical stress, and molecular structure of human stratum corneum ex vivo," *J. Biophoton.* **8**(3), 217–225, Wiley Online Library (2015).

67. S.H. Chung, A.E. Cerussi, S.I. Merritt, J. Ruth, and B.J. Tromberg, "Non-invasive tissue temperature measurements based on quantitative diffuse optical spectroscopy (DOS) of water," *Phys. Med. Biol.* **55**(13), 3753, IOP Publishing (2010).

68. K.A. Martin, "Direct measurement of moisture in skin by NIR spectroscopy," *J. Soc. Cosmet. Chem.* **44**, 249 (1993).

69. H. Arimoto and M. Egawa, "Non-contact skin moisture measurement based on near-infrared spectroscopy," *Appl. Spectrosc.* **58**(12), 1439–1446, SAGE Publications (2004).

70. R.R. Anderson and J.A. Parrish, "Optical properties of human skin," in *The Science of Photomedicine*, pp. 147–194, Springer (1982).

71. A.N. Bashkatov, E.A. Genina, V.I. Kochubey, and V.V. Tuchin, "Optical properties of human skin, subcutaneous and mucous tissues in the wavelength range from 400 to 2000 nm," *J. Phys. D. Appl. Phys.* **38**(15), 2543, IOP Publishing (2005).

72. T.L. Troy and S.N. Thennadil, "Optical properties of human skin in the near infrared wavelength range of 1000 to 2200 nm," *J. Biomed. Opt.* **6**(2), 167–177, International Society for Optics and Photonics (2001).

73. T. Lister, P.A. Wright, and P.H. Chappell, "Optical properties of human skin," *J. Biomed. Opt.* **17**(9), 90901, International Society for Optics and Photonics (2012).

74. M.E. Darvin, C. Sandhagen, W. Koecher, et al., "Comparison of two methods for noninvasive determination of carotenoids in human and animal skin: Raman spectroscopy versus reflection spectroscopy," *J. Biophoton.* **5**(7), 550–558, Wiley Online Library (2012).

75. G. Zonios, A. Dimou, I. Bassukas, et al., "Melanin absorption spectroscopy: New method for noninvasive skin investigation and melanoma detection," *J. Biomed. Opt.* **13**(1), 14017, International Society for Optics and Photonics (2008).

76. J.J. Riesz, *The Spectroscopic Properties of Melanin*, University of Queensland (2007).

77. E. Kaxiras, A. Tsolakidis, G. Zonios, and S. Meng, "Structural model of eumelanin," *Phys. Rev. Lett.* **97**(21), 218102, APS (2006).

78. M. Doi and S. Tominaga, "Spectral estimation of human skin color using the Kubelka-Munk theory," in *Color Imaging VIII: Processing, Hardcopy, and Applications* **5008**, pp. 221–228, International Society for Optics and Photonics (2003).

79. M.E. Darvin, I. Gersonde, S. Ey, et al., "Noninvasive detection of beta-carotene and lycopene in human skin using Raman spectroscopy," *Laser Phys. Lawrence* **14**(2), 231–233, Interperiodica (2004).

80. J.B. Dawson, D.J. Barker, D.J. Ellis, et al., "A theoretical and experimental study of light absorption and scattering by in vivo skin," *Phys. Med. Biol.* **25**(4), 695, IOP Publishing (1980).

81. B. Stam, M.J.C. Van Gemert, T.G. Van Leeuwen, and M.C.G. Aalders, "3D finite compartment modeling of formation and healing of bruises may identify methods for age determination of bruises," *Med. Biol. Eng. Comput.* **48**(9), 911–921, Springer (2010).

82. L.O. Svaasand, L.T. Norvang, E.J. Fiskerstrand, et al., "Tissue parameters determining the visual appearance of normal skin and port-wine stains," *Lasers Med. Sci.* **10**(1), 55–65, Springer (1995).

83. W. Verkruysse, R. Zhang, B. Choi, et al., "A library based fitting method for visual reflectance spectroscopy of human skin," *Phys. Med. Biol.* **50**(1), 57, IOP Publishing (2004).

84. T. Hegyi, I.M. Hiatt, I.M. Gertner, R. Zanni, and T. Tolentino, "Transcutaneous bilirubinometry II. Dermal bilirubin kinetics during phototherapy," *Pediatr. Res.* **17**(11), 888–891, Nature Publishing Group (1983).

85. C. Choe, J. Ri, J. Schleusener, J. Lademann, and M.E. Darvin, "The non-homogenous distribution and aggregation of carotenoids in the stratum corneum correlates with the organization of intercellular lipids in vivo," *Exp. Dermatol.* **28**(11), 1237–1243, Wiley Online Library (2019).

86. V.V. Tuchin, *Handbook of Optical Sensing of Glucose in Biological Fluids and Tissues*, CRC press (2008).

87. D. Zhu, K.V. Larin, Q. Luo, and V.V. Tuchin, "Recent progress in tissue optical clearing," *Laser Photon. Rev.* **7**(5), 732–757, Wiley Online Library (2013).

88. R.R. Anderson and J.A. Parrish, "The optics of human skin," *J. Invest. Dermatol.* **77**(1), 13–19 (1981).

89. H. Takiwaki, "Measurement of skin color: Practical application and theoretical considerations," *J. Med. Investig.* **44**, 121–126 (1998).

90. J.R. Mourant, J.P. Freyer, A.H. Hielscher, et al., "Mechanisms of light scattering from biological cells relevant to noninvasive optical-tissue diagnostics," *Appl. Opt.* **37**(16), 3586–3593, Optical Society of America (1998).

91. A.N. Bashkatov, E.A. Genina, V.I. Kochubey, et al., "Optical properties of melanin in the skin and skinlike phantoms," in *Controlling Tissue Optical Properties: Applications in Clinical Study* **4162**, pp. 219–226, International Society for Optics and Photonics (2000).

92. E.V. Salomatina, B. Jiang, J. Novak, and A.N. Yaroslavsky, "Optical properties of normal and cancerous human skin in the visible and near-infrared spectral range," *J. Biomed. Opt.* **11**(6), 64026, International Society for Optics and Photonics (2006).

93. M.J.C. Van Gemert, S.L. Jacques, H. Sterenborg, and W.M. Star, "Skin optics," *IEEE Trans. Biomed. Eng.* **36**(12), 1146–1154, IEEE (1989).

94. V.V. Tuchin, L. Wang, and D.A. Zimnyakov, *Optical Polarization in Biomedical Applications*, Springer Science & Business Media (2006).

95. D.W. Leonard and K.M. Meek, "Refractive indices of the collagen fibrils and extrafibrillar material of the corneal stroma," *Biophys. J.* **72**(3), 1382–1387, Elsevier (1997).

96. R. Drezek, A. Dunn, and R. Richards-Kortum, "Light scattering from cells: Finite-difference time-domain simulations and goniometric measurements," *Appl. Opt.* **38**(16), 3651–3661, Optical Society of America (1999).

97. M. Friebel and M. Meinke, "Model function to calculate the refractive index of native hemoglobin in the wavelength range of 250–1100 nm dependent on concentration," *Appl. Opt.* **45**(12), 2838–2842, Optical Society of America (2006).

98. E.N. Lazareva and V.V. Tuchin, "Measurement of refractive index of hemoglobin in the visible/NIR spectral range," *J. Biomed. Opt.* **23**(3), 35004, International Society for Optics and Photonics (2018).

99. A.N. Bashkatov, E.A. Genina, and V.V. Tuchin, "Optical properties of skin, subcutaneous, and muscle tissues: A review," *J. Innov. Opt. Health Sci.* **4**(01), 9–38, World Scientific (2011).

100. S.V. Patwardhan, A.P. Dhawan, and P.A. Relue, "Monte Carlo simulation of light-tissue interaction: Three-dimensional simulation for trans-illumination-based imaging of skin lesions," *IEEE Trans. Biomed. Eng.* **52**(7), 1227–1236, IEEE (2005).

101. S.L. Jacques, C.A. Alter, and S.A. Prahl, "Angular dependence of HeNe laser light scattering by human dermis," *Lasers Life. Sci.* **1**(4), 309–333 (1987).

102. W.A.G. Bruls and J.C. Van Der Leun, "Forward scattering properties of human epidermal layers," *Photochem. Photobiol.* **40**(2), 231–242, Wiley Online Library (1984).

103. S.A. Prahl, "Light transport in tissue." Doctoral dissertation, The University of Texas at Austin (1988).

104. S.L. Jacques, "The role of skin optics in diagnostic and therapeutic uses of lasers," in *Lasers in Dermatology*, pp. 1–21, Springer (1991).

105. J.F. Beek, P. Blokland, P. Posthumus, et al., "In vitro double-integrating-sphere optical properties of tissues between 630 and 1064 nm," *Phys. Med. Biol.* **42**(11), 2255, IOP Publishing (1997).

106. Y. Du, X.H. Hu, M. Cariveau, et al., "Optical properties of porcine skin dermis between 900 nm and 1500 nm," *Phys. Med. Biol.* **46**(1), 167, IOP Publishing (2001).

107. X. Ma, J.Q. Lu, H. Ding, and X.-H. Hu, "Bulk optical parameters of porcine skin dermis at eight wavelengths from 325 to 1557 nm," *Opt. Lett.* **30**(4), 412–414, Optical Society of America (2005).

108. F.A. Duck, *Physical Properties of Tissues: A Comprehensive Reference Book*, Academic Press (2013).

109. G.J. Tearney, M.E. Brezinski, J.F. Southern, et al., "Determination of the refractive index of highly scattering human tissue by optical coherence tomography," *Opt. Lett.* **20**(21), 2258–2260, Optical Society of America (1995).

110. A.R. Knuettel and M. Boehlau-Godau, "Spatially confined and temporally resolved refractive index and scattering evaluation in human skin performed with optical coherence tomography," *J. Biomed. Opt.* **5**(1), 83–93, International Society for Optics and Photonics (2000).

111. S.A. Alexandrov, A.V. Zvyagin, K.D. Silva, and D.D. Sampson, "Bifocal optical coherence refractometry of turbid media," *Opt. Lett.* **28**(2), 117–119, Optical Society of America (2003).

112. A.V. Zvyagin, K.D. Silva, S.A. Alexandrov, et al., "Refractive index tomography of turbid media by bifocal optical coherence refractometry," *Opt. Express* **11**(25), 3503–3517, Optical Society of America (2003).

113. M. Sand, T. Gambichler, G. Moussa, et al., "Evaluation of the epidermal refractive index measured by optical coherence tomography," *Ski. Res. Technol.* **12**(2), 114–118, Wiley Online Library (2006).

114. A.R. Knuettel, S.M. Bonev, and W. Knaak, "New method for evaluation of in vivo scattering and refractive index properties obtained with optical coherence tomography," *J. Biomed. Opt.* **9**(2), 265–274, International Society for Optics and Photonics (2004).

115. A. Roggan, K. Dorschel, O. Minet, D. Wolff, and G. Muller, "The optical properties of biological tissue in the near infrared wavelength range," in *Laser-induced Interstitial Therapy*, pp. 10–44, SPIE Press (1995).

116. K. Walkley, "Bound water in stratum corneum measured by differential scanning calorimetry," *J. Invest. Dermatol.* **59**(3), 225–227, Elsevier (1972).

117. T. Inoue, K. Tsujii, K. Okamoto, and K. Toda, "Differential scanning calorimetric studies on the melting behavior of water in stratum corneum," *J. Invest. Dermatol.* **86**(6), 689–693 (1986).

118. J.J. Bulgin and L.J. Vinson, "The use of differential thermal analysis to study the bound water in stratum corneum membranes," *Biochim. Biophys. Acta Gen. Subj.* **136**(3), 551–560, Elsevier (1967).

119. H. Tagami, M. Ohi, K. Iwatsuki, et al., "Evaluation of the skin surface hydration in vivo by electrical measurement," *J. Invest. Dermatol.* **75**(6), 500–507, Elsevier (1980).

120. C.M. Lee and H.I. Maibach, "Bioengineering analysis of water hydration: An overview," *Exog. Dermatol.* **1**(6), 269–275, Karger Publishers (2002).

121. E. Xhauflaire-Uhoda, G. Loussouarn, C. Haubrechts, D. Saint Léger, and G.E. Piérard, "Skin capacitance imaging and corneosurfametry. A comparative assessment of the impact of surfactants on stratum corneum," *Contact Dermatitis* **54**(5), 249–253, Wiley Online Library (2006).

122. P. Clarys, R. Clijsen, J. Taeymans, and A.O. Barel, "Hydration measurements of the stratum corneum: Comparison between the capacitance method (digital version of the C orneometer CM 825®) and the impedance method (S kicon-200 EX®)," *Ski. Res. Technol.* **18**(3), 316–323, Wiley Online Library (2012).

123. A.O. Barel and P. Clarys, "In vitro calibration of the capacitance method (Corneometer CM 825) and conductance method (Skicon-200) for the evaluation of the hydration state of the skin," *Ski. Res. Technol.* **3**(2), 107–113, Wiley Online Library (1997).

124. M. Gniadecka, "Potential for high-frequency ultrasonography, nuclear magnetic resonance, and Raman spectroscopy for skin studies," *Ski. Res. Technol.* **3**(3), 139–146, Wiley Online Library (1997).

125. T. Yamamura and T. Tezuka, "The water-holding capacity of the stratum corneum measured by 1 H-NMR," *J. Invest. Dermatol.* **93**(1), 160–164, Elsevier (1989).

126. M. Takenouchi, H. Suzuki, and H. Tagami, "Hydration characteristics of pathologic stratum corneum: Evaluation of bound water," *J. Invest. Dermatol.* **87**(5), 574–576, Elsevier (1986).

127. A.A. Gurjarpadhye, W.C. Vogt, Y. Liu, and C.G. Rylander, "Effect of localized mechanical indentation on skin water content evaluated using OCT," *Int. J. Biomed. Imaging* **2011**, 817250-1-8 Hindawi (2011).

128. P.L. Walling and J.M. Dabney, "Moisture in skin by near-infrared reflectance spectroscopy," *J. Soc. Cosmet. Chem.* **40**(3), 151–171, Citeseer (1989).

129. R.O. Potts, D.B. Guzek, R.R. Harris, and J.E. McKie, "A noninvasive, in vivo technique to quantitatively measure water concentration of the stratum corneum using attenuated total-reflectance infrared spectroscopy," *Arch. Dermatol. Res.* **277**(6), 489–495, Springer (1985).

130. D. Bommannan, R.O. Potts, and R.H. Guy, "Examination of stratum corneum barrier function in vivo by infrared spectroscopy," *J. Invest. Dermatol.* **95**(4), 403–408 (1990).

131. K. Wichrowski, G. Sore, and A. Khaiat, "Use of infrared spectroscopy for in vivo measurement of the stratum corneum moisturization after application of cosmetic preparations," *Int. J. Cosmet. Sci.* **17**(1), 1–11, Wiley Online Library (1995).

132. F. Pirot, Y.N. Kalia, A.L. Stinchcomb, et al., "Characterization of the permeability barrier of human skin in vivo," *Proc. Natl. Acad. Sci. U.S.A.* **94**(4), 1562–1567, National Acad. Sciences (1997).

133. G.W. Lucassen, G.N.A. Van Veen, and J.A.J. Jansen, "Band analysis of hydrated human skin stratum corneum attenuated total reflectance Fourier transform infrared spectra in vivo," *J. Biomed. Opt.* **3**(3), 267–281, International Society for Optics and Photonics (1998).

134. J.R. Hansen and W. Yellin, "NMR and intrared spectroscopic studies of stratum corneum hydration," in *Water Structure at the Water-Polymer Interface*, pp. 19–28, Springer (1972).

135. H. Ramløv and D.S. Friis (eds.), *Antifreeze Proteins*, vol. **2**, Springer (2020).

136. S.L. Zhang, C.L. Meyers, K. Subramanyan, and T.M. Hancewicz, "Near infrared imaging for measuring and visualizing skin hydration. A comparison with visual assessment and electrical methods," *J. Biomed. Opt.* **10**(3), 31107, International Society for Optics and Photonics (2005).

137. T. Von Zglinicki, M. Lindberg, G.M. Roomans, and B. Forslind, "Water and ion distribution profiles in human skin," *Acta Derm. Venereol.* **73**(5), 340 (1993).

138. R.R. Warner, M.C. Myers, and D.A. Taylor, "Electron probe analysis of human skin: Determination of the water concentration profile," *J. Invest. Dermatol.* **90**(2), 218–224 (1988).

139. A.Y. Sdobnov, M.E. Darvin, J. Schleusener, J. Lademann, and V.V. Tuchin, "Hydrogen bound water profiles in the skin influenced by optical clearing molecular agents: Quantitative analysis using confocal Raman microscopy," *J. Biophoton.* **12**(5), e201800283-1–11. (2019) [doi:10.1002/jbio.201800283].

140. H. Nakazawa, N. Ohta, and I. Hatta, "A possible regulation mechanism of water content in human stratum corneum via intercellular lipid matrix," *Chem. Phys. Lipids* **165**(2), 238–243, Elsevier (2012).

141. Y.T. Kelman, S. Asraf, N. Ozana, N. Shabairou, and Z. Zalevsky, "Optical tissue probing: Human skin hydration detection by speckle patterns analysis," *Biomed. Opt. Express* **10**(9), 4874–4883, Optical Society of America (2019).

142. H. Iwasaki, K. Miyazawa, and S. Nakauchi, "Visualization of the human face skin moisturizing ability by spectroscopic imaging using two near-infrared bands," in Spectral Imaging: 8th International Symposium on Multispectral Color Science **6062**, p. 606203, International Society for Optics and Photonics (2006).

143. T. Yu, X. Wen, Q. Luo, D. Zhu, and V.V. Tuchin, "Quantitative analysis of dehydration in porcine skin for assessing mechanism of optical clearing," *J. Biomed. Opt.* **16**(9), 95002, International Society for Optics and Photonics (2011).

144. B.P. Yakimov, D.A. Davydov, V.V. Fadeev, G.S. Budylin, and E.A. Shirshin, "Comparative analysis of the methods for quantitative determination of water content in skin from diffuse reflectance spectroscopy data," *Quantum Electron.* **50**(1), 41, IOP Publishing (2020).

145. C.V. Raman and K.S. Krishnan, "A new type of secondary radiation," *Nature* **121**(3048), 501–502, Nature Publishing Group (1928).

146. H. Abramczyk, *Introduction to Laser Spectroscopy*, Elsevier (2005).

147. K. Kong, C. Kendall, N. Stone, and I. Notingher, "Raman spectroscopy for medical diagnostics: From in-vitro biofluid assays to in-vivo cancer detection," *Adv. Drug Deliv. Rev.* **89**, 121–134, Elsevier (2015).

148. C. Krafft, M. Schmitt, I.W. Schie, et al., "Label-free molecular imaging of biological cells and tissues by linear and nonlinear Raman spectroscopic approaches," *Angew. Chemie Int. Ed.* **56**(16), 4392–4430, Wiley Online Library (2017).

149. M. Jermyn, J. Desroches, K. Aubertin, et al., "A review of Raman spectroscopy advances with an emphasis on clinical translation challenges in oncology," *Phys. Med. Biol.* **61**(23), R370, IOP Publishing (2016).

150. A. Quatela, L. Miloudi, A. Tfayli, and A. Baillet-Guffroy, "In vivo Raman microspectroscopy: Intra- and intersubject variability of stratum corneum spectral markers," *Skin Pharmacol. Physiol.* **29**(2), 102–109, Karger Publishers (2016).

151. A. Huizinga, A.C.C. Bot, F.F.M. de Mul, G.F.J.M. Vrensen, and J. Greve, "Local variation in absolute water content of human and rabbit eye lenses measured by Raman microspectroscopy," *Exp. Eye Res.* **48**(4), 487–496, Academic Press (1989).

152. N.J. Bauer, J.P. Wicksted, F.H. Jongsma, et al., "Noninvasive assessment of the hydration gradient across the cornea using confocal Raman spectroscopy," *Invest. Ophthalmol. Vis. Sci.* **39**(5), 831–835, The Association for Research in Vision and Ophthalmology (1998).

153. P.J. Caspers, H.A. Bruining, G.J. Puppels, G.W. Lucassen, and E.A. Carter, "In vivo confocal Raman microspectroscopy of the skin: Noninvasive determination of molecular concentration profiles," *J. Invest. Dermatol.* **116**(3), 434–442, Elsevier (2001).

154. M. Gniadecka, O.F. Nielsen, D.H. Christensen, and H.C. Wulf, "Structure of water, proteins, and lipids in intact human skin, hair, and nail," *J. Invest. Dermatol.* **110**(4), 393–398, Elsevier (1998).

155. M. Egawa and H. Tagami, "Comparison of the depth profiles of water and water-binding substances in the stratum corneum determined in vivo by Raman spectroscopy between the cheek and volar forearm skin: Effects of age, seasonal changes and artificial forced hydration," *Br. J. Dermatol.* **158**(2), 251–260, Wiley Online Library (2008).

156. E. Boireau-Adamezyk, A. Baillet-Guffroy, and G.N. Stamatas, "Mobility of water molecules in the stratum corneum: Effects of age and chronic exposure to the environment," *J. Invest. Dermatol.* **134**(7), 2046–2049 (2014).

157. C. Choe, S. Choe, J. Schleusener, J. Lademann, and M.E. Darvin, "Modified normalization method in in vivo stratum corneum analysis using confocal Raman microscopy

to compensate nonhomogeneous distribution of keratin," *J. Raman Spectrosc.* **50**(7), 945–957, Wiley Online Library (2019).

158. M.E. Darvin, C.-S. Choe, J. Schleusener, and J. Lademann, "Non-invasive depth profiling of the stratum corneum in vivo using confocal Raman microscopy considering the non-homogeneous distribution of keratin," *Biomed. Opt. Express* **10**(6), 3092–3103, Optical Society of America (2019).

159. Q. Sun, "The Raman OH stretching bands of liquid water," *Vib. Spectrosc.* **51**(2), 213–217, Elsevier (2009).

160. Q. Sun, "Local statistical interpretation for water structure," *Chem. Phys. Lett.* **568**, 90–94, Elsevier (2013).

161. Y. Maeda and H. Kitano, "The structure of water in polymer systems as revealed by Raman spectroscopy," *Spectrochim. Acta Part A Mol. Biomol. Spectrosc.* **51**(14), 2433–2446, Elsevier (1995).

162. C. Choe, J. Schleusener, J. Lademann, and M.E. Darvin, "Keratin-water-NMF interaction as a three layer model in the human stratum corneum using in vivo confocal Raman microscopy," *Sci. Rep.* **7**(1), 1–13, Nature Publishing Group (2017).

163. C. Choe, J. Schleusener, S. Choe, J. Lademann, and M.E. Darvin, "A modification for the calculation of water depth profiles in oil-treated skin by in vivo confocal Raman microscopy," *J. Biophoton.* **13**(1), e201960106, Wiley Online Library (2020).

164. G. Puppels, P. Caspers, and C. Nico, "Comment on Choe et al. 'A modification for the calculation of water depth profiles in oil-treated skin by in vivo Raman microscopy,'" *J. Biophoton.* **13**(6), e202000043-1–3 (2020).

165. M.E. Darvin, C. Choe, J. Schleusener, S. Choe, and J. Lademann, "Response to comment by Puppels et al. on 'A modification for the calculation of water depth profiles in oil-treated skin by in vivo Raman microscopy'," *J. Biophoton.* **13**(6), e202000093-1–3 (2020).

166. Q. Zhang, K.L. Andrew Chan, G. Zhang, et al., "Raman microspectroscopic and dynamic vapor sorption characterization of hydration in collagen and dermal tissue," *Biopolymers* **95**(9), 607–615, Wiley Online Library (2011).

167. T.T. Nguyen, T. Happillon, J. Feru, et al., "Raman comparison of skin dermis of different ages: focus on spectral markers of collagen hydration," *J. Raman Spectrosc.* **44**(9), 1230–1237, Wiley Online Library (2013).

168. S. Mangelsdorf, T. Vergou, W. Sterry, J. Lademann, and A. Patzelt, "Comparative study of hair follicle morphology in eight mammalian species and humans," *Ski. Res. Technol.* **20**(2), 147–154, Wiley Online Library (2014).

169. S. Carvalho, N. Gueiral, E. Nogueira, et al., "Glucose diffusion in colorectal mucosa: A comparative study between normal and cancer tissues," *J. Biomed. Opt.* **22**(9), 91506, International Society for Optics and Photonics (2017).

170. B. Schulz, D. Chan, J. Bäckström, and M. Rübhausen, "Spectroscopic ellipsometry on biological materials – investigation of hydration dynamics and structural properties," *Thin Solid Films* **455**, 731–734, Elsevier (2004).

171. C. Choe, J. Schleusener, S. Choe, et al., "Stratum corneum occlusion induces water transformation towards lower bonding state: A molecular level in vivo study by confocal Raman microspectroscopy," *Int. J. Cosmet. Sci.* **42**(5), 482–493 (2020).

172. I. Carneiro, S. Carvalho, R. Henrique, L. Oliveira, and V.V. Tuchin, "Simple multimodal optical technique for evaluation of free/bound water and dispersion of human liver tissue," *J. Biomed. Opt.* **22**(12), 125002, International Society for Optics and Photonics (2017).

173. И.Ю. Янина, И. Шлойзенер, Ю. Ладеманн, В.В. Тучин, and М.Е. Дарвин, "Исследование Эффективности Оптического Просветления Кожи Растворами Глицерина Методом Конфокальной Микроспектроскопии Комбинационного Рассеяния Света," *Журнал Технической Физики* **128**(6), 753 (2020) [doi:10.21883/os.2020.06.49407.52-20].

174. J. Schleusener, J. Lademann, and M.E. Darvin, "Depth-dependent autofluorescence photobleaching using 325, 473, 633, and 785 nm of porcine ear skin ex vivo," *J. Biomed. Opt.* **22**(9), 91503, International Society for Optics and Photonics (2017).

175. E.N. Lazareva, P.A. Dyachenko, A.B. Bucharskaya, N.A. Navolokin, and V.V. Tuchin, "Estimation of dehydration of skin by refractometric method using optical clearing agents," *J. Biomed. Photonics Eng.* **5**(2), 20305 (2019).

176. R. Barer, "Spectrophotometry of clarified cell suspensions," *Science* **121**(3151), 709–715, JSTOR (1955).

177. S.H. Chung, A.E. Cerussi, C. Klifa, et al., "In vivo water state measurements in breast cancer using broadband diffuse optical spectroscopy," *Phys. Med. Biol.* **53**(23), 6713, IOP Publishing (2008).

178. E.M. Barroso, R.W.H. Smits, T.C. Bakker Schut, et al., "Discrimination between oral cancer and healthy tissue based on water content determined by Raman spectroscopy," *Anal. Chem.* **87**(4), 2419–2426, ACS Publications (2015).

179. V.D. Genin, D.K. Tuchina, A.J. Sadeq, et al., "Ex vivo investigation of glycerol diffusion in skin tissue," *J. Biomed. Photonics Eng.* **2**(1), 010303-1–5 (2016).

180. A.S. Kolesnikov, E.A. Kolesnikova, K.N. Kolesnikova, et al., "THz monitoring of the dehydration of biological tissues affected by hyperosmotic agents," *Phys. wave Phenom.* **22**(3), 169–176, Springer (2014).

181. S.R. Utz, V.V. Tuchin, and E.M. Galkina, "The dynamics of some human skin biophysical parameters in the process of optical clearing after hyperosmotic solutions topical application," *Vestn. Dermatol. Venerol.* **91**(4), 60–68 (2015).

182. V.D. Genin, E.A. Genina, S.V. Kapralov, et al., "Optical clearing of the gastric mucosa using 40%-glucose solution," *J. Biomed. Photon. Eng.* **5**(3), 30302 (2019).

183. L.M. Oliveira, M.I. Carvalho, E.M. Nogueira, and V.V. Tuchin, "Skeletal muscle dispersion (400–1000 nm) and kinetics at optical clearing," *J. Biophoton.* **11**(1), e201700094, Wiley Online Library (2018).

9

Optical and mechanical properties of cartilage during optical clearing

Yulia M. Alexandrovskaya, Olga I. Baum, Vladimir Yu. Zaitsev, Alexander A. Sovetsky,
Alexander L. Matveyev, Lev A. Matveev, Kirill V. Larin, Emil N. Sobol, and Valery V. Tuchin

CONTENTS

Introduction

Cartilage is a connective tissue with a load-bearing function and is present all over the human body [1, 2]. Evolutionary cartilage is a more ancient form of skeletal tissue. The development of the bone skeleton of the embryo originates from the hyaline cartilage through endochondral ossification [1]. Among all the different tissue types, cartilage is unique due to the absence of blood vessels in a healthy state. This property makes it both compliant and more complex to handle, study, and treat. On one hand, the relatively stationary structure of the cartilaginous matrix enables one to obtain its content, mechanical, optical, physicochemical, and other properties with high accuracy and to more reliably transmit the results of *ex vivo* studies to *in vivo* applications. On the other hand, the commonly known mechanisms of living tissue metabolism, such as oxygen and glucose supply, immune response, and regeneration, can be only very roughly considered regarding cartilage and need to be kept in mind when some diagnostic or analytic methods are brought from other fields where they have shown high efficacy. Particularly, liquid exchange within cartilage has been shown to proceed through pores of the submicron level [3–5]. The size of pores in healthy hyaline-type

articular cartilage shown by different methods does not exceed 5–15 nm. Liquid diffusion through such small pores requires a significantly longer time in comparison to blood flow; therefore, the metabolism rate is rather low. Alternations in this fragile balance of exchange between cartilage and external synovial fluid lead to the development of pathology.

Nontraumatic disruptions of natural cartilaginous functions are always somewhat connected with metabolic disorders being the most evident cause of decease [2, 6]. These pathologies, such as osteoarthritis, develop slowly but present a significant challenge in terms of diagnosing them at an early stage and treating them. Cartilage pathologies affect about one third of adults aged ≥45, and among adults with arthritis, 27% report severe joint pain [7]. Chronic cartilaginous diseases hold strong leadership positions among those decreasing quality of life to a great extent, especially for elderly people, and billions of dollars are being spent to overcome its consequences.

In addition to nontrivial metabolic properties, cartilaginous tissue presents a challenge of *in vivo* availability. Naturally, joint hyaline cartilage is situated at bone boundaries covered from two sides, such as the femoral, knee, elbow, interphalangeal, ankle, and facet (back spine) joints, which allow for the smooth movement of corresponding bones; the thyroid,

TABLE 9.1

Selected data on diagnostics and treatment of cartilage by means of optical methods at certain wavelengths

Wavelength, nm	Technique description	References
420–720	Spectral-derived quantification of color-changing for degraded tissue, increasing of "redness" due to bone hemoglobin absorption at 550 nm with decreasing cartilage thickness	[21]
	Polarized reflectance maps with green transmission filter (560 nm) of superficial collagen network and subsurface chondrocyte organization	[22]
780	Quantitative assessment of second harmonic generation (SHG) intensity and formulation of criteria to distinguish between collagen I and II	[10]
800	SHG of collagen matrix architecture and two-photon fluorescence of cell lysis debris, zonal structure of collagen, and signs of calcification	[23]
850	OCT-based displacement measurements under compressive load, identification of two mechanically different subzones of radial zone of articular cartilage	[24]
980	Ear cartilage reshaping in animals with water cooling	[25]
1064	Ear cartilage reshaping in patients	[26]
~1300	2D-OCT roughness parameters for surface integrity quantification	[27]
	Polarization-sensitive OCT analysis of depth-dependent collagen orientation	[28,29]
	Quantitative assessment of mechanical stiffness drop for enzymatically degraded tissue with OCT-based air-jet indentation	[30]
	OCE-based depth-resolved imaging of laser irradiation-induced strain-evolution and heat-induced ensembles of pores in bulk matrix, quantification of their mean parameters	[31,32, 33]
	OCE-based evaluation of slow postirradiation strain in laser-reshaped cartilaginous implants	[34]
	OCE-based revealing of thermal-stresses and strains in laser-assisted modification of cartilage	[35]
1320	Stabilization of costal cartilage warping	[19]
1450	Thermoforming of tracheal cartilage	[36]
1540–1560	Laser correction of ear protrusions avoiding skin burn	[37]
	Laser correction of nasal septum shape	[15]
	Laser reshaping of costal cartilage	[17]
	Laser-induced regeneration of spine disks and joint cartilage	[13,14]
2100	Reshaping of human nasal septum	[38]

cricoid, and trachea cartilages, which maintain the larynx anatomy in a compact manner; and intervertebral discs, which provide mobility and flexibility of the spine core. In this way, methods of cartilage medical analysis are strongly limited to those allowing for its examination in a noncontact way, such as ultrasound, radiography, and magnetic resonance imaging (MRI), or those specially equipped to minimize any traumatic effect, such as arthroscopic observation. However, the resolution and precision of standard imaging do not always satisfy the criteria of efficacy when it comes to early cartilaginous disorders – that is, exactly where the effectiveness of the diagnosis would allow minimizing the possible negative consequences of the disease.

Thus, optical imaging techniques are being developed to enhance the informational content of diagnostic feedback. Visualization of cartilage with optical coherence tomography (OCT) opened new perspectives for the evaluation of early osteoarthritis, allowing for detection of its surface irregularities and systematic quantitative classification of small lesions [8, 9], precise measurements of its thickness and grade of fibrillation [9], the obtaining of backscatter and extinction coefficients sensitive to structural changes [9], and detailed investigation of extracellular matrix organization [8]. In particular, the detection of tissue deterioration with the polarization-sensitive OCT is capable of distinguishing between collagen II and I types, which indicates the regions of "normal" hyaline-type cartilage and fibrocartilage of altered structure,

correspondingly [10]. In this regard, new emerging modalities such as OCT-based elastography (OCE) have great potential to give further insights in mechanical alterations of cartilage caused by pathological processes [9]. The recent investigation of breast tissue and model animal tumors by means of OCE revealed that mechanical stiffness can be a reliable indicator of pathology comparable in information value and accuracy with staining-based biopsy [11, 12]. Meanwhile, in comparison to conventionally used histology, it is a rapid and noninvasive technique.

The application of light-based techniques to cartilage is not limited to potentials of early diagnostics of osteoarthritis. Due to its less susceptibility to drugs and extremely slow restoration after common surgery, the light-based technologies of cartilage treatment based on near infrared (NIR) laser action were outlined in the recent decades and showed inspiring results for laser-assisted regeneration of intervertebral discs and knee joint cartilage [13, 14]. The radiation is transmitted through a thin optical fiber, which makes the procedure less invasive than surgical intervention. Another promising potential of the technique of laser action on cartilage is its nondestructive reshaping, enabling the correction, for example, of the nasal septum [15] or ear shape [16] and obtain load-bearing implants of costal cartilage of semicircle geometry [17, 18] or stabilize its undesirable spontaneous warping [19]. The effect originates from the thermal-induced relaxation of internal stresses [20]. The completeness of stress relaxation depends on the efficiency

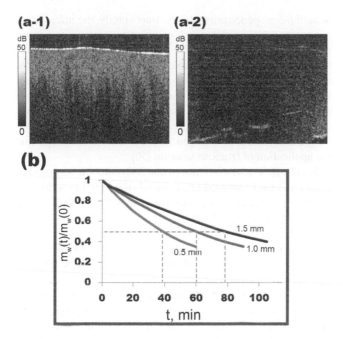

FIGURE 9.1 Cartilage drying: (a) OCT structural images of (a-1) wet and (a-2) dried cartilage (the residual amount of water is about 4%), the panel size is 2 × 4 mm; (b) the kinetics of cartilage air drying depending on the tissue thickness (0.5–1.5 mm); Y-axis indicates the relative water mass (m_w) at t and 0 moments, the dashed green line shows the time moments when the initial water content decreases by 50%.

Cartilage optics at optical clearing

Basic concept

Water occupies about 75% of the total cartilage mass [1, 3, 4]. Though it does not have strong absorbers in the range of 400–2000 nm, the high scattering strongly limits light penetration depth in cartilage. The enhancement of its transparency during several minutes of air drying due to evaporation of subsurface water can already be seen on visual inspection. For instrumentation analysis, the completely dried cartilage can be much more transparent than the liquid-saturated tissue. Figure 9.1(a) shows the comparative OCT visualization of cartilage with (a-1) ~75 and (a-2) ~4% of water content. The subsurface level of dried tissue is almost transparent; the characteristic granular patterns and surface cannot be seen.

As shown in [39], drying of cartilage at room temperature is completely reversible in terms of its optical properties. However, the drying technique is hardly applicable to *in vivo* tissue analysis as well as to *ex vivo* specimen examination as it dramatically transforms the whole tissue structure and disturbs cell functioning. Moreover, rather a long time is required to remove water. Figure 9.1(b) shows the drying kinetics for cartilage of three different thicknesses. From 40 to 80 min is needed to decrease the water content within a 0.5–1.5 mm tissue layer by 50% (Figure 9.1(b)). The tissue heating with moderate intensity of NIR laser can slightly increase its transparency due to two codirectional processes: surface water evaporation and decrease of water absorption with growing temperature [40].

The opacity of cartilage is caused by its heterogeneous structure and light scattering by collagen, minor proteins, and cells. Optical properties of cartilage were established for numerous wavelengths [41], including the infrared range [42]. It must be noted that each parameter strongly depends on the tissue type, age, presence or absence of degradation signs, water content, and technique of measurement. An important property of cartilage is structural anisotropy, which determines the anisotropy of optical parameters [43, 44].

The replacement of interstitial water with OCA can be a more rapid and effective solution for enhancing light penetration through cartilage. To date, different OCAs were tested to increase cartilage transparency [40, 44–48]. Among the nonreactive chemicals, two types of agents can be outlined: (1) sugar alcohols provoking pronounced shrinkage, such as glycerol, ethylene and polyethylene glycols, sorbitol, and others; and (2) agents with mild action, such as sugars or some radiopaque substances. The application of each type depends on the concrete purpose and whether the cartilage structure needs to be preserved or not. The ability of glycerol to partially dissolve collagen was experimentally shown [49]. Further, some applications of OCA for cartilage OC have been investigated.

of heat delivery, i.e., the delivery of laser radiation throughout the whole thickness of cartilage, including the deep layers.

Some demonstrative examples of recent advances in cartilage diagnostics and modification by means of light-based methods according to certain wavelengths are summarized in Table 9.1 [21–38]. One can see that the ranges of visible and NIR light related to the first and second transparency windows bring reliable information about cartilage's fine structure in a less subjective and sometimes even quantitative manner, which is unachievable for more approximate methods such as MRI and radiography. Meanwhile, growing water absorption with increasing wavelength is utilized to generate heat-induced effects, still not compromising matrix structure, but enabling cartilage properties to be temporarily modified to allow reshaping and moderate laser-assisted regeneration.

Due to the high scattering of light within cartilage, the problem of adequate optical feedback from deeper tissue layers usually occurs. The general limitation of optical imaging as well as NIR laser-induced heating is the comparatively low penetration depth of radiation, not exceeding 1–2 mm for most of the used wavelengths. Fortunately, the thickness of human joint cartilage is usually somewhat close to that achievable for optical techniques – about 1.5–2.0 mm. Some thicker types of cartilage, for example, costal cartilage with dimensions equal to those of ribs, present a more challenging task for light-based techniques.

Thus, OC of cartilage may significantly contribute to the already existing methods of its optical diagnostics and treatment as well as broaden the useful optical range involving the wavelengths with higher interstitial scattering. Further, the recent results on cartilage OC are summarized and discussed.

Visible range

In [45], the OC of bovine osteochondral samples harvested from femoral condyles of bovine knee with gradually increasing fructose concentrations was performed. Interestingly, the OC process was combined with a multitude of labeling

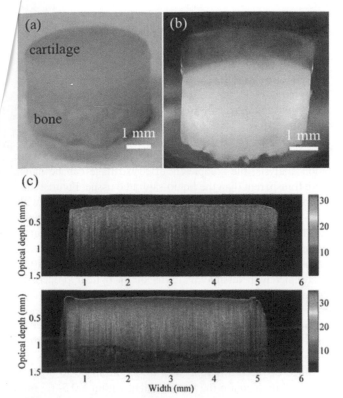

FIGURE 9.2 Articular cartilage OC visualization: (a) natural appearance of the sample before OC, (b) after the saturation with the iohexol solution, (c) OCT image of the cartilage before (top) and after 15 min of OC (bottom). Average thickness of the cartilage measured before OC was 0.9 mm. Color bar indicates OCT signal in dB. Reprinted with permission from [47].

techniques typical for microscopic observation of cartilage cells and matrix proteins. The OC ability of fructose solution was proven to be effective for cartilage, but not for osteochondral bone. The application of OCA allowed for the visualization of the fluorescent markers for cell membranes and nuclei as deep as the working distance of the confocal microscope objective. Through the reversible OC process, the cell volume did not change [45]. The immunolabeling for collagen II and VI could not be observed at 200 μm depth, indicating the limit

of antibodies' penetration ability. Importantly, the microstructure of cartilage components revealed by scanning electron microscopy was not altered by the OC process, which makes the sugar solutions very suitable to be used as OCA for microscopic observation of deep tissue layers, including fluorescent techniques with markers of small molecular weight that can effectively penetrate dense tissue matrix. In particular, intercellular chondrocyte connections in cartilage were successfully nondestructively imaged with confocal microscopy and the application of fructose solution [50].

In [47], 39% iohexol (Omnipaque 350) solution was used to reduce the backscattered OCT signal from a 0.9 mm layer of articular cartilage, allowing for imaging of the cartilage–bone interface (Figure 9.2(a,b,c)).

The backscattered OCT signal was maximally reduced at approximately 50 min of OC; however, the bone surface could already be visualized after 15 min of OCA application (Figure 9.2(c)). The measured refractive indices of cartilage before and after 100 min of OC were 1.386 ± 0.008 and 1.504 ± 0.016 respectively. The calculation given on the basis of Gladstone-Dale law for refractive index of composite media [47] showed that about 50% of the initial volume of interstitial fluid was replaced with iohexol. Meanwhile, as solution of iohexol possesses low osmolality (1–3 times that of blood plasma), no dramatic shrinkage was observed during the study – a decrease of less than 15%–20% of the total cartilage thickness. The return to normal opacity was observed after 10 min immersion in PBS [47]. In contrast to iohexol, the immersion in polyethylene glycol (PG) gave a decrease of about 60% of the initial cartilage thickness [46]. The saturation OC time was 30 min, whereas 5 min of application was enough to visualize cartilage–bone interface. Thus, as PG disrupts cartilage structure, it is more suitable when only bone imaging is needed. For evaluation of both cartilage and bone structure, it is better to use iohexol.

On the basis of data discussed in [40, 48], the effect of several common OCAs can be described for costal cartilage (Figure 9.3(a,b)). In comparison to articular cartilage, which has a mean thickness of about 1 mm, costal cartilage grafts used for implantation are 1.5–3.0 times thicker [17, 18], which makes their deep layers less achievable for optical radiation.

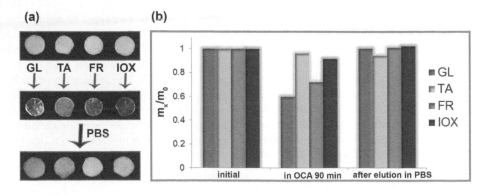

FIGURE 9.3 Demonstration of effect of 35% glycerol (GL), 8.3% tartaric acid (TA), 35% fructose (FR), and 39% iohexol (IOX) on 1.5 mm costal cartilage samples: (a) visualization of OC ability at 50 min of immersion in corresponding OCAs and restoration of the initial structure after immersion in PBS; (b) mass dynamics of cartilage under the corresponding OCAs effect after 90 min of immersion in OCA and after elution in PSB in relation to initial mass value (m_0).

In [40], fructose and glycerol solutions of different concentrations were used. The pronounced OC effect in the visible range was reported for both OCAs, while fructose application resulted in less intensive shrinkage. When glycerol is applied in concentration higher than 45%–50%, the reduction of optical transmittance can be observed due to the deformation processes caused by dehydration and shrinkage [40]. The noticeable drop in mass value for samples immersed in 35% glycerol and 50% fructose was observed to be 40% and 30% respectively, which can be attributed to OCA-induced dehydration (Figure 9.3(b)). Along with the possibility of effective OC in the visible range, a solution of iohexol does not provoke such dramatic dehydration and cartilage shrinkage (Figure 9.3(a)) [48]. The increase of the tissue transmittance intensity probing with green light after 90 min immersion in 40% iohexol [48] was comparable to that of 20% fructose and 35% glycerol [40], while the observed macroscopic deformation was minimal (Figure 9.3(a)).

Tartaric acid as a potential OCA appeared to be ineffective for cartilage (Figure 9.3(a, b)) leading to irreversible alterations in its structure and a noticeable mass drop that was not restored with immersion in PBS.

NIR range

Cartilage OC in the NIR range involves some additional challenges connected with the increasing absorption of OH-containing molecules, the presence of overtones, and combination bands with corresponding absorption effects, which may to a great extent compromise the OC ability of a substance. From 1000 to 2000 nm, the water absorption spectrum has peaks at 1450 and 1930 nm, attributed to the first overtone of symmetric stretch vibration and the combination band of OH bond, respectively [51]. Organic molecules, for example, sugars [52], also have peaks within the NIR range. Therefore, the intensity of corresponding peaks of absorption should be considered. A spectrum recorded after the immersion of ear cartilage in iohexol showed three additional unidentified peaks at 1240, 1300, and 1360 nm, which appeared as a result of the interaction of iohexol with cartilage; and one at ~1200 nm, related to iohexol absorption [44], while in the visible range, the OC process did not change the transmittance spectrum. In [48], heat-induced decomposition of iohexol molecules was detected by Raman spectroscopy when cartilage samples immersed in iohexol were subjected to moderate laser exposure with a wavelength of 1560 nm, a duration of 15 s, and a power density higher than 350 W/cm². It was shown that saturation with OCA caused an increase in maximal temperature at laser heating of 15% and resulted in thermal decomposition of a substance [48]. Although the OC ability of iohexol in the NIR range remains quite effective (resulting in up to a twofold increase in tissue transmittance), the additional heating of tissue caused by iohexol absorption threatens tissue integrity and the chemical stability of OCA itself [48].

At ~1500 nm, glycerol and fructose showed remarkably high OC ability with the transmittance increase in 2.5 and >10 times correspondingly [40], though at saturation point showing substantial shrinkage and dehydration (Figure 9.4(b)). The temperature growth for 1560 nm laser irradiation was observed to be 17% and 28% for fructose and glycerol, respectively (Figure 9.4(a)).

Since all considered agents cause dehydration in the amount of at least a few percent, an increase in the absorption of NIR radiation can be associated either with absorption by the OCAs themselves, or with a more compact packing of water molecules under their action. Some explanations for particular mechanisms of osmotic dehydration can be outlined by analyzing the local dynamics of the tissue's mechanical properties under the action of hyperosmotic OCAs, which are discussed in the next section.

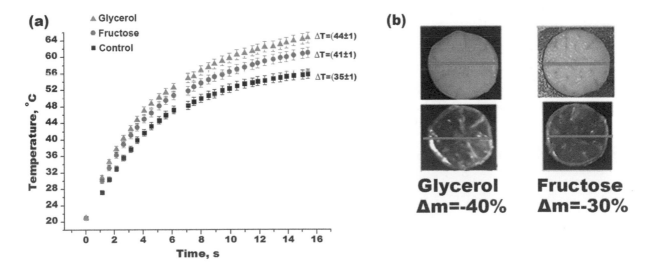

FIGURE 9.4 (a) Kinetics of the temperature maximum on the cartilage surface in the course of its laser irradiation with λ=1.56 μm. The samples are saturated with glycerol and fructose solutions of 35% wt. concentration: (b) Cartilage samples before (upper) and after saturation with glycerol and fructose (lower), red lines are of equal length on upper and lower pictures, Δm indicates the mass loss. Cartilage thickness is 1 mm. Reprinted with permission from [40].

Cartilage mechanics at optical clearing

Osmotic stress

Many commonly used OCAs are nonreactive chemicals – biocompatible substances already used in medical diagnostics and pharmaceutics, such as glycerol, sugars, sugar alcohols, etc. The radiopaque substance iohexol used in current studies is approved by the Food and Drug Administration for intravenous injection to enhance the contrast of radio diagnostics. Thus, the possible toxicity of these OCAs is minimal.

However, the safety of OCA injection can be compromised by osmotic stress caused by a sudden change in solute concentration in tissue bulk and around cells. Even if the concentration gradients are not critical for particular type of cells, the intensive replacement of water in the tissue matrix may result in the significant alteration of bulk mechanics and the generation of high gradients of local strain, affecting its function. Therefore, close attention is paid to the study of tissue mechanics under OC and identifying acceptable thresholds of mechanical load created by OCA administration. One should note that common measurements of elasticity parameters before and after OCA application do not show the magnitude of osmotic stress in the course of diffusion, which can be critical for tissue safety. A noncontact instrumentation allowing for continuous monitoring of osmotic effects has been lacking. After a discussion of established practices, some novel approaches to overcoming the previously mentioned limitation are discussed.

Compression tests

In [45], the mechanical testing of cartilage at OC with fructose solution was performed using an unconfined compression technique with a servoelectric materials testing system. Samples were compressed with 0.05 strain increments at a physiological strain of 0.179/s coupled with a 10-min hold to ensure stress relaxation. The instantaneous and equilibrium moduli were obtained for control, cleared, and uncleared cartilage discs, showing a significant increase in both moduli for cleared tissue: from 2.6 to 3.3 MPa and from 0.4 to 0.8 correspondingly (Figure 9.5). The same measurements after elution of OCA showed some decrease in relation to the initial values: about 10% and 50% for instantaneous and equilibrium moduli respectively, which is considered by the authors as a result of the proteolytic process of natural tissue degradation during the time of experiment and not as the effect of OC.

Several explanations of the observed OCA-induced increase in tissue stiffness is proposed in [45]: (1) the higher viscosity of OCA (fructose solution) contributes to increased sustainability of bulk pores to liquid outflow under compression; (2) the OCA-induced breakage of intermolecular hydrogen bonds results in the increased contribution of covalent collagen bonds in total tissue stiffness. Other results of the study, such as the preservation of cell dimensions in the course of OC, show that the first is the more likely explanation. Notice that no substantial shrinkage of or alterations in tissue microstructure were observed with the application of fructose solution as OCA in this study [45].

Phase-stabilized swept source OCE measurements

Stiffness measurements for a nasal septum were performed in [53] under the action of 20% glucose (Figure 9.6). In this work, for the first time, routine compression testing was supplemented with phase-stabilized swept source OCE with an air-pulse delivery system capable of delivering a short duration (≤1 ms) focused air-pulse to induce a localized displacement in

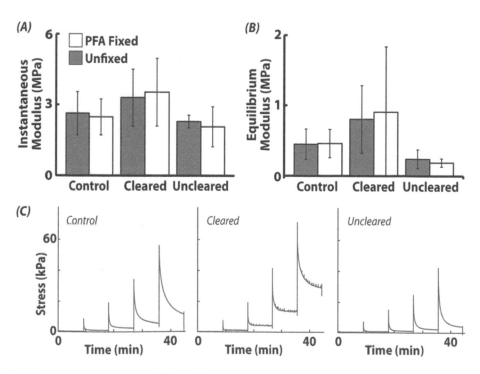

FIGURE 9.5 Mechanical compression testing of cartilage stiffness under OC: (a) instantaneous modulus for 4% paraformaldehyde (PFA) fixed and unfixed samples, (b) equilibrium modulus for the same groups of samples, (c) corresponding stress relaxation curves. Reprinted with permission from [45].

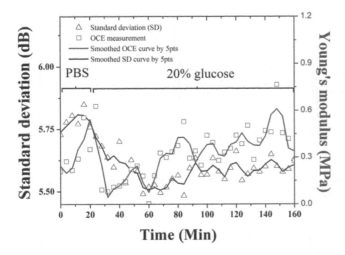

FIGURE 9.6 Elasticity of cartilage measured by OCE and uniaxial mechanical testing. The cartilage sample was immersed in 1X PBS for 20 min, then in 20% glucose for 140 min. Reprinted with permission from [53].

sample, which propagated as an elastic wave that was imaged by OCT. By synchronizing the air-pulse with the OCT imaging, two-dimensional depth-resolved elasticity maps were generated. The quantification of Young's modulus based on elastic wave group velocity evaluation from displacement profiles is given in detail in [53].

It was shown that glucose solution reduces the refractive index mismatches between collagen and interstitial fluid in comparison to PBS, indicating the course of the OC process [53]. The Young's modulus deviations in the process of cartilage OC were obtained from OCE measurements. Interestingly, the dynamics of the modulus changed direction within the time intervals of diffusion; for the first 10–15 min of immersion, the tissue stiffness decreased and then gradually began to increase towards the initial value for the remaining 110 min. The authors suggest that the gradual increase in cartilage stiffness reflects the reversibility of the OC process when glucose is forced out of the extracellular matrix by compression. The correlation between OCE- derived stiffness and that obtained from standard compression tests was established; however, the absolute values of moduli were between two and three times lower for OCE [53], which was attributed to the anisotropy of cartilage biomechanical properties and the different geometry of measurements in the two methods.

Meanwhile, for the cleared cartilage, no substantial increase in the Young's modulus from its initial value was detected in [53] compared to the results obtained in [45]. The possible explanation may be the different time of diffusion; in [53], the total duration of immersion was only 120 min, whereas in [45] the samples were left overnight in OCA solution. Probably, the results of [45] show moduli at an equilibrium concentration of OCA, and in [53] the initial dynamics of osmotic-mediated tissue mechanics is shown.

Perspectives of noncontact strain mapping by phase-resolved OCE

Basic principles of phase-resolved OCE

Compared to elastic moduli evaluation either by compression tests or by OCE, direct noncontact strain mapping can

be a more rapid and demonstrative technique for monitoring the effects of the hyperosmotic action of OCA on tissue mechanics.

Coherent optical techniques for studying the modification of biological tissues impregnated by osmotically active OCAs have already been widely used. In particular, the observation of statistical properties (such as the evolution of characteristic correlational spatiotemporal scales) of speckle patterns in samples illuminated by coherent light during impregnation by OCAs was discussed over two decades ago [54].

However, although the penetration of osmotically active OCAs in the tissue is a microscopic stochastic process that can be described by diffusion equations, the propagation of the OCA front into the tissue bulk is accompanied by the creation of quite regular strains corresponding to either shrinking of the tissue or, on the contrary, a kind of swelling. Although the idea of measuring microscopic deformation and strain in tissues using OCT was also proposed over two decades ago in the seminal paper by Schmitt [55], practically operable OCT-based methods of strain mapping have only been demonstrated in the past few years.

In this section, we describe the first application of one such recently developed OCT-based elastographic technique [33, 56, 57] to visualize and quantify the strains of osmotic origin in cartilaginous samples impregnated by OCA (glycerol). Application of this new OCT-based elastographic technique opened previously inaccessible prospects for studying in detail the spatiotemporal dynamics of osmotic strain in biological tissues.

In phase-resolved OCT, the phase variation Φ of the backscattered signal is proportional to the axial displacements U of scatterers:

$$U = \frac{\lambda_0 \Phi}{4\pi n},$$ (9.1)

where λ_0 is the optical wavelength in vacuum, and n is the refractive index of the material. It can be shown that phase variations in the OCT signal can be more tolerant to strain-induced decorrelation [56, 57]. Consequently, even for "typical" OCT systems (i.e. without the need for a super-broadband spectrum to reduce the deformation-induced decorrelation), phase-resolved OCE of the displacements of scatterers can be done much more reliably in comparison with the correlational speckle tracking. This is an important advantage of phase-resolved approaches to the estimation of strains. The estimation of the axial gradient of phase variations $\Phi(z)$ and, therefore, $U(z)$ makes it possible to estimate local axial strains ε by finding the axial derivative

$$\varepsilon = \frac{\partial U}{\partial z}.$$ (9.2)

It is important to emphasize that in OCT images, both the displacements and distances are naturally measured in pixels (the physical size of which is dependent on the refractive index). Therefore, when the gradients are calculated via Equation (9.2), the refractive index is cancelled and the strain estimation is correctly found independently of the refractive index. This feature of the phase-resolved method of strain estimation is important for studying the action of OCAs, penetration of which may noticeably change the refractive index.

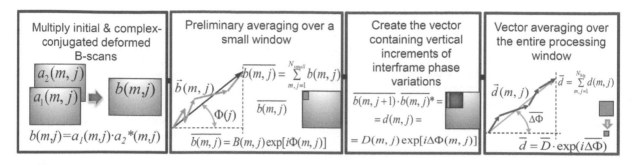

FIGURE 9.7 Schematic view of the evaluation of the axial phase-gradient in the vector approach adapted for laterally inhomogeneous strains with nonhorizontal phase-variation isolines. All intermediate transformations are performed with complex-valued signals, and the sought phase gradient is singled out at the last stage. See reference [61] for more detail.

Another important issue is that, for unambiguous estimation of interframe phase variation $\Phi(z)$ and, therefore, displacements $U(z)$, the condition $U(z) < \lambda / 4$ should be fulfilled. For larger displacements, the phase wrapping occurs because of the 2π-periodicity of the optical-wave phase. To exclude this phase-variation ambiguity in estimations of displacements, the conventional approach is to ensure that the interframe displacements do not exceed $\pm \lambda / 4$ [58]. However, in the method of local-strain estimation used, this limitation is eliminated, so that the displacements may be supra-wavelength and the method is operable until the *phase difference* between neighboring vertical pixels becomes wrapped; at the same time the total interframe phase variation in each of the pixels may be many times wrapped.

A detailed discussion of axial strain estimation based on phase-resolved data for compared deformed and reference OCT scans was presented in [59]. In that paper, the least-square

method (including the improved version with amplitude weighting) was used to estimate the local slope of function $\Phi(z)$. Here we will use another method of finding phase gradients. It was proposed in [60, 61] and was called the "vector method," because it operates with complex-valued OCT signals $a\exp(i\phi)$ considering them as vectors in the complex plane. The sought phase gradient is singled out at the very last stage of the processing procedure. The main steps of the vector method are schematically shown in Figure 9.7.

This method is especially computationally efficient and also enables amplitude weighting to suppress contributions of small-amplitude noisy pixels. The increased tolerance of the method to various noises (including decorrelation ones) makes this method operable under elevated interframe strains (up to ~10^{-2}). These features make the vector method especially suitable for mapping aperiodic strains, for example, those related to deformations of osmotic pressure origin, for which enhancement of signal-to-noise ratio (SNR) via conventional periodic averaging is impossible.

For performing observation of aperiodic strains on fairly large time intervals (up to tens of minutes), sparsing of the recorded sequence of OCT frames is also used as discussed in [62]. On the one hand, this is useful to reduce the volume of data recorded during such long intervals, and on the other hand, this is favorable for enhancing SNR in mapping slowly varying strains. Note that besides "instantaneous" interframe strains (in fact reflecting the strain rate) the total cumulative strain for the entire observation interval can also be mapped.

This technique is used to obtain the below-presented examples demonstrating the spatiotemporal dynamics of strain in cartilaginous samples during the process of OC. To the best of our knowledge, such direct visualization of osmotic-origin strains is made here for the first time. In the color-coded graphs shown below, the negative strain (tissue shrinking) corresponds to the blue color and positive strains corresponding to dilatation (swelling) of the tissue are shown by the red color.

Brief description of a typical experiment on OCA-induced strain mapping

Samples of costal porcine cartilage were taken from local butcher immediately after the slaughter and were stored frozen at −15°C. Before the experiment, the samples were thawed to room temperature in saline solution (0.9% NaCl) or PSB to recover their mechanical properties. It has been shown that such freezing and thawing weakly affect the biomechanical

FIGURE 9.8 Schematic view of OCE setup for visualization of osmotic strains caused by tissue impregnation by OCA.

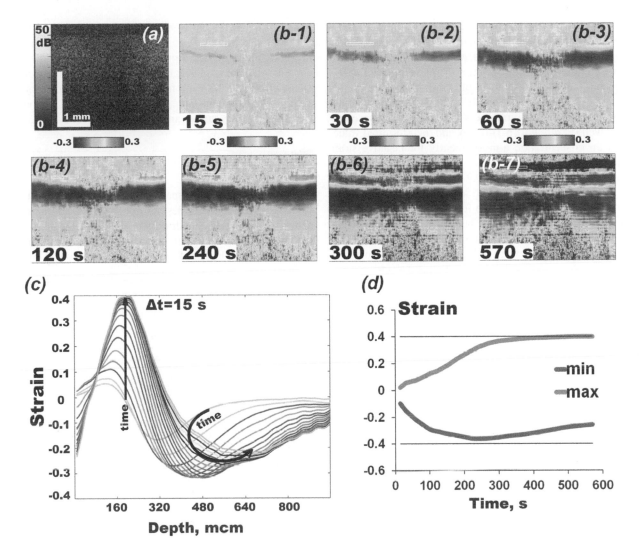

FIGURE 9.9 OCE visualization of strains during cartilage OC with 100% glycerol: (a) OCT structural image of cartilage under glycerol layer; (b-1)–(b-7) cumulative strain maps at certain time intervals from the start of glycerol diffusion; (c) time evolution of the strain profile at the cumulative strain map with a time step of 15 sec between the subsequent curves and 10 min total observation interval; (d) time evolution of strain minimum and maximum for panel (c).

properties of cartilaginous tissue. Cylindrical cuts of cartilage with d≈10 mm and thickness of 2 mm were prepared using a metal punch tool. The tissue was placed in the OCT setup (Figure 9.8) and at the start of OCT recording, a 1 mm layer of glycerol solution was dropped onto the cartilage surface.

To monitor slow strains, a custom-made spectral-domain OCT setup was used, operating at a central wavelength of 1300 nm (with ~90 nm spectral width), 20 kHz rate of obtaining spectral fringes, and 20 Hz rate of acquiring B-scans, covering 4 mm laterally and 2 mm in depth (in air). Besides structural images, it allowed for obtaining depth- and laterally resolved 2D maps of interframe and cumulative strains using estimation of local axial gradients of interframe phase variations [35].

OCA-induced strain sign-changing and concentration dependence of strain

Figure 9.9 shows cumulative strain evolution with time of cartilage immersion in 100% glycerol. Just a few seconds after the application of the glycerol, shrinkage of the subsurface layer of tissue can be clearly seen (Figure 9.9(b-1)–(b-3), (c)). At about 2 min of immersion, the maximum negative strain goes deeper, while the strain with a positive sign appears near the surface (Figure 9.9(b-4)) and gradually increases up to the end of OCT recording at 10 min (Figure 9.9(b-4)–(b-7)). The time evolution of strain along the cartilage depth is shown in Figure 9.9(c), and the corresponding kinetics of strain minimum and maximum are given in Figure 9.9(d). As one can see, glycerol acts here as an extremely hygroscopic agent causing pronounced water desorption which results in substantial tissue shrinkage (40% for 10 min). The observed positive-signed strain can be of more complex nature connected with the solute concentration and tissue composition. Note that the strain minimum and maximum are of the same amplitude (Figure 9.9(d)) indicating a kind of "compensational" process.

Therefore, the injection of highly hygroscopic OCA, such as glycerol, presents a complex multistage process in which the value of deformation is split due to simultaneous dehydration and swelling during diffusion of the agent.

The concentration dependence of the osmotic strain is shown in Figure 9.10. The decrease in glycerol concentration leads to a subsequent reduction in resulted strain. The amplitude difference becomes less sharp and more extended in depth from 100% to 50% glycerol, as shown on the "waterfall" view of cumulative strain evolution maps (Figure 9.10(a-1)–(a-2)), and the total amplitude of strain drops almost ten times for a 25% solution (Figure 9.10(a-3)). The same splitting on minima and maxima persists for smaller concentrations as well (Figure 9.10(b-1)-(b-3)), and their terminal values depend on concentration almost linearly for negative strain (Figure 9.10c).

The data shown in Figures 9.9 and 9.10 demonstrate the capacity of phase-resolved OCE to monitor spatiotemporal dynamics of osmotic strain. The found dependences must be studied in more detail with the help of a phenomenological description of the observed phenomena of diffusion and dehydration. However, for now, it can be declared that in the course of OC with hyperosmotic OCA, such sign-changing processes with high amplitudes of positive and negative strain may significantly affect the local tissue integrity, whatever the initial and final value of its elastic modulus. The development of local osmotic strain monitoring methodology is important for optimizing OCA concentration, especially when it is intended to be used for *in vivo* studies and medical diagnostic applications.

Conclusion

OC of cartilage is a prospective emerging technique that is opening up new possibilities not only for medical diagnostics and treatment of subchondral tissues but also for understanding natural mechanisms in terms of their development and functioning. For the last decade certain progress was made in

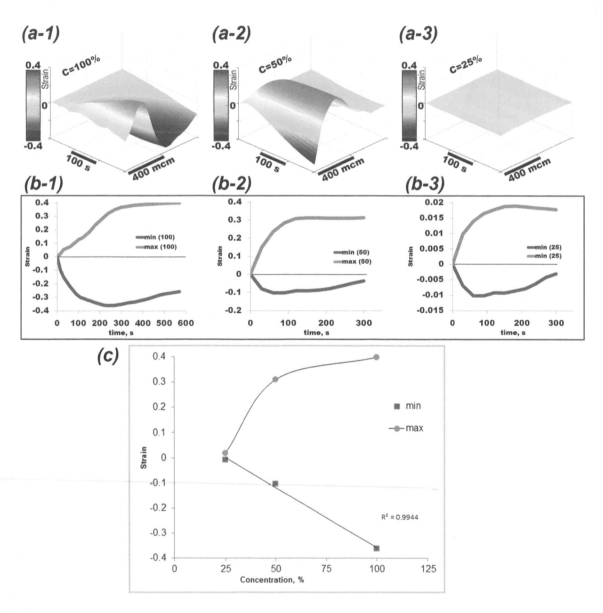

FIGURE 9.10 Dependence of strain evolution in cartilage on initial glycerol concentration: (a-1)-(a-3) the "waterfall" view of cumulative strain evolution for cartilage immersion in 100, 50, and 25% glycerol; (b-1)–(b-3) the corresponding kinetics of maximal and minimal tstrain registered at the time of observation; (c) the concentration dependence of cumulative strain minima and maxima.

finding effective OCAs for cartilage OC in visible and NIR spectral ranges, optimization of their concentration, estimation of diffusion kinetics parameters, and degree of reversibility. The future prospects are connected with regulation of osmotic and mechanical stresses for improving safety through better control of rapid and complex processes on the OCA-tissue interface.

Acknowledgment

This work was supported by the Ministry of Science and Higher Education within the State assignment to FSRC "Crystallography and Photonics" RAS in part of cartilage modification and impregnation by OCAs. The development of OCE algorithms for osmotic strain mapping was supported by the State assigment to IAP RAS (contract No 0030-2021-0007). V.V.T. was supported by a grant of the Government of the Russian Federation No 075-15-2021-615.

REFERENCES

1. S. Grässel, A. Aszódi (eds.), *Cartilage. Volume 1: Physiology and Development*, Springer, Cham (2016).
2. K.A. Athanasiou, E.M. Darling, J.C. Hu, G.D. DuRaine, A.Hari Reddi, *Articular Cartilage*, Taylor and Francis Group, CRC press, Boca Raton, FL (2013).
3. Freeman M.A.R. (ed.), *Physiochemical Properties of Articular Cartilage*, Cambridge University Press, Cambridge, UK (1979).
4. Maroudas, P. Bullough, "Permeability of articular cartilage," *Nature* 219, 1260–1261 (1968).
5. D. Majda, A. Bhattarai, J. Riikonen, et al., "New approach for determining cartilage pore size distribution: NaCl-thermoporometry," *Micropor. Mesopor. Mat.* 241, 238–45 (2017).
6. S. Grässel, A. Aszódi (eds.), *Cartilage. Volume 2: Pathophysiology*, Springer, Cham (2017).
7. J. Hootman, C. Helmick, K. Barbour, K. Theis, M. Boring, "Updated projected prevalence of self-reported doctor-diagnosed arthritis and arthritis-attributable activity limitation among US adults, 2015–2040," *Arthritis Rheumatol.* 68, 1582–7 (2016).
8. H. Jahr, N. Brill, S. Nebelung, "Detecting early stage osteoarthritis by optical coherence tomography?" *Biomarkers* 20(8), 590–596 (2015).
9. S.J. Matcher, "What can biophotonics tell us about the 3D microstructure of articular cartilage?" *Quantum Imaging Med. Surg.* 5, 143–58 (2015).
10. P.J. Su, W.L. Chen, T.H. Li, "The discrimination of type I and type II collagen and the label-free imaging of engineered cartilage tissue," *Biomaterials* 31(36), 9415–21 (2010).
11. E.V. Gubarkova, A.A. Sovetsky, V.Yu. Zaitsev, et al., "OCT-elastography-based optical biopsy for breast cancer delineation and express assessment of morphological/molecular subtypes," *Biomed. Opt. Express* 10(5), 2244 (2019).
12. M.A. Sirotkina, E.V. Gubarkova, A.A. Sovetsky, et al., "In vivo assessment of functional and morphological alterations in tumors under treatment using OCT-angiography combined with OCT-elastography," *Biomed. Opt. Express* 11(3), 1365 (2020).

13. E. Sobol, O. Baum, A. Shekhter, A. Guller, A. Baskov, "Laser-induced regeneration of cartilage," *J. Biomed. Opt.* 16(8), 080902 (2011).
14. H. Jelinkova (ed.), *Lasers in Orthopaedic Surgery*, Woodhead Publishing Limited, Oxford (2013).
15. B.F. Wong, J. Ilgner (eds.), *Cartilage Reshaping of the Nose*, Springer, New York (2016).
16. F.M. Leclère, P.M. Vogt, V. Casoli, S. Vlachos, S. Mordon, "Laser-assisted cartilage reshaping for protruding ears: A review of the clinical applications," *Laryngoscope* 125, 2067–2071 (2015).
17. O.I. Baum, Yu.M. Soshnikova, E.N. Sobol et al., "Laser reshaping of costal cartilage for transplantation," *Lasers Surg. Med.* 43, 511 (2011).
18. O.I. Baum, Yu.M. Alexandrovskaya, V.M. Svistushkin, S,V. Starostina, E.N. Sobol, "New clinical application of laser correction of cartilage shape for implantation in otolaryngology," *Laser Phys. Lett.* 16, 035603 (2019).
19. A. Foulad, P. Ghasri, R. Garg, B. Wong, "Stabilization of costal cartilage graft warping using infrared laser irradiation in a porcine model," *Arch. Facial. Plast. Surg.* 12(6), 405–11 (2010).
20. E. Sobol, A. Sviridov, A. Omelchenko, "Laser reshaping of cartilage," *Biotechnol. Genet. Eng. Rev.* 17, 553–78 (2000).
21. J. Kinnunen, J.S. Jurvelin, J. Mäkitalo, M. Hauta-Kasari, P. Vahimaa, S. Saarakkala, "Optical spectral imaging of degeneration of articular cartilage," *J. Biomed. Opt.* 15(4), 046024 (2010).
22. R.N. Huynh, B. Pesante, G. Nehmetallah, C.B. Raub, "Polarized reflectance from articular cartilage depends upon superficial zone collagen network microstructure," *Biomed. Opt. Express* 10, 5518–5534 (2019).
23. J.C. Mansfield, C.P. Winlove, "A multi-modal multiphoton investigation of microstructure in the deep zone and calcified cartilage," *J. Anat.* 220(4), 405–16 (2012).
24. M. Ravanfar, G. Yao, "Simultaneous tractography and elastography imaging of the zone-specific structural and mechanical responses in articular cartilage under compressive loading," *Biomed. Opt. Express* 10, 3241–3256 (2019).
25. A. El Kharbotly, T. El Tayeb, Y. Mostafa, I. Hesham, "Diode laser (980nm) cartilage reshaping," *Proc. SPIE* 7883, 788325 (2011).
26. F.M. Leclère, S. Mordon, J. Alcolea, P. Martinez-Carpio, M. Vélez, M. Trelles, "1064-nm Nd: YAG laser-assisted cartilage reshaping for treating ear protrusions," *Laryngoscope* 125(11), 2461–7 (2015).
27. N. Brill, J. Riedel, B. Rath, et al., "Optical coherence tomography-based parameterization and quantification of articular cartilage surface integrity," *Biomed. Opt. Express* 6, 2398–2411 (2015).
28. N. Ugryumova, J. Jacobs, M. Bonesi, S.J. Matcher, "Novel optical imaging technique to determine the 3-D orientation of collagen fibers in cartilage: Variable-incidence angle polarization-sensitive optical coherence tomography," *Osteoarthritis Cartilage* 17(1), 33–42 (2009).
29. D.K. Kasaragod, Z. Lu, J. Jacobs, S.J. Matcher, "Experimental validation of an extended Jones matrix calculus model to study the 3D structural orientation of the collagen fibers in articular cartilage using polarization-sensitive optical coherence tomography," *Biomed. Opt. Express* 3, 378–387 (2012).

30. Y.-P. Huang, S.-Z. Wang, S. Saarakkala, Y.-P. Zheng, "Quantification of stiffness change in degenerated articular cartilage using optical coherence tomography-based air-jet indentation," *Connective Tissue Res.* 52(5), 433–443 (2011).

31. V.Y. Zaitsev, A.L. Matveyev, L.A. Matveev, et al., "Optical coherence tomography for visualizing transient strains and measuring large deformations in laser-induced tissue reshaping," *Laser Phys. Lett.* 13(11), 115603 (2016).

32. V.Y. Zaitsev, A.L. Matveyev, L.A. Matveev, et al., "Optical coherence elastography for strain dynamics measurements in laser correction of cornea shape," *J. Biophoton.* 10(11), 1450–1463 (2017).

33. V.Yu. Zaitsev, A.L. Matveyev, L.A. Matveev, et al., "Revealing structural modifications in thermomechanical reshaping of collagenous tissues using optical coherence elastography," *J. Biophoton.* 12(3), e201800250 (2019).

34. Yu.M. Alexandrovskaya, O.I. Baum, A.A. Sovetsky, et al., "Observation of internal stress relaxation in laser-reshaped cartilaginous implants using OCT-based strain mapping," *Laser Phys. Lett.*, In Print, (2020).

35. O.I. Baum, V.Y. Zaitsev, A.V. Yuzhakov et al., "Interplay of temperature, thermal-stresses and strains in laser-assisted modification of collagenous tissues: Speckle-contrast and OCT-based studies," *J. Biophoton.* 13(1), e201900199 (2020).

36. Y. Chae, D. Protsenko, P.K. Holden, C. Chlebicki, B.J.F. Wong, "Thermoforming of tracheal cartilage: Viability, shape change, and mechanical behavior," *Lasers Surg. Med.* 40(8), 550–561 (2008).

37. F.M. Leclère, I. Petropoulos, S. Mordon, "Laser-assisted cartilage reshaping (LACR) for treating ear protrusions: A clinical study in 24 patients," *Aesthetic Plast. Surg.* 34(2), 141–6 (2010).

38. A. Sviridov, E. Sobol, N. Jones, et al., "The effect of holmium laser radiation on stress, temperature and structure of cartilage," *Laser Med. Sci.* 13, 73–77 (1998).

39. V.N. Bagratashvili, E.N. Sobol, A.P. Sviridov, V.K. Popov, A.I. Omelchenko, S.M. Howdle, "Thermal and diffusion processes in laser-induced stress relaxation and reshaping of cartilage," *J. Biomech.* 30(8), 813–817 (1997).

40. Yu.M. Alexandrovskaya, K. Sadovnikov, A. Sharov et al., "Controlling the near infrared transparency of costal cartilage by impregnation with clearing agents and magnetite nanoparticles," *J. Biophotonics* 11(2), e201700105 (2017).

41. D.W. Ebert, C.J. Roberts, S.K. Farrar, W.M. Johnston, A.S. Litsky, A.L. Bertone, "Articular cartilage optical properties in the spectral range 300–850 nm," *J. Biomed. Opt.* 3(3), 326 (1998).

42. A.V. Yuzhakov, A.P. Sviridov, E.M. Shcherbakov, O.I. Baum, E.N. Sobol, "Optical properties of costal cartilage and their variation in the process of non-destructive action of laser radiation with the wavelength 1.56 μm," *Quantum Electron.* 44(1), 65–68 (2014).

43. Yu.M. Soshnikova, M.M. Keselman, O.I. Baum, E.V. Shults, M.V. Obrezkova, V.V. Lunin, E.N. Sobol, "Effect of anisotropy and drying of costal cartilage on its optical transmittance in laser reshaping of implants with 1, 2, and 3 mm in thickness," *Lasers Surg. Med.* 48(9), 887–892 (2016).

44. G. Simonenko, "Refractive index anisotropy and diffusion rate in cartilage tissue," *J. Biomed. Photon. Eng.* 3(3), 030302 (2017).

45. C.P. Neu, T. Novak, K.F. Gilliland, P. Marshall, S. Calve, "Optical clearing in collagen- and proteoglycan-rich osteochondral tissues," *Osteoarthritis Cartilage* 23(3), 405–13 (2015).

46. A. Bykov, T. Hautala, M. Kinnunen, et al., "Optical clearing of articular cartilage: A comparison of clearing agents," *Proc. SPIE* 9540, 95400A (2015).

47. A. Bykov, T. Hautala, M. Kinnunen, et al., "Imaging of subchondral bone by optical coherence tomography upon optical clearing of articular cartilage," *J. Biophoton.* 9(3), 270–275 (2016).

48. Yu.M. Alexandrovskaya, E.G. Evtushenko, M.V. Obrezkova, V.V. Tuchin, E.N. Sobol, "Control of optical transparency and infrared laser heating of costal cartilage via injection of iohexol," *J. Biophoton.* 11(12), e201800195 (2018).

49. A.T. Yeh, B. Choi, J.S. Nelson, and B.J. Tromberg, "Reversible dissociation of collagen in tissues," *J. Invest. Dermatol.* 121(6), 1332–1335 (2003).

50. S. Calve, A. Ready, C. Huppenbauer, R. Main, C.P. Neu, "Optical clearing in dense connective tissues to visualize cellular connectivity in situ," *Plos One* 10(1), e0116662 (2015).

51. K. Buijs, G.R. Choppin, "Near-infrared studies of the structure of water. I. Pure water," *J. Chem. Phys.* 39, 2035 (1963).

52. K. Izutsu, Y. Hiyama, C. Yomota, T. Kawanishi, "Near-infrared analysis of hydrogen-bonding in glass- and rubber-state amorphous saccharide solid," *AAPS Pharm. Sci. Technol.* 10(2), 524–529 (2009).

53. C.-H. Liu, M. Singh, J. Li et al., "Quantitative assessment of hyaline cartilage elasticity during optical clearing using optical coherence elastography," *Modern Technol. Med.* 7(1), 44–51 (2015).

54. V.V. Tuchin, "Coherent optical techniques for the analysis of tissue structure and dynamics," *J. Biomed. Opt.* 4(1), 106–24 (1999).

55. J. Schmitt, "OCT elastography: Imaging microscopic deformation and strain of tissue," *Opt. Express* 3(6), 199–211 (1998).

56. V.Yu. Zaitsev, I.A. Vitkin, L.A. Matveev, V.M. Gelikonov, A.L. Matveyev, G.V. Gelikonov, "Recent trends in multimodal optical coherence tomography. II. The correlation-stability approach in OCT elastography and methods for visualization of microcirculation," *Radiophys. Quantum Electron.* 57(3), 210–225 (2014).

57. V.Yu. Zaitsev, A.L. Matveyev, L.A. Matveev, G.V. Gelikonov, V.M. Gelikonov, A. Vitkin, "Deformation-induced speckle-pattern evolution and feasibility of correlational speckle tracking in optical coherence elastography," *J. Biomed. Opt.* 20(7), 075006-1-12 (2015).

58. H.H. Müller, L. Ptaszynski, K. Schlott, et al., "Imaging thermal expansion and retinal tissue changes during photocoagulation by high speed OCT," *Biomed. Opt. Express* 3(5):1025–1046 (2012).

59. B.F. Kennedy, S.H. Koh, R.A. McLaughlin, K.M. Kennedy, P.R.T. Munro, D.D. Sampson, "Strain estimation in phase-sensitive optical coherence elastography," *Biomed. Opt. Express* 3(8), 1865–1879 (2012).

60. V.Yu. Zaitsev, A.L. Matveyev, L.A. Matveev, G.V. Gelikonov, A.A. Sovetsky, A. Vitkin, "Optimized phase gradient measurements and phase-amplitude interplay in optical coherence elastography," *J. Biomed. Opt.* 21, 116005 (2016).

61. A.L. Matveyev, L.A. Matveev, A.A. Sovetsky, G.V. Gelikonov, A.A. Moiseev, V.Yu. Zaitsev, "Vector method for strain estimation in phase-sensitive optical coherence elastography," *Laser Phys. Lett.* 15, 065603 (2018).

62. V.Yu. Zaitsev, L.A. Matveev, A.L. Matveyev et al., "Optimization of phase-resolved optical coherence elastography for highly-sensitive monitoring of slow-rate strains," *Laser Phys. Lett.* 16(6), 065601 (2019).

10

Compression optical clearing

Olga A. Zyuryukina and Yury P. Sinichkin

CONTENTS

Introduction

Studies of the effect of external mechanical compression on the diffuse reflectance spectra of biological tissue *in vivo* are of interest for a number of reasons. Many of the first studies showed that the method of diffusion reflective spectroscopy has great potential in the field of noninvasive diagnosis and monitoring of human conditions *in vivo*. The spectral composition of radiation diffusely reflected by biological tissue carries information about its morphological and functional state, its structure, the quantity of blood vessels and their blood supply, the spatial distribution of chromophores inside the biological tissue and their concentration, and the intensity of metabolic processes occurring in the biological tissue [1]. Therefore, changes in diffuse reflection of biological tissues under conditions of their compression are of interest. Secondly, the key element in such *in vivo* studies is the fiberoptic sensor, which is in contact with the surface of the studied biological tissue, creating uncontrolled pressure on the surface of the biological tissue by the sensor. As a result, the diffuse reflection spectrum can change uncontrollably depending on the magnitude of the mechanical compression, which is a source of errors in the analysis of the spectra. Thirdly, with local mechanical

compression created by the end of the fiberoptic sensor (the area of the mechanical compression being of the order of several mm²⁾, at the place of application of compression in the biological tissue a gradient of the refractive index is created, and this volume of tissue fulfills the role of lenses for probing light propagating in tissue. When mechanical compression is applied to a relatively large area of the surface of the biological tissue (of the order of several cm²), the compressed volume of the biological tissue is a spatially homogeneous medium without lens effects, in which, in contrast to the region of local compression, the light propagation pattern changes, which can manifest itself in a change in the shape of the diffuse reflection spectrum. Fourth, as a result of the application of external compression, the scattering properties of biological tissue are reduced; therefore, this compression method can be used for controlling the optical parameters of biological tissues, an alternative to the widely used method based on the introduction of chemical agents in biological tissue [2–5]. Fifth, at present, external mechanical compression is used as a method that allows one to increase the resolution and image contrast in optical coherent tomography of biological tissues [6]. Further, the method of compression optical clearing of biological tissues can be implemented in the instrument version; in particular, a

DOI: 10.1201/9781003025252-11

device for bleaching biological tissues (tissue optical clearing device, TOCD) has been developed [7]. Finally, mechanical compression of the skin allows one to evaluate the content of chromophores in it, the absorption of which under normal conditions is veiled by the absorption of other chromophores. Thus, squeezing blood from the compression application area allows one to evaluate the content of carotenoids in the skin by diffuse reflection spectra [8] and the content of melanin in the skin by fluorescence spectra [1].

The method of immersion clearing of biological tissues is currently used quite widely. In the processes of the clearing of biological tissues, an important role is played by water transport. Since the result of external compression is a change in the water content in the area of compression application, the compression method for controlling the optical parameters of biological tissues is a promising method. Moreover, it has a number of advantages compared to the immersion method, since mechanical compression of biological tissue is less invasive and safe; in contrast to the immersion method, the barrier functions of the *stratum corneum* and the entire epidermis as a whole are preserved, and the compression method should have greater speed, greater controllability, and repeatability of application results.

Despite the rather large volume of publications related to the mechanical compression of biological tissues as a result, various forms were discovered due to the presence of external compressors (local or nonlocal) and different detection geometries of reflected biological tissue light (using optical sensor or without it) [9]. In addition, the effect of external compression on the optical properties of biological tissue is inertial, and therefore, the measurement results depend on the delay time between compression and contact and measurements [10]; this issue is not well understood.

Finally, it is important to know the mechanism of changes in the optical properties of biological tissues under external compression, the dynamics of changes in the structure and component composition of biological tissues under external pressure, and the effect of such changes on the optical properties of biological tissues and, as a consequence, on diffuse reflectance spectra of biological tissues.

The effect of external mechanical compression on the optical properties of biological tissues

The effect of increasing the depth of penetration of laser radiation into biological tissue by applying external local mechanical pressure on it was demonstrated almost 40 years ago [11]. Since then, a rather large number of publications have appeared related to studies of the effect of external mechanical compression of biological tissues on their optical properties (absorption and scattering).

The first studies [12, 13] were carried out on *ex vivo* biological tissue samples. The authors noted that external compression changes the optical properties of biological tissue samples, which was reflected in the change in diffuse reflection and light transmission of biological tissue samples. So, in [12], studies of the effect of compression on soft tissue samples (human skin, sclera, and aorta of a bull, sclera of a pig) showed

a decrease in diffuse reflection coefficient and an increase in transmittance in the spectral region of 400–1800 nm. The decrease in reflection is due to a decrease in the thickness of tissue samples (up to 72%) and a decrease in the weight of tissue samples (decreases to 40%), while the effects of pressure on the optical properties of biological tissues increased with increasing pressure. It was concluded that compression increases the absorption and scattering coefficient of biological tissue, and a possible mechanism of effect is an increase in the concentration of scatterers.

Similar results were obtained in [13]. The authors investigated the effect of mechanical compression on pig skin samples *ex vivo*, and it was found that by mechanical compression of the samples, there is an increase in the transmission of light through them, and the effect is irreversible.

Possible mechanisms responsible for the change in optical properties are associated by the authors with a change in the thickness or weight of tissues and are directly related to their dehydration. Compression reduces the thickness of the sample and causes fluid to leak out of the samples, while changes in tissue thickness and water content affect the optical properties of the samples in different ways.

Under compression conditions, the density of the absorbing or scattering centers increases, which is the result of a decrease in the space between the cellular components. Since the scattering or absorption cross-sections remain constant or slightly decrease, as a result of compression, the absorption and scattering coefficients increase. An increase in the volume concentration of water due to a decrease in the thickness of the tissue may also be the reason for the increase in the absorption coefficient when applying compression.

The absorption coefficients for compressed samples are large, even if they have similar thicknesses with samples without compression. This is due to the fact that the physical structures of uncompressed samples (even if they are dehydrated) are looser than the structures of compressed samples, which are denser and more compact.

Changes in the physical structure of soft tissue are one of the causes of changes in the optical properties of tissue under compression. Changes will be irreversible if the structure does not return to its original state after compression. If the thickness of the tissue does not return to its original state after compression is removed, this may be an explanation of the fact that changes in optical properties as a result of compression are irreversible.

The authors of [7, 14] also investigated the diffuse reflection of light and its transmission by porcine skin samples *ex vivo* under conditions of mechanical compression. It has been suggested that to explain the results obtained, it is not enough to reduce the thickness of the sample, and it is necessary to involve other effects, including dynamic ones.

Mechanical compression also changes the optical properties of biological tissues *in vivo*, which is accompanied by spectral changes both in diffuse light reflection by biological tissues [10, 14–22, 23] and in autofluorescence of biological tissues [22, 24–26]. The authors of [16, 24] were among the first researchers to note the effect of compression on the diffuse reflection and autofluorescence spectra of human skin *in vivo*. It was noted that the pressure exerted on the skin reduces the depth of the dip in the green region of the spectrum, which is an indicator of the presence of blood (hemoglobin) in the biological tissue, as a result of which the reflection coefficient of the biological tissue in this region increases and reduces the reflection coefficient of the skin in the yellow-red region of the spectrum. In this case, an isosbestic point was found in the diffuse reflection spectrum at a wavelength of the order of 600 nm, the reflection coefficient at which practically did not change in the presence or absence of compression [15]. External compression also led to an increase in the intensity of autofluorescence of the skin in the short-wave region of the visible range of the spectrum [24].

The effect of compression on the *in vivo* diffuse reflection spectra of the skin in the range 1100–1700 nm was studied in [16], where it was shown that with increasing pressure of the fiberoptic sensor on the skin surface, diffuse reflection decreases. The compression was increased by pressing the sensor end into the skin. It was found that, as the contact increases, the diffuse reflection initially fluctuates significantly, and after a certain contact time, the fluctuations stabilize. The authors associated this behavior of the spectra with a change in the internal structure of skin tissue. Based on the biomechanical structure of human skin and subcutaneous tissue, it can be considered that subcutaneous fat and muscle tissue have good elasticity and, to some extent, resistance to external compression. Therefore, when measuring the diffuse reflection of biological tissue under compression conditions, we can assume that the skin and subcutaneous tissue change accordingly, affecting the stability of the measured spectra. However, as the contact time increases, the structure of the biological tissue stabilizes, and the degree of its influence on the spectra weakens. The authors determined the optimal contact state and the optimal measurement time at which the influence of pressure on diffuse reflection measurements is minimal.

A similar problem was solved by the authors of [21] in the visible range of the skin reflectance spectrum. The effect of the pressure of a fiberoptic sensor on the surface of the skin *in vivo* on diffuse reflectance spectra in the range 345–1000 nm was studied. The authors found that with increasing compression, the diffuse reflection spectra behave differently. In one case, an increase in pressure led to a decrease in diffuse reflectance in the wavelength range of 530–680 nm ("parallel pattern"). In another case, with increasing pressure, the spectra "revolved" around the isosbestic point of 590 nm, so that the spectra below this wavelength increased and the diffuse reflectance decreased ("pivot pattern").

The influence of external mechanical pressure on the diffuse reflectance spectra in the near-infrared region of soft tissues was studied in [27–29]. The effect of static and dynamic (changing) pressure on the diffuse reflectance of soft tissues (*in vitro* fat and muscles and *in vivo* skin of different parts of the palm and wrist – over the muscle, over the vein, and over the bone) was studied. It was found that in the case of skin samples, the applied pressure leads to a decrease in diffusely reflectance and scattering, while the concentration of chromophores (water, hemoglobin, lipids) and, accordingly, absorption, increased. The authors suggested that the pressure-induced spectral changes strongly depend on the location of the site and the underlying tissues. In addition, the same contact pressure causes larger changes in the spectra of the near-infrared spectral range than in the spectral range from 650 to 900 nm.

In [22], the effects of pressure of a fiberoptic sensor on the diffuse reflectance and autofluorescence spectra of human skin (neck, finger, and forearm) *in vivo* were studied. The effects of short-term (less than 2 s) and long-term (more than 30 s) mechanical effects on spectral measurements were studied. It was found that at high pressure, significant spectral changes occur during prolonged compression, while the pressure of the sensor affects not only the optical but also the physiological parameters of the skin tissue.

The effect of compression on the fluorescence of biological tissues is often controversial. So, in [24] there was an increase in the intensity of autofluorescence of human skin *in vivo* in the case of skin compression, while the effect of external pressure on cervical fluorescence is negligible [25, 26].

The effect of external mechanical compression on the physiological properties of biological tissue

The authors of [22] believe that changes in the spectra are due to changes in physiological characteristics. Moreover, changes in the spectra are specific for skin of different morphology. After applying compression, an increase in the concentration of hemoglobin in the skin of the neck is noted, while the content of hemoglobin in the skin of the finger and forearm decreases. Hemoglobin in the skin of the neck decreases only at a sufficiently high pressure (0.77×10^5 Pa), in contrast to the skin of the finger and forearm, where the concentration of hemoglobin decreases even at low pressure. With increasing pressure, the hemoglobin content continues to decrease, while the degree of hemoglobin saturation with oxygen also decreases, and after 60 s, hemoglobin in the tissues is completely absent. The authors believe that such differences in the behavior of the spectra are due to the morphology of the samples. The skin of the neck lies on a more muscular and elastic tissue compared to the finger and forearm. This tissue does not compress quickly, so the neck requires more pressure to achieve the same effect as for the finger or forearm. Due to the massiveness and elasticity of the neck, the compressed tissue shows a temporary increase in blood content, which subsequently decreases.

Compared to absorption, the reduced scattering coefficient varies to a less extent. With increasing pressure, it increases for the forearm, decreases for the skin of the neck, and changes little for the finger. The authors hypothesize that a decrease in the scattering coefficient with increasing pressure on the skin of the neck is due to the displacement of water in the epidermis and dermis. In the case of the forearm, the presence of bone prevents the transition of the dermis to the hypodermis, contributing to an increase in depth penetration of the sensor into the dermis. As a result, total scattering increases as the sensor reaches the dermis with highly scattering collagen. This also confirms the increase in collagen fluorescence in the case of the forearm.

In [18], the effects of the pressure of a fiberoptic sensor on the physiological characteristics of mouse muscle tissue *in vivo* were studied. It was found that with increasing pressure, the diameters of blood vessels and the degree of oxygenation of hemoglobin decrease, while the reduced scattering coefficient at a wavelength of 700 nm increases. The authors suggest that pressure compresses the blood vessels, decreasing the flow of blood, which leads to a decrease in the flow of oxygenated blood coming into the tissue. The pressure of the sensor can also increase the density of the scatterers per unit volume, which may be associated with an increase in the reduced scattering coefficient. The authors note that hemoglobin and myoglobin make the predominant contribution to muscle absorption.

The authors of [19] also suggested that changes in the physiological parameters of the tissue, such as changes in blood volume, oxygenation of blood hemoglobin, and tissue metabolism, can make a large contribution to the spectral changes in diffuse reflection and fluorescence.

In [17], lung tissue was subjected to repeated compression cycles. When squeezing tissue, a decrease in scattering, as well as in hemoglobin content and saturation of the tissue with oxygen, was noted. The authors also note the effect of hyperemia during repeated cycles. It is concluded that significant changes in the physiological characteristics of tissue induced by compression can be used in the optical imaging of lung tissue.

The influence of the pressure of a fiberoptic sensor on the spectral measurements of diffusely reflected light from the mucosa of the inner side of the lower lip of a person was studied *in vivo* by the authors of [10]. An object was probed by linearly polarized light within 100–200 μm of the depth of the surface layer. The difference polarization spectrum (with an average sounding depth of 100 μm) and the cross-polarized component (average sounding depth of 200 μm) were measured. Several parameters were calculated from the measurement data, including hemoglobin content, blood oxygen saturation, blood vessel diameters, and total scattering intensity. It was found that when applying compression, the difference signal significantly increases and the cross-polarized component significantly decreases. A decrease in the hemoglobin content and the degree of hemoglobin oxygenation during pressure application was noted.

The authors of [27–29] suggested that the pressure-induced spectral changes strongly depend on the location of the site and what the underlying tissues are, which provides additional information for improving the sensitivity and specificity of soft tissue classification. In addition, the same contact pressure causes larger changes in the spectra of the near infrared spectral range than in the spectral range from 650 to 900 nm.

Similar findings were reported in a study of blood glucose using optical coherence tomography [30].

Currently, external mechanical compression is used as a method to increase the resolution and contrast of the image in optical coherence tomography of biological tissues, [6, 30, 31]. So, in [6], the effectiveness of mechanical compression of biological tissues was studied to improve the differentiation of pathological changes in the structure of biological tissue using the method of optical coherence tomography. Experiments were completed on the effect of compression on *ex vivo* images of the rectum, and the effectiveness of compression was shown to differentiate inflammation and rectal carcinoma when diagnosed by *ex vivo* OCT. Registration of OCT images in two orthogonal polarizations allows the retrieval of additional information about the structure of the investigated tissue. It was shown in [31] that mechanical compression increases the contrast of the epidermis–dermis transition and decreases the contrast of the *stratum corneum*–epidermis transition, which, according to the authors, is associated with structural and functional changes in the tissue caused by compression.

The method of compression optical clearing of biological tissues can be implemented in the instrument version; in particular, a device for clearing biological tissues (tissue optical clearing device, TOCD) has been developed [7, 32–34]. It was noted above that localized compression is an effective method for increasing the resolution and contrast of structures underneath tissues subjected to compression [7]. The method was implemented in a device for optical tissue enlightenment (TOCD), which contained ensembles of mechanical pins or lenses that mechanically acted on biological tissue in certain areas. One scheme ensured the transport of water and blood from the zones of exposure to the pins, providing zones of dehydration of reduced thickness with possibly altered optical properties. The second scheme made it possible to increase the depth of light penetration into biological tissue due to the microlens system.

In a device [34] based on a fiberoptic sensor (diaphragm-based fiberoptic interferometric pressure sensor, DFPI), pressure is used to stabilize the diffuse reflectance spectra of mucous tissues and allows rapid screening of oral and cervical malignancies.

As a result of compression, water is squeezed out of the biological tissue, which determines the absorption of the biological tissue in the near infrared (IR) spectrum region and the scattering properties of the biological tissue both in the near infrared and in the visible range of the spectrum, as well as the blood contained in the blood vessels of the compressed biological tissue. Under normal conditions, the hemoglobin contained in the biological tissue of the blood is one of the main chromophores, and its absorption spectrum veils the absorption spectra of other, less absorbing chromophores. The imposition of external compression leads to the displacement of blood from the biological tissue, and its absorption in the short-wavelength region of the visible spectrum decreases, as a result of which it becomes possible to evaluate the content of other chromophores in the biological tissue from diffuse reflection or autofluorescence spectra. Thus, the authors of [8] used mechanical compression to determine carotenoids in human skin *in vivo*, and the possibility of determining the content of melanin pigment in human skin from the autofluorescence spectra of the skin under external compression was shown in [1, 15].

Water transport in biological tissues under external mechanical compression

Due to the compactness of the hydrogen bond between H_2O and the protein matrix, the water inside the biological tissue can be divided into two groups: the main (free) water and the bound water. They have various mechanical and spectral properties. Basic water can easily be transported inside the skin. Bound water is very tightly affiliated and difficult to separate from the solid matrix.

The complex structure of biological tissues and the related variations in the refractive index of their components make such media highly scattering in the visible and near-infrared regions of the spectrum. Skin tissue is not an exception, and, according to the model of a two-phase nonlinear mixture [35], it can be represented as an incompressible elastic solid matrix, formed mainly from elastic fibers and cells, filled with water. The solid matrix provides water resistance to spreading. This model best represents the composition and mechanical properties of the skin [32, 35].

The mobility of water within a solid matrix consisting of elastic fibers and cells is a decisive factor in the nonlinear mechanical response. This mobility is determined by the time-dependent behavior of the solid matrix and the state of the water. The ratio of skin deformation to relaxation time when applying pressure depends on the elastic properties of collagen fibers and the amount and viscosity of the fluid in the skin. Since bound water is closely tied to a solid matrix, the ratio of free and bound water reflects the deformation of the matrix and the movement of free water at a certain pressure.

The dynamics of the volume fraction of water during compression of tissue samples was analyzed by the authors of [36] on the basis of experiments with *ex vivo* pig skin performed by the OCT method. The analysis was carried out under the assumption that the skin can be represented as a two-component mixture of water and protein. The results were compared with the results of weight measurements of the samples during compression and during the drying of the samples. It was found that localized mechanical compression of skin samples reduces tissue thickness (more than two times), reduces the water content in the samples (volume fraction decreased by more than three times), and increases the refractive index of samples (from 1.39 to 1.50). It is noted that compression leads to the same effects as in the case of dehydration of samples during their air immersion (drying). Local mechanical compression not only reduces scattering in the skin due to a decrease in the content of unbound water in the dermal collagen matrix, as a result of which the protein structure becomes more tightly packed, leading to better matching of the refractive indices of the components of the biological tissue, but also reduces the absorption of the samples.

The correlation of the increase in local mechanical pressure with a decrease in the reduced refractive index during the compression time is noted by the authors of [14]. Pressing the light source into the biological tissue increased the transmission of light through the tissue, while the relative transmittance increases as a function of compressive strain (relative change in thickness). The reduced scattering coefficient decreased by 12.0% with a relative decrease in thickness of 0.44 and by 35.6% with a strain of 0.71. An increase in strain to 0.45 lowered the reduced scattering coefficient by ~15%.

Detailed studies of the effect of the pressure of a fiberoptic sensor on the optical properties of biological tissue are given in [37], with an emphasis on changes in water content. The objects of research were *in vitro* samples of pig skin and *in vivo* human skin. The results of the studies showed that the contact pressure of the sensor when reaching a certain threshold and duration can significantly affect the reflectance spectrum. Since a biological tissue with a high water content is largely incompressible, the authors propose a mechanism for changing the optical characteristics of biological tissue during its compression, which is associated with the deformation of the biological tissue and the displacement of water from the compressed tissue region. The amount of water displaced from the localized pressure region is the main factor in the variation of the reflectance spectra. The simulation of such variations was carried out on the basis of a two-phase skin model, according to which the skin is a combination of an incompressible elastic solid matrix formed mainly of elastic fibers and cells, which play a major role in the nonlinear mechanical response of biological tissue to external influences, and fill the solid matrix of the liquid. The results showed that changes in free and bound water in the tissue are associated with pressure in a nonlinear manner. In the case of pig skin *in vitro*, the application of compression leads to the displacement of free water from the compression area, while the movement of bound water is difficult. At a pressure of about 400 kPa, the content of bound water decreases by 30%, which, according to the authors, is associated with tissue deformation. It is shown that temporary changes in the content of free and bound water in the *ex vivo* pig skin occur over a period of about 6 minutes after compression, after which the process stabilizes, and the stabilization time depends on the magnitude of the applied pressure. In the early stages of tissue deformation at a certain pressure, the main water leaves the compression area at a high speed due to the high stress inside the tissue. Due to the relaxation of the stress of biological tissues, the deformation and migration rate of the main water gradually decreases. After a certain time, depending on the amount of compression, the deformation of the tissue and the transport of the main water become constant.

Despite the fact that tissue deformation under compression makes a significant contribution to the effect of optical enlightenment, this contribution is not enough for the experimentally observed increase in light transmission by compressed media. As a result of the studies [7], it was found that the increase in transmittance continues to increase even if the tissue thickness remains constant. Mechanical clearing can be a function of the relative change in tissue thickness (compression strain), and not just an absolute decrease in tissue thickness due to applied pressure.

This suggests that the mechanisms of optical clearing can be more than a simple decrease in thickness. The authors of [38] note that dehydration is an important mechanism of optical clearing using chemical agents, and analyze how dehydration can contribute to a decrease in light scattering. An explanation of the optical clearing mechanism gives an expression relating the reduced scattering coefficient to the volume fraction of scattering particles. The parabolic dependence of the reduced scattering coefficient on the packing density of scattering particles shows that the effects of dehydration modify the packing density of scattering particles. The relationship between refractive indices and changes in the shape or structure of scattering particles affects the scattering cross section. According to authors, the clearing of biological tissue is inversely proportional to its state of hydration. These results are in good agreement with the correlation between water losses and the clearing ability of chemical agents for samples of muscle tissue of a pig and a stomach noted by the authors of [39, 40].

Dehydration can reduce light scattering by increasing the volume fraction of scattering particles. Dehydration of biological tissue makes collagen fibrils and organelles more closely packed, but does not cause significant changes in their size. Since the natural density of collagen fibrils is about 65% and increases due to dehydration to 90%, such structural changes can give a 60% decrease in the reduced scattering coefficient under the assumption that the refractive index of the base substance does not change. A contribution to optical clearing can also come from a decrease in the mismatch of refractive indices between fibrils and interfibrillar space caused by an increase in the concentration of proteoglycans (or exogenous immersion substances) [41].

Quantitative measurements of dehydration of biological tissue as a result of external compression are very difficult, especially if biological tissue is studied *in vivo*. If *ex vivo* dehydration of samples of biological tissues can be estimated by their weight measurements during drying or compression, then *in vivo* such measurements are not possible, and the only measured parameter in spectral experiments is the diffuse reflectance of biological tissues. Water makes a significant contribution to the formation of the diffuse reflectance spectrum of biological tissue. The change in the water content in the biological tissue is clearly manifested in the diffuse reflectance spectra as a result of changes in the absorbing and scattering properties of the tissue. Tissue dehydration reduces its absorption in the infrared region and alters tissue scattering in the visible and infrared wavelength ranges. The effect of external mechanical compression on the water content in human skin tissue *in vivo* was considered in [42, 43]. This paper presents the results of a study of the effect of dehydration of *ex vivo* samples of cow muscle tissue during drying and compression on their diffuse reflectance spectra in order to identify correlations between diffuse reflectance spectra of biological tissues and their dehydration.

Mechanical optical clearing is achieved due to local tissue compression and is presumably explained by the displacement of intercellular water and blood, which leads to local tissue dehydration and a change in the scattering and absorbing properties of the tissue [36, 38]. As the tissue is compressed, water and blood move from areas undergoing high compression deformation (relative change in thickness), which reduces the local volume fraction of water and, therefore, the mismatch of the refractive indices between the tissue components and the reduction of scattering and absorption coefficients [10].

The effect of external mechanical compression of the skin on the spectrum of its diffuse reflectance (model)

Skin tissue model

To assess the changes in the optical and physiological properties of the skin *in vivo* under external mechanical compression and their influence on the diffuse reflectance spectra of the skin, the diffuse reflectance spectra of the skin model were calculated taking into account changes in its absorbing and scattering properties that are manifested when external compression is applied.

The calculation of the diffuse reflection spectrum of skin tissue was carried out on the basis of a simplified model of skin tissue, according to which the skin is presented in the form of a homogeneous semi-infinite scattering medium with absorption. Such a model of skin tissue for the analysis of diffuse reflectance spectra of the skin is quite acceptable, since the scattering properties of the skin are determined mainly by water; although the water content in the skin has a certain gradient as it penetrates the skin (from 15% in the *stratum corneum* to 70% at the depth of the dermis), if we take into account the thickness of the layers, we can consider its distribution in the skin to be uniform. The same applies to the absorption of the skin in the near infrared range, since in this range the main chromophore is water. As for the blue-green spectral range, here the papillary dermis blood (superficial vascular plexus) plays a decisive role in the absorption of light, the content of which determines the behavior of the diffuse reflectance spectrum of the skin when pressure is applied. Blood of a rather narrow layer of the superficial vascular plexus without loss of generality was evenly distributed throughout the entire volume of the skin, participating in the formation of back diffuse light scattering.

The optical properties of such a medium are characterized by the spectral dependences of the absorption coefficient $\mu_a(\lambda)$ and reduced scattering coefficient $\mu'_s(\lambda)$, which is associated with the scattering coefficient $\mu_s(\lambda)$ and scattering anisotropy factor $g(\lambda)$ by a simple relation:

$$\mu'_s(\lambda) = \mu_s(\lambda)(1 - g(\lambda)).$$

In the diffusion approximation of the theory of radiation transfer, the authors of [44] obtained an analytical expression for the spectral dependence of the coefficient of diffuse reflection of light by such a medium in the form:

$$R = \frac{a'}{2}\left(1 + \exp\left(-\frac{4}{3}A\sqrt{3(1-a')}\right)\right)\frac{1}{1+\sqrt{3(1-a')}}, \quad (10.1)$$

where $a' = \dfrac{\mu'_s}{\mu_a + \mu'_s}$, coefficient A introduced to take into account the Fresnel refection at the air–medium interface. For the skin $A = 3.72$.

The value A has a spectral dependence, which can be determined as follows.

Absorption spectrum $\mu_a(\lambda)$ of the model medium is determined by three main chromophores (melanin and hemoglobin in visible spectral range and water in near IR spectral range) with their relative contributions.

$$\mu_a(\lambda) = C_{hem}(C_{oxy}\,\mu_{oxy}(\lambda) + C_{deoxy}\,\mu_{deoxy}(\lambda))$$
$$+ C_{mel}\,\mu_{mel}(\lambda) + C_{water}\,\mu_{water}(\lambda), \quad (10.2)$$

In Equation (10.2), the spectral dependences of the absorption of oxy- and deoxygenated hemoglobin (in cm^{-1}) were taken in the form of the following equation [45]:

$$\mu_{oxy,deoxy}(\lambda) = \frac{2,303 \times 150 \times \varepsilon_{oxy,deoxy}(\lambda)}{66500}, \quad (10.3)$$

where $\varepsilon_{oxy,deoxy}(\lambda)$ – the spectra of molar extinction coefficients for the two forms of hemoglobin, respectively; C_{oxy}, C_{deoxy} determine the degree of oxygenation of hemoglobin ($Y = C_{oxy}/(C_{oxy} + C_{deoxy})$). In the calculation, the ratio of hemoglobin forms was determined as $C_{oxy} = 0.7$ and $C_{deoxy} = 0.3$.

The pigment eumelanin was taken as epidermal melanin [45], the extinction spectrum of which can be approximated by the following formula:

$$\mu_{mel}(\lambda) = 544.936 \times \exp\left(-\frac{\lambda}{106.207}\right). \quad (10.4)$$

The absorption spectrum $\mu_{water}(\lambda)$ of water was taken from [46]. The scattering of a model medium, as was said earlier, is determined by the spectral dependence of the dermis reduced scattering coefficient, which can be represented as a superposition of Rayleigh scattering and Mie scattering [47]:

$$\mu'_S(\lambda) = 0.55 \times 10^{12} \times \lambda^{-4} + 36.85 \times \lambda^{-0.22}. \quad (10.5)$$

In the calculation of the diffuse reflection coefficient in formula (10.1), scattering was taken into account as $C_{scat}\mu'_S$, where C_{scat} is a certain coefficient, and the change in the scattering properties of the model medium was taken into account by varying this coefficient C_{scat}.

The skin reflectance spectrum calculation algorithm

The calculation consisted in determining the values of the coefficients C_{hem}, C_{mel}, C_{water}, as well as the spectrum of the reduced scattering coefficient (more precisely, the value of the coefficient C_{scat}), at which the calculated diffuse reflectance spectrum of the skin was as close as possible to the experimentally obtained spectrum using the integrating sphere technique. It was found that the maximum coincidence of the spectra was obtained for the following values of the coefficients that determine the relative contribution of chromophores to skin absorption: $C_{hem} = 0.0024$; $C_{mel} = 0.14$; $C_{water} = 1$, $C_{scat} = 0.5$.

The influence of the water content on the diffuse reflectance spectrum of the model medium was determined by varying the coefficient C_{water} in Equation (10.2), and the value $C_{water} = 1$ is conventionally taken as water content in normal skin. The effect of scattering on the diffuse reflectance spectrum was determined by varying the scattering properties of the model medium (the value of the coefficient C_{scat}). In this case, the spectral distribution of the reduced scattering coefficient, in which the calculated diffuse reflectance spectrum of the skin was as close as possible to the experimentally obtained spectrum, was considered the spectrum of the reduced scattering coefficient of normal skin.

Calculation results

The results of calculating the diffuse reflectance spectra of a model medium with different scattering and different absorption (water concentration in the NIR range of the spectrum and hemoglobin in the visible range of the spectrum) are presented below.

Figure 10.1 shows the diffuse reflectance spectra of the model medium, calculated for different values of the reduced scattering coefficient of the medium. The relative changes in the scattering properties of the model medium for which the spectral dependences are obtained are shown in Table 10.1.

Figure 10.1 shows that with increasing scattering, the reflection coefficient of the model medium increases in the entire spectrum from 400 to 2000 nm. The scattering properties of biological tissue are determined by its water content, while a decrease in water content leads to a decrease in scattering and, as a consequence, should lead to a decrease in reflection coefficient.

The kinetics of the reflectance spectrum of a model medium caused by a change in its scattering properties can be seen in Figure 10.2, where the dependences of the reflectance coefficient at fixed wavelengths on the magnitude of the reduced scattering coefficient are given. The selected wavelengths were 970 nm, 1190 nm, and 1450 nm where water has local absorption maxima; 1070 nm and 1250 nm where local water

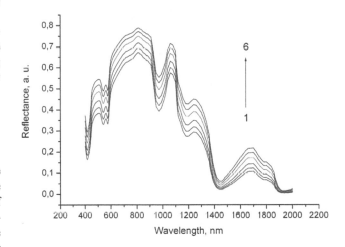

FIGURE 10.1 Diffuse reflectance spectra of a model medium with different scattering properties (the values of reduced scattering coefficient are given in the Table 10.1).

TABLE 10.1

The values of the reduced scattering coefficient (in arbitrary units) of the model medium for which the diffuse reflection spectra are shown in Figure 10.1.

Spectrum #	Reduced scattering coefficient, arb. unit
1	0.4
2	0.6
3	0.8
4	1.0 (norm). In Figure 10.1 it is highlighted in red
5	1.2
6	1.5

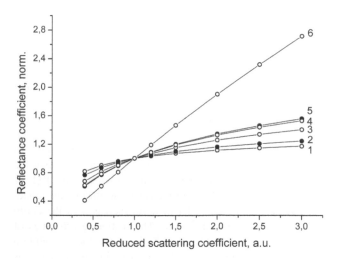

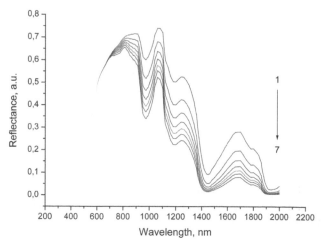

FIGURE 10.2 The dependence of the normalized reflectance coefficients of the model medium on the fixed wavelengths on the medium scattering properties. $1 - \lambda = 810$ nm, $2 - \lambda = 1070$ nm, $3 - \lambda = 970$ nm, $4 - \lambda = 1250$ nm, $5 - \lambda = 1190$ nm, $6 - \lambda = 1450$ nm.

FIGURE 10.3 Diffuse reflectance spectra of a model medium with different water content (the values of water concentration are given in the Table 10.3).

absorption minima take place; and a wavelength of 810 nm where water absorption is minimal.

For clarity, the values of the reduced scattering coefficient are expressed in arbitrary units, assuming the coefficient value for normal skin tissue to be unity. The actual values of the reduced scattering coefficient for skin tissue without compression at the selected wavelengths are given in Table 10.2.

The effect of absorption due to the water present in the model medium on the diffuse reflectance spectrum of the model medium can be seen from Figure 10.3, which shows the diffuse reflectance spectra of the model medium at different water contents. The relative contributions of water to the absorption of the model medium for which the spectral dependences are obtained are given in Table 10.3.

The kinetics of the reflectance spectrum of the model medium caused by a change in the water content can be seen from Figure 10.4, which shows the dependences of the reflectance coefficient at fixed wavelengths on the concentration of water in the model medium. The same wavelengths were taken, for which the dependences of the reflectance coefficient on the value of the reduced scattering coefficient were calculated.

In the visible spectrum, the diffuse reflection spectra of the skin are affected by the absorption of blood (hemoglobin) contained in the skin. The diffuse reflectance spectra of the model sample at different blood concentrations (hemoglobin)

are shown in Figure 10.5, and the relative contributions of hemoglobin to the absorption of the model medium for which the spectral dependences are obtained are given in Table 10.4.

Thus, the water content in the model sample affects the reflectance coefficient in the near IR region differently: a decrease in the water content in the medium leads to a decrease in its absorption, which is reflected in an increase in the reflectance coefficient; on the other hand, a decrease in water content leads to a decrease in the scattering properties of the medium, which leads to a decrease in reflectance coefficient. These results can be the basis for the analysis of experimentally obtained spectra of skin reflectance in the near infrared range.

In the visible range of the spectrum, skin compression leads to a decrease in the absorption of biological tissue due to blood (hemoglobin) and, as a result, to an increase in reflection coefficient.

Effect of external mechanical compression on optical and physiological properties of the skin (experiment)

Experimental setup and the object of study

The experimental setup is shown schematically in Figure 10.6 and included light sources HL-2000 (Ocean Optics, USA),

TABLE 10.2

Optical parameters of the skin without compression at selected wavelengths.

Wavelength, nm	Reduced scattering coefficient, cm⁻¹	Absorption coefficient, cm⁻¹
810	9.72	0.074
970	8.74	0.475
1070	8.36	0.136
1190	8.03	1.041
1250	7.90	0.886
1450	7.55	28.60

TABLE 10.3

Water concentrations (in arbitrary units) in a model medium for which diffuse reflectance spectra are calculated.

Spectrum #	Water concentration, arb. units
1	0.2
2	0.4
3	0.6
4	0.8
5	1.0 (normal). In Figure 10.3 it is highlighted in red
6	1.2
7	1.5

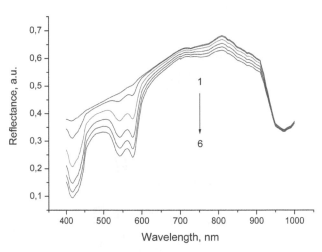

FIGURE 10.5 The dependence of the normalized reflectance coefficients of the model medium on the fixed wavelengths on the medium hemoglobin content; the spectrum of the skin with normal hemoglobin concentration is highlighted in red (the values of hemoglobin concentrations are given in the Table 10.4).

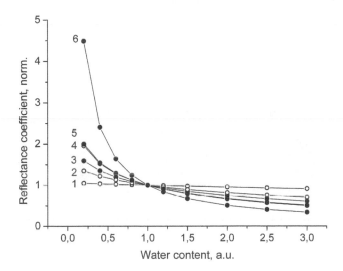

FIGURE 10.4 The dependence of the normalized reflectance coefficients of the model medium on the fixed wavelengths on the medium water content. $1 - \lambda = 810$ нм, $2 - \lambda = 1070$ нм, $3 - \lambda = 970$ нм, $4 - \lambda = 1250$ нм, $5 - \lambda = 1190$ нм, $6 - \lambda = 1450$ нм.

TABLE 10.4

Hemoglobin concentrations (in arbitrary units) in a model medium for which diffuse reflection spectra of the skin were obtained.

Spectrum #	Hemoglobin concentration, arb. units
1	0.0002
2	0.00096
3	0.0024 (norm). In Figure 10.5 it is highlighted in red
4	0.00384
5	0.00576
6	0.00768

fiberoptic sensors, and two fiberoptic spectrometers USB4000 (Ocean Optics, USA) (spectrum registration range 400–1000 nm) and NIR Quest512-2.2 (Ocean Optics, USA) (spectrum registration range 900–2200 nm), associated with personal computers, and provided registration of spectra of diffuse reflected skin light in the spectral range 400–2000 nm.

In this work, we used fiberoptic sensors of our own design and two Ocean Optics fiberoptic sensors (R400-7-VIS/NIR and R600-7-VIS-125F) to measure the diffuse reflectance spectra of skin *in vivo*.

In the first case, the sensor design included a half-ring with a radius of 40 mm with two optical fibers fixed in it (core diameter 400 μm, numerical aperture 0.2) to supply radiation to the skin surface and collect the light reflected from skin. A collimating lens was located at the end of the fiber leading light normal to the skin surface, as a result of which the irradiation spot diameter of the skin surface was 6 mm.

The receiving fiber was fixed at an angle of 30° relative to the illuminating fiber at a distance from the skin surface so that the size of the skin area from which radiation diffusely reflected by the skin was collected was two times the size of the skin irradiation spot. This was done to minimize losses

in the detected light of the long-wavelength part of the spectrum [9]. When registering the skin diffuse reflectance spectra under conditions of its external mechanical compression, a thin quartz glass with a diameter of 30 mm was placed between the half-ring and the skin surface, on which a pressure p was applied in the range from 0 to 10^6 Pa (Figure 10.7).

In the second case, sensors were used, one of which contained seven optical fibers with a diameter of 400 μm located in a steel tip with a diameter of 6.3 mm, and other contained seven optical fibers with a diameter of 600 μm located in a steel tip with a diameter of 3.2 mm. The sensors used quartz-polymer fibers with a numerical aperture of 0.22, which corresponds to an exit beam cone angle of 24.8°. Six optical fibers arranged around the circumference served to supply radiation to the object of study, and the central optical fiber served to detect light diffusely reflected by biological tissue. The receiving fiber was connected with USB4000 and NIR Quest512-2.2 fiberoptic spectrometers, which, in turn, were connected to a personal computer.

The sensors were mounted in special holders that provided the necessary area for applying pressure on the skin. In turn, the sensors with holders were mounted on a half-ring with a radius of 40 mm and provided external pressure in the range from 0 to 10^6 Pa (Figure 10.8).

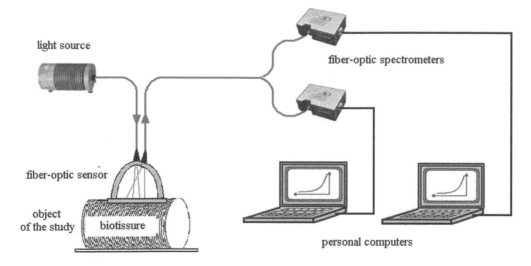

FIGURE 10.6 Experimental setup.

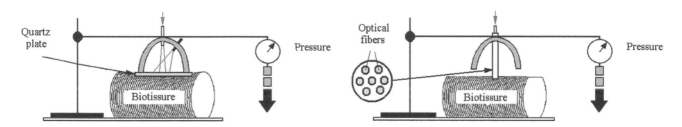

FIGURE 10.7 Scheme of the device for registration of the skin diffuse reflectance spectra under conditions of its external mechanical compression. Detector 30 mm.

FIGURE 10.8 Scheme of the device for registration of the skin diffuse reflectance spectra under conditions of its external mechanical compression. Detectors 3.2, 6.3, 10.0, 13.8, and 15.8 mm.

The size of the pressure application region was provided either by the ends of the sensors (3.2 mm, 6.3 mm) or by the holders (10 mm, 13.8 mm, and 15.8 mm).

External compression affects the blood content in the skin tissue, which can be estimated by the magnitude of the dip in the diffuse reflection spectrum in the spectral region 545–575 nm (erythema degree), and the water content, which determines the scattering of the skin in the entire visible range and its absorption in the range of over 800 nm. In the spectral region of 700–800 nm, the effect of water absorption on the diffuse reflectance spectrum is minimal compared with the scattering effect, and we can assume that changes in the diffuse reflectance spectrum of the skin in this range are primarily due to the scattering properties of the skin. This is taken into account when analyzing changes in reflectance spectra in the region above 800 nm, in which water determines both the absorbing and scattering properties of the skin.

The object of the study was the skin of the inner side of the human forearm *in vivo*. Diffuse reflectance spectra of human skin were measured on 10 volunteers aged 20–65 years with III and IV skin types; one volunteer had V skin type, according to Fitzpatrick. On each volunteer, diffuse reflectance spectra were measured at different values of applied pressure and different sizes of compression areas, resulting in a total of about 40 measurements.

The design of the sensor mounting in the form of a half-ring, when it was fixed along the forearm, allowed minimizing

changes in the measurement geometry, which resulted in a high degree of reproducibility of the results (more than 95%).

The main factor determining systematic measurement errors is the difference in the skins of different people and different skin areas of one volunteer. The discrepancies in the reflectance of the skin, obtained from different parts of the skin and with the same parts of the skin surface in different people, reached 10%–15%. Therefore, diffuse reflectance spectra of the skin under external compression for each volunteer were recorded from the same forearm site. Accuracy of spectral measurements during external compression was within 5%. For different volunteers, there were some differences in the temporary change in the spectra; however, general patterns persisted.

To determine the degree of dehydration of *ex vivo* samples of cow muscle tissue under the influence of external mechanical pressure, a special nozzle was made on the sensor with a cylindrical cell (Figure 10.9). The slots in the bottom and walls of the cuvette served to divert fluid squeezed out of the tissue under the influence of external mechanical compression. The nozzle and the cuvette were simulated using the program Autodesk Fusion 360 and printed on a 3D printer in black PLA plastic. Samples of muscle tissue were cut in the form of cylinders with a diameter of ~20 mm and a thickness of ~25 mm, which were placed in a cuvette and subjected to external pressure using a nozzle.

After applying pressure for a time of about 6 minutes, reflectance spectra were recorded with a time step of 5 s, after which

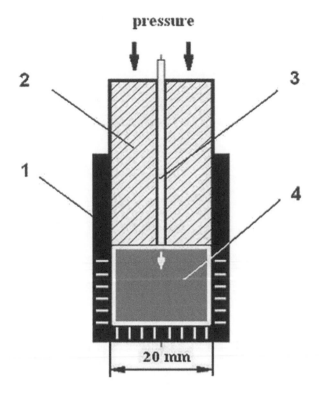

FIGURE 10.9 Scheme nozzles for squeezing fluid from muscle tissue samples. 1 – cuvette, 2 – nozzle, 3 – fiberoptic sensor, 4 – tissue sample.

the external cylinder with a sample of biological tissue was weighed. As a result of squeezing the sample in the cylinder, water came out of it (and from the external cylinder, respectively), which did not participate in weight measurements. The measurement results were averaged.

Experimental results

Temporal kinetics of *in vivo* *diffuse reflectance spectra of the skin in the visible spectrum*

Figures 10.10 and 10.11 show the temporal dynamics of the diffuse reflectance spectra of the skin after applying external

pressure of different values: $p = 13.9$ kPa (Figure 10.10(a)) and $p = 110$ kPa (Figure 10.11(a)) and after its removal (Figure 10.10(b) and Figure 10.11(b)).

Within 4–6 min, a change in the reflectance coefficient occurs in the entire spectral region, and a change in the spectral range of 500–600 nm, which practically disappears 5 min after compression of 100 kPa.

After removal of the external compression, a sharp decrease in the reflection coefficient of the skin occurs over the entire spectral range within a few seconds, while a dip again forms in the region of 500–600 nm. The restoration of the reflectance spectrum to its original state takes about 50 min.

This can be seen from Figures 10.12 and 10.13, which show the temporary changes in the reflectance of the skin of a volunteer at two wavelengths (545 and 800 nm) under conditions of application and removal at different pressures.

Common to all diffuse reflection spectra of the skin in the range 400–1000 nm, measured using different sensors, is the different behavior of the reflectance spectrum in the region of 500–600 nm and in the spectral region above 600 nm.

Temporal kinetics of *in vivo* *diffuse reflectance spectra of the skin in the spectral range 500–600 nm*

Different behaviors of skin diffuse reflectance are due to the different nature of the changes in the scattering and absorption properties of biological tissue under external mechanical compression. As is known, the main chromophores of skin tissue, which determine the spectrum of diffuse reflectance of the skin in the visible range of the spectrum, are the pigment of melanin and hemoglobin in the blood located in the papillary dermis. Obviously, under the influence of compression, the melanin content in the skin does not change, while the blood content in the skin can change, especially when high pressures are applied to the skin (about 100 kPa). This is evidenced by the behavior of the dip in the spectrum of skin reflectance in the spectral region of 500–600 nm, due to the absorption of hemoglobin.

Two factors affect the behavior of the reflection spectrum of the skin in the spectral region of 500–600 nm. The first factor

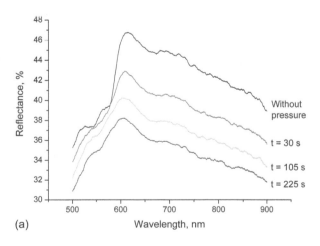

(a)

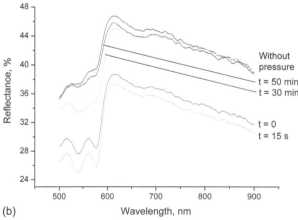

(b)

FIGURE 10.10 Temporal changes in the reflectance spectrum of the skin *in vivo* with applying external compression (a) and after its removal (b). $p = 13.9$ kPa.

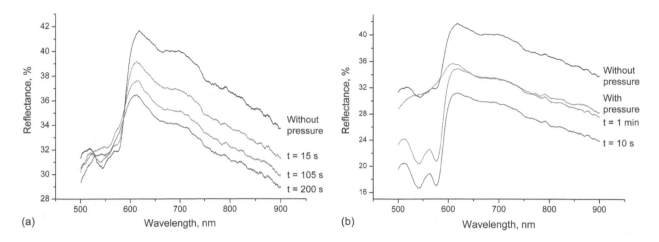

FIGURE 10.11 Temporal changes in the reflectance spectrum of the skin *in vivo* with applying external compression (a) and after its removal (b). $p = 110$ kPa.

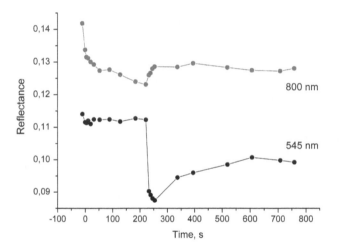

FIGURE 10.12 Temporary changes in skin *in vivo* reflectance on two wavelengths with the application of external compression (time interval 0–250 s) and after its removal (time interval over 250 s). Skin type V. $p = 100$ kPa.

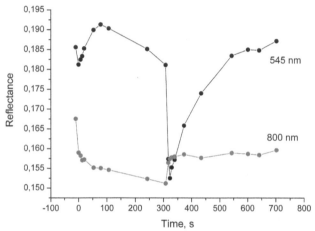

FIGURE 10.13 Temporary changes in skin *in vivo* reflectance on two wavelengths with applying external compression (time interval 0–310 s) and after its removal (time interval over 310 s). $p = 60$ kPa.

is the blood supply of the papillary dermis (erythema index) under normal conditions. In cases when the manifestation of hemoglobin absorption in the skin reflection spectrum without compression is small, during the skin compression process, a decrease in the reflectance coefficient is observed in the entire spectral range of 400–1000 nm (Figure 10.14).

The diffuse reflection spectra for a volunteer with type V skin behave similarly (Figure 10.15). For this type of skin, there is practically no dip in the spectral range 500–600 nm. When applying compression, the reflection practically does not change.

Quantitatively, the blood content is estimated by the severity of erythema: the greater the blood supply of the dermis, the greater the redness of the skin (the stronger the dip in the blue-green region of the spectrum) and the higher the degree of erythema. For skin type V, strong melanin pigmentation of the skin veils erythema, and it becomes less pronounced, as is seen in Figure 10.15.

Such a behavior of the spectra, when a decrease in the reflection coefficient in the entire spectral range of 400–1000 nm is

noted during compression, was called by the authors of [21] a "parallel pattern."

The second factor affecting the behavior of the spectra is the amount of external compression. A decrease in the reflectance coefficient in the entire spectral range of 400–1000 nm is observed with a large value of compression.

When the absorption spectrum of hemoglobin is clearly visible in the skin reflection spectrum without compression or the compression is not large enough, the behavior of the dip in the skin reflection spectrum in the region of 500–600 nm is different: when applying compression, an increase in the reflection coefficient is observed with a subsequent decrease. This case, shown in Figure 10.11, is noted in the literature [21] as a "pivot pattern."

A sharp increase in the reflectance coefficient over the entire range, while in the 500–600 nm range, the gap again forms for a time within 30–50 min depending on the applied pressure and the size of the compression application area.

Thus, temporary changes in reflectance coefficients in the region of 500–600 nm (hemoglobin absorption region) are sensitive to the amount of blood in the skin in the area of applied compression and the magnitude of this compression.

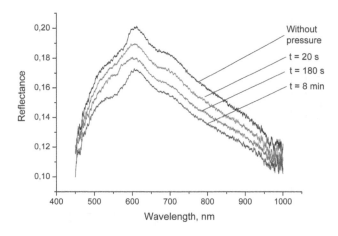

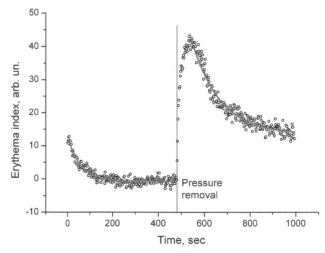

FIGURE 10.14 Temporal changes in the *in vivo* human skin reflectance with the application of external compression. $p = 100$ kPa.

FIGURE 10.16 Kinetics of erythema index of the skin with the application of external compression (time interval 0–480 s) and after its removal (time interval above 480 s). Pressure 200 kPa. Solid lines – approximation of experimental data by exponential function.

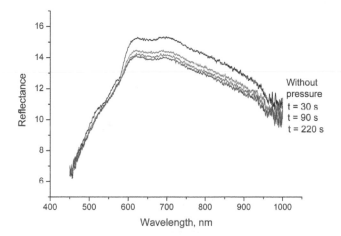

FIGURE 10.15 Temporal changes in the *in vivo* human skin reflectance with the application of external compression. Skin type V. $p = 100$ kPa.

Instead of analyzing the effect of amount of blood content in the skin on its value of diffuse reflectance, it is much more efficient to analyze the temporal behavior of the erythema index of the skin in the process of compression.

Kinetics of changes in blood content and the degree of oxygenation of hemoglobin in the skin tissue during compression

A quantitative assessment of the blood content (hemoglobin) and the degree of oxygenation of blood hemoglobin in the skin tissue was carried out according to the diffuse reflectance spectra of the skin in the visible range using pigmentation indices [1]. In particular, the quantitative blood (hemoglobin) content in skin tissues *in vivo* allows us to estimate the erythema index, which gives the results of a comparison of the optical density (OD) of the skin in the region of 510–610 nm, where characteristic hemoglobin absorption bands are present:

$$E = 100 \left[\begin{array}{c} OD_{560} + 1.5 \left(OD_{545} + OD_{575} \right) \\ - 2.0 \left(OD_{510} + OD_{610} \right) \end{array} \right], \quad (10.6)$$

where the indices determine the wavelengths in nm. An alternative approximation widely used in transmission oximetry is based on a comparison of the OD value at isosbestic points, i.e., at such wavelengths where the absorption of hemoglobin does not depend on the state of its oxygenation. In the spectral region of 500–600 nm, there are five such isosbestic points: 502, 529, 545, 570, and 584 nm. Differences in OD values between two isosbestic points will be proportional to the hemoglobin content in the sample and are independent of the state of oxygenation. As a result, the degree of oxygenation of hemoglobin in the blood can be calculated according to the following expression [48, 49]:

$$Y = \alpha \left(\left(\frac{OD_{570} - OD_{570}}{13} - \frac{OD_{570} - OD_{545}}{12} \right) \frac{1}{H} + \beta \right), \quad (10.7)$$

where the hemoglobin index is as follows:

$$H = \frac{OD_{545} - OD_{529}}{16} - \frac{OD_{570} - OD_{545}}{25}, \quad (10.8)$$

α and β – correction factors (for our experimental setup $\alpha=31$, $\beta=1$).

Figure 10.16 shows the kinetics of changes in the skin erythema index when applying external compression of different values and after its removal. It is seen that in the case of external compression, the erythema index decreases, which indicates a decrease in the blood content in the skin tissue, and the temporary change in the skin erythema index is well approximated by an exponential function.

After time determining the transport of blood from the compression area, the content of the blood remaining in the compression area of the skin stops changing (the erythema index stabilizes). The average time of displacement of blood from the compression area depends on the applied pressure (18.19 ± 10.34 s for compression $p = 41.6$ kPa to 58.16 ± 16.6 s for compression $p = 76.2$ kPa). As the area of pressure application

TABLE 10.5

The times of displacement of blood from the compression area for different sensors. Compression p = 55.5 kPa

Sensor, mm	Time, s
3.2	3.7 ± 1.7
6.3	6.7 ± 2.1
10.0	13.0 ± 3.8
13.8	18.5 ± 2.2
15.8	23.0 ± 2.6
30.0	30.0 ± 8.2

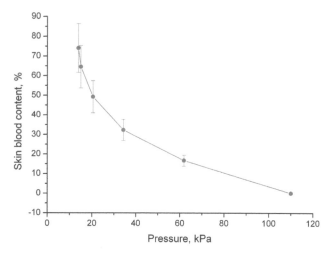

FIGURE 10.18 Blood (hemoglobin) content in the skin depending on the applied compression.

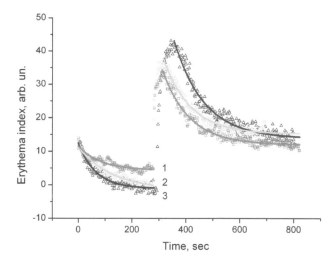

FIGURE 10.17 Kinetics of erythema index of the skin with the application of external compression of different values. Solid lines – approximation of experimental data by exponential function.

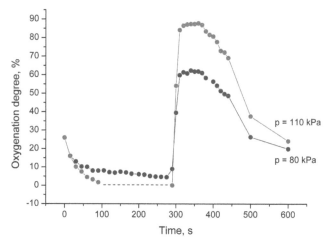

FIGURE 10.19 Kinetics of the degree of oxygenation of hemoglobin of blood contained in the skin with the application of external compression (time interval 0–290 s) and after its removal (time interval above 290 s). The dashed line shows the time range when there is no blood in the skin.

increases, the time of blood displacement also increases. This can be seen in Table. 10.5, which shows the average time of blood displacement when using different sensors. The values of the times are in good agreement with the average linear velocity of capillary blood flow in humans (0.5–1 mm/s).

The amount of blood displaced from the compression area depends on the amount of compression (Figure 10.17), while with compression over 110 kPa, the blood is completely displaced (Figure 10.18). When external compression is removed, there is a sharp increase in the blood content in the volume of skin tissue that has been compressed. This happens over a period of about 30 s. Then the erythema index values return to their original state after 30–50 min, depending on the magnitude of the compression, and the relaxation of the erythema index is well described by the exponential function.

Of particular interest is the behavior of the degree of oxygenation of hemoglobin of blood contained in the volume of skin tissue subjected to mechanical compression. Figure 10.19 shows the kinetics of changes in the degree of oxygenation under mechanical compression and after its removal. After applying compression, a decrease in blood content was accompanied by a decrease in the degree of oxygenation of the hemoglobin contained in it. The removal of compression led to a sharp increase (two or three times) in the degree of oxygenation.

It should be noted that in the diffuse reflectance spectra of the skin, mainly blood in the superficial vascular plexus appears, and the degree of hemoglobin oxygenation is a parameter that depends on the ratio of arterial and venous blood in the probed volume of skin tissue. If in the normal state the degree of oxygenation was about 32%, then a possible reason for the increase in the degree of oxygenation after removal of compression may be a sharp injection of arterial blood into the volume of skin tissue that was compressed, since the valves of the veins do not allow squeezed venous blood to return to the compression area after it has been removed [50].

Temporal kinetics of in vivo *diffuse reflectance spectra of the skin in the spectral range 600–800 nm*

In the spectral range of 600–2000 nm the spectrum of diffuse reflectance of the skin under conditions under compression is

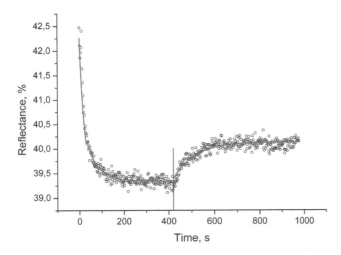

FIGURE 10.20 Kinetics of reflectance coefficient of the skin at the wavelength 800 nm with the application of external compression (time interval 0–420 s) and after its removal (time interval above 420 s). $p = 100$ kPa. Solid lines – approximation of experimental data by exponential function. Vertical line – compression removal time.

TABLE 10.6

Fast and slow relaxation times for reflectance coefficients averaged over wavelengths at different compression values

Compression, кPa	Fast relaxation, s	Slow relaxation, s
27.7	12.72 ± 6.20	160.93 ± 96.51
41.6	8.28 ± 3.83	105.56 ± 44.68
55.5	5.18 ± 1.16	82.50 ± 46.50
63.9	4.46 ± 2.11	69.75 ± 38.36
80.5	3.60 ± 1.84	66.12 ± 36.37

significantly affected by the amount of water it contains; in the 600–800 nm range, water determines the scattering properties of the skin, and in a spectral range above 800 nm, in addition to the scattering properties, the water content determines the absorbing properties of the skin.

Figure 10.20 shows the kinetics of the reflectance coefficients of the skin at a wavelength of 800 nm when compression is applied and removed. A distinctive feature of the behavior of the reflectance coefficient after compression is its monotonic decrease according to the two-exponential law. The characteristic decay times of exponential functions are in the order of few seconds (fast relaxation) and several minutes (slow relaxation). Table 10.6 shows the times of the fast and slow relaxation for different values of compression averaged over the reflectance coefficient at different wavelengths.

The two-exponential time dependence of the reflectance coefficients of the skin during compression can be due to the presence of both free and accessible water in the skin. At the initial application of pressure, a sharp deformation and compression of the collagen matrix occurs (according to the two-phase model), which is accompanied by the removal of free water from the compression region (quick relaxation). Subsequently, the rate of water flowing from the skin area tissue subject to compression decreases and a process of slow dehydration of this volume of the skin (slow relaxation) occurs,

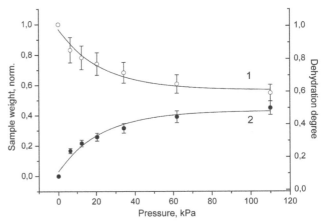

FIGURE 10.21 Temporary changes in weight (1) and dehydration degree (2) of muscle samples during their compression.

in which bound water is involved. The characteristic times of an exponential decrease in reflectance coefficients in the spectral region of 600–900 nm after compression is nothing more than the times of dehydration of a tissue sample.

In the spectral region of 600–800 nm, the diffuse reflectance spectrum of the skin depends only on the amount of water contained in the skin. Since water absorption in this spectral region can be neglected, it can be concluded that the amount of water contained in the skin determines only the scattering properties of the skin. Compression of the skin leads to a decrease in the water content in the skin, or its dehydration.

Following the authors of [51], the degree of dehydration of biological tissue changes from zero to the maximum degree without any dimension can be defined as

$$H_D(t) = \frac{W(t=0) - W_0}{W(t=0)} \left(1 - \exp\left(-t / \tau_D\right)\right), \quad (10.9)$$

where $W(t = 0)$ is the initial weight of the sample and the difference $(W(t = 0) W_0)$ shows the amount of water that left the biological tissue as a result of dehydration, while the time dependence of the degree of dehydration H_D can be written as follows:

$$H_D(t) = A_D \left(1 - \exp\left(-t / \tau_D\right)\right), \quad (10.10)$$

where A_D is the dimensionless parameter, characterizes the maximum degree of dehydration, τ_D is the dehydration time constant characterizing the speed of the process.

Compression of biological tissue samples should lead to a change in their physical – including optical – parameters, which are caused by a number of factors, including deformation of the samples, a change in their density, and a change in the amount of water contained in them. Information on the weight of the samples and the degree of their dehydration as a result of compression obtained, taking into account relations of Equations (10.9) and (10.10), is shown in Figure 10.21.

Figure 10.21 shows that at a pressure on the samples of the order of 110 kPa, the weight of the samples decreases by 45%. This decrease in sample weight is due to the displacement of water from the samples and their dehydration. The maximum

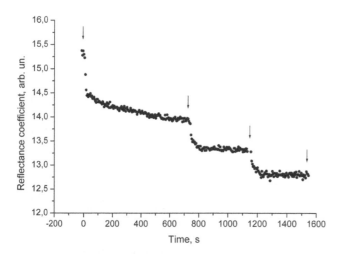

FIGURE 10.22 Temporary changes in diffuse reflectance of tissue sample under compression of different values. λ = 900 nm. The arrows indicate times of competition changes.

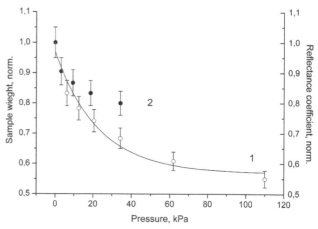

FIGURE 10.23 Temporal changes in weight (1) and diffuse reflectance coefficient (2) of muscle tissue samples depending of applied compression. λ = 900 nm

degree of dehydration of the samples during compression, as can be seen from Figure 10.21, was about 50%.

Figure 10.22 shows temporary changes in the coefficient of diffuse reflectance of a tissue sample at a wavelength of 900 nm with a sequential increase in applied compression (times of compression change are shown by arrows), which is well approximated by the exponential dependence (Figure 10.23).

Figure 10.23 also provides information on the state of dehydration of tissue samples under compression, depending on the weight of the samples with compression applied. The results show that compression of the samples leads to a decrease in the diffuse reflectance.

This fact may be explained by the difference in the physical structures of the samples during dehydration [38]. Compressed samples are very dense and compact, while compaction of scattering particles leads to a change in the volume fraction of scattering particles φ_S and, as a consequence, to a change in the reduced scattering coefficient μ_S', which are related by the relationship [38, 41]:

$$\mu_S' = \frac{\varphi_S(1-\varphi_S)}{V_P}\sigma_S', \qquad (10.11)$$

where V_P – is the volume of the scattering particle and σ_S' – is the cross-section of the reduced scattering, depending on the wavelength of light, the radius of the scattering particle, and the refractive indices of the scattering particle and the environment. This dependence has the form of a parabola with a maximum μ_S' at $\varphi_S = 0.5$ and vanishing at = 0 and $\varphi_S = 1$, that is, when the medium becomes single-phase.

The data given in Figure 10.23 show that the behavior of the coefficient of diffuse reflectance of a tissue sample correlates with the process of reducing the weight of the sample during compression. Both processes are exponential in nature, with a decrease in the reflectance coefficient due to dehydration of the sample. The relative decrease in the reflectance coefficient differs in value from the relative decrease in the

weight of the sample, which is due to the fact that the diffuse reflectance of the sample depends on its scattering and absorbing properties [44], that is, it depends on the wavelength of the probe light. An increase in the wavelength can give a quantitative coincidence between the dependences shown in Figure 10.23.

Temporal kinetics of *in vivo diffuse reflectance spectra of the skin in the NIR*

In the spectral region of 900–2200 nm, the diffuse reflectance spectra of the skin were recorded during external compression using Ocean Optics fiberoptic sensors (R400-7-VIS/NIR and R600-7-VIS-125F) with nozzles providing a smaller area compression overlay.

Figure 10.24 shows the diffuse reflectance spectra of the skin of the human forearm *in vivo* with detecting reflected light using a 10 mm sensor when applying external pressure has of different valued.

Thus, in the near infrared region of the spectrum, the application of external compression to the skin shows a tendency to decrease the reflectance coefficient of the skin in the entire spectral range. This is also seen in Figure 10.25, which shows the temporary changes in the skin reflectance coefficients at a wavelength of 1070 nm. The temporal kinetics of reflectance coefficients at other wavelengths has a similar character.

After removing the compression, the spectra are restored to their original position in almost the same time period as in the visible range (about 30–50 min).

Changing the size of the area of compression application does not fundamentally change the character of changes in the diffuse reflectance spectra of the skin *in vivo*.

It is seen that an increase in compression leads to a decrease in the reflectance coefficient of the skin. In the case of using a 10 mm sensor, a decrease in the reflectance coefficient during compression amounted to 10.36% at a pressure of 28 kPa, 12.89% at a pressure of 50 kPa, 16.36% at a pressure of 100 kPa, and 21.86% at a pressure of 143 kPa.

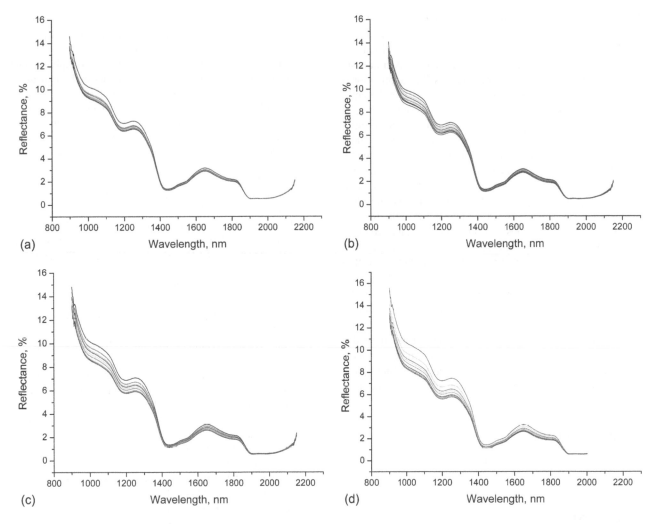

FIGURE 10.24 Temporal changes in diffuse reflectance spectrum of the human skin *in vivo* with applied external mechanical compressions of different value: 28 kPa (a), 50 kPa (b), 100 kPa (c), и 143 kPa (d).

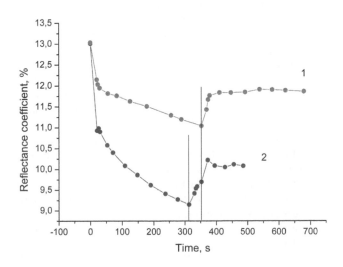

FIGURE 10.25 Temporal changes in diffuse reflectance spectrum of the human skin *in vivo* at wavelength 1070 nm with applied external mechanical compressions of values 45 kPa (1) and 108 kPa (2). Vertical lines – compression removal times.

Despite some differences in the dynamics of the spectra under external compression due to the morphological characteristics of the skin of volunteers, the overall result is a decrease in the scattering coefficient over time after applying compression.

Unlike the spectral range 600–800 nm, where the water present in the skin determines the scattering properties of the skin, in the spectral region above 900 nm, the water determines not only the scattering of the skin, but also its absorption. Since a decrease in the absorption of water in the skin resulting from its displacement from the compression region leads to an increase in the reflectance coefficient, the experimental results allow us to conclude that the effect of reducing skin scattering due to the displacement of water from the compression region prevails over the effect of reducing the absorption of skin tissue, caused by the same process water displacement from biological tissue in the same area of external compression.

Summary

The application of external mechanical pressure of up to 105 Pa on the human skin *in vivo* leads to a decrease in the skin diffuse reflectance coefficient in the spectral range 400–2000 nm. Changes in skin geometry (its thickness and density) and its physiological parameters (blood and water content, blood oxygen saturation), and resulting changes in optical skin parameters are the reasons for the spectral changes. The application of compression leads to a decrease in skin scattering in the entire spectrum range 400–2000 nm, and the process of reducing the scattering properties prevails in the formation of the value of diffuse reflectance of the skin in NIR spectral region. Blood is completely expelled from the skin at a pressure of about 100 kPa. A decrease in blood content is accompanied by a decrease in the degree of oxygenation of hemoglobin in the skin blood. After releasing the compression for a few seconds, the blood content and its degree of oxygenation significantly increase (between two and three times and between three and five times, respectively) compared to normal skin without compression. This fact can be used as the basis for a method of controlling the degree of hemoglobin oxygenation in skin blood. The times of changes in the optical and physiological parameters of the skin as a result of external compression and restoration of their initial values after its removal were estimated. The research results contribute to the further development of the method of optical clearing of biological tissues, which is used to control the optical parameters of tissues to increase the efficiency of diagnosis and treatment of various diseases by optical methods.

Acknowledgments

This work was supported by the Russian Foundation for Basic Research (project No. 19-32-90177).

REFERENCES

1. Yu.P. Sinichkin, N. Kollias, G. Zonios, S.R. Utz, V.V. Tuchin, "Back reflectance and fluorescence spectroscopy of the human skin in vivo," in V.V. Tuchin (ed.), *Handbook on Optical Biomedical Diagnostics and Imaging*, SPIE Press, Bellingham, 725–785 (2002).
2. V.V. Tuchin, I.L. Maksimova, D.A. Zimnyakov, I.L.Kon, A.H.Mavlyutov, A.A. Mishin, "Light propagation in tissues with controlled optical properties," *J. Biomed. Opt.* **2**(4), 401–417 (1997).
3. O. Vargas, E. K Chan., J.K. Barton, H.G. Rylander, A.J. Welch, "Use of an agent to reduce scattering in skin," *Laser Surg. Med.* **24**(2), 133–141 (1999).
4. V.V. Tuchin, "Optical clearing of tissue and blood using immersion method," *J Phys D.* **38**, 2497–2518, (2005).
5. E.A. Genina, A.N. Bashkatov, Yu.P. Sinichkin, I.Yu. Yanina, V.V. Tuchin, "Optical clearing of tissues: Benefits for biology, medical diagnostics and phototherapy," *J. Biomed. Photon. Eng.* **1**(1), 22–58 (2015).
6. P.D. Agrba, M.Y. Kirillin, A.I. Abelevich, E.V. Zagainova, V.A. Kamensky, "Compression as a method for increasing the informativity of optical coherence tomography of biotissues," *Opt. Spectrosc.* **107**(6), 901–906 (2009).
7. A. Izquierdo-Roman, W.C. Vogt, L. Hyacinth, C.G. Rylander, "Mechanical tissue optical clearing technique increases imaging resolution and contrast through ex vivo porcine skin," *Lasers Surg. Med.* **43**, 814–823 (2011).

8. I.V. Ermakov, W. Gellermann, "Dermal carotenoid measurements via pressure mediated reflection spectroscopy," *J. Biophoton.* **5**(7), 559–570 (2012).
9. L.E. Dolotov, Yu.P. Sinichkin, "Features of applying fiber-optic sensors in spectral measurements of biological tissues," *Opt. Spectrosc.* **115**(2), 187–192 (2013).
10. S. Ruderman, A.J. Gomes, V. Stoyneva, et al., "Analysis of pressure, angle and temporal effects on tissue optical properties from polarization-gated spectroscopic probe measurements," *Biomed. Opt. Express* **1**, 489–499 (2010).
11. G.A. Askar'yan, "Enhancement of transmission of laser and other radiation by soft turbid physical and biological media," *Sov. J. Quantum Electron.* **12**(7), 877–880 (1982).
12. E.K. Chan, B. Sorg, D. Protsenko, M. O'Neil, M. Motamedi, A.J. Welch, "Effects of compression on soft tissue optical properties," *IEEE J. Sel. Top. Quantum Electron.* **2**(4), 943–950 (1996).
13. H. Shangguan, S.A. Prahl, S.L. Jacques, L.W. Casperson, "Pressure effects on soft tissues monitored by changes in tissue optical properties," *Proc. SPIE.* **3254**, 366–371 (1998).
14. W.C. Vogt, A. Izquierdo-Roman, B. Nichols, L. Lim, J.W. Tunnell, C.G. Rylander, "Effects of mechanical indentation on diffuse reflectance spectra, light transmission, and intrinsic optical properties in ex vivo porcine skin," *Lasers Surg. Med.* **44**, 303–309 (2012).
15. Yu.P. Sinichkin, S.R. Uts, E.A. Pilipenko, "Spectroscopy of human skin in vivo: 1. Reflection spectra," *Opt. Spectrosc.* **80**(2), 228–234 (1996).
16. W. Chen, R. Liu, K. Xu, R.K. Wang "Influence of contact state on NIR diffuse reflectance spectroscopy in vivo," *J. Phys. D: Appl. Phys.* **38**, 2691 (2005).
17. S.A. Carp, T. Kauffman, Q. Fang et al., "Compression-induced changes in the physiological state of the breast as observed through frequency domain photon migration measurements," *J. Biomed. Opt.* **11**(6), 064016 (2006).
18. R. Reif, M.S. Amorosino, K.W. Calabro, O.A.' Amar, S.K. Singh, I.J. Bigio, "Analysis of changes in reflectance measurements on biological tissues subjected to different probe pressures," *J. Biomed. Opt.* **13**(1), 010502 (2008).
19. Y. Ti, W.C. Lin, "Effects of probe contact pressure on in vivo optical spectroscopy," *Opt. Express* **16**(6), 4250–4262 (2008).
20. A. Cerussi, S. Siavoshi, A.Durkin, et al., "Effect of contact force on breast tissue optical property measurements using a broadband diffuse optical spectroscopy handheld probe," *Appl. Opt.* **48**, 4270–4277 (2009).
21. J.A. Delgado Atencio, E.E. Orozco Guillén, S. Vázquezy Montiel, et al., "Influence of probe pressure on human skin diffuse reflectance spectroscopy measurements," *Opt. Mem. Neural Networks* **18**(1), 6–14 (2009).
22. L. Lim, B. Nichols, N. Rajaram, J.W. Tunnell, "Probe pressure effects on human skin diffuse reflectance and fluorescence spectroscopy measurements," *Biomed. Opt.* **16**(1), 011012-1-9 (2011).
23. Xu.U. Zhang, D.J. Faber, T.G. van Leeuwen, H.J.C.M. Sterenborg, "Effect of probe pressure on skin tissue optical properties measurements using multi-diameter single fiber reflectance spectroscopy," *J. Phys.: Photon.* **2**, 034008 (2020).

24. Yu.P. Sinichkin, S.R. Uts, I.V. Meglinskii, E.A. Pilipenko, "Spectroscopy of human skin in vivo: II. Fluorescence spectra," *Opt. Spectrosc.* **80**(3), 383–389 (1996).

25. A. Nath, K. Rivoire, S.Chang, et al., "Effect of probe pressure on cervical fluorescence spectroscopy measurements," *J. Biomed. Opt.* **9**(3), 523–533 (2004).

26. K. Rivoire, A. Nath, D. Cox, E.N. Atkinson, R. Richards-Kortum, M. Follen, "The effects of repeated spectroscopic pressure measurements on fluorescence intensity in the cervix," *Am. J. Obstet. Gynecol.* **191**(5), 1606–1617 (2004).

27. B. Cugmas, M. Bürmena, F. Pernuša, B. Likar, "Analysis of soft tissue near infrared spectra under dynamic pressure effects," *Proc. SPIE*, **8220**, 822007 (2012).

28. B. Cugmas, M. Bürmena, V. Bregar, F. Pernuša, B. Likar, "Pressure-induced near infrared spectra response as a valuable source of information for soft tissue classification," *J. Biomed. Opt.* **18**(4), 047002 (2013).

29. B. Cugmas, M. Bürmena, V. Bregar, F. Pernuša, B. Likar, "Impact of contact pressure–induced spectral changes on soft-tissue classification in diffuse reflectance spectroscopy: Problems and solutions," *J. Biomed. Opt.* **19**(3), 037002 (2014).

30. V.V. Sapozhnikova, R.V. Kuranov, I. Cicenaite, R.O. Esenaliev, D.S. Prough, "Effect on blood glucose monitoring of skin pressure exerted by an optical coherence tomography probe," *J. Biomed. Opt.* **13**(2), 0211122008 (2008).

31. M.Y. Kirillin, P.D. Agrba, V.A. Kamensky, "In vivo study of the effect of mechanical compression on formation of OCT images of human skin," *J. Biophoton.* **3**(12), 752–758 (2010).

32. C.G. Rylander, T.E. Milner, S.A. Baranov, J.S. Nelson, "Mechanical tissue optical clearing devices: Enhancement of light penetration in ex vivo porcine skin and adipose tissue," *Lasers Surg. Med.* **40**(10), 688–694 (2008).

33. C. Drew, T.E. Milner, C.G. Rylander, "Mechanical tissue optical clearing devices: Evaluation of enhanced light penetration in skin using optical coherence tomography," *J. Biomed. Opt.* **14**(6), 064019 (2009).

34. B. Yu, A. Shah, V.K. Nagarajan, D.G. Ferris, "Diffuse reflectance spectroscopy of epithelial tissue with a smart fiber-optic probe," *Biomed. Opt. Express* **5**(3), 675–689 (2014).

35. C.W.J. Oomens, D.H. Vancampen, H.J. Grootenboer, "A mixture approach to the mechanics of skin," *J. Biomech.* **20**(9), 877–885 (1987).

36. A.A. Gurjarpadhye, W.C. Vogt, Ya. Liu, C.G. Rylander, "Effect of localized mechanical indentation on skin water content evaluated using OCT," *J. Biomed. Imaging* **2011**, 817250 (2011).

37. C. Li, J. Jiang, K. Xu, "The variations of water in human tissue under certain compression: Studied with diffuse reflectance spectroscopy," *J. Innov. Opt. Health Sci.* **6**(1), 1350005 (2013).

38. C.G. Rylander, O.F. Stumpp, T.E. Milner, et al., "Dehydration mechanism of optical clearing in tissue," *J. Biomed. Opt.* **11**(P), 041117 (2006).

39. X. Xu, R.K. Wang, "Synergistic effect of hyperosmotic agents of dimethyl sulfoxide and glycerol on optical clearing of gastric tissue studied with near infrared spectroscopy," *Phys. Med. Biol.* **49**(3), 457–468 (2004).

40. X. Xu, R.K. Wang, "The role of water desorption on optical clearing of biotissue: Studied with near infrared reflectance spectroscopy," *Med. Phys.* **30**(6), 1246–1253 (2003).

41. M.R. Hardisty, D.F. Kienle, T.L. Kuhl, S.M. Stover, D.P. Fyhrie, "Strain-induced optical changes in demineralized bone," *J. Biomed. Opt.* 19(3), 035001 (2014).

42. I.A. Nakhaeva, O.A. Zyuryukina, M.R. Mohammed, Yu.P. Sinichkin, "The effect of external mechanical compression on in vivo water content in human skin," *Opt. Spectrosc.* **118**(5), 834–840 (2015).

43. O.A. Zyuryukina, Yu. P. Sinichkin, Dynamics of optical and physiological properties of human skin in vivo during its compression," *Opt. Spectrosc.* **127**(3), 498–506 (2019).

44. T.J. Farrell, M.S. Patterson, B. Wilson, "A diffuse theory model of spatially resolved, steady-state diffuse reflectance for the noninvasive determination of tissue optical properties in vivo," *Med. Phys.* 19, 879–888 (1992).

45. S. Prah, 1 Oregon medical laser center. URL: http://omlc.ogi.edu.

46. R. Nachabe, J.W.van der Hoorn, R. van de Molengraaf, et al., "Validation of interventional fiber optic spectroscopy with MR spectroscopy, MAS-NMR spectroscopy, high-performance thin-layer chromatography, and histopathology for accurate hepatic fat quantification," *Invest. Radiol.* **47**, 209–216 (2012).

47. A.N. Bashkatov, E.A. Genina, V.I. Kochubey, V.V. Tuchin, "Optical properties of human skin, subcutaneous and mucous tissues in the wavelength range from 400 to 2000 nm," *J. Phys. D: Appl. Phys.* **38**, 2543–2555 (2005).

48. J.W. Feather M. Haijzadeh, J.B. Dawson, et al., "A portable scanning reflectance spectrophotometer using visible wavelengths for rapid measurement of skin pigments," *Phys. Med. Biol.* **34**, 807–820 (1989).

49. M. Haijzahen, J.W. Feather, J. Dawson, "B. An investigation of factors affecting the accuracy of in vivo measurements of skin pigments by reflectance spectroscopy," *Phys. Med. Biol.* **35**, 1301–1315 (1990).

50. I.A. Nakhaeva, M.R. Mohammed, O.A. Zyuryukina, Yu.P. Sinichkin, "The effect of external mechanical compression on in vivo optical properties of human skin," *Opt. Spectrosc.* **117**(3), 506–512. (2014).

51. E.A. Genina, A.N. Bashkatov, A.A. Korobko, et al., "Optical clearing of human skin: Comparative study of permeability and dehydration of intact and photothermally perforated skin," *J Biomed. Opt.* **13**(2), 021102 (2008).

Part II

Tissue optical clearing method for biology (3D imaging)

11

Optical clearing for multiscale tissues and the quantitative evaluation of clearing methods in mouse organs

Tingting Yu, Jianyi Xu, and Dan Zhu

CONTENTS

Introduction

The mapping of three-dimensional (3D) structures through intact tissue is essential in many biomedical researches, such as reconstructing neuronal connectivity of the brain in neuroscience and understanding the morphogenesis of organs in developmental biology [1–4]. To this end, multiple approaches could be used to reconstruct the 3D information. Making thin sections is the most common method, using either traditional histological sectioning or automated serial sectioning techniques [5, 6].

The optical sectioning technique provides an alternative method for 3D reconstruction of tissue structures [7, 8], such as single-photon (confocal) microscopy [8], multiphoton microscopy [7, 9], and light-sheet microscopy [10–12]. However, the optical imaging depth suffers from strong light scattering due to the inherent heterogeneity of biological tissues as well as light absorption due to endogenous pigments. The tissue optical clearing technique [13, 14] has been proposed to improve imaging depth for the 3D visualization of large tissues.

Over the past decade, a variety of optical clearing methods have been developed to achieve deep imaging. Tissue optical clearing has two different major approaches with different final goals: *in vitro* methods, mostly used for obtaining the intact structures of tissues [15–22]; and *in vivo* methods, mostly used for monitoring the intravital function

or structure [23–27]. The latter one has some evident limitations and difficulties connected with the toxicity of optical clearing agents and short exposure times (from minutes to a few hours only). Both approaches can use similar optical clearing agents and technologies for agent delivery. The well-developed delivery technologies for some *in vivo* approaches could be of interest to *in vitro* 3D imaging of tissues, and, vice versa, some novel *in vitro* technologies to *in vivo* applications. The two approaches can enrich each other. Tissue optical clearing methods enhance the applicability of optical imaging in the biomedical field to a high degree, and in particular they offer significant tools for neural imaging with high resolution and have accelerated the development of neuroscience. This chapter will focus on *in vitro* optical clearing methods.

For the end-users, an ideal optical clearing method should provide not only high transparency, but also good fluorescence preservation, fine morphology maintenance, and wide applicability for different kinds of samples. Apart from the above properties, clearing speed, easy handling, safety, and cost are also taken into account. However, it is impractical to achieve all of these characteristics in one method. In addition, most *in vitro* optical clearing methods were developed for specific tissue types. Only a few methods [22, 28–31] have demonstrated clearing and imaging performance on some other organs. In practical applications, it is necessary but difficult to make an appropriate choice from the numerous clearing methods to meet the need of particular experiments.

On the one hand, some good reviews have provided good references for researchers in different fields. Tuchin and his collaborators have given several good reviews on the optical clearing of tissues and blood from basic principles and methods for various applications [13, 14, 32–36]. Zhu et al. also gave an excellent review to introduce progress in tissue optical clearing, including the mechanisms, enhancing methods, and applications for *in vitro* tissue microstructure in 3D and *in vivo* vascular function imaging [37]. Richardson and Lichtman presented a review on the clearing methods [1], described the physical mechanisms, and classified them into solvent-based clearing and aqueous-based clearing. Susaki and Ueda reviewed recent methodological achievements in whole-body and whole-organ clearing and imaging, as well as advances in cellular labeling and image informatics [2]. Tainaka et al. focused mainly on the chemical principles in current tissue-clearing and staining technologies in order to better understand the underlying mechanisms [38]. Additionally, Silvestri's review proposed a taxonomy of tissue optical clearing methods in the view of microscopists [39]. Recently, Yu et al. reviewed optical clearing methods based on the sample features in terms of size and age from the perspective of end-users.

On the other hand, some comparison work has been done. For instance, Kiefer et al. [40] applied ten clearing methods to brain, lung, heart, kidney, muscle, and embryo, providing qualitative data on clearing efficacy, maintenance of tissue integrity, duration of the protocols, and some other handling properties, but there was a lack of quantitative data on fluorescence preservation and imaging depth. Yu et al. made a systematically quantitative assessment of different methods on different organs, making the assessment more detailed and more specific for researchers to choose the most suitable clearing protocols.

This chapter will first review the up to date *in vitro* tissue optical clearing techniques in the view of sample features based on Yu et al.'s review, and then will exhibit the results of the comprehensive evaluation of the performances of some typical clearing methods in various intact mouse organs shown in the literature [41].

A brief review of optical clearing methods for multiscale biological tissues

The following describes the latest clearing methods suitable for multiscale tissues, from small tissue blocks, embryos, or neonatal samples to intact and adult organs, and further to whole bodies. The combinations of these methods and suitable choices for immunostaining and lipophilic dyes are also discussed.

Small sample/neonatal sample/embryo clearing

FocusClear was first introduced by Chiang's group, and its content was selected from the group consisting of dimethyl sulfoxide (DMSO), diatrizoate acid, glucamine, and many other chemical agents [42]. The Chiang group have applied it to clear the Drosophila brain to study the kinds of circuit of the fly related to olfactory representation [15], early memory trace [43], long-term memory formation [44], visual information processing [45], and auditory processing pathways [46]. As a mature commercial clearing reagent, FocusClear has been used by researchers for various species, such as plant (e.g., maize [47]), zebrafish [48], insect (e.g., Drosophila [49]), and mouse (e.g., brain, intestine, skin, islet, lymph node) [50–53], but mainly for small samples.

Soon afterward, graded sucrose solutions were exploited to study the spatial distributions of neurons and vessels in tissue slabs (1–2 mm) of the murine cortex [54]. Ke et al. tested various sugar solutions and found fructose solution to be a better clearing agent with high solubility in water and high refractive index (RI) of saturated solution, termed SeeDB [18, 55–57]. It showed good transparency and fine morphology preservation for whole mouse embryos, neonatal brains, and adult brain slices (Figure 11.1). They demonstrated the applications of SeeDB in developmental studies and microcircuit reconstruction in the brain, including visualizing long-range callosal axon projections and describing the wiring diagram of sister mitral cells in mouse olfactory bulb. To further achieve super-resolution imaging of thick tissues, Ke et al. developed SeeDB2 [58] by using iohexol solutions instead of fructose solutions and introducing saponin and Tris-EDTA buffer. It enabled large-scale synaptic mapping of thick fly and mouse brain samples in combination with super-resolution microscopy.

A drawback of SeeDB is the extremely high viscosity of the saturated fructose solution. To solve this problem, FRUIT [60] has been proposed, which utilized a cocktail of fructose and urea while retaining the advantages of original SeeDB. Sorbitol has also been combined with urea to produce ScaleS, which presented good structural stability [61]. Its fine preservation of samples' ultrastructure potentially allowed it to be combined with serial sectioning imaging techniques, such as serial two-photon tomography (STP).

Another clearing agent suitable for embryos and neonatal brains is *Clear^{T/T2}* [19], which is detergent- and solvent-free. This agent not only provides a fast clearing procedure and minimal tissue volume changes, but also has fine compatibility with multiple fluorescent probes, including lipophilic dyes, tracers, immunostaining, and fluorescent-protein labeling. Based on *Clear^{T2}*, Yu et al. developed a rapid and versatile clearing agent using a combination of triethanolamine and formamide, termed RTF [59], which achieved better clearing capability. Costantini et al. also introduced a versatile and simple clearing method based on 2,2'-thiodi-ethanol (TDE), and demonstrated hippocampus imaging with two-photon serial sectioning [21]. With the low viscosity and tunable refractive index of the water-based solution in varying ratios, TDE was tested on both the mouse brain and the formalin-fixed human dysplastic brain tissues with immunostaining. Economo et al. used DMSO/sorbitol to clear mouse brain tissues and combined it with STP to conduct whole-brain imaging, which realized tracing of neuronal axons through the whole brain [62]. Recently, Zhu et al. developed FOCM, which consists of DMSO, urea,

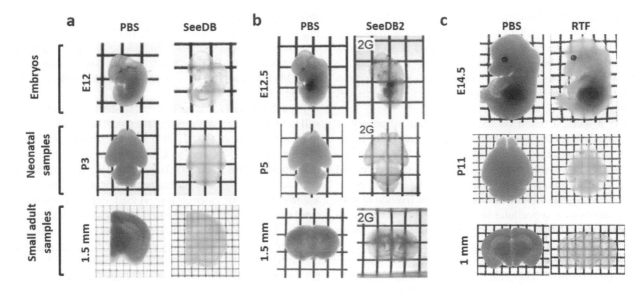

FIGURE 11.1 Transparency images of mouse embryos, neonatal brains, and adult mouse brain slices obtained with different methods (SeeDB, SeeDB2, and RTF). Images adapted from refserences [18, 58, 59].

D-sorbitol, and glycerol and can rapidly clarify 300-*μm* thick brain slices within 2 min [63].

RapiClear (Sunjin Lab, Taiwan) is a commercial water-soluble clearing agent for enhanced visualization of both fluorescence and nonfluorescence-labeled biological specimens. The product includes several versions with different refractive indexes. Though its composition is undisclosed, RapiClear provides an alternative method for viewing cell morphology in the tissues of mammals, plants, insects, and biomaterials by simply applying it in the mounting procedure, as claimed by SunJin Lab [64].

Intact and adult organ clearing

For large or intact organ clearing, Spalteholz first developed a clearing method using organic solvents a century ago [65], which damaged the superficial tissue areas and could not preserve fluorescence. Recently, Dodt's group has done much work in solvent-based clearing to achieve deep imaging of fluorescently labeled tissues [11, 12, 17, 66–71]. These approaches commonly include two steps, involving dehydration with lipid solvation and clearing by refractive index matching. At first, alcohol and BABB (benzyl alcohol and benzyl benzoate) were adopted for tissue clearing [11, 12, 66–68], which could rapidly decrease the fluorescent signal. Further screening found that tetrahydrofuran (THF) combined with BABB performed better in fluorescence preservation [17]. To achieve better transparency, dibenzyl ether (DBE) was used as a substitute for BABB combined with THF and formed a new protocol [69, 70], termed 3DISCO. Benefiting from the high transparency, the typical neuronal structure and vasculature of the whole mouse brain and spinal cord segment were obtained with 3DISCO combined with light-sheet fluorescence microscopy (LSFM). In addition, differing from other published solvent-based clearing methods based on BABB, FluoClearBABB [72] used 1-propanol or tert-butanol as the dehydration agent and

possessed the advantage of allowing long-term storage with no obvious fluorescent decrease. Combined with LSFM, it could also achieve brain-wide imaging. Ethanol-ECi was also proposed to clear various tissue organs, including bone, heart, and kidney, by using pH-adjusted ethanol for dehydration and nontoxic ethyl cinnamate for refractive index matching [73]. Recently, the Dodt group introduced a clearing protocol named sDISCO [74], achieving improved tissue transparency and longer clearing times with higher fluorescence preservation by the addition of the antioxidant propyl gallate in purified DBE or BABB.

Most methods were developed with the principle of preserving endogenous fluorescence, such as green fluorescent protein (GFP). With a distinct view, Renier et al. tried to visualize these proteins via whole-mount immunolabeling combining with 3DISCO and introduced iDISCO [75] for volume imaging of large intact tissues, including adult organs of rodents, such as brain and kidney, and even human embryonic and fetal specimens [76]. A lot of unprecedented molecularly labeled structures within multiple large and complex tissues were demonstrated with iDISCO. In the follow-up studies, the same group improved the original iDISCO protocol to reduce sample shrinkage and better preserve tissue morphology, called iDISCO+ [77].

However, the clearing and labeling of large adult human tissues or organs remained challenging. Roebroeck et al. reported a simple, fast, and low-cost method for the cytoarchitectonic labeling of human brain tissues, MASH, which was based on methanol pretreatment and hydrogen peroxide bleaching [78]. MASH was suitable for large formalin-fixed adult brain samples (4–5 cm thick) and compatible with small organic dyes, such as acridine orange, methylene blue, methyl green, or neutral red. Recently, Ertürk's group published SHANEL [79] to deeply label intact human organs based on CHAPS-mediated tissue permeabilization by forming small micelles, and achieved clearing by a combination of ethanol for dehydration, DCM (dichloromethane) for delipidation, and BABB for RI matching. This technique enabled deep molecular labeling and clearing of centimeters-sized human organs, such as the human brain and kidney [79].

As well as the solvent-based methods described above, a number of aqueous-based clearing methods for large tissues are constantly emerging.

Sca*l*e, a method based on the serendipitous discovery of urea's clearing ability, can render whole mouse brain transparent while preserving fluorescent signals [16]. With the customed long-working-distance objective lenses corrected for the clearing solution (Sca*l*eView, Olympus, Japan), Sca*l*e allows the imaging of intact brains and reconstruction of neuronal projections at subcellular resolution in combination with two-photon microscopy. However, the samples cleared with Sca*l*e are highly fragile and difficult to handle, with obvious tissue swelling.

By comprehensive chemical screening corresponding to the components of Sca*l*eA2, including polyhydric alcohols, detergents, and small hydrophilic molecules, Susaki et al. developed an efficient and scalable urea-based clearing protocol for whole-brain imaging, termed CUBIC [28, 80, 81], which involved a series of immersion steps over about two weeks. The CUBIC protocol could achieve good transparency in a reasonable time and was applicable to various fluorescent protein-labeled brains and immunostained tissues. Apart from rodent brains, it has also been used to reconstitute 3D images of nuclear-stained marmoset brain with LSFM. Following CUBIC, a series of modified protocols have been proposed. For instance, CUBIC-X uses imidazole and antipyrine reagents to expand the sample, thereby achieving efficient clearing of the whole brain, and was used to construct the 3D whole-brain atlas at single-cell-resolution [82]. Recently, the Ueda group updated the cocktails and described a series of CUBIC protocols (I–IV) [83] which enable the 3D imaging of mouse organs; mouse body, including bone; and even large primate and human tissues.

Additionally, Gomez-Gaviro et al. have optimized the CUBIC protocol by adjusting an incubation time for chicken embryo imaging studies [84]. Chen et al. developed a urea-based amino-sugar mixture including Meglumine, 1,3-Dimethyl-2-imidazolidinone, urea and Triton X-100, named UbasM, which can enable simple, volumetric imaging of whole adult mouse organs [85]. Li et al. also developed an easy-to-use method, C_e3D, by mixing N-methylacetamide, Histodenz with Triton X-100, which generates excellent tissue transparency for most organs [86].

Distinct from the solvent-based and urea-based methods mentioned above, CLARITY is a pioneer method based on tissue transformation utilizing hydrogel embedding [20, 87]. After hydrogel-tissue hybridization for stabilizing tissue structure, the samples were firstly treated with a customed electrophoretic tissue clearing (ETC) setup to actively remove lipids using strong ionic detergents (sodium dodecyl sulfate, SDS), then finally immersed in a refractive index matching solution (e.g., FocusClear). This method can render the tissues not only optically transparent but also macromolecule permeable, which enables multi-round molecular phenotyping in intact tissues. A piece of commercial equipment, named X-CLARITY (Logos

Biosystems, South Korea), has been manufactured based on this method. Further, advanced CLARITY [88] was proposed to process multiple samples simultaneously with multiplexed ETC setup. CLARITY-optimized light-sheet microscopy (COLM) [88] was also developed for fast data collection with high resolution.

Though CLARITY is an active lipid-removal method, it is still limited to a slow speed and by the risk of damaging samples. Epp et al. optimized the CLARITY procedure by combination of temperatures (37°C and 55°C) for clearing whole brains and other intact organs faster [50]. Bastrup et al. optimized the ETC chamber to rapidly clear the samples without any tissue damage [89]. Stochastic electrotransport (SE) was presented for the rapid clearing and staining of intact tissues with nondestructive process by using a rotational electric field [90]. It can achieve complete clearing of mouse organs within 1–3 days and staining with nuclear dyes, proteins, and antibodies within 1 day, but it requires complex customized devices. The MAP protocol [91] expands the tissues four- to five-fold for super-resolution imaging through hydrogel embedding, tissue denaturation, and expansion, which is also compatible with SE. Meanwhile, Chung's group introduced SWITCH [92] for molecular labeling of both animal and large human samples with no need for any special equipment. The SWITCH-processed samples can be rapidly cleared at elevated temperatures (e.g., 60–80°C), with adult mouse brain, rat brain, human and marmoset samples being demonstrated. They also demonstrated a minimum of 22 rounds of molecular labeling for a postmortem human tissue. However, they could not preserve endogenous fluorescence due to the harsh clearing procedure (e.g., too high a temperature). The clearing capability of several methods on various tissues and organs were demonstrated in Figure 11.2.

PACT [29] is a passive protocol based on CLARITY [20] by just incubation with no use of ETC set-up. PACT's effectiveness was tested on various tissues, including the brain, spinal cord, some internal organs (kidney, heart, lung, intestine, et al.), and even human basal-cell carcinoma tissues [29]. The authors further investigated its compatibility with fluorescent proteins, immunostaining, and smFISH, as well as the enhanced penetrating capability of immunolabels throughout the tissue blocks. PACT has been modified to expand tissues for better separation of compact structures (ePACT) [93]. It has also been optimized to clear difficult-to-image tissues such as bone by introduction of EDTA (PACT-deCAL) [93] or the combination of EDTA with amino alcohol (Bone CLARITY) [94] and lipid droplet–rich tissues by lipase assist [95]. Woo et al. proposed a faster method by adding α-thioglycerol to the original PACT clearing solution, called mPACT [31]. Its clearing process could also be accelerated with an appropriately elevated temperature [96]. Researchers have also made efforts to simplify the PACT clearing procedure. For instance, Sung et al. described a portable method, SCM, that could generate a tissue–hydrogel hybrid by increasing initiator concentration rather than utilizing vacuum pumping and nitrogen backfilling [97]. The Gentleman group even recommended the omission of the acrylamide-hydrogel as long as the tissue is fully fixed [98] and introduced FASTClear [99], which is a quite userfriendly protocol for human brain tissues with a maximum of 3 mm in thickness, and they further developed OPTIClear [100],

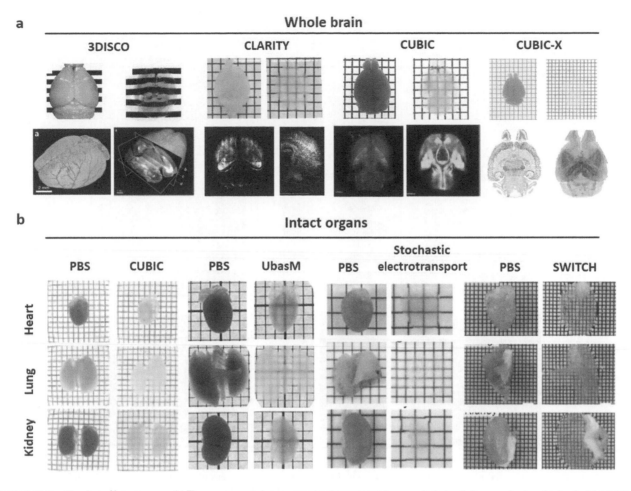

FIGURE 11.2 Clearing of intact organs. (a) Transparency and representative fluorescence images of whole adult mouse brains obtained with 3DISCO, CLARITY, CUBIC, and CUBIC-X. (b) Transparency images of other intact internal organs (heart, lung, and kidney) obtained with different methods (CUBIC, UbasM, Stochastic electrotransport, SWITCH). Adapted from references [28, 70, 80, 82, 85, 88, 90, 92].

a method optimized for fresh and archival human brain tissues, including formalin-fixed paraffin-embedded materials.

As is known, most clearing technologies chiefly focused on interrogating proteins over large volumes. To further access the biological information contained in RNA of large intact tissues, EDC-CLARITY was developed to retain RNAs by using carbodiimide-based chemistry [101]. This method enables multiplexed, volumetric visualization of diverse coding and noncoding RNAs in a variety of intact tissues, including mouse brain and human lobe resections. In recent years, Chung's group proposed SHIELD [102], which stabilized the tissue structure by polyepoxides and degreased by SDS immersion or combined with SE, followed by refractive index matching reagents. It can preserve protein fluorescence and antigenicity, transcription products, and tissue structures under harsh conditions, and has been used for studying human brain tissues and metastasis of breast cancer cells in mouse kidney [102]. The latest eFLASH method [103] first fixed tissue structures based on SHIELD, using SE for lipid removal and 3D labeling. By adjusting the pH of the labeling solution and the concentration of sodium deoxycholate (SDC) to control the binding affinity of the probe-target, it can quickly realize the complete and uniform labeling of various tissues, including mouse organs, marmoset brain tissues, and brain organoids.

For plant tissues, PEA-CLARITY was introduced by Palmer et al. to clear whole plant organs, and could improve the imaging depth and aid antibody penetration by the introduction of plant enzymes [104]. In addition, the Scale-like method and ClearSee, two urea-based methods, were developed by different groups especially for plant tissues [105, 106]. ClearSee can rapidly render plant tissues transparent, diminish chlorophyll autofluorescence, and preserve the fluorescence of fluorescent proteins (FPs). Combining with confocal and two-photon microscopy, the pistil, phloem, leaf, root tips, and whole seedling have been better visualized.

Whole-body clearing

Meanwhile, methodologies for whole-body clearing and imaging are constantly emerging. These whole-body clearing methods provide advanced opportunities for organ- and organism-scale histological analysis of multicellular systems [107].

The CB-Perfusion [28, 81] protocol enabled whole-body clearing and imaging of adult mice by a combination of direct transcardial perfusion of CUBIC cocktail and subsequent immersion over two weeks (Figure 11.3). By using LSFM, the single-cell resolution images were acquired by dissecting an adult body into four parts because adult mouse bodies

Whole-body clearing

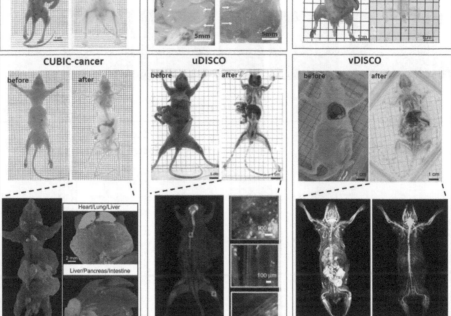

FIGURE 11.3 Transparency images of whole bodies obtained with CB-Perfusion (before: 8-week-old, after: 19-week-old), CUBIC-cancer (6-week-old), PARS (4~12-week-old), PEGASOS (2-month-old), uDISCO (12-week-old), and vDISCO (3~4-month-old), and the representative fluorescence images with CUBIC-cancer, uDISCO and vDISCO. Images adapted from references [22, 28, 29, 108, 109, 110].

could not fit into the existing setup. Then, Ueda's group also described an optimized protocol (CUBIC-cancer) by using a more efficient delipidation cocktail and higher refractive index matching solution for the detection of cancer metastasis in a whole-animal context at single-cell resolution [108]. Further, they developed a highly optimized 3D staining imaging pipeline, CUBIC-HV. It could uniformly label and clear whole adult mouse brains, an adult marmoset brain hemisphere, a tissue block of a postmortem adult human cerebellum, and an entire infant marmoset body, with dozens of antibodies and cell-impermeant nuclear stains [107].

Recently, Zhu's group developed MACS [111] based on MXDA and sorbitol, achieving ultrafast clearing speed, robust lipophilic dye compatibility, and fine morphology maintenance. It could clear the whole brain of an adult mouse in 2.5 days by convenient incubation, which was more than four times faster than the commonly used aqueous clearing methods, such as CUBIC; and allowed 3D imaging and visualization of various organs, such as rodent brain, spinal cord, spleen, and intestine.

PARS [29, 31, 93] was developed from the basic principle of CLARITY. It is another method using the intrinsic circulatory system (or cerebrospinal fluid route) to deliver the solutions and facilitate fast whole-body clearing and labeling. By optimizing the hydrogel embedding, clearing, and imaging agents,

PARS performed all the steps in situ prior to tissue extraction, including clearing and labeling. PARS can render intact whole organisms transparent (Figure 11.3) for imaging while preserving fluorescent signals and tissue architectures.

Thereafter, ACT-PRESTO [112] was reported by optimizing the conditions for tissue–hydrogel polymerization and electrophoresis in the original CLARITY protocol. It could clear various whole organisms, including mouse, zebrafish, and Xenopus, within hours to days with fine architecture and fluorescence preservation. The penetration of macromolecules deep into thick tissues, such as antibodies, could also be expedited by applying pressure in ACT-PRESTO. It has been applied for the clearing and imaging results of large volume tissues, such as rat or rabbit brains and human spinal cord segments.

As well as aqueous-based clearing methods, some solvent-based methods also achieved the transparency of whole bodies and demonstrated whole-body imaging. The latest PEGASOS [109] by Zhao's group introduced EDTA and Quadrol for decalcification and decolorization, used Quadrol to adjust the pH of tert-butanol for dehydration, and introduced PEGMMA500 and benzyl benzoate to match the refractive index, which could effectively clear both soft and hard tissues and turn the whole adult mouse body transparent (Figure 11.3). Ertürk's group developed uDISCO [22], a new method for shrinkage-mediated imaging of entire organs and even rodent bodies. By using tert-butanol as dehydration reagent and a mixture of BABB and diphenyl ether (DPE) with antioxidant α-tocopherol (Vitamin E) as refractive index matching agent, uDISCO can preserve fluorescent proteins over months and be notably better than 3DISCO. Except for the endogenous protein, it is also compatible with diverse labeling methods, including virus labeling and immunostaining. uDISCO can effectively clear not only the soft organs, such as the brain, spinal cord, heart, lung, and kidney, but also hard tissues such as calcified bone and even archival human tissue. The neuronal connections of the whole mouse were reconstructed with uDISCO, as shown in Figure 11.3.

The Zhu group proposed a modified clearing method, named FDISCO, by adjusting the temperature and pH of clearing reagents [113]. This method retained the advantages of 3DISCO, such as high clearing capability, meanwhile effectively overcoming the problem of fluorescence quenching. FDISCO had been applied to high-resolution 3D imaging and reconstruction of neuronal and vascular networks in various tissues and organs, and showed a potential for use in the imaging of fluorescent proteins with low expression [113]. The group also optimized uDISCO to develop a-uDISCO [114], increasing the fluorescence signal by about two times.

Recently, the Ertürk group developed a nanobody-based whole-body immunolabeling technology, named vDISCO [110], to boost the intensity of fluorescent proteins. It could enhance fluorescent signals more than 100 times and thereby allowed panoptic imaging of the whole body at subcellular resolution. vDISCO had been applied to visualize whole-body neuronal projections (Figure 11.3) and assess central nervous system trauma effects on neuronal projections and inflammatory processes in the entire mouse body [110]. Pan et al.

applied the vDISCO method to image cancer metastasis at single-cell resolution and to reveal the efficacy of therapeutic antibody targeting in entire mice [115].

There is no doubt that these methods developed for whole organisms can be applied to clear intact organs or even smaller samples. Figure 11.4 provides a rough guide for end-users to choose suitable clearing protocols according to the samples' size and age. Those methods are also marked with different colors based on the main chemical principles and clearing mechanisms [116].

Combination of different methods

The clearing agents and protocols mentioned above can potentially be shuffled and combined to develop better protocols or satisfy certain applications, based on their chemical compositions and putative functions in the clearing process. Many researchers have explored combining strategies for specific purposes, such as the combinations of TDE, SeeDB, Sca*le*CUBIC-R2, or RapiClear with CLARITY for economical refractive index matching alternatives [21, 112, 117]; fructose in SeeDB with urea in Sca*le* for low viscosity of clearing solutions (FRUIT [60]); RapiClear with Sca*le* for better transparency [64]; tert-butanol in FluoClearBABB with 3DISCO (uDISCO [22]) for better fluorescence preservation; SDS in CLARITY with detergents in iDISCO for immunostaining improvement (FASTClear [99]); and amino alcohol in CUBIC with CLARITY for decolorization of bone tissues (Bone CLARITY [94]). Recently, Miyawaki et al. [118] reported a method for 3D visualization of cerebral vascular networks, SeeNet, by combining hybrid hydrogel polymerization and SDC delipidation with Sca*le*CUBIC-R2 incubation for RI matching before imaging.

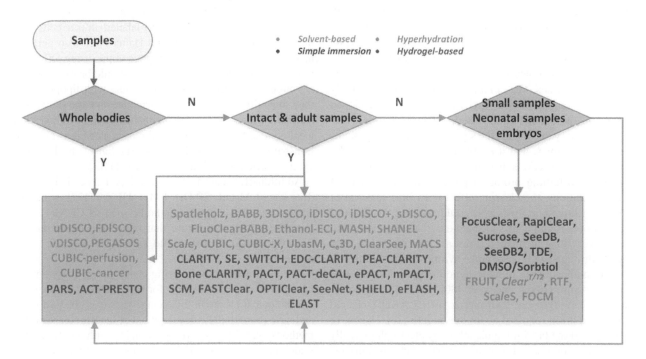

FIGURE 11.4 Gross flow chart for choosing the clearing protocols based on the size and age of samples. Font colors for different methodologies. Adapted from reference [116].

Though the combinations of different clearing agents or protocols showed effectiveness in certain experiments, it is still crucially important to discover and develop other reagents with better clearing performance, along with exploring fundamental mechanisms of tissue clearing in order to better understand how it happens. Table 11.1 listed and compared most of the reported *in vitro* tissue clearing methods.

Methods suitable for immunostaining and lipophilic dyes

In real applications, various probes might be used for fluorescent labeling of specific targets, including endogenous fluorescent proteins, immunofluorescent labels, chemical dyes, and so on [119]. The compatibility of the clearing methods and the probes mentioned above have been listed in Table 11.1.

Except for the transgenic labeling, immunostaining plays an important role for labeling specific molecules, structures, and cell types inside tissues. Compatibility with immunostaining is critical for the clearing methodologies and has been demonstrated in most methods. For some clearing protocols, the immunolabeling steps have been performed prior to clearing (e.g., FocusClear [15, 43, 44, 46], *Clear*^{T2} [19], SeeDB [18], iDISCO [75, 77], vDISCO [110], etc.) or final refractive index matching solutions (e.g., CUBIC [28, 80, 107], PACT [29, 93], CLARITY [20, 87, 88, 103], ScaleS [60], etc). It should be noted that it is quite challenging to get the molecular labels into deep tissues by relying on passive diffusion.

For the immunostaining protocols performed prior to clearing, MeOH, DMSO with hydrogen peroxide, repeated freeze–thaw procedures, or higher incubation temperatures are used for permeabilization in conventional whole-mount immunostaining protocols [75, 120–123]. Immunostaining results of various organs (e.g., mouse hemispheres, whole mouse embryos, brains, kidneys) have been demonstrated with iDISCO [75], which used dehydration–rehydration, an increase in incubation temperature, and higher-concentration DMSO (relative to previous whole-mount immunohistochemistry) to permeabilize the samples, but the technique is still limited by long incubation times. The immunolabeling of the whole body remained challenging. Recently, vDISCO [110] was developed as a pressure-driven, nanobody-based, whole-body immunolabeling technology to enhance the signal of fluorescent proteins more than 100 times, and it enabled visualization of whole-body neuronal projections in adult mice for the first time.

For immunostaining protocols performed before final RI matching solutions, the clearing effect with lipid removal or formation of large pore hydrogels could enhance the penetration of antibodies and hence accelerate the labeling protocols. However, very long incubation times are still needed to fully penetrate through the large samples within weeks. For example, the immunostaining of adult hypothalamus after CUBIC treatment required about one week [80], and the immunohistochemistry for tyrosine hydroxylase in CLARITY-cleared intact mouse brain required about one month [20]. Hence, Li et al. proposed the fast immunolabeling by electrophoretically driven infiltration [124]. Later on, stochastic electrotransport [90] was developed to overcome the slow penetration speed and uneven antibody distribution. The effectiveness of this

acceleration on CLARITY and CUBIC has been demonstrated, but it was not successful when combined with iDISCO. SWITCH [92] is another method developed for increasing antibody penetration and yielding a more uniform fluorescent signal distribution. It enables more than 20 rounds of relabeling of a postmortem human tissue for multiplexed imaging with precise coregistration. ACT-PRESTO [112] was also used to expedite antibody penetration into large samples by applying pressure. Chung's group reported a rapid method, eFLASH, that enables complete and uniform labeling of organ-scale tissue within one day by dynamically modulating chemical transport and reaction kinetics to establish system-wide uniform labeling conditions [103]. Additionally, a combination strategy has also been utilized; for example, Liu et al. [99] have combined CLARITY with iDISCO to improve immunolabeling based on delipidation and detergent permeabilization. Recently, Ueda's group experimentally evaluated broad 3D staining conditions by using an artificial tissue-mimicking material and then proposed a superior 3D staining protocol, CUBIC-HV, based on the combination of optimized conditions [107]. And the Chung group developed ELAST, a technology that transforms tissues into elastic hydrogels, which are highly stretchable and compressible, enabling reversible shape transformation and faster delivery of probes into intact tissue specimens via mechanical thinning [125].

Lipophilic dyes are commonly used to visualize the neuronal projections in the brain [18, 19, 60, 61] and vasculatures [126]. Taking DiI as an example, this dye was not compatible with many clearing protocols, such as 3DISCO, uDISCO, CUBIC, PACT, CLARITY, and so on. These incompatible clearing methods usually contained solvent agents or high-concentration detergents, such as Triton X-100 or SDS. FocusClear, *Clear*$^{T/T2}$, SeeDB, SeeDB2, FRUIT, and ScaleS allow imaging of DiI-labeled samples. For example, Kuwajima et al. labeled the retinal axon with DiI and cleared it with *Clear*$^{T/T2}$, Ke et al. obtained stacked images of DiI-labeled mitral cells in the olfactory bulb with SeeDB, and Hama et al. also performed neural tract tracing in mouse hippocampus using DiI in combination with ScaleS clarification. It should be mentioned that although these methods are compatible with lipophilic dyes, they show limited clearing performance on large samples such as whole-brain. The recently published MACS method showed ideal compatibility with lipophilic dyes as well as high tissue transparency, allowing the reconstruction of the DiI-labeled vascular structures of various organs [111].

Additionally, Jensen et al. developed modified lipophilic dyes (e.g., CM-DiI, SP-DiI, FM 1-43FX) [127], which were compatible with clearing methods, such as CLARITY. It is expected in the near future that more labeling markers will be developed to be resistant to various harsh tissue clearing processes.

The numerous tissue clearing techniques that are emerging have provided powerful tools for better understanding of various biological events in different tissues over the past decade, and have promoted the development of the life sciences. These methods were developed for samples with multiple spatial scales and ages, from small tissue blocks, embryos, and neonatal samples to intact adult organs, and further to whole rodent bodies. The classification of the clearing methods based on the

TABLE 11.1

Comparison of *in vitro* tissue optical clearing methods. Adapted from reference [116]

Method	Time to clear	Fluorescent labels		IM	Final RI	Morphology alternation	Complexity
		FPs	LDs				
Small samples/neonatal samples/embryos							
FocusClear	hours–days	yes	yes	yes	1.45	Shrinkage	Incubation
Sucrose	1 day	yes	no	yes	1.44	Shrinkage	Incubation
SeeDB	days	yes	yes	yes	1.50	no	Incubation
SeeDB2	days	yes	yes	yes	1.52	no	Incubation
FRUIT	days	yes	yes	N.T.	1.48	Slight expansion	Incubation
ScaleS	days	yes	yes	yes	1.44	Transient shrinkage / expansion	Incubation
Clear$^{T/T2}$	hours–days	yes	yes	yes	1.44	no	Incubation
RTF	hours–days	yes	yes	yes	1.46	no	Incubation
TDE	days–weeks	yes	no	yes	1.42	no	Incubation
DMSO /D-sorbitol	days	yes	no	N.T.	1.47	no	Incubation
FOCM	minutes–days	yes	yes	yes	1.49	no	Incubation
RapiClear	days–weeks	yes	N.T.	N.T.	1.47–1.55	no	Incubation
Large and adult samples							
BABB	days	yes	no	yes	1.55	Shrinkage	Incubation
3DISCO	hours–days	yes	no	yes	1.56	Shrinkage	Incubation
sDISCO	hours–days	yes	no	yes	1.56	Shrinkage	Incubation
iDISCO	hours–days	yes	no	yes	1.56	Shrinkage	Incubation
iDISCO+	days	yes	no	yes	1.56	Shrinkage	Incubation
MASH	days	no	no	yes	1.56	Shrinkage	Incubation
SHANEL	months	no	no	yes	1.56	Shrinkage	Perfusion, Incubation
FluoClearBABB	hours–days	yes	no	N.T.	1.56	Shrinkage	Incubation
Ethanol-ECi	hours–days	yes	no	yes	1.56	Shrinkage	Incubation
CLARITY	days	yes	no	yes	1.45	Transient expansion	Embedding, ETC
SeeNet	days–weeks	yes	no	yes	1.45	Transient expansion	Perfusion, Embedding, Incubation
PACT	days–weeks	yes	no	yes	1.42–1.48	Slight expansion	Embedding, Incubation
SCM	days–weeks	yes	no	yes	1.46	expansion	Embedding, Incubation
FASTClear	weeks	no	no	yes	1.42/1.56	no or shrinkage	Incubation
OPTIClear	days–months	yes	yes	yes	1.47–1.48	no	Embedding, Incubation
Stochastic electrotransport	days	yes	no	yes	1.46	Transient expansion	Embedding, ETC
SWITCH	days	no	yes	yes	1.47	Slight expansion	Embedding
SHIELD	weeks	yes	no	yes	1.46	Expansion	Embedding, ETC
eFLASH	days–weeks	yes	no	yes	1.46	Expansion	Embedding, ETC
ELAST	weeks	yes	no	yes	1.46–1.52	no	Embedding, Incubation
EDC-CLARITY	days	yes	no	yes	1.45	Transient expansion	Embedding, ETC
Bone CLARITY	weeks	yes	no	yes	1.47	no	Embedding, Incubation
PEA-CLARITY	weeks	yes	no	yes	1.33/1.45/1.46	no	Embedding, Incubation
Scale	weeks–months	yes	N.T.	yes	1.38	Expansion	Incubation
CUBIC	days–weeks	yes	no	yes	1.45	Transient expansion	Incubation
CUBIC-X	days–weeks	yes	no	yes	1.47	Expansion	Incubation
UbasM	days–weeks	yes	no	yes	1.47–1.48	Transient expansion	Incubation
C_e3D	weeks	yes	no	yes		Slight shrinkage	Incubation
ClearSee	days–weeks	yes	yes	N.T.	1.41	no	Incubation
Whole bodies							
PARS	days–weeks	yes	no	yes	1.42–1.48	no	Perfusion, Embedding

(Continued)

TABLE 11.1 (CONTINUED)

Comparison of *in vitro* tissue optical clearing methods. Adapted from reference [116]

Method	Time to clear	Fluorescent labels		IM	Final RI	Morphology alternation	Complexity
		FPs	LDs				
CB-Perfusion	days–weeks	yes	no	yes	1.45	Transient expansion	Perfusion, Incubation
CUBIC-cancer	days-weeks	yes	no	yes	1.44–1.52	Transient expansion	Perfusion, Incubation
CUBIC-HV	weeks	yes	no	yes	1.52	Transient expansion	Perfusion, Incubation
MACS	hours–days	yes	yes	yes	1.51	no	Incubation
ACT-PRESTO	hours–days	yes	no	yes	1.43–1.48	Expansion	Embedding, ETC
uDISCO	hours–days	yes	no	yes	1.56	Shrinkage	Incubation
FDISCO	hours–days	yes	no	yes	1.56	Shrinkage	Incubation
vDISCO	days–weeks	yes	no	yes	1.56	Shrinkage	Perfusion, Incubation
PEGASOS	hours–days	yes	no	yes	1.54	Shrinkage	Perfusion, Incubation

*Abbreviations: FPs, fluorescent proteins; RI, refractive index; LDs, lipophilic dyes; IM, immunostaining; Embedding, hydrogel embedding; ETC, electrophoretic tissue clearing.

sample features in terms of size and age would, from the perspective of the end-users, provide a general guide for researchers to choose suitable clearing methods.

Quantitative assessment of optical clearing methods in various intact mouse organs

To evaluate and compare the performances of clearing methods for different organs, Xu et al. [41] applied seven clearing methods, including 3DISCO, uDISCO, SeeDB, FRUIT, ScaleS, CUBIC, and PACT, to the intact brain, heart, liver, lung, kidney, spleen, stomach, small intestine, skin, and muscle. Then the clearing efficiency, sample deformation, fluorescence preservation, and imaging depth of these methods were quantitatively evaluated. Finally, based on the systemic evaluation of various parameters described above, the appropriate clearing method for specific organs, such as the kidney or intestine, was screened out. This section will examine the main results of the comprehensive evaluation of the performances of some typical clearing methods in various intact mouse organs.

Clearing time and transparency

Several typical optical clearing methods, including both solvent-based methods (3DISCO and uDISCO) and aqueous-based clearing methods (SeeDB, FRUIT, ScaleS, CUBIC, and PACT), were used to clear the adult mouse organs and tissues, which were divided into solid organs (brain, heart, liver, lung, kidney, and spleen), hollow organs (stomach and small intestine), and muscle and skin [41]. Figure 11.5 showed the clearing protocols of each method for solid organs and muscles.

For solid organs and muscles, the clearing protocols of aqueous clearing methods were conducted according to the published literature. However, the dehydration time of 3DISCO and uDISCO for solid organs (except brains and livers) and muscles were shortened to take account of the tradeoff between clearing efficiency and poor fluorescence

retention. For the hollow organs and skin, half of the clearing time specified in Figure 11.5 was used.

The transparency of all samples cleared with different clearing protocols are shown in Figure 11.6 and the quantitative data obtained by using a USAF 1951 Resolution Target are shown in Table 11.2. For solid organs and skeletal muscle, 3DISCO and uDISCO could render them transparent in the shortest time, except spleen, which contained too much residual blood. SeeDB, FRUIT, and ScaleS could only achieve weak transparency on them even when treated for a long time, while CUBIC and PACT could provide better transparency than other aqueous clearing methods, but took more clearing time. For hollow organs, all aqueous clearing methods provided better transparency compared to most solid organs, even with the shortened processing time, owing to their thinness. Because of the special physiological structure of stomach, the transparency of stomach fundus and stomach body was evaluated respectively, as shown in Table 11.2 [41].

As mentioned, for the samples containing residual blood, such as spleens, the transparency cleared with 3DISCO and uDISCO was rather limited due to the aggregation of heme. By contrast, Quadrol in ScaleCUBIC-1 reagent could remove heme effectively, and the SDS in PACT could also remove blood in samples. Thus, if there is much residual blood in the samples, it is suggested that researchers use CUBIC and PACT, but not SeeDB, FRUIT, or ScaleS, which have weak clearing capability for the large-volume solid organs and skeletal muscles. The hollow organs are easy to clear owing to their thin structures; however, if stomachs need to be cleared completely, 3DISCO and uDISCO are recommended [41].

Morphological retention

Size changes in organs and tissues after clearing were shown in Figure 11.7. For each organ, the size changes in tissues treated with different clearing protocols are different due to the different reagent compositions and clearing principles of each method. Furthermore, for each method, the

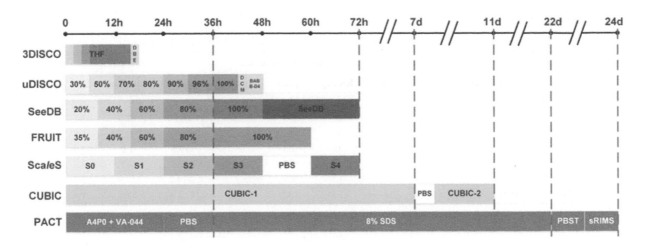

FIGURE 11.5 Clearing protocols of different methods for solid organs and muscles. Adapted from reference [41].

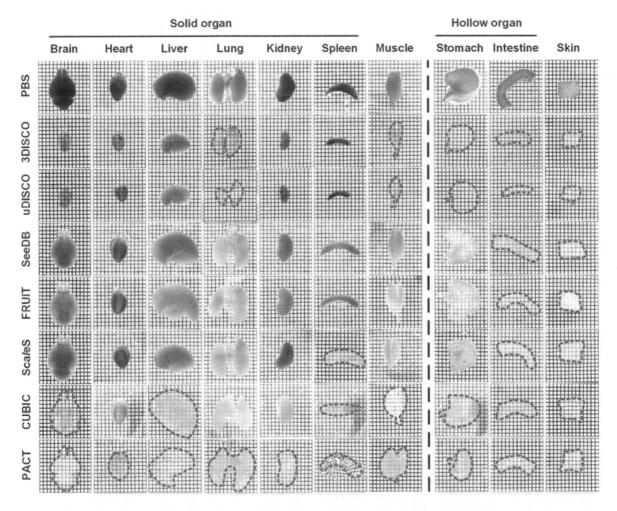

FIGURE 11.6 Comparison of the transparency of organs and tissues treated with different clearing methods. The red dotted line was used to outline the edge of cleared samples. Grid size, 1.45 mm × 1.45 mm. Reprinted with permission from reference [41].

size changes in different tissues are also different because of the differences in tissue structures and components [41]. For example, PACT-treated liver expanded by 40.4%, while skin after PACT shrank by 3.4%. The solvent-based clearing methods, such as 3DISCO and uDISCO, could cause severe shrinkage, which mainly resulted from harsh dehydration, but would be beneficial for large-volume imaging [22]. By contrast, most of the aqueous clearing methods could maintain the morphology of different tissues well, except for PACT, which induced serious expansion for most organs. The rSBR of images obtained by shortened protocol was slightly higher than that by long-time protocol [41].

TABLE 11.2

Quantitative assessment of tissue transparency with various methods. Data from Reference [41]

| | Mean Line – pairs per mm | | | | | | | | | |
	Brain	Heart	Liver	Lung	Kidney	Spleen	Muscle	Stomach[a]	Intestine	Skin
3DISCO	43.6	36.4	27.7	39.5	48.6	6.3	32.1	37.7/37.7	40.4	47.1
uDISCO	52.8	45.4	10.1	57	54.9	8.3	48.9	27.7/27.7	47.5	57
SeeDB	0	0	0	0	0	0	0	0.0/11.7	1.5	1.8
FRUIT	0	0	0	0	0	1.3	0	0.0/9.5	1.9	2.6
Sca*l*eS	0	0	0	0	0	8.2	0	0.0/7.4	2.7	3.3
CUBIC	18	4.9	2	1.6	3.5	9.7	3.7	1.2/14.2	17.9	3.3
PACT	19.6	9.5	2	1.8	3.4	16.6	10.8	2.0/12.86	17.3	1.4

[a] The transparency of stomach fundus and body were evaluated respectively.

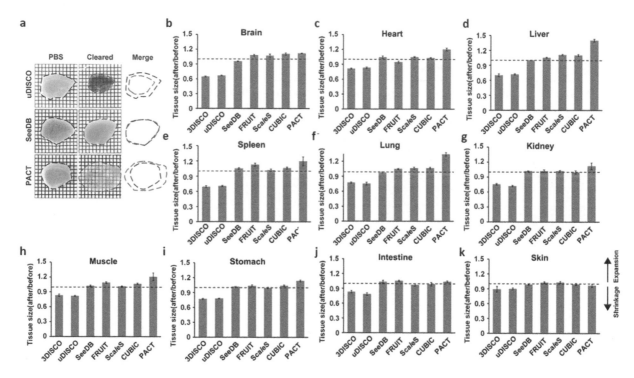

FIGURE 11.7 Quantitative comparison of sample deformation after clearing. (a) The borders of livers before and after clearing are outlined with blue and red dotted lines, respectively. (b–k) Quantification of the linear change for different organs and tissues. Adapted from reference [41].

Fluorescence preservation

The assessment of fluorescence preserving capability is an essential reference for the choice of clearing methods. Kiefer et al. only focused on the preservation of mOrange2 fluorescence protein and provided some qualitative data on fluorescence imaging [40]. Xu et al. investigated the compatibility of each method with genetic labeling and various chemical fluorescent dyes, including endogenous GFP (*Cx3cr1*-GFP line mice), DyLight 594-conjugated LEL, Alexa Fluor 647-conjugated α-BTX, and DAPI, by imaging the samples before and after clearing [41]. The fluorescence-preserving capability was expressed with the relative change of relative signal to noise ratio (rSBR) (Figure 11.8), which takes into account the influence on both signal and background.

For DAPI, SeeDB and Sca*l*eS maintained the rSBR at 68.3% and 68.7%, while other methods, such as CUBIC and PACT,

decreased the rSBR to less than 50.0% and even quenched fluorescence substantially. For DyLight 594-conjugated LEL, SeeDB showed the best fluorescence-preserving capability with a rSBR of up to 71.8%, which was higher than 3DISCO (51.8%) and PACT (45.4%) and the other methods (close to 30.0% or less). For Alexa Fluor 647-conjugated α-BTX, most of these methods demonstrated poor fluorescence preservation, especially CUBIC and PACT, which was thought to be due to the breaking of binding bond [41].

For endogenous GFP, FRUIT could maintain the rSBR at 86.4%, followed by Sca*l*eS and SeeDB with 68.6% and 62.2%, respectively, while CUBIC and PACT only maintained similar levels to 3DISCO and uDISCO (47.2% and 37.9%). It should be noted that the PACT clearing time used for the intestines in Figure 11.8 was 10 days, which was designed for the hollow organs with half of the clearing time used for solid organs. The rSBR of images obtained by shortened protocol was slightly

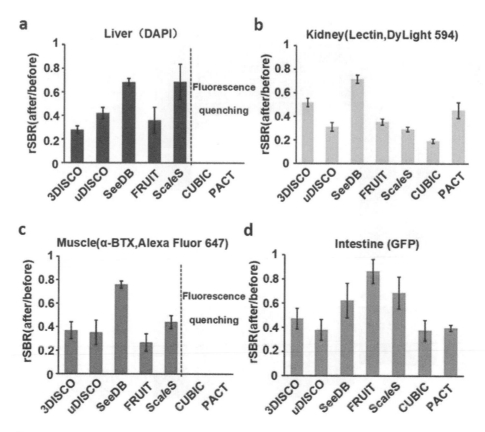

FIGURE 11.8 Quantification of fluorescence preservation for various clearing methods in liver, kidney, and gastrocnemius labeled respectively with DAPI, LEL, and α-BTX, as well as small intestine expressing GFP. Adapted from reference [41].

higher than that by long-time protocol [41]. The results of the treatment of 1 mm-thick brain slices (*Thy1*-GFP-M) using a 3.5-day PACT protocol showed that the total fluorescence intensity of signal after clearing was maintained to a relatively high level compared to that before clearing, while the calculated rSBR was low, indicating a decreased image quality after clearing [41]. It was supposed that the decreased image quality was due to PACT-induced sample expansion and background fluorescence enhancement. These results showed that each clearing method has its applicability for certain types of fluorescent labeling and imaging.

Imaging depth

The imaging depth is determined by tissue transparency and fluorescence intensity, and can directly reflect the clearing performance. To quantitatively compare the imaging depth of various methods, the kidney sections (solid organ) stained with LEL and small intestines (hollow organ) expressing endogenous GFP were imaged with confocal microscopy. Figure 11.9 showed the reconstructed XZ views of image stacks and the representative images at the deepest level where the signal was still available, and the standardized imaging depth based on deformation parameters.

For the kidney labeled with LEL, the imaging depth was greatly improved to ~1300 μm by 3DISCO and uDISCO. PACT increased the imaging depth on kidney slices up to ~680 μm, while the other aqueous clearing methods could only achieve

about 100–300 μm (Figure 11.9a and 9b). For small intestine, the difference in imaging depth achieved by all these clearing methods was less obvious than that of kidneys, and aqueous clearing methods provided similar values of imaging depth except SeeDB (Figure 11.9c and 9d). As 3DISCO and uDISCO are organic solvent-based methods with excellent clearing efficiency, their "imaging depth" for the intestines was determined by the thickness of the intestine walls and was not considered to be actual imaging depth [41].

It should be mentioned that the transparency of the kidneys after CUBIC and PACT were similar (Table 11.2), but imaging depth for CUBIC was less than that of PACT. This was due to the poor fluorescence retention of CUBIC for DyLight 594-conjugated LEL. By contrast, for small intestines (GFP), the transparency and fluorescence preservation of CUBIC and PACT were similar, so their imaging depth was at the same level [41].

To select a best-fit optical clearing method for specific organs, a comprehensive consideration from different viewpoints is required. For solid organs and muscles with large volumes and dense structures, 3DISCO, uDISCO, and PACT were preferred to clear the whole samples rather than other aqueous clearing methods. PACT was not considered due to complex processes and serious expansion, unless there was residual blood in samples. Thus, considering both transparency and fluorescence preservation, 3DISCO was selected to clear the whole kidney labeled with LEL. For hollow organs and skin, most methods can clear them effectively, so the methods that

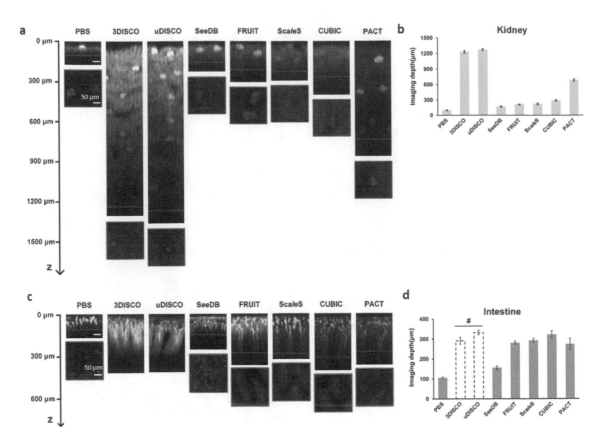

FIGURE 11.9 Comparison of fluorescence imaging depth for kidney sections (LEL) and the small intestine inner wall (GFP) treated with different methods. (a,c) XZ views of z-stacks and representative images at the indicated depth where the clear signal could be achieved for kidney (a) and small intestine (c). (b,d) Quantification of imaging depth. Adapted from reference [41].

can preserve fluorescence better were given priority. Thus, FRUIT was selected to clear small intestines of *Cx3cr1*-GFP mouse labeled with LEL, because it could not only provide good transparency, but also preserve GFP more stably [41].

Overall, the quantitative evaluation of several typical clearing methods on various organs and tissues provided important references for the selection of appropriate optical clearing methods in related research.

Summary

In recent years, tens of tissue optical clearing methods have emerged, providing multiple options for researchers to perform 3D visualization of various tissues and organs. However, for end-users, it is still difficult to select a best-fit optical clearing method from the numerous methods in practical applications.

This chapter reviews the latest *in vitro* clearing methods from the perspective of the end-users for multiscale tissues, from small tissue blocks, embryos, or neonatal samples to intact and adult organs, and further to whole rodent bodies. Then, taking some typical clearing methods (i.e., 3DISCO, uDISCO, SeeDB, FRUIT, ScaleS, CUBIC and PACT) as examples, the quantitative evaluation of their performances in various intact mouse tissues and organs (i.e., intact brain, heart, liver, lung, kidney, spleen, stomach, small intestine, skin, and muscle) were exhibited, including the clearing efficiency, sample deformation, fluorescence preservation, and imaging depth of these methods.

Either the brief review of clearing methods for multiscale tissues or the quantitative evaluation in various organs is expected to be helpful for researchers in selecting the most suitable clearing method to meet their specific needs.

Acknowledgments

T.Y. and D.Z. were supported by the National Key Research and Development Program of China grant number 2017YFA0700501, and NSFC grant numbers 61860206009, 81701354, 81961138015.

REFERENCES

1. D.S. Richardson, and J.W. Lichtman, "Clarifying tissue clearing," *Cell* **162**(2), 246–257 (2015).
2. E.A. Susaki, and H.R. Ueda, "Whole-body and whole-organ clearing and imaging techniques with single-cell resolution: Toward organism-level systems biology in mammals," *Cell Chem. Biol.* **23**(1), 137–157 (2016).
3. S.Y. Kim, K. Chung, and K. Deisseroth, "Light microscopy mapping of connections in the intact brain," *Trends Cognit. Sci.* **17**(12), 596–599 (2013).
4. A. Miyawaki, "Brain clearing for connectomics," *Microscopy* **64**(1), 5–8 (2015).

5. T. Ragan, L.R. Kadiri, K.U. Venkataraju, et al., "Serial two-photon tomography for automated ex vivo mouse brain imaging," *Nat. Methods* **9**(3), 255–258 (2012).

6. H. Gong, D. Xu, J. Yuan, et al., "High-throughput dual-colour precision imaging for brain-wide connectome with cytoarchitectonic landmarks at the cellular level," *Nat. Commun.* **7**, 12142 (2016).

7. F. Helmchen, and W. Denk, "Deep tissue two-photon microscopy," *Nat. Methods* **2**(12), 932–940 (2005).

8. J.A. Conchello, and J.W. Lichtman, "Optical sectioning microscopy," *Nat. Methods* **2**(12), 920–931 (2005).

9. W.R. Zipfel, R.M. Williams, and W.W. Webb, "Nonlinear magic: multiphoton microscopy in the biosciences," *Nat. Biotechnol.* **21**(11), 1369–1377 (2003).

10. J. Mertz, "Optical sectioning microscopy with planar or structured illumination," *Nat. Methods* **8**(10), 811–819 (2011).

11. H.-U. Dodt, U. Leischner, A. Schierloh, et al., "Ultramicroscopy: Three-dimensional visualization of neuronal networks in the whole mouse brain," *Nat. Methods* **4**(4), 331–336 (2007).

12. K. Becker, N. Jahrling, E.R. Kramer, F. Schnorrer, and H.-U. Dodt, "Ultramicroscopy: 3D reconstruction of large microscopical specimens," *J. Biophoton.* **1**(1), 36–42 (2008).

13. V.V. Tuchin, "Optical clearing of tissues and blood using the immersion method," *J Phys D Appl Phys* **38**(3), 2497–2518 (2005).

14. V.V. Tuchin, "Optical immersion as a new tool to control optical properties of tissues and blood," *Laser Phys.* **15**(8), 1109–1136 (2005).

15. H.H. Lin, J.S. Lai, A.L. Chin, Y.C. Chen, and A.S. Chiang, "A map of olfactory representation in the Drosophila mushroom body," *Cell* **128**(6), 1205–1217 (2007).

16. H. Hama, H. Kurokawa, H. Kawano, et al., "Scale: a chemical approach for fluorescence imaging and reconstruction of transparent mouse brain," *Nat. Neurosci.* **14**(11), 1481–1488 (2011).

17. A. Erturk, C.P. Mauch, F. Hellal, et al., "Three-dimensional imaging of the unsectioned adult spinal cord to assess axon regeneration and glial responses after injury," *Nat. Med.* **18**(1), 166–171 (2012).

18. M.T. Ke, S. Fujimoto, and T. Imai, "SeeDB: A simple and morphology-preserving optical clearing agent for neuronal circuit reconstruction," *Nat. Neurosci.* **16**(8), 1154–1161 (2013).

19. T. Kuwajima, A.A. Sitko, P. Bhansali, C. Jurgens, W. Guido, and C. Mason, "ClearT: a detergent- and solvent-free clearing method for neuronal and non-neuronal tissue," *Development* **140**(6), 1364–1368 (2013).

20. K. Chung, J. Wallace, S.Y. Kim, et al., "Structural and molecular interrogation of intact biological systems," *Nature* **497**(7449), 332–337 (2013).

21. I. Costantini, J.P. Ghobril, A.P. Di Giovanna, et al., "A versatile clearing agent for multi-modal brain imaging," *Sci. Rep.* **5**, 9808 (2015).

22. C. Pan, R. Cai, F.P. Quacquarelli, et al., "Shrinkage-mediated imaging of entire organs and organisms using uDISCO," *Nat. Methods* **13**(10), 859–867 (2016).

23. D. Zhu, J. Zhang, H. Cui, Z. Mao, P. Li, and Q. Luo, "Short-term and long-term effects of optical clearing agents on blood vessels in chick chorioallantoic membrane," *J. Biomed. Opt.* **13**(2), 021106 (2008).

24. J. Wang, D. Zhu, M. Chen, and X. Liu, "Assessment of optical clearing induced improvement of laser speckle contrast imaging," *J. Innov. Opt. Heal. Sci.* **03**(03), 159–167 (2010).

25. D. Zhu, J. Wang, Z. Zhi, X. Wen, and Q. Luo, "Imaging dermal blood flow through the intact rat skin with an optical clearing method," *J. Biomed. Opt.* **15**(2), 026008 (2010).

26. J. Wang, R. Shi, and D. Zhu, "Switchable skin window induced by optical clearing method for dermal blood flow imaging," *J. Biomed. Opt.* **18**(6), 061209 (2013).

27. R. Shi, M. Chen, V.V. Tuchin, and D. Zhu, "Accessing to arteriovenous blood flow dynamics response using combined laser speckle contrast imaging and skin optical clearing," *Biomed. Opt. Express* **6**(6), 1977–1989 (2015).

28. K. Tainaka, S.I. Kubota, T.Q. Suyama, et al., "Whole-body imaging with single-cell resolution by tissue decolorization," *Cell* **159**(4), 911–924 (2014).

29. B. Yang, J.B. Treweek, R.P. Kulkarni, et al., "Single-cell phenotyping within transparent intact tissue through whole-body clearing," *Cell* **158**(4), 945–958 (2014).

30. H. Lee, J.H. Park, I. Seo, S.H. Park, and S. Kim, "Improved application of the electrophoretic tissue clearing technology, CLARITY, to intact solid organs including brain, pancreas, liver, kidney, lung, and intestine," *BMC Dev. Biol.* **14**, 48 (2014).

31. J. Woo, M. Lee, J.M. Seo, H.S. Park, and Y.E. Cho, "Optimization of the optical transparency of rodent tissues by modified PACT-based passive clearing," *Exp. Mol. Med.* **48**(12), e274 (2016).

32. V.V. Tuchin, *Optical Clearing of Tissues and Blood*, SPIE Press, Bellingham, WA (2006).

33. V.V. Tuchin, "A clear vision for laser diagnostics (review)," *IEEE J. Sel. Top. Quantum Electron.* **13**(6), 1621–1628 (2007).

34. E.A. Genina, A.N. Bashkatov, and V.V. Tuchin, "Tissue optical immersion clearing," *Expert Rev. Med. Devices* **7**(6), 825–842 (2010).

35. K.V. Larin, M.G. Ghosn, A.N. Bashkatov, E.A. Genina, N.A. Trunina, and V.V. Tuchin, "Optical clearing for OCT image enhancement and in-depth monitoring of molecular diffusion," *IEEE J. Sel. Top. Quantum Electron.* **18**(3), 1244–1259 (2012).

36. E.A. Genina, A.N. Bashkatov, Yu.P. Sinichkin, I.Yu. Yanina, and V.V. Tuchin, "Optical clearing of biological tissues: prospects of application in medical diagnostics and phototherapy," *J. Biomed. Photon. Eng.* **1**(1), 22–58 (2015).

37. D. Zhu, K.V. Larin, Q. Luo, and V.V. Tuchin, "Recent progress in tissue optical clearing," *Laser Photonics. Rev.* **7**(5), 732–757 (2013).

38. K. Tainaka, A. Kuno, S.I. Kubota, T. Murakami, and H.R. Ueda, "Chemical principles in tissue clearing and staining protocols for whole-body cell profiling," *Annu. Rev. Cell Dev. Biol.* **32**, 713–741 (2016).

39. L. Silvestri, I. Costantini, L. Sacconi, and F.S. Pavone, "Clearing of fixed tissue: A review from a microscopist's perspective," *J. Biomed. Opt.* **21**(8), 081205 (2016).

40. M. Orlich, and F. Kiefer, "A qualitative comparison of ten tissue clearing techniques," *Histol. Histopathol.* **33**(2), 181–199 (2018).

41. J. Xu, Y. Ma, T. Yu, and D. Zhu, "Quantitative assessment of optical clearing methods in various intact mouse organs," *J. Biophoton.* **12**(2), e201800134 (2019).

42. A.S. Chiang, "Aqueous tissue clearing solution," United States Patent Number US 6472216 B1. (2002).

43. Y. Wang, A. Mamiya, A.S. Chiang, and Y. Zhong, "Imaging of an early memory trace in the Drosophila mushroom body," *J. Neurosci.* **28**(17), 4368–4376 (2008).

44. C.C. Chen, J.K. Wu, H.W. Lin, et al., "Visualizing long-term memory formation in two neurons of the Drosophila brain," *Science* **335**(6069), 678–685 (2012).

45. C.Y. Lin, C.C. Chuang, T.E. Hua, et al., "A comprehensive wiring diagram of the protocerebral bridge for visual information processing in the Drosophila brain," *Cell Rep.* **3**(5), 1739–1753 (2013).

46. J.S. Lai, S.J. Lo, B.J. Dickson, and A.S. Chiang, "Auditory circuit in the Drosophila brain," *Proc. Natl. Acad. Sci. U. S. A.* **109**(7), 2607–2612 (2012).

47. B.H. Lee, R. Johnston, Y. Yang, et al., "Studies of aberrant phyllotaxyl mutants of maize indicate complex interactions between auxin and cytokinin signaling in the shoot apical meristem," *Plant Physiol.* **150**(1), 205–216 (2009).

48. H. Diekmann, P. Kalbhen, and D. Fischer, "Characterization of optic nerve regeneration using transgenic zebrafish," *Front. Cell. Neurosci.* **9**, 118 (2015).

49. P.Y. Hsiao, C.L. Tsai, M.C. Chen, Y.Y. Lin, S.D. Yang, and A.S. Chiang, "Non-invasive manipulation of Drosophila behavior by two-photon excited red-activatable channelrhodopsin," *Biomed. Opt. Express* **6**(11), 4344–4352 (2015).

50. J.R. Epp, Y. Niibori, H.L. Liz Hsiang, et al., "Optimization of CLARITY for clearing whole-brain and other intact organs(1,2,3)," *eNeuro* **2**(3), 1–15 (2015).

51. A.J. Moy, B.V. Capulong, R.B. Saager, et al., "Optical properties of mouse brain tissue after optical clearing with FocusClear," *J. Biomed. Opt.* **20**(9), 95010 (2015).

52. J.H. Juang, C.H. Kuo, S.J. Peng, and S.C. Tang, "3-D imaging reveals participation of donor islet schwann cells and pericytes in islet transplantation and graft neurovascular regeneration," *EBioMedicine* **2**(2), 109–119 (2015).

53. E. Song, H. Seo, K. Choe, et al., "Optical clearing based cellular-level 3D visualization of intact lymph node cortex," *Biomed. Opt. Express* **6**(10), 4154–4164 (2015).

54. P.S. Tsai, J.P. Kaufhold, P. Blinder, et al., "Correlations of neuronal and microvascular densities in murine cortex revealed by direct counting and colocalization of nuclei and vessels," *J. Neurosci.* **29**(46), 14553–14570 (2009).

55. M.T. Ke, and T. Imai, "3D fluorescence imaging of the brain using an optical clearing agent, SeeDB," *Jikken-Igaku* **32**, 449–455 (2014).

56. M.T. Ke, and T. Imai, "Optical clearing of fixed brain samples using SeeDB," *Curr. Protoc. Neurosci.* **66**, 2–22 (2014).

57. M.T. Ke, "Optical clearing and deep-tissue fluorescence imaging using fructose," Ph.D. thesis, Kyoto University (2014).

58. M.T. Ke, Y. Nakai, S. Fujimoto, et al., "Super-resolution mapping of neuronal circuitry with an index-optimized clearing agent," *Cell Rep.* **14**(11), 2718–2732 (2016).

59. T. Yu, J. Zhu, Y. Li, et al., "RTF: A rapid and versatile tissue optical clearing method," *Sci. Rep.* **8**(1), 1964 (2018).

60. B. Hou, D. Zhang, S. Zhao, et al., "Scalable and DiI-compatible optical clearance of the mammalian brain," *Front. Neuroanat.* **9**, 19 (2015).

61. H. Hama, H. Hioki, K. Namiki, et al., "ScaleS: An optical clearing palette for biological imaging," *Nat. Neurosci.* **18**(10), 1518–1529 (2015).

62. M.N. Economo, N.G. Clack, L.D. Lavis, et al., "A platform for brain-wide imaging and reconstruction of individual neurons," *Elife* **5**, e10566 (2016).

63. X. Zhu, L. Huang, Y. Zheng, et al., "Ultrafast optical clearing method for three-dimensional imaging with cellular resolution," *Proc. Natl. Acad. Sci. U.S.A.* **116**(23), 11480–11489 (2019).

64. "http://www.sunjinlab.com/index.php."

65. W. Spalteholz, *Über das Durchsichtigmachen von menschlichen und tierischen Präparaten und seine theoretischen Bedingungen*, Leipzig, S. Hirzel (1914).

66. N. Jährling, K. Becker, E.R. Kramer, and H.-U. Dodt, "3D-Visualization of nerve fiber bundles by ultramicroscopy," *Med. Laser Appl.* **23**(4), 209–215 (2008).

67. N. Jahrling, K. Becker, and H.-U. Dodt, "3D-reconstruction of blood vessels by ultramicroscopy," *Organogenesis* **5**(4), 227–230 (2009).

68. N. Jahrling, K. Becker, C. Schonbauer, F. Schnorrer, and H.-U. Dodt, "Three-dimensional reconstruction and segmentation of intact Drosophila by ultramicroscopy," *Front. Syst. Neurosci.* **4**, 1 (2010).

69. K. Becker, N. Jahrling, S. Saghafi, R. Weiler, and H.-U. Dodt, "Chemical clearing and dehydration of GFP expressing mouse brains," *PLoS One* **7**(3), e33916 (2012).

70. A. Erturk, K. Becker, N. Jahrling, et al., "Three-dimensional imaging of solvent-cleared organs using 3DISCO," *Nat. Protoc.* **7**(11), 1983–1995 (2012).

71. P.J. Keller, and H.-U. Dodt, "Light sheet microscopy of living or cleared specimens," *Curr. Opin. Neurobiol.* **22**(1), 138–143 (2012).

72. M.K. Schwarz, A. Scherbarth, R. Sprengel, J. Engelhardt, P. Theer, and G. Giese, "Fluorescent-protein stabilization and high-resolution imaging of cleared, intact mouse brains," *PLoS One* **10**(5), e0124650 (2015).

73. A. Klingberg, A. Hasenberg, I. Ludwig-Portugall, et al., "Fully automated evaluation of total glomerular number and capillary tuft size in nephritic kidneys using light-sheet microscopy," *J. Am. Soc. Nephrol.* **28**(2), 452–459 (2017).

74. C. Hahn, K. Becker, S. Saghafi, et al., "High-resolution imaging of fluorescent whole mouse brains using stabilised organic media (sDISCO)," *J. Biophoton.* **12**(8), e201800368 (2019).

75. N. Renier, Z. Wu, D.J. Simon, J. Yang, P. Ariel, and M. Tessier-Lavigne, "iDISCO: A simple, rapid method to immunolabel large tissue samples for volume imaging," *Cell* **159**(4), 896–910 (2014).

76. M. Belle, D. Godefroy, G. Couly, et al., "Tridimensional visualization and analysis of early human development," *Cell* **169**(1), 161–173 e112 (2017).

77. N. Renier, E.L. Adams, C. Kirst, et al., "Mapping of brain activity by automated volume analysis of immediate early genes," *Cell* **165**(7), 1789–1802 (2016).

78. S. Hildebrand, A. Schueth, A. Herrler, R. Galuske, and A. Roebroeck, "Scalable labeling for cytoarchitectonic characterization of large optically cleared human neocortex samples," *Sci. Rep.* **9**(1), 10880 (2019).

79. S. Zhao, M.I. Todorov, R. Cai, et al., "Cellular and molecular probing of intact human organs," *Cell* **180**(4), 796–812 e719 (2020).

80. E.A. Susaki, K. Tainaka, D. Perrin, et al., "Whole-brain imaging with single-cell resolution using chemical cocktails and computational analysis," *Cell* **157**(3), 726–739 (2014).

81. E.A. Susaki, K. Tainaka, D. Perrin, H. Yukinaga, A. Kuno, and H.R. Ueda, "Advanced CUBIC protocols for whole-brain and whole-body clearing and imaging," *Nat. Protoc.* **10**(11), 1709–1727 (2015).

82. T.C. Murakami, T. Mano, S. Saikawa, et al., "A three-dimensional single-cell-resolution whole-brain atlas using CUBIC-X expansion microscopy and tissue clearing," *Nat. Neurosci.* **21**(4), 625–637 (2018).

83. K. Tainaka, T.C. Murakami, E.A. Susaki, et al., "Chemical landscape for tissue clearing based on hydrophilic reagents," *Cell Rep.* **24**(8), 2196–2210 e2199 (2018).

84. M.V. Gomez-Gaviro, E. Balaban, D. Bocancea, et al., "Optimized CUBIC protocol for 3D imaging of chicken embryos at single-cell resolution," *Development* **114**(11), 2092–2097 (2017).

85. L. Chen, G. Li, Y. Li, et al., "UbasM: An effective balanced optical clearing method for intact biomedical imaging," *Sci. Rep.* **7**(1), 12218 (2017).

86. W. Li, R.N. Germain, and M.Y. Gerner, "Multiplex, quantitative cellular analysis in large tissue volumes with clearing-enhanced 3D microscopy (Ce3D)," *Proc. Natl. Acad. Sci. U.S.A.* **114**(35), E7321–E7330 (2017).

87. K. Chung, and K. Deisseroth, "CLARITY for mapping the nervous system," *Nat. Methods* **10**(6), 508–513 (2013).

88. R. Tomer, L. Ye, B. Hsueh, and K. Deisseroth, "Advanced CLARITY for rapid and high-resolution imaging of intact tissues," *Nat. Protoc.* **9**(7), 1682–1697 (2014).

89. J. Bastrup, and P.H. Larsen, "Optimized CLARITY technique detects reduced parvalbumin density in a genetic model of schizophrenia," *J. Neurosci. Methods* **283**, 23–32 (2017).

90. S.Y. Kim, J.H. Cho, E. Murray, et al., "Stochastic electrotransport selectively enhances the transport of highly electromobile molecules," *Proc. Natl. Acad. Sci. U.S.A.* **112**(46), E6274–6283 (2015).

91. T. Ku, J. Swaney, J.Y. Park, et al., "Multiplexed and scalable super-resolution imaging of three-dimensional protein localization in size-adjustable tissues," *Nat. Biotechnol.* **34**(9), 973–981 (2016).

92. E. Murray, J.H. Cho, D. Goodwin, et al., "Simple, scalable proteomic imaging for high-dimensional profiling of intact systems," *Cell* **163**(6), 1500–1514 (2015).

93. J.B. Treweek, K.Y. Chan, N.C. Flytzanis, et al., "Whole-body tissue stabilization and selective extractions via tissue-hydrogel hybrids for high-resolution intact circuit mapping and phenotyping," *Nat. Protoc.* **10**(11), 1860–1896 (2015).

94. A. Greenbaum, K.Y. Chan, T. Dobreva, et al., "Bone CLARITY: Clearing, imaging, and computational analysis of osteoprogenitors within intact bone marrow," *Sci. Transl. Med.* **9**(387), (2017).

95. M. Lai, X. Li, J. Li, Y. Hu, D.M. Czajkowsky, and Z. Shao, "Improved clearing of lipid droplet-rich tissues for three-dimensional structural elucidation," *Acta Biochim. Biophys. Sin.* **49**(5), 465–467 (2017).

96. T. Yu, Y. Qi, J. Zhu, et al., "Elevated-temperature-induced acceleration of PACT clearing process of mouse brain tissue," *Sci. Rep.* **7**, 38848 (2017).

97. K. Sung, Y. Ding, J. Ma, et al., "Simplified three-dimensional tissue clearing and incorporation of colorimetric phenotyping," *Sci. Rep.* **6**, 30736 (2016).

98. H.M. Lai, A.K. Liu, W.L. Ng, et al., "Rationalisation and validation of an acrylamide-free procedure in three-dimensional histological imaging," *PLoS One* **11**(6), e0158628 (2016).

99. A.K.L. Liu, H.M. Lai, R.C. Chang, and S.M. Gentleman, "Free of acrylamide sodium dodecyl sulphate (SDS)-based tissue clearing (FASTClear): A novel protocol of tissue clearing for three-dimensional visualization of human brain tissues," *Neuropathol. Appl. Neurobiol.* **43**(4), 346–351 (2017).

100. H.M. Lai, A.K.L. Liu, H.H.M. Ng, et al., "Next generation histology methods for three-dimensional imaging of fresh and archival human brain tissues," *Nat. Commun.* **9**(1), 1066 (2018).

101. E.L. Sylwestrak, P. Rajasethupathy, M.A. Wright, A. Jaffe, and K. Deisseroth, "Multiplexed intact-tissue transcriptional analysis at cellular resolution," *Cell* **164**(4), 792–804 (2016).

102. Y.G. Park, C.H. Sohn, R. Chen, et al., "Protection of tissue physicochemical properties using polyfunctional cross-linkers," *Nat. Biotechnol.* **37**(1), 73–83 (2018).

103. D.H. Yun, Y.-G. Park, J.H. Cho, et al., "Ultrafast immunostaining of organ-scale tissues for scalable proteomic phenotyping," (2019). https://doi.org/10.1101/660373

104. W.M. Palmer, A.P. Martin, J.R. Flynn, et al., "PEA-CLARITY: 3D molecular imaging of whole plant organs," *Sci. Rep.* **5**, 13492 (2015).

105. D. Kurihara, Y. Mizuta, Y. Sato, and T. Higashiyama, "ClearSee: A rapid optical clearing reagent for whole-plant fluorescence imaging," *Development* **142**(23), 4168–4179 (2015).

106. C.A. Warner, M.L. Biedrzycki, S.S. Jacobs, R.J. Wisser, J.L. Caplan, and D.J. Sherrier, "An optical clearing technique for plant tissues allowing deep imaging and compatible with fluorescence microscopy," *Plant Physiol.* **166**(4), 1684–1687 (2014).

107. E.A. Susaki, C. Shimizu, A. Kuno, et al., "Versatile whole-organ/body staining and imaging based on electrolyte-gel properties of biological tissues," *Nat. Commun.* **11**(1), 1982 (2020).

108. S.I. Kubota, K. Takahashi, J. Nishida, et al., "Whole-body profiling of cancer metastasis with single-cell resolution," *Cell Rep.* **20**(1), 236–250 (2017).

109. D. Jing, S. Zhang, W. Luo, et al., "Tissue clearing of both hard and soft tissue organs with the PEGASOS method," *Cell Res.* **28**(8), 803–818 (2018).

110. R. Cai, C. Pan, A. Ghasemigharagoz, et al., "Panoptic imaging of transparent mice reveals whole-body neuronal projections and skull-meninges connections," *Nat. Neurosci.* **22**(2), 317–327 (2019).

111. J. Zhu, T. Yu, Y. Li, et al., "MACS: Rapid aqueous clearing system for 3D mapping of intact organs," *Adv. Sci.* **7**(8), 1903185 (2020).

112. E. Lee, J. Choi, Y. Jo, et al., "ACT-PRESTO: Rapid and consistent tissue clearing and labeling method for 3-dimensional (3D) imaging," *Sci. Rep.* **6**, 18631 (2016).

113. Y. Qi, T. Yu, J. Xu, et al., "FDISCO: Advanced solvent-based clearing method for imaging whole organs," *Sci. Adv.* **5**(1), eaau8355 (2019).

114. Y. Li, J. Xu, P. Wan, T. Yu, and D. Zhu, "Optimization of GFP fluorescence preservation by a modified uDISCO clearing protocol," *Front. Neuroanat.* **12**, 67 (2018).

115. C. Pan, O. Schoppe, A. Parra-Damas, et al., "Deep learning reveals cancer metastasis and therapeutic antibody targeting in the entire body," *Cell* **179**(7), 1661–1676 e1619 (2019).

116. T. Yu, Y. Qi, H. Gong, Q. Luo, and D. Zhu, "Optical clearing for multiscale biological tissues," *J. Biophoton.* **11**(2), e201700187 (2018).

117. L. Ye, W.E. Allen, K.R. Thompson, et al., "Wiring and molecular features of prefrontal ensembles representing distinct experiences," *Cell* **165**(7), 1776–1788 (2016).

118. T. Miyawaki, S. Morikawa, E.A. Susaki, et al., "Visualization and molecular characterization of whole-brain vascular networks with capillary resolution," *Nat. Commun.* **11**(1), 1104 (2020).

119. V. Marx, "Optimizing probes to image cleared tissue," *Nat. Methods* **13**(3), 205–209 (2016).

120. R.V. Sillitoe, and R. Hawkes, "Whole-mount immunohisto-chemistry: A high-throughput screen for patterning defects in the mouse cerebellum," *J. Histochem. Cytochem.* **50**(2), 235–244 (2002).

121. T. Yokomizo, T. Yamada-Inagawa, A.D. Yzaguirre, M.J. Chen, N.A. Speck, and E. Dzierzak, "Whole-mount three-dimensional imaging of internally localized immunostained cells within mouse embryos," *Nat. Protoc.* **7**(3), 421–431 (2012).

122. J.A. Gleave, J.P. Lerch, R.M. Henkelman, and B.J. Nieman, "A method for 3D immunostaining and optical imaging of the mouse brain demonstrated in neural progenitor cells," *PLoS One* **8**(8), e72039 (2013).

123. M. Belle, D. Godefroy, C. Dominici, et al., "A simple method for 3D analysis of immunolabeled axonal tracts in a transparent nervous system," *Cell Rep.* **9**(4), 1191–1201 (2014).

124. J. Li, D.M. Czajkowsky, X. Li, and Z. Shao, "Fast immuno-labeling by electrophoretically driven infiltration for intact tissue imaging," *Sci. Rep.* **5**, 10640 (2015).

125. T. Ku, W. Guan, N.B. Evans, et al., "Elasticizing tissues for reversible shape transformation and accelerated molecular labeling," *Nat. Methods* **17**(6), 609–613 (2020).

126. Y. Li, Y. Song, L. Zhao, G. Gaidosh, A.M. Laties, and R. Wen, "Direct labeling and visualization of blood vessels with lipophilic carbocyanine dye DiI," *Nat. Protoc.* **3**(11), 1703–1708 (2008).

127. K.H. Jensen and R.W. Berg, "CLARITY-compatible lipophilic dyes for electrode marking and neuronal tracing," *Sci. Rep.* **6**, 32674 (2016).

12

Ultrafast aqueous clearing methods for 3D imaging

Tingting Yu, Jingtan Zhu, and Dan Zhu

CONTENTS

Introduction

In the past decade, various tissue optical clearing methods have been developed for imaging deeper in biological tissues, providing powerful tools for visualizing three-dimensional structures in large-volume tissues [1–3]. Principally, they are divided into solvent-based clearing methods and aqueous-based clearing methods [4, 5]. Solvent-based methods can achieve a high level of tissue transparency and substantial tissue shrinkage within a short time, but are faced with the problem of quenching endogenous fluorescence signals, such

as 3DISCO [6]. In recent years, some advanced solvent-based methods, such as uDISCO [7], PEGASOS [8], FDISCO [9], sDISCO [10], and vDISCO [11], were proposed to overcome the problem. Simultaneously, some aqueous-based clearing methods were introduced for good fluorescence preservation and scalable tissue size maintenance. However, the optical clearing of many aqueous-based methods was generally slow and required a long incubation time for sufficient transparency, particularly for large tissues. For instance, it took nearly 1 month to transparentize the adult mouse brain by Scale [12]; CUBIC and its variants required about 1–2 weeks or more for

DOI: 10.1201/9781003025252-14

clearing [13–15]; and CLARITY, utilizing an electrophoretic tissue clearing (ETC) system, also took at least 1–2 weeks to clear the entire mouse brain [16, 17]. The slow clearing process prevented their application for large-volume tissues or whole organs, as well as for tissue slices that required fast processing.

To address this issue, many research groups have worked to develop rapid tissue clearing methods; some shortened the clearing time by screening different chemical mixtures, such as FOCM [18], *Clear*$^{T/T2}$ [19], RTF [20], Sca*l*eSQ [21], and MACS [22], and others used methods involving modifying the clearing conditions, such as ACT method [23]. These methods successfully achieved rapid clearing performance for multiscale tissues, from hundreds-micron-thick tissue slices and several-millimeter-thick tissue blocks to adult whole organs and even whole bodies, providing critical and rapid techniques for highly efficient research.

This chapter will introduce some ultrafast aqueous clearing protocols reported in recent years from the angle of their applicability in tissues at different scales.

Rapid and simple optical clearing methods for hundreds-micron-thick tissue sections

A rapid and simple clearing protocol would benefit for quick data acquisition in given experiments for the clearing of serial hundreds-micron-thick tissue sections. This section will introduce two clearing methods suitable for thin tissue sections and describe their performance in three-dimensional imaging at cellular resolution.

Rapid optical clearing method for brain sections based on sugar/sugar-alcohol

Sugar and sugar-alcohol had presented good clearing potential on skin tissues [24–26] and had also been used as clearing agents for brain tissues [27, 28]. To clear the thin brain sections, Yu et al. developed a rapid clearing method based on pure sugar/sugar-alcohol solution [29]. After screening out several sugars and sugar-alcohols such as glycerol, sorbitol, sucrose, and fructose, their clearing processes were recorded and compared, the fluorescence preservation and imaging depth improvement were also quantitatively analyzed, and then the preservation of structures was further investigated.

Sugar/sugar-alcohol solution renders brain sections transparent rapidly

Several sugars and sugar-alcohols commonly used in skin optical clearing were preselected due to their high clearing potential. During the clearing processes with different agents, the white-light images of thin brain sections (100-μm-thick) above the central area of 1951 USAF were recorded, as shown in Figure 12.1. The results showed that the striatum regions

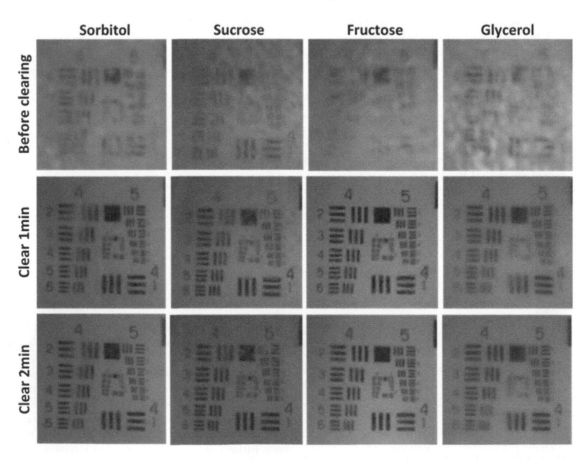

FIGURE 12.1 White-light images of brain sections before and after clearing with sorbitol, sucrose, fructose, and glycerol. Reprinted with permission from reference [29].

of the brain with abundant nerve fibers become transparent within 2 minutes by all tested agents. After treatment, the quantification based on the resolution of 1951 USAF target measurement showed that sorbitol (16.31 ± 1.12 μm), sucrose (16.95 ± 1.12 μm) and fructose (14.81 ± 1.21 μm) could achieve better transparency than glycerol (30.18 ± 1.97 μm), and fructose presented the best clearing effect [29].

Fluorescence preservation and imaging depth improvement with sugar/sugar-alcohol solutions

Figure 12.2a–b showed the quantification of fluorescence preservation of endogenous GFP and exogenous dye prodium iodide (PI) after clearing with different agents. The results showed that sorbitol and sucrose could enhance the mean GFP fluorescence as well as $Clear^{T2}$, while fructose kept the value almost as it was before clearing. For PI, the mean fluorescence intensity increases by 1.5, 2.3, and 3 times after clearing with sorbitol, sucrose, and fructose, respectively. In general, sorbitol and sucrose evidently increase both GFP and PI fluorescence intensity. Fructose can greatly increase the fluorescence intensity of PI while slightly decreasing the fluorescence intensity of GFP.

Then, to assess the imaging depth by different agents, confocal fluorescence images of *Thy1*-EGFP-M brain slices were obtained before and after clearing [29]. Figure 12.2c demonstrated the maximum intensity projections of the image stacks of samples cleared by sorbitol, indicating increased imaging depth after clearing. The imaging depth by the agents was obtained based on the contrast. As shown in Figure 12.2d, the imaging depth has been increased almost 3 fold for sorbitol, sucrose, and fructose, which is significantly higher than $Clear^{T2}$. In a word, treatment with the three clearing agents can significantly improve the optical imaging depth.

Size and morphology maintenance after clearing with sugar/sugar-alcohol solutions

As shown in Figure 12.3a, the contraction rates of samples after clearing with sorbitol, sucrose, and fructose were measured. The results demonstrated that there was no significant difference between the horizontal and longitudinal directions, which determined the basic structure preservation. To further evaluate the cell morphology preservation, a pyramidal neuron in the cortex was imaged before and after clearing with sorbitol (Figure 12.3b). The neuronal morphology remained intact, as seen from the images, and the soma, dendrite, and axon were preserved well. Sucrose and fructose showed fine structure preservation, as did sorbitol [29].

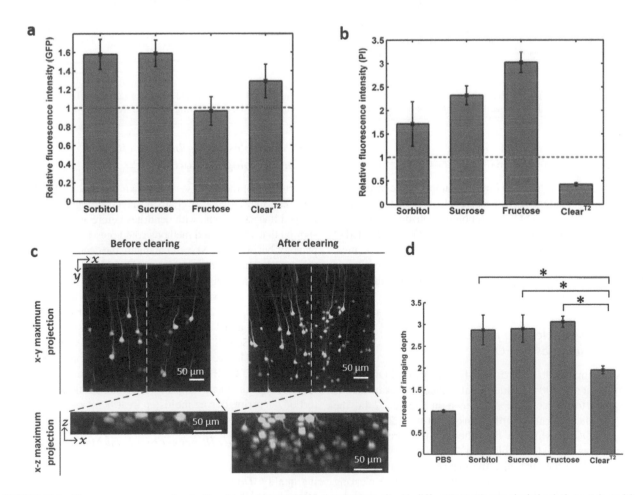

FIGURE 12.2 Fluorescence preservation (a–b) and improvement of imaging depth (c–d) with different sugar/sugar-alcohol solutions. Adapted from reference [29].

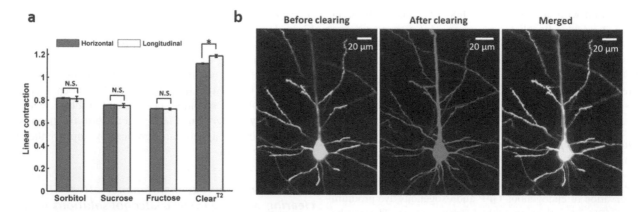

FIGURE 12.3 Size and morphology maintenance after clearing. (a) Contraction rates of brain slices in two directions. (b) Typical pyramidal neuron before and after sorbitol clearing. Adapted from reference [29].

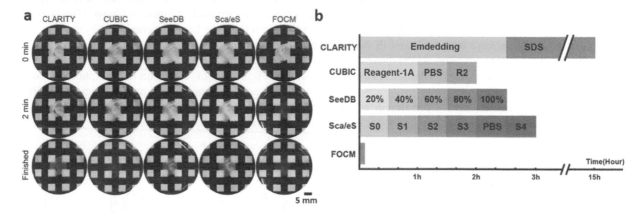

FIGURE 12.4 Ultrafast tissue clearing process of 300-μm-thick brain sections with FOCM. (a) Bright-field images of samples at 0 min, 2 min, and the end of clearing with different methods. (b) Comparison of schedules and clearing time with other clearing protocols. Adapted from reference [18].

FOCM: Ultrafast optical clearing method for brain sections

To overcome the limitation of time consumption in conventional optical clearing methods, Zhu et al. [18] also reported an ultrafast optical clearing method, FOCM, combining dehydration of water-miscible polar solvent and hydration of urea. FOCM consisted of four commonly used reagents: dimethyl sulfoxide (DMSO), urea, D-sorbitol, and glycerol. It could render a 300-μm-thick brain section transparent within 2 min, and it also showed high fluorescence preservation and involved a simple protocol.

Ultrafast tissue clearing process of thin brain sections by FOCM

With the aim of ultrafast tissue slice clearing, FOCM was implemented to 300-μm-thick brain slices to examine its clearing performance [18]. As shown in Figure 12.4a, the half brain slices turned highly transparent after 2-min incubation in FOCM, while for the other clearing protocols, the brain samples were still turbid after 2 mins of treatment. The comparison of schedules and the clearing time of different clearing protocols are shown in Figure 12.4b. The clearing process of FOCM reduced the clearing time by ~450 times versus CLARITY, ~60 times versus CUBIC, ~90 times versus ScaleS, and ~75 times versus SeeDB.

FOCM enables negligible size change and morphology distortion

The outlines of the brain section drawn before and after FOCM treatment showed good overlap, as shown in Figure 12.5a. The merged images of soma with dendrites before and after clearing showed that all the tested methods could preserve the soma morphology, but only FOCM could maintain the dendrites, indicating the excellent preserving capability of FOCM for fine structures.

The quantifications of linear expansion by different clearing methods were compared in Figure 12.5b. The results showed that there was only 2.12% tissue expansion for FOCM, while 22.0%, 15.9%, and 24.5% expansion occurred with CLARITY, CUBIC, and ScaleS, respectively, and 13.5% shrinkage with SeeDB [18]. Hence, the morphology distortion in dendrites in CLARITY, CUBIC, ScaleS, and SeeDB might be tissue expansion or shrinkage and the unavoidable human-made distortion occurring during multistep treatment.

FOCM improves fluorescence imaging quality

To examine the improvement of FOCM in fluorescence imaging quality, *Thy1*-EGFP-M mice were imaged with confocal microscopy [18]. Figure 12.6 showed that FOCM could significantly increase the imaging depth. Through the 200-μm-deep imaging volume, in PBS, the noise started to increase

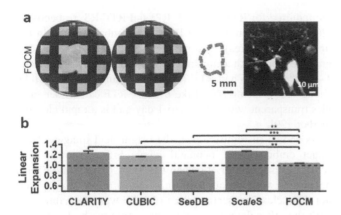

FIGURE 12.5 FOCM enables excellent tissue morphology preservation. (a) The outlines of brain sample (300-*μm*-thick, *Thy1*-EGFP-M) and the neurons imaging before and after clearing by FOCM. (b) Sample morphology changes after clearing with different methods. Adapted from reference [18].

at the depth of 60 *μm* and details were severely lost at a depth of 120–200 *μm*, whilst after FOCM clearing, the details of dendrites and cell bodies remained through the whole volume.

The image calculation showed that before clearing, the signal-to-background ratio (SBR) declined rapidly from 56.9 ± 2.6 to 15.2 ± 2.7 at a depth of 60–120 *μm*. By contrast, the SBR maintained at 50–80 through the 200-*μm* volume after FOCM clearing (Figure 12.6c). The above results showed that the FOCM fast clearing method could improve imaging depth by nearly three times, with SBR increasing at deep depth.

Raid optical clearing methods for embryos and millimeter-thick adult tissue blocks

For the clearing of embryos and millimeter-thick adult tissue blocks, several rapid clearing methods have been proposed. This section will introduce three typical clearing methods suitable for this type of tissue, including *Clear*$^{T/T2}$ [19], RTF [20], and Sca*l*eSQ [21].

A rapid clearing method based on formamide or formamide/polyethylene glycol

After observing the clearing process of 20-*μm*-thick cryosections of embryonic mouse brain during the *in situ* hybridization experiment, Kuwajima et al. [19] found the clearing potential of formamide, and proposed *Clear*^{T} and *Clear*^{T2} based on formamide and formamide/polyethylene glycol (PEG). The two clearing protocols could rapidly render embryos and neonatal brain sections transparent with minimal volume changes. *Clear*$^{T/T2}$ had been used to enhance visualization of cell morphology and connections in neuronal and nonneuronal tissue.

Clear$^{T/T2}$: Detergent- and solvent-free rapid tissue clearing methods

As the detergent- and solvent-free clearing methods, *Clear*^{T} and *Clear*^{T2} utilized formamide or formamide/PEG as the main clearing agents. For the *Clear*^{T} clearing procedure, the samples were incubated in 20%, 40%, 80%, and 95% formamide solutions (in PBS) gradually. For the *Clear*^{T2} clearing procedure, the samples were incubated in a 25% formamide/10% PEG solution followed by a 50% formamide/20% PEG solution. For both the *Clear*^{T} and *Clear*^{T2} methods, the incubation times in each step vary according to tissue thickness for the desired transparency.

The intact embryos, embryonic heads, and postnatal dissected brains have been cleared successfully by *Clear*^{T}. After gradually incubated in 20%, 40%, 80%, and 95% formamide for 30 min, 30 min, 2 h, and 30 min, followed by incubation in 95% formamide overnight, the *Clear*^{T} procedure achieved rapid clearing of the intact embryo (E14.5) and postnatal day 0 (P0) brain. Compared with Sca*l*eA2, *Clear*^{T} rendered embryonic brains transparent significantly faster. In addition, the clearing was reversible with PBS. The cleared embryonic head turned opaque after treated with PBS solution for 30 min, and the recovered samples could be stored for at least 1 month. This reversible effect facilitates long-term tissue storage in the case of unavoidably delayed research or repetitive observation.

Although the original *Clear*^{T} protocol effectively cleared the E14.5 embryo (actin-GFP), it reduced GFP fluorescence, and while the 50% formamide maintained fluorescence, it failed to clear embryos. Hence, *Clear*^{T2} was proposed to clear

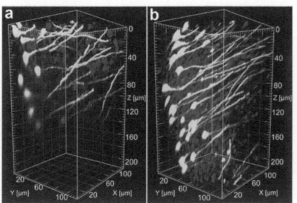

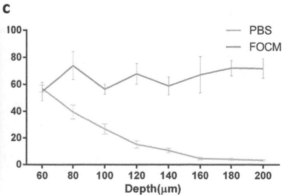

FIGURE 12.6 FOCM improves the fluorescence imaging quality. (a–b) 3D reconstruction of neurons in *Thy1*-EGFP-M mouse with DAPI staining before and after FOCM clearing. (c) SBR analysis from the depth of 60 *μm* to 200 *μm*. Adapted from reference [18].

the samples and maintain fluorescence. To stabilize the GFP fluorescence, PEG was mixed with formamide as the clearing agent in *Clear^{T2}*. With a graded series of formamide/PEG solutions, *Clear^{T2}* cleared embryos and maintained fluorescence. The transparency of embryonic heads and P0 brain sections treated with *Clear^T* is better than with *Clear^{T2}*. For both *Clear^T* and *Clear^{T2}*, the outline drawings of before and after clearing overlapped well, indicating minimal size changes.

Clear^T could achieve better transparency than *Clear^{T2}* and was compatible with dye tracers such as DiI and CTB, but it could not maintain the fluorescent signal of genetically encoded proteins. Therefore, tissue samples labeled with DiI or CTB alone are best cleared by *Clear^T*, and *Clear^{T2}* is suitable for transgenic labeled samples and immunostaining labeling, as well as dye tracers [19].

RTF: A rapid and versatile clearing method based on *Clear^{T2}*

As described before, *Clear^{T2}* was a relatively simple and rapid clearing method for preserving fluorescent signals of lipophilic dyes and immunostaining, as well as fluorescent proteins with no use of solvents or detergents [19]. Nevertheless, the transparency of tissues treated with *Clear^{T2}* was not sufficient, especially for adult brain blocks.

RTF [20] was a modified clearing method based on *Clear^{T2}*. It could achieve better transparency in both developing and adult brain tissues while retaining clearing rapidity.

Improved transparency of tissues by rapid RTF clearing

The RTF clearing reagents were composed of formamide, triethanolamine, and distilled water [20]. RTF-R1 refers to 30% triethanolamine, 40% formamide, 30% water; RTF-R2 refers to 60% triethanolamine, 25% formamide, 15% water; RTF-R3 refers to 70% triethanolamine, 15% formamide, 15% water. For the RTF clearing protocol, the samples were sequentially incubated in RTF-R1, RTF-R2, and RTF-R3. The incubation time in each solution depends on tissue type and thickness. For whole embryos or heads (E11-E15), 2–3 h, 2–3 h, and 5–14 h

are required for RTF-R1, RTF-R2, and RTF-R3, respectively [20]. For intact brains (E16-P12), the incubation time in the final solution increases to overnight or longer [20]. For adult brain sections with a thickness of 800–1500 *μm*, 30–50 min, 1–1.5 h, and 1–1.5 h in RTF-R1, RTF-R2, and RTF-R3 will be enough for clearing [20]. Like *Clear^{T2}*, RTF can render samples transparent within hours to 1 day and is a rapid clearing method [20].

As shown in Figure 12.7a, the bright-field images of the intact embryos and neonatal whole brains showed that RTF achieved better transparency than *Clear^{T2}*. RTF also performed well on adult brain sections while *Clear^{T2}* did not. The clearing performance of RTF was also compared with the other detergent- and solvent-free methods, such as SeeDB, ScaleSQ(0), and FRUIT. Figure 12.7b illustrated the clearing procedures of different clearing methods and the transparency of 1-mm-thick adult brain slices with each method. Overall, RTF showed similar transparency as SeeDB, ScaleSQ(0), and FRUIT, but required less incubation time. After the clearing procedure, the RTF-treated adult brain section was close to its original size (0.98 ± 0.05), and the size change showed no significant difference with *Clear^{T2}* (0.95 ± 0.03). In addition, RTF clearing is also reversible, as is *Clear^{T/T2}*.

RTF enables visualization of axons in intact embryos and neurons in embryonic brain

RTF is compatible with various fluorescent labeling methods, including endogenous fluorescent proteins, immunostaining, and DiI-labeling. Figure 12.8 showed the applications of RTF on 3D visualization of nerves and neurons in intact mouse embryo and embryonic brain. The whole-mount embryo (E12.5) was immunostained with antibody to neurofilament (anti-NF), then cleared with RTF and imaged with confocal microscopy. As shown in Figure 12.8a, the RTF-cleared embryo showed a more complete view of the axon tracts and arbors in the nervous system. The enlarged images of the whisker pad and forelimb indicated that more information deep in RTF-cleared tissues could be obtained to demonstrate the projection and innervation details.

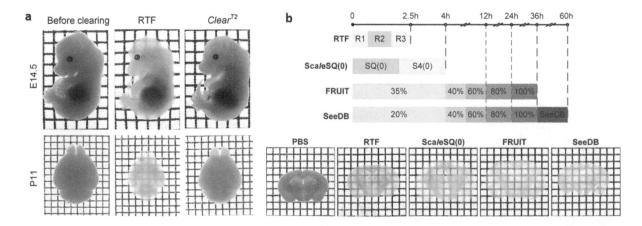

FIGURE 12.7 Rapid clearing using RTF. (a) Intact embryos (E14.5) and P11 whole brains cleared with RTF and *Clear^{T2}*. (b) Comparison of clearing performance on 1-mm-thick adult brain slices with other detergent- and solvent-free clearing methods. Grid size, 1.45 mm × 1.45 mm. Adapted from reference [20].

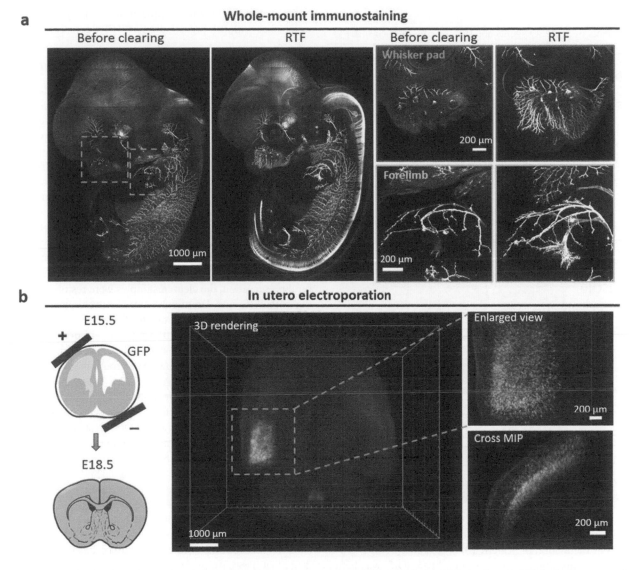

FIGURE 12.8 Visualization of axons in whole-mount embryo and neurons in embryonic brain by RTF clearing. (a) E12.5 whole mouse embryo immunostained with anti-NF. (b) After IUE at E15.5, The whole brain harvested at E18.5 after IUE at E15.5. Adapted from reference [20].

Figure 12.8b demonstrated the reconstruction of endogenous fluorescent proteins in the whole embryonic brain by RTF. The whole brains harvested at E18.5 after in utero electroporation (IUE) were cleared and imaged with Ultramicroscope. The reconstruction and resampling images clearly demonstrated the 3D distribution of transfected neurons in whole brain.

Overall, RTF demonstrated better tissue transparency and fluorescence preservation than *Clear^{T2}*, which was thought to be due to RTF clearing agents' higher refractive index (RI = 1.46 for RTF versus 1.41 for *Clear^{T2}*) and the alkalinity of triethanolamine in RTF solutions, respectively [20]. RTF is expected to be applied for observing neural development during different stages, such as the full-scale information of neuron localization and neuronal migration.

ScaleSQ: Rapid clearing of thick brain blocks

As a pioneering method utilizing hyperhydration of urea, Scale could obtain good transparency of tissues, but required weeks or even months to clear a whole mouse brain and caused obvious swelling and fragility. To address this issue, Miyawaki's group [21] further proposed ScaleS by introducing agents such as sorbitol and DMSO, and designed different experimental protocols for certain purposes, such as standard ScaleS, AbScale, ChemScale, and ScaleSQ. Among these protocols, ScaleSQ was a quick version of ScaleS aiming for rapid clearing of 1–2 mm-thick brain slices.

Rapid clearing process of ScaleSQ

ScaleSQ includes the ScaleSQ(0) protocol and ScaleSQ(5) protocol. The ingredients of ScaleSQ solutions and the corresponding clearing protocols are listed in Table 12.1. It should be noted that ScaleSQ solutions must be kept constantly at temperatures above 30°C due to the precipitating effect of high-concentration urea at lower temperatures [21].

ScaleSQ was developed as a rapid clearing protocol for applications requiring fast processing, such as high-throughput mapping of serial brain sections requiring brain slice thicknesses of 1–2 mm or less. Figure 12.9 showed the transmission

TABLE 12.1

Ingredients of ScaleSQ agents and the clearing protocols [21]

Reagents	Ingredients	ScaleSQ protocol
ScaleSQ(0) reagent	22.5% D-sorbitol; 9.1M urea	Clearing: ScaleSQ(0/5) (~2 h, 37°C)
ScaleSQ(5) reagent	22.5% D-sorbitol; 9.1M urea; 5% Triton X-100	Mounting: ScaleS4(0) (~2 h, RT)
ScaleS4(0) reagent	40% D-sorbitol; 10% glycerol; 4M urea; 15%–25% DMSO	

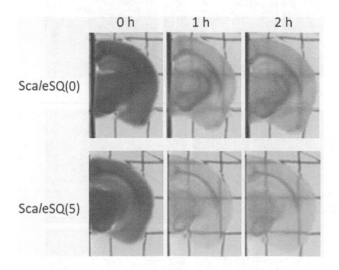

FIGURE 12.9   Transmission images of 1-mm-thick adult brain slices during incubation in ScaleSQ(0) or ScaleSQ(5). Adapted from reference [21].

images of 1-mm-thick brain slices of adult *Thy1*-EYFP-H mice after incubation in ScaleSQ(0) or ScaleSQ(5) for 1–2 h at 37°C. The transparency achieved with ScaleSQ(5) was better than ScaleSQ(0). This is because ScaleSQ(5) had the addition

of 5% Triton X-100, which allowed the complete clarification of the slice with full preservation of YFP fluorescence.

Preservation of fluorescence and ultrastructure

After incubation in ScaleSQ(0) or ScaleSQ(5) for 1–2 h, the brain slices were imaged with a fluorescence stereomicroscope. Then the cleared slices were incubated in ScaleS4(0) for another 2 h and imaged with light sheet microscopy. Figure 12.10 shows the 3D reconstruction of YFP-expressing neurons in the indicated cortical regions. For ScaleSQ(0)-cleared samples, the light-sheet imaging revealed that some opacity remained in the slice, while ScaleSQ(5) achieved better image quality with increased transparency.

Except for the clearing speed, ScaleS enabled the ultrastructural preservation for electron microscopy (EM), which had not been well addressed by other protocols. To investigate the ability of ScaleSQ(0) and ScaleSQ(5) in terms of ultrastructural preservation, the cleared samples were restored by washing with PBS, and 1-mm-cube samples were excised for the preparation of ultrathin sections for EM.

Figure 12.11 shows the transmission electron microscopy (TEM) observation of brain samples restored from ScaleSQ(0) and ScaleSQ(5). For the ScaleSQ(0) group, the excitatory synapse could be clearly observed by EM. ScaleSQ(0) treatment accurately preserved anatomical ultrastructure, and the substantial membrane integrity in ScaleSQ(0)-treated samples was confirmed by EM analysis. For ScaleSQ(5)-treated samples, there was some deterioration in ultrastructural preservation due to the use of a detergent, Triton X-100, and the excitatory synapse could also be observed by EM. The EM imaging and the transparency observation demonstrated the ultrastructural preservation/clearing trade-off.

In summary, these clearing methods, including *Clear^{T/T2}*, RTF, and ScaleSQ, are merely suitable for the clearing of embryos and millimeter-thick adult tissue blocks. There is no doubt that they have made certain contributions to the research fields of development and neuroscience. However, the clearing

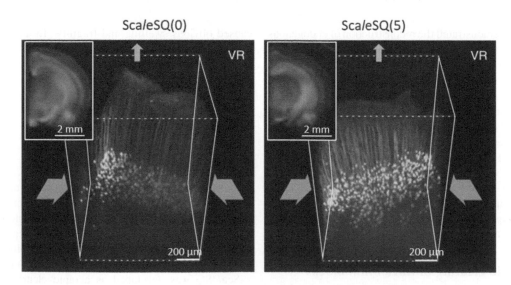

FIGURE 12.10   Fluorescence images and 3D reconstruction of neurons in cortex of *Thy1*-EYFP-H brain sections after clearing with ScaleSQ(0) or ScaleSQ(5). Adapted from reference [21].

Restored from Sca/eSQ(0)   Restored from Sca/eSQ(5)

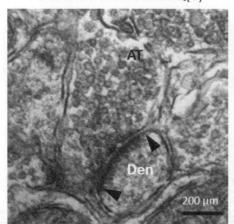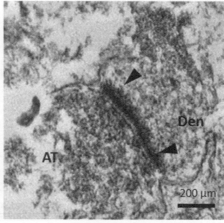

FIGURE 12.11   TEM observation of brain samples restored from Sca/eSQ(0) or Sca/eSQ(5). Reprinted with permission from reference [21].

capability of these methods on intact organs, such as the whole mouse brain, is still limited. The next section will introduce some rapid clearing methods for intact organs or even whole bodies.

Rapid clearing methods for 3D mapping of intact organs

Up to now, a variety of aqueous clearing methods applicable to intact organs or even entire bodies have been reported, such as the well-known CLARITY-series and CUBIC-series techniques, which are gel-embedding-based and hyperhydration-based clearing methods, respectively. Both of these suffered from a long clearing time for large-volume tissues. This section will introduce a rapid gel-embedding-based clearing method, ACT, and will then describe the advanced hyperhydration-based clearing system MACS. As different types of methods, both ACT and MACS performed excellently in terms of clearing speed for intact organs.

ACT: A rapid and scalable clearing method for large samples

The hydrogel-based clearing methods include a protein and acrylamide crosslinking step that selectively immobilizes proteins and other macromolecules, resulting in a tissue-embedded hydrogel. Lipid components are selectively removed either by passive diffusion [30, 31] or actively by electrophoresis [16, 17]. Even with active electrophoresis, the entire clearing process was still slow. Taking CLARITY as an example, clearing the entire mouse brain required at least 1–2 weeks [16]. To overcome this problem, Lee et al. proposed the rapid ACT method [23] to clear large tissues by modifying conditions for tissue–hydrogel polymerization and electrophoresis. It could clear large organs or whole bodies within 1 day while preserving tissue architecture and fluorescent protein signals.

ACT clearing system

ACT is an optimized clearing protocol based on CLARITY, which works by modifying conditions for tissue–hydrogel

polymerization and electrophoresis [23]. Like the original CLARITY, for ACT, the samples were first fixed with paraformaldehyde (PFA) followed by acrylamide infusion without bis-acrylamide. Clearing using ACT results in less protein–acrylamide crosslinking compared to that of CLARITY, resulting in a higher porosity hydrogel to allow rapid extraction of lipids and better diffusion of macromolecules [23]. After polymerization, the samples were actively cleared with an electrophoresis step. Figure 12.12a shows a diagram of the ETC system of ACT.

To address the problem of inconsistency during ETC, a modified version of the ETC chamber system using a platinum plate was designed to generate a dense regular current [23]. The dimensions of the ACT–ETC chamber, containing the electrode area, electrode–electrode distance, and inner-chamber dimensions, were different from the original CLARITY-ETC chamber; the detailed values are listed in Figure 12.12b. With this design of ETC chamber, the clearing solution was maintained without changing pH (Figure 12.12c) or color during the extended ETC period. Moreover, the temperature controlling panel greatly reduced heat generation. In addition, the long ETC chamber allowed all air bubbles to float to the top where they were removed through the top outlet of the chamber. ACT could rapidly obtain cleared tissue without tissue surface burning, collapse, or protein loss [23].

Comparison of ACT with other methods

To compare the efficacy of ACT with that of other clearing methods, including SeeDB, Sca/eA2, CUBIC, BABB, iDISCO, CLARITY, and PACT, 1-mm-thick brain sections were used. For 1-mm-thick brain sections, 2 hours of ACT clearing were sufficient to achieve nearly complete optical transparency, while the other methods required 1–3 days to achieve comparable optical transparency (Figure 12.13c-e). The ACT-processed samples expanded about 80% in size, like the other gel-embedding methods such as CLARITY and PACT, and returned to their original size after in mounting media such as RIMS (Figure 12.13f). These results showed that the ACT method was a speedy clearing process with recovery to the original size.

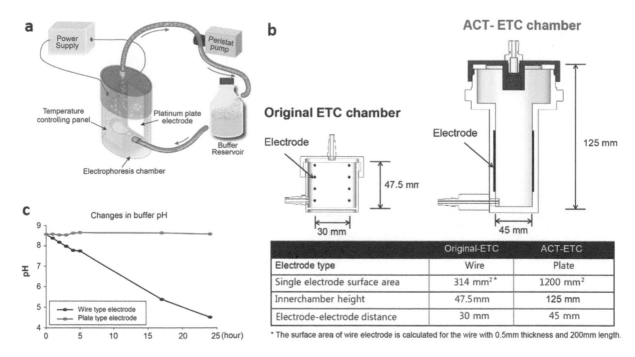

FIGURE 12.12 ACT clearing system and comparison of ETC setup with original CLARITY. (a) Diagram of the ACT-ETC system. (b) Dimensions of the ACT-ETC chamber and the original CLARITY-ETC chamber. (c) Changes in pH of the clearing buffers during ETC. Adapted from reference [23].

Scalability of ACT for clearing of various whole organs and whole bodies

For a mouse brain, 6 h of ETC made it nearly completely transparent, and ACT could also be used to clear larger brains within a reasonable time, as shown in Figure 12.14. To sufficiently clear an adult rat brain and a rabbit brain hemisphere, which are about 4 times and 23 times larger than an adult mouse brain, respectively, 15 h and 50 h were required. Human spinal cord tissue (1.3 cm diameter) fixed with formalin was rendered transparent after 100 h of ACT clearing. It should be noted that the time mentioned here did not include the time for fixation and RI matching in RIMS or CUBIC-mount solution, which took several hours to days depending on tissue size [23].

The short processing time of ACT makes it applicable to much larger specimens, such as the whole body (Figure 12.15). The whole mouse body (3 weeks old) with skin removed could be cleared by ACT within 24 h followed by 3 days incubation in CUBIC-mount solution. As shown in Figure 12.15 with ACT, substantial transparency was achieved in many other organisms, such as zebrafish, rat embryos, chicken, *Xenopus*, and small octopus.

In a word, ACT is scalable for large organs or even whole bodies of adult animals.

Ultrafast aqueous clearing system for 3D mapping of intact organs

For the clearing of large tissues, methods based on hyperhydration of urea and urea-like chemicals play an important role. The most representative method is CUBIC and its series protocols [13–15, 32], which could efficiently achieve high transparency in mouse organs, mouse body, and even large primate and

human tissues. However, they generally required 1–2 weeks or more for sufficient clearing. In addition, few clearing methods are compatible with widely used lipophilic dyes while maintaining high clearing performance due to the use of organic solvents or high concentration detergents. Recently, Zhu et al. developed MACS [22], a rapid and highly efficient aqueous clearing method based on m-xylylenediamine (MXDA). MACS achieved high transparency in intact organs and rodent bodies in a fairly short time (e.g., 2.5 days for an adult whole brain) only by simple incubation, and showed ideal compatibility with multiple probes, especially for lipophilic dyes.

Clearing protocol of MACS

MACS clearing cocktails mainly consist of MXDA and sorbitol. The former is a colorless and water-miscible liquid with a high RI, and it has two NH_2 groups similar to urea, indicating good potential in tissue clearing. The addition of sorbitol provided increased performance in both clearing effect and fluorescence preservation. To achieve high transparency, MACS was designed as a tri-step protocol with three cocktails: MACS-R0 (20 vol% MXDA and 15% w/v sorbitol in distilled water, RI = 1.40), MACS-R1 (40 vol% MXDA and 30% w/v sorbitol in PBS, RI = 1.48), and MACS-R2 (40 vol% MXDA and 50% w/v sorbitol in distilled water, RI = 1.51).

The incubation time for each cocktail varied for different kinds of tissues and organs, as shown in Figure 12.16. With MACS, an intact mouse embryo could be cleared within hours. For an intact adult brain, MACS could render it highly transparent within only 2.5 days. Notably, for hard tissues or whole bodies, EDTA incubation at 37°C was introduced for decalcification prior to MACS-R0 incubation. In practical experiments,

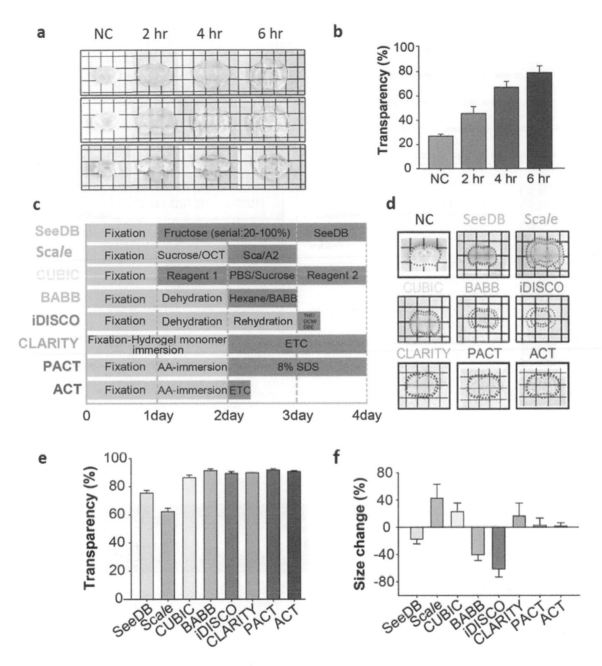

FIGURE 12.13 Clearing performance of ACT and comparison with other clearing methods. (a–b) The ACT-cleared whole-brains were cut into 2-mm-thick coronal slices to observe and measure transparency over time. Grid size, 5 mm × 5 mm. (c) Comparison of procedures, processing times of different methods for 1-mm thick brain blocks. (d) Bright-field images of cleared brain blocks. Dotted green lines indicate original sizes of blocks and red lines mark sizes after clearing. (e) Normalized transparency of cleared slices. (f) Size changes after clearing. Adapted from reference [23].

the time of each step could be adjusted based on the real-time clearing performance.

Comparison of MACS with other clearing protocols

The clearing performance of MACS was compared with some typical clearing protocols, including SeeDB2, Sca/eS, CUBIC-L/R, PACT, and uDISCO. As shown in Figure 12.17a, MACS took only 6 h to clear 2-mm-thick brain blocks, and the achieved transparency was equivalent to CUBIC-L/R, which took about 60 h. For an adult whole mouse brain, MACS took

only 2.5 days to achieve a high level of transparency, while it took 10 days for CUBIC-L/R and 20 days for PACT. The transparency achieved by SeeDB2 and Sca/eS was limited even with long clearing time.

Figure 12.17c was the bar graph for the clearing time for the whole brain of different methods, showing that MACS was much faster than other available clearing methods. The transmittance of the whole brain showed the equivalent transparency between MACS and CUBIC-L/R (Figure 12.17b). In addition, the quantification of linear expansion induced by each method indicated that MACS nearly maintained the

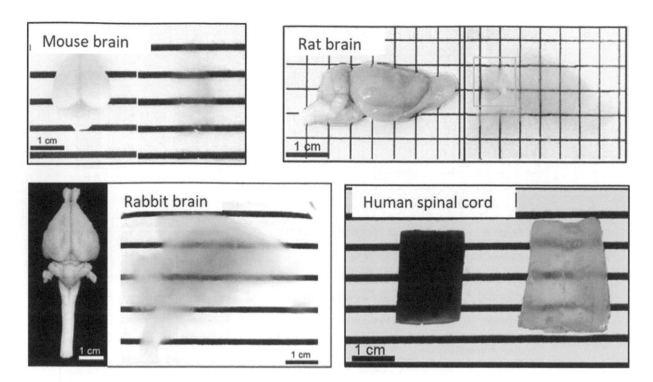

FIGURE 12.14   Whole organ clearing of different species by ACT. Adapted from reference [23].

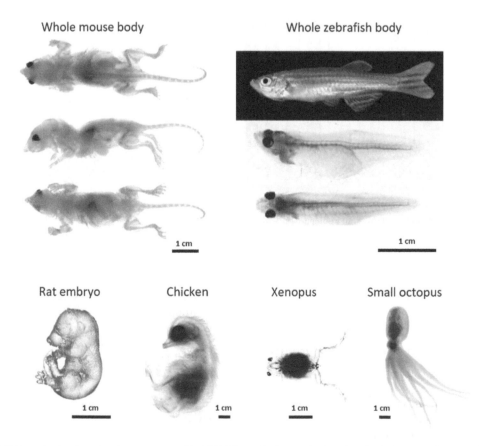

FIGURE 12.15   Whole-body clearing of various organisms by ACT followed by immersing in refractive index matching solution, including a whole mouse body (3 weeks old), whole zebrafish body, whole rat embryo body (E18), chicken, *Xenopus*, and small octopus. Adapted from reference [23].

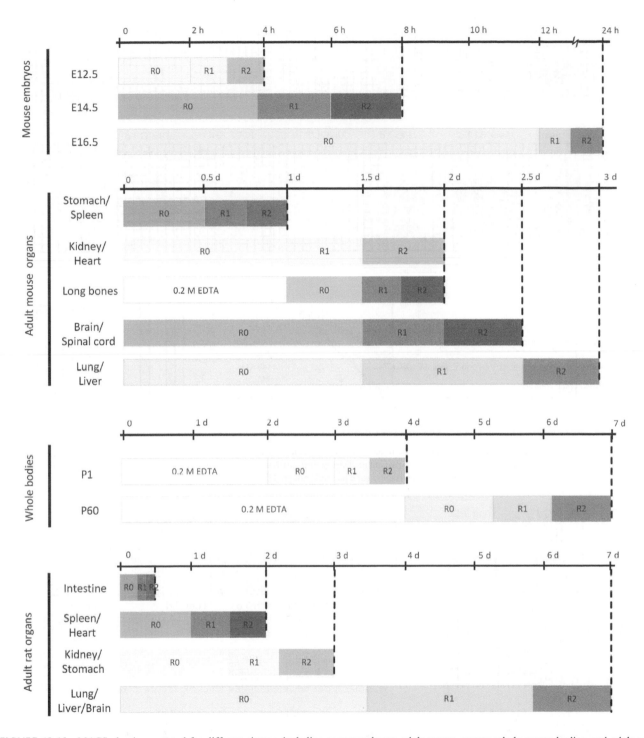

FIGURE 12.16 MACS clearing protocol for different tissues, including mouse embryos, adult mouse organs, whole mouse bodies, and adult rat organs. Adapted from reference [22].

sample size after transient expansion, while the other tested methods showed either obvious tissue expansion or substantial shrinkage (Figure 12.17e).

Applicability of MACS for various whole organs and whole bodies

MACS can be applied to various whole organs and whole bodies. Figure 12.18 shows bright-field images of different tissues

before and after clearing by MACS. The images showed that MACS could efficiently clear both soft mouse internal organs such as heart, lung, and spleen, and hard bones such as femur. For larger rat organs, MACS also showed good clearing performance with the brain, heart, kidney, liver, intestine, stomach, lung, and spleen, they turned optically transparent in a reasonable time. For whole bodies, MACS demonstrated superior ability in clearing mouse embryos and pups, and it also achieved high level of transparency in the adult P60 mouse

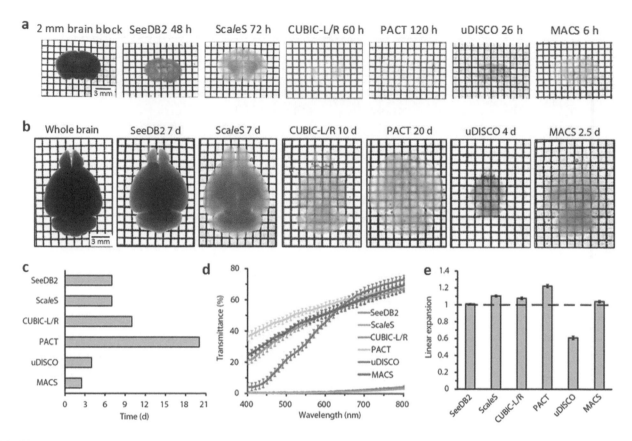

FIGURE 12.17 Rapid clearing of MACS for multiscale tissues. (a-b) Bright-field images of 2-mm-thick mouse brain slices and whole adult mouse brains cleared by different methods. (c) Quantitative comparison of clearing time. (d) Transmittance curves of cleared whole brains treated by different methods. (e) Linear expansion of whole brains after clearing by each method. Adapted from reference [22].

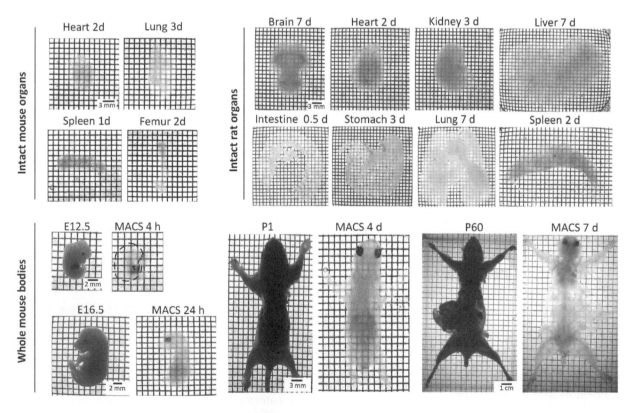

FIGURE 12.18 Clearing performance of MACS for various organs and whole bodies, including intact mouse organs, intact rat internal organs, whole embryo bodies, P1 whole mouse body, and P60 whole mouse body. Adapted from reference [22].

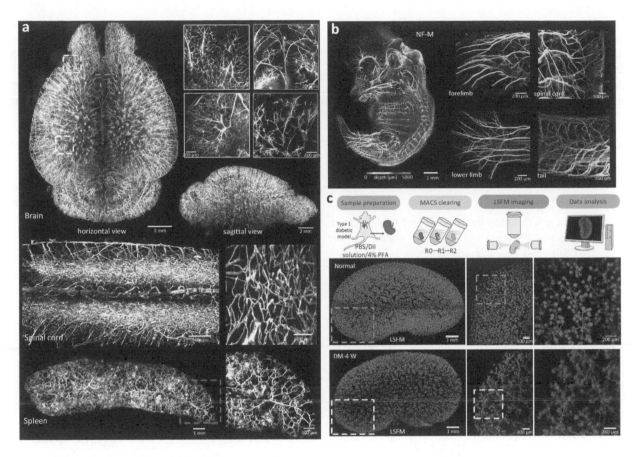

FIGURE 12.19 Applications of MACS in various tissues. (a) DiI-labeled vasculatures in an entire adult mouse brain, spinal cord, and spleen cleared by MACS. (b) 3D reconstruction of whole embryo (E14.5) labeled for neurofilament (NF-M). (c) 3D pathology of glomeruli in normal and diabetic kidneys cleared by MACS. Adapted from reference [22].

body (Figure 12.18). These demonstrated that MACS is a rapid clearing method with wide applicability.

Application of MACS in 3D visualization of neuronal and vascular structures

MACS could preserve the signals of multiple fluorescent probes, including endogenous fluorescent proteins, immunostaining, and, in particular, lipophilic dyes. With good fluorescence preserving capability, MACS enabled 3D reconstruction of neuronal and vascular structures in various intact tissues.

After being labeled with DiI perfusion, the intact organs were dissected and cleared with MACS, and then were imaged with Ultramicroscope. Figure 12.19a demonstrated the 3D reconstruction of the DiI-labeled blood vasculature in various tissues, including whole mouse brain, spinal cord, and spleen. As seen from the high-magnification images in the cortex, middle of the brain, cerebellum, and hippocampus, as well as the sagittal view of the reconstructed brain, the surrounded capillaries were clearly visualized. The intact E14.5 embryo immunostained with anti-neurofilament antibody was also imaged after MACS clearing, as shown in Figure 12.19b. The fine labeling and detection of nerve fibers allowed observation of details of the nerve innervation in the forelimb, spinal cord, lower limb, and tail in the embryo body.

With its excellent compatibility with lipophilic dyes, MACS enabled the analysis of 3D pathology of DiI-labeled glomeruli in normal and diabetic kidneys. Figure 12.19c shows the experimental workflow for 3D pathology of glomeruli: the mouse kidneys were first labeled by DiI perfusion and cleared with MACS before undergoing light-sheet imaging. The images demonstrated that the kidney of a type 1 diabetic model at 4 weeks (DM-4 W) presented a different glomerulus distribution from that of normal kidney, as well as an abnormal glomerulus structures observed from the high-magnification light-sheet images.

Overall, MACS is a rapid, highly efficient clearing method with robust compatibility. MACS could provide a valuable alternative to 3D analysis of large volume tissues and facilitate diagnostic studies for pathological diseases, and it could even contribute to 3D pathology of human clinical samples in the future.

Summary

The long incubation time needed for many aqueous-based clearing methods greatly limits their wide applications. To shorten the clearing time, different methods were developed by either screening different chemical mixtures or modifying the clearing conditions. This chapter introduces these ultrafast aqueous clearing protocols from the angle of their applicability in tissues with different scales, such as the sugar/sugar-alcohol based methods and FOCM for hundreds-micron-thick tissue slices; Clear$^{T/T2}$, RTF, and ScaleSQ for several-millimeter-thick tissue blocks; and ACT and MACS for whole adult organs and even whole bodies. These ultrafast clearing methods provide alternatives for those who require speedy clearing. They are expected to play important roles in the high-efficient 3D mapping of various biological tissues.

Acknowledgments

T.Y. and D.Z. were supported by the National Key Research and Development Program of China grant number 2017YFA0700501, and NSFC grant numbers 61860206009, 81701354, 81961138015.

REFERENCES

1. E.A. Susaki, and H.R. Ueda, "Whole-body and whole-organ clearing and imaging techniques with single-cell resolution: Toward organism-level systems biology in mammals," *Cell Chem. Biol.* **23**(1), 137–157 (2016).
2. D. Zhu, K.V. Larin, Q. Luo, and V.V. Tuchin, "Recent progress in tissue optical clearing," *Laser Photonics Rev.* **7**(5), 732–757 (2013).
3. L. Silvestri, I. Costantini, L. Sacconi, and F.S. Pavone, "Clearing of fixed tissue: A review from a microscopist's perspective," *J. Biomed. Opt.* **21**(8), 081205 (2016).
4. D.S. Richardson, and J.W. Lichtman, "Clarifying tissue clearing," *Cell* **162**(2), 246–257 (2015).
5. K. Tainaka, A. Kuno, S.I. Kubota, T. Murakami, and H.R. Ueda, "Chemical principles in tissue clearing and staining protocols for whole-body cell profiling," *Annu. Rev. Cell Dev. Biol.* **32**, 713–741 (2016).
6. A. Erturk, K. Becker, N. Jahrling, et al., "Three-dimensional imaging of solvent-cleared organs using 3DISCO," *Nat. Protoc.* **7**(11), 1983–1995 (2012).
7. C. Pan, R. Cai, F.P. Quacquarelli, et al., "Shrinkage-mediated imaging of entire organs and organisms using uDISCO," *Nat. Methods* **13**(10), 859–867 (2016).
8. D. Jing, S. Zhang, W. Luo, et al., "Tissue clearing of both hard and soft tissue organs with the PEGASOS method," *Cell Res.* **28**(8), 803–818 (2018).
9. Y. Qi, T. Yu, J. Xu, et al., "FDISCO: advanced solvent-based clearing method for imaging whole organs," *Sci. Adv.* **5**(1), eaau8355 (2019).
10. C. Hahn, K. Becker, S. Saghafi, et al., "High-resolution imaging of fluorescent whole mouse brains using stabilised organic media (sDISCO)," *J. Biophoton.* **12**(8), e201800368 (2019).
11. R. Cai, C. Pan, A. Ghasemigharagoz, et al., "Panoptic imaging of transparent mice reveals whole-body neuronal projections and skull-meninges connections," *Nat. Neurosci.* **22**(2), 317–327 (2019).
12. H. Hama, H. Kurokawa, H. Kawano, et al., "Scale: A chemical approach for fluorescence imaging and reconstruction of transparent mouse brain," *Nat. Neurosci.* **14**(11), 1481–1488 (2011).
13. E.A. Susaki, K. Tainaka, D. Perrin, et al., "Whole-brain imaging with single-cell resolution using chemical cocktails and computational analysis," *Cell* **157**(3), 726–739 (2014).
14. K. Tainaka, S.I. Kubota, T.Q. Suyama, et al., "Whole-body imaging with single-cell resolution by tissue decolorization," *Cell* **159**(4), 911–924 (2014).
15. K. Tainaka, T.C. Murakami, E.A. Susaki, et al., "Chemical landscape for tissue clearing based on hydrophilic reagents," *Cell Rep.* **24**(8), 2196–2210 e2199 (2018).
16. K. Chung, J. Wallace, S.Y. Kim, et al., "Structural and molecular interrogation of intact biological systems," *Nature* **497**(7449), 332–337 (2013).
17. R. Tomer, L. Ye, B. Hsueh, and K. Deisseroth, "Advanced CLARITY for rapid and high-resolution imaging of intact tissues," *Nat. Protoc.* **9**(7), 1682–1697 (2014).
18. X. Zhu, L. Huang, Y. Zheng, et al., "Ultrafast optical clearing method for three-dimensional imaging with cellular resolution," *Proc. Natl. Acad. Sci. U. S. A.* **116**(23), 11480–11489 (2019).
19. T. Kuwajima, A.A. Sitko, P. Bhansali, C. Jurgens, W. Guido, and C. Mason, "ClearT: A detergent- and solvent-free clearing method for neuronal and non-neuronal tissue," *Development* **140**(6), 1364–1368 (2013).
20. T. Yu, J. Zhu, Y. Li, et al., "RTF: A rapid and versatile tissue optical clearing method," *Sci. Rep.* **8**(1), 1964 (2018).
21. H. Hama, H. Hioki, K. Namiki, et al., "ScaleS: An optical clearing palette for biological imaging," *Nat. Neurosci.* **18**(10), 1518–1529 (2015).
22. J. Zhu, T. Yu, Y. Li, et al., "MACS: Rapid aqueous clearing system for 3D mapping of intact organs," *Adv. Sci.* **7**(8), 1903185 (2020).
23. E. Lee, J. Choi, Y. Jo, et al., "ACT-PRESTO: Rapid and consistent tissue clearing and labeling method for 3-dimensional (3D) imaging," *Sci. Rep.* **6**, 18631 (2016).
24. Z. Mao, D. Zhu, Y. Hu, X. Wen, and Z. Han, "Influence of alcohols on the optical clearing effect of skin in vitro," *J. Biomed. Opt.* **13**(2), 021104 (2008).
25. J. Wang, N. Ma, R. Shi, Y. Zhang, T.T. Yu, and D. Zhu, "Sugar-induced skin optical clearing: From molecular dynamics simulation to experimental demonstration," *IEEE J. Sel. Top. Quantum Electron.* **20**(2), 7101007 (2014).
26. W. Feng, R. Shi, N. Ma, D.K. Tuchina, V.V. Tuchin, and D. Zhu, "Skin optical clearing potential of disaccharides," *J. Biomed. Opt.* **21**(8), 081207 (2016).
27. M.T. Ke, S. Fujimoto, and T. Imai, "SeeDB: A simple and morphology-preserving optical clearing agent for neuronal circuit reconstruction," *Nat. Neurosci.* **16**(8), 1154–1161 (2013).
28. P.S. Tsai, J.P. Kaufhold, P. Blinder, et al., "Correlations of neuronal and microvascular densities in murine cortex revealed by direct counting and colocalization of nuclei and vessels," *J. Neurosci.* **29**(46), 14553–14570 (2009).

29. T. Yu, Y. Qi, J. Wang, et al., "Rapid and prodium iodide-compatible optical clearing method for brain tissue based on sugar/sugar-alcohol," *J. Biomed. Opt.* **21**(8), 081203 (2016).

30. B. Yang, J.B. Treweek, R.P. Kulkarni, et al., "Single-cell phenotyping within transparent intact tissue through whole-body clearing," *Cell* **158**(4), 945–958 (2014).

31. J.B. Treweek, K.Y. Chan, N.C. Flytzanis, et al., "Whole-body tissue stabilization and selective extractions via tissue-hydrogel hybrids for high-resolution intact circuit mapping and phenotyping," *Nat. Protoc.* **10**(11), 1860–1896 (2015).

32. S.I. Kubota, K. Takahashi, J. Nishida, et al., "Whole-body profiling of cancer metastasis with single-cell resolution," *Cell Rep.* **20**(1), 236–250 (2017).

13

Challenges and opportunities in hydrophilic tissue clearing methods

Etsuo A. Susaki and Hiroki R. Ueda

CONTENTS

Introduction

Biological tissue is composed of three-dimensional (3D) structures. Its organization is complicated, and the components may be rare or unevenly distributed. Preparing sections for observing internal structures has long been the gold standard of histology and pathology. However, such 2D-based inspections have difficulty detecting these exact features of biological tissues. Accurate capture of the original features thus requires 3D observation and analysis.

The earliest tissue clearing technology developed by a German anatomist, Spalteholz, over 100 years ago was an attempt to break through the limitations of such classical histology and pathology [1]. Prompted by recent optical microscope technologies and the growing need for 3D observation of the whole brain in neuroscience, modern tissue clearing methods were rapidly developed in the 2000s. In particular, a study published by Dodt and colleagues in 2007 set the course for a research scheme that combines efficient tissue clearing with a light microscope for 3D imaging, including light-sheet microscopy [2].

Current tissue clearing methods are roughly classified into

1. methods using organic solvents (hydrophobic methods);

2. methods using water-soluble compounds (hydrophilic methods);

3. methods of embedding tissue in an artificial gel (hydrogel-based method).

Among these, various hydrophilic tissue clearing methods have already been developed and tested in a wide range of biological applications due to their high histocompatibility, safety, simplicity, and applicability to fluorescent proteins. This chapter mainly discusses tissue clearing chemistry of hydrophilic reagents and their recent applications in biomedical research.

Brief history of the development of hydrophilic tissue clearing methods

Although much shorter than the history of organic solvent-based reagents derived from Spalteholz, hydrophilic tissue clarification reagents also have a long quarter-century history. Especially in the early stages, single-agent use of various chemicals was intensively tested. In the 1990s, Tuchin and colleagues investigated how to suppress light scattering in biological tissue samples from a physical perspective. Assuming that the refractive index match of a solvent with a biological sample, such as human sclera and skin, can suppress

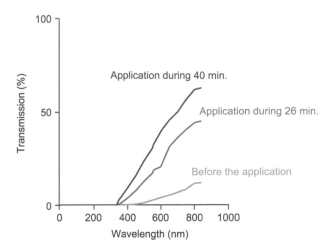

FIGURE 13.1 An early tissue clearing experiment with hydrophilic reagent. The light transmission in human sclera was increased by application of Trazograph, an X-ray contrast agent composed of diatrizoate meglumine and diatrizoate sodium solution. The graph was recaptured from Figure 2 of Bakutkin et al. Proc. SPIE (1995) [3].

light scattering and render the sample translucent, they tested a range of chemicals including X-ray contrast agents (e.g., Trazograph, meglumine sodium amidotrizoate), alcohols (e.g., polyethylene glycol, glycerol, propylene glycol), sugars (e.g., glucose, dextran), and water-soluble organic solvents [dimethyl sulfoxide (DMSO)] [3–9] (Figure 13.1). Similarly, several groups also tested related agents including sugars (e.g., glucose), polyol (e.g., glycerol, propylene glycol, ethylene glycol, butanediol), and DMSO for optical clearing of biological samples [10–17].

Following these pioneering studies, the development of modern clearing reagents with 3D tissue imaging by fluorescence microscopy began to be considered. The FocusClear reagent developed by Chiang and colleagues is one of the earliest hydrophilic clearing methods developed by combining an X-ray contrast agent and a detergent and was applied to insect brain imaging [18, 19]. However, the work by Dodt and colleagues, who reported excellent examples of tissue clearing (by BABB) and the use of light-sheet microscopy [2], prompted the development of numerous subsequent tissue clearing methods.

In 2011, Miyawaki and colleagues serendipitously found urea to be a novel category of tissue clearing agents and developed a series of Sca*l*e reagents to clear whole mouse brains [20]. In contrast to most organic solvent-based clearing reagents, Sca*l*e was able to preserve fluorescent protein signals more efficiently. Sca*l*e's ability was epoch-making because, at that time, organic solvent reagents were the only way to efficiently clear whole mammalian tissues. Besides, this technique brought a new idea to the tissue clearing process – hyperhydration – that contributes to some internal noncovalent bonds inside tissue and leads to sample swelling. Later, they developed a new reagent, Sca*l*eS, which combines urea and sugar (sorbitol) to improve cleaning performance while suppressing excessive swelling [21].

Since the development of the aforementioned studies, the trend in exploring new tissue clearing compounds and

developing more generalized clearing protocols has accelerated. Imai and colleagues developed SeeDB, a clearing protocol that uses fructose to maintain tissue integrity, and showed its application in neural circuit analysis [22]. They recently developed SeeDB2 using iohexol (Histodentz™), an X-ray contrast agent, and made it possible to apply it to superresolution imaging by making the refractive index in the optical path almost uniform [23]. Mason and colleagues developed *Clear*^{T} by combining polyethylene glycol and formamide and applied it to mouse embryo imaging [24]. Ueda and colleagues constructed a comprehensive chemical profiling system to identify better compounds with potent tissue-clearing activity. They found that a range of amino alcohols had both delipidation and decolorization activity, and used them to design a new clearing protocol, CUBIC (Clear, Unobstructed Brain/Body Imaging Cocktails and Computational Analysis) [25, 26]. The first-generation CUBIC reagents also contained urea, as used in the Sca*l*e recipe. However, they eventually screened more than 1600 compounds and developed urea-free second-generation reagents. Their screening also identified a series of aromatic amides with strong refractive index-matching properties [27].

Many additional studies since 2015 have further demonstrated a wide range of hydrophilic tissue clearing methods and their applications. Some of these methods have achieved comparable clearing capabilities to organic solvent reagents without losing the biocompatibility, biosafety, and protein storage benefits. Therefore, hydrophilic tissue clearing methods have been successful in offering opportunities for various biomedical applications. The timeline for the development of these methods is summarized in Figure 13.2.

Chemistry of hydrophilic tissue clearing methods

The roles of tissue clearing reagents can be categorized into 1) reduction of light scattering by delipidation, decalcification, and RI matching; and 2) reduction of light absorption by decolorization [28–30]. Most recent works have attempted to combine chemicals with distinct functions to develop a competent tissue clearing protocol. However, how can we estimate the contribution of each compound to the clearing process? One useful approach is to comprehensively and quantitatively measure the potency of a chemical in terms of various aspects of the tissue clearing process. Ueda and colleagues developed a novel chemical screening and profiling system to discover the chemical structures and properties required for hydrophilic tissue clearing reagents. They first screened and found first candidates for potent tissue clearing, including amino alcohols, out of a small library of 40 chemicals [26]. Ueda and colleagues recently extended this approach to a comprehensive assessment of the chemical properties required for tissue clearing, including delipidation, decolorization, decalcification, pH, refractive index (RI), and fluorescent protein quenching [27]. They performed chemical profiling of >1600 chemical candidates ("CUBIC chemical profiling") and finally developed improved tissue clearing reagents. Their CUBIC chemical profiling also provided in-depth information on the chemical profiles that contribute to each tissue clearing process. Based

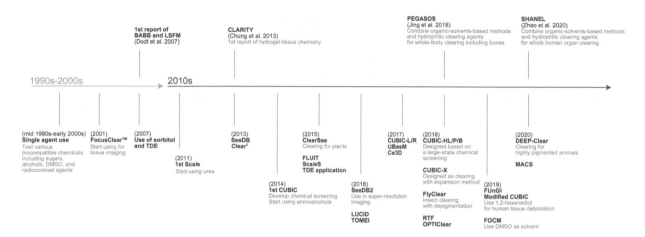

FIGURE 13.2 History of hydrophilic tissue clearing reagents. Most early studies used a single chemical to test for changes in the optical properties of biological samples (such as skin). The latest clearing reagents are designed particularly for 3D tissue imaging by fluorescence microscopy. Various functions, such as the ability to preserve tissue components and fluorescent proteins and compatibility with dyes and antibodies, were considered. The combination of multiple compounds with different functions is important for achieving the various capabilities.

on these insights, this section discusses the role of compounds used in recent hydrophilic tissue clearing methods.

Delipidation

The role of delipidation is to remove lipids, a strong light scatterer in the biological tissues, and to promote the penetration of compounds in reagents (so-called permeabilization). In most cases, tissue samples for clearing are fixed with a cross-linker such as paraformaldehyde. The side chains of basic amino acids such as lysine in proteins are crosslinked. On the other hand, the lipids react less with such fixatives and are efficiently removed by delipidation. The degree of delipidation has been shown to correlate with final transparency after RI matching [27] (Figure 13.3). Therefore, this step is especially important in applications where high sample transparency is required. Several ionic and nonionic detergents and some alcohols and amines are used for recent hydrophilic clearing protocols (Figure 13.4).

Detergents are the first choice as delipidation compounds. Many hydrophilic tissue clearing methods use detergents such as Triton X-100, saponin, SDS, or sodium deoxycholate from low (0.1%~) to high (~15%) concentrations [20, 21, 23, 26, 27, 31–36]. However, they generally have a large molecular size and it is difficult to get them to penetrate tissues. To overcome this issue, a new clearing method (SHANEL) employed CHAPS, an ionic detergent, to form smaller micelles [37]. SHANEL can thus be applied to large lipid-rich samples such as human and porcine tissues. A combination with urea [26, 27, 35]) or electric field-promoted penetration [38, 39] has also been proposed to improve the tissue permeability of concentrated detergents. Perfusion of the whole animal body with a clearing reagent is also an option [25, 40].

In addition to detergents, alcohols and amines also have delipidation ability [26, 27, 41]. In the CUBIC chemical profiling, aliphatic amines, amino alcohols, and amino ethers were the most effective groups for lipid solubility (Figure 13.5). Aliphatic amines have a significantly better delipidation

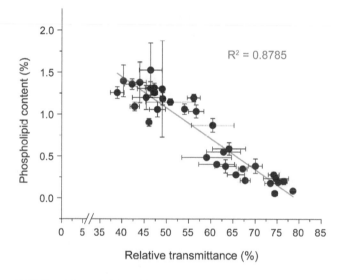

FIGURE 13.3 Delipidation efficiency and tissue transparency. From the >1600 chemical library, some highly lipid-solving chemicals were selected for use in the delipidation of mouse brain hemispheres. The delipidated brain hemispheres were subjected to the measurement of residual phospholipid content and light transmittance after RI matching. Data are means ± SDs (n = 2). The plot was linearly fitted ($R^2 = 0.8785$). Adapted from Tainaka et al. (2018) [27].

efficiency than amino alcohols and detergents. In addition, the logP value (octanol/water partition ratio, an index of a chemical's lipophilicity) was found to be the most influential property for a chemical's delipidation activity (Figure 13.6). Therefore, it was suggested that the relatively hydrophobic and uncharged amine derivatives have a high permeability in tissues and exhibit a remarkable delipidation activity. Based on these findings, they chose N-butyldiethanolamine and 1,3-bis(aminomethyl)cyclohexane for the component in animal and human tissue delipidation reagents (CUBIC-L and -HL). In their recent study, they also applied a glycol (1,2-Hexanediol) as an alternative human tissue delipidation agent [41]. Quadrol used in ScaleCUBIC-1 has a relatively low logP value and requires urea for its uptake into tissues.

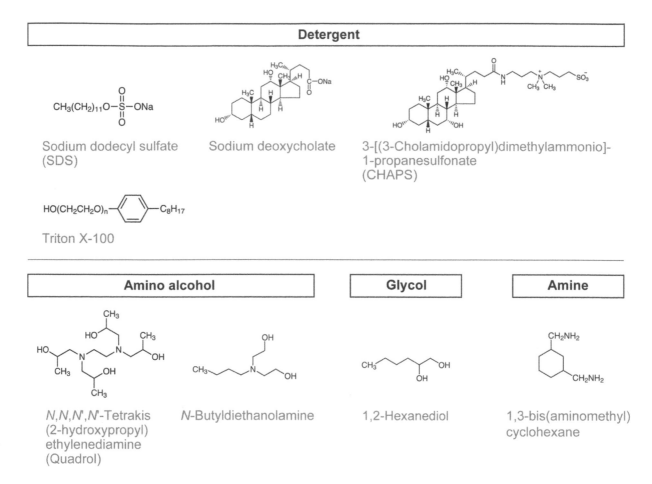

FIGURE 13.4 Representative delipidation compounds used in modern tissue clearing methods. Images of the chemical structures are provided by Tokyo Chemical Industry.

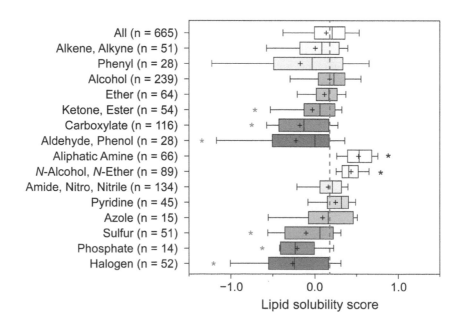

FIGURE 13.5 Delipidation ability of each chemical category. The library of >1600 chemicals was screened for lipid solubilization ability. The box plots show the distribution of lipid solubility scores classified with the chemical category (normalized by the mean of alcohol group, dot line). Aliphatic amines, amino alcohols, and amino ethers showed the most effective functions. The box plots indicate the 25th–75th percentile (boxes), 10th–90th percentile (whiskers), median (vertical lines), and mean (cross). Asterisks indicate the groups showing statistically significant values. Adapted from Tainaka et al. (2018) [27].

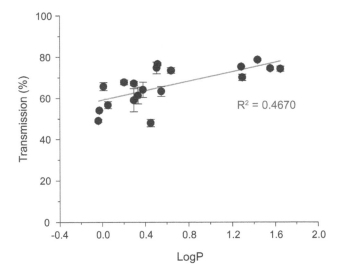

FIGURE 13.6 Correlation of logP values and final tissue transparency. Multivariate linear regression analysis of lipid solubility scores and features of the screened chemicals revealed logP to be the most influential property. The plot shows the correlation between lipid solubility score and logP of the highly lipid-soluble amine group. The final transparency of chemically treated brains was measured after delipidation with each amine and RI matching. The plot was linearly fitted ($R^2 = 0.4670$). Data are means + SDs (n = 2). Adapted from Tainaka et al. (2018) [27].

It is considered essential to adjust the pH of delipidation reagents from neutral to alkaline, as many biological materials tend to be insoluble in acids [42]. The CUBIC chemical profiling also found that a basic condition (pH >10) was significantly superior to an acidic one (pH <6) in terms of delipidation ability [27]. Therefore, CUBIC's delipidation reagents (Sca*le*CUBIC-1, CUBIC-L, CUBIC-HL) are adjusted to be alkaline.

Decolorization

Decolorization is intended to remove endogenous compounds with light absorption properties (heme and other pigments) and autofluorescence (lipofuscin and chlorophyll). Given the applications of tissue clearing in a wide range of biomedical studies, the discovery of heme-removing compounds with minimal effects on endogenous proteins was a breakthrough. In 2014, Ueda and colleagues accidentally discovered that one of their Sca*le*CUBIC agents Quadrol, a type of amino alcohol, has the unique property of mild heme removal from tissue samples, potentially via interfering heme's coordinate bond [25]. Several tissue clearing protocols since this study have employed the same chemicals for decolorization purposes [43–45]. Toward human tissue clearing, Ertürk and colleagues selected other amino alcohol *N*-methyldiethanolamine (NMDEA). They used it in their SHANEL protocol [37]. *m*-Xylylenediamine (MXDA), an aromatic amine, was recently reported as a strong clearing and decolorization agent [46].

Comprehensive chemical profiling [27] (Figure 13.7) demonstrated high decolorization potential for agents classified as cationic detergents, phenyls, amines, amino alcohols, azoles, and pyridines. This profile has some overlap with that of delipidating capacity. Of the agents tested, *N*-alkylimidazoles were the most effective decoloring agents. They also performed a multivariate linear regression model analysis on amine groups to investigate decolorization factors in compounds with high delipidation capacity. Multiple polar groups (e.g., $-NH_2$ and $-OH$) and longer chain lengths between polar groups (C–C length ≥3) were associated with efficient decolorization capabilities. They finally chose an amine [1,2-bis(aminomethyl) cyclohexane] and amino alcohol (N-butyldiethanolamine) as delipidation/decolorization agents and an N-alkylimidazole (1-methylimidazole) as a decolorizing accelerator. Since

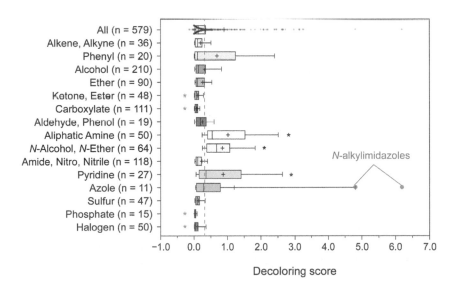

FIGURE 13.7 Decoloring ability of each chemical category. The library of >1600 chemicals was screened for decoloring ability. The box plots show the distribution of decoloring scores classified with the chemical category (normalized by the mean of alcohol group, dot line). Amines, amino alcohols, pyridines, and azoles showed significantly high decoloring functions. In particular, *N*-alkylimidazoles showed the most prominent ability. The values of two representatives *N*-alkylimidazoles (1-ethylimidazole and 1-(3-aminopropyl)imidazole) are indicated in the plot. The box plots represent the 25th–75th percentile (boxes), 10th–90th percentile (whiskers), median (vertical lines), and mean (cross). Asterisks indicate the groups showing statistically significant values. Adapted from Tainaka et al. (2018) [27].

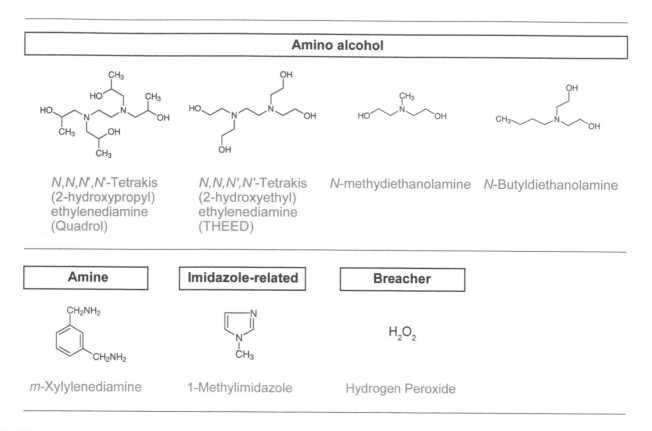

FIGURE 13.8 Representative decolorization compounds used in modern tissue clearing methods. Images of the chemical structures are provided by Tokyo Chemical Industry.

N-alkylimidazoles have a structure similar to that of the histidine residue (heme ligand of hemoglobin), the compounds potentially contribute to highly efficient heme elution through competitive inhibition of a noncovalent binding.

Removing other natural pigments from biological tissue remained more challenging. Melanin, pterins, ommochromes, and carotenoids are included in such pigments. Several studies have tried to develop hydrophilic tissue clearing methods with pigment breaching. Most of these studies primarily targeted melanin and applied hydrogen peroxide (H_2O_2) to directly degrade the pigment structure. For example, the breaching and optical clearing were achieved in the whole beetle and fish bodies [47, 48] or mammalian whole eyes [49]. In the case of decolorization and clearing of entire crustacean bodies, additional decalcification with EDTA was adopted [50]. However, the H_2O_2-based pigment breaching can generally cause loss of fluorescent protein signals. Most 3D imaging applications require the recovery of labeling signals by postimmunostaining/ISH.

The combination of H_2O_2 and other hydrophilic chemicals has been proposed as providing a more efficient decolorization procedure. Dodt and colleagues developed FlyClear, which applies other amino alcohol N,N,N′,N′-Tetrakis(2-hydroxy ethyl)ethylenediamine (THEED), with a structure similar to Quadrol, to depigmentation of the fly body after permeabilization with cold acetone [35]. Later, they improved the protocol by adding H_2O_2 treatment [51]. This advanced protocol, DEEP-Clear, successfully removed various types of body pigments from the bodies of worm, squid, zebrafish, and tetrapods, while preserving tissue integrity for immunolabeling, 5-ethynyl-2′-deoxyuridine (EdU) labeling, and RNA in situ

hybridization (ISH). Figure 13.8 summarizes the chemicals used for decolorization and depigmentation purposes.

Suppression of chlorophyll autofluorescence was another challenge for plant clearing. ClearSee was developed to meet the needs of plant tissue clearing and 3D imaging while maintaining fluorescent protein signals [33]. Higashiyama and colleagues conducted a chemical screen inspired by CUBIC's research and discovered several detergents with chlorophyll-removing activity. They finally applied sodium deoxycholate to the final recipe and successfully obtained 3D images of fixed Arabidopsis leaves, roots, and pistil. Matsunaga and colleagues also succeeded in suppressing chlorophyll autofluorescence by a mixed fixative of acetic acid and ethanol, followed by thiodiethanol-based optical purification (TOMEI) [52]. Here, the choice of fixative also appears to be another critical depigmenting factor. The clearing of an entire beetle was successful in combination with a mixed formaldehyde and ethanol fixative [47].

Decalcification

Since the bone matrix contains calcium phosphate (hydroxyapatite) as the main component and strongly scatters light, decalcification is essential for clearing bone tissue. Removal of bone tissue hydroxyapatite is promoted under strongly acidic conditions, especially below pH 2.0. This is probably because the pKa1 value of phosphate is 2.12, which is related to the equilibrium shift of phosphate [27]. However, the acid treatment is unsuitable for clearing to observe biological tissue. Neutral EDTA solutions are widely used as calcium chelating agents in nearly all clearing methods reported to date [43–45, 53].

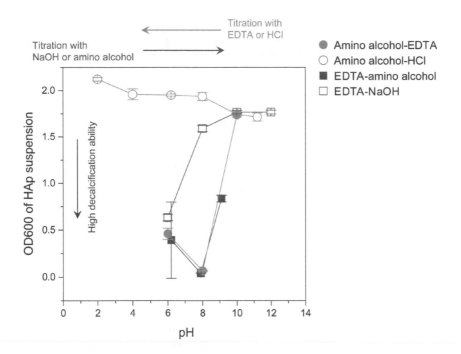

FIGURE 13.9 pH and organic base dependence of decalcification ability of EDTA. The decalcification capacity of 10 wt% amino alcohol (*N*-butyldiethanolamine) titrated with EDTA or HCl, and 10 wt% EDTA titrated with *N*-butyldiethanolamine or NaOH is shown. The decalcification capacity was measured by the decrease of OD 600 of hydroxyapatite (HAp) suspension. EDTA's decalcification activity was enhanced by adjusting the pH using amino alcohol (organic base) rather than HCl or NaOH (inorganic acid or base). The activity of the EDTA-organic base mix was maximized at pH 8. Data are means ± SDs (n = 3). Adapted from Tainaka et al. (2018) [27].

CUBIC chemical profiling also confirmed the utility of EDTA [27]. Structural analogs of EDTA, CyDTA, and PDTA had comparable effects, suggesting that the common chemical properties of these compounds are essential for decalcification. They further found that the 10 wt% EDTA-Na solution's decalcification efficiency was high in the neutral range and dropped significantly at pH 8.0. Interestingly, adjusting the pH using organic bases such as amino alcohols and amines, rather than inorganic bases such as NaOH, could enhance EDTA's activity of dissolving HAp (Figure 13.9). Their profiling finally revealed that the combination of EDTA and imidazole showed the highest decalcification activity. This recipe was adopted as the CUBIC-B demineralization reagent. The pH dependence shown in Figure 13.9 suggests a significant contribution of EDTA and imidazole that also promotes protonation of PO_4^{2-} in addition to the chelation of calcium, and leads to HAp dissolution.

Refractive index matching

The delipidation and decalcification steps can change the composition of biological tissues and alternate the refractive index (RI) of the entire sample. These steps can contribute to improving tissue clearing efficiency. However, RI adjustment of the sample and surrounding solvent should be placed in the final step of tissue clearing, as it is of paramount importance for the completion of the process. Thus, the process contributes to tissue clearing by highly suppressing light scattering inside the tissue samples. Most recent hydrophilic clearing methods have adjusted the RI of their final clearing reagents around 1.4–1.5 (Figure 13.10). Note that this step is achieved by other mechanisms, not just by matching RI values (Figure 13.11). Swelling the sample is a form of this final clearing step, as described below. Other mechanisms, such as changes in inner scattering structures (e.g., collagen) and birefringence, have also been suggested [54]. For example, the clearing capability of sugars and polyols, such as glucose, fructose, and sorbitol, may be partly attributable to their interaction between collagen helices, fibrillogenesis inhibition, and potential of collagen solubilization, which were correlated with optical clearing degree [55–57].

One form of RI matching is to use high RI solvents. Since the RIs of lipids (~1.45) and proteins (~1.54) are higher than that of water (1.33), RI matching can be achieved by exchanging a solvent (e.g., phosphate buffer) for a reagent with the RI of around 1.5 [28–30]. Compounds with high RI, or those that can increase the RI of a solution due to their high solubility, are used as compositions of such RI matching reagents. Note that merely a high RI is not enough; a particular type of biocompatibility is also required [27–30]. For example, increasing the reagent's RI with inorganic compounds (e.g., zinc iodide, tungstic acid) did not contribute to tissue clearing [22] (Susaki et al., unpublished observation). Alcohols, sugars, water-soluble aromatic compounds, polyethers, X-ray contrast agents, and some amines and amides may have been serendipitously selected as compounds with such dissolving properties, which were also proposed as "RI mixing effect" [29, 30].

Polyalcohols, including the sugar group (e.g., glucose, fructose, sucrose, sorbitol, xylitol) and polyol group (e.g., propylene glycol, glycerol), were applied in early studies [5, 7, 56] and have been used in multiple modern protocols [20–22, 26, 31, 33, 46, 58–63]. Amino alcohols, such as

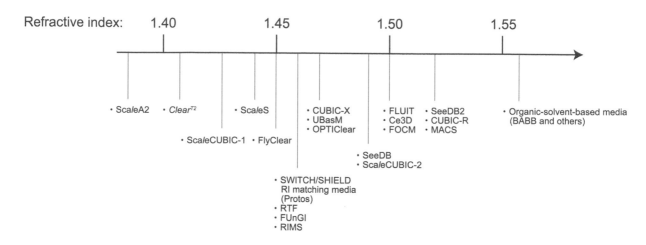

FIGURE 13.10 The range of RI values of hydrophilic RI-matching reagents. Most RI values were adjusted to 1.4–1.5. Organic solvents have high RI values (1.5–1.56), which explains some of their superior clearing capabilities.

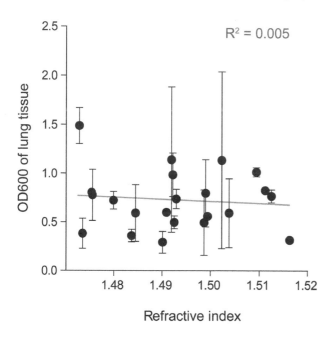

FIGURE 13.11 No correlation between the reagent's clearing efficiency and its RI value in high RI range. Twenty-two extremely water-soluble RI-matching candidates (RI range: 1.47–1.52) were selected from the CUBIC chemical screening. Their clearing efficiency was examined by the OD600 of the mouse lungs cleared by the chemical. The scatterplot shows that their clearing capacities are not solely dependent on their RI values. The plot was linearly fitted ($R^2 = 0.005$). Data are means ± SDs (n = 2). Adapted from Tainaka et al. (2018) [27].

Quadrol and triethanolamine, were identified as the superior compatible compound of glycerol used in Sca*l*e. They have been used in Sca*l*eCUBIC and RTF (Rapid clearing method based on Triethanolamine and Formamide) reagents [26, 64]. Meglumine is also a nitrogen-containing polyol that has been used since the beginning of the development of hydrophilic tissue clearing methods (as a component of Trazographs) [3, 4, 6]. Modern protocols such as UBasM, OPTIClear, and FlyClear have also adopted this chemical [31, 34, 35]. Thiodiethanol was first reported as an clearing reagent by Hell and colleagues [65] and adopted by several recent methods [34, 52, 63, 66, 67]. Clear*T* used polyethylene glycol (PEG)

as one of its compositions [24]. Polyethylene glycol methacrylate (PEGMMA) was used in the PEGASOS method as an RI matching agent mixed with organic solvents [45]. Various X-ray contrast agents have also been frequently used. Tuchin and colleagues used Trazograph (containing diatrizoate) in their early studies [3, 4, 6]. Recently, FlyClear also used this compound [35]. Several recent methods, such as RIMS, SeeDB2, Ce3D, and OPTIClear, used iohexol [23, 32, 34, 40]. In the SWITCH study, iohexol and another X-ray contrast agent, iodixanol, were mixed with meglumine [68]. The water-soluble organic solvent DMSO is also occasionally used [5, 21, 59]. It was also used as a solvent instead of water in ultrafast optical clearing method (FOCM) [61].

CUBIC chemical profiling by Ueda and colleagues indicated that aromatic chemical groups (phenyl, pyridine, and azole groups) exhibited higher RI values in solution. Furthermore, they revealed that the amide group contains chemicals with a particularly strong tissue-clearing ability compared to other groups. Finally, they proposed a CUBIC-R recipe by mixing the pyrazolone group (antipyrine) with the aromatic amide group (N-methylnicotinamide or nicotinamide). These compounds can be a low-cost alternative to X-ray contrast agents, which are also classified as aromatic amides. Consistent with their insights, several other amines and amides (e.g., 1,3-Dimethyl-2-imidazolidinone, N-Methylacetamide, m-Xylylenediamine) have been independently adopted by other RI-matching reagents [31, 32, 46]. Figure 13.12 summarizes various RI-matching chemicals used in hydrophilic clearing reagents.

Another form of RI matching is sample swelling. Recently, Ueda, Susaki, and colleagues discovered that fixed and delipidated tissues for clearing purposes exhibit repeated and reversible swelling–shrinkage behavior, indicating that biological tissue can be defined as a certain type of gel [29, 30, 69, 70]. They found that the fixed and delipidated tissue behaves like a cross-linked polypeptide gel [69, 70]. In particular, the swelling–shrinkage property of tissue can be mimicked by a fixed gelatin gel [69, 70] (Figure 13.13). This was also consistent with the previous findings by Toyoichi Tanaka, who had reported that some natural polymers could be defined as gels [71, 72]. As explained below, some hydrophilic tissue clearing chemicals

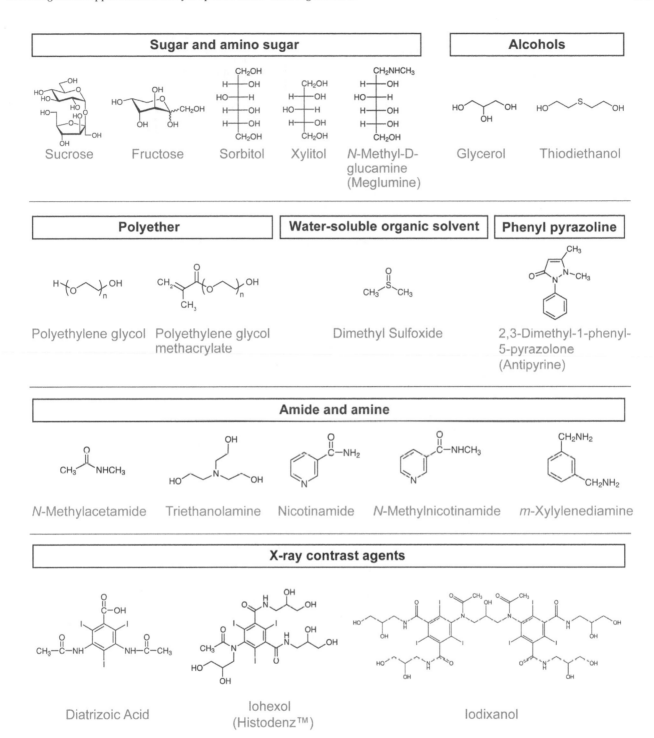

FIGURE 13.12 Representative RI-matching compounds used in modern tissue clearing methods. A part of the chemical structure images are provided by Tokyo Chemical Industry.

and alkaline-adjusted reagents have the tissue gel swelling effect. Apparently, swelling contributes to RI matching primarily by relaxing the light-scattering structure and reducing the average RI by reducing the density of tissue gel polymer per volume of RI matching by swelling is one of the features of hydrophilic tissue clearing methods, as opposed to organic solvent-based methods, which shrink the sample by dehydration.

Hyperhydration with urea and urea-related compounds may contribute to RI matching in this form. Urea was first adopted in the Sca*l*e method [20] and then used in a wide range of protocols including Sca*l*eCUBIC, FRUIT, UBasM, FUnGI, ClearSee, FryClear, FOCM, and SUT (Scheme Update on tissue Transparency) [26, 31, 33, 35, 58, 60, 61, 73]. Urea has high hydration energy due to bearing both hydrogen donor and acceptor groups and the strong synergistic effect of the two NH_2 groups [42, 74]. Formamide, a urea-related denaturant used in the *Clear^T* and RTF protocols [24, 64], may also contribute to hyperhydration. However, the swelling effect of

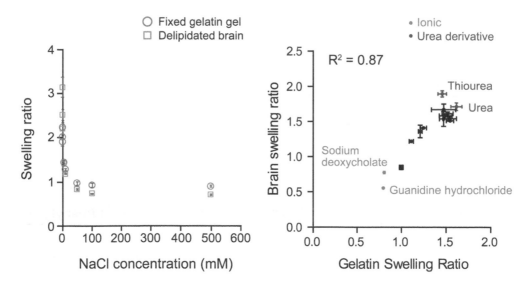

FIGURE 13.13 Similarity in the swelling–shrinkage behaviors of fixed gelatin gel and the delipidated brain. (Left) Ion strength–dependent shrinkage curves of the fixed gelatin gels and delipidated mouse brain hemispheres. Similar shrinkage patterns were observed between these gels. Data are mean ± SDs (n = 3). Adapted from Susaki et al. (2020) [70]. (Right) Correlation of the swelling ratio between the fixed gelatin gels (n = 3) and delipidated mouse brain hemispheres (n = 2) treated with PBS, water, or 12 chemicals used in [26]. Urea-related and ionic chemicals are manifested. A high similarity between these gels was observed. The plot was linearly fitted (R^2 = 0.87). Data are means ± SDs (n = 2 for brains, n = 3 for gelatin). Adapted from Murakami et al. (2018) [69].

formamide seems significantly weaker due to the single NH_2. The sample size did not change after clearing with these protocols. 1,3-Dimethyl-2-imidazolidinone was found to enhance tissue clearing in UBasM recipe, potentially working as urea-like, moderate hyperhydration agent [31].

Hyperhydration with urea and related compounds also plausibly increase the internal osmotic pressure, which induces the influx of outer molecules, including water and other ingredients, and promotes their incorporation [42]. This mechanism may explain the success of several protocols such as Sca*l*eCUBIC-2, Sca*l*eS, FRUIT, and FUnGI, which combined urea with other RI matching compounds such as sugars [21, 26, 58, 60]. In the first Sca*l*e design, glycerol was added to prevent excess hydration and tissue expansion [20]. The final osmotic pressure is balanced in these reagents, which allows clearing while maintaining the tissue size.

Ueda and colleagues exploited gelatin gel as a tissue-mimicking gel to comprehensively screen compounds with tissue swelling activity. They first identified 11 compounds that showed more vigorous swelling activity than urea. Then, they found that the combination of imidazole and antipyrine exhibited the most potent swelling activity. They finally proposed a hydrophilic clearing method with expansion (CUBIC-X) that simultaneously achieves tissue swelling and RI matching [69]. CUBIC-X can swell tissues up to 10 times its original volume ratio. They used CUBIC-X to realize a concept similar to high-resolution imaging with artificial gel-induced tissue swelling (Expansion Microscopy [75, 76]). CUBIC-X contributed to the construction of mouse whole-brain single-cell resolution atlas (CUBIC-Atlas).

In a tissue sample that have undergone clearing, carboxyl groups are the primary residues of ionization because the amino groups are crosslinked by fixatives such as formalin. Therefore, alkaline pH can induce tissue gel swelling due to

the progressed ionization of the carboxyl groups. To design a RI matching reagent with a moderate swelling capacity, CUBIC-R contains antipyrine having both high RI and tissue swelling properties, with the addition of amino alcohol to adjust the pH to weak alkaline range [27, 69]. Hyperhydration and swelling compounds are summarized in Figure 13.14.

Features of modern hydrophilic tissue clearing methods

The primary advantage of hydrophilic tissue clearing methods is their high safety. In the practical aspect of the experiment and disposal procedure, hydrophilic tissue clearing methods are easy to handle and safe. It is not necessary to use the fume hood, which is an essential part of organic solvent–based methods. The high compatibility of hydrophilic tissue clearing methods with microscopy instruments allows standard microscopes to be used without special care, and organ-scale data acquisition to be applied with complex setups such as automated sectioning tomography [59, 62].

High biocompatibility is the second advantage of hydrophilic tissue clearing methods. As introduced above, Tuchin and colleagues initially thought that hydrophilic clearing agents could be used for medical purposes by applying them to living tissue. They measured in vivo optical reflectance and transmittance in rabbit eyes and the human skins administered with glucose, Trazograph, glycerol, and DMSO [5]. Welch and colleagues also tested in vivo clearing by injecting glycerol underneath hamster skin [10]. Recently, the optical clearing window of a living mouse skull was tested by using several biocompatible compounds such as EDTA, urea, and glycerol [77, 78]. Iodixanol was applied to increase the 3D imaging quality of living cultured cells, organoids, zebrafish embryos,

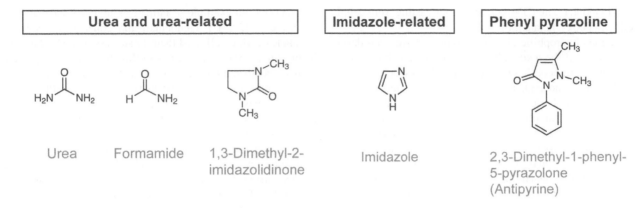

FIGURE 13.14 Representative hyperhydration and swelling compounds used in modern tissue clearing methods. Images of the chemical structures are provided by Tokyo Chemical Industry.

and planarian flatworms by RI adjustment [79]. Glycerol administration to animals may have a similar effect [80].

Due to their high biocompatibility, hydrophilic tissue clearing methods have also provided an alternative direction to overcome the lower ability of organic solvent–based methods to preserve fluorescent protein (FP) signals. Miyawaki and colleagues developed the first Scale protocol specifically to address this shortcoming of organic solvent reagents [20]. Most of the following hydrophilic clearing reagents also paid attention to preserving the FP signals. Thus, the degree of FP signal retention is one of the critical indicators in the development of hydrophilic tissue clearing methods.

Various applications have been reported based on the FP preservation capacity. Imai and colleagues demonstrated high-resolution imaging of multicolor FP expression (Tetbow) in the mouse brain using the SeeDB2 method [23, 81]. Ueda and colleagues first succeeded in comparing neural circuit function in the brains of multiple mice labeled with the Arc-dVenus transgene [26, 82]. They performed an omics-type clustering analysis of whole-brain neural activities under the administration of MK-801 (NMDA receptor antagonist) or saline [83]. CUBIC has been exploited in whole-brain neural circuit tracing by rabies virus and AAV tools [70, 84]. CUBIC also contributed to the detection of systemic metastasis of FP-labeled cancer cells and the effects of anticancer drug administration in a mouse model [85]. Similarly, FP-labeled cancer tissue was subjected to clearing and multicolor imaging using the FUnGI method [60].

The flexibility of protocol design according to the target tissue and experimental purpose is another advantage of the hydrophilic tissue clearing methods. A wide range of species, from mammals to colored arthropods, mollusks, and plants, could be cleared [33, 35, 47, 50–52]. In addition to FPs, multiple dyes, antibodies, and nuclear stains are used. The rational design of tissue clearing reagents based on chemical profiling is a case where this advantage was maximized [27].

Some hydrophilic tissue clearing methods minimized delipidation to avoid adverse effects on sample morphology and composition. Lipophilic dyes can be used in such lipid-preserving protocols [21, 22, 31, 34, 46, 58, 64] because strong delipidation with detergent resulted in signal loss. These protocols have also enabled high-resolution imaging of anatomical

structures. ScaleS was designed to achieve brain circuit analysis by combining mesoscopic and microscopic observation with two-photon and electron microscopies, respectively [21]. SeeDB2 matched the RI of cleared tissue with that of immersion oil (RI = 1.518) to minimize light scattering and spherical aberration. This reagent enabled volumetric superresolution imaging of mouse and fly brains and revealed synaptic connectivity [23]. Ce3D was initially used for immunostaining-based 3D histocytometry, a quantitative analysis of composition and tissue distribution of multiple cell populations in lymphoid tissues [32]. Later on, fluorescence in situ hybridization (FISH) of Ce3D-cleared samples was also demonstrated [86].

On the other hand, organ- or body-scale 3D imaging by light-sheet microscope (LSFM) requires a protocol with strong clearing ability. CUBIC is a representative that has been designed for a high-throughput imaging of large cleared specimens using LSFM [25, 26]. Delipidation helps to render the sample highly transparent after RI matching. Moreover, a recently developed CUBIC expansion protocol, CUBIC-X, based on hydrophilic reagents intentionally swells the sample to improve clearing efficiency and image resolution [27, 69] without using any exogenous hydrogels that were used in the hydrogel-based expansion microscopy (ExM) protocols [75, 76]. Another new protocol, called *m*-xylylenediamine (MXDA)-based aqueous clearing system (MACS), achieved rapid and efficient clearing without detergent. Samples cleared by MACS could be applied for lipophilic dye labeling and whole-brain LSFM imaging [46]. The combined structure of MXDA's aromatic and amine residues may provide multiple functions in the clearing procedure, such as high penetration efficiency and RI. Hydrophilic clearing methods are generally less efficient than organic solvent reagents. However, these ingenuities have made it possible to obtain comparable sample transparency.

While most tissue specimens used for tissue clearing are fixed with formalin or paraformaldehyde, hydrophilic tissue clearing reagents can also be applied to other fixation methods using artificial gels or epoxy compounds. RI matching reagents for samples prepared by hydrogel–tissue chemistry [87] are based on polyols or X-ray contrast agents [39, 40, 68]. In the original protocols, these samples were delipidated with SDS [39, 40]. However, the applicability of ScaleCUBIC-1 was also tested [26]. Samples prepared with acrylamide and glycidyl methacrylate (GMA) for vascular casting were delipidated with deoxycholate [36].

Numerous hydrophilic clearing protocols have been developed because of the many advantages discussed above. Since each of these protocols has been designed for a specific experimental purpose and application, users need to understand their characteristics and select an appropriate protocol. The current situation may be complicated for researchers who have begun to consider the use of tissue clearing. Some reports on method-by-method comparisons may be helpful for selection. For example, the FP signal preservation capabilities of representative hydrophilic clearing methods have been systematically compared in the literature [21, 23, 31, 61, 88–90].

Opportunities for hydrophilic tissue clearing methods in histological and pathological applications

This section discusses recent advances in hydrophilic tissue clearing methods in histology and pathology as representative and promising applications.

Human 3D histology and pathology

Most tissue clearing methods have been validated for compatibility with tissue staining. Therefore, 3D human clinical pathology is an important area where these modern tissue clearing methods can contribute. Several organic solvent clearing methods have intensively tested human 3D histology examinations [37, 91, 92]. However, hydrophilic clearing methods also show promise in similar applications. Ueda, Tainaka, and colleagues designed a highly efficient clearing reagent for human tissues based on their chemical profiling

and reported an example of human tissue block clearing and 3D imaging [27, 41]. Gentleman and colleagues developed OPTIClear, which allowed them to successfully 3D stain and image human brain tissue a few millimeters thick. They quantitatively analyzed the 3D shape of catecholamine neurons in the midbrain and the 3D distribution of magnocellular neurons in the basal forebrain. Visvader and colleagues tested the FUnGI method on whole-mount immunostaining of human breast tissue from healthy women and the masses of breast tumor xenografts from patients [60]. Hildebrand, Galuske, and colleagues updated the FRUIT recipe [58] by adding sucrose and increasing the concentration of 1-thioglycerol up to 20%. The hFRUIT protocol with improved clearing efficiencies allowed partial LSFM imaging of the human brain labeled with lipophilic dyes [93].

3D observation of pathological specimens has been shown to improve diagnostic sensitivity and objectivity. Ueda, Morii, and colleagues compared conventional 2D cross-sectional examinations with 3D observations on lymph nodes dissected from human colorectal cancer patients (Figure 13.15). Their 3D observation succeeded in increasing the detection sensitivity of micrometastatic cancer cells from 85% of conventional 2D observation to 100% [94]. Liu and colleagues stained prostate cancer biopsies with eosin and a nuclear stain DRAQ5 and transparentized them using CLARITY. The 3D fluorescence images were converted to pseudotones of hematoxylin and eosin (H & E) staining and diagnosed by two independent pathologists. Interestingly, using some 2D images extracted from the data, their diagnosis following the Gleason score varied between 3 and 4. However, when examining the 3D images, both pathologists rated a score of 3. These results suggest that while 2D images do not provide sufficient information to assess atypicality of structure, 3D evaluation enables stable pathological judgment by providing continuous structural information [95].

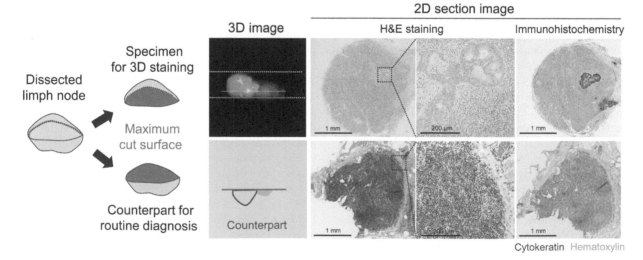

FIGURE 13.15 Example of 3D pathology to identify micrometastasis of colon cancer with hydrophilic tissue clearing method. The lymph nodes dissected from human colorectal cancer patients were half-cut and subjected to 1) 3D staining, CUBIC clearing, and light-sheet microscopy imaging, followed by a post-hoc 2D sectioning examination (position indicated by the orange line in the 3D image); 2) conventional 2D-based routine diagnosis. Cancer cell micrometastasis was occasionally detected by the 3D examination (arrowhead) but not by the 2D diagnosis. The sensitivity of micrometastatic cancer cell detection was 100% for the 3D imaging and 85% for the 2D diagnosis. Adapted from Nojima et al. (2017) [94].

These examples provide evidence that 3D pathology is potentially useful when 1) consensus among pathologists cannot be obtained due to difficulty in diagnosis, and 2) high diagnostic accuracy is required to determine the means of treatment (e.g., drug use or need for surgery). It is noted that several studies have tested sample deparaffinization before clearing and, conversely, paraffin embedding after clearing [94, 96, 97]. These favorable combinations with conventional pathological methods are vital for disseminating 3D pathology. In this regard, hydrophilic tissue clearing methods are potentially advantageous because of their safety and convenience.

3D pathology will also help ensure the teaching data necessary for training supervised machine learning classifiers. 3D imaging can collect significantly large amounts of data per specimen. Platforms for 3D microscope image analysis using machine learning are gradually being prepared, including Cell Profiler (https://cellprofiler.org), ilastik (https://www.ilastik .org), and Weka (Https://imagej.net/Trainable_Weka_Segm entation) [98–100]. Unlike CT and MRI, 3D pathology applications to clinical specimens and machine-learning-based computer diagnostics are still in their infancy. However, the current situation may help expand the use of tissue clearing and 3D imaging in future clinical computational pathology.

Design of versatile and robust 3D staining methods

Recent developments in advanced genetic tools and 3D staining methods have further expanded the use of tissue clearing and 3D imaging in biomedical applications. Improving efficiency in 3D staining is an urgent task, particularly considering the use of tissue clearing in human clinical research and examination, as discussed in the previous section.

Since a combination of tissue clearing and whole-mount staining using frog embryos was reported [101], attempts to stain and visualize 3D tissue samples have progressed almost parallel to the biological application of tissue clearing. Most clearing methods reported recently tested their applicability to probing the sample with dyes and antibodies. iDISCO, developed by Tessier-Lavigne and colleagues, is among the successful 3D staining and clearing methods, which can stain whole mouse embryos with various antibodies and entire mouse brains with c-Fos antibody [102, 103]. This protocol was initially proposed for organic solvent–based tissue clearing. However, a combinatory use of iDISCO with hydrophilic tissue clearing methods (e.g., CUBIC) was recently reported [104]. Among the hydrophilic clearing methods, Miyawaki and colleagues created ChemScale and AbScale, where urea may potentially help deeper penetration of dyes and antibodies. They used the methods for 3D immunolabeling of Aβ plaques in the hemisphere of a mouse model of Alzheimer's disease [21].

However, the staining conditions have been empirically determined due to the alchemistic feature of histological techniques. This limits the design of ideal 3D staining protocols. In particular, the probe penetration problem during 3D staining is significant and has never been entirely resolved. Even small molecule dyes can be resistant to dispersion in 3D tissue, suggesting a complex physicochemical environment in the staining system rather than simple reasons such as molecular weight.

Logical and evidence-based designs of 3D staining methods based on physicochemical principles have recently been attempted. Chung and colleagues have developed a stochastic electrotransport (SE) method that used a rotating electric field to enhance the dispersion of electromobility probes in a target sample [38]. SE supported homogeneous 3D staining throughout the whole mouse brain using a nuclear stain, lectin, and an antihistone antibody. They also devised a SWITCH method that controls antibody diffusion and staining steps with ionic detergent SDS [68]. In the latest study, they have developed a combination method named eFLASH, which realized the SE and SWITCH procedures simultaneously [105]. This method used an SE buffer containing deoxycholate and D-sorbitol. SE induced the degradation of these compounds and gradually switched the initial probe unbinding condition (high concentration of deoxycholate in alkaline pH) to the binding condition (decomposed deoxycholate and neutralized pH). eFLASH was able to use 3D immunostaining of mouse hemispheres and part of the marmoset cortical tissue using antibodies commonly used in neuroscience. This method also preserved the FP signal during staining. However, these approaches are specifically designed for samples fixed with polyacrylamide or epoxy compounds. Formalin-fixed samples may require special consideration. A device dedicated to electrophoresis is also required.

Susaki, Ueda, and colleagues recently attempted to develop a simple diffusion-based 3D staining method [70]. They first clarified the physicochemical properties of biological tissues as an electrolyte gel, mainly composed of the crosslinked polypeptides. Based on this insight, they extensively searched for optimal 3D staining conditions by using tissue-mimicking artificial gelatin gels and simulation of the diffusion–reaction scheme. They finally combined the identified requirements and successfully designed an ideal 3D staining protocol in a bottom-up manner. The new method, named CUBIC-HistoVIsion (CUBIC-HV), provided whole-organ and whole-body staining and imaging of large-volume samples, such as a mouse brain, a marmoset hemisphere, an infant marmoset body, and a human pathological specimen block (Figure 13.16). Importantly, applying these 3D staining requirements to iDISCO and AbScale significantly improved their staining efficiency, suggesting the development strategy based on tissue's physicochemical properties is common to other histological methods. However, it is of note that CUBIC-HV does not support multiplex labeling. Also, the simple diffusion strategy may potentially cause inhomogeneous staining in large 3D samples.

Opportunities for hydrophilic tissue clearing methods in organ- and organism-wide whole-cell profiling

Whole-cell profiling of whole organs and organisms has been achieved by combining advanced genetic and histological cell labeling tools, efficient tissue clearing, and rapid 3D imaging. This section focuses on one implementation form of this concept. First, an organ-scale single-cell resolution atlas is created as a reference by detecting cell locations and converting

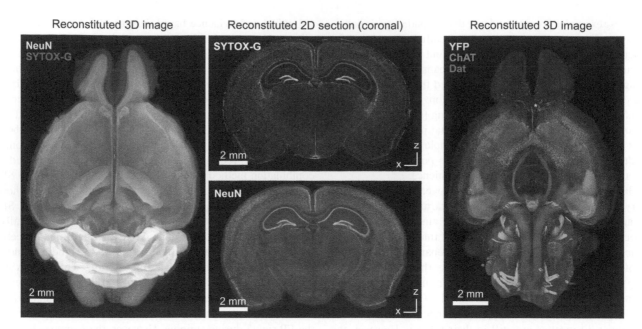

FIGURE 13.16 Versatile whole-organ 3D staining and imaging with hydrophilic tissue clearing method. (Left and middle) Whole-organ staining and 3D imaging of adult mouse brain. The sample was stained with an anti-NeuN antibody and a nuclear stain SYTOX-G. The middle part of the image was reconstituted as coronal sections to show the internal staining results. (Right) Whole-organ staining and 3D imaging of adult Thy1-YFP-H Tg mouse brain. The sample was stained with anti-ChAT and anti-Dat antibodies. Both of the FP transgene signals and the immunostained signals were observable. The reconstituted 3D images are views from the ventral side. Adapted from Susaki et al. (2020) [70].

them to coordinate data. Then, information about cell types, activities, and connections is added to this *blank map*. Nuclear labeling, high-resolution imaging, and computational detection of all cells in the organs is the first part. Low-magnification whole-organ imaging of many samples and quantitative comparison analysis of the labeled cells on the standard reference data is the key to the second part.

Ueda and colleagues have already demonstrated a primitive form of this concept. They compared the whole-brain neural activity of multiple Arc-dVenus transgenic mice [82]. In this Tg strain, the destabilized Venus transgene expression is induced by the promoter of the immediate-early gene Arc during neural excitation. They successfully compared the whole-brain activity of two mice with and without light stimulation [26]. Later, they collected whole-brain activity images of 20 mice under eight different experimental conditions (drug-treated or nontreated groups × four time points). After computational detection of the labeled cells, a clustering analysis, as in the case of transcriptome data processing, was applied. These experiments successfully extracted brain regions showing different activity patterns associated with specific experimental conditions [83]. The following sessions discuss the more recent development of this research scheme.

Whole-organ atlas with single-cell resolution

A 3D organ atlas with anatomical information serves as a reference for multiple sample registration and analysis. Waxholm Space Atlas [106] and Allen Brain Atlas (ABA) [107] were used for organ-scale whole-cell profiling frameworks [26, 103, 108]. They are composed of volume-based regional data, and experts have determined the region boundaries. These features

restrict end-users from editing the content to incorporate their findings. In addition, previous ABA versions consisted of a series of 2D cross-sectional images, resulting in shaggy region boundaries (the latest version has remedied this problem [109]).

Point-based cell coordinate data describe another possible form of 3D organ atlas. Such cell-based atlases can reflect the architectures of organs and organisms by the basic unit of life and can be edited by users according to research-driven discoveries on different cell types, states, and connections. Ueda and colleagues tried to actually create a digitized whole mouse brain single-cell resolution atlas containing coordinates of all cells (about 100 million) with ABA's region information [69, 110]. To this end, they have developed a new CUBIC-X protocol for achieving simultaneous swelling and clearing of samples to improve transparency and image resolution, based on the idea of expansion microscopy [75]. They also built a custom light-sheet microscope equipped with a high-resolution objective (10× /NA0.6) to perform whole-brain imaging of swollen samples at subcellular resolution. They finally collected approximately 1.3 million images (~14 TB) of all cell nuclei in the mouse brain prepared with CUBIC-X and propidium iodide staining and calculated the cell nuclear coordinates by using a GPU-based nucleus detection algorithm. The final coordinate data to which the ABA region information was transferred constituted CUBIC-Atlas, the world's first whole-cell single-cell resolution atlas (Figure 13.17).

CUBIC-Atlas works as a whole-organ single-cell resolution reference for multisample analyses. To prove the concept, the whole-brain neural activity data in Tatsuki et al. (2016) [83] was reanalyzed using CUBIC-Atlas. Probabilistic mapping and annotation techniques were used to map labeled cells from different brains to the closest coordinates within the

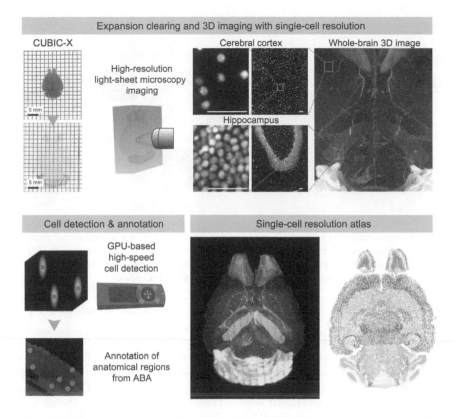

FIGURE 13.17 Construction of the CUBIC single-cell resolution atlas. Whole-organ data resources, consisting of all cell coordinates, facilitates whole cell profiling at organ scale. Ueda and colleagues tried to develop CUBIC-Atlas, the world's first whole-cell single-cell resolution atlas. The clearing and expansion protocol CUBIC-X was designed to obtain whole-organ single-cell resolution data (1.3 million images, approximately 14 TB) using a specialized light-sheet microscope equipped with a high-resolution objective lens. The cell coordinates were calculated with a GPU-based algorithm and annotated from Allen Brain Atlas (ABA). Finally, the whole-organ single-cell resolution atlas of the adult mouse brain was constructed. Scale bars in the microscopy images were 50 μm, using a pre-CUBIC-X brain as a baseline. Adapted from Murakami et al. (2018) [69].

single-cell resolution atlas. This whole-brain analysis revealed that the upper and lower regions of the granule cell layer of the dentate gyrus are distinct subregions belonging to different neural activity clusters [69].

Although the CUBIC-Atlas was previously prepared only for the whole mouse brain, in principle, it can also be extended to various whole organs and the whole animal body. A recent light-sheet microscope system for faster and higher resolution imaging, named MOVIE, can accelerate this direction [110]. The system delivers between five and eight times faster imaging than the previous CUBIC-X whole-brain imaging system [69] (several hours vs. 3–4 days per whole mouse brain) by integrating 1) continuous acquisition (MOVIE-scan), 2) real-time autofocusing (MOVIE-focus), and 3) skipping of blank images (MOVIE-skip). MOVIE-scan was particularly effective in reducing the imaging time. The prevailing light-sheet microscopies employ a stop-and-exposure operation where each image at a specific z-position is acquired after moving and stopping the stage. The total imaging duration mainly depends on this positioning time. To circumvent this constraint, MOVIE-scan realized a combination of uniform stage movement and synchronized continuous image acquisition. MOVIE-skip also contributed to shorter imaging times by skipping areas outside the sample. Frequent focal plane adjustments with MOVIE-focus ensured the image quality of the entire organ sample.

Associated with the MOVIE system development, an improved, highly parallelized nuclear detection algorithm was also investigated. This algorithm achieved >90% accuracy (F-score) to identify whole-cell coordinates throughout the adult mouse brain. The CUBIC atlas of an 8-week-old C57BL/6N mouse brain was recounted by the refined algorithm. The accuracy of the detection was improved to identify the cells in adult mouse brain with the number of approximately 1×10^8 cells [110].

In contrast to the volume-based atlas, the point-based atlas can provide a comparison analysis platform and an editable information resource across all organs and species. Biologically relevant cell coordinates, rather than arbitrary region boundaries, provide versatile functionality. The hydrophilic tissue clearing method, along with compatible whole organ nuclear staining, has contributed significantly to the development of this new resource for whole-cell profiling.

Whole-organ comparison analysis of multiple samples

Comprehensive mapping and comparing cell types, cell functions, and cell connections throughout the organ allow for the extraction of the functional regions and essential cellular components in an unbiased manner. With the advent of modern tissue clearing methods and light-sheet microscopy

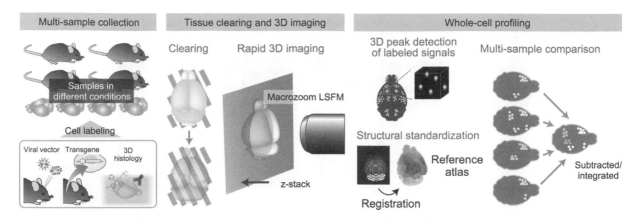

FIGURE 13.18 Scheme of organ-wide whole-cell profiling with tissue clearing. This figure outlines the organ-scale multisample comparison analysis framework described in this section. The collected, prelabeled multisamples are then subjected to tissue clearing and 3D imaging (typically, a large-field, relatively low-resolution imaging meets the requirements for this purpose). After calculating the detected cell coordinates and standardizing the structure, the multisample data can be compared directly on a unified platform such as CUBIC-Atlas.

applications, an organ-scale multisample comparison has become possible, avoiding the laborious preparation of consecutive 2D physical slices and their image acquisition (Figure 13.18). Similar to the CUBIC demonstration that profiled 20 whole-brain neural activities as described above [69, 83], one of the organic solvent–based methods, iDISCO+, has realized whole-brain c-Fos immunostaining and signal mapping (ClearMap) to analyze brain-scale neural activities [103]. Uchida and colleagues applied CLARITY to trace rabies virus–labeled neural circuits. They successfully performed a whole-brain analysis of dopaminergic neural circuits by clearing and imaging a total of 77 brains [111]. This type of analysis has been spreading by using various tissue clearing approaches. However, given the scalability of the comparison analysis, which requires processing large numbers of samples, hydrophilic clearing methods seem advantageous due to their simplicity, reproducibility, and compatibility with multiple labeling tools.

The sample should be prelabeled with cell type or activity markers, depending on the experimental purposes. Advanced genetic tools and histological staining methods facilitate organ-scale cell labeling. For example, AAV derivatives discovered by Cre recombination-based AAV targeted evolution (CREATE) allows systemic gene delivery to the specific organs (e.g., brain labeling by AAV-PHP.eB) via peripheral injection of the virus [112, 113]. Whole-organ/body histological labeling and imaging have also been recently developed [70]. This study realized whole-organ/body nuclear staining to obtain the global structural information of the sample needed to register the data to the reference.

The labeled sample is then subjected to clearing and 3D imaging. In most cases of multiple sample data collection, large-field, relatively low-resolution imaging can be applied to meet imaging throughput and data size suppression requirements. Employing a high-speed macro zoom light-sheet microscopy such as mesoSPIM [114], which can image a whole mouse brain within 7–8 min with a resolution of 6.5 μm^3, improves the feasibility of organ-scale whole-cell profiling. Generally, the data size collected by such a macro zoom light-sheet microscopy only reaches several tens of gigabytes

at most, which can be sufficiently processed by a desktop workstation.

When cells labeled with specific cell types or activity markers are sparsely distributed in the sample, low-resolution microscopy can provide sufficient image data quality for subsequent computational cell detection and analysis [70, 103]. However, applications for axonal tracing, identification of dense cell populations, and detection of intracellular organelles require the use of higher resolution microscopes. The MOVIE system with a high resolution (10× /NA0.6) objective lens was used for accurate cell detection and counting of whole-brain samples labeled by AAV-PHP.eB and cleared by CUBIC [110]. The hydrophilic clearing methods also facilitated an alternative whole-organ high-resolution imaging with automated serial sectioning and imaging systems. Chandrashekar, Myers, and colleagues took advantage of delipidation by ScaleCUBIC-1 and RI-matching by the mixture of DMSO and D-sorbitol for their serial two-photon tomography. This combination significantly increased the acquisition depth from the cut surface (> 200 μm) and allowed the imaging of fine axon structures in the entire mouse brain volume. The system was finally used for collecting neural projections sparsely labeled by FP-harboring AAV [59, 115]. Gong and colleagues used a mixture of sugar and urea for their wide-field large-volume sectioning tomography system to increase imaging depth (8×) and shorten acquisition time (1/2×) [62]. It is of note that these sectioning systems may have lower scalability in imaging of multiple samples due to slower throughput (several days per whole mouse brain imaging) compared to light-sheet microscopy.

To quantitatively analyze the obtained multisample data on the single-cell resolution atlas, the coordinates of labeled cells should first be calculated. Then, the nuclear stain data is registered and aligned to the reference atlas to calculate the registration matrix (standardization). This matrix is applied to the cell coordinates for standardization. This operation allows a comparison analysis of many whole-organ samples on the same platform. Future software developments for data visualization and quantitative analysis will help end-users incorporate this organ-wide whole-cell profiling scheme into their studies.

Conclusions and perspectives

Recently developed tissue clearing technologies, especially hydrophilic reagents and protocols, offer a wide range of opportunities for various biomedical applications. The hydrophilic tissue clearing methods utilize the intrinsic properties of the tissue itself to achieve the clearing state. This approach is particularly contrasted with the hydrogel–tissue chemistry techniques, where artificial gels and strong fixatives were used to artificially add new properties for the specific clearing and labeling purposes [39, 68, 116, 117]. Contributions of many developers and large-scale profiling have revealed the chemical categories of hydrophilic clearing agents and their representative compounds. These chemicals were characterized according to the properties required for the clearing processes. These accumulated insights enable flexible protocol design of the methods.

This chapter focuses on two noteworthy applications. Human 3D histopathology is a potential implementation of tissue clearing technology in society. Hydrophilic tissue clearing methods are highly compatible with conventional histopathological methods. In addition, the safety of reagents and ease of protocol open up possibilities for automation in routine diagnostics. Whole-organ/body-scale whole-cell profiling is also a promising application to take advantage of the scalability of the hydrophilic clearing method. Again, the simplicity of the methods is crucial for handling large numbers of samples. Extending this approach to other model organisms such as rats and marmosets may enable a wide range of applications in diverse basic research, drug discovery, and toxicity assessments. Improving throughput at every step of clearing, staining, imaging, and data analysis will be essential.

Converting all human cells into digitized data is one of the ultimate goals of tissue clearing technology. It requires attempts to evolve more efficient technologies by linking different clearing strategies. Zhao and colleagues reported a PEGASOS method combining organic solvent reagents and water-soluble compounds, demonstrating systematic clearing of animals, including bones [45]. vDISCO and uDISCO reported by Ertürk and colleagues used some water-soluble decalcifying and decolorizing agents. They recently reported the SHANEL method, which employs CHAPS and NMDEA to clear the entire human brain and kidney [37]. Chemicals with high delipidation and RI matching capabilities found in CUBIC chemical profiling [27] may contain agents that are better suited for clearing human tissues. Further refinements of the clearing methods contribute to whole-cell profiling of the human body, providing an approach that complements sequence-based cell cataloging such as the Human Cell Atlas project (https://www.humancellatlas.org/).

REFERENCES

1. W. Spalteholz. *Über das Durchsichtigmachen von menschlichen und tierischen Präparaten.* S. Hirzel, Leipzig (1914).
2. H.U. Dodt, U. Leischner, A. Schierloh, et al. "Ultramicroscopy: Three-dimensional visualization of neuronal networks in the whole mouse brain". *Nat. Methods.* 4(4), 331–336 (2007).
3. V.V. Bakutkin, I.L. Maksimova, T.N. Semyonova, V.V. Tuchin, and I.L. Kon. "Controlling optical properties of sclera". *Proc. SPIE.* 2393 (1995).
4. D.A. Zimnyakov, V.V. Tuchin, A.A. Michin, I.L. Kon, and A.N. Serov. "In-vitro human sclera structure analysis using tissue optical immersion effect". *Proc. SPIE.* 2673 (1996).
5. V.V. Tuchin, A.N. Bashkatov, E.A. Genina, et al. "Optics of living tissues with controlled scattering properties". *Proc. SPIE.* 3863, 10–21 (1999).
6. V.V. Tuchin, I.L. Maksimova, D.A. Zimnyakov, et al. "Light propagation in tissues with controlled optical properties". *J. Biomed. Opt.* 2(4), 401–417 (1997).
7. A.N. Bashkatov, V.V. Tuchin, E.A. Genina, et al. "Human sclera dynamic spectra: In-vitro and in-vivo measurements". *Proc. SPIE.* 3591 (1999).
8. V.V. Tuchin, X. Xu, and R.K. Wang. "Dynamic optical coherence tomography in studies of optical clearing, sedimentation, and aggregation of immersed blood". *Appl. Opt.* 41(1), 258–271 (2002).
9. X. Xu, R.K. Wang, J.B. Elder, and V.V. Tuchin. "Effect of dextran-induced changes in refractive index and aggregation on optical properties of whole blood". *Phys. Med. Biol.* 48(9), 1205 (2003).
10. O. Vargas, E.K. Chan, J.K. Barton, H.G. Rylander, and A.J. Welch. "Use of an agent to reduce scattering in skin". *Lasers Surg. Med.* 24(2), 133–141 (1999).
11. G. Vargas, K.F. Chan, S.L. Thomsen, and A.J. Welch. "Use of osmotically active agents to alter optical properties of tissue: Effects on the detected fluorescence signal measured through skin". *Lasers Surg. Med.* 29(3), 213–220 (2001).
12. H. Liu, B. Beauvoit, M. Kimura, and B. Chance. "Dependence of tissue optical properties on solute-induced changes in refractive index and osmolarity". *J. Biomed. Opt.* 1(2), 200–211 (1996).
13. B. Chance, H. Liu, T. Kitai, and Y. Zhang. "Effects of solutes on optical properties of biological materials: Models, cells, and tissues". *Anal. Biochem.* 227(2), 351–362 (1995).
14. R.K. Wang, X. Xu, V.V. Tuchin, and J.B. Elder. "Concurrent enhancement of imaging depth and contrast for optical coherence tomography by hyperosmotic agents". *J. Opt. Soc. Am. B.* 18(7), 948–953 (2001).
15. X. Xu and R.K. Wang. "The role of water desorption on optical clearing of biotissue: Studied with near infrared reflectance spectroscopy". *Med. Phys.* 30(6), 1246–1253 (2003).
16. J. Jiang and R.K. Wang. "Comparing the synergistic effects of oleic acid and dimethyl sulfoxide as vehicles for optical clearing of skin tissue in vitro". *Phys. Med. Biol.* 49(23), 5283–5294 (2004).
17. B. Choi, L. Tsu, E. Chen, et al. "Determination of chemical agent optical clearing potential using in vitro human skin". *Lasers Surg. Med.* 36(2), 72–75 (2005).
18. A. Chiang, Y. Liu, S. Chiu, et al. "Three-dimensional mapping of brain neuropils in the cockroach, Diploptera punctata". *J. Comp. Neurol.* 440(1), 1–11 (2001).
19. Y.-C. Liu and A.-S. Chiang. "High-resolution confocal imaging and three-dimensional rendering". *Methods.* 30(1), 86–93 (2003).
20. H. Hama, H. Kurokawa, H. Kawano, et al. "Scale: A chemical approach for fluorescence imaging and reconstruction of transparent mouse brain". *Nat. Neurosci.* 14(11), 1481–1488 (2011).

21. H. Hama, H. Hioki, K. Namiki, et al. "ScaleS: An optical clearing palette for biological imaging". *Nat. Neurosci.* 18(10), 1518–1529 (2015).

22. M.-T. Ke, S. Fujimoto, and T. Imai. "SeeDB: a simple and morphology-preserving optical clearing agent for neuronal circuit reconstruction". *Nat. Neurosci.* 16(8), 1154–1161 (2013).

23. M.-T. Ke, Y. Nakai, S. Fujimoto, et al. "Super-resolution mapping of neuronal circuitry with an index-optimized clearing agent". *Cell Rep.* 14(11), 2718–2732 (2016).

24. T. Kuwajima, A.A. Sitko, P. Bhansali, et al. "ClearT: A detergent- and solvent-free clearing method for neuronal and non-neuronal tissue". *Development.* 140(6), 1364–1368 (2013).

25. K. Tainaka, S.I. Kubota, T.Q. Suyama, et al. "Whole-body imaging with single-cell resolution by tissue decolorization". *Cell.* 159(4), 911–924 (2014).

26. E.A. Susaki, K. Tainaka, D. Perrin, et al. "Whole-brain imaging with single-cell resolution using chemical cocktails and computational analysis". *Cell.* 157(3), 726–739 (2014).

27. K. Tainaka, T.C. Murakami, E.A. Susaki, et al. "Chemical landscape for tissue clearing based on hydrophilic reagents". *Cell Rep.* 24(8), 2196–2210 (2018).

28. E.A. Susaki and H.R. Ueda. "Whole-body and whole-organ clearing and imaging techniques with single-cell resolution: Toward organism-level systems biology in mammals". *Cell Chem. Biol.* 23(1), 137–157 (2016).

29. H.R. Ueda, A. Erturk, K. Chung, et al. "Tissue clearing and its applications in neuroscience". *Nat. Rev. Neurosci.* 21(2), 61–79 (2020).

30. H.R. Ueda, H.U. Dodt, P. Osten, et al. "Whole-brain profiling of cells and circuits in mammals by tissue clearing and light-sheet microscopy". *Neuron.* 106(3), 369–387 (2020).

31. L. Chen, G. Li, Y. Li, et al. "UbasM: An effective balanced optical clearing method for intact biomedical imaging". *Sci. Rep.* 7(1), 12218 (2017).

32. W. Li, R.N. Germain, and M.Y. Gerner. "Multiplex, quantitative cellular analysis in large tissue volumes with clearing-enhanced 3D microscopy (Ce3D)". *Proc. Natl. Acad. Sci. U.S.A.* 114(35), E7321–E7330 (2017).

33. D. Kurihara, Y. Mizuta, Y. Sato, and T. Higashiyama. "ClearSee: A rapid optical clearing reagent for whole-plant fluorescence imaging". *Development.* 142(23), 4165–4179 (2015).

34. H.M. Lai, A.K.L. Liu, H.H.M. Ng, et al. "Next generation histology methods for three-dimensional imaging of fresh and archival human brain tissues". *Nat. Commun.* 9(1), 1066 (2018).

35. M. Pende, K. Becker, M. Wanis, et al. "High-resolution ultramicroscopy of the developing and adult nervous system in optically cleared Drosophila melanogaster". *Nat. Commun.* 9(1), 4731 (2018).

36. T. Miyawaki, S. Morikawa, E.A. Susaki, et al. "Visualization and molecular characterization of whole-brain vascular networks with capillary resolution". *Nat. Commun.* 11(1), 1104 (2020).

37. S. Zhao, M.I. Todorov, R. Cai, et al. "Cellular and molecular probing of intact human organs". *Cell.* 180(4), 796–812 e19 (2020).

38. S.Y. Kim, J.H. Cho, E. Murray, et al. "Stochastic electrotransport selectively enhances the transport of highly electromobile molecules". *Proc. Natl. Acad. Sci. U.S.A.* 112(46), E6274–83 (2015).

39. K. Chung, J. Wallace, S.Y. Kim, et al. "Structural and molecular interrogation of intact biological systems". *Nature.* 497(7449), 332–337 (2013).

40. B. Yang, J.B. Treweek, R.P. Kulkarni, et al. "Single-cell phenotyping within transparent intact tissue through whole-body clearing". *Cell.* 158(4), 945–958 (2014).

41. M. Inoue, R. Saito, A. Kakita, and K. Tainaka. "Rapid chemical clearing of white matter in the post-mortem human brain by 1,2-hexanediol delipidation". *Bioorg. Med. Chem. Lett.* 29(15), 1886–1890 (2019).

42. K. Tainaka, A. Kuno, S.I. Kubota, T. Murakami, and H.R. Ueda. "Chemical principles in tissue clearing and staining protocols for whole-body cell profiling". *Annu. Rev. Cell Dev. Biol.* 32, 713–741 (2016).

43. R. Cai, C. Pan, A. Ghasemigharagoz, et al. "Panoptic imaging of transparent mice reveals whole-body neuronal projections and skull-meninges connections". *Nat. Neurosci.* 22(2), 317–327 (2019).

44. J.B. Treweek, K.Y. Chan, N.C. Flytzanis, et al. "Whole-body tissue stabilization and selective extractions via tissue-hydrogel hybrids for high-resolution intact circuit mapping and phenotyping". *Nat. Protoc.* 10(11), 1860–1896 (2015).

45. D. Jing, S. Zhang, W. Luo, et al. "Tissue clearing of both hard and soft tissue organs with the PEGASOS method". *Cell Res.* 28(8), 803–818 (2018).

46. J. Zhu, T. Yu, Y. Li, et al. "MACS: Rapid aqueous clearing system for 3D mapping of intact organs". *Adv. Sci. Lett.* 7, 1903185 (2020).

47. M. Kuroda and S. Kuroda. "Whole-body clearing of beetles by successive treatment with hydrogen peroxide and CUBIC reagents". *Entomol. Sci.* 12, 15 (2020).

48. K. Futami, O. Furukawa, M. Maita, and T. Katagiri. "Application of hydrogen peroxide- melanin bleaching and fluorescent nuclear staining for whole-body clearing and imaging in fish". *Fish Pathol.* 54(4), 101–103 (2020).

49. Y. Ye, T.A.D. Duong, K. Saito, et al. "Visualization of the retina in intact eyes of mice and ferrets using a tissue clearing method". *Transl. Vis. Sci. Technol.* 9(3), 1 (2020).

50. A. Konno and S. Okazaki. "Aqueous-based tissue clearing in crustaceans". *Zoological Lett.* 4, 13 (2018).

51. M. Pende, K. Vadiwala, H. Schmidbaur, et al. "A versatile depigmentation, clearing, and labeling method for exploring nervous system diversity". *Sci Adv.* 6(22), eaba0365 (2020).

52. J. Hasegawa, Y. Sakamoto, S. Nakagami, et al. "Three-dimensional imaging of plant organs using a simple and rapid transparency technique". *Plant Cell Physiol.* 57(3), 462–472 (2016).

53. A. Greenbaum, K.Y. Chan, T. Dobreva, et al. "Bone CLARITY: Clearing, imaging, and computational analysis of osteoprogenitors within intact bone marrow". *Sci. Transl. Med.* 9(387), eaah6518 (2017).

54. Q. Xie, N. Zeng, Y. Huang, V.V. Tuchin, and H. Ma. "Study on the tissue clearing process using different agents by Mueller matrix microscope". *Biomed. Opt. Express.* 10(7), 3269–3280 (2019).

55. N. Kuznetsova, S.L. Chi, and S. Leikin. "Sugars and polyols inhibit fibrillogenesis of type I collagen by disrupting hydrogen-bonded water bridges between the helices". *Biochemistry.* 37(34), 11888–11895 (1998).

56. J. Hirshburg, B. Choi, J.S. Nelson, and A.T. Yeh. "Correlation between collagen solubility and skin optical clearing using sugars". *Lasers Surg. Med.* 39(2), 140–144 (2007).

57. A.T. Yeh and J. Hirshburg. "Molecular interactions of exogenous chemical agents with collagen – implications for tissue optical clearing". *J. Biomed. Opt.* 11(1), 014003 (2006).

58. B. Hou, D. Zhang, S. Zhao, et al. "Scalable and DiI-compatible optical clearance of the mammalian brain". *Front. Neuroanat.* 9, 19 (2015).

59. M.N. Economo, N.G. Clack, L.D. Lavis, et al. "A platform for brain-wide imaging and reconstruction of individual neurons". *Elife.* 5, e10566 (2016).

60. A.C. Rios, B.D. Capaldo, F. Vaillant, et al. "Intraclonal plasticity in mammary tumors revealed through large-scale single-cell resolution 3D imaging". *Cancer Cell.* 35(4), 618–632 e6 (2019).

61. X. Zhu, L. Huang, Y. Zheng, et al. "Ultrafast optical clearing method for three-dimensional imaging with cellular resolution". *Proc. Natl. Acad. Sci. U.S.A.* 116(23), 11480–11489 (2019).

62. H. Wu, X. Yang, S. Chen, et al. "On-line optical clearing method for whole-brain imaging in mice". *Biomed. Opt. Express.* 10(5), 2612–2622 (2019).

63. H. Mizutani, S. Ono, T. Ushiku, et al. "Transparency-enhancing technology allows three-dimensional assessment of gastrointestinal mucosa: A porcine model". *Pathol. Int.* 68(2), 102–108 (2018).

64. T. Yu, J. Zhu, Y. Li, et al. "RTF: A rapid and versatile tissue optical clearing method". *Sci. Rep.* 8(1), 1964 (2018).

65. T. Staudt, M.C. Lang, R. Medda, J. Engelhardt, and S.W. Hell. "2,2′-Thiodiethanol: a new water soluble mounting medium for high resolution optical microscopy". *Microsc. Res. Tech.* 70(1), 1–9 (2007).

66. I. Costantini, J.P. Ghobril, A.P. Di Giovanna, et al. "A versatile clearing agent for multi-modal brain imaging". *Sci. Rep.* 5, 9808 (2015).

67. Y. Aoyagi, R. Kawakami, H. Osanai, T. Hibi, and T. Nemoto. "A rapid optical clearing protocol using 2,2′-thiodiethanol for microscopic observation of fixed mouse brain". *PLoS One.* 10(1), e0116280 (2015).

68. E. Murray, J.H. Cho, D. Goodwin, et al. "Simple, scalable proteomic imaging for high-dimensional profiling of intact systems". *Cell.* 163(6), 1500–1514 (2015).

69. T.C. Murakami, T. Mano, S. Saikawa, et al. "A three-dimensional single-cell-resolution whole-brain atlas using CUBIC-X expansion microscopy and tissue clearing". *Nat. Neurosci.* 21(4), 625–637 (2018).

70. E.A. Susaki, C. Shimizu, A. Kuno, et al. "Versatile whole-organ/body staining and imaging based on electrolyte-gel properties of biological tissues". *Nat. Commun.* 11(1), 1982 (2020).

71. T. Amiya and T. Tanaka. "Phase transitions in crosslinked gels of natural polymers". *Macromolecules.* 20(5), 1162–1164 (1987).

72. M. Shibayama and T. Tanaka. "Volume phase transition and related phenomena of polymer gels". In: *Responsive Gels: Volume Transitions I*, Springer, Berlin, 1–62 (1993).

73. Z. Wang, J. Zhang, G. Fan, et al. "Imaging transparent intact cardiac tissue with single-cell resolution". *Biomed. Opt. Express.* 9(2), 423–436 (2018).

74. P. Jedlovszky and A. Idrissi. "Hydration free energy difference of acetone, acetamide, and urea". *J. Chem. Phys.* 129(16), 164501 (2008).

75. F. Chen, P.W. Tillberg, and E.S. Boyden. "Expansion microscopy". *Science.* 347(6221), 543–548 (2015).

76. T. Ku, J. Swaney, J.Y. Park, et al. "Multiplexed and scalable super-resolution imaging of three-dimensional protein localization in size-adjustable tissues". *Nat. Biotechnol.* 34(9), 973–981 (2016).

77. Y.-J. Zhao, T.-T. Yu, C. Zhang, et al. "Skull optical clearing window for in vivo imaging of the mouse cortex at synaptic resolution". *Light Sci Appl.* 7, 17153 (2018).

78. C. Zhang, W. Feng, Y. Zhao, et al. "A large, switchable optical clearing skull window for cerebrovascular imaging". *Theranostics.* 8(10), 2696–2708 (2018).

79. T. Boothe, L. Hilbert, M. Heide, et al. "A tunable refractive index matching medium for live imaging cells, tissues and model organisms". *Elife.* 6, e27240 (2017).

80. K. Iijima, T. Oshima, R. Kawakami, and T. Nemoto. "Optical clearing of living brains with MAGICAL to extend in vivo imaging". *Science* 24(1), 101888 (2020).

81. R. Sakaguchi, M.N. Leiwe, and T. Imai. "Bright multicolor labeling of neuronal circuits with fluorescent proteins and chemical tags". *Elife.* 7 (2018).

82. M. Eguchi and S. Yamaguchi. "In vivo and in vitro visualization of gene expression dynamics over extensive areas of the brain". *Neuroimage.* 44(4), 1274–1283 (2009).

83. F. Tatsuki, G.A. Sunagawa, S. Shi, et al. "Involvement of Ca(2+)-dependent hyperpolarization in sleep duration in mammals". *Neuron.* 90(1), 70–85 (2016).

84. L. Wang, S. Gillis-Smith, Y. Peng, et al. "The coding of valence and identity in the mammalian taste system". *Nature.* 558(7708), 127–131 (2018).

85. S.I. Kubota, K. Takahashi, J. Nishida, et al. "Whole-body profiling of cancer metastasis with single-cell resolution". *Cell Rep.* 20(1), 236–250 (2017).

86. W. Li, R.N. Germain, and M.Y. Gerner. "High-dimensional cell-level analysis of tissues with Ce3D multiplex volume imaging". *Nat. Protoc.* 14(6), 1708–1733 (2019).

87. V. Gradinaru, J. Treweek, K. Overton, and K. Deisseroth. "Hydrogel-tissue chemistry: Principles and applications". *Annu. Rev. Biophys.* 47, 355–376 (2018).

88. P. Matryba, A. Sosnowska, A. Wolny, et al. "Systematic evaluation of chemically distinct tissue optical clearing techniques in murine lymph nodes". *J. Immunol.* 204(5), 1395–1407 (2020).

89. J. Xu, Y. Ma, T. Yu, and D. Zhu. "Quantitative assessment of optical clearing methods in various intact mouse organs". *J. Biophotonics.* 12(2), e201800134 (2019).

90. P. Wan, J. Zhu, J. Xu, et al. "Evaluation of seven optical clearing methods in mouse brain". *Neurophotonics.* 5(3), 035007 (2018).

91. M. Belle, D. Godefroy, G. Couly, et al. "Tridimensional visualization and analysis of early human development". *Cell.* 169(1), 161–173 e12 (2017).

92. S. Hildebrand, A. Schueth, A. Herrler, R. Galuske, and A. Roebroeck. "Scalable labeling for cytoarchitectonic characterization of large optically cleared human neocortex samples". *Sci. Rep.* 9(1), 10880 (2019).

93. S. Hildebrand, A. Schueth, K. von Wangenheim, et al. "hFRUIT: An optimized agent for optical clearing of DiI-stained adult human brain tissue". *Sci. Rep.* 10(1), 9950 (2020).

94. S. Nojima, E.A. Susaki, K. Yoshida, et al. "CUBIC pathology: Three-dimensional imaging for pathological diagnosis". *Sci. Rep.* 7(1), 9269 (2017).

95. A.K. Glaser, N.P. Reder, Y. Chen, et al. "Light-sheet microscopy for slide-free non-destructive pathology of large clinical specimens". *Nat. Biomed. Eng.* 1(7), 0084 (2017).

96. Y. Chen, Q. Shen, S.L. White, et al. "Three-dimensional imaging and quantitative analysis in CLARITY processed breast cancer tissues". *Sci. Rep.* 9(1), 5624 (2019).

97. M.E. van Royen, E.I. Verhoef, C.F. Kweldam, et al. "Three-dimensional microscopic analysis of clinical prostate specimens". *Histopathology.* 69(6), 985–992 (2016).

98. C. McQuin, A. Goodman, V. Chernyshev, et al. "CellProfiler 3.0: Next-generation image processing for biology". *PLoS Biol.* 16(7), e2005970 (2018).

99. Sommer, C.; Straehle, C.; Kothe, U.; Hamprecht, F.A.,. "Ilastik: Interactive learning and segmentation toolkit". *Proc. IEEE Int. Symp. Biomed. Imaging.* 230–233 (2011).

100. I. Arganda-Carreras, V. Kaynig, C. Rueden, et al. "Trainable Weka segmentation: A machine learning tool for microscopy pixel classification". *Bioinformatics.* 33(15), 2424–2426 (2017).

101. J.A. Dent, A.G. Polson, and M.W. Klymkowsky. "A whole-mount immunocytochemical analysis of the expression of the intermediate filament protein vimentin in Xenopus". *Development.* 105(1), 61–74 (1989).

102. N. Renier, Z. Wu, D.J. Simon, et al. "iDISCO: A simple, rapid method to immunolabel large tissue samples for volume imaging". *Cell.* 159(4), 896–910 (2014).

103. N. Renier, E.L. Adams, C. Kirst, et al. "Mapping of brain activity by automated volume analysis of immediate early genes". *Cell.* 165(7), 1789–1802 (2016).

104. J. McKey, L.A. Cameron, D. Lewis, I.S. Batchvarov, and B. Capel. "Combined iDISCO and CUBIC tissue clearing and lightsheet microscopy for in toto analysis of the adult mouse ovary†". *Biol. Reprod.* 102(5), 1080–1089 (2020).

105. D.H. Yun, Y.-G. Park, J.H. Cho, et al. "Ultrafast immunostaining of organ-scale tissues for scalable proteomic phenotyping". bioRxiv. 660373 (2019).

106. G.A. Johnson, A. Badea, J. Brandenburg, et al. "Waxholm space: An image-based reference for coordinating mouse brain research". *Neuroimage.* 53(2), 365–372 (2010).

107. E.S. Lein, M.J. Hawrylycz, N. Ao, et al. "Genome-wide atlas of gene expression in the adult mouse brain". *Nature.* 445(7124), 168–176 (2007).

108. E.A. Susaki, K. Tainaka, D. Perrin, et al. "Advanced CUBIC protocols for whole-brain and whole-body clearing and imaging". *Nat. Protoc.* 10(11), 1709–1727 (2015).

109. Q. Wang, S.-L. Ding, Y. Li, et al. "The Allen mouse brain common coordinate framework: A 3D reference atlas". *Cell.* 181(4), 936–953 .e20 (2020).

110. K. Matsumoto, T.T. Mitani, S.A. Horiguchi, et al. "Advanced CUBIC tissue clearing for whole-organ cell profiling". *Nat. Protoc.* 14(12), 3506–3537 (2019).

111. W. Menegas, J.F. Bergan, S.K. Ogawa, et al. "Dopamine neurons projecting to the posterior striatum form an anatomically distinct subclass". *Elife.* 4, 230–233 (2015).

112. B.E. Deverman, P.L. Pravdo, B.P. Simpson, et al. "Cre-dependent selection yields AAV variants for widespread gene transfer to the adult brain". *Nat. Biotechnol.* 34(2), 204–209 (2016).

113. K.Y. Chan, M.J. Jang, B.B. Yoo, et al. "Engineered AAVs for efficient noninvasive gene delivery to the central and peripheral nervous systems". *Nat. Neurosci.* 20(8), 1172–1179 (2017).

114. F.F. Voigt, D. Kirschenbaum, E. Platonova, et al. "The mesoSPIM initiative: Open-source light-sheet microscopes for imaging cleared tissue". *Nat. Methods.* 16(11), 1105–1108 (2019).

115. M.N. Economo, S. Viswanathan, B. Tasic, et al. "Distinct descending motor cortex pathways and their roles in movement". *Nature.* 563(7729), 79–84 (2018).

116. Y.G. Park, C.H. Sohn, R. Chen, et al. "Protection of tissue physicochemical properties using polyfunctional cross-linkers". *Nat. Biotechnol.* (2018).

117. T. Ku, W. Guan, N.B. Evans, et al. "Elasticizing tissues for reversible shape transformation and accelerated molecular labeling". *Nat. Methods.* 17(6), 609–613 (2020).

14

Combination of tissue optical clearing and 3D fluorescence microscopy for high-throughput imaging of entire organs and organisms

Peng Fei and Chunyu Fang

CONTENTS

Introduction

Biologists have always wanted to observe various organs and tissues to study the structure, function, and dysfunction of their cell components. In the past, this often required tissue extraction and histological preparation to gain access. Traditional optical microscopy techniques use a linear (single-photon) absorption process to produce contrast, but they are limited to high-resolution imaging near the tissue surface (less than 100 μm) because of the strong and multiple light scattering at greater depths. Scattering particularly severely affects the signal intensity in confocal microscopes. Confocal microscopes can achieve high three-dimensional resolution and optical sectioning by pinholes which reject all light that does not appear to come from the focus.

In the past two decades, a new optical microscope technology has been developed that uses nonlinear light-mass interactions to generate signal contrast. Nonlinear optical microscopy technology has a special function that reduces its sensitivity to scattering, so it is very suitable for high-resolution imaging in tissues. In particular, the two-photon excited fluorescence laser scanning microscope (2PLSM), combined with *in vivo* fluorescence labeling technology, has opened up a rapidly expanding field for imaging research of intact tissues and living animals.

Although 2PLSM is currently a mature technology that uses turn-key laser sources and commercial microscope systems, it is still important to understand its basic principles and key technologies, especially when optimizing the microscope system to achieve a large imaging depth. In this review, we discuss the physical principles, especially focusing on the imaging system parameters that are important for deep imaging, and summarize the technical issues related to the application of 2PLSM in high-resolution imaging of live animals.

Principle of traditional TPEM

In optical microscopy, one can distinguish between linear and nonlinear excitation. Traditional techniques, including confocal microscopy, generate contrast from light–matter interactions, in which the elementary process involves a single photon and which therefore depend linearly on the incident light intensity. Nonlinear techniques are fundamentally different in that they use "higher-order" light–matter interactions involving multiple photons for contrast generation. The nonlinear nature of these interactions leads to qualitatively new imaging properties.

Several different nonlinear processes can occur when light interacts with matter (Figure 14.1). Most widely used in biological imaging is fluorescence excitation by two-photon absorption [1]. Two photons that arrive "simultaneously" (within ~0.5 fs) at a molecule combine their energies to promote the molecule to an excited state, which then proceeds along the normal fluorescence-emission (or photochemical-reaction) pathway [1, 2]. Similarly, three or more photons can combine to cause excitation.

The efficiency of multiphoton absorption depends on the physical properties of the molecule (the "multiphoton absorption cross-section") [3, 4], and on the spatial and temporal

DOI: 10.1201/9781003025252-16

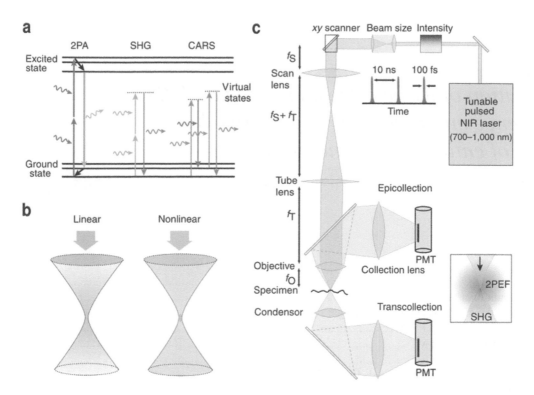

FIGURE 14.1 Traditional two-photon nolinear optical microscopy [15]. (**a**) Jablonski diagram, illustrating two-photon absorption (2PA), second-harmonic generation (SHG), and coherent anti-Stokes Raman scattering (CARS). Note that in second-harmonic generation and Raman scattering, no actual electronic excitation takes place. (**b**) Spatial confinement of signal generation with nonlinear excitation. Visible ("blue-ish") light is used for excitation in single-photon microscopy, whereas near-infrared ("red-ish") light is used in TPEM. In single-photon microscopy, an entire cone of fluorescence light (green) is generated, whereas nonlinear signal production is localized to the vicinity of the focal spot. (**c**) Generic nonlinear laser-scanning microscope. A laser source provides near-infrared ultrashort pulses; intensity and beam size are adjusted before coupling the laser beam to the microscope. The focal lengths of the scan lens (f_S), the tube lens (f_T), and the objective (f_O) are indicated. Two-photon excited fluorescence (2PEF), which is isotropically emitted (inset), can be collected in epi- and/or transcollection mode, using whole-area detection by photomultiplier tubes (PMTs). Forward-directed optical-harmonic and Raman signals are detected in transcollection mode in transparent samples. For *in vivo* experiments, epicollection is used exclusively.

distribution of the excitation light. Most nonlinear processes have in common that the transition probabilities are extremely low at "normal" light intensities. To generate sufficient signal, excitation light has to be concentrated in space and time. High spatial densities are (cheaply) generated by focusing a laser beam through a high numerical aperture (NA) objective. Concentration in the time domain requires the use of (expensive) lasers that emit "ultrashort" pulses (less than a picosecond long) with correspondingly high peak intensities. For laser pulses of width τ occurring at a rate f_R, the signal is enhanced by a factor of $1/\left(\tau f_R\right)^{n-1}$ compared to continuous-wave illumination, where n is the number of photons involved in the elementary process. Lasers typically used in TPEM provide 100-fs pulses at about 100 MHz, with a "two-photon advantage" of about 10 [5].

Multiphoton absorption is but one of several possible nonlinear interactions [5]. Another is optical-harmonic generation, in which two or more photons are "simultaneously" scattered, generating a single photon of exactly twice (thrice, and so on) the incoming quantum energy (Figure 14.1a). Harmonic generation requires no actual absorption but is enhanced near a resonance, albeit at the expense of parasitic absorption [6]. It also differs from multiphoton absorption in that it is a coherent – that is, phase-preserving – process, which causes speckles,

possible cancellation, predominantly forward-directed emission, and supralinear dependence on the chromophore density. In practice, only second-harmonic [6] and third-harmonic [7, 8] generation have been used. Second-harmonic (but not third-harmonic) generation depends on the absence of inversion symmetry, which not only requires that individual molecules are inversion-asymmetric (as most biological molecules are) but also that they are spatially ordered. Second-harmonic generation has, therefore, been useful for investigating ordered structural protein assemblies such as collagen fibers [9] or microtubules [10]. Similarly, dyes that are incorporated preferentially in one leaflet of the plasma membrane can be used to detect membrane voltage [11, 12]. A further process used for microscopy is coherent anti-Stokes Raman scattering (Figure 14.1a), which is sensitive to molecular vibration states and can be used to detect the presence of specific chemical bond types [13, 14].

All nonlinear microscopy techniques require expensive pulsed laser systems to achieve sufficient excitation rates. Two major advantages make the investment worthwhile. First, because multiple excitation photons combine their quantum energies in nonlinear microscopy, the photons generated (or the transitions excited) have higher energies than the excitation light, making emission "bluer" than the excitation, which is different from traditional fluorescence. For commonly used fluorescent markers, multiphoton absorption occurs in the near-infrared wavelength range (700–1000 nm), whereas emission occurs in the visible spectral range. Near-infrared light not only penetrates deeper into scattering tissue but is also generally less phototoxic owing to the lack of significant endogenous (one-photon) absorbers in most tissues.

The second major advantage of two-photon absorption and, in fact, of all nonlinear contrast mechanisms, is that the signal (S) depends supralinearly ($S \propto I^n$) on the density of photons, that is, the light intensity (I). As a consequence, when focusing the laser beam through a microscope objective, multiphoton absorption is spatially confined to the perifocal region (Figure 14.1b). The absence of multiphoton absorption in out-of-focus planes contrasts with confocal microscopy, where (single-photon) absorption occurs within the entire excitation light cone. The lack of out-of-focus excitation in nonlinear microscopy further reduces photodamage and thus increases tissue viability, which is crucial for long-term imaging [16]. Localization of excitation also provides excitation-based three-dimensional resolution with no need for spatially resolved detection through a confocal pinhole. By the same token, multiphoton absorption allows highly localized photo-manipulations, such as photobleaching and photolytic release of caged compounds, within femtoliter volumes [17–19].

Localization of excitation is maintained even in strongly scattering tissue because the density of scattered excitation photons generally is too low to generate a significant signal, making nonlinear microscopy far less sensitive to light scattering than traditional microscopy (Figure 14.2). This is of paramount importance for deep imaging, because it means that all fluorescence photons are known to originate from near the focus and thus can provide useful signal. However, the deep imaging described above often only involves samples at the micron level, and imaging depths of 1 mm are currently extremely difficult to achieve and need to be combined with other advanced imaging techniques. These additional add-ons raise the cost of the device while also placing higher demands on the construction of the optical path. In the following subsection, the implications and limitations of conventional two-photon excitation microscopy due to the effects of sample scattering are detailed.

Limitations of traditional TPEM application by light scattering

In most biological tissues, absorption of light is negligible compared to scattering, particularly in the near-infrared wavelength range. Scattering is the deflection of a light "ray" from its original direction; if the photon energy stays unchanged, it is termed *elastic*. Elastic scattering depends on refractive index inhomogeneities, which are present even in glass but are much stronger in tissue because cells are a heterogeneous mixture of molecules and supramolecular structures with varying molecular polarizabilities. The strength of scattering is described by the *mean free path* (l_s), the average distance between scattering events. The likelihood and angular distribution of scattering depend on refractive index variation, object size, and wavelength λ. For very small objects (such as isolated atoms or molecules in a gas) scattering is nearly isotropic and strongly wavelength-dependent ($\propto \lambda^{-4}$; *Rayleigh scattering*). For objects comparable in size to the wavelength (as in cells), scattering is directed mostly in the forward direction. This can be quantified by the anisotropy parameter (g) or by the *transport mean free path*, $l_t = l_s / (1 - g)$, which is the distance after which "direction memory" is lost. Measurements in brain gray matter yielded values for ls of 50–100 μm at 630 nm in extracted tissue and of about 200 μm at 800 nm *in vivo* [20, 21]. Scattering decreases with wavelength, albeit less than expected for Rayleigh scattering [21]. The anisotropy parameter generally is high (≈ 0.9) in brain tissue [22]. In nonlinear optical microscopy, only ballistic (nonscattered) photons contribute to signal generation in the focal volume. The ballistic power follows a Lambert-Beer–like exponential decline with imaging depth z

$$P_{ball} = P_0 e^{-z/l_s} \tag{14.1}$$

with length constant l_s and surface power P_0. Because of the quadratic intensity-dependence in TPEM, the fluorescence signal declines as

$$F_{2PE} \propto \left(e^{-z/l_s}\right)^2 = e^{-2z/l_s} \tag{14.2}$$

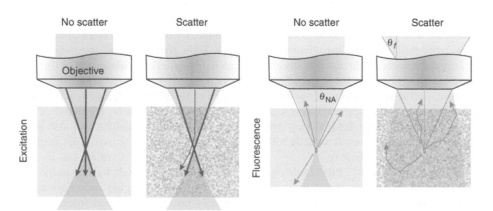

FIGURE 14.2 Signal generation and fluorescence collection in clear tissue (no scatter) and in scattering tissue (scatter).

Conversely, we need an exponentially increasing laser power entering the surface ($P_0 \propto e^{z/l}$) to maintain the same ballistic intensity at the focus. The reduction in focal intensity depends on l_s and not l_t, because even a small deflection from the original path causes a ray to miss the focus (Figure 14.2). In contrast, the forward-directed angular distribution of scattering is important for the calculation of the near-surface background. Both l_s and g not only depend on tissue type but can also change substantially with age and upon removal of the tissue from the animal [21]. Surface scattering can become important if the beam crosses between media with substantially different refractive indices, for example when imaging through the skull. Scattering of fluorescence photons is important for the detection process. Because of the short mean free path of visible light, the ballistic fraction becomes quickly irrelevant with increasing focal depth. For sufficient depth, multiply scattered fluorescence light leaves the sample from a diffusely radiating region on the surface, which has a full-width-at-half-maximum intensity (FWHM) of about 1.5 × the focal depth, independent of the scattering length [23].

The maximum achievable imaging depth is proportional to the scattering mean-free-path and depends logarithmically on available laser power, two-photon advantage, and collection efficiency [21]. With a laser oscillator providing ~100-fs pulses, the maximum depth usually is limited by the available average power (1 W average power allows imaging depths of about 600–800 μm in the neocortex) [20, 24–27]. When a regenerative amplifier is used, it is possible to image deeper (up to 1 mm in the neocortex) [28]. Eventually, however, fluorescence generated near the sample surface becomes a limiting factor. The resulting contrast reduction might be impossible to overcome in samples with a wide fluorophore distribution as, for example, in transgenic mice with extensive GFP expression. The achievable imaging depth also strongly depends on other tissue properties such as microvasculature organization, cell body arrangement, and collagen or myelin content, which will more or less degrade the laser focus and limit signal generation deep inside the tissue. If deeper structures need to be reached and if less-scattered longer wavelengths cannot be used, mechanical penetration or removal [29] of overlying tissue may be necessary. Efforts in this direction have been made using very narrow objective lenses made from gradient-index (GRIN) material [30–32].

In clear tissue, all excitation light reaches the focus, but in scattering tissue, scattering (even by a small angle) causes light rays to miss the focus and be lost to signal generation. This leads to a roughly exponential decrease in excitation with depth. In clear tissue only fluorescence light rays initially emitted into the collection cone, determined by the objective's NA, can be detected, but in scattering tissue, fluorescence light is (multiply) scattered and may even "turn around." Fluorescence light apparently originates from a large field of view but a larger fraction than in the nonscattering case is actually within the angular acceptance range θ_f of the objective.

Besides normal imaging of brain slices, TPEM is also used for high-resolution imaging in various organs of living animals. This chapter shows additional aspects that are important for *in vivo* imaging, using mainly examples in the intact brain.

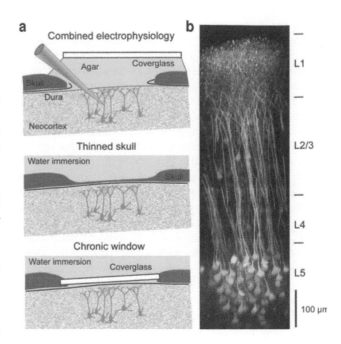

FIGURE 14.3 *In vivo* two-photon imaging in the intact neocortex. (a) Different types of brain access. Open cranial window with the dura mater removed so that micropipettes for cell labeling and electrophysiological recordings can be inserted (top). Pulsation of the exposed brain is reduced by covering the craniotomy with agar and a cover glass. Thinned-skull (20–40 μm thickness) preparation (middle). Cellular structures are either prelabeled (for example, with fluorescent proteins in transgenic mice) or stained through a tiny hole lateral to the thinned area. Chronically implanted glass window replacing the skull (bottom). Agar is used underneath the window for stabilization. (b) Example of deep two-photon imaging in mouse neocortex. Maximum-intensity side projection of a fluorescence image stack, obtained in a transgenic mouse expressing Clomeleon, a genetically encoded chloride indicator, under the control of the Thy1-promoter, preferentially in deep layer 5 (L5) pyramidal cells. Data were taken with a 10 W pumped Ti: sapphire oscillator using a 40×, NA 0.8 water-immersion lens (Zeiss). Note that nearly the entire depth of the neocortex can be imaged.

For short-term studies of the brain, a cranial window is opened above the area of interest, which also provides access for recording electrodes (Figure 14.3a). Other organs are surgically exposed in a similar manner or, as in the case of kidney, are exteriorized [33]. A general problem for *in vivo* imaging is motion induced by heartbeat and breathing. Therefore, tissue pulsation should be dampened as much as possible, for example, by covering or embedding the exposed organ with agar. Tight control of the anesthesia or artificial respiration can help to alleviate pulsation. For time-lapse imaging of cell structures, we recommend acquiring small subvolumes (image stacks of 10–20 focal planes) in anticipation of lateral or focal drift, which can then be corrected offline using correlation methods. Image acquisition can also be synchronized to the heartbeat by triggering individual frames using a simultaneously recorded electrocardiogram, to ensure that all images have the same phase relationship to the heart beat [34].

For long-term imaging over days to months, animals are multiply reanesthetized and image stacks of the same subvolume are acquired. In the brain, these experiments are performed either through the thinned skull [35–38] or through a chronically implanted glass window [39–41] (Figure 14.3a). In cancer research, a dorsal skin-fold chamber with a glass window is used for imaging implanted tumors [42]. To repeatedly find the same structure (for example, a cell, a blood vessel, an amyloid plaque, or even a synapse), stable anatomical landmarks, such as the surface blood vessel pattern, can be used.

Functional imaging from neuronal dendrites in the intact neocortex has, so far, still required the introduction of synthetic calcium indicators via electrodes [24–26, 43–45], which has the advantage of providing additional *in vivo* intracellular electrophysiological data. The combination of two-photon optophysiology with electrophysiology [46] will be crucial for studying synaptic integration and single-cell computation in living animals. Great hope still rests on the *in vivo* application of genetically encoded functional indicators [47–51].

However, for classical *in vivo* neurological observations with windows opened in the mouse brain, the presence of the cranium greatly increases the scattering and absorption of light during propagation. Figure 14.3a demonstrates the application of imaging on a thinner skull and the replacement of the skull with a thin slide by a surgical approach. However, we are currently unable to be clear whether this surgical procedure has an impact on neural activity in the brain of mice – it is common sense to infer that opening a window on the brain and thinning or even removing the skull would necessarily have a dramatic impact on the actions of the organism, making it an important obstacle to the application of two-photon excitation microscopy to *in vivo* imaging observations. This obstacle is also driving the development of tissue optical clearing techniques, making skull transparency an important and much needed developing direction.

Tissue optical clearing with TPEM

As mentioned earlier, conventional two-photon excitation imaging is affected by the scattering of light in the sample, even though the use of long wavelengths (second or even third harmonics) can enhance the penetration of the excitation light, resulting in a higher light intensity and less scattering of the excitation light irradiated on the sample, yet it remains powerless when faced with, for example, a centimeter-scale sample of a mouse brain. Sequential two-photon excitation (STP) in combination with extremely high-precision mechanical slices allows high-resolution imaging of intact mouse brains with submicron accuracy, but it poses the problem of long imaging times.

So the idea of combining tissue optical clearing techniques with the already well-established two-photon excitation imaging came naturally. As described in the previous sections, existing tissue optical removal techniques have made it possible to transparent the intact mouse brain, nervous system, and even the whole mouse, at which point, in combination with two-photon excitation imaging, the imaging depth of the sample will be greatly increased, limited only by the working distance of the excitation and detection mirrors, and in combination with the electromechanical displacement stage, stitching imaging in a single plane will be possible, and in the horizontal plane will no longer be limited by the sample size.

Taking the 2019 paper *Reconstruction of 1,000 Projection Neurons Reveals New Cell Types and Organization of Long-Range Connectivity in the Mouse Brain* by Janelia Farm's Jayaram Chandrashekar research group published in *Cell* as an example [52], the researchers wanted to achieve an ideal submicron level resolution while observing information on the distribution of locations such as different brain regions through which long-distance projection neurons pass throughout the brain, which challenged both the resolution and sample size of the imaging. Using a well-established high-resolution two-photon excitation imaging system combined with the CUBIC tissue optical clearing method, the researchers achieved a 3D resolution of $0.3 \times 0.3 \times 1 \ \mu m^3$ to acquire a complete three-dimensional data of the whole mouse brain, and based on this data completed the localization and tracking of the specific labeled nerves. Figure 14.4 shows a schematic of the imaging with the results.

In addition to tissue optical clearing techniques combined with STP, researchers are now exploring ways to make the skull transparent. This is because, for *in vivo* imaging, both the chemical immersion and rinsing required by current tissue optical scavenging techniques result in the inactivation of the sample, resulting in the death of the organism itself, which in turn defeats the purpose and significance of *in vivo* observations. Therefore, it is a compromise solution for less impactful cranial hyaline. At the same time, more advanced tissue optical clearing technologies that are completely harmless to the organism itself are also being explored and researched in full swing. It is believed that in the near future, with the flourishing development of tissue optical clearing technology, two-photon excitation microscopy *in vivo* imaging and its combination will no longer be just a dream, and many life phenomena and problems in the organism, such as the propagation and regulation of calcium signals, will see greater breakthroughs.

Tissue optical clearing with LSFM

In the previous subsection, this book focused on the principles of traditional two-photon excitation microscopy methods and the problems faced by two-photon excitation microscopy methods when faced with the problem of light scattering and absorption in biological sample tissues. It concludes with a presentation on how tissue optical scavenging techniques can help solve problems and greatly enhance the scale of 3D imaging, opening a new door for more and broader research into biomedical problems.

This section will detail an emerging three-dimensional microscopic imaging method developed in recent years – light-sheet fluorescence microscopy. Similar to the structure of the previous section, the principle of the light-sheet fluorescence microscopy method is first described. Unlike the two-photon excitation microscopy method, due to the principle limitation of the light-sheet fluorescence microscopy method itself, it is only able to detect the signal at a depth of about 100 μm without the tissue optical clearing technique. Only when the two are combined can the advantages of light-sheet fluorescence microscopy imaging be fully reflected. This section will

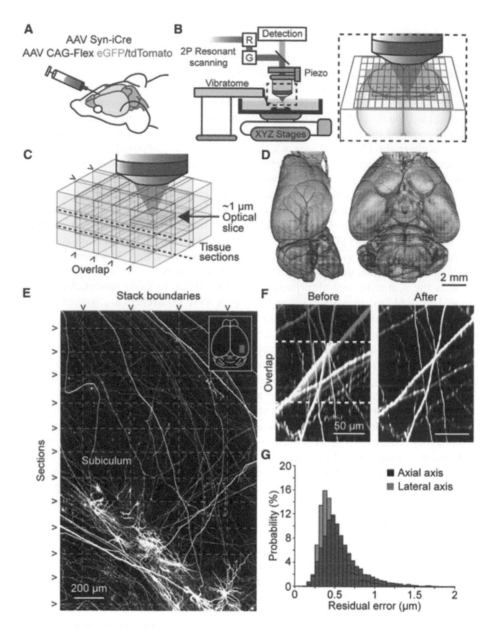

FIGURE 14.4 Imaging pipeline. (a) Animals were injected in targeted brain areas with a combination of a low-titer AAV Syn-iCre and a high-titer AAV CAG-Flex-(eGFP/tdTomato). (b) Two-photon microscope with an integrated vibratome. Inset: sequential imaging of partially overlapping image stacks. (c) Image stacks overlapped in x, y, and z. (d) Rendered brain volume after stitching. (e) Horizontal maximum intensity projection through a $1300 \times 2000 \times 3600$ mm³ volume of the motor cortex containing labeled somata and neurites. Horizontal dashed lines mark physical tissue sections; vertical dashed line represents stack boundaries. Dashed box is region shown in (f). (f) Example of boundary region between two adjacent image stacks before (left) and after stitching (right). Dashed line indicates overlap region. (g) Residual stitching error in the lateral and axial directions.

present several of the more cutting-edge light-sheet imaging methods available today, demonstrating the different scales of research that have been conducted when they are combined with tissue optical clearing techniques, fully demonstrating the convenience that these two techniques provide to life science research today.

Fluorescence microscopy in concert with genetically encoded reporters is a powerful tool for biological imaging over space and time. Classical approaches have taken us so far and continue to be useful, but the pursuit of new biological insights often requires higher spatiotemporal resolution in ever-larger, intact samples and, crucially, with a gentle touch, such that biological processes continue unhindered. LSFM is making strides in each of these areas and is so named to reflect the mode of illumination: a sheet of light illuminates planes in the sample sequentially to deliver volumetric imaging. LSFM was developed as a response to inadequate four-dimensional (4D; x, y, z, and t) microscopic imaging strategies in developmental and cell biology, which overexpose the sample and poorly temporally resolve its processes. It is LSFM's fundamental combination of optical sectioning and parallelization (Figure 14.5) that allows long-term biological studies with minimal phototoxicity and rapid acquisition.

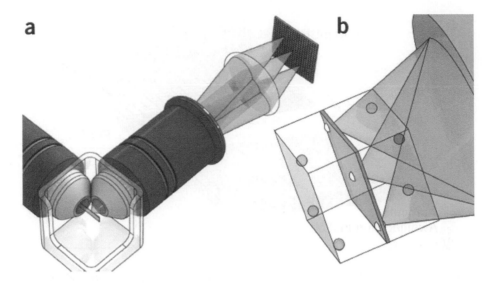

FIGURE 14.5   Light-sheet fluorescence microscopy [65]. (a) The archetypal light-sheet microscope: paired, orthogonal optical paths provide plane-wise illumination and wide-field fluorescence detection. (b) Optical sectioning by selective illumination of a single plane.

Principle of conventional LSFM

The starting point for the design of light-sheet systems should be a specific biological question or application. Indeed, LSFM is particularly suited for construction around the sample, because the decoupled illumination and detection paths of LSFM provide endless scope for customization and because the microscope may be arbitrarily arranged in space. Although commercial systems perform a crucial role in multiuser environments, the most exacting applications require custom solutions. The power of the custom approach becomes particularly apparent when the application pushes the limits of the technology, for example for high-speed *in toto* imaging of neural activity [53], cardiac dynamics [54, 55], gastrulation in whole embryos [56, 57], and physiologically representative subcellular imaging [58]. Likewise, where concessions must be made to sample-mounting protocols to allow normal development of physically sensitive embryos [59–62], geometric flexibility is crucial, whereas for behavioral studies the ability to spatially [63] or spectrally [64] avoid visually evoked responses may prove invaluable.

To understand how LSFM elegantly sidesteps many of the issues that plague conventional microscopies, consider the question: how does microscopy allow us to visualize biological tissues in 4D (x, y, z, and t)? A cursory appreciation of optics is sufficient to understand that out-of-focus objects appear blurred. This is why epifluorescence microscopy, which captures volumes as 2D projections, can only achieve high-contrast imaging in thin samples. The acquisition of images without somehow being able to discriminate based on depth reduces biological systems, which are three-dimensional without exception, to a planar representation. Just as tissue can be mechanically sectioned, sectioning can be achieved noninvasively by optical sectioning, which point-scanning confocal and multiphoton microscopies achieve by the removal of the out-of-focus signal and by confining excitation to the focal volume, respectively. However, in each case, the serial nature of the acquisition process limits the speed with which volumetric data can be collected. Also, crucially, regions that do not direct contribute to the useful signal are exposed repeatedly, which leads to photodamage.

The manner in which LSFM overcomes these limitations is remarkably simple. Taking a wide-field microscope as its basis, the sample is illuminated from the side with a sheet of light, ensuring that signal arises only from in-focus regions (Figure 14.5), thereby reducing the total exposure. A camera collects the resulting fluorescence signal, sequentially imaging the volume as 2D optical sections, thus parallelizing the imaging process within each plane. As such, the dwell time at each point is orders of magnitude higher than that in the point-scanned case, which allows for commensurate reductions in peak light intensity. Because the peak intensity and total power delivered will each have a bearing on photodamage rates, the combination of intrinsic sectioning (entire illuminated volume contributes to useable signal) and parallelization (plane-wise acquisition, millisecond exposure times) allows for gentle and rapid imaging. It is this combination of speed, 3D resolving power, and low phototoxicity that makes LSFM such an attractive imaging tool to confront a range of biological questions.

A historical perspective is useful in understanding some of the most fundamental choices that need to be made in building or choosing a light-sheet microscope. The first, at least in a form that would be recognizable to a modern user, was developed a little over a decade ago and called *selective-plane-illumination microscopy* (SPIM). SPIM (Figure 14.6) illuminates the sample with a static 2D light sheet focused by a cylindrical lens, and its use demonstrated for the first time that long-term fluorescence imaging of entire developing embryos could be achieved without unduly impairing their health. As a testament to the strength of SPIM, this fully parallelized scheme (simultaneous whole-plane illumination and detection) has yet to be improved upon for its use in long-term developmental imaging. However, a drawback of illuminating an entire plane at once from the side is the presence of striped artifacts in the

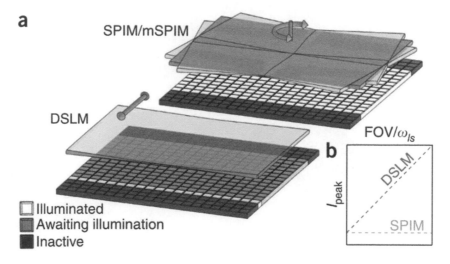

FIGURE 14.6 Parallelization of light-sheet generation. (a) SPIM illuminates and captures fluorescence from the entire FOV simultaneously, whereas mSPIM reduces striped artifacts by pivoting the light sheet about its center. DSLM produces a virtual light sheet by time-sharing the beam, with fluorescence arising only from the illuminated strip at any given time. (b) To maintain identical SNRs, DSLM requires higher peak intensities (I_{peak}) as the FOV size increases (along the scanning axis) relative to the light-sheet thickness, ω_{ls}.

resulting data, which arise from the refraction, scattering, and absorption of coherent light within tissue. In relatively transparent samples (such as zebrafish embryos), the effects are minimal, but in optically dense samples (such as fruit flies), they may be more severe. A later variant, multidirectional SPIM (mSPIM) (Figure 14.6) resonantly pivots the light sheet about its focus, illuminating more uniformly and thus reducing the stripe artifacts [66].

Digitally scanned light-sheet microscopy (DSLM) (Figure 14.6) provides the counter to the full parallelization of SPIM, sweeping out a virtual light sheet by scanning a Gaussian beam through the sample [67]. Because only part of the plane is illuminated at a given time point (i.e., the pixel dwell time decreases), the peak laser power delivered to the sample must increase proportionally to maintain the SNR, thus increasing the chances of fluorophore saturation and rates of photodamage. In optically dense samples, however, DSLM is superior in reducing striping, and the increase in intensity may be manageable even if not desirable (Figure 14.6b). All light-sheet microscopes are ultimately based on either the SPIM or DSLM architecture, and the choice of whether to scan or not can be crucial in balancing photodamage, imaging speed, and quality.

The lateral and axial resolution in light-sheet microscopy is determined slightly differently from that of conventional techniques. The product of the illumination and detection PSFs determines the axial resolution. Thinner, high-NA light sheets, therefore, provide superior axial resolution; however, diffraction dictates that a tightly focused (high-NA) Gaussian light sheet diverges rapidly away from the focus (Figure 14.7). Generally, the region over which the light sheet spreads by no more than $\sqrt{2}$ times the thickness at waist is taken to demarcate the area that is useful for imaging. For a focused Gaussian beam, the resulting light sheet length, z_{ls}, and the thickness, ω_{ls}, are given as:

$$z_{ls} = \frac{2\lambda}{\pi NA^2} \qquad (14.3)$$

$$\omega_{ls} = \frac{2\lambda}{\pi NA} \qquad (14.4)$$

The NA dependence demonstrates that there are diminishing returns on decreasing the light-sheet thickness in terms of the achievable FOV, and so typically some trade-off between usable FOV and axial resolving ability has to be made. Early implementations of LSFM focused on whole-embryo imaging for good reason: isotropic, subcellular resolution requires an ultrathin (high-NA) light sheet, which severely limits FOV.

The theoretically achievable lateral resolution is simply that of a wide-field microscope, governed by the wavelength and NA of the objective lens used for detection. For low magnification–high NA detection lenses, undersampling is frequently employed to sacrifice resolution for field of view (FOV), and sCMOS (scientific complementary metal-oxide-semiconductor) cameras are generally favored since they deliver ~4-16× larger FOVs (by area) than typical EMCCD (electron-multiplying charge couple device) cameras for equivalent spatial sampling. For cases in which excellent light sensitivity is required, EMCCDs may offer a superior solution, notably for superresolution and multiphoton implementations. Moreover, under light-starved conditions, a high-detection NA is favorable, as the collection efficiency scales with NA [2]. However, another consequence of high NA is that the DOF of the objective lens will be small, and so only a thin slice of the sample remains in focus. Usually the light sheet has to be thicker to cover the FOV, which compromises the sectioning ability and lowers the contrast.

Three challenges and corresponding methods for LSFM

The first challenge for achieving high-resolution imaging in LSFM is largely a geometric issue. High numerical aperture (NA) detection optics are favorable for light-collection

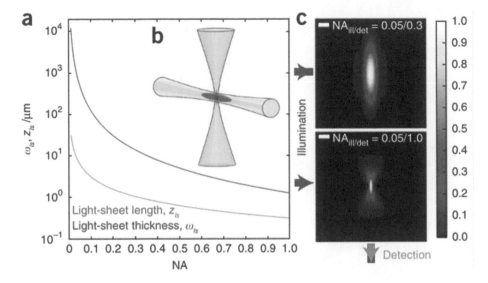

FIGURE 14.7 Spatial resolution in light-sheet fluorescence microscopy. (a) Interplay between light-sheet thickness (ω_{ls}) and length (z_{ls}). (b) Overlap of illumination (blue) and detection (green) PSFs yields the system PSF. (c) Influence of the detection NA on the system PSF, displayed as a summed projection orthogonal to the illumination and detection axes. The color scale defines the normalized intensity of the system PSF.

efficiency and lateral resolution, whereas high-NA illumination produces thinner light sheets, yielding superior axial resolution and sectioning ability. Unfortunately, concurrent high NA in both pathways is sterically constrained as high-NA lenses are, by necessity, bulky. Although the size of the objective lens, which dictates the achievable NA, requires little consideration in an "epi" configuration, trying to position high-NA (>0.9) water-immersion lenses such that their orthogonally oriented foci overlap is a fruitless task. Regardless, for many applications, the light-sheet NA (which governs the light-sheet length and thickness) may be much lower, for example, 0.06 > NA > 0.02 is typical to cover a field of view (FOV) of 50–500 μm (λ_{ill} = 488 nm). This permits the use of ultralong working-distance, low-NA lenses, substantially relaxing the constraints on the choice of objective lens for detection. A summary of objective lenses that are typically used for illumination and detection in LSFM is given as a guide to what can be orthogonally coaligned. It is worth noting that some particularly advanced systems have used custom-designed objectives [53, 68]; however, the cost and complexity may be prohibitive for the majority of microscopists.

In principle, LSFM provides an ideal platform for far-field superresolution imaging. In a live-cell context, in particular, the LSFM approach is beneficial relative to near-field techniques (for example, total internal reflection fluorescence microscopy), which image molecules located within one wavelength of the coverslip. Localization-based techniques exploit the photophysics of molecular probes and allow spatial resolution of tens of nanometers. However, in thick, living samples, the indiscriminate nature of the illumination and the numerous exposures required to stochastically construct the image lead to photodamage and out-of-focus signal. However, all the aforementioned have a precondition: the tissue optical clearing. Without the clearing techniques, because LSFM uses normal lasers, the scattering and attenuation caused by biological samples prevent the propagation of excitation light. By confining the illumination to a thin plane, both can be ameliorated,

but it still penetrates for a depth of only about 100 μm. At the same time, the emission fluorescence is also unable to spread to the superficial layer and be collected by detection objective lens.

In spite of the mutual exclusivity between high detection and illumination NA, a number of localization-based superresolution light-sheet fluorescence microscopes have been reported. The individual molecule localization selective-plane illumination microscope (IML-SPIM) provides a prime example with a high NA_{det} (1.1) limiting NA_{ill} (0.3) [69]. Although axial localization is achieved through depth-dependent astigmatism [70], the relatively thick light sheet compromises sectioning. Naturally, as one ventures to smaller feature sizes or exceptionally low light levels, the required increase in detection NA eventually becomes a bottleneck, and alternative geometries have to be sought.

To overcome this limitation, Gebhardt et al. used a pair of vertically opposed objective lenses with a 45° mirrored cantilever to redirect a thin light sheet (full-width at half maximum [FWHM] thickness of 1 μm) horizontally into the detection plane (Figure 14.8a), permitting the use of an ultrahigh-NA detection objective lens (NA_{det} of 1.4 in oil) [71]. This reflected light-sheet microscopy (RLSM) provides an additional benefit, as the inverted imaging geometry facilitates construction around conventional microscope bodies. Single-objective SPIM (soSPIM) launches the light sheet from the detection objective lens ($NA_{det,max}$ of 1.4 in oil) (Figure 14.8b). In this case, a microcavity mirror coupled to the sample support is used to horizontally reflect the light sheet and deliver similar sectioning to that by RLSM [72]. Because the illumination and detection planes are coupled in soSPIM, the production of volumetric data is more complex and requires a combination of scanning and refocusing. To change the axial position of the light sheet, the beam is swept laterally across the mirror while the detection plane remains coaligned by translating the objective lens. Naturally, this is accompanied by a shift in the light-sheet waist across the FOV. To compensate, an electrically

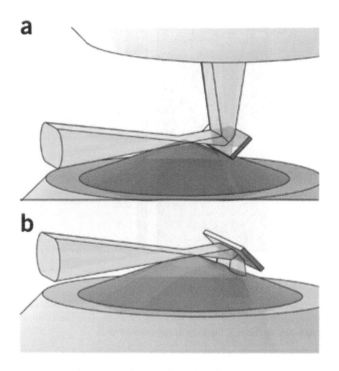

FIGURE 14.8 Reflected light-sheet geometries. (a) RLSM – the light sheet is launched from an opposing objective. (**b**) soSPIM – the light sheet is launched from the detection objective.

tunable lens repositions the illumination focus. Li et al. provided another single-objective variant, termed *axial-plane optical microscopy* (APOM) [73]. The light sheet is delivered in an epi configuration via an ultrahigh-NA (1.4 in oil) objective lens to illuminate a single axial plane. A remotely situated objective and 45°-tilted mirror subsequently serve to rotate and reimage the axial plane onto a camera. Experiments and simulations confirmed that APOM achieved a resolution comparable to that of conventional epifluorescence microscopy, although aberrations were incurred toward the periphery of the FOV.

An elegant solution that avoids a reflective element is the so-called π-SPIM (Figure 14.9), which uses a pair of nonorthogonal objectives (90 < θ < 180) and oblique illumination to relax mechanical constraints somewhat [74], in essence forming a two-objective variant of a highly inclined and laminated optical sheet (HILO) microscope [75]. Although the obliquity sacrifices some of the illumination objective NA, the geometry allows a combined illumination and detection NA near the theoretical maximum, using only off-the-shelf components (NA_{ill} = 0.71; NA_{det} = 1.1 (water); fill factor = 0.986).

The second challenge in achieving high-resolution imaging with LSFM is to maintain high axial resolution over a large FOV. Because the overall or system point-spread function (PSF) arises from the overlap of illumination and detection PSFs, isotropic resolution is achievable, in principle, by sufficient axial confinement of the light sheet [76]. Maintaining a thin light sheet across a FOV > 10 μm is problematic, however, as a high-NA Gaussian light sheet spreads rapidly away from the focus. Consequently, the most common light-sheet variants fail to achieve a truly isotropic PSF, and axial resolution is typically no better than ~1 μm (or between two and three times worse than lateral resolution). Although this is perfectly adequate for cellular resolution, it may be limiting for cases in which subcellular resolution is required across a large FOV.

A number of solutions to this problem, using nondiffracting beam modes, have emerged, the most common being the Bessel beam, whose cross-section consists of a narrow central core surrounded by a series of rings of diminishing intensity (Figure 14.10). These beams are governed by diffraction like any other beam, but they maintain an invariant profile over many times the Rayleigh range of a Gaussian beam of equal NA. Planchon et al. used a moderate NA (0.8) in both the illumination and detection pathways to deliver ~300 nm isotropic resolution over a FOV spanning 40 μm along the propagation axis of a scanned Bessel beam [77] (Figure 14.11). Bessel beams have been shown to offer measured improvements in a turbid medium, penetrating 1.55× further into human skin

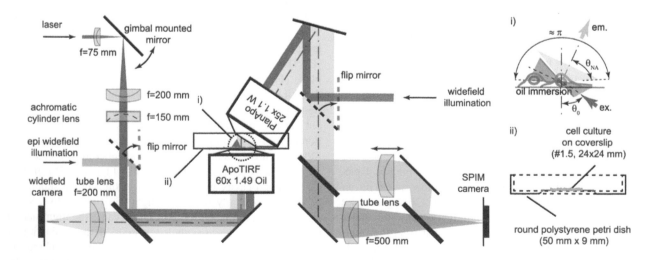

FIGURE 14.9 πSPIM setup. Schematic dual-color πSPIM setup with an oblique light sheet produced by off-center passage of the beam through the illumination lens (1.49 ApoTIRF 60×), and the detection lens (1.1 W, 25×) arranged orthogonally to the oblique light sheet. (i) Close-up of the focal region showing the angular range of the complementing illumination and emission cones. (ii) Mid-plane cross-section of the glass-bottom dish used for sample mounting.

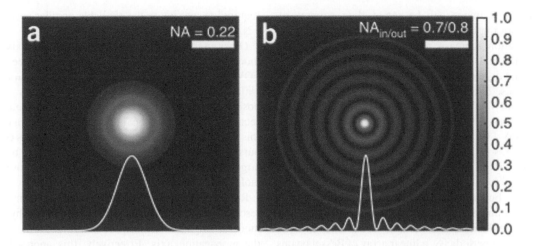

FIGURE 14.10 Gaussian and Bessel beams for light-sheet generation. (a,b) Gaussian (a) and Bessel (b) beams produce a light sheet with equivalent length (~6 μm; λ_{ill} = 488 nm; refractive index, n_{imm} = 1.33) despite the disparity in NA. Only a single NA is necessary to define the Gaussian beam, whereas the Bessel beam features two corresponding values that define the inner and outer NA of the annular spectrum of the beam. Although the Bessel beam features a much smaller central beam lobe only, ~11% of the total Bessel beam irradiance is contained therein, and only ~30% is contained within the Gaussian beam waist. The color scale shows the normalized intensity of each of the beams. Scale bars, 1 μm.

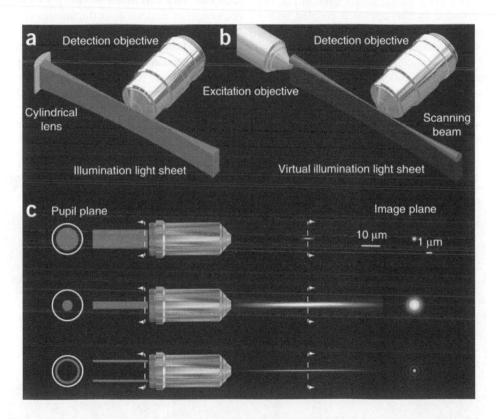

FIGURE 14.11 Plane illumination microscopy with scanned Bessel beams [82]. (a,b) The light sheet of conventional plane illumination can be created by a sheet-like nonscanning line focus (a) or by scanning a focused Gaussian beam to produce a virtual light sheet (b). (c) Comparison of Gaussian beams at 0.5 (top) and 0.15 (middle) excitation NA and a Bessel beam with 0.5 NA_{OD} and 0.46 NA_{ID} (bottom). The Bessel beam has a central peak slightly narrower than even the high-NA Gaussian beam, but it is at the same time as long as the low-NA Gaussian beam. However, it also has strong side lobes, which must be taken into account when it is used for plane illumination.

tissue than Gaussian light sheets [78]. Although impressive, the contrast-limiting effect of the concentric side lobes and associated additional photon load discussed earlier are similarly applicable. This may be ameliorated by sectioning the Bessel beam in a manner such that the ring structure is suppressed above and below the imaging plane while maintaining the self-reconstruction ability, although this broadens the resulting light sheet [79]. However, the beam side lobes illuminate out-of-focus regions, which compromises sectioning ability, degrades contrast, and unnecessarily exposes the sample. By combining with optical-sectioning structured illumination (OS-SIM) or two-photon excitation [58, 80], this out-of-focus contribution to the image can be removed or suppressed, respectively, with the caveat that the additional exposures or pulsed illumination increase rates of photobleaching [80]. Additionally, the OS-SIM algorithm produced reconstruction artifacts and discarded a large portion of the useful signal [77]. Fei et al. used the electronic rolling shutter of sCMOS and the confocal line detection effect helped to reject the effect of side lobes, which achieved subcellular resolution at 1-mm range FOV [81]. Gao et al. later adapted the technique to utilize super-resolution structured-illumination algorithms (SR-SIM) to provide an improvement in resolution of between 1.5-fold and 1.9-fold, an increased SNR, and a more judicious use of the photon budget [80]. By using multiple Bessel beams in parallel, a commensurate reduction in peak intensity was possible, further reducing photodamage.

Given the correct periodicity, a linear array of Bessel beams may interfere destructively, such that the rings are somewhat suppressed, producing an optical lattice. This realization led to the development of lattice light-sheet microscopy (LLSM), which is capable of delivering ultrathin (FWHM 1 μm) light sheets in a highly parallelized manner [68] (Figure 14.12). The optical efficiency afforded by this approach coupled with a NA_{det} of 1.1 allowed the electron-multiplying charge couple device (EMCCD) to be replaced with a faster scientific complementary metal-oxide semiconductor (sCMOS) camera. Along with low magnification (25×), the result is a larger FOV than previous Bessel beam implementations [77, 80] (~80 × 80 μm) and a plane-wise imaging rate as high as 200 or 1,000 fps [68] for multi- and single-color imaging, respectively. With the increased detection NA, custom illumination optics were

required to maximize the available angular space, delivering a $NA_{ill,max}$ of 0.65.

Other pseudo nondiffracting beams also exist. The Airy mode has been shown to yield thinner light sheets over larger FOVs than Gaussian or Bessel beams of comparable NA [83]. Unfortunately, to allow the beam side lobes to contribute positively to image formation, the data must be deconvolved, requiring that the Airy beam side lobes remain in focus. In turn this has limited the detection NA to 0.4, which is counter to the pursuit of high resolution. For now, the Airy beam remains largely a curiosity, and future studies in a more demanding biological context are welcomed.

Rather than use complicated beam shaping, high axial resolution and large FOVs can be achieved by sweeping a moderate-high NA Gaussian beam through the sample along the propagation axis. Effectively, this approach shares the focus temporally between different focal depths to produce a long and thin light sheet, while sacrificing some (1D) parallelization relative to the analogous SPIM- or DSLM-based approach. Dean, Fiolka, and Fei et al. have used ultrasonic lenses or voice coil motor to sweep a focused beam through the sample to achieve sheet thicknesses (FWHM) of 465 nm and 1.7 μm over FOVs of 50 μm and 3.3 mm, respectively [83–85]. Dean, Fiolka [84], and Fei [85] adopted confocal line detection (CLD), which effectively captures the 2D image line-wise to remove out-of-plane contributions from the beam tails (Figure 14.13a). Relative to DSLM, the 2D-scan or sweep process and associated decrease in dwell time required higher intensities still, which spurred the development of axially swept light-sheet microscopy (ASLM). In 2019, the new

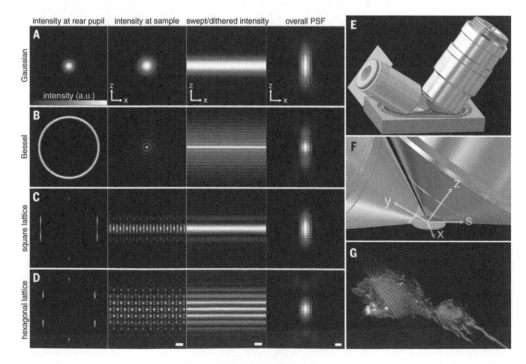

FIGURE 14.12 Lattice light-sheet microscopy. (A–D) The traditional Gaussian light-sheet (DSLM), a Bessel beam with side lobes, the square lattice in (C) optimizes the confinement of the excitation to the central plane, and the hexagonal lattice in (D) optimizes the axial resolution as defined by the overall PSF of the microscope. (E–F) Model showing the core of microscope, with orthogonal excitation (left) and detection (right) objectives dipped in a media-filled bath. (G) Representation of a lattice light sheet (blue-green) intersecting a cell (gray) to produce fluorescence (orange) in a single plane. The cell is swept through the light sheet to generate a 3D image.

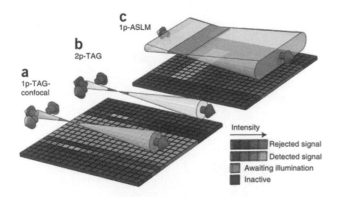

FIGURE 14.13 Axially swept light-sheet geometries. (a,b) 2D virtual light-sheet production (axially swept or laterally scanned) using ultrasonic (tunable acoustic gradient; TAG) lenses by one-photon excitation with confocal line detection (CLD) to remove undesired signal (a) or two-photon excitation to suppress undesired signal (b). (c) 1D virtual light-sheet production (axially swept) with CLD to remove undesired signal or with sequential acquisition of images at different beam waist positions and subsequent image stitching.

ASLM represented by ctASLM [86] and mesoSPIM [88] further applied this technology to whole mouse brain imaging, bringing new opportunities for the imaging of large-scale biological samples.

In one sense, ASLM is akin to SPIM, producing a short depth-of-field (DOF) light sheet with cylindrical optics (Figure 14.13c). ASLM sweeps the short light sheet through the sample using a remotely situated objective lens and swept piezo mirror for aberration-free refocusing [88]. Like DSLM, this approach also produces a virtual light sheet in one dimension. ASLM produces ultrathin light sheets over large FOVs by using CLD to remove fluorescence from the beam tails, delivering comparable resolution to LLSM without the need for complicated processing or reconstruction. Illuminating out-of-focus regions is, however, more wasteful with the photon budget.

The stability and structure of the Bessel beam makes it an ideal candidate for confocal treatment, as the beam penetrates further into tissue without being unduly perturbed. Fahrbach et al. explored the depth-dependent attenuation of signal for Gaussian and Bessel beams, demonstrating that the decay in signal with depth is more severe for the former [89]. It is also worth bearing in mind that the spatial filter is only 1D and so will not remove light that is scattered along the slit axis. Silvestri et al. noted that this causes an increase in background signal at depth while the useful signal decreases [90]. CLD requires no additional exposures or postprocessing steps and has become a widespread and powerful tool for scanned LSFM systems. Recent developments include beam multiplexing to utilize the twinned rolling shutters of the current generation of sCMOS cameras, which delivers higher imaging speed [91] as well as more complex multiview implementations [53, 92].

Relative to conventional microscopies, LSFM performs remarkably well at low-to-moderate NA (owing to both general efficiency and axial resolving power derived from overlap of orthogonal illumination and detection PSFs). Correspondingly, samples can be positioned far from mechanically small, long-working-distance objective lenses to allow easy manipulation

of, and unparalleled optical access to, the sample. Even so, the speed of LSFM is crucial in allowing multiview imaging; conventional methods are simply too slow to take advantage of the paradigm, even when geometry allows it.

The first aspect of multiview imaging concerns improved axial resolution. Because axial resolution is typically lower than lateral, imaging from two directions (separated by 90°) will produce two data sets, which, taken together, sample all axes with the best possible resolution. The different viewing angles can subsequently be combined to produce a single data set with improved axial resolution [93]. A number of multiview registration [94] and deconvolution algorithms exist, with the most powerful ones capable of performing real-time processing [95] and requiring fewer views [96]. Two views are insufficient to provide truly isotropic resolution; however, additional views reduce the temporal resolution, unnecessarily expose the sample, and encode increasing amounts of redundant information [97].

Large and opaque samples may additionally benefit from the superior sample coverage offered by multiview imaging. Full optical coverage can be achieved, even on distal sides of the embryo, by sequentially recording image stacks from different viewing angles and computationally fusing them to produce a single high-resolution data set. For example, consider an embryo with a degree of rotational symmetry, such as the ellipsoidal Drosophila embryo or the spherical zebrafish embryo. In addition to the small-scale imaging of embryos, multiview imaging is more widely used in large-scale imaging. Due to the weak penetration of the Gaussian light sheet we mentioned, with multiview imaging, it can completely penetrate the sample deep layer, so only needs to penetrate half of the depth to complete the imaging. With the deconvolution algorithm [98], the original details of the sample can be completely restored. As shown in Figure 14.14, the light sheet and imaging optics overlap to provide good optical coverage of a quarter of the embryo; therefore, by taking four views that are spaced by 90° each, the entire sample can be covered. Acquisitions from a few closely spaced angles can also help when the sample exhibits a complex geometry that may be changing during a time-lapse experiment, as these views increase the chance to capture a critical event from the best possible angle.

Multidirectional SPIM (mSPIM) was the first technique to add a second illumination lens, which effectively provides two views of the sample as it is sequentially illuminated from either side [66]. Although synchronous double-sided illumination is possible, this results in a loss of contrast as the light sheet spreads toward the opposing side of the embryo. Because a rudimentary fusion of the two views can be achieved by stitching together the good half of each image, only half of the FOV needs to be covered by each light sheet. As such, each light sheet may be made thinner by a factor of $\sqrt{2}$ without compromising the FOV. Nevertheless, full optical coverage still requires at least one rotation, as half of the embryo remains inaccessible to the single detection path.

The latest IsoView allows simultaneous illumination and detection in all four paths, and eliminates crosstalk either spatially, by using phase-shifted confocal line detection; or spectrally, by switching between colors in the orthogonal pathways [53]. Both modes require scanning all four objectives

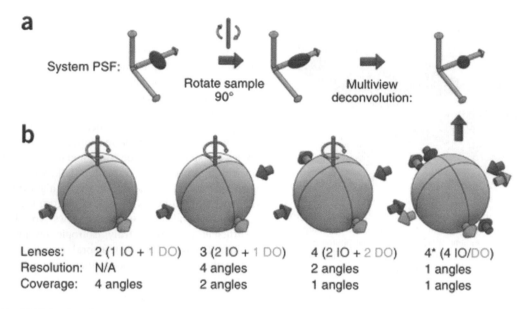

FIGURE 14.14 Multiview imaging. (a) Improved axial resolution can be achieved by reconstructing images taken from different angles, either achieved by sample rotation (two, three, or four lenses) or by using all of the lenses for illumination and detection (4×). (b) Optical coverage arises from the overlap of efficiently illuminated and detected quadrants. The (minimum) number of imaging angles to provide full optical coverage and improved axial resolution is given for each case.

to refocus the corresponding detection plane, with the result being that the beam waist of each light sheet is translated an equal distance. Correspondingly, the light sheet is more weakly focused to span the additional distance, compromising sectioning somewhat. Nevertheless, multiview deconvolution produced data sets with a spatial resolution of 2.5 μm or better, even in the center of the Drosophila embryo.

Though not specific to LSFM, refractive-index matching by chemical clearing of tissue finds a natural home in this context, allowing for exceptionally large, fixed samples to be imaged with microscopic resolution and in a reasonable period of time. Nevertheless, the transition to larger sample sizes does provide some unique optical challenges, and although clearing makes even deep tissues accessible, without corresponding changes to the optical components, they remain tantalizingly out of reach. Dodt et al. reported an ultramicroscope that uses low magnification and NA optics to image cleared mouse brains over centimeter-sized FOVs [99] naturally, with some sacrifice to spatial resolution. The generation of thinner, less-divergent light sheets benefits subcellular and macroscopic LSFM imaging alike. Saghafi et al. were able to shape the illuminating light sheet using several aspheric and cylindrical lenses in series to produce light sheets with a 4-μm thickness at the beam waist and with little divergence over several millimeters [100]. Others have used binary-pupil phase masks to achieve similar results [101, 102]. Tomer et al. adopted a different approach to imaging optically cleared tissues in the CLARITY optimized light-sheet microscope (COLM) [103], which tiles the acquisition process to cover large FOVs. The superior collection efficiency afforded by high-NA optics compensates somewhat for the additional exposures by making better use of the available light, whereas the relatively high magnification and NA affords submicron resolution. To compensate for misalignment caused by residual refractive index

inhomogeneities deep inside tissue, an autocalibration routine adjusts the light-sheet position such that the two planes maintain coalignment throughout the volume.

The third challenge is to achieve high imaging throughput while maintaining high 3D resolution. The aforementioned methods such as ASLM and multiview light sheet imaging all need to improve the conventional light sheet imaging system, such as scanning with a galvanometer, using the CLD effect to reject the side lobes of the Bessel beam or reject the part out of Gaussian light sheet's Rayleigh range. As for multiview imaging, a rotating motor can be used to rotate the sample and acquire information from multiple angles, and then an algorithm can be used to fuse it to a whole 3D volume.

The main problem of these methods is to reduce the imaging frame rate, which in turn reduces the imaging throughput. ASLM is affected by the response and movement time of the galvanometer or ETL, and due to the use of sCMOS electronic rolling shutter, the maximum frame rate of the camera will be reduced by half, which greatly limits the total optical throughput of the imaging system. Similarly, multiview imaging cannot complete the three-dimensional scanning of the sample at a fixed angle of view. The more precise the angle divided during the acquisition process, the higher the resolution and the longer the acquisition time. The time required for four angles is four times that of single-sided illumination, while the time required for eight angles will be further doubled.

The key to the improvement is to combine light sheet fluorescence microscopy with the resolution enhancement algorithm, achieving that under the combination of a thicker light sheet excitation and a lower magnification detection objective, and acquiring the raw data with low three-dimensional resolution with extremely high throughput. After acquisition, the three-dimensional resolution enhancement algorithm can be used to enhance 3D resolution of raw data to obtain high-resolution

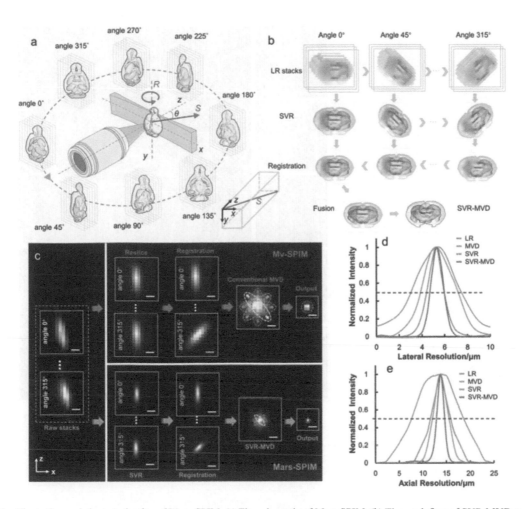

FIGURE 14.15 Illustration and characterization of Mars-SPIM. (a) The schematic of Mars-SPIM. (b) The work flow of SVR-MVD procedure which can *in toto* reconstruct the whole brain at isotropically enhanced resolution. (c) The resolution comparison between single-view raw image, SVR only, MVD only, and SVR-MVD, via resolving subresolution fluorescent beads (≈500 nm diameter). Images x–z show the lateral and axial extents of the resolved beads (red circles). (d, e) The intensity plots of the linecuts through the resolved beads along the lateral and axial directions, respectively. The SVR-MVD shows an obviously highest isotropic resolution at ≈1.4 μm, which is compared to ≈4.2 (lateral) and 12 μm (axial) in raw image. Scale bars: 5 μm in (c).

data comparable to data acquired by the combination of a thinner light sheet excitation and a higher magnification detection objective. The most commonly used algorithm is deconvolution. If the point spread function (PSF) of the optical imaging system is known, the PSF can be used to complete the deconvolution operation. The commonly used algorithms are the Lucy-Richardson algorithm and Wiener filtering. Deconvolution is simple and convenient, but the disadvantage is that the effect is limited. For images that are too blurry, it is difficult to achieve the desired resolution enhancement effect. In the research work of Fei et al. [98], the deconvolution algorithm and the subvoxel-resolving (SVR) algorithm are combined to achieve better enhancement effects. With the multiangle imaging at four times, a complete mouse brain was observed within half an hour, and achieved a three-dimensional resolution of approximately 1.5 μm (Figure 14.15), overcoming the contradiction between high throughput and high resolution.

Several other commonly used algorithms in superresolution research, such as SIM, require the synthesis of multiple raw images and therefore cannot meet the requirement of increasing the acquisition throughput. Compression sensing, which is also based on a single raw image, is also favored by researchers. Compressed sensing is based on sparse raw signals, allowing image acquisition at sampling rates lower than the Nyquist sampling frequency, while for dense signals, images can be converted to sparse domain by Fourier transform, etc. to recover, with equally impressive effects. Therefore, researchers reduced the imaging time in two ways: the first is by directly reducing the exposure time during the acquisition of single-frame images to obtain the original image with low signal-to-noise ratio and then using the compressed sensing to recover the results of high SNR [104, 105]; the second is to use the same thicker light sheet and low-magnification detection objective lens, and after obtaining the blurred image, use the compressed sensing to enhance the 3D resolution [81]. Both methods can increase the imaging throughput, and due to the more pronounced effect of 3D resolution, the second method can increase the throughput up to 64 times, achieving whole mouse brain data acquisition within minutes, which can be recovered to ~1 μm isotropic resolution (Figure 14.16).

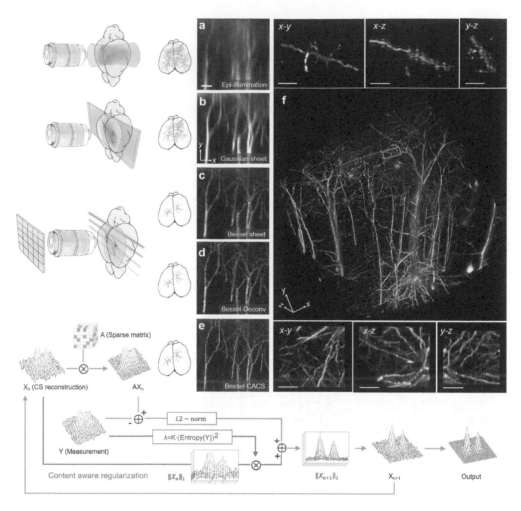

FIGURE 14.16 CACS Bessel plane illumination microscopy and other imaging modes. a, Epi-illumination mode (left), and maximum-intensity-projection (MIP) in the *y–z* plane from a 3D image stack of a cleared transgenic mouse brain tissue with GFP labelled to the neurons (right, 3.2 × /0.28 objective); b, Gaussian sheet mode; *c–e*, Raw, deconvolution and CACS modes under synchronized Bessel plane illumination and how CACS computation being further applied to super-resolve the image (e). Scale bar, 20 μm. f, Volume rendering of GFP-labelled neurons in a cleared mouse cortex tissue by 3.2 × Bessel-CACS mode. Insets: MIPs along orthogonal axes of the cubical volume of interest (boxes). Scale bar, 10 μm.

Conclusions and outlook

This chapter respectively introduces the principles of traditional TPEM and LSFM and the problems of light scattering and absorption when imaging biological tissues. The main strength of TPEM and Bessel light sheet fluorescence microscopy techniques is the ability to maintain resolution and contrast within scattering tissue. The long wavelength and concentric circle structure formed by interference ensure that when they penetrate thick biological tissues, they can better avoid the influence of absorption and keep their shape unchanged. Even so, the penetration depth is still unsatisfactory when facing opaque biological samples.

The emergence and development of tissue optical clearing technology successfully extended the imaging size to several millimeters, even centimeters. For the study of *in vitro* imaging, this combination provides great convenience for the whole brain nerve observation and other fields. At the same time, LSFM can also use the two-photon long wavelength as an illumination laser source, which can further enhance the penetration depth. With the current variety of new modified light sheets, such as multiview imaging and ASLM, it can achieve high-throughput, high-resolution imaging, which can image a whole mouse brain in several hours with cellular 3D resolution. At the same time, for small samples at the single cell level, submicron three-dimensional resolution can also be observed.

Even though there is currently a shortage of applications for *in vivo* imaging, we still believe that as tissue optical clearing technology continues to evolve, new harmless and rapid clearing techniques will inevitably emerge in the future that can make tissues transparent without harming the activity of the organisms themselves. A new revolution will also take place in the field of live imaging at that time.

Acknowledgments

The authors thank Shangbang Gao, Luoying Zhang, Haohong Li, Man Jiang, and Bo Xiong for discussions and comments on the chapter and Yunyun Han for the discussion on the imaging and visualization. The Mars-SPIM and CACS Bessel projects were supported by the National Key R&D program of China (2017YFA0700501, D.Z. and P.F.), the National Natural Science Foundation of China (21874052 for P.F., 61860206009 for D.Z., 81873793 for W.M.), the Innovation Fund of WNLO (P.F. and D.Z.), and the Junior Thousand Talents Program of China (P.F.).

REFERENCES

1. Denk, W., Piston, D.W. and Webb, W.W. *Two-Photon Molecular Excitation in Laser-Scanning Microscopy[M]*, Springer US, 445–458 (1995).
2. Göppert-Mayer, M. *Über Elementarakte mit zwei Quantensprüngen.* Annalen Der Physik **401**(3), 273–294 (1931).
3. Xu, C. and Webb, W.W. "Measurement of two-photon excitation cross sections of molecular fluorophores with data from 690 to 1050 nm," *Journal of the Optical Society of America B* **13**, 481–491 (1996).
4. Zipfel, W.R., Williams, R.M. and Webb, W.W. "Nonlinear magic: Multiphoton microscopy in the biosciences," *Nature Biotechnology* **21**, 1369–1377 (2003).
5. Mertz, J. "Nonlinear microscopy: New techniques and applications," *Current Opinion in Neurobiology* **14**, 610–616 (2004).
6. Campagnola, P.J. and Loew, L.M. "Second-harmonic imaging microscopy for visualizing biomolecular arrays in cells, tissues and organisms," *Nature Biotechnology* **21**, 1356–1360 (2003).
7. Squier, J., Muller, M., Brakenhoff, G.J. and Wilson, K.R. "Third harmonic generation microscopy," *Optics Express* **3**, 315–324 (1998).
8. Oron, D., Yelin, D., Tal, A.E. et al. "Depth-resolved structural imaging by third-harmonic generation microscopy," *Journal of Structural Biology* **147**, 3–11 (2004).
9. Mohler, W., Millard, A.C. and Campagnola, P.J. "Second harmonic generation imaging of endogenous structural proteins," *Methods* **29**, 0–109 (2003).
10. Daniel A. Dombeck, Karl A. Kasischke, Harshad D. Vishwasrao, Martin Ingelsson, Bradley T. Hyman, and Watt W. Webb. "Uniform polarity microtubule assemblies imaged in native brain tissue by second-harmonic generation microscopy," *Proceedings of the National Academy of Sciences of the United States of America*, **100**(12), 7081–7086 (2003).
11. Bouevitch, O., Lewis, A., et al. "Probing membrane potential with nonlinear optics," *Biophysical Journal* **65**(2), 672–679 (1993).
12. Moreaux, L. "Membrane imaging by simultaneous second-harmonic generation and two-photon microscopy," *Optics Letters* **25**, 320–322 (2000).
13. Cheng, J.X., Volkmer, A. and Xie, X.S. "Theoretical and experimental characterization of coherent anti-stokes Raman scattering microscopy," *Journal of the Optical Society of America B*, **19**(6), 1363–1375 (2002).
14. Wang, H., Fu, Y., Zickmund, P., Shi, R. and Cheng, J.X. "Coherent anti-stokes Raman scattering imaging of axonal myelin in live spinal tissues," *Biophysical Journal* **89**, 581–591 (2005).
15. Helmchen, F. and Denk, W. "Deep tissue two-photon microscopy," *Nature Methods* **2**, 932–940 (2005).
16. Squirrell, J.M., Wokosin, D.L., White, J.G. and Bavister, B.D. "Long-term two-photon fluorescence imaging of mammalian embryos without compromising viability," *Nature Biotechnology* **17**, 763–767 (1999).
17. Denk, W. "Two-photon scanning photochemical microscopy: mapping ligand-gated ion channel distributions," *Proceedings of the National Academy of Sciences of the United States of America* **91**, 6629–6633 (1994).
18. Svoboda K. and Tank D.W. "Direct measurement of coupling between dendritic spines and shafts," *Science* **272**(5262), 716–719 (1996).
19. Matsuzaki, M., Ellis-Davies, G., Nemoto, T. et al. "Dendritic spine geometry is critical for AMPA receptor expression in hippocampal CA1 pyramidal neurons," *Nature Neuroscience* **4**, 1086 (2001).
20. Kleinfeld, D., Mitra, P.P., Helmchen, F. and Denk, W. "Fluctuations and stimulus-induced changes in blood flow observed in individual capillaries in layers 2 through 4 of rat neocortex," *Proceedings of the National Academy of Sciences* **95**(26), 15741–15746 (1998).
21. Oheim, M., Beaurepaire, E., Chaigneau, E., Mertz, J. & Charpak, S.. "Two-photon microscopy in brain tissue: Parameters influencing the imaging depth," *Journal of Neuroscience Methods* **111**(1), 29–37 (2001).
22. Yaroslavsky, A.N., Schulze, P.C., Yaroslavsky, I.V., Schober, R. and Schwarzmaier, H.J. "Optical properties of selected native and coagulated human brain tissues in vitro in the visible and near infrared spectral range," *Physics in Medicine & Biology* **47**, 2059–2073 (2002).
23. Beaurepaire, E. and Mertz, J. "Epifluorescence collection in two-photon microscopy," *Appl Opt* **41**, 5376–5370 (2002).
24. Svoboda, K., Denk, W., Kleinfeld, D. and Tank, D.W. "In vivo dendritic calcium dynamics in neocortical pyramidal neurons," *Nature* **385**(6612), 161–165 (1997).
25. Helmchen, F., Svoboda, K., Denk, W. and Tank, D.W. "In vivo dendritic calcium dynamics in deep-layer cortical pyramidal neurons," *Nature Neuroscience* **2**, 989–996 (1999).
26. Svoboda, K., Helmchen, F., Denk, W. and Tank, D.W. "Spread of dendritic excitation in layer 2/3 pyramidal neurons in rat barrel cortex in vivo," *Nature Neuroscience* **2**, 65–73 (1999).
27. Nimmerjahn, A., Kirchhoff, F., Kerr, J.N.D. and Helmchen, F. "Sulforhodamine 101 as a specific marker of astroglia in the neocortex in vivo," *Nature Methods* **1**, 31–37 (2004).
28. Theer, P., Hasan, M.T. and Denk, W. "Two-photon imaging to a depth of 1000 microm in living brains by use of a Ti:Al2O3 regenerative amplifier," *Optics Letters* **28**, 1022 (2003).
29. Mizrahi, A. "High-resolution in vivo imaging of hippocampal dendrites and spines," *Journal of Neuroscience* **24**, 3147–3151 (2004).
30. Jung, J.C. and Schnitzer, M.J. "Multiphoton endoscopy," *Optics Letters* **28**, 902–904 (2003).
31. Jung, J.C., Mehta, A.D., Aksay, E., Stepnoski, R. and Schnitzer, M.J. "In vivo mammalian brain imaging using one- and two-photon fluorescence microendoscopy," (2004).
32. Levene, M.J. "In vivo multiphoton microscopy of deep brain tissue," *Journal of Neurophysiology* **91**, 1908–1912 (2004).

33. Molitoris, B.A. "Intravital multiphoton microscopy of dynamic renal processes," *American Journal of Physiology Renal Physiology* **288**, F1084–F1089 (2005).
34. Nimmerjahn, A., Kirchhoff, F. and Helmchen, F. "Resting microglial cells are highly dynamic surveillants of brain parenchyma in vivo," *Science* **308**, 1314–1318 (2005).
35. Christie, R.H., Bacskai, B.J., Zipfel, W.R., Williams, R.M. and Hyman, B.T. "Growth arrest of individual senile plaques in a model of Alzheimer's disease observed by in vivo multiphoton microscopy," *Journal of Neuroscience* **21**, 858–864 (2001).
36. Grutzendler, J., Kasthuri, N. and Can, W.-B. "Long-term dendritic spine stability in the adult cortex," *Nature* **420**, 812–816 (2002).
37. Yoder, E.J. and Kleinfeld, D. "Cortical imaging through the intact mouse skull using two-photon excitation laser scanning microscopy," *Microscopy Research and Technique* **56**(4), 304–305 (2002).
38. Zuo, Y., Lin, A., Chang, P. and Gan, W.B. "Development of long-term dendritic spine stability in diverse regions of cerebral cortex," *Neuron* **46**(2), 181–189 (2002).
39. Trachtenberg, J.T. et al. "Long-term in vivo imaging of experience-dependent synaptic plasticity in adult cortex," *Nature* **420**, 788–794 (2002).
40. Holtmaat, A.J.G.D., Trachtenberg, J.T., Wilbrecht, L., Shepherd, G.M. and Svoboda, K. "Transient and persistent dendritic spines in the neocortex in vivo," *Neuron* **45**(2), 279–291 (2005).
41. Majewska, A. and Sur, M. "Motility of dendritic spines in visual cortex in vivo: Changes during the critical period and effects of visual deprivation," *Proceedings of the National Academy of Sciences of the U S A* **100**, 16024–16029 (2004).
42. Jain, R., Munn, L. and Fukumura, D. "Dissecting tumour pathophysiology using intravital microscopy," *Nat Rev Cancer* **2**, 266–276 (2002).
43. Charpak, S., Mertz, J. and Beaurepaire, E. "Odor-evoked calcium signals in dendrites of rat mitral cells," *Proceedings of the National Academy of Sciences of the United States of America* **98**, 1230–1234 (2001).
44. Waters, J., Larkum, M., Sakmann, B. and Helmchen, F. "Supralinear Ca2+ influx into dendritic tufts of layer 2/3 neocortical pyramidal neurons in vitro and in vivo," *Journal of Neuroscience* **23**, 8558–8567 (2003).
45. Waters, J. and Helmchen, F. "Boosting of action potential backpropagation by neocortical network activity in vivo," *The Journal of Neuroscience* **24**(49) 11127–11136 (2004).
46. Margrie, T.W., Meyer, A.H., Caputi, A., Monyer, H. and Brecht, M. "Targeted whole-cell recordings in the mammalian brain in vivo," *Neuron* **39**, 911–918 (2003).
47. Jing, W.W., Wong, A.M., Flores, J., Vosshall, L.B. and Axel, R. "Two-photon calcium imaging reveals an odor-evoked map of activity in the fly brain," *Cell* **112**, 271–282 (2003).
48. Holtmaat, A., Trachtenberg, J.T., Wilbrecht, L., Shepherd, G.M. and Svoboda, K. "Transient and persistent dendritic spines in the neo cortex in vivo," *Neuron* **45**(2), 279–291 (2005).
49. Gero Miesenböck. "Genetic methods for illuminating the function of neural circuits," *Current Opinion in Neurobiology* **14**, 395–402 (2004).
50. Kerr, R., Lev-Ram, V., Baird, G. Vincent, P, & Schafer, W. "Optical imaging of calcium transients in neurons and pharyngeal muscle of c. elegans," *Neuron* **26**(3), 583–594 (2000).
51. Reiff, D.F. "In vivo performance of genetically encoded indicators of neural activity in flies," *Journal of Neuroscience* **25**, 4766–4778 (2005).
52. Winnubst, J., Bas, E., Ferreira, T.A., Wu, Z. and Chandrashekar, J. "Reconstruction of 1,000 projection neurons reveals new cell types and organization of long-range connectivity in the mouse brain," *Cell* **179**(1), 268–281 (2019).
53. Chhetri, R.K., Amat, F., Wan, Y. et al. "Whole-animal functional and developmental imaging with isotropic spatial resolution," *Nature Methods* **12**, 1171–1178 (2015).
54. Arrenberg, A.B., Stainier, D.Y.R., Baier, H. and Huisken, J. "Optogenetic control of cardiac function," *Science* **330**, 971–974 (2010).
55. Mickoleit, M., Schmid, B., Weber, M. et al. "High-resolution reconstruction of the beating zebrafish heart," *Nature Methods* **11**, 919–922 (2014).
56. Schmid, B., Shah, G., Scherf, N. et al. "High-speed panoramic light-sheet microscopy reveals global endodermal cell dynamics," *Nature Communications* **4**, 1–10 (2013).
57. Rauzi, M., Krzic, U., Saunders, T. et al. "Embryo-scale tissue mechanics during Drosophila gastrulation movements," *Nature Communications* **6**, 8677 (2015).
58. Erik, S.W., Meghan, K.D., Kevin M.D. et al. "Quantitative multiscale cell imaging in controlled 3D microenvironments," *Developmental Cell* **36**, 462–475 (2016).
59. Strnad, P., Gunther, S., Reichmann, J. et al. "Inverted light-sheet microscope for imaging mouse pre-implantation development," *Nature Methods* **13**, 139–142 (2016).
60. Wu, Y., Ghitani, A., Christensen, R. et al. "Inverted selective plane illumination microscopy (iSPIM) enables coupled cell identity lineaging and neurodevelopmental imaging in Caenorhabditis elegans," *Proceedings of the National Academy of Sciences of the United States of America* **108**(43), 17708–17713 (2011). doi:10.1073/pnas.1108494108
61. Kaufmann, A., Mickoleit, M., Weber, M. and Huisken, J. "Multilayer mounting enables long-term imaging of zebrafish development in a light sheet microscope," *Development* **139**, 3242–3247 (2012).
62. Wu, Y., Wawrzusin, P., Senseney, J. et al. "Spatially isotropic four-dimensional imaging with dual-view plane illumination microscopy," *Nature Biotechnology* **31**, 1032–1038 (2013).
63. Vladimirov, N., Mu, Y., Kawashima, T. et al. "Light-sheet functional imaging in fictively behaving zebrafish" *Nature Methods* **11**, 883–884 (2014).
64. Ahrens, M., Orger, M., Robson, D. et al. "Whole-brain functional imaging with two-photon light-sheet microscopy," *Nature Methods* **12**, 379–380 (2015).
65. Power, R. and Huisken, J. "A guide to light-sheet fluorescence microscopy for multiscale imaging," *Nature Methods* **14**, 360–373 (2017).
66. Huisken, J. and Stainier, D.Y. "Even fluorescence excitation by multidirectional selective plane illumination microscopy (mSPIM)," *Optics Letters* **32**, 2608–2610 (2007).
67. Keller, P.J., Schmidt, A.D., Wittbrodt, J. and Stelzer, E.H.K. "Reconstruction of zebrafish early embryonic development by scanned light sheet microscopy," *Science* **322**, p.1065–1069 (2008).
68. Chen, B.C., Legant, W.R., Wang, K. et al. "Lattice light-sheet microscopy: imaging molecules to embryos at high spatiotemporal resolution," *Science* **346**, 1257998 (2014).

69. Cella Zanacchi, F., Lavagnino, Z., Perrone Donnorso, M. et al. "Live-cell 3D super-resolution imaging in thick biological samples," *Nature Methods* **8**, 1047–1049 (2011).

70. Huang, B., Jones, S.A., Brandenburg, B. and Zhuang, X. "Whole-cell 3D STORM reveals interactions between cellular structures with nanometer-scale resolution," *Nature Methods* **5**, 1047–1052 (2008).

71. Gebhardt, J., Suter, D., Roy, R. et al. "Single-molecule imaging of transcription factor binding to DNA in live mammalian cells," *Nature Methods* **10**, 421–426 (2013).

72. Galland, R., Grenci, G., Aravind, A. et al. "3D high- and super-resolution imaging using single-objective SPIM," *Nature Methods* **12**, 641–644 (2015).

73. Li, T., Ota, S., Kim, J. et al. "Axial plane optical microscopy," *Scientific Reports* **4**, 7253 (2014).

74. Theer, P., Dragneva, D. and Knop, M. "πSPIM: High NA high resolution isotropic light-sheet imaging in cell culture dishes," *Scientific Reports* **6**, 32880 (2016).

75. Tokunaga, M., Imamoto, N. and Sakata-Sogawa, K. "Highly inclined thin illumination enables clear single-molecule imaging in cells," *Nature Methods* **5**, 159–161 (2008).

76. Gao, L. "Optimization of the excitation light sheet in selective plane illumination microscopy," *Biomedical Optics Express* **21**, 1571–1572 (2015).

77. Planchon, T., Gao, L., Milkie, D. et al. "Rapid three-dimensional isotropic imaging of living cells using Bessel beam plane illumination," *Nature Methods* **8**, 417–423 (2011).

78. Fahrbach, F.O., Simon, P. and Rohrbach, A. "Microscopy with self-reconstructing beams," *Nature Photonics* **4**, 780–785 (2010).

79. Fahrbach, F.O., Gurchenkov, V., Alessandri, K., Nassoy, P. and Rohrbach, A. "Self-reconstructing sectioned Bessel beams offer submicron optical sectioning for large fields of view in light-sheet microscopy," *Optics Express* **21**, 11425 (2013).

80. Gao, L., Shao, L., Higgins, C., Poulton, J, Peifer, M, & Davidson, M., et al. "Noninvasive imaging beyond the diffraction limit of 3d dynamics in thickly fluorescent specimens," *Cell* **151**(6), 1370–1385 (2012).

81. Fang, C., Yu, T., Chu, T. et al. "Minutes-timescale 3D isotropic imaging of entire organs at subcellular resolution by content-aware compressed-sensing light-sheet microscopy," *Nature Communications* **12**, 107 (2021).

82. Gao, L., Shao, L., Chen, B.C. et al. "3D live fluorescence imaging of cellular dynamics using Bessel beam plane illumination microscopy," *Nature Protocols* **9**, 1083–1101 (2014).

83. Vettenburg, T., Dalgarno, H., Nylk, J. et al. "Light-sheet microscopy using an Airy beam," *Nat Methods* **11**, 541–544 (2014).

84. Dean, K.M. and Fiolka, R. "Uniform and scalable light-sheets generated by extended focusing," *Optics Express* **22**, 26141 (2014).

85. Ping, J., Zhao, F., Nie, J., Yu, T. and Fei, P. "Propagating-path uniformly scanned light sheet excitation microscopy for isotropic volumetric imaging of large specimens," *Journal of Biomedical Optics* **24**, 1 (2019).

86. Chakraborty, T., Driscoll, M.K., Jeffery, E. et al. "Light-sheet microscopy of cleared tissues with isotropic, subcellular resolution," *Nature Methods* **16**, 1109–1113 (2019).

87. Voigt, F.F., Kirschenbaum, D., Platonova, E. et al. "The meso-SPIM initiative: Open-source light-sheet microscopes for imaging cleared tissue," *Nature Methods* **16**, 1105–1108 (2019).

88. Dean, K.M., Roudot, P., Welf, E.S., Danuser, G., and Fiolka, R. "Deconvolution-free subcellular imaging with axially swept light sheet microscopy," *Biophysical Journal* **108**(12), 2807–2815 (2015).

89. Fahrbach, F.O. and Rohrbach, A. "Propagation stability of self-reconstructing Bessel beams enables contrast-enhanced imaging in thick media," *Nature Communications* **3**, 632 (2012).

90. Silvestri, L., Bria, A., Sacconi, L., Iannello, G. and Pavone, F.S. "Confocal light sheet microscopy: Micron-scale neuroanatomy of the entire mouse brain," *Optics Express* **20**, 20582 (2012).

91. Z. Yang, L. Mei, F. Xia, Q. Luo, L. Fu, and H. Gong. "Dual-slit confocal light sheet microscopy for in vivo whole-brain imaging of zebrafish," *Biomedical Optics Express* **6**, 1797–1811 (2015).

92. Medeiros, G., Norlin, N., Gunther, S. et al. "Confocal multiview light-sheet microscopy," *Nature Communications* **6**, 8881 (2015).

93. Verveer, P., Swoger, J., Pampaloni, F. et al. "High-resolution three-dimensional imaging of large specimens with light sheet-based microscopy," *Nature Methods* **4**, 311–313 (2007).

94. Preibisch, S., Saalfeld, S., Schindelin, J. and Tomancak, P. "Software for bead-based registration of selective plane illumination microscopy data," *Nature Methods* **7**, 418–419 (2010).

95. Benjamin, S. and Jan, H. "Real-time multi-view deconvolution," *Bioinformatics* **31**(20), 3398–3400 (2015).

96. Preibisch, S., Amat, F., Stamataki, E. et al. "Efficient Bayesian-based multiview deconvolution," *Nature Methods* **11**, 645–648 (2014).

97. Swoger, J., Verveer, P., Greger, K., Huisken, J. and Stelzer, E.H.K. "Multi-view image fusion improves resolution in three-dimensional microscopy," *Optics Express* **15**, 8029–8042 (2007).

98. Nie, J., Liu, S., Yu, T. et al. "Fast, 3D isotropic imaging of whole mouse brain using multiangle-resolved subvoxel SPIM," *Advanced Science* **7**(3), 1901891 (2020).

99. Dodt, H., Leischner, U., Schierloh, A. et al. "Ultramicroscopy: Three-dimensional visualization of neuronal networks in the whole mouse brain," *Nature Methods* **4**, 331–336 (2007).

100. Saghafi, S., Becker, K., Hahn, C. and Dodt, H.U. "3D-ultramicroscopy utilizing aspheric optics," *Journal of Biophotonics* **7**(1–2), (2014).

101. Wilding, D., Pozzi, P., Soloviev, O. et al. "Pupil filters for extending the field-of-view in light-sheet microscopy," *Optics Letters* **41**, 1205 (2016).

102. Golub, I., Chebbi, B. and Golub, J. "Toward the optical 'magic carpet': Reducing the divergence of a light sheet below the diffraction limit," *Optics Letters* **40**, 5121 (2015).

103. Tomer, R., Ye, L., Hsueh, B. and Deisseroth, K. "Advanced CLARITY for rapid and high-resolution imaging of intact tissues," *Nature Protocols* **9**, 1682–1697 (2014).

104. Woringer, M., Darzacq, X., Zimmer, C. and Mir, M. "Faster and less phototoxic 3D fluorescence microscopy using a versatile compressed sensing scheme," *Optics Express* **25**, 13668–13683 (2017).

105. Bai, C., Liu, C., Jia, H. et al. "Compressed blind deconvolution and denoising for complementary beam subtraction light-sheet fluorescence microscopy," *IEEE Transactions on Biomedical Engineering*, 2979–2989 (2019).

15

Endogenous fluorescence preservation from solvent-based optical clearing

Tingting Yu, Yisong Qi, and Dan Zhu

CONTENTS

Introduction

Recently, tissue optical clearing techniques have emerged to reduce scattering and improve light penetration depth by introducing various chemical agents and tools [1–5]. A number of clearing methods have been developed [1, 3, 6, 7], including aqueous-based clearing methods, such as CLARITY [8], PACT-PARS [9], CUBIC [10, 11], Sca*l*eS [12], OPTIClear [13], and C$_e$3D [14]; and organic solvent–based clearing methods, such as BABB [15], 3DISCO [16, 17], iDISCO [18], uDISCO [19], FDISCO [20], sDISCO [21], FluoClearBABB [22],

Ethanol-ECi [23], and PEGASOS [24]. Each kind of method has its advantages: the former can preserve protein-based fluorescence, while the latter can provide a favorable speed of clearing [25]. These methods provide essential tools for obtaining high-resolution 3D images of intact tissues and elucidating many biological events [7].

The organic solvent–based clearing protocols, such as 3DISCO, can achieve the highest level of tissue transparency and shrinkage [3, 6], thereby facilitating the imaging of large samples, such as the brain, spinal cord, lung tumors, mammary glands, immune organs, and embryos [16, 18]. However,

DOI: 10.1201/9781003025252-17

3DISCO results in a rapid decline in endogenous fluorescence signals during the tissue clearing and storage procedure (with a half-life of approximately 1–2 d), which has impeded the application of 3DISCO, especially for large samples such as the whole brain [16, 19]. Hence, the loss of endogenous fluorescence remains a significant concern for solvent-based clearing methods.

To address this issue, researchers have made efforts from different aspects, involving introducing a resin formulation during archiving of cleared samples, screening the fluorescence-friendly compounds (e.g., dehydration and refractive index matching reagents as well as antioxidants), and adjusting the conditions (e.g., pH and temperature) in the clearing and storage process. These works provided valuable alternatives for the long-term preservation of endogenous fluorescence. This chapter will introduce some representative works from the above aspects in the following sections.

Introducing a resin for archiving of cleared samples

Another method is to introduce a transparent solid resin to prevent fluorescence of cleared samples from fading, photobleaching, and mechanical damage without impairing their transparency. Becker et al. developed a solid resin formulation

that can maintain the specimens' transparency and provide a constant high level of fluorescence.

Screening of the resin and resin embedding of the specimens

First, to find the optimal resin formulation, various resins such as acrylic-, polyester-, and epoxy-based resins and different amine-based curing agents were screened for their ability to preserve tissue transparency and fluorescence of cleared samples. The optimal formulation contained 11.5 ml (at 50°C) of the epoxy resin D.E.R. 332 (bisphenol A diglycidyl ether), 3.5 ml of the flexibilizer D.E.R. 736 (polypropylene glycol diglycidyl ether), and 3 ml of the cyclic allopathic amine epoxy curing reagent isophorone diamine (IPDA, 5-Amino-1,3,3-trimethylcyclohexane-methylamine). The mixture should be carefully mixed and degassed in a vacuum chamber at about 100 mbar for about 1 h to achieve homogeneous optical properties.

The resin embedding was carried out in a cubic mold fabricated from Silastic E-RTV silicon rubber, as shown in Figure 15.1. Its side length was up to the size of the samples, e.g., 15 mm for mouse brains or 10 mm for mouse hippocampi.

Before embedding in the cubic mold, the specimens should be preincubated in a small resin mixture for about 15 min after gently removing the clearing solution from the specimens' surface by a sheet of tissue paper. The molds were first filled about one third with the degassed resin mixture and the specimens

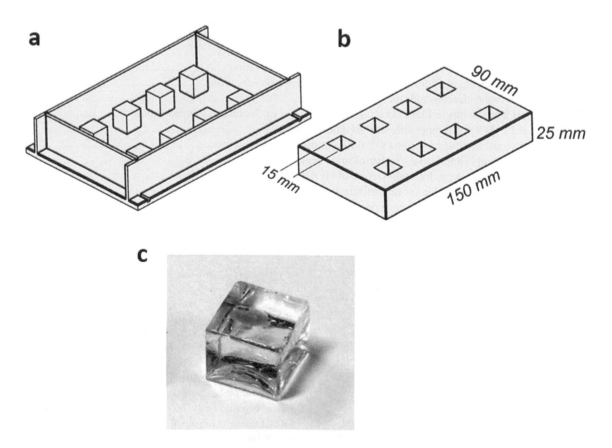

FIGURE 15.1 Fabrication of molds for resin embedding from Silastic E-RTV silicone rubber. (a) Casting frame made from acrylic glass. (b) Silicon rubber mold. (c) Cured resin block with embedded cleared mouse brain hemisphere. Reprinted with permission from reference [26].

embedded, and then they were filled up with resin mixture. The orientation of the specimens was adjusted by using an injection cannula. The resin cubes were cured at room temperature for at least 2 days in the dark to finish the embedding.

Resin embedding enables fluorescence preservation with repetitive long-term illumination

Green fluorescent protein (GFP) exhibits a decay in fluorescence activity during illumination. As shown in Figure 15.2, the thin layer in the mouse hippocampal region (*Thy1*-EGFP-M) was illuminated with an optical section exposed for 2 h constantly to a high-power light sheet, and the illumination was repeated on the following 2 days. The fluorescence within the cleared hemisphere incubated in DBE (control group) dropped severely during illumination, and the signal was lost after the second illumination period. In contrast, the fluorescence could be preserved largely by resin-embedding under the same experimental conditions (Figure 15.2).

The fluorescence intensity within the corresponding regions of interest was used to quantify the resin-mediated fluorescence protection and signal maintenance (Figure 15.3). The resin-embedded and control samples were illuminated as has been described, and images were taken every 6 min during three successive light exposures (day 1–day 3) of 120 min duration [26]. The plots of relative fluorescence intensity at day 1 and day 3 over time demonstrated that photobleaching is much lower in the resin-embedded sample than in the control and is limited to the initial illumination phase (Figure 15.3b).

Resin embedding enables fluorescence preservation during long-term storage

Due to the fluorescence stabilizing effect of resin embedding, long-term storage of cleared GFP-expressing samples becomes possible, e.g., for archiving purposes or repetitive investigations. An embedded mouse hippocampus remained fluorescent over several months, without relevant loss in signal quality. As shown in Figure 15.4, the resin-embedded hippocampus was still fluorescent even after more than 2 months, while the control sample without embedding lost fluorescence completely within a few days.

The transparent solid resin protects DBE-cleared samples from fluorescence fading caused by high-energy illumination and long-term storage. As is known, peroxides and aldehydes in DBE are generated continuously in the presence of oxygen or water, and can be enhanced by light [27]. Peroxides can bind covalently to GFP, quenching fluorescence permanently [28].

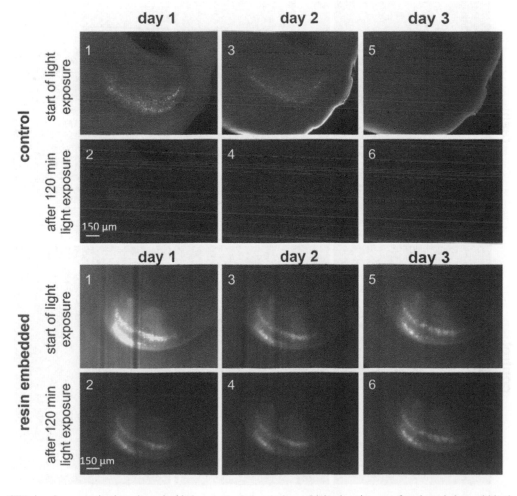

FIGURE 15.2   GFP signal preservation in resin-embedded mouse brain hemispheres. Light-sheet images of a selected plane within tissues in a control and a resin-embedded brain hemisphere before and after 120 min of constant high power illumination. Adapted from reference [26].

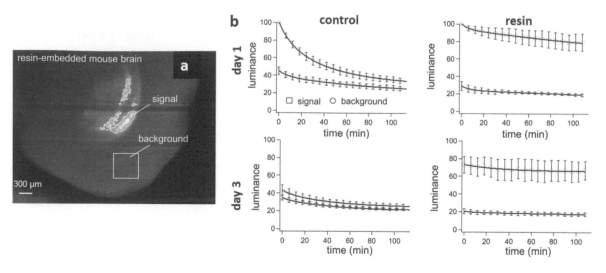

FIGURE 15.3 Quantification of fluorescence preservation during repetitive long-term illumination. Adapted from reference [26].

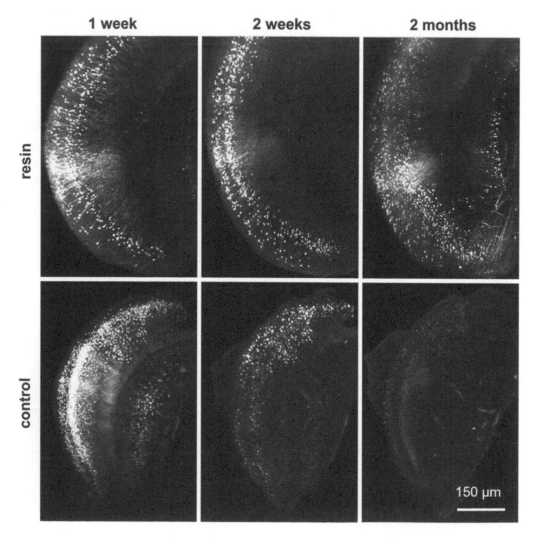

FIGURE 15.4 Fluorescence preservation by resin embedding during long-term storage. Adapted from reference [26].

It is speculated that resin embedding could prevent access by these molecules. Besides light-sheet imaging of large samples, this resin formulation is potentially to have various applications, such as the production of transparent cell cultures or slices, which could be embedded for microscopy utilizing high illumination intensities and for long-term storage.

Screening fluorescence-friendly organic compounds

As described above, embedding specimens in a transparent solid resin [26] enabled long-term archiving but required a complicated embedding procedure and incurred a risk of low imaging quality associated with reduced transparency. Some researchers developed fluorescence-friendly clearing methods by introducing new dehydrating and refractive index matching agents as well as antioxidants. For instance, Schwarz et al. proposed using 1-propanol and tert-butanol in combination with pH control to produce FluoClearBABB [22]. Pan et al. developed the uDISCO method to maintain endogenous fluorescence signals for several months by introducing (diphenyl ether) DPE to clearing combined with alpha-tocopherol [19]. Hahn et al. described sDISCO by dissolving the antioxidant propyl gallate in DBE or BABB to suppress generation of

peroxides and aldehydes, thus preserving fluorescence signals [21]. Recently, Zhao's group also proposed PEGASOS, in which polyethylene glycol (PEG) component clearing agents efficiently protect endogenous fluorescence [24].

FluoClearBABB improves endogenous fluorescence

Schwarz et al. [22] described a clearing method, FluoClearBABB, which mainly differs from the other BABB protocols in the use of 1-propanol or tert-butanol during dehydration and a basic pH throughout the procedure.

Hydrocarbon content of alcoholic component contributes to fluorescence preservation

As shown in Figure 15.5a, when the typical BABB clearing protocol (E-BABB, ethanol for dehydration, and BABB for clearing) using ethanol as dehydration agent was applied to the GFP expressed in *E. coli* cells, the majority of fluorescence loss was observed after dehydration (96.5% loss at 80% ethanol). When methanol was used as the dehydration agent, the loss of fluorescence was more rapid and severe (99.7% loss at 80% methanol) than ethanol [22]. While using 1-propanol and tert-butanol for dehydration, more fluorescence was retained (24.2% with 1-propanol, and 79.2% with tert-butanol) [22],

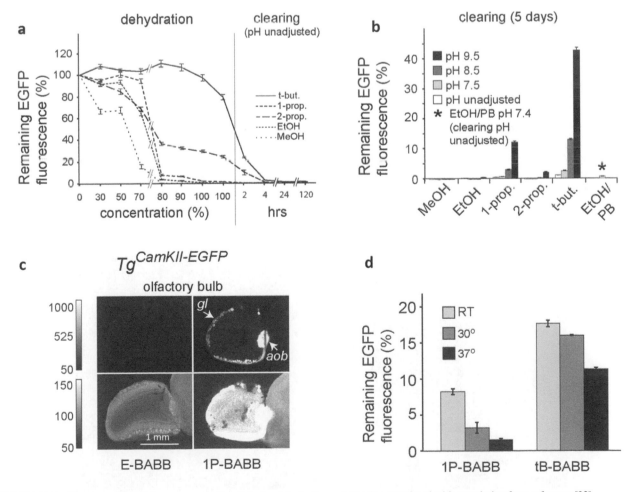

FIGURE 15.5 Changes of EGFP fluorescence in *E.coli* during dehydration and clearing. Reprinted with permission from reference [22].

indicating that more hydrocarbon content of the alcoholic component of dehydration solution contributed to more retained fluorescence.

Nevertheless, even when using tert-butanol for dehydration, most of the fluorescence was still lost during the BABB procedure (Figure 15.5b). Schwarz et al. [22] tried to increase the pH of both the dehydration and clearing solutions by adding triethylamine, which would not cause precipitation in the solutions. The effect of basic pH during clearing was proved to be useful for retaining EGFP fluorescence, which was most pronounced for both 1-propanol and tert-butanol dehydrated samples. It should be noted that the fluorescence was quite stable at high pH up to 9.5, while increasing the pH to 10.5 did not further improve the fluorescence preservation.

The optimization of dehydration and clearing parameters was also effective to improve EGFP stability in mouse brain tissues. As shown in Figure 15.5c, the EGFP fluorescence was observed to be weak in the olfactory bulb of mouse hemisphere cleared with E-BABB but significant for the one cleared with 1P-BABB (1-propanol for dehydration and BABB for clearing).

The experiment on EGFP expressed in *E. coli* cells showed that the fluorescence dropped severely at 37°C while it reduced only slightly at 30°C (Figure 15.5d). In addition, the clearing effect was found to be more efficient at higher temperatures for older animals [22]. Hence, to provide the optimal fluorescence preservation in combination with excellent clearing, especially for large samples such as adult mouse brains, tB-BABB (tert-butanol for dehydration and BABB for clearing) at 30°C was chosen to be the most reliable protocol.

FluoClearBABB enables long-term preservation of GFP fluorescence

FluoClearBABB enables long-term preservation of GFP fluorescence; for example, the level of fluorescence was unchanged after 68 days in 1P-BABB clearing solution [22]. Moreover, the fluorescence intensity of the neurons in mouse brain stored at 4°C after tB-BABB/30°C clearing was stable after 8 months, with only a minor loss of signal intensity over time as shown in Figure 15.6. Even repetitive imaging of the same sample region did not cause significant photobleaching, indicating the excellent fluorescence preserving capability of FluoClearBABB clearing protocols.

uDISCO based on diphenyl ether preserves endogenous fluorescence for months

To overcome the fluorescence quenching of 3DISCO, Pan et al. searched for organic compounds that preserved fluorescence while providing a potent tissue-clearing effect, resulting in a tissue clearing method, uDISCO, based on DPE [19].

uDISCO maintains endogenous GFP signal for months

uDISCO reveals fluorescence notably better and maintains it several times longer within *Thy1*-EGFP-M mouse brains compared with 3DISCO [19] (Figure 15.7). The quantitative fluorescence signal of uDISCO-cleared samples showed some decrease within the first few weeks and stabilized after about 1 month. The signal at 3 months after clearing was equivalent to that at 1 month. As DPE has a melting point of 26°C, it was mixed with benzyl alcohol (–15°C melting point) and benzyl benzoate (18 °C melting point) to obtain a mixture that is liquid at room temperature, named BABB-D [19]. Additionally, uDISCO utilizes the antioxidant α-tocopherol (Vitamin E) to scavenge peroxides and tert-butanol, a dehydrating reagent more stable than tetrahydrofuran (THF) used in 3DISCO. With the resulting protocol, the fine structures of labeled neurons in the lipid-dense brain and spinal cord can be detected over weeks to months.

uDISCO shows better fluorescence signal quality after whole-body clearing than 3DISCO

For system-level interrogation, uDISCO was applied to the whole body of an adult *Thy1*-EGFP-M mouse [19]. Figure 15.8a and 8b showed the images of cleared mouse body with skull and vertebra removed by a standard fluorescence stereomicroscope. For uDISCO-cleared whole body, the individual neurons and their extensions throughout the brain, spinal cord, and limbs in the intact mouse could

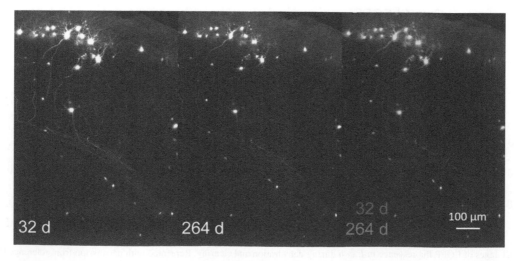

FIGURE 15.6 Long-term preservation of fluorescence. Adapted from reference [22].

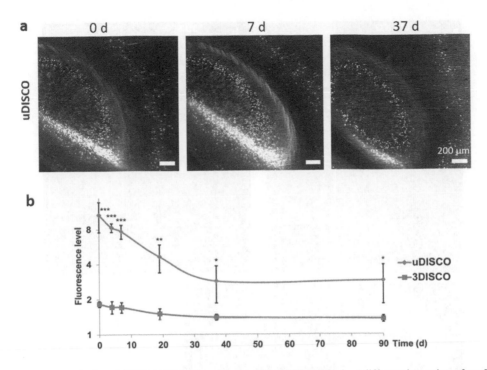

FIGURE 15.7 Preservation of GFP after uDISCO. (a) Light-sheet images of *Thy1*-EGFP-M brain at different time points after uDISCO clearing. (b) Quantifications of fluorescence-level after uDISCO and 3DISCO over time. Adapted from reference [19].

be readily identified, while for 3DISCO-cleared samples, the preservation of endogenous GFP signal was too weak to be detected with fluorescence stereomicroscope.

Figure 15.8c–d show the representative light-sheet images of the mouse brains at similar depth after whole-body uDISCO and 3DISCO clearing. The signal intensity (Figure 15.8e–f) in different brain regions between the two clearing protocols showed that uDISCO obtained a better fluorescence signal and revealed more structural details.

Propyl gallate in sDISCO stabilizes fluorescence of samples immersed in DBE or BABB

The solvent-based clearing methods generally used DBE or BABB for refractive index matching after dehydration. However, both substances continuously form peroxides or aldehydes, which severely quench the fluorescence of the proteins [28]. Becker et al. had tried to remove these compounds before application and achieved improvements in EGFP fluorescence intensity and stability [10]. However, both peroxides and aldehydes can regenerate by even trace amounts of oxygen, water, or by illumination. To overcome this problem, Hahn et al. reported the sDISCO method [21] by addition of the antioxidant propyl gallate, which efficiently suppresses the reaccumulation of peroxides and aldehydes and protects EGFP fluorescence in specimens immersed in DBE or BABB.

Propyl gallate prevents generation of peroxides and aldehydes and stabilizes fluorescence

Hahn et al. [21] tested several antioxidants, including propyl gallate, butylhydroxyltoluol, butylhydroxylanisol, and alpha-tocopherol. Among them, propyl gallate showed the highest capacity to eliminate peroxides from DBE. The addition of

propyl gallate could eliminate peroxides (10 mg/L) within 1 minute, while the other tested antioxidants could not do so in a comparable time.

To further test the ability of propyl gallate to suppress the generation of peroxides and aldehydes in purified DBE, Quantofix-25 peroxide test strips were used for assessing peroxide level, and Brady's test was used for the aldehydes [21]. Without propyl gallate, peroxide levels rose quickly and aldehydes accumulated with a short delay. By contrast, with addition of 0.2% (w/v) propyl gallate, no peroxides or aldehydes could be detected even after 1 month under air bubbling or after 1 year in a specimen container stored at 4°C. They also demonstrated that 0.4% propyl gallate did not interfere with the transparency of the specimens and is compatible with both GFP and YFP fluorescence, allowing for high-resolution imaging [21]. Hence, for sDISCO, 0.4% propyl gallate was added to DBE.

To quantify the fluorescence preserving ability of propyl gallate over time, the mouse hemispheres (*Thy1*-EGFP-M) were cleared either in DBE containing propyl gallate (stabilized DBE) or in DBE without propyl gallate as a control [21]. The fluorescence images were taken by Ultramicroscopy immediately after clearing and again after 1 month. As shown in Figure 15.9a, the signal was almost gone within the non-stabilized hemispheres (3DISCO), and the signal was still bright within the hemisphere in stabilized DBE (sDISCO) (Figure 15.9a). The signal-to-noise ratio (SNR) dropped to 0.24 ± 0.05 without stabilization, whereas it kept stable at 1.04 ± 0.15 with stabilization [21].

sDISCO protects fluorescence better than uDISCO and FluoClearBABB

To compare the signal stabilization performance of sDISCO with uDISCO and FluoClearBABB, the hemispheres were

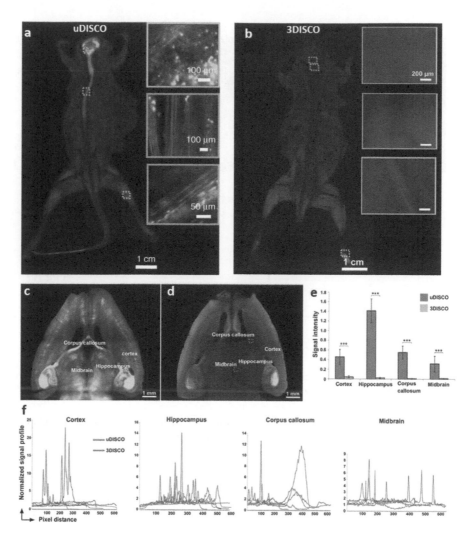

FIGURE 15.8 Whole-body clearing and imaging of adult mice. (a–b) Adult GFP-M mouse cleared with uDISCO and 3DISCO, respectively. (c–f) Representative light-sheet images of mouse brains after clearing with uDISCO and 3DISCO, respectively, and the signal intensity and profiles at different brain regions. Adapted from reference [19].

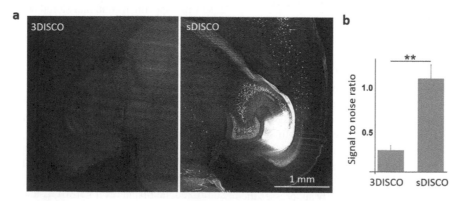

FIGURE 15.9 Comparison of signal preservation in purified but nonstabilized DBE (3DISCO) and stabilized DBE (sDISCO) in hemispheres of the same mouse brain (*Thy1*-EGFP-M). (a) Fluorescence images after immersing in nonstabilized DBE or stabilized DBE for 1 month. (b) Quantification of the signal-to-noise ratio. Adapted from reference [21].

cleared by each method and imaged by Ultramicroscopy [21]. Equivalent sections were illuminated continuously for 14 min with maximal laser power, an image was recorded every 2 min, and the ratio of fluorescence after illumination to fluorescence in the beginning of illumination was determined as

Measurement 1. After 1 week, the experiment was repeated as Measurement 2. As shown in Figure 15.10, after 14 min, the sDISCO signal was reduced to $65.6 \pm 5.2\%$ during the first measurement, and to $51.2 \pm 6.0\%$ during the second measurement [21]. In contrast, FluoClearBABB and uDISCO showed

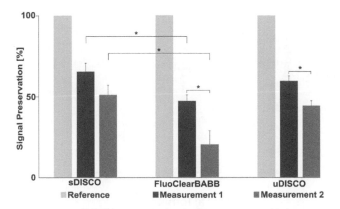

FIGURE 15.10 Signal preservation of YFP from brain hemispheres (*Thy1*-EYFP-H) in organic media of sDISCO, FluoClearBABB, and uDISCO. Reprinted with permission from reference [21].

significantly higher decay of relative signal preservation in both two measures. It was concluded that, in stabilized media, fluorescence was better preserved during illumination than in FluoClearBABB and uDISCO.

PEG component in PEGASOS clearing agents efficiently protects endogenous fluorescence

Since most tissue clearing methods enabled high tissue transparency and 3D visualization of soft organs, hard tissues such as bones and teeth are difficult to clear. In addition, as mentioned above, the preservation of endogenous fluorescence remains an important concern. The Zhao group developed a polyethylene glycol (PEG)-assisted solvent system, named PEGASOS, which could render most types of tissues transparent and preserve endogenous fluorescence [24].

PEGASOS retains most of the GFP and tdTomato fluorescence

For the quantitative evaluation of the influence of the clearing process on endogenous fluorescence, intestine samples were harvested from either adult CAG-EGFP or Tie2-Cre:: Ai14 mice and were then processed with PEGASOS for soft tissue organs. To evaluate the maximum effects, each step in the PEGASOS procedure lasted for 24 h. Fluorescence intensity after PFA fixation was defined as the original intensity. As

shown in Figure 15.11, both GFP and tdTomato fluorescence retained about 70% of the original intensity at the end of the PEGASOS process. In addition, the fluorescence intensity of the samples 1 week after PEGASOS clearing was maintained well. It should be noted that the increase in fluorescence intensity during the delipidation and dehydration steps might be due to the increase of autofluorescence [24].

PEGASOS clearing medium enables long-term fluorescence preservation

The preservation of endogenous fluorescence in different clearing media was evaluated and compared (Figure 15.12). Figure 15.12c demonstrated the representative images of EGFP and tdTomato-labeled intestine tissues placed in various clearing media at different time points. The quantitative results showed that for the BB-PEG (benzyl benzoate-PEG) medium, the endogenous fluorescence intensity of both EGFP and tdTomato increased in the first week and maintained at the same level even 1 month later (Figure 15.12a, b). In contrast, after 1 week of treatment with the uDISCO or FluoClearBABB clearing media, GFP and tdTomato fluorescence intensities were reduced by around 50%.

PEG component in clearing medium contributes to fluorescence preservation

Further, the effective component that contributed to the improved fluorescence preserving capability was investigated. As shown in Figure 15.13a, the GFP fluorescence was well preserved in BB-PEG medium, but was rapidly quenched at day 1 in benzyl benzoate medium, indicating that the protection of fluorescence by the BB-PEG medium can be attributed to the presence of PEG. This might be because PEGs have amphiphilic characteristics and can form an amenable interface to protect endogenous proteins from denaturation [24].

As is known, PEG is a large chemical family with many different forms. Among the tested PEGs, the modified PEGs such as PEGMMA and PEG diacrylate (PEGDA) provided the best protection for endogenous fluorescence, while for the unmodified PEG200, PEG400, and PEG1000, less preservation effect was observed (Figure 15.13b). This result indicated that the modification groups on PEGs might contribute to the stronger effects of fluorescence protection.

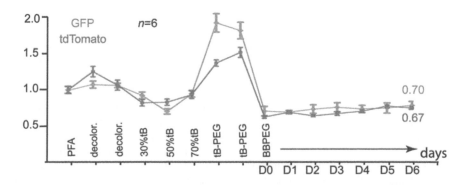

FIGURE 15.11 Fluorescence intensity changes of GFP and tdTomato during PEGASOS. Adapted from reference [24].

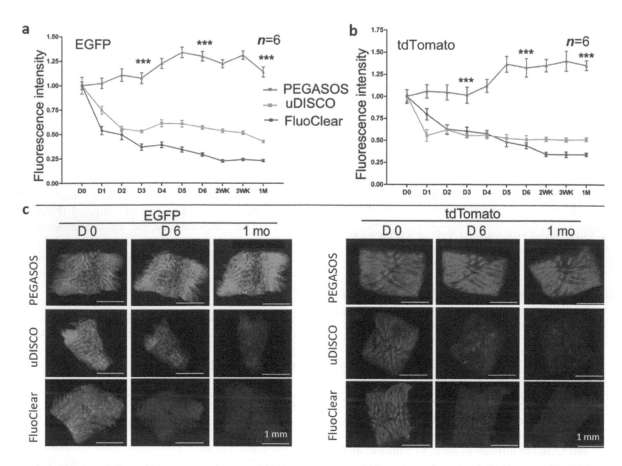

FIGURE 15.12 Fluorescence changes of GFP and tdTomato within different clearing media. (a–b) Fluorescence intensity of the samples along with time after being placed into the clearing media. (c) Representative images of intestine tissues placed in different media at various time points. Adapted from reference [24].

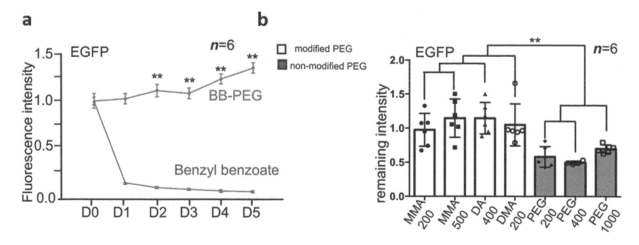

FIGURE 15.13 (a) Quantitative comparison between fluorescence intensity of PEGASOS cleared samples placed in BB-PEG medium or benzyl benzoate. (b) Dehydrated samples were placed in BB-PEG clearing medium of different formulations (BB (75% v/v) + various PEGs (25% v/v)). Adapted from reference [24].

Adjusting the conditions in the clearing and storage process

Except for the screening of fluorescence-friendly chemicals, the other solution is to adjust the conditions during the clearing and storage. The Zhu group has proposed an advanced method named FDISCO (DISCO with superior fluorescence preserving capability) [20]. FDISCO can achieve a high level of preservation of endogenous fluorescence. They also developed a-uDISCO by introducing the pH adjustment to optimize uDISCO, achieving better fluorescence preservation [29]. Both protocols provide alternatives methods for imaging of fluorescent proteins (FPs) with low expression.

Temperature and pH adjustments in FDISCO achieves a high level of fluorescence preservation

To overcome the fluorescence quenching problem, Qi et al. [20] searched for an optimized protocol that can better preserve fluorescence based on 3DISCO. As described in previous literature, 3DISCO involved dehydration with THF solutions (50 vol%, 70 vol%, 80 vol%, and 100 vol%), and refractive index matching with DBE, which were both conducted at room temperature [16]. Inspired by the effect of temperature and pH on GFP stability reported in previous studies, Qi et al. proposed modifying these conditions during clearing and storage, aiming for better endogenous fluorescence preservation [20].

Development of FDISCO by temperature and pH adjustments

To investigate the effects of the temperature and pH of the clearing agents on GFP, Qi et al. started the experiments by *in vitro* assay of recombinant EGFP expressed in *E.coli*. They dissolved EGFP in THF solutions under different pH (adjusted to 9.0 with triethylamine) and temperature conditions, including 4°C/pH 9.0, 4°C, 25°C (used in 3DISCO), and 37°C (Figure 15.14a–c).

The fluorescence intensity of the EGFP/THF solutions over time and fluorescence spectra showed that the 4°C/pH 9.0 group had the best EGFP fluorescence preservation. These results indicated that the EGFP fluorescence was quenched more slowly at lower temperatures and that the alkaline pH also contributed to fluorescence preservation. These results indicated that less EGFP was denatured under the lower temperature and alkaline pH condition by THF.

Then, the effect of the temperature and pH of THF on EGFP-labeled mouse brain sections (*Thy1*-EGFP-M) was studied. The fluorescence images and quantification showed that the lower temperature and alkaline clearing condition also led to greater EGFP fluorescence in brain tissues (Figure 15.14d, e). Furthermore, after clearing, the samples were stored in pure DBE at 4°C and 25°C, and it was found that at 4°C, the EGFP fluorescence signals were maintained for weeks and that the storage time was significantly longer than that at 25°C (Figure 15.14f). These results indicated that the lower temperature was beneficial for EGFP stability during long-term storage in DBE.

Hence, 4°C/pH 9.0 and 4°C were selected as suitable clearing conditions for THF and DBE, respectively, resulting in an optimized protocol with improved fluorescence preservation named FDISCO.

FDISCO preserves the fluorescence signals of different fluorescent proteins

Besides EGFP, FDISCO was proved to be effective for the imaging of other FPs, such as EYFP (*Thy1*-EYFP-H mouse) and tdTomato (Sst-IRES-Cre:: Ai14 mouse) [20]. FDISCO demonstrated better fluorescence preservation than 3DISCO, similar to the results for EGFP (Figure 15.15). In particular, FDISCO was suitable for tdTomato fluorescence, which is incompatible with uDISCO clearing. Moreover, the endogenous fluorescence signals could still be detected after 28 d of storage (Figure 15.15d, e).

For larger samples such as whole brains, longer incubation times in clearing solution could result in more quenching of fluorescence signals. By observing the neuronal structures in different regions of the whole brain (Figure 15.16a, b), FDISCO-treated samples revealed more neuronal details and higher levels of fluorescence signals compared to 3DISCO or uDISCO. At each time point, the fluorescence level of FDISCO samples was significantly higher than those of 3DISCO and uDISCO samples (Figure 15.16d). Notably, FDISCO could maintain EGFP fluorescence at a fairly high level over 1 year, which allowed repeated imaging of the cleared samples (Figure 15.16c, d). For mouse brains expressing EYFP (*Thy1*-EYFP-H) and tdTomato (*Sst*-IRES-Cre:: Ai14), FDISCO also demonstrated good imaging quality and achieved fine reconstruction of neurons in different brain regions at single-cell resolution.

Temperature adjustment is applicable to other clearing protocols

Moreover, the low-temperature adjustment strategy proved to be effective for fluorescence preservation with other clearing protocols[20]. The results showed that the EGFP and EYFP fluorescence intensities after clearing with ethanol and BABB at 4°C were higher than for clearing at 25°C (Figure 15.17a, b). The lower temperature of the storage procedure also contributed to better fluorescence maintenance of cleared tissues for uDISCO clearing (Figure 15.17c, d).

a-uDISCO based on pH adjustment for optimization of GFP fluorescence preservation

In addition, Li et al. applied pH adjustment to uDISCO clearing protocol and developed a-uDISCO method for better GFP fluorescence preservation [29].

Determination of optimal pH value for the optimization of clearing protocol

To identify the optimal pH value, both the dehydration and the refractive index matching solutions (i.e., tert-butanol solutions and BABB-D4) were adjusted using triethylamine. The pH ranges of 7.5–8.0, 9.0–9.5, and 10.5–11.0 were investigated. The images of cleared brain sections (*Thy1*-EGFP-M) were acquired by fluorescence stereomicroscope on day 0 and day 3 after the onset of clearing, as shown in Figure 15.18a. The fluorescence decay during storage was shown in Figure 15.18c. The results showed that pH values in the range of 9.0–9.5 and 10.5–11.0 provided better fluorescence preservation than the other groups. The pH ranging from 9.0 to 9.5 was selected as the optimal values for uDISCO due to the addition of less triethylamine (which can decrease transparency) than that of 10.5–11.0.

a-uDISCO achieves improved fluorescence preservation

To assess the GFP fluorescence-preserving capability of a-uDISCO, the fluorescence signals in brain sections and hemispheres by either a-uDISCO or uDISCO were compared [29].

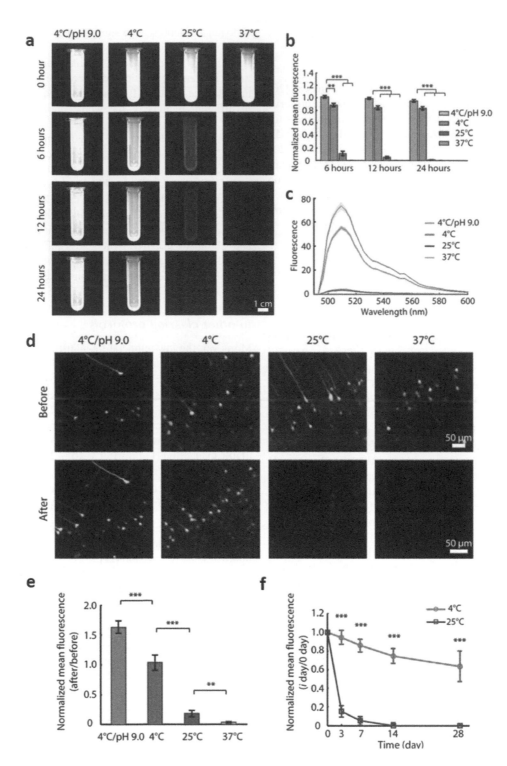

FIGURE 15.14 Development of FDISCO by temperature and pH adjustments. (a–b) Fluorescence changes of recombinant EGFP dissolved in 30% THF over time under the indicated conditions. (c) Emission spectra of EGFP/THF solutions at 24 h. (d–e) Confocal fluorescence images and normalized mean fluorescence intensity of EGFP under the indicated conditions. (f) Quantification of the fluorescence level of cleared samples over time stored in DBE at 4° and 25°C. Adapted from reference [20].

The maximum intensity projection (MIP) images before and after clearing and the relative mean fluorescence intensities are shown in Figure 15.19. The results showed that a-uDISCO obtained significantly higher fluorescence intensity (~136%) than uDISCO (~73%) immediately after clearing. After immersion in BABB-D4 for 3 days, less than 7% of the fluorescence intensity was retained in original uDISCO cleared tissues, whereas for a-uDISCO, ~80% of the mean fluorescence intensity had survived after immersion in pH-adjusted BABB-D4 (pH = 9.0–9.5) for 3 days.

Due to the difference between the clearing protocol for different sized tissues, the fluorescence preservation of a-uDISCO

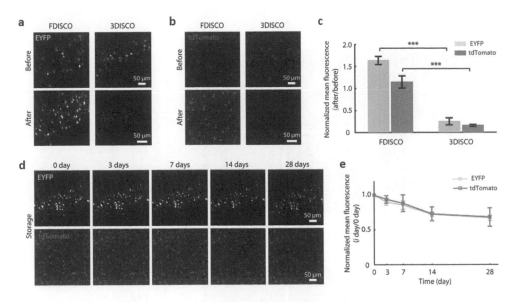

FIGURE 15.15 Fluorescence images of EYFP (a) and tdTomato (b) in 1-mm-thick brain slices before and after FDISCO or 3DISCO clearing. (c) Quantification of EYFP and tdTomato fluorescence after the clearing. (d–e) EYFP and tdTomato images of FDISCO-cleared brain slices over time (d) and quantification data (e). Adapted from reference [20].

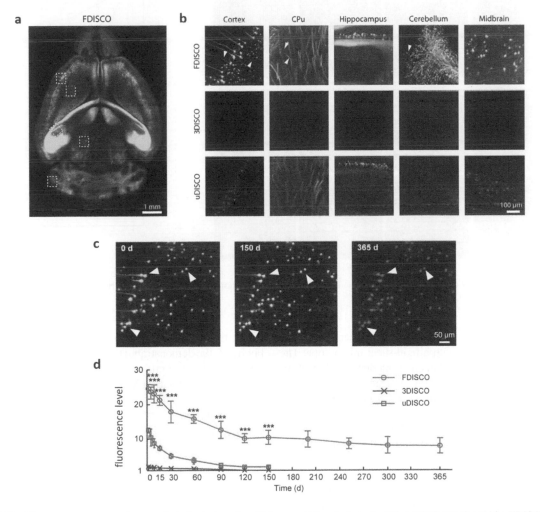

FIGURE 15.16 Fluorescence preservation on whole-brain imaging. (a) Image of the whole brain (*Thy1*-EGFP-M) cleared by FDISCO. (b) High-magnification images of the cleared brains immediately after FDISCO, 3DISCO, and uDISCO clearing. (c) Images of cortical neurons in the FDISCO-cleared brain taken at 0, 150, and 365 days after clearing, respectively. (d) Fluorescence level quantification of cleared brains over time after FDISCO, 3DISCO, and uDISCO clearing. Adapted from references [20, 30].

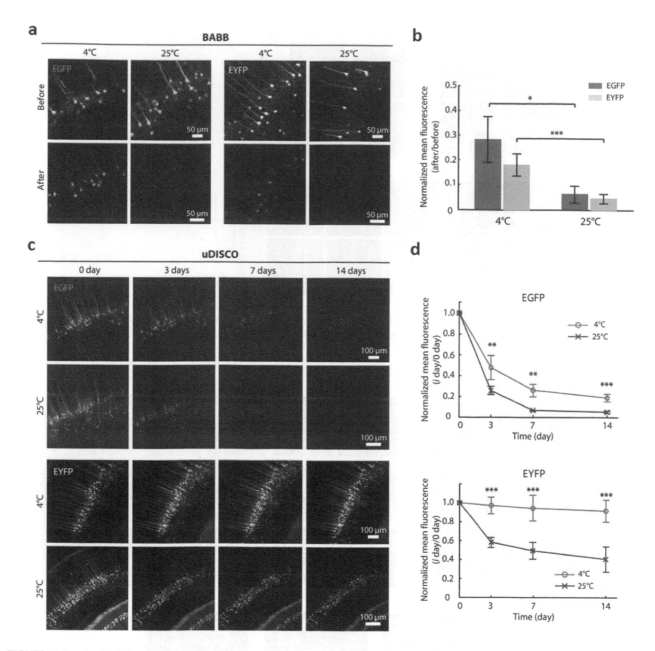

FIGURE 15.17 Applicability of low temperature for increasing the endogenous fluorescence preservation of BABB (a–b) and uDISCO (c–d) methods. Reprinted with permission from reference [20].

in hemispheres was also assessed. The cleared hemispheres of mouse brains were imaged with a light-sheet microscope. The fluorescence images were obtained immediately after clearing with either a-uDISCO or uDISCO. Transverse projections of hemispheres and high-magnification images were obtained at the cortex, hippocampus, and superior colliculus, as shown in Figure 15.19c. The results indicated that a-uDISCO fluorescence was notably brighter than that of uDISCO and allowed the visualization of finer structures, such as dendrites and axons.

Moreover, the high-magnification images of cleared brains at the indicated time points after clearing were used to evaluate the fluorescence changes during long-term storage. The hippocampus's fluorescence images obtained at the four time points qualitatively showed that fluorescence was maintained for a

longer time in a-uDISCO than in uDISCO (Figure 15.19d). Figure 15.19e demonstrated that the fluorescence levels at each time point were significantly higher for a-uDISCO than the original protocol during long-term storage.

Summary

The loss of endogenous fluorescence remains a major concern for solvent-based clearing methods. This chapter discusses several strategies for preserving endogenous fluorescence in solvent-based clearing methods by describing several typical experiments, including introducing a resin formulation during archiving of cleared samples; screening

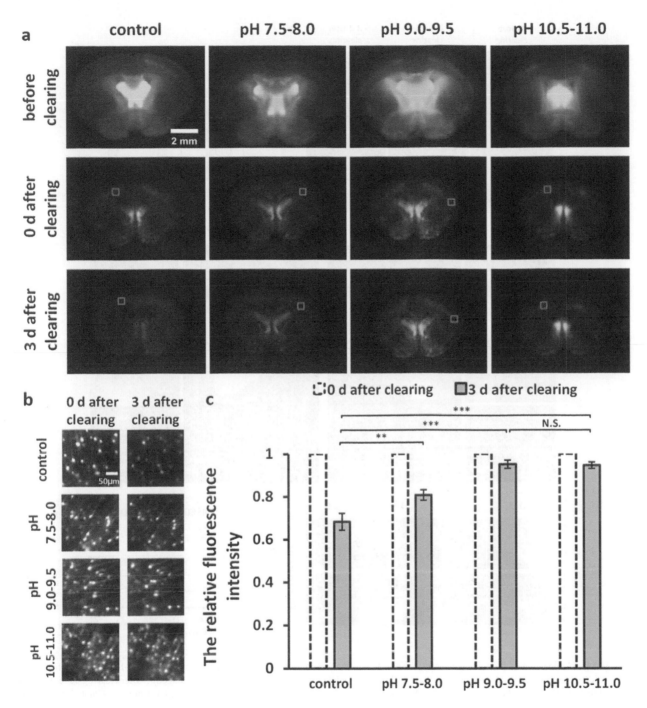

FIGURE 15.18 Screening of optimal pH to optimize uDISCO for better GFP fluorescence preservation. (a) Images of 1-mm-thick brain sections were obtained with a fluorescence stereomicroscope before and after clearing under different pH conditions. Green represents GFP signals. (b) Cropped images of the cortex areas indicated in (a) with red boxes. (c) Relative mean fluorescence intensity of cortex areas of cleared brain slices. Reprinted with permission from reference [29].

fluorescence-friendly compounds such as FluoClearBABB, uDISCO, sDISCO, and PEGASOS; and adjusting the conditions in the clearing and storage process such as FDISCO and a-uDISCO. These methods provided effective and valuable alternatives for the long-term preservation of endogenous fluorescence, facilitating their wide applications in biomedical studies.

Acknowledgments

T.Y. and D.Z. were supported by the National Key Research and Development Program of China grant number 2017YFA0700501, and NSFC grant numbers 61860206009, 81701354, 81961138015.

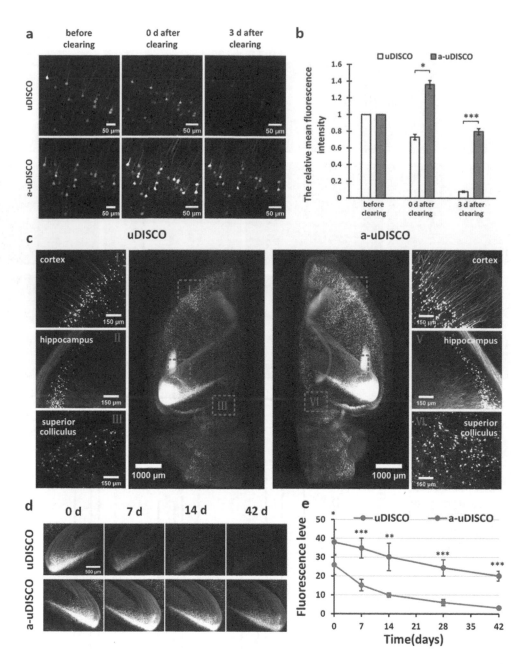

FIGURE 15.19 Comparison of fluorescence preservation in 1-mm-thick brain sections and hemispheres of mice (*Thy1*-EGFP-M). (a–b) Maximum intensity projections of the confocal image stacks of cortical neurons and the fluorescence intensity at day 0 and day 3 after clearing. (c–e) Light-sheet imaging of left and right hemispheres cleared with uDISCO and a-uDISCO, and fluorescence-level during long-term storage. Adapted from reference [29].

REFERENCES

1. E.A. Susaki, and H.R. Ueda, "Whole-body and whole-organ clearing and imaging techniques with single-cell resolution: Toward organism-level systems biology in mammals," *Cell Chem. Biol.* **23**(1), 137–157 (2016).

2. L. Silvestri, I. Costantini, L. Sacconi, and F.S. Pavone, "Clearing of fixed tissue: A review from a microscopist's perspective," *J. Biomed. Opt.* **21**(8), 081205 (2016).

3. D.S. Richardson, and J.W. Lichtman, "Clarifying tissue clearing," *Cell* **162**(2), 246–257 (2015).

4. V.V. Tuchin, *Optical Clearing of Tissues and Blood*, SPIE Press, Bellingham, WA (2005).

5. D. Zhu, K.V. Larin, Q. Luo, and V.V. Tuchin, "Recent progress in tissue optical clearing," *Laser Photonics Rev.* **7**(5), 732–757 (2013).

6. K. Tainaka, A. Kuno, S.I. Kubota, T. Murakami, and H.R. Ueda, "Chemical principles in tissue clearing and staining protocols for whole-body cell profiling," *Annu. Rev. Cell Dev. Biol.* **32**, 713–741 (2016).

7. T. Yu, Y. Qi, H. Gong, Q. Luo, and D. Zhu, "Optical clearing for multiscale biological tissues," *J. Biophoton.* **11**(2), e201700187 (2018).

8. K. Chung, J. Wallace, S.Y. Kim, et al., "Structural and molecular interrogation of intact biological systems," *Nature* **497**(7449), 332–337 (2013).

9. B. Yang, J.B. Treweek, R.P. Kulkarni, et al., "Single-cell phenotyping within transparent intact tissue through whole-body clearing," *Cell* **158**(4), 945–958 (2014).

10. E.A. Susaki, K. Tainaka, D. Perrin, et al., "Whole-brain imaging with single-cell resolution using chemical cocktails and computational analysis," *Cell* **157**(3), 726–739 (2014).

11. K. Tainaka, S.I. Kubota, T.Q. Suyama, et al., "Whole-body imaging with single-cell resolution by tissue decolorization," *Cell* **159**(4), 911–924 (2014).

12. H. Hama, H. Hioki, K. Namiki, et al., "ScaleS: An optical clearing palette for biological imaging," *Nat. Neurosci.* **18**(10), 1518–1529 (2015).

13. H.M. Lai, A.K.L. Liu, H.H.M. Ng, et al., "Next generation histology methods for three-dimensional imaging of fresh and archival human brain tissues," *Nat. Commun.* **9**(1), 1066 (2018).

14. W. Li, R.N. Germain, and M.Y. Gerner, "Multiplex, quantitative cellular analysis in large tissue volumes with clearing-enhanced 3D microscopy (Ce3D)," *Proc. Natl. Acad. Sci. U. S. A.* **114**(35), E7321–E7330 (2017).

15. H.U. Dodt, U. Leischner, A. Schierloh, et al., "Ultramicroscopy: Three-dimensional visualization of neuronal networks in the whole mouse brain," *Nat. Methods* **4**(4), 331–336 (2007).

16. A. Erturk, K. Becker, N. Jahrling, et al., "Three-dimensional imaging of solvent-cleared organs using 3DISCO," *Nat. Protoc.* **7**(11), 1983–1995 (2012).

17. K. Becker, N. Jahrling, S. Saghafi, R. Weiler, and H.-U. Dodt, "Chemical clearing and dehydration of GFP expressing mouse brains," *PLoS One* **7**(3), e33916 (2012).

18. N. Renier, Z. Wu, D.J. Simon, J. Yang, P. Ariel, and M. Tessier-Lavigne, "iDISCO: A simple, rapid method to immunolabel large tissue samples for volume imaging," *Cell* **159**(4), 896–910 (2014).

19. C. Pan, R. Cai, F.P. Quacquarelli, et al., "Shrinkage-mediated imaging of entire organs and organisms using uDISCO," *Nat. Methods* **13**(10), 859–867 (2016).

20. Y. Qi, T. Yu, J. Xu, et al., "FDISCO: Advanced solvent-based clearing method for imaging whole organs," *Sci. Adv.* **5**(1), eaau8355 (2019).

21. C. Hahn, K. Becker, S. Saghafi, et al., "High-resolution imaging of fluorescent whole mouse brains using stabilised organic media (sDISCO)," *J. Biophoton.* **12**(8), e201800368 (2019).

22. M.K. Schwarz, A. Scherbarth, R. Sprengel, J. Engelhardt, P. Theer, and G. Giese, "Fluorescent-protein stabilization and high-resolution imaging of cleared, intact mouse brains," *PLoS One* **10**(5), e0124650 (2015).

23. A. Klingberg, A. Hasenberg, I. Ludwig-Portugall, et al., "Fully automated evaluation of total glomerular number and capillary tuft size in nephritic kidneys using light-sheet microscopy," *J. Am. Soc. Nephrol.* **28**(2), 452–459 (2017).

24. D. Jing, S. Zhang, W. Luo, et al., "Tissue clearing of both hard and soft tissue organs with the PEGASOS method," *Cell Res.* **28**(8), 803–818 (2018).

25. P. Ariel, "A beginner's guide to tissue clearing," *Int. J. Biochem. Cell Biol.* **84**, 35–39 (2017).

26. K. Becker, C.M. Hahn, S. Saghafi, N. Jahrling, M. Wanis, and H.-U. Dodt, "Reduction of photo bleaching and long term archiving of chemically cleared GFP-expressing mouse brains," *PLoS One* **9**(12), e114149 (2014).

27. F. Eichel, and D. Othmer, "Benzaldehyde by autoxidation by dibenzyl ether," *Ind. Eng. Chem.* **41**(11), 2623–2626 (2002).

28. A.A. Alnuami, B. Zeedi, S.M. Qadri, and S.S. Ashraf, "Oxyradical-induced GFP damage and loss of fluorescence," *Int. J. Biol. Macromol.* **43**(2), 182–186 (2008).

29. Y. Li, J. Xu, P. Wan, T. Yu, and D. Zhu, "Optimization of GFP fluorescence preservation by a modified uDISCO clearing protocol," *Front. Neuroanat.* **12**, 67 (2018).

30. Y. Qi, "FDISCO: An innovative tissue optical clearing method for whole-organ imaging," Ph.D. thesis, Huazhong University of Science and Technology (2019).

16

Progress in ex situ tissue optical clearing – shifting immuno-oncology to the third dimension

Paweł Matryba, Leszek Kaczmarek, and Jakub Gołąb

CONTENTS

Introduction

Tissue optical clearing (TOC) relies on the application of the selected optical clearing agents (OCAs) that enhance the optical properties of the studied tissue by either decreasing its light absorption and/or scattering coefficient. This, in turn, limits the number of interactions between photons and the sample, thus improving its penetration by laser light of the microscope setup [1]. The concept of TOC covers two distinct fields of research, namely in vivo [2, 3] and ex situ TOC [4], that currently rely on different OCAs and aims of their application. In vivo TOC utilizes OCAs that fall into three general categories of chemicals [5]: sugars (i.e., glucose, sucrose, fructose); alcohols (i.e., glycerol, sorbitol, propylene glycol); and electrolyte solutions that are widely used as a X-ray contrast agents (i.e., Trazograph™, Hypaque™, Verografin™), with still new potential OCAs, such as Omnipaque™ [6], emerging. As mentioned, these are mild chemicals that in principle do not contribute to any long-lasting effects on immersed tissue but transiently improve its imaging contrast and/or depth of the imaging. Although such treatment does not always guarantee excellent imaging conditions and generation of histology-like data, it can serve as a valuable "window" to study dynamic changes in organs such as brain [7] and skin [8, 9] in vivo. Moreover, it holds great promise in elucidating patterns of kinetics of OCAs distribution between normal vs. pathological, carcinomatous tissues and thus contributes to faster,

almost noninvasive diagnosis of potentially malignant tissue [10, 11]. Researchers interested in getting in-depth knowledge on in vivo TOC are highly recommended to refer to a recent comprehensive review on this topic by Oliveira and Tuchin [5], while we continue discussion on the ex situ TOC.

Ex situ TOC refers to all studies in which OCAs are applied on isolated and fixed organs. Incidentally, this type of tissue clearing is sometimes called "ex vivo" in the literature to underline the opposition to "in vivo" clearing; however this is misleading, as "ex vivo" implies studies on unfixed tissues/organs to study how a particular physical/chemical factor affects the living tissue under conditions that are usually intolerable for a living organism. The main idea behind ex situ TOC is different from in vivo TOC, and aims to allow researchers to study tissues with resolution similar to the one offered by classical histology with a three-dimensional view, often spanning the entire organ of interest. Similarly to in vivo TOC, in ex situ TOC, one can readily distinguish different chemicals that this time can be classified into four basic groups [12], i.e., organic solvents, hyperhydrating solutions, high-refractive index aqueous solutions, and tissue transforming techniques. As reviewed recently [4], all these groups influence rodent organs in distinct ways, and thus possess either advantageous or disadvantageous features depending on the type of organ studied and biological question to be answered.

Here, we aim to succinctly highlight the most characteristic features of each ex situ TOC group; focus on current limitations

DOI: 10.1201/9781003025252-18

and emerging solutions; and present relevant applications in the field of basic immunology and immuno-oncology research.

Overview of ex situ TOC methods

In the following section, a description of the major ex situ TOC approaches is provided, with their current taxonomy summarized in Figure 16.1.

Organic solvents

Historically, organic solvents were the first OCAs, proposed as early as 1911 by Spalteholz [13]. Rediscovered 100 years later in the 2010s, these agents are currently widely used due to the exceptionally high refractive index (RI) easily achieved by these solutions. Such high RI (1.54–1.56) matches that of proteins (>1.50) and, in general, guarantees the highest transparency of the resultant sample [14]. It should be underlined, however, that the light transmittance measured for brains cleared with organic solvents is the highest vs. other clearing methods starting from 550–600 nm [15, 16], as these possess an amber-like color that limits transmittance in shorter wavelengths. This might explain why multiplex detection is rather avoided with organic solvents, with dyes usually excited at ~650 nm (recently in the field of TOC, application of dyes that are excited in the near-infrared region has also emerged [17]). However, excitation in the spectrum of green fluorescent protein (GFP) is not useless, as it detects autofluorescent signals that can help position the target of interest within the context of the tissue or even serve as a reference for Allen Brain Atlas annotation in the case of this organ [18].

Rapid decay of the fluorescent signals of proteinaceous fluorophores and toxicity were perceived as major weaknesses of organic solvents, with both of these being addressed in the following protocols [19]. As GFP/YFP (YFP – yellow fluorescent protein/RFP – red fluorescent protein), etc. (hereafter abbreviated as XFP for simplicity, if applicable) were suspected to require water molecules to sustain their function [1], it was initially believed that quenching of proteinaceous fluorophores would remain an inherent limitation of organic solvents that, immiscible in water, need the tissue to be completely dehydrated before the RI matching step. This belief, however, was proved misleading by several research groups that independently presented that fluorescence might be greatly stabilized simply by using slightly alkaline (pH 9.0–9.5) chemicals. This was validated for both dehydrating agents (*tert*-butanol in the case of FluoClearBABB [20], a-uDISCO [21], and PEGASOS [22]; THF in FDISCO [15]; ethanol in ECi [23]; and 1-propanol in 2nd generation of ECi [24]) and RI matching organic solvents (BABB-D in a-uDISCO [21]). Besides alkalization, the introduction of peroxide scavengers to both dehydrating alcohols and RI matching organic solvents might be beneficial for XFP preservation. Apparently, the first such attempt was presented by Pan et al. [25], who added alpha-tocopherol to the final RI matching solution, and this approach has recently been exploited by Hahn et al. [26] In this study led by Dodt, a pioneer who combined organic solvent-based TOC and ultramicroscopy [27], researchers showed that these are both aldehydes and peroxides that contribute to rapid quenching of XFP fluorescent signals. Thus, their protocol relies first on the elimination of aldehydes and peroxides by column chromatography with basic activated aluminum oxide from THF and DBE (dehydrating and RI-matching solution, respectively). Second, as peroxides tend to develop with time and contribute to the

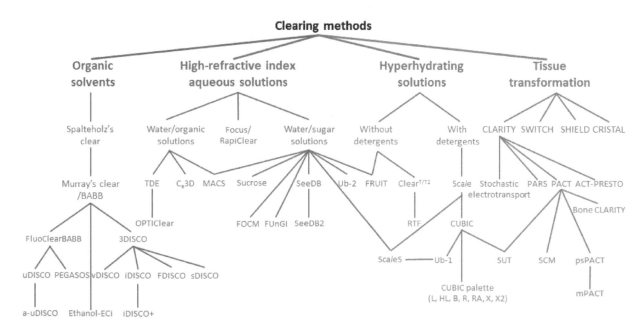

FIGURE 16.1 Updated taxonomy of the major ex situ TOC methods. Although TOC methods still fall into four main chemical groups, newer, advanced protocols start to merge and include mechanisms observed as advantageous for clearing process from distinct categories, i.e. MACS that combine high-RI organic compounds (MDXA) with sugar (sorbitol); or the recent FOCM and FUnGI, which incorporated the addition of hyperhydrating and denaturing urea. Adapted under the terms of the Creative Commons Attribution – Noncommercial License from reference [4].

generation of new aldehydes, a set of promising compounds that are known to prevent peroxide generation was tested, revealing propyl gallate to be the most effective eliminator of peroxides from DBE (significantly more effective than, e.g. alpha-tocopherol). Importantly, the resultant protocol, named sDISCO, greatly stabilizes the XFP signal, so that successful imaging can be performed 1 year after the clearing (and possibly later, but no further data regarding stability was provided). Notably, all of the steps of the animal perfusion, fixation, and TOC were performed at 4°C (as in FDISCO protocol) and with slightly alkaline PBS (pH 8.3), both of which conditions were reported to stabilize signal from XFPs.

In order to overcome the toxicity of organic solvents serving as RI-matching solutions, which are dangerous for both human health (potential carcinogens) and a vast majority of standard microscopic lenses and other parts of the imaging apparatus, ethyl cinnamate (ECi, approved food flavor) was introduced by Klingberg et al. [23] to first study kidney, bone, and heart, the application of which was further extended by Masselink et al. [24] on such distinct samples as *D. melanogaster*, *X. laevis*, and cerebral organoids, as well as adult and larvae of axolotl.

To summarize, the main drawbacks of organic solvent–based TOC, i.e. preservation of XFPs signals and toxicity, have already been overcome, leaving researchers with an easy-to-apply, rapid, cost-effective, and efficient TOC method. However, solvent-based TOC is not yet "one size fits all," and success with applying this set of methodology still depends on reflection on three additional factors: 1) the type of endogenous chromophores and/or light-absorbing molecules, 2) compatibility with immunohistochemistry and lipophilic stains and 3) effective resolution required to accurately answer the biological question.

Endogenous chromophores such as heme, melanin, lipofuscin, and riboflavin absorb light in its visible spectrum and as such, might severely limit the quality of the acquired data [28]. Although pioneer solvent-based TOC methods, like a BABB/ Murray's clear, could not eliminate these light-absorbing molecules, novel ideas and progress in other chemical groups of TOC have now allowed researchers to profit from both the remarkable transparency achievable with organic solvents and the removal of endogenous chromophores (mainly the most abundant heme – present in both hemoglobin and myoglobin proteins). This obstacle might be overcome either by the introduction of a bleaching step or heme elution. A bleaching step with 5–10-times diluted 30% H_2O_2 in methanol was first introduced in the widely utilized iDISCO protocol [29] and later optimized with the addition of sucrose specifically for clearing and decolorization of mouse immune organs (the spleen is exceptionally difficult to decolorize, followed by subcutaneous and mesenteric lymph nodes and thymus) and presented under the name ImmuView [30]. It should be noted, however, that the harsh treatment of tissue during iDISCO might lead to the loss of antigenicity of particular epitopes and/or reactivity of clones of antibodies, i.e. CD31 [31, 32], which therefore must be carefully screened during preliminary experiments. A constantly updated list of already validated antibodies can be found at https://idisco.info/validated-antibodies/. Decolorization with Quadrol (*N,N,N',N'*-tetrakis(2-hydroxypropyl)ethylenediamine), a heme-releasing aminoalcohol [33], was first reported in

CUBIC (one of the pioneer methods from the hyperhydrating solutions group) and later incorporated as additional step before dehydration in case of solvent-based TOC [17, 22]. Notably, such an application has so far been reserved for studies employing so-called whole-body clearing (which aim at making transparent the majority of, if not all, rodent organs) to ensure efficient penetration of organs by Quadrol during the perfusion step.

The other issue worth briefly mentioning is the possible incompatibility of immunohistochemistry with organic solvents even without harsh methanol/H_2O_2 pretreatment [34, 35]. Although this problem might be attributed to any group of TOC methods, loss of antigenicity seems to be more pronounced in case of solvents i.e., 3DISCO [34, 35]. However, very recent articles also show that a combination of, first, tissue incubation in CUBIC, followed by a solvent-based TOC, might significantly enhance immunostaining efficacy either by decreasing background signals or enhancing antibody penetration [36–39].

Furthermore, the excellent transparency of these protocols comes at the expense of, at least partially, lipid solvation which makes them incompatible with lipophilic stains that are widely utilized for neuronal and vascular tracing, for example DiI [40]. Last but not least, the dehydration step causes anticipated tissue shrinkage of the specimen that ranges from ~10% in case of skin to as much as ~35% shrinkage of brain tissue [41]. This might be especially advantageous when a large volume of tissues, even whole bodies, need to be screened, as elegantly presented by Pan et al. [42], who applied solvent-based TOC to quantitively approach dynamics of cancer metastasis using various cancer models. Moreover, by applying deep-learning algorithms, this team revealed the efficacy of antibody-drug targeting of metastases in the entire body.

Hyperhydrating solutions

As tissues are composed mainly of water (RI = 1.33) and proteins (RI > ~1.50), this natural RI mismatch can be overcome either by removing water (as in the case of organic solvent TOC) or tissue hyperhydration. Hyperhydration homogenizes tissue composition by decreasing its overall RI and partial dilution of light absorbers/scatterers. Hyperhydration solution TOC are represented by i.e. Sca*le* [43], CUBIC [44], UbasM [40] (which utilizes a high concentration of mild detergents, such as Triton X-100) and Clear^T/Clear^{T2} [45], RTF [46], and FRUIT [47] (in which detergents are almost absent). In general, the first group is capable of clearing large tissue blocks, organs, and even whole rodent bodies, while detergent-free solutions are much more widely applied to small samples such as spheroids [48] or embryos [46] at early stages of development. Undoubtedly, the most frequently applied, out of hyperhydrating solutions, is CUBIC, so far successfully used to study almost every murine [44] and rat [49] organ. What makes CUBIC protocols so widely utilized are two easy-to-prepare, inexpensive, and nontoxic solutions (designated as CUBIC-R1 and -R2), which, thanks to the already mentioned Quadrol, also effectively decolorize the organs. Although in principle CUBIC-R1 was developed to perform delipidation (15 wt% Triton X-100), decolorization (25 wt% Quadrol), and hyperhydration (25 wt% urea), with

CUBIC-R2 serving as high RI solution for RI-matching (50 wt% sucrose), we have recently presented that in the case of murine lymph node clearing, immersion in solely CUBIC-R1 is sufficient to perform successful, high resolution light-sheet imaging [35]. Based on the available literature, treatment with CUBIC-R1 rather preserves antigenicity of the majority of epitopes and allows for deeper antibody penetration due to "loosening" of tissue structure and its partial delipidation. Although images of traceable structures, such as neuronal projections or vasculature, are usually acquired with success, CUBIC-R1 can lead to distortion or even complete loss of signal from immune cell markers (e.g. CD4, CD8, B220) [34, 35]. Importantly, concentration of the paraformaldehyde (PFA) used during the fixation step can exert tremendous influence on the quality of the final image following immunohistochemistry, with 4% PFA preserving immune cell markers more efficiently than 1% PFA, which further confirms the possible negative impact of CUBIC-R1 on tissue protein content or at least their structure [35] (Figure 16.2). Thus, if the quality of the final image seems to be compromised, it is advised to perform immunolabeling before CUBIC TOC, as long as the penetration of the full tissue thickness can be achieved.

Further imperfections of CUBIC protocols are their incompatibility with DiI labeling [40] (observed again due to high concentration of Triton X-100 but overcame in UbasM technique) and moderate stabilization of XFPs fluorescence. It is important to note that while CUBIC properly stabilizes XFPs fluorescence and quenching of XFPs to the level at which these are undetectable is improbable [50], samples left in CUBIC-R1 will gradually decrease signal intensity over time [35, 50]. Thus, it is pivotal to perform imaging during the same time-points after the clearing, if a fluorescence intensity is measured for quantification.

Moreover, the original CUBIC solutions could not transparentize bones [51] (making actual whole-body imaging inaccessible), and Triton X-100 was not an ideal delipidating agent. Thus, the recent efforts of Tainaka et al. [52] led to the development of "second generation" CUBIC solutions that are now capable of more efficient delipidation, decolorization, decalcification, and RI-matching, allowing actual whole-body clearing to be performed for the first time.

Application of new delipidating (10 w%/10 w% N-butyldiethanolamine/Triton X-100, aka CUBIC-L) and RI-matching (45 w%/30 w% antipyrine/nicotinamide) solutions allowed the same research group to study whole-body/organ cancer metastases in a number of widely used murine models (Panc-1, SUIT-2, B16F10, to name just a few), with spectacular resolution, allowing for detection down to a level

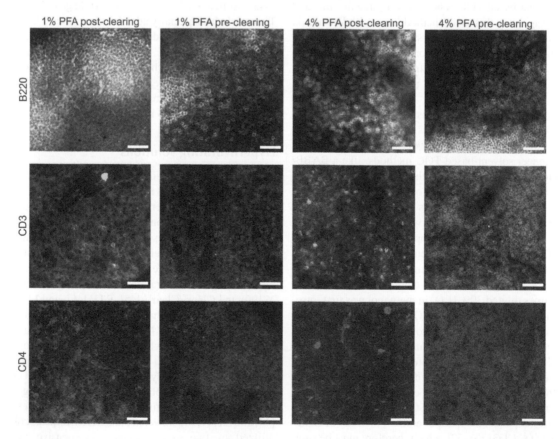

FIGURE 16.2 Efficiency of immunostaining performed on CUBIC-cleared samples might be compromised. CUBIC was already successfully applied in >150 studies, but some of them report compromised retention of protein content [77], which might lead to a poor/absent immunostaining signal [34]. We have recently found that if the precise morphology of cells needs to be retained for further analysis, immunostaining performed before clearing might be beneficial in the case of particular epitopes [35]. Moreover, fixation of tissue with 4% PFA, when compared with 1% PFA (regularly used during fixation of immune organs), may further increase the quality of the resultant image. Postclearing, immunostaining was performed after TOC; preclearing, immunostaining was performed before TOC. Scale bar, 50 mm. Reproduced under AAI guidelines from reference [35].

of a single cancer cell. Further increase in delipidation efficacy was presented by Inoue et al. [53], who introduced 1,2-hexanediol as a potent agent for blocks of lipid-rich human white matter specimens.

Perhaps the most prominent limitation of the current CUBIC pipeline is still a relatively long processing time. Although hands-on time is short, the whole-body clearing can take up to 2–3 weeks in case of adult animals. This was very recently overcome by Zhu et al. [54], who introduced a new TOC solution that is based on m-xylylenediamine (MDXA). Analogously to urea, MDXA presents hyperhydrating capability due to the presence of two NH_2 groups; however, its water solution achieves a much higher RI (up to RI = 1.57), similar to that of organic solvents. The group prepared three solutions which gradually increased concentration of MDXA and sorbitol, termed MACS-R0, -R1, and -R2 (MXDA-based Aqueous Clearing System), which first assure that the entire organ/organism is well penetrated with diluted solution (R0) and later homogenize tissue RI (R1) and finally match RI for the imaging (R2, RI = 1.51, as the final solution guarantees almost intact tissue size, MACS should be perceived as a high-refractive index organic (MDXA) and sugar (sorbitol) aqueous solution described in detail in the next section). Such high RI of the final of MACS solutions efficiently transparentize the entire body of adult mouse, even bones, without the need to perform prior decalcification and thus opens the possibility for further utilization for whole-body imaging. Moreover, it was observed that MACS decolorizes heme-rich tissues, possibly in a mechanism of Fe release. A number of the performed quantifications confirmed that MACS is a rapid TOC protocol (rendering mouse brain transparent in as little as 2.5 days), entirely compatible with lipophilic dyes, a good stabilizer of XFPs fluorescence (e.g. EYFP, EGFP, tdTomato), and an efficient tissue-decolorizing agent applicable to both embryos and whole organs/organism. However, it also relies on toxic reagent, as MDXA is fatal if inhaled (category 2), labeled as toxic and corrosive by GHS, and assigned to health hazard category 4 by NFPA and 3 by HIMS, (meaning "deadly" and "serious hazard", respectively).

High-refractive index aqueous solutions

Although vastly heterogenous from a chemical perspective, the protocols that fall into this TOC group, which consist of one-two solutions or gradient solutions of sugar, are mainly applicable to small pieces of tissue (e.g. 1 mm-thick brain slices, lymph nodes, organoids), and achieve final RI of ~1.46–1.50. This group of reagents is represented by sucrose and glycerol solutions, SeeDB [55], SeeDB2 [56], FRUIT [47] (being at the border of hyperhydrating solutions due to urea and high-RI aqueous solutions due to gradient of fructose), second solution of UbasM [40], FocusClear™ [57], 2-2′ thiodiethanol [58], C_e3D [34] (clearing-enhanced 3D microscopy), the recently presented FOCM [59] and FUnGI [60, 61], and the already described MACS [54].

Gradients of sucrose and fructose (SeeDB) were the first to reveal the potential of high-RI aqueous solutions to render mouse brain transparent. Although SeeDB is a poor TOC method for whole-organ clearing, it is still successfully applied to a variety of thick tissue sections imaged with confocal microscopy [62, 63]. The second generation of SeeDB, SeeDB2, was also invented for imaging brain tissue sections, but this time with superresolution (achieving lateral resolution of 50–150 nm). Utilization of a patent-protected FocusClear was at first vast, ranging from mammals [57] and insects [64] to plants [65] and even biomaterials [66], and serving as mounting medium for CLARITY-cleared samples, but it is now limited by its relatively high price. Similarly to CUBIC, UbasM protocol relies on two solutions made of urea, with additional chemicals introduced to prevent tissue swelling and to replace a high concentration of Triton X-100, which (in comparison to CUBIC) makes UbasM-treated samples compatible with DiI labeling. Recently, a simple incubation in 75% and 88% v/v glycerol solutions were shown to be highly efficient for both spheroids [67] and murine lymph nodes [35], respectively. Unfortunately, long-term storage of lymph nodes in 75% glycerol leads to tissue deformation and quenching of XFPs, EGFP, and DsRed in particular [35, 68].

Recent efforts in the development of high-RI aqueous solutions aimed to generate protocols that would render samples transparent rapidly and preserve their endogenous fluorescence. These criteria are fulfilled by two chemically similar methods – FUnGI [61] and FOCM [59] – and. distinct from them, the C_e3D method. FUnGI, consisting of 50% v/v glycerol, 2.5 M fructose, 2.5 M urea, 10.6 mM Tris Base, and 1 mM EDTA, was successfully applied to a number of organoids and small murine organs, rendering them transparent within as little as 2–4 hours. Notably, FUnGI-cleared samples can be stored in this reagent at −20°C for at least 18 months and thawed without any significant loss of fluorescent signal. On the other hand, FOCM is made of urea, d-sorbitol, and glycerol dissolved in DMSO, the ratio of which depends on the size of the organ to be cleared, and it was to be proved useful for TOC of both hemisphere and thick brain slices. In case of the latter, FOCM achieves spectacular clearing time, requiring as little as 2 minutes to render a 300-μm-thick slice transparent and ready for confocal imaging. Similarly to FUnGI, FOCM neither quenches XFPs nor causes tissue distortion. C_e3D [34] is another, very simple, "one-pot" TOC method that relies on N-methylacetamide and Histodenz and guarantees robust tissue transmittance of a variety of murine organs. Excellent retention of both fluorescence of endogenous XFPs and the epitopes, as confirmed by multiplex immunohistochemistry, with rapid clearing taking 1 day to complete and the possibility of detecting RNA in clarified samples [69], makes C_e3D a promising candidate for widely utilized TOC method. However, two possible limitations of C_e3D exist: inability of tissue decolorization and minor (10%–20%) shrinkage of tissue volume [34, 35, 70] that might impede object segmentation, especially in case of densely packed immune cells.

Tissue transforming methods

The CLARITY protocol [71] (Clear Lipid-exchanged Acrylamide-hybridized Rigid Imaging/Immunostaining/ In situ hybridization-compatible Tissue-hYdrogel), in which acrylamide/bisacrylamide solution creates a hydrogel mesh inside the tissue, laid the foundation for a new branch of methods that transform tissues into rigid tissue–hydrogel matrices. Since the first report of relatively complex protocol of CLARITY in 2013, which relied on electrophoretic tissue clearing for lipid extraction with a strong detergent, as well as the performance of sodium dodecyl sulfate at high temperatures, which often led to tissue damage, a plethora of advanced CLARITY protocols and modifications have been developed [72, 73]. These modifications include titration of monomer solutions, their concentration [74] and composition [75, 76], resigning from electrophoretic delipidation [77]; the development of new, stable devices for stochastic electrotransport [78] to improve both clearing and labeling efficiency; and the introduction of new, inexpensive RI matching solutions [75, 77], with new polymers still being designed and tested for possibly better mechanistic properties for the TOC process [79].

Currently, apparently the most advanced techniques in the field of tissue transformation come from Chung's and Boyden's laboratories. A novel tissue preservation method called SHIELD [80] (stabilization to harsh conditions via intramolecular epoxide linkages to prevent degradation) utilizes polyglycerol 3-polyglycidyl ether as a resin acting in a tissue as a cross-linker that forms additional intramolecular bonds and protects tertiary structure of proteins. Briefly, a solution that contains polyglycerol 3-polyglycidyl ether is first infused along with PFA during the perfusion step of the animal; next, selected organs are additionally immersed in a similar solution but one devoid of PFA, and finally, a process of crosslinking is initiated at 37°C. SHIELD ensures excellent preservation of tissue structure upon TOC, antigenicity (with more than 50 antibodies validated in original article), and fluorescence of XFPs even under harsh conditions e.g. 24 hours of incubation at 70°C. Another issue optimized by Chung's group is the uniform immunolabeling of large organs. SWITCH [81] was the first protocol developed for that purpose that relied on two solutions – SWITCH-off, which inhibits binding of antibodies with the tissue, thus allowing for their unlimited spread, followed by SWITCH-on, which restores the tissue's and the antibodies' reactivity. Unfortunately, one should be bear in mind that the process of tissue immersion with SWITCH-on also takes time, which means that there is a gradient of tissue-probe reactivity from the border to the core of the specimen, potentially still leading to uneven labeling. The most recent idea to overcome the issue of uneven immunohistochemistry of large specimens, such as mouse brain hemisphere, was presented in a protocol named eFLASH [82] (electrophoretically driven fast labeling using affinity sweeping in hydrogel). The idea behind eFLASH is to gradually increase the affinity between probes and their targets by balancing them with a concentration of bile salts, e.g. sodium deoxycholate, and pH. Yun et al. [82] observed that the affinity of various antibodies rises as the concentration of bile salts decreases and pH falls (from basic to neutral). Thus, since the start of the protocol, the probes can penetrate fixed and delipidated samples in an unbiased manner without the risk of a huge depletion of antibodies at the sample's border. Such a protocol not only provides an opportunity to perform unbiased immunolabeling of large tissue volumes, but also greatly decreases the amount of probes needed, which are otherwise significantly depleted before reaching

the core of the specimen (if their targets are highly abundant in a particular tissue). When tested on a SHIELD-fixed, delipidated tissues with immunolabeling performed using the previously described technique of the stochastic electrotransport, eFLASH collectively allowed for the uniform labeling of murine brain with as little as 3–5 µg of antibody, depending on the target. Boyden's group renders tissues transparent via intensive expansion of the transformed samples [83]. Development of protocols that rely on gel anchoring, followed by digestion of proteins and immersion (expansion) in water, results in 4.5× linear tissue expansion (or even ≈16–22×, if the process of gel anchoring and expansion is repeated) and >99% composition of the resultant tissue of water. However, it should be noted that the final tissue transparency is rather a side-product, as this team aims at superresolution microscopy, e.g. capturing the morphology of dendritic spines with conventional spinning disk systems, rather than at imaging exceptionally large volumes of optically cleared specimens.

To summarize, tissue transforming techniques offer plenty of novel methodological advances that make such TOC convenient, but could be implemented with other groups of TOC to strengthen the applicability of the latter. Although fixation with the SHIELD methodology seems advantageous for the majority of TOC techniques, its application is rather limited, perhaps due to the doubts of researchers regarding work with potentially toxic resins. The commercialization of easy to use setups for SHIELD, stochastic electrotransport, or eFLASH could possibly widen implementation of these tools in research groups composed mainly of biologists.

Application of TOC to immune organs

Although TOC has already been successfully applied to literally every murine organ, the immune organs appear to be extraordinarily challenging to clear and image. This is either because of a high abundance of hemoglobin (i.e. spleen and bone marrow), or extremely high cellularity (i.e. lymph nodes), which hinders quantitative measurements. Here, applications of TOC to primary and secondary immune organs, along with their limitations and perspectives, are highlighted.

Bone and bone marrow

Having less dense distribution of cells, as compared with other lymphoid organs, bone marrow (BM) offers favorable whole-organ imaging and segmentation conditions. Due to heterogeneity of bone tissue (hard, solid minerals and soft marrow [84]) and relatively high RI [85] of its components (e.g. apatite ~1.62 [86], lipids ~1.45 [87] and collagen type I ~1.43 [88]), initially only organic solvents could effectively match the RI of bones. One of the first applications of TOC, to half bone and BM plugs, was presented by Acar et al. [89], who, by using Murray's clear (dehydration in methanol followed by RI-matching in BABB), visualized possible distinct niches in BM that are occupied by populations of either dividing or nondividing hematopoietic stem cells (HSCs). During the selection of an appropriate TOC approach, the authors observed that harsh, SDS-mediated lipid removal during either electrophoretic (CLARITY) or passive (PACT) protocols did not offer optimal clearing conditions and, moreover, resulted in the destruction of several cell surface epitopes. Although expected, it was also experimentally proven that other TOC methods that are characterized by rather low RI (e.g. CUBIC, Focus Clear, and Sca*le*A2) were not effective for bone/BM clearing. Finally, besides Murray's clear, the 3DISCO technique (dehydration with THF followed by RI-matching in dibenzyl ether) was applied by Acar et al. [89] during experiments in which XFPs were visualized, as it led to better stabilization of DsRed and tdTomato signal, but resulted in worse clearing efficiency otherwise. Similar results were obtained by Berke et al. [90], who performed a side-by-side comparison of eight TOC methods. In addition, this group showed that it is not only the value of RI that influences the efficacy of bone TOC, but also the mechanism of clearing (i.e. dehydration vs. hyperhydration) as 97% TDE (RI ~1.47) outperformed other high-RI solutions, such as SeeDB (RI ~1.50). The detailed protocol for solvent-based TOC, bone/BM imaging, and segmentation was recently described by Gorelashvili et al. [91]. Using this pipeline and in vivo two-photon imaging, the authors could provide strong evidence for a revised model of megakaryopoiesis, in which megakaryocytes are located at the BM sinusoids and are being replenished by the precursory cells that originate not from the periostic but rather from the sinusoidal niche [92]. Using an ethanol-ECi protocol (called "simpleCLEAR"), Grüneboom et al. [93] discovered a new type of vessels that originate in BM, travel perpendicularly through the cortical bone, and finally connect to periosteal circulation. These structures, named transcortical vessels (TCVs), are highly abundant in both murine and human bones, and express either arterial or venous markers. Interestingly, this group found TCVs to exhibit significant remodeling capability, i.e., new TCVs develop within weeks in a model of chronic (but not acute) arthritis (thus potentially contributing to enhanced efflux of leukocytes to joints) and decrease in number upon irradiation and aging. A variation of another solvent-based TOC, iDISCO, was recently presented as the BoneClear [94] method. The major novelty introduced in BoneClear is to combine the iDISCO scaffold with prior bone decalcification with EDTA. BoneClear-processed murine bones exhibit unprecedented transparency and excellent immunostaining (as verified with nine primary antibodies) and XFP preservation capabilities. As a proof of concept, the authors applied BoneClear and visualized the entire innervation of femur and hindpaw, followed by the process of re-innervation upon injury and neuropathy in chemotherapy-induced murine model (treated with Paclitaxel).

It should be underlined, however, that the successful application of organic solvents to bone/BM imaging seems to heavily rely on the experience of a particular research group. As already noted, Acar et al. [89] reported that3DISCO stabilized the signal from XFPs better than Murray's clear, while Berke et al. [90] has shown that it is 3DISCO that leads to the most dramatic loss of XFPs when compared to other solvents. Moreover, Coutu et al. [95] screened more than ten clearing/mounting media specifically for the purpose of BM imaging and reported "efficient bone and marrow clearing"

with ethanol-ECi, but also "high background fluorescence and staining artefacts, massive tissue distortion, and shrinkage," while the same protocol in the hands of the Gunzer group [93] allowed them to obtain compelling imaging conditions and discover, as mentioned above, CTVs. As the advanced protocols from all other chemical groups of TOC already achieve optimal and similar bone clearing (i.e. Bone CLARITY [96], a modified CLARITY approach, in which an additional step of EDTA-mediated decalcification and decolorization with Quadrol was introduced, and CUBIC-B [52], which relies on EDTA and imidazole cocktail for the efficient decalcification), it is now even more difficult to select the best method [97]. Most probably, such a method simply does not exist, and the choice should be made upon initial screening to verify suitability with particular bones/epitopes/XFPs. Nonetheless, both Bone CLARITY and CUBIC were already applied to bone clearing (besides the proof of concept studies performed by the inventors) and facilitated description of the dynamics of HIV-1-mediated spread and infection of human CD4- and CD68-expressing macrophages in humanized mice [98] and near homogenous attachment of long-term HSCs to the vascular endothelial cadherin positive cells [99], respectively. Undoubtedly, progress not only in TOC approaches but also in new light-sheet microscopy setups will greatly add to high-throughput and precise imaging of bone and BM [100].

TOC of the isolated BM seems to be relatively easier than that of the whole bone [101]. The first such technique was implemented in a study published as early as 2013, in which Nombela-Arrieta et al. [102] applied FocusClear to visualize distribution of hematopoietic stem and progenitor cells in the BM of murine femoral bones. Interestingly, they found that although these cells are preferentially located in endosteal zones, in close contact with the microvessels, they exhibit a strong hypoxic phenotype defined by e.g. expression of HIF-1α. The same group has recently presented optimized multiscale 3D quantitative microscopy of BM cellular components using RapiClear (RI = 1.52) as an OCA [103]. The experimental workflow, in which BMs are imaged with confocal microscopy, first using low magnification (10–20×) to define regions of interest and then higher magnification (40–93×) to achieve subcellular resolution, results in a mosaic image that is finally converted into single volumetric reconstruction, as presented in Figure 16.3. Using such an approach and flow cytometry, Gomariz et al. [103] found tremendous discrepancies between these methods, when stromal cells within the entire murine femurs were to be counted. 3D quantitative microscopy revealed ~30× greater numbers of both sinusoidal endothelial cells and CXCL12-abundant reticular cells, a fundamental knowledge which greatly supports the need for the introduction of histocytometry approaches for the reevaluation of cellular composition of tissues. A similar histocytometric approach to BM imaging was presented by Coutu et al. [95], who selected TDE as an optimal TOC chemical. In addition to the TOC pipeline, the authors developed the software x-dimensional image analysis toolbox (XiT) for high-throughput image analysis, which allowed them to elucidate relationships between hematopoietic cells, bone matrix, and marrow Schwann cells.

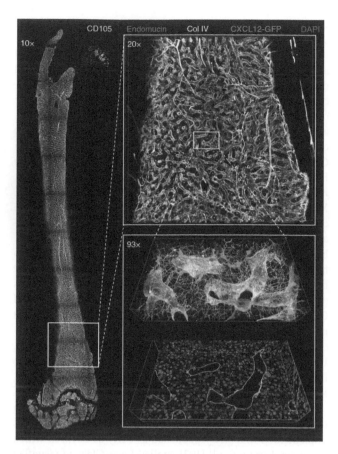

FIGURE 16.3 Multiscale 3D quantitative microscopy of BM cellular components. The figure presents the resultant mosaic image of BM subjected to TOC with RapiClear. In this approach, tissue is first imaged under low magnification settings to get an overview of tissue structure and properly select unbiased regions of interest, which are further studied under higher magnification settings. Reproduced under the terms of the CC-BY Creative Commons Attribution 4.0 International License from reference [103].

Spleen

Although extremely rich in light-absorbing pigments, especially heme, spleen has already been optically cleared by a number of laboratories, using every chemically distinct TOC group [22, 44, 74, 75, 104, 105]. It should be noted, however, that in the majority of cases, only macrophotographies were provided without fluorescent imaging to assess whether the residual pigments do not compromise usefulness of the particular TOC method [22, 40, 41, 75, 106]. Thus far, application of BABB as a RI-matching solution to thick spleen segments has facilitated precise annotation of HSC localization and the discovery of new migratory capacity of neutrophils [107]. Using ImmuView (an iDISCO-based protocol, the chemical composition of which has already been discussed here), Ding et al. [30] presented a compelling characterization of the sympathetic and parasympathetic innervations of the spleen and other major immune organs (subcutaneous and mesenteric lymph nodes, Peyer's patches, thymus) and proved >99% of synaptophysin positive (pan-neural marker) neural fibers to be of the sympathetic (TH positive) type (Figure 16.4). 3D visualization of spleen neuroanatomy was also assessed by

Murray et al. [108], who, by applying the CLARITY protocol, quantified interactions between TH-positive axon endings and populations of choline acetyltransferase expressing CD3+ T- and B220+ B-cells. Lastly, CUBIC was successfully applied to the spleen by Kieffer et al. [109] to aid tracking of HIV-1 and HIV-1 infected CD3+ T-cells spread, similarly to the already outlined, bone-oriented experiment [98].

Lymph nodes

Similarly to spleen and BM, the possibility of visualizing LNs in 3D was appreciated by several research groups that had already utilized every chemical group of TOC methods for that purpose [68, 69, 110, 111]. Although small and relatively pigment-free, successful imaging of LNs characterized by immense cellular density still requires the prudent selection of TOC. We have recently performed a profound evaluation of >10 protocols that span the entire chemical spectrum of TOC methods [35]. After side-by-side comparison of the resultant LN transparency, size change, compatibility with proteinaceous fluorophores (GFP and RFP, in particular), immunostaining, and H&E staining, it should be concluded that both CUBIC-R1 (first reagent of classical CUBIC protocol) and C$_e$3D guarantee optimal TOC conditions for murine LNs. Nevertheless, one of the first studies in which murine LNs were subjected to solvent-based TOC was presented by Woodruff et al. [112] who, by using a model of influenza vaccination, revealed rapid repositioning of LN-resident dendritic cells (DCs) from the T-cell cortex to the medullary interfollicular region. Interestingly, the resident DCs captured virus effectively (represented by expression of CXCL10) and activated viral-specific CD4+ T-cells (represented by expression of CD69, CD44, and CXCR3) even before the expected time of arrival of migratory DCs from skin. BABB-mediated RI-matching was also a core technique utilized by Kumar et al. [113, 114] to study lymphoid remodeling upon viral infection with special emphasis on number and volume of B-cell follicles and length of high endothelial venules (HEV). The dynamics of tumor-induced lymphangiogenesis in 4T1 (breast cancer) and B16F10 (melanoma) murine models was also genuinely revealed by Commerford et al. [115] who used BABB clearing. Although these studies were very informative, it should be reemphasized that dehydration observed in case of solvent-based TOC always leads to severe tissue shrinkage (even 50%–60% of original LN area), thus precluding further direct quotation of such results to define e.g. total length of HEV network under physiological conditions, as unfortunately still is the case [116]. Moreover, tissue shrinkage is perceived to be advantageous when large tissue volumes need to be imaged and the targets of imaging are positioned relatively loosely, i.e. neuronal projections, vasculature, or sites of metastasis. Thus, in a majority of studies trying to quantitatively elucidate immune responses in 3D fashion by inspecting changes in numbers of immune cells, organic solvents will significantly hinder analysis. This issue was already reported by Cabeza-Cabrerizo et al. [117] who, after applying uDISCO to mesenteric LNs and spleen, were unable to study distribution and clustering of DCs in this organs, and thus limited their analysis to nonlymphoid tissue, lung, and small intestine in

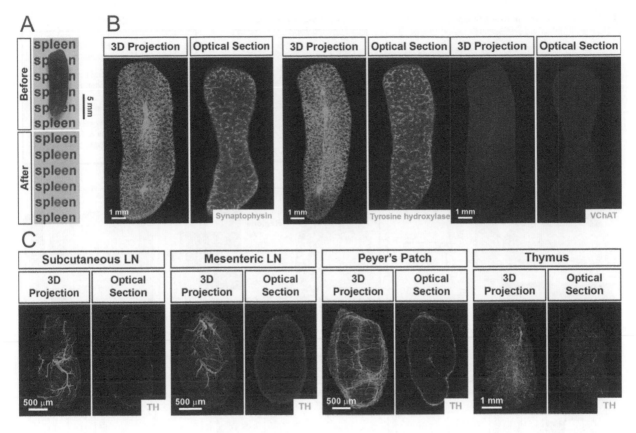

FIGURE 16.4 ImmuView as a method for unbiased inspection of neuronal markers in immune organs. (A) ImmuView, a modified iDISCO protocol, achieves unprecedented transparency of the spleen. (B) This approach allows a view of the entire neuronal network of this heme-rich organ to be obtained, revealing that the spleen is full of panicle-shaped sympathetic fibers (synaptophysin, TH), while it lacks parasympathetic ones (VChAT). (C) When applied to other immune organs, ImmuView substantiated histological observations that although both lymph nodes and thymus contain dozens of sympathetic fibers (similarly to the spleen), these are located either on the organ's surface or run along the vasculature. Reproduced under the terms of the CC-BY Creative Commons Attribution 4.0 International License from reference [30].

particular. Meager but consistent tissue shrinkage (between 25% and 10% volume reduction depending on organ, with lung and bone retaining their original size), seems to be also the only weakness of an otherwise excellent C$_e$3D method. C$_e$3D, applicable to the number of organs but developed with a view to enhance the histocytometry approach, was already applied to generate high-quality imaging data and develop advanced MATLAB toolbox for spatial cellular analysis, and called Histo-Cytometric Multidimensional Analysis Pipeline [118] (CytoMAP, Figure 16.5).

Another possible challenge that arises during imaging of the cleared LNs is the opacity of the surrounding adipose tissue, which is not cleared effectively with majority of TOC protocols (with the exception of organic solvents) and autofluorescent signal coming from the fibrous capsule (especially in the ~488 nm channel). The first issue should be addressed after LN fixation by gentle removal of the surrounding tissue, preferably under the stereomicroscope. The second limitation can be overcome by slicing LN into thick (200–500 μm) slices and starting the z-stack from the cutting plane. Thus, it is important to first decide whether true whole-mount imaging is indeed indispensable for answering a particular hypothesis. Imaging LN sections is definitely a more straightforward approach as these usually do not require power ramping nor do they present with a significant

autofluorescent signal of the fibrous capsule. In addition, it greatly expedites immunostaining and overcomes the potential problem of uneven labeling. Studies performed with TOC applied to thick LN slices have already contributed to a number of discoveries of particular importance [110, 119]. For instance, Mondor et al. [120] presented a detailed scenario of LN vascular network remodeling in which it is orchestrated by the clonal proliferation of HEVs, while Dubey et al. [119] deciphered crosstalk between fibroblastic reticular and B-cells that promotes both de novo follicle formation and lymphangiogenesis.

TOC in cancer research

The possibility of tracing vasculature over long distances and performing fast inspection of entire organs or even murine bodies with cellular resolution prompted researchers to apply TOC in cancer research both at basic science and clinical levels. Although as early as in 2006, a case report by Dickie et al. [121] presented that highly detailed imaging of cleared, carbon perfused arteries of orthotopically implanted intracranial tumor is feasible with Murray's clear, the first advanced study deciphering intratumoral vasculature comes from Dobosz et al. [122]. Using HER-2 overexpressing human breast cancer cell line (KPL-4) to generate tumor xenograft, injection

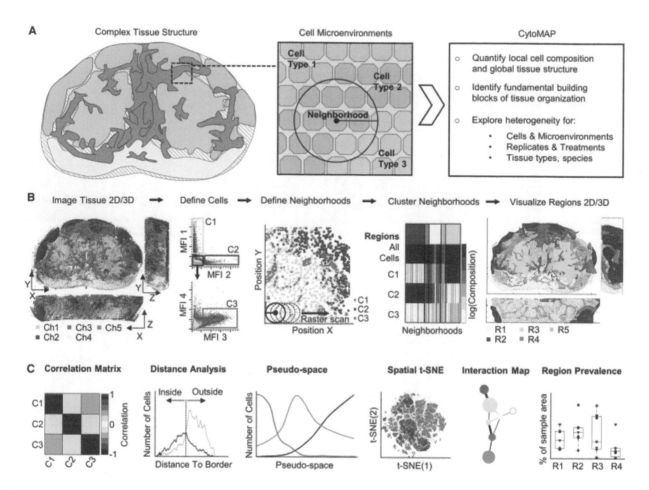

FIGURE 16.5 Scheme representing CytoMAP pipeline. (A) CytoMAP is a tool for highly accurate quantitative analyses to reveal how local cell microenvironments form global tissue structure and to study both intra- and intersample tissue heterogeneity. (B) After acquisition of high-resolution data, hierarchical gating of cell objects is performed and passed into CytoMAP for segmentation using clustering algorithms that ultimately allows exploration of detected clusters in 2D or 3D space. (C) In CytoMAP, several tools for data quantification were included, e.g. analysis of spatial correlations between different cell types, investigation of distance relationships of cells with architectural landmarks, analysis of neighborhood heterogeneity within individual tissues or across multiple samples, and quantitative visualization of tissue architecture. Adapted under the terms of the CC-BY Creative Commons Attribution 4.0 International License from reference [118].

of lectin-Alexa Fluor 647, and BABB-mediated TOC, the researchers obtained a precise view of tumor vasculature and intratumoral distribution of Alexa Fluor 750-trastuzumab (a humanized monoclonal antibody that has a high affinity for the extracellular domain of HER2). While the distribution of trastuzumab was mosaic, the observed heterogeneity relied vastly on the composition of tumor tissue (with necrotic areas marginally penetrated when compared with soft, highly vascularized tumor compartments). Hence, the developed pipeline was employed to study the effect of administration of bevacizumab (humanized monoclonal antibody inhibiting growth of blood vessels) on trastuzumab distribution. Strikingly, intratumoral vasculature (number of vessel segments, branching points, vessel volume) was already significantly diminished 1 day after injection and strengthened over the entire period of treatment. This effect, however, was observed only in the case of vessels located at the tumor periphery and on vessels with a diameter between 10 and 30 μm and significantly decreased penetration of the HER2-targeting antibody. In a similarly designed study, Pöschinger et al. [123] applied TOC to verify the sensitivity of the dynamic contrast-enhanced

microcomputed tomography as a tool for noninvasive assessment of efficacy of antiangiogenic compounds. Interestingly, the data obtained with subcellular resolution in case of TOC and light-sheet imaging was highly correlated by the results of in vivo micro-computed tomography. Relation between tumor and vasculature was also studied by Lagerweij et al. [32] in an orthotopic glioblastoma xenograft models. Using CLARITY and iDISCO on 3–5 mm-thick brain slices (to assure unbiased penetration of the antibodies) they could visualize migration patterns of glioblastoma tumor cells in relation to microvasculature. It turned out that in more vascularized, gray matter, tumor cells migrate in proximity of vessels, while being distant from vasculature in white matter, observation of which suggests migration behavior through tracts. The value of 3D imaging of glioblastoma was also presented by Yang et al. [124] who provided precise characterization of its microenvironment in terms of cell stemness, immune infiltration, and microvasculature. Angiotropism in brain metastases of skin melanoma was also recently presented by Rodewald et al. [125] in 3 mm-thick biopsies of eight patients, which were successfully cleared with CLARITY protocol. Cell migration of tumor cells was also extensively studied by Hume et al. [126], who engineered adipose tissue in a collagen structure to recapitulate breast tissue environment. In this model, CUBIC was applied to first assure uncompromised structure of the generated tissue (using second harmonic generation imaging to detect collagen fibers), and second, it allowed the migratory behavior of tdTomato[+] MDA-MB-231 breast cancer cell line seeded in such scaffolds to be described. This approach was further utilized to demonstrate effect of several chemotherapeutics on migratory tumor cell behavior (represented by e.g. migration distance or number of migratory cells change upon treatment [127]).

Another possible application of TOC to cancer research is in deciphering tumor microenvironment (TME [128]). For instance, Cuccarese et al. [129] reported a significant heterogeneity in the density and infiltration pattern of tumor-associated macrophages in pulmonary tumor model, while Messal et al. [130] precisely analyzed morphological changes during the initial steps of the development of pancreatic cancer. 3D observations revealed two types of neoplastic growth of pancreatic ducts – exophytic and endophytic – in the case of small and large ducts, respectively, showing that epithelial tumorigenesis might be determined by the tension imbalance. To perform successful, multiplex immunostaining of TME, Lee et al. [131] proposed a 3T approach (transparent tumor tomography) that relies on TOC of 400 μm-thick sections with 80% D-fructose solution. This protocol allowed them to present spatial distribution of several markers important in immune TME, such as HER2, Ki-67, CD45, CD31, PD-L1 [131], and spatial pharmacokinetics and distribution of one of the checkpoint inhibitors, anti-PD-L1 antibody, along with its ligand in murine lung and mammary carcinomas [132]. The same method was recently used to examine immune infiltrates of human head and neck tumor biopsies [133]. The TME of breast cancer models was also reinvestigated in pilot studies by several research groups that used TOC of every chemical group, i.e. CUBIC and SeeDB [134], FUnGI [60, 135], CLARITY [136], and uDISCO [137]. Besides fluorophore-conjugated trastuzumab and anti-PD-L1 antibodies, other therapeutic agents might be visualized in TME to understand their pharmacokinetics and local distribution and identify relevant biological barriers. In a series of studies, the group led by Chan tested compatibility and optimized TOC (mainly CLARITY approach) to visualize 3D distribution of nanoparticles, a promising group of carriers for antitumor therapeutic agents [138–140]. Based on their initial observations, although CLARITY could efficiently clear several organs of interest, the electrophoresis-driven approach led to substantial, ~30% loss of nanoparticles [138]. Thus, the group optimized the technique in terms of temperature and duration of clearing and resignation from active lipid removal [139] which, coupled with visualization of gold nanoparticles with light scattering [141], allowed them to perform unbiased, artificial intelligence–based profound analysis of interactions between nanoparticles and micrometastases [142] in isolated organs. Micrometastases were already studied in the context of whole mouse body at single-cell resolution with two chemically distinct TOC approaches. First, Kubota et al. [143] described two new CUBIC solutions (CUBIC-L and -R for delipidation and RI matching, respectively) and reported that RI = 1.52 of

final RI-matching solution guarantees optimal tissue transmittance. Using this approach, whole-body distribution of several, widely used cancer models were screened down to a single cell level, and quantitative evaluation of few therapeutic agents (i.e. doxorubicin, fluorouracil, cyclophosphamide) on the number and volume of tumor foci was performed. Importantly, the organs cleared with these CUBIC solutions remain compatible with further classical histological analysis, and thus it readily bridges the gap between in vivo live inspection of tumor bioluminescence and 2D histology [143]. Recently, Pan et al. [42] exploited the same idea with vDISCO technology, in which fluorescence of the tumor cells was greatly enhanced using Atto-conjugated antibodies. Such a bright signal allowed the performance of screening of metastases and visualization of tumor-targeting drugs in an automatic, AI-based way with the developed algorithms. Interestingly, the group presented that antibody-based drugs may miss as many as 23% of micrometastatic clusters of tumor cells. It should be mentioned, however, that the majority of the approved anticancer nanoparticle-based therapeutic and experimental formulations are lipid-based [144] (e.g. liposomes, solid lipid nanoparticles, and nanostructed lipid carriers) which are effectively removed with the currently most often used TOC techniques – CUBIC, CLARITY, and DISCO, in particular. To overcome this limitation, Syed et al. [140] developed a tag that, when conjugated with therapeutic liposomes, remains crosslinked to the tissue, even after the liposome undergoes efficient removal. A combination of such approach with the discussed CUBIC or DISCO whole-body clearing might even further extend the applicability of these techniques for preclinical screening of the efficacy of antitumor targeting by emerging compounds.

Summary

Although valuable for examining literally every biological specimen [4], TOC opens critically important new avenues for immuno-oncology at both basic science and clinical levels. Along with the development of advanced quantitative tools, such as CytoMAP [118], it will be shortly possible to assess the crosstalk between immune and cancer cells during every step of tumorigenesis, extract distinct microenvironmental niches, create their mathematical models, and finally reveal their key components to be targeted during cancer therapy. Moreover, as presented, it could be valuable to evaluate tissue sections of cancer patients in 3D and, possibly, create a new generation of more reliable cancer grading systems. This, however, will require much effort to firstly describe new TOC methods that (1) are rapid and further compatible with standard histopathological pipelines and (2) guarantee comparable results, independently of staff experience; and secondly, to build standardized, easy-to-operate light-sheet fluorescence microscopes.

Acknowledgments

This work was supported by NCN Grant UMO-2019/35/N/NZ6/04043 to PM.

REFERENCES

1. D.S. Richardson, J.W. Lichtman, "Clarifying tissue clearing," *Cell* **162**(2), 246–257 (2015).
2. E.A. Genina, A.N. Bashkatov, Y.P. Sinichkin, I.Y. Yanina, V.V. Tuchin, "Optical clearing of biological tissues: Prospects of application in medical diagnostics and phototherapy," *J. Biomed. Photon. Eng.* **1**(1), 22–58 (2015).
3. A.N. Bashkatov, K.V. Berezin, K.N. Dvoretskiy, et al., "Measurement of tissue optical properties in the context of tissue optical clearing," *J. Biomed. Opt.* **23**(9), 091416 (2018).
4. P. Matryba, L. Kaczmarek, J. Gołąb, "Advances in ex situ tissue optical clearing," *Laser Photon. Rev.* **13**(8), 1800292 (2019).
5. L.M.C. Oliveira, V.V. Tuchin, *The Optical Clearing Method: A New Tool for Clinical Practice and Biomedical Engineering*, SpringerBriefs in Physics, Springer, Berlin (2019).
6. A. Sdobnov, M.E. Darvin, J. Lademann, V. Tuchin, "A comparative study of ex vivo skin optical clearing using two-photon microscopy," *J. Biophoton.* **10**(9), 1115–1123 (2017).
7. C. Zhang, W. Feng, Y. Zhao, et al., "A large, switchable optical clearing skull window for cerebrovascular imaging," *Theranostics* **8**(10), 2696–2708 (2018).
8. X. Yang, Y. Liu, D. Zhu, R. Shi, Q. Luo, "Dynamic monitoring of optical clearing of skin using photoacoustic microscopy and ultrasonography," *Opt. Express* **22**(1), 1094–1104 (2014).
9. R. Shi, M. Chen, V.V. Tuchin, D. Zhu, "Accessing to arteriovenous blood flow dynamics response using combined laser speckle contrast imaging and skin optical clearing," *Biomed. Opt. Express* **6**(6), 1977–1989 (2015).
10. L. Pires, V. Demidov, I.A. Vitkin, et al., "Optical clearing of melanoma in vivo: Characterization by diffuse reflectance spectroscopy and optical coherence tomography," *J. Biomed. Opt.* **21**(8), 081210 (2016).
11. G. Terentyuk, E. Panfilova, V. Khanadeev, et al., "Gold nanorods with a hematoporphyrin-loaded silica shell for dual-modality photodynamic and photothermal treatment of tumors in vivo," *Nano Res.* **7**(3), 325–337 (2014).
12. L. Silvestri, I. Costantini, L. Sacconi, F.S. Pavone, "Clearing of fixed tissue: A review from a microscopist's perspective," *J. Biomed. Opt.* **21**(8), 081205 (2016).
13. W. Spalteholz. *Über das Durchsichtigmachen von menschlichen und tierischen Präparaten und seine theoretischen Bedingungen*, S. Hirzel, Leipzig (1911).
14. M. Orlich, F. Kiefer, "A qualitative comparison of ten tissue clearing techniques," *Histol. Histopathol.* **33**(2), 181–199 (2018).
15. Y. Qi, T. Yu, J. Xu, et al., "FDISCO: Advanced solvent-based clearing method for imaging whole organs," *Sci. Adv.* **5**(1), eaau8355 (2019).
16. M. Stefaniuk, E.J. Gualda, M. Pawlowska, et al., "Light-sheet microscopy imaging of a whole cleared rat brain with Thy1-GFP transgene," *Sci. Rep.* **6**, 28209 (2016).
17. R. Cai, C. Pan, A. Ghasemigharagoz, et al., "Panoptic imaging of transparent mice reveals whole-body neuronal projections and skull-meninges connections," *Nat. Neurosci.* **22**(2), 317–327 (2019).
18. N. Renier, E.L. Adams, C. Kirst, et al., "Mapping of brain activity by automated volume analysis of immediate early genes," *Cell* **165**(7), 1789–1802 (2016).
19. A. Ertürk, K. Becker, N. Jährling, et al., "Three-dimensional imaging of solvent-cleared organs using 3DISCO," *Nat. Protoc.* **7**(11), 1983–1995 (2012).
20. M.K. Schwarz, A. Scherbarth, R. Sprengel, et al., "Fluorescent-protein stabilization and high-resolution imaging of cleared, intact mouse brains," *PLoS ONE* **10**(5), e0124650 (2015).
21. Y. Li, J. Xu, P. Wan, T. Yu, D. Zhu, "Optimization of GFP fluorescence preservation by a modified uDISCO clearing protocol," *Front. Neuroanat.* **12**, 67 (2018).
22. D. Jing, S. Zhang, W. Luo, et al., "Tissue clearing of both hard and soft tissue organs with the PEGASOS method," *Cell Res.* **28**(8), 803–818 (2018).
23. A. Klingberg, A. Hasenberg, I. Ludwig-Portugall, et al., "Fully automated evaluation of total glomerular number and capillary tuft size in nephritic kidneys using lightsheet microscopy," *J. Am. Soc. Nephrol.* **28**(2), 452–459 (2017).
24. W. Masselink, D. Reumann, P. Murawala, et al., "Broad applicability of a streamlined ethyl cinnamate-based clearing procedure," *Development* **146**(3), dev166884 (2019).
25. C. Pan, R. Cai, F.P. Quacquarelli, et al., "Shrinkage-mediated imaging of entire organs and organisms using uDISCO," *Nat. Methods* **13**(10), 859–867 (2016).
26. C. Hahn, K. Becker, S. Saghafi, et al., "High-resolution imaging of fluorescent whole mouse brains using stabilised organic media (sDISCO)," *J. Biophoton.* **12**(8), e201800368 (2019).
27. H.-U. Dodt, U. Leischner, A. Schierloh, et al., "Ultramicroscopy: Three-dimensional visualization of neuronal networks in the whole mouse brain," *Nat. Methods* **4**(4), 331–336 (2007).
28. K. Tainaka, A. Kuno, S.I. Kubota, T. Murakami, H.R. Ueda, "Chemical principles in tissue clearing and staining protocols for whole-body cell profiling," *Annu. Rev. Cell Dev. Biol.* **32**, 713–741 (2016).
29. N. Renier, Z. Wu, D.J. Simon, et al., "iDISCO: A simple, rapid method to immunolabel large tissue samples for volume imaging," *Cell* **159**(4), 896–910 (2014).
30. X. Ding, H. Wang, X. Qian, et al., "Panicle-shaped sympathetic architecture in the spleen parenchyma modulates antibacterial innate immunity," *Cell Rep.* **27**(13), 3799–3807 (2019).
31. Y. Cao, H. Wang, Q. Wang, X. Han, W. Zeng, "Three-dimensional volume fluorescence-imaging of vascular plasticity in adipose tissues," *Mol. Metab.* **14**, 71–81 (2018).
32. T. Lagerweij, S.A. Dusoswa, A. Negrean, et al., "Optical clearing and fluorescence deep-tissue imaging for 3D quantitative analysis of the brain tumor microenvironment," *Angiogenesis* **20**(4), 533–546 (2017).
33. K. Tainaka, S.I. Kubota, T.Q. Suyama, et al., "Whole-body imaging with single-cell resolution by tissue decolorization," *Cell* **159**(4), 911–924 (2014).
34. W. Li, R.N. Germain, M.Y. Gerner, "Multiplex, quantitative cellular analysis in large tissue volumes with clearing-enhanced 3D microscopy (Ce3D)," *Proc. Natl. Acad. Sci.* **114**(35), E7321–E7330 (2017).

35. P. Matryba, A. Sosnowska, A. Wolny, et al., "Systematic evaluation of chemically distinct tissue optical clearing techniques in murine lymph nodes," *J. Immunol.* **204**(5), 1395–1407 (2020).

36. Y. Li, J. Xu, J. Zhu, T. Yu, D. Zhu, "Three-dimensional visualization of intramuscular innervation in intact adult skeletal muscle by a modified iDISCO method," *Neurophotonics* **7**(1), 015003 (2020).

37. J. McKey, L.A. Cameron, D. Lewis, I.S. Batchvarov, B. Capel, "Combined iDISCO and CUBIC tissue clearing and lightsheet microscopy for in toto analysis of the adult mouse ovary," *Biol. Reprod.* **102**(5), 1080–1089 (2020).

38. M. Lesage, M. Thomas, J. Bugeon, et al., "C-Eci: A cubic-Eci combined clearing method For 3D follicular content analysis in the fish ovary†," *Biol. Reprod.* **103**, 1099–1109 (2020).

39. P. Matryba, A. Wolny, M. Pawłowska, et al., "Tissue clearing-based method for unobstructed 3D imaging of mouse penis with subcellular resolution," *J. Biophoton.* **13**, e202000072, (2020).

40. L. Chen, G. Li, Y. Li, et al., "UbasM: An effective balanced optical clearing method for intact biomedical imaging," *Sci. Rep.* **7**(1), 12218 (2017).

41. J. Xu, Y. Ma, T. Yu, D. Zhu, "Quantitative assessment of optical clearing methods in various intact mouse organs," *J. Biophoton.* **12**(2), e201800134 (2019).

42. C. Pan, O. Schoppe, A. Parra-Damas, et al., "Deep learning reveals cancer metastasis and therapeutic antibody targeting in the entire body," *Cell* **179**(7), 1661–1676 (2019).

43. H. Hama, H. Kurokawa, H. Kawano, et al., "Scale: A chemical approach for fluorescence imaging and reconstruction of transparent mouse brain," *Nat. Neurosci.* **14**(11), 1481–1488 (2011).

44. E.A. Susaki, K. Tainaka, D. Perrin, et al., "Advanced CUBIC protocols for whole-brain and whole-body clearing and imaging," *Nat. Protoc.* **10**(11), 1709–1727 (2015).

45. T. Kuwajima, A.A. Sitko, P. Bhansali, et al., "ClearT: A detergent- and solvent-free clearing method for neuronal and non-neuronal tissue," *Development* **140**(6), 1364–1368 (2013).

46. T. Yu, J. Zhu, Y. Li, et al., "RTF: a rapid and versatile tissue optical clearing method," *Sci. Rep.* **8**(1), 1964 (2018).

47. B. Hou, D. Zhang, S. Zhao, et al., "Scalable and DiI-compatible optical clearance of the mammalian brain," *Front. Neuroanat.* **9**, 19 (2015).

48. E.C. Costa, A.F. Moreira, D. de Melo-Diogo, I.J. Correia, "ClearT immersion optical clearing method for intact 3D spheroids imaging through confocal laser scanning microscopy," *Opt. Laser Technol.* **106**, 94–99 (2018).

49. P. Matryba, L. Bozycki, M. Pawłowska, L. Kaczmarek, M. Stefaniuk, "Optimized perfusion-based CUBIC protocol for the efficient whole-body clearing and imaging of rat organs," *J. Biophoton.* **11**(5), e201700248 (2017).

50. P. Wan, J. Zhu, J. Xu, et al., "Evaluation of seven optical clearing methods in mouse brain," *Neurophotonics* **5**(3), 035007 (2018).

51. L. Bozycki, K. Łukasiewicz, P. Matryba, S. Pikula, "Whole-body clearing, staining and screening of calcium deposits in the mdx mouse model of Duchenne muscular dystrophy," *Skelet. Muscle* **8**(1), 21 (2018).

52. K. Tainaka, T.C. Murakami, E.A. Susaki, et al., "Chemical landscape for tissue clearing based on hydrophilic reagents," *Cell Rep.* **24**(8), 2196–2210 (2018).

53. M. Inoue, R. Saito, A. Kakita, K. Tainaka, "Rapid chemical clearing of white matter in the post-mortem human brain by 1,2-hexanediol delipidation," *Bioorg. Med. Chem. Lett.* **29**(15), 1886–1890 (2019).

54. J. Zhu, T. Yu, Y. Li, et al., "MACS: Rapid aqueous clearing system for 3D mapping of intact organs," *Adv. Sci.* **7**, 1903185 (2020).

55. M.-T. Ke, S. Fujimoto, T. Imai, "SeeDB: A simple and morphology-preserving optical clearing agent for neuronal circuit reconstruction," *Nat. Neurosci.* **16**(8), 1154–1161 (2013).

56. M.-T. Ke, Y. Nakai, S. Fujimoto, et al., "Super-resolution mapping of neuronal circuitry with an index-optimized clearing agent," *Cell Rep.* **14**(11), 2718–2732 (2016).

57. A.J. Moy, B.V. Capulong, R.B. Saager, et al., "Optical properties of mouse brain tissue after optical clearing with FocusClear™," *J. Biomed. Opt.* **20**(9), 95010 (2015).

58. I. Costantini, J.-P. Ghobril, A.P. Di Giovanna, et al., "A versatile clearing agent for multi-modal brain imaging," *Sci. Rep.* **5**, 9808 (2015).

59. X. Zhu, L. Huang, Y. Zheng, et al., "Ultrafast optical clearing method for three-dimensional imaging with cellular resolution," *Proc. Natl. Acad. Sci.* **116**(23), 11480–11489 (2019).

60. A.C. Rios, B.D. Capaldo, F. Vaillant, et al., "Intraclonal plasticity in mammary tumors revealed through large-scale single-cell resolution 3D imaging," *Cancer Cell* **35**(6), 953 (2019).

61. J.F. Dekkers, M. Alieva, L.M. Wellens, et al., "High-resolution 3D imaging of fixed and cleared organoids," *Nat. Protoc.* **14**(6), 1756–1771 (2019).

62. H. Morales-Navarrete, H. Nonaka, A. Scholich, et al., "Liquid-crystal organization of liver tissue," *eLife* **8**, e44860 (2018).

63. K. Meyer, H. Morales-Navarrete, S. Seifert, et al., "Bile canaliculi remodeling activates YAP via the actin cytoskeleton during liver regeneration," *Mol. Syst. Biol.* **16**(2), e8985 (2020).

64. C.-L. Wu, T.-F. Fu, Y.-Y. Chou, S.-R. Yeh, "A single pair of neurons modulates egg-laying decisions in Drosophila," *PLOS ONE* **10**(3), e0121335 (2015).

65. B. Lee, R. Johnston, Y. Yang, et al., "Studies of aberrant phyllotaxy1 mutants of maize indicate complex interactions between auxin and cytokinin signaling in the shoot apical meristem," *Plant Physiol.* **150**(1), 205–216 (2009).

66. S.-C. Tang, Y.-Y. Fu, W.-F. Lo, T.-E. Hua, H.-Y. Tuan, "Vascular labeling of luminescent gold nanorods enables 3-D microscopy of mouse intestinal capillaries," *ACS Nano* **4**(10), 6278–6284 (2010).

67. E. Nürnberg, M. Vitacolonna, J. Klicks, et al., "Routine optical clearing of 3D-cell cultures: simplicity forward," *Front. Mol. Biosci.* **7**, 20 (2020).

68. E. Song, H. Seo, K. Choe, et al., "Optical clearing based cellular-level 3D visualization of intact lymph node cortex," *Biomed. Opt. Express* **6**(10), 4154–4164 (2015).

69. W. Li, R.N. Germain, M.Y. Gerner, "High-dimensional cell-level analysis of tissues with Ce3D multiplex volume imaging," *Nat. Protoc.* **14**(6), 1708 (2019).

70. G.D.P. Bossolani, I. Pintelon, J.D. Detrez, et al., "Comparative analysis reveals Ce3D as optimal clearing method for in toto imaging of the mouse intestine," *Neurogastroent. Motil.* **31**(5), e13560 (2019).

71. K. Chung, J. Wallace, S.-Y. Kim, et al., "Structural and molecular interrogation of intact biological systems," *Nature* **497**(7449), 332–337 (2013).

72. K.H.R. Jensen, R.W. Berg, "Advances and perspectives in tissue clearing using CLARITY," *J. Chem. Neuroanat.* **86**, 19–34 (2017).

73. H. Du, P. Hou, W. Zhang, Q. Li, "Advances in CLARITY-based tissue clearing and imaging," *Exp. Ther. Med.* **16**(3), 1567–1576 (2018).

74. J.R. Epp, Y. Niibori, H.-L. Liz Hsiang, et al., "Optimization of CLARITY for clearing whole-brain and other intact organs," *eNeuro* **2**(3), ENEURO.0022–15.2015. (2015).

75. J. Woo, M. Lee, J.M. Seo, H.S. Park, Y.E. Cho, "Optimization of the optical transparency of rodent tissues by modified PACT-based passive clearing," *Exp. Mol. Med.* **48**(12), e274 (2016).

76. T. Koda, S. Dohi, H. Tachi, et al., "One-shot preparation of polyacrylamide/poly(sodium styrenesulfonate) double-network hydrogels for rapid optical tissue clearing," *ACS Omega* **4**(25), 21083–21090 (2019).

77. B. Yang, J.B. Treweek, R.P. Kulkarni, et al., "Single-cell phenotyping within transparent intact tissue through whole-body clearing," *Cell* **158**(4), 945–958 (2014).

78. S.-Y. Kim, J.H. Cho, E. Murray, et al., "Stochastic electrotransport selectively enhances the transport of highly electromobile molecules," *Proc. Natl. Acad. Sci.* **112**(46), E6274–E6283 (2015).

79. Y. Ono, I. Nakase, A. Matsumoto, C. Kojima, "Rapid optical tissue clearing using poly(acrylamide-co-styrenesulfonate) hydrogels for three-dimensional imaging," *J. Biomed. Mater. Res. Part B Appl. Biomater.* **107B**, 2297–2304 (2019).

80. Y.-G. Park, C.H. Sohn, R. Chen, et al., "Protection of tissue physicochemical properties using polyfunctional crosslinkers," *Nat. Biotechnol.* **37**(1), 73–83 (2019).

81. E. Murray, J.H. Cho, D. Goodwin, et al., "Simple, scalable proteomic imaging for high-dimensional profiling of intact systems," *Cell* **163**(6), 1500–1514 (2015).

82. D.H. Yun, Y.-G. Park, J.H. Cho, et al., "Ultrafast immunostaining of organ-scale tissues for scalable proteomic phenotyping," *bioRxiv* 660373 (2019). DOI: 10.1101/660373

83. S. Alon, G.H. Huynh, E.S. Boyden, "Expansion microscopy: Enabling single cell analysis in intact biological systems," *The FEBS Journal* **286**(8), 1482–1494 (2019).

84. E.A. Genina, A.N. Bashkatov, V.V. Tuchin, "Optical clearing of cranial bone," *Adv. Opt. Technol.* **2008**, 1–8 (2008).

85. A. Ascenzi, C. Fabry, "Technique for dissection and measurement of refractive index of osteones," *J. Biophys. Biochem. Cytol.* **6**(1), 139–142 (1959).

86. F. Duck, *Physical Properties of Tissues*, Elsevier, San Diego (1990).

87. V. Tuchin, *Optical Clearing of Tissues and Blood*, SPIE – The International Society for Optical Engineering, Bellingham, Washington (2005).

88. X.J. Wang, T.E. Milner, M.C. Chang, J.S. Nelson, "Group refractive index measurement of dry and hydrated type I collagen films using optical low-coherence reflectometry," *J. Biomed. Opt.* **1**(2), 212–216 (1996).

89. M. Acar, K.S. Kocherlakota, M.M. Murphy, et al., "Deep imaging of bone marrow shows non-dividing stem cells are mainly perisinusoidal," *Nature* **526**(7571), 126–130 (2015).

90. I.M. Berke, J.P. Miola, M.A. David, M.K. Smith, C. Price, "Seeing through musculoskeletal tissues: Improving in situ imaging of bone and the lacunar canalicular system through optical clearing," *PLoS ONE* **11**(3), e0150268 (2016).

91. M.G. Gorelashvili, K.G. Heinze, D. Stegner, "Optical clearing of murine bones to study megakaryocytes in intact bone marrow using light-sheet fluorescence microscopy," *Methods Mol. Biol.* **1812**, 233–253 (2018).

92. D. Stegner, J.M.M. vanEeuwijk, O. Angay, et al., "Thrombopoiesis is spatially regulated by the bone marrow vasculature," *Nat. Commun.* **8**(1), 127 (2017).

93. A. Grüneboom, I. Hawwari, D. Weidner, et al., "A network of trans-cortical capillaries as mainstay for blood circulation in long bones," *Nat. Metab.* **1**(2), 236 (2019).

94. Q. Wang, K. Liu, L. Yang, H. Wang, J. Yang, "BoneClear: Whole-tissue immunolabeling of the intact mouse bones for 3D imaging of neural anatomy and pathology," *Cell Res.* **29**(10), 870–872 (2019).

95. D.L. Coutu, K.D. Kokkaliaris, L. Kunz, T. Schroeder, "Multicolor quantitative confocal imaging cytometry," *Nat. Methods* **15**(1), 39–46 (2018).

96. A. Greenbaum, K.Y. Chan, T. Dobreva, et al., "Bone CLARITY: Clearing, imaging, and computational analysis of osteoprogenitors within intact bone marrow," *Sci. Transl. Med.* **9**(387), eaah6518 (2017).

97. D. Jing, Y. Yi, W. Luo, et al., "Tissue clearing and its application to bone and dental tissues," *J. Dent. Res.* **98**(6), 621–631 (2019).

98. M.S. Ladinsky, W. Khamaikawin, Y. Jung, et al., "Mechanisms of virus dissemination in bone marrow of HIV-1–infected humanized BLT mice," *eLife* **8**, e46916 (2019).

99. J.Y. Chen, M. Miyanishi, S.K. Wang, et al., "Hoxb5 marks long-term haematopoietic stem cells revealing a homogenous perivascular niche," *Nature* **530**(7589), 223–227 (2016).

100. T. Chakraborty, M.K. Driscoll, E. Jeffery, et al., "Light-sheet microscopy of cleared tissues with isotropic, subcellular resolution," *Nat. Methods* **16**(11), 1109–1113 (2019).

101. A. Gomariz, S. Isringhausen, P.M. Helbling, C. Nombela-Arrieta, "Imaging and spatial analysis of hematopoietic stem cell niches," *Ann. N. Y. Acad. Sci.* **1466**, 5–16 (2019).

102. C. Nombela-Arrieta, G. Pivarnik, B. Winkel, et al., "Quantitative imaging of haematopoietic stem and progenitor cell localization and hypoxic status in the bone marrow microenvironment," *Nat. Cell Biol.* **15**(5), 533–543 (2013).

103. A. Gomariz, P.M. Helbling, S. Isringhausen, et al., "Quantitative spatial analysis of haematopoiesis-regulating stromal cells in the bone marrow microenvironment by 3D microscopy," *Nat. Commun.* **9**(1), 1–15 (2018).

104. C.M. McErlean, J.K.R. Boult, D.J. Collins, et al., "Detecting microvascular changes in the mouse spleen using optical computed tomography," *Microvasc. Res.* **101**, 96–102 (2015).

105. J.D. Ventura, J. Beloor, E. Allen, et al., "Longitudinal bioluminescent imaging of HIV-1 infection during antiretroviral therapy and treatment interruption in humanized mice," *PLoS Pathog.* **15**(12), e1008161 (2019).

106. E. Lee, J. Choi, Y. Jo, et al., "ACT-PRESTO: rapid and consistent tissue clearing and labeling method for 3-dimensional (3D) imaging," *Sci. Rep.* **6**, 18631 (2016).

107. M. Casanova-Acebes, J.A. Nicolás-Ávila, J.L. Li, et al., "Neutrophils instruct homeostatic and pathological states in naive tissues," *J. Exp. Med.* **215**(11), 2778–2795 (2018).

108. K. Murray, D.R. Godinez, I. Brust-Mascher, et al., "Neuroanatomy of the spleen: Mapping the relationship between sympathetic neurons and lymphocytes," *PLOS ONE* **12**(7), e0182416 (2017).

109. C. Kieffer, M.S. Ladinsky, A. Ninh, R.P. Galimidi, P.J. Bjorkman, "Longitudinal imaging of HIV-1 spread in humanized mice with parallel 3D immunofluorescence and electron tomography," *elife* **6**, e23282 (2017).

110. L.K. Dubey, P. Karempudi, S.A. Luther, B. Ludewig, N.L. Harris, "Interactions between fibroblastic reticular cells and B cells promote mesenteric lymph node lymphangiogenesis," *Nat. Commun.* **8**(1), 367 (2017).

111. J. Abe, A.J. Ozga, J. Swoger, et al., "Light sheet fluorescence microscopy for in situ cell interaction analysis in mouse lymph nodes," *J. Immunol. Methods* **431**, 1–10 (2016).

112. M.C. Woodruff, C.N. Herndon, B.A. Heesters, M.C. Carroll, "Contextual analysis of immunological response through whole-organ fluorescent imaging," *Lymphat. Res. Biol.* **11**(3), 121–127 (2013).

113. V. Kumar, E. Scandella, R. Danuser, et al., "Global lymphoid tissue remodeling during a viral infection is orchestrated by a B cell-lymphotoxin-dependent pathway," *Blood* **115**(23), 4725–4733 (2010).

114. V. Kumar, S. Chyou, J.V. Stein, T.T. Lu, "Optical projection tomography reveals dynamics of HEV growth after immunization with protein plus CFA and features shared with HEVs in acute autoinflammatory lymphadenopathy," *Front. Immunol.* **3**, 282 (2012).

115. C.D. Commerford, L.C. Dieterich, Y. He, et al., "Mechanisms of tumor-induced lymphovascular niche formation in draining lymph nodes," *Cell Rep.* **25**(13), 3554–3563 (2018).

116. A. Ager, "High endothelial venules and other blood vessels: Critical regulators of lymphoid organ development and function," *Front. Immunol.* **8**, 45 (2017).

117. M. Cabeza-Cabrerizo, J. van Blijswijk, S. Wienert, et al., "Tissue clonality of dendritic cell subsets and emergency DCpoiesis revealed by multicolor fate mapping of DC progenitors," *Sci. Immunol.* **4**(33), eaaw1941 (2019).

118. C.R. Stoltzfus, J. Filipek, B.H. Gern, et al., "CytoMAP: A spatial analysis toolbox reveals features of myeloid cell organization in lymphoid tissues," *Cell Rep.* **31**(3), 107523 (2020).

119. L.K. Dubey, L. Lebon, I. Mosconi, et al., "Lymphotoxin-dependent B Cell-FRC crosstalk promotes de novo follicle formation and antibody production following intestinal helminth infection," *Cell Rep.* **15**(7), 1527–1541 (2016).

120. I. Mondor, A. Jorquera, C. Sene, et al., "Clonal proliferation and stochastic pruning orchestrate lymph node vasculature remodeling," *Immunity* **45**(4), 877–888 (2016).

121. R. Dickie, R.M. Bachoo, M.A. Rupnick, et al., "Three-dimensional visualization of microvessel architecture of whole-mount tissue by confocal microscopy," *Microvasc. Res.* **72**(1–2), 20–26 (2006).

122. M. Dobosz, V. Ntziachristos, W. Scheuer, S. Strobel, "Multispectral fluorescence ultramicroscopy: Three-dimensional visualization and automatic quantification of tumor morphology, drug penetration, and antiangiogenic treatment response," *Neoplasia* **16**(1), 1–13 (2014).

123. T. Pöschinger, A. Renner, F. Eisa, et al., "Dynamic contrast-enhanced micro-computed tomography correlates with 3-dimensional fluorescence ultramicroscopy in antiangiogenic therapy of breast cancer xenografts," *Invest. Radiol.* **49**(7), 445–456 (2014).

124. R. Yang, J. Guo, Z. Lin, et al., "The combination of two-dimensional and three-dimensional analysis methods contributes to the understanding of glioblastoma spatial heterogeneity," *J. Biophoton.* **13**(2), e201900196 (2020).

125. A.-K. Rodewald, E.J. Rushing, D. Kirschenbaum, et al., "Eight autopsy cases of melanoma brain metastases showing angiotropism and pericytic mimicry. Implications for extravascular migratory metastasis," *J. Cutan. Pathol.* **46**(8), 570–578 (2019).

126. R.D. Hume, L. Berry, S. Reichelt, et al., "An engineered human adipose/collagen model for in vitro breast cancer cell migration studies," *Tissue Eng. Part A* **24**(17–18), 1309–1319 (2018).

127. R.D. Hume, S. Pensa, E.J. Brown, et al., "Tumour cell invasiveness and response to chemotherapeutics in adipocyte invested 3D engineered anisotropic collagen scaffolds," *Sci. Rep.* **8**(1), 1–15 (2018).

128. N.S. Joshi, E.H. Akama-Garren, Y. Lu, et al., "Regulatory T cells in tumor-associated tertiary lymphoid structures suppress anti-tumor T cell responses," *Immunity* **43**(3), 579–590 (2015).

129. M.F. Cuccarese, J.M. Dubach, C. Pfirschke, et al., "Heterogeneity of macrophage infiltration and therapeutic response in lung carcinoma revealed by 3D organ imaging," *Nat. Commun.* **8**, 14293 (2017).

130. H.A. Messal, S. Alt, R.M.M. Ferreira, et al., "Tissue curvature and apicobasal mechanical tension imbalance instruct cancer morphogenesis," *Nature* **566**(7742), 126–130 (2019).

131. S.S.-Y. Lee, V.P. Bindokas, S.J. Kron, "Multiplex three-dimensional optical mapping of tumor immune microenvironment," *Sci. Rep.* **7**(1), 17031 (2017).

132. S.S.-Y. Lee, V.P. Bindokas, S.J. Kron, "Multiplex three-dimensional mapping of macromolecular drug distribution in the tumor microenvironment," *Mol. Cancer Ther.* **18**(1), 213–226 (2019).

133. S.S.-Y. Lee, V.P. Bindokas, M.W. Lingen, S.J. Kron, "Nondestructive, multiplex three-dimensional mapping of immune infiltrates in core needle biopsy," *Lab. Invest.* **99**(9), 1400–1413 (2019).

134. B. Lloyd-Lewis, F.M. Davis, O.B. Harris, et al., "Imaging the mammary gland and mammary tumours in 3D: Optical tissue clearing and immunofluorescence methods," *Breast Cancer Res.* **18**(1), 127 (2016).

135. J.F. Dekkers, M. Alieva, L.M. Wellens, et al., "High-resolution 3D imaging of fixed and cleared organoids," *Nat. Protoc.* **14**(6), 1756–1771 (2019).

136. Y. Chen, Q. Shen, S.L. White, et al., "Three-dimensional imaging and quantitative analysis in CLARITY processed breast cancer tissues," *Sci. Rep.* **9**(1), 1–13 (2019).

137. E. Lagoutte, C. Villeneuve, V. Fraisier, et al., "A new analytical pipeline for the study of the onset of mammary gland oncogenesis based on mammary organoid transplantation and organ clearing," *bioRxiv* 860270 (2019). DOI: https://doi.org/10.1101/860270

138. S. Sindhwani, A.M. Syed, S. Wilhelm, et al., "Three-dimensional optical mapping of nanoparticle distribution in intact tissues," *ACS Nano* **10**(5), 5468–5478 (2016).

139. S. Sindhwani, A.M. Syed, S. Wilhelm, W.C.W. Chan, "Exploring passive clearing for 3D optical imaging of nanoparticles in intact tissues," *Bioconjugate Chem.* **28**(1), 253–259 (2017).

140. A.M. Syed, P. MacMillan, J. Ngai, et al., "Liposome imaging in optically cleared tissues," *Nano Lett.* **20**(2), 1362–1369 (2020).

141. A.M. Syed, S. Sindhwani, S. Wilhelm, et al., "Three-dimensional imaging of transparent tissues via metal nanoparticle labeling," *J. Am. Chem. Soc.* **139**(29), 9961–9971 (2017).

142. B.R. Kingston, A.M. Syed, J. Ngai, S. Sindhwani, W.C.W. Chan, "Assessing micrometastases as a target for nanoparticles using 3D microscopy and machine learning," *Proc. Nat. Acad. Sci.* **116**(30), 14937–14946 (2019).

143. S.I. Kubota, K. Takahashi, J. Nishida, et al., "Whole-body profiling of cancer metastasis with single-cell resolution," *Cell Rep.* **20**(1), 236–250 (2017).

144. B. García-Pinel, C. Porras-Alcalá, A. Ortega-Rodríguez, et al., "Lipid-based nanoparticles: Application and recent advances in cancer treatment," *Nanomaterials* **9**(4), 638 (2019).

Part III

Towards *in vivo* tissue optical clearing

17

In vivo *skin optical clearing methods for blood flow and cell imaging*

Dongyu Li, Wei Feng, Rui Shi, and Dan Zhu

CONTENTS

Introduction

With the development of optical engineering, various noninvasive and minimally invasive optical imaging modalities have witnessed widespread exciting applications in biomedical diagnostics, including but not limited to laser speckle contrast imaging (LSCI) [1, 2], hyperspectral imaging (HSI) [3, 4], photoacoustic imaging [5, 6], and confocal imaging [7, 8]. So far, interest in using optical methods for physiological-condition monitoring and pathological diagnostics and therapies has been increasing because of their simplicity, safety, low cost, and contrast and resolution features in contrast to conventional X-ray computed tomography, ultrasound imaging, and some nuclear medicine imaging [9, 10].

Unfortunately, the main limitation of the majority of optical imaging technologies, especially optical imaging, is the strong scattering in superficial tissue, which causes dramatic decreases in spatial resolution, imaging contrast, and penetration depth [9, 10]. Among tissues and organs, skin is known as the largest organ of the body, as well as the most important route for foreign compounds entering the body. More importantly, skin is the first barrier layer for accessing *in vivo* physiological or pathological characteristics via noninvasive optical imaging [11, 12]. Therefore, in recent years, the skin optical clearing technique, especially *in vivo* skin optical clearing, has attracted a lot of attention.

Generally, skin roughly consists of three major layers, which from the outmost, are the epidermis, dermis, and subcutaneous adipose tissue. The dermis is rich in collagen and is thus the main scattering source, and thus it is the target for treatment with optical clearing agents (OCAs) to make the skin transparent. Usually, high refractive agents such as alcohol and sugars with hydroxyl groups are chosen as skin OCAs because they can disassociate dermic collagen by forming hydrogen bonds and bridges with the collagens. Thus, the more hydroxyl groups the agents have, the higher their ability to destabilize higher-order structures of collagen by disrupting water bridges and the hydration shells, and the higher potential they have to make the skin transparent to light [13–19].

Since the dermis is under the epidermis, which contains a highly keratinized layer named the *stratum corneum* (SC), it is not easy for topically applied OCAs to penetrate into the dermis. Thus, intradermal injection of OCAs has been used to improve the optical clearing of *in vivo* skin effectively [20]. However, intradermal injection of OCAs into the dermis can induce visible skin lesions [21], leading to less satisfaction for *in vivo* structural and functional imaging. Therefore, various methods for OCA penetration enhancement have been developed and used together with the *in vivo* skin optical clearing technique.

In this chapter, we will firstly introduce the enhancement methods that help topical applied OCAs penetrate into the dermis. Then, we will demonstrate in detail the current achievements of the *in vivo* skin optical clearing technique used for different skins, including dorsal skin, ear skin, and footpad skin in rodents. Meanwhile, skin optical clearing window–assisted optical imaging for vascular and cellular structure and functions will be introduced as well.

Enhancement method of skin optical clearing

Compared with immersion-based optical *ex vivo* clearing treatment, *in vivo* skin optical clearing might be much more difficult because that the barrier function of SC and compact epidermis restricts the penetration of OCAs into dermis. For *in vivo* topical application of OCAs, in order to develop effective and safe ways to accelerate the permeability of skin, many methods have been proposed, such as physical enhancement, chemical enhancement, and combinations of both.

Physical enhancement methods

Laser and intense pulsed light (IPL) have been used as common tools for clinical cosmetic therapy, and also have been used to effectively improve drug or OCA transepidermal delivery. Stumpp et al. [22] used a 980 nm diode laser to irradiate hamster and rat skin with artificial absorption substrates on the surface. The laser irradiation was strongly absorbed and provided precise heating to breach the SC diffusion barrier. Then, glycerol was applied to the skin surface, followed by optical coherence tomography (OCT) imaging for 30–45 min. Results indicated that there was an improvement in depth penetration of up to 42% for the ability to measure an OCT signal.

Tuchin et al. [23] found that skin permeability could be effectively enhanced by creating an island of limited thermal damage using a flash-lamp IPL system. In their study, the mask with a pattern of carbon-black absorbing centers was applied to the skin surface to absorb light strongly and to create a lattice of islets of damage. Fresh rat skin, farm pig skin, and Yucatan pig skin *ex vivo*, and human skin samples *ex vivo* and *in vivo* were used to verify the optical clearing efficacy of glycerol at different concentrations, 40% glucose as well as 60% propylene glycol. Transmittance was measured dynamically to record the clearing process. After 1–1.5 h of OCA application, a lattice of islands of limited thermal damage was produced on the skin surface, allowing for substantially enhanced penetration rate of topically applied OCAs and enhanced optical transmittance.

Further, different light sources at different dose and pulse modes were applied to topically irradiate rat skin *in vivo*, and glycerol was subsequently applied to the dorsal skin [15]. Reflectance of the treated skin area was detected to evaluate optical clearing efficacy. Results showed that the relative decrease of reflectance for irradiated skin samples was eight to nine times more than that of nonirradiated skin samples (as Figure 17.1 shows).

In addition to light irradiation, other mechanical options were also utilized to breach the SC barrier function, including ultrasound application [24], sandpaper grinding [25], microneedle rolling [26], and tape stripping [27].

Xu et al. [24, 28] successfully enhanced the permeation of OCAs into the skin dermis by ultrasound. Ultrasound treatment was performed for about 15 min on the *in vivo* human skin followed by occlusion with OCAs on the surface for 60 min. The ultrasound was performed in a pulsed mode at the frequency of 4 MHz to minimize thermal effects. The OCAs used include glycerol, propylene glycol, butanediol, butanol,

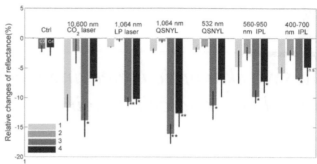

FIGURE 17.1 Relative changes in reflectance at 575 nm (1 and 3) and 615 nm (2 and 4) of *in vivo* rat skin. Ctrl: skin treated by glycerol. For experimental groups, 1, 2-skin treated by light irradiation; 3, 4-skin treated by light irradiation and glycerol. Data are shown as Mean ± SD (n = 6). *$P < 0.05$; **$P < 0.01$; n.s., nonsignificant difference compared to control group. QSNYL: Q-switched Nd: YAG laser. Reprinted from reference [15].

and polyethylene glycol (PEG200, PEG400). Compared with OCA application alone, the light penetration depth of OCT imaging for the ultrasound-OCA treated group was significantly increased.

Stumpp et al. [25] used a piece of fine 220-grit sandpaper to gently rub glycerol or dextrose solution into *in vivo* hamster skin after hair depilation. The manual rubbing with OCAs on the skin surface was continued for 2–4 min in different directions and performed gently to avoid injury to the skin. From the quantitative analysis of OCT signals, the light penetration depth with rubbing was obviously increased by 36%–43% compared to a control group to which only OCAs were applied.

Yoon et al. [29] utilized a microneedle roller (192 needles on a cylindrical surface, with 70 μm diameter and 500 μm height for each needle) to roll on the *ex vivo* porcine skin samples and then applied glycerol. All the skin samples were placed over a customized modulation transfer function target (MTFT) to acquire cross-polarized images. The contrast and relative contrast of the MTFT were quantitatively calculated to evaluate the enhanced efficacy of the microneedle roller in skin optical clearing. They found the contrast was increased by approximately two-fold for the microneedle roller treated samples compared with the control samples 30 min after glycerol application, as Figure 17.2 shows.

Tape stripping is another common method used for SC removing [27]. Tape stripping was usually performed gently and repeated for about 30 strippings to make the skin surface glisten but not to damage it. In addition, medical-grade cyanoacrylate adhesive bonded on the slide surface was also used to remove the SC. The slide with cyanoacrylate was applied to skin with moderate pressure for appropriate 3 min and then released, keeping contact for 2 min more. When the slide was peeled, the SC was successfully adhered to the substrate and removed from skin. This process should be repeated for a few times to make the skin glisten. OCAs were subsequently applied to the treated skin area by a low-pressure transdermal application device which also infiltrated the OCAs permeation. Tattoos in the skin dermis with removed SC could be seen much more clearly than those with nonremoved SC [30].

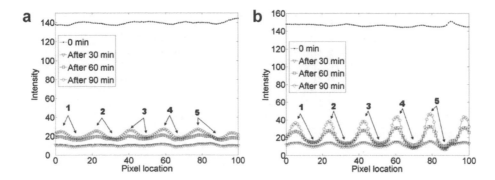

FIGURE 17.2 Modulation transfer functions (MTFs) on the (a) control and (b) test skin sample. Contrasts of the MTFs were calculated at five locations, as indicated in both figures. Reprinted from reference [29].

One of the mechanisms of skin optical clearing is tissue dehydration. A mechanical facility called a tissue optical clearing device (TOCD) was invented by Rylander et al. [31] to displace water in tissue and reduce light scattering. The TOCD consisted of an array of pins fixed on the inner chamber surface and two brims to interface with skin surface. When the device was connected to a vacuum pump (750 mm Hg), the brims formed an airtight seal, and the pins induced spatially localized tissue compression. OCT imaging demonstrated the light penetration depth at spatial positions was increased up to three-fold and the contrast of the epidermal–dermal junction was also increased.

Chemical enhancement methods

In addition to physical methods, chemical penetration enhancers used in medicine and cosmetics have also been introduced into skin optical clearing techniques [32], including Azone [33], Oleic acid (OA) [34], Dimethyl sulfoxside (DMSO) [35], propylene glycol (PG) [36], and Thiazone [19], to accelerate the permeability of OCAs into skin to improve the skin optical clearing efficacy.

Xu et al. [33] evaluated the effect of Azone as a penetration enhancer to elevate OCA permeation. 40% glycerol was mixed with Azone with different concentrations, respectively. Sixty minutes after the skin samples had been treated by glycerol-2%-Azone or glycerol-5%-Azone mixture, the light transmittance at 1276 nm was increased by 31% and 41%, respectively, while the transmittance was increased by 25% after treatment with glycerol alone. Meanwhile, the diffuse reflectance at 1066 nm was decreased by 21% and 29% after treatment with glycerol-2%-Azone or glycerol-5%-Azone mixture, respectively, while glycerol alone only led to a 17% reduction. It is worth mentioning that, compared with 80% glycerol alone, the optical clearing effect of 40%-glycerol-5%-Azone was even higher, which indicated that OCA permeation was markedly enhanced by Azone.

Dimethyl sulfoxide (DMSO) was used not only as hyperosmotic agent to induce optical clearing efficacy but also as an OCA penetration enhancer. Previous studies have showed that the mixed solution of glycerol and DMSO was able to play a synergistic effect on the optical clearing of gastric and skin tissue, due to the carrier effect of DMSO [35]. However, DMSO involves potential toxicity and possible side effects for *in vivo*

skin, which limits its usage to improve *in vivo* skin optical clearing.

Oleic acid (OA) is a monounsaturated fatty acid and has also been widely used as a kind of safe transdermal enhancer in the field of drug delivery. In the field of skin optical clearing, the synergistic effect of OA as a promoter to facilitate OCA permeation into skin tissue has been studied [34]. Mixtures of OA-PG and DMSO-PG were respectively applied to fresh porcine skin *in vitro*. From the results of near infrared spectroscopy, mass loss measured, and skin permeability assessment, it was speculated that OA had a similar enhancing effect to DMSO. OA has been considered to be a safe enhancer, so it could also be an optimal enhancer for skin optical clearing.

Thiazone, as an innovative penetration enhancer in the pharmaceutical and cosmetics industries, has a 2.99 times greater skin penetration effect than that of Azone. Compared with DMSO, Azone, or PG, it was found that Thiazone possessed the highest penetration enhancing effect for *in vitro* skin. Zhu et al. [19] calculated the reduced scattering coefficient and absorption coefficient of *in vitro* rat skin sample treated with PEG-400 mixed Thiazone. Figure 17.3 shows that the mixture induced almost no influence on the absorption coefficient but decreased the reduced scattering coefficient significantly, while saline did not. It was found that with the addition of Thiazone, the speed and the efficacy of optical clearing were higher than that without Thiazone, suggesting that Thiazone indeed enhances the penetration of PEG-400.

Combinations of physical and chemical enhancement methods

A solo physical breaching or chemical enhancer was effective to induce skin optical clearing efficacy. Their combination would show much greater potential to overcome the SC barrier function and facilitate OCA penetration into deep skin.

Ultrasound is usually utilized to combine with Thiazone to help OCAs penetrate into skin. Zhong et al. [37] applied a mixed solution of PEG-400 and 0.25% Thiazone to the surface of *in vitro* porcine skin samples or *in vivo* human skin. It was shown that the diffuse reflectance decreased by approximately 33.7, 3.3, and 2.7 times in 0.25% Thiazone PEG-400 solution with ultrasound treatment compared with that of the samples in control, PEG-400, and 0.25% Thiazone PEG-400 solution, respectively. At 60 min, the decrease in diffuse reflectance of

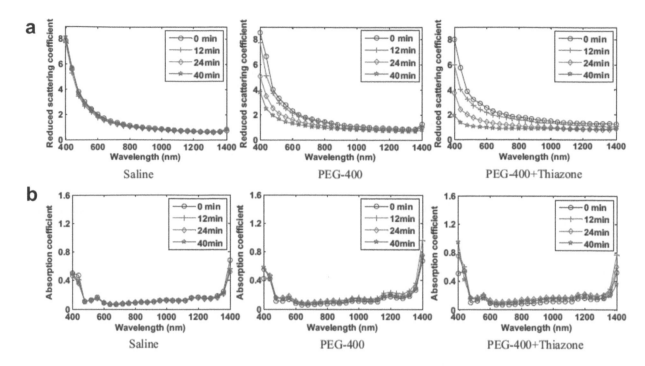

FIGURE 17.3 Typical spectra of (a) reduced scattering and (b) absorption of *in vitro* skin sample at initial state and different times (12, 24, and 40 min) after the application of the chemical agents (saline, PEG-400, and a mixture of PEG-400 and Thiazone). Reprinted from reference [19].

samples in 0.25% Thiazone PEG-400 solution with ultrasound treatment is about 2.2- and 1.2-fold compared with that of the samples in PEG-400 and 0.25% Thiazone PEG-400 solution, respectively, at 540 nm. Meanwhile, the OCT imaging depth increased by 41.3%, which was a significant difference compared with control samples.

The associative action of tape stripping with chemical enhancers, *i.e.*, Thiazone, Azone, and PG, has been performed over *in vivo* rat dorsal skins [38]. Tape stripping was implemented onto the epithelium surface three or four times to loosen the skin SC and then a mixed solution of enhancer/PEG400 was applied to the stripped skin area. Diffuse reflectance was measured, and it was found that chemical enhancers combined with the physical stripping method were able to facilitate more PEG400 penetration into the skin and decreased reflectance. The changes caused by the four kinds of agents vary; the largest changes in reflectance were induced by PEG-400/Thiazone, then PEG-400/Azone second, PEG-400/PG third, and pure PEG-400, as Figure 17.4 shows.

Wen et al. [36] demonstrated that significant optical clearing of *in vivo* rat skin was achieved within 15 min by topical application of an optical clearing agent PEG-400, a chemical enhancer (Thiazone or propanediol), and physical massage, as Figure 17.5 shows. The OCT signal falls as a function of depth in the dermis. As optical clearing develops, the superficial layers of the dermis become lighter, which means less scattering is occurring, and hence photons can travel deeper to carry out imaging. Further comparisons demonstrated that only when all three components (PEG-400, Thiazone, and massage) were applied together could a 15 min treatment achieve a three-fold increase in the OCT reflectance from a 300 μm depth and 31% enhancement in image depth.

In conclusion, topical application of the physical and chemical enhancement or the combinations of both with OCAs onto

skin *in vitro* can reduce scattering dramatically and, on skin *in vivo*, can make the skin transparent. These enhancements enable us to observe dermal blood vessels and image dermal blood flow through intact skin using diverse optical imaging techniques.

In vivo skin optical clearing windows for vascular and cellular imaging

Some peripheral vascular diseases are tightly coupled with abnormal structure and function of cutaneous microcirculation [39–41]. In addition, abundant microvascular networks in the dermis also permit the skin to become a target tissue for establishing tumor models [42]. Furthermore, skin also allows the realization of immune response monitoring. Thus, the ability to image the structure and function of dermal blood vessels and cells is of considerable importance for investigating changes in cutaneous microenvironment, and tracking tumor angiogenesis, development, and interventional treatment.

With the developments of lasers and optics technology, various optical imaging approaches have been becoming the primary tools for intravital imaging over the last decades. The advantages in terms of superior spatiotemporal resolution, multichannel parallel detection, and time-lapse dynamic monitoring permit the monitoring of blood flow dynamics [43–47], tracking of angiogenesis and tumor growth, and interventional treatment [48–50], *etc.* for living small animals. However, when the optical imaging technique is used for skin, it always suffers from high scattering of skin, leading to unsatisfactory imaging quality.

In previous studies, a skin chamber window achieved through a surgical operation was developed to resolve this problem and has been used for addressing vascular and

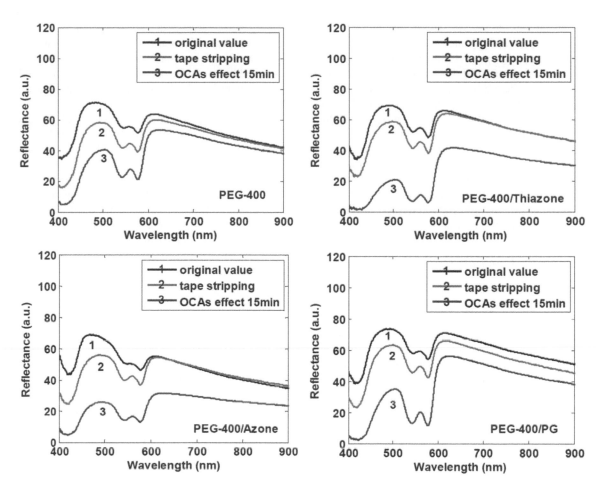

FIGURE 17.4 Diffuse reflectance of *in vivo* rat skin before and after treatment of tape stripping and chemical enhancer/OCA solution. Reprinted from reference [38].

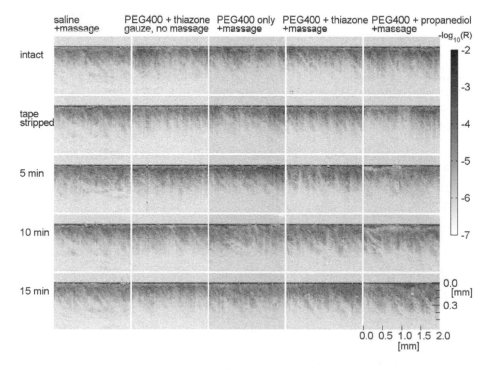

FIGURE 17.5 Computer-flattened optical coherence tomography (OCT) images of rat dorsal skin during topical application of different optical clearing agents. Reprinted from reference [36].

cellular events [51–53]. However, it is well known that surgical operations are always accompanied by bleeding and changes in the normal physiological environment during the process of model establishment [53, 54]. Fortunately, the *in vivo* skin optical clearing technique [55–61] provides an alternative solution by establishing a surgery-free optical clearing skin window through topical application of OCAs on the skin, which demonstrated its powerful capacity in optimizing the imaging performances for structural and functional visualization. Up to now, various skin OCAs have been developed for different kinds of skin, including dorsal skin [19, 36, 62–65], ear skin [66, 67], and footpad skin [68, 69] in rodents, to help visualize blood vessel diameter, blood flow distribution, and cells with higher resolution and in deeper tissue.

Dorsal skin optical clearing window

Dorsal skin has a large area and provides abundant microvasculature to be observed, and thus is the most targeted place for the development of *in vivo* skin optical clearing agents.

Zhu et al. [19] developed a skin optical clearing window to make the rat dorsal skin transparent by topically applying PEG-400 and Thiazone after tape stripping three or four times, as shown in Figure 17.6. In the beginning, the dermal blood vessel is invisible to the naked eye. The image quality becomes better after topical application of the mixture, and more details of the dermal blood vessels can be seen. After 12 min, even the small branch vessels can be distinguished clearly. Moreover, this situation will be allowed to continue.

After 40 min of treatment with the mixture, saline solution was applied onto the area of interest. Immediately, the skin recovered to its initial state and the blood vessels are out of view. In addition, the dorsal skin optical clearing window also proved suitable for LSCI to monitor distribution of dermal blood flow.

Wang et al. further evaluated the switchability and safety of the dorsal skin clearing window established by PEG-400 and Thiazone [70, 71]. In their study, the rat received a topical application of warm OCA (37°C), and the OCA was removed from the skin 30 min later with lukewarm water (37°C). The OCA was repeatedly applied on the same area of rat skin in vivo over the following 3 days. White-field images and laser speckle contrast images were recorded to evaluate the capability of repeated optical clearing. The results demonstrated that on the second or the third day, the initial state of the rat skin was almost the same as the intact skin, with no wrinkles or edema occurring in the skin. Besides, it was difficult to image blood flow because the skin was turbid, while after topical application of OCA on the same place, the skin became transparent again, and even the small branches of dermal blood vessels could be clearly observed. The details of dermal blood-flow distribution information were available to be acquired by the LSCI technique with high resolution. With treatment with saline solution, the skin recovered to its initial state, and the blood vessel structure and flow information were concealed by the high scattering of skin once again without any obvious difference (Figure 17.7). In addition, quantitative analysis showed that, after repeated application of OCA on rat skin, the

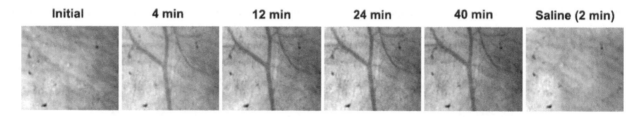

FIGURE 17.6 Photographs of in vivo rat skin at initial state, 4, 12, 24, 40 min after treatment of mixed solution of PEG-400 and Thiazone, and 2 min after treatment of saline. Reprinted from reference [19].

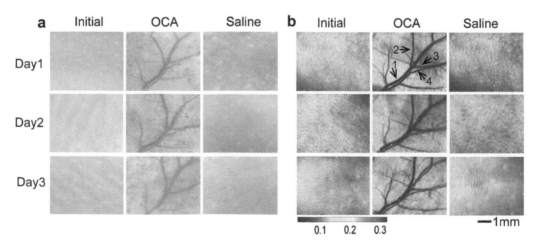

FIGURE 17.7 Typical visual photos (a) and corresponding laser speckle temporal contrast images (b) of *in vivo* rat skin. The arrows 1–4 indicate the blood vessels that were used for diameter measuring. Reprinted from reference [71].

diameters of dermal blood vessels measured over the following two days were essentially unchanged from that measured on the first day.

Further, the effects of optical clearing treatment on the microstructure of skin were investigated, as Figure 17.8 shows. It can be seen that after topical application of the mixture for 30 min on the third day, the epidermis of skin became extremely thin, the stratum corneum was destroyed, and the arrangement of subepidermal collagen became loose. However, after 2 weeks, the stratum corneum of rat skin recovers to normal. On the 28th day after treatment, the skin recovered to its initial state, and the development of skin was not affected by the optical clearing agent.

The combination of a dorsal skin optical clearing window and LSCI make it possible to access dermal arteriovenous blood flow dynamic response through the intact skin. Shi et al. [62] introduced topical application of PEG-400 and Thiazone on mice dorsal skin to make it transparent, and used LSCI to monitor blood flow response to Noradrenaline (NA) in blood vessels. With the assistance of the dorsal skin optical clearing window, they performed dynamic monitoring of arterial and venous blood flow response to NA using LSCI. Figure 17.9 shows the typical white-light images, blood flow velocity maps, and profiles of flow velocity values along the horizontal white line at status of turbid skin; with OCA treatment for 5 min; and 5, 10, 15, 25, and 30 min after NA injection through the OCA-treated skin, respectively. The results demonstrate that the cutaneous microvasculature and blood flow can be detected by LSCI with superior resolution and contrast after topical treatment of OCA to skin for 5 min. From the profiles of the flow velocity, it could be found that NA injection firstly induces a significant increase in the flow velocity, which attains the maximum response amplitude in about 10 min, before gradually decreasing and finally recovering to the base line in about 25 min.

Further quantification demonstrates that arteries and veins have different vasoconstrictive responses to NA. As shown in Figure 17.10, the NA-induced decrease in vascular diameter

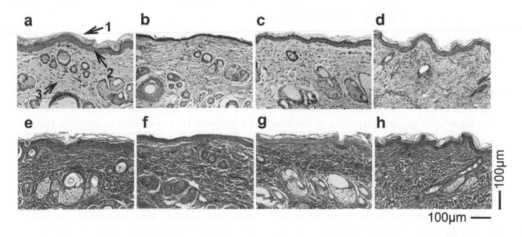

FIGURE 17.8 H&E stains (top row) and Masson stains (bottom row) of tissue sections of rat skin. (a) and (e) control skin sections; (b) and (f), (c) and (g), and (d) and (h) skin sections on the first day, the 14th day, and the 28th day after treatment of OCA for three times. The arrows 1–3 point out the *stratum corneum*, epidermis, and dermis, respectively. Reprinted from reference [71].

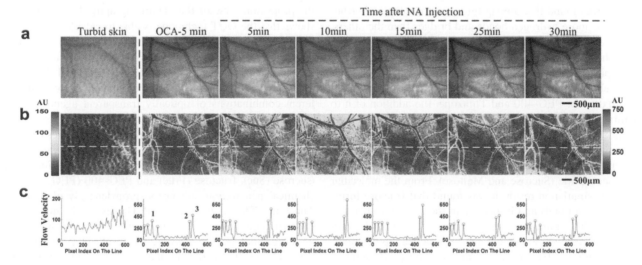

FIGURE 17.9 Monitoring cutaneous blood flow dynamics at status of turbid skin; OCA-5 min; and 5, 10, 15, 25, 30 min after NA injection accompanied with OCA-treatment: white-light images (a); blood flow velocity maps (b); profiles of the flow velocity along the horizontal white line (c), respectively. Bar = 500 μm. Reprinted from reference [62].

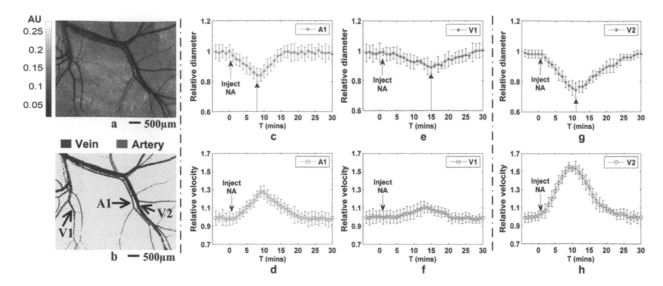

FIGURE 17.10 Speckle contrast image (a); arteries–veins separation image (b); NA injection-induced relative changes in vascular diameter and flow velocity in artery A1 (c, d); vein V1 (e, f); and vein V2 (g, h) accompanied by OCA treatment. The injection time is set to be 0, which is consistent with the arrow shown in (c–h). Bar = 500 *μm*. Reprinted from reference [62].

and increase in flow velocity in artery (marked by A1) are smaller than those in the parallel vein (marked by V2), while a smaller vein (marked by V1) showed weaker diameter decrease and flow velocity increase after NA injection. In addition, the recovery of vascular diameter is quicker than that of the flow velocity for the artery A1. Whereas the recovery of the vascular diameter is slower than that of the flow velocity for both veins V1 and V2.

Other than alcohols like glycerol and PEG-400, sugars with hydroxyl groups also attract attention in the development of novel *in vivo* skin optical clearing methods. Wang et al. [72] studied the optical clearing effect of three monosaccharides (fructose, glucose, and ribose) on rat skin by molecular dynamics simulation and in vitro experiment. Finally, fructose was selected for *in vivo* experiments, since it showed best potential in *in vitro* experiments, and its feasibility was verified. In their study, it was found that among three monosaccharides, fructose forms more hydrogen bonds and bridges than the others, and its solution provided the best optical clearing results when a rat skin sample was immersed for 24 min. For the further *in vivo* topical treatment, as Figure 17.11 shows, although fructose itself does not have a comparable optical clearing effect to a mixture of PEG-400 and Thiazone, the addition of it to the mixture can obviously make rat dorsal skin more transparent and help access to higher resolution and contrast of blood flow images captured by LSCI (Table 17.1). Feng et al. [73] further compared optical clearing ability between fructose and two disaccharide (Sucrose and Maltose). From the molecular dynamics simulation result, it was found that fructose forms less hydrogen bonds and bridges than the two disaccharides. However, the saturation concentration of maltose is very low, which means it is not suitable for practical applications. Finally, saturated fructose (6 M) and sucrose (3 M) solution was used for *in vivo* dorsal skin optical clearing, and showed similar efficacy. Therefore, sucrose offers another choice as an effective OCA besides fructose.

In addition to white-field imaging and LSCI, skin optical clearing windows are also beneficial for improving the performance of OCT for skin tissue and vascular imaging. Li et al. [74] topically applied a mixture of fructose, PEG-400, and Thiazone to rat dorsal skin to enhance imaging depth, resolution, and contrast of OCT structural and angiographic images. The layered structures of turbid skin are quite vague and difficult to distinguish. In contrast, after optical clearing, the layered structures and features are clearly identified, including the epidermis, dermis, hypodermis, and muscles. The epidermis–dermis and hypodermis–muscle boundaries are visualized clearly, and the dermis exhibits a decreased OCT scattering intensity. The OCT angiograms also demonstrate that the hypodermal blood vessels present improved visibility with optical clearing treatment, due to the decreased scattering attenuation in the dermis, which can be also observed in the projection view of the 3D angiography. Their study also shows that the OCT technique is able to calculate the reduction of scattering coefficient and refractive-index mismatching caused by OCA in each layer of skin, suggesting it can be an effective method to evaluate the efficacy of optical clearing.

Shi et al. [75] applied OCT angiography to evaluate six different combinations of optically transparent agents induced changes in imaging performances (*i.e.*, improvements in signal intensity and imaging depth), skin optical properties (changes of scattering and refractive index mismatching extent), and permeability rate. In their study, three optical clearing agents (sucrose (Suc), fructose (Fruc) and PEG-400 (PEG)) and two chemical penetration enhancers (propylene glycol (PG) and Thiazone (Thiaz)) were used. PEG was firstly mixed with the two penetration enhancers, respectively, and then mixed with Fruc and Suc, respectively. The results demonstrated that PEG+Thiaz+Suc has the optimal capacity to enhance imaging performance and decrease the scattering and refractive index mismatching, since Thiaz is superior to PG, and Suc is superior to Fruc (Figure 17.12).

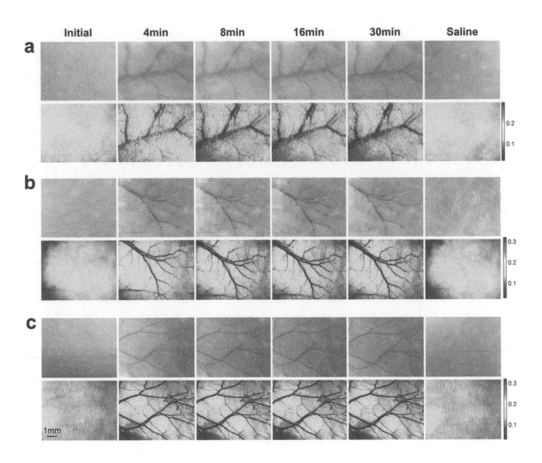

FIGURE 17.11 Typical white-light images (top row) and laser speckle temporal contrast images (bottom row) of rat skin *in vivo* before and after topical application of (a) fructose, (b) the mixture of PEG-400 and Thiazone (PT), and (c) the mixture of fructose, PEG-400 and Thiazone (FPT). Reprinted from reference [72].

TABLE 17.1

Contrast-to-noise ratio of laser speckle contrast images after treatment with the three solutions for 30 min. Reprinted with permission from reference [72]

	Contrast-to-noise ratio		
Vessel diameter	**4 weeks**	**PT**	**FPT**
80–300 μm	4.6 ± 2.0	4.6 ± 2.0	14.0 ± 3.0
30–80 μm	-	4.3 ± 1.6	7.4 ± 2.0

Combined with the skin optical clearing technique, OCT angiography is also able to perform cutaneous hemodynamic analysis with high spatiotemporal resolution and contrast. NA-induced changes in vascular diameter and flow flux can be estimated over time. The flow flux is measured based on the signal strength of OCT angiograms and normalized with the baseline. Figure 17.13 shows that After NE injection, the diameter exhibits a gradual decrease of up to 32.8% at the time instant ~14 min, then returns to ~96% of the baseline in 25 min. Along with the dynamic change in diameter, the flow flux exhibits a gradual increase of up to ~30% at the time instant ~15 min, and then returns to ~105% of the baseline after 25 min.

Acoustic resolution photoacoustic microscopy can offer deep imaging depth and high spatial resolution, and combines the advantages of optical imaging and acoustic imaging. It is a noninvasive imaging technique, which is very important for *in vivo* imaging. It can be used to image single cells without markers, to image microvessels with high resolution, to analyze the components of different tissues, and to detect the blood parameters with high specificity. Liu et al. [76] combined the *In vivo* skin optical clearing method and photoacoustic microscopy to image subcutaneous vessels, and the results showed that OCA can enhance photoacoustic amplitude and retain spatial resolution, as shown in Figure 17.14.

Ear skin optical clearing window

Understanding of gene expression, physiological function, and treatment efficacy is crucial for tumor study. Thus, the ability to monitor tumor growth, transfer, angiogenesis, and blood vessel recruitment is of considerable importance [77]. The specific microenvironment and semitransparent property of mice ear permits it to become an available window for continuously studying the angiogenic activities of tumors [77]. Unfortunately, the residual scattering of the ear skin still influences the optical imaging quality. Therefore, it is of great biological and medical significance to further decrease the scattering of the ear skin in mice. The *in vivo* tissue optical clearing technique presents a new opportunity to enhance the imaging contrast or imaging depth of optical methods.

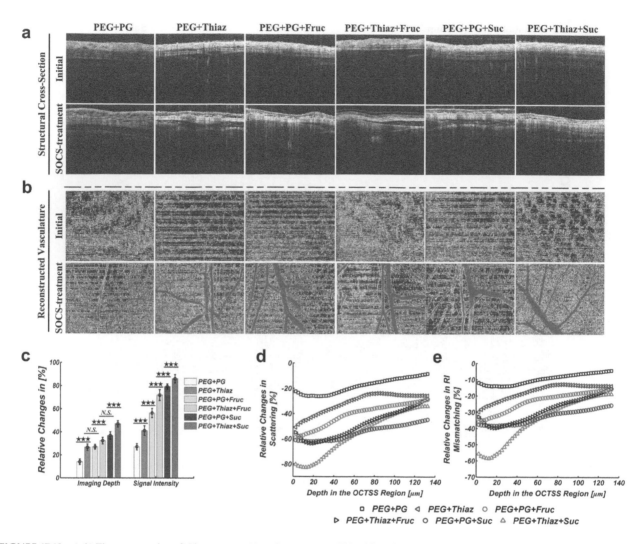

FIGURE 17.12 (a-b) The cross-section of skin structure (a), and reconstructed blood flow distribution information (b), before or after different SOCA treatments. (c–e) Relative changes of imaging depth and signal intensity (c), scattering coefficient (d) and RI mismatching (e) induced by different SOCA treatments. ★★★ refers to (p < 0.001), and N.S. refers to No Significance. Reprinted from reference [75].

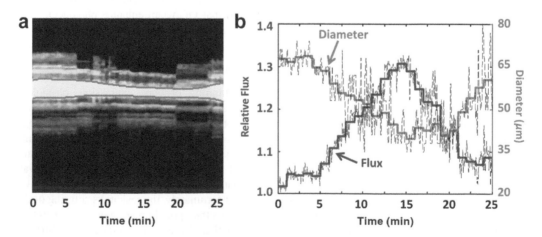

FIGURE 17.13 Blood flow dynamic responses to vasoactive drug monitored by Angio-OCT. (a) The M-mode (line of motion over time) Angio-OCT image of the time-lapse blood flow dynamic responses (the fitted red curves enhance the visualization of response changes) and (b) the dynamic changes in vessel diameter and normalized blood flow flux. Reprinted from reference [74].

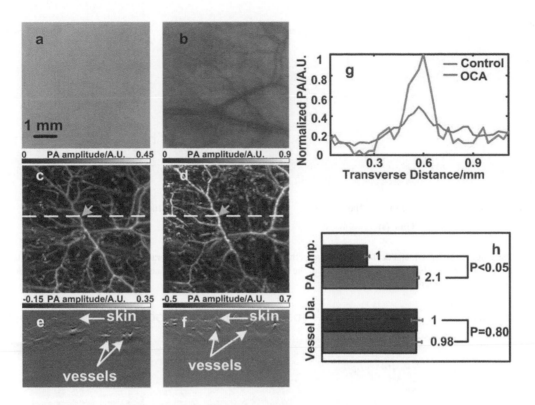

FIGURE 17.14 Comparative images and statistical analysis for *in vivo* imaging with a mixture of PEG-400 and Thiazone. (a) and (b) are photographs of skin before and after immersion. (c) and (d) are photoacoustic maximum amplitude projection (MAP) images projected along the depth direction. (e) and (f) are B-scan images denoted by the dashed lines in (c) and (d). The horizontal axis is the scan direction, and the vertical axis is the depth. (g) shows transverse plots of the vessel indicated by the green arrows in (c) and (d). (h) shows the statistical results for ten randomly selected vessels. The scale bar is the same for (a)–(f). Reprinted from reference [76].

Wang et al. [67] established an ear skin optical clearing window on mice by topically applying an ear skin optical clearing agent (ESOCA), which was a mixture of PEG-400, fructose, and Thiazone, and found that the transmittance spectrum of the mice ear increased significantly within 5 min. In addition, the relative change at 633 nm was also calculated, and it was apparent that compared to the initial value, the transmittance had increased by 111.0 ± 8.2 % after topical application of OCA for 10 min. Furthermore, LSCI was performed to monitor blood vessel structure and flow distribution in ear skin. After topical application of ESOCA on the ear skin for 5 min, the mouse ear becomes more transparent and even the small branches of cutaneous blood vessels can be clearly observed in white-light images. The LSCI shows that it is still difficult to distinguish the blood flow information of vessel-3 at initial state. After treatment with ESOCA for 10 min, the mouse ear becomes entirely transparent, and both the structure and flow distribution information of all the blood vessels can be detected with higher contrast. Thus it can be concluded that smaller branch blood vessels can be distinguished and no blood flow is blocked after application of ESOCA.

They further investigated the safety of ESOCA by quantitatively evaluating its effect on the diameter of the cutaneous blood vessels. The results show that after topical application of ESOCA on ear skin, the diameters of cutaneous blood vessels are essentially unchanged compared with the initial measurements of diameter. There are also no significant changes in vessel diameters between 5 min, 10 min, and initial. This means that the ESOCA has no obvious side effects on cutaneous blood vessels.

Ding et al. [66] compared the efficacy of ESOCA to that of glycerol. The transmittance of the ear skin increased significantly in the ESOCA treated group after 10 min of topical treatment, while glycerol only created a slight enhancement. Quantitative analysis showed that there were significant differences between the intact and ESOCA-treated ear skin (p<0.01) at 535 nm, 635 nm, and 665 nm, but no differences between the intact and glycerol-treated or saline-washed skin. They next performed confocal *in vivo* flow cytometry (IVFC) in the ear blood vessels to monitor the velocity and number of labeled red blood cells. With the assistance of an ear skin optical clearing window, the peak intensity, peak number, and signal-to-noise ratio of IVFC at deeper depth were significantly increased compared to the glycerol-treated group (Figure 17.15).

Footpad skin optical clearing window

Comparing with the dorsal skin or the ear skin, footpad skin is more convenient for intravital optical imaging due to its characteristic feature of hairlessness [78]. As a classic physiological site [79], the footpad has been widely used to study the immune function of immonocytes, *e.g.* tracking of cellular recruitment and migration at the single-cell level using microscopical imaging [79–82]. Unfortunately, both imaging contrast and imaging depth suffered from its strong scattering,

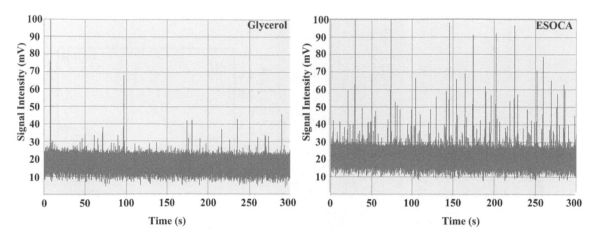

FIGURE 17.15 IVFC signals of 1,1'-Dioctadecyl-3,3,3',3'-Tetramethylindodicarbocyanine Perchlorate (DiD)-labeled red blood cells (left: treated by glycerol; right: treated by ESOCA). The signals were recorded at the depth of 180 μm for 6 min. Each peak represented a DiD-labeled cell traversing the excitation slit. Reprinted from reference [66].

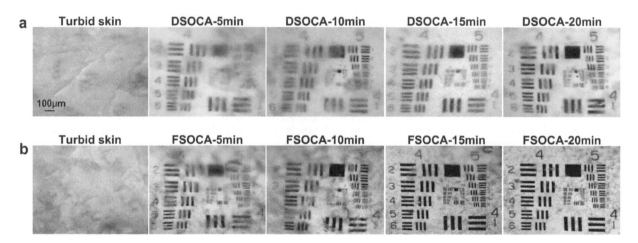

FIGURE 17.16 DSOCA- and FSOCA-induced time-dependent *in vitro* footpad skin optical clearing. Reprinted from Ref. [68].

and thus the investigations of immunological events were usually limited to the superficial tissue only [83, 84]. The novel *in vivo* skin optical clearing technique provides a new idea for deep-tissue optical monitoring vascular and cellular dynamics with high resolution.

Shi et al. [68] reported a footpad skin optical clearing agent (FSOCA), which consisted of fructose, ethanol, DMSO, PEG-400, and Thiazone, to established a switchable optical clearing window on mice footpad. In their study, an *ex vivo* experiment was first performed for comparison between DSOCA (a mixture of PEG-400 and Thiazone) and FSOCA. Footpad skin samples were taken and immersed in two OCAs, respectively, followed by being put on a 1951 United States Air Force (USAF) resolution test target. In can be seen in Figure 17.16 that the footpad skin is turbid initially and the USAF target was completely concealed. After the footpad skin samples were immersed into DSOCA and FSOCA, respectively, the USAF target beneath the footpad skin samples can gradually be seen. However, the footpad skin becomes more transparent after treatment by FSOCA at each time point than the DSOCA does, which allows the USAF target to be distinguished much more clearly. The result indicates that the resolution can

achieve 13.72 ± 2.47 μm and 7.06 ± 0.61 μm after treatment by DSOCA and FSOCA, respectively, for 20 min.

Figure 17.17 shows the typical results on footpad skin *in vivo* in conditions of intact treatment with FSOCA for 5 min, 15 min, and 20 min, and recovery with physiological saline. It can be seen that, initially, the footpad skin is turbid, and it is difficult to distinguish the blood vessels and blood flow distribution. However, after treatment with FSOCA for 5 min, the cutaneous blood vessels and blood flow distribution can be observed. Even the blood flow of the small branch vessels can be monitored by the LSCI technique with superior imaging resolution and contrast after FSOCA treatment for 15 min. In addition, after treatment with physiological saline, the footpad skin can recover to the turbid status, and the blood vessels and the blood flow distribution can only be distinguished hazily.

In addition, FSOCA could also optimize the performance of *in vivo* cell imaging. Fluorescence-labeled monocytes were imaged with confocal microscopy. Figure 17.18 shows that through the intact footpad skin, only sparse monocytes can be distinguished, and the fluorescence signal intensity is relatively weak. However, for transparent footpad skin,

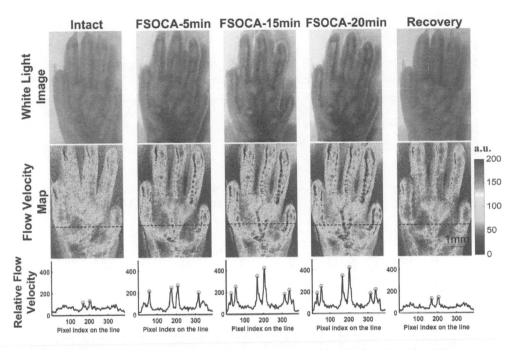

FIGURE 17.17 White-light images (top row), relative blood flow velocity maps (middle row) and profiles of blood-flow index along the dashed line (bottom row) through the intact, FSOCA-5 min, FSOCA-15 min, FSOCA-20 min, and recovered footpad skin. The circles indicate the blood vessels of interest. Reprinted from reference [68].

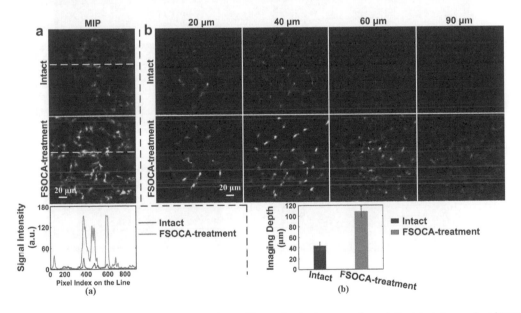

FIGURE 17.18 (a) Maximum intensity projection (MIP) maps imaged by confocal microscopy through the intact (top row) and transparent footpad skin (middle row), and profiles of fluorescence signal intensity along the dashed line (bottom row). (b) Depth-resolved fluorescence images (the first two rows) and the imaging depth (the third row) before and after FSOCA treatment. Data are expressed as mean ± SEM. Reprinted from reference [68].

a remarkable increase could be observed in the number of monocytes, and fluorescence signal intensity increases by 3.5 times on average compared to the turbid footpad skin. In addition, the improvement in imaging depth was also evaluated, suggesting the imaging depth was only 40 μm through intact footpad, while the cells can still be distinguished with acceptable fluorescence signal intensity when the imaging depth reaches 90 μm through the transparent footpad skin; that is, the imaging depth is improved by 1.3-fold on average.

The safety of FSOCA was evaluated by dynamically monitoring vascular and cellular changes in repeatedly optical clearing treated footpad skin over the course of 3 days. Figure 17.19a–d show the distribution information of blood vessels and blood flow of footpad skin before and after repeated FSOCA treatment in vivo at 0, 4, 24, 48, and 72 h, indicating that the distribution of blood vessels can be imaged clearly every time after treatment, and repeatedly being treated with FSOCA does not influence blood flow velocity or vascular diameter. Other than vascular monitoring, the distribution of monocytes in skin was

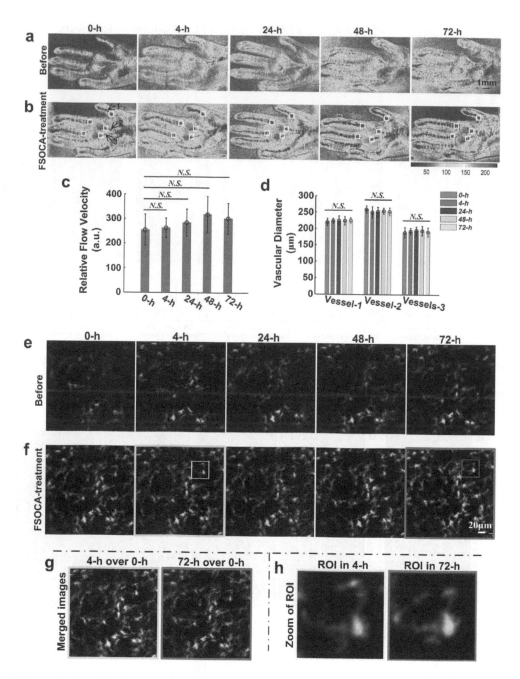

FIGURE 17.19 Safety evaluation of FSOCA. (a–b) Repeated blood-flow imaging *in vivo* before (a) and after (b) FSOCA treatment. (c) The averaged relative flow velocity from the five white rectangles in (b). (d) The diameters of blood vessels pointed with the three black arrows in (b). Data are expressed as mean ± SEM. N.S. refers to no significance, Mann-Whitney test. (e–f) Repeated cell imaging *in vivo* before (a) and after (b) FSOCA treatment. (c) Merged images of 4 h (cyan border) or 72 h (magenta border) together with that at 0 h. (d) Zoom of ROI in 4 h (yellow rectangle) or 72 h (red rectangle). Reprinted from reference [68].

also studied. Since monocytes are sensitive to inflammation, if FSOCA causes any change to the microenvironment of the skin, the monocytes will respond to this and it will be reflected in changes in morphology and distribution. However, as shown in Figure 17.19e–h, a negligible difference can be observed within the period. Furthermore, the results of H&E staining show that no inflammation cells infiltrate into the epidermis and dermis. Such evidence suggests that repeated FSOCA treatment does not induce any observable histopathologic or microstructural changes in footpad skin.

In conclusion, DSOCA, ESOCA, and FSOCA were developed to make the dorsal skin, ear skin, and footpad skin transparent by topical application *in vivo*, providing a safe optical clearing skin window for blood vessel and cell imaging. It not only allows the cutaneous vascular structure and blood flow to be monitored by LSCI and OCT with higher contrast, but also permits the IVFC or LSCM to count or image fluorescent cells much deeper and with higher fluorescence signal intensity, thus having great significance in biological research.

Summary

The *in vivo* tissue optical clearing technique provide a non-invasive skin window for optical visualization of structure and dynamic functions in dermis. By topically applying this technique using alcohols and sugars, the scattering of skin can be significantly reduced. To overcome the limitations to the penetration of OCAs into the dermis, physical and chemical methods or a combination of these are usually introduced to promote the permeability of OCAs. Targeting different types of skin, various formulas of skin OCA have been developed, especially for dorsal skin, ear skin, and footpad skin, allowing optical imaging systems to perform vascular and cellular monitoring in certain areas with promising imaging depth and resolution. The optical clearing skin window not only makes it possible to track microvasculature and blood flow distribution with higher resolution and cells in much deeper skin, but also decreases the risk of phototoxicity due to the reduced light power needed. In addition, the novel *in vivo* skin optical clearing methods have been proved to hardly have any impact on cutaneous vasculature or microenvironment.

Combining this switchable skin optical clearing window with modern intravital imaging techniques will deliver new insights for investigating blood-flow dynamics and cellular immune function *in vivo*, which will be significant for developing and evaluating the outcomes of some new therapeutics for some vascular and immunological diseases.

Acknowledgments

This chapter was supported by the National Science Foundation of China (Grant Nos. 61860206009, 81171376) and China Postdoctoral Science Foundation funded project (No. BX20190131, 2019M662633).

REFERENCES

1. M. Draijer, E. Hondebrink, T. van Leeuwen, and W. Steenbergen, "Review of laser speckle contrast techniques for visualizing tissue perfusion," *Lasers Med Sci* **24**(4), 639–651 (2009).
2. D.A. Boas, and A.K. Dunn, "Laser speckle contrast imaging in biomedical optics," *J. Biomed. Opt.* **15**(1), 011109 (2010).
3. L. Khaodhiar, T. Dinh, K.T. Schomacker et al., "The use of medical hyperspectral technology to evaluate microcirculatory changes in diabetic foot ulcers and to predict clinical outcomes," *Diabetes Care* **30**(4), 903–910 (2007).
4. D. Yudovsky, A. Nouvong, K. Schomacker, and L. Pilon, "Monitoring temporal development and healing of diabetic foot ulceration using hyperspectral imaging," *J. Biophoton.* **4**(7–8), 565–576 (2011).
5. H.F. Zhang, K. Maslov, G. Stoica, and L.V. Wang, "Functional photoacoustic microscopy for high-resolution and noninvasive in vivo imaging," *Nat. Biotechnol.* **24**(7), 848–851 (2006).
6. J.J. Yao, and L.H.V. Wang, "Transverse flow imaging based on photoacoustic Doppler bandwidth broadening," *J. Biomed. Opt.* **15**(2), 021304 (2010).
7. D. Semwogerere, and E.R. Weeks, "Confocal microscopy," *Encycl. Biomater. Biomed. Eng.* **10**, 1081 (2005).
8. R. Dickie, R.M. Bachoo, M.A. Rupnick et al., "Three-dimensional visualization of microvessel architecture of whole-mount tissue by confocal microscopy," *Microvasc. Res.* **72**(1–2), 20–26 (2006).
9. V.V. Tuchin, *Tissue Optics: Light Scattering Methods and Instruments for Medical Diagnostics*, Chapter 13, 3rd edition, SPIE Press, Bellingham, WA, 661–709 (2015).
10. V.V. Tuchin, *Tissue Optics: Light Scattering Methods and Instruments for Medical Diagnostics*, Chapter 14, 3rd edition, SPIE Press, Bellingham, WA, 711–747 (2015).
11. M. Roustit, and J.L. Cracowski, "Non-invasive assessment of skin microvascular function in humans: An insight into methods," *Microcirculation* **19**(1), 47–64 (2012).
12. G. Mahe, A. Humeau-Heurtier, S. Durand, G. Leftheriotis, and P. Abraham, "Assessment of skin microvascular function and dysfunction with laser speckle contrast imaging," *Circ. Cardiovasc. Imag.* **5**(1), 155–163 (2012).
13. X. Wen, V.V. Tuchin, Q. Luo, and D. Zhu, "Controlling the scattering of intralipid by using optical clearing agents," *Phys. Med. Biol.* **54**(22), 6917–6930 (2009).
14. J. Wang, N. Ma, R. Shi et al., "Sugar-induced skin optical clearing: From molecular dynamics simulation to experimental demonstration," *IEEE J. Sel. Top. Quantum Electron.* **20**(2), 256–262 (2014).
15. C.H. Liu, Z.W. Zhi, V.V. Tuchin, Q.M. Luo, and D. Zhu, "Enhancement of skin optical clearing efficacy using photo-irradiation," *Laser. Surg. Med.* **42**(2), 132–140 (2010).
16. J. Hirshburg, B. Choi, J.S. Nelson, and A.T. Yeh, "Collagen solubility correlates with skin optical clearing," *J. Biomed. Opt.* **11**(4), 040501 (2006).
17. O. Stumpp, A. Welch, T. Milner, and J. Neev, "Enhancement of transepidermal skin clearing agent delivery using a 980 nm diode laser," *Laser. Surg. Med.* **37**(4), 278–285 (2005).
18. V.V. Tuchin, G.B. Altshuler, A.A. Gavrilova, A.B. Pravdin, and I.V. Yaroslavsky, "Optical clearing of skin using flash lamp-induced enhancement of epidermal permeability," *Laser. Surg. Med.* **38**(9), 824–836 (2006).
19. D. Zhu, J. Wang, Z. Zhi, X. Wen, and Q. Luo, "Imaging dermal blood flow through the intact rat skin with an optical clearing method," *J. Biomed. Opt.* **15**(2), 026008 (2010).
20. E.I. Galanzha, V.V. Tuchin, A.V. Solovieva et al., "Skin backreflectance and microvascular system functioning at the action of osmotic agents," *J. Phys. D: Appl. Phys.* **36**(14), 1739 (2003).
21. Z. Mao, Z. Han, X. Wen, Q. Luo, and D. Zhu, "Influence of glycerol with different concentrations on skin optical clearing and morphological changes in vivo," *Photon. Optoelectron. Meet.* **7278**, 72781T–72781T-72787 (2008).
22. O.F. Stumpp, A.J. Welch, T.E. Milner, and J. Neev, "Enhancement of transepidermal skin clearing agent delivery using a 980 nm diode laser," *Laser. Surg. Med.* **37**(4), 278–285 (2005).
23. V.V. Tuchin, G.B. Altshuler, A.A. Gavrilova et al., "Optical clearing of skin using flash lamp–induced enhancement of epidermal permeability," *Laser. Surg. Med.* **38**(9), 824–836 (2006).
24. X. Xu, and Q. Zhu, "Sonophoretic delivery for contrast and depth improvement in skin optical coherence tomography," *IEEE J. Sel. Top. Quantum Electron.* **14**(1), 56–61 (2008).

25. O. Stumpp, B. Chen, and A.J. Welch, "Using sandpaper for noninvasive transepidermal optical skin clearing agent delivery," *J. Biomed. Opt.* **11**(4), 041118 (2006).

26. T. Son, J. Yoon, C.Y. Ko et al., "Contrast enhancement of laser speckle skin image: Use of optical clearing agent in conjunction with micro-needling," *Proc. SPIE* 6842, 68420E (2008).

27. R.J. Mcnichols, M.A. Fox, A. Gowda et al., "Temporary dermal scatter reduction: Quantitative assessment and implications for improved laser tattoo removal," *Laser. Surg. Med.* **36**(4), 289–296 (2005).

28. X. Xu, Q. Zhu, and C. Sun, "Assessment of the effects of ultrasound-mediated alcohols on skin optical clearing," *J. Biomed. Opt.* **14**(3), 034042 (2009).

29. J. Yoon, T. Son, E.-h. Choi et al., "Enhancement of optical skin clearing efficacy using a microneedle roller," *J. Biomed. Opt.* **13**(2), 021103 (2008).

30. M.A. Fox, D.G. Diven, S. Karen et al., "Dermal scatter reduction in human skin: A method using controlled application of glycerol," *Laser. Surg. Med.* **41**(4), 251–255 (2009).

31. C.G. Rylander, T.E. Milner, S.A. Baranov, and J.S. Nelson, "Mechanical tissue optical clearing devices: Enhancement of light penetration and heating of ex-vivo porcine skin and adipose tissue," *Laser. Surg. Med.* **40**(10), 688–694 (2008).

32. A.C. Williams, and B.W. Barry, "Penetration enhancers," *Adv. Drug Del. Rev.* **56**(5), 603–618 (2004).

33. X. Xu, and Q. Zhu, "Evaluation of skin optical clearing enhancement with Azone as a penetration enhancer," *Opt. Commun.* **279**(1), 223–228 (2007).

34. J. Jiang, and R.K. Wang, "Comparing the synergistic effects of oleic acid and dimethyl sulfoxide as vehicles for optical clearing of skin tissue in vitro," *Phys. Med. Biol.* **49**(23), 5283–5294 (2004).

35. K. Moulton, F. Lovell, E. Williams et al., "Use of glycerol as an optical clearing agent for enhancing photonic transference and detection of Salmonella typhimurium through porcine skin," *J. Biomed. Opt.* **11**(5), 054027 (2006).

36. X. Wen, S.L. Jacques, V.V. Tuchin, and D. Zhu, "Enhanced optical clearing of skin in vivo and optical coherence tomography in-depth imaging," *J. Biomed. Opt.* **17**(6), 066022 (2012).

37. H.Q. Zhong, Z.Y. Guo, H.J. Wei et al., "Synergistic effect of ultrasound and thiazone-PEG 400 on human skin optical clearing in vivo," *Photochem. Photobiol.* **86**(3), 732–737 (2010).

38. J. Wang, X. Zhou, S. Duan, Z. Chen, and D. Zhu, Improvement of in vivo rat skin optical clearing with chemical penetration enhancers, *Proc. SPIE* **7883**, 78830Y-1 (2011).

39. H.A. Struijker-Boudier, A.E. Rosei, P. Bruneval et al., "Evaluation of the microcirculation in hypertension and cardiovascular disease," *Eur. Heart. J.* **28**(23), 2834–2840 (2007).

40. F. Quondamatteo, "Skin and diabetes mellitus: What do we know?" *Cell Tissue Res.* **355**(1), 1–21 (2014).

41. M. Rossi, A. Carpi, F. Galetta, F. Franzoni, and G. Santoro, "The investigation of skin blood flowmotion: a new approach to study the microcirculatory impairment in vascular diseases?" *Biomed. Pharmacother.* **60**(8), 437–442 (2006).

42. L.A. Holowatz, C.S. Thompson-Torgerson, and W.L. Kenney, "The human cutaneous circulation as a model of generalized microvascular function," *J. Appl. Physiol.* **105**(1), 370–372 (2008).

43. S.M. Daly, and M.J. Leahy, "'Go with the flow': A review of methods and advancements in blood flow imaging," *J. Biophoton.* **6**(3), 217–255 (2013).

44. D.D. Postnov, V.V. Tuchin, and O. Sosnovtseva, "Estimation of vessel diameter and blood flow dynamics from laser speckle images," *Biomed. Opt. Express* **7**(7), 2759–2768 (2016).

45. Y. Liu, X. Yang, H. Gong et al., "Assessing the effects of norepinephrine on single cerebral microvessels using optical-resolution photoacoustic microscope," *J. Biomed. Opt.* **18**(7), 076007 (2013).

46. A.Y. Neganova, D.D. Postnov, J.C. Jacobsen, and O. Sosnovtseva, "Laser speckle analysis of retinal vascular dynamics," *Biomed. Opt. Express* **7**(4), 1375–1384 (2016).

47. S. Huang, M. Shen, D. Zhu et al., "In vivo imaging of retinal hemodynamics with OCT angiography and doppler OCT," *Biomed. Opt. Express* **7**(2), 663–676 (2016).

48. B.J. Vakoc, R.M. Lanning, J.A. Tyrrell et al., "Three-dimensional microscopy of the tumor microenvironment in vivo using optical frequency domain imaging," *Nat. Med.* **15**(10), 1219–1223 (2009).

49. Y. Zhao, A.J. Bower, B.W. Graf, M.D. Boppart, and S.A. Boppart, "Imaging and tracking of bone marrow-derived immune and stem cells," *Methods Mol. Biol.* **1052**(75), 1–20 (2013).

50. J.A. Lee, R.T. Kozikowski, and B.S. Sorg, "In vivo microscopy of microvessel oxygenation and network connections," *Microvasc. Res.* **98**, 29–39 (2014).

51. G.E. Koehl, A. Gaumann, and E.K. Geissler, "Intravital microscopy of tumor angiogenesis and regression in the dorsal skin fold chamber: Mechanistic insights and preclinical testing of therapeutic strategies," *Clin. Exp. Metastasis* **26**(4), 329–344 (2009).

52. G.M. Palmer, A.N. Fontanella, S. Shan et al., "In vivo optical molecular imaging and analysis in mice using dorsal window chamber models applied to hypoxia, vasculature and fluorescent reporters," *Nat. Protoc.* **6**(9), 1355–1366 (2011).

53. N. Duansak, and S. Chatpun, "Intravital microscopy in a dorsal skinfold chamber: Hemodynamics, tumor angiogenesis and inflammation," *Int. J. Biomed. Res.* **4**(2), 65–73 (2013).

54. A.L.B. Seynhaeve, and T.L.M.T. Hagen, *High-Resolution Intravital Microscopy of Tumor Angiogenesis*, Humana Press, New York (2016).

55. M. Kinnunen, A.V. Bykov, J. Tuorila et al., "Optical clearing at cellular level," *J. Biomed. Opt.* **19**(7), 71409 (2014).

56. J. Wang, Y. Zhang, P. Li, and Q. Luo, "Review: Tissue optical clearing window for blood flow monitoring," *IEEE J. Sel. Top. Quantum Electron.* **20**(2), 92–103 (2014).

57. D. Zhu, K.V. Larin, Q. Luo, and V.V. Tuchin, "Recent progress in tissue optical clearing," *Laser Photon. Rev.* **7**(5), 732–757 (2013).

58. E.A. Genina, A.N. Bashkatov, Yu.P. Sinichkin, Yu. Yanina, and V.V.Tuchin, "Optical clearing of biological tissues: Prospects of application in medical diagnostics and phototherapy," *J. Biomed. Photon. Eng.* **1**(1), 22–58 (2015).

59. D.S. Richardson, and J.W. Lichtman, "Clarifying tissue clearing," *Cell* **162**(2), 246–257 (2015).

60. E. Song, H. Seo, K. Choe et al., "Optical clearing based cellular-level 3D visualization of intact lymph node cortex," *Biomed. Opt. Express* **6**(10), 4154–4164 (2015).

61. X. Yang, Z. Yang, Z. Kai, and Y. Zhao, "Skull optical clearing solution for enhancing ultrasonic and photoacoustic imaging," *IEEE Trans. Med. Imaging* **35**(8), 1903–1906 (2016).

62. R. Shi, M. Chen, V.V. Tuchin, and D. Zhu, "Accessing to arteriovenous blood flow dynamics response using combined laser speckle contrast imaging and skin optical clearing," *Biomed. Opt. Express* **6**(6), 1977–1989 (2015).

63. L. Guo, R. Shi, C. Zhang et al., "Optical coherence tomography angiography offers comprehensive evaluation of skin optical clearing in vivo by quantifying optical properties and blood flow imaging simultaneously," *J. Biomed. Opt.* **21**(8), 081202 (2016).

64. Y.Y. Bai, X. Gao, Y.C. Wang et al., "Image-guided pro-angiogenic therapy in diabetic stroke mouse models using a multi-modal nanoprobe," *Theranostics* **4**(8), 787–797 (2014).

65. X.W. Zhu, G. Perry, M.A. Smith, and X.L. Wang, "Abnormal mitochondrial dynamics in the pathogenesis of Alzheimer's disease," *J. Alzheim. Dis.* **33**(Supplement 1), S253–S262 (2013).

66. Y. Ding, J. Wang, Z. Fan et al., "Signal and depth enhancement for in vivo flow cytometer measurement of ear skin by optical clearing agents," *Biomed. Opt. Express* **4**(11), 2518–2526 (2013).

67. W. Jing, S. Rui, Z. Yang, and Z. Dan, "Ear skin optical clearing for improving blood flow imaging," *Photon. Laser. Med.* **2**(1), 37–44 (2013).

68. R. Shi, W. Feng, C. Zhang, Z. Zhang, and D. Zhu, "FSOCA-induced switchable footpad skin optical clearing window for blood flow and cell imaging in vivo," *J. Biophoton.* **10**(12), 1647–1656. (2017).

69. R. Shi, W. Feng, C. Zhang et al., "In vivo imaging the motility of monocyte/macrophage during inflammation in diabetic mice," *J. Biophoton.* **11**(5), e201700205 (2018).

70. J. Wang, and D. Zhu, "Preliminary investigations of rat skin after topical application of optical clearing agent," *Proc. SPIE* **7999**(1), 2121–2127 (2011).

71. J. Wang, R. Shi, and D. Zhu, "Switchable skin window induced by optical clearing method for dermal blood flow imaging," *J. Biomed. Opt.* **18**(6), 061209 (2013).

72. W. Jing, M. Ning, S. Rui et al., "Sugar-induced skin optical clearing: From molecular dynamics simulation to experimental demonstration," *IEEE J. Sel. Top. Quantum Electron.* **20**(2), 256–262 (2014).

73. W. Feng, R. Shi, N. Ma et al., "Skin optical clearing potential of disaccharides," *J. Biomed. Opt.* **21**(8), 081207 (2016).

74. L. Guo, R. Shi, C. Zhang et al., "Optical coherence tomography angiography offers comprehensive evaluation of skin optical clearing in vivo by quantifying optical properties and blood flow imaging simultaneously," *J. Biomed. Opt.* **21**(8), 081202 (2016).

75. R. Shi, L. Guo, C. Zhang et al., "A useful way to develop effective in vivo skin optical clearing agents," *J. Biophoton.* **10**(6–7), 887–895 (2016).

76. Y. Liu, X. Yang, D. Zhu, R. Shi, and Q. Luo, "Optical clearing agents improve photoacoustic imaging in the optical diffusive regime," *Opt. Letter.* **38**(20), 4236–4239 (2013).

77. A. Pourtiermanzanedo, C. Vercamer, E.V. Belle et al., "Expression of an Ets-1 dominant-negative mutant perturbs normal and tumor angiogenesis in a mouse ear model," *Oncogene* **22**(12), 1795–1806 (2003).

78. Q. Wang, H. Ilves, P. Chu et al., "Delivery and inhibition of reporter genes by small interfering RNAs in a mouse skin model," *J. Invest. Dermatol.* **127**(11), 2577–2584 (2007).

79. B.H. Zinselmeyer, J.N. Lynch, X. Zhang, T. Aoshi, and M.J. Miller, "Video-rate two-photon imaging of mouse footpad: A promising model for studying leukocyte recruitment dynamics during inflammation," *Inflamm. Res.* **57**(3), 93–96 (2008).

80. F. Liu, L. Zhang, R.M. Hoffman, and M. Zhao, "Vessel destruction by tumor-targeting Salmonella typhimurium A1-R is enhanced by high tumor vascularity," *Cell Cycle* **9**(22), 4518–4524 (2010).

81. T.R. Mempel, M.L. Scimone, J.R. Mora, and U.H. von Andrian, "In vivo imaging of leukocyte trafficking in blood vessels and tissues," *Curr. Opin. Immunol.* **16**(4), 406–417 (2004).

82. B. Wang, B.H. Zinselmeyer, J.R. McDole, P.A. Gieselman, and M.J. Miller, "Non-invasive imaging of leukocyte homing and migration in vivo," *J. Visual. Express* (46), e2062 (2010).

83. R.N. Germain, E.A. Robey, and M.D. Cahalan, "A decade of imaging cellular motility and interaction dynamics in the immune system," *Science* **336**(6089), 1676–1681 (2012).

84. B. Roediger, L.G. Ng, A.L. Smith, B. Fazekas de St Groth, and W. Weninger, "Visualizing dendritic cell migration within the skin," *Histochem. Cell Biol.* **130**(6), 1131–1146 (2008).

18

In vivo *skull optical clearing for imaging cortical neuron and vascular structure and function*

Dongyu Li, Yanjie Zhao, Chao Zhang, and Dan Zhu

CONTENTS

Introduction

In vivo observation of brain structure and activities is of vital significance to better understand not only normal brain physiology but also dysfunctions of vasculature and neural networks related to various brain diseases [1–5]. Modern optical imaging technology has attracted much attention in biomedical investigation due to its low invasiveness and high resolution, and because it is able to obtain biological structure and function *in vivo*, it thus plays an important role in the field of brain science [6–9]. For instance, laser speckle contrast imaging (LSCI) has been widely applied to map changes in cortical blood flow (CBF), which has provided valuable insights into many aspects of cortical function [3, 10]. Hyperspectral imaging (HSI) [11, 12] has been used for monitoring oxygen saturation (SO_2) in cortical tissues and blood vessels [13, 14]. Optical coherent tomography (OCT) has been used to perform three-dimensional (3D) volumetric imaging of the internal microstructure in biological tissue over 1–2 mm imaging scale with micron-meter resolution [15, 16]. Nonlinear optical microscopy has emerged and has drawn much attention in

brain scientific research because it provides *in vivo* deep-tissue neuron-vasculature observation in cortical [17–25], and has been applied to study specific diseases and behavior [26–28].

However, the turbid skull above the cortex is a barrier to noninvasive cerebral optical imaging. In order to overcome the scattering of the intact skull, the previous investigations were performed with several types of skull window based on craniotomy, including the open-skull glass window, the thinned-skull cranial window, and their variants [29–34]. The open-skull glass window is to remove a section of the skull and replace it with a glass coverslip. The window is convenient for repeat imaging during a month. However, the operation may cause an inflammatory response, which would last for 3 weeks, making it hard to visualize the cortex right after the surgery [35]. The thinned-skull window is to thin the skull down to at most 25 μm, where the skull will be transparent enough for imaging. It tend to be a minimally invasive method. Nevertheless, the grinding of the high-speed drill will greatly heat up the part of the skull involved. Besides, due to the regrowth of skull, for repeat observation, one has to repeatedly thin the skull over and over again, which makes it relatively inconvenient to use [31]. Thus, despite their frequent use in cortical imaging, these two types of cranial windows present problems that are difficult to solve: the associated inflammatory response, as well as the complexities in surgical procedures and high skill requirements for laboratory personnel. As a consequence, it is urgent to develop a safe and easy-to-handle skull window technique for cortical neuroangiography.

The tissue optical clearing technique can reduce the scattering of tissue and has great potential for solving this problem. In recent years, the *in vivo* skull optical clearing technique has been developed and made promising progress. Just by applying reagents to the skull, scattering can be significantly reduced and the skull rendered "transparent" to light [36, 37]. With the novel skull optical clearing window, many optical imaging systems have been used to observe the structures and functions of cortical vasculature and neuron systems with high resolution [38–41]. According to different requirements, different methods of *in vivo* skull optical clearing have been developed for high-resolution imaging of spinous processes, switchable cranial window establishment, and long-term cortical monitoring, respectively [36, 37, 42–44].

In this chapter, the development of the *in vivo* skull optical clearing technique as well as its applications in cortical neurovascular imaging are introduced. The advantages and limitations of each method are also demonstrated.

SOCS: Providing a surgery-free skull window

In the last decades, the tissue optical clearing technique has allowed great progress in *ex vivo* cortical imaging [35, 45, 46]. Combined with various optical imaging systems, it is possible to visualize brain, spinal cord, etc. at high resolution and obtain 3D imaging of neuron-vasculature [47, 48]. However, early research in tissue optical clearing techniques were limited to soft tissue. As for the skull, only some *in vitro* attempts were made to reduce scattering [49].

The main components of the skull are calcium hydroxyapatite (16%), collagen (16%), lipids (54%), and water (14%) [49]. Aiming to reduce the scattering of each component in skull, Wang et al. [42] developed a skull optical clearing solution (SOCS) that consisted of laurinol, weak alkaline substances, EDTA, dimethyl sulfoxide, sorbitol, alcohol, and glucose, and they tested its compatibility of *in vivo* application. The results

demonstrated that the novel skull optical clearing method provides an innovative transparent cranial window for accessing cortical structural and functional information without craniotomy. In the formular, the sorbitol was used to dissolve the collagen, the laurinol was used to dissolve the lipid, and the EDTA was used to dissolve the calcium hydroxyapatite. They also found that the application of weak alkaline substances could enhance the efficacy of skull optical clearing.

In the *in vivo* experiment, the scalp was removed carefully, and the skull was topical treated with SOCS for 25 min. The results showed that the intact skull was almost turbid and the cortical blood vessels were hardly visible, but that the cortical microvessels become visible gradually with prolonged action time of SOCS. Finally, they directly removed the skull for comparison, and found that the minimum resolution diameter of microvessels through exposed cortex and transparent skull after treatment of SOCS for 25 min was 12.8 ± 0.9 and 14.4 ± 0.8 μm, respectively. They also performed LSCI of cortical microvessels for optical clearing efficiency evaluation. The result showed that the blood flow in some large vessels can be measured hazily through the intact skull (Figure 18.1a), while the blood flow distribution of cortical microvessels in details can be distinguished clearly through the transparent skull (Figure 18.1b) and exposed cortex (Figure 18.1c).

SOCS could improve not only the quality of white-light imaging and LSCI, but also of photoacoustic imaging. Yang et al. [50] investigated the difference of the cortical vascular photoacoustic imaging through the skull before and after the treatment with SOCS *in vivo*. As shown in Figure 18.2, after topical treatment of the skull with SOCS for 25 min, the photoacoustic signal amplitude of the same vessels in the cortex is significantly enhanced, and concealed vessels in deep tissue can be observed clearly through the transparent skull window compared with the untreated turbid skull. With the optical clearing skull window, the photoacoustic signal increased 2.59-fold without affecting the diameter of the vessels.

SOCS provides a surgery-free skull window for cortical vascular imaging for the first time, proving the possibility of *in vivo* skull optical clearing. Despite the novelty and advantages, SOCS still has limitations. The current resolution may not be able to visualize the fine structure of neurons in cortical. Moreover, the SOCS-induced skull window cannot be used for long-term observation. Furthermore, the safety of SOCS was not investigated in the reports.

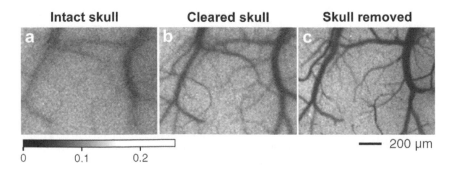

FIGURE 18.1 Laser speckle temporal contrast images. (a–c) demonstrate the speckle contrast images from intact skull (a), the transparent skull after SOCS treatment for 25 min without (b), and with exposed cortex (c). Reprinted from reference [42].

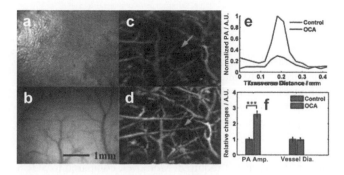

FIGURE 18.2 Typical images of cortical blood vessels, (a) reflection images obtained from untreated skull, (b) reflection images obtained from the transparent skull after SOCS treatment for 25 min, (c) photoacoustic images obtained from intact skull, (d) photoacoustic images obtained from the transparent skull after SOCS treatment for 25 min, (e) typical photoacoustic signal of the same vessel denoted by green arrows in (c) and (d), (f) statistical analysis of amplitude of photoacoustic signal and diameter of vessels before and after SOCS treatment. Reprinted from reference [50].

"Transparent skull": For chronic imaging

In biomedical research, sometimes months of continuous observation are required to monitor structural and hemodynamic changes during the occurrence and development of some cerebrovascular diseases. The open skull glass window could be used for 1–2 months, but longer would be impossible due to the regrown of skull. The thinned skull window required repeated thinning over a long time, which means it is not easy to handle.

Silasi et al. [43] reported a intact skull chronic window with clear dental cement and demonstrated its feasibility and stability for bi-hemispheric wide-field imaging of spontaneous activity in awake mice. Seed pixel functional connectivity (correlation) maps generated from spontaneous activity in awake mice show similar patterns of connectivity to the anesthetized state, indicating that the preparation facilitates functional cortical mesoscopic mapping in different brain states.

In their study, lidocaine (0.1 ml, 0.2%) was injected under the scalp, and mice also received a 0.5 ml subcutaneous injection of a saline solution containing burprenorphine (2 mg/ml), atropine (3 μg/ml), and glucose (20 mM). A skin flap extending over both hemispheres approximately 8 mm in diameter (3 mm anterior to bregma to posterior end of skull and down lateral to eye level) was cut and removed, and sterile cotton tips were used to gently wipe off any fascia or connective tissue on the skull surface. The clear version of C&B-Metabond dental cement was prepared by mixing 1 scoop of C&B Metabond powder, 6 drops of C&B Metabond Quick Base, and 1 drop of C&B Universal catalyst in a ceramic or glass dish. When the cement was partially dry (1–2 min), it was applied directly on the intact skull, and a cover glass was gently placed on the intact skull. The mixture remains transparent after it solidifies with veins and arteries observed at the end of the procedure. Figure 18.3 shows the procedure for the chronic window implant.

For longitudinal studies, intermittent imaging in awake mice is typically performed for up to 2 months. The fluorescence of GcaMP was used for mapping neuron activity. Figure 18.4 shows that, during the 2 months, the imaging contrast stays almost the same, and it is unnecessary to increase excitation light power or camera exposure to achieve a constant level of fluorescence across the imaging sessions, suggesting that both optical transparency and GCaMP6 expression in the cortex are stable over the long-term.

Steinzeig et al. [44] developed another chronic skull window, named transparent skull (TS), for the investigation of visual system plasticity. The procedure for transparent skull surgery is shown in Figure 18.5. For transparent skull establishment, a thin layer of cyanoacrylate glue was first applied to the exposed skull for 15 min. Then, a nail-polish-like consistency of acryl powder dissolved in methyl methacrylate liquid was applied to the skull. The first layer was allowed to dry for about 40 min before the second layer was applied. The layers were left to dry overnight. The next day, the acrylic layer was polished only in the region of interest (ROI) with fine acryl polishers. A metal bar holder was first glued to the skull

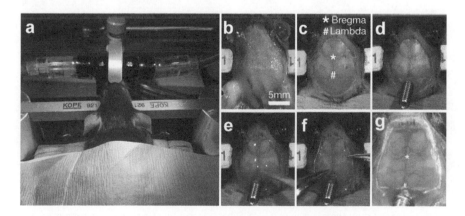

FIGURE 18.3 Surgical procedure for chronic window implant. Anesthetized mice are stabilized in a stereotaxic frame (A), the skin between the ears and eyes is shaved (B), and the skin covering the occipital, parietal, and frontal bones is cut away (C). The underlying skull is cleared of fascia and a setscrew is attached with dental cement over the occipital bone (D). A thick layer of dental cement is applied over the exposed (intact) skull (E) and a piece of cover glass is carefully lowered over the cement (F). In contrast to the opaque nature of the dry skull (C, D), once the chronic preparation dries, surface vessels are clearly visible through the intact skull (G). Reprinted from reference [43].

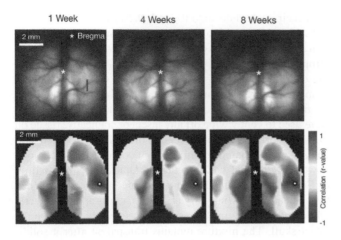

1 Week   4 Weeks   8 Weeks

FIGURE 18.4 Transcranial windows maintain optical transparency for at least 8 weeks. Surface vasculature remained clearly visible, and spontaneous activity recordings showed similar correlation maps for at least 8 weeks after chronic window implantation. Reprinted from reference [43].

surface and then covered with a mixture of cyanoacrylate glue and dental cement. Finally, transparent nail polish was added to the inside of the metal holder above the ROI.

It is worth mentioning that the transparent skull window has accessibility for injections. To test the possibility of

injections through the transparent skull, a small hole was drilled above the ROI, 5 μL of saline was injected into the brain parenchyma, and the intrinsic signal was measured. The results demonstrated that saline injection through the transparent skull into the brain parenchyma caused no disruption of the intrinsic signal quality after the procedure (Figure 18.6).

Despite the relatively low resolution of the two methods above, they can be used for continuous observation for 2 months, indicating they could be used in conditions where high resolution is not required. Moreover, compared to SOCS and the skull-clearing method developed by Silasi et al., Steinzeig's procedure is relatively time-consuming.

SOCW: Optical clearing skull window for synaptic imaging

Zhao et al. [36] developed a skull optical clearing window (SOCW), through which synaptic resolution could be achieved in cortex with TPLSM. This technique allowed them to repeatedly image neurons, microglia, and microvasculature of mice. It was also applied to study the plasticity of dendritic spines in critical periods and to visualize dendrites and microglia after laser ablation.

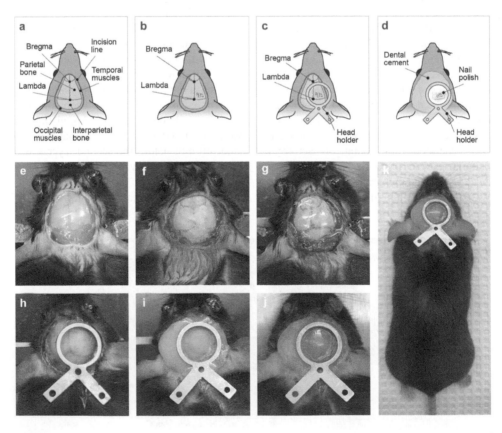

FIGURE 18.5 Procedure for transparent skull surgery in mice. (A–D) Schematic view of experimental preparations: (A) Dorsal view on the mouse head with important landmarks and indication of the incision area; (B) Skull after clearing with cyanoacrylate glue obtained increased transparency; (C) Head holder implantation scheme; (D) Nail polish and dental cement application scheme. (E–J) Photographs of the TS surgery main steps (see the protocol for detailed description): (E) Skull after scalp incision; (F) Cleaned bone surface; (G) Skull covered with cyanoacrylate glue and acryl; (H) Head holder mounting; (I) Head holder fixation with dental cement; (J) Nail polish added to the window. (K) Overview of a mouse after the TS surgery. Reprinted from reference [44].

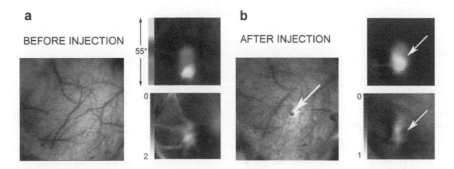

FIGURE 18.6 Transparent skull preparation allows injections through the skull. (a) Images of the intact transparent skull. Left: surface vasculature photograph. Right: representative optical imaging data (color images – phase map of retinotopy. Color bar represents the phase; b/w images – magnitude map of intrinsic signal). (b) Corresponding images after saline injection through the transparent skull. The injection site is indicated with arrows. Reprinted from reference [44].

Methods

The optical clearing agents used in the study include collagenase, EDTA disodium, and glycerol, since the main components of the inorganic matrix and organic matrix are calcium hydroxyapatite and collagen, respectively. As the mice grow older, the ratio of inorganic matrix to organic matrix increases. Therefore, for infantile mice (<P20), collagenase was used to dissolve collagenous fiber; for older mice (>P20), EDTA disodium was used to chelate calcium ions (decalcification). In addition, glycerol was used to match the refractive index.

The skull clearing procedure consists of two steps. For the first step, for mice aged P15–P20, the intact skull is topically treated with 10% collagenase for 5–10min; for mice aged P21–P30, the reagent is replaced by 10% EDTA disodium. The thickness of the skull increases as mice age; thus, for mice older than P30, the skull has to be thinned to approximately 100 μm before clearing and then treated with 10% EDTA disodium for 5–10 min. For the second step, the first reagent above the skull is removed and 80% glycerol is dropped onto the skull. The skull could return to the opaque state very quickly after treatment with PBS (phosphate-buffered saline), and retreatment with optical clearing agent (OCAs) made the skull re-clear very quickly. If the imaging interval was beyond one day, the time for clearing was slightly longer than the first time. Figure 18.7 is the schematic diagram of the SOCW technique.

In vivo two-photon microscopy through SOCW

The authors acquired *in vivo* images of YFP-expressing dendritic spines/EGFP-expressing microglia/FITC-dextran-labeled cerebral vasculature by a two-photon microscope (FV1200; Olympus, Tokyo, Japan) with a Ti:sapphire laser. The image quality was considerably improved by SOCW and it was sufficient for imaging the dendritic spines through the cleared skull. Besides, significant increases in fluorescence intensity and imaging depth were enhanced after clearing (Figure 18.8a), and the mean depth increased approximately twofold (Figure 18.8b). Through SOCW, good visibility of dendrites, microglia, and blood vessels up to 250 μm below the pial surface could be obtained (Figure 18.8c–e). In addition, such results also suggested that the SOCW technique is compatible with various fluorescent proteins and dyes, including GFP, YFP, and tetramethylrhodamine, etc.

Dynamic synaptic monitoring

SOCW was applied to dynamically monitor the plasticity of dendritic protrusions of juvenile mice (P19) over 60 min. The results demonstrated that dendritic protrusions exhibited strong motility. The dendritic branches were studded with numerous spines (56.6 ± 1.4 spine per 100 μm), including mushroom-type, stubby, and thin spines, as well as long filopodia-like protrusions. The spines were observed to have changed within 1 hour, including the appearance and disappearance of spines (Figure 18.9a–c) and changes in shape (Figure 18.9f), which points to changes in the wiring of neuronal circuits. It was counted that spine formation and elimination were 3.20 ± 0.21% and 1.83 ± 0.52%, respectively. Compared with dendritic spines, filopodia exhibited higher motility (Figure 18.9d–e). In addition, filopodia can even convert into spine-like protrusions, demonstrating that filopodia are likely to be the precursors of dendritic spines. These results suggested that the plasticity of dendritic protrusions during the third week is very intense.

Repeated observation through SOCW

The capability of SOCW-assisted repeated observation of the cortex was evaluated. The *in vivo* short-term visualization showed the SOCW allowed researchers to repeatedly perform high-resolution two-photon microscopy for 3 days without inducing morphology changes in dentrites (Figure 18.10a–b) or microglia (Figure 18.10c). In addition, they also imaged the FITC-dextran-filled cerebral vasculature of C57BL/6 mice at 0 and 21 days after clearing and found that the distribution of cerebral vasculature was nearly the same (Figure 18.10d), indicating its safety.

Further safety evaluation was carried by observing microglia 2 days after clearing and visualizing the expression of the GFAP in astrocytes 10 days after clearing. Figure 18.10e–f shows that both microglia and astrocyte GFAP expression remained consistent with that under the contralateral side, suggesting they were in a nonactive state.

In conclusion, the SOCW technique can render mouse skull highly transparent, especially in the case of young mice, making it possible to visualize neural spines in cortex. With SOCW, repeated dynamic observation of cortical neuron and vasculature can be realized without causing side effects. However,

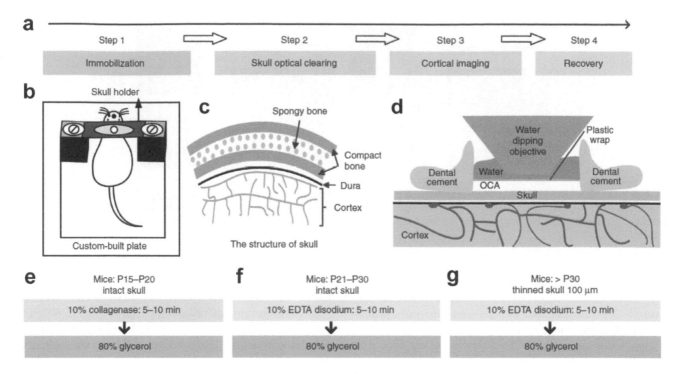

FIGURE 18.7 Schematic diagram of SOCW technique for cortical imaging. (a) Four steps: immobilization, skull optical clearing, cortical imaging, and recovery. (b) A custom-built head immobilization device consisting of a skull-holder and a custom-built plate is used to reduce motion artifact during imaging. (c) Anatomical structure of mouse skull. (d) Schematic of the SOCW. A layer of plastic wrap is placed over the cleared skull to separate the water-immersion objective from the OCA. (e–g) The skull optical clearing methods for mice aged e P15–P20, f P21–P30, g older than P30, which include two steps, the first step is different for mice of different ages. Step 1: For mice aged P15–P20, the intact skull was topically treated with 10% collagenase (w v⁻¹) for 5–10 min; for mice aged P21–P30, the intact skull was topically treated with 10% EDTA disodium (w v⁻¹) for 5–10 min; for mice older than P30, the thinned skull (100 μm) was treated with 10% EDTA disodium (w v⁻¹) for 5–10 min. Step 2: 80% glycerol (v v⁻¹) was dropped onto the cleared skull. Reprinted from reference [36].

for mice older than 30 days, the skull should be thinned to 100 μm before optical clearing treatment, which makes SOCW minimally invasive to adult mice. In addition, long-term multi-repeated (more than 3×) opening of the optical clearing skull window with SOCW technique is lacking, which might be another limitation for further applications.

USOCA: Switchable optical clearing skull window for long-term cortical observation

Other than "permanent skull window", Zhang et al. [37] developed a switchable optical clearing skull window by applying a urea-based skull optical clearing agent (USOCA), which could be used to realize repeated cortical vascular imaging for as long as 5 months.

Methods

USOCA consists of solution 1 (S1) and solution 2 (S2). S1 is a saturated urea in 75% (vol/vol) ethanol solution. The volume–mass ratio of ethanol and urea is about 10:3. S2 is a high-concentration sodium dodecylbenzenesulfonate (SDBS) solution that is prepared by mixing 0.7 M NaOH solution with dodecylbenzenesulfonic acid at a volume–mass ratio of 24:5, and pH is kept at 7.2–8. During the optical clearing procedure,

a holder was glued onto the exposed skull, and the mouse was immobilized in a custom-built plate. Then, S1 was topically applied to the skull for about 10 min with a swab gently rubbing the skull to make it turn transparent. Finally, S1 was removed and S2 was applied to the skull for about 5 min for the skull to be effectively cleared. Figure 18.11 is the preparation of USOCA and the experimental procedure.

As shown in Figure 18.12, the white-light cortical vascular image through the transparent skull is similar to that for the exposed cortex. Besides, when the imaging experiment is over, the optical clearing skull window can be "closed" by scrubbing with phosphate-buffered saline, drying, and skin suturing.

Two-photon laser scanning microscopy enhancement using USOCA

Two-photon laser scanning microscopy (TPLSM) was performed to evaluate the enhancement of signal intensity, imaging contrast, and imaging depth due to skull optical clearing (Figure 18.13). The z-axis average signal intensity maps indicated that the imaging depth was also greatly enhanced. Before clearing, the imaging depth was about 120 μm, and only a few vessels (67.5 ± 7.9 μm) could be barely detected. However, with the optical clearing skull window, the imaging depth increased to about 300 μm, and both large vessels and microvessels (4.5 ± 0.31 μm) could be distinguished.

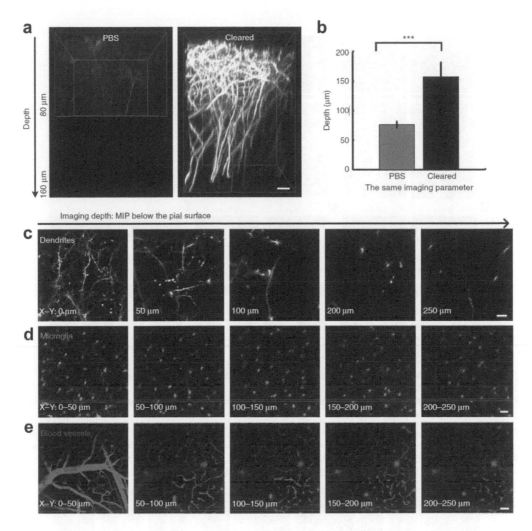

FIGURE 18.8 Imaging depth through the SOCW. (a) Orthogonal (x–z) projections of dendrites through the intact skull, before and after skull optical clearing, demonstrating that the depth is obviously enhanced after clearing (the imaging parameter and data processing were the same). Scale bar = 10 μm. (b) The depth when imaging the dendrites of Thy1-YFP neurons, before and after skull optical clearing (P30, n=10 mice; statistical method: one-way analysis of variance (ANOVA); P0.001). (c–e) Imaging depth through the SOCW after optimizing the imaging parameters (P30, n = 6 mice). (c) Dendrites and spines of Thy1-YFP neurons at different depths through the SOCW. Scale bar = 10 μm. (d) Maximum z-axis projections across 50 μm of microglia through the SOCW. Scale bar = 25 μm. (e) Maximum z-axis projections across 50 μm of FITC-dextran-filled cerebral vasculature through the SOCW. Scale bar = 50 μm. Reprinted from reference [36].

USOCA for different ages and long-term repeated observation

The authors then used a LSCI/HSI dual-model optical imaging system to monitor CBF and SO₂, respectively. The results demonstrated that the USOCA was compatible to different ages of mice, from 2 months old to 8 months old. Cortical vessels were not visible on the white-light and SO₂ maps of the untreated skull, where laser speckle temporal contrast analyzing method could not distinguish cortical vessels clearly. For 2-month-old and 4-month-old mice, topical PBS treatment could slightly increase the imaging quality, but it was useless for 6-month-old and 8-month-old mice. However, with the application of USOCA, abundant cortical vessels were clearly observed, and the contrast and resolution of the CBF and SO₂ maps were greatly enhanced and were almost comparable to those obtained in all four age groups with the skull removal.

To evaluate the ability of USOCA-based optical clearing skull window for long-term observation, the authors performed optical clearing in the same area of the skull every other day for a week and once a week as the mice aged from 2 to 7 months (Figure 18.14). The long-term repeated clearing and imaging results demonstrated that the optical clearing skull window was efficiently and repeatedly established once a month, and the vascular network was maintained relatively well, with only a few vessels appearing and "disappearing" over the 6 months, which was a normal phenomenon of cerebral angiogenesis [51].

USOCA for large-field visualization of middle cerebral artery occlusion

USOCA was used for bi-hemispheric skull optical clearing, and LSCI/HSI was performed to monitor CBF and SO₂

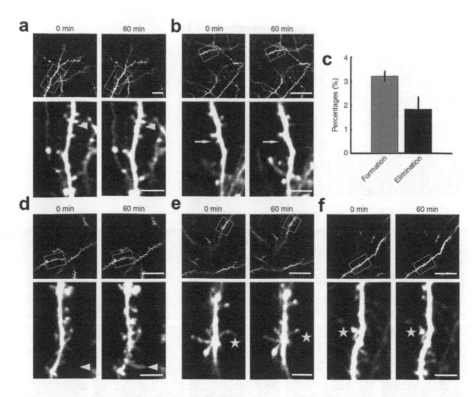

FIGURE 18.9 Dynamical monitoring of the plasticity of dendritic protrusions in infantile mice (P19) through the SOCW. Time-lapse images of dendritic branches over an hour (60 min). (a and b) Dendritic spines can (a) appear and (b) disappear within an hour. (c) Percentage of spines formed and eliminated within an hour, according to the SOCW technique (n = 6 mice). (d and e) Filopodia can also (d) appear and (e) convert into a spine-like protrusion within an hour. (f) The morphology of dendritic spines can change within an hour. The triangle, arrow, and asterisk show appearance, disappearance, and morphological changes, respectively. Scale bar = 25 μm (above) and 5 μm (below). Reprinted from reference [36].

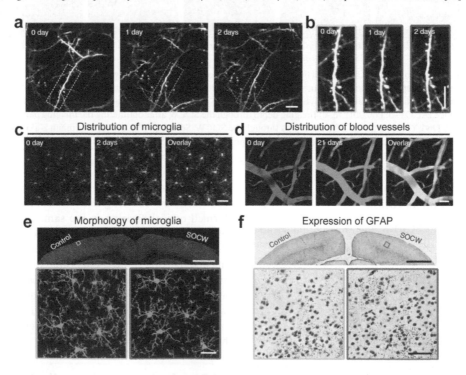

FIGURE 18.10 (a and b) Repeated imaging of the dendrites b and spines c of Thy1-YFP neurons obtained over a 1-day interval (P28–P30, n = 10 mice). Scale bar = 10 μm. (c) Distribution of microglia through the SOCW. After the SOCW technique, microglia remain at the same position (P28, n = 4 mice). Scale bar = 25 μm. (d) Maximum projections (z-axis) across 10–40 μm of FITC-dextran-filled cerebral vasculature under the SOCW obtained 0 (P28, n = 4 mice) and 21 days after the treatment. We observe that the cerebrovascular morphology remains nearly unchanged 21 days after forming the SOCW. Scale bar = 25 μm. (e) Histological images of microglia in the case of the SOCW technique; microglia in both the treated and control sides appear normal (P30, n = 3 mice). Scale bar = 1 mm (above) and 25 μm (below). (f) GFAP expression under the SOCW technique. The treated and control hemispheres show similar levels of GFAP expression (P38, n = 3 mice). Scale bar = 1 mm (above) and 50 μm (below). Reprinted from reference [36].

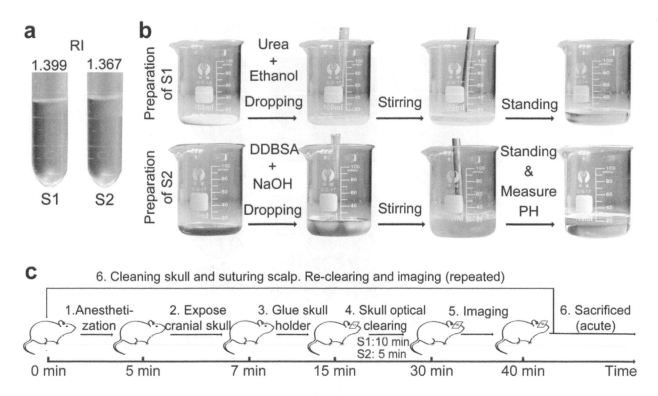

FIGURE 18.11 Preparation of USOCA and the experimental procedure. (A) Photograph of USOCA. (B) Steps of the preparation procedures for S1 and S2. (C) Steps in the experimental procedure. Reprinted from reference [37].

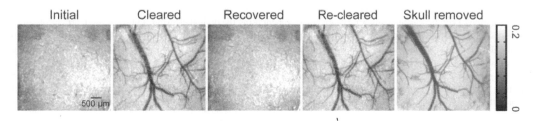

FIGURE 18.12 Typical white-field cortical images under various conditions. Reprinted from Ref [37].

changes after middle cerebral artery occlusion (MCAO) surgery [52]. With the establishment of the large optical clearing skull window, nearly all the cerebral blood vessels in bi-hemispheres were visible, and SO$_2$ and CBF of both hemispheres were obtained. When MCAO surgery was performed on mice, the cortical hemodynamic changes induced by this acute ischemia was clearly monitored (Figure 18.15a).

Over time, both CBF and SO$_2$ at the surgery-operated side decreased radically. As shown in Figure 18.15b, CBF of the middle cerebral artery (MCA) on the operated side decreased ~60% of its initial value 10 min after MCAO surgery, and then dropped to ~0 at 30 min after MCAO. In addition, CBF of the veins on the operated side decreased ~40% 10 min after surgery, and then decreased by ~60% of the initial value 30 min after surgery. The changes in SO$_2$ showed almost the same pattern as CBF. At the opposite site, CBF increased dramatically. In arteries, CBF increased to almost three times of its initial value, and in veins, it increased by ~50%. However, the SO$_2$ of both arteries and veins in the off-surgery site was observed to increase only slightly (Figure 18.15c).

USOCA for visualization of cortical vascular leakage

The blood–brain barrier (BBB) plays a key role in the health of the central nervous system. It is necessary to *in vivo* monitor the BBB permeability for assessing drug release with high resolution. Feng et al. [38] combined spectral imaging with an USOCA-based optical clearing skull window to *in vivo* dynamically monitor BBB opening caused by 5-aminolevulinic acid (5-ALA)-mediated photodynamic therapy (PDT), in which the Evans blue dye (EBd) acted as an indicator of the BBB permeability.

In their study, a spectral imaging system was used to obtain the reflectance spectrum of EBd and calculate the corresponding absorption spectrum of EBd at various concentrations. Thus, the absorption of EBd at 620 nm was used to calibrate its concentration. Thus, with the spectral imaging system, the distribution of EBd both inside and outside microvasculature could be observed with accurate concentration during the PDT induced cerebrovascular leakage process (Figures 18.16 and 18.17).

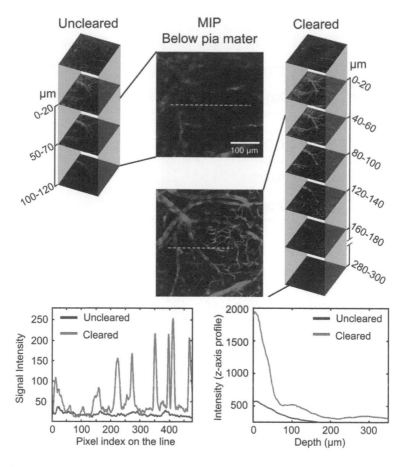

FIGURE 18.13 Evaluation of cerebrovascular imaging depth enhancement using the optical clearing skull window. Cortical vessels (red) and second harmonic generation (SHG) signals of skull (navy blue) at particular imaging depths were detected using TPLSM before and after clearing. Maximum intensity projection (MIP) maps of cortical vessels below pia mater were obtained in the same area before and after clearing. Images (1024 × 1024 pixels, 0.78 μm/pixel) show maximum projections across 0–120 μm (uncleared) and 0–300 μm (cleared) below the pial surface; the z-step was 2 μm, the dwell time was 6.1 μs/pixel and the average power increased linearly from the top to the bottom (20–150 mW, excitation: 880 nm, detection: 595 ± 25 nm, objective: 16 × W 0.8 NA, zoom factor: 1). At the bottom, the left-hand plot is the signal intensity along the dashed lines in the MIP images. The right-hand plot is the signal intensity (z-axis profile) as a function of imaging depth before and after skull clearing. Reprinted from reference [37].

Safety evaluation of USOCA

Like safety evaluation of Zhao et al. to SOCW [36], Zhang et al. [37] also measured the performance of microglia with and without USOCA treatment. The size, shape, and location of somata remained stable during 60 min of observation. Two days after skull optical clearing, microglia in the superficial layers had neither assumed amoeboid shapes nor extension toward the pial surface. Besides, there was no obvious difference in GFAP expression in astrocytes at the optical clearing skull window 10 days after skull optical clearing.

Furthermore, in addition, the body weights of the mice had not changed dramatically 28 days after a 3-day optical clearing treatment, and there was no significant difference between the control and experimental groups with respect to organ-to-body weight ratio, and the hematoxylin and eosin (H&E) staining results demonstrated that there are no obvious differences between liver (Figure 18.18a) and kidney (Figure 18.18b) slices of experimental and control groups, and no cellular aggregation or infiltration phenomena were present.

So far, *in vivo* tissue optical clearing technique provides a skull window for cortical optical imaging with a lot of advantages, including safety, non/minimized invasiveness, high resolution, switchability, and capability for long-term observation. However, the optical clearing methods discussed above still suffer from limited efficiency in reducing the scattering of skull, and thus deep-cortical imaging has not been realized.

Combination of SOCW and USCOA: Syncretic *in vivo* skull optical clearing method for deep-cortical imaging

SOCW allows two-photon imaging to achieve synaptic resolution, but for adult mice, it is still necessary to thin the skull to 100 μm before topical application of OCA. USOCA provides a large, switchable optical clearing skull window by simply treating it with two kinds of solution, but the two-photon imaging depth is still limited to about 300 μm, which is much less than that through an open-skull glass window. Is it possible to

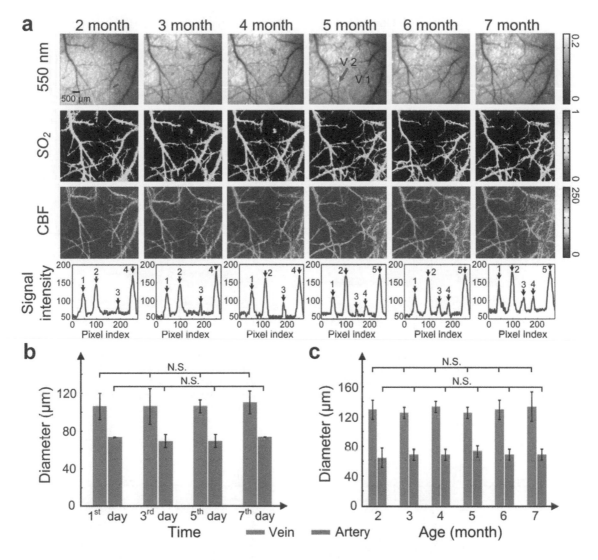

FIGURE 18.14 (a) Optical clearing skull window was established and cleared once a month over 6 months on the same mouse to image cortical vessels. Red and blue arrows indicate newly formed vessels and vessels that "disappeared," respectively. The color bar in the CBF images represents the relative blood flow velocity. The color bar in the SO₂ images represents the oxyhemoglobin saturation. (b) Diameter of vein and artery in short-term clearing. N.S.: no significant difference. (c) Diameter of vein (blue short line in (a), top left panel) and artery (red short line in (a), top left panel) in long-term clearing. Reprinted from reference [37].

combine SOCW and USOCA to further improve the imaging quality?

Chen et al. [53] proved the feasibility of combining the two methods both theoretically and experimentally. In their study, they unraveled the molecular process underlying two skull optical clearing methods by label-free hyperspectral stimulated Raman scattering (SRS) microscopy, thereby discovering the optimal clearing strategy to turn a turbid skull into a transparent skull window. Using SRS, one could directly 3D-visualize the effects of different OCAs on different components of the skull.

The main components of SOCW are collagenase and EDTA. After treatment of 10% collagenase on the surface of the skull for 5–10 min, the collagen fibers on the surface were depolymerized effectively (Figure 19.19a–b). However, the collagenase was not able to degrade hydroxyapatite and collagens embedded in hydroxyapatite. On the contrary, a large number of cavities formed around the bone lacunae only after 5–10 min of EDTA immersion (Figure 18.19a, c), and the thickness of the hydroxyapatite in the skull was significantly reduced from the original position of 15 to 40 μm. This implies that EDTA can not only disintegrate hydroxyapatite but also homogenizes the proteins and realizes great skull transparency. Meanwhile, the collagen layers maintained their original thickness and thus actually preserved the integrity of the skull.

The main components of USOCA are ethanol solution of urea (S1) and NaOH solution of dodecylbenzenesulfonate (S2). After treated with USOCA for 15 min, the proteins in both the surface and deep zones of the skull were depolymerized and redistributed homogeneously (Figure 18.19d–e). The results also demonstrated that USOCA has a greater ability to dissolve the proteins that does collagenase.

Since the USOCA and EDTA of SOCW make skull transparent via different mechanisms, it is possible to combine

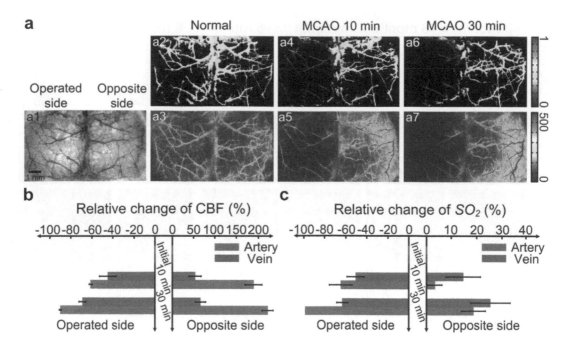

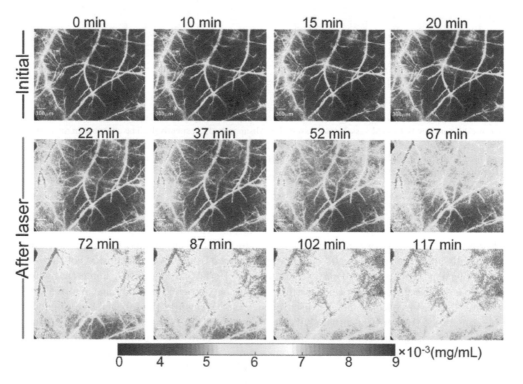

FIGURE 18.15 Establishment of a large optical clearing skull window over bi-hemispheres and monitoring the cortical hemodynamic changes induced by MCAO. (a, a1) White-light image of bi-hemispheric cortical vessels; rectangles are areas with arteries (red rectangles) and veins (blue rectangles) chosen for analysis. (a2–a3) SO_2 and CBF maps, respectively, of the bi-hemispheres in the normal state. (a4–a5) SO_2 and CBF maps, respectively, 10 min after MCAO. (a6–a7) SO2 and CBF maps, respectively, 30 min after MCAO. (b–c) Bar graphs showing relative changes in SO_2 and CBF. The color bar in the first row in (a) represents the oxyhemoglobin saturation. The color bar in the second row in (a) represents the relative blood flow velocity. Reprinted from reference [37].

FIGURE 18.16 Typical Evans blue dye (EBd) concentration maps before and after the laser irradiation (initial and after laser) obtained using the spectral imaging method for evaluating the blood–brain barrier (BBB) permeability (color bar represents the EBd concentration). Reprinted from reference [38].

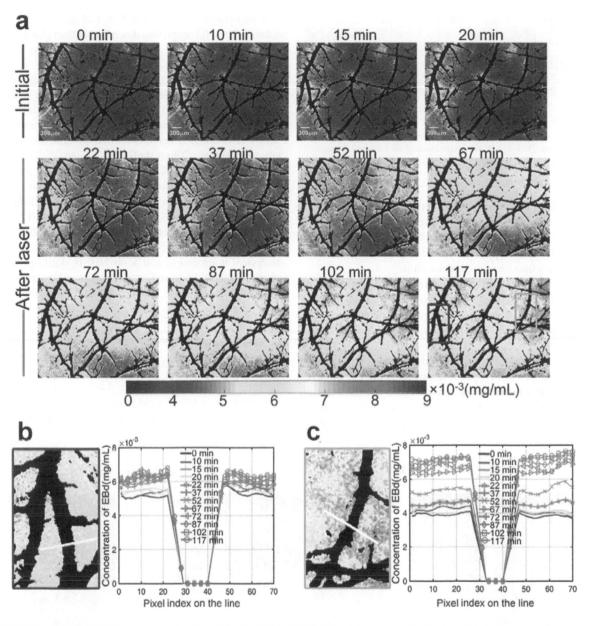

FIGURE 18.17 Distribution of the extravascular Evans blue dye (EBd). (A) Typical spatiotemporal distribution of the EBd maps at the extravascular positions (color bar represents the EBd concentration). (B and C) Magnified views of the blue and green boxes in (A) at 117 min and the profiles of the EBd concentration maps along the green lines of the magnified views at the corresponding time points. Reprinted from reference [38].

the OCAs of two methods (Figure 18.19f). Chen et al. [53] then treated the mouse skull with saline, USOCA, EDTA, and glycerol sequentially. After being topically treated with USOCA (25 min), 10% EDTA (25 min), and 80% glycerol *in vivo*, more cortical blood vessels and their detailed branches were visually observable. In addition, safety assessment of this method was performed. The result of H&E staining of a brain slice showed no difference between the untreated and treated regions. It indicates that this optical clearing technique does not affect the microenvironment. Finally, three-photon deep-tissue vascular microscopy was performed in the 2-month-old mouse through the optical clearing skull window established by the syncretic method, and Texas red-labeled blood vessels down to a depth of 850 μm were still visible (Figure 18.20).

USOCA together with SOCW allows three-photon microscopy to achieve deep tissue imaging in cortex. However, although safety is proved with *ex vivo* experiments, the skull becomes soft after treatment with all the OCAs, and thus the capability of the skull window to realize repeated imaging remains unclear.

VNSOCA: Skull optical clearing window for the region of visible to NIR-II light

The skull optical clearing techniques above mainly focus on decreasing the cranial scattering, which is indeed effective in the region of visible (Vis) light and near infrared (NIR)-I

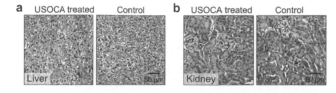

FIGURE 18.18 (a) H&E staining (liver) of the experimental and control groups 28 days after establishment of the USOCA-based window. (b) H&E staining (kidney) of the experimental and control groups 28 days after establishment of the USOCA-based window. Reprinted from reference [37].

light, since the main limitation of optical penetration here is scattering. However, as for the optical clearing in the NIR-II region, where the original scattering of skull is much less than it is at shorter wavelengths, the absorption of the skull window should be taken into serious consideration.

Li et al. [54] quantitatively measured the transmittance of USOCA in Vis-NIR-II region, and found it has nonnegligible absorption in NIR-II region due to the main component, water (Figure 18.21a–c), which limited its feasibility in NIR-II excited third harmonic generation (THG) imaging (Figure 18.21d–e). To overcome this obstacle, the authors developed a Vis-NIR-II compatible skull optical clearing agent (VNSOCA) based on USOCA by replacing water with D_2O. As shown in Figure 18.21, both S1 and S2 of VNSOCA demonstrate much higher transmittance in the NIR-II region that that of USOCA.

As a consequence, when VSOCA was used as an imaging medium, the THG signal was much higher than that of USOCA.

As shown in Figure 18.22, in the application of NIR-II excited deep-tissue vascular THG microscopy in cerebral cortex, the vis-NIR-II skull optical clearing window could remarkably increase the signal intensity as well as imaging depth without craniotomy. Compared with turbid skull, the transparent skull window promised an increased NIR-II excited THG imaging depth of more than three times, which is close even to that without skull [55].

In summary, Vis-NIR-II skull optical clearing agents were developed with D_2O instead of water, which could effectively make the skull transparent from visible to NIR-II region. Compared to previous USOCA, NIR-II excited THG signal intensity could be enhanced several times through each agent of VNSOCA. The combination of NIR-II fs excited nonlinear optical microscopy, and a vis-NIR-II skull optical clearing window could be used to observe vascular structures and activities in deep tissue, and holds great potential for deep-tissue observation and manipulation.

Summary and prospect

The recently developed *in vivo* skull optical clearing technique provides a safe optical window on mice without craniotomy [36, 37]. With the novel skull optical clearing window,

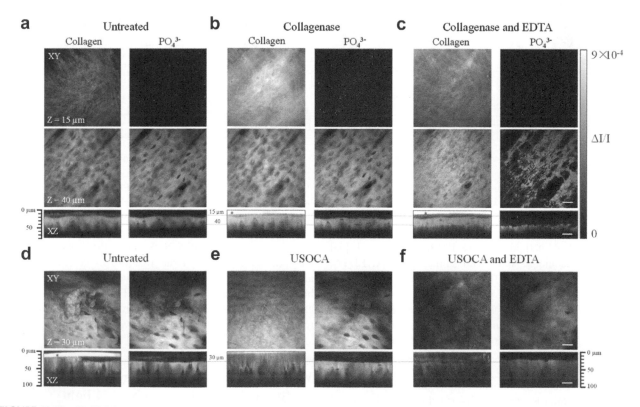

FIGURE 18.19 3D SRS images of skull optical clearing process. (a–c) Microstructural images of collagenous fibers and hydroxyapatite in a piece of parietal bone from an 18-day-old mouse. The skull was untreated (a), treated with 10% collagenase for 5–10 min (b), and treated with 10% EDTA for 5–10 min (c), successively. (d–f) Microstructural variations of collagenous fibers and hydroxyapatite in parietal bones from 2-month-old mice. The bone before treatment (d), treatment of USOCA for 15 min (e), and then treatment of 10% EDTA for 5–10 min on another bone (f). Reprinted from reference [53].

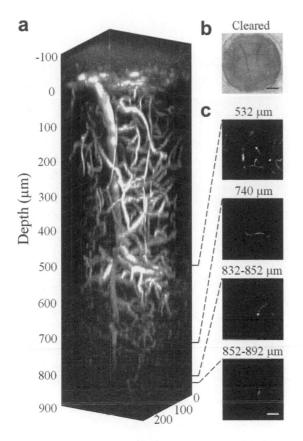

various optical imaging techniques, including LSCI, HSI, fluorescence imaging, and multiphoton microscopy, have been applied to observe cortical neuron, microglia, and vascular structures and functions [36, 37, 39–41]. Different skull optical clearing methods could be chosen for different conditions. Clear dental cement–based and acryl-based transparent skulls provide a long-term window for investigating neural signaling and brain development with relatively low-resolution requirements. Collagenase and EDTA-based SOCW are transparent enough for observing neuroplastic development at synaptic level in juvenile mice, but it is necessary to slightly thin the skull for adult mice for synaptic resolution. Urea and sodium dodecylbenzenesulfonate–based USOCA can produce a switchable skull window for at most 8-month-old mice, and can be applied to repeatedly monitor the same mice from 2 months old to 7 months old. For one-off deep cortical visualization, one can combine SOCW with USOCA. In addition, replacing water with D_2O in the clearing agents could enhance the feasibility of the NIR-II-based optical imaging technique.

Despite the achievements, there is still much room for improvement in *in vivo* skull optical clearing techniques. On one hand, it is necessary to develop an optical clearing skull window combined with the advantages of high transparency, high imaging resolution, long-term repeated observation, safety, and convenience. On the other hand, current methods focus on mouse skull. In the future, methods could be developed to make thicker skulls transparent. *In vivo* skull clearing methods for larger animals, such as rats, rabbits, and even primates, will provide more support for biomedical research.

FIGURE 18.20 Three-photon excitation imaging of mouse vasculature through the transparent skull window. (a) 3D reconstruction of the vasculature of the mouse brain with Texas Red-labeled (BALB/c mouse, 26 g, 8 weeks old; similar results for n = 3). (b) Photographed image of the mouse skull after optical clearing. (c) Three-photon excitation images at various depths. Scale bars, 1 mm (b) and 50 *μm* (c). Reprinted from reference [53].

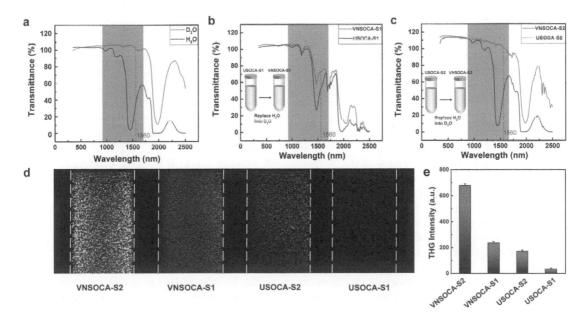

FIGURE 18.21 (a) Transmission spectra of D_2O (red line) and H_2O (blue line). (b) Transmission spectra of S1 of VNSOCA (red line) and USOCA (blue line). (c) Transmission spectra of S2 of VNSOCA (red line) and USOCA (blue line). (d) THG images of glass capillaries filled with DCCN nanocrystal dispersion with the 25× objective immersed in S1 and S2 of VNSOCA and USOCA respectively. The dashed lines represent the edges of the capillaries. Green represents the THG signals. (e) The THG intensities of DCCN nanocrystal dispersion filled in capillaries with the objective immersed in S1 and S2 of VNSOCA and USOCA respectively. Reprinted from reference [54].

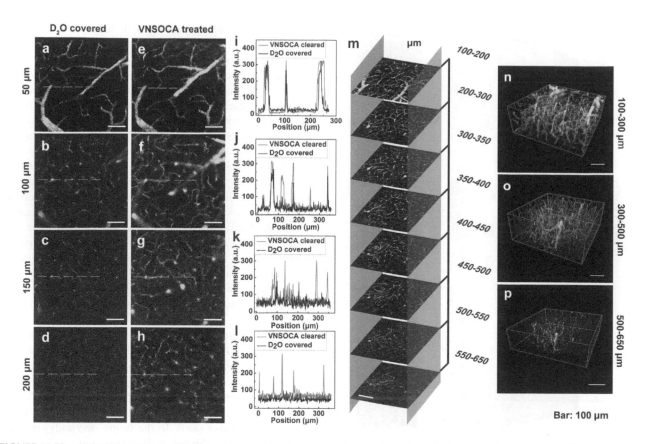

FIGURE 18.22 (a–d) THG scanning microscopy at different depth using the 25× objective without skull clearing. (e–h) THG scanning microscopy at different depth using the 25× objective with skull clearing. (i–l) Intensity profiles along the white dashed lines across the vasculature in (a–h), respectively. (m) THG imaging of cortical vasculature at particular imaging depths. (n–p) 3D reconstruction of vasculature in certain volumes. Reprinted from reference [54].

Acknowledgments

This chapter was supported by the National Science Foundation of China (Grant Nos. 61860206009, 81870934) and China Postdoctoral Science Foundation funded project (Grant Nos. BX20190131, 2019M662633).

REFERENCES

1. P. McGonigle, "Animal models of CNS disorders," *Biochem. Pharmacol.* **87**(1), 140–149 (2014).
2. M. Wiesmann, B. Zinnhardt, D. Reinhardt et al., "A specific dietary intervention to restore brain structure and function after ischemic stroke," *Theranostics* **7**(2), 493–512 (2017).
3. M. Draijer, E. Hondebrink, T. van Leeuwen, and W. Steenbergen, "Review of laser speckle contrast techniques for visualizing tissue perfusion," *Lasers Med. Sci.* **24**(4), 639–651 (2009).
4. A. Devor, S. Sakadzic, V.J. Srinivasan et al., "Frontiers in optical imaging of cerebral blood flow and metabolism," *J. Cereb. Blood Flow Metab.* **32**(7), 1259–1276 (2012).
5. J. Yao, L. Wang, J.M. Yang et al., "High-speed label-free functional photoacoustic microscopy of mouse brain in action," *Nat. Methods* **12**(5), 407–410 (2015).
6. C.W.T. Leung, Y.N. Hong, S.J. Chen et al., "A photostable AIE luminogen for specific mitochondrial imaging and tracking," *J. Am. Chem. Soc.* **135**(1), 62–65 (2013).
7. K.J. Zuzak, M.D. Schaeberle, E.N. Lewis, and I.W. Levin, "Visible reflectance hyperspectral imaging: characterization of a noninvasive, in vivo system for determining tissue perfusion," *Anal. Chem.* **74**(9), 2021–2028 (2002).
8. C. Yue, C. Zhang, G. Alfranca et al., "Near-infrared light triggered ROS-activated theranostic platform based on Ce6-CPT-UCNPs for simultaneous fluorescence imaging and chemo-photodynamic combined therapy," *Theranostics* **6**(4), 456–469 (2016).
9. S. Wang, J. Lin, T. Wang, X. Chen, and P. Huang, "Recent advances in photoacoustic imaging for deep-tissue biomedical applications," *Theranostics* **6**(13), 2394–2413 (2016).
10. P. Li, S. Ni, L. Zhang, S. Zeng, and Q. Luo, "Imaging cerebral blood flow through the intact rat skull with temporal laser speckle imaging," *Opt. Lett.* **31**(12), 1824–1826 (2006).
11. W.J. Jeffcoate, D.J. Clark, N. Savic, P.I. Rodmell, and F.L. Game, "Use of HSI to measure oxygen saturation in the lower limb and its correlation with healing of foot ulcers in diabetes," *Diabet. Med.* **32**(6), 798–802 (2015).
12. K. Radrich, and V. Ntziachristos, "Quantitative multi-spectral oxygen saturation measurements independent of tissue optical properties," *J. Biophoton.* **9**(1–2), 83–99 (2016).

13. D. Ravì, H. Fabelo, G.M. Callic, and G.Z. Yang, "Manifold embedding and semantic segmentation for intraoperative guidance with hyperspectral brain imaging," *IEEE Trans. Med. Imaging* **36**(9), 1845–1857.

14. L.C. Cancio, A.I. Batchinsky, J.R. Mansfield et al., "Hyperspectral imaging: A new approach to the diagnosis of hemorrhagic shock," *J. Trauma Acute Care Surg.* **60**(5), 1087–1095 (2006).

15. Y.X. Cheng, L. Guo, C. Pan et al., "Statistical analysis of motion contrast in optical coherence tomography angiography," *J. Biomed. Opt.* **20**(11), 116004 (2015).

16. C.L. Chen, and R.K. Wang, "Optical coherence tomography based angiography [Invited]," *Biomed. Opt. Express* **8**(2), 1056–1082 (2017).

17. H. Liu, X. Deng, S. Tong et al., "In vivo deep-brain structural and hemodynamic multiphoton microscopy enabled by quantum dots," *Nano Lett.* **19**(8), 5260–5265 (2019).

18. D. Kobat, N.G. Horton, and C. Xu, "In vivo two-photon microscopy to 1.6-mm depth in mouse cortex," *J Biomed Opt* **16**(10), 106014 (2011).

19. D. Kobat, M.E. Durst, N. Nishimura et al., "Deep tissue multiphoton microscopy using longer wavelength excitation," *Opt. Express* **17**(16), 13354–13364 (2009).

20. S. Wang, W. Xi, F. Cai et al., "Three-photon luminescence of gold nanorods and its applications for high contrast tissue and deep in vivo brain imaging," *Theranostics* **5**(3), 251–266 (2015).

21. H. Zhang, N. Alifu, T. Jiang et al., "Biocompatible aggregation-induced emission nanoparticles with red emission for in vivo three-photon brain vascular imaging," *Journal of Materials Chemistry B* **5**(15), 2757–2762 (2017).

22. Y. Wang, X. Han, W. Xi et al., "Bright AIE nanoparticles with F127 encapsulation for deep-tissue three-photon intravital brain angiography," *Advanced Healthcare Materials* **6**(21), 1700685 (2017).

23. N.G. Horton, K. Wang, D. Kobat et al., "In vivo three-photon microscopy of subcortical structures within an intact mouse brain," *Nat Photonics* **7**(3), 205–209 (2013).

24. T. Wang, D.G. Ouzounov, C. Wu et al., "Three-photon imaging of mouse brain structure and function through the intact skull," *Nat. Methods* **15**(10), 789–792 (2018).

25. R. Prevedel, A.J. Verhoef, A.J. Pernia-Andrade et al., "Fast volumetric calcium imaging across multiple cortical layers using sculpted light," *Nat. Methods* **13**(12), 1021–1028 (2016).

26. J.J. Iliff, M.H. Wang, Y.H. Liao et al., "A paravascular pathway facilitates CSF flow through the brain parenchyma and the clearance of interstitial solutes, including amyloid beta," *Sci. Transl. Med.* **4**(147), 147ra111–147ra111 (2012).

27. M.A. Busche, G. Eichhoff, H. Adelsberger et al., "Clusters of hyperactive neurons near amyloid plaques in a mouse model of Alzheimer's disease," *Science* **321**(5896), 1686–1689 (2008).

28. J. Kain, C. Stokes, Q. Gaudry et al., "Leg-tracking and automated behavioural classification in Drosophila," *Nat. Commun.* **4**, 1910 (2013).

29. A. Holtmaat, T. Bonhoeffer, D.K. Chow et al., "Long-term, high-resolution imaging in the mouse neocortex through a chronic cranial window," *Nat. Protoc.* **4**(8), 1128–1144 (2009).

30. J.E. Levasseur, E.P. Wei, A.J. Raper, H.A. Kontos, and J.L. Patterson, "Detailed description of a cranial window technique for acute and chronic experiments," *Stroke* **6**(3), 308–317 (1975).

31. G. Yang, F. Pan, C.N. Parkhurst, J. Grutzendler, and W.B. Gan, "Thinned-skull cranial window technique for long-term imaging of the cortex in live mice," *Nat. Protoc.* **5**(2), 201–208 (2010).

32. X.Z. Yu, and Y. Zuo, "Two-photon in vivo imaging of dendritic spines in the mouse cortex using a thinned-skull preparation," *J. Visualized Exp.* (87), e51520 (2014).

33. G.J. Goldey, D.K. Roumis, L.L. Glickfeld et al., "Removable cranial windows for long-term imaging in awake mice," *Nat. Protoc.* **9**(11), 2515–2538 (2014).

34. T.H. Kim, Y. Zhang, J. Lecoq et al., "Long-term optical access to an estimated one million neurons in the live mouse cortex," *Cell Rep.* **17**(12), 3385–3394 (2016).

35. H.T. Xu, F. Pan, G. Yang, and W.B. Gan, "Choice of cranial window type for in vivo imaging affects dendritic spine turnover in the cortex," *Nat. Neurosci.* **10**(5), 549–551 (2007).

36. Y.J. Zhao, T.T. Yu, C. Zhang et al., "Skull optical clearing window for in vivo imaging of the mouse cortex at synaptic resolution," *Light Sci. Appl.* **7**(2), 17153 (2018).

37. C. Zhang, W. Feng, Y.J. Zhao et al., "A large, switchable optical clearing skull window for cerebrovascular imaging," *Theranostics* **8**(10), 2696–2708 (2018).

38. W. Feng, C. Zhang, T.T. Yu, O. Semyachkina-Glushkovskaya, and D. Zhu, "In vivo monitoring blood-brain barrier permeability using spectral imaging through optical clearing skull window," *J. Biophoton.* **12**(4), e201800330 (2019).

39. C. Zhang, W. Feng, E. Vodovozova et al., "Photodynamic opening of the blood-brain barrier to high weight molecules and liposomes through an optical clearing skull window," *Biomed. Opt. Express* **9**(10), 4850–4862 (2018).

40. C. Zhang, W. Feng, Y. Li et al., "Age differences in photodynamic therapy-mediated opening of the blood-brain barrier through the optical clearing skull window in mice," *Lasers Surg. Med.* **51**(7), 625–633 (2019).

41. W. Feng, S. Liu, C. Zhang et al., "Comparison of cerebral and cutaneous microvascular dysfunction with the development of type 1 diabetes," *Theranostics* **9**(20), 5854–5868 (2019).

42. J. Wang, Y. Zhang, T. Xu, Q. Luo, and D. Zhu, "An innovative transparent cranial window based on skull optical clearing," *Laser Phys. Lett.* **9**(6), 469–473 (2012).

43. G. Silasi, D.S. Xiao, M.P. Vanni, A.C.N. Chen, and T.H. Murphy, "Intact skull chronic windows for mesoscopic wide-field imaging in awake mice," *J. Neurosci. Methods* **267**, 141–149 (2016).

44. A. Steinzeig, D. Molotkov, and E. Castren, "Chronic imaging through 'transparent skull' in mice," *PLoS One* **12**(8), e0181788 (2017).

45. V.V. Tuchin, I.L. Maksimova, D.A. Zimnyakov et al., "Light propagation in tissues with controlled optical properties," *J. Biomed. Opt.* **2**(4), 401–417 (1997).

46. D. Zhu, K.V. Larin, Q. Luo, and V.V. Tuchin, "Recent progress in tissue optical clearing," *Laser Photonics Rev.* **7**(5), 732–757 (2013).

47. Y. Sato, T. Miyawaki, A. Ouchi et al., "Quick visualization of neurons in brain tissues using an optical clearing technique," *Anat. Sci. Int.* **94**(2), 199–208 (2019).

48. A. Erturk, C.P. Mauch, F. Hellal et al., "Three-dimensional imaging of the unsectioned adult spinal cord to assess axon regeneration and glial responses after injury," *Nat. Med.* **18**(1), 166–171 (2012).

49. E.A. Genina, A.N. Bashkatov, and V.V. Tuchin, "Optical clearing of cranial bone," *Adv. Opt. Technol.* **2008**, 267867 (2008).

50. X.Q. Yang, Y. Zhang, K. Zhao et al., "Skull optical clearing solution for enhancing ultrasonic and photoacoustic imaging," *IEEE Trans. Med. Imaging* **35**(8), 1903–1906 (2016).

51. I.M. Wittko-Schneider, F.T. Schneider, and K.H. Plate, "Cerebral angiogenesis during development: Who is conducting the orchestra?" in *Cerebral Angiogenesis*, pp. 3–20, Springer (2014).

52. O. Engel, S. Kolodziej, U. Dirnagl, and V. Prinz, "Modeling stroke in mice-middle cerebral artery occlusion with the filament model," *J. Visualized Exp.* 47, e2423 (2011).

53. Y. Chen, S. Liu, H. Liu et al., "Coherent Raman scattering unravelling mechanisms underlying skull optical clearing for through-skull brain imaging," *Anal. Chem.* **91**(15), 9371–9375 (2019).

54. D.Y. Li, Z. Zheng, T.T. Yu et al., "Vis-NIR-II skull optical clearing window for in vivo cortical vasculature imaging and targeted manipulation," *J. Biomed. Opt.*, 202000142 (2020).

55. Z. Zheng, D. Li, Z. Liu et al., "Aggregation-induced nonlinear optical effects of AIEgen nanocrystals for ultradeep in vivo bioimaging," *Advanced Materials* **31**(34), 1904799 (2019).

19

In vivo *skin optical clearing in humans*

Elina A. Genina, Alexey N. Bashkatov, Vladimir P. Zharov, and Valery V. Tuchin

CONTENTS

Introduction

During the last 30 years, interest in the development and application of optical methods for clinical functional imaging of physiological conditions and neoplasms hidden under tissue layers, aimed at diagnostics and therapy of cancer and other diseases, has been continually growing. It is initiated by the unique informativity, relative simplicity, safety, and sufficiently low cost of optical instruments. However, the main limitation of optical diagnostic methods, including optical coherence tomography (OCT), spectroscopic methods, laser speckle contrast imaging, photoacoustics, etc., is the strong scattering of light in tissues and blood that significantly reduces the light penetration depth and, therefore, decreases contrast and spatial resolution [1].

The difficulty of increasing the informativeness and accuracy of the optical methods of diagnostics and the effectiveness of laser therapy can be successfully overcome through the use of tissue optical clearing techniques with biocompatible optical clearing agents (OCAs) and methods for their incorporation into biological tissues [2–6].

One of the important problems in optical diagnostics and treatment is the reduction of skin scattering, aimed at imaging the inhomogeneities hidden inside or under the skin or blood microvessels. However, it is rather difficult to obtain a satisfactory optical clearing of skin using noninvasive or minimally invasive means, since the skin epidermis is a natural barrier that impedes OCA penetration into the dermis [7].

Beneficial results in skin optical clearing were obtained for many optical imaging techniques; however, most investigations were carried out for *in vitro* or *ex vivo* samples or for animal models and were not realized for humans *in vivo*.

In this chapter, we briefly review the structure and optical properties of human skin, discuss features and possibilities of *in vivo* skin optical clearing in humans using chemical and physical enhancers of epidermal permeability, and examine applications of the approaches used to some optical diagnostic methods, such as OCT, reflectance and transmittance spectroscopy, fluorescence spectroscopy, Raman spectroscopy and microscopy, laser speckle contrast imaging, hyperspectral imaging, and photoacoustic imaging.

Skin structure and optical properties

Skin is a complex heterogeneous tissue in which the content of absorbers and scatterers is distributed differently depending on the depth [8, 9]. In humans, the thicknesses of the main skin layers are as follows: i) the epidermis is the outer layer with a thickness of 100–150 μm, bloodless, being a stratified squamous keratinizing epithelium; ii) the dermis is a dense fiber-elastic tissue 1–4 mm thick, containing a vascular network; and iii) the hypodermis is subcutaneous fatty tissue with a thickness of 1–6 mm, depending on the area of the body. The randomly inhomogeneous distribution of water and various chromophores and pigments in the skin produces the variation in the average optical parameters of its layers. However, there are areas in the skin where the gradient of changes in the structure and the number of chromophores in depth is approximately zero [10]. This allows the main layers of the skin to be subdivided into sublayers, in accordance with the physiological nature and physical and optical properties of the skin components and the content of pigments.

The epidermis can be subdivided into the two sublayers: nonliving and living epidermis, which in turn can be subdivided into five sublayers, differing in the number of cell rows and the shape of cells: i) superficial *stratum corneum* (SC); ii) *stratum lucidum*; iii) *stratum granulosum*; iv) *stratum spinosum*; and v) *stratum basale*, which is the "germ" layer of the epidermis. The predominant cells are the keratinocytes.

The nonliving epidermis or the *stratum corneum* is approximately 10–20 μm thick and consists of dead flat cells (corneocytes) approximately 40 μm in diameter and 0.5 μm thick [7]. The membranes of corneocytes are strongly keratinized, cell organelles and cytoplasm disappear during keratinization, and a densely packed layer is formed, with a high content of lipids (~20%) and proteins (~60%) and a relatively low content of water (~20%) [8–11]. Lipids fill the intercellular space. The tile-like packing of cells and lipid–protein intercellular matrix of the *stratum corneum* provides a natural skin barrier for the penetration of exogenous substances into the body [7].

The thickness of the living epidermis is 50–120 μm [7, 12]. This layer contains most of the skin pigments, mainly melanin. Melanin is formed in melanocytes occurring in the *stratum basale*, which contain a large number of structural organelles called melanosomes, which in turn are granules filled with pigment [13]. When the granules mature, they leave the melanocytes and invade keratinocytes, where melanin accumulates [13–15]. There are two types of this pigment: the red/yellow phaeomelanin and a brown/black eumelanin [13]. The ratio between the concentration of phaeomelanin and eumelanin present in human skin varies from individual to individual, with much overlap between skin types. In accordance with the Fitzpatrick Skin Type Classification Scale [16], the skin subdivides into the six phototypes (from pale white to deeply pigmented dark brown or black skin). The ratio between the concentrations of pheomelanin and eumelanin found in human skin can range from 0.049 to 0.36 [17]. The melanin absorption level depends on how many melanosomes per unit volume are in the epidermis. The volume fraction of the epidermis occupied by melanosomes varies from 1.3% (type I) to 43% (type VI) [18]. Analysis of melanosome size revealed a significant and progressive variation in size with ethnicity, African skin having the largest melanosomes followed in turn by Indian, Mexican, Chinese, and European [19]. In addition to keratinocytes and melanocytes, Langerhans cells, Merkel cells, etc. are found in the epidermis.

The water content in the epidermis is unevenly distributed and sharply increases during the transition from the *stratum corneum* to the living epidermis, reaching ~50% in the *stratum granulosum* and 65%–70% in the area of the *stratum spinosum* and basal lamina [7]. A basal lamina separates the epidermis from the dermis.

The dermis is the main bulk of the skin and without sharp boundaries passes into the subcutaneous fatty tissue. The dermis is arbitrarily divided into two anatomical regions. The thinner of these, the so-called papillary dermis, is the outer part of the connective tissue of the dermis that forms under the epidermis. The papillary dermis contains the lymphatic plexuses and blood vessels. Its thickness is about 100 μm. The reticular dermis constitutes its main part (approximately 1–2 mm) and lies under the papillary dermis. It is characterized as a relatively acellular, fairly dense connective tissue [8].

Dermis is a vascularized tissue. Blood vessels are located in the dermis in the form of a superficial vascular plexus and a deep venous plexus (on the border with the subcutaneous tissue) [8].

The main absorbers in the visible spectral range are the blood hemoglobin, β-carotene (0.22–0.63 nmol/g [20]), and bilirubin. In the IR spectral range, absorption properties of skin dermis are defined by absorption of water. The scattering properties of the dermis are mainly defined by the fibrous structure of the tissue, where collagen fibrils are packed in collagen bundles and have lamellae structure. The light scatters on both single fibrils and scattering centers, which are formed of by the interlacement the collagen fibrils and bundles. Papillary and reticular dermis differs by the size of collagen fibers and blood content. The small size of the collagen fibers in the papillary dermis (diameter less than visible light wavelength) makes this layer highly backscattering. Within the reticular dermis, the large size of collagen fiber bundles causes highly forward-directed scattering. Thus, any light that reaches this layer is passed on deeper into the skin and contributes to some extent to the spectrum reflected from the skin [21]. The blood volume fraction in the skin varies from 0.2% [22] to 4% [23]. The volume fraction of water in the dermis is estimated as 65%–75% [24–27].

The hypodermis is formed by the aggregation of adipocytes (fat cells) containing stored lipids presented by triglycerides in the form of a number of small droplets or a single big drop in each cell [28, 29]. The mean diameter varies from 50 μm [28] to 120 μm [29]. In the spaces between the cells, there are blood capillaries (arterial and venous plexus), nerves, and reticular fibrils connecting each cell and providing metabolic activity of fat tissue [28, 29]. Absorption of the human adipose tissue is defined by absorption of hemoglobin, lipids, and water (~10.9 ± 1.4%) [30]. The main scatterers of adipose tissue are spherical droplets of lipids, which are uniformly distributed within adipocytes.

Knowledge of optical properties of skin is of great importance for interpretation and quantification of the diagnostic data, and for prediction of light distribution and absorbed energy for therapeutic and surgical use [12, 31–38]. The difficulties involved in determining the optical properties of tissue *in vivo* are well known: advanced instrumentation is required (time or phase resolved spectroscopy) and the volume of tissue probed is hard to define [32]. However, it has been shown that the differential pathlength of light in tissue measured *post-mortem* is close to that measured *in vivo* [39]. Tissue optical properties are usually described in terms of the following parameters: the absorption coefficient μ_a, scattering coefficient μ_s, scattering anisotropy factor g, and transport scattering coefficient ($\mu'_s = \mu_s(1-g)$). Optical measurements of tissue, such as reflectance, are the result of the combined effect of these properties.

The absorption and transport scattering coefficients of Caucasian and Negroid skin (epidermis and dermis), and subdermal fat at the selected wavelengths in the range between 620 and 1000 nm are presented in Figure 19.1 (a). The absorption coefficient for Negroid skin is an order of magnitude greater than Caucasian at visible wavelengths, reducing steadily with increasing wavelength. The scattering spectra show no significant differences between the Negroid and Caucasian skin types [32].

Figure 19.1(b) shows the five experimental reflectance spectra for two lightly (type II) and three darkly pigmented (type VI) skin types. It is clear that in the spectral range shorter than ~900 nm, the reflectance is defined mostly by blood and melanin content, and at longer than 900 nm the main role is the water content.

Optical coherence tomography (OCT) allows one to evaluate the attenuation coefficient ($\mu_t = \mu_a + \mu_s$) of skin *in vivo* within the limits of an OCT penetration depth of 0.2–3 mm depending on the probing wavelength. Melanin granules are the major backreflecting particles in OCT. Figure 19.1 (c) demonstrates OCT images of II and VI skin phototypes. It is clear that epidermis with high melanin concentration looks brighter. Increasing brightness of the basal and suprabasal layers with increasing skin phototypes is confirmed with laser scanning confocal microscopy [40]. Combination of absorption and a local variation of the refractive index in epidermis caused by melanin influences local variations of a backscattering capability of the epidermis that leads to significant deviations in OCT A-scans of the volunteers with dark skin. For the volunteers with light skin, such deviations were not observed [41].

Refractive indices of skin and its components is an important focus of interest in tissue optics because the index of refraction determines light reflection and refraction at the interfaces between air and skin, detecting fiber and skin, and skin layers; it also strongly influences light propagation and distribution within skin [35]. For different components of skin, values of refractive index in the visible and NIR range can be estimated as follows: SC of human skin, 1.5–1.55 [42]; epidermis of human skin, 1.34–1.39 [42, 43]; dermis of human skin, 1.4–1.41 [43]; human subcutaneous fat, 1.44–1.455 [44, 45]; blood, 1.4 [45]; melanin, 1.6–1.7 [1, 35]; collagen fibers, 1.474 [1].

Mechanisms of skin optical clearing

For living unfixed tissues, the optical clearing phenomenon is based mostly on i) matching of refractive indices of tissue components, ii) dehydration of tissue, and iii) dissociation of its collagen, which breaks down collagen fibers into thinner ones when tissue is immersed into a highly concentrated OCA [2–4, 46]. For fibrous tissues, such as dermis, both processes – namely the water loss and the diffusion of the hyperosmotic agent into tissue – occur simultaneously, but the clearing

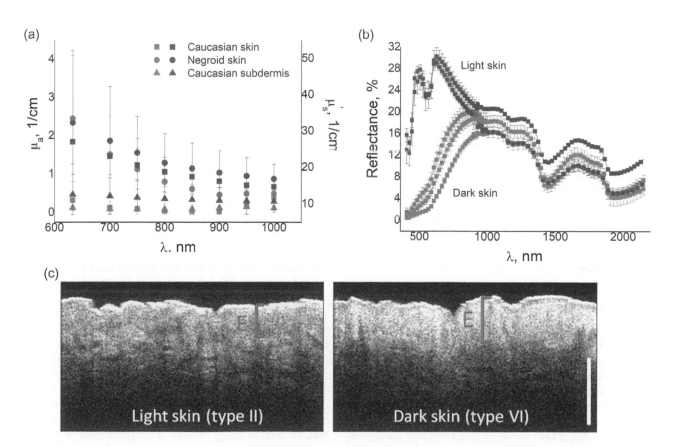

FIGURE 19.1 a) Absorption (red) and transport scattering (blue) coefficients of Caucasian and Negroid skin (epidermis and dermis), and Caucasian subdermal tissue. Data adapted from reference [32]; b) backr eflectance spectra of light (blue) and dark (red) skin; c) OCT images of the light and dark skin. E with a bracket marks epidermal layer. Bar corresponds to 300 μm.

effect of the interstitial fluid replacement with the OCA solution prevails over all other mechanisms. It was shown, for example in references [47–49], that in total, the transparency of tissue samples *ex vivo* stimulated with evaporation of water increased. But this effect could be caused predominantly by thinning of the samples. Experiment with skin samples *ex vivo* showed that simultaneously with the increase in collimated transmission during a sample drying, a decrease in sample thickness and an increase in the light attenuation coefficient were observed [50]. This fact can probably be explained by an increase in the concentration of tissue scatterers per unit volume during dehydration [47], which led to an increase in the value of the light attenuation coefficient. In contrast to natural drying in air, at immersion clearing, a part of the OCA penetrates into the interstitial space and mixes with the interstitial fluid. Thus, a partial matching effect between refractive indices of tissue scatterers and interstitial fluid occurs.

The OCA-induced destabilization of the collagen structure can lead to the additional reduction of optical scattering in tissue due to the decrease in the basic scatterer size [51–54]. For skin *ex vivo* and *in vitro*, reduction of scattering coefficient can be extremely high [53, 55–58].

General principles of optical clearing mechanisms for non-living and living tissues are the same; however, there are significant differences. In *in vivo* studies, the need to significantly reduce the time of the treatment (not more than dozens of minutes) and use only biologically compatible agents is also complicated by the need to account for a physiological response of living tissue to the action of the agent and its redistribution (washing out) with a possible quick exit from the area of interest.

Besides, in living tissues, the refractive index is a function of the physiological or pathological condition of the tissue. Depending on the specific features of the tissue condition, the refractive index of scatterers and/or the base substance can change (increase or decrease), and correspondingly the light scattering will change [1]. In addition, the introduction of some OCAs, in particular, glycerol and glucose, into a tissue affects the condition of microcirculation in the tissue, causing a transient stasis of the microvessels [58–64]. In the case of transepidermal delivery of hydrophilic OCAs into the skin, the diffusion is inhibited by the existence of the epidermis lipid barrier.

Transepidermal delivery: Physical and chemical enhancers

Significant optical clearing effect in skin *in vivo* was reported in Refs. [65, 66]. The human skin reflectance was reduced up to about 4-fold within 1 hour of treatment under action of 40% glucose solution [65]. Decreasing in reduced scattering coefficient of human skin by 3.8-fold was achieved during 50% propylene glycol action [66]. However, these treatments were invasive as the OCAs were injected under the skin. The authors noted that at low concentrations, OCAs did not provide a sufficiently effective optical clearing, and at high concentrations it caused edema, chemical burns, partial necrosis, and scarring [60, 65].

To provide noninvasiveness, a topical OCA application to skin is needed; however, the SC is a major barrier for penetration of any chemicals including OCA (especially hydrophilic ones) into skin.

Such biocompatible agents as ethanol, DMSO, propylene glycol, azone and thiazone, liquid paraffin, and fatty acids were used as penetration enhancers of OCA diffusivity through the SC *in vivo* [67–82]. Ethanol has the ability to modify the skin barrier property as well as enhance pore transport [67, 83, 84]. Also, application of ethanol leads to the delipidation of skin surfaces, which facilitates OCA penetration.

Also, it was found that the permeability enhancing mechanism for DMSO acts in two ways. First, DMSO can induce the transition of ceramide bilayers from the gel phase to the liquid crystalline phase. Second, DMSO can weaken the lipid bilayers in the SC that consist of a high amount of ceramides [85]. The use of a mixture of DMSO and ethanol as a solvent improves skin permeability to a greater extent than more concentrated solutions of each of these agents [86].

In reference [87] it was shown that mixtures of OCAs with propylene glycol can enhance the effectiveness of optical clearing, while the clearing effect induced by propylene glycol itself can be lower than that induced by the other OCAs. The mechanism of the effect of enhancing the permeability of the SC under the action of propylene glycol is the solvation of keratin upon replacement of water in the binding hydrogen groups and the inclusion of propylene glycol in the polar heads of the lipid bilayer [68].

Thiazone is an innovative agent; it has a ~3-fold stronger effect in terms of increasing skin permeability than azone [3]. It is also more effective than propylene glycol [3, 87].

Fatty acids affect the skin by disrupting the organization of SC lipids, which increases their fluidity [88, 89]. Liquid paraffin is a kind of highly refined mineral oil used in cosmetics and pharmaceutical excipients to retain water in skin and as a solvent [75, 90].

Besides chemical enhancers, physical methods have also been used for advanced optical clearing. Large numbers of methods such as SC removal [91–94], low intensive and high intensive laser irradiation of skin surface [95, 96], photothermal and mechanical microperforation [97–99], laser ablation [100, 101], or fractional microablation [102] of SC, iontophoresis [103], and sonophoresis [41, 104] were suggested for reduction of the skin barrier function.

Partial removal of SC provides enhancement of penetration of OCAs through epidermis and can be made using tape stripping [91], medical grade cyanoacrylate adhesive [92], rubbing with sandpaper [93], or microdermabrasion [94], which are minimally invasive procedures. In Reference [41] it was shown that removing the SC layer about 8 ± 4 μm increased OCT-signal amplitude by about 10%–15% at depths of 400–800 μm.

Measurements of reflection spectra before and after irradiation demonstrated that Q-switched and long-pulse Nd:YAG laser radiation induced an 8–9-fold improvement in transepidermal penetration of OCA compared to nonirradiated skin [96]. In another study, Stumpp et al. [100] showed that laser radiation at the wavelength of 980 nm provided heating of the SC and caused a disruption in the function of the protective barrier.

The integrity of the SC of the epidermis was violated using a flash-lamp [98] or mechanical device – a microneedle roller [99]. These devices were used to create transdermal microchannels in skin samples to improve the penetration of OCAs into the skin. Also microchannels with different shapes and depths can be created using laser fractional microablation [102]. However, optical or mechanical injury of epidermis *in vivo* can impair the optical clearing of the skin. In Refs. [101, 102] it was shown that surface or fractional ablation of skin led to local edema of the affected region, which increased the scattering coefficient.

Electrophoresis can be applied not only with polar electrically charged drugs. The dependence of electroosmotic diffusion on the pH of the solution and the content of NaCl ions in it was studied in [103]. For the study, a solution containing 0.07M NaCl and 0.13M D-mannitol in 5 mM citric acid buffer with pH = 6 was used. Compared with passive diffusion, after switching on the current, the volumetric flux of the agent gradually increased by approximately 20 times, reaching a constant value.

The use of ultrasound (US) with a frequency of 1–1.5 MHz (sonophoresis) as a physical enhancer allows deeper penetration of OCAs [105] into the skin. This mode of US induces cavitation effects, which consist in the formation of oscillating microbubbles in skin and results in pushing substances deeper into the skin because of pressure difference, the disturbance of SC cell ordering, and heating of the tissue [105–107]. For example, a 60% glycerol solution and a 60% PEG-400 solution under the action of US demonstrated a 1.5-fold and 1.6-fold excess of the efficiency of skin optical clearing, respectively, compared with the action of only glycerol and PEG-400 [108]. Besides, it was shown that US alone can increase light-depth penetration for OCT up to ~1.5-fold during 2 min [41]. Apparently, local cavitation of microbubbles in the skin leads to a change in the ordering of the structure of collagen fibers in the dermis and, consequently, to a decrease in the volume fraction of scatterers [105].

Significant results in optical clearing were achieved when multimodal treatments were used, for example, chemical and physical penetration enhancers [109–112] or combinations of several physical treatments [41, 92, 102, 113–115].

Examples of human skin optical clearing *in vivo* applications

Optical coherence tomography

Optical clearing allows one to improve imaging contrast and increase optical probing depth of OCT. There are many examples of the use of the optical clearing technique for *in vivo* human skin OCT study. In Reference [116], the authors investigated the effect of glycerol and ultrasound gel on optical clearing of the skin. For both agents, an increase in the optical probing depth from 1 to 1.5 mm was observed. Both agents helped to reduce the level of intensity of artifacts in the OCT images. Topical application of a 50% aqueous solution of propylene glycol also resulted in an increase in probing depth and an improvement in visualization of subsurface blood vessels

on OCT images (820 nm) [66]. Also, the possibility of observation of subepidermal cavity, malignant melanoma, superficial vein of the human finger, and control of the human skin scattering properties under the topical application of OCAs was shown in Refs. [117–120]. He et al. [121] demonstrated that a 50% glycerol solution applied to the surface of human skin *in vivo* improved the registration of glucose in blood using OCT.

In reference [80], salicylic acid (SA) was used in combination with a chemical diffusion enhancer, azone, to increase the permeability of the epidermis. It was found that the combination of 1% SA and 1% azone gave the greatest synergistic effect in increasing the OCT probing depth (1310 nm) from 0.34 ± 0.04 mm to 0.45 ± 0.03 mm (the probing depth was determined as 1/e of the light penetration depth).

In reference [110], it was shown that after complex processing with thiazone, PEG-400, and US, the observation depth increased by 41.3% in comparison with the nontreated control samples.

Optical clearing of dark skin is a serious problem because of light absorption in pigmented epidermis. Reference [41] developed optical clearing protocol for highly pigmented skin. The application of oleic acid (OA) as an lipophilic OCA in combination with microdermabrasion and sonophoresis for optical clearing of epidermis compared with separate use was studied for both light and dark skin using OCT (930 nm) (see Figure 19.2). Sequent application of microdermabrasion, OA, and US (MOUSE – Microdermabrasion, Oleic acid, UltraSound Effect) for light skin (type II) allowed increasing OCT signal amplitude up to 3.3-fold with more than twice improved depth penetration during 30 min. For dark skin (type VI) at the depth of 400 μm, the increase in signal amplitude compared to initial values was about 34% after MOUSE [41].

Reflectance, transmittance, and fluorescence spectroscopy

In vivo investigations of human skin with diffuse reflectance (DR) and fluorescence spectroscopy are widely used. They give unique quantitative information about skin endogenous and exogenous chromophores. Optical clearing allows for increasing accuracy of the evaluations [113, 118, 122–130].

Experiments with laboratory animals showed decreasing in DR of skin [58, 59, 131–133] and increasing in fluorescence signal from low-level light sources under skin [123, 126, 130] under action of different OCAs.

The results of *in vivo* investigation of human skin clearing caused by matching effect after a 40% glucose solution subcutaneous injection was reported for the first time in Reference [65]. An almost four-fold reduction in DR at the wavelength 650 nm was observed during about an hour, and then recovery of skin optical properties took place (see Figure 19.3).

Zhong et al. [110] proposed a new physical method in combination with mixed solution of thiazone and PEG-400 penetration into tissue to assess the skin optical clearing. The authors showed that the DR at 540 nm of human skin *in vivo* at 60 min after being treated by PEG-400 with 0.25% thiazone and 5-min ultrasound decreased by approximately 2.22-fold.

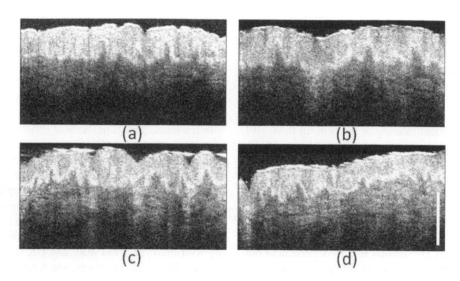

FIGURE 19.2 Optical clearing of dark skin: a) intact skin; b) skin after microdermabrasion and 30-min oleic acid application; c) skin after oleic acid application and 5-min sonophoresis (common duration of the procedure was 30 min); d) skin after MOUSE protocol (microdermabrasion, oleic acid application, and 5-min sonophoresis). Bar corresponds to 300 *μm*.

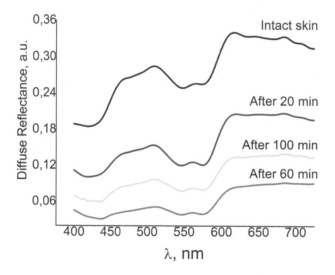

FIGURE 19.3 Optical diffuse reflectance spectra of human skin *in vivo* measured before and after subcutaneous injection of 40% glucose solution (adapted from reference [65]).

Zhao et al. [80] studied the change in DR induced by SA alone and SA combined with azone at different combinations on human skin *in vivo*. The results showed that the desmolytic ability of SA at a relatively low concentration on human skin could be significantly enhanced by azone.

A computational optical clearing (COC) method for evaluation of the chromophore concentration in tissue was developed in reference [129]. It was shown that COC enhances the ability of low-cost DR spectroscopy for *in vivo* detection of dermal beta-carotene in humans.

Multimodal fibered optical spectroscopy is a point optical biopsy technique combining spatially resolved DR and autofluorescence intensity spectra measurements. It is a valuable tool for *in vivo* noninvasive investigation of skin optical properties in the UV-visible spectral range, of particular interest for diagnosing and studying skin carcinogenesis stages [134].

However, it provides bulk fluorescence signals from which the individual endogenous fluorophore contributions need to be disentangled. Skin optical clearing allows for increasing tissue transparency, thus providing access to more accurate in-depth information. In reference [135], optical clearing was jointed to the multimodal fiber optical spectroscopy for improvement of the sensitivity of the approach. Authors studied the time behavior of skin intrinsic fluorophores during optical clearing process at the topical application of OCA (mixture of 45% PEG-400 with 5% polypropylene glycol and 50% sucrose) to *ex vivo* human skin strip lying on a fluorescent gel substrate. The analysis of the estimated abundance matrix allowed one to identify the relative contributions of two sources related to the skin intrinsic fluorescence of collagen/elastin (in dermis) and flavins (in epidermis) and one source related to the chlorine e6 fluorescence arising from gel substrate.

Transillumination is one of the existing methods for optical imaging of the internal structure of biological tissue [136, 137]. However, strong light scattering in tissues causes image blurring. A new method for enhancement of optical imaging of proximal interphalangeal joints in humans using optical clearing of skin was suggested in Reference [138]. In particular, it was found that glycerol application to the human skin during 60 min caused an increase in contrast in 1.5- and 1.7-fold for 820 nm and 904 nm, respectively.

Raman spectroscopy

A Raman spectrum represents a molecular fingerprint of tissue and provides quantitative information regarding its chemical composition. Raman spectroscopy (RS) and microscopy (RM) are widely used in dermatology and skin physiology for *in vivo* human skin analysis [20, 139–143]. The penetration of OCAs through the skin was also investigated [71, 144–146]. The changes in hydrogen-bound water molecule types during optical clearing measured using RM allowed the dehydration mechanism of the skin optical clearing to be studied [147]. Enejder et al. [148] measured RS for different glucose concentrations in nondiabetic volunteers using an oral glucose tolerance protocol.

Due to the reduction of elastic light scattering at tissue optical clearing, interaction of a probing laser beam with the target molecules is more effective. OCAs can increase the signal-to-noise ratio and significantly improve the Raman signal as well as reduce the systematic error caused by misdetection of surface and subsurface spectra. The long-term application of OCAs can lead to additional decrease in the Raman signal, which should be taken into consideration in analysis of deep-located tissue regions [4, 149].

For example, the use of glycerol allowed one to increase the probing depth in skin from 50 to 400 μm for *in vivo* determination of the composition of various types of tattoo ink based on the analysis of Raman spectra that was suggested by Darvin et al. [150].

Photoacoustic diagnostics

Photoacoustic microscopy (PAM) includes optical irradiation and acoustical detection of a target hidden inside tissue. In the optical component, the lateral resolution is determined by the diffraction-limited optical focusing. As photons travel in tissue, the focusing capability degrades due to optical scattering [151]. It was shown that optical clearing using glycerol is an efficient way to enhance the sensitivity and resolution of PAM [151].

Clinical applications of photoacoustic flow cytometry (PAFC) for detection of circulating tumor cells in deep blood vessels are also hindered by laser beam scattering, resulting in loss of PAFC sensitivity and resolution. Refs. [152, 153]

demonstrated first applications of optical clearing to enhance the sensitivity of PA detection methods for *in vivo* flow cytometry and to improve imaging of sentinel lymph nodes, respectively.

Figure 19.4 shows an example of the use of skin optical clearing in human *in vivo* to minimize light scattering and thus increase optical resolution and sensitivity of PAFC presented in reference [115]. Authors used alcohol skin cleaning, microdermabrasion, and glycerol application enhanced by massage and sonophoresis.

Laser speckle contrast imaging

The laser speckle contrast imaging (LSCI) technique provides images of the two-dimensional blood flow distribution at high spatial and temporal resolutions. This approach is widely used for the study of changes in cutaneous microvascular blood flow under action of different stimuli [154, 155]. However, high scattering of skin limits the penetration depth of light and reduces the imaging contrast, the spatial resolution, and the probing depth of the method [156, 157]. The scientific group of Zhu et al. [53, 73, 81, 158–160] developed an optical clearing method using multicomponent OCAs with chemical enhancers, allowing for imaging of mice dermal blood flow through the skin at high contrast and resolution. Thus, more data related to the dermal blood vessels or blood flow can be obtained.

In Refs. [131, 132] blood microcirculation in the underlying tissues at topical application of OCA in rats *in vivo* was studied. The results showed that applying OCAs led to a reduction in blood flow velocity at the insignificant vasodilatation.

Agafonov at al. [161] investigated the applicability of the LSCI technique for the study of nail-bed microcirculation

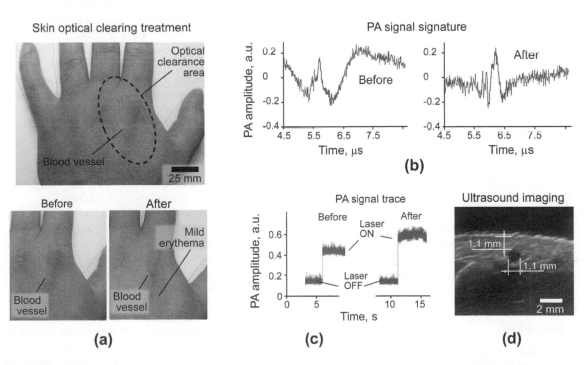

FIGURE 19.4 Optical clearing of human skin for PAFC. a) Visual contrast of the vein. b) Typical changes in photoacoustic signal waveform. c) Typical PAFC traces for a vein in human hand before and after optical clearing. d) US characterization of the selected vein (Reprinted with permission from [115] © The Optical Society).

parameters in humans. Aqueous 50% glycerol solution with DMSO allowed increasing the accuracy of the estimation of absolute values of blood flow with and without occlusion.

Thus, the combination of LSCI and OCA is promising approach for obtaining important information about the state of blood microcirculation in humans at different diseases.

Hyperspectral imaging

Hyperspectral imaging (HSI) is a hybrid modality that combines imaging and spectroscopy. By collecting spectral information at each pixel of a 2D detector array, HSI generates a 3D dataset of spatial and spectral information. As an emerging imaging modality for medical applications, HSI offers great potential for noninvasive disease diagnosis and surgical guidance [162–166].

HSI have good temporal resolution, which allow one to noninvasively monitor wide-field peripheral microcirculatory blood flow and blood oxygen in real time [167]. However, there is a technical challenge to study targeted skin vasculature directly because of high scattering of skin.

HSI combined with the skin optical clearing technique provided a possible way to noninvasively monitor hemodynamics of cutaneous microvessels in mice model *in vivo* [168]. To realize visualization of the skin microvascular dysfunction of type 1 diabetic mice, a combination of LSCI and HSI to simultaneously monitor the noradrenaline (NE)-induced responses of vascular blood flow and blood oxygen with the development of diabetes through optical clearing skin window was realized in Reference [168]. Thus, the *in vivo* skin optical clearing method is prospective for the use in humans because it can provide a feasible solution to realize visualization of cutaneous microvessels for monitoring microvascular reactivity under pathological conditions.

Conclusion

We reviewed the specific features and methods for the optical clearing of skin. The impact of the OCA on a tissue allows for efficient control of the optical properties, particularly, the reduction of tissue scattering coefficient, which facilitates the increase of efficiency of optical imaging (optical biopsy) in medical applications. The main focus was on the human skin *in vivo*. The differences in optical clearing of lightly and highly pigmented skin were shown. Application of *in vivo* skin optical clearing to the optical diagnostic methods, such as OCT, reflectance and transmittance spectroscopy, fluorescence spectroscopy, Raman spectroscopy and microscopy, laser speckle contrast imaging, hyperspectral imaging, and photoacoustic imaging was also reviewed.

Acknowledgments

The reported study was partly funded by RFBR and INSF, project number 20-52-56005 (E.A.G and A.N.B.); V.V.T was supported by the Government of the Russian Federation (grant #14.Z50.31.0044).

REFERENCES

1. V.V. Tuchin, *Tissue Optics: Light Scattering Methods and Instruments for Medical Diagnostics*, 3rd ed., **PM254**, SPIE Press, Bellingham (2015).
2. E.A. Genina, A.N. Bashkatov, and V.V. Tuchin, "Tissue optical immersion clearing," *Expert Rev. Med. Dev.* **7**, 825–842 (2010).
3. D. Zhu, K. Larin, Q. Luo, and V.V. Tuchin, "Recent progress in tissue optical clearing", *Laser Photon. Rev.* **7**(5), 732–757 (2013).
4. A. Yu. Sdobnov, M.E. Darvin, E.A. Genina, A.N. Bashkatov, J. Lademann, and V.V. Tuchin, "Recent progress in tissue optical clearing for spectroscopic application," *Spectrochim. Acta Part A: Molec. Biomolec. Spectr.* **197**, 216–229 (2018).
5. I. Costantini, R. Cicchi, L. Silvestri, F. Vanzi, and F.S. Pavone, "In-vivo and ex-vivo optical clearing methods for biological tissues: Review," *Biomed. Opt. Express* **10**, 5251–67 (2019).
6. M.V. Gómez-Gaviro, D. Sanderson, J. Ripoll, and M. Desco, "Biomedical applications of tissue clearing and three-dimensional imaging in health and disease," *iScience* **23**, 101432 (2020).
7. H. Schaefer and T.E. Redelmeier, *Skin Barrier. Principles of Percutaneous Absorption*, Karger, Basel (1996).
8. G.F. Odland, "Structure of the skin," in *Physiology, Biochemistry, and Molecular Biology of the Skin*, Vol. 1, L.A. Goldsmith (ed.), Oxford University Press, Oxford, 3–62 (1991).
9. T.J. Ryan, *Cutaneous Circulation. Physiology, Biochemestry, and Molecular Biology of the Skin*, Vol. 2, L.A. Goldsmith (ed.), Oxford University Press, Oxford, 1019–1084 (1991).
10. J.S. Stenn, *The Skin. Cell and Tissue Biology*, L. Weiss (ed.), Urban and Shwarzenberg, Baltimore, 541–572 (1988).
11. M.A. Farage, K.W. Miller, P. Elsner, and H.I. Maibach, "Structural characteristics of the aging skin: A review," *J. Toxicology (Cutaneous Ocul. Toxicol.)* **26**, 343–357 (2007).
12. X. Chen, W. Lin, C. Wang, et al., "In vivo real-time imaging of cutaneous hemoglobin concentration, oxygen saturation, scattering properties, melanin content, and epidermal thickness with visible spatially modulated light," *Biomed. Opt. Express* **8**(12), 5468–5482 (2017).
13. M.R. Chedekel, *Photophysics and Photochemistry of Melanin. Melanin: Its Role in Human Photoprotection*, L. Zeise, M.R. Chedekel, and T.B. Fitzpatrick (eds.), Valdenmar, Overland Park, 11–22 (1995).
14. I.A. Menon, and H.F. Haerman, "Mechanisms of action of melanin," *Br. J. Dermatol.* **51**, 109–112 (1977).
15. N. Kollias, and A. Baqer, "On the assessment of melanin in human skin in vivo," *J. Photochem. Photobiol.* **43**(1), 49–54 (1986).
16. Australian Radiation Protection and Nuclear Safety Agency. *Fitzpatrick Skin Phototype*. [Online]. Available: https://www.arpansa.gov.au/sites/g/files/net3086/f/legacy/pubs/RadiationProtection/FitzpatrickSkinType.pdf.
17. D. Parsad, K. Wakamatsu, A.J. Kanwar, B. Kumar, and S. Ito, "Eumelanin and phaeomelanin contents of depigmented and repigmented skin in vitiligo patients," *Br. J. Dermatol.* **149**, 624–626 (2003).

18. S.L. Jacques, *Origins of Tissue Optical Properties in the UVA, Visible and NIR Regions. Advances in Optical Imaging and Photon Migration*, Vol. 2, R.R. Alfano, and J.G. Fujimoto (eds.), OSA TOPS: Optical Society of America, Washington, 364–371 (1996).

19. S. Alaluf, D. Atkins, K. Barrett, M. Blount, N. Carter and A. Heath, "Ethnic variation in melanin content and composition in photoexposed and photoprotected human skin," *Pigment Cell Res.* **15**, 112–118 (2002).

20. M.E. Darvin, I. Gersonde, M. Meinke, W. Sterry, and J. Lademann, "Non-invasive in vivo determination of the carotenoids beta-carotene and lycopene concentrations in the human skin using the Raman spectroscopic method," *J. Phys. D: Appl. Phys.* **38**, 2696–2700 (2005).

21. E. Claridge, S. Cotton, P. Hall, and M. Moncrieff, "From colour to tissue histology: Physics-based interpretation of images of pigmented skin lesions," *Med. Image Anal.* **7**, 489–502 (2003).

22. M.J.C. van Gemert, S.L. Jacques, H.J.C.M. Sterenborg, and W.M. Star, "Skin optics," *IEEE Trans. Biomed. Eng.* **36**(12), 1146–1154 (1989).

23. I.V. Meglinski, and S.J. Matcher, "Quantitative assessment of skin layers absorption and skin refectance spectra simulation in visible and near-infrared spectral region," *Physiol. Meas.* **23**, 741–753 (2002).

24. T.L. Troy, and S.N. Thennadil, "Optical properties of human skin in the near infrared wavelength range of 1000 to 2200 nm," *J. Biomed. Opt.* **6**, 167–176 (2001).

25. G. Altshuler, M. Smirnov, and I. Yaroslavsky, "Lattice of optical islets: A novel treatment modality in photomedicine," *J. Phys. D: Appl. Phys.* **38**, 2732–2747 (2005).

26. S.L. Jacques, "Optical assessment of cutaneous blood volume depends on the vessel's size distribution: A computer simulation study," *J. Biophoton.* **3**, 75–81 (2010).

27. R.F. Reinoso, B.A. Telfer, and M. Rowland, "Tissue water content in rats measured by desiccation," *J. Pharmacol. Toxicol. Methods* **38**, 87–92 (1997).

28. Y. Taton, *Obesity, Pathophysiology, Diagnostics, Therapy*, Medical Press, Warsaw (1981).

29. D.Y. Shurigin, P.O. Vyazitskiy, and K.A. Sidorov, *Obesity*, Medicine, Leningrad (1975).

30. G, Nishimura, I. Kida, and M. Tamura, "Characterization of optical parameters with a human forearm at the region from 1.15 to 1.52 μm using diffuse reflectance measurements," *Phys. Med. Biol.* **51**, 2997–3011 (2006).

31. R. Marchesini, C. Clemente, E. Pignoli, and M. Brambilla, "Optical properties of in vitro epidermis and their possible relationship with optical properties of in vivo skin," *J. Photochem. Photobiol. B*, **16**, 127–140 (1992).

32. C.R. Simpson, M. Kohl, M. Essenpreis, and M. Cope, "Near-infrared optical properties of ex vivo human skin and subcutaneous tissues measured using the Monte Carlo inversion technique," *Phys. Med. Biol.* **43**, 2465–2478 (1998).

33. A.N. Bashkatov, E.A. Genina, V.I. Kochubey, and V.V. Tuchin, "Optical properties of human skin, subcutaneous and mucous tissues in the wavelength range from 400 to 2000 nm," *J. Phys. D: Appl. Phys.* **38**, 2543–2555 (2005).

34. E. Salomatina, B. Jiang, J. Novak, and A.N. Yaroslavsky, "Optical properties of normal and cancerous human skin in the visible and nearinfrared spectral range," *J. Biomed. Opt.* **11**, 064026 (2006).

35. A.N. Bashkatov, E.A. Genina, and V.V. Tuchin, "Optical properties of skin, subcutaneous and muscle tissues: A review," *J. Innov. Opt. Health Sci.* **4**(1), 9–38 (2011).

36. D. Yudovsky and L. Pilon, "Retrieving skin properties from *in vivo* spectral reflectance measurements," *J. Biophoton.* **4**(5), 305–314 (2011).

37. S.L. Jacques, "Quick analysis of optical spectra to quantify epidermal melanin and papillary dermal blood content of skin," *J. Biophoton.* **8**(4), 309–316 (2015).

38. X.U. Zhang, P. van der Zee, I. Atzeni, D.J. Faber, T.G. van Leeuwen, and H.J.C.M. Sterenborg, "Multidiameter single-fiber reflectance spectroscopy of heavily pigmented skin: Modeling the inhomogeneous distribution of melanin," *J. Biomed. Opt.* **24**(12), 127001 (2019).

39. D.T. Delpy, M. Cope, P. van der Zee, S. Arridge, S. Wray, and J. Wyatt, "Estimation of optical pathlength through tissue from direct time of flight measurement," *Phys. Med. Biol.* **33** 1433–1442 (1988).

40. C. Antoniou, J. Lademann, H. Richter, et al., "Analysis of the melanin distribution in different ethnic groups by *in vivo* laser scanning microscopy," *Laser Phys. Lett.* **6**(5), 393–398 (2009).

41. E.A. Genina, Yu.I. Surkov, I.A. Serebryakova, A.N. Bashkatov, V.V. Tuchin, and V.P. Zharov, "Rapid ultrasound optical clearing of human light and dark skin," *IEEE Trans. Med. Imag.* **39**(10), 3198–3206 (2020).

42. G.J. Tearney, M.E. Brezinski, J.F. Southern, B.E. Bouma, M.R. Hee, and J.G. Fujimoto, "Determination of the refractive index of highly scattering human tissue by optical coherence tomography," *Opt. Lett.* **20**, 2258–2260 (1995).

43. M. Sand, T. Gambichler, G. Moussa, et al., "Evaluation of the epidermal refractive index by optical coherence tomography," *Skin Res. Technol.* **12**, 114–118 (2005).

44. A. Roggan, K. Dörschel, O. Minet et al., "The optical properties of biological tissue in the near infrared wavelength range: Review and measurements," in *Laser-Induced Interstitial Thermotherapy*, **PM25**, G. Müller, A. Roggan (eds.), SPIE Press, Bellingham, 10–44 (1995).

45. F.P. Bolin, L.E. Preuss, R.C. Taylor, and R.J. Ference, "Refractive index of some mammalian tissues using a fiber optic cladding method," *Appl. Opt.* **28**, 2297–2303 (1989).

46. E.A. Genina, A.N. Bashkatov, Yu.P. Sinichkin, I.Yu. Yanina, and V.V. Tuchin, "Optical clearing of biological tissues: Prospects of application in medical diagnostics and phototherapy," *J. Biomed. Photon. Eng.* **1**(1), 22–58 (2015).

47. C.G. Rylander, O.F. Stumpp, T.E. Milner, et al., "Dehydration mechanism of optical clearing in tissue," *J. Biomed. Opt.*, **11**(4), 041117 (2006).

48. L. Oliveira, A. Lage, M.P. Clemente, and V. Tuchin, "Rat muscle opacity decrease due to the osmosis of a simple mixture," *J. Biomed. Opt.* **15**(5), 055004 (2010).

49. Y. Tanaka, A. Kubota, M. Yamato, T. Okano, and K. Nishida, "Irreversible optical clearing of sclera by dehydration and cross-linking," *Biomaterials* **32**, 1080–1090 (2011).

50. S.M. Zaytsev, A.N. Bashkatov, W. Blondel, M. Amouroux, V.V. Tuchin, and E.A. Genina, "Impact of ex vivo skin dehydration on collimated transmittance spectra kinetics," *Proc. SPIE* **11845**, 118450M (2021).

51. J. Hirshburg, B. Choi, J.S. Nelson, and A.T. Yeh, "Correlation between collagen solubility and skin optical clearing using sugars", *Lasers Surg. Med.* **39**, 140–144 (2007).

52. J.M. Hirshburg, K.M. Ravikumar, W. Hwang, and A.T. Yeh, "Molecular basis for optical clearing of collagenous tissues", *J. Biomed. Opt.* **15**(5), 055002 (2010).

53. J. Wang, N. Ma, R. Shi, Y. Zhang, T. Yu, and D. Zhu, "Sugar-induced skin optical clearing: From molecular dynamics simulation to experimental demonstration", *IEEE J. Sel. Top. Quantum Electron.* **20**(2), 7101007 (2014).

54. A.N. Bashkatov, K.V. Berezin, K.N. Dvoretskiy, et al., "Measurement of tissue optical properties in the context of tissue optical clearing," *J. Biomed. Opt.* **23**(9), 091416 (2018).

55. B. Choi, L. Tsu, E. Chen, et al., "Determination of chemical agent optical clearing potential using in vitro human skin", *Lasers Surg. Med.* **36**, 72–75 (2005).

56. V.D. Genin, D.K. Tuchina, A.J. Sadeq, E.A. Genina, V.V. Tuchin, and A.N. Bashkatov, "*Ex vivo* investigation of glycerol diffusion in skin tissue," *J. Biomed. Photon. Eng.* **2**(1), 010303 (2016).

57. D K. Tuchina, V.D. Genin, A.N. Bashkatov, E.A. Genina, and V.V. Tuchin, "Optical clearing of skin tissue ex vivo with polyethylene glycol," *Opt. Spectrosc.* **120**(1), 28–37 (2016).

58. E.A. Genina, A.N. Bashkatov, Yu.P. Sinichkin, and V.V. Tuchin, "Optical clearing of skin under action of glycerol: Ex vivo and in vivo investigations", *Opt. Spectrosc.* **109**(2), 225–231 (2010).

59. E.I. Galanzha, V.V. Tuchin, A.V. Solovieva, T.V. Stepanova, Q. Luo, and H. Cheng, J., "Skin backreflectance and microvascular system functioning at the action of osmotic agents", *Phys. D: Appl. Phys.* **36**, 1739–1746 (2003).

60. G. Vargas, A. Readinger, S.S. Dosier, and A.J. Welch, "Morphological changes in blood vessels produced by hyperosmotic agents and measured by optical coherence tomography", *Photochem. Photobiol.* **77**, 541–549 (2003).

61. H. Cheng, Q. Luo, S. Zeng, S. Chen, W. Luo, and H. Gong, "Hyperosmotic chemical agent's effect on *in vivo* cerebral blood flow revealed by laser speckle", *Appl. Opt.* **43**, 5772–5777 (2004).

62. A.N. Bashkatov, A.N. Korolevich, V.V. Tuchin, et al., "*In vivo* investigation of human skin optical clearing and blood microcirculation under the action of glucose solution," *Asian J. Phys.* **15**(1), 1–14 (2006).

63. D. Zhu, J. Zhang, H. Cui, Z. Mao, P. Li, and Q. Luo, "Short-term and long-term effects of optical clearing agents on blood vessels in chick chorioallantoic membrane", *J. Biomed. Opt.* **13**(2), 021106 (2008).

64. R.T. Zaman, A.B. Parthasarathy, G. Vargas, et al., "Perfusion in hamster skin treated with glycerol," *Lasers Surg. Med.* **41**, 492–503 (2009).

65. V.V. Tuchin, A.N. Bashkatov, É.A. Genina, Yu.P. Sinichkin, and N.A. Lakodina, "*In vivo* investigation of the immersion-liquid-induced human skin clearing dynamics," *Tech. Phys. Lett.* **27**(6), 489–490 (2001).

66. R.K. Wang, and V.V. Tuchin, "Enhance light penetration in tissue for high resolution optical imaging techniques by the use of biocompatible chemical agents," *J. X-Ray Sci. Technol.* **10**, 167–176 (2002).

67. M. Sznitowska, "The influence of ethanol on permeation behavior of the porous pathway in the stratum corneum," *Int. J. Pharmacol.* **137**, 137–140 (1996).

68. A.C. Williams and B.W. Barry, "Penetration enhancers," *Adv. Drug Deliv. Rev.* **56**, 603–618 (2004).

69. H. Trommer and R.H.H. Neubert, "Overcoming the stratum corneum: The modulation of skin penetration," *Skin Pharmacol. Physiol.* **19**, 106–121 (2006).

70. M.E. Lane, "Skin penetration enhancers," *Int. J. Pharm.* **447**, 12–21 (2013).

71. P.J. Caspers, A.C. Williams, E.A. Carter, et al., "Monitoring the penetration enhancer dimethyl sulfoxide in human stratum corneum *in vivo* by confocal Raman spectroscopy," *Pharm. Res.* **19**(10), 1577–1580 (2002).

72. X. Wen, S.L. Jacques, V.V. Tuchin, and D. Zhu, "Enhanced optical clearing of skin in vivo and optical coherence tomography in-depth imaging," *J. Biomed. Opt.* **17**, 066022 (2012).

73. D. Zhu, J. Wang, Z. Zhi, X. Wen, and Q. Luo, "Imaging dermal blood flow through the intact rat skin with an optical clearing method," *J. Biomed. Opt.* **15**(2), 026008 (2010).

74. K. Chen, Y. Liang, and Y. Zhang, "Study on reflection of human skin with liquid paraffin as the penetration enhancer by spectroscopy," *J. Biomed. Opt.* **18**(10), 105001 (2013).

75. J. Wang, Y. Liang, S. Zhang, Y. Zhou, H. Ni, and Y. Li, "Evaluation of optical clearing with the combined liquid paraffin and glycerol mixture," *Biomed. Opt. Express* **2**(8), 2329–2338 (2011).

76. D. Abookasis and T. Moshe, "Reconstruction enhancement of hidden objects using multiple speckle contrast projections and optical clearing agents," *Opt. Commun.* **300**, 58–64 (2013).

77. Y. Liu, X. Yang, D. Zhu, and Q. Luo, "Optical clearing agents improve photoacoustic imaging in the optical diffusive regime," *Opt. Lett.* **38**(20), 4236–4239 (2013).

78. Z. Deng, L. Jing, N. Wu, et al., "Viscous optical clearing agent for *in vivo* optical imaging," *J. Biomed. Opt.* **19**(7), 076019 (2014).

79. L. Guo, R. Shi, C. Zhang, D. Zhu, Z. Ding, and P. Li, "Optical coherence tomography angiography offers comprehensive evaluation of skin optical clearing *in vivo* by quantifying optical properties and blood flow imaging simultaneously," *J. Biomed. Opt.* **21**(8), 081202 (2016).

80. Q. Zhao, C. Dai, S. Fan, J. Lv, and L. Nie, "Synergistic efficacy of salicylic acid with a penetration enhancer on human skin monitored by OCT and diffuse reflectance spectroscopy," *Sci. Rep.* **6**, 34954 (2016).

81. R. Shi, W. Feng, C. Zhang, Z. Zhang, and D. Zhu, "FSOCA-induced switchable footpad skin optical clearing window for blood flow and cell imaging *in vivo*," *J. Biophoton.* **10**(12), 1647–1656 (2017).

82. W. Feng, R. Shi, C. Zhang, S. Liu, T. Yu, and D. Zhu, "Vizualization of skin microvascular dysfunction of type I diabetic mice using *in vivo* skin optical clearing method," *J. Biomed. Opt.* **24**(3), 031003 (2019).

83. T. Kurihara-Bergstrom, K. Knutson, L.J. de Noble, and C.Y. Goates, "Percutaneous absorption enhancement of an ionic molecule by ethanol–water system in human skin," *Pharm. Res.* **7**, 762–766 (1990).

84. E.A. Genina, A.N. Bashkatov, and V.V. Tuchin, "Effect of ethanol on the transport of methylene blue through stratum corneum," *Med. Laser Appl.* **23**, 31–38 (2008).

85. R. Notman, W.K. Den Otter, M.G. Noro, W.J. Briels, and J. Anwar, "The permeability enhancing mechanism of DMSO in ceramide bilayers simulated bymolecular dynamics," *Biophys. J.* **93**, 2056–2068 (2007).

86. J.-M. Andanson, K.L.A. Chan, and S.G. Kazarian, "High-throughput spectroscopic imaging applied to permeation through the skin," *Appl. Spectrosc.* **63**(5), 512–517 (2009).

87. Z. Zhi, Z. Han, Q. Luo, and D. Zhu, "Improve optical clearing of skin in vitro with propylene glycol as a penetration enhancer," *J. Innov. Opt. Health Sci.* **2**, 269–278 (2009).

88. J. Jiang and R.K. Wang, "Comparing the synergistic effects of oleic acid and dimethyl sulfoxide as vehicles for optical clearing of skin tissue in vitro," *Phys. Med. Biol.* **49**, 5283–5294 (2004).

89. J. Jiang and R.K. Wang, "How different molarities of oleic acid as enhancer exert its effect on optical clearing of skin tissue in vitro," *J. X-Ray Sci. Technol.* **13**, 149–159 (2005).

90. F. Sharif, E. Crushell, K. O'Driscoll, and B. Bourke, "Liquid paraffin: A reappraisal of its role in the treatment of constipation," *Arch. Dis. Child.* **85**(2), 121–124 (2001).

91. H.-J. Weigmann, J. Lademann, S. Schanzer, et al., "Correlation of the local distribution of topically applied substances inside the stratum corneum determined by tape stripping to differences in bioavailability," *Skin Pharmacol. Appl. Ski. Physiol.* **14**(Supplement 1), 98–102 (2001).

92. R.J. McNichols, M.A. Fox, A. Gowda, S. Tuya, B. Bell, and M. Motamedi, "Temporary dermal scatter reduction: Quantitative assessment and implications for improved laser tattoo removal," *Lasers. Surg. Med.* **36**(4), 289–296 (2005).

93. O. Stumpp, B. Chen, and A.J. Welch, "Using sandpaper for noninvasive transepidermal optical skin clearing agent delivery," *J. Biomed. Opt.* **11**(4), 041118 (2006).

94. W.-R. Lee, R.-Y. Tsai, C.-L. Fang, C.-J. Liu, C.-H. Hu, and J.-Y. Fang, "Microdermabrasion as a novel tool to enhance drug delivery via the skin: An animal study," *Dermatol. Surg.* **32**, 1013–1022 (2006).

95. C. Tse, M.J. Zohdy, J.Y. Ye, and M. O'Donnell, "Penetration and precision of subsurface photodistribution in porcine skin tissue with infrared femtosecond laser pulses," *IEEE Trans. Biomed. Eng.* **55**(3), 1211–1218 (2008).

96. C. Liu, Z. Zhi, V.V. Tuchin, Q. Luo, and D. Zhu, "Enhancement of skin optical clearing efficacy using photo-irradiation," *Lasers Surg. Med.* **42**, 132–140 (2010).

97. A.N. Bashkatov, E.A. Genina, A.A. Gavrilova, et al., "What exactly causes increase in skin transparency: Water replacement or dehydration?" *Lasers Surg. Med.* **38**(Suppl. 18), 84 (2006).

98. V.V. Tuchin, G.B. Altshuler, A.A. Gavrilova, et al., "Optical clearing of skin using flashlamp-induced enhancement of epidermal permeability," *Lasers Surg. Med.* **38**(9), 824–836 (2006).

99. J. Yoon, T. Son, E. Choi, B. Choi, J.S. Nelson, and B. Jung, "Enhancement of optical clearing efficacy using a microneedle roller," *J. Biomed. Opt.* **13**(2), 021103 (2008).

100. O.F. Stumpp, A.J. Welch, T.E. Milner, and J. Neev, "Enhancement of transepidermal skin clearing agent delivery using a 980 nm diode laser," *Laser Surg. Med.* **37**(4), 278–285 (2005).

101. N.S. Ksenofontova, E.A. Genina, A.N. Bashkatov, G.S. Terentyuk, and V.V. Tuchin, "OCT study of skin optical clearing with preliminary laser ablation of epidermis," *J. Biomed. Photon. Eng.* **3**(2), 020307 (2017).

102. E.A. Genina, A.N. Bashkatov, G.S. Terentyuk, and V.V. Tuchin, "Integrated effects of fractional laser microablation and sonophoresis on skin immersion optical clearing *in vivo*, *J. Biophoton.* **13**(7), e202000101 (2020).

103. A.K. Nugroho, G.L. Li, M. Danhof, and J.A. Bouwstra, "Transdermal iontophoresis of rotigotine across human stratum corneum in vitro: Influence of pH and NaCl concentration," *Pharm. Res.* **21**, 844–850 (2004).

104. X. Xu and Q. Zhu, "Sonophoretic delivery for contrast and depth improvement in skin optical coherence tomography," *IEEE J. Sel. Top. Quantum Electron.* **14**(1), 56–61 (2008).

105. B.E. Polat, D. Hart, R. Langer, and D. Blankschtein, "Ultrasound-mediated transdermal drug delivery: Mechanisms, scope, and emerging trends," *J. Controlled Release* **152**, 330–348 (2011).

106. X. Xu and C. Sun, "Ultrasound enhanced skin optical clearing: Microstructural changes," *J. Innov. Opt. Health Sci.* **3**(3), 189–194 (2010).

107. A. Tezel and S. Mitragotri, "Interaction of inertial cavitation bubbles with stratum corneum lipid bilayers during low frequency sonophoresis," *Biophys. J.* **85**, 3502–3512 (2003).

108. X. Xu, Q. Zhu, and C. Sun, "Assessment of the effects of ultrasound-mediated alcohols on skin optical clearing," *J. Biomed. Opt.* **14**, 034042 (2009).

109. X. Xu, Q. Zhu, and C. Sun, "Combined effect of ultrasound-SLS on skin optical clearing," *IEEE Photon. Technol. Lett.* **20**(24), 2117–2119 (2008).

110. H. Zhong, Z. Guo, H. Wei, et al., "Synergistic effect of ultrasound and Thiazone: PEG 400 on human skin optical clearing *in vivo*," *Photochem. Photobiol.* **86**, 732–737 (2010).

111. E.A. Genina, A.N. Bashkatov, E.A. Kolesnikova, M.V. Basco, G.S. Terentyuk, and V.V. Tuchin, "Optical coherence tomography monitoring of enhanced skin optical clearing in rats *in vivo*," *J. Biomed. Opt.* **19**(2), 021109 (2014).

112. S.V. Zaitsev, Y.I. Svenskaya, E.V. Lengert, et al., "Optimized skin optical clearing for optical coherence tomography monitoring of encapsulated drug delivery through the hair follicles," *J. Biophoton.* **13**(4), e201960020 (2020).

113. M.A. Fox, D.G. Diven, K. Sra, et al., "Dermal scatter reduction in human skin: A method using controlled application of glycerol," *Lasers Surg. Med.* **41**, 251–255 (2009).

114. J. Yoon, D. Park, T. Son, J. Seo, J.S. Nelson, and B. Jung, "A physical method to enhance transdermal delivery of a tissue optical clearing agent: Combination of microneedling and sonophoresis," *Lasers Surg. Med.* **42**, 412–417 (2010).

115. Yu.A. Menyaev, D.A. Nedosekin, M. Sarimollaoglu, M.A. Juratli, E.I. Galanzha, V.V. Tuchin, and V.P. Zharov, "Optical clearing in photoacoustic flow cytometry," *Biomed. Opt. Express* **4**(12), 3030–3041 (2013).

116. Y.M. Liew, R.A. McLaughlin, F.M. Wood, and D.D. Sampson, "Reduction of image artifacts in three-dimensional optical coherence tomography of skin *in vivo*," *J. Biomed. Opt.*, **16**(11), 116018 (2011).

117. V.V. Tuchin, *Optical Clearing of Tissues and Blood*, PM154, SPIE Press, Bellingham (2006).

118. L. Pires, V. Demidov, I.A. Vitkin, V. Bagnato, C. Kurachi, and B.C.Wilson, "Optical clearing of melanoma in vivo: Characterization by diffuse reflectance spectroscopy and optical coherence tomography," *J. Biomed. Opt.* **21**, 081210 (2016).

119. S.G. Proskurin, "Optical coherence-domain imaging of subcutaneous human blood vessels *in vivo*," *Adv. Life Sci.* **1**(2), 40–44 (2011).

120. N. Bosschaart, D.J. Faber, T.G. van Leeuwen, and M.C.G. Aalders, "*In vivo* low-coherence spectroscopic measurements of local hemoglobin absorption spectra in human skin," *J. Biomed. Opt.*, **16**(10), 100504 (2011).

121. R. He, H. Wei, H. Gu, et al., "Effects of optical clearing agents on noninvasive blood glucose monitoring with optical coherence tomography: A pilot study," *J. Biomed. Opt.* **17**(10), 101513 (2012).

122. Y. He, R.K. Wang, and D. Xing, "Enhanced sensitivity and spatial resolution for in vivo imaging with low-level light-emitting probes by use of biocompatible chemical agents," *Opt. Lett.* **28**(21), 2076–2078 (2003).

123. E.D. Jansen, P.M. Pickett, M.A. Mackanos, and J. Virostko, "Effect of optical tissue clearing on spatial resolution and sensitivity of bioluminescence imaging," *J. Biomed. Opt.* **11**(4), 041119 (2006).

124. A.A. Gavrilova, V.V. Tuchin, A.B. Pravdin, I.V. Yaroslavsky, and G.B. Altshuler, "Skin spectrophotometry under the islet photothermal effect on the epidermal permeability," *Opt. Spectros.* **104**(1), 140–146 (2008).

125. A.N. Bashkatov, E.A. Genina, V.V. Tuchin, and G.B. Altshuler, "Skin optical clearing for improvement of laser tattoo removal," *Laser Phys.* **19**(6), 1312–1322 (2009).

126. S. Karma, J. Homan, C. Stoianovic, and B. Choi, "Enhanced fluorescence imaging with DMSO-mediated optical clearing," *J. Innov. Opt. Health Sci.* **3**(3), 153–158 (2010).

127. C. Liu, R. Shi, M. Chen, and D. Zhu, "Quantitative evaluation of enhanced laser tattoo removal by skin optical clearing," *J. Innov. Opt. Health Sci.* **8**(3), 1541007 (2015).

128. L. Pires, V. Demidov, B.C. Wilson, et al., "Dual-agent photodynamic therapy with optical clearing eradicates pigmented melanoma in preclinical tumor models," *Cancers* **12**, 1956 (2020).

129. A. Morovati, M.A. Ansari, and V.V. Tuchin, "In vivo detection of human cutaneous beta-carotene using computational optical clearing," *J. Biophoton.* **13**(1), e202000124 (2020).

130. S.M. Zaytsev, W. Blondel, M. Amouroux, et al., "Optical spectroscopy as an effective tool for skin cancer features analysis: Applicability investigation," *Proc. SPIE* **11457**, 1145706 (2020).

131. A.N. Bashkatov, E.A. Genina, I.V. Korovina, V.I. Kochubey, Yu.P. Sinichkin, and V.V. Tuchin, "In vivo and in vitro study of control of rat skin optical properties by acting of osmotical liquid," *Proc. SPIE* **4224**, 300–311 (2000).

132. Z. Mao, X. Wen, J. Wang, and D. Zhu, "The biocompatibility of the dermal injection of glycerol in vivo to achieve optical clearing," Proc. SPIE 7519, 75191 (N) (2009).

133. D.K. Tuchina, P.A. Timoshina, V.V. Tuchin, A.N. Bashkatov, and E.A. Genina, "Kinetics of rat skin optical clearing at topical application of 40%-glucose: Ex vivo and in vivo studies," *IEEE J. Sel. Top. Quant. Electron.* **25**(1), 7200508 (2019).

134. M. Amouroux, G. Diaz-Ayil,W. Blondel, G. Bourg-Heckly, A. Leroux, and F. Guillemin, "Classification of ultraviolet irradiated mouse skin histological stages by bimodal spectroscopy: Multiple excitation autofluorescence and diffuse reflectance," *J. Biomed. Opt.* **14**, 014011 (2009).

135. P. Rakotomanga, C. Soussen, G. Khairallah, et al., "Source separation approach for the analysis of spatially resolved multiply excited autofluorescence spectra during optical clearing of *ex vivo* skin," *Biomed. Opt. Express* **10**(7), 3410–3424 (2019).

136. T. Barozzino and M. Sgro, "Transillumination of the neonatal skull: Seeing the light," *Canadian Med. Assoc. J.* **167**(11), 1271–1272 (2002).

137. U. Zabarylo and O. Minet, "Pseudo colour visualization of fused multispectral laser scattering images for optical diagnosis of rheumatoid arthritis," *Laser Phys. Lett.* 7(1), 73–77 (2010).

138. E.A. Kolesnikova, A.S. Kolesnikov, U. Zabarylo, et al. "Optical clearing of human skin for the enhancement of optical imaging of proximal interphalangeal joints," *Proc. SPIE* **9031**, 90310C (2014).

139. L. Chrit, C. Hadjur, S. Morel et al., "*In vivo* chemical investigation of human skin using a confocal Raman fiber optic microprobe," *J. Biomed. Opt.* **10**(4), 044007 (2005).

140. L. Binder, S. SheikhRezaei, A. Baierl, L. Gruber, M. Wolzt, and C. Valenta, "Confocal Raman spectroscopy: In vivo measurement of physiological skin parameters - a pilot study," *J. Dermatol. Sci.* **88**(3), 280–288 (2017).

141. C. Choe, J. Schleusener, J. Lademann, and M.E. Darvin, "Keratin-water-NMF interaction as a three layer model in the human stratum corneum using *in vivo* confocal Raman microspecroscopy," *Sci. Rep.* **7**, 15900 (2017).

142. C. Choe, J. Schleusener, J. Lademann, and M.E. Darvin, "Age related depth profiles of human stratum corneum barrier-related molecular parameters by confocal Raman microscopy *in vivo*," *Mech. Ageing Dev.* **172**, 6–12 (2018).

143. C. Choe, J. Schleusener, J. Lademann, and M.E. Darvin, "Human skin in vivo has a higher skin barrier function than porcine skin ex vivo: Comprehensive Raman microscopic study of the stratum corneum," *J. Biophotonics*, **11**, e201700355 (2018).

144. E. Ciampi, M. van Ginkel, P.J. McDonald, et al., "Dynamics *in vivo* mapping of model moisturiser ingress into human skin by GARfield MRI," *NMR Biomed.* **24**, 135–144 (2011).

145. C. Choe, J. Lademann, and M.E. Darvin, "Confocal Raman microscopy for investigating the penetration of various oils into the human skin *in vivo*," *J. Dermatol. Sci.* **79**, 171–178 (2015).

146. C. Choe, J. Lademann, and M.E. Darvin, "Analysis of human and porcine skin in vivo/ex vivo for penetration of selected oils by confocal Raman microscopy," *Skin Pharmacol. Physiol.* **28**(6), 318–330 (2015).

147. A.Y. Sdobnov, M.E. Darvin, J. Schleusener, J. Lademann, and V.V. Tuchin, "Hydrogen bound water profiles in the skin influenced by optical clearing molecular agents – quantitative analysis using confocal Raman microscopy," *J. Biophoton.* **11**, e201800283, (2018)

148. A.M.K. Enejder, T.G. Scecina, J. Oh, et al., "Raman spectroscopy for noninvasive glucose measurements," *J. Biomed. Opt.* **10**, 031114 (2005).

149. A.Yu. Sdobnov, J. Lademann, M.E. Darvin, and V.V. Tuchin, "Methods for optical skin clearing in molecular optical imaging in dermatology," Biochemistry (Moscow) **84**(Suppl. 1), S144–S158 (2019).

150. M.E. Darvin, J. Schleusener, F. Parenz, et al., "Confocal Raman microscopy combined with optical clearing for identification of inks in multicolored tattooed skin in vivo," *Analyst*, **143**, 49904999 (2018)

151. Y. Zhou, J. Yao, and L.V. Wang, "Optical clearing-aided photoacoustic microscopy with enhanced resolution and imaging depth," *Opt. Lett.* **38**(14), 2592–2595 (2013).

152. V.P. Zharov, E.I. Galanzha, E.V. Shashkov, N.G. Khlebtsov, and V.V. Tuchin, "In vivo photoacoustic flow cytometry for monitoring of circulating single cancer cells and contrast agents," *Opt. Lett.* **31**(24), 3623–3625 (2006).

153. E.I. Galanzha, M.S. Kokoska, E.V. Shashkov, J.W. Kim, V.V. Tuchin, and V.P. Zharov, "In vivo fiber-based multi-color photoacoustic detection and photothermal purging of metastasis in sentinel lymph nodes targeted by nanoparticles," *J. Biophoton.* **2**(8–9), 528–539 (2009).

154. A. Khalil, A. Humeau-Heurtier, G. Mahe, and P. Abraham, "Laser speckle contrast imaging: Age-related changes in microvascular blood flow and correlation with pulse-wave velocity in healthy subjects," *J. Biomed. Opt.* **20**(5), 051010 (2015).

155. M. Ogami, R. Kulkarni, H. Wang, R. Reif, and R.K. Wang, "Laser speckle contrast imaging of skin blood perfusion responses induced by laser coagulation," *Quant. Electron.* **44**(8), 746–750 (2014).

156. D. Briers, D.D. Duncan, E. Hirst, et al., "Laser speckle contrast imaging: Theoretical and practical limitations," *J. Biomed. Opt.* **18**(6), 066018 (2013).

157. M.A. Davis, S.M.S. Kazmi, and A.K. Dunn, "Imaging depth and multiple scattering in laser speckle contrast imaging," *J. Biomed. Opt.* **19**(8), 086001 (2014).

158. R. Shi, M. Chen, V.V. Tuchin, and D. Zhu, "Accessing to arteriovenous blood flow dynamics response using combined laser speckle contrast imaging and skin optical clearing," *Biomed. Opt. Express* **6**(6), 1977–1989 (2015).

159. J. Wang, R. Shi, Y. Zhang, and D. Zhu, "Ear skin optical clearing for improving blood flow imaging," *Photon. Lasers Med.* **2**(1), 37–44 (2013).

160. Y. Ding, J. Wang, Z. Fan, et al., "Signal and depth enhancement for *in vivo* flow cytometer measurement of ear skin by optical clearing agents," *Biomed. Opt. Express* **4**(11), 2518–2526 (2013).

161. D.N. Agafonov, P.A. Timoshina, M.A. Vilensky, I.V. Fedosov, and V.V. Tuchin, "The study of nail bed microcirculation by laser speckle imaging technique," *Izv. Saratov Univ. (N.S.), Ser. Phys.* **11**(2), 14–19 (2011) (in Russian).

162. G. Lu and B. Fei, "Medical hyperspectral imaging: A review," *J. Biomed. Opt.* **19**(1), 010901 (2014).

163. S. Miclos, S.V. Parasca, M.A. Calin, D. Savastru, and D. Manea, "Algorithm for mapping cutaneous tissue oxygen concentration using hyperspectral imaging," *Biomed. Opt. Express* **6**(9), 3420–3430 (2015).

164. G.V.G. Baranoski, A. Dey, and T.F. Chen, "Assessment the sensitivity of human skin hyperspectral responses to increasing anemia severity levels," *J. Biomed. Opt.* **20**(9), 095002 (2015).

165. I.A. Bratchenko, V.P. Sherendak, O.O. Myakinin, et al., "*In vivo* hyperspectral imaging of skin malignant and benign tumors in visible spectrum," *J. Biomed. Photon. Eng.* **4**(1), 010301 (2018).

166. E.J.M. Baltussen, E.N.D. Kok, S.G.B. Konig, et al., "Hyperspectral imaging for tissue classification, a way toward smart laparoscopic colorectal surgery," *J. Biomed. Opt.* **24**(1), 016002 (2019).

167. E. Zherebtsov, V. Dremin, A. Popov, et al., "Hyperspectral imaging of human skin aided by artificial neural networks," *Biomed. Opt. Express* **10**(7), 3545–3559 (2019).

168. W. Feng, R. Shi, C. Zhang, T. Yu, and D. Zhu, "Lookup-table-based inverse model for mapping oxygen concentration of cutaneous microvessels using hyperspectral imaging," *Opt. Express* **25**(4), 3481–3495 (2017).

20

Optical clearing of blood and tissues using blood components

Olga S. Zhernovaya, Elina A. Genina, Valery V. Tuchin, and Alexey N. Bashkatov

CONTENTS

Introduction

Strong light scattering in biological tissues is determined by refractive index mismatch between its components, such as collagen fibers, cellular organelles, interstitial fluid, etc. Scattering prevents the obtaining of good contrast and well-resolved images of microstructure of tissues by optical imaging techniques, such as optical coherence tomography (OCT), visible and near-infrared spectroscopy, fluorescence spectroscopy, microscopy, and others. The optical clearing method is a possible solution for improving imaging quality and increasing imaging depth for optical imaging techniques. Light scattering in blood is mainly caused by the refractive index mismatch between erythrocytes (red blood cells, RBCs) and blood plasma, as erythrocytes have a higher refractive index than blood plasma. Application of biocompatible optical clearing agents (OCAs) decreases light scattering in blood which leads to an increase in optical penetration depth and enhancement of quality of imaging. Determination of optimal type and concentration of optical clearing agents for optical clearing of blood is required for effective and nondestructive usage of the optical clearing method in *in vivo* applications.

The optical clearing method has been successfully applied to many biological tissues, such as skin, sclera, and muscle [1–9]. The optical clearing method also showed applicability for improving light transport in whole blood, including in *in vivo* applications [10–18]. A number of optical clearing agents, such as dextrans, glycerol, glucose, fructose, hemoglobin, and others demonstrated a potential to be used for optical clearing of blood. The mechanism of optical clearing of blood depends on the properties of a particular OCA. The optical clearing agents can cause aggregation of erythrocytes, refractive index matching between RBCs and blood plasma, alterations of morphology of erythrocytes, and some other effects which change the scattering properties of blood and allow the optical clearing effect in blood to be achieved.

Optical properties of blood

Whole blood consists of plasma and formed elements: erythrocytes (red blood cells, RBCs), leukocytes (white blood cells), and thrombocytes (platelets). Erythrocytes are the predominant group of blood cells. Blood plasma makes up about 55% of whole blood volume and is composed of water (about 90%) and dissolved proteins, glucose, hormones, and other substances. In norm, hematocrit (volume fraction of erythrocytes) in adults is 41%–53% for males and 36%–46% for females [19–21]. An erythrocyte contains about 70% of water, 25% of hemoglobin, and 5% of lipids and glucose. Hemoglobin in erythrocytes is contained in the membrane, which is formed by mostly lipids and proteins. In norm, an erythrocyte is formed as a biconcave disc with the diameter of about 8 μm [20].

DOI: 10.1201/9781003025252-23

Blood is a highly scattering and absorbing medium in the visible and near infrared regions. Scattering properties of RBCs, the main scatterers in blood, are determined by their size, shape, volume, and mass. Optical properties of RBCs also depend on concentration of hemoglobin in erythrocyte. Absorption properties of blood in the visible range are mainly determined by the absorption bands of hemoglobin. The absorption and scattering spectra of blood and anisotropy factor in visible and near infrared regions are represented in Figure 20.1 [21].

The absorption spectra of blood have characteristic absorption peaks of hemoglobin in the visible range at about 410, 540, and 575 nm (for oxygenated blood), 430 and 555 nm (for deoxygenated blood), and isobestic point at 805 nm. In the near infrared region, the absorption spectra of blood are determined by the absorption bands of water with the maxima at 1450 and 1930 nm. The main source of light scattering in blood is the refractive index mismatch between erythrocytes and blood plasma. The scattering properties of a single RBC depend on its refractive index, which is mainly determined by refractive index of hemoglobin in RBCs. In norm, the concentration of hemoglobin is 120–175 g/L in blood and 310–370 g/L inside the erythrocyte [19]. Refractive index of erythrocytes is much higher than plasma; in the range of 800–1000 nm, the refractive index of hemoglobin with the concentration of 287 g/L is 1.40, while the refractive index of blood plasma is 1.33–1.34 in the same wavelength region [11, 22–24]. Scattering properties of blood mainly depend on hematocrit (the volume fraction of erythrocytes), but there are several factors which can influence the scattering properties of blood, such as aggregation of erythrocytes, sedimentation of blood, hemolysis, and deformation of the RBCs under various conditions. Spectrally, it is clearly seen that in the range from 250 to 630 nm, blood scattering spectrum has dips in the area of hemoglobin absorption bands, and in the range from 630 to 2000 nm, the scattering coefficient monotonically decreases as wavelength increases. The wavelength dependence of scattering anisotropy factor has similar behavior. It is clearly seen that in the range 250–630 nm, the dependence increases but has dips in the area of hemoglobin absorption bands, and in the range from 630 to 2500 nm, the anisotropy factor is slightly decreased in the area of water absorption bands with maxima 1445 and 1930 nm.

Optical clearing of blood by dextrans

Dextrans are considered to be promising agents for optical clearing of blood due to their biocompatibility, relatively high refractive index, and ability to induce aggregation of blood. The influence of dextrans with different molecular weights on the optical properties of blood and their applicability for optical clearing of blood were investigated in several studies [10–12]. Addition of dextran to blood in comparison to intravenous contrast agents and their influence on blood transmittance was studied by OCT *in vitro* in circulating blood by Brezinski et al. [10]. Both dextran and intravenous contrast agents were found to significantly increase the light penetration in blood. Application of dextran and intravenous contrast to blood led to the increase of the OCT signal intensity to 69% and 45%, respectively, while the control measurements with saline did not provide any significant effect of reduction in scattering [10]. The increase of the OCT signal intensity after the addition of intravenous contrast agent to blood was suggested to be due to the decrease in the volume of erythrocyte caused by the addition of the agent. The effect of the increased light penetration of blood after application of dextran was mainly attributed to refractive index matching between RBCs and ground matter.

The results of OCT study of optical clearing of blood by several clearing agents, including dextrans with different molecular weights, were presented by Tuchin *et al.* [11, 12]. It was found that the mechanism and capability of dextrans to improve light transport in blood depends on the molecular weight of the dextrans and their concentration (Figure 20.2).

Application of dextran with high molecular weight (Dx500 – dextran with molecular weight of 473,000) demonstrated better optical clearing efficiency than dextran with low molecular weight [11, 12]. This effect was explained by the assumption that dextran with high molecular weight not only had a greater ability to match the refractive indices of erythrocytes and blood plasma due to the high refractive index of dextrans, but also had a greater ability to induce aggregation of blood, which resulted in enhancement of transmittance of blood. Application of dextran with low molecular weights (Dx10 – dextran with molecular weight of 10,500) improved light transport only due to refractive index matching between RBCs and ground matter. It was also found that high concentration of

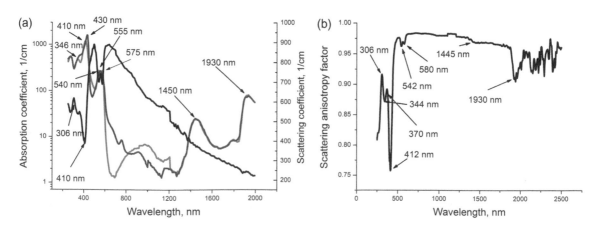

FIGURE 20.1 Optical properties of blood: (a) absorption coefficient μ_a of oxygenated (red line) and deoxygenated (blue line) blood with the hematocrit of 45%, scattering coefficient μ_s (black line) and (b) anisotropy factor g. Tabulated data from Ref. [21].

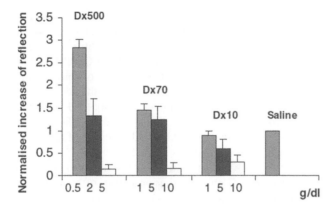

FIGURE 20.2 The effects of dextrans compared to that of the saline control on light transmission after 10 min of blood sedimentation. Reproduced from Ref. [12] by permission of IOP Publishing. All rights reserved.

dextrans Dx500 and Dx70 (dextran with molecular weight of 65,500) provided a noticeable optical clearing effect at the initial phase of the application, which was believed to be due to refractive index matching caused by dextrans.

Optical clearing of blood by glucose and fructose

Glucose is widely used as an optical clearing agent for improvement of light penetration in biological tissues for various optical techniques [4, 11, 25–31]. The refractive index of highly concentrated glucose and fructose solutions is higher than that of blood plasma; therefore, glucose and fructose can be used for enhancement of light transport by refractive index matching between erythrocytes and blood plasma.

Theoretical modeling of optical properties of blood upon addition of glucose solutions with different concentrations was performed by Bashkatov et al. [18]. The influence of glucose on the scattering coefficient of blood was estimated for the wavelength range of 400–1000 nm, and it was found that the scattering coefficient of blood decreases at the increase of the concentration of glucose. It was also found that the scattering coefficient of blood nonlinearly depends on the concentration of glucose. The minimum of the scattering coefficient for most wavelengths was observed at the concentration of glucose about 0.6–0.7 g/mL.

The experimental estimation of the optimal concentration of glucose and fructose required for blood optical clearing was performed *in vitro* using the OCT system working at 930 nm [32]. The decrease of total attenuation coefficient of blood samples with glucose and fructose with the concentrations in the range of 50–400 g/L was observed for all samples, but the most noticeable decrease (about 1.3×) of the total attenuation coefficient was observed for the blood samples with the concentration of glucose and fructose of 400 g/L. A shrinking of erythrocytes and partial hemolysis subsequently took place at high concentrations glucose or fructose solutions added to the blood, of the order of 200–400 g/L, while at their low concentrations of 50 g/L did not lead to the destruction of erythrocytes. The observed shrinking of erythrocytes and hemolysis at high concentrations of glucose and fructose was attributed to the increase in osmolarity

of ground matter due to the addition of high concentrations of glucose and fructose. Thereby, the decrease of scattering of blood after the addition of glucose and fructose solutions can be explained not only by the refractive index matching between RBCs and ground matter, but also by the changes in morphology of RBCs at high concentrations of glucose and fructose. Hemolysis also causes the increase of refractive index of ground matter due to release of hemoglobin in blood plasma, which also leads to the decrease of scattering in blood.

Application of optical clearing agents which change blood osmolarity, such as glucose and fructose, leads to changes in the morphology of erythrocytes, which can be a possible method to change optical properties of blood and increase the transmittance of blood. In *in vivo* conditions, the RBCs which were shrunken due to application of an optical clearing agent are expected to return to their normal shape. However, the lysis of some erythrocytes caused by addition of the solutions which change blood osmolarity may prevent the using of these solutions for *in vivo* applications.

Optical clearing of blood by hemoglobin solutions

Application of hemoglobin, albumin, or multimolecular solutions, such as a whole blood, as optical clearing agents for optical clearing of tissues and blood is promising due to their biocompatibility and high refractive indices. Addition of a highly concentrated hemoglobin solution to blood in order to achieve the optical clearing effect was firstly proposed by Tuchin *et al.* [13], where the Mie-based theoretical analysis of the scattering coefficient of blood at local hemolysis was presented. It was reported that the scattering coefficient of blood decreased upon increase of the degree of hemolysis (Figure 20.3). Popescu et al. [33] have proven this concept experimentally by studying the dynamics and morphology of erythrocytes by Hilbert phase microscopy.

In the spectral range of 400–500 nm, the scattering coefficient of blood decreases to 30% upon raise of the hemolysis up to 20% (Figure 20.3). In the spectral range of 500–1000 nm, the 40% decrease in the scattering coefficient was observed at the increase of the degree of hemolysis up to 20%. The reduction of the scattering coefficient was supposed to be attributed to the increase of refractive index of blood plasma after release

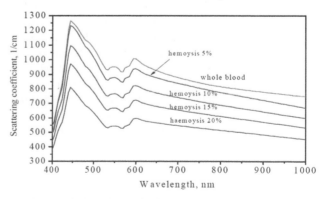

FIGURE 20.3 Scattering coefficient of blood in the spectral range of 400–1000 nm at several degrees of the hemolysis. Reprinted with permission from Ref. [13] © The Optical Society.

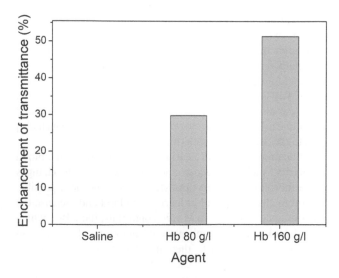

FIGURE 20.4 Enhancement of transmittance after addition of Hb solutions with the concentrations of 80 and 160 g/L. Reprinted from Ref. [14].

of the hemoglobin due to hemolysis, and refractive index matching between erythrocytes and blood plasma, which led to decrease of scattering. Thus, it was theoretically proved that the method of optical clearing of blood by hemoglobin solutions can enhance light transport in blood.

The optical clearing of blood by hemoglobin solutions was experimentally studied *in vitro* by using of an OCT system working at 930 nm [14]. The total attenuation coefficient and enhancement of transmission of blood after addition of hemoglobin solutions were derived from the OCT measurements for blood samples after addition of hemoglobin solutions with the concentrations of 80 and 160 g/L, which had high refractive indices (1.362 and 1.390, respectively). Addition of hemoglobin solutions to blood led to a noticeable decrease of total attenuation coefficient and to the enhancement of transmittance of blood (Figure 20.4). The total attenuation coefficient of blood mixed with the Hb solutions with the concentrations of 80 and 160 g/L decreased by a factor of 1.3 and 1.5, respectively, compared to the control sample. The enhancement of transmittance for blood mixed with hemoglobin solutions was 30% for the 80 g/L and 51% for 160 g/L concentrations, respectively.

After addition of Hb solutions, the refractive index of ground matter became higher due to high refractive indices of hemoglobin solutions. The high refractive indices of added Hb solutions provided the effect of optical clearing by matching of refractive index of RBCs and blood plasma. The microscopic images of blood samples obtained by smear microscopy did not show any aggregation or considerable deformation of the form of the erythrocytes after addition of Hb solutions compared to the blood samples with saline. Hence, it was assumed that the main cause of the observed optical clearing effect in blood after addition of Hb solutions was the refractive index matching between blood plasma and RBCs.

Optical clearing of blood by PEG, PPG, and PG

Among the various substances which can be potentially used for optical clearing of blood, polyethylene glycol (PEG), propylene

glycol (PG) and polypropylene glycol (PPG) can be appropriate agents due to their high refractive index (about 1.4, which is comparable to the refractive index of RBCs) and ability to change the aggregation properties of blood. Previously, PEG, PPG, and PG were successfully applied for optical clearing of some biological tissues: the application of PEG and PEG-thiazone mixture improved light penetration in human skin [34]; PPG- and PEG-based polymer mixtures was also found to reduce light scattering in dermis [35]; and PG was used for enhancement of imaging of skin and gastrointestinal tissues [3, 25, 36].

The *in vitro* OCT experimental study of the influence of PEG, PPG, and PG on scattering properties of blood showed that addition of these agents caused considerable improvement of light transport in blood [14]. For added PEG, PG, and PPG, the enhancement of transmittance of blood samples was found to be 94%, 148%, and 162%, respectively (Figure 20.5). The smear microscopy study showed that the addition of PEG caused shrinking and elongating of the erythrocytes. It was also found that PG possibly induced partial hemolysis of erythrocytes, because after addition of PG, erythrocytes became much thinner. Application of PEG and PPG induced aggregation of blood and led to the formation of big aggregates of erythrocytes, which resulted in changes in the scattering properties of blood and decrease in the total attenuation coefficient of blood after addition of these agents.

The effect of the optical clearing of blood after addition of PEG and PPG was supposed to be mainly caused by aggregation of erythrocytes. The adhesion (or aggregation) of RBCs possibly led to a decrease of scattering due to the "sieve" effect, when light travels in blood without interaction with scattering elements [37]. High refractive indices of PEG and PPG increase the refractive index of ground matter, which also makes an impact on overall optical clearing effect. The addition of PG causes the lysis of erythrocytes, which can probably provide a decrease in scattering along with refractive index matching between erythrocytes and blood plasma after release of hemoglobin. The addition of saline to the blood samples with PEG or PPG led to recovery of the erythrocytes, which was observed for some cells; therefore, the aggregation of erythrocytes caused by PEG and PPG can be reversible.

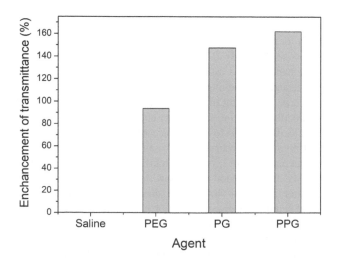

FIGURE 20.5 Enhancement of transmittance after addition of saline, PEG, PG and PPG to blood. Reprinted and adapted from Ref. [14].

Optical clearing of blood in blood vessels

Optical imaging systems such as optical coherence tomography (OCT) and integrated optical and acoustic imaging systems can be applied for imaging of blood vessels and intravascular imaging [15–17, 38]. OCT allows more precise morphological information of atherosclerotic plaques and stent lesions to be obtained than is possible with other conventional intracoronary imaging techniques. The frequency-domain OCT (FD-OCT) imaging systems and combined OCT-ultrasound systems have been used for intravascular imaging using a catheter inserted into the vessel of interest [15, 17]. Application of optical imaging methods for intravascular imaging has a major limitation: strong scattering of blood, which causes signal attenuation. Using optical clearing agents for imaging of blood vessels was proven to be a promising method for improving imaging quality.

For intracoronary imaging, the FD-OCT imaging systems require an injection of contrast media [15]. The increased amount of contrast media used for OCT imaging may lead to impairment of renal function. In order to reduce the risk of that, Ozaki et al. [15] suggested the use of low-molecular-weight dextran L (LMD-L) as a promising alternative flushing agent for intracoronary FD-OCT imaging, which can replace or reduce the use of contrast media. The comparative quantitative and qualitative study of using contrast media or LMD-L for FD-OCT imaging demonstrated that that the image quality and lumen measurement of FD-OCT imaging with LMD-L were comparable to those with contrast media. In addition, the volumes of LMD-L and contrast media required for FD-OCT imaging were the same. The obtained results demonstrated that contrast media can be replaced by LMD-L as a flushing agent for intracoronary FD-OCT imaging.

Li et al. [17] experimentally investigated the application of three chemicals (dextran, mannitol, and iohexol) as the flushing agents for *in vivo* imaging of rabbit abdominal aortas by the integrated intravascular ultrasound and optical coherence tomography (IVUS-OCT) system (Figure 20.6). Dextran was

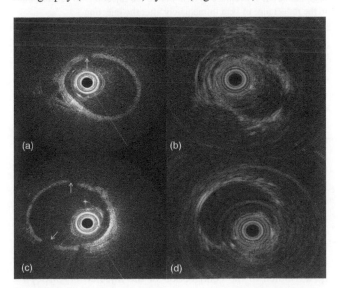

FIGURE 20.6   The images obtained with alternative flushing agents: (a, b) dextran (c, d) and mannitol. (a, c) are the OCT images of rabbit artery; (b, d) are the corresponding IVUS images. Arrows denote areas where the artery wall is not visible in the image. Reprinted from Ref. [17].

found to be the most effective flushing agent for intravascular imaging among these three chemicals, since dextran improved both IVUS and OCT signals, especially when used in high concentrations. The enhancement of OCT imaging quality was attributed not only to displacing of blood by the injected chemicals, but also to refractive index matching in blood between erythrocytes and blood plasma, since dextran, mannitol, and iohexol have higher refractive indices than blood plasma.

Recently, experiments on mice demonstrated that injection of saline in a tail vein of a mouse and consequent partial dilution of blood did not lead to a noticeable enhancement of the OCT probing depth when imaging through skin [14]. The injection of fructose intradermally without injection of optical clearing agent in the vein also did not provide a noticeable increase of OCT imaging depth [14]. In order to increase the OCT probing depth, it seems to be necessary to use optical clearing agents, which reduce light scattering in blood by refractive index matching and/or inducing aggregation of erythrocytes.

The enhancement of light transport after intravenous injection of PEG-300 or hemoglobin solution as the optical clearing agents was studied by OCT in *in vivo* experiments on two living mice [16]. The results of the study demonstrated that the optical clearing effect can be noticed immediately after the injections of PEG-300 or hemoglobin solution in the tail veins of the mice. The injections of the optical clearing agents in the blood vessels provided a rapid enhancement of light transport which allowed the borders of the veins and tissues surrounding the veins to be observed.

In some cases, the injection of an optical clearing agent into the blood vessel does not lead to rapid enhancement of light transport in the area of interest. It was recently reported that for the OCT imaging performed through skin, the combination of intradermal injections and intravenous injections of optical clearing agents can provide fast enhancement of OCT imaging and increase OCT probing depth [16]. The injection of glucose in skin was previously reported to be more effective for the immediate reduction of light scattering compared to the topical application of the clearing agent [26]. A significant decrease in skin reflectance was observed immediately after injection of glucose into the skin [26, 27].

In the *in vitro* experiments with mice, fructose solution with the concentration of 400 g/L was used as an optical clearing agent, and was injected in the skin in order to reduce scattering of the tissues near the tail vein of a mouse, while PEG-300 was injected into the tail veins [16].

The intradermal injection of fructose solution (400 g/L) in combination with the intravenous injection of PEG-300 led to a rapid optical clearing effect, whereas the injection of PEG-300 in the vein did not did not immediately provide any rapid enhancement of optical clearing after the injection. The combination of injections of the optical clearing agents intradermally and intravenously allowed the vein borders and tissues lying below the vein to be observed immediately after the injections. The injection of an optical clearing agent in skin can be used for enhancement of OCT imaging when the injection of the clearing agent into the blood vessels does not provide any noticeable and rapid optical clearing effect.

Overall, the use of optical clearing agents demonstrated the applicability for enhancement of light transport in highly scattering media for *in vivo* optical imaging techniques. The optical clearing agents which induce aggregation of erythrocytes and/or refractive index matching proved to be effective for application in OCT imaging and combined OCT and ultrasound imaging systems. The aggregation of blood is expected to be reversible in *in vivo* cases as erythrocytes are able to disaggregate in the flowing blood due to shear stress. The method of optical clearing demonstrated the potential to improve the quality of *in vivo* OCT imaging and increase the imaging depth in endoscopic mode and also when imaging through skin.

In vivo optical clearing of skin by blood and hemoglobin

The efficiency of application of hemoglobin solutions for the enhancement of transmittance of blood was recently theoretically [13] and experimentally proved [14, 16]. Application of blood and hemoglobin solutions as optical clearing agents can also be a potentially effective method for improvement of light transport in other biological tissues, such as skin. Since highly concentrated hemoglobin solutions and whole blood have high refractive indices, they can possibly provide an optical clearing effect in skin due to refractive index matching between collagen fibrils and ground matter (interstitial fluid).

The optical clearing effect upon application of hemoglobin and whole blood as the optical clearing agents was experimentally studied by *in vivo* measurements of optical reflectance in the near infrared (900–2000 nm) and visible (400–1000 nm) regions using two different fiberoptic spectrometers: USB4000-Vis-NIR and NIRQuest-512-2.2 (Ocean Optics, USA). The results of the measurements are presented at Figure 20.7 and Figure 20.8. From the figures, it is clearly seen that the intradermal injection of hemoglobin solution into rat skin (within rat dorsal area) caused the decrease of reflectance both in the visible and in infrared regions. In the near infrared range, the skin reflectance reaches its minimum after 5 min after injection of hemoglobin solution (Figure 20.7a). The reflectance drops more than two-fold after 5 min at the wavelengths of 1000 nm and 1300 nm. For the region of shorter wavelengths (Figure 20.8), a bigger decrease in skin reflectance was observed at 600 nm [almost three-fold in 5 min after the injection of hemoglobin solution (Figure 20.8a)]. The reflectance decreases from 16.1% to 8.4% at the wavelength of 805 nm and from 13.8% to 6.4% at 930 nm.

The main cause of the reflectance decrease after injection of hemoglobin solution was probably the refractive index matching between refractive indices of collagen fibrils and ground matter. This matching occurred rather fast after the injection of hemoglobin solution in skin, since it can be seen that the reflectance drops in the 5 min after injection. The refractive index of collagen fibrils is 1.41–1.47 [39], while the refractive index of interstitial fluid is lower compared to that of collagen fibers and estimated as 1.33–1.35 [24]. The mismatch of refractive indices of collagen fibrils and interstitial fluid is considered to be the main cause of scattering of skin. Since hemoglobin solution with the concentration of 120 g/L has a rather high refractive index (about 1.367–1.384 in the wavelength range used in this study), therefore, the observed optical clearing effect can be attributed to increase of refractive index

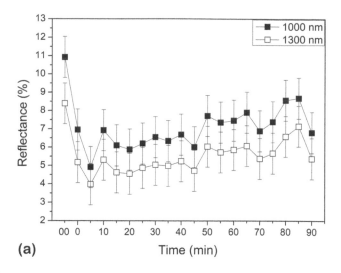

(a)

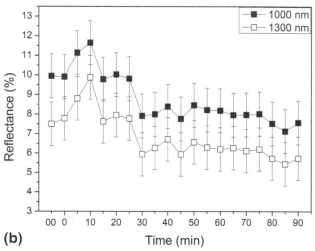

(b)

FIGURE 20.7 Temporal optical reflectance of rat skin after injection of hemoglobin (120 g/L) (a) and blood (2:1, in heparin) (b) solutions measured at 1000 and 1300 nm by the fiber optic spectrometer #1. "00" corresponds to the reflectance of skin before the injection; "0" corresponds to the reflectance of skin immediately after the injection.

of the ground matter as a results of the injection of hemoglobin, which caused a decrease in scattering.

The injection of blood solution into the skin also led to a decrease in reflectance, but after a longer time (30–40 min) (Figures 20.7b and Figure 20.8b). Moreover, the relative drop in skin reflectance after blood injection was less than that after the injection of hemoglobin. This fact can be explained by the assumption that since the refractive index of blood diluted in heparin is 1.343 (at 589 nm), which is lesser than the refractive index of the hemoglobin solution (1.384), the hemoglobin solution therefore provides better matching of refractive indices between tissue components than the blood solution.

It is also can be noted that in the first 10–15 min after the injection of blood solution, the reflectance of skin is higher than it was initially (Figure 20.7b and Figure 20.8b). This can be attributed to the contribution of the injected erythrocytes, which increase the overall scattering of skin. The reflectance of the skin started to decrease after about 30 min after the injection of blood solution. This time can be associated with the lysis of erythrocytes, which result in the decrease of the amount

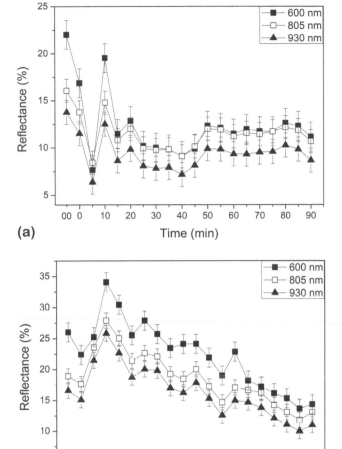

(a)

(b)

FIGURE 20.8 Temporal optical reflectance of rat skin after injection of hemoglobin (120 g/L) (a) and blood (2:1, in heparin) (b) solutions measured at 600, 805 and 930 nm by the fiber optic spectrometer #2. "00" corresponds to the reflectance of skin before the injection; "0" corresponds to the reflectance of skin immediately after the injection.

of scatterers (erythrocytes) and release of hemoglobin out of the erythrocytes, which leads to an overall decrease in scattering.

Overall, the injection of nonscattering hemoglobin solution caused a more noticeable and immediate optical clearing effect in skin compared to that after injection of blood solution. In case of using the whole blood solution, a period of time of about 10–15 min after the injection is needed for the lysis of RBCs to occur in order to produce free hemoglobin, which further provides refractive index matching.

In vitro optical clearing of gastric wall mucosa by hemoglobin

The contactless laser coagulation and ablation of pathologic formations in different organs is widely used in modern clinics [5, 6, 24, 40]. The minor invasiveness of the procedure reduces the risk of postoperative complications. The sources used for this purpose include, particularly, the ytterbium (1075 nm), CO2 (9.4–10.6 μm), Nd:YAG (1064 nm), diode (980, 808,

810 nm), and other lasers. The main advantage of lasers generating at the wavelengths that coincide with the absorption bands of water is the small depth of light penetration into tissues, which prevents damage to the adjacent healthy tissue underlying the affected area. However, cheaper diode lasers have recently become more and more widely used. The radiation from infrared diode lasers in the range 800–1100 nm penetrates deep enough, since in this range, the absorption of such tissue components as hemoglobin, melanin, proteins, and water is relatively small [24]. Thus, the danger of tissue damage or perforation arises. To prevent these complications, the authors of some papers [41–43] propose increasing the thickness of the irradiated object. For example, in order to protect the mucosa of stomach and intestinal wall in the course of laser endoscopic resection, the layers of mucosa and submucosa were separated by certain spacing, filled with glycerol, sodium hyaluronate, or hydrogel. An alternative possible approach is to optimize the laser's impact by varying the optical parameters of the tissue. The control of optical parameters can be implemented both by enhancing the absorption properties of the object itself and by reducing the scattering in the tissues adjacent to the lesion focus. In the first case, on the one hand, the laser beam penetration depth is reduced and, on the other hand, the fraction of energy absorbed in the lesion focus is essentially increased, thus improving the laser coagulation efficiency. In the second case, the precision of laser radiation focusing is facilitated.

The absorption and scattering properties of the stomach wall mucosa are well-studied [44]. It has been shown that the use of biocompatible immersion agents, such as solutions of glycerol, propylene glycol, etc., results in efficient optical clearing (i.e., light scattering reduction) of the gastric tissues in the near IR spectral region [45–49]. Earlier, we proposed using hemoglobin in order to control the scattering properties of blood by creating a local hemolysis site in a blood vessel [13]. In this section, we propose using blood hemoglobin as an absorbing agent, and demonstrated the effect of aqueous hemoglobin solution on the optical properties of gastric wall mucosa in order to improve the conditions for laser coagulation in the visible and near IR spectral ranges.

Figure 20.9 presents the spectra of the absorption coefficient and the reduced scattering coefficient of the gastric wall mucosa before and after the injection of aqueous hemoglobin solution with concentration 70 g/L. The averaging was executed over five samples of the tissue. From Figure 20.9(a), it follows that in the spectral range from 350 to 1250 nm, one can observe an essential increase in the absorption coefficient of the mucosa (by 2–4.5× depending on the wavelength). In the spectral range 1250–2500 nm, the increase is expressed essentially weaker. Such behavior of the absorption coefficient is related to the characteristic absorption of hemoglobin in the visible wavelength range. From Figure 20.9(b), it follows that the injection of the aqueous solution of hemoglobin has practically no effect on the scattering characteristics of the mucosa.

The result of estimating the depth of light penetration into the tissue $\left(\delta = 1 / \sqrt{3\mu_a(\mu_a + \mu'_s)} \right)$ is shown in Figure 20.10. The expression used is applicable to the case when the tissue surface is uniformly illuminated by the radiation from a point source, located at some distance from the surface. This corresponds to the conditions of real laser surgery of gastric wall, since in this case the illuminating probe is introduced directly into the

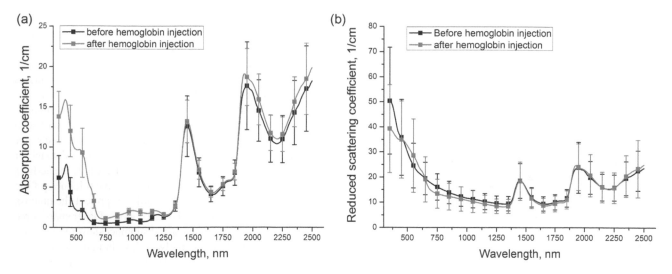

FIGURE 20.9 Optical properties of human gastric wall mucosa measured before (black line) and after the injection of aqueous hemoglobin solution (red line): (a) absorption coefficient spectra; (b) reduced scattering coefficient spectra. Vertical bars show the standard deviation. Adapted from Ref. [40].

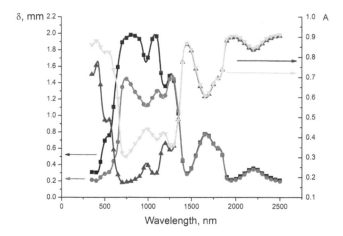

FIGURE 20.10 Wavelength dependence of the depth of radiation penetration into the human gastric wall mucosa (δ) (black line, before the hemoglobin injection; red line, after the hemoglobin injection), calculated from the experimental data of Figures 20.9 (a) and (b). Results of Monte Carlo simulation of the fraction of photons absorbed in the mucosa (A) versus the wavelength (blue line, before the hemoglobin injection; green line, after the hemoglobin injection). Adapted from Ref. [40].

stomach cavity without contact with the mucosa surface. The depth of light penetration into the mucosa was calculated using the values of the absorption coefficient (see Figure 20.9a) and the reduced scattering coefficient (see Figure 20.9b). Another important parameter is the fraction of the incident radiation energy absorbed in the mucosa. This parameter was calculated in the course of Monte Carlo modeling with algorithm presented in [50], and the result is presented in Figure 20.10.

In this figure, one can clearly see that the depth of laser radiation penetration essentially decreases in the visible range of wavelengths (at the wavelength of Nd:YAG laser radiation (1064 nm) the penetration depth decreases by nearly 60%, whereas the fraction of absorbed energy at this wavelength increases by nearly 90%). The computer simulation allows the evaluation of the change of the absorbed energy fraction at the wavelengths of diode-based laser systems used for the

photodestruction of tissue neoplasms. Thus, for the wavelength 810 nm, the injection of hemoglobin aqueous solution leads to the increase of the absorbed energy fraction by nearly 80%, with an almost 50% reduction in the laser radiation penetration depth. For the wavelength 970 nm, the increase of the absorbed energy fraction amounts to nearly 65%, with a decrease of about 50% in the laser radiation penetration depth.

Conclusions

Light scattering in biological tissues and blood can be effectively reduced by application of biocompatible optical clearing agents. The method of immersion optical clearing showed applicability for decreasing light scattering and enhancing light transport in tissues and blood. Various biocompatible optical clearing agents can be effectively applied for enhancement of light transport in blood and, therefore, for improving image quality and increasing imaging depth for optical imaging techniques.

The effect of optical clearing of blood can be caused by several factors, such as refractive index matching between erythrocytes and blood plasma after addition of optical clearing agents, alterations in size and form of erythrocytes, hemolysis, changes in aggregation properties of blood, and other factors. Refractive index matching between blood plasma and erythrocytes is considered to be one of the main causes of the improvement of light transport in blood. Along with refractive index matching, several other effects can influence the transmittance of blood after addition of clearing agents. For example, the optical clearing agents which induce aggregation of blood were experimentally proved to induce a noticeable optical clearing effect. Deformation of RBCs and variations in their form and size also change the scattering properties of blood. The partial lysis of RBCs can be a possible reason for the enhancement of transmittance of blood due to reduction of scattering between blood plasma and RBCs, followed by destruction of the RBCs. The release of hemoglobin from erythrocytes also increases the refractive index of blood plasma and thereby improves light transport in blood.

The experiments carried out on animals demonstrate the possibility of using the optical clearing method for *in vivo* applications. The optical clearing method has the potential to increase the probing depth of optical imaging techniques not only in endoscopic mode, but also when imaging through skin. For more effective *in vivo* optical clearing in blood vessels through skin, the injection of optical clearing agent into the vein can be made along with injection of optical clearing agent in skin above the vein in order to reduce light scattering in skin and achieve the optical clearing effect more rapidly and effectively.

Due to the high refractive index of highly concentrated hemoglobin solutions and blood, these substances can also be potentially applied as optical clearing agents for enhancement of light transport in biological tissues. The *in vivo* experiments carried out in animals demonstrate that the injection of highly concentrated hemoglobin solution into skin caused an immediate and considerable drop in optical reflectance of skin; therefore, the application of hemoglobin solutions can be a promising method to achieve a fast and efficient optical clearing effect in skin and other biological tissues for *in vivo* applications of optical imaging techniques.

Acknowledgments

E.A.G. and A.N.B. were supported by RFBR grant number 18-52-16025. V.V.T. was supported by grant number 13.2251.21.0009 of the Ministry of Science and Higher Education of the Russian Federation.

REFERENCES

1. V.V. Tuchin, I.L. Maksimova, D.A. Zimnyakov, I.L. Kon, A.H. Mavlyutov, and A.A. Mishin, "Light propagation in tissues with controlled optical properties," *Journal of Biomedical Optics* **2**(4), 401–417 (1997).

2. G. Vargas, E.K. Chan, J.K. Barton, H.G. Rylander, and A.J. Welch, "Use of an agent to reduce scattering in skin," *Lasers in Surgery and Medicine* **24**(2), 133–141 (1999).

3. R.K. Wang, X. Xu, V.V. Tuchin, and J.B. Elder, "Concurrent enhancement of imaging depth and contrast for optical coherence tomography by hyperosmotic agents," *Journal of the Optical Society of America B* **18**(7), 948–953 (2001).

4. A.N. Bashkatov, E.A. Genina, Y.P. Sinichkin, V.I. Kochubey, N.A. Lakodina, and V.V. Tuchin, "Glucose and mannitol diffusion in human dura mater," *Biophysical Journal* **85**(5), 3310–3318 (2003).

5. E.A. Genina, A.N. Bashkatov, and V.V. Tuchin, "Tissue optical immersion clearing," *Expert Review of Medical Devices* **7**(6), 825–842 (2010).

6. D. Zhu, K.V. Larin, Q. Luo, and V.V. Tuchin, "Recent progress in tissue optical clearing," *Laser & Photonics Reviews* **7**(5), 732–757 (2013).

7. X. Wen, Z. Mao, Z. Han, V.V. Tuchin, and D. Zhu, "In vivo skin optical clearing by glycerol solutions: Mechanism," *Journal of Biophotonics* **3**(1–2), 44–52 (2010).

8. L.M. Oliveira, M.I. Carvalho, E.M. Nogueira, and V.V. Tuchin, "Skeletal muscle dispersion (400–1000 nm) and kinetics at optical clearing," *Journal of Biophotonics* **11**(1), e201700094 (2018).

9. X. Xu, and R.K. Wang, "The role of water desorption on optical clearing of biotissue: Studied with near infrared reflectance spectroscopy," *Medical Physics* **30**(6), 1246–1253 (2003).

10. M. Brezinski, K. Saunders, C. Jesser, X. Li, and J. Fujimoto, "Index matching to improve optical coherence tomography imaging through blood," *Circulation* **103**(15), 1999–2003 (2001).

11. V.V. Tuchin, X. Xu, and R.K. Wang, "Dynamic optical coherence tomography in studies of optical clearing, sedimentation, and aggregation of immersed blood," *Applied Optics* **41**(1), 258–271 (2002).

12. X. Xu, R.K. Wang, J.B. Elder, and V.V. Tuchin, "Effect of dextran-induced changes in refractive index and aggregation on optical properties of whole blood," *Physics in Medicine and Biology* **48**(9), 1205–1221 (2003).

13. V. Tuchin, D. Zhestkov, A. Bashkatov, and E. Genina, "Theoretical study of immersion optical clearing of blood in vessels at local hemolysis," *Optics Express* **12**(13), 2966–2971 (2004).

14. O. Zhernovaya, V.V. Tuchin, and M.J. Leahy, "Blood optical clearing studied by optical coherence tomography," *Journal of Biomedical Optics* **18**(2), 026014 (2013).

15. Y. Ozaki, H. Kitabata, H. Tsujioka et al., "Comparison of contrast media and low-molecular-weight dextran for frequency-domain optical coherence tomography," *Circulation Journal* **76**(4), 922–927 (2012).

16. O. Zhernovaya, V.V. Tuchin, and M.J. Leahy, "Enhancement of OCT imaging by blood optical clearing in vessels: A feasibility study," *Photonics & Lasers in Medicine* **5**(2), 151–159 (2016).

17. J. Li, H. Minami, E. Steward et al., "Optimal flushing agents for integrated optical and acoustic imaging systems," *Journal of Biomedical Optics* **20**(5), 056005 (2015).

18. A.N. Bashkatov, D.M. Zhestkov, É.A. Genina, and V.V. Tuchin, "Immersion clearing of human blood in the visible and near-infrared spectral regions," *Optics and Spectroscopy* **98**(4), 638–646 (2005).

19. A. Kratz, M. Ferraro, P.M. Sluss, and K.B. Lewandrowski, "Normal reference laboratory values," *New England Journal of Medicine* **351**(15), 1548–1563 (2004).

20. R. Munker, E. Hiller, J. Glass, and R. Paquette, Eds., *Modern Hematology*, 2nd ed., Humana Press Inc., Totowa, NJ (2007).

21. N. Bosschaart, G.J. Edelman, M.C.G. Aalders, T.G. van Leeuwen, and D.J. Faber, "A literature review and novel theoretical approach on the optical properties of whole blood," *Lasers in Medical Science* **29**(2), 453–479 (2014).

22. M. Friebel, and M. Meinke, "Model function to calculate the refractive index of native hemoglobin in the wavelength range of 250–1100 nm dependent on concentration," *Appl. Opt.* **45**(12), 2838–2842 (2006).

23. S. Cheng, H.Y. Shen, G. Zhang, C.H. Huang, and X.J. Huang, "Measurement of the refractive index of biotissue at four laser wavelengths," *Proceedings of SPIE* **4916**, 172–176 (2002).

24. V.V. Tuchin, *Tissue Optics: Light Scattering Methods and Instruments for Medical Diagnostics*, 3rd ed., SPIE Press, Bellingham, WA (2015).

25. X. Guo, Z. Guo, H. Wei et al., "In vivo comparison of the optical clearing efficacy of optical clearing agents in human skin by quantifying permeability using optical coherence tomography," *Photochemistry and Photobiology* **87**(3), 734–740 (2011).

26. A.N. Bashkatov, A.N. Korolevich, V.V. Tuchin et al., "*In vivo* investigation of human skin optical clearing and blood microcirculation under the action of glucose solution," *Asian Journal of Physics* **15**(1), 1–14 (2006).

27. E.I. Galanzha, V.V. Tuchin, A.V. Solovieva, T.V. Stepanova, Q. Luo, and H. Cheng, "Skin backreflectance and microvascular system functioning at the action of osmotic agents," *Journal of Physics D: Applied Physics* **36**(14), 1739–1746 (2003).

28. R.V. Kuranov, V.V. Sapozhnikova, D.S. Prough, I. Cicenaite, and R.O. Esenaliev, "In vivo study of glucose-induced changes in skin properties assessed with optical coherence tomography," *Physics in Medicine and Biology* **51**(16), 3885–3900 (2006).

29. L.M. Oliveira, M.I. Carvalho, E.M. Nogueira, and V.V. Tuchin, "The characteristic time of glucose diffusion measured for muscle tissue at optical clearing," *Laser Physics* **23**(7), 075606 (2013).

30. N. Sudheendran, M. Mohamed, M.G. Ghosh, V.V. Tuchin, and K.V. Larin, "Assessment of tissue optical clearing as a function of glucose concentration using optical coherence tomography," *Journal of Innovative Optical Health Sciences* **3**(3), 169–176 (2010).

31. G. Vargas, K.F. Chan, S.L. Thomsen, and A.J. Welch, "Use of osmotically active agents to alter optical properties of tissue: Effects on the detected fluorescence signal measured through skin," *Lasers in Surgery and Medicine* **29**(3), 213–220 (2001).

32. O.S. Zhernovaya, E. Jonathan, V.V. Tuchin, and M.J. Leahy, "Study of optical clearing of blood by immersion method," *Proceedings of SPIE* 7898, 78981B (2011).

33. G. Popescu, T. Ikeda, C.A. Best, K. Badizadegan, R.R. Dasari, and M.S. Feld, "Erythrocyte structure and dynamics quantified by Hilbert phase microscopy," *Journal of Biomedical Optics* **10**(6), 060503 (2005).

34. H. Zhong, Z. Guo, H. Wei et al., "Synergistic effect of ultrasound and thiazone–PEG 400 on human skin optical clearing in vivo," *Photochemistry and Photobiology* **86**(3), 732–737 (2010).

35. M.H. Khan, B. Choi, S. Chess, K.M. Kelly, J. McCullough, and J.S. Nelson, "Optical clearing of in vivo human skin: Implications for light-based diagnostic imaging and therapeutics," *Lasers in Surgery and Medicine* **34**(2), 83–85 (2004).

36. R.K. Wang, and J.B. Elder, "Propylene glycol as a contrasting agent for optical coherence tomography to image gastrointestinal tissues," *Lasers in Surgery and Medicine* **30**(3), 201–208 (2002).

37. L. Northam, and G.V.G. Baranoski, "A novel first principles approach for the estimation of the sieve factor of blood samples," *Optics Express* **18**(7), 7456–7469 (2010).

38. I.-K. Jang, G.J. Tearney, B. MacNeill et al., "In vivo characterization of coronary atherosclerotic plaque by use of optical coherence tomography," *Circulation* **111**(12), 1551–1555 (2005).

39. D.W. Leonard, and K.M. Meek, "Refractive indices of the collagen fibrils and extrafibrillar material of the corneal stroma," *Biophysical Journal* **72**(3), 1382–1387 (1997).

40. A.N. Bashkatov, E.A. Genina, V.A. Grishaev, S.V. Kapralov, V.I. Kochubey, and V.V. Tuchin, "Study of the changes of gastric wall mucosa optical properties under the impact of aqueous solutions of haemoglobin and glucose for improving conditions of the laser coagulation," *Journal of Biomedical Photonics & Engineering* **3**(4), 040304 (2017).

41. T. Uraoka, T. Fujii, Y. Saito et al., "Effectiveness of glycerol as a submucosal injection for EMR," *Gastrointestinal Endoscopy* **61**(6), 736–740 (2005).

42. H. Yamamoto, H. Kawata, K. Sunada et al., "Success rate of curative endoscopic mucosal resection with circumferential mucosal incision assisted by submucosal injection of sodium hyaluronate," *Gastrointestinal Endoscopy* **56**(4), 507–512 (2002).

43. I. Kumano, M. Ishihara, S. Nakamura et al., "Endoscopic submucosal dissection for pig esophagus by using photo-crosslinkable chitosan hydrogel as submucosal fluid cushion," *Gastrointestinal Endoscopy* **75**(4), 841–848 (2012).

44. A.N. Bashkatov, E.A. Genina, and V.V. Tuchin, "Tissue optical properties," in *Handbook of Biomedical Optics* D.A. Boas, C. Pitris, and N. Ramanujam, Eds., pp. 67–100, Taylor & Francis Group, LLC, CRC Press (2011).

45. R.K. Wang, X. Xiangqun, H. Yonghong, and J.B. Elder, "Investigation of optical clearing of gastric tissue immersed with hyperosmotic agents," *IEEE Journal of Selected Topics in Quantum Electronics* **9**(2), 234–242 (2003).

46. X. Xu, R. Wang, and J.B. Elder, "Optical clearing effect on gastric tissues immersed with biocompatible chemical agents investigated by near infrared reflectance spectroscopy," *Journal of Physics D: Applied Physics* **36**(14), 1707–1713 (2003).

47. Y.H. He, and R.K. Wang, "Dynamic optical clearing effect of tissue impregnated with hyperosmotic agents and studied with optical coherence tomography," *Journal of Biomedical Optics* **9**(1), 200–206 (2004).

48. X. Xu, and R.K. Wang, "Synergistic effect of hyperosmotic agents of dimethyl sulfoxide and glycerol on optical clearing of gastric tissue studied with near infrared spectroscopy," *Physics in Medicine and Biology* **49**(3), 457–468 (2004).

49. H. Xiong, Z. Guo, C. Zeng, L. Wang, Y. He, and S. Liu, "Application of hyperosmotic agent to determine gastric cancer with optical coherence tomography ex vivo in mice," *Journal of Biomedical Optics* **14**(2), 1–5 (2009).

50. L. Wang, S.L. Jacques, and L. Zheng, "MCML: Monte Carlo modeling of light transport in multi-layered tissues," *Computer Methods and Programs in Biomedicine* **47**(2), 131–146 (1995).

21

Blood and lymph flow imaging at optical clearing

**Polina A. Dyachenko (Timoshina), Arkady S. Abdurashitov,
Oxana V. Semyachkina-Glushkovskaya, and Valery V. Tuchin**

CONTENTS

Introduction

As known, mammals have two specialized vascular circulatory systems: the circulatory vascular network and the lymphatic vascular network [1–4]. The lymphatic system is represented by lymphatic capillaries, lymphatic vessels, and lymph nodes. Lymphatic capillaries absorb excess tissue fluid and fine solid particles. The lymph formed in them flows through the lymphatic vessels, which merge with each other and form several large vessels that flow into the veins in the chest area. Lymph nodes are located along the lymphatic vessels. These are small pink bean-like formations that function as biological filters; they trap particles collected in the lymph and destroy microorganisms. Lymph nodes are also included in the immune system; because lymphocytes are formed in them, antibodies are produced [5].

The circulatory system consists of the heart and blood vessels – arteries, veins and capillaries. Arteries branch many times into smaller ones and form blood capillaries, in which the exchange of substances between blood and body tissues takes place. The capillaries have extremely thin walls; from one layer of epithelial tissue, part of the blood plasma seeps through them, replenishing the amount of tissue fluid, nutrients, oxygen, carbon dioxide, and other substances. The circulatory and lymphatic systems are closely related. The liquid enters the tissues only through the arteries in the blood, and flows out of the tissues in two ways: through the veins in the blood and through the lymphatic vessels in the form of lymph.

Near the heart, blood and lymph flow merge again. This is also important because in the intestine, some nutrients do not enter the bloodstream, but pass into the lymph [4].

In this regard, monitoring the state of both the blood circulatory and lymphatic systems is one of the important problems of modern medical diagnostics. This is due to the fact that many diseases, such as diseases of the cardiovascular system, atherosclerosis, *diabetes mellitus*, chronic venous insufficiency, and others, cause functional and morphological changes in the microvasculature. Damage at the level of microcirculation/macrocirculation is the basis for the development of stress-induced diseases, such as gastrointestinal hemorrhages, arterial hypertension, hemorrhagic pancreatitis, myocardial infarction, strokes, etc. [6–12]. It is also known that lymphatic vessels are active participants in basic physiological and pathophysiological processes. Until recently, lymphatic dysfunction was mainly associated with primary and secondary lymphedema. Surprisingly, however, lymphatic vascular defects have been found in conditions such as obesity, cardiovascular disease, inflammation, hypertension, atherosclerosis, Crohn's disease, glaucoma, and various neurological disorders such as Alzheimer's disease [4].

Currently, the most effective diagnostic methods for determining the physiological parameters of blood flow and microcirculation are the methods based on dynamic light scattering [13], as well as on principles of OCT, the so called Doppler OCT (DOCT) [14–17]. One of the prospective methods for the assessment of blood flow is laser speckle contrast imaging (LSCI) [17–23]. However, these methods require an increase in a sufficiently shallow sounding depth of biological tissues, and an increase in the spatial resolution and contrast of images of the structural components of tissues. The method of optical clearing (OC), which reduces the efficiency of scattering from static scatterers, allows one to extend the applicability of the optical methods to larger imbedding depths of the vessels.

Methods of optical imaging

Optical coherence tomography

OCT is a widespread and well-developed optical tool for nondestructive imaging with high spatial and temporal resolution [24]. From the OCT signal, both structural (skin layers boundaries, vessel geometry, etc.) [25–27] and functional (blood flow velocity, perfusion, oxygenation, etc.) data can be extracted simultaneously [28–32].

The main principles of this technique are low-coherence interferometry and the confocal detection scheme. The first is achieved by utilizing a broad-band light source like superluminescent diodes (SLD) or tunable lasers (swept-sources). Such sources enable a coherence gain property of the OCT – the ability to reconstruct a depth-resolved (topographic) signal, raised from the small low-coherence volumes into which sample can be divided. The quantitative characteristics of these volumes are defined by the spectral properties of the source, like frequency bandwidth for SLD or tuning range for a swept-source one [33]. All of the modern OCT systems are optical

fiber-based, this means that the illumination and detection is performed confocaly through the fiber. This is similar to conjugate diaphragms in traditional confocal microscopy that reduces the amount of unwanted backscattered light reaching the detector plane, thus improving image contrast. Usually scanning along the (x, y) plane is performed point-wisely using galvo-mirrors. The OCT signal at each point of a sample is usually called an A-scan. A stack of A-scans along the x-axis is called B-scan, while several B-scans taken along the y-axis forms a C-scan. These are the generally accepted terms. There are a modifications of the OCT that required scanning along one axis (line-field OCT) [34–36] or full-field variants [37].

By the detection principle, OCTs are divided into two major domains: *temporal* and *spectral*. Their respective optical schemes are illustrated in Figure 21.1. Let us briefly introduce the concept and the pros and cons of both setups.

Time-domain optical coherence tomography (TD-OCT)

From the hardware perspective, TD-OCT is a reflection mode Michelson interferometer (Figure 21.1a) in which a coherent superposition between a sample and a reference wave is recorded by a photosensitive element. By physically moving the reference mirror along its axis, a relative time delay between sample and reference arm changes, and thus scanning is performed along the sample's depth z. The obvious and main disadvantage of such a detection scheme is the mechanically movable parts in the reference arm, which greatly reduce acquisition speed and repeatability of the TD-OCT setup.

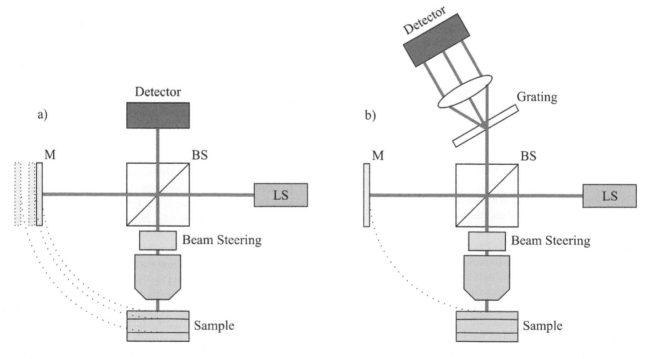

FIGURE 21.1 Principle optical schemes of the temporal (a) and spectral (b) domain OCT setups: M – reference mirror, BS – beam splitter, and LS – light source.

Spectral domain optical coherence tomography (SD-OCT)

To overcome the mechanical scanning problem, a spectral domain approach was developed. In this setup, prior to the registration process a coherent superposition is decomposed down to fundamental optical frequencies either by a diffraction grating (parallel detection) or via a tunable laser source (sequential detection). Although the detection in a spectral domain significantly increases the acquisition speed, it comes with a cost of limited imaging depth due to discretization process of the spectrum and finite resolving power of the diffraction grating or finite spectral width of the swept-source laser line ("fall-off" effect [37]).

The vast majority of commercially available OCT systems are spectral domain ones, so it is relevant to describe a basic mathematical model of signal formation in SD-OCT as well as important properties such as *transverse* and *longitudinal* resolution and their dependencies of the light source's spectral characteristics.

Basic mathematical model of OCT signal formation in spectral domain

The SD-OCT signal can be described as follows: let us assume that the plane wave (single spectral component of the arbitrary source $S(k)$, where k is the wavenumber) approaches the specimen. Each reflective layer within the specimen contributes a complex factor $a_j \exp\{-ink\Delta z_j\}$ to the resulted complex amplitude distribution, where Δz_j is the additional geometrical path difference that light needs to travel inside the sample with respect to reference mirror position. A simplified drawing of such a process is presented in Figure 21.2.

The cumulative signal from all layers at a particular spectral component is given as:

$$I(k) = S(k)\left|\sum_j a_j \exp^{-ikn\Delta z_j} + \exp^{-ikz_{ref}}\right|^2, \quad (21.1)$$

where n is the refractive index of the specimen and z_{ref} is the absolute position of the reference mirror. The beam splitter ratio as well as the mirror's reflectivity coefficients and total energy losses inside the optical elements are excluded from this equation because they are just constant values that are not important for the understanding of the underlying physics behind the OCT signal formation. In Equation 21.1 inside the parenthesis is the Fourier decomposition of the sample's complex reflectivity a_j, thus an inverse Fourier transform of the Equation 21.1 should be taken to recover a depth-dependent complex OCT signal:

$$U(z) = FFT^{-1}\{I(k)\}. \quad (21.2)$$

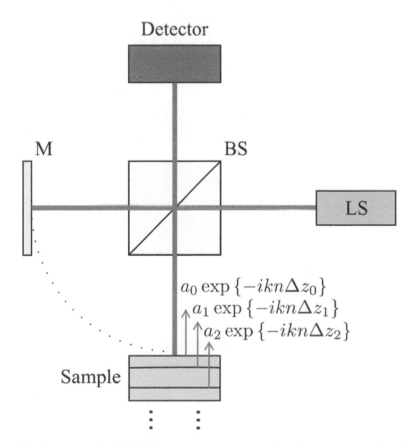

FIGURE 21.2 Illustration of the OCT signal formation. Light from the source (LS) is split into two equal beams by the beam splitter (BS). One part is directed to the reference mirror M, and another is guided to the sample. Coherent superposition of the reference beam and all elementary waves reflected from the sample is recorded at the detector plane.

$|U(z)|^2$ is carrying out the structural or morphological information about the specimen, more precisely about the depth-dependent refractive index changes. Both Equations 21.1 and 21.2 describe the ideal OCT signal in a spectral domain prior to the registration process. Besides the scattering and absorption coefficients of the sample, the registration procedure by itself limits the imaging depth of the SD-OCT. Any spectroscopic device operates with a finite spectral resolution σ, which is the minimum distance between two adjacent wavelengths that device could resolve. The spectrum of the coherence superposition of reference and sample arm is recorded on a CCD (charge couple device), which typically has a square shaped pixel with a finite area. This two-step registration process could be described as convolutions in a spectral domain of an ideal OCT signal with a spectral resolution curve of the spectroscopic device and a pixel-shape curve of a CCD array and is known as "roll-off" effect in the SD-OCT. By assuming a Gaussian resolution curve and a square pixel shape, Equation 21.2 could be rewritten as follows (according to a Fourier theorem, convolution in a spectral domain is equal to multiplication in a space domain with the Fourier transforms of corresponding functions):

$$U(z) = FFT^{-1}\{I(k)\} \times FFT\left\{\frac{1}{\sqrt{2\pi\sigma^2}}\exp\left(\frac{-k}{2\sigma^2}\right)\right\}$$

$$\times FFT\left\{\frac{1}{\Delta k}rect\left(\frac{k}{\Delta k}\right)\right\}. \tag{21.3}$$

From the physical point of view, the detection process has the effect of vanishing high frequency oscillations in the spectrum of the OCT signal. High frequency oscillations come from the interference between sample wave with a large additional propagation distance Δz and a reference field, thus the "fall-off" effect "suppresses" the intensity of deep reflective layers and reduces the imaging depth of the SD-OCT system as illustrated in Figure 21.3. This "roll-off" problem could be partially overcome by using a swept-source laser instead of SLD and diffraction grating. The spectral width of a sweepable laser line is usually much smaller than the resolving power of a spectroscopic device and thus the "roll-off" effect could be significantly reduced [38].

Resolving power of the OCT

In traditional microscopy, transverse and longitudinal resolutions are linked together through the numerical aperture of the

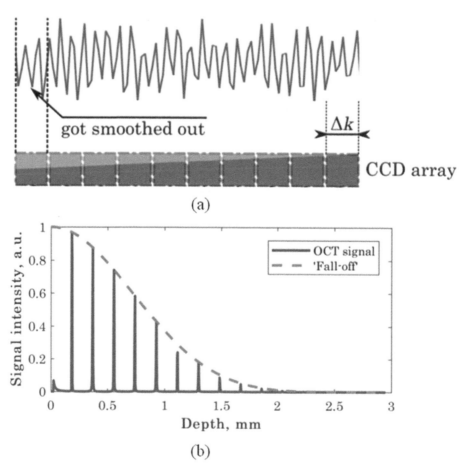

FIGURE 21.3 Graphical representation of the "roll-off" effect in the SD-OCT systems: high frequencies spectrum oscillations are averaged other the CCD pixel area and get smoothed out by the resolution curve of the spectroscopic device (a); ideal OCT signal in a space domain (b) got modulated by a "roll-off" curve which is a product of second and third factors in Equation 21.3 thus reducing the imaging depth of an SD-OCT setup (adapted from [38]).

imaging lens. Transverse resolution of the OCT system is diffraction limited and guided by the numerical aperture of the lens used to focus light on the sample. The general equation is:

$$\Delta(x,y) \approx 0.61 \frac{\lambda_0}{NA}, \qquad (21.4)$$

where λ_0 is the central wavelength of the light source and NA is the numerical aperture of the lens. Longitudinal resolution is determined by the spectral characteristics of the source or its temporal coherence length. Considering Gaussian spectral shape, the longitudinal resolution is expressed as follows:

$$l_c \approx \frac{\lambda_0^2}{\Delta \lambda}, \qquad (21.5)$$

where $\Delta \lambda$ is the spectral bandwidth of the light source.

Usually, OCT systems have a focusing lens with relatively low numerical aperture, and thus the depth of focus (DOF $\approx 2\lambda_0 / NA^2$) of this lens is much larger compared to the temporal coherence of the source. This property of OCT to unbound longitudinal resolution from the numerical aperture of the lens is extremely important and widely used to image relatively thick specimens with a high transverse resolution without any focusing corrections.

Functional OCT

It was mentioned earlier that a reconstructed complex OCT signal can be used to obtain functional data such as blood-flow velocity and perfusion. Modern commercial spectral domain OCT systems operate on 100–250 kHz (x, y) scanning frequencies, while some custom-made devices can perform even faster than that [39, 40]. Such high scanning frequencies open the possibility of detecting small relative phase changes of the light reflected from the neighbor spatial locations of the specimen, due to the Doppler shift. The Doppler frequency shift occurs when a photon is scattering on a moving particle such as red blood cell (RBC). Frequency shift is proportional to the relative phase different between adjacent A-scans as follows [41]:

$$\Delta f = \frac{\Delta \varphi}{2\pi \cdot dt} = \frac{2V \cos(\theta)}{\lambda_0}, \qquad (21.6)$$

where Δf is the Doppler frequency shift, $\Delta \varphi$ is the phase difference between adjacent A-scans, dt is the A-scan period, θ is the angle between mean velocity vector and illumination light beam as illustrated in Figure 21.4, and V is the magnitude of the velocity vector.

As well as a scattering induced frequency shift, a time-variant speckle dynamic could also be used for blood-flow monitoring [42]. Speckles arise from the interference of multiple secondary wavelets with random amplitude and phase distribution. These wavelets originate from the coherent light scattering on the micro-inhomogeneities in the sample bulk. If the spatial distribution of such inhomogeneities does not change with time, the phase relations between secondary wavelets will be preserved and the intensity distribution in the

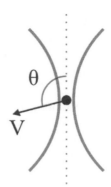

FIGURE 21.4 Illustration of Doppler angle.

speckle pattern will be still. If the phase relations are non-constant in time either because of Doppler shift or some mechanical strains, intensity distribution of the speckle pattern will be time-dependent and proportional to these phase changes. Usually a variance of time-dependent speckle intensity is calculated to estimate the magnitude of motion inside a region of a sample, but more sophisticated algorithms based on variance of the complex amplitude or phase-variance approaches have been developed [42].

Lymph fluid, due to its relatively high transparency in a broad spectral range and low concentration of lymphocytes, is challenging to image using the OCT system. Besides that, some lymphatic vessels (like meningeal) are fairly small, down to an order of tenths of microns [43], thus requiring a high NA objective and dense lateral scanning. One way to overcome such concerns is to introduce some sort of contrast agent into the lymph fluid to increase the scattering signal from these regions. Composite gold nanorods (GNRs) are particles that are widely used for such purposes [44, 45]. By adjusting their length and diameter, the plasmon resonance can be precisely tuned to match the central wavelength of the OCT's light source. Advanced methods of OCT signal processing based on intensity and phase analysis as well as sophisticated protocols for spectroscopic OCT were developed recently to compute the distribution of GNRs within tissues and enhance OCT signal from the small blood and lymphatic vessels [46, 47]. For intensity and phase angiography approaches, GNRs serves the same role as red blood cell (RBCs) in a regular blood vessel – time-varying inhomogeneity with random intensities and phases of the scattered light. There are some label-free approaches to lymphatic vessels segmentation using OCT, but they use a point scanning technique (M-mode OCT) and cannot provide wide-field spatial distribution of the vessels [48].

Laser speckle contrast imaging

Laser speckle contrast imaging (LSCI) or laser speckle contrast analysis (LASCA) is a simple yet powerful optical tool for noncontact optical monitoring of blood flow. It relies on the numerical analysis of the laser speckle pattern formed by the coherent light scattering out of dynamic objects. Speckle patterns can be divided to subjective and objective. An objective pattern forms on a screen or other photosensitive materials and can be observed by the optical system. Their mean size is entirely dependent on object surface roughness, illuminated area A, and distance to the screen L:

$$\mu_{speckle} = \frac{\lambda A}{L}. \qquad (21.7)$$

In contrast, subjective speckles form by the optical system and are registered by the CCD or eye retina. In this case, mean speckle size is independent of surface microstructure and is determined only by the properties of the imaging system:

$$\mu_{speckle} \approx \frac{\lambda(1+M)}{NA}, \qquad (21.8)$$

where M is the magnification of the optical system.

If dynamic subjective speckles are registered by the CCD device, with finite exposure local speckle blurring may occur. The reason for such blurring is that the exposure time of the light sensor is much larger compared to the characteristic time period τ_c of the speckle intensity fluctuations due to the moving particles. A quantity called speckle contrast is calculated to numerically characterize the local blurring of the speckle pattern [49]:

$$K = \frac{\sigma}{I}, \qquad (21.9)$$

where σ is the standard deviation of speckle's intensity and I is its mean value.

In order to obtain blood-flow data from the speckle-contrast images, a mathematical relation between K and τ_c should be derived. Pioneering development was done by Fercher and Briers in 1981 [49]. In the assumption of Lorenz distribution of particle's velocity vectors, Equation 21.9 could be expanded as follows:

$$K = \frac{\sigma}{I} = \sqrt{\beta\left\{\frac{\tau_c}{T} + \frac{\tau_c^2}{2T^2}\exp\left[\left(\frac{-2T}{\tau_c}\right)\right]-1\right\}}, \quad (21.10)$$

where T is the exposure time of a light sensor and β is the setup specific constant that includes ration of pixel-speckle size, a fraction of Doppler-shifted photons, polarization ratio, and coherent properties of the source. From the Equation 21.10, a characteristic time τ_c can be calculated. A well-known equation could be used to recover velocity value from the τ_c if precise control of experimental condition was carried out [50]:

$$\tau_c = \frac{\lambda}{2NA \cdot V}. \qquad (21.11)$$

It would be useful to describe briefly the most common approaches for blood-flow characterization by the means of time-variant speckles:

1. LASCA: spatial laser speckle contrast calculated within one recorded frame using sliding window with typical size of 5×5 or 7×7 pixels and generally depending on a speckle size [51];

2. LSI (laser speckle imaging): statistical properties of time-variant speckles calculated pixel-wisely within a few laser speckle images of the sample [52, 53];

3. LSPI (laser speckle perfusion imaging): speckle contrast calculated using a specially developed spatio-temporal algorithm [54, 55];

4. MESI (multi-exposure speckle imaging): a series of speckle images taken at different exposures to determine the velocity distribution maps with a high dynamic range [56].

Optical clearing for assessment of blood and lymph microcirculation

The term *optical clearing* (OC) groups together a broad range of chemical, hardware, and software tools for the probing depth increase of different optical methods. For instance, wavefront shaping techniques (adaptive optics) can be used to eliminate the effect of light scattering by superficial layers of tissue, thus allowing one to investigate underlying layers [57]. This is a promising tool, but it requires advanced hardware with a combination of mathematically complex iterative algorithms to calculate the correct wavefront shape. Thus this method in its current state can only be used *ex vivo* for thin tissues slices or for semitransparent tissues and in combination with immersion optical clearing [58].

On the other hand, immersion optical clearing agents (OCAs) can be used to match the refractive index of the intracellular fluid and the cell membrane to minimize the scattering effect and/or reduce the water concentration to minimize absorption in the IR region, which is important for multiphoton and many other optical methods [59]. An encouraging aspect of immersion OC is the fact that these OCAs can be also used as immersion liquids for high NA objectives allowing for high-resolution *in vivo* imaging. Companies like Olympus and Zeiss already have a commercially available objective lens specifically designed for working with glycerol as an immersion fluid. These objectives usually come with a rotating ring that can be used to adjust the position of the lenses inside the objective to obtain optimal performance with a particular OCA as an immersion liquid.

There are some fascinating *in vitro* OC methods like CLARITY [60], in which a whole organ is chemically transformed into the completely transparent nanoporous hydrogel-hybridized form (crosslinked to a three-dimensional network of hydrophilic polymers), allowing high-resolution whole-organ tomography (see Chapters 3, 11–16). However, these technologies are not suitable for *in vivo* studies of blood and lymph vessel networks.

The applicability of the optical methods for blood and lymph-flow imaging is mainly dependent on the embedding depth of the vessels under study, being more accurate for superficial vessels. The OC, which reduces the efficiency of scattering from static scatterers, allows one to extend the applicability of the method to larger embedding depths of the vessels.

For example, the authors of Reference [61] measured the temporal optical reflectance spectra of rat skin and mesentery *in vivo* impregnated with osmotic liquids (glycerol or glucose) or distillate water to analyze the state of blood microcirculation using reflectance spectra and digital transmission microscopy during the application of these agents. Glucose and glycerol solutions were injected slowly into the dermal layer of skin (approximately 0.1 ml) in a shaved area of rat pad. A 75% solution of glycerol and different concentrations of glucose solution – 20%, 25%, 30%, 35%, and 40% – were used. To study blood microvessels *in vivo*, these solutions were applied topically to rat mesenteric microvessels using a micropipette. The dose of each agent administered was 10 *μl*.

In these experiments, an injection of 75% glycerol into the skin resulted in a significant, stable OC effect (reduction in skin reflectivity across the entire spectrum), but this effect was slightly less than with a 40%-glucose injection. At the same time, a virtual window of transparency in the skin with a diameter of 3–4 mm was formed in the injection area, which was kept transparent for an hour. Through this window, small microvessels of the skin were clearly visible.

The application of these OCAs directly to the rat mesenteric microvessels caused a short-term slowdown and local stagnation in various microvessels (arterioles, venules, capillaries) and expansion of microvessels directly in the area of OCA application. A decrease in the osmolarity of the OCA led to significantly smaller changes in microcirculation; for example, a 20%-glucose solution caused only short-term and insignificant changes in microcirculation. Assessment of the effect of OCA on blood flow is of great importance for the further application of this combined technology for imaging blood hemodynamics at great depths.

A combination of LSCI with immersion OC demonstrates a high efficiency for evaluation of microhemodynamics of different tissues. In [62], this combination increased the resolution, contrast, and sensitivity of the LSCI method, making it possible to rebuild artery–vein separations and quantitatively assess the blood-flow dynamical changes in terms of flow velocity and vascular diameter at a single artery or vein level. Cheng et al. used LSCI to evaluate quantitatively OCA-induced changes in blood flow [63]. They found that after removing the rabbit skull and a small area of dura mater, the application of glycerol on the dura mater around the exposed cortex decreases the cerebral blood flow by 20%–30% immediately. Zhu et al. [64] applied the LSCI to investigate both the

short-term and long-term effects of glycerol and glucose on blood vessels in chick chorioallantoic membrane (CAM). Mao et al. [65] studied 30% glycerol solution's effect on the dermal blood vessels of the rat skin by LSCI.

The effect of exogenous glucose on the mouse cerebral microvascular network dynamics at topical application of 20% and 45%-glucose aqueous solutions was studied using LSCI [66]. Figure 21.5 shows typical speckle contrast images after trepanation skull for intact brain in 1 min after brain treatment by 20% glucose and in 1 min after 45% glucose application. It is clear that image contrast is increased and more small vessels can be visible with better contrast, and the changes are more obvious for the higher concentration of glucose solution.

The effects of glucose solution as an OCA with different concentrations on cortical vascular network and blood flow were quantitatively estimated in comparison with the intact state. The results show that exogenous glucose solution provides immediate sagittal vein expansion at 10% at the application of 20% glucose, 15% at the application of 45% glucose, and corresponding blood flow decrease on 14% and 18%, respectively [66].

The impact of 40%-glucose aqueous solution on blood microcirculation and diameter of blood vessels in hypodermis in rats *in vivo* studied by LSCI can be evaluated from the histograms for blood velocity (Figure 21.6a) and diameter of blood vessels (Figure 21.6b) before and after application of the OCA for each animal [67].

The averaged values and standard deviations of blood velocity for ten laboratory rats before and after glucose impact were 1.32 ± 0.34 mm/sec and 0.59 ± 0.19 mm/sec, respectively. There was a statistically significant difference between these values (p<0.001). The mean value and standard deviation of vessel diameter before and after glucose impact were (166 ± 25) *μm* and (180 ± 16) *μm*, respectively. There was no significant difference between these parameters (p>0.05). The results have shown that applying a 40%-glucose solution has led to reduction of blood flow velocity (about 2.2-fold) at the insignificant vasodilatation (increase of microvessel size by 8%). The solution attracts water from the surrounding tissues, thereby increasing the diameter of the vessels and accordingly slowing down blood flow.

The recent studies using LSCI were focused on vascular permeability at application of a specific OCA, *i.e.*,

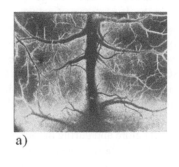
a)

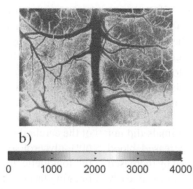
b)

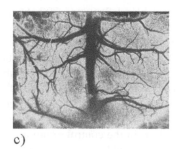
c)

| 0 | 1000 | 2000 | 3000 | 4000 |

FIGURE 21.5 Speckle contrast images of brain vessels after skull trepanation with velocity (a.u.) color bar under the action of the OCA: initial (a), 1 min after application of 20% glucose (b), 1 min after application of 45% glucose (c) (adapted from [66]).

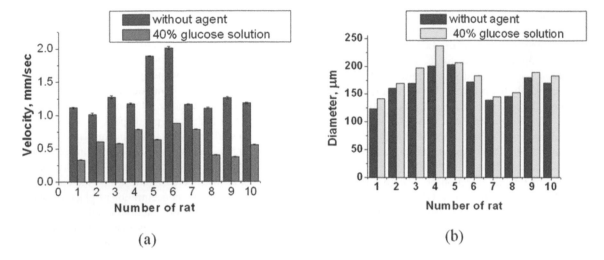

FIGURE 21.6 Variation of velocity of blood (a) and diameter of blood vessels (b) in hypodermis of rats *in vivo* without and with 40% glucose action (adapted from [67]).

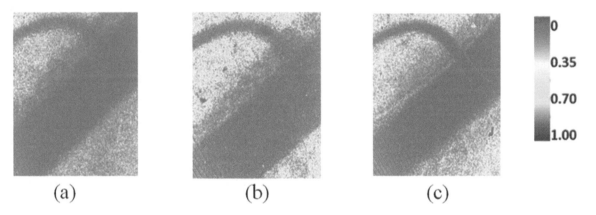

FIGURE 21.7 Contrast images of blood vessels of pancreas after topical application of aqueous solution of 70% Omnipaque™-300: without agent (a), in 1 min after application (b), in 5 min after application (c) (adapted from [68]).

Omnipaque™-300, namely, on the analysis of blood flow in pancreatic vessels. Omnipaque™-300 is a radiopaque agent for X-ray image contrasting with iohexol as an active substance [68]. Its aqueous solutions were also used. The solution (0.5 ml) was topically applied with a pipette to the pancreas surface of rat under anesthesia. The calibration of the LSCI system made it possible to quantify changes of the blood flow. Application of 100% Omnipaque™-300 demonstrates a 65% increase of blood flow, and 70% Omnipaque™-300 gives a 50% increase in blood flow in the group of diabetic animals (experiments were conducted 16 days after the injection of alloxan). Figure 21.7 shows the distribution of contrast of the speckle images when pancreas was subjected topically to an aqueous 70% solution of Omnipaque™-300. It is evident that use of the OCA improved the speckle contrast images of studied vessels to the 5th min of its application.

Blood flow in the control group of healthy animals did not show any noticeable change (Figure 21.8). Increased blood flow after application of the OCA could be caused by increased vascular endothelial permeability in diabetes even at early stages of the disease. In both cases to the 10th min, the blood-flow velocity was completely restored. The difference found

in the behavior of the response of blood vessels to the action of agents for animals in the control and diabetic groups can be associated with significant differences in the permeability of normal and glycated tissues for OCA molecules.

Thus, application of OC could be used not only for getting better speckle images of blood flow distribution or decrease of scattering of tissue in the living organ at surgical procedures, but also for monitoring complications related to changes in vascular endothelial permeability in the development of pathologies.

Recently, the application of OCAs to improve the imaging quality of cerebral blood flow of newborn mice was described [69]. A 60%-glycerol solution and a 70%-Omnipaque™-300 solution in water/DMSO (25%/5%) were chosen as the OCAs. LSCI was used for the imaging of cerebral blood flow in newborn mouse brains during topical OCA application in the area of the fontanelle. Figure 21.9 shows the speckle contrast images of the cerebral blood vessel before and during the action of the 60%-glycerol solution on the newborn mouse skin surface in the fontanelle area. It can be seen that the quality of vascular imaging has been improved significantly.

Also, it was shown that the Omnipaque/DMSO solution gives a better result of OC than aqueous 70% Omnipaque

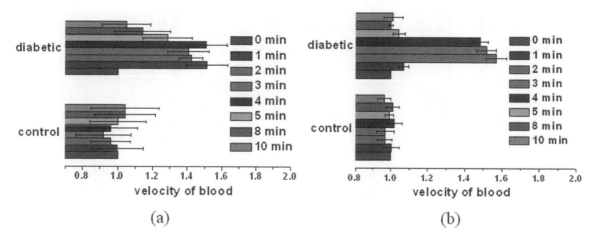

FIGURE 21.8   Velocity of blood flow in the rat pancreas of control and diabetic groups when topically exposed to an aqueous 70% (a) and 100% (b) solutions of Omnipaque™-300.

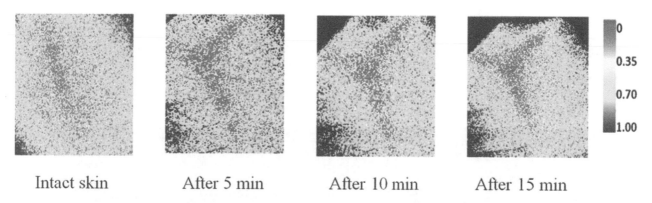

Intact skin   After 5 min   After 10 min   After 15 min

FIGURE 21.9   LSCI of cerebral vessels at application of 60%-glycerol solution to skin surface of the newborn mouse in the fontanelle area (adapted from [69]).

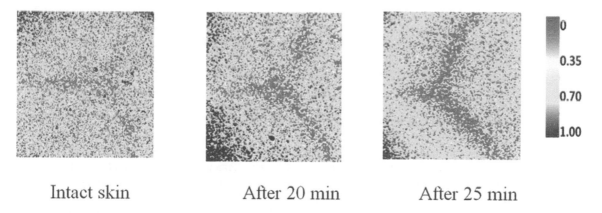

Intact skin   After 20 min   After 25 min

FIGURE 21.10   LSCI of cerebral vessels at application of aqueous Omnipaque (70%) solution in water (25%) and DMSO (5%) on skin surface of the newborn mouse in the fontanelle area (adapted from [69]).

solution (Figure 21.10). All the OCAs used have demonstrated effective skin clearing of newborn mice in the fontanelle area. The OC is more efficient if an aqueous 60%-glycerol solution is applied. However, glycerol solutions caused a reduction in the cerebral blood flow of up to 12%.

These results demonstrate the effectiveness of glycerol and Omnipaque solutions as OCAs for the investigation of cerebral blood flow in newborn mice without scalp removal and skull thinning.

To improve visualization of vascular foot of newborn rats, OC studies were conducted. Figure 21.11 shows speckle images of the paw vessels when exposed to an aqueous 60%-glycerol solution, obtained by recording the transmitted light through the paw.

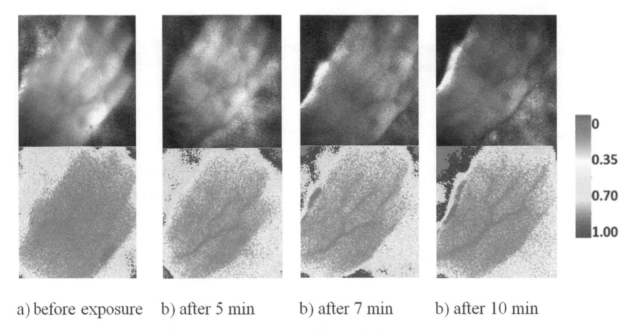

a) before exposure   b) after 5 min   b) after 7 min   b) after 10 min

FIGURE 21.11 Images of the paw vessels of the newborn rat in coherent light (633 nm) (upper row, black-white) and the calculated distribution of speckle contrast (1–0) (color) when exposed to an aqueous solution of 60% glycerol: before exposure (a), after 5 min (b), 7 min (c), and 10 min (d) of glycerol application.

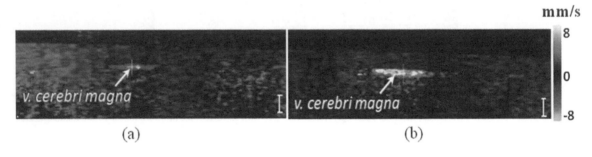

FIGURE 21.12 DOCT images of the brain through cranial bone of a rat from group I: without preliminary OC (a), 1 hour after the application of a multicomponent OCA (ethanol – 45%, dehydrated glycerol – 25%, PEG-300 – 10%, and water – 20%) (b). Vertical segments correspond to an optical thickness of 500 μm. The scale of the distribution of blood flow velocity is shown on the right side of the images (adapted from [71]).

The images obtained demonstrate the effectiveness of the usage of OCAs for assessing the microcirculation of blood in tissues. This contributes to the improvement of visualization of blood vessels by any optical method, including LSCI.

A number of studies have shown that DOCT is an effective optical method for analyzing blood flow in healthy and diseased tissues [13–16, 70]. However, the use of OC can significantly increase the resolution and imaging depth when visualizing blood vessels in the skin or in the brain cortex through the intact cranial bone. This can increase the efficiency of early diagnosis of various pathologies or diseases, for example, stroke and other cerebral hemodynamic disorders. OC of the rat skull with an OCA containing ethanol and thiazone as enhancers of tissue permeability allowed for the measurement of cerebral blood flow by DOCT without removal of the cranial bone [71]. Figure 21.12 shows DOCT images of the brain through the cranial bone of a rat without preliminary OC (a) and 1 hour after the application of a multicomponent OCA (ethanol – 45%, dehydrated glycerol – 25%, PEG-300 – 10%, and water – 20%) (b)[71]. OC made it possible to increase the

resolution of the DOCT image of the blood vessel and estimate blood velocity. A similar result was obtained when using a solution containing thiazone.

The use of the OC also increases the efficiency of photoacoustic (PA) diagnostic methods and the image quality of deep blood vessels and small vessels [72]. For the first time, OC was demonstrated as a great tool for improving the PA signal from melanoma cells in deep blood vessels [73]. Authors demonstrated biocompatible and rapid OC of skin to minimize light scattering and thus increase the optical resolution and sensitivity of PA flow cytometry (PAFC). An OC effect was achieved in 20 min by sequent skin cleaning, microdermabrasion, and glycerol application enhanced by massage and sonophoresis. Using a 0.8 mm mouse skin layer over a glass tube taken as an *in vitro* blood vessel phantom, authors demonstrated a 1.6-fold decrease in laser spot blurring. As a result, the peak rate for B16F10 melanoma cells in blood flow increased 1.7-fold. The feasibility of PA contrast enhancement of human hand veins with the improvement of PA signal amplitude up to 40% was also demonstrated.

The OC effect remained for an hour during PAFC measurements. OCAs containing several biocompatible agents, such as PEG-400, fructose, and thiazone, can be used to improve signal quality [74].

A high-resolution imaging of the microstructure, lymphatic system, and blood vessel network is required. However, the lymphatic system, due to relatively low scattering properties of the lymph, looks almost transparent and thus it is difficult to image it optically. Paper [75] introduces a novel diagnostic and therapeutic platform for *in vivo* noninvasive detection and treatment of metastases in sentinel lymph nodes (SLNs) at single-cell level using an integrated system of multicolor lymph PAFC. The authors presented the results of microscopic examination of lymph nodes *ex vivo*. For this study, lymph nodes were excised from the mouse without fat and surrounding tissues and placed in a well of a microscope slide (8 mm in diameter and 2 mm in depth) to obtain optical images at the different magnifications. To decrease beam blurring due to light scattering in lymph nodes, the nodes were embedded for 5 min in 40% glucose, 100% DMSO, or 80% glycerol as hyperosmotic OCAs. Comparisons of optical images with different OCAs and physiological solution (control) revealed that glycerol had a maximal OC. The treatment of SLNs with 80% glycerol permitted the label-free imaging of a fresh node at the cellular level, including localization of immune-related, metastatic, and other cells (e.g., lymphocytes, macrophages, dendritic cells, and melanoma cells) as well as surrounding microstructures (e.g., afferent lymph vessels, subcapsular and transverse sinuses, the medulla, a reticular meshwork, follicles, and the venous vessels).

Furthermore, in Refs. [72, 76] the feasibility of the OC approach to improve PA imaging contrast of lymphatics was discussed. Such OCAs as PEG-400 and glycerol significantly improved the PA amplitude and image quality of deep-sealed blood vessels and shallow vessels, respectively [77].

The first demonstration of successful application of OCT for imaging of the lymphatic vessels in the meninges after opening of the blood–brain barrier (BBB) is described in paper [78], which might be a new useful strategy for noninvasive analysis of lymphatic drainage in daily clinical practice. Also, the capability of photodynamic treatment and OCT for the assessment of glymphatics, a network of lymphatic vessels in the central nervous system, is explored by Semyachkina-Glushkovskaya et al. [79–82]. Figure 21.13 shows the OCT image of deep lymphatic node in the brain [79].

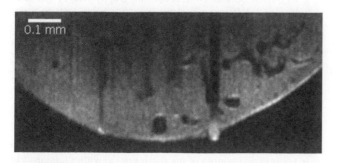

FIGURE 21.13 OCT image of deep lymphatic node in the brain (adapted from [79]).

The method of free labeling optical microangiography (OMAG) based on OCT technology was invented and well developed by Wang et al. [83–92]. The 3D information about the microcirculation that supplies the lymph nodes and efferent lymphatic vessels has been found [83–85]. To distinguish a lymphatic vessel, especially lymphatic capillaries, from background tissue, ultra-high resolution is required. An ultrahigh resolution OMAG system was used for simultaneous 3D imaging of microstructure, lymphatic, and blood vessels [86–88]. Authors developed an automatic segmentation algorithm for the extraction of lymphatic vessels from the OCT microstructure image, based on the fact that the lymph fluid is optically transparent, thus giving optical signals near the system noise floor. This method has been used in many studies to visualize blood perfusion and build microvasculature maps [89] of various living tissues, such as retina [90], cerebral [91], renal [92], and skin [88] microcirculation.

Vakoc et al. [93] applied a new method and instrumentation for advanced OCT technology, termed optical frequency domain imaging (OFDI) [94], for measurements of tumor angiogenesis, lymphangiogenesis, tissue viability, and both vascular and cellular responses to therapy.

Kalchenko et al. [95] used a long-exposure laser speckle imaging (LSI) approach for noninvasive *in vivo* imaging of blood and lymph vessels. They presented the images of lymphatic vessels of the mouse external ear, obtained *in vivo* using fluorescent intravital microscopy mode, where LSI image is superimposed with a mask of lymph vessels derived from the fluorescent intravital microscopy image. Authors demonstrated that a long-exposure-time LSI imaging approach allows for visualizing not only the blood vessels but also lymphatic vessels of the mouse ear *in vivo*.

Summary

This chapter presented the latest advances in the use of noninvasive optical methods for assessing blood flow and lymph flow in combination with the OC method. Laser speckle imaging, photoacoustic flow cytometry, and OCT have recently become the most widely used and beneficial optical technologies in biology and medical practice. Due to the numerous experimental data, these technologies, and all their modifications, are characterized as promising for medical research. Despite this, all the methods require an increase in a sufficiently small depth of probing of biological tissues, and an increase in the spatial resolution and contrast of images of the structural components of biological tissues. The chapter presented various studies that have aimed to advance existing optical methods in order to increase spatial resolution and probing depth. One of the promising and actively developing methods is the tissue OC. The use of OC allows for expanding the scope of optical methods to greater probing depths for quantification of properties of vessels and cell streams. Also, the use of OCAs allows for monitoring of complications associated with changes in vascular endothelial permeability caused by development of pathologies. Microvascular reactions to external influence from an OCA can be a potential biomarker for early prevention of the onset and development of diseases such as diabetes, cancer, pancreatic necrosis, etc.

Acknowledgments

P.A.D. and V.V.T. were supported by RF Governmental Grant 14.Z50.31.0044. O.V.S.-G. was supported by grants of RF Government 075-15-2019-1885, RSF 20-15-00090, RFBR 19-515-55016 China-a, and RFBR 20-015-00308-a.

REFERENCES

1. T.V. Petrova and G.Y. Koh, "Organ-specific lymphatic vasculature: From development to pathophysiology," *J. Exp. Med.* **215**(1), 35–49 (2018).
2. T.V. Petrova, T. Karpanen, C. Norrmén et al., "Defective valves and abnormal mural cell recruitment underlie lymphatic vascular failure in lymphedema distichiasis," *Nat. Med.* **10**(9), 974–981 (2004).
3. T. Tammela and K. Alitalo, "Lymphangiogenesis: Molecular mechanisms and future promise," *Cell* **140**(4), 460–476 (2010).
4. G. Oliver , J. Kipnis, G.J.Randolph, and N.L. Harvey, "The lymphatic vasculature in the 21st century: Novel functional roles in homeostasis and disease," *Cell* **182**(2), 270–296 (2020).
5. T. Barrett, P.L. Choyke, and H. Kobayashi, "Imaging of the lymphatic system: New horizons," *Contrast Media Mol. Imaging* **1**(6), 230–245 (2006).
6. M.J. Leahy, F.F. de Mul, G.E. Nilsson, and R. Maniewski, "Principles and practice of the laser-Doppler perfusion technique," *Technol. Health Care* **7**(2–3), 143–162 (1999).
7. K. Yaoeda, M. Shirakashi, S. Funaki, H. Funaki, T. Nakatsue, and H. Abe, "Measurement of microcirculation in the optic nerve head by laser speckle flowgraphy and scanning laser Doppler flowmetry," *Am. J. Ophthalmol.* **129**(6), 734–739 (2000).
8. S. Hanazawa, R.L. Prewitt, and J.K. Terzis, "The effect of pentoxifylline on ischemia and reperfusion injury in the rat cremaster muscle," *J. Reconstr. Microsurg.* **10**(01), 21–26 (1994).
9. J.B. Dixon, D. Zawieja, A. Gashev, and G. Coté, "Measuring microlymphatic flow using fast video microscopy," *J. Biomed. Opt.* **10**(6), 64016 (2005).
10. H.H. Lipowsky, S. Usami, and S. Chien, "In vivo measurements of 'apparent viscosity' and microvessel hematocrit in the mesentery of the cat," *Microvasc. Res.* **19**(3), 297–319 (1980).
11. Z. Chen, T.E. Milner, D. Dave, and J.S. Nelson, "Optical Doppler tomographic imaging of fluid flow velocity in highly scattering media," *Opt. Lett.* **22**(1), 64–66 (1997).
12. P.R. Schvartzman and R.D. White, "Magnetic resonance imaging," in *E.J. Topol Textbook of Cardiovascular Medicine*, 2nd ed., Lippincott Williams Wilkins, Philadelphia, PA, 1213–1256 (2002).
13. V.V. Tuchin (Ed.), *Handbook of Optical Biomedical Diagnostics*, vol. 2, *Methods*, 2nd ed., SPIE Press, Bellingham (2016), DOI:10.1117/3.2219608.
14. V. Doblhoff-Dier, L. Schmetterer, W. Vilser, G. Garhöfer, M. Gröschl, R.A. Leitgeb, and R.M. Werkmeister, "Measurement of the total retinal blood flow using dual beam Fourier-domain Doppler optical coherence tomography with orthogonal detection planes," *Biomed. Opt. Express* **5**(2), 630–642 (2014).
15. Y. Huang, Z. Ibrahim, D. Tong et al., "Microvascular anastomosis guidance and evaluation using real-time three-dimensional Fourier-domain Doppler optical coherence tomography," *J. Biomed. Opt.* **18**(11), 111404 (2013).
16. Z. Chen, T.E. Milner, X. Wang, S. Srinivas, and J.S. Nelson, "Optical Doppler tomography: Imaging in vivo blood flow dynamics following pharmacological intervention and photodynamic therapy," *Photochem. Photobiol.* **67**(1), 56–60 (1998).
17. J.D. Briers, "Laser Doppler, speckle and related techniques for blood perfusion mapping and imaging.," *Physiol. Meas.* **22**(4), R35–R66 (2001), DOI:10.1088/0967-3334/22/4/201
18. K. Basak, M. Manjunatha, and P.K. Dutta, "Review of laser speckle-based analysis in medical imaging," *Med. Biol. Eng. Comput.* **50**(6), 547–558 (2012).
19. D.J. Briers, D.A.Zimnyakov, O.V. Ushakova, and V.V. Tuchin, "Speckle technologies for monitoring and imaging tissues and tissue-like phantoms," in: *Handbook of Optical Biomedical Diagnostics*, 2nd ed., vol. 2: Methods, 429–496, SPIE Press, Bellingham (2016), DOI:10.1117/3.2219608.ch8
20. I. Sigal, R. Gad, A.M. Caravaca-Aguirre, Y. Atchia, D.B. Conkey, R. Piestun, and O. Levi, "Laser speckle contrast imaging with extended depth of field for in-vivo tissue imaging," *Biomed. Opt. Express* **5**(1), 123–135 (2014).
21. J.C. Ramirez-San-Juan, R. Ramos-Garcia, I. Guizar-Iturbide, G. Martinez-Niconoff, and B. Choi, "Impact of velocity distribution assumption on simplified laser speckle imaging equation," *Opt. Express* **16**(5), 3197–3203 (2008).
22. A.K. Dunn, "Laser speckle contrast imaging of cerebral blood flow," *Ann. Biomed. Eng.* **40**(2), 367–377 (2012), DOI:10.1007/s10439-011-0469-0.
23. P. Li, S. Ni, L. Zhang, S. Zeng, and Q. Luo, "Imaging cerebral blood flow through the intact rat skull with temporal laser speckle imaging," *Opt. Lett.* **31**(12), 1824–1826 (2006), DOI:10.1364/OL.31.001824.
24. W. Drexler and J.G. Fujimoto, "Optical coherence tomography," in: *Optical Coherence Tomography: Technology and Applications*, 2nd ed., Springer, Berlin (2015), DOI:10.1007/978-3-319-06419-2.
25. J. Welzel, "Optical coherence tomography in dermatology: A review," *Skin Res. Technol. Rev. Artic.* **7**(1), 1–9 (2001).
26. E. Sattler, R. Kästle, and J. Welzel, "Optical coherence tomography in dermatology," *J. Biomed. Opt.* **18**(6), 061224 (2013), DOI:10.1117/1.JBO.18.6.061224.
27. E. Jonathan, J. Enfield, and M.J. Leahy, "Correlation mapping method for generating microcirculation morphology from optical coherence tomography (OCT) intensity images," *J. Biophoton.* **4**(9), 583–587 (2011), DOI:10.1002/jbio.201000103.
28. M. Bonesi, S. Matcher, and I. Meglinski, "Doppler optical coherence tomography in cardiovascular applications," *Laser Phys.* **20**(6), 1491–1499 (2010).
29. Y. Jia, E. Wei, X.Wang et al., "Optical coherence tomography angiography of optic disc perfusion in glaucoma," *Ophthalmology* **121**(7), 1322–1332 (2014), DOI:10.1016/j.ophtha.2014.01.021.
30. J. Qin, J. Jiang, L.An, D. Gareau, and R.K. Wang, "In vivo volumetric imaging of microcirculation within human skin under psoriatic conditions using optical microangiography," *Lasers Surg. Med.* **43**(2), 122–129 (2011).

31. J. Yi, W.Liu, S.Chen et al., "Visible light optical coherence tomography measure retinal oxygen metabolic response to systemic oxygenation (Conference Presentation)," *Proc. SPIE* **9697**, 96970I (2016).

32. A.F. Fercher, W. Drexler, C.K. Hitzenberger, and T. Lasser, "Optical coherence tomography – Principles and applications," *Rep. Prog. Phys.* **66**(2), 239–303 (2003), DOI:10.1088/0034-4885/66/2/204.

33. S.-W. Lee and B.-M. Kim, "Line-field optical coherence tomography using frequency-sweeping source," *IEEE J. Sel. Top. Quantum Electron.* **14**(1), 50–55 (2008).

34. A. Dubois, O. Levecq, H. Azimani, et al., "Line-field confocal optical coherence tomography for high-resolution non-invasive imaging of skin tumors," *J. Biomed. Opt.* **23**(10), 106007 (2018).

35. Y. Nakamura, S. Makita, M.Yamanari, M. Itoh, T.Yatagai, and Y. Yasuno, "High-speed three-dimensional human retinal imaging by line-field spectral domain optical coherence tomography," *Opt. Express* **15**(12), 7103–7116 (2007).

36. A. Dubois, K.Grieve, G. Moneron, R. Lecaque, L. Vabre, and C. Boccara, "Ultrahigh-resolution full-field optical coherence tomography," *Appl. Opt.* **43**(14), 2874–2883 (2004), DOI:10.1364/AO.43.002874.

37. M. Hagen-Eggert, P. Koch, and G. Hüttmann, "Analysis of the signal fall-off in spectral domain optical coherence tomography systems," *Proc. SPIE* **8213**, 82131K (2012).

38. J. Xu, S. Song, S. Men, and R.K. Wang, "Long ranging swept-source optical coherence tomography-based angiography outperforms its spectral-domain counterpart in imaging human skin microcirculations," *J. Biomed. Opt.* **22**(11) 116007 (2017).

39. V.X.D. Yang, M.L. Gordon, B. Qi et al., "High speed, wide velocity dynamic range Doppler optical coherence tomography (Part I): System design, signal processing, and performance," *Opt. Express* **11**(7), 794–809 (2003).

40. O.P. Kocaoglu, T.L. Turner, Z. Liu, and D.T. Miller, "Adaptive optics optical coherence tomography at 1 MHz," *Biomed. Opt. Express* **5**(12), 4186–4200 (2014).

41. G. Liu and Z. Chen, *Phase-Resolved Doppler Optical Coherence Tomography*, INTECH Open Access, London (2012)

42. U. Baran and R.K. Wang, "Review of optical coherence tomography based angiography in neuroscience," *Neurophotonics* **3**(1), 010902 (2016), DOI:10.1117/1.NPh.3.1.010902.

43. G. Schwalbe, "Der Arachnoidalraum, ein Lymphraum und Sein Zusammenhang mit dem Perichorioidalraum," *Zentralbl. Med. Wiss.* **7**, 465–467 (1869).

44. L. Tong, Q. Wei, A. Wei, and J.-X. Cheng, "Gold nanorods as contrast agents for biological imaging: Optical properties, surface conjugation and photothermal effects," *Photochem. Photobiol.* **85**(1), 21–32 (2009).

45. O. Semyachkina-Glushkovskaya,V. Chehonin, E. Borisova et al., "Photodynamic opening of the blood-brain barrier and pathways of brain clearing," *J. Biophoton.* **11**(8), e201700287 (2018), DOI:10.1002/jbio.201700287.

46. A.L. Oldenburg, M.N. Hansen, A. Wei, and S.A. Boppart, "Plasmon-resonant gold nanorods provide spectroscopic OCT contrast in excised human breast tumors," *Proc. SPIE* **6867**, 68670E (2008).

47. E.D. SoRelle, O.Liba, Z. Hussain, M.Gambhir, and A. de la Zerda, "Biofunctionalization of large gold nanorods realizes ultrahigh-sensitivity optical imaging agents," *Langmuir* **31**(45), 12339–12347 (2015).

48. C. Blatter, E F.J. Meijer, A.S. Nam, D. Jones, B.E. Bouma, T.P. Padera, B.J. Vakoc, "In vivo label-free measurement of lymph flow velocity and volumetric flow rates using Doppler optical coherence tomography," *Sci. Rep.* **6**(1), 1–10 (2016).

49. A.F. Fercher and J.D. Briers, "Flow visualization by means of single-exposure speckle photography," *Opt. Commun.* **37**(5), 326–330 (1981), DOI:10.1016/0030-4018(81)90428-4.

50. D.D. Duncan and S.J. Kirkpatrick, "Can laser speckle flowmetry be made a quantitative tool?" *J. Opt. Soc. Am. A* **25**(8), 2088–2094 (2008), DOI:10.1364/JOSAA.25.002088.

51. J.D. Briers and S. Webster, "Laser speckle contrast analysis (LASCA): A nonscanning, full-field technique for monitoring capillary blood flow," *J. Biomed. Opt.* **1**(2), 174–179 (1996).

52. H. Cheng, Q. Luo, S. Zeng, S. Chen, J.Cen, and H. Gong, "Modified laser speckle imaging method with improved spatial resolution," *J. Biomed. Opt.* **8**(3), 559–565 (2003).

53. A.K. Dunn, H. Bolay, M.A. Moskowitz, and D.A. Boas, "Dynamic imaging of cerebral blood flow using laser speckle," *J. Cereb. Blood Flow Metab.* **21**(3), 195–201 (2001).

54. K.R. Forrester, C. Stewart, J. Tulip, C. Leonard, and R.C. Bray, "Comparison of laser speckle and laser Doppler perfusion imaging: Measurement in human skin and rabbit articular tissue," *Med. Biol. Eng. Comput.* **40**(6), 687–697 (2002).

55. K.R. Forrester, J. Tulip, C. Leonard, C. Stewart, and R.C. Bray, "A laser speckle imaging technique for measuring tissue perfusion," *IEEE Trans. Bio Med. Eng.* **51**(11), 2074–2084 (2004).

56. A.B. Parthasarathy, W.J. Tom, A. Gopal, X. Zhang, and A.K. Dunn, "Robust flow measurement with multi-exposure speckle imaging," *Opt. Express* **16**(3), 1975–1989 (2008).

57. H. Yu, J. Park, K. Lee, J. Yoon, K. Kim, S. Lee, and Y. Park, "Recent advances in wavefront shaping techniques for biomedical applications," *Curr. Appl. Phys.* **15**(5), 632–641 (2015).

58. H. Yu, P. Lee, Y.Ju Jo, K.R. Lee, V.V. Tuchin, Y. Jeong, and Y.K. Park, "Collaborative effects of wavefront shaping and optical clearing agent in optical coherence tomography," *J. Biomed. Opt.* **21**(12), 121510 (2016), DOI:10.1117/1.JBO.21.12.121510.

59. E.A. Calle, S. Vesuna, S. Dimitrievska et al., "The use of optical clearing and multiphoton microscopy for investigation of three-dimensional tissue-engineered constructs," *Tissue Eng. C* **20**(7), 570–577 (2014).

60. K. Chung, J. Wallace, S.-Y. Kim et al., "Structural and molecular interrogation of intact biological systems," *Nature* **497**(7449), 332–337 (2013).

61. E.I. Galanzha, V.V. Tuchin, A.V. Solovieva, T.V. Stepanova, Q. Luo, and H. Cheng, "Skin backreflectance and microvascular system functioning at the action of osmotic agents," *J. Phys. D* **36**(14), 1739–1746 (2003).

62. R. Shi, M. Chen, V.V. Tuchin, and D. Zhu, "Accessing to arteriovenous blood flow dynamics response using combined laser speckle contrast imaging and skin optical clearing," *Biomed. Opt. Express* **6**(6), 1977–1989 (2015), DOI:10.1364/BOE.6.001977.

63. H. Cheng, Q. Luo, S. Zeng, S. Chen, W. Luo, and H. Gong, "Hyperosmotic chemical agent's effect on in vivo cerebral blood flow revealed by laser speckle," *Appl. Opt.* **43**(31), 5772–5777 (2004).

64. D. Zhu, K. Larin, Q. Luo, and V.V. Tuchin, "Recent progress in tissue optical clearing," *Laser Photonics Rev.* **7**(5), 732–757 (2013).

65. Z. Mao, Z. Han, X. Wen, Q. Luo, and D. Zhu, "Influence of glycerol with different concentrations on skin optical clearing and morphological changes in vivo," *Proc. SPIE* **7278**, 72781T (2009).

66. P.A. Timoshina, R. Shi, Y. Zhang et al., "Comparison of cerebral microcirculation of alloxan diabetes and healthy mice using laser speckle contrast imaging," *Proc. SPIE* **9448**, 94480B (2015).

67. D.K. Tuchina, P.A. Timoshina, V.V. Tuchin, A.N. Bashkatov, and E.A. Genina, "Kinetics of rat skin optical clearing at topical application of 40% glucose: Ex vivo and in vivo studies," *IEEE J. Sel. Top. Quantum Electron.* **25**(1), 1–8 (2018).

68. P.A. Timoshina, A.B. Bucharskaya, D.A. Alexandrov, and V.V. Tuchin, "Study of blood microcirculation of pancreas in rats with alloxan diabetes by laser speckle contrast imaging," *J. Biomed. Photon. Eng.* **3**(2), 20301 (2017).

69. P.A. Timoshina, E.M. Zinchenko, D.K. Tuchina, M.M. Sagatova, O.V. Semyachkina-Glushkovskaya, and V.V. Tuchin, "Laser speckle contrast imaging of cerebral blood flow of newborn mice at optical clearing," *Proc. SPIE* **10336**, 1033610 (2017).

70. V.J. Srinivasan, J.Y. Jiang, M.A. Yaseen et al., "Rapid volumetric angiography of cortical microvasculature with optical coherence tomography," *Opt. Lett.* **35**(1), 43–45 (2010).

71. E.A. Genina, A.N. Bashkatov, O.V. Semyachkina-Glushkovskaya, and V.V. Tuchin, "Optical clearing of cranial bone by multicomponent immersion solutions and cerebral venous blood flow visualization," *Izv. Saratov Univ.(NS), Ser. Phys.* **17**(2), 98–110 (2017).

72. R.W. Sun, V.V. Tuchin, V.P. Zharov, E.I. Galanzha, and G.T. Richter, "Current status, pitfalls and future directions in the diagnosis and therapy of lymphatic malformation," *J. Biophoton.* **11**(8), e201700124 (2018), DOI:10.1002/jbio.201700124

73. Y.A. Menyaev, D.A. Nedosekin, M. Sarimollaoglu, M.A. Juratli, E.I. Galanzha, V.V. Tuchin, and V.P. Zharov, "Optical clearing in photoacoustic flow cytometry," *Biomed. Opt. Express* **4**(12), 3030 (2013).

74. Y. Ding, J. Wang, Z. Fan et al., "Signal and depth enhancement for in vivo flow cytometer measurement of ear skin by optical clearing agents," *Biomed. Opt. Express* **4**(11), 2518–2526 (2013).

75. E.I. Galanzha, M.S. Kokoska, E.V. Shashkov, J.-W. Kim, V.V. Tuchin, and V.P. Zharov, "In vivo fiber-based multicolor photoacoustic detection and photothermal purging of metastasis in sentinel lymph nodes targeted by nanoparticles," *J. Biophoton.* **2**(8–9), 528–539 (2009).

76. V.V. Tuchin, V.P. Zharov, and E.I. Galanzha, "Biophotonics for lymphatic theranostics in animals and humans," *J. Biophoton.* **11**(8), 15750479 (2018), DOI:10.1002/jbio.201811001

77. Yong Zhou, Junjie Yao, and Lihong V. Wang, "Optical clearing-aided photoacoustic microscopy with enhanced resolution and imaging depth," *Opt. Lett.* **38**(14), 2592–2595 (2013).

78. O. Semyachkina-Glushkovskaya, A. Abdurashitov, A. Dubrovsky et al., "Application of optical coherence tomography for in vivo monitoring of the meningeal lymphatic vessels during opening of blood–brain barrier: Mechanisms of brain clearing," *J. Biomed. Opt.* **22**(12), 121719 (2017), DOI:10.1117/1.JBO.22.12.121719

79. A. Abdurashitov, V. Tuchin, and O. Semyachkina-Glushkovskaya, "Photodynamic therapy of brain tumors and novel optical coherence tomography strategies for in vivo monitoring of cerebral fluid dynamics," *J. Innov. Opt. Health Sci.* **13**(2), 2030004 (2020), DOI:10.1142/S1793545820300049

80. E. Zinchenko, N. Navolokin, A. Shirokov et al., "Pilot study of transcranial photobiomodulation of lymphatic clearance of beta-amyloid from the mouse brain: Breakthrough strategies for nonpharmacologic therapy of Alzheimer's disease," *Biomed. Opt. Express* **10**(8), 4003–4017 (2019), DOI:10.1364/BOE.10.004003

81. O. Semyachkina-Glushkovskaya, A. Abdurashitov, M. Klimova et al., "Photostimulation of cerebral and peripheral lymphatic functions" *Transl. Biophoton.* **2**(1–2), e201900036 (2020), DOI:10.1002/tbio.201900036

82. O. Semyachkina-Glushkovskaya, A. Abdurashitov, A. Dubrovsky et al., "Photobiomodulation of lymphatic drainage and clearance: Perspective strategy for augmentation of meningeal lymphatic functions," *Biomed. Opt. Express* **11**(2), 725–734 (2020), DOI:10.1364/BOE.383390

83. Y. Jung, Z. Zhi, and R.K. Wang, "Three-dimensional optical imaging of microvascular networks within intact lymph node in vivo," *J. Biomed. Opt.* **15**(5), 50501 (2010).

84. R.K. Wang, "Three-dimensional optical micro-angiography maps directional blood perfusion deep within microcirculation tissue beds in vivo," *Phys. Med. Biol.* **52**(23), N531 (2007).

85. R.K. Wang, S.L. Jacques, Z. Ma, S. Hurst, S.R. Hanson, and A. Gruber, "Three dimensional optical angiography," *Opt. Express* **15**(7), 4083 (2007), DOI:10.1364/OE.15.004083

86. Z. Zhi, Y. Jung, and R.K. Wang, "Label-free 3D imaging of microstructure, blood, and lymphatic vessels within tissue beds in vivo," *Opt. Lett.* **37**(5), 812–814 (2012).

87. R.K. Wang, L. An, P. Francis, and D.J. Wilson, "Depth-resolved imaging of capillary networks in retina and choroid using ultrahigh sensitive optical microangiography," *Opt. Lett.* **35**(9), 1467–1469 (2010).

88. L. An, J. Qin, and R.K. Wang, "Ultrahigh sensitive optical microangiography for in vivo imaging of microcirculations within human skin tissue beds.," *Opt. Express* **18**(8), 8220–8228 (2010), DOI:10.1364/OE.18.008220

89. S. Yousefi, J.Qin, Z.Zhi, and R.K. Wang, "Label-free optical lymphangiography: Development of an automatic segmentation method applied to optical coherence tomography to visualize lymphatic vessels using Hessian filters," *J. Biomed. Opt.* **18**(8), 086004 (2013), DOI:10.1117/1.JBO.18.8.086004

90. L. An, T. Shen, and R.K. Wang, "Using ultrahigh sensitive optical microangiography to achieve comprehensive depth resolved microvasculature mapping for human retina," *J. Biomed. Opt.* **16**(10) 106013 (2011).

91. Y. Jia, P. Li, and R.K. Wang, "Optical microangiography provides an ability to monitor responses of cerebral microcirculation to hypoxia and hyperoxia in mice," *J. Biomed. Opt.* **16**(9), 96019 (2011).

92. Z. Zhi, Y. Jung, Y. Jia, L. An, and R.K. Wang, "Highly sensitive imaging of renal microcirculation in vivo using ultrahigh sensitive optical microangiography," *Biomed. Opt. Express* **2**(5), 1059–1068 (2011).

93. B.J. Vakoc, R.M. Lanning, J.A. Tyrrell et al., "Three-dimensional microscopy of the tumor microenvironment in vivo using optical frequency domain imaging," *Nat. Med.* **15**(10), 1219–1223 (2009).

94. S.-H. Yun, G.J. Tearney, J.F. de Boer, N. Iftimia, and B.E. Bouma, "High-speed optical frequency-domain imaging," *Opt. Express* **11**(22), 2953–2963 (2003).

95. V. Kalchenko, Y. Kuznetsov, I. Meglinski, and A. Harmelin, "Label free in vivo laser speckle imaging of blood and lymph vessels," *J. Biomed. Opt.* **17**(5), 50502 (2012).

Part IV

Combination of tissue optical clearing and optical imaging/spectroscopy for diagnostics

22

Optical clearing aided photoacoustic imaging in vivo

Yong Zhou and Lihong V. Wang

CONTENTS

Introduction

Combining the advantages of optical excitation and of acoustic detection, photoacoustic imaging (PAI) has recently shown its unparalleled strength in providing structural [1], molecular [2], functional [3–5], and metabolic information [6]. Based on its focusing mechanism, PAI can be classified into two different implementations: optical-resolution PAI (OR-PAI) and acoustic-resolution PAI (AR-PAI) [1]. In OR-PAI, the lateral resolution is determined by the diffraction-limited optical focusing. As photons travel in tissue, the focusing capability degrades due to optical scattering. In addition, the maximum penetration depth of OR-PAI in biological tissue is mainly limited by the optical transport mean free path, affected by both absorption and scattering. Since the tissue scattering is typically an order of magnitude larger than the absorption, it is the scattering that predominantly limits the penetration depth of OR-PAI. In AR-PAI, the resolution is determined by acoustic detection, so the optical scattering does not have a significant impact on the resolution. However, it does limit the maximum penetration depth of AR-PAI for the same reason as in OR-PAI [7].

To reduce the scattering, tissue optical clearing (TOC) techniques have been widely used in many high-resolution optical imaging modalities, such as laser speckle contrast imaging and optical coherence tomography [8]. The basic mechanism of TOC is to diffuse a high-refractive-index optical clearing agent (OCA) into the tissue, which reduces the refractive index mismatch between intracellular components and extracellular fluids and thus decreases the scattering. In addition, tissue dehydration caused by OCAs has also been used to explain TOC [8]: on one hand, it increases refractive index matching due to decreased volume fraction of free water in the interstitial fluid; on the other hand, it increases packing of scatters, which may engender spatial correlations between scatterings, both of which reduce scattering further.

In this chapter, we discuss the state-of-the-art methods of using optical clearing in photoacoustic imaging *in vivo*. We first introduce how optical clearing has been applied to enhance the imaging depth and lateral resolution of OR-PAI. Then we discuss how optical clearing improves penetration of AR-PAI. Finally, we summarize the application of optical clearing in photoacoustic imaging and suggest future research directions.

Optical clearing–aided optical resolution photoacoustic imaging

In 2013, Zhou et al. first applied optical clearing in a typical OR-PAI implementation [9], as shown in Figure 22.1. Briefly, a tunable dye laser (CBR-D, Sirah, GmbH.) pumped by a Nd:YLF laser (INNOSAB, Edgewave, GmbH.) was used as the light source. After being focused by a condenser lens, filtered by a pinhole, and reflected by a mirror, the laser beam was finally focused by an objective lens into the sample. Ultrasonic detection was achieved by a wide-band ultrasonic transducer (V214-BC, Panametrics-NDT Inc.), which was placed confocally with the optical objective lens. Each laser pulse yielded a one-dimensional depth-resolved PA image (A-line) by recording the time course of a photoacoustic signal. A three-dimensional (3D) image was obtained by raster scanning the sample and piecing together A-lines.

They first demonstrated the efficiency of optical clearing in a phantom experiment. A U.S. penny [Figure 22.2(a)] covered by a piece of freshly harvested mouse skin was imaged at 570 nm. The phantom was immersed in the glycerol–water solution. As shown in the baseline image [Figure 22.2(b)], there were almost no PA signals from the coin because of the strong scattering of the mouse tissue. During optical clearing, the PA signals from the coin became stronger and stronger, and more

DOI: 10.1201/9781003025252-26

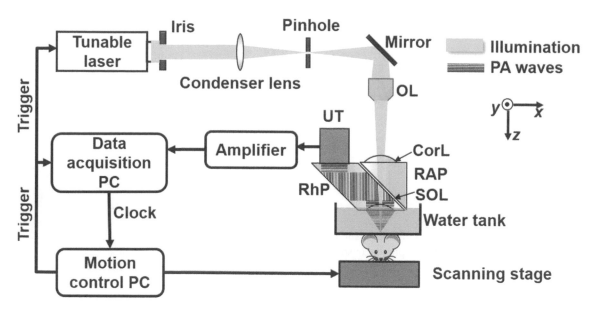

FIGURE 22.1 System schematic used in the experiment. CorL, correction lens; OL, objective lens; RAP, right-angle prism; RhP, rhomboid prism; SOL, silicone oil layer; UT, ultrasonic transducer. Reprinted with permission from reference [9] © The Optical Society.

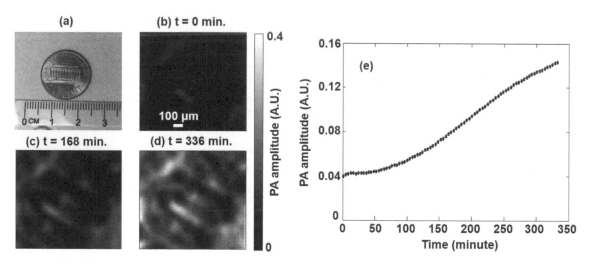

FIGURE 22.2 Optical clearing aided PAI of a coin covered by mouse skin. (a) Photograph of the one-cent coin used in the experiment. (b)–(d) PA images of the coin at different times during optical clearing. (e) Time course of the PA amplitude averaged over the whole image during optical clearing. The higher PA amplitude indicates that the optical clearing increased penetration depth. Reprinted with permission from reference [9] © The Optical Society.

features became resolvable, as shown from Figure 22.2(c)–(d). Figure 22.2(e) shows the average total PA signal amplitude from the coin versus time; after 250 min, the amplitude increased by more than three-fold over the baseline value. Note that the PA signals were still increasing at the point when they stopped the experiment. This phantom experiment clearly illustrates that the scattering of the tissue was reduced by optical clearing, and thus the penetration depth was enhanced.

Next, they used a nude mouse in an *in vivo* experiment. Because the diffusion of glycerol across the epidermal layer is very slow, they directly injected the glycerol–water solution into the mouse scalp to create a local optical clearing window. Figures 22.3(a) and (b) show the maximum amplitude projection (MAP) of the PA images of blood vessels in the mouse scalp before and after optical clearing, respectively. For

a better visualization of the deeper vessels in the scalp, most of the superficial capillaries seen in Figure 22.3(b) were digitally removed, with the result shown in Figure 22.3(c). From Figures 22.3(a)–(c), a clear PA signal amplitude enhancement after optical clearing can be observed. Three representative lines were taken from Figures 22.3(a)–(c) for further comparison, as shown in Figures 22.3(d)–(f). The averaged PA amplitude had increased by about eight times after optical clearing.

The optical clearing effect was quantified along the depth direction, as shown in Figures 22.3(g)–(h). To quantify the total signal enhancement after optical clearing, the averaged PA signal amplitude along the depth was calculated and is shown in Figure 22.3(i). Based on Beer's law, the fitted attenuation coefficients before and after optical clearing were 57.4 cm^{-1} and 103.1 cm^{-1}, respectively. At first glance, the

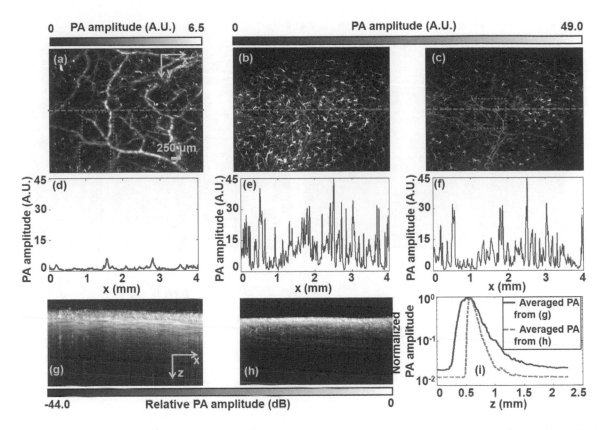

FIGURE 22.3 *In vivo* PAI of mouse scalp facilitated by optical clearing. (a) PAI image before optical clearing. (b) PAI image with capillaries after optical clearing. (c) PAI image with parts of capillary layer digitally removed after optical clearing. (d–f) PA amplitude along the red dashed line in (a–c). Maximum amplitude projection image along the *y* axis (g) before and (h) after optical clearing. (i) PA amplitude averaged along the *x* direction of (g) (blue line) and (h) (red dashed line). Adapted with permission from reference [9] © The Optical Society.

attenuation appears to have increased after optical clearing. However, since dehydration due to optical clearing caused the total thickness of the tissue to decrease by about 2.57 times [Figures 22.3(g)–(h)], the scatter and absorber number density should increase by about 2.57 times. If both the scattering and absorption cross sections had remained constant after optical clearing, the attenuation coefficient should have increased by 2.57 times rather than the 1.80 times observed in the experiment. Thus, the scattering or absorption cross section was reduced after optical clearing. Because optical clearing did not affect the absorption cross section, it was concluded that the 0.77 times less increase in the attenuation coefficient resulted from the decrease in the scattering cross section.

The enhanced PA detection sensitivity after optical clearing enabled better capillary imaging [Figure 22.3(b)]. Figure 22.4(a) and (b) are close-up images of the red-dashed-square areas in Figures 22.3(a) and (b), respectively. The large vessels in the periphery of both images indicate that they are from the same area in the mouse scalp. As shown in Figure 22.4(a), very few capillaries can be detected before optical clearing. However, after optical clearing, more capillaries can be distinctly observed, as shown in Figure 22.4(b). As shown in Figures 22.4(c)–(d), the vessel density increased by about ten times after clearing, and the averaged PA signal amplitude from individual capillaries increased by a factor of 22. The much denser capillaries and stronger PA amplitudes

in Figure 22.4(d) indicate that, after optical clearing, the local laser fluence at the capillaries was increased. There were two possible reasons for the increased local laser fluence: (1) enhanced focusing on these capillaries due to the dehydration-induced shrinkage of the tissue, and (2) reduced scattering loss. As shown in Figure 22.3(i), because of dehydration, the averaged thickness of the tissue decreased from about 0.88 mm to 0.37 mm. Assuming the light beam intensity had a Gaussian shape, it was estimated that the average fluence increase caused by shrinkage was about five times. Therefore, it can be concluded that the remaining 4.4-fold increase in the PA signal amplitude was due to the decreased scattering coefficient. Because the background noise remained constant during the experiment, the noise-equivalent sensitivity depended only on the PA signal amplitude. Therefore, sensitivity should also increase 4.4 times due to the decreased scattering coefficient.

Finally, the lateral resolution enhancement after optical clearing was analyzed. By zooming in on the deeper vessels in Figures 22.3(a)–(b), it can be seen that the vessels before optical clearing were much more blurred than those after optical clearing, as shown in Figures 22.5(a) and (b). Again, the averaged PA amplitude was increased by about five times [Figures 22.5(c)–(d)]. More importantly, the lateral resolution improvement was analyzed by comparing the vessel diameters. As shown in Figure 22.5(e), a representative vessel had nominal diameters of 75 μm and 30 μm before and after clearing,

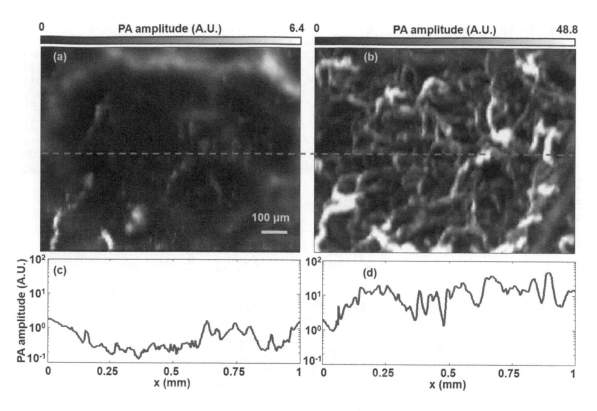

FIGURE 22.4 Optical clearing reveals more capillaries. (a–b) Close-ups of PAM images in the red dashed boxes in Figures 22.3(a–b). (c–d) PA amplitudes along the red dashed lines in (a–b). Reprinted with permission from reference [9] © The Optical Society.

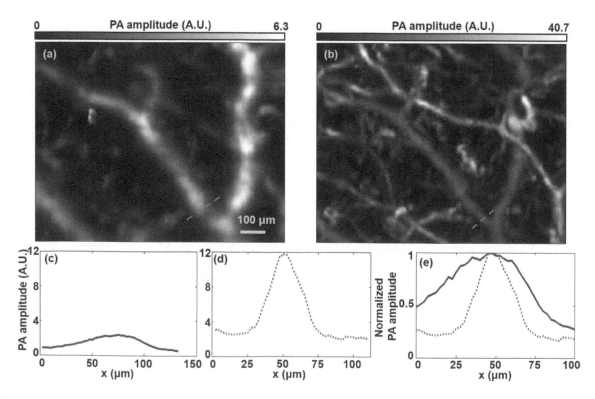

FIGURE 22.5 Optical clearing improves the lateral resolution for deep vessels. (a) Close-up of PAM image in the green dashed box in Figure 22.3(a). (b) Close-up of PAM image in the green dashed box in Figure 22.3(c). (c–d) PA amplitudes along the red dashed lines in (a–b). (e) Combined curves in (c) and (d) after normalization. Reprinted with permission from reference [9] © The Optical Society.

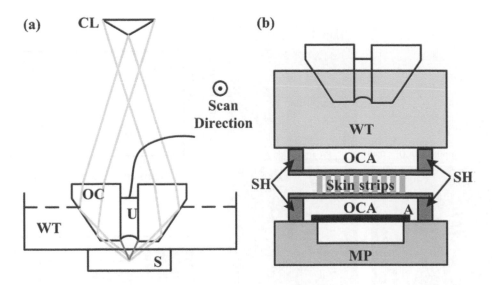

FIGURE 22.6 (a) Schematic of the imaging head. (b) Diagram of the sample setup. A, absorber; CL, conical lens; MP, metal plate; OC, optical condenser; S, sample; SH, sample holder; U, ultrasonic transducer; WT, water tank. Reprinted with permission from reference [10] © The Optical Society.

respectively. The measured vessel diameter decreased at least by 2.5 times after optical clearing. Because the measured vessel diameter was the convolution of the actual lateral resolution with the true vessel diameter, the actual resolution improvement should be better than 2.5 times.

Optical clearing–aided acoustic resolution photoacoustic imaging

Liu et al. did a very comprehensive test of the use of optical clearing in AR-PAI [10]. A typical dark-field illumination AR-PAI was used, as shown in Figure 22.6(a). Briefly, a conical lens and an optical condenser were used to provide the dark-field illumination. A focused ultrasonic transducer (center frequency, 50 MHz; NA, 0.47; V30011, Olympus, USA) was placed in the center of the optical condenser to detect the induced ultrasonic waves. Between the sample and the transducer, the ultrasonic wave was coupled by a water tank with a bottom window sealed with a polyethylene membrane. As shown in Figure 22.6(b), the rat skin specimens were immersed in two layers of OCA. A black taper was used to measure the amplitude, while some carbon fibers with a diameter of 6 μm were used to measure the spatial resolution.

Five commonly used OCAs (glycerol, PEG-400, DMSO, glucose (40%), and oleic acid) were first characterized on AR-PAI *ex vivo*. As shown in Figure 22.7, DMSO, PEG-400, and glycerol enhanced the photoacoustic signal, while glucose and oleic acid caused subtle decrease in the amplitude similar to PBS. Meanwhile, no obvious changes were found in the lateral resolution after OCAs were applied.

Based on the *ex vivo* experiments, PEG-400, glycerol, and DMSO were selected for *in vivo* experiments. Because PEG 400 does not easily penetrate the dermis, thiazone was chosen as a penetration enhancer for the *in vivo* experiment. To further improve dermal penetration, massaging was applied with OCAs. The experimental results from PEG-400, glycerol, and DMSO are shown in Figure 22.8, Figure 22.9, and Figure 22.10, respectively.

As shown in Figure 22.8(a) and (b), the mixture of PEG-400 and thiazone produced an obvious optical clearing effect. As shown in Figure 22.8(c), (d), and (g), although the photoacoustic amplitude of subcutaneous vessels is greatly improved, the measured vascular diameters are the same, indicating there are no lateral resolution changes. Based on the B-scan images shown in Figure 22.8(e) and (f), the blood vessels seem to be closer to the skin surface, probably due to the dehydration effect. A statistical analysis from ten randomly chosen vessels confirms that there is significant photoacoustic amplitude improvement but no big changes in lateral resolution, as shown in Figure 22.8(h).

As shown in Figure 22.9(a) and (b), the glycerol produced very little optical clearing effect visually. However, as shown in Figure 22.9(c), (d), and (g), the photoacoustic amplitude of subcutaneous vessels is greatly improved. Different from the mixture of PEG-400 and thiazone, there was no obvious dehydration effect, as shown in Figure 22.9(e) and (f).

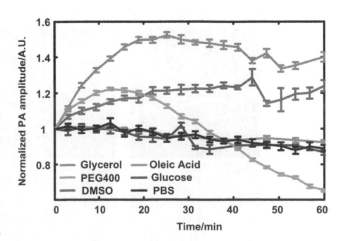

FIGURE 22.7 Changes in the photoacoustic amplitude with different OCAs. Not all OCAs enhanced the amplitude. Reprinted with permission from reference [10] © The Optical Society.

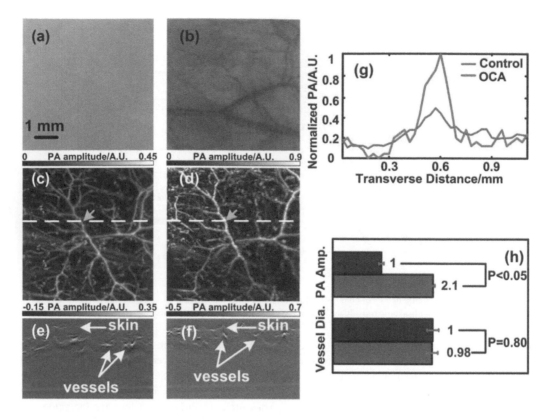

FIGURE 22.8 Comparative images and statistical analysis for *in vivo* imaging with a mixture of PEG-400 and thiazone. (a) and (b) are photographs of skin before and after immersion. (c) and (d) are photoacoustic maximum amplitude projection (MAP) images projected along the depth direction. (e) and (f) are B-scan images denoted by the dashed lines in (c) and (d). The horizontal axis is the scan direction, and the vertical axis is the depth. (g) shows transverse plots of the vessel indicated by the green arrows in (c) and (d). (h) shows the statistical results for ten randomly selected vessels. The scale bar is the same for (a)–(f). Reprinted with permission from reference [10] © The Optical Society.

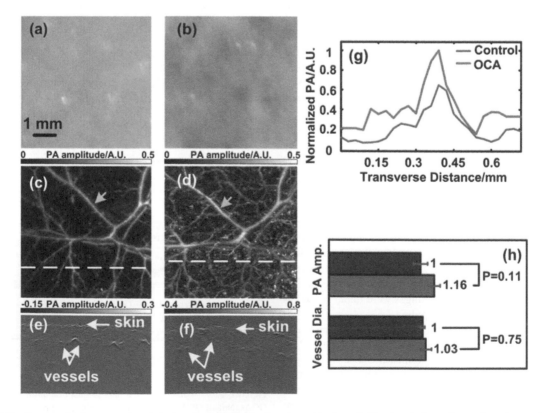

FIGURE 22.9 Comparative images and statistical analysis for *in vivo* imaging with glycerol. The descriptions of the subgraphs are the same as in Figure 22.8. Reprinted with permission from reference [10] © The Optical Society.

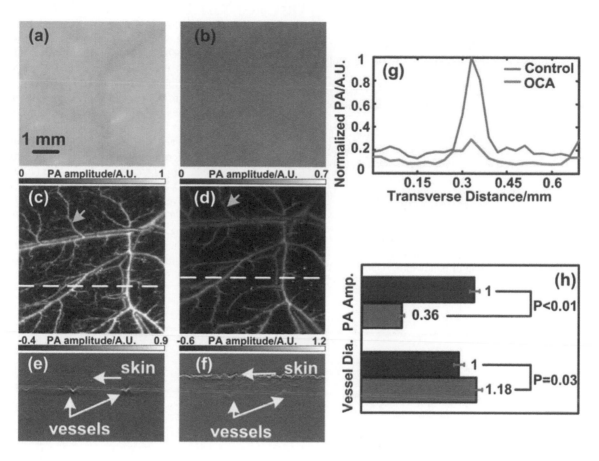

FIGURE 22.10 Comparative images and statistical analysis for *in vivo* imaging with DMSO. The descriptions of the subgraphs are the same as in Figure 22.8. Reprinted with permission from reference [10] © The Optical Society.

A statistical analysis from ten randomly chosen vessels shows that there is about 10% photoacoustic amplitude improvement but no big changes in lateral resolution, as shown in Figure 22.9(h).

Surprisingly, the *in vivo* experiment from DMSO presents contradictory results compared to *ex vivo*. As shown in Figure 22.10(a) and (b), there is no obvious optical clearing effect with DMSO. The photoacoustic amplitude of subcutaneous vessels greatly decreased after applying DMSO, as shown in Figure 22.10(c), (d), and (g). As shown in Figure 22.10(e) and (f), the subcutaneous blood vessels are deeper from the skin surface, which is probably from the swelling caused by stimulation. Based on the statistical result in Figure 22.10(h), the photoacoustic signal decreased to only 36% of the original ones. The physiological response of applying DMSO (such as swelling) probably caused the decrease of photoacoustic signals. Thus, DMSO may not be appropriate for AR-PAI as an OCA.

optical agents are beneficial in photoacoustic imaging, such as DMSO, glucose, and oleic acid. Glycerol has been proven to be able to improve photoacoustic amplitude in both OR-PAI and AR-PAI.

Because photoacoustic imaging performance is determined by both optical excitation and acoustic detection, a clearing agent that can reduce both optical and acoustic scattering is desired. So far, a majority of studies have focused on the optical side, and few works have been carried out from the acoustic perspective. It has been shown that a common optical clearing agent will introduce more acoustic attenuation due to dehydration, thus deteriorating the imaging performance. In 2016, Yang et al. proposed a new optical clearing agent that can be used to increase not only the transmittance of light, but also that of ultrasound in the skull in AR-PAI [11]. However, no obvious resolution improvement was observed. In the future, an optical clearing agent that can also reduce acoustic scattering in the skull would be very helpful for brain imaging with photoacoustic imaging techniques.

Conclusions

In summary, optical clearing has shown great capability in reducing optical scattering, thus improving photoacoustic signals in both OR-PAI and AR-PAI. In addition, because the lateral resolution is determined by optical focus in OR-PAI, utilizing optical clearing can also lead to an increase of lateral resolution in OR-PAI. Finally, it has been found that not all

Acknowledgments

This work was supported in part by National Institutes of Health grants R01 CA186567 (NIH Director's Transformative Research Award), R01 NS102213, U01 NS099717, R35 CA220436 (Outstanding Investigator Award), and R01 EB028277.

REFERENCES

1. L.V. Wang and S. Hu, "Photoacoustic tomography: In vivo imaging from organelles to organs," *Science* 335(6075), 1458–1462 (2012).

2. Y. Zhou, C. Zhang, D.-K. Yao, and L.V. Wang, "Photoacoustic microscopy of bilirubin in tissue phantoms," *J. Biomed. Opt.* 17(12), 126019 (2012).

3. S.L. Chen, T. Ling, S.W. Huang, H. Won Baac, and L.J. Guo, "Photoacoustic correlation spectroscopy and its application to low-speed flow measurement," *Opt. Lett.* 35(8), 1200–1202 (2010)

4. J. Yao and L.V. Wang, "Photoacoustic tomography: fundamentals, advances and prospects," *Contrast Media Mol. Imag.* 6(5), 332–345 (2011).

5. J. Yao, K.I. Maslov, Y. Shi, L.A. Taber, and L.V. Wang, "In vivo photoacoustic imaging of transverse blood flow by using Doppler broadening of bandwidth," *Opt. Lett.* 35(9), 1419–142 (2010).

6. J. Yao, K.I. Maslov, Y. Zhang, Y. Xia, and L.V. Wang, "Label-free oxygen-metabolic photoacoustic microscopy in vivo," *J. Biomed. Opt.* 16(7), 076003 (2011).

7. L.V. Wang and H. Wu, *Biomedical Optics: Principles and Imaging*, Wiley, Hoboken, NJ (2007).

8. D. Zhu, K.V. Larin, Q. Luo, and V.V. Tuchin, "Recent progress in tissue optical clearing," *Laser Photonics Rev.* 1–26 (2013).

9. Y. Zhou, J. Yao, and L.V. Wang, "Optical clearing-aided photoacoustic microscopy with enhanced resolution and imaging depth," *Opt. Lett.* 38(14), 2592–2595 (2013).

10. Y. Liu, X. Yang, D. Zhu, R. Shi, and Q. Luo, "Optical clearing agents improve photoacoustic imaging in the optical diffusive regime," *Opt. Lett.* 38(20), 4236–4239 (2013).

11. X. Yang, Y. Zhang, K. Zhao et al., "Skull optical clearing solution for enhancing ultrasonic and photoacoustic imaging," *IEEE Trans. Med. Imag.* 35(8), 1903–1906 (2016).

23

Enhancement of contrast in photoacoustic – fluorescence tomography and cytometry using optical clearing and contrast agents

Julijana Cvjetinovic, Daniil V. Nozdriukhin, Maksim Mokrousov, Alexander Novikov, Marina V. Novoselova, Valery V. Tuchin, and Dmitry A. Gorin

CONTENTS

Introduction

A vast amount of imaging techniques are successfully used in biomedical research to provide anatomical, functional, and molecular information about organisms *in vivo* at different scales and accuracy levels. All biomedical visualization modalities have their own features affecting imaging performance, the information provided, safety, and usability. The crucial thing is obtaining enough image contrast to make an accurate diagnosis and optimize therapy, as well as monitor already diagnosed and/or treated diseases. Accordingly, an urgent task is to clearly define the border normal/pathological tissue, which is undoubtedly important not only for hematological/oncological, but also for inflammatory, infectious, and degenerative diseases.

Photoacoustic (PA) imaging is one of the most promising methods with great potential to become a universal diagnostic tool in medicine since it possesses high contrast and excellent resolution [1–5]. It is based on the PA effect, which implies irradiation of tissue with pulsed light, whereby the light energy is absorbed by chromophores and converted into heat [6]. Such a process leads to thermoelastic oscillation of the irradiated object and, hence, the subsequent generation of a pressure wave which can be detected using a conventional ultrasound transducer. After the signal receiving and the reconstruction routine, images can be obtained and post-processed. The sound waves exhibit significantly less scattering and dissipation during propagation in a tissue than photons, so deeper tissue penetration could be achieved. As a noninvasive, easy-to-use, and highly sensitive method, PA is beneficial for both preclinical and clinical research, providing structural and functional information about biological systems based on their light-absorption coefficient [7–10, 14–16]. This information is provided by either endogenous or exogenous contrast media. Photoacoustic imaging can also be used to characterize non-human natural objects, such as unicellular diatom algae, which contain chromophores, such as chlorophylls and carotenoids, and therefore provide a strong PA response [10].

Endogenous chromophores such as hemoglobin, melanin, collagen, and lipids are strong light absorbers demonstrated as photoacoustic contrast agents [11–13]. However, their application field is quite limited due to the finite amount of the natural contrast substances types. To further improve the quality of imaging in specific tasks, exogenous contrast agents should be employed; many diseases cannot be identified using only intrinsic chromophores.

On the other hand, fluorescent tomography has drawn particular attention due to its extremely high sensitivity and a large number of fluorescent probes that can be readily used to highlight poorly contrasting tissues further [14]. Moreover, reflection-mode fluorescent molecular technology can be applied for intraoperative imaging [14, 15]. Nevertheless, the main challenge of this method is its low spatial resolution [16]. Since photoacoustic imaging has lower sensitivity than fluorescence imaging, but higher spatial resolution and penetration depth, they can be combined to overcome their limitations in clinical applications [16].

Multimodal visualization, which refers to the simultaneous acquiring of signals for more than one imaging technique – the combination of several imaging types – of chemical markers, biomolecules, and cellular compartments has emerged as a new correlative imaging approach with an outstanding opportunity for biomedical applications [17]. Several imaging methods, such as computed tomography (CT), PAI, fluorescence imaging, magnetic resonance imaging (MRI), surface-enhanced Raman scattering (SERS), and positron emission tomography (PET), have already been used together to combine the strengths and eliminate the drawbacks of each method and to enhance image resolution [18–20]. Marker-based multimodal imaging was demonstrated by using the fluorescence imaging (FLI)/SERS approach [21–23], and CT/MRI methods [24]. FLI/MRI/PAI hybrid imaging mode was used in the *in vivo* studies [25–27]. A combination of PET/MRI has been reported using nanoparticles instead of conventional positron-emitting radionuclides [28]. A single-photon emission computed tomography (SPECT)/CT pair was applied to assess the biodistribution of radio-labeled nanocarriers [29, 30]. SPECT/MRI/CT imaging modality has been shown to be useful in imaging the kidneys and bladder [31].

One of the main limitations of photoacoustic imaging is limited *in vivo* penetration depth due to strong light scattering in biological medium. This problem can be solved using two different approaches. The first and most widely used implies the application of different exogenous contrast agents aimed at increasing the sensitivity, reliability, imaging contrast, and resolution of medical imaging methods, as well as penetrability to tissue. The second is based on using high-refractive-index optical clearing agents (OCAs). In return, optical clearing techniques can be chemically based, involving treating the tissue with chemical reagents (glycerol, polyethylene glycol (PEG)-400, polypropylene glycol (PPG)-425, dimethyl sulfoxide (DMSO), hyaluronic acid (HA), etc.) [32–36]; and physically based, by altering the tissue properties with physical impact (ultrasound application, fractional laser resurfacing, grinding, mechanical compression, etc.). The tissue optical clearing (TOC) method will be described in detail on the following pages.

To the best of our knowledge, even though there are many review articles devoted to the analysis of contrast agents and tissue optical clearing, these two approaches have not been systematized simultaneously. So, we decided to fill this gap and provide an overview of recent progress in the development of exogenous photoacoustic and fluorescent contrast agents, as well as the most commonly used tissue optical clearing routes. Moreover, we analyzed a novel and promising method for overcoming conventional depth limit *in vivo* based on a combination of contrast agents and optical clearing.

Contrast and contrast-enhancing methods

The crucial role in biomedical imaging is played by contrast – the relation of the image background and the difference between the image and adjacent background. The contrast is forming according to the signal-to-noise ratio (SNR) of the detector. It could be caused by absorption, fluorescence, scattering, phase shift, reflectance, and other optical processes guided by light–specimen interactions.

The optical contrast of the image can be described with the basic Weber's law:

$$K = (I - I_b)/I_b, \qquad (23.1)$$

where I is the intensity of the object of interest and I_b is the intensity of the background.

In an ideal case, the light absorption of the surrounding tissue is low, while the absorption of the region of interest is drastically higher to form sufficient image contrast. However, in most cases, they are more or less equal, which tends to the implementation of the contrast-enhancing techniques. The contrast-enhancing techniques could be software-based, which could reveal new crucial information from the image on the post-processing step; and hardware-based, which require the tuning of a study protocol and will elucidate the peculiarities of studying the object. Besides, hardware-based methods could be separated into two principal categories:

- Reduction of the background signal (optical clearing)
- Increasing SNR or K (contrast agents' utilization).

Photoacoustic contrast agents

One of the greatest advantages of photoacoustic imaging is the ability to use intrinsic chromophores to visualize blood vessel structure, blood flow, and thermal burns; to perform blood oxygenation mapping and functional brain imaging, and to detect and characterize primary melanoma and metastatic melanoma cells [37–39]. However, most diseases cannot be detected with endogenous PAI molecules, leading to the development of new exogenous contrast agents that can provide sufficient PA signal even at low concentrations. Contrast agents can improve not only image resolution but also penetration depth, and can help in targeted imaging [38, 40]. The ideal contrast agent must meet specific criteria [37, 38, 40, 41]. A contrast agent must consist of a signaling compound, which produces the response for imaging, and a targeting moiety that enables the readout of a specific biological entity or process. The signaling compound is required to have a high molar extinction coefficient to maximize the amount of absorbed light, since photoacoustic image contrast is absorption-based. The signaling compound should preferably have high absorption in the near-infrared (NIR) biological transparency window, from ~620 nm to ~950 nm, where absorption and scattering of tissue are quite low, which increases the penetration depth [42]. Moreover, the compound must have high photostability to ensure that spectral features are not changed by light irradiation; low quantum yield, to maximize the nonradiative conversion of light energy to heat; efficient conversion of heat energy to produce acoustic waves [43]; and sharp peaks in its absorption spectrum to ensure unambiguous identification by different image processing techniques, e.g., spectral unmixing, even at low molar concentrations [44].

The targeted contrast agent must overpass circulatory and cellular barriers. First, contrast agents in the bloodstream will interact with biomolecules [45, 46]. These interactions, mainly adsorption onto the surface of the contrast agent, affect immune responses and target binding [46, 47]. The extravasation from the circulatory system to extravascular targets requires a maximum size of the contrast agent to be <10 nm. It has been reported that contrast agents with a size of up to 100 nm have been found to pass through the leaky vessels in tumors – the enhanced permeability and retention (EPR) effect [48]. After overcoming the circulatory barrier, the contrast agent can overpass the cellular barrier by active targeting to the cell surface receptors [49], transporters [50], or metabolic enzymes [51], or by passive uptake through diffusion or endocytosis. To facilitate active targeting, the biological target should be (over)expressed at an early stage of the disease and found at low levels in off-target tissues. Small molecules, peptides, adhirons [52], antibodies [53], aptamers [54], and proteins can be used for active targeting. Biocompatibility needs to be carefully considered in terms of minimizing toxicity and immune response, as well as optimizing clearance from the circulatory system and penetration into target tissues.

The field of development of new active contrast agents that generate a robust photoacoustic response is rapidly evolving, given their importance for assessing the morphological structure and function of tissues, as well as for the diagnosis, monitoring, and treatment of diseases.

Considering the importance of image contrast enhancement in photoacoustic research, great efforts are being made to develop new exogenous agents that can be classified into inorganic and organic materials by their chemical structure. The classification of PA contrast agents is shown schematically in Figure 23.1. The group of inorganic PA contrast agents includes metallic nanoparticles, carbon-based nanomaterials, and transition metal chalcogenides. On the other hand, small organic molecules and semiconductor polymer nanoparticles belong to the group of organic materials.

Endogenous contrast agents

One of the most significant advantages of photoacoustic imaging is the ability to use intrinsic chromophores, such as oxy- and deoxyhemoglobin, melanin, collagen, lipids, and water, as contrast agents. Hemoglobin is a dominant absorber in blood in the visible range, so it can be exploited to obtain information about the structure of blood vessels [55], total hemoglobin concentration [56], microvascular blood flow [57], oxygen saturation [58], and metabolic rate of oxygen [59]. The vascular structure can be easily visualized by photoacoustic tomography (PAT) due to the very high absorption of both oxygenated and deoxygenated hemoglobin in the wavelength range 650–900 nm. Another essential capability of PAT is a measurement of oxygen saturation of hemoglobin, a parameter that indicates the level of tumor progression [60].

Melanin, a natural pigment found in skin, hair, and eyes, has a strong optical absorption throughout the optical window. By using melanin as a contrast agent, photoacoustic imaging is capable of detecting an early-stage melanoma [59, 61] as well as metastatic melanoma cells [62]. An early-stage melanoma in mouse ear was detected by measuring the metabolic rate

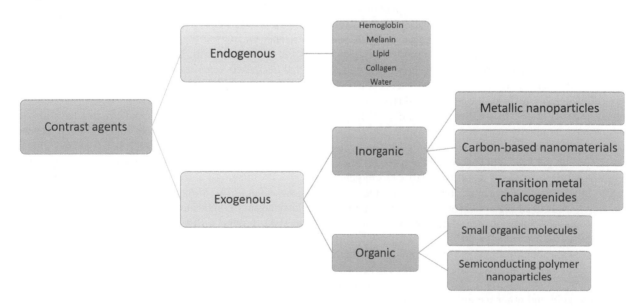

FIGURE 23.1 Classification of the PA contrast agents.

of oxygen as an indicator of tissue viability and functionality [59]. Since the optical penetration depth in areas rich in melanin can be significantly reduced, melanin-based nanomaterials are being developed to improve the performance of PA imaging in such tissues [63].

Lipids and water are also important indicators of various diseases. For example, lipid deposition can be identified by using wavelengths of 930 nm [64] and 1210 nm [65]. Recently, Kosik et al. demonstrated a new fast, low-cost, and safe method of visualizing whole breast tumors based on their lipid profile by using an intraoperative photoacoustic screening (iPAS) soft-tissue scanner [64]. According to their study, this method provides volumetric visualization of breast tumors with similar performance to dynamic contrast-enhanced magnetic resonance imaging (DCE-MRI), without injecting exogenous contrast agents. Pleitez and co-workers showed that mid-infrared optoacoustic microscopy (MiROM) could be used for label-free imaging of carbohydrates (1085–1000 cm^{-1}), lipids (around 2850 cm^{-1}), and proteins (at 1550 cm^{-1}) in cells and tissues [66]. Besides, MiROM with high contrast, image quality, sensitivity, and specificity can also observe nucleic acids and water. Water content in the tissue can be detected and measured with laser-based PAT at the 975 nm wavelength [67].

Even though these naturally occurring contrast agents are safe and reveal many physiological characteristics of the organism, their use is very often limited. Therefore, external contrast agents must be administered to obtain a sufficient photoacoustic response and to detect specific diseases *in vivo*.

Exogenous contrast agents

Metallic nanoparticles

Metallic nanoparticles are currently the most abundant and up-and-coming class of photoacoustic contrast agents due to their high absorption coefficients; photothermal efficiency; chemical, physical, and biochemical flexibility; tunable surface properties; and size/morphology-dependent optical characteristics [37, 38, 40, 41]. Additionally, they come in a variety of shapes and sizes. However, potential problems that still need to be addressed include biocompatibility, biodegradability, reproducibility, purification, quantification, and homogeneity [38].

In recent years, gold nanostructures have been widely used as photoacoustic contrast agents due to their excellent optical absorption, which arises from the localized surface plasmon resonance (LSPR) effect. In addition, gold nanoparticles are biocompatible, chemically inert, and have customizable physical and chemical properties that make them good candidates for a variety of biomedical applications. Collective oscillations of electrons on the surface of metal nanoparticles are generated by light illumination of nanoparticles at a specific wavelength, which leads to strong optical absorption, conversion to heat, thermoelastic expansion, and detection of the acoustic wave by ultrasound transducers [40]. LSPR is dependent on particle size, shape, composition, and environment, and therefore the optical properties of gold nanostructures can be easily altered by changing some of these parameters [68]. Due to this phenomenon, gold and other metallic nanomaterials can also be used for photothermal therapy (PTT) [69, 70].

Gold nanoparticles of various shapes and sizes, including nanospheres, nanorods, nanoshells, nanoprisms, nanocages, nanostars, nanodisks, nanovesicles, nanotripods, nanoplates, nanowreaths, rhombic dodecahedra, and bipyramids, are ideal candidates for different applications [37, 41, 69, 71]. Among them, gold nanospheres and nanorods are most often used as photoacoustic contrast agents. Although the absorption peak of the gold nanosphere is at approximately 520 nm [72], it can be shifted to 600 nm by adjusting their size [73] or even to 800 nm by coating hollow nanospheres of 40–50 nm in size with PEG [74]. Moreover, the results showed that PEGylated gold nanospheres provide a higher photoacoustic response than blood when examining the vasculature of the mouse brain.

Gold nanorods have attracted the attention of researchers due to the strong LSPR effect created along the longitudinal direction and the fact that their size can be smaller than, for example, spherical gold nanoshells [75]. Their distribution in the tissue can be detected by using a laser PA imaging system at low concentrations, which is very promising for new diagnostic modalities when taking into account their ability to attach to vesicles and target cancer cells. The PA signal can be amplified using gold nanorods encapsulated in a 20 nm silica shell [76] by about 3.8 times, which can be explained by changes in the interfacial heat conductivity between gold particles thermally insulated from the surrounding aqueous matrix. Due to the relatively low thermal conductivity of silica, the particles better retain the heat generated during the laser pulse, which results in a larger temperature difference and, therefore, a stronger photoacoustic signal. The longitudinal peak shifts to the NIR region, depending on the thickness of the silica shells. Gold nanorods coated with silica shell of 75 nm thickness have an absorption peak at 806 nm. A study by Jokerst et al. [77] showed that silica-coated gold nanorods have tremendous potential for labeling mesenchymal stem cells and visualizing them in the musculature of living mice, increasing the PA signal by a factor of four. The silica coating increased the cellular uptake of gold particles by more than five times, without adversely affecting cell metabolism, proliferation, or pluripotency. Kim et al. [78] have developed hybrid nanoparticles consisting of silver-coated gold nanorods as theranostic agents for *in vivo* photoacoustic monitoring of localized release of Ag^+ ions with high antibacterial potential. The PA signal decreased when the nanorods were coated with silver, but it was restored and enhanced by etching the silver shell with a ferricyanide solution.

Several other forms of gold nanoparticles have also been used successfully as PA contrast agents. Dual plasmonic gold nanostars with tunable plasmon bands coated with silica proved their ability to serve as a theranostic nanoagents for the *in vivo* PA imaging-guided PTT upon illumination with 1064 nm laser light [79]. Moreover, a nanoprobe based on Fe_3O_4/gold nanostars composite nanoparticles modified with folic acid (FA) through a PEG spacer was applied for the multimodal MR/CT/PA imaging-guided PTT of tumors [80]. These nanoparticles have proven to be very useful for multimodal imaging due to their superior properties, such as colloidal stability, hemocompatibility, cytocompatibility, X-ray attenuation, intense absorption in the NIR region, and ultrahigh r_2 relaxivity. Biocompatible PEGylated gold nanoprisms with an absorption peak at 830 nm were exploited for the enhancement of PA signal to visualize tumor angiogenesis in gastrointestinal cancer cells [81]. The surface plasmon band peak of gold nanoprisms can be further shifted to 980 nm by antibody- and (RGD)-peptide functionalization [82]. These antibody/peptide nanoprism probes showed their imaging and therapeutic potential for *in situ* photoacoustic imaging, angiography, and localized hyperthermia treatment. Branched nanoporous gold shell generates strong surface Raman scattering (SERS) as well as PA signals accompanied by high photothermal conversion efficiency [83]. Luke et al. [84] developed silica-coated gold nanoplates that provide a strong photoacoustic response from the sentinel lymph node (SLN) when irradiated by a 1064 nm laser. In view of their high photothermal stability, low toxicity, and long-term retention in the lymph node, they may be well suited as a contrast agent for PA-based SLN mapping. PEGylated triangular gold nanoplates conjugated with anti-epidermal growth factor receptor (EGFR)-peptide was used for computed tomography and photoacoustic imaging–guided photothermal therapy of non-small-cell lung cancer [85]. These therapeutic agents, having strong NIR absorption, high photothermal conversion efficiency, and good biocompatibility, have shown increased tumor accumulation and therapeutic efficacy after laser irradiation at 680 nm. Liang and co-workers [86] introduced liposome-coated gold nanocages conjugated with the dendritic cells antibody aCD11c for targeted delivery of adjuvant monophosphoryl lipid A (MPLA) and melanoma antigen peptide TRP2. Fluorescence and photoacoustic imaging were applied *in vivo* to visualize the accumulation and migration of particles to regional lymph nodes (RLN). Results proved that this nanovaccine is a promising candidate for antitumor immunotherapy as well as *in vivo* tracking.

Silver nanoparticles represent another class of potential photoacoustic contrast agents because they also exhibit the LSPR phenomenon [87] and are a more potent light absorber than gold. However, some challenges, such as their reactivity and potential cytotoxicity, must be addressed to increase the effectiveness of silver nanomaterials *in vivo* [88]. Despite the fact that silver is used as an antimicrobial agent and in wound healing, questions about its toxicity remain [89, 90]. A system consisting of silica core covered with a porous silver layer demonstrated a strong photoacoustic response that decreases with decreasing concentration of the particles [91]. PEGylated form of this silica–silver system showed no toxicity at concentrations of silver up to 2 mg/ml. The functionality of such a platform could be improved by encapsulating drugs and therapeutic agents or attaching antibodies to the surface. Silver nanoplates can be functionalized by using more biocompatible reagents and readily used for various biomedical applications [92]. Antibody EGFR (a-EGFR) functionalized silver nanoplates were employed as contrast agents for the ultrasound and PA imaging of an orthotopic pancreatic tumor in a mouse. Their nontoxicity was confirmed at concentrations up to 1 mg/ml.

Recently, Silvestri and co-workers designed a hybrid nanosystem based on silica, eumelanin, and silver [93]. The obtained self-structured nanoplatform demonstrated high PA contrast, stability to aggregation, efficient uptake by human pancreatic cancer cells, biocompatibility, and low cytotoxicity.

This study revealed not only the excellent potential of these particles as contrast agents for photoacoustics but also the detection of rhodamine isothiocyanate–conjugated particles using fluorescence imaging. The intrinsic contrast of eumelanin was enhanced by using silver nanoparticles.

Iron oxide nanoparticles Fe_3O_4, as FDA-approved nanoprobes with excellent magnetic properties, are widely used for various imaging applications [94]. Their unique features make them a superb contrast medium for MRI [94, 95]. Superparamagnetic iron oxide nanoparticles (SPIONs) can also be employed in photoacoustic imaging due to their biocompatibility, biodegradability, efficient contrast properties, facile surface functionalization, rapid response to the magnetic field, and optical stability [94–96]. For example, Alwi et al. [97] showed that silica-coated SPIONs have strong potential to be employed as PA contrast agents under 1064 nm laser excitation. They found that the maximum depth characterization of such nanoparticles was up to 24 mm. In contrast, the minimum detectable concentrations were ~0.17 and ~0.23 mg/ml at depths of 5 mm and 10 mm inside a tissue-like Intralipid solution, respectively. Armanetti et al. developed a dual MR/PA imaging nanosystem based on bio-conjugated near-infrared light-absorbing magnetic nanoparticles [98]. Iron oxide clusters encapsulated into carboxy-terminated poly (D, L-lactide-co-glycolide)-block-poly(ethylene glycol) (PLGA-b-PEG-COOH) nanoparticles with an average hydrodynamic diameter of 40 nm showed outstanding PA performance at the μM level when irradiated in the NIR region from 680 to 970 nm. A strong superparamagnetic behavior, detection at 1.5 T, and good nuclear relaxivity values make them suitable as MRI contrast agents as well.

Xi et al. employed iron oxide nanoparticles labeled with a NIR-absorbing dye (NIR 830) and conjugated to amino-terminal fragments of urokinase plasminogen activator (uPA) for targeted imaging of breast cancer tissue containing the uPA receptor [99]. The photoacoustic signal from targeted nanoparticles injected in the orthotopic mouse mammary tumor model was three times higher compared to nontargeted particles and ten times higher than the signal from control mice. Kanazaki and co-workers [100] examined nanosystems based on 20-nm, 50-nm, and 100-nm iron oxide nanoparticles (IONPs) conjugated with three different anti-HER2 targeting moieties (whole IgG, single-chain fragment variable (scFv), and peptide) for HER2-targeted PA tumor imaging. According to the results, 20-nm anti-HER2 scFv-conjugated IONPs can be considered as the most suitable contrast agent for PA imaging since they exhibited the highest affinity to HER2 and significantly higher tumor uptake and tumor-to-blood ratios when compared to other particles. Moreover, they were able to achieve selective visualization of HER2-positive tumors in PA studies.

Gold–iron oxide Janus magnetic–plasmonic nanoparticles, developed by Reguera et al., have proven to be attractive candidates for multimodal imaging [101]. Janus nanoparticles with two distinct sides having different surface features are composed of an iron oxide nanosphere and a branched gold nanostar. Their magnetic and plasmonic properties have been extensively evaluated to determine their applicability as multimodal contrast agents for MRI, CT, PA, and optical imaging under bright- and dark-field illumination, transmission electron microscopy, and SERS imaging. The results have proven that they can also be exploited in biomedical applications such as photo- and magnetothermal therapies or targeted and controlled drug delivery, owing to their unique superparamagnetic behavior and intense plasmonic responses in the NIR, as well as easy surface functionalization. Table 23.1 summarizes the essential parameters and application of the aforementioned PA contrast agents.

Carbon materials

Other promising PA-contrast agents are carbon nanomaterials. They have a broad light absorption in the whole optical spectra, but the phenomena of the absorption vary on the conformation of the carbon nanostructure. Carbon structures could be separated into three categories depends on their dimensionality [37]:

1. 0D – carbon dots, nanodiamonds, and fullerenes
2. 1D – single- and multiwalled carbon nanotubes (SWCNT and MWCNT)
3. 2D – graphene

Fullerenes, an allotrope of carbon whose molecule consists of carbon atoms connected by single and double bonds to form a closed or partially closed mesh with fused rings of between five and seven atoms, act mostly like molecules. The absorption could be described in terms of HOMO–LUMO transitions (highest occupied molecular orbital and lowest unoccupied molecular orbital) of the one electron after the proper wavelength light absorption. Due to weak luminescence, they could be great PA-contrast agents. The fullerenes mostly absorb at 300–700 nm [102].

The absorption mechanism of carbon dots – nanoparticles consisting of carbon with surface passivation and certain surface states – is still under consideration. At least a couple related to two peaks of absorption was pronounced. The first one, in the short wavelength range (220–250 nm), could be assigned to $\pi–\pi^*$ transition of aromatic –C=C–, and –C–C– bonds in the sp^2 hybridized domain of graphitic core. The second one, in the longer range (300–450 nm), could be assigned to $n–\pi^*$ transition of –C=O, C–N or –C–OH bonds in the sp^3 hybridized domains, which originated from carboxyl (–COOH) or amine (–NH$_2$) groups exists on the surface of dots [103]. The carbon dots mostly absorb at 300–600 nm [104–106].

The nanodiamonds are carbon nanoparticles in which all atoms are sp3- hybridized. Their optical features include strong absorption and tiny luminescence simultaneously, caused by local defects in their microcrystalline structure generally induced by vacancy centers – inclusions of another atom such as N, Si, Ge, Ni. Nanodiamonds effectively absorb light at 450–900 nm, depending on the fabrication process [107–109].

Carbon nanotubes' absorption properties imposed by their semiconducting bandgap may vary depending on the nanotube diameter and the chiral properties. They also display minute luminescence being isolated in micelles, but their photoacoustic response is stable enough to provide *in vivo* contrast. The

carbon nanotubes absorb in a broad range of 300–1800 nm, depending on the fabrication process [110–112].

The pure graphene was pronounced as a gapless semiconductor and shown weak absorption features in the NIR [113]. However, due to the abundance of surface heterogeneities, graphene oxide (GO) and reduced graphene oxide (rGO) show strong absorption in the NIR. The exact origins of this behavior are still not clearly defined. Still, it is considered as the electronic transition at the boundaries between two phases: the oxidized graphene and nonoxidized sp2 graphene domains. Optical features and facile functionalization turns GO/rGO to promising PA contrast agent [114].

Despite all of the advantages of carbon materials, there are two crucial shortcomings: low molar-extinction coefficient compared to metallic materials, and the smooth decreasing absorption curve in NIR regions, which makes carbon structures unable for multiplexing in PA studies. However, at least the low molar absorption issue could be overcome by increasing the local concentration of carbon structures by depositing them on particles/isolating them in micelles or by decorating them with light-absorbing substances to improve their properties (metal–carbon and dye–carbon combinations).

Toumia et al. developed graphene-coated PVA microbubbles with high NIR absorption and PA response and performed an *in vivo* photoacoustic imaging utilizing such contrast agents [115]. Song et al. manufactured carbon nanotube rings (CNTR) coated with gold nanoparticles (CNTR@AuNPs) and successfully demonstrated drastically enhanced Raman and PA signal (2 orders higher) in comparison with uncoated CNTR [116]. Gold-coated nanodiamonds, studied by Orlanducci, showed enhanced absorption and PA signal intensity [117]. Yashchenok et al. demonstrated that microcapsules made of Au-coated SWCNTs provide a two-fold higher PA signal than the whole blood [118].

The fullerene derivatives consisting of a new type of NIR dye bound with fullerene have shown excellent NIR absorption with several prominent peaks, thus enabling multiplexing [102].

cRGD-functionalized SWCNTs were used in the PA imaging of tumors and provided an eight times higher signal than bare SWCNTs. Then such construction was coated with ICG dye or QSY21 (an optical dye and quencher) and showed an increase in the absorption spectrum of the particles for order and two-orders high PA intensity in comparison with bare SWCNTs [119]. Also, SWCNTs could be functionalized for better biocompatibility and uptake, for example, with hyaluronic acid-5β-cholanic acid conjugated with folic acid [120]. Noble-metal NP-coated SWCNTs possess an increased NIR absorption, enhanced biocompatibility, and high PA signal. For example, Au-coated CNTs were developed, and their absorbance intensity was 85 times higher than that of bare SWCNTs at the same concentration [121]. The photothermal conversion further was increased by intercalating such an arrangement with gold nanoparticles. Reduced graphene oxide – rGO – has a larger surface area in comparison to CNTs and could be better dispersed in biological conditions. ICG-PDA-RGO composite sheets shown perfect PA signal formation by quenched ICG and high absorption at 750–800 nm [122]. rGO-coated Au nanorods were reported to have an enhanced

TABLE 23.1

The main properties of metallic nanoparticles and fluorescent dyes used as photoacoustic contrast agents

Contrast agents	Type	Size (nm)	λ (nm)	Functionalization/coating	Applications	Ref
Gold nanoparticles	Nanosphere	40–50	800	PEG	Mouse brain vasculature	[74]
	Nanorods	50 × 15	760	PEG	Tissue contrast	[75]
		Silica shell thickness 0, 6, 20, 75 nm	786,790,802, 806	PEG, silica	Contrast amplifier	[76]
		Length: 82.99 ± 3.86 Width: 64.20 ± 3.48	676	Silica	Imaging and quantitation of mesenchymal stem cells in rodent muscle tissue	[77]
		Length: 50–55 Width: 12–14 Shell thickness of Ag: 10	750	Silver	Imaging of bacterial-induced infections and antibacterial activity	[78]
	Nanostars	Silica thickness: 10	700, 1000–1200	Silica	Tumor theranostics, photothermal therapy	[79]
		9.3 nm	810	Folic acid-modified Fe_3O_4	MR/CT/PA imaging and PTT of tumors	[80]
	Nanoprisms	Thickness: 10 Edge lengths: 120	830, 530	PEG	MSOT imaging of gastrointestinal cancer	[81]
		110	980, 530	Antibody- and peptide (RGD)	Imaging, angiography, and localized hyperthermia	[82]
	Nanoshell	60	~800–900	Redox-active polymer nanoparticles	Tumor contrast, PTT	[83]
	Nanoplates	96 ± 5	945, 1035	Silica	SLN imaging	[84]
		77.9 ± 7.0	665	Anti-EGFR peptide	CT/PA imaging-guided PTT of non-small cell lung cancer	[85]
	Nanocages	~50–60	~800	Poly(vinyl pyrrolidone), liposome coated, antibody aCD11c modified	Antitumor immunotherapy, *in vivo* tracking	[86]
Silver nanoparticles	Silica nanospheres coated with a porous silver layer	180–520	490	Silica	Image-guided therapy	[91]
	Nanoplates	Edge length: 25.3, 60.9, 128.0, 218.6 Thickness: 10.4, 12.5, 18.0, 25.6	550, 720, 900, 1080	PEG, antibody to EGFR	Orthotopic pancreatic cancer imaging	[92]
	Hybrid silica, eumelanin, silver nanostructure	35, 234	710	Rhodamine B isothiocyanate	Multimodal imaging in cancer diagnosis and therapy	[93]
Magnetite nanoparticles	SPIONs	Core: 8 Coating: 3	NIR	Silica	Tissue contrast	[97]
	Fluorescent nano bioreactor magnetite nanoshells	41	494, 516 975	Fluorescein 488-NHS	MR/PA imaging and cell-tracking	[98]
	receptor-targeted nanoparticles	10	730	NIR-830 dye, urokinase plasminogen activator receptor	Breast cancer contrast	[99]
	antibody conjugated nanoparticles	20, 50, 100	710, 797	Anti-HER2 antibody	Imaging of HER2-expressing tumors	[100]
	Janus plasmonic-magnetic gold–iron oxide nanoparticles	20–50	700–800	Raman reporter 4-mercaptobenzoic acid (4-MBA)	Multimodal imaging: MRI, CT, PA, optical imaging, TEM and SERS	[101]

(Continued)

TABLE 23.1 (CONTINUED)

The main properties of metallic nanoparticles and fluorescent dyes used as photoacoustic contrast agents

Contrast agents	Type	Size (nm)	λ (nm)	Functionalization/coating	Applications	Ref
Heptamethine cyanine dyes	Indocyanine green (ICG)	<2	780	/	Sentinel lymph node mapping	[149]
	CDnir7	<2	806	/	Macrophages and inflammation visualization	[238]
	IRDye800cw	<2	774	2-deoxy glucose	Cancer imaging with neutropilin-1 receptor conjugation	[50, 239]
	IRDye800	<2	792	Peptide (c(KRGDf))	Brain tumor imaging	[240]
	IC-5-T	<2	830	/	Tumor imaging	[161]
	IC7-1-Bu	<2	823	/	Tumor imaging	[241]
	IR780	<2	780	Caspase inhibitor	Tumor imaging	[242]
	L^1, L^2 probes	<2	776	N-methyl-D-aspartate receptor (NMDAR) antagonists	Selective NMDAR imaging	[243]
Azo dyes	Methylene blue	<2	664	/	Sentinel lymph node identification	[244–247]
	Evans blue	<2	620	/	Sentinel lymph node imaging	[248]
Others (e.g. Naphthalocyanine dyes,	Alexa Fluor 750	<2	750	/ Peptide; Peptide with black hole quencher Herceptin	Cancer imaging with HER2 conjugation	[249]; [250]; [251; 252]
Pyrrolopyrrole cyanine dyes)	SNARF-5F carboxylic acid (Fluorone dye)	<2	564; 532	/	pH imaging	[253, 254]
	CF-750	<2	755	/	Epidermal growth factor detection	[255]
	ATTO740	<2	740	Peptide	Prostate cancer imaging	[256]
	SiNc (Naphthalocyanine dye)	**<2**	770	/	Cancer imaging	[257]
	Octabutoxy Naphthalocyanine	**<2**	900	PEG, Sn(IV) chloride	Vascular imaging	[258]

PA signal [123]. Here, Au increases the optical absorption efficiency of rGO, and due to the interaction between AuNP and rGO, the photocurrent of rGO gets enhanced, which increases its photothermal properties. Thus, intercalating carbon structures with noble-metal nanoparticles increase two parameters: the absorption itself and photothermal conversion ratio, which highly increase the PA response.

Semiconducting structures

Among the fluorescent/photoacoustic contrast agent family, there is a group of particles based on semiconducting structures: quantum dots, transition metal chalcogenides, MXenes, and semiconducting polymeric particles.

Quantum dots (QDs) were initially engineered as a fluorescent probe with a mechanism of light emission a bit different from typical bulk semiconductors, and a bandgap depends on the type of materials used and on the size of a quantum dot. Due to the small size (smaller than exciton Bohr radius) of QDs, the conduction and valence bands turn to quantized energy levels with quantization according to Pauli's principle. The larger the dot, the lower energy it can emit due to the more closely located energy levels in which the exciton could be trapped. The electron from a "valence" energy level could be excited with any energy, which allows it to enter the conduction band so that QDs can absorb the light in a broad range. After the excitation and relaxation process, an electron recombines with the hole with the emission of the photon. Due to the aforementioned structure peculiarities, the emission contour is narrow: around 30–50 nm on FWHM. Still, the quantum yield of QDs is limited (20%–50%), and the absorbed light energy, mainly spent on nonradiative relaxation, leads to the photoacoustic response of such structures by mechanism "excitation – non-radiative relaxation – heating – compression wave generation". Another advantage of QDs is their relatively high absorption cross-section in comparison to fluorescent dyes and endogenous tissue background (10^{-14} cm^2 for QDs in contrast to 10^{-16} cm^2 for dyes and 10^{-16}–10^{-18} cm^2 for tissue). Also, such a combination of properties makes QDs good dual-modal contrast agents because the PA response mainly following the absorption curve and can be monitored separately from the fluorescent response [124, 125].

The application range of the QDs is quite vast. Bi_2O_2Se quantum dots demonstrated an excellent photothermal conversion ratio of 37.5% with good photothermal stability. Being passively accumulated in tumors, they enable efficient PA imaging [125]. MoO_{3-x} QDs were demonstrated as nice NIR-absorbers with photoacoustic, photodynamic, and photothermal modalities for theranostic tumors [126]. Ag_2S and PbS QDs enables NIR-II fluorescence and photoacoustic imaging *in vitro* and *in vivo* with low to moderate toxicity [127–129]. CdSe, CdTe, and InAs QDs showed themselves to be perfect contrast agents for fluorescent imaging [130–133].

The next type of inorganic contrast agents is related to chalcogenides – chemical compounds containing at least one chalcogen (group 16 in the periodic table) anion and one electropositive cation – and MXenes, which are 2D inorganic compounds made of several atomic-thick layers of transition metal carbides, nitrides, or carbonitrides. For example, good

PA responses were shown by the PEGylated tungsten disulfide (WS_2-PEG) and TiS_2 nanosheets [134, 135]. CuS and Ag_2S quantum dots, which are chalcogenides structures, were described before. Among MXenes, 2D Ti_3C_2- and Nb_2C- based structures [136, 137] have huge potential. Also, $Ti_3C_2T_x$ have shown broad absorption at 300–600 nm and stable and robust photoluminescence at 400–700 nm, which means they could be used for cell staining [138].

The next type is semiconducting polymer nanoparticles (SPN) or polymer dots. Semiconducting polymers (SP) represent organic π-conjugated units on their polymer backbones. In such a configuration, electrons become delocalized and can diffuse by tunneling, hopping, or other related mechanisms along polymer chains and quickly reach the neighboring acceptors. Commonly, SPs are direct bandgap semiconductors with a donor–acceptor mechanism of copolymer conjugation, and the bandgap depends on the molecular structure of the polymer. High absorptivity and amplified energy transfer lead to diverse applications of the particles made of these polymers in the biomedical field.

SPNs have a large and tunable absorption coefficient, controllable size, and high photostability.

Perylene-diimide (PDI) is a widely used semiconducting polymer that was applied to PA imaging due to its strong NIR absorption and excellent biocompatibility, high photothermal stability, and good light-to-heat conversion [139]. Also, it can be easily modified. PDI-based nanoparticles were developed for PA imaging of tumors.

Recent studies showed that semiconducting polymer nanoparticles could effectively convert light energy into thermoelastic oscillations, allowing the development of responsive PA-imaging probes. The combination of SPNs and dyes leads to a series of superior properties. The PA contrast agent consists of degradable carried semiconducting oligomer amphiphile, and the ROS-inert dye NIR775 via a self-assembly method was made for sensing ClO– [140]. NIR775 remained intact, while the semiconducting oligomer amphiphile degraded in the presence of ClO–, giving rise to the ratiometric PA signal.

Various polymer dots with absorption in the 250–600 nm range and emission in the 400–850 nm range were synthesized by nanoprecipitation and miniemulsion method and utilized for *in vivo* and *in vitro* studies [141, 142].

Fluorescent dyes

The chemical basis of small-molecule NIR dyes [143] represents a series of conjugated double bonds and/or (mostly aromatic) ring systems. In these dyes, less energy is needed for excitation because, in these highly conjugated ring systems, electrons are delocalized. The energy gap between the ground and excited states results in a distinct absorption peak. Relaxation of the excited electrons, depending on the lifetime and electronic configuration of the excited states, can occur in radiative (fluorescence or bioluminescence) or nonradiative ways [144].

One of the main classes of NIR dyes used for photoacoustic imaging is heptamethine cyanine dyes [145, 146]. Indocyanine green (ICG), a US Food and Drug Administration–approved dye with low toxicity, belongs to this group. ICG has an absorption maximum at 780 nm in aqueous solution and relatively low quantum yield for fluorescence [147] (0.027). ICG exhibits high plasma protein binding, and because of this, it is used clinically as a blood-flow contrast agent [148]. ICG was used for identification of the closest "sentinel" lymph node in the lymphatic drainage from a tumor [149] by photoacoustic imaging, which is very important for cancer staging. The hydrophobicity of ICG, as well as other small-molecule dyes, can lead to large concentration- and environment-dependent changes in optical properties [150] and photoinstability.

Another main class of NIR dyes used for molecular PAI is squaraines. These dyes are characterized by an electron-deficient central four-membered ring that is often conjugated to two electron-donating groups to generate a donor-acceptor-donor form, which increases the delocalization of the electrons. The optical properties can be tuned by variation of the donor strengths to produce sharp absorption bands in the NIR range [151]. Despite their good optical properties and photostability, squaraines are often limited by insolubility, aggregation, and chemical reactivity [152, 153]. In PA imaging, they have been used only in conjunction with phospholipids, albumin, or polymers owing to their low solubility [154–156].

Other dye classes with fewer examples of use as PA contrast agents are rhodamines and azo dyes. Rhodamines are characterized by a xanthene core unit with dialkylamino groups and have good photostability [157], but few have been reported in the NIR range. The azo dyes methylene blue (664 nm) and Evans blue (620 nm) have been used for PA imaging but are limited by their relatively blue-shifted absorption peaks [158, 159] and potential to cause photodamage under intense irradiation [160].

A tendency toward aggregation and photobleaching are the common issues in the use of small-molecule dyes along with a low molar extinction coefficient ($<3 \times 10^5$ M^{-1}cm^{-1}) and low solubility. These problems can be solved by attaching hydrophilic groups, introducing triplet-state quenchers [161], and integrating stabilizing groups [162]. The spectral range can be changed by varying the size of the conjugated system and attaching the electron-donating or electron-withdrawing groups. The chemical flexibility can allow the creation of contrast agents that can partially compensate for their low molar extinction coefficient by minimizing the influence of the background signal from endogenous chromophores. Small-molecule dyes, despite their photophysical limitations, are good candidates for molecular PA imaging in preclinical extravascular studies due to their small size and biocompatibility.

There are multiple examples of fluorescence quenching by excited-state electron transfer from inter- and intramolecular amines [163–167], and the reversal of this quenching by electronic mediation has been observed. It was shown that the fluorescence of bis[[[2-(dimethylamino)ethyl]methylamino]methyl]anthracene is dramatically enhanced by amine chelation with zinc chloride [168]. The intramolecular quenching of a fluorophore by a nitroxide radical was observed, and the direct relationship between the loss of the radical and increased fluorescence yield was demonstrated [169]. The fluorescence is quenched electronically and is then restored by complexation or free-radical reaction in both cases [170].

The self-aggregation of the dyes becomes significant and is caused by self-assembly of the dyes at the substrate's surface [171]. This results in a clustering of the dye molecules, substantial quenching (up to 90%) of the fluorescence, and low brightness of the imaging probe. The ordered clusters of molecular dyes that possess different properties from that of constituting organic molecules are known as J-aggregates. These aggregates were discovered by Jelley and hence the name J-aggregates [172, 173]. The experimental data, collected over the years, started to show that such self-association of dyes in solution or at the solid–liquid interfaces leads to significant changes in the energies of their electronic transitions is a frequently observed phenomenon. [174]

H- and J-type aggregates are traditionally classified on the basis of the observed spectral shift of the absorption maximum relative to the respective monomer absorption band (hypsochromic for H-type and bathochromic for J-type). J-aggregates are fluorescent, and their fluorescence quantum yield often surpasses that of the monomeric dyes. On the contrary, H-aggregates have highly quenched fluorescence. This effect was observed for multiple dimer aggregates of fluorophores, including fluorescein, eosin, thionine, methylene blue, and certain cyanine dyes, and the nonradiative character of the excited state became accepted as a general feature of H-aggregates [175–177].

Forster and Kasha theories could explain the nonfluorescent nature of H-type aggregates [178, 179]. A blue-shifted band absorption spectrum can be observed in the UV/Vis because two exciton states arise in the case of face-to-face-stacked dimer aggregates, and only the transition to the higher energy exciton state is allowed. Rapid internal conversion of the excited state into the lower energy state leads to the quenching of fluorescence due to the decreased transition probability for a radiative process between these two states. A few exceptions have been reported at low temperatures in frozen solutions [180] and for dye aggregates embedded in Langmuir–Blodgett layers. [181] There are examples of tethered mero- and hemicyanine chromophores, which can yield fluorescent H-type aggregates.[182, 183] In these cases, the UV/Vis absorption spectra are quite different from those of conventional cyanine dye H-type dimers.

The ICG is widely used in medical diagnosis, and it has potential applications in photodynamic therapy [184]. At high dye concentrations (C= 1.3×10^{-3} mol/L and 1.6×10^{-3} mol/L) [150, 185], the formation of a J-aggregate-like [172, 186] red-shifted narrow absorption band within about two weeks was observed [150, 185]. J-aggregation was identified as a thread-like attachment of molecules to a high degree of aggregation [187, 188]. A dipolar coupling of adjacent molecules leads to the red absorption shift [189], and the extended coherence length of the Frenkel excitons causes the line narrowing [190, 191]. J-aggregation is common for many cyanine dyes [187, 188].

An overview of metal nanoparticles and fluorescent dyes is presented in Table 23.1.

Contrast agents based on plasmonic–fluorophore interfaces

Another promising combination is fluorophore–metal particle conjugates. Such types of materials could provide both photoacoustic and fluorescent signals depending on conformation.

Fe- and Cu- based nanoparticles are mostly pronounced as fluorescence quenchers, while Au and Ag could even enhance the fluorescent response (SEF – surface-enhanced fluorescence) by the local field if they are separated from fluorophore at a ~5 nm distance.

Fluorophores are, probably, the most popular biosensing substances. However, they have significant drawbacks: a tiny optical cross-section and low photostability. The first one limits the detection of individual fluorophores in the media. The second one tends to decay of brightness, which is necessary for detection and quantification. SEF was firstly observed in the 1980s by the groups of Gramila, Nitzan, and Reigler [192–194], and occurs when a fluorophore is placed near the surface of a plasmonic metal nanoparticle, enhancing the fluorophore emission intensity by orders of magnitude with its high electromagnetic field. This process could be assigned to two phenomena: the localization of the incident light due to the significant absorption and scattering cross-sections of the plasmonic particle, and the decrease in the fluorophore's fluorescence lifetime that increases the frequency of excited state–ground state transitions.

The most noticeable effect occurs when the plasmon resonance of the metal nanoparticle is spectrally matched by the absorbance/emission of the fluorophore placed near the surface [195]. It was demonstrated that SEF is also very sensitive to the size and shape of the plasmonic particle.

Fu and Lakowicz demonstrated that covalently linking Cy5 by nucleotides to gold nanorods at an 8 nm distance can increase the optical signal of the dye and therefore improve signal sensitivity, achieving a 40 times higher fluorescence emission rate with a 7 times shorter fluorescence decay rate [196]. Tam et al. demonstrated a 50 times molecular fluorescence enhancement for ICG when in proximity to gold nanoshells [197]. Kim et al. manufactured complicated multimodal nanospheres with magnetite-conjugated Raman labeling compounds and Ag shell coated with silica-conjugated rhodamine-B-isothiocyanate (RITC) for SEF. Then, cell labeling and sorting were performed using these particles [198].

If the fluorophore is located on the surface of the metallic particle, it exhibits fluorescence quenching due to the suppression of quantum efficiency by the nonradiative process. Yamaoka et al. demonstrated a drastically enhanced PA response from a mixture of 20 nm gold nanoparticles with fluorescent dye Rhodamine B, physically adsorbed on their surface. The low specific heat of AuNP (and, hence, high light-to-heat conversion ratio) together with the particular absorption wavelength of Rhodamine B selectively evokes a two-photon absorption at excitation pulse energy 0.31 mJ/pulse [199]. Novoselova et al. have shown that gold nanorods and ICG inserted in a spherical polyelectrolytes matrix produce higher signal than bare gold nanorods at 820 nm excitation and 15 µJ/pulse. Ultrasharp peaks in the PA spectrum of such particles were considered as a nonlinear PA effect and could not be achieved on gold nanorods or ICG separately [200]. Zhang et al. fabricated fluorescent nanodiamonds – gold nanoparticles conjugates which successfully demonstrated a PA signal rise in comparison to gold nanoparticles or fluorescent nanodiamonds themselves, which were used in this experiment. Negatively charged nitrogen vacancies in nanodiamonds provide red fluorescence emission (600–800 nm) under yellow light excitation. Gold nanoparticles were bound to nanodiamonds to form photoacoustic-active clusters with 4-fold and 9-fold suppressed fluorescence under 530 and 565 nm excitation, respectively, which is guided by efficient energy transfer within the conjugates between surface plasmon resonance of AuNP and nitrogen centers of nanodiamonds [107].

Tissue optical clearing

Tissue optical clearing is a phenomenon of increasing the light penetration depth by decreasing light scattering in biological tissues and cells [34], which leads to improved optical image contrast and spatial resolution. It is used mostly for the visible and NIR spectral ranges where biological tissues are low absorbing (the so-called second near-infrared window is located in this region [201]), but highly scattering media. TOC finds applications in biological imaging of *in vivo*, *ex vivo*, and *in vitro* samples via optical coherence tomography, second harmonic generation microscopy, confocal reflectance microscopy, two-photon microscopy, and light-sheet microscopy [202] as well as in medicine for diagnostics, therapy, and laser surgery [203]. There are various methods and approaches of TOC with their advantages, drawbacks, and specifics of implementation.

The main reason for light scattering in biological tissues and cells is refractive indices mismatching of its structural components and surrounding media or interstitial liquid cells [34]. Thus, the most popular TOC approach includes the immersion of samples into special agents known as optical-clearing agents (OCAs). In general, OCA techniques involve the impregnation of OCAs into the extracellular space and lead to the matching of refractive indices of tissue components and surrounding media and make the tissue more transparent [34]. Another approach is utilizing ultrasound transducers that can induce cavitation bubbles [204] and change the scattering properties of the tissue from Rayleigh to Mie scattering [205] or can create an optical waveguide in tissue by acoustic or pressure waves, resulting in refractive index contrast that allows steering the trajectory of light [206]. Other approaches, such as squeezing (compressing) or stretching, extrusion, or flattening, can also be used to increase light transmission through soft tissue but have specific application conditions and are thus less common. The scheme of TOC approaches is shown in Figure 23.2.

Tissue optical clearing method using immersion of tissues into optical-clearing agents

A TOC method based on immersion of tissues into OCAs was developed by Tuchin and coworkers [207, 208]. The main mechanisms of TOC are:

- Matching of refractive indices of tissue components
- Tissue dehydration
- Dissociation of tissue collagen

The tissue structure is densely packed with various components with higher refractive indices (i.e., collagen, elastic

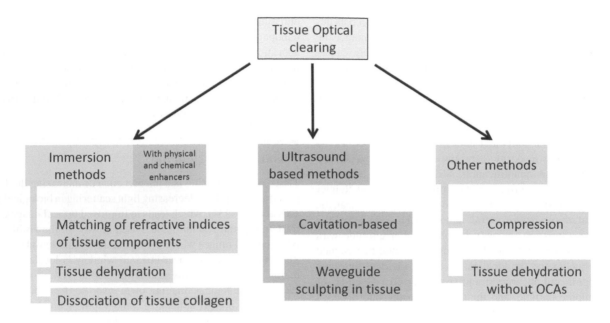

FIGURE 23.2 A scheme representing different approaches for TOC.

fibers, cells, cell compartments), while the surrounding media has a lower refractive index (i.e., interstitial fluid or cytoplasm) [34]. This heterogeneity of biological tissue makes light travel at different speeds and angles, resulting in strong light scattering [209, 210]. The primary mechanism of OCAs in TOC is based on the fact that most OCAs usually have a high refractive index in comparison to the extracellular fluid. Thus, penetration of OCAs with higher refractive indices and higher osmolality into the extracellular space matches the refractive indices of tissue components and surrounding media. It leads to a significant reduction in scattering and improvement in optical imaging depth and contrast [34].

Another important mechanism of TOC can be tissue dehydration. Application of hyperosmotic OCA leads to drawing of water from cells and/or collagen fibers out of the tissue by causing the water flux from the interstitial space to the tissue surface or out of the tissue and increasing the osmolality of interstitial fluid. It gives an additional refractive index matching effect and makes the tissue denser, reducing the tissue thickness. It improves the optical transmittance and reduces the diffuse reflectance of the tissue [34].

The molecular mechanism of the TOC is related to the dissociation of tissue collagen. It is known that the major scattering centers in tissues are collagen fibers that have complex self-assembled structures, and the primary bonding force between collagen triple helices is hydrogen bonding. Thus, OCAs with multiple hydroxyl groups and strong electronegativity destabilize the higher-order structure of collagen and dissociate it. However, this effect can be relatively easily reversed [34, 211].

The TOC efficacy depends on treatment time, or how long the tissue is immersed in OCA [34]. Optical clearing of skin is more difficult compared to the soft tissues because of the limitation of OCAs' penetration into the dermis due to the barrier function of the stratum corneum. The skin application of OCAs at low concentrations is not sufficient, while a high concentration of OCAs could cause unnecessary damage to the skin, inducing edema, suppuration, or even scarring [34, 212]. Therefore, effective and safe enhancement methods of TOC can be considered to breach the stratum corneum's integrity and accelerate OCA permeability. These methods can be divided into chemical enhancers, physical methods, and their combinations. Physical or mechanical methods of enhancement include the application of ultrasound, light irradiation, light fractional ablation, sandpaper grinding, microneedle rolling, microdermabrasion, and mechanical compressing. Chemical-penetration enhancers used in medicine and cosmetics such as Azone, Oleic acid, DMSO, Thiazone, propylene glycol, DMSO, and linoleic acid can accelerate the permeability of OCA into the skin and improve the efficacy of TOC [34, 208]. Skin optical clearing efficacy can be improved by using the synergistic combination of ultrasound and some chemical enhancers such as sodium lauryl sulfate or Thiazone [213, 214]. Tape stripping can also be used in combination with chemical enhancers such as Thiazone, Azone, and propylene glycol [215].

As tissues have different structures and components, the TOC process with OCA can also be different. Moreover, there are differences between *in vivo*, *ex vivo*, and *in vitro* mechanisms of TOC, leading to different efficacy of optical clearing. *In vivo* optical clearing is still challenging, and most of the examples in literature target the skin [202]. Before *in vivo* application of OCA or its chemical enhancers to the tissues, it is necessary to be mindful of safety concerns as there can be side effects on the skin, blood vessels, organs, or body [34]. It has been shown that the topical application of OCAs can cause an inflammation in a short time; however, the skin morphology and microstructure can be recovered after several days. Also, the application of OCA can induce occlusion of blood vessels and affect blood flow [216]. Therefore, biocompatible OCAs are required. Besides, for *in vivo* investigations, the metabolism must be considered that can take away some OCA and decrease its local concentration [34]. An overview of OCAs can be found in [203].

TOC techniques can significantly improve the penetration depth and image contrast of optical microscopy. In this case, most techniques include several steps. The combination of optical clearing and labeling techniques is very significant in three-dimensional reconstructions of tissue and organ in life-science. An overview of the optical clearing techniques used for imaging in microscopy, including the ability for fluorescence preservation, can be found in [209] and [210]. It must be noted that high tissue transparency was obtained only *in vitro* with a treatment time up to several days or weeks [209]. The treatment may include additional steps with tissue fixation and delipidation that are entirely not applicable for *in vivo* study [204].

Combination of optical clearing and contrast agents

As we mentioned earlier, the main problem of PA imaging – penetration depth – can be overcome with OCAs, which diffuse into tissues and partially replace water [34, 203, 207, 208, 217]. A decrease in the difference between the refractive indices of tissues and surrounding media leads to a reduction of the light scattering and, as a consequence, to an improvement in the PA imaging depth [34, 203, 208, 218, 219].

The most widely used OCAs are glycerol, polyethylene glycol, propylene glycol, butylene glycol, mannitol, sorbitol, xylitol, glucose, dextrose, fructose, sucrose, ribose, saccharose, verografin, and trasograph [34, 203, 207, 208, 220–229]. They can be used together with chemical enhancers, such as DMSO, ethanol, oleic acid, sodium lauryl sulfate, Azone, and Thiazone (benzisothiazol-3(2H)-one-2-butyl-1,1-dioxide), or physical enhancers [34, 203, 208, 223, 230]. Glycerol is the most popular and safest agent with a long history of use in various fields [34, 203, 208, 218, 226, 230, 231]. Youn et al. showed that 70% glycerol is the optimal concentration to maximize the tissue optical clearing effect for topical applications by increasing the light penetration depth in tissue and inducing collagen dissociation [232].

According to the study by Zhou and co-workers [218], the penetration depth and lateral resolution of optical-resolution photoacoustic microscopy (OR-PAM) were significantly enhanced by using an 88% glycerol–water solution at 570 nm. To accelerate optical clearing, the authors directly injected the glycerol solution into the tissue, although this may lead to bleeding and edema. Since high concentrations of glycerol can potentially block blood flow [233], some other agents, such as PEG-400, can be used instead, as suggested by Liu et al. [223]. They applied several types of OCAs to rat dorsal skin and showed that a mixture of PEG-400 and thiazone caused severe dehydration and improved the imaging capabilities of acoustic resolution photoacoustic microscopy (AR-PAM) to detect deep blood vessels in the skin. On the other hand, glycerol was found to be better for imaging shallow blood vessels.

Yang et al. investigated the changes in the skin during optical clearing by using a combination of photoacoustic microscopy and ultrasonography [234]. For this study, they used glycerol and PEG-400 and revealed that glycerol has a higher clearing potential than PEG-400 *ex vivo*. Still, both of them caused similar skin dehydration and increased PA signal *in vivo*. Furthermore, they were able to describe the temporal relationship between optical clearing and dehydration quantitatively. A forward photoacoustic signal for glycerol was enhanced throughout the entire imaging procedure, while the response from PEG-400 was sustained only for the first 20 min. According to the shift in the photoacoustic arrival time, the glycerol diffusion process lasted about 45 min, and for PEG-400, it lasted 12 min. Using these techniques, they successfully monitored dermal changes depending on various parameters. Liopo and co-workers proposed a method for enhanced optical clearing of skin based on the treatment of PEG-400 and PPG-425, as well as their mixture with hyaluronic acid [35]. Pretreatment of pig skin for about 30 min with 0.5% HA before injection of the mixture of PPG-425 and PEG-400 resulted in a ~47-fold increase in transmission of red and near-infrared light and an improvement in PA contrast.

Numerous studies confirm that the application of optical cleaning improves the sensitivity and image quality of photoacoustic flow cytometry (PAFC). PAFC, in combination with photothermal and optical clearing methods, was used to characterize the kinetics of ICG and single cancer cells labeled with gold nanorods and ICG [219]. The contrast of blood microvessels in the mouse ear was improved by using glycerol as OCA. In addition, the shapes of leukocytes and erythrocytes in the capillaries could be easily distinguished. The PA response from cancer cells labeled with gold nanorods was between five and seven times greater than from cells labeled with ICG and comparable to the signal from ICG alone in the bloodstream. It is estimated that PAFC can detect one cancer cell in the background of 10^7–10^8 healthy blood cells in small animals. Galanzha et al. proved the feasibility of using an integrated system of multicolor photoacoustic lymph flow cytometry, PA-lymphography, absorption image cytometry, and photothermal therapy for the detection and treatment of metastases in SLN at a single-cell level, as well as counting of disseminated tumor cells (DTCs) in prenodal lymphatics [235]. PEG-coated magnetic nanoparticles with a Fe_2O_3 core (MNPs) and golden carbon nanotubes (GNTs) were used as PA lymphographic contrast agents. High-amplitude PA signals were obtained with a PA contrast of 55 for GNTs and 14 for MNPs at 850 nm and 639 nm. To reduce the scattering, they immersed the lymph nodes in 40% glucose, 100% DMSO, and 80% glycerol as OCAs to reduce the scattering. The results showed that the treatment with glycerol was the most effective, since it allowed the label-free imaging of a fresh node at the cellular level and localization of immune-related, metastatic, and other cells, such as lymphocytes, macrophages, dendritic cells, and melanoma cells, as well as surrounding microstructures. They also tested the possibility of PA technology to detect nonpigmented metastasis by performing dual-labeling of breast cancer cells with GNT-folate-fluorescein isothiocyanate (FITC) conjugates. Menyaev et al. confirmed that optical clearing of skin with glycerol enhanced by massage and sonophoresis resulted in a 1.6-fold increase in PA signal amplitude from blood [219]. Furthermore, the peak rate for B16F10 melanoma cells increased by 1.7 times. They successfully demonstrated the high sensitivity of PAFC label-free detection of melanoma in a phantom model and observed an increase in PA signal from human hand veins up to 40%.

Our group recently demonstrated a novel approach to increasing penetration depth by integrating PA imaging and TOC into a single procedure [236]. The idea was to use OCA not only for optical clearing but also as an acoustic coupling medium. We tested several acoustic coupling agents, including water, ultrasound gel, a mixture of 30% ultrasound gel and 70% glycerol and 99.4% glycerol by applying them between the surface of the mouse skin and the transducer of raster-scanning optoacoustic mesoscope (RSOM; iThera Medical GmbH, Germany) under 532 nm laser illumination. As expected, the application of a mixture of glycerol and ultrasound gel provided 1.5–2 times greater penetration depth and a significant improvement in the visualization of blood vessels during 60 min, without spending time on pretreatment with these agents, as shown in Figure 23.3. The optimal imaging time is 5–45 min after OCA is applied to the skin.

We observed an interesting phenomenon upon intradermal injection of 70% glycerol solution in mouse limb. Shortly after injection (5–15 min), the image quality and penetration depth significantly improved, allowing us to visualize small vessels. However, during the diffusion of glycerol into the tissue, relatively deep and large vessels were also visible against the background of the superficial vasculature, as demonstrated in Figure 23.4. We assume that this phenomenon is based on the slowing/stopping of local blood flow and optical clearing of blood cells inside the vessels.

Additionally, we demonstrated quick RSOM lymphography by using a combination of label-free PA angiography, which exploits hemoglobin as a contrast agent, with PA lymphography based on the labeling of lymphatic vessels with exogenous PA contrast agents. In that way, we were able to distinguish blood and lymphatic vessels. We have also proved the potential for PA-controlled drug delivery to lymphatic vessels by coadministration of PA contrast dye Cy5 and multilayer nanocomposites, consisting of poly-L-arginine hydrochloride (PARG), dextran sulfate sodium salts (DEX), ICG, and gold nanorods adsorbed

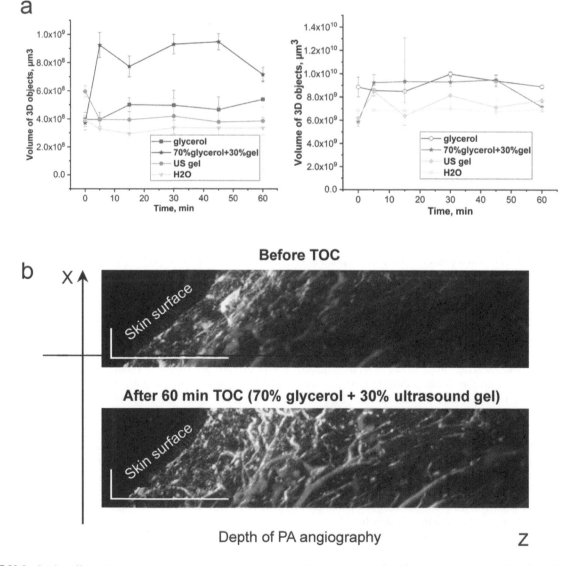

FIGURE 23.3 In vivo effects of RSOM & TOC at the topical application of different OCAs on the skin surface of mouse limb during RSOM imaging procedure. (a) Comparative quantitative analysis (volume of 3D objects) for large (red color signal, left) and small (green color signal, right) vessels at the application of water, ultrasound (US) gel, pure glycerol, and mixture of 70% glycerol + 30% ultrasound gel. (b) PA images (XZ scans) of the same limb area before (conventional RSOM imaging) and after 70% glycerol + 30% ultrasound gel TOC (RSOM & TOC imaging). Scale bar = 0.5 mm. Reprinted with permission from Ref. [236].

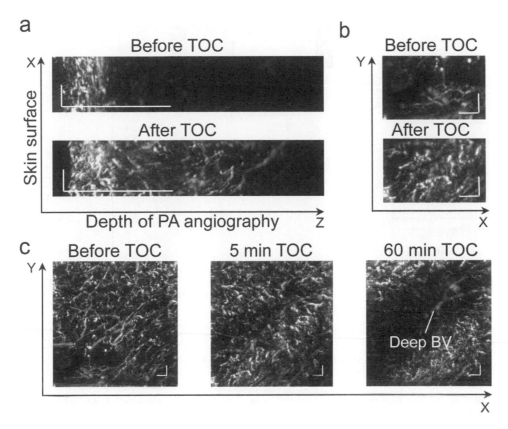

FIGURE 23.4 (a) Increasing depth of detectable blood vessels in 5 min after intradermal injection of 70 % glycerol. (b, c) Effects of injectable TOC on RSOM imaging of the same limb area over 60 min of monitoring: conventional RSOM images of vessels before TOC; improvement in mapping of superficial small vessels in 5 min after glycerol injection due to skin clearing (bottom in "b" and middle in "c") and "open a window" for imaging of deep and large vessels in 60 min after injection as a result of double clearing of skin and superficial vessels (right in "c"). Scale bar=0.5 mm. Reprinted with permission from Ref. [236].

on the surface of vaterite particles. The capsules loaded with bovine serum albumin (BSA) labeled with ICG accumulated only in the areas of local lymph nodes, demonstrating its potential for targeted drug delivery to the lymphatic system. Our results pave the way for in-depth, high-quality 3D PA angio- and lymphography, as well as for PA-guided lymphatic drug delivery using a novel integrated RSOM and TOC approach.

Ultrasound-based optical clearing

While the use of ultrasound for TOC is not widely exploited, the light penetration depth can also be increased by implementing an ultrasound. There are two proposed approaches that utilize ultrasound for this purpose. The first one uses ultrasound to induce cavitation activity of microbubbles locally. It can lead to a change in the ordering of the collagen fiber structure in the dermis with a subsequent decrease in the volume fraction of scatterers [204]. Besides, cavitation microbubbles can lead to a change of scattering regime in tissue from Rayleigh to Mie scattering (see Figure 23.5, left), so the light is scattered predominantly in the forward direction [205].

In another approach, an ultrasound is used to create virtual optical waveguides in tissue. In this case, acoustic waves create local pressure differences that result in refractive index contrasts, thereby forming an optical waveguide, as illustrated in Figure 23.6 [206]. The application of this technology was demonstrated in confining light through mouse brain tissue.

Other optical clearing approaches

External compression can significantly change the optical and scattering properties of the tissue. During the compression, water moves out from the intercellular space, and the protein concentration in it increases, resulting in the reduction of scattering [203]. Physical methods such as extrusion or flattening or air-drying methods can also be used for TOC, inducing the mechanism of tissue dehydration discussed above but without OCA treatment, showing similar efficacy but being more time consuming compared to the case with the application of OCAs [34, 237].

Summary

In conclusion, it should be noted that a wide range of strategies is currently being developed to improve image contrast, which will help researchers to investigate various diseases more deeply using photoacoustic and fluorescence methods. Although different types of contrast agents have quite impressive properties, many of them also exhibit essential issues that need to be addressed, such as elimination from the circulatory system, relatively low biocompatibility, poor targeting, the need for a high concentration of agents, suboptimal tissue biodistribution, lack of appropriate pharmacokinetic assessment, etc.

This chapter discussed two different approaches to improving contrast, namely the use of contrast media and tissue optical

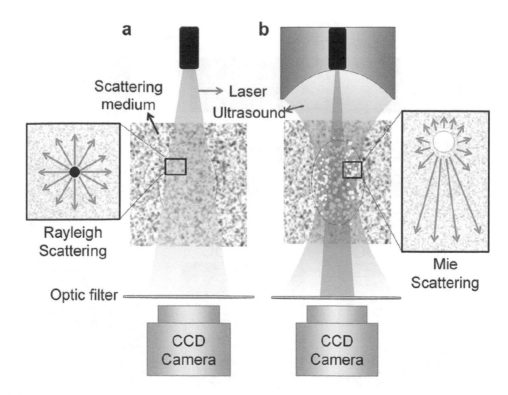

FIGURE 23.5 Scheme of the mechanism of increased light penetration due to air bubbles in a medium that were induced by ultrasound. Reprinted with permission from Ref. [205].

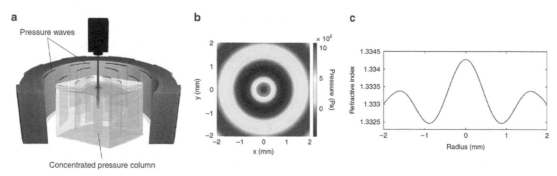

FIGURE 23.6 a) Schematic illustration of acousto-optic waveguide formation in a tissue; b) Finite-elements method simulation of standing ultrasonic wave in an infinite cylindrical transducer placed in water; and c) refractive index profile at a radial cross-section of the cylindrical waveguide. Reprinted with permission from Ref. [206].

clearing by chemical and physical means. Besides that, a novel approach to increasing the depth of penetration by integrating PA and TOC images into a single procedure was elucidated. It was shown that OCA could be simultaneously exploited as an acoustic coupling medium. The use of a mixture of glycerol and ultrasound gel provided 1.5–2 times greater penetration depth and significantly improved the visualization of blood vessels within 60 min, without wasting time on pretreatment with these agents.

We believe that the combination of contrast agents and both optical clearing strategies – chemical and physical – will significantly advance efforts to bring biomedical imaging closer to more effective clinical use.

Acknowledgment

This work was supported by the Russian Foundation for Basic Research (RFBR grant 18-29-08046), the Government of the Russian Federation (grant 14.Z50.31.0044 to support scientific research projects implemented under the supervision of leading scientists at Russian institutions and Russian institutions of higher education) and Saratov State University.

REFERENCES

1. V. Ntziachristos and D. Razansky, "Molecular imaging by means of multispectral optoacoustic tomography (MSOT)," *Chem. Rev.* **110**(5), 2783–2794 (2010) [doi:10.1021/cr9002566].

2. L.V. Wang and S. Hu, "Photoacoustic tomography: In vivo imaging from organelles to organs," *Science* **335**(6075), 1458–1462, American Association for the Advancement of Science (2012) [doi:10.1126/science.1216210].

3. L.V. Wang and J. Yao, "A practical guide to photoacoustic tomography in the life sciences," *Nat. Methods* **13**(8), 627–638, Nature Publishing Group (2016) [doi:10.1038/nmeth.3925].

4. M. Xu and L.V. Wang, "Photoacoustic imaging in bio-medicine photoacoustic imaging in biomedicine," *Rev. Sci. Instrum.* 041101(4), 41101–41122 (2006) [doi:10.1063/1.2195024].

5. S. Manohar and D. Razansky, "Photoacoustics: A historical review," *Adv. Opt. Photonics* **8**(4), 586 (2016) [doi:10.1364/AOP.8.000586].

6. S. Mallidi, G.P. Luke, and S. Emelianov, "Photoacoustic imaging in cancer detection, diagnosis, and treatment guidance," *Trends Biotechnol.* **29**(5), 213–221 (2011) [doi:10.1016/j.tibtech.2011.01.006].

7. A. Taruttis, G.M. Van Dam, and V. Ntziachristos, "Mesoscopic and macroscopic optoacoustic imaging of cancer," *Cancer Res.* **75**(8), 1548–1559, American Association for Cancer Research Inc. (2015) [doi:10.1158/0008-5472.CAN-14-2522].

8. A.B.E. Attia, G. Balasundaram, M. Moothanchery, et al., "A review of clinical photoacoustic imaging: Current and future trends," *Photoacoustics* **16**, 100144, Elsevier GmbH (2019) [doi:10.1016/j.pacs.2019.100144].

9. J. Kim, J.Y. Kim, S. Jeon, et al., "Super-resolution local-ization photoacoustic microscopy using intrinsic red blood cells as contrast absorbers," *Light Sci. Appl.* **8**(1), 103, Springer (2019) [doi:10.1038/s41377-019-0220-4].

10. J. Cvjetinovic, A.I. Salimon, M.V. Novoselova, et al., "Photoacoustic and fluorescence lifetime imaging of dia-toms," *Photoacoustics* **18**, 100171, Elsevier GmbH (2020) [doi:10.1016/j.pacs.2020.100171].

11. M. Martinho Costa, A. Shah, I. Rivens, et al., "Quantitative photoacoustic imaging study of tumours in vivo: Baseline variations in quantitative measurements," *Photoacoustics* **13**, 53–65, Elsevier GmbH (2019) [doi:10.1016/j.pacs.2018.12.002].

12. Q. Fan, K. Cheng, X. Hu, et al., "Transferring biomarker into molecular probe: Melanin nanoparticle as a naturally active platform for multimodality imaging," *J. Am. Chem. Soc.* **136**(43), 15185–15194, American Chemical Society (2014) [doi:10.1021/ja505412p].

13. P. Wang, P. Wang, H. W. Wang, and J.-X. Cheng, "Mapping lipid and collagen by multispectral photoacoustic imag-ing of chemical bond vibration," *J. Biomed. Opt.* **17**(9), 0960101, SPIE-Intl Soc Optical Eng (2012) [doi:10.1117/1.jbo.17.9.096010].

14. J. Ge, S.J. Erickson, and A. Godavarty, "Fluorescence tomo-graphic imaging using a handheld-probe-based optical imager: Extensive phantom studies," *Appl. Opt.* **48**(33), 6408–6416, OSA – The Optical Society (2009) [doi:10.1364/AO.48.006408].

15. A. Godavarty, A.B. Thompson, R. Roy, et al., "Diagnostic imaging of breast cancer using fluorescence-enhanced opti-cal tomography: Phantom studies," *J. Biomed. Opt.* **9**(3), 488, SPIE-Intl Soc Optical Eng (2004) [doi:10.1117/1.1691027].

16. B. Wang, Q. Zhao, N.M. Barkey, D.L. Morse, and H. Jiang, "Photoacoustic tomography and fluorescence molecular tomography: A comparative study based on indocyanine green," *Med. Phys.* **39**(5), 2512–2517 (2012) [doi:10.1118/1.3700401].

17. L. Martí-Bonmatí, R. Sopena, P. Bartumeus, and P. Sopena, "Multimodality imaging techniques," *Contrast Media Mol. Imaging* **5**(4), 180–189 (2010) [doi:10.1002/cmmi.393].

18. S. Park, U. Jung, S. Lee, D. Lee, and C. Kim, "Contrast-enhanced dual mode imaging: Photoacoustic imaging plus more," *Biomed. Eng. Lett.* **7**(2), 121–133, Springer Verlag (2017) [doi:10.1007/s13534-016-0006-z].

19. U. Bagci, J.K. Udupa, N. Mendhiratta, et al., "Joint segmen-tation of anatomical and functional images: Applications in quantification of lesions from PET, PET-CT, MRI-PET, and MRI-PET-CT images," *Med. Image Anal.* **17**(8), 929–945 (2013) [doi:10.1016/j.media.2013.05.004].

20. R. Cicchi, A. Cosci, S. Rossari, et al., "Combined fluo-rescence-Raman spectroscopic setup for the diagnosis of melanocytic lesions," *J. Biophoton.* **7**(1–2), 86–95 (2014) [doi:10.1002/jbio.201200230].

21. Z. Liu, B. Ye, M. Jin, et al., "Dye-free near-infrared sur-face-enhanced Raman scattering nanoprobes for bioimag-ing and high-performance photothermal cancer therapy," *Nanoscale* **7**(15), 6754–6761, Royal Society of Chemistry (2015) [doi:10.1039/c5nr01055a].

22. S. Pal, A. Ray, C. Andreou, et al., "DNA-enabled ratio-nal design of fluorescence-Raman bimodal nanoprobes for cancer imaging and therapy," *Nat. Commun.* **10**(1), 1926, Nature Publishing Group (2019) [doi:10.1038/s41467-019-09173-2].

23. N. Zhang, S. Ye, Z. Wang, R. Li, and M. Wang, "A dual-signal twinkling probe for fluorescence-SERS dual spec-trum imaging and detection of miRNA in single living cell via absolute value coupling of reciprocal signals," *ACS Sens.* **4**(4), 924–930, American Chemical Society (2019) [doi:10.1021/acssensors.9b00031].

24. Y. Xi, J. Zhao, J.R. Bennett, et al., "Simultaneous CT-MRI reconstruction for constrained imaging geometries using structural coupling and compressive sensing," *IEEE Trans. Bio Med. Eng.* **63**(6), 1301–1309, IEEE Computer Society [doi:10.1109/TBME.2015.2487779].

25. M.F. Kircher, A. De La Zerda, J.V. Jokerst, et al., "A brain tumor molecular imaging strategy using a new triple-modality MRI-photoacoustic-Raman nanoparticle," *Nat. Med.* **18**(5), 829–834 (2012) [doi:10.1038/nm.2721].

26. Y. Liu, N. Kang, J. Lv, et al., "Deep photoacoustic/lumi-nescence/magnetic resonance multimodal imaging in living subjects using high-efficiency upconversion nanocompos-ites," *Adv. Mater.* **28**(30), 6411–6419, Wiley-VCH Verlag (2016) [doi:10.1002/adma.201506460].

27. C. Qin, K. Cheng, K. Chen, et al., "Tyrosinase as a multi-functional reporter gene for photoacoustic/MRI/PET triple modality molecular imaging," *Sci. Rep.* **3**(1), 1490 (2013) [doi:10.1038/srep01490].

28. J. Garcia, T. Tang, and A.Y. Louie, "Nanoparticle-based multimodal PET/MRI probes," *Nanomedicine* **10**(8), 1343–1359, Future Medicine Ltd. (2015) [doi:10.2217/nnm.14.224].

29. S.R. Cherry, "Multimodality imaging: Beyond PET/CT and SPECT/CT," *Semin. Nucl. Med.* **39**(5), 348–353 (2009) [doi:10.1053/j.semnuclmed.2009.03.001].

30. Y. Xing, J. Zhao, P.S. Conti, and K. Chen, "Radiolabeled nanoparticles for multimodality tumor imaging," *Theranostics* **4**(3), 290–306 (2014) [doi:10.7150/thno.7341].

31. Å.A. Barrefelt, T.B. Brismar, G. Egri, et al., "Multimodality imaging using SPECT/CT and MRI and ligand functional-ized 99mTc-labeled magnetic microbubbles," *EJNMMI Res.* **3**(1), 1–14, Springer Verlag (2013) [doi:10.1186/2191-219X-3-12].

32. R. Lin, J. Chen, H. Wang, et al., "Longitudinal label-free optical-resolution photoacoustic microscopy of tumor angiogenesis in vivo," *Quantum Imaging Med. Surg.* **5**(1), 23–239 (2015) [doi:10.3978/j.issn.2223-4292.2014.11.08].

33. X. Yang, Y. Zhang, K. Zhao, et al., "Skull optical clearing solution for enhancing ultrasonic and photoacoustic imaging," *IEEE Trans. Med. Imaging* **35**(8), 1903–1906 (2016) [doi:10.1109/TMI.2016.2528284].

34. D. Zhu, K.V. Larin, Q. Luo, and V.V. Tuchin, "Recent progress in tissue optical clearing," *Laser Photonics Rev.* **7**(5), 732–757 (2013) [doi:10.1002/lpor.201200056].

35. A. Liopo, R. Su, D.A. Tsyboulski, and A.A. Oraevsky, "Optical clearing of skin enhanced with hyaluronic acid for increased contrast of optoacoustic imaging," *J. Biomed. Opt.* **21**(8), 081208 (2016) [doi:10.1117/1.jbo.21.8.081208].

36. A.K. Bui, R.A. McClure, J. Chang, et al., "Revisiting optical clearing with dimethyl sulfoxide DMSO," *Lasers Surg. Med.* **41**(2), 142–148 (2009) [doi:10.1002/lsm.20742].

37. M. Maturi, E. Locatelli, I. Monaco, and M. Comes Franchini, "Current concepts in nanostructured contrast media development for in vivo photoacoustic imaging," *Biomater. Sci.* **7**(5), 1746–1775, Royal Society of Chemistry [doi:10.1039/c8bm01444b].

38. J. Weber, P.C. Beard, and S.E. Bohndiek, "Contrast agents for molecular photoacoustic imaging," *Nat. Methods* **13**(8), 639–650 (2016) [doi:10.1038/nmeth.3929].

39. P.K. Upputuri and M. Pramanik, "Recent advances in photoacoustic contrast agents for in vivo imaging," *Wiley Interdiscip. Rev. Nanomed. Nanobiotechnology* **14**(4), e1618, 1–23 (2020) [doi:10.1002/wnan.1618].

40. P.K. Upputuri and M. Pramanik, "Photoacoustic imaging in the second near-infrared window: A review," *J. Biomed. Opt.* **24**(04), 1 (2019) [doi:10.1117/1.jbo.24.4.040901].

41. Q. Fu, R. Zhu, J. Song, H. Yang, and X. Chen, "Photoacoustic imaging: Contrast agents and their biomedical applications," *Adv. Mater.* **31**(6), 1–31 (2019) [doi:10.1002/adma.201805875].

42. R. Weissleder, "A clearer vision for in vivo imaging: Progress continues in the development of smaller, more penetrable probes for biological imaging," *Nat. Biotechnol.* **19**(4), 316–317 (2001) [doi:10.1038/86684].

43. P. Beard, "Biomedical photoacoustic imaging," *Interface Focus* **1**(4), 602–631 (2011) [doi:10.1098/rsfs.2011.0028].

44. B. Cox, J.G. Laufer, S.R. Arridge, and P.C. Beard, "Quantitative spectroscopic photoacoustic imaging: A review," *J. Biomed. Opt.* **17**(6), 061202 (2012) [doi:10.1117/1.jbo.17.6.061202].

45. S. Tenzer, D. Docter, J. Kuharev, et al., "Rapid formation of plasma protein corona critically affects nanoparticle pathophysiology," *Nat. Nanotechnol.* **8**(10), 772–781, Nature Publishing Group (2013) [doi:10.1038/nnano.2013.181].

46. I. Lynch and K.A. Dawson, "Protein-nanoparticle interactions," *Nano Today* **3**(1–2), 40–47 (2008) [doi:10.1016/S1748-0132(08)70014-8].

47. M. Lundqvist, "Nanoparticles: Tracking protein corona over time," *Nat. Nanotechnol.* **8**(10), 701–702, Nature Publishing Group (2013) [doi:10.1038/nnano.2013.196].

48. R.A. Petros and J.M. Desimone, "Strategies in the design of nanoparticles for therapeutic applications," *Nat. Rev. Drug Discov.* **9**(8), 615–627 (2010) [doi:10.1038/nrd2591].

49. L. Nie, S. Wang, X. Wang, et al., "In vivo volumetric photoacoustic molecular angiography and therapeutic monitoring with targeted plasmonic nanostars," *Small* **10**(8), 1585–1593, Wiley-VCH Verlag (2014) [doi:10.1002/smll.201302924].

50. M.R. Chatni, J. Xia, R. Sohn, et al., " Tumor glucose metabolism imaged in vivo in small animals with whole-body photoacoustic computed tomography ," *J. Biomed. Opt.* **17**(7), 0760121, SPIE-Intl Soc Optical Eng (2012) [doi:10.1117/1.jbo.17.7.076012].

51. A. Dragulescu-Andrasi, S.R. Kothapalli, G.A. Tikhomirov, J. Rao, and S.S. Gambhir, "Activatable oligomerizable imaging agents for photoacoustic imaging of furin-like activity in living subjects," *J. Am. Chem. Soc.* **135**(30), 11015–11022 (2013) [doi:10.1021/ja4010078].

52. C. Tiede, A.A.S. Tang, S.E. Deacon, et al., "Adhiron: A stable and versatile peptide display scaffold for molecular recognition applications," *Protein Eng. Des. Sel.* **27**(5), 145–155, Oxford University Press (2014) [doi:10.1093/protein/gzu007].

53. J. Löfblom, J. Feldwisch, V. Tolmachev, et al., "Affibody molecules: Engineered proteins for therapeutic, diagnostic and biotechnological applications," *FEBS Lett.* **584**(12), 2670–2680 (2010) [doi:10.1016/j.febslet.2010.04.014].

54. H. Sun, X. Zhu, P.Y. Lu, et al., "Oligonucleotide aptamers: New tools for targeted cancer therapy," *Mol. Ther. Nucleic Acids* **3**, e182, Nature Publishing Group (2014) [doi:10.1038/mtna.2014.32].

55. S. Hu and L.V. Wang, "Photoacoustic imaging and characterization of the microvasculature," *J. Biomed. Opt.* **15**(1), 011101, SPIE-Intl Soc Optical Eng (2010) [doi:10.1117/1.3281673].

56. X. Wang, X. Xie, G. Ku, L.V. Wang, and G. Stoica, "Noninvasive imaging of hemoglobin concentration and oxygenation in the rat brain using high-resolution photoacoustic tomography," *J. Biomed. Opt.* **11**(2), 024015, SPIE-Intl Soc Optical Eng (2006) [doi:10.1117/1.2192804].

57. H. Fang, K. Maslov, and L.V. Wang, "Photoacoustic Doppler effect from flowing small light-absorbing particles," *Phys. Rev. Lett.* **99**(18), 184501 (2007) [doi:10.1103/PhysRevLett.99.184501].

58. H.F. Zhang, K. Maslov, G. Stoica, and L.V. Wang, "Functional photoacoustic microscopy for high-resolution and noninvasive in vivo imaging," *Nat. Biotechnol.* **24**(7), 848–851 (2006) [doi:10.1038/nbt1220].

59. J. Yao, K.I. Maslov, Y. Zhang, Y. Xia, and L.V. Wang, "Label-free oxygen-metabolic photoacoustic microscopy in vivo," *J. Biomed. Opt.* **16**(7), 076003 (2011) [doi:10.1117/1.3594786].

60. M. Li, Y. Tang, and J. Yao, "Photoacoustic tomography of blood oxygenation: A mini review," *Photoacoustics* **10**(December 2017), 65–73, Elsevier (2018) [doi:10.1016/j.pacs.2018.05.001].

61. J.-T. Oh, M.-L. Li, H.F. Zhang, et al., "Three-dimensional imaging of skin melanoma in vivo by dual-wavelength photoacoustic microscopy," *J. Biomed. Opt.* **11**(3), 034032, SPIE-Intl Soc Optical Eng (2006) [doi:10.1117/1.2210907].

62. I. Stoffels, S. Morscher, I. Helfrich, et al., "Metastatic status of sentinel lymph nodes in melanoma determined noninvasively with multispectral optoacoustic imaging," *Sci. Transl. Med.* **7**(317), 317ra199–317ra199 (2015) [doi:10.1126/scitranslmed.aad1278].

63. D.L. Longo, R. Stefania, S. Aime, and A. Oraevsky, "Melanin-based contrast agents for biomedical optoacoustic imaging and theranostic applications," *Int. J. Mol. Sci.* **18**(8) (2017) [doi:10.3390/ijms18081719].

64. I. Kosik, M. Brackstone, A. Kornecki, et al., "Lipid-weighted intraoperative photoacoustic tomography of breast tumors: Volumetric comparison to preoperative MRI," *Photoacoustics* **18**, 100165, Elsevier GmbH (2020) [doi:10.1016/j.pacs.2020.100165].

65. G.S. Sangha, E.H. Phillips, and C.J. Goergen, "In vivo photoacoustic lipid imaging in mice using the second near-infrared window," *Biomed. Opt. Express* **8**(2), 736, The Optical Society (2017) [doi:10.1364/boe.8.000736].

66. M.A. Pleitez, A.A. Khan, A. Soldà, et al., "Label-free metabolic imaging by mid-infrared optoacoustic microscopy in living cells," *Nat. Biotechnol.* **38**(3), 293–296, Springer (2020) [doi:10.1038/s41587-019-0359-9].

67. Z. Xu, C. Li, and L.V. Wang, "Photoacoustic tomography of water in phantoms and tissue," *J. Biomed. Opt.* **15**(3), 036019, SPIE-Intl Soc Optical Eng (2010) [doi:10.1117/1.3443793].

68. E. Ringe, J.M. McMahon, K. Sohn, et al., "Unraveling the effects of size, composition, and substrate on the localized surface plasmon resonance frequencies of gold and silver nanocubes: A systematic single-particle approach," *J. Phys. Chem. C* **114**(29), 12511–12516 (2010) [doi:10.1021/jp104366r].

69. W. Li and X. Chen, "Gold nanoparticles for photoacoustic imaging," *Nanomedicine* **10**(2), 299–320 (2015) [doi:10.2217/nnm.14.169].

70. D. Kumar, R.K. Soni, and D.P. Ghai, "Pulsed photoacoustic and photothermal response of gold nanoparticles," *Nanotechnology* **31**(3), 035704, Institute of Physics Publishing 1–12 (2020) [doi:10.1088/1361-6528/ab47ae].

71. P. Huang, J. Lin, W. Li, et al., "Biodegradable gold nanovesicles with an ultrastrong plasmonic coupling effect for photoacoustic imaging and photothermal therapy," *Angew. Chem.* **125**(52), 14208–14214, Wiley (2013) [doi:10.1002/ange.201308986].

72. P.K. Jain, K.S. Lee, I.H. El-Sayed, and M.A. El-Sayed, "Calculated absorption and scattering properties of gold nanoparticles of different size, shape, and composition: Applications in biological imaging and biomedicine," *J. Phys. Chem. B* **110**(14), 7238–7248 (2006) [doi:10.1021/jp057170o].

73. S. Link and M.A. El-Sayed, "Spectral properties and relaxation dynamics of surface plasmon electronic oscillations in gold and silver nanodots and nanorods," *J. Phys. Chem. B* **103**(40), 8410–8426, American Chemical Society (1999) [doi:10.1021/jp9917648].

74. W. Lu, Q. Huang, G. Ku, et al., "Photoacoustic imaging of living mouse brain vasculature using hollow gold nanospheres," *Biomaterials* **31**(9), 2617–2626 (2010) [doi:10.1016/j.biomaterials.2009.12.007].

75. M. Eghtedari, A. Oraevsky, J.A. Copland, et al., "High sensitivity of in vivo detection of gold nanorods using a laser optoacoustic imaging system," *Nano Lett.* **7**(7), 1914–1918 (2007) [doi:10.1021/nl070557d].

76. Y.S. Chen, W. Frey, S. Kim, et al., "Silica-coated gold nanorods as photoacoustic signal nanoamplifiers," *Nano Lett.* **11**(2), 348–354 (2011) [doi:10.1021/nl1042006].

77. J.V. Jokerst, M. Thangaraj, and S.S. Gambhir, "Photoacoustic imaging of mesenchymal stem cells in living mice via silica-coated gold nanorods," *Proc SPIE* **8943**, 89431O, (2014) [doi:10.1117/12.2036786].

78. T. Kim, Q. Zhang, J. Li, L. Zhang, and J.V. Jokerst, "A gold/silver hybrid nanoparticle for treatment and photoacoustic imaging of bacterial infection," *ACS Nano* **12**(6), 5615–5625 (2018) [doi:10.1021/acsnano.8b01362].

79. V. Raghavan, C. O'Flatharta, R. Dwyer, et al., "Dual plasmonic gold nanostars for photoacoustic imaging and photothermal therapy," *Nanomedicine* **12**(5), 457–471, Future Medicine Ltd. (2017) [doi:10.2217/nnm-2016-0318].

80. Y. Hu, R. Wang, S. Wang, et al., "Multifunctional Fe 3 O 4 @ Au core/shell nanostars: A unique platform for multimode imaging and photothermal therapy of tumors," *Sci. Rep.* **6**(June), 1–12, Nature Publishing Group (2016) [doi:10.1038/srep28325].

81. C. Bao, N. Beziere, P. Del Pino, et al., "Gold nanoprisms as optoacoustic signal nanoamplifiers for in vivo bioimaging of gastrointestinal cancers," *Small* **9**(1), 68–74 (2013) [doi:10.1002/smll.201201779].

82. C. Bao, J. Conde, F. Pan, et al., "Gold nanoprisms as a hybrid in vivo cancer theranostic platform for in situ photoacoustic imaging, angiography, and localized hyperthermia," *Nano Res.* **9**(4), 1043–1056 (2016) [doi:10.1007/s12274-016-0996-y].

83. J. Song, X. Yang, Z. Yang, et al., "Rational design of branched nanoporous gold nanoshells with enhanced physico-optical properties for optical imaging and cancer therapy," *ACS Nano* **11**(6), 6102–6113 (2017) [doi:10.1021/acsnano.7b02048].

84. G.P. Luke, A. Bashyam, K.A. Homan, et al., "Silica-coated gold nanoplates as stable photoacoustic contrast agents for sentinel lymph node imaging," *Nanotechnology* **24**(45) (2013) [doi:10.1088/0957-4484/24/45/455101].

85. Y. Zhao, W. Liu, Y. Tian, et al., "Anti-EGFR peptide-conjugated triangular gold nanoplates for computed tomography/photoacoustic imaging-guided photothermal therapy of non-small cell lung cancer," *ACS Appl. Mater. Interfaces* **10**(20), 16992–17003 (2018) [doi:10.1021/acsami.7b19013].

86. R. Liang, J. Xie, J. Li, et al., "Liposomes-coated gold nanocages with antigens and adjuvants targeted delivery to dendritic cells for enhancing antitumor immune response," *Biomaterials* **149**, 41–50, Elsevier Ltd (2017) [doi:10.1016/j.biomaterials.2017.09.029].

87. S. Farooq, F. Dias Nunes, and R.E. de Araujo, "Optical properties of silver nanoplates and perspectives for biomedical applications," *Photonics Nanostruct. Fundam. Appl.* **31**, 160–167, Elsevier B.V. (2018) [doi:10.1016/j.photonics.2018.07.001].

88. Y. Huang, S. He, W. Cao, K. Cai, and X.J. Liang, "Biomedical nanomaterials for imaging-guided cancer therapy," *Nanoscale* **4**(20), 6135–6149 (2012) [doi:10.1039/c2nr31715j].

89. J.Y. Maillard and P. Hartemann, "Silver as an antimicrobial: Facts and gaps in knowledge," *Crit. Rev. Microbiol.* **39**(4), 373–383 (2013) [doi:10.3109/1040841X.2012.713323].

90. S.-U. Victor and V.-B. José Roberto, "Gold and silver Nanotechology on medicine," *J. Chem. Biochem.* **3**(1), 21–33 (2015) [doi:10.15640/jcb.v3n1a2].

91. K. Homan, J. Shah, S. Gomez et al., "Silver nanosystems for photoacoustic imaging and image-guided therapy," *J. Biomed. Opt.* **15**(2), 021316, SPIE-Intl Soc Optical Eng (2010) [doi:10.1117/1.3365937].

92. K.A. Homan, M. Souza, R. Truby, et al., "Silver nanoplate contrast agents for in vivo molecular photoacoustic imaging," *ACS Nano* **6**(1), 641–650 (2012) [doi:10.1021/nn204100n].

93. B. Silvestri, P. Armanetti, G. Sanità, et al., "Silver-nanoparticles as plasmon-resonant enhancers for eumelanin's photoacoustic signal in a self-structured hybrid nanoprobe," *Mater. Sci. Eng. C Mater. Biol. Appl.* **102**(February), 788–797, Elsevier (2019) [doi:10.1016/j.msec.2019.04.066].

94. Y. Javed, K. Akhtar, H. Anwar, and Y. Jamil, "MRI based on iron oxide nanoparticles contrast agents: Effect of oxidation state and architecture," *J. Nanoparticle Res.* **19**(11) (2017), 366, 1–25 [doi:10.1007/s11051-017-4045-x].

95. Z.R. Stephen, F.M. Kievit, and M. Zhang, "Magnetite nanoparticles for medical MR imaging," *Mater. Today* **14**(7–8), 330–338, Elsevier B.V. (2011) [doi:10.1016/S1369-7021(11)70163-8].

96. H. Chen, Z. Yuan, and C. Wu, "Nanoparticle probes for structural and functional photoacoustic molecular tomography," *BioMed Res. Int.* **2015**(1), 757101, 1–11 (2015) [doi:10.1155/2015/757101].

97. R. Alwi, S. Telenkov, A. Mandelis, et al., "Silica-coated super paramagnetic iron oxide nanoparticles (SPION) as biocompatible contrast agent in biomedical photoacoustics," *Biomed. Opt. Express* **3**(10), 2500 (2012) [doi:10.1364/boe.3.002500].

98. P. Armanetti, A. Flori, C. Avigo, et al., "Spectroscopic and photoacoustic characterization of encapsulated iron oxide super-paramagnetic nanoparticles as a new multiplatform contrast agent," *Spectrochim. Acta A Mol. Biomol. Spectrosc.* **199**, 248–253, Elsevier B.V. (2018) [doi:10.1016/j.saa.2018.03.025].

99. L. Xi, S.R. Grobmyer, G. Zhou, et al., "Molecular photoacoustic tomography of breast cancer using receptor targeted magnetic iron oxide nanoparticles as contrast agents," *J. Biophoton.* **7**(6), 401–409 (2014) [doi:10.1002/jbio.201200155].

100. K. Kanazaki, K. Sano, A. Makino, et al., "Development of anti-HER2 fragment antibody conjugated to iron oxide nanoparticles for in vivo HER2-targeted photoacoustic tumor imaging," *Nanomed. Nanotechnol. Biol. Med.* **11**(8), 2051–2060, Elsevier Inc. (2015) [doi:10.1016/j.nano.2015.07.007].

101. J. Reguera, D. Jiménez De Aberasturi, M. Henriksen-Lacey, et al., "Janus plasmonic-magnetic gold-iron oxide nanoparticles as contrast agents for multimodal imaging," *Nanoscale* **9**(27), 9467–9480, Royal Society of Chemistry (2017) [doi:10.1039/c7nr01406f].

102. Z.H. Zhao, D. Wang, H. Gao, et al., "Photoacoustic effect of near-infrared absorbing fullerene derivatives with click moieties," *Dyes Pigment* **164**, 182–187, Elsevier Ltd (2019) [doi:10.1016/j.dyepig.2019.01.022].

103. F. Yan, Z. Sun, H. Zhang, et al., "The fluorescence mechanism of carbon dots, and methods for tuning their emission color: A review," *Microchim. Acta* **186**(8), 583, Springer-Verlag Wien (2019) [doi:10.1007/s00604-019-3688-y].

104. Y. Song, S. Zhu, and B. Yang, "Bioimaging based on fluorescent carbon dots," *RSC Adv.* **4**(52), 27184–27200, Royal Society of Chemistry (2014) [doi:10.1039/c3ra47994c].

105. Q. Jia, X. Zheng, J. Ge, et al., "Synthesis of carbon dots from Hypocrella Bambusae for bimodel fluorescence/photoacoustic imaging-guided synergistic photodynamic/photothermal therapy of cancer," *J. Colloid Interface Sci.* **526**, 302–311, Academic Press Inc. (2018) [doi:10.1016/j.jcis.2018.05.005].

106. J. Ge, Q. Jia, W. Liu, et al., "Red-emissive carbon dots for fluorescent, photoacoustic, and thermal theranostics in living mice," *Adv. Mater.* **27**(28), 4169–4177, Wiley-VCH Verlag (2015) [doi:10.1002/adma.201500323].

107. B. Zhang, C.-Y. Fang, C.-C. Chang, et al., "Photoacoustic emission from fluorescent nanodiamonds enhanced with gold nanoparticles," *Biomed. Opt. Express* **3**(7), 1662, The Optical Society (2012) [doi:10.1364/boe.3.001662].

108. T. Zhang, H. Cui, C.Y. Fang, et al., "Targeted nanodiamonds as phenotype-specific photoacoustic contrast agents for breast cancer," *Nanomedicine* **10**(4), 573–587 (2015) [doi:10.2217/nnm.14.141].

109. L.O. Usoltseva, D.S. Volkov, D.A. Nedosekin, et al., "Absorption spectra of nanodiamond aqueous dispersions by optical absorption and optoacoustic spectroscopies," *Photoacoustics* **12**(October), 55–66, Elsevier (2018) [doi:10.1016/j.pacs.2018.10.003].

110. L. Xie, G. Wang, H. Zhou, et al., "Functional long circulating single walled carbon nanotubes for fluorescent/photoacoustic imaging-guided enhanced phototherapy," *Biomaterials* **103**, 219–228, Elsevier Ltd (2016) [doi:10.1016/j.biomaterials.2016.06.058].

111. A. De La Zerda, C. Zavaleta, S. Keren, et al., "Carbon nanotubes as photoacoustic molecular imaging agents in living mice," *Nat. Nanotechnol.* **3**(9), 557–562 (2008) [doi:10.1038/nnano.2008.231].

112. P.A. Obraztsov, S.V. Garnov, E.D. Obraztsova, et al., "Passive mode-locking of diode-pumped YAG:Nd solid state laser operated at $\lambda = 1.32$ μm using carbon nanotubes as saturable absorber," *J. Nanoelectron. Optoelectron.* **4**(2), 227–231 (2009) [doi:10.1166/jno.2009.1037].

113. M. Noroozi, A. Zakaria, S. Radiman, and Z.A. Wahab, "Environmental synthesis of few layers graphene sheets using ultrasonic exfoliation with enhanced electrical and thermal properties," *PLOS ONE* **11**(4), e0152699, Public Library of Science (2016) [doi:10.1371/journal.pone.0152699].

114. G. Lalwani, X. Cai, L. Nie, L.V. Wang, and B. Sitharaman, "Graphene-based contrast agents for photoacoustic and thermoacoustic tomography," *Photoacoustics* **1**(3–4), 62–67, Elsevier GmbH (2013) [doi:10.1016/j.pacs.2013.10.001].

115. Y. Toumia, F. Domenici, S. Orlanducci, et al., "Graphene meets microbubbles: A superior contrast agent for photoacoustic imaging," *ACS Appl. Mater. Interfaces* **8**(25), 16465–16475, American Chemical Society (2016) [doi:10.1021/acsami.6b04184].

116. J. Song, F. Wang, X. Yang, et al., "Gold nanoparticle coated carbon nanotube ring with enhanced Raman scattering and photothermal conversion property for theranostic applications," *J. Am. Chem. Soc.* **138**(22), 7005–7015, American Chemical Society (2016) [doi:10.1021/jacs.5b13475].

117. S. Orlanducci, "Gold-decorated nanodiamonds: Powerful multifunctional materials for sensing, imaging, diagnostics, and therapy," *Eur. J. Inorg. Chem.* **48**(48), 5138–5145, Wiley-VCH Verlag (2018) [doi:10.1002/ejic.201800793].

118. A.M. Yashchenok, J. Jose, P. Trochet, G.B. Sukhorukov, and D.A. Gorin, "Multifunctional polyelectrolyte microcapsules as a contrast agent for photoacoustic imaging in blood," *J. Biophoton.* **9**(8), 792–799, Wiley-VCH Verlag (2016) [doi:10.1002/jbio.201500293].

119. A. De La Zerda, S. Bodapati, R. Teed, et al., "Family of enhanced photoacoustic imaging agents for high-sensitivity and multiplexing studies in living mice," *ACS Nano* **6**(6), 4694–4701 (2012) [doi:10.1021/nn204352r].

120. G. Wang, F. Zhang, R. Tian, et al., "Nanotubes-embedded indocyanine green-hyaluronic acid nanoparticles for photoacoustic-imaging-guided phototherapy," *ACS Appl. Mater. Interfaces* **8**(8), 5608–5617, American Chemical Society (2016) [doi:10.1021/acsami.5b12400].

121. J.W. Kim, E.I. Galanzha, E.V. Shashkov, H.M. Moon, and V.P. Zharov, "Golden carbon nanotubes as multimodal photoacoustic and photothermal high-contrast molecular agents," *Nat. Nanotechnol.* **4**(10), 688–694, Nature Publishing Group (2009) [doi:10.1038/nnano.2009.231].

122. D. Hu, J. Zhang, G. Gao, et al., "Indocyanine green-loaded polydopamine-reduced graphene oxide nanocomposites with amplifying photoacoustic and photothermal effects for cancer theranostics," *Theranostics* **6**(7), 1043–1052, Ivyspring International Publisher (2016) [doi:10.7150/thno.14566].

123. H. Moon, H. Kim, D. Kumar, et al., "Amplified photoacoustic performance and enhanced photothermal stability of reduced graphene oxide coated gold nanorods for sensitive photoacoustic imaging," *ACS Nano* **9**(3), 2711–2719, American Chemical Society (2015) [doi:10.1021/nn506516p].

124. J.D. Schiffman and R.G. Balakrishna, "Quantum dots as fluorescent probes: Synthesis, surface chemistry, energy transfer mechanisms, and applications," *Sens Actuators B* **258**, 1191–1214, Elsevier B.V. (2018) [doi:10.1016/j.snb.2017.11.189].

125. H. Xie, M. Liu, B. You, et al., "Biodegradable Bi2O2Se quantum dots for photoacoustic imaging-guided cancer photothermal therapy," *Small* **16**(1), 1–11 (2020) [doi:10.1002/smll.201905208].

126. D. Ding, W. Guo, C. Guo, et al., "MoO3-x quantum dots for photoacoustic imaging guided photothermal/photodynamic cancer treatment," *Nanoscale* **9**(5), 2020–2029, Royal Society of Chemistry (2017) [doi:10.1039/c6nr09046j].

127. D.H. Zhao, X.Q. Yang, X.L. Hou, et al., "In situ aqueous synthesis of genetically engineered polypeptide-capped Ag2S quantum dots for second near-infrared fluorescence/photoacoustic imaging and photothermal therapy," *J. Mater. Chem. B* **7**(15), 2484–2492 (2019) [doi:10.1039/c8tb03043j].

128. M. Vijaya Bharathi, S. Maiti, B. Sarkar, K. Ghosh, and P. Paira, "Water-mediated green synthesis of PBS quantum dot and its glutathione and biotin conjugates for non-invasive live cell imaging," *R. Soc. Open Sci.* **5**(3), 171614, 1–9 (2018) [doi:10.1098/rsos.171614].

129. X. Shi, S. Chen, M.Y. Luo, et al., "Zn-doping enhances the photoluminescence and stability of PbS quantum dots for in vivo high-resolution imaging in the NIR-II window," *Nano Res.* **13**(8), 2239–2245 (2020) [doi:10.1007/s12274-020-2843-4].

130. O.T. Bruns, T.S. Bischof, D.K. Harris, et al., "Next-generation in vivo optical imaging with short-wave infrared quantum dots," *Nat. Biomed. Eng.* **1**(4), 0056, 1–11 (2017) [doi:10.1038/s41551-017-0056].

131. C.T. Matea, T. Mocan, F. Tabaran, et al., "Quantum dots in imaging, drug delivery and sensor applications," *Int. J. Nanomedicine* **12**, 5421–5431 (2017) [doi:10.2147/IJN.S138624].

132. K.H. Lee, "Quantum dots: A quantum jump for molecular imaging?" *J. Nucl. Med.* **48**(9), 1408–1410 (2007) [doi:10.2967/jnumed.107.042069].

133. K.J. McHugh, L. Jing, A.M. Behrens, et al., "Biocompatible semiconductor quantum dots as cancer imaging agents," *Adv. Mater.* **30**(18), 1706356, Wiley-VCH Verlag (2018) [doi:10.1002/adma.201706356].

134. L. Cheng, J. Liu, X. Gu, et al., "Pegylated WS2 nanosheets as a multifunctional theranostic agent for in vivo dual-modal CT/photoacoustic imaging guided photothermal therapy," *Adv. Mater.* **26**(12), 1886–1893 (2014) [doi:10.1002/adma.201304497].

135. X. Qian, S. Shen, T. Liu, L. Cheng, and Z. Liu, "Two-dimensional TiS2 nanosheets for in vivo photoacoustic imaging and photothermal cancer therapy," *Nanoscale* **7**(14), 6380–6387, Royal Society of Chemistry (2015) [doi:10.1039/c5nr00893j].

136. Z. Li, H. Zhang, J. Han, et al., "Surface nanopore engineering of 2D MXenes for targeted and synergistic multitherapies of hepatocellular carcinoma," *Adv. Mater.* **30**(25), 1–11 (2018) [doi:10.1002/adma.201706981].

137. H. Lin, S. Gao, C. Dai, Y. Chen, and J. Shi, "A two-dimensional biodegradable niobium carbide (MXene) for photothermal tumor eradication in NIR-I and NIR-II bio-windows," *J. Am. Chem. Soc.* **139**(45), 16235–16247 (2017) [doi:10.1021/jacs.7b07818].

138. L. Zhou, F. Wu, J. Yu, et al., "Titanium carbide (Ti3C2Tx) MXene: A novel precursor to amphiphilic carbide-derived graphene quantum dots for fluorescent ink, light-emitting composite and bioimaging," *Carbon* **118**(Y), 50–57, Elsevier Ltd (2017) [doi:10.1016/j.carbon.2017.03.023].

139. Z. Yang, R. Tian, J. Wu, et al., "Impact of semiconducting perylene diimide nanoparticle size on lymph node mapping and cancer imaging," *ACS Nano* **11**(4), 4247–4255 (2017) [doi:10.1021/acsnano.7b01261].

140. C. Yin, X. Zhen, Q. Fan, W. Huang, and K. Pu, "Degradable semiconducting oligomer amphiphile for ratiometric photoacoustic imaging of hypochlorite," *ACS Nano* **11**(4), 4174–4182 (2017) [doi:10.1021/acsnano.7b01092].

141. Y.H. Chan and P.J. Wu, "Semiconducting polymer nanoparticles as fluorescent probes for biological imaging and sensing," *Part. Part. Syst. Charact.* **32**(1), 11–28, Wiley-VCH Verlag (2015) [doi:10.1002/ppsc.201400123].

142. C. Wu and D.T. Chiu, "Highly fluorescent semiconducting polymer dots for biology and medicine," *Angew. Chem. Int. Ed.* **52**(11), 3086–3109 (2013) [doi:10.1002/anie.201205133].

143. S. Luo, E. Zhang, Y. Su, T. Cheng, and C. Shi, "A review of NIR dyes in cancer targeting and imaging," *Biomaterials* **32**(29), 7127–7138, Elsevier Ltd (2011) [doi:10.1016/j.biomaterials.2011.06.024].

144. H.H. Jaffé and A.L. Miller, "The fates of electronic excitation energy," *J. Chem. Educ.* **43**(9), 469–473 (1966) [doi:10.1021/ed043p469].

145. Y. Lin, R. Weissleder, and C.H. Tung, "Novel near-infrared cyanine fluorochromes: Synthesis, properties, and bioconjugation," *Bioconjug. Chem.* **13**(3), 605–610 (2002) [doi:10.1021/bc0155723].

146. F. Song, X. Peng, E. Lu, et al., "Syntheses, spectral properties and photostabilities of novel water-soluble near-infrared cyanine dyes," *J. Photochem. Photobiol. A* **168**(1–2), 53–57 (2004) [doi:10.1016/j.jphotochem.2004.05.012].

147. R. Philip, A. Penzkofer, W. Bäumler, R.M. Szeimies, and C. Abels, "Absorption and fluorescence spectroscopic investigation of indocyanine green," *J. Photochem. Photobiol. A* **96**(1–3), 137–148 (1996) [doi:10.1016/1010-6030(95)04292-X].

148. W.M. Kuebler, "How NIR is the future in blood flow monitoring?" *J. Appl. Physiol. (1985)* **104**(4), 905–906 (2008) [doi:10.1152/japplphysiol.00106.2008].

149. C. Kim, K.H. Song, F. Gao, and L.V. Wang, "Sentinel lymph nodes and lymphatic vessels: Noninvasive dual-modality in vivo mapping by using indocyanine green in rats – Volumetric spectroscopic photoacoustic imaging and planar fluorescence imaging," *Radiology* **255**(2), 442–450 (2010) [doi:10.1148/radiol.10090281].

150. M.L.J. Landsman, G. Kwant, G.A. Mook, and W.G. Zijlstra, "Light absorbing properties, stability, and spectral stabilization of indocyanine green," *J. Appl. Physiol.* **40**(4), 575–583 (1976) [doi:10.1152/jappl.1976.40.4.575].

151. K. Umezawa, D. Citterio, and K. Suzuki, "Water-soluble NIR fluorescent probes based on squaraine and their application for protein labeling," *Anal. Sci.* **24**(2), 213–217 (2008) [doi:10.2116/analsci.24.213].

152. S. Sreejith, P. Carol, P. Chithra, and A. Ajayaghosh, "Squaraine dyes: A mine of molecular materials," *J. Mater. Chem.* **18**(3), 264–274 (2008) [doi:10.1039/b707734c].

153. S.H. Kim, J.H. Kim, J.Z. Cui, et al., "Absorption spectra, aggregation and photofading behaviour of near-infrared absorbing squarylium dyes containing perimidine moiety," *Dyes Pigment* **55**(1), 1–7 (2002) [doi:10.1016/S0143-7208(02)00051-7].

154. D. Zhang, Y.X. Zhao, Z.Y. Qiao, et al., "Nano-confined squaraine dye assemblies: New photoacoustic and near-infrared fluorescence dual-modular imaging probes in vivo," *Bioconjug. Chem.* **25**(11), 2021–2029 (2014) [doi:10.1021/bc5003983].

155. F.F. An, Z.J. Deng, J. Ye, et al., "Aggregation-induced near-infrared absorption of squaraine dye in an albumin nanocomplex for photoacoustic tomography in vivo," *ACS Appl. Mater. Interfaces* **6**(20), 17985–17992 (2014) [doi:10.1021/am504816h].

156. S. Sreejith, J. Joseph, M. Lin, et al., "Near-infrared squaraine dye encapsulated micelles for in vivo fluorescence and photoacoustic bimodal imaging," *ACS Nano* **9**(6), 5695–5704 (2015) [doi:10.1021/acsnano.5b02172].

157. M. Beija, C.A.M. Afonso, and J.M.G. Martinho, "Synthesis and applications of rhodamine derivatives as fluorescent probes," *Chem. Soc. Rev.* **38**(8), 2410–2433 (2009) [doi:10.1039/b901612k].

158. E. Morgounova, Q. Shao, B.J. Hackel, D.D. Thomas, and S. Ashkenazi, "Photoacoustic lifetime contrast between methylene blue monomers and self-quenched dimers as a model for dual-labeled activatable probes," *J. Biomed. Opt.* **18**(5) (2013) [doi:10.1117/1.jbo.18.5.056004], [056004].

159. J. Yao, K. Maslov, S. Hu, and L.V. Wang, "Evans blue dye-enhanced capillary-resolution photoacoustic microscopy in vivo," *J. Biomed. Opt.* **14**(5), 054049 (2009) [doi:10.1117/1.3251044].

160. D. Gabrielli, E. Belisle, D. Severino, A.J. Kowaltowski, and M.S. Baptista, "Binding, aggregation and photochemical properties of methylene blue in mitochondrial suspensions," *Photochem. Photobiol.* **79**(3), 227 (2004) [doi:10.1562/be-03-27.1].

161. S. Onoe, T. Temma, K. Kanazaki, M. Ono, and H. Saji, "Development of photostabilized asymmetrical cyanine dyes for in vivo photoacoustic imaging of tumors," *J. Biomed. Opt.* **20**(9), 96006 (2015) [doi:10.1117/1.jbo.20.9.096006].

162. G.A. Reynolds and K.H. Drexhage, "Stable heptamethine pyrylium dyes that absorb in the infrared," *J. Org. Chem.* **42**(5), 885–888 (1977) [doi:10.1021/jo00425a027].

163. G.J. Kavarnos and N.J. Turro, "Photosensitization by reversible electron transfer: Theories, experimental evidence, and examples," *Chem. Rev.* **86**(2), 401–449 (1986) [doi:10.1021/cr00072a005].

164. A.R. McIntosh, A. Siemiarczuk, J.R. Bolton, et al., "Intramolecular Photochemicak electron transfer. 1. ERP and optical absorption evidence for stabilized charge," *J. Phys. Chem. A* **111**(25), 7215–7223 (1983).

165. T. Okada, N. Mataga, Y. Sakata, and S. Misumi, "Ultrafast intersystem crossing of intramolecular heteroexcimers," *J. Photochem.* **17**(1), 130–131 (1981) [doi:10.1016/0047-2670(81)85252-5].

166. Y. Wang, M.C. Crawford, and K.B. Eisenthal, "Picosecond laser studies of intramolecular excited-state charge-transfer dynamics and small-chain relaxation," *J. Am. Chem. Soc.* **104**(22), 5874–5878 (1982) [doi:10.1021/ja00386a004].

167. P. Vanderauwera, F.C. DeSchryver, A. Weller, M.A. Winnik, and K.A. Zachariasse, "Intramolecular exciplex formation and fluorescence quenching as a function of chain length in ω-dimethylaminoalkyl esters of 2-anthracenecarboxylic acid," *J. Phys. Chem.* **88**(14), 2964–2970 (1984) [doi:10.1021/j150658a010].

168. M.E. Huston, K.W. Haider, and A.W. Czarnik, "Chelation enhanced fluorescence in 9,10-bis[[(2-(dimethylamino)ethyl)methylamino]methyl]anthracene," *J. Am. Chem. Soc.* **110**(13), 4460–4462 (1988) [doi:10.1021/ja00221a083].

169. N.V. Blough and D.J. Simpson, "Chemically mediated fluorescence yield switching in nitroxide–fluorophore adducts: Optical sensors of radical/redox reactions," *J. Am. Chem. Soc.* **110**(6), 1915–1917 (1988) [doi:10.1021/ja00214a041].

170. C. Munkholm, D. Parkinson, and D.R. Walt, "Intramolecular fluorescence self-quenching of fluoresceinamine," *J. Am. Chem. Soc.* **112**(7), 2608–2612 (1990).

171. S.R. Mujumdar, R.B. Mujumdar, C.M. Grant, and A.S. Waggoner, "Cyanine-labeling reagents: Sulfobenzindocyanine succinimidyl esters," *Bioconjug. Chem.* **7**(3), 356–362 (1996) [doi:10.1021/bc960021b].

172. E.E. Jelley, "Spectral absorption and fluorescence of dyes in the molecular state," *Nature* **138**(3502), 1009–1010 (1936) [doi:10.1038/1381009a0].

173. E.E. Jelley, "Molecular, nematic and crystal states of I: I-diethyl--Cyanine chloride," *Nature* **159**(4036), 631–632 (1937) [doi:10.1038/159333a0].

174. J.L. Bricks, Y.L. Slominskii, I.D. Panas, and A.P. Demchenko, "Fluorescent J-aggregates of cyanine dyes: Basic research and applications review," *Methods Appl. Fluoresc.* **6**(1), 012001, 1–31 IOP Publishing (2018) [doi:10.1088/2050-6120/aa8d0d].

175. U. Rösch, S. Yao, R. Wortmann, and F. Würthner, "Fluorescent H-aggregates of merocyanine dyes," *Angew. Chem.* **45**(42), 7026–7030 (2006) [doi:10.1002/ange.200602286].

176. E. Rabinowitch and L.F. Epstein, "Polymerization of dyestuffs in solution. Thionine and methylene blue 1," *J. Am. Chem. Soc.* **63**(1), 69–78 (1941) [doi:10.1021/ja01846a011].

177. K. Bergmann and C.T. O'Konski, "A spectroscopic study of methylene blue monomer, dimer, and complexes with montmorillonite," *J. Phys. Chem.* **67**(10), 2169–2177 (1963) [doi:10.1021/j100804a048].

178. M. Kasha, H.R. Rawls, and M.A. El-Bayoumi, "The exciton model in molecular spectroscopy," *Pure Appl. Chem.* **11**(3–4), 371–392 (1965) [doi:10.1351/pac196511030371].

179. T. Förster, "Energy migration and fluorescence. 1946," *J. Biomed. Opt.* **17**(1), 011002 (2012) [doi:10.1117/1.jbo.17.1.011002].

180. M. Van Der Auweraer, G. Biesmans, and F.C. De Schryver, "On the photophysical properties of aggregates of 3-(2-phenyl)-indolocarbocyanines," *Chem. Phys.* **119**(2–3), 355–375 (1988) [doi:10.1016/0301-0104(88)87196-9].

181. M. Van Der Auweraer, B. Verschuere, and F.C. De Schryver, "Absorption and fluorescence properties of rhodamine b derivatives forming Langmuir-Blodgett films," *Langmuir* **4**(3), 583–588 (1988) [doi:10.1021/la00081a016].

182. L. Lu, R.J. Lachicotte, T.L. Penner, J. Perlstein, and D.G. Whitten, "Exciton and charge-transfer interactions in nonconjugated merocyanine dye dimers: Novel solvatochromic behavior for tethered bichromophores and excimers," *J. Am. Chem. Soc.* **121**(36), 8146–8156 (1999) [doi:10.1021/ja983778h].

183. S. Zeena and K.G. Thomas, "Conformational switching and exciton interactions in hemicyanine-based bichromophores," *J. Am. Chem. Soc.* **123**(32), 7859–7865 (2001) [doi:10.1021/ja010199v].

184. Y. Gu, J.H. Li, and Z.H. Gou, "Selective protection of normal hepatocytes by indocyanine green in photodynamic therapy for the hepatoma of rat," in International Conference on Photodynamic Therapy and Laser Medicine, Beijing, China, **1616**, 266–274 (1991) [doi:10.1117/12.137021].

185. K.J. Baker, "Binding of sulfobromophthalein (BSP) sodium and indocyanine green (ICG) by plasma a1 lipoproteins.," *Proc. Soc. Exp. Biol. Med.* **122**(4), 957–963 (1966) [doi:10.3181/00379727-122-31299].

186. G. Scheibe, "Über die Veränderlichkeit der Absorptionsspektren in Lösungen und die Nebenvalenzen als ihre Ursache," *Angew. Chem.* **50**(11), 212–219 (1937) [doi:10.1002/ange.19370501103].

187. A.H. Herz, "Aggregation of sensitizing dyes in solution and their adsorption onto silver halides," *Adv. Colloid Interface Sci.* **8**(4), 237–298 (1977) [doi:10.1016/0001-8686(77)80011-0].

188. D. Möbius, "Scheibe aggregates," *Adv. Mater.* **7**(5), 437–444 (1995) [doi:10.1002/adma.19950070503].

189. H. Fidder, J. Knoester, and D.A. Wiersma, "Optical properties of disordered molecular aggregates: A numerical study," *J. Chem. Phys.* **95**(11), 7880–7890 (1991) [doi:10.1063/1.461317].

190. E.W. Knapp, "Lineshapes of molecular aggregates, exchange narrowing and intersite correlation," *Chem. Phys.* **85**(1), 73–82 (1984) [doi:10.1016/S0301-0104(84)85174-5].

191. J. Knoester, "Nonlinear optical line shapes of disordered molecular aggregates: Motional narrowing and the effect of intersite correlations," *J. Chem. Phys.* **99**(11), 8466–8479 (1993) [doi:10.1063/1.465623].

192. D.A. Weitz, S. Garoff, J.I. Gersten, and A. Nitzan, "The enhancement of Raman scattering, resonance Raman scattering, and fluorescence from molecules adsorbed on a rough silver surface," *J. Chem. Phys.* **78**(9), 5324–5338 (1983) [doi:10.1063/1.445486].

193. F.R. Aussenegg, A. Leitner, M.E. Lippitsch, H. Reinisch, and M. Riegler, "Novel aspects of fluorescence lifetime for molecules positioned close to metal surfaces," *Surf. Sci.* **189–190**(C), 935–945 (1987) [doi:10.1016/S0039-6028(87)80531-9].

194. D.A. Weitz, J.I. Gersten, S. Garoff, C.D. Hanson, and T.J. Gramila, "Fluorescent lifetimes of molecules on silver-island films," *Opt. Lett.* **7**(2), 89 (1982) [doi:10.1364/ol.7.000089].

195. M. Swierczewska, S. Lee, and X. Chen, "The design and application of fluorophore-gold nanoparticle activatable probes," *Phys. Chem. Chem. Phys.* **13**(21), 9929–9941 (2011) [doi:10.1039/c0cp02967j].

196. Y. Fu and J.R. Lakowicz, "Enhanced fluorescence of Cy5-labeled oligonucleotides Near silver island films: A distance effect study using single molecule spectroscopy," *J. Phys. Chem. B* **110**(45), 22557–22562 (2006) [doi:10.1021/jp060402e].

197. F. Tam, G.P. Goodrich, B.R. Johnson, and N.J. Halas, "Plasmonic enhancement of molecular fluorescence," *Nano Lett.* **7**(2), 496–501 (2007) [doi:10.1021/nl062901x].

198. H.M. Kim, D.M. Kim, C. Jeong, et al., "Assembly of plasmonic and magnetic nanoparticles with fluorescent silica shell layer for tri-functional SERS-magnetic-fluorescence probes and its bioapplications," *Sci. Rep.* **8**(1), 1–10 (2018) [doi:10.1038/s41598-018-32044-7].

199. Y. Yamaoka and T. Takamatsu, "Enhancement of multiphoton excitation-induced photoacoustic signals by using gold nanoparticles surrounded by fluorescent dyes," *Photons Plus Ultrasound Imaging Sens.* **7177**, 71772A (2009) [doi:10.1117/12.808367].

200. M.V. Novoselova, D.N. Bratashov, M. Sarimollaoglu, et al., "Photoacoustic and fluorescent effects in multilayer plasmon-dye interfaces," *J. Biophoton.* **12**(4), e201800265 (2019) [doi:10.1002/jbio.201800265].

201. A.M. Smith, M.C. Mancini, and S. Nie, "Bioimaging: Second window for in vivo imaging," *Nat. Nanotechnol.* **4**(11), 710–711, Nature Publishing Group (2009) [doi:10.1038/nnano.2009.326].

202. I. Costantini, R. Cicchi, L. Silvestri, F. Vanzi, and F.S. Pavone, "In-vivo and ex-vivo optical clearing methods for biological tissues: Review," *Biomed. Opt. Express* **10**(10), 5251–5267 (2019) [doi:10.1364/BOE.10.005251].

203. E.A. Genina, A.N. Bashkatov, Y.P. Sinichkin, I.Y. Yanina, and V.V. Tuchin, "Optical clearing of biological tissues: Prospects of application in medical diagnostics and phototherapy," *J. Biomed. Photonics Eng.* **1**(1), 22–58 (2015) [doi:10.18287/jbpe-2015-1-1-22].

204. E.A. Genina, Y.I. Surkov, I.A. Serebryakova, et al., "Rapid ultrasound optical clearing of human light and dark skin," *IEEE Trans. Med. Imaging*, **39**(10), 3198–3206, Institute of Electrical and Electronics Engineers (IEEE) (2020) [doi:10.1109/tmi.2020.2989079].

205. H. Kim and J.H. Chang, "Increased light penetration due to ultrasound-induced air bubbles in optical scattering media," *Sci. Rep.* **7**(1), 16105, Nature Publishing Group (2017) [doi:10.1038/s41598-017-16444-9].

206. M. Chamanzar, M.G. Scopelliti, J. Bloch, et al., "Ultrasonic sculpting of virtual optical waveguides in tissue," *Nat. Commun.* **10**(1), 92, Nature Publishing Group (2019) [doi:10.1038/s41467-018-07856-w].

207. V.V. Tuchin, I.L. Maksimova, D.A. Zimnyakov, et al., "Light propagation in tissues with controlled optical properties," *Proc. SPIE* **2925**(4), 118–142 (1996) [doi:10.1117/12.281502].

208. E.A. Genina, A.N. Bashkatov, and V.V. Tuchin, "Tissue optical immersion clearing," *Expert Rev. Med. Devices* **7**(6), 825–842 (2010) [doi:10.1586/erd.10.50].

209. D.S. Richardson and J.W. Lichtman, "Clarifying tissue clearing," *Cell* **162**(2), 246–257, Cell Press (2015) [doi:10.1016/j.cell.2015.06.067].

210. E.C. Costa, D.N. Silva, A.F. Moreira, and I.J. Correia, "Optical clearing methods: An overview of the techniques used for the imaging of 3D spheroids," *Biotechnol. Bioeng.* **116**(10), 2742–2763, John Wiley and Sons Inc. (2019) [doi:10.1002/bit.27105].

211. A.T. Yeh and J. Hirshburg, "Molecular interactions of exogenous chemical agents with collagen: Implications for tissue optical clearing," *J. Biomed. Opt.* **11**(1), 014003, SPIE-Intl Soc Optical Eng (2006) [doi:10.1117/1.2166381].

212. M.H. Khan, S. Chess, B. Choi, K.M. Kelly, and J.S. Nelson, "Can topically applied optical clearing agents increase the epidermal damage threshold and enhance therapeutic efficacy?" *Lasers Surg. Med.* **35**(2), 93–95 (2004) [doi:10.1002/lsm.20078].

213. X. Xu, Q. Zhu, and C. Sun, "Combined effect of ultrasound-SLS on skin optical clearing," *IEEE Photon. Technol. Lett.* **20**(24), 2117–2119 (2008) [doi:10.1109/LPT.2008.2006987].

214. H. Zhong, Z. Guo, H. Wei, et al., "Synergistic effect of ultrasound and thiazone-PEG 400 on human skin optical clearing in vivo," *Photochem. Photobiol.* **86**(3), 732–737 (2010) [doi:10.1111/j.1751-1097.2010.00710.x].

215. J. Wang, X. Zhou, S. Duan, Z. Chen, and D. Zhu, "Improvement of in vivo rat skin optical clearing with chemical penetration enhancers," *Proc SPIE* **7883**, 78830Y (2011) [doi:10.1117/12.874859].

216. G. Vargas, A. Readinger, S.S. Dozier, and A.J. Welch, "Morphological changes in blood vessels produced by hyperosmotic agents and measured by optical coherence tomography," *Photochem. Photobiol.* **77**(5), 541–549, American Society for Photobiology (2007) [doi:10.1562/0031-8655(2003)0770541mcibvp2.0.co2].

217. T. Yu, Y. Qi, H. Gong, Q. Luo, and D. Zhu, "Optical clearing for multiscale biological tissues," *J. Biophoton.* **11**(2), e201700187, Wiley-VCH Verlag (2018) [doi:10.1002/jbio.201700187].

218. Y. Zhou, J. Yao, and L.V. Wang, "Optical clearing-aided photoacoustic microscopy with enhanced resolution and imaging depth," *Opt. Lett.* **38**(14), 2592, The Optical Society (2013) [doi:10.1364/ol.38.002592].

219. V.P. Zharov, E.I. Galanzha, E.V. Shashkov, N.G. Khlebtsov, and V.V. Tuchin, "In vivo photoacoustic flow cytometry for monitoring of circulating single cancer cells and contrast agents," *Opt. Lett.* **31**(24), 3623 (2006) [doi:10.1364/ol.31.003623].

220. B. Choi, L. Tsu, E. Chen, et al., "Case report: Determination of chemical agent optical clearing potential using in vitro human skin," *Lasers Surg. Med.* **36**(2), 72–75 (2005) [doi:10.1002/lsm.20116].

221. M.G. Ghosn, N. Sudheendran, M. Wendt, et al., "Monitoring of glucose permeability in monkey skin in vivo using Optical Coherence Tomography," *J. Biophoton.* **3**(1–2), 25–33 (2010) [doi:10.1002/jbio.200910075].

222. J.M. Hirshburg, K.M. Ravikumar, W. Hwang, and A.T. Yeh, "Molecular basis for optical clearing of collagenous tissues," *J. Biomed. Opt.* **15**(5), 055002, SPIE-Intl Soc Optical Eng (2010) [doi:10.1117/1.3484748].

223. Y. Liu, X. Yang, D. Zhu, R. Shi, and Q. Luo, "Optical clearing agents improve photoacoustic imaging in the optical diffusive regime," *Opt. Lett.* **38**(20), 4236 (2013) [doi:10.1364/ol.38.004236].

224. E.A. Genina, A.N. Bashkatov, A.A. Korobko, et al., "Optical clearing of human skin: Comparative study of permeability and dehydration of intact and photothermally perforated skin," *J. Biomed. Opt.* **13**(2), 021102, SPIE-Intl Soc Optical Eng (2008) [doi:10.1117/1.2899149].

225. A.N. Bashkatov, E.A. Genina, V.I. Kochubey, V.V. Tuchin, and Y.P. Sinichkin, "Influence of osmotically active chemical agents on the transport of light in scleral tissue," *Proc SPIE* **3726**, 403 (1999) [doi:10.1117/12.341424].

226. X. Weny, Z. Maoy, Z. Han, V.V. Tuchin, and D. Zhu, "In vivo skin optical clearing by glycerol solutions: Mechanism," *J. Biophoton.* **3**(1–2), 44–52 (2010) [doi:10.1002/jbio.200910080].

227. V. Hovhannisyan, P.-S. Hu, S.-J. Chen, C.-S. Kim, and C.-Y. Dong, "Elucidation of the mechanisms of optical clearing in collagen tissue with multiphoton imaging," *J. Biomed. Opt.* **18**(4), 046004, SPIE-Intl Soc Optical Eng (2013) [doi:10.1117/1.jbo.18.4.046004].

228. A.N. Bashkatov, E.A. Genina, Y.P. Sinichkin, et al., "Glucose and mannitol diffusion in human dura mater," *Biophys. J.* **85**(5), 3310–3318, Biophysical Society (2003) [doi:10.1016/S0006-3495(03)74750-X].

229. Z. Zhu, G. Wu, H. Wei, et al., "Investigation of the permeability and optical clearing ability of different analytes in human normal and cancerous breast tissues by spectral domain OCT," *J. Biophoton.* **5**(7), 536–543 (2012) [doi:10.1002/jbio.201100106].

230. J. Jiang and R.K. Wang, "Comparing the synergistic effects of oleic acid and dimethyl sulfoxide as vehicles for optical clearing of skin tissue in vitro," *Phys. Med. Biol.* **49**(23), 5283–5294 (2004) [doi:10.1088/0031-9155/49/23/006].

231. Y.A. Menyaev, D.A. Nedosekin, M. Sarimollaoglu, et al., "Optical clearing in photoacoustic flow cytometry," *Biomed. Opt. Express* **4**(12), 3030 (2013) [doi:10.1364/boe.4.003030].

232. E. Youn, T. Son, H.-S. Kim, and B. Jung, "Determination of optimal glycerol concentration for optical tissue clearing," *Proc SPIE* **8207**, 82070J (2012) [doi:10.1117/12.909790].

233. D. Zhu, J. Zhang, H. Cui, et al., "Short-term and long-term effects of optical clearing agents on blood vessels in chick chorioallantoic membrane," *J. Biomed. Opt.* **13**(2), 021106, SPIE-Intl Soc Optical Eng (2008) [doi:10.1117/1.2907169].

234. X. Yang, Y. Liu, D. Zhu, R. Shi, and Q. Luo, "Dynamic monitoring of optical clearing of skin using photoacoustic microscopy and ultrasonography," *Opt. Express* **22**(1), 1094 (2014) [doi:10.1364/oe.22.001094].

235. E.I. Galanzha, M.S. Kokoska, E.V. Shashkov, et al., "In vivo fiber-based multicolor photoacoustic detection and photothermal purging of metastasis in sentinel lymph nodes targeted by nanoparticles," *J. Biophoton.* **2**(8–9), 528–539 (2009) [doi:10.1002/jbio.200910046].

236. M.V. Novoselova, T.O. Abakumova, B.N. Khlebtsov, et al., "Optical clearing for photoacoustic lympho- and angiography beyond conventional depth limit in vivo," *Photoacoustics* **20**(June), 100186, Elsevier (2020) [doi:10.1016/j.pacs.2020.100186].

237. C.G. Rylander, O.F. Stumpp, T.E. Milner, et al., "Dehydration mechanism of optical clearing in tissue," *J. Biomed. Opt.* **11**(4), 041117, SPIE-Intl Soc Optical Eng (2006) [doi:10.1117/1.2343208].

238. N.Y. Kang, S.J. Park, X. Wei Emmiline Ang, et al., "A macrophage uptaking near-infrared chemical probe CDnir7 for in vivo imaging of inflammation," *Chem. Commun.* **50**(50), 6589–6591 (2014) [doi:10.1039/c4cc02038c].

239. J.L. Kovar, W. Volcheck, E. Sevick-Muraca, M.A. Simpson, and D.M. Olive, "Characterization and performance of a near-infrared 2-deoxyglucose optical imaging agent for mouse cancer models," *Anal. Biochem.* **384**(2), 254–262, Elsevier Inc. (2009) [doi:10.1016/j.ab.2008.09.050].

240. M.L. Li, J.T. Oh, X. Xie, et al., "Simultaneous molecular and hypoxia imaging of brain tumors in vivo using spectroscopic photoacoustic tomography," *Proc. IEEE* **96**(3), 481–489 (2008) [doi:10.1109/JPROC.2007.913515].

241. T. Temma, S. Onoe, K. Kanazaki, M. Ono, and H. Saji, "Preclinical evaluation of a novel cyanine dye for tumor imaging with in vivo photoacoustic imaging," *J. Biomed. Opt.* **19**(9), 90501 (2014) [doi:10.1117/1.jbo.19.9.090501].

242. Q. Yang, H. Cui, S. Cai, X. Yang, and M.L. Forrest, "In vivo photoacoustic imaging of chemotherapy-induced apoptosis in squamous cell carcinoma using a near-infrared caspase-9 probe," *J. Biomed. Opt.* **16**(11), 116026 (2011) [doi:10.1117/1.3650240].

243. N. Sim, S. Gottschalk, R. Pal, et al., "Wavelength-dependent optoacoustic imaging probes for NMDA receptor visualisation," *Chem. Commun.* **51**(82), 15149–15152, Royal Society of Chemistry (2015) [doi:10.1039/c5cc06277b].

244. K.H. Song, E.W. Stein, J.A. Margenthaler, and L.V. Wang, "Noninvasive photoacoustic identification of sentinel lymph nodes containing methylene blue in vivo in a rat model," *J. Biomed. Opt.* **13**(5), 54033 (2008) [doi:10.1117/1.2976427].

245. C. Kim, T.N. Erpelding, L. Jankovic, and L.V. Wang, "Performance benchmarks of an array-based hand-held photoacoustic probe adapted from a clinical ultrasound system for non-invasive sentinel lymph node imaging," *Philos. Trans. R. Soc. A* **369**(1955), 4644–4650 (2011) [doi:10.1098/rsta.2010.0353].

246. Q. Shao, E. Morgounova, C. Jiang, et al., "In vivo photoacoustic lifetime imaging of tumor hypoxia in small animals," *J. Biomed. Opt.* **18**(7), 76019 (2013) [doi:10.1117/1.jbo.18.7.076019].

247. S. Ashkenazi, "Photoacoustic lifetime imaging of dissolved oxygen using methylene blue," *J. Biomed. Opt.* **15**(4), 40501 (2010) [doi:10.1117/1.3465548].

248. L. Song, C. Kim, K. Maslov, K.K. Shung, and L.V. Wang, "High-speed dynamic 3D photoacoustic imaging of sentinel lymph node in a murine model using an ultrasound array," *Med. Phys.* **36**(8), 3724–3729 (2009) [doi:10.1118/1.3168598].

249. D. Razansky, C. Vinegoni, and V. Ntziachristos, "Multispectral photoacoustic imaging of fluorochromes in small animals," *Opt. Lett.* **32**(19), 2891 (2007) [doi:10.1364/ol.32.002891].

250. K.M. Stantz, M. Cao, B. Liu, K.D. Miller, and L. Guo, "Molecular imaging of neutropilin-1 receptor using photoacoustic spectroscopy in breast tumors," *Photons Plus Ultrasound Imaging Sens.* **7564**, 75641O (2010) [doi:10.1117/12.842271].

251. J. Levi, S.R. Kothapalli, S. Bohndiek, et al., "Molecular photoacoustic imaging of follicular thyroid carcinoma," *Clin. Cancer Res.* **19**(6), 1494–1502 (2013) [doi:10.1158/1078-0432.CCR-12-3061].

252. S. Bhattacharyya, S. Wang, D. Reinecke, et al., "Synthesis and evaluation of near-infrared (NIR) dye-Herceptin conjugates as photoacoustic computed tomography (PCT) probes for HER2 expression in breast cancer," *Bioconjug. Chem.* **19**(6), 1186–1193 (2008) [doi:10.1021/bc700482u].

253. M.R. Chatni, J. Yao, A. Danielli, et al., "Functional photoacoustic microscopy of pH," *J. Biomed. Opt.* **16**(10) 100503 (2011).

254. T.D. Horvath, G. Kim, R. Kopelman, and S. Ashkenazi, "Ratiometric photoacoustic sensing of pH using a 'sonophore', " *Analyst* **133**(6), 747–749 (2008) [doi:10.1039/b800116b].

255. S.V. Hudson, J.S. Huang, W. Yin, et al., "Targeted noninvasive imaging of EGFR-expressing orthotopic pancreatic cancer using multispectral optoacoustic tomography," *Cancer Res.* **74**(21), 6271–6279 (2014) [doi:10.1158/0008-5472.CAN-14-1656].

256. J. Levi, A. Sathirachinda, and S.S. Gambhir, "A high-affinity, high-stability photoacoustic agent for imaging gastrin-releasing peptide receptor in prostate cancer," *Clin. Cancer Res.* **20**(14), 3721–3729 (2014) [doi:10.1158/1078-0432.CCR-13-3405].

257. N. Beziere and V. Ntziachristos, "Optoacoustic imaging of naphthalocyanine: Potential for contrast enhancement and therapy monitoring," *J. Nucl. Med.* **56**(2), 323–328 (2015) [doi:10.2967/jnumed.114.147157].

258. H. Huang, D. Wang, Y. Zhang, et al., "Axial pegylation of tin octabutoxy Naphthalocyanine extends blood circulation for photoacoustic vascular imaging," *Bioconjug. Chem.* **27**(7), 1574–1578 (2016) [doi:10.1021/acs.bioconjchem.6b00280].

24

Tissue optical clearing in the terahertz range

Olga A. Smolyanskaya, Kirill I. Zaytsev, Irina N. Dolganova, Guzel R. Musina,
Daria K. Tuchina, Maxim M. Nazarov, Alexander P. Shkurinov, and Valery V. Tuchin

CONTENTS

Introduction

During the past few decades, rapid progress in terahertz (THz) technology has been observed [1–9], which has stimulated the development of THz biomedical applications [10], with an emphasis on

- label-free medical diagnosis of malignant and benign neoplasms [11–18];
- evaluation of scar treatment strategies [19, 20];
- sensing of glycated tissues and blood [21–23];
- studying hydration [24], viability [25], and injuries [26, 27] of tissues.

The majority of these applications rely on the strong sensitivity of THz-waves to the content and state (free or bound) of tissue water [10, 28]. In this way, water is used as an endogenous marker of pathological tissues. Despite the attractiveness of THz technologies in the aforementioned applications, their transfer to clinical practice is strongly limited by the small depth of THz-wave penetration in tissues (either tissue *in vivo* or freshly-excised tissue *ex vivo*). The typical penetration depth ranges from tens to hundreds of micrometers, depending on the tissue type and the considered electromagnetic frequency and, thus, allows one to probe only the superficial tissue properties [10, 11].

In order to reduce the THz-radiation absorption by tissues and, thus, to improve the THz-wave penetration depth, several approaches have already been developed based on different physical principles, among them

- freezing of tissues [29, 30];
- dehydration of tissues [31, 32];
- paraffin-embedding of tissues [15];
- lyophilization of tissues [33];
- compression optical clearing of tissues [34, 35];
- immersion optical clearing (IOC) of tissues [36–47].

Such techniques as tissue freezing, dehydration, paraffin-embedding, and lyophilization are rather time-consuming and require complex preparations. Moreover, they can be predominantly used only in *ex vivo* applications of THz technology. Some of them lead to structural changes in tissues as a result of their long-term exposure. From Reference [34], we notice that a compression of tissues can be used to change their optical properties at THz frequencies. However, a compression-driven reduction of THz-wave absorption in tissues seem to be rather small, while this tissue clearing modality remains quite novel and deserves further comprehensive studies in order to make objective conclusions. Among all the aforementioned techniques, IOC of tissues remains the most attractive for THz-wave penetration depth enhancement [10, 11]. First, IOC was introduced in the visible and near-infrared ranges [45–47], and then, it was transferred to the THz [36–44, 48].

DOI: 10.1201/9781003025252-28

IOC relies on application of specific hyperosmotic agents that interact with tissues (*ex vivo* or *in vivo*) and change their optical properties (either reversibly or irreversibly). In the visible range, a hyperosmotic agent allows for changing a dielectric contrast between scatterers in tissues, thus reducing their electromagnetic-wave extinction coefficient. In turn, in the THz range, where the Mie scattering in tissues is usually assumed to be much weaker and the effective medium theory is applied to describe the THz-wave tissue interactions [10, 11], a hyperosmotic agent partially substitutes tissue water, thus decreasing an effective refractive index and absorption coefficient of tissues. In this way, for the THz applications, IOC agents should be characterized by hyperosmotic status, low THz-wave absorption, and high diffusion coefficient. Nevertheless, a lack of data about the THz optical properties of various IOC hyperosmotic agents, and about a rate of their diffusion in various tissue types, does not allow for selecting the optimal ones for applications in the THz frequency range.

THz dielectric permeability of water and biological tissues with and without application of hyperosmotic immersion agents

THz radiation is sensitive to the content of free and bound water in tissues. A hyperosmotic immersion agent applied to the tissues induced there a flow of free and weakly bound water from the tissue (inter- and intracellular fluid) and diffuses inside the tissue by replacing free water in the inter- and intracellular spaces. These two facts can help to enhance the THz-wave penetration depth, increase the contrast of the THz images, and expand the diagnostic capabilities associated with the properties of both bound water and unbound water–related features of tissues.

It is known that the frequency-dependent THz refractive index and absorption coefficient of the skin, as well as muscle tissue and blood, are similar to that of liquid water; therefore, we can conclude that a significant part of the water in tissues mentioned above is located there in a free state. After dehydrating agents are applied to tissues, mainly bound water will remain.

Water can be classified as a free (bulk) water (it forms no strong bonds with biological molecules) and bound water (it surrounds biological molecules and interacts with them) [1, 2]. With a strict analysis, it is necessary to separate strongly bound and weakly bound water [49]. Strongly or weakly bound water, as well as free (unbound) water, makes valuable, but different contributions to the THz dielectric response of biological tissues [10, 49, 50]. Reorientation time of hydrogen-bonded water molecules changes in the vicinity of a biomolecule [51]. This encourages scientists to use THz spectroscopy for studying different forms of water in biological media, including blood, tissues, and solutions [10, 50, 52]. The transition of water from its free to a bound state and back is the reason for a change in the THz response in biological objects [53]. An important feature of bound water is its much lower absorption at THz frequencies, as is evident from Figure 24.1.

Sensitivity to the presence of free water has a range from 0.01 to 1.00 THz, since the characteristic frequency of the peak

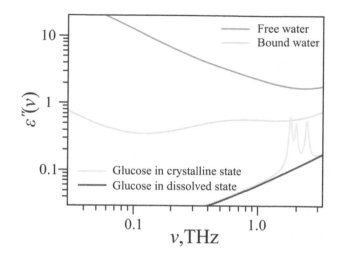

FIGURE 24.1 Imaginary part of the complex dielectric permittivity of bound water and other glucose solution components [28].

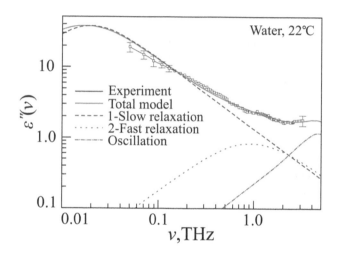

FIGURE 24.2 Imaginary part of the complex dielectric permittivity of water, where contribution of each term of the dielectric response model (see Equation 24.1) is highlighted.

absorption of free water corresponds to 20 GHz (Figure 24.2), while its spectral width extends by 2–3 octaves in both directions from this peak [54]. Therefore, to increase the sensitivity of the THz spectroscopy, it is necessary to expand the dynamic range of the experimental system towards low frequencies.

The frequency dependence of the THz complex dielectric permittivity of water, biological liquids, and tissues is described by relaxation models, for example:

$$\varepsilon_{water} = \varepsilon_\infty + \frac{\Delta\varepsilon_1}{1+i\omega\tau_1} + \frac{\Delta\varepsilon_2}{1+i\omega\tau_2} + \frac{A_1}{\omega_{01}^2 - \omega^2 + i\gamma_{01}\omega} + \frac{i\delta}{\omega\varepsilon_0} \quad (24.1)$$

with very similar parameters [49–51, 54]. The physical processes related to presented parameters in Equation 24.1 are the following: τ_1 is assigned to the cooperative reorientation time of hydrogen-bonded (HB) bulk water molecules involving HB switching events [55, 56]. The whole first term is responsible for the so-called "slow relaxation" process. The "fast relaxation" process can be explained by the two-fraction model of water, where a part of water is classified as non-hydrogen-bonded water isolated from the HB network, and τ_2 is assigned to the collisional relaxation of non-hydrogen-bonded water [50]. The noticeable sensitivity to the "fast relaxation" value is found in conductive ionic solutions [57, 58], or in water–alcohol mixtures [59]. Resonance processes described by the third term correspond to several known vibrational modes in the THz frequency range [60, 61], while, in our case, the only significant vibrational mode is located around 5 THz. The last term contains resistivity δ, it becomes nonzero for ionic solutions, in particular for NaCl solution.

All known processes in polar solutions below tens of THz ranges are very broadband, so it essential to combine several experimental techniques, each for its own spectral range, to obtain spectra of many octaves.

The relaxation dielectric response of tissues and solutions is characterized by the absence of clearly defined spectral features in the 0.1–3.0 THz range. At the same time, the sensitivity of radiation to water content makes the THz spectroscopy and imaging attractive medical diagnostic tools [62–64].

In the skin, the water is represented not in a pure state, but with the dissolved proteins, sugars, and other elements. While conducting THz-diagnostics of skin, the main contribution to the THz response is determined by free water; the lesser contribution by bound water molecules around the dissolved elements; and the neglected contribution by the dissolved substances themselves (see Figure 24.3a and Figure 24.1). The most significant changes occur in the low-frequency region of the THz spectrum, where dispersion and absorption are increased in free water, but not in bound water. Therefore, it is desirable to optimize THz time-domain spectrometers (THz-TDS) for operation in the low-frequency range of 0.1–0.5 THz (see Figure 24.3-b).

In healthy tissues, protein of different structural organization (tertiary and quaternary structure of proteins) can form hydrogen bonds with surrounding water molecules [14, 15], so a hydrating shell around the protein is formed, and the mobility of surrounding water molecules is reduced. In malignant tissues, proteins are less bound to neighboring water molecules, and these tissues contain more free water than bound water in the intercellular space.

The balance between free and bound water in soft tissues can be estimated by studying the kinetics of IOC hyperosmotic agent penetration in tissues to provide the most effective optical clearing of tissues. However, such a study in the THz frequency range was carried out only in work [65] for normal and pathological tissues. Tissue dehydration occurs due to free water migration from tissue and its substitution by a hyperosmotic immersion agent. The process is accompanied by reduced light scattering due to the refractive index matching of scatterers (e.g. collagen fibrils in soft tissues) and intercellular fluid, which now contains a larger amount of agent and a smaller amount of water; and the free space between the scatterers is reduced (tissue shrinkage). For the THz range, dehydration is accompanied by reduced absorption, due to a lower water content. The kinetics of the dehydration processes and agent substitution is determined by the characteristic diffusion time τ_f (or the diffusion coefficient D_f) [66]. Such kinetics characterizes properly the tissue under study and the agent applied to it. In particular, in other frequency ranges, it was shown that the diffusion rate can be used as a parameter for differentiation between healthy and pathological tissues (cancerous and diabetic) with high accuracy and sensitivity [67]. For diabetes, the glycation of proteins (e.g. albumin in the blood) leads to a reduction of molecule penetration into tissues [67]. The experimentally measured kinetics of agent penetration has demonstrated a higher content of free water in the cancerous mucous membrane of the colon than in a healthy one; a method for the quantitative determination of free and bound water in soft tissues has been proposed in Reference [68]. Therefore, THz radiation is promising for work with dehydration, as it is highly sensitive to the presence of free water.

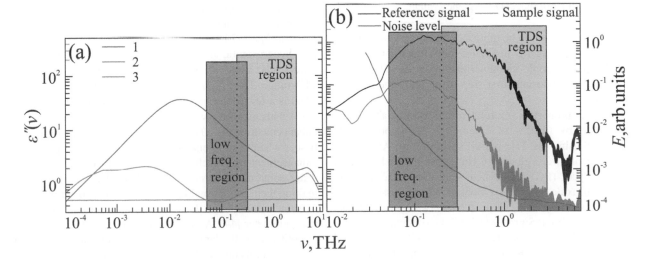

FIGURE 24.3 Imaginary part ε'' of the dielectric permittivity ε for the three components of the aqueous solution (a): free water (blue line) (1); bound water [61](red line) (2); solution spectrum (pink line) (3); dynamic range of the used THz spectrometer (b): spectra of reference (black line) and sample (red line) signals $E(\nu)$, while the latter is passed through a 500-μm-thick layer of water; blue line defines a noise level. The low-frequency part of the THz spectrum is highlighted in dark gray; the traditional range for THz–TDS is highlighted in gray [52].

Hyperosmotic agents for THz immersion optical clearing of tissue

The problem of selecting the optimal IOC hyperosmotic agents for applications in THz biophotonics still remains unaddressed, since quite a limited number of such agents were systematically studied in the THz range [36–42]. There are several parameters which can be considered when selecting the optimal IOC hyperosmotic agents. Among them are the following:

- The key parameter is a THz-wave absorption coefficient of an agent, as compared to that of liquid water. By substituting the main THz-wave absorber in a tissue (i.e. water), an agent reduces its THz-wave absorption. In this way, the ratio between the THz-wave absorption coefficient of a liquid water and of an IOC hyperosmotic agent can be considered as a first (raw) assumption for achievable enhancement of the IOC-induced enhancement of THz-wave penetration depth. It is worth noting the that THz-wave absorption coefficient seems to be a general parameter for IOC hyperosmotic agent selection, which should not strongly depend on the considered tissue type.

- Another important parameter is a diffusion coefficient of an agent in tissues. It defines the rate of diffusion process and is proportional to the total duration of the IOC procedure. The latter might be of crucial importance for *in vivo* THz probing. For the selected agent, the diffusion coefficient varies over different tissue types, as well as over healthy and pathological tissues. Thus, measurements of diffusion coefficients should be performed independently for different tissues.

- Finally, toxicity of IOC hyperosmotic agents must be very important, especially for the *in vivo* applications of IOC in the THz frequency range.

Despite this, it appears to be quite a daunting task to compare IOC hyperosmotic agents in such a multidimensional space, and their spectroscopic characterization in the THz range seems to be an important first step to be performed in this subsection.

THz optical properties – refractive index n and absorption coefficient α (by field) – of common hyperosmotic agents were studied using the THz pulsed spectroscopy. Among the analyzed agents were glucose ("Hungrana KFT," Hungary), propylene glycol (PG) ("Chemical Line Co. Ltd.," Russia), dimethyl sulfoxide (DMSO) ("SpektrChem," Russia), polyethylene glycol (PEG) 200 ("Nizhnekamskneftekhim," Russia), PEG 300 ("Sigma-Aldrich," Germany), PEG 400 ("Nizhnekamskneftekhim," Russia), PEG 600 ("Norchem," Russia), fructose ("PanReac," Spain), sucrose ("PanReac," Spain), glycerol ("SpektrChem", Russia), Dextran 40 ("AppliChem," Germany), and Dextran 70 ("AppliChem," Germany). For their experimental characterization, an original

transmission-mode THz–TDS with a vacuum THz-beam path, equipped with a cuvette for measurements of liquid samples, and software for the inverse spectroscopy problem solving were applied. A detailed description of this experimental setup and the signal processing routine can be found in Refs. [6, 41–44, 69, 70]. Along with the pure agents, their aqueous solutions with different concentrations were evaluated. Aqueous solutions of agents were prepared in a laboratory. For this purpose, the volume/volume method (for liquid agents) and the weight/volume method (for powders) were used, along with a deionized water and an analytical laboratory weight ("AND HR-250AZG"), featuring a weight limit of 210 mg and a resolution of 0.1 mg.

In Figure 24.4, the THz–TDS data for the selected agents and their aqueous solutions are presented [44]. Optical properties of glycerol and PEG 200 are shown for a pure agent and its 20%–80% aqueous solutions. In turn, since obtaining aqueous solution of glucose with concentration above 50% is a daunting task, their optical properties are shown for a limited range of concentrations. It should be noticed that THz optical properties of agents and their solution are compared with that of a deionized water, shown in gray.

From Figure 24.4, one can see that refractive index n and absorption coefficient α of all agents is less than that of a liquid water, which makes them potentially attractive for the tissue IOC in the THz range. Such a reduction in THz optical properties of aqueous solutions is due to partial substitution of highly polar water molecules, which feature high refractive index and THz-wave absorption in a liquid state, by less polar molecules of an agent, possessing much lower refractive index and absorption coefficient. Furthermore, some water molecules in a solution can be bound by hydrophilic parts of agent molecules, thus leading to an additional reduction in its THz optical constants [10, 28]. Nevertheless, the detailed analysis of the THz dielectric response of aqueous solution of IOC hyperosmotic agents, including hydration of agent molecules, formation of macromolecular complex, and their evolution in time, as well as development of related physical and mathematical model of a complex dielectric permittivity, deserve further comprehensive studies.

From Figure 24.4, one can see that, among the considered IOC hyperosmotic agents, glycerol and PEG 200 possess the lowest THz-wave absorption in the considered 0.20–0.25 THz range. Their THz-wave absorption (~60–70 cm^{-1} @ 1.0 THz) is a few times smaller than that of a liquid water (~210 cm^{-1}). This yields a raw estimation of the maximal achievable enhancement of the THz-wave penetration depth in hydrated tissues when these agents are applied. In fact, the penetration enhancement can reach ~3.0 times. In turn, glucose and its water solutions possess much higher THz-wave absorption, making them suboptimal for tissue IOC in THz biophotonics.

As mentioned above, the diffusion coefficient is also an important parameter [44]. Among the considered set of IOC hyperosmotic agents, the largest diffusion coefficient is observed for 100% glycerol and its 70% solution (see Figure 24.5). Thus, the use of glycerol might provide the fastest IOC of tissues. For rat brain tissues *ex vivo* with a 500 μm thickness, the total duration of the IOC procedure can be as small as 20–30 min. Thus,

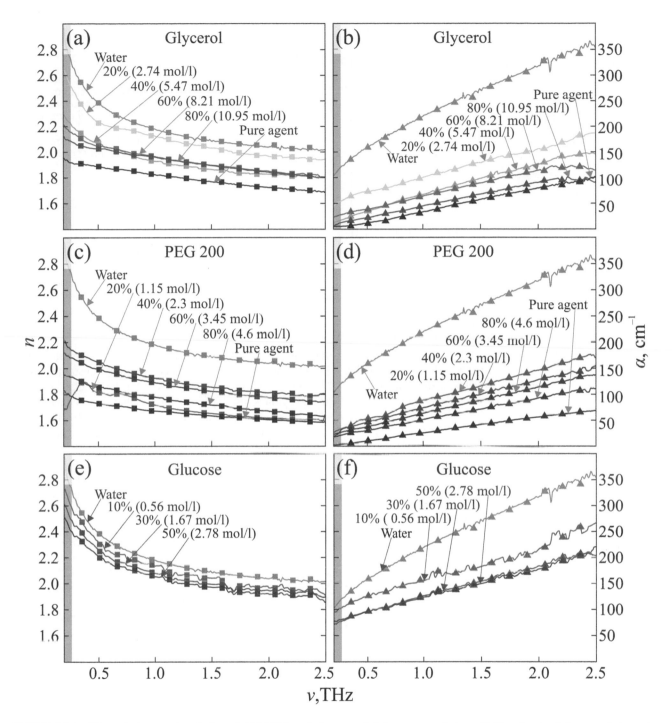

FIGURE 24.4 THz optical properties of glycerol, PEG 200, and glucose and their aqueous solutions: (a),(b) refractive index *n* and absorption coefficient α (by field) of glycerol and its aqueous solutions; (c),(d),(e),(f) equal data set for PEG 200 and glucose. Here, the concentration presented as percentages (20%–80% for glycerol and PEG 200; 10%–50% for glucose), while molar concentrations are defined in brackets, as an amount of hyperosmotic agent (by volume) solute in deionized water. A yellow-colored area at low frequencies defines the spectral range, where distortions of experimental data due to the THz-beam diffraction at the cuvette aperture might be expected.

given the properties of absorption and distribution of agents in biological tissue, glycerol appears to be the optimal agent for the tissue IOC in the THz range. Finally, in the sub-THz frequency range, IOC hyperosmotic agents possess lower THz-wave, and there is no rapid increase in refractive index.

For THz spectroscopy, 100% glycerol is of interest as a water-free agent. The spectrum of glycerol at low frequencies

and up to THz frequency range was studied, for example, in Reference [71]. The THz dielectric response of glycerol can be described as

$$\varepsilon_{\text{glycerol}}(\nu) = 2.6 + \frac{45}{\left(1 + i5200\nu\right)^{0.7}} + \frac{1.2}{1 + i0.6\nu}, \quad (24.2)$$

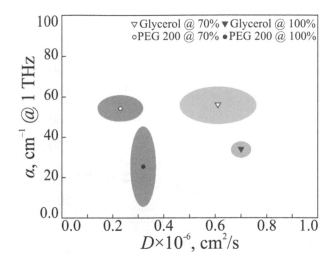

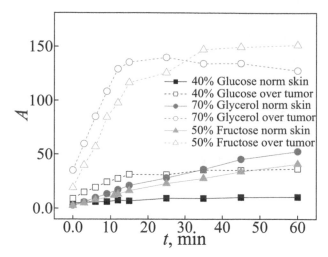

FIGURE 24.5   A comparison of the IOC hyperosmotic agents, both pure and their 70% aqueous solutions, using a 2D nomogram, which accounts for their THz-wave absorption coefficient α (by field) at 1.0 THz and binary diffusion coefficients D. Here, markers show the mean values, while the colored areas behind represent the measurement errors.

FIGURE 24.7   Time-dependent of the THz pulse amplitude for samples of healthy rat skin and a skin above the tumor during the IOC procedure with solutions of 40% glucose, 70% glycerol, and 50% fructose, 30% alcohol, and 20% saline [65].

where ν is in [THz], while that of glycerol aqueous solution can be described by Looyenga's effective medium model

$$\varepsilon_{\text{solution}}\left(\nu, C_f, C_b\right) = \left(\begin{array}{c} \left(1 - C_f - C_b\right)\varepsilon_{\text{glycerol}}(\nu)^{\frac{1}{3}} + C_f \varepsilon_{\text{water}}(\nu)^{\frac{1}{3}} \\ + C_b \varepsilon_{\text{bound}}(\nu)^{\frac{1}{3}} \end{array} \right)^3 ,$$

(24.3)

where C_f and C_b are volume fractions of free and bound water, considering that some of the water is bound

$$\varepsilon_{\text{bound}}(\nu) = 3 + \frac{45}{1 + i1785\nu} + \frac{0.885}{1 + i1.558\nu} + \frac{791}{1109} \frac{1}{39.5\nu^2 + i211\nu}.$$

(24.4)

In this way, it is possible to describe the THz spectrum of the agent while increasing the water concentration in a solution.

Also, using the effective medium theory, one can characterize a tissue response where glycerol partially substitutes water. The THz spectrum of "dry" tissue can be roughly approximated by the known spectrum of dry albumin. The data from skin clearing are presented in Figures 24.6 and 24.7. This is an *in vitro* measurement of the transmission of skin samples, on both sides of which an IOC agent is delivered [65].

Another *in vivo* method of skin diagnostics in the THz frequency range is to measure the concentration of water that the agent pulled from the tissue at a fixed time and fixed volume of the agent. The outlet velocity of water correlates with the condition of the tissue site (for example, pathology), and the high sensitivity of THz to the presence of water allows measuring water presence in agent, for example glycerol, in an amount of several percent by volume (Figure 24.8). The concentration of water in such modified solutions can be also estimated with a high precision using optical refractometry.

It is worth making a comparison of the clearing abilities of IOC hyperosmotic agents with formalin [72]. In this case, the

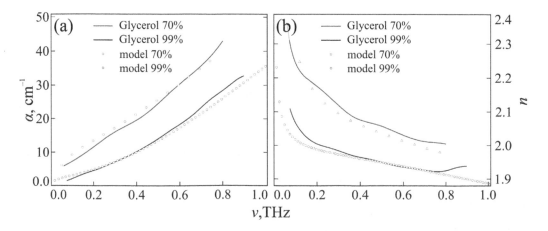

FIGURE 24.6   THz dielectric response of pure (99%) glycerol and 70% glycerol solution, as compared with its model (see Equations (24.2)–(24.4)) with 20% and 10% of free and bound water, respectively.

optical clearing takes 24–72 hours in muscle tissue (for example, for 0.2 THz, the refractive index decreases from 2.5 to 2.2). The traditionally used IOC hyperosmotic agents, such as glycerol, PEG, and PG [45,73], possess significant advantages over formalin; namely, clearing takes only 8–10 min and leads to a 2–5 times decrease in the THz-wave absorption by muscle tissues in the frequency range of 0.2–1.0 THz.

When optical clearing, the Fresnel reflection at the tissue–air interface significantly reduced. It is due to decrease in the refractive index from 3.1–2.6 to 2.1–1.9 at 0.2–1.0 THz, which is a side benefit of optical clearing. To obtain flat interface of a soft tissue, dielectric substrate is usually used as a second media, and depending on substrate material ($n_{polymer}<n_{water}$; $n_{crystal}>n_{water}$), contrast in THz range may be reduced or enhanced for dehydrated region. On the other hand, for THz spectroscopy realized in reflection mode, which is interesting for *in vivo* diagnostics, a refractive index matching of tissue layers may lead to a loss of useful signal. However in practice, it is not achievable, and refractive index mismatch remains sufficiently high.

The average diffusion coefficient D_f has been calculated using Equation (24.2), where

$$D_f = \frac{d^2}{\tau_f}, \qquad (24.5)$$

the characteristic diffusion time for the molecular flows (water/agent) τ_f was determined from the experimental kinetic curves for tissue slices of thickness d. At the characteristic diffusion time τ_f, the kinetic curves become saturated. It was found that optical clearing efficiency of glycerol and PEG is better than that of PG. In these measurements, the agent penetrated a 0.7-mm-thick layer of tissue from above, while the reflection was measured from the opposite side of a tissue layer; therefore, the thickness of the tissue also affects the characteristic time of "clearing."

When dehydrating agents of 100% glycerol, 40% aqueous glucose solution, PEG-600, or PG were applied to normal and cancerous skin of rats, PEG-600 provided the highest efficiency of the THz IOC of tissues [37–39]. Optical clearing of a cancerous tissue occurs faster than that of a healthy tissue; in addition, optical clearing is faster for *in vivo* studies (13–17 min) compared with *in vitro* experiments (20–80 min). However, in *in vivo* conditions, the subsequent rehydration process occurs due to the physiological response of living tissue to dehydration and osmotic stress.

It is known that the clearing time for healthy and pathologic regions of the tissue differs by 20%–40% (the pathologic tissue is cleared faster). Therefore, by choosing the time for imaging to be a half of the clearing saturation time for the healthy tissue, it can be possible to get considerable contrast improvement for selecting the region with pathology in the terahertz imaging.

In Reference [40], Smolyanskaya et al. showed that a deeper penetration of THz-waves in the mouse skin *in vitro* treated with glycerol can be associated with a decrease in the ratio of free and bound water in a sample. The data obtained experimentally by nuclear magnetic resonance (NMR) showed that the total amount of water in percent and the hydration

coefficient of the mouse skin sample decreased with glycerol treatment. The best results were achieved with glycerol concentrations of 70% and 100%, and the total amount of water decreased by 28.8% and 25.3%, respectively.

Yamaguchi et al. proposed the following equation for estimating the water content at the site of tissue measurement [74]:

$$n_{tissue} = \frac{A}{100} n_{water} + \frac{(100-A)}{100} n_{other}, \qquad (24.6)$$

where n_{tissue}, n_{water}, and n_{other} stand for the refractive index, water, and other biological components in a tissue, respectively; A is the percentage of water in the volume, where the THz-waves interact with a tissue.

The optical properties of freshly extracted normal and tumor tissue of the brain are similar to each other; the refractive index at 1.0 THz is about 2.1–2.2 [74]. The refractive index decreases sharply under the action of the dehydration process in both normal and tumor tissues; it reaches the values of 1.5 for normal tissue and 1.8 for tumor. Thus, the dehydration process results in a greater difference in optical properties of normal and tumor tissue in the THz range.

Figure 24.9 represents the THz images registered immediately after application of glycerol over stratum corneum *in vitro*, and after 3, 10, and 30 min after application [48]. Results show the change in normalized intensity of the THz signal with respect to the initial state of every sample. For glycerol, a change is clearly visible between Figures 24.9 (a) and (d). The images show how the area of interaction is increasing and becoming darker.

Other approaches to tissue dehydration

Introduction

There are other approaches for tissue dehydration, such as *in vivo* compression optical clearing, and *in vitro* approaches for

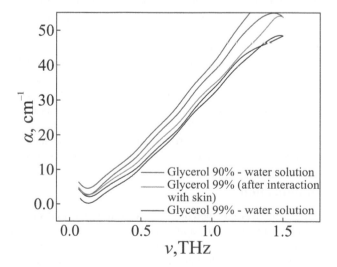

FIGURE 24.8 Absorption coefficient spectra for 99% (black) and 90% glycerol (blue), as well as glycerol applied to the skin of patients (red) [65].

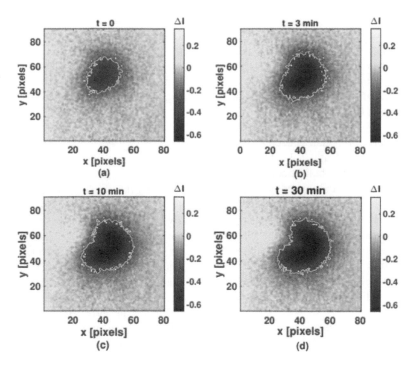

FIGURE 24.9 THz images registered (a) immediately after application of glycerol over stratum corneum in vitro, and after (b) 3, (c) 10, and (d) 30 min after application (Reprinted with permission from reference [48], © The Optical Society).

tissue samples such as freezing, heating, gelatin or paraffin embedding, or drying at low or high temperature (lyophilization) also belong to a promising group of optical clearing methods. However, these approaches have their drawbacks. For example, freezing can initiate water crystallization in the intra- and extracellular spaces and destroy the structure of biological structures inside the sample.

During lyophilization, tissue rapidly freezes in cold and low-pressure environments, resulting in sublimation of water (at solid-to-gas transition 90%–99% of water is removed), while the structure and molecular integrity are preserved. Lyophilized, frozen, or gelatin/paraffin embedded samples provide repeatability over a long period, and they are also convenient in experimental processing (cutting, mounting in holder, etc.) and at measuring in the setup [15, 75, 76]. Unfortunately, these methods are unusable *in vivo*.

Mechanical shrinkage (squeezing, compression) or stretching of soft tissues causes a temporary displacement of free water (intercellular fluid and blood in tissue) in force initiation site, and also provides a dense packing of tissue components (collagen fibers, etc.) [45, 46, 77]. This method is applicable *in vivo*, but to ensure a comfortable and repeatable mechanical action, it is necessary to develop special tissue mounts [78].

Lyophilization

Lyophilization is a method of gentle dehydration of a substance. This approach also found application in THz spectroscopy and imaging. For example, in [79], lyophilization of tissues was performed, which made it possible to remove a large amount of water while maintaining the freshness of the sample. In addition, lyophilized tissue samples are easy to handle, and their texture and size do not change over time, which allows consistent and stable measurements in the THz frequency range.

Comparison of lyophilized and fresh tissue shows that lyophilization can be one of the ways to overcome the problems of tissue dehydration while maintaining the cellular structure.

Tissue lyophilization has advantages over other dehydration approaches. First of all, a biological sample, dried and pressed into a "pellet" (also called a *tablet*), is conveniently mounted in the spectrometer holder; its thickness is uniform and convenient to measure, and therefore the optical constants are well calculated in a wide range of frequencies, THz radiation is weakly attenuated in the sample, and, certainly, the sample in this case retains its molecular and structural properties.

In Reference [80], the human blood plasma from the diabetic and control groups was lyophilized and pressed into a tablet, after which its refractive index was reconstructed as a function of the lateral coordinated using the holographic imaging. Different physiological reasons may influence the color of blood plasma pellets in visible, for example, hemolysis or different lipid concentrations (see Figure 24.10); increased triglyceride concentration may cause turbidity, which also can be visible. Analyzing the frequency-dependent character of

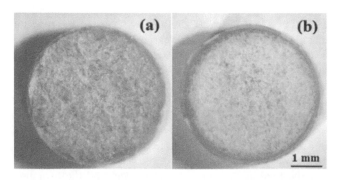

FIGURE 24.10 Photographs of nondiabetic (a) and diabetic (b) pellets.

refractive index in the range 0.2–1.2 THz, sufficient differences between non-diabetic and diabetic plasma pellets were demonstrated. Both diabetic and nondiabetic plasma pellets could contain glycated and nonglycated proteins in different concentrations. These factors could lead to inhomogeneity of pellets.

In Reference [33], Png et al. reported that samples of lyophilized tissue, such as the kidney, colon, and stomach of laboratory animals (rats), retain their thickness and volume after lyophilization. If the tissue was dried with nitrogen, then protein denaturation and tissue necrosis occur. Therefore, drying should be done by lyophilization at moderately low temperatures and low pressure for 1–2 days.

Reference [79] is devoted to the study of lyophilized solutions of bovine serum albumin (BSA) and lyophilized solutions of monoclonal antibodies (mAb) mixed with excipients: polysorbate-80, sucrose, trehalose, and arginine. Experiments were conducted for a variable temperature at indicating the transition temperature to the glassy state using THz–TDS. The characteristics of the structural dynamic properties of protein solutions with excipients were studied. THz–TDS measurements of the solutions demonstrated two different processes of glass transition. The transition with a lower temperature (217 K, 262 K, 219 K, 205 K) is associated with the beginning of the local protein movement due to secondary relaxation, while the transition with a higher temperature (298 K, 318 K, 336 K, 339 K) characterizes the beginning of α-relaxation. The absorption in the terahertz frequency range of some solutions at high temperatures (298 K, 318 K, 336 K, 339 K) does not increase. This can be related to an increase in protein aggregation, which prevents a further increase in the mobility of the protein matrix with the excipient at high temperatures.

THz optical properties of frozen tissues

One of the natural and most common factors that affect the state of water is temperature. Tissue freezing allows the effect of optical clearing to be achieved by decreasing water absorption just because "slow relaxation" in Equation 24.1 decreases at low temperatures. In Reference [29], the potential of this approach was revealed in practice: pork tissue was frozen to a temperature below -33°C. As a result, the authors achieved good differentiation of striated muscle and adipose tissue in the THz range due to the difference in refractive index and absorption coefficient.

This technique is also applicable to study pathology [30]. It was described that in the case of malignant melanoma of the oral cavity using THz imaging at room temperature (20°C) and a temperature below zero (–20°C), the THz images correlated well with histological images. THz spectroscopic studies of frozen tissue at –20°C showed better contrast due to the low content of liquid water.

THz optical properties of dehydrated tissues

It is worth noting that tissue dehydration can also occur naturally without the use of hyperosmotic agents. Natural dehydration processes in three types of biological tissues, including cattle, sheep, and pig, were studied using the THz digital holographic imaging system [32]. By comparing THz absorption spectra of these specimens and the temporal evolution of their relative variations, dehydration features of adipose and muscle tissues are precisely distinguished. The adipose tissue presents an exponential decay of THz absorbance with increasing drying time, and muscle tissue shows a more complex process due to the change in morphology. These data make it possible to characterize the fundamental features of natural tissue dehydration as an independent method of optical clearing, as well as an integral part of most other methods. To optimize the quality of measured THz images, the reconstruction of optical constants were also carried out using THz digital holographic imaging system and an image reconstruction algorithm based on the angular spectrum theory. In Reference [31], an experimental model was proposed for assessing natural dehydration, taking into account the biological and molecular features of various tissues.

THz optical properties of paraffin-embedded tissues

The need for dehydration appears not only with the use of THz technology, but also in the process of routine histological examination. To preserve morphological properties and manufacture high-quality histological slides, biological tissues go through several stages of dehydration using alcohol of various concentrations and subsequent fixation in paraffin. This method of dehydration finds its application in the practice of THz research. In particular, using THz–TDS, dielectric properties of paraffin-embedded brain gliomas and normal brain tissues were measured [15]. As a result, spectral differences between gliomas and normal brain tissue were obtained. Compared to normal brain tissue, gliomas of the brain fixed in paraffin have higher refractive index and absorption coefficient.

In reference [72], Emma MacPherson et al. reported that the refractive index and absorption coefficient of muscle and adipose tissues were reduced during the substitution of intercellular fluid by formalin. Fat cells in adipose tissue contain much less water, about 10%–30%; therefore, THz optical properties of adipose tissue are not as dramatically reduced as that of muscle tissue.

In addition, THz pulse imaging of cutaneous malignant melanoma dehydrated with ethanol and fixed in paraffin was performed [81]. First, tissue images were obtained based on information about the time-domain amplitude of THz electric field. Then, the areas of normal and cancerous tissues were determined using the mathematical morphology of multiscale, multi-azimuthal, and multistructural elements. The physical meaning of the image was analyzed by calculating the refractive index and absorption coefficient of cutaneous malignant melanoma in different areas. The refractive index of both normal and cancerous tissues showed an abnormal dispersion. The refractive index of cancerous tissues ranges from 0.2 to 0.7 THz, while in healthy and adipose tissues, no pronounced changes were observed. The absorption of cancerous tissues was higher, with a maximum of 0.37 THz. Thus, tumor and healthy skin tissue can be differentiated based on the difference in their THz optical characteristics.

In reference [82], freshly extracted and paraffin-embedded normal and tumor tissues of the brain were studied.

If the water in the tissue sample is replaced by paraffin, then n_{water} in the equation (Figure 24.6) will be replaced by a refractive index of paraffin $n_{paraffin}$. In this paper, the percentage of water in the tissue was estimated using the obtained $n_{paraffin}$ value and by subtracting the $n_{tissue-paraffin}$ from the n_{tissue}. The estimated percentage of water, A, is approximately 5% higher for tumor tissue compared to healthy tissue.

THz optical properties of mechanically compressed tissues

In some tissues, the redistribution of water, providing optical clearing, can be achieved by mechanical compression of tissues. This technique is most relevant for the skin: experimental models have been developed and tested that evaluate the effect of compression on the dielectric properties in the THz range. Study has shown that for some tissues, in particular, the skin, a similar approach can be quite effective [34, 35].

From data in reference [34], it follows that a compression of tissues can be used to change their optical properties at THz frequency range. However, a compression-driven reduction of THz-wave absorption in tissues seem to be rather small, while this tissue-clearing modality still remains quite novel and deserves further comprehensive studies in order to make objective conclusions (see Figure 24.11).

Discussion

IOC of tissues can serve as a tool for increasing the informative features of THz spectra and images of healthy and pathological tissues in diagnostic applications. In particular, one of the socially important diseases for which THz diagnostics can be applied is diabetes mellitus. The fundamental basis for the use of THz technologies in the diagnosis of diabetes is the peculiarities of water distribution in the tissues; there is an increase in the water content in individual compartments of the body as a result of both the hyperosmotic effect of glucose and the accumulation of glycolysis products [83].

It was shown that THz time-domain spectroscopy can be used to noninvasively assess blood glucose levels, while THz technologies are not inferior in terms of the analysis accuracy [21, 22]. However, it is important not only to assess blood glucose level as a key marker of this pathology, but also to study the secondary lesions of tissues and organs and their biochemical restructuring. It was shown that some metabolic effects of these secondary pathological changes can be detected using THz imaging. For example, metabolic and functional changes in human foot tissues caused by biochemical molecular rearrangement as a result of diabetes mellitus are clearly observed at various stages of their progression using THz imaging [21–23]. IOC techniques can significantly improve the efficiency of THz diagnostics in this area; for example, due to biochemical remodeling, diabetically modified tissues have a lower degree

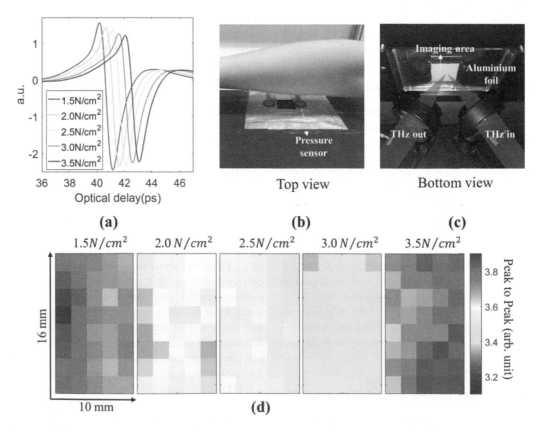

FIGURE 24.11 (a) THz pulse response of skin under different pressures, shifted for clear display. Photographs of the experimental setup: (b) top view, just before placing the arm on the quartz window and (c) bottom view during measurement (skin is contacting the quartz window) (d) THz images of skin under different pressures. (Reprinted with permission from reference [34], © The Optical Society).

of optical clearing under the action of optical clearing agents, as compared to healthy tissues [10, 11].

One of the most intensively studied areas of targeted application of THz technology is oncology. Since the penetrating ability of THz radiation is relatively low, diagnostic approaches based on its application can be most effectively used to study superficial tissues directly during surgical access [11]. A considerable amount of experimental data has been accumulated regarding the optical properties of healthy tissues and pathologies of the skin in the THz frequency range, including the primary tumors of the skin [11, 12]. In such applications, IOC techniques might allow for visualizing subcutaneously located tumor nodes using THz spectroscopy and imaging [36].

As part of the surgical intervention, direct access can be provided to almost any tissue. This opens great opportunities for the development of intraoperative THz diagnostics [10, 11]. For example, it has been shown that it is possible to differentiate between intact (healthy) tissues and tumor tissue with different localization and nosology using THz technologies, including neoplasms of the brain [14–18], breast [84–86], and colon [87–90]. In such applications, IOC procedures can be adjusted to increase the effectiveness and information content of intraoperative THz diagnostics. In addition, the discussed intraoperative diagnostic approach can be extended to non-tumor pathology. For example, THz spectroscopy has been shown to differentiate cirrhotic altered tissue from intact liver tissue [91]. It is curious that this difference is not solely due to the difference in the water content in the studied tissues. Apparently, other molecular factors make a significant contribution to the formation of differences in the properties of normal and pathologically altered tissues in the THz range, and this pattern can be characteristic not only for liver cirrhosis, but also for many other pathologies. The solvation features of biological molecules, which arise as a result of various biochemical rearrangements, probably play their role in determining the differing properties of normal and pathological cells and tissues [92].

Prospective measurements be aimed at enhancement the biophysical and mathematical methodology binding interactions between normal/pathological tissues and immersion agents to improve THz biomedical imaging. A precise and predicted no-effect concentration of immersion agent must be defined for *in vivo* application. Many soft and hard tissues and body fluids (blood etc.) remain unstudied in optical clearing for more effective THz diagnostic and therapeutic applications. The proportion by volume of tissue components, as well as changes in the real and imaginary parts of the THz dielectric permittivity and Fresnel reflection at the interfaces inside the tissue under the influence of immersion agents, also need to be clarified.

Conclusions

In this chapter, recent developments in the area of biological tissue IOC in the THz ranges were discussed. Increasing attention has been paid to this modality of tissue optical clearing, since it holds strong potential in different branches of THz biophotonics. Indeed, IOC opens novel opportunities for

unmasking non-water-related molecular and structural features of tissues at THz frequencies, as well as for improving the depth of THz-wave penetration in tissues. Along with tissue IOC, other approaches to reduce the impact of liquid water on the THz dielectric response of tissues, such as dehydration, lyophilization, paraffin-embedding, freezing, and compression of tissues, were also overviewed in this work. Strong demand for optical clearing of tissues at THz frequencies would obviously push further developments in the considered area of THz biophotonics, thus bringing THz technologies much closer to clinical practice.

Acknowledgment

This chapter was supported by **the** Russian Foundation for Basic Research (RFBR), Project # 17-00-00275 K (17-00-00186, 17-00-00270, 17-00-00272). Development and assembling of the experimental setup and cuvette for the THz dielectric spectroscopy of liquids, as well as processing the data of THz measurements by G.R. Musina, G.A. Komandin, and D.K. Tuchina, were supported by the Russian Science Foundation (RSF), Project # 18-12-00328. A comparison of IOC agents using original nomogram-based technique by I.N. Dolganova was supported by the Russian Federation President Grant for Young Russian Scientists # 2541.2019.8.

REFERENCES

1. S. Atakaramians, S. Afshar, T.M. Monro, and D. Abbot, "Terahertz dielectric waveguides," *Adv. Opt. Photonics* **5**(2), 169–215 (2013).

2. G.A. Komandin, S.V. Chuchupal, S.P. Lebedev et al., "BWO generators for terahertz dielectric measurements," *IEEE Trans. Terahertz Sci. Technol.* **3**(4), 440–444 (2013).

3. S. Lepeshov, A. Gorodetsky, A. Krasnok, E. Rafailov, and P. Belov, "Enhancement of terahertz photoconductive antenna operation by optical nanoantennas," *Laser Photonics Rev.* **11**(1), 1600199 (2017).

4. H. Guerboukha, K. Nallappan, and M. Skorobogatiy, "Toward real-time terahertz imaging," *Adv. Opt. Photon.* **10**(4), 843–938 (2018).

5. A.E. Yachmenev, D.V. Lavrukhin, I.A. Glinskiy et al., "Metallic and dielectric metasurfaces in photoconductive terahertz devices: A review," *Opt. Eng.* **59**(6), 061608 (2019).

6. G.A. Komandin, V.B. Anzin, V.E. Ulitko et al., "Optical cryostat with sample rotating unit for polarization-sensitive terahertz and infrared spectroscopy," *Opt. Eng.* **59**(6), 061603 (2019).

7. M. Ryzhii, T. Otsuji, V. Ryzhii et al., "Concepts of infrared and terahertz photodetectors based on vertical graphene van der Waals and HgTe-CdHgTe heterostructures," *Opto-Electronics Rev.* **27**(2), 219–223 (2019).

8. A.E. Yachmenev, S.S. Pushkarev, R.R. Reznik, R.A. Khabibullin, and D.S. Ponomarev, "Arsenides- and related III–V materials-based multilayered structures for terahertz applications: Various designs and growth technology," *Prog. Cryst. Growth Charact. Mater.* **66**(2), 100485 (2020).

9. M.S. Islam, C. Cordeiro, M. Franco, J. Sultana, A. Cruz, and D. Abbott, "Terahertz optical fibers," *Opt. Express* **28**(11), 16089–16117 (2020).

10. O.A. Smolyanskaya, N.V. Chernomyrdin, A.A. Konovko, et al., "Terahertz biophotonics as a tool for studies of dielectric and spectral properties of biological tissues and liquids," *Prog. Quantum Electron.* **62**, 1–77 (2018).

11. K.I. Zaytsev, I.N. Dolganova, N.V. Chernomyrdin et al., "The progress and perspectives of terahertz technology for diagnosis of neoplasms: A review," *J. Opt.* **22**(1), 013001 (2019).

12. K.I. Zaytsev, I.N. Dolganova, N.V. Chernomyrdin et al., "In vivo terahertz spectroscopy of pigmentary skin nevi: Pilot study of non-invasive early diagnosis of dysplasia," *Appl. Phys. Lett.* **106**(5), 53702 (2015).

13. D. Li, Z. Yang, A. Fu et al., "Detecting melanoma with a terahertz spectroscopy imaging technique," *Spectrochim. Acta Part A Mol. Biomol. Spectrosc.* **234**, 118229 (2020).

14. S.J. Oh, S.-H. Kim, Y.B. Ji et al., "Study of freshly excised brain tissues using terahertz imaging," *Biomed. Opt. Express* **5**(8), 2837–2842 (2014).

15. K. Meng, T.-N. Chen, L.-G. Zhu et al., "Terahertz pulsed spectroscopy of paraffin-embedded brain glioma," *J. Biomed. Opt.* **19**(7), 077001 (2014).

16. Y.B. Ji, S.J. Oh, S.G. Kang et al., "Terahertz reflectometry imaging for low and high grade gliomas," *Sci. Rep.* **6**, 36040 (2016).

17. A.A. Gavdush, N.V. Chernomyrdin, K.M. Malakhov et al., "Terahertz spectroscopy of gelatin-embedded human brain gliomas of different grades: A road toward intraoperative THz diagnosis," *J. Biomed. Opt.* **24**(2), 27001 (2019).

18. L. Wu, D. Xu, Y. Wang et al., "Study of in vivo brain glioma in a mouse model using continuous-wave terahertz reflection imaging," *Biomed. Opt. Express* **10**(8), 3953–3962 (2019).

19. S. Fan, B. Ung, E. Parrott, V. Wallace, and E. Pickwell-MacPherson, "In vivo terahertz reflection imaging of human scars during and after the healing process," *J. Biophoton.* **10**(9), 1143–1151 (2017).

20. J. Wang, Q. Sun, R. Stantchev, T.-W. Chiu, A. Ahuja, and E. Pickwell-MacPherson, "In vivo terahertz imaging to evaluate scar treatment strategies: Silicone gel sheeting," *Biomed. Opt. Express* **10**(7), 3584–3590 (2019).

21. O.A. Smolyanskaya, E.N. Lazareva, S.S. Nalegaev et al., "Multimodal optical diagnostics of glycated biological tissues," *Biochemistry* **84**(1), 124–143 (2019).

22. O. Cherkasova, M. Nazarov, and A. Shkurinov, "Noninvasive blood glucose monitoring in the terahertz frequency range," *Opt. Quantum Electron.* **48**(3), 217 (2016).

23. G.G. Hernandez-Cardoso, S.C. Rojas-Landeros, M. Alfaro-Gomez et al., "Terahertz imaging for early screening of diabetic foot syndrome: A proof of concept," *Sci. Rep.* **7**, 42124 (2017).

24. A.V. Shchepetilnikov A.M. Zarezin, V.M. Muravev, P.A. Gusikhin, and I.V. Kukushkin, "Quantitative analysis of water content and distribution in plants using terahertz imaging," *Opt. Eng.* **59**(6), 61617 (2020).

25. N. Bajwa, J. Au, R. Jarrahy et al., "Non-invasive terahertz imaging of tissue water content for flap viability assessment," *Biomed. Opt. Express* **8**(1), 460–474 (2017).

26. H. Zhao, Y. Wang, L. Chen et al., "High-sensitivity terahertz imaging of traumatic brain injury in a rat model," *J. Biomed. Opt.* **23**(3), 36015 (2018).

27. Y. Cao, P. Huang, J. Chen, W. Ge, D. Hou, and G. Zhang, "Qualitative and quantitative detection of liver injury with terahertz time-domain spectroscopy," *Biomed. Opt. Express* **11**(2), 982–993 (2020).

28. O.P. Cherkasova, M.M. Nazarov, M. Konnikova, and A.P. Shkurinov, "THz spectroscopy of bound water in glucose: Direct measurements from crystalline to dissolved state," *J. Infrared Millimeter Terahertz Waves* **41**, 1057–1068 (2020).

29. H. Hoshina, A. Hayashi, N. Miyoshi, F. Miyamaru, and C. Otani, "Terahertz pulsed imaging of frozen biological tissues," *Appl. Phys. Lett.* **94**(12), 123901 (2009).

30. Y.C. Sim, K.M. Ahn, J.Y. Park, C.-S. Park, and J.-H. Son, "Temperature-dependent terahertz imaging of excised oral malignant melanoma," *IEEE J. Biomed. Heal. Inf.* **17**(4), 779–784 (2013).

31. Y. He, K. Liu, C. Au, Q. Sun, E. Parrott, and E. PickWell-MacPherson, "Determination of terahertz permittivity of dehydrated biological samples," *Phys. Med. Biol.* **62**(23), 8882–8893 (2017).

32. L. Guo, X. Wang, P. Han et al., "Observation of dehydration dynamics in biological tissues with terahertz digital holography," *Appl. Opt.* **56**(13), F173–F178 (2017).

33. G.M. Png, J.W. Choi, B.W.H. Ng, S.P. Mickan, D. Abbott, and X.C. Zhang, "The impact of hydration changes in fresh bio-tissue on THz spectroscopic measurements.," *Phys. Med. Biol.* **53**(13), 3501–3517 (2008).

34. J. Wang, R. Stantchev, Q. Sun, T.-W. Chiu, A.T. Ahuja, and E. MacPherson, "THz in vivo measurements: The effects of pressure on skin reflectivity," *Biomed. Opt. Express* **9**(12), 6467–6476 (2018).

35. Q. Sun, E. Parrott, Y. He, and E. Pickwell-MacPherson, "In vivo THz imaging of human skin: accounting for occlusion effects," *J. Biophotonics* **11**(2), e201700111 (2018).

36. S.J. Oh, S.-H. Kim, K. Jeong et al., "Measurement depth enhancement in terahertz imaging of biological tissues," *Opt. Express* **21**(18), 21299– 21305 (2013).

37. A.S. Kolesnikov, E.A. Kolesnikova, K.N. Kolesnikova et al., "THz monitoring of the dehydration of biological tissues affected by hyperosmotic agents," *Phys. Wave Phenom.* **22**(3), 169–176 (2014).

38. A.S. Kolesnikov, E.A. Kolesnikova, D.K. Tuchina et al., "In-vitro terahertz spectroscopy of rat skin under the action of dehydrating agents," *Proc. SPIE* **9031**, 90310D (2014).

39. A.S. Kolesnikov, E.A. Kolesnikova, A.P. Popov, M.M. Nazarov, A.P. Shkurinov, and V.V. Tuchin, "In vitro terahertz monitoring of muscle tissue dehydration under the action of hyperosmotic agents," *Quantum Electron.* **44**(7), 633–640 (2014).

40. O.A. Smolyanskaya, I.J. Schelkanova, M.S. Kulya et al., "Glycerol dehydration of native and diabetic animal tissues studied by THz-TDS and NMR methods," *Biomed. Opt. Express* **9**(3), 1198–1215 (2018).

41. G.R. Musina, I.N. Dolganova, K.M. Malakhov et al., "Terahertz spectroscopy of immersion optical clearing agents: DMSO, PG, EG, PEG," *Proc SPIE* **10800**, 108000F (2018).

42. G.R. Musina, A.A. Gavdush, D.K. Tuchina et al., "A comparison of terahertz optical constants and diffusion coefficients of tissue immersion optical clearing agents," *Proc. SPIE* **11065**, 110651Z (2019).

43. G.R. Musina, A.A. Gavdush, N.V. Chernomyrdin et al., "Optical properties of hyperosmotic agents for immersion clearing of tissues in the terahertz range," *Opt. Spectrosc.* **128**(7), 1026–1035 (2020).

44. G.R. Musina, I.N. Dolganova, N.V. Chernomyrdin, A.A. Gavdush, V.E. Ulitko, O.P. Cherkasova, D.K. Tuchina, P.V. Nikitin, A.I. Alekseeva, N. Bal, G.A. Komandin, V. Kurlov, V.V. Tuchin, and K.I. Zaytsev, "Optimal hyperosmotic agents for tissue immersion optical clearing in terahertz biophotonics," *Journal of Biophotonics* 13(12), e202000297 (2020), DOI: 10.1002/jbio.202000297.

45. V.V. Tuchin, *Tissue Optics: Light Scattering Methods and Instruments for Medical Diagnostics*, 3rd ed. Bellingham, SPIE Press, 2015, 988 p.

46. E.A. Genina, A.N. Bashkatov, Yu.P. Sinichkin, I.Yu. Yanina, and V.V. Tuchin, "Optical clearing of biological tissues: Prospects of application in medical diagnostics and phototherapy," *J. Biomed. Photon. Eng.* **1**(1), 22–58 (2015).

47. A.N. Bashkatov, K.V. Berezin, K.N. Dvoretskiy et al., "Measurement of tissue optical properties in the context of tissue optical clearing," *J. Biomed. Opt.* **23**(9), 91416 (2018).

48. D.I. Ramos-Soto, A.K. Singh, E. Saucedo-Casas, E. Castro-Camus, and M. Alfaro-Gomez, "Visualization of moisturizer effects in stratum corneum in vitro using THz spectroscopic imaging," *Appl. Opt.* **58**(24), 6581–6585 (2019).

49. A. Charkhesht, C.K. Regmi, K.R. Mitchell-Koch, S. Cheng, and N.Q. Vinh, "High-precision megahertz-to-terahertz dielectric spectroscopy of protein collective motions and hydration dynamics," *J. Phys. Chem. B* **122**(24), 6341–6350 (2018).

50. M.M. Nazarov, O.P. Cherkasova, and A.P. Shkurinov, "Study of the dielectric function of aqueous solutions of glucose and albumin by THz time-domain spectroscopy," *Quantum Electron.* **46**(6), 488–495 (2016).

51. B. Born and M. Havenith, "Terahertz dance of proteins and sugars with water," *J. Infrared Millimeter Terahertz Waves* **30**(12), 1245–1254 (2009).

52. M.M. Nazarov, O.P. Cherkasova, E.N. Lazareva et al., "A complex study of the peculiarities of blood serum absorption of rats with experimental liver cancer," *Opt. Spectrosc.* **126**(6), 721–729 (2019).

53. U. Heugen, G. Schwaab, E. Brundermann et al., "Solute-induced retardation of water dynamics probed directly by terahertz spectroscopy," *Proc. Natl. Acad. Sci.* **103**(33), 12301–12306 (2006).

54. L. Comez, M. Paolantoni, P. Sassi, S. Corezzi, A. Morresi, and D. Fioretto, "Molecular properties of aqueous solutions: A focus on the collective dynamics of hydration water," *Soft Matter* **12**(25), 5501–5514 (2016).

55. M. Nagai, H. Yada, T. Arikawa, and K. Tanaka, "Terahertz time-domain attenuated total reflection spectroscopy in water and biological solution," *Int. J. Infrared Millimeter Waves* **27**(4), 505–515 (2006).

56. H. Yada, M. Nagai, and K. Tanaka, "Origin of the fast relaxation component of water and heavy water revealed by terahertz time-domain attenuated total reflection spectroscopy," *Chem. Phys. Lett.* **464**(4–6), 166–170 (2008).

57. N. Penkov, V. Yashin, Jr. E. Fesenko, A. Manokhin, and E. Fesenko, "A study of the effect of a protein on the structure of water in solution using terahertz time-domain spectroscopy," *Appl. Spectrosc.* **72**(2), 257–267 (2018).

58. P.U. Jepsen and H. Merbold, "Terahertz reflection spectroscopy of aqueous NaCl and LiCl solutions," *J. Infrared Millimeter Terahertz Waves* **31**(4), 430–440 (2010).

59. S. Sarkar, D. Saha, S. Banerjee, A. Mukherjee, and P. Mandal, "Broadband terahertz dielectric spectroscopy of alcohols," *Chem. Phys. Lett.* **678**, 65–71 (2017).

60. K. Shiraga, A. Adachi, M. Nakamura, T. Tajima, K. Ajito, and Y. Ogawa, "Characterization of the hydrogen-bond network of water around sucrose and trehalose: microwave and terahertz spectroscopic study," *J. Chem. Phys.* **146**(10), 105102 (2017).

61. T. Fukasawa, T. Sato, J. Watanabe, Y. Hama, W. Kunz, and R. Buchner, "Relation between dielectric and low-frequency Raman spectra of hydrogen-bond liquids," *Phys. Rev. Lett.* **95**(19), 197802 (2005).

62. S. Yamaguchi, Y. Fukushi, O. Kubota, T. Itsuji, T. Ouchi, and S. Yamamoto, "Brain tumor imaging of rat fresh tissue using terahertz spectroscopy," *Sci. Rep.* **6**, 30124 (2016).

63. M. Borovkova, M. Khodzitsky, P. Demchenko, O. Cherkasova, A. Popov, and I. Meglinski, "Terahertz time-domain spectroscopy for non-invasive assessment of water content in biological samples," *Biomed. Opt. Express* **9**(5), 2266–2276 (2018).

64. N. V Chernomyrdin, A.A. Gavdush, S.-I. Beshplav et al., "In vitro terahertz spectroscopy of gelatin-embedded human brain tumors: A pilot study," *Proc. SPIE* **10716**, 107160S (2018).

65. E.N. Lazareva, M.M. Nazarov, A.B. Bucharskaya, V.V. Tuchin, and A.P. Shkurinov, "THz properties of rat skin and extracts at exposure to hyperosmotic solutions (Conference Presentation)," *Proc. SPIE* **11363**, 1136328 (2020).

66. L.M. Zurk, B. Orlowski, D.P. Winebrenner, E.I. Thorsos, M.R. Leahy-Hoppa, and I.M. Hayden, "Terahertz scattering from granular material," *J. Opt. Soc. Am. B* 24(9), 2238–2243 (2007).

67. D.K. Tuchina, R. Shi, A.N. Bashkatov et al., "Ex vivo optical measurements of glucose diffusion kinetics in native and diabetic mouse skin," *J. Biophoton.* **8**(4), 332–346 (2015).

68. L. Oliveira, M.I. Carvalho, E. Nogueira, and V.V. Tuchin, "Optical clearing mechanisms characterization in muscle," *J. Innov. Opt. Health Sci.* **9**(5), 1650035 (2016).

69. D.V. Lavrukhin, A.E. Yachmenev, A.Yu. Pavlov et al., "Shaping the spectrum of terahertz photoconductive antenna by frequency-dependent impedance modulation," *Semicond. Sci. Technol.* **34**(3), 34005 (2019).

70. A.A. Gavdush, V.E. Ulitko, G.R. Musina et al., "A method for reconstruction of terahertz dielectric response of thin liquid samples," *Proc. SPIE* **11060**, 110601G (2019).

71. U. Schneider, P. Lunkenheimer, R. Brand, and A. Loidl, "Dielectric and far-infrared spectroscopy of glycerol," *J. Non-Cryst. Solids* **235**, 173–179. (1998).

72. Y. Sun, B.M. Fischer, and E. Pickwell-MacPherson, "Effects of formalin fixing on the THz properties of biological tissues," *J. Biomed. Opt.* **14**(6), 6401–6417 (2009).

73. V.V. Tuchin, *Optical Clearing of Tissues and Blood*, SPIE Press, Bellingham, WA (2006).

74. S. Yamaguchi, Y. Fukushi, O. Itsuji, T. Ouchi, and S. Yamamoto, "Origin and quantification of differences between normal and tumor tissues observed by terahertz spectroscopy," *Phys. Med. Biol.* **61**, 6808–6820 (2016).

75. S. Fan, B. Ung, E. Parrott, and E. Pickwell-MacPherson, "Gelatin embedding: A novel way to preserve biological samples for terahertz imaging and spectroscopy," *Phys. Med. Biol.* **60**(7), 2703–2713 (2015).

76. Y. He, B.S.-Y. Ung, E. Parrott, A.T. Ahuja, and E. Pickwell-MacPherson, "Freeze-thaw hysteresis effects in terahertz imaging of biomedical tissues," *Biomed. Opt. Express* **7**(11), 4711–4717 (2016).

77. L. Oliveira and V.V. Tuchin, *The Optical Clearing Method: A New Tool for Clinical Practice and Biomedical Engineering*, Springer, Basel, 2019, 177 p.

78. C.G. Rylander, T.E. Milner, S.A. Baranov, and J.S. Nelson, "Mechanical tissue optical clearing devices: Enhancement of light penetration in ex vivo porcine skin and adipose tissue," *Lasers Surg. Med.* **40**(10), 688–694 (2008).

79. T.A. Shmool, P.J. Woodhams, M. Leutzsch et al., "Observation of high-temperature macromolecular confinement in lyophilised protein formulations using terahertz spectroscopy," *Int. J. Pharm. X* **1**, 100022 (2019).

80. M. Kulya, E. Odlyanitskiy, Q. Cassar et al., "Fast terahertz spectroscopic holographic assessment of optical properties of diabetic blood plasma," *Int. J. Infrared Millimeter Waves* **41**(9), 1041–1056 (2020).

81. J. Li, Y. Xie, and P. Sun, "Edge detection on terahertz pulse imaging of dehydrated cutaneous malignant melanoma embedded in paraffin," *Front. Optoelectron.* **12**(3), 317–323 (2019).

82. J.H. Son, "Terahertz electromagnetic interactions with biological matter and their applications," *J. Appl. Phys.* **105**(10), 102033 (2009).

83. A. Caduff, P. Ben Ishai, and Y. Feldman, "Continuous noninvasive glucose monitoring: Water as a relevant marker of glucose uptake in vivo," *Biophys. Rev.* **11**(6), 1017–1035 (2019).

84. A.J. Fitzgerald, V.P. Wallace, M. Jimenez-Linan et al., "Terahertz pulsed imaging of human breast tumors," *Radiology* **239**(2), 533–540 (2006).

85. P.C. Ashworth, E. Pickwell-MacPherson, E. Provenzano et al., "Terahertz pulsed spectroscopy of freshly excised human breast cancer," *Opt. Express* **17**(15), 12444–12454 (2009).

86. B.C.Q. Truong, H.D. Tuan, A.J. Fitzgerald, V.P. Wallace, and H.T. Nguyen, "A dielectric model of human breast tissue in terahertz regime," *IEEE Trans. Biomed. Eng.* **62**(2), 699–707 (2014).

87. C.B. Reid, A. Fitzgerald, G. Reese et al., "Terahertz pulsed imaging of freshly excised human colonic tissues," *Phys. Med. Biol.* **56**(14), 4333–4353 (2011).

88. P. Doradla, K. Alavi, C.S. Joseph, and R.H. Giles, "Detection of colon cancer by continuous-wave terahertz polarization imaging technique," *J. Biomed. Opt.* **18**(9), 090504 (2013).

89. D. Hou, X. Li, J. Cai et al., "Terahertz spectroscopic investigation of human gastric normal and tumor tissues," *Phys. Med. Biol.* **59**(18), 5423–5440 (2014).

90. P. Doradla, C. Joseph, and R.H. Giles, "Terahertz endoscopic imaging for colorectal cancer detection: Current status and future perspectives," *World J. Gastrointest. Endosc.* **9**(8), 346–358 (2017).

91. S. Sy, S. Huang, Y.-X. Wang et al., "Terahertz spectroscopy of liver cirrhosis: investigating the origin of contrast," *Phys. Med. Biol.* **55**(24), 7587–7596 (2010).

92. K. Shiraga, T. Suzuki, N. Kondo, and Y. Ogawa, "Hydration and hydrogen bond network of water around hydrophobic surface investigated by terahertz spectroscopy," *J. Chem. Phys.* **141**(23), 235103 (2014).

25

Magnetic resonance imaging study of diamagnetic and paramagnetic agents for optical clearing of tumor-specific fluorescent signal in vivo

Alexei A. Bogdanov Jr., Natalia I. Kazachkina, Victoria V. Zherdeva, Irina G. Meerovich, Daria K. Tuchina, Ilya D. Solovyev, Alexander P. Savitsky, and Valery V. Tuchin

CONTENTS

Introduction

The task of efficient delivery of light into deep subcutaneous layers of biological tissue has been long recognized as a complicated task due to the presence of multiple light scattering layers and inclusions [1]. However, there is a multitude of *in vivo* diagnostic applications that could greatly benefit from improvements in the efficacy of light delivery (reviewed in [2–4]). They range from Raman spectroscopy–enhanced diagnostics [5] to photodynamic therapy [6]. The anatomy and structure of skin and underlying tissues such as subcutaneous fat and connective tissue such as fascia and muscle fibers create multiple barriers to light penetration deeper into the body due to strong scattering and absorption (reviewed in [7]). It is well established that the stratum corneum and epidermis contain scatterers that are particularly strong, while deeper layers of the skin, such as the dermis and hypoderm, have very large scattering coefficients which exceed the absorption coefficients of these tissues by 10–100 times [8, 9]. Therefore, optical clearing directed at better light delivery into the tissue is expected to be critical for improving *in vivo* imaging techniques due to its ability to decrease scattering of light in multimodal settings [10, 11]. Optical clearing of skin surface is known as one of effective approaches for increasing depth penetration and improving quality of optical diffusion tomography and multiphoton tomography. For example, it has been shown recently that the use of an X-ray contrast agent (Omnipaque®) shows promise as optical clearing agent for fluorescence diffusion tomography due to its ability to decrease background fluorescence of the upper layers of the skin [12].

It is believed that OC compositions work because of three main possible mechanisms that are acting in concert with each other *in vitro*, which primarily includes clearing during the immersion in OC solutions. The same factors may potentially be at work *in vivo*. The first of those mechanisms is based on equalizing the difference between the values of refractive indices between various components of biological tissues and tissue fluids due to the penetration of OC into the tissues [13]. The second effect is due to the dehydration of tissues caused by hyperosmotic effects of the components of OC compositions [14, 15]. The third hypothetic mechanism is based on reversible dissociation of collagen fibers due to OC components interfering with hydrogen bonds (which keep the collagen structure together), leading to lowering of light scattering of collagen by reducing size of scattering particles [16, 17]. Out of all available optical clearing compositions tested so far, those containing glycerol and/or DMSO are used in

DOI: 10.1201/9781003025252-29

research most frequently. However, it should be noted that in the majority of cases it is difficult to identify a clear-cut correlation between the properties of a given OC composition and efficacy of OC [18]. This is just one of the reasons why OC components require to be studied individually and in combination, which in the end helps to improve the properties of these OC. The OC with intended use for potential *in vivo* applications should be studied with special care since OC selected by using immersion methods *in vitro* are not necessarily biocompatible. The possible steps to take along the lines of required testing should include the following: 1) selecting OC compositions with low toxicity and high biocompatibility that allow tissue architecture preservation; 2) analyzing the effects of OC on various tissue components individually in realistic model systems; 3) investigating cumulative OC effects in living systems (advanced experimental animal models), ideally carrying grafts of human tissues including the skin (adoptive xenotransplantation) using SCID and/or humanized immunodeficient rodent SCID strains that can harbor functional human xenografts.

Furthermore, in addition to key optical imaging modalities, it is important to deploy an available imaging approach that would allow the investigation of tissue properties in a parallel or independent experiment. Those imaging modalities should have 1) a sufficiently wide longitudinal (time) window for an appropriate number of data point acquisitions during the imaging study, i.e. sufficient spatiotemporal resolution (OC effects are transient); 2) the ability to capture changes in tissue (micro) environment in a sufficiently large volume of tissue; and 3), the spatial resolution of the alternative imaging modality that is to be coupled with optical imaging should be sufficient for meaningful interpretation (i.e. in comparatively similar range). However, the role of multimodality imaging in the assessment of OC properties and effects has only recently began to emerge [19, 20]. In the current study we examined the ability of magnetic resonance imaging (MRI) to provide insights into the *in vivo* effects of OC compositions. In this chapter, we review the results of two OC types, i.e. diamagnetic and paramagnetic, both of which may be used separately or in combination to provide optical clearing effects. The reason behind the ability of MRI to provide means for detecting the effects of OC on the skin as well as a relatively thin layer of subdermal space is in the ability of water-miscible and biocompatible compounds to alter proton relaxation times of water in the skin and underlying tissue, either directly by shortening of relaxation times by OC components, or indirectly due to changing the tissue properties such as local water diffusion rates or tissue hydration that affects the MR signal.

Materials and methods

Optical clearing solutions

We used topical applications of a diamagnetic mixture of glycerol/DMSO/water (70% glycerol, 5% DMSO, 25% water, ACS for analysis grade or similar), which was prepared by mixing, equilibrated for 1 h and used within 24 h for experiments *in vitro* and *in vivo*. Gadovist® (gadobutrol, 1.0 mmol/ml, Bayer HealthCare Pharmaceuticals, Germany) was used nondiluted as a paramagnetic OC solution.

Animal model

Animal studies were approved by the Ethics Committee of the Saratov State Medical University (Protocol No 8, April 10, 2018). Optical measurements were performed using athymic *nu/nu* mice (Research Center of Biotechnology of the RAS, Moscow, Russia). An animal model of cancer was obtained using a group of mice (n = 7, male) that were inoculated subcutaneously with epithelioid human carcinoma (HEp2 or A549) tumor cell suspension (10^6 cells in 100 µl of DPBS) in the right flank. Cells were expressing a far-red cell marker protein Tag RFP [21, 22].

Optical imaging

For low resolution imaging, UVP iBox studio (Analytik Jena AG) was used for locating subcutaneous fluorescent tumor nodules. Fluorescence measurements were performed in anesthetized animals using an inverted microscope (Eclipse TE 2000 U, Nikon) equipped with DCS-120 Confocal Scanning System (Becker & Hickl GmbH, Berlin, Germany) setup, the WL-SC-480-6 supercontinuum laser, acousto–optic tunable filter AOTF-V1-D-FDS-SM (FIANIUM, UK), and HPM-100-40 hybrid detector (Becker & Hickl GmbH, Berlin, Germany). The signal was acquired using DCS-120 system with a beam splitter quartz plate and longpass HQ 550LP Chroma and bandpass 580BP40 Omega filters. TagRFP fluorescence was excited *in vivo* at 540 nm. To study the OC effect, the OC mixtures were applied on the skin surface above the tumors for 15 min. Thereafter, the excess of the mixture was carefully absorbed using a cotton swab and the images were recorded.

Magnetic resonance imaging

MRI was performed at 1T (M3, Aspect Imaging, Shoham, Israel) with animals maintained at 37°C under gas anesthesia (1.5–2% isoflurane in oxygen) using body RF coil. Initially, pre-contrast images were obtained (T2w FSE pulse sequence: TR/TE 4000/42, FOV 40x40, 128x240 matrix, ETL 8, NEX 4) selecting 4–6 1 mm-thick tomographic slices on scout images. Anesthesia-mediated breath frequency was kept at 30–40 b/m, and built-in respiratory triggering was used during the acquisitions to minimize motion artifacts. In some T2w FSE experiments, an inversion prepulse TI=120 ms was used for fat suppression. Diffusion-weighted pulse sequences (DWI) were applied in some experiments: TR/TE=1200/52.5 ms, b=133.8, 407.1, 760.3 and 975.6 s/mm², δ = 2 ms. Scanning time was approximately 5 min. 3D Gradient-echo (GRE) T1-weighted sequences (TR/TE= 10/2.9 ms, FA 40, or TR/TE= 60/2.9 ms, FA 20, NEX 7, FOV 40 × 40 mm, 256 × 256 imaging matrix) were used in experiments involving gadobutrol topical application. Scanning times were 6 min (short TR) and 24 min (long TR), respectively. The application of diamagnetic OC composition (n = 5) and gadobutrol (n =3) was performed in separate groups of animals using the same conditions as described above. The cradle with the coil was placed back in the bore of the magnet and series of T2w FSE or 3D GRE images were acquired. Signal intensity changes were determined by 16-bit TIFF image analysis using Fiji/ImageJ. ADC values were calculated by fitting $ln(S/So)$ dependence vs. b factor, where S/So is a ratio of diffusion-weighted to non-diffusion-weighted MR signals.

Statistical analysis

The differences in fluorescence intensity before and after OC were considered statistically significant ($p<0.05$) according to one-tailed nonparametric Mann–Whitney–Wilcoxon (MWW) t-test paired observation. MRI intensity data were analyzed using one-tailed nonparametric (unpaired) MWW t-test, n = 6–12 data points per group.

Results and discussion

Fluorescence and MRI measurements. Optical clearing of skin using a diamagnetic glycerol and DMSO containing mixture

Subcutaneous tumors expressing red fluorescent TagRFP with an emission maximum of 584 nm [23] were used as a source of fluorescence in animal experiments. Cells expressing TagRFP can be imaged *in vivo* at 2–3 weeks after tumor inoculation into the flank of the animals [24]. The location of tumors was first identified by using a CCD camera with imaging at low resolution to verify the location, and then the animals were placed in

a special cassette for macroscopic imaging of tumor nodules in more detail using the setup involving excitation with a tunable laser source and data collection using a time-correlated single photon counting detector. The animals were imaged to measure fluorescence intensity changes before and after the topical application and after the removal of the excess of glycerol/DMSO/water OC compositions. As shown in Figure 25.1 fluorescence intensity measurements provide semiquantitative assessment of OC efficacy; a comparison of the images of two closely located tumor nodules before and after OC indicate that fluorescence of both nodules increased after the application. These differences become especially apparent after the removal of the excess of OC composition (Figure 25.1 A, B, and C) and the observed differences could be compared quantitatively by measuring fluorescence intensity change along the trace shown in Figure 25.1 C. The representative trace plot shows that the tumor nodule that appeared only weakly fluorescent before OC showed a strong increase in fluorescence intensity (Figure 25.1).

Fluorescence intensity measurements allowed the enhancement of the background vs. enhancement of true sources to be compared, i.e. small areas occupied by red fluorescent cells within the area they occupy under the skin of mice. To assess the effectiveness of measurements performed over the skin surface of tumors in mice, the measurements were performed over time to determine fluorescence intensity of TagRFP before and after the optical clearing during tumor progressions. We determined that in those areas, the mean photon counts were increased by a factor that varied between 1 and 3 after a 15-minute optical clearing using a mixture containing glycerol

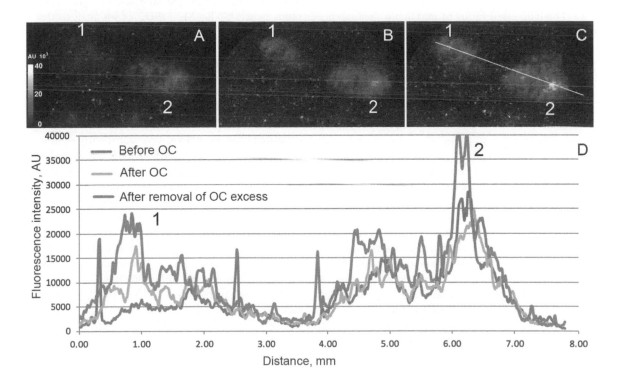

FIGURE 25.1 TagRFP red fluorescence imaging of two closely located subcutaneous Hep2 tumor nodules before (A) and after OC application (B) and after the removal of the excess of OC (C). The representative combination of fluorescence intensity profile plots corresponding to images A–C is shown in Figure 1D with two distinct areas (1 and 2) of regional fluorescence enhancement clearly visible. Nodules 1 and 2 and corresponding intensity peaks are numbered on the images accordingly. Pseudo color scale is shown in panel A.

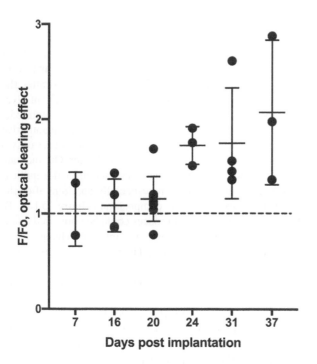

FIGURE 25.2 TagRFP fluorescence intensity change post implantation, showing the dependence of fluorescence intensity ratio changes (expressed as F_i/F_o, where F_o is the initial maximum fluorescence intensity of given ROI before OC). In 19 ROI out of 22 total (n = 3/animals) a fluorescence intensity increase (i.e. $F_i/F_o>1$) was detected indicating optical clearing effect. Individual observations for each ROI (black circles) and mean ± SD values are shown. The values plotted above the dashed line $F_i/F_o = 1$ indicate the OC effect.

and DMSO. The calculated ratios of maximum fluorescence intensity show that an optical clearing effect ($F_i/F_o>1$) was observed in 19 regions of interest (ROI) over a total of 22 total cases in three animals (Figure 25.2). The OC effect showed a tendency to increase over the course of tumor growth. This observed dependence on the time post-cell implantation may be a consequence of the increase in the density of fluorescent cells within the field of view and their relative proximity to the surface of the skin.

A variety of contrast-enhanced [25, 26] and noncontrast MRI techniques are available for measurements of small blood volume changes in the tissue [27–29]. As one of the potential options compatible with diamagnetic OC solutions low field proton (^{1}H) magnetic resonance imaging (MRI at 1T) T2-weighted pulse sequences were chosen as means to investigate the potential changes of skin/subcutaneous tissue properties before and after OC composition was applied. For MRI in mouse models of cancer, glycerol/DMSO OC composition was used and fast spin echo (FSE) pulse sequences with T2-weighting [30] were acquired that enable imaging of MR signal change resulting from transverse relaxation rate changes. In this case, the MR signal may be strongly affected by the factors promoting changes in the content of blood (local blood volume), vessel diameter, and blood oxygenation of the tissue [31, 32]. The influence of T2* effects that take place during the change of e.g. capillary blood vessel diameters in epidermis that cause local field inhomogeneities, and T2* relaxation effects related to differences in magnetic susceptibility were

eliminated by using a 180° pulse of spin-echo sequence [33]. In T2 imaging strength of the MR signal is judged by hypointensity of images, since the signal intensity (SI)

$$SI = k[H](1 - e^{-TR/T1}) \bullet e^{-TE/T2},$$

where H is the proton density and SI is determined by the factor $e^{-TE/T2}$ if TR> T1.

To apply FSE with MR signal change occurring primarily due to T2 relaxation, one should wait for most of the longitudinal magnetization to recover, and then apply a series of 180-degree radiofrequency pulses to collect series of echoes that emphasize T2 decay of magnetization (nuclear spin dephasing). With longer echo times, the differences in T2 relaxation properties of different tissues become more apparent. MR images acquired over time using T2w fast spin-echo (FSE) MRI pulse sequences with approximately 10–15 min delays between the acquisitions in the same animals shortly after optical imaging (Figure 25.3 A–D) showed significant quantitative differences between normalized MR signal intensities of axial peripheral tissue/skin slices before and after OC mixture applications in five animals with implanted ectopic subcutaneous tumor (Figure 25.3).

The application of OC composition onto the skin resulted in peripheral low-level "darkening" T2-weighted MR signal of the subcutaneous space below the treated skin, i.e. at the tumor periphery (Figure 25.3 A–D). Even though the differences in signal intensity caused by OC application were difficult to appreciate visually, the changes in MR signal were measurable by using local region-of-interest analysis. The differences in MR signal allowed the measurement of signal-to-noise (SNR) ratios (where SNR= SI_{mean}/SD_{noise}), an analysis that reflected normalized MR signal change differences which were negative in value after OC treatment. This indicated, as expected, the increase of the ROI-specific mean T2w MR signal. The obtained results obtained in a group of five animals with two different tumor types expressing TagRFP are summarized in Figure 25.3E. In all animals, SNR decreased after OC and in three out of five of the peripheral, subcutaneous tumors, the measured changes in SNR were statistically significant.

The measured change of SNR on T2-weighted MR images could potentially be a direct consequence of OC dilution in the skin, in subcutaneous tissue layers, and in the lymph of the subdermal extracellular space. To investigate whether dilution is causing the change in SNR, we conducted phantom experiments. After diluting OC and applying the same T2w FSE pulse sequences, we observed no statistically significant differences in normalized T2w MR signal in the case of signal intensities measured in 100% OC, in OC mixtures at various dilutions, and in pure water. Thus, this control MRI experiment showed that the observed differences in T2w MR signal intensity were not due to OC dilution when glycerol or DMSO concentrations decrease after penetrating the tissue.

Diffusion-sensitive MR pulse sequences may be applied to monitor H_2O diffusion changes in tumor tissue vs. normal tissue before and after application of OC composition. Diffusion-weighted MR signal changes should be derived from the images obtained at $b >100$ s/mm² in soft tissues [34]. The obtained apparent diffusion coefficients (ADC) were

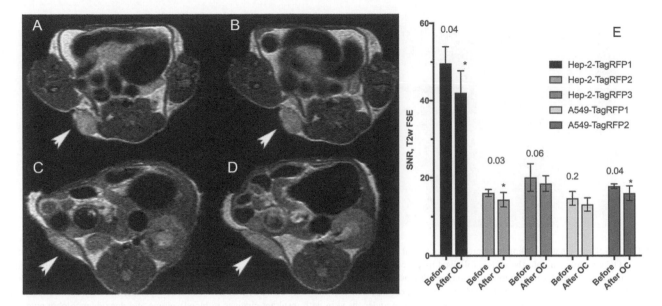

FIGURE 25.3 Imaging and SNR changes analysis in ectopic tumor-bearing animals. 1T MR images obtained before (A, C) and after OC (B, D) using T2-weighted fast-spin-echo pulse sequences at 1T. A- T2w FSE (TR/TE=4000/40 ms, 7NEX, 0.7 mm slice thickness, with external respiratory gating) of Hep2-TagRFP (A, B) or A549-TagRFP (C, D) expressing tumor-bearing mice (arrows). Images were analyzed for calculations of tissue MR signal intensity. E shows the mean SNR measured at 1T before and after application of glycerol/DMSO/water OC in 5 animals (3 – Hep2-TagRFP tumors and 2 with A549-TagRFP tumors). Mean normalized MR signal intensity was measured using T2w FSE pulse sequence (TR/TE 4000/39.8) in 6–10 0.7 mm thick tomographic slices; $SNR_i = SI_i/SD_{noise}$, p-values (MWW test, n = 6–12 data points) are shown above the bars. Data shown as mean ± SD. Asterisk indicates statistically significant changes of SNR (p<0.05).

compared by assuming monoexponential dependence on high b factor values and calculated using the fitting the data using standard equation: $SI = S_0 e^{-bD}$, where SI is the signal intensity with DW; S_0 is the signal intensity without DW; b is the attenuation coefficient (s/mm²); and D (ADC) is the diffusion rate constant for the given voxel (mm²/s).

Overall, the ADC values were larger in the tumors than in the normal muscle (Figure 25.4). This observation was expected since H_2O diffusion in cancer tissue as a rule is faster than in the normal muscle [35]. Also as expected, most of the skin/subdermal ROI showed an increase of ADC after OC, both in tumors as well as normal muscle (Figure 25.4). In the case of tumors, those changes were in most cases statistically significant. The penetration of OC components into the skin and possibly subdermal layer should induce the gradients of water diffusion because of water structuring and creation of differentially mobile water pools in the tissue.

Therefore, based on the results obtained without and with diffusion sensitization, the following explanation of observed MR signal differences could be put forward (Figure 25.5). Glycerol and DMSO are known to alter skin water content transiently due to the penetration into the skin and sequestration of water in dermis/epidermis [14, 36]. According to molecular dynamic simulations, higher glycerol content results in more extensive hydrogen-bond networks since the hydrogen bonding lifetime shows a tendency to increase as glycerol concentration increases while water molecules show faster reorientational motions than glycerol [37]. The consequence is shorter proton relaxation times of tissue/glycerol structured water and hydroxyls of glycerol molecules. One cannot exclude a potentially more complex interaction of both DMSO and glycerol with the tissue resulting in transient water depletion

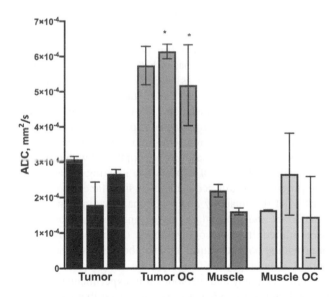

FIGURE 25.4 Apparent diffusion coefficient values in the tissues obtained by using diffusion weighted (DWI) MRI pulse sequence application at TR/TE = 1200/52.5 ms and b = 407.12, 757.87, and 975.6 s/mm² (three data points per each ROI). The increase of ADC after OC application in tumors was statistically significant, marked by asterisks (n = 3/ tissue).

of the deeper subdermal layers and relative increase in blood volume (capillary density) in the peripheral tissue. The consequence of that would be an increase in deoxyhemoglobin content, and this effect may contribute to a paramagnetic effect and resultant shortening of water proton relaxation times. The same effect, i.e. relative increase of blood volume and local tissue concentration of paramagnetic deoxyhemoglobin, may

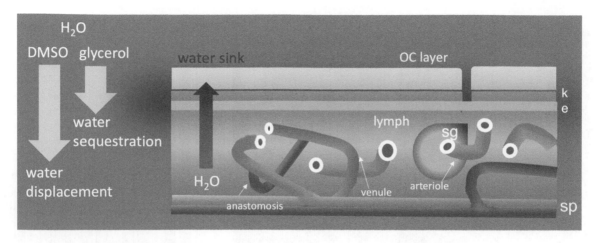

FIGURE 25.5 A schematic showing sources of potential effects caused by application of OC composition on the skin surface. The application of a layer of OC composition onto the skin is followed by penetration of the components through keratinized (k) and epidermal (e) layers and interaction with water of extracellular fluid/lymph that may result in water sequestration, displacement from hydration layers of macromolecules, and facilitation of water movement towards the layer of OC (water sink) on the surface of the skin. Those effects and transient hyperthermia may cause expansion (dilation) of peripheral vasculature of superficial plexus (sp) under the skin with resultant changes in blood volume and deoxyhemoglobin content. Sg – sebaceous gland.

be caused by also by mild local hyperthermia due to OC interfering with normal heat exchange which may be followed by vasodilation.

The use of MR contrast agents for OC

The development of various cross-overs of imaging modalities providing anatomical and functional data (with the notable exception of radioisotope-based SPECT and PET combined with CT and MRI, such as SPECT/CT, PET/CT, and PET/MRI), is still at relatively early stages of conceptualization and preclinical translation [38–40]. However, such new multimodality hybrids may have important advantages over traditional *in vivo* imaging. One important and early recognized advantage of optical imaging combined with MRI is its safety due to the lack of exposure to ionizing radiation [41]. The progress towards acquisitions of simultaneous or back-to back MRI (for anatomy and functional imaging) and optical (functional and molecular) images and data sets has so far been limited. It was recognized early on that MRI is capable of providing both structural and metabolic information, while optical tomography delivers specific and quantitative data reflecting anatomy and metabolism in experimental oncology [42, 43]. Several previous reports suggested that multimodality imaging integrating, for example, described multispectral frequency domain diffuse optical tomography with MRI, by which MRI-generated maps may be used to instruct the algorithms involved in optical reconstructions. This approach afforded an estimation absorption coefficient of an inclusion in a mouse-size phantom with 11% error [44]. Furthermore, the application of fluorescence lifetime imaging mode enabled improvements in MRI-assisted tumor identification in the liver of experimental animals [45]. These achievements, though incremental in nature, also spurred activity in multimodality optical/MRI contrast agent development [46, 47]. Nevertheless, the limitations in efficiency of light delivery to deeper tissues and resultant complexity of imaging data interpretation

are major obstacles on the road to comprehensive integration of optical imaging with other modalities. The use of optical clearing compositions with paramagnetic properties that can be detected by MRI may be viewed as a measure directed at improving the prospects of optical and MR image integration.

Paramagnetic clinical MR contrast agents (MR CA) are highly soluble and thermodynamically stable low molecular weight chelates of Gd(III), which have been in clinical use for contrast agent-assisted MRI since the 1980s due to their ability to shorten relaxation rates of water protons (relaxivity) [48–50]. For *in vivo* IV use, the high-relaxivity Gd(III)-based MR CA are supplied formulated at high concentrations (at 0.5M or 1M). These MR CA are hyperosmolar (1300–1600 mOsm) and usually viscous (2–5 Pa·s) solutions that do not absorb light in the visible/near infrared ranges of the spectrum [51]. In this regard, Gd(III)-based compounds have advantages as potential OC agents over experimental iron (III)- and manganese (II)-containing low-molecular-weight experimental paramagnetic contrast agents [52, 53] because the majority of chelated iron (III) and manganese (II), i.e. high relaxivity complexes absorb light in the visible range of the spectrum. The main rationale behind the idea to assess the OC properties of clinical MR CA was in enabling multimodal imaging *in vivo*. If in similarity with X-ray contrast agents, local application of paramagnetic gadolinium-based contrast agents leads to optical clearing of the skin, it could be also feasible to perform contrast-enhanced MR imaging to determine the time course of MRI contrast agent penetration into the skin. It would be also feasible to collect information contained in MR images that reflects local anatomy and is useful for correcting optical imaging results to account for the influence of fine structure of peripheral layers of normal subcutaneous tissue, or a peripheral lesion.

The initial optical clearing feasibility work using MR CA was conducted by optimization of contrast agent use with the following tests *in vitro* involving mouse skin samples: 1) measurements of skin thickness with OCT, and 2) measurements of a gain in transmittance. These experiments showed, in the case of gadobutrol (Gadovist®), a 12-fold increase in transmittance of light in a broad spectral range during the period of 10 min after application of MR CA on the skin *ex vivo* [19]. The OCT images showed an increase of light beam probing depth with a better contrast of the skin sample images as a result of MR contrast agent mediated optical clearing. In addition, optical coherence tomography reveled improved images reflecting skin microarchitecture, which were obtained after gadobutrol application (Figure 25.6A). After immersion of the skin in MR Cas, the OCT scans showed less scattering. Overall, results obtained by using gadobutrol indicated that this highly concentrated CA was superior to other CAs tested, if compared by collimated transmittance increase after application onto the skin samples. For example, gadobutrol mediated transmittance (Figure 25.6B) was five times higher than that of GdDTPA (Magnevist®) or GdDOTA (Dotarem®) over a period of 1 h post treatment [19].

Reliable MR imaging of various phase and tissue boundaries such as skin–air interfaces may be complicated, and the resultant images are generally prone to boundary artefacts [54, 55]. In addition, high concentrations of gadolinium-based MR CA generally cause strong magnetic susceptibility artefacts [56]. Due to the nature of these artifacts, the detection of locally applied concentrated MR CA is uncomplicated. However, gathering quantitative useful information reflecting the volume and penetration of MR CA into the tissue by measuring MR signal intensity changes over time strongly depends on chosen pulse sequence. This is why the experiments were initially conducted by following Gd-based contrast resulting from the influence of high Gd(III) magnetic susceptibility ($\Delta\chi$) on SNR.

Fast MR image acquisition mode using GRE pulse sequence with high T1-weighting (short TR and minimal TE) resulted in local image distortion by gadobutrol (at 1 M gadolinium) due to water proton relaxation signal affected by local susceptibility with low SNR due to high image noise (Figure 25.7). T1-weighted GRE with higher TR/TE ratio resulted in slower MRI acquisition, high T1 weighting, less distortion, higher image contrast, and higher SNR due to lower noise (Figure 25.7). During the course of gradient echo (GRE) T1 weighted pulse sequences acquisition, it became very apparent that by measuring signal loss, one could follow the fate of the contrast agent on the skin. This can be achieved by using either MRI highly sensitive to local magnetic susceptibility change ($\Delta\chi$), or less sensitive which is associated with less signal loss. In the latter case, the overall image distortion due to high $\Delta\chi$ is less pronounced. Indeed with the MR signal loss in the case of a solution high magnetic susceptibility causes magnetic field inhomogeneity that results in dephasing of nuclear spins during data acquisition [57]. This leads to a local loss of signal, which is the proportional to voxel size and TE. As any change of MR signal local signal loss can be quantified. Figure 25.7 illustrates the distribution of contrast agent on the surface of the skin and peripheral tissue layers. Skin signal intensity drops after the application of gadobutrol (i.e. the contrast agent showing the highest efficacy of optical clearing *in vivo*). The observed effect of gadobutrol is still detectable and the signal loss is still present on MR images even at 40 min after CA application.

Signal loss due to 100% gadobutrol application onto the skin surface above ectopic tumors in live mice was measurable with both TR = 10 and TR = 60 ms (Figure 25.8 A and B). It is apparent that GRE acquisitions at shorter TR (Figure 25.8A)

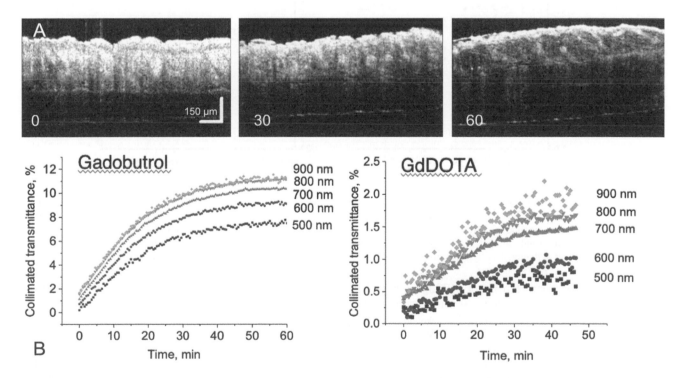

FIGURE 25.6 A – Optical coherence tomography B-scans of albino mouse skin samples *ex vivo* before (0 min), after 30 and 60 min of immersion in gadobutrol (Gadovist®); B – time dependence of collimated transmittance of albino mouse skin samples during the immersion in gadobutrol and GdDOTA (Dotarem®).

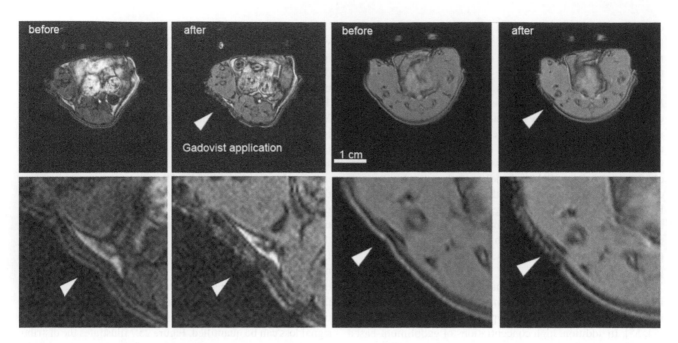

FIGURE 25.7 The dependence of signal contrast on gradient echo MRI pulse sequence parameters caused by 100% gadobutrol application onto the skin of nu/nu mouse. The first two upper row images show the results obtained on axial images using T1w-GRE (TR/TE= 10/2.9 ms, flip angle 40, NEX 8, FOV 40 × 40 mm). The second two were obtained using T1w-GRE with TR/TE= 60/2.9 ms, flip angle 20, NEX 7, FOV 40 × 40 mm. The lower row show enlarged ROI before and after the application of gadobutrol. The area showing the loss of signal is pointed by arrowheads.

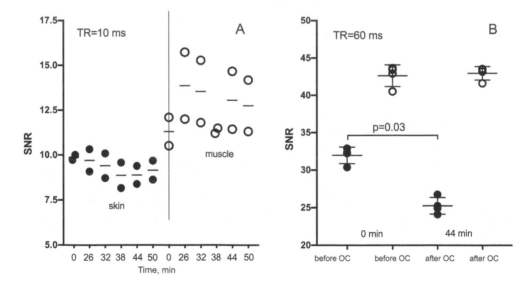

FIGURE 25.8 Time-dependent MR signal intensity changes (expressed as SNR change) in the skin/peripheral tumor tissue (closed circles) and contralateral normal skeletal muscle (open circles) measured before and after application of 100% gadobutrol on the surface of the skin above ectopic Hep2-RFP tumors grown in athymic mice. A – data obtained using T1w-GRE pulse sequence (TR/TE = 10/2.9 ms, flip angle 40, NEX 8, FOV 40 × 40 mm), data shown as mean and actual measured SNR (n = 2, average of 3 slices). B – data obtained using T1w-GRE with TR/TE = 60/2.9 ms, flip angle 20, NEX 7, FOV 40 × 40 mm, data shown as mean ± SD and actual measured SN, n = 3 at two time points (before OC and after OC).

showed a time-dependent decrease of MR signal, reflecting gadobutrol-mediated signal loss in the skin and peripheral tissue (a 2–3 mm thick layer). Over time there was also a small increase in the average SNR observed in the contralateral normal muscle, indicating that the contrast agent may have traversed the skin barrier and was enhancing contralateral skeletal muscle due to the dilution of MR CA in the lymph/extracellular fluid. With GRE acquisitions at longer TR = 60 (Figure 25.8B), the time resolution of imaging was insufficient

to track the dynamics of signal change. However, the MR signal intensity loss in the skin (closed circles) is apparent at the endpoint (44 min) after the completion of OC skin treatment, and this change was statistically significant (p<0.05). The OC amplification apparently resulted in no change of SNR in the muscle (open circles) using this pulse sequence parameter. It is possible that unlike extreme T1-weighting scenario at TR = 60, the enhancement of the contralateral muscle due to transfer the gadobutrol through the skin was either very low, or this effect was measurable only at the early time points in the beginning of the experiment because of the short half-life of gadobutrol in mice (about 15–20 min after intravenous administration [58]). In effort to improve the observed OC effect of gadobutrol, an attempt to obtain a system similar to glycerol/DMSO mixture was made by replacing glycerol (the main component of the OC composition) by gadobutrol for the purpose of improving both its transdermal transport and optical clearing. In experimental setup essentially identical to the one used for assessing OC effect of glycerol and DMSO in mouse ectopic

tumors (see Figure 25.1), a consistent increase of fluorescence intensity of TagRFP tumor marker was observed over time during the period of 1 h after OC application (Figure 25.9A). Fluorescence intensity quantification showed that OC mixtures containing DMSO resulted in faster clearing, but the overall fluorescence intensity increase effect was similar in terms of red fluorescent protein intensity increase over the period of 1 h. Both CCD measurements and more sophisticated photon counting using a DCS-120 detector showed similar trends in terms of intensity change.

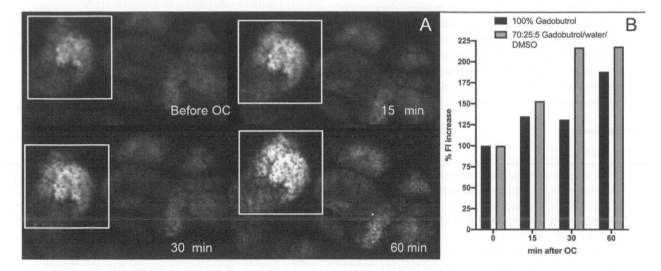

FIGURE 25.9 OC effect of a mixture of gadobutrol/water/DMSO (75:25:5 by volume) after application on the surface of the skin of athymic mice bearing superficial Hep2-TagFRP tumors before, and at 15, 30, and 60 min after application (A). B – Comparative fluorescence intensity measurements of TagRFP emission achieved by using a photon integration mode of DCS-120 detector after OC with 100% gadobutrol (black columns) and gadobutrol/water/DMSO (75:25:5) (gray columns).

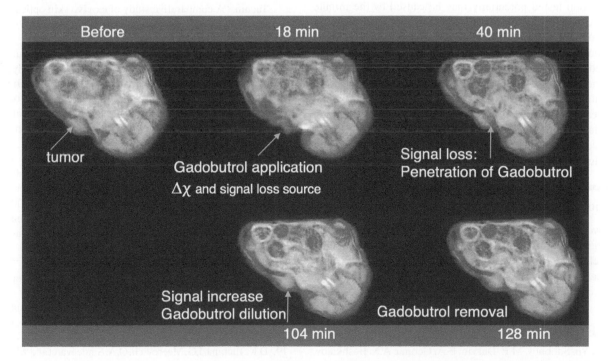

FIGURE 25.10 Pseudo-color Z-stack image averages (n = 4, 1 mm slices) obtained using T1-weighted pulse sequence (TR 60/TE 3, FA FOV 40 × 40 mm, 256 × 256 matrix, NEX 5, FA 20°. At 18 min, there is local signal loss in the area above the tumor due to the application of an OC composition consisting of gadobutrol/water/DMSO (70:25:5%).

The effect of OC composition containing gadobutrol and DMSO on MR effect was studied in more detail by applying a 3D gradient echo pulse sequence to measure the signal in the skin and underlying tissue. The local changes of MR signal i.e., its decrease followed by a subsequent increase became more apparent after collecting four consecutive tomographic image slices and generating a z-stack with MR signal averaging (Figure 25.10). The averaged z-stack images allow more precise tracking of the signal change over time *in vivo*; as expected, the initial decrease in MR signal intensity due to the magnetic susceptibility effect was followed by a signal loss in the interstitium, suggesting that some of the locally applied gadobutrol was absorbing through the skin, potentially aided by DMSO at approximately 40 min after application. After additional 60 min, the increase in MR signal intensity was observed presumably due to dilution in the interstitium, and soon after the CA started to disappear from the tumor due to removal from the imaged tissue volume, and this effect resulted in the loss of MR signal intensity.

Conclusions

In summary, magnetic resonance imaging is complementary to optical imaging of fluorescent protein marker in that it enables tracking the fate of OC compositions on the skin surface and underlying peripheral tissue. Both diamagnetic and paramagnetic solutions with OC properties induce changes in local water proton relaxation properties, and resultant changes in MR signal-to-noise ratios can be followed over time to improve OC protocols. The ability of MRI to resolve kinetics of local changes due to OC-mediated changes in relaxivity of the tissue is strictly pulse sequence–dependent. It is likely that T2w MR hypointensity increases in the case of diamagnetic OC composition tested potentially may be caused by the minute increase of deoxyhemoglobin content due to skin permeability and water immobilization and/or displacement. The clinical MR contrast agent with the highest concentration (gadobutrol) resulted in an OC effect which was readily detectable by optical coherence tomography, collimated transmittance increase, and fluorescence intensity measurements *in vivo*. MR imaging of topical gadobutrol application suggests transient penetration into the skin, which is enhanced in the presence of DMSO and results in transient changes in tissue signal after applying T1-weighted gradient-echo pulse sequences.

Acknowledgments

This work was supported by the Government of the Russian Federation (grant no. 14.W03.31.0023 to support scientific research projects implemented under the supervision of leading scientists at Russian institutions and Russian institutions of higher education).

REFERENCES

1. A.Yu. Sdobnov, M.E. Darvin, E.A. Genina, A.N. Bashkatov, J. Lademann, and V.V. Tuchin, "Recent progress in tissue optical clearing for spectroscopic application," *Spectrochim Acta A Mol. Biomol. Spectrosc.* 197, 216–229 (2018).

2. C. Bremer, V. Ntziachristos, and R. Weissleder, "Optical-based molecular imaging: Contrast agents and potential medical applications," *Eur. Radiol.* 13 (2), 231–243 (2003).

3. T.Hamaoka, K.K. McCully, M. Niwayama, and B. Chance, "The use of muscle near-infrared spectroscopy in sport, health and medical sciences: Recent developments," *Philos. Trans. A Math. Phys. Eng. Sci.* 369 (1955), 4591–4604 (2011).

4. A. Zelmer and T.H. Ward, "Noninvasive fluorescence imaging of small animals," *J. Microsc.* 252 (1), 8–15 (2013).

5. N.K. Das, Y. Dai, P. Liu, et al., "Raman plus X: Biomedical applications of multimodal Raman spectroscopy," *Sensors* 17 (7), 1592-1-20 (2017); 10.3390/s17071592.

6. S. Mallidi, S. Anbil, A.L. Bulin, G. Obaid, M. Ichikawa, and T. Hasan, "Beyond the barriers of light penetration: Strategies, perspectives and possibilities for photodynamic therapy," *Theranostics* 6 (13), 2458–2487 (2016).

7. A. Yu., Sdobnov, J. Lademann, M.E. Darvin, and V.V. Tuchin, "Methods for optical skin clearing in molecular optical imaging in dermatology," *Biochemistry* 84 (Suppl 1), S144–S158 (2019).

8. E. Gratton, "Applied physics. Deeper tissue imaging with total detection," *Science* 331 (6020), 1016–1017 (2011).

9. V.V. Tuchin, *Tissue Optics: Light Scattering Methods and Instruments for Medical Diagnostics.* 3rd ed., PM254, Bellingham, WA, SPIE Press, 2015.

10. M. Oldham, H. Sakhalkar, T. Oliver, et al. "Three-dimensional imaging of xenograft tumors using optical computed and emission tomography," *Med. Phys.* 33 (9), 3193–3202 (2006).

11. M. Oldham, H. Sakhalkar, T. Oliver, G. Allan Johnson, and M. Dewhirst, "Optical clearing of unsectioned specimens for three-dimensional imaging via optical transmission and emission tomography," *J. Biomed. Opt.* 13 (2), 021113 (2008).

12. A. Yu. Sdobnov, M.E. Darvin, J. Lademann, and V.V. Tuchin, "A comparative study of ex vivo skin optical clearing using two-photon microscopy," *J. Biophot.* 10(9), 1115–1123 (2017).

13. L.M. Oliveira, M.I. Carvalho, E.M. Nogueira, and V.V. Tuchin, "Diffusion characteristics of ethylene glycol in skeletal muscle," *J. Biomed. Opt.* 20 (5), 051019 (2015).

14. X. Xu and R.K. Wang, "Synergistic effect of hyperosmotic agents of dimethyl sulfoxide and glycerol on optical clearing of gastric tissue studied with near infrared spectroscopy," *Phys. Med. Biol.* 49(3), 457–468 (2004).

15. E.A. Genina, A.N. Bashkatov, A.A. Korobko, et al., "Optical clearing of human skin: Comparative study of permeability and dehydration of intact and photothermally perforated skin," *J. Biomed. Opt.* 13 (2), 021102 (2008).

16. J.M. Hirshburg, K.M. Ravikumar, W. Hwang, and A.T. Yeh, "Molecular basis for optical clearing of collagenous tissues," *J. Biomed. Opt.* 15 (5), 055002 (2010).

17. E.A. Genina, A.N. Bashkatov, and V.V. Tuchin, "Tissue optical immersion clearing," *Expert Rev. Med. Devices* 7 (6), 825–842 (2010).

18. D. Zhu, K.V. Larin, Q. Luo, and V.V. Tuchin, "Recent progress in tissue optical clearing," *Laser Photonics Rev.* 7 (5), 732–757 (2013).

19. D.K. Tuchina, I.G. Meerovich, O.A. Sindeeva, et al., "Magnetic resonance contrast agents in optical clearing: Prospects for multimodal tissue imaging," *J. Biophoton.* 13, e201960249 (2020); https://doi.org/10.1002/jbio.201960249

20. A.A. Bogdanov Jr, V.V. Tuchin, I.G. Meerovich, et al., "Towards registration of optical and MR signal changes in subcutaneous tumor volume in vivo after optical skin clearing," *Proc. SPIE.* 11239, 112390N-1-8 (2020); DOI: 10.1117/12.2545312

21. A.L. Rusanov, T.V. Ivashina, L.M. Vinokurov, et al., "Lifetime imaging of FRET between red fluorescent proteins," *J. Biophotonics* 3 (12), 774–783 (2010).

22. A.P. Savitsky, A.L. Rusanov, V.V. Zherdeva, T.V. Gorodnicheva, M.G. Khrenova, and A.V. Nemukhin, "FLIM-FRET imaging of caspase-3 activity in live cells using pair of red fluorescent proteins," *Theranostics* 2 (2), 215–226 (2012).

23. E.M. Merzlyak, J. Goedhart, D. Shcherbo, et al., "Bright monomeric red fluorescent protein with an extended fluorescence lifetime," *Nat. Methods* 4 (7), 555–557 (2007).

24. V. Zherdeva, N.I. Kazachkina, V. Shcheslavskiy, and A.P. Savitsky, "Long-term fluorescence lifetime imaging of a genetically encoded sensor for caspase-3 activity in mouse tumor xenografts," *J. Biomed. Opt.* 23(3), 035002 (2018).

25. S. Huang, C.T. Farrar, G. Dai, et al., "Dynamic monitoring of blood–brain barrier integrity using water exchange index (WEI) during mannitol and CO_2 challenges in mouse brain," *NMR Biomed.* 26 (4), 376–385 (2013).

26. Y.R. Kim, M.D. Savellano, D.H. Savellano, R. Weissleder, and A. Bogdanov, Jr., "Measurement of tumor interstitial volume fraction: Method and implication for drug delivery," *Magn. Reson. Med.* 52 (3), 485–494 (2004).

27. R. Abramovitch, D. Frenkiel, and M. Neeman, "Analysis of subcutaneous angiogenesis by gradient echo magnetic resonance imaging," *Magn. Reson. Med.* 39 (5), 813–824 (1998).

28. M. Neeman, J.M. Provenzale, and M.W. Dewhirst, "Magnetic resonance imaging applications in the evaluation of tumor angiogenesis," *Semin. Rad. Oncol.* 11 (1), 70–82 (2001).

29. J. Hua, P. Liu, T. Kim, et al., "MRI techniques to measure arterial and venous cerebral blood volume," *Neuroimage* 187, 17–31 (2019).

30. J. Hennig, A. Nauerth, and H. Friedburg, "RARE imaging: A fast imaging method for clinical MR," *Magn. Reson. Med.* 3 (6), 823–833 (1986).

31. I. Kida, T. Yamamoto, and M. Tamura, "Interpretation of BOLD MRI signals in rat brain using simultaneously measured near-infrared spectrophotometric information," *NMR Biomed.* 9 (8), 333–338 (1996).

32. J. Dennie, J.B. Mandeville, J.L. Boxerman, S.D. Packard, B.R. Rosen, and R.M Weisskoff, "NMR imaging of changes in vascular morphology due to tumor angiogenesis," *Magn. Reson. Med.* 40(6), 793–799 (1998).

33. G.B. Chavhan, P.S. Babyn, B. Thomas, M.M. Shroff, and E.M. Haacke, "Principles, techniques, and applications of T2*-based MR imaging and its special applications," *Radiographics* 29 (5),1433–1449 (2009).

34. A.R. Padhani, G. Liu, D.M. Koh, et al., "Diffusion-weighted magnetic resonance imaging as a cancer biomarker: Consensus and recommendations," *Neoplasia* 11 (2), 102–125 (2009).

35. D.M. Koh and D.J. Collins, "Diffusion-weighted MRI in the body: Applications and challenges in oncology," *AJR Am .J. Roentgenol.* 188 (6), 1622–1635 (2007).

36. A. Ross and J.N. Kearney, "The measurement of water activity in allogeneic skin grafts preserved using high concentration glycerol or propylene glycol," *Cell Tissue Bank* 5 (1), 37–44 (2004).

37. C. Chen, W.Z. Li, Y.C. Song, and J. Yang, "Hydrogen bonding analysis of glycerol aqueous solutions: A molecular dynamics simulation study," *J. Mol. Liq.* 146 (1–2), 23–28 (2009).

38. B.J. Pichler, H.F. Wehrl, and M.S. Judenhofer, "Latest advances in molecular imaging instrumentation," *J. Nucl. Med.* 49 (Suppl 2), 5S–23S (2008).

39. M. Estorch and I. Carrio, "Future challenges of multimodality imaging," *Recent Results Cancer Res.* 187, 403–415 (2013).

40. M.B. Syed, A.J. Fletcher, R.O. Forsythe, et al., "Emerging techniques in atherosclerosis imaging," *Br. J. Radiol.* 92 (1103), 20180309 (2019).

41. M. Nahrendorf, D.E. Sosnovik, and R. Weissleder, "MR-optical imaging of cardiovascular molecular targets," *Basic Res. Cardiol.* 103 (2), 87–94 (2008).

42. G. Gulsen, H. Yu, J. Wang, et al., "Congruent MRI and near-infrared spectroscopy for functional and structural imaging of tumors," *Technol. Cancer Res. Treat.* 1 (6), 497–505 (2002).

43. S. Merritt, F. Bevilacqua, A.J. Durkin, et al., "Coregistration of diffuse optical spectroscopy and magnetic resonance imaging in a rat tumor model," *Appl Opt* 42 (16), 2951–2959 (2003).

44. G. Gulsen, O. Birgul, M.B. Unlu, R. Shafiiha, and O. Nalcioglu, "Combined diffuse optical tomography (DOT) and MRI system for cancer imaging in small animals," *Technol. Cancer. Res. Treat.* 5 (4), 351–363 (2006).

45. A. Erten, D. Hall, C. Hoh, et al., "Enhancing magnetic resonance imaging tumor detection with fluorescence intensity and lifetime imaging," *J. Biomed. Opt.* 15 (6), 066012 (2010).

46. M. Longmire, P.L. Choyke, and H. Kobayashi, "Dendrimer-based contrast agents for molecular imaging," *Curr. Top. Med. Chem.* 8 (14), 1180–1186 (2008).

47. A. Bumb, C.A. Regino, M.R. Perkins, et al., "Preparation and characterization of a magnetic and optical dual-modality molecular probe," *Nanotechnology* 21 (17), 175704 (2010).

48. P. Caravan, J.J. Ellison, T.J. McMurry, and R.B. Lauffer, "Gadolinium(III) chelates as MRI contrast agents: Structure, dynamics, and applications," *Chem. Rev.* 99 (9), 2293–2352 (1999).

49. A.E. Merbach and E. Toth, *The Chemistry of Contrast Agents in Medical Magnetic Resonance Imaging*, A.E. Merbach and E. Toth (Eds.). 2nd ed., Chichester UK, Wiley, 2013.

50. E. Boros, E.M. Gale, and P. Caravan, "MR imaging probes: Design and applications," *Dalton Trans.* 44 (11), 4804–4818 (2015).

51. ACR Manual on Contrast Media, *ACR Committee on Drugs and Contrast Media*, American College of Radiology, ISBN: 978-1-55903-012-0, 2020, 129 p.; https://www.acr.org/-/media/ACR/Files/Clinical-Resources/Contrast_Media.pdf.

52. E.M. Gale, I.P. Atanasova, F. Blasi, I. Ay, and P. Caravan, "A manganese alternative to gadolinium for MRI contrast," *J. Am. Chem. Soc.* 137 (49), 15548–15557 (2015).

53. J. Wahsner, E.M. Gale, A. Rodriguez-Rodriguez, and P. Caravan, "Chemistry of MRI contrast agents: Current challenges and new frontiers," *Chem. Rev.* 119 (2), 957–1057 (2019).

54. S. Khanna and J.V. Crues, 3rd, "Complexities of MRI and false positive findings," *Ann. NY Acad. Sci.* 1154, 239–258 (2009).

55. S. Matsuoka, A.R. Hunsaker, R.R. Gill, et al., "Functional MR imaging of the lung," *Magn. Reson. Imaging Clin. N. Am.* 16(2), 275–289 (2008).

56. T.P. Roberts and D. Mikulis, "Neuro MR: principles," *J. Magn. Reson. Imaging* 26 (4), 823–837 (2007).

57. V. Kuperman, "Image artifacts," Chapter 6, in *Magnetic Resonance Imaging. Physical Principles and Applications*, I. Mayergoyz (Ed.), Academic Press, San Diego CA, 2000.

58. H. Vogler, J. Platzek, G. Schuhmann-Giampieri, et al., "Pre-clinical evaluation of gadobutrol: A new, neutral, extracellular contrast agent for magnetic resonance imaging," *Eur. J. Radiol.* 21 (1), 1–10 (1995).

26

Use of optical clearing and index matching agents to enhance the imaging of caries, lesions, and internal structures in teeth using optical coherence tomography and SWIR imaging

Daniel Fried

CONTENTS

Introduction

Studies have shown that various agents can be applied to tooth surfaces to enhance the imaging of caries lesions (tooth decay) and internal structures such as the dentinal enamel junction (DEJ), cracks, root canals and the pulp chamber. The refractive indices (RI) of enamel, dentin, and cementum are quite high at 1.63 and 1.54, and 1.58, respectively, and surface scattering and specular reflection can interfere with optical penetration into the tooth. Moreover, dental hard tissues are porous structures, and that porosity increases markedly with demineralization and hypomineralization. Filling the porous structure of subsurface caries lesions with a high RI fluid enables better optical penetration through such lesions to better view the underlying tooth structure for better assessment of lesion severity. For example, it has been demonstrated that optical clearing agents can be used to increase the visibility of deeply penetrating occlusal lesion images that have reached the underlying dentin and spread laterally under the enamel in optical coherence tomography (OCT) images that cannot be detected with dental radiographs. High refractive index fluids can also increase the contrast of caries lesions. In addition, the loss of the mobile water in sound enamel and dentin or displacement

by another fluid profoundly changes light scattering in these tissues and optical penetration. The use of exogenous agents to enhance dental imaging is of increasing interest to the dental community with the introduction of new near-infrared (NIR) and short wavelength infrared (SWIR) transillumination and reflectance imaging methods along with OCT that exploit the high transparency of dental enamel at longer wavelengths beyond the visible range.

Optical properties of dental hard tissues from 400 to 2300 nm

Light attenuation in enamel and dentin are dominated by scattering in the visible and near-infrared (NIR) ranges [1]. The scattering coefficient of enamel decreases with increasing wavelength from 400 cm^{-1} in the near-UV [1] to as low a value as 2–3 cm^{-1} at 1310 nm in the short wavelength infrared (SWIR) ranges [2, 3] and continues to decrease beyond 1700 nm [4] (see Figure 26.1). The strong wavelength dependence of enamel suggests that scattering is dominated by the small submicron enamel crystals that act as Rayleigh type

DOI: 10.1201/9781003025252-30

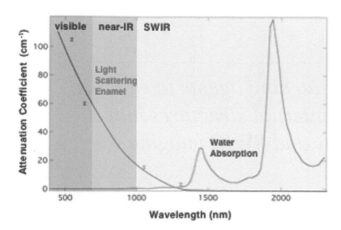

FIGURE 26.1 Light attenuation in enamel (red trace) due to light scattering and light attenuation due to absorption by water (blue trace) from [3, 7]. The visible, NIR (700–100 nm) and SWIR (1000–2500 nm) regions are also indicated by color shading.

scatterers [1]. In contrast, light scattering in dentin remains high across the visible and NIR ranges due to the dentinal tubule structures a few microns in diameter that act as cylindrical Mie scatterers. The scattering of dentin is highly anisotropic and quite variable with position in the tooth and exceeds 100–200 cm^{-1} across the visible and NIR [5] ranges. It has been suggested that the tubules can act as waveguides to guide visible light [6].

The absorption coefficient of enamel is too low to be measured in the visible range while that of dentin is essentially wavelength independent with a value of approximately 4 cm^{-1} above 400 nm [5]. At wavelengths beyond 1400 nm, water absorption increases [7] (see Figure 26.1) and becomes an important source of attenuation in enamel and dentin and contributes to the high contrast of caries lesions on both crown and root surfaces.

The demineralization of enamel and dentin occurs due to organic acids from cariogenic bacteria in the plaque or biofilm on the tooth surface. Highly porous demineralized areas of coronal caries appear whiter and are more opaque. Pores are formed in the tooth structure that increase the magnitude of light scattering by orders of magnitude. Demineralized tissues appear whiter than the surrounding sound tissues due to the increased diffuse reflectance, and clinicians refer to early caries lesions as white spot lesions [8, 9]. Hypomineralized areas of the tooth formed due to developmental defects and fluorosis are also highly porous and appear similar to demineralization due to caries. The earliest attempts to measure the optical properties of dental caries were limited to measurements of backscattered light from optically thick, multilayered sections of simulated caries lesions [10–12]. The microstructure of enamel and dentin caries lesions is complex, consisting of various turbid and transparent zones [13, 14]. During remineralization, pores and tubules are filled with mineral, and those areas are typically more transparent. There have been some attempts to measure and simulate light scattering in artificial and natural lesions [8, 12, 15]. Lesion areas are highly porous, allowing index matching fluids to imbibe into the lesion, effectively eliminating the lesion from an optical standpoint,

i.e., the lesion becomes more transparent than the sound tissue. One approach is to assume that carious lesions will be filled with saliva, and that measurement in water replicates the conditions that will be encountered clinically. Darling et al. [16] measured the optical properties of lesion areas imbibed in water at 1300 nm and compared the increase in the magnitude of the scattering coefficient with mineral loss measured using microradiography. The attenuation in the lesion exceeds 100 cm^{-1}, a factor of 50 times higher than that of sound enamel at 2–3 cm^{-1}, and is a similar magnitude to that of the dentin. Moreover, it is hypothesized that water absorption is a major contributor to the contrast between sound and demineralized enamel beyond 1300 nm since deeply penetrating photons in sound enamel are likely absorbed by water. A comparison of three reflectance images of the occlusal surface of an extracted tooth with demineralization in the fissures are shown in the visible range, 1300 nm and 1450 nm in Figure 26.2 (top). Demineralization should appear whiter in reflectance, and the highest contrast of demineralization is at 1450 nm, while at visible wavelengths only the dark stains in the fissure can be seen. Multispectral comparisons of lesion contrast provide insight into the mechanism responsible for higher contrast at longer SWIR wavelengths. In a recent study [4], reflectance images of demineralization on tooth enamel surfaces were acquired at wavelengths near 1450, 1860, 1880, and 1950 nm. The magnitude of water absorption is similar at 1450 and 1880 nm but varies markedly between 1860, 1880, and 1950 nm. The highest contrast was at 1950 nm; however, the markedly higher contrast at 1880 compared to 1450 nm and similar contrast between 1860 and 1880 nm suggests that the enamel scattering coefficient continues to decrease beyond 1300 nm and that reduced light scattering in sound enamel is most responsible for the higher lesion contrast at longer SWIR wavelengths. Longer wavelength studies of the contrast of demineralization and calculus on root surfaces suggest that the water absorption is a greater contributor to the contrast between sound and demineralized dentin, since the contrast no longer increases beyond 1400 nm, unlike for enamel [17]. Figure 26.3 shows images of the root surface of an extracted tooth at visible, 850, 1300, and 1500 nm. Note how dark the tooth root appears due to the higher water absorption at 1500 nm, while images beyond 1400 nm all appear similar to the 1500 nm image in Figure 26.3 [17]. Tooth surfaces are often heavily stained, and stains accumulate within the porous structure of caries lesions where they can profoundly change the lesion's optical appearance. The visual diagnosis of lesion activity is based on the lesion texture and color [18]. However, lesion color and texture are not reliable indicators of caries presence and activity, and clinicians often misdiagnose lesions based on the presence of stain, particularly in the pits and fissures of the occlusal surface. A major advantage in imaging in the SWIR is that the stains that are common on tooth occlusal surfaces do not interfere at wavelengths beyond 1150 nm since none of the known chromophores absorb light at longer wavelengths [19]. This can be seen in Figures 26.2 and 26.3, where in the visible images the lesion areas appear darker than the sound tooth structure due to absorption of stain, while the stain does not interfere at longer wavelengths. The photon energy is not sufficient for electronic excitation of the chromophores [20, 21].

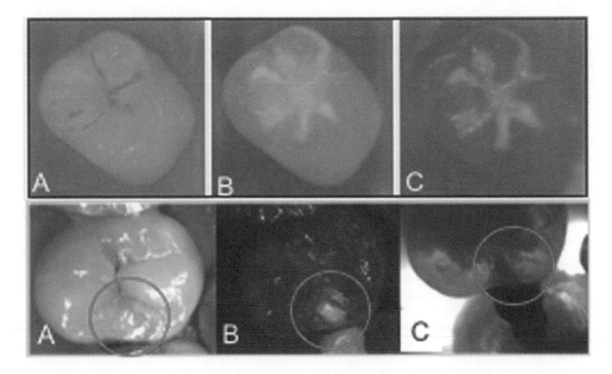

FIGURE 26.2 (top) NIR reflectance images of a tooth with stained fissures and demineralization at visible, 1300 nm, and 1450 nm. Reprinted with permission from reference [19]. (bottom) *In vivo* images of a tooth with an approximal lesion and a stained fissure on the occlusal surface that was misdiagnosed as having a lesion. (A) Visible reflectance, (B) SWIR reflectance (1500–1750 nm), and (C) occlusal transillumination (1300 nm) images are shown. The SWIR images confirm that an approximal lesion is present but indicate that there is no demineralization present on the occlusal surface. Reprinted with permission from reference [72].

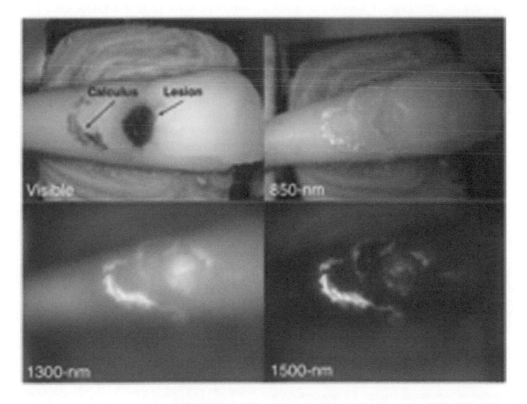

FIGURE 26.3 NIR Reflectance images of the root surface of a tooth showing dental calculus and root caries at visible, 850, 1300, and 1500–1750 nm wavelengths. Reprinted with permission from reference [17].

It is necessary to use SWIR wavelengths greater than 1150 nm to avoid significant interference from stains when measuring lesion contrast in reflectance and transillumination modalities [19]. Therefore, stains can be easily differentiated from actual demineralization in the SWIR range, which is not possible at visible and NIR wavelengths. Chung et al. [22] demonstrated that absorption due to stains contributed more to the lesion contrast than increased scattering due to demineralization at visible wavelengths [23]. Since it is impractical to remove stains from the deep grooves and fissures on tooth occlusal surfaces, lack of interference from stains at longer SWIR wavelengths is a significant advantage for transillumination and optical coherence tomography.

Influence of hydration

Teeth are highly porous with large fractions of mobile water, and the state of hydration of the tooth profoundly influences the lesion contrast and optical penetration in enamel, dentin, and cementum. The strong influence of hydration can be seen in transillumination measurements through thick tooth sections with lesions at 1300 nm; the contrast decreased markedly if the samples were left out of water overnight and allowed to dehydrate [22]. This effect can clearly be seen for a 5 mm-thick section in Figure 26.4. The mean contrast was 0.53 (0.097) fully hydrated and almost completely opaque 0.016(0.039) when dry for n = 6 samples, 5 mm thick [22]. The lesion contrast completely recovered after the samples were immersed in water overnight and rehydrated. Hydration studies involving the transillumination of the tooth sections indicated that loss of mobile water resulted in a significant reduction in sound enamel transparency and loss of lesion contrast. It is likely that the loss of mobile water leaves pores that act as scattering sites. In another study, it was found that rapid heating with a carbon dioxide laser can dehydrate dental hard tissues; the attenuation coefficient of enamel at 1310 nm

increased significantly from 2.12 ± 0.82 to 5.08 ± 0.98 with loss of mobile water due to heating [24, 25]. These results indicate that it is extremely important to ensure that teeth are kept well hydrated for SWIR imaging studies, including optical coherence tomography [26]. It has been well established that a layer of water or saliva influences the contrast and visibility of early white spot lesions. Lesions are visible with higher contrast if the tooth surface is air-dried, and clinicians are taught that if the lesion is only visible when the tooth is dry, it is shallow and superficial. The rate of water loss from lesions has also been correlated with lesion activity; arrested lesions are less permeable to water due to the highly mineralized outer layer covering the lesion, and both optical and thermal imaging [27, 28] have been used to monitor optical changes associated with water loss. More recently, OCT and SWIR imaging have been used to assess lesion activity, and those methods are discussed in a following section.

Refractive index (RI), polarized light, and imbibing agents

Tooth enamel has a very high refractive index, resulting in high reflectivity from the enamel surface and high refraction and internal reflection that render photographing and imaging teeth more challenging. The refractive indices (RI) of enamel, dentin, and cementum are quite high at 1.63 and 1.54, and 1.58, respectively [29]. Optical coherence tomography has also been used to measure the refractive index in dental hard tissues [30]. The visibility of scattering structures on highly reflective surfaces such as teeth can be enhanced by the use of crossed polarizers to remove the glare from the surface [31, 32]. The contrast between sound and demineralized enamel can be further enhanced by depolarization of the scattered light in the area of demineralized enamel [33, 34].

Polarized light microscopy has been used to study teeth for more than 150 years [35]. Tooth enamel and dentin are birefringent due to both mineral and organic components in these tissues. Different zones of caries lesions have been identified and studied under polarized light [36–38]. Darling et al. [39] and others [40] developed quantitative methods to measure the mineral loss by imbibing different fluids of varying refractive index into tooth sections and measuring the birefringence. Quinoline has a refractive index of 1.63, the same as dental enamel, and it is common practice to immerse tooth sections in quinoline to study birefringence changes in the various zones of caries lesions.

Dental caries and demineralization and remineralization

During the past century, the nature of dental caries (tooth decay) in the US has changed markedly due to the introduction of fluoride to the drinking water, the advent of fluoride dentifrices and rinses, and improved dental hygiene. In spite of these advances, dental decay continues to be the leading cause of tooth loss in the US [41–43]. Root caries is an increasing problem with our aging population. Today the majority of newly discovered caries lesions are localized to the occlusal

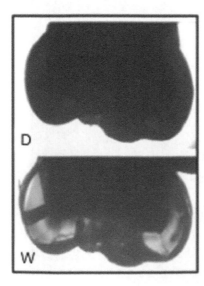

FIGURE 26.4 Transillumination images through a 5 mm-thick tooth section at 1300 nm (W) and dry (D). Reprinted with permission from reference [22].

pits and fissures of the posterior dentition (occlusal caries lesions) and the proximal contact sites between teeth (approximal or interproximal caries lesions). In the caries process, demineralization occurs as organic acids generated by bacterial plaque that diffuse through the porous enamel of the tooth dissolve the mineral. If the decay process is not arrested, the demineralization spreads through the enamel and reaches the dentin, where it rapidly accelerates due to the markedly higher solubility and permeability of dentin. The lesion spreads throughout the underlying dentin to encompass a large area, resulting in loss of integrity of the tissue and cavitation. When the mineral content falls below a certain level the tissue collapses forming an obvious cavity, and it is this process that led to use of "cavity" to describe a caries lesion. Caries lesions are usually not detected until after the lesions have progressed to the point at which surgical intervention and restoration are necessary, often resulting in the loss of healthy tissue structure and weakening of the tooth. Therefore, new diagnostic tools are needed for the detection and characterization of caries lesions in the early stages of development.

Caries detection and diagnostics

Conventional methods of caries detection and diagnostics

Caries lesions are routinely detected in the US using visual/tactile (explorer) methods coupled with radiography. These diagnostic and treatment paradigms were developed long ago and were adequate for large, cavitated lesions; however, they do not have sufficient sensitivity or specificity for the diagnosis of the early noncavitating caries lesions prevalent today. Radiographic methods do not have the sensitivity for early lesions, particularly occlusal lesions, and by the time the lesions are radiolucent, they have often spread extensively throughout the underlying dentin, at which point surgical intervention becomes necessary [44]. At that stage in the decay process, it is far too late for preventive and conservative intervention, and a large portion of carious and healthy tissue will need to be removed, often compromising the mechanical integrity of the tooth. If carious lesions are detected early enough, it is likely that they can be arrested/reversed by nonsurgical means through fluoride therapy, antibacterial therapy, or dietary changes [45].

Accurate determination of the degree of lesion activity and severity is of paramount importance for the effective employment of the treatment strategies mentioned above. Since optical diagnostic tools exploit changes in the light scattering of the lesion, they have great potential to assess whether or not the caries lesion is active and expanding, or whether the lesion has been arrested and is undergoing remineralization.

The most difficult to detect and the most common early enamel lesions are occlusal lesions. Occlusal lesions constitute 80% of the new lesions found today [46]. In the conventional method of occlusal caries detection, the clinician probes areas in the dentition that appear suspicious upon an initial visual inspection with the dental explorer [47]. If the probed area is soft and provides some resistance upon retraction of the

instrument, the site is deemed to be carious. Studies suggest that the use of the dental explorer to probe for caries may actually promote or accelerate lesion formation [44, 48]. Thus, the use of a blunt explorer or none at all has been recommended by leading cariologists [49–51]. Clinicians base their diagnosis of occlusal lesions and treatment planning on the pit and fissure color and texture which is strongly influenced by the degree of staining. This can be misleading because lesion color does not provide sufficient information about the state of the lesion, i.e., whether it is progressing or arrested. Moreover, pigmentation can be due to staining from diet and other environmental factors and not from infection by microorganisms [52].

Optical transillumination

Optical transillumination was used extensively before the discovery of X-rays for detection of dental caries. The development of high-intensity fiberoptic light sources a few decades ago revived interest in optical transillumination for the detection of approximal lesions [53–58]. During fiberoptic transillumination, carious lesion appears dark because of decreased transmission due to increased scattering and absorption by the lesion; however operation in the visible range where light scattering in enamel is high greatly limits performance.

Almost 20 years ago, it was discovered that enamel was almost completely transparent at 1300 nm [2, 59].

The first NIR and SWIR transillumination studies involved proximal transillumination with a similar geometry to radiographs [60], but soon it was discovered the tooth can be imaged from the occlusal surface after shining light at and below the gum line, which we call occlusal transillumination [20, 61]. Approximal lesions can be imaged by occlusal transillumination of the proximal contact points between teeth by directing SWIR light below the crown while imaging the occlusal surface [59–61]. The first clinical study in 2009 demonstrated that approximal lesions appearing on radiographs could be detected *in vivo* with SWIR imaging proximal and occlusal transillumination with similar sensitivity [61].

SWIR transillumination from the occlusal surface can also be used to image occlusal lesions with high contrast, and multiple studies have shown that occlusal caries lesions can be imaged with high contrast *in vivo* and that SWIR occlusal transillumination is an excellent screening tool for occlusal lesions. [20, 34, 61–65].

In a recent clinical study [66], SWIR transillumination and SWIR reflectance were used to screen premolar teeth scheduled for extraction for all coronal caries lesions, both approximal and occlusal, and the diagnostic performance of SWIR imaging was compared with radiography. SWIR imaging was shown to be significantly more sensitive than radiography for the detection of lesions on both occlusal and proximal tooth surfaces *in vivo*. *In vivo* SWIR transillumination (1300 nm) and SWIR reflectance (1600 nm) measurements of a tooth with caries are shown in Figure 26.2 (bottom) [65]. There are now several NIR imaging systems available for caries detection, and they employ both occlusal transillumination and reflectance; however, current commercial systems are only available at wavelengths less than 1000 nm, due to the high cost of sensors at longer wavelengths [67–69].

Reflectance measurements

Early enamel white spot lesions can be discriminated from sound enamel by visual observation or by visible-light diffuse reflectance imaging [9, 10]. However, color, in addition to the intensity of the reflected light, plays a large role in detecting those changes. Moreover, such changes are difficult to quantify, and the color of sound tooth structure varies markedly due to stains. In a recent study of natural lesions on the occlusal surfaces of extracted teeth, the image contrast was actually negative as opposed to being positive in visible reflectance measurements, indicating that absorption due to stains contributed more than increased scattering due to demineralization to the lesion contrast [22]. This renders the method useless in areas that are subject to heavy staining, namely the occlusal surfaces where most lesions are likely to develop. In fact, visible light reflectance was proposed three decades ago for use in monitoring early demineralization on tooth surfaces, but it has proven to be unsuccessful due to the problems indicated above [9]. In the early 1980s, ten Bosch et al. [8, 9] introduced an optical monitor that used optical fibers for visible light reflectance measurements on tooth surfaces.

In contrast to transillumination, which initially focused on approximal lesions, the first studies utilizing SWIR wavelengths for reflectance measurements targeted occlusal caries lesions. The contrast between sound and demineralized enamel is greatest in the SWIR due to the minimal scattering of sound enamel, and this can be exploited for reflectance imaging of early demineralization [16]. Wu et al. [70] reported that the contrast between early demineralization was significantly higher at 1310 nm than in the visible range. Zakian acquired hyperspectral reflectance images of occlusal caries lesions and demonstrated that multiwavelength images could be used to aid diagnosis [21]. The highest contrast is achieved at longer SWIR wavelengths coincident with higher water absorption [71]. Water in the underlying dentin and surrounding sound enamel absorbs the deeply penetrating light and reduces the reflectivity in sound areas. In turn, this results in higher contrast between sound and demineralized enamel. Hyperspectral reflectance measurements by Zakian show that the tooth appears darker with increasing wavelength [21]. The performance of SWIR reflectance is even more dramatic for natural lesions on the occlusal surfaces of extracted teeth that are typically stained, as can be seen in the images presented in Figure 26.2. The contrast is significantly higher for wavelengths greater than 1600 nm than at other wavelengths, and stains interfere significantly at wavelengths less than 1000 nm [19].

The first clinical study using SWIR reflectance utilized the wavelength range of 1500–1700 nm, and the diagnostic performance was higher than radiography and other NIR imaging modalities for the detection of proximal and occlusal lesions [66]. *In vivo* images of a tooth with an approximal lesion and a stained fissure on the occlusal surface that was misdiagnosed as having a lesion are also shown in Figure 26.2. The SWIR images confirm that an approximal lesion is present but indicate that there is no demineralization present on the tooth occlusal surface [72].

Recent imaging studies at wavelengths from 400 to 2350 nm show that high-contrast images of root caries can be acquired at wavelengths beyond 1400 nm [17]. Stains interfere significantly at wavelengths less than 1150 nm, yielding nondiagnostic contrast for root caries lesions. Significantly higher (P<0.05) lesion contrast was measured at wavelengths greater than 1400 nm, where water absorption is high. There was no further increase in contrast beyond 1400 nm as was observed for enamel, which suggests that the scattering coefficient of dentin does not decrease further with increasing wavelength. Therefore, it is likely that the high absorption of water at longer wavelengths reduced the light scattering from the surrounding and underlying normal dentin, thus increasing the lesion contrast. Reflectance images of the root surface of a tooth showing dental calculus and root caries at visible, 850, 1300, and 1500–1750 nm wavelengths are shown in Figure 26.3.

Recently, with the introduction of commercial NIR transillumination systems, contrast agents have been investigated for enhancing dental imaging. Li et al. have investigated the use of injecting indocyanine green (ICG) [73–76] in live rodents, and Abdelaziz et al. [77] have used externally applied ICG gels and liquids to determine if human approximal lesions are cavitated or not in NIR images.

Reflectance measurements for lesion activity assessment

Optical changes due to the loss of water from porous lesions can be exploited to assess lesion severity with fluorescence, thermal, and SWIR imaging [27, 78, 79]. Since arrested lesions are less permeable to water due to the highly mineralized surface layer, changes in the rate of water loss can be related to changes in lesion structure, porosity, and activity. The highly mineralized surface layer can be detected with OCT and can serve as an indicator that the lesion has been repaired or remineralized [80–82]. When lesions become arrested by mineral deposition in the outer layers of the lesion, the diffusion of fluids into the lesion is inhibited. Therefore, the lesion permeability reflects the degree of lesion activity, and it can be indirectly measured via changes in SWIR reflectance during dehydration [83]. The lesion permeability monitored with SWIR reflectance measurements at 1450 nm is extremely sensitive to the transparent surface zone thickness [83].

Optical coherence tomography

Optical coherence tomography (OCT) has great potential for imaging teeth in the SWIR due to the high transparency of enamel. OCT has been under development for dental applications for almost 25 years, although there are no dedicated OCT systems commercially available for dentists. Light scattering in sound enamel significantly decreases beyond 1300 nm, which may be advantageous for achieving greater imaging depths for optical coherence tomography. The first images of the soft and hard tissue structures of the oral cavity were acquired by Colston et al. [84, 85]. Feldchtein et al. [86] presented high-resolution dual wavelength 830- and 1280-nm images of dental hard tissues, enamel, and dentin caries and restorations *in vivo*. OCT can be used to measure the reflectivity within dental hard tissues to a depth of more than 5 mm in enamel and 1–2 mm in dentin. OCT is valuable for obtaining high-resolution images of lesion structure and severity to aid diagnosis. The advantages of using cross-polarization OCT (CP-OCT) and polarization sensitive (PS-OCT) to monitor demineralization and remineralization has been demonstrated in several studies utilizing various lesion models and natural lesions [23, 33, 82, 87–89]. The ability of PS-OCT to monitor the formation of a distinct transparent surface zone due to remineralization has also been demonstrated [80–82]. Many clinicians are primarily interested in knowing how deep the occlusal lesions have actually penetrated into the tooth so that they can decide whether a restoration is necessary. The identification of occlusal lesions penetrating to dentin is poor, with an accuracy of ~50% [90, 91]. OCT is ideally suited for monitoring and improving the diagnosis of hidden subsurface lesions. Typically, lesions spread laterally under the enamel upon contacting the more soluble and softer dentin. Therefore, OCT can be used to determine if occlusal lesions have penetrated to the underlying dentin by detecting the lateral spread across the dentinal–enamel junction [65].

Since OCT can be used to image deep into the tooth, it can be potentially used to monitor erosion and wear by monitoring the remaining enamel thickness. Studies have shown that OCT can be used to measure the enamel thickness [92]. Multiple clinical [93] and *in vitro* studies have shown that [94–96] erosion and wear can be monitored using optical coherence tomography.

Transillumination and reflectance are better suited for rapid screening for the detection of caries lesions, while OCT is better suited for acquiring tomographic images of the lesion showing structure and depth penetration that can aid in making an accurate diagnosis. Since the devices utilize similar InGaAs detectors and SWIR light sources, it is likely that the technology will merge in the future, leading to a single device that can be used for both screening and the acquisition of high-resolution images.

Caries lesions and subsurface structures from tooth occlusal surfaces

In vitro and *in vivo* studies have demonstrated that OCT can be used to determine if occlusal lesions have penetrated to the underlying dentin [65, 72] by detecting the lateral spread at the dentinal–enamel junction (DEJ). As mentioned earlier, most newly discovered lesions are in the occlusal surfaces, and radiographs do not perform well in detecting the penetration of such lesions through the enamel and into the dentin before surgical intervention is required; new methods are thus needed to detect these hidden occlusal lesions. Demineralization strongly scatters light and blocks optical penetration both into and out of the tooth. Even though the optical penetration of SWIR light can easily exceed 7 mm through sound enamel to image lesions on proximal surfaces with high contrast [59], the large increase in light scattering due to demineralization [16] typically limits optical penetration in highly scattering lesions (also dentin and bone) to 1–2 mm, thus cutting off the OCT signal before it reaches the dentinal–enamel junction (DEJ). The demineralization typically is highly localized to the pit and fissures and does not spread laterally until it reaches the DEJ, where the

FIGURE 26.5 A diagram of the optical paths considered for the detection of dentinal lesions from the occlusal surface using OCT. The path at the center of the fissure (red) is blocked due to strong scattering by the demineralized enamel, while the paths through the adjacent sound transparent enamel (green) can reach the underlying dentin where the lesion has spread laterally. The occurrence of strong subsurface reflections near the DEJ adjacent to fissures with demineralization indicates the existence of severe occlusal lesions that have penetrated to the underlying dentin and may need surgical intervention.

more soluble dentin is more easily demineralized. Therefore, most lesions extend laterally along the DEJ upon reaching the underlying dentin. OCT can detect that lateral spread since it can reach those lesion areas near the DEJ around the periphery of the lesion without having to penetrate through the demineralized, highly scattering enamel. Therefore the existence of very strong scattering/reflectivity from the surface of a pit and fissure that blocks optical penetration accompanied by strong reflectance peripheral to the fissure located below the surface at the position of the DEJ indicates the presence of an occlusal lesion that is severe enough to reach the dentin and spread significantly. The concept is illustrated in the diagram shown in Figure 26.5. Such an indication suggests that surgical intervention may be necessary, while the lack of spread suggests that the lesion is confined to the enamel and a small area of dentin and may be more successfully treated with remineralization therapy. A recent OCT image of an extracted tooth with an occlusal lesion taken using a new OCT prototype imaging system [97] that penetrates into the dentin and spreads laterally in the is shown in Figure 26.6 along with the matching microCT image dentin. The same approach can be used to image deep interproximal lesions from the tooth occlusal surface under the sound enamel [72, 98].

In the first clinical study using OCT to test this approach, 12 out of 14 lesions scheduled for surgical intervention examined *in vivo* using OCT exhibited increased reflectivity below

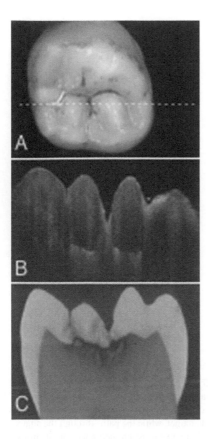

FIGURE 26.6 (A) Color, (B) OCT, and (C) microCT images of an extracted tooth with a large occlusal lesion that has spread extensively in the underlying dentin. There is demineralization in all three fissures blocking optical penetration at the center of each fissure. Strong subsurface reflections adjacent to the left and center fissures in the OCT image indicate that the lesions have penetrated to the dentin in those areas but not for the fissure on the right, where the microCT confirms that the demineralization is localized to the surface.

lesion body to better reach the true depth of the lesion. Even though such penetration is anticipated to lower the contrast of the lesion near the tooth surface, it is also expected to increase the optical penetration to deeper layers in the lesion.

The first study using high refractive index fluids to enhance the imaging of caries lesions with OCT was carried out 15 years ago by Jones et al. [26], and propylene glycol and glycerol were used to demonstrate that the application of high refractive index liquids increased penetration depth in sound enamel to a greater degree than dried or water-moistened samples. The results also demonstrated that image contrast between sound and carious enamel is dependent on the viscosity of the liquid and the degree of porosity of the carious lesion. In later studies by Kang et al. [99, 100] on extracted human teeth with occlusal lesions, OCT images were acquired after the application of water, glycerol, BABB (33% benzyl alcohol + 67% benzyl benzoate) and a Cargille liquid (Cedar Grove, NJ) (hydrogenated terphenyl 1-bromo-naphthalene) with a refractive index of 1.61. The purpose of these later studies focused on enhancing the detection of the penetration of occlusal lesions into the underlying dentin. The intensity ratio between the reflection at the surface and subsurface reflection from the lesion at the DEJ was measured at the same position for the different agents, and the contrast ratio was calculated and compared. Figure 26.7 shows OCT b-scans and a-scans at the indicated position (red arrows) acquired for one extracted tooth from that study for the four liquids; note how more and more of the body of the lesion at the center of the fissure becomes progressively visible along with the subsurface lesion areas in the dentin peripheral to the fissure. The intensity ratio is plotted in Figure 26.8 for the ten samples, and there was a significant increase in the integrated reflectivity and the contrast ratio of the subsurface lesion area for the higher refractive index liquids.

Those studies showed that optical clearing agents and image analysis methods (edge detection) can be used to increase the optical penetration and the visibility of subsurface lesions and the DEJ under sound and demineralized enamel in OCT images. The use of optical clearing agents significantly increases the visibility of subsurface lesions located under sound enamel peripheral to the pits and fissures in the occlusal surface. In addition, the visibility of the DEJ is also increased, which is potentially valuable for measuring the remaining enamel thickness for monitoring tooth wear and erosion. In a later study by Kang et al. [101, 102], it was demonstrated that a polysiloxane dental impression material (VPS) can be used to further improve the visibility of subsurface lesions and the DEJ in OCT images. OCT images are shown in Figures 26.9 and 26.10 of two extracted teeth from that study before and after application of VPS to the surfaces. The first tooth had a lesion that penetrated to the underlying dentin, while the demineralization was localized to the fissure surface for the second tooth. The DEJ is more visible after application of the VPS in Figure 26.10. Improving the visibility of the DEJ is important since the distance from the surface of the tooth to the DEJ is the remaining enamel thickness, which can be used to monitor tooth erosion and wear. The excellent performance of the VPS impression material is particularly exciting because these impression materials have been used clinically for many years and they are transparent, odorless, and

the DEJ, which suggested that the lesions had spread to the dentin; considering that none of the lesions were visible on radiographs, this is a remarkable improvement in sensitivity over radiography and other existing technology [65]. In a subsequent clinical study using a high-speed swept source OCT, it was demonstrated that 3D images of hidden occlusal lesions showing the lateral spread at the DEJ in multiple directions could be rapidly acquired [72]. In contrast to fluorescence detection methods such as the Diagnodent, OCT has the added advantage of producing an image showing the spread of demineralization as opposed to a single reading that does not indicate the depth or area of demineralization, and stains do not absorb NIR light, so there is minimal interference from stain.

Optical clearing or index matching agents can be used to increase the optical penetration into the tooth and increase the contrast of the deeper demineralization near the DEJ and also allow better resolution of the lesion depth. A liquid on the tooth surface with a refractive index closer to that of tooth enamel (1.63) reduces the surface scattering and specular reflectance, increasing the quality of the OCT images. In addition, filling the pores of the lesion greatly reduces the scattering in the lesion, allowing better light penetration through the

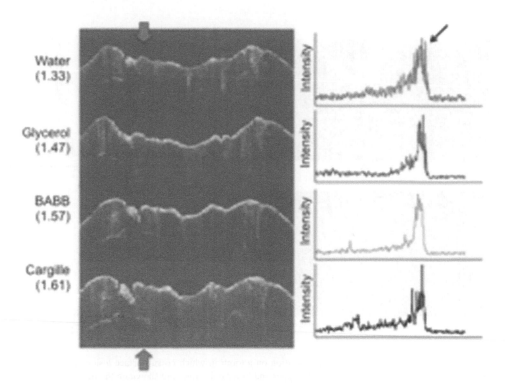

FIGURE 26.7 (Left) CP-OCT B-scans taken at the position of the dotted line in Figure 26.1 for the four liquids of varying refractive index (RI), water, glycerol, BABB, and Cargille liquid. CP-OCT A-scans extracted at the position of the upper and lower red arrows are shown on the left. The position of the two peaks analyzed is shown by the arrow for both the B-scan and A-scan images. Reprinted with permission from reference [100].

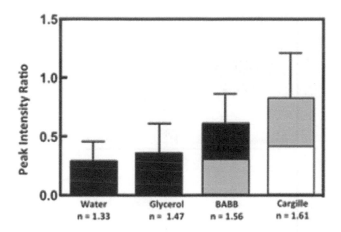

FIGURE 26.8 Mean ± S.D. of the peak intensity ratios of each of the ten liquids for ten samples. The values of the refractive index are also listed for each liquid. Groups with the same color are statistically similar (P>0.05). Reprinted with permission from Reference [99].

tasteless. Therefore, they can be used immediately as optical clearing agents for clinical OCT imaging. Moreover, since the impression material hardens in place it is easy to apply to upper (maxillary) teeth as well by holding in place for 10–15 seconds while it hardens as is currently done now with an impression tray. Polysiloxanes are used for intraocular lenses and dental impression materials, and the refractive index (RI) can be varied from 1.4 to 1.6 with RI increasing with the phenyl concentration [103, 104]. Kang used a common dental transparent vinyl polysiloxane (VPS) impression material (RI = 1.4) which showed a remarkable improvement in the visibility of subsurface structures around the fissure. The higher performance of the lower refractive index VPS impression material (RI = 1.4) suggests that index matching is not the most important criterion for increasing optical penetration. Viscosity and surface tension are also important factors because they influence the capillary penetration of the agent into the lesion pores and the narrow fissures. Vinyl polysiloxane impression materials are designed for intimate contact with tooth surfaces for accurate impressions; therefore they are hydrophilic and have a low enough viscosity to penetrate tooth fissures.

Carneiro et al. [105] investigated the use of silver nanoparticles to improve the contrast between sound and demineralized enamel in OCT images. The nanoparticles appeared to act as a contrast agent, increasing the visibility of lesion areas.

OCT Imaging of root surfaces

Optical clearing agents can also be used to increase the performance of OCT for the diagnosis of root caries and other defects on root surfaces due to the high permeability of the dentinal tubule network. The application of fluids with a higher refractive index than water greatly improves optical penetration of both sound and demineralized root surfaces. We suspect these agents act as optical clearing agents by filling the pores and tubules structures to reduce internal light scattering, and they act as index matching agents by reducing scattering and reflection at root surfaces. The diagnosis of root caries and root fractures is of increasing importance due to our aging population. In a recent study, 20 teeth with suspected root caries and dental calculus were imaged with optical coherence tomography (OCT) with and without the

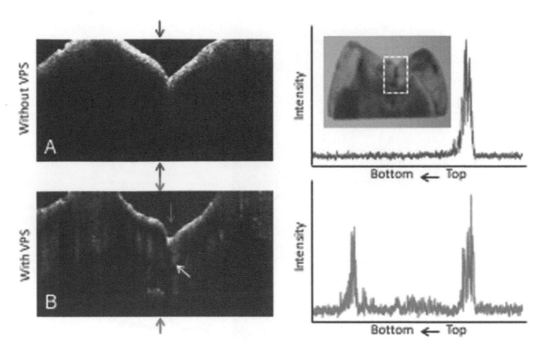

FIGURE 26.9   (Left) OCT b-scans taken at the same fixed position on a tooth in which no subsurface lesion was present, acquired without (A) and with (B) VPS. OCT a-scans extracted at the position of the upper and lower red arrows are shown on the right. Reprinted with permission from reference [102].

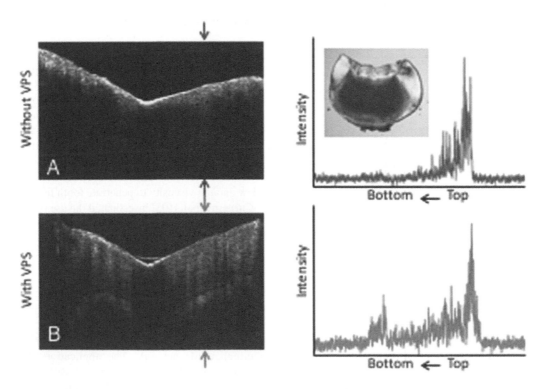

FIGURE 26.10   (A) Polarized and non-polarized (B) light microscopy images acquired of tooth sections matching the position of the OCT scans shown in Figures 26.2 and 26.4. Reprinted with permission from [102].

addition of water (W), glycerol (G), and propylene glycol (PG) [106]. The reflectance was monitored in sound and demineralized areas before and after application of W, G, and PG. The mean values ± standard deviation for the attenuation coefficient, $1/e^2$ optical penetration depth, and depth weighted reflectivity were calculated for a-scans positioned near the most severe area of each lesion for the twenty lesions. The optical penetration depth was significantly higher for PG and G (p <0.05) compared to wet and dry surfaces. However, the optical penetration was statistically similar for PG and

G (p > 0.05) [106]. The visibility of subsurface root canals was also measured before and after application of the respective fluids. Propylene glycol and glycerol have similar refractive indices of 1.43 and 1.47 respectively, but very different viscosities. Glycerol has a much higher viscosity. The performance of glycerol was slightly higher than propylene glycol, but that difference was not significant. Glycerol is well suited for use in dental imaging due to its high biocompatibility. It is a common additive for food and pharmaceuticals and is used topically, orally, and intravenously for various medical applications. Its high viscosity, low vapor pressure, and miscibility in water are also important advantages for OCT imaging. The low vapor pressure is useful for *in vitro* OCT imaging of tooth samples over an extended period of time where it is important to avoid water loss due to evaporation. The optical penetration depth increases by almost a factor of two over wet lesions and by more than a factor of five over the dry lesion if glycerol is applied. The large increase in optical penetration or decrease in optical attenuation with the addition of optical clearing agents clearly indicates that they can be used to increase the visibility of deeper regions in root caries lesions.

The identification of root fractures is also a significant problem in dentistry, and OCT can potentially be used for imaging tooth roots using endoscopic scanning systems that can be inserted into the root canal [107]. The optical penetration through the root dentin can be greatly improved with an optical clearing agent, and one can easily envision filling the root with glycerol to improve imaging. The root surfaces of the same 20 teeth were also scanned at areas closer to the apical tip of the root where the root canal was visible well below the surface, and those are shown in Figure 26.11. The line profile used for comparison is shown in gray, and the position of the a-scan chosen for further analysis is indicated by the arrows. Note that on some regions of the scan, both sides of the root canal are visible to the left of the arrows, particularly in the image with glycerol. The mean integrated reflectivity

of the root canal wall (weighted and unweighted with respect to the depth in the tooth) and the ratio of the reflectivity of the root surface and the wall of the root canal were calculated for the 20 samples. Since the distance from the root surface to the root canal varied markedly over the 20 teeth, it was necessary to factor in the depth or distance of the root canal from the surface in the comparison. The ratio of the root canal to surface intensity represents the visibility of the root canal wall, and it increases by a factor of three from wet to dry and by another factor of two from wet to G and PG. The bar chart in Figure 26.12 shows the mean ratios for the four conditions. The integrated reflectivity from the root canal wall also increased significantly progressing from dry to wet to G and PG [106].

Conclusions

One can easily envision how optical clearing agents can be utilized clinically for dental imaging, particularly an agent such as VPS that works well on both mandibular and maxillary occlusal surfaces. After an initial scan of a suspect fissure using OCT, those areas can be rescanned after application of the optical clearing agent to enhance the ability to detect the subsurface lesion either by revealing a subsurface reflection that was not visible without enhancement by the agent or by increasing the intensity of a very weak reflection for additional confirmation. It is also important to note how the speed and performance of OCT has improved over the past two decades. The OCT image shown in Figure 26.6B was acquired with a MEMS-based 3D SS-OCT system with a scanning range of 8 mm and scanning rate of 100 kHz, while the OCT images acquired in Figures 26.7, 26.9, and 26.10 were taken with a first-generation time domain OCT system with a scanning range of 6 mm and a scanning rate of only 100 Hz. Newer systems are 1000 times faster, with longer scanning range and higher sensitivity. Clinical OCT systems should soon be available for dentistry as manufacturing costs also decrease.

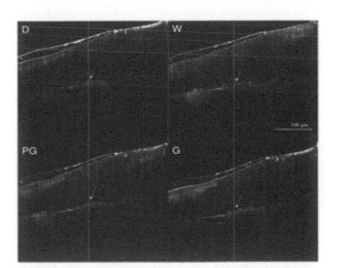

FIGURE 26.11 OCT b-scan images of a sound area of the root showing the subsurface root canal: (D) dry, (W) wet, (PG) propylene glycol, (G) glycerol. Note that the line profile used for data collection and comparison is indicated by the vertical line. The position of the root canal is indicated by an arrow. Reprinted with permission from reference [106].

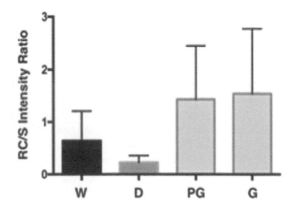

FIGURE 26.12 Chart of the mean ratio of the reflectance of the root canal over the reflectance from the root surface. Repeated wet and dry scans were averaged with (W) wet, (D) dry, (PG) propylene glycol, and (G) glycerol. Similar colors indicate that the values are not statistically different (p>0.05). Reprinted with permission from reference [106].

Acknowledgments

The author would like to acknowledge the contributions of his students and collaborators who contributed much of the work at UCSF involving the use of optical clearing agents described in this chapter, including Robert Jones, Hobin Kang, Vincent Yang, and Marwa Abdel Aziz. In addition, the author would like to acknowledge the support of the NIH/NIDCR for that work over the past 20 years.

REFERENCES

1. D. Spitzer, and J.J. ten Bosch, "The absorption and scattering of light in bovine and human dental enamel" *Calcif Tiss Res*, 17, 129–137 (1975).
2. R.S. Jones, and D. Fried, "Attenuation of 1310-nm and 1550-nm laser light through sound dental enamel" *Proc SPIE*, 46101, 87–190 (2002).
3. D. Fried, R.E. Glena, J.D. Featherstone, and W. Seka, "Nature of light scattering in dental enamel and dentin at visible and near-infrared wavelengths" *Appl Opt*, 34(7), 1278–1285 (1995).
4. K.H. Chan, and D. Fried, "Multispectral cross-polarization reflectance measurements suggest high contrast of demineralization on tooth surfaces at wavelengths beyond 1300-nm due to reduced light scattering in sound enamel" *J Biomed Opt*, 23(6), 060501 (2018).
5. J.J. ten Bosch, and J.R. Zijp, *Optical Properties of Dentin in Dentine and Dentine Research in the Oral Cavity*. IRL Press, Oxford, UK (1987).
6. G.B. Altshuler, "Optical model of the tissues of the human tooth" *J Opt Technol*, 62(8), 516–521 (1995).
7. G.M. Hale, and M.R. Querry, "Optical constants of water in the 200-nm to 200-μm wavelength region" *Appl Opt*, 12(3), 555–563 (1973).
8. J. Brinkman, J.J. ten Bosch, and P.C.F. Borsboom, "Optical quantification of natural caries in smooth surfaces of extracted teeth" *Caries Res*, 22(5), (257–262) (1988).
9. J.J. ten Bosch, H.C. van der Mei, and P.C.F. Borsboom, "Optical monitor of in vitro caries. A comparison with chemical and microradiographic determination of mineral loss in early lesions" *Caries Res*, 18(6), 540–547 (1984).
10. B. Angmar-Mansson, and J.J. ten Bosch, "Optical methods for the detection and quantification of caries" *Adv Dent Res*, 1(1), 14–20 (1987).
11. J.J. ten Bosch, and J.C. Coops, "Tooth color and reflectance as related to light scattering and enamel hardness" *J Dent Res*, 74(1), 374–380 (1995).
12. C.C. Ko, D. Tantbirojn, T. Wang, and W.H. Douglas, "Optical scattering power for characterization of mineral loss" *J Dent Res*, 79(8), 1584–1589 (2000).
13. H.M. Theuns, R.P. Shellis, A. Groeneveld, J.W.E.v. Dijk, and D.F.G. Poole, "Relationships between birefringence and mineral content in artificial caries lesions in enamel" *Caries Res*, 27(1), 9–14 (1993).
14. J.S. Wefel, B.H. Clarkson, and J.R. Heilman, "Natural root caries: A histologic and microradiographic evaluation" *J Oral Pathol*, 14(8), 615–623 (1985).
15. J. Vaarkamp, J.J. ten Bosch, and E.H. Verdonschot, "Light propagation through teeth containing simulated caries lesions" *Phys Med Biol*, 40(8), 1375–1387 (1995).
16. C.L. Darling, G.D. Huynh, and D. Fried, "Light scattering properties of natural and artificially demineralized dental enamel at 1310-nm" *J Biomed Opt*, 11(3), 034023 (2006).
17. V.B. Yang, D.A. Curtis, and D. Fried, "Cross-polarization reflectance imaging of root caries and dental calculus at wavelengths from 400–2350-nm" *J Biophoton*, 11(11), e201800113 (2018).
18. N. Pitts, *Detection, Assessment, Diagnosis and Monitoring of Caries*. Karger, Basel (2009).
19. C. Ng, E.C. Almaz, J.C. Simon, D. Fried, and C.L. Darling, "Near-infrared imaging of demineralization on the occlusal surfaces of teeth without the interference of stains" *J Biomed Opt*, 24(3), 036002 (2019).
20. C. Buhler, P. Ngaotheppitak, and D. Fried, "Imaging of occlusal dental caries (decay) with near-IR light at 1310-nm" *Opt Express*, 13(2), 573–582 (2005).
21. C. Zakian, I. Pretty, and R. Ellwood, "Near-infrared hyperspectral imaging of teeth for dental caries detection" *J Biomed Opt*, 14(6), 064047 (2009).
22. S. Chung, D. Fried, M. Staninec, and C.L. Darling, "Multispectral near-IR reflectance and transillumination imaging of teeth" *Biomed Opt Express*, 2(10), 2804–2814 (2011).
23. S.L. Chong, C.L. Darling, and D. Fried, "Nondestructive measurement of the inhibition of demineralization on smooth surfaces using polarization-sensitive optical coherence tomography" *Lasers Surg Med*, 39(5), 422–427 (2007).
24. C.L. Darling, and D. Fried, "Real-time near IR (1310 nm) imaging of CO_2 laser ablation of enamel" *Opt Express*, 16(4), 2685–2693 (2008).
25. L.H. Maung, C. Lee, and D. Fried, "Near-IR Imaging of thermal changes in enamel during laser ablation" *Proc. SPIE* 7549, 754902 (2010).
26. R.S. Jones, and D. Fried, "The effect of high index liquids on PS-OCT imaging of dental caries" *Proc. SPIE*, 5687, 34–41 (2005).
27. K. Kaneko, K. Matsuyama, and S. Nakashima, "Quantification of early carious enamel lesions by using an infrared camera" in *Early Detection of Dental Caries II Proceedings of the 4th Annual Indiana Conference*. Indiana: Indiana University, 483–499 (1999).
28. M. Ando, G.K. Stookey, and D.T. Zero, "Ability of quantitative light-induced fluorescence (QLF) to assess the activity of white spot lesions during dehydration" *Am J Dent*, 19(1), 15–18 (2006).
29. F.C. Besic, and M.R. Wiemann, "Dispersion staining, dispersion, and refractive indices in early enamel caries" *J Dent Res*, 51(4), 973–985 (1971).
30. Z. Meng, X.S. Yao, H. Yao, Y. Liang, T. Liu, Y. Li, G. Wang, and S. Lan, "Measurement of the refractive index of human teeth by optical coherence tomography" *J Biomed Opt*, 14(3), 034010 (2009).
31. P.E. Benson, A. Ali Shah, and D. Robert Willmot, "Polarized versus nonpolarized digital images for the measurement of demineralization surrounding orthodontic brackets" *Angle Orthod*, 78(2), 288–293 (2008).

32. M.J. Everett, B.W. Colston, U.S. Sathyam, L.B.D. Silva, D. Fried, and J.D.B. Featherstone, "Non-invasive diagnosis of early caries with polarization sensitive optical coherence tomography (PS-OCT)" *Proc SPIE*, 3593, 177–183 (1999).

33. D. Fried, J. Xie, S. Shafi, J.D.B. Featherstone, T. Breunig, and C.Q. Lee, "Early detection of dental caries and lesion progression with polarization sensitive optical coherence tomography" *J Biomed Opt*, 7(4), 618–627 (2002).

34. D. Fried, J.D.B. Featherstone, C.L. Darling, R.S. Jones, P. Ngaotheppitak, and C.M. Buehler, "Early caries imaging and monitoring with Near-IR light" *Dent Clin N Am*, 49(4), 771–794 (2005).

35. W.J. Schmidt, and A. Keil, *Polarizing Microscopy of Dental Tissues*. Pergamon Press, New York (1971).

36. K. Ekstrand, L. Holmen, and K. Qvortrup, "A polarized light and scanning electron microscopic study of human fissure and lingual enamel of unerupted mandibular third molars" *Caries Res*, 33(1), 41–49 (1999).

37. L. Holmen, A. Thylstrup, B. Ogaard, and F. Kragh, "A polarized light microscopic study of progressive stages of enamel caries in vivo" *Caries Res*, 19(4), 348–354 (1985).

38. W.F. Waggoner, and J.J. Crall, "A study of the carious lesion utilizing radiography, polarized light microscopy, and scanning electron microscopy" *Quintessence Int Dent Dig*, 15(11), 1163–1174 (1984).

39. A.I. Darling, "Studies of the early lesion of enamel caries with transmitted light, polarised light and radiography" *Br Dent J*, 101, 289–287 and 328–341 (1956).

40. R.C.G. De Medeiros, J.D. Soares, and F.B. De Sousa, "Natural enamel caries in polarized light microscopy: Differences in histopathological features derived from a qualitative versus a quantitative approach to interpret enamel birefringence" *J Microsc*, 246(2), 177–189.

41. H.H. Chauncey, R.L. Glass, and J.E. Alman, "Dental caries, principal cause of tooth extraction in a sample of US male adults" *Caries Res*, 23(3), 200–205 (1989).

42. L.M. Kaste, R.H. Selwitz, R.J. Oldakowski, J.A. Brunelle, D.M. Winn, and L.J. Brown, "Coronal caries in the primary and permanent dentition of children and adolescents 1–17 years of age: United States, 1988–1991" *J Dent Res*, 75, 631–641 (1996).

43. D.M. Winn, J.A. Brunelle, R.H. Selwitz, L.M. Kaste, R.J. Oldakowski, A. Kingman, and L.J. Brown, "Coronal and root caries in the dentition of adults in the United States, 1988–1991" *J Dent Res*, 75, 642–651 (1996).

44. J.M. ten Cate, and J.P. van Amerongen, "Caries diagnosis: Conventional methods" *Early Detection of Dental Caries*. Proceedings of the 1st Annual Indiana Conference, Indiana University, 27–37 (1996).

45. NIH, "Diagnosis and management of dental caries throughout life" *NIH Consensus Statement*, 18(1) (2001).

46. D.J. Dodds, "Dental caries diagnosis: Toward the 21st century" *Nat Med*, 2, 281 (1996).

47. J.P. Carlos, and J.A. Brunelle, *Oral Health Surveys of the NIDR*. Diagnostic Criteria and Procedures U.S. Department of Health and Human Services, Bethesda (1991).

48. K. Ekstrand, V. Qvist, and A. Thylstrup, "Light microscope study of the effect of probing in occlusal surfaces" *Caries Res*, 21(4), 368–374 (1987).

49. E.A.M. Kidd, D.N.J. Ricketts, and N.B. Pitts, "Occlusal caries diagnosis: A changing challenge for clinicians and epidemiologists" *J Dent Res*, 21(6), 3232–3331 (1993).

50. A. Lussi, A. Firestone, V. Schoenberg, P. Hotz, and H. Stich, " In vivo diagnosis of fissure caries using a new electrical resistance monitor" *Caries Res*, 29(2), 81–87 (1991).

51. A. Lussi, S. Imwinkelreid, N.B. Pitts, C. Longbottom, and E. Reich, "Performance and reproducibility of a laser fluorescence system for detection of occlusal caries in vitro" *Caries Res*, 33(4), 261–266 (1999).

52. K. Ekstrand, D.N.J. Ricketts, E.A.M. Kidd, V. Qvist, and A. Thylstrup, "Reproducibility and accuracy of three methods for assessment of demineralization depth on the occlusal surface" *Caries Res*, 31, 224–231 (1997).

53. C.M. Pine, "Fiber-optic transillumination (FOTI) in caries diagnosis" in *Early Detection of Dental Caries. Proceedings of the 1st Annual Indiana Conference*. Indiana: Indiana University, 51–66 (1996).

54. J. Peltola, and J. Wolf, "Fiber optics transillumination in caries diagnosis" *Proc Finn Dent Soc*, 77(4), 240–244 (1981).

55. J. Barenie, G. Leske, and L.W. Ripa, "The use of fiber optic transillumination for the detection of proximal caries" *Oral Surg Oral Med Oral Pathol*, 36(6), 891–897 (1973).

56. R.D. Holt, and M.R. Azeevedo, "Fiber optic transillumination and radiographs in diagnosis of approximal caries in primary teeth" *Commun Dent Health*, 6(3), 239–247 (1989).

57. C.M. Mitropoulis, "The use of fiber optic transillumination in the diagnosis of posterior approximal caries in clinical trials" *Caries Res*, 19(4), 379–384 (1985).

58. A. Schneiderman, M. Elbaum, T. Schultz, S. Keem, M. Greenebaum, and J. Driller, "Assessment of dental caries with digital imaging fiber-optic transillumination (DIFOTI): In vitro study" *Caries Res*, 31(2), 103–110 (1997).

59. G. Jones, R.S. Jones, and D. Fried, "Transillumination of interproximal caries lesions with 830-nm light" *Proc SPIE*, 5313, 17–22 (2004).

60. R. Jones, G. Huynh, G. Jones, and D. Fried, "Near-infrared transillumination at 1310-nm for the imaging of early dental decay" *Opt Express*, 11(18), 2259–2265 (2003).

61. M. Staninec, C. Lee, C.L. Darling, and D. Fried, "In vivo near-IR imaging of approximal dental decay at 1,310 nm" *Lasers Surg Med*, 42(4), 292–298 (2010).

62. K. Hirasuna, D. Fried, and C.L. Darling, "Near-IR imaging of developmental defects in dental enamel" *J Biomed Opt*, 13(4), 1–7, 044011 (2008).

63. C. Lee, D. Lee, C.L. Darling, and D. Fried, "Nondestructive assessment of the severity of occlusal caries lesions with near-infrared imaging at 1310 nm" *J Biomed Opt*, 15(4), 047011 (2010).

64. L. Karlsson, A.M.A. Maia, B.B.C. Kyotoku, S. Tranaeus, A.S.L. Gomes, and W. Margulis, "Near-infrared transillumination of teeth: Measurement of a system performance" *J Biomed Opt*, 15(3), 036001 (2010).

65. M. Staninec, S.M. Douglas, C.L. Darling, K. Chan, H. Kang, R.C. Lee, and D. Fried, "Nondestructive clinical assessment of occlusal caries lesions using Near-IR imaging methods" *Lasers Surg Med*, 43(10), 951–959 (2011).

66. J.C. Simon, S.A. Lucas, R.C. Lee, M. Staninec, H. Tom, K.H. Chan, C.L. Darling, and D. Fried, "Near-IR transillumination and reflectance imaging at 1300-nm and 1500–1700-nm for in vivo caries detection" *Lasers Surg Med*, 48(6), 828–836 (2016).

67. M. Abdelaziz, and I. Krejci, "DIAGNOcam: A near infrared digital imaging transillumination (NIDIT) technology" *Int J Esthet Dent*, 10(1), 158–165 (2015).

68. M. Abdelaziz, I. Krejci, T. Perneger, A. Feilzer, and L. Vazquez, "Near infrared transillumination compared with radiography to detect and monitor proximal caries: A clinical retrospective study" *J Dent*, 70, 40–45 (2018).

69. J. Kuhnisch, F. Sochtig, V. Pitchika, R. Laubender, K.W. Neuhaus, A. Lussi, and R. Hickel, "In vivo validation of near-infrared light transillumination for interproximal dentin caries detection" *Clin Oral Investig*, 20(4), 821–829 (2015).

70. J. Wu, and D. Fried, "High contrast near-infrared polarized reflectance images of demineralization on tooth buccal and occlusal surfaces at lambda = 1310-nm" *Lasers Surg Med*, 41(3), 208–213 (2009).

71. W.A. Fried, K.H. Chan, D. Fried, and C.L. Darling, "High contrast reflectance imaging of simulated lesions on tooth occlusal surfaces at near-IR wavelengths" *Lasers Surg Med*, 45(8), 533–541 (2013).

72. J.C. Simon, H. Kang, M. Staninec, A.T. Jang, K.H. Chan, C.L. Darling, R.C. Lee, and D. Fried, "Near-IR and CP-OCT imaging of suspected occlusal caries lesions" *Lasers Surg Med*, 49(3), 215–224 (2017).

73. Z. Li, T. Hartzler, A. Ramos, M.L. Osborn, Y. Li, S. Yao, and J. Xu, "Optimal imaging windows of indocyanine green-assisted near-infrared dental imaging with rat model and its comparison to X-ray imaging" *J Biophoton*, 8(1–2), e201960232 (2020).

74. Z. Li, S. Yao, and J. Xu, "Indocyanine-green-assisted near-infrared dental imaging – The feasibility of in vivo imaging and the optimization of imaging conditions" *Sci Rep*, 9(1), 8238 (2019).

75. Z. Li, W. Zaid, T. Hartzler, A. Ramos, M.L. Osborn, Y. Li, S. Yao, and J. Xu, "Indocyanine green-assisted dental imaging in the first and second near-infrared windows as compared with X-ray imaging" *Ann N Y Acad Sci*, 1448(1), 42–51 (2019).

76. Z. Li, S. Yao, J. Xu, Y. Wu, C. Li, and Z. He, "Endoscopic near-infrared dental imaging with indocyanine green: A pilot study" *Ann N Y Acad Sci*, 1421(1), 88–96 (2018).

77. M. Abdelaziz, I. Krejci, and D. Fried, "Enhancing the detection of proximal cavities on near infrared transillumination images with Indocyanine Green (ICG) as a contrast medium: In vitro proof of concept studies" *J Dent*, 91, 103222 (2019).

78. C.M. Zakian, A.M. Taylor, R.P. Ellwood, and I.A. Pretty, "Occlusal caries detection by using thermal imaging" *J Dent*, 38(10), 788–795 (2010).

79. P. Usenik, M. Burmen, A. Fidler, F. Pernus, and B. Likar, "Near-infrared hyperspectral imaging of water evaporation dynamics for early detection of incipient caries" *J Dent*, 42(10), 1242–1247 (2014).

80. H. Kang, C.L. Darling, and D. Fried, "Nondestructive monitoring of the repair of enamel artificial lesions by an acidic remineralization model using polarization-sensitive optical coherence tomography" *Dent Mater*, 28(5), 488–494 (2012).

81. R.S. Jones, and D. Fried, "Remineralization of enamel caries can decrease optical reflectivity" *J Dent Res*, 85(9), 804–808 (2006).

82. R.S. Jones, C.L. Darling, J.D. Featherstone, and D. Fried, "Imaging artificial caries on the occlusal surfaces with polarization-sensitive optical coherence tomography" *Caries Res*, 40(2), 81–89 (2006).

83. R.C. Lee, C.L. Darling, and D. Fried, "Assessment of remineralization via measurement of dehydration rates with thermal and near-IR reflectance imaging" *J Dent*, 43(8), 1032–1042 (2015).

84. B. Colston, M. Everett, L. Da Silva, L. Otis, P. Stroeve, and H. Nathel, "Imaging of hard and soft tissue structure in the oral cavity by optical coherence tomography" *Appl Opt*, 37(19), 3582–3585 (1998).

85. B.W. Colston, U.S. Sathyam, L.B. DaSilva, M.J. Everett, P. Stroeve, and L. Otis, "Dental OCT" *Opt Express*, 3(3), 230–238 (1998).

86. F.I. Feldchtein, G.V. Gelikonov, V.M. Gelikonov, R.R. Iksanov, R.V. Kuranov, A.M. Sergeev, N.D. Gladkova, M.N. Ourutina, J.A. Warren, and D.H. Reitze, "In vivo OCT imaging of hard and soft tissue of the oral cavity" *Opt Express*, 3(3), 239–251 (1998).

87. R.S. Jones, C.L. Darling, J.D.B. Featherstone, and D. Fried, "Remineralization of in vitro dental caries assessed with polarization sensitive optical coherence tomography" *J Biomed Opt*, 11(1), 014016 (2006).

88. K.H. Chan, A.C. Chan, W.A. Fried, J.C. Simon, C.L. Darling, and D. Fried, "Use of 2D images of depth and integrated reflectivity to represent the severity of demineralization in cross-polarization optical coherence tomography" *J Biophoton*, 8(1–2), 36–45 (2015).

89. R.C. Lee, H. Kang, C.L. Darling, and D. Fried, "Automated assessment of the remineralization of artificial enamel lesions with polarization-sensitive optical coherence tomography" *Biomed Opt Express*, 5(9), 2950–2962 (2014).

90. J.D. Bader, and D.A. Shugars, "The evidence supporting alternative management strategies for early occlusal caries and suspected occlusal dentinal caries" *J Evid Based Dent Pract*, 6(1), 91–100 (2006).

91. J.D. Bader, D.A. Shugars, and A.J. Bonito, "A systematic review of the performance of methods for identifying carious lesions" *J Public Health Dent*, 62(4), 201–213 (2002).

92. A. Algarni, H. Kang, D. Fried, G.J. Eckert, and A.T. Hara, "Enamel thickness determination by optical coherence tomography: In vitro validation" *Caries Res*, 50(4), 400–406 (2016).

93. C.H. Wilder-Smith, P. Wilder-Smith, H. Kawakami-Wong, J. Voronets, K. Osann, and A. Lussi, "Quantification of dental erosions in patients with GERD using optical coherence tomography before and after double-blind, randomized treatment with esomeprazole or placebo" *Am J Gastroenterol*, 104(11), 2788–2795 (2009).

94. M.A. Alghilan, F. Lippert, J.A. Platt, G.J. Eckert, C. Gonzalez-Cabezas, D. Fried, and A.T. Hara, "In vitro longitudinal evaluation of enamel wear by cross-polarization optical coherence tomography" *Dent Mater*, 35(10), 1464–1470 (2019).

95. M.A. Alghilan, F. Lippert, J.A. Platt, G.J. Eckert, C. Gonzalez-Cabezas, D. Fried, and A.T. Hara, "Impact of surface micromorphology and demineralization severity on enamel loss measurements by cross-polarization optical coherence tomography" *J Dent*, 81, 52–58 (2019).

96. J. Seeliger, M. Machoy, R. Koprowski, K. Safranow, T. Gedrange, and K. Wozniak, "Enamel thickness before and after orthodontic treatment analysed in optical coherence tomography" *BioMed Res Int*, 2017, 8390575 (2017).

97. A.F. Zuluaga, V. Yang, J. Jabbour, T. Ford, N. Kemp, and D. Fried, "Real-time visualization of hidden occlusal and approximal lesions with an OCT dental handpiece" *Proc. SPIE*, 10857, 108570E (2019).

98. P. Ngaotheppitak, C.L. Darling, D. Fried, J. Bush, and S. Bell, "PS-OCT of occlusal and interproximal caries lesions viewed from occlusal surfaces" *Proc SPIE*, 6137, 6137L (2006).

99. H. Kang, C.L. Darling, and D. Fried, "Enhancing the detection of hidden occlusal caries lesions with OCT using high index liquids" *Proc SPIE*, 8929, 89290 (2014).

100. H. Kang, C.L. Darling, and D. Fried, "Enhanced detection of dentinal lesions in OCT images using the RKT transformation" *Proc SPIE*, 9306, 93060P (2015).

101. H. Kang, C.L. Darling, and D. Fried, "Enhancement of OCT images with vinyl polysiloxane (VPS)" *Proc. SPIE*, 9692, 9692Y (2016).

102. H. Kang, C.L. Darling, and D. Fried, "Use of an optical clearing agent to enhance the visibility of subsurface structures and lesions from tooth occlusal surfaces" *J Biomed Opt*, 21(8), 081206 (2016).

103. P. Chevalier, and D. Ou, "High refractive index polysiloxanes and their preparation" Patent US7429638B2 2008.).

104. X. Hao, J.L. Jeffery, T.P.T. Le, G. McFarland, G. Johnson, R.J. Mulder, Q. Garrett, F. Manns, D. Nankivil, E. Arrieta, A. Ho, J.M. Parel, and T.C. Hughes, "High refractive index polysiloxane as injectable, in situ curable accommodating intraocular lens" *Biomaterials*, 33(23), 5659–5671 (2012).

105. A.F. Carneiro, C.B.O. Mota, A.F. Sousa, E.J. Da Silva, A.F. Da Silva, M.E.M. Gerbi, and A.S. Gomes, "Silver nanoparticles as optical clearing agent enhancers to improve caries diagnostic by optical coherence tomography" *Proc. SPIE*, 10507, 1050719 (2018).

106. V.B. Yang, D.A. Curtis, and D. Fried, "Use of optical clearing agents for imaging root surfaces with optical coherence tomography" *IEEE J Sel Top Quantum Electron*, 25(1), 1–7 (2018).

107. H. Shemesh, G. van Soest, M.K. Wu, and P.R. Wesselink, "Diagnosis of vertical root fractures with optical coherence tomography" *J Endod*, 34(6), 739–742 (2008).

27

Optical clearing of adipose tissue

Irina Yu. Yanina, Yohei Tanikawa, Daria K. Tuchina, Polina A. Dyachenko (Timoshina),
Yasunobu Iga, Shinichi Takimoto, Elina A. Genina, Alexey N. Bashkatov, Georgy S. Terentyuk,
Nikita A. Navolokin, Alla B. Bucharskaya, Galina N. Maslyakova, and Valery V. Tuchin

CONTENTS

Introduction

Structure of adipose tissue

Mammalian adipose tissue (AT), including humans, is mainly composed of fatty lobules, which are minimal essential units [1–4]. A fat lobule with the size of millimeters consists of 10^2 to 10^3 adipocytes. Every lobule is anatomically separated by fibrous septa, where blood microcirculation supplies nutrition to adipocytes [5]. In addition, the septa functionally support the lobules, provide resistance to the enlarging lobules, and offer a scaffold to which blood vessels, nerves, and lymphatics are attached [6–8].

White adipose tissue (WAT) is neither uniform nor inflexible because it undergoes constant remodeling, adapting in size and number of adipocytes to changes in nutrient availability and hormonal environment. Fat depots from different areas of the body display distinct structural and functional properties and have disparate roles in pathology. Anatomically, WAT is located at visceral, subcutaneous, intermuscular, and intramuscular sites in mammals [9–11]. The two major types of WAT are visceral fat, localized within the abdominal cavity and mediastinum, and subcutaneous fat in the hypodermis. Subcutaneous WAT is found in the abdomen, hip, thigh, and gluteal locations.

Thickness of subcutaneous AT varies from 6.64 ± 3.16 mm (for men) to 34.02 ± 5.42 mm for women with cellulite. Thickness of subcutaneous AT for women without cellulite is 8.34 ± 2.44 mm [12]. Ho et al. [13] shows that the thickness of subcutaneous layer for women varies from 6.8 ± 2.3 mm to 13.6 ± 4.7 mm. The influence of ageing on subcutaneous fat thickness was investigated by Petrofsky et al. [7]. They found that for the younger group of volunteers (average age of 25.7 ± 2.9 years) the thickness is 8.2 ± 1.6 mm, whereas for the older group of volunteers (average age of 66.9 ± 18.3 years) the thickness is 7.1 ± 1.1 mm.

DOI: 10.1201/9781003025252-31

The composition of AT is known to vary with changes in the adipose mass. During expansion of the adipose mass observed during growth or with the development of obesity, the lipid-to-protein ratio increases and the relative water content decreases [14, 15]. For example, Geri et al. [16] showed that with weight increase, water content in pig AT decreases from 13% to 9.2%, whereas lipid content increases from 73.2% to 75.7%. Salans et al. [17] found that lipid content in adipocyte cells increases from 0.44 ± 0.02 μg lipid/cell for non-obese humans to 0.88 ± 0.02 μg lipid/cell for obese humans (subcutaneous AT in the gluteal, abdomen, and triceps areas); and from 0.39 ± 0.02 μg lipid/cell for non-obese humans to 0.81 ± 0.02 μg lipid/cell for obese humans (deep AT, including preperitoneal, mesentery, and omentum).

Unilocular adipocytes are usually round in shape. Adipocyte diameter changes during development [18]. In humans, they are from 40 to 50 μm in the fetus, from 50 to 80 μm in newborns [19–23], from 90 to 130 μm in children, from 50 to 200 μm in normal adults, and from 90 to 270 μm in obese adults [15, 17, 24–30]. The maximum size is limited by oxygen diffusion and by interaction with the extracellular matrix. Fang et al. [31] demonstrated that the distribution of the fat cells is essentially bimodal, with two separate populations of smaller and larger cells. Fraction of the small cells is $(14.9 \pm 2.4)\%$ (subcutaneous fat, healthy volunteers), $(31.4 \pm 5.0)\%$ (subcutaneous fat, patients with type 2 diabetes mellitus), $(14.0 \pm 3.5)\%$ (omentum, healthy volunteers), $(30.5 \pm 5.3)\%$ (omentum, patients with type 2 diabetes mellitus), $(14.2 \pm 2.0)\%$ (mesentery, healthy volunteers), and $(23.6 \pm 4.9)\%$ (mesentery, patients with type 2 diabetes mellitus). They also found that mean adipocyte size is (105.7 ± 3.2) μm for subcutaneous fat of healthy volunteers, (91.2 ± 4.5) μm for subcutaneous fat of patients with type 2 diabetes mellitus, (95.1 ± 5.4) μm for omentum of healthy volunteers, (82.4 ± 4.9) μm for omentum of patients with type 2 diabetes mellitus, (91.9 ± 4.0) μm for mesentery of healthy volunteers, and (81.5 ± 4.2) μm for mesentery of patients with type 2 diabetes mellitus. The mean size of small cell fraction is (42.1 ± 1.7) μm (subcutaneous fat, healthy volunteers), (39.0 ± 2.6) μm (subcutaneous fat, patients with type 2 diabetes mellitus), (43.1 ± 3.4) μm (omentum, healthy volunteers), (33.9 ± 1.8) μm (omentum, patients with type 2 diabetes mellitus), (41.6 ± 1.5) μm (mesentery, healthy volunteers), and (32.9 ± 2.0) μm (mesentery, patients with type 2 diabetes mellitus) [31]. The mean size of large cell fraction is (116.9 ± 2.7) μm (subcutaneous fat, healthy volunteers), (114.4 ± 3.3) μm (subcutaneous fat, patients with type 2 diabetes mellitus), (104.0 ± 5.1) μm (omentum, healthy volunteers), (103.6 ± 3.4) μm (omentum, patients with type 2 diabetes mellitus), (100.1 ± 4.1) μm (mesentery, healthy volunteers), and (97.0 ± 2.9) μm (mesentery, patients with type 2 diabetes mellitus) [31]. McLaughlin et al. [32] showed that the size of the small subcutaneous adipose cells is (67 ± 8) μm (for males) and (72 ± 10) μm (for females). Bjorntorp and Sjostrom [33] reported that the mean size of human adipocyte cells is (134.0 ± 2.2) μm in femoral area, (126.8 ± 2.7) μm in the abdominal area, and (131.8 ± 2.0) μm in the gluteal area. For pig AT, the adipocyte size increases from 70 μm to 80 μm with weight increase [16]. Heredia et al. [34] investigated subcutaneous adipocytes in rats and found that mean cell size is

(55.84 ± 2.95) μm (for 6-month animals), (67.65 ± 5.56) μm (for 14-month animals), and (58.04 ± 3.85) μm (for 20-month animals).

In the space between the cells, there are blood capillaries (arterial and venous plexus), nerves, and reticular fibrils connecting each cell and providing metabolic activity of AT [29, 30, 35]. Typical blood vessel sizes in subcutaneous AT vary from 25 to 120 μm [36–38]. Blood volume fraction in human subcutaneous fat is 5% [38, 39] with oxygen saturation 90.8% [40]. For swine AT, Nachabe et al. [41] found the blood volume fraction to be $(0.8 \pm 0.1)\%$ with oxygen saturation $(96.1 \pm 2.3)\%$.

Adipocytes plays a paramount role in the homeostatic control of whole body metabolism [42–44]. Their principal function is to monitor energy balance by storing triacylglycerol in periods of energy excess and mobilizing it during energy deprivation. Besides the classical function of storing fat, adipocytes secrete numerous lipid and protein factors. In total, they are considered to constitute a major endocrine organ. They have a profound impact on the metabolism of other tissues, the regulation of appetite, insulin sensitivity, immunological response, and vascular disease. The hormones and adipokines, as well as other biologically active agents released from fat cells, affect many physiological and pathological processes [3]. Visceral obesity correlates with an increased risk of insulin resistance and cardiovascular disease, while an increase of subcutaneous fat is associated with favorable plasma lipid profiles [17, 25]. Visceral adipocytes show higher lipogenic and lipolytic activities and produce more pro-inflammatory cytokines, while subcutaneous adipocytes are the main source of leptin and adiponectin. Moreover, AT is associated with skeletal muscles (intramyocellular and intermuscular fat), and with the epicardium is believed to provide fuel for skeletal and cardiac muscle contraction. However, increased mass of either epicardial or intermuscular AT correlates with cardiovascular risk, while the presence of the intramyocellular fat is a risk factor for the development of insulin resistance. WAT is unique in that it can account for as little as 3% of total body weight in elite athletes or as much as 70% in the morbidly obese [26]. With the development of obesity, WAT undergoes a process of tissue remodeling in which adipocytes increase in both number (hyperplasia) and size (hypertrophy). Metabolic derangements associated with obesity, including type 2 diabetes, occur when WAT growth through hyperplasia and hypertrophy cannot keep pace with the energy storage needs associated with chronic energy excess.

Although the major cells of AT are the adipocyte, this is not the only cell type presented in AT, nor is it the most abundant [45]. Other cell types described include stem cells, preadipocytes, macrophages, neutrophils, lymphocytes, and endothelial cells.

Characterization of fat in terms of localization and innervation was the first possible upon application of solvent-based tissue optical clearing (TOC) performed by some groups [46–49]. The most detailed 3D view on fat under general and cold-exposure conditions has recently been presented by Chi et al. [48, 50], who, by building upon iDISCO (immunolabeling-enabled 3DISCO), described a novel TOC method focused on AT, termed Adipo-Clear. Although consisting of numerous steps, this approach is straightforward, does not require any specialized instruments, and can be completed within 12 days for the entire posterior subcutaneous fat depot (2 days for delipidation and clearing followed by 10 days for staining with primary and secondary antibodies). Accompanied by LSFM (Ultramicroscope II, LaVisionBiotec), it revealed striking differences in the distribution of uncoupling protein 1 (a hallmark protein of brown fat) expressing cells within posterior subcutaneous fat depot, which was further correlated with asymmetric prevalence of sympathetic innervation. A relevant advantage that further underscores Adipo–Clear utilization for fat imaging is postdelipidation autoflorescent signal, which reflects perilipin staining and therefore might be used to detect tissue architecture and precisely delineate cells contours. This is especially important as Adipo-Clear requires 8–10 days for immunostaining. It should be also noted that tissue treatment with methanol, a key Adipo-Clear reagent, might impair the structure of particular epitopes, as it was already shown for CD31 and CD105, common vascular epitopes. This encouraged Cao et al. [51] to search for alternatives and resulted in the development of a modified iDISCO technique for TOC, in which methanol-based permeabilization is replaced by detergent-based steps. In addition, the authors reported the compatibility of such an approach with an intravenous injection of Alexa dye-conjugated Isolectin to effectively label endothelial cells within less than 24 h. Co-staining with CD31 and imaging of *Tek-Cre; Rosa26-LSL-tdTomato* mice (in which vasculature is labeled with *tdTomato*) revealed that >99.9% of CD31 cells were also Isolectin positive, while only 89.4% of Isolectin positive cells were also *tdTomato* positive. This means that Isolectin is a highly specific and sensitive marker of endothelial cells and, interestingly, exposes imperfection of Cre-activity in a fraction of endothelial cells. Finally, the aforementioned approach allowed the authors to describe events of vascular plasticity in WAT of both wild-type and obese mice strains in a response to a cold challenge, and they proved this plasticity to be catecholamine-mediated.

Optical properties of adipose tissue

Optical spectroscopy techniques have the unique advantage to serve as a promising tissue analysis tool because they provide accurate quantitative information based on their difference in light absorption and scattering properties across visible-near infrared (NIR) range. Absorption of the AT is defined in this spectral range mainly by absorption of blood hemoglobin, lipids, collagen, and water [8, 14, 16, 38–41, 52–55]. Optical properties of blood were recently well investigated and summarized in reference [56] (for hematocrit 45%) and presented in Figure 27.1. The main absorption bands of blood in the visible are at the wavelengths 415, 542, and 576 nm (oxygenated hemoglobin), and 433, 556, and 758 nm (deoxygenated hemoglobin) [54, 56].

Data on lipid content in AT can be found in Refs. [16, 40, 52, 57–62]. Dev et al. [52] demonstrated that for beige AT, the lipid fraction is 0.628 ± 0.076, falling in between the lipid fraction of WAT (0.7 ± 0.095) and brown AT (0.518 ± 0.112). Lanka et al. [40] show that lipid concentration in AT is (895.2 ± 203.5) mg/cm^3. Woodard and White [57] found the lipid content to be (74.27 ± 12.95)%. Torricelli et al. [58] showed that lipid content in abdominal AT is (77.7 ± 17.1)%. For subcutaneous swine, AT Nachabe et al. [59] found the lipid content to be (89.6 ± 5.0)% and for visceral swine AT as (74.0 ± 4.1)%. At the same time, Zamora-Rojas et al. [60], with reference to Aberle et al. [61], Geri et al. [16], and Lebret and Mourot [62], reported that for pig AT, lipid content is 10.23%. Figure 27.2 shows absorption spectrum of pork lard (Figure 27.2a) [63] and mice liver fat (Figure 27.2 b) [64]. Absorption bands of lipids have maxima at 760, 930, 1042, 1211, 1393, 1414, 1720, 1760, and 2142 nm [41, 54, 63, 64].

Water content in AT was measured in Refs. [14, 40, 57–59, 65–67]. DiGirolamo and Owens [14] investigated water content in rat AT and found that intracellular water varies from 5% to 7% in small fat cells and from 1% to 3% in large ones. Lanka et al. [40] found water concentration in adult abdominal AT to be (183.2 ± 106.9) mg/cm^3. Woodard and White [57] reported that water content in AT is (21.0 ± 9.6)%. Torricelli et al. [58] measured water content in human abdominal AT as (21.7 ± 16.8)%. For swine, the water volume fraction was found to be 0.122 ± 0.05 for subcutaneous AT, whereas for visceral AT it was 0.249 ± 0.09 [59]. Ballard et al. [65] measured the water content in rat AT as (4.7 ± 0.2)%, whereas Reinoso et al. [66] found the water content to be (17.5 ± 2.0)%. Nishimura et al. [67] measured water content in human subcutaneous AT as (10.9 ± 1.4)%.

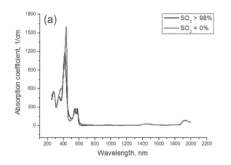
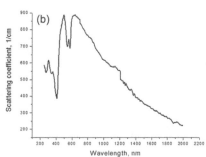
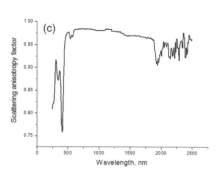

FIGURE 27.1 Spectral properties of blood (Hct = 45%): absorption coefficient μ_a, black line SO$_2$ > 98% and red line SO$_2$ = 0% (a); scattering coefficient μ_s (b); scattering anisotropy factor g (c) [56].

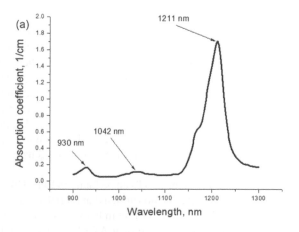

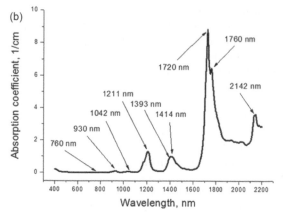

FIGURE 27.2 Absorption spectrum of pork lard (a) [63] and mice liver fat (b) [64].

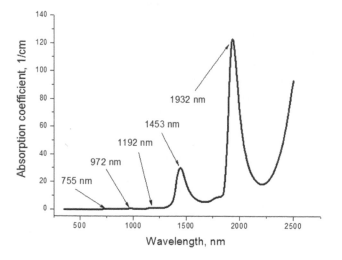

FIGURE 27.3 Absorption spectrum of liquid water. Summarized from [68–85].

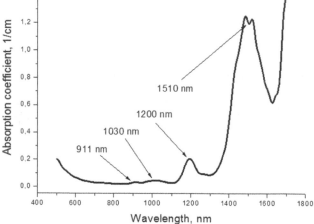

FIGURE 27.4 Absorption spectrum of collagen [88]. Concentration is 120 mg/cm³.

Optical properties of water were investigated in the Refs. [68–85], and they are summarized in Figure 27.3. It is clear that in the visible and NIR, water has absorption bands with maxima at 755, 972, 1192, 1453, and 1932 nm [54].

Collagen and other proteins form the main carcasses of ATs. Lanka et al. [40] found collagen content in adult abdomen AT to be (68.1 ± 56.8) mg/cm³ or (4.8 ± 3.6)% taking into account that collagen mass density at physiological hydration is 1.41 g/cm³ [86] or 1.42 g/cm³ [87]. Woodard and White [57] reported that collagen content in AT is (4.4 ± 3.5)%. The main absorption bands of collagen have maximum at 911, 1030, 1200, and 1510 nm [54]. Figure 27.4 shows the absorption spectrum of collagen solution for concentration 120 mg/cm³ [88] or ~8.5%.

The main scatterers of AT are spherical droplets of lipids, which are uniformly distributed within adipocytes at physiological temperatures. The AT optical properties have been measured with integrating sphere technique in the visible and NIR ranges [60, 89–94], and the results are summarized in Table 27.1. Diffuse reflectance spectroscopy was used for measurement AT optical properties [8, 59, 95–98], and the results are summarized in Table 27.2. Table 27.3 presented optical properties of AT obtained with the time-domain technique [40, 58, 99].

Refractive indices of human and animal AT are presented in Chapter 7.

Different approaches to optical clearing of adipose tissue

Tissue optical clearing (TOC) is a technique the goal of which is to transform tissue into optically transparent specimen not only for *in vitro* but also for *ex vivo* or *in vivo* biomedical examination, through application of different optical clearing agents (OCAs), such as glycerol, iohexol (Omnipaque®), propylene glycol, polyethylene glycol, dimethyl sulfoxide (DMSO), glucose, and fructose solutions [55, 100–112].

Despite the large number of publications, the problem of finding the most effective OCA, and exposure time for different tissues as well as application of physical and chemical enhancers of its penetration, is still ongoing. Not enough attention has been paid to study the effect of OCAs on the parameters and components of the tissue *in vivo* and on the safety of topical application of OCAs to the tissue. It was shown that prolonged use of highly concentrated OCAs can have a negative effect on the skin (local hemostasis, dehydration, etc.) and subcutaneous tissues. Some OCAs serve as penetration enhancers and can cause a change in the skin morphology due

TABLE 27.1

Optical properties of human and animal adipose tissue (AT) measured *in vitro* and *ex vivo* (rms values are given in parentheses). IS – single integrating sphere; DIS – double integrating sphere; IMC – inverse Monte Carlo method; IAD – inverse adding-doubling method.

AT	λ, nm	μ_a, cm⁻¹	μ_s, cm⁻¹	g	μ_s', cm⁻¹	Remarks
Pig subcutaneous	1150	0.46	179.4	0.952	8.66	DIS, IAD; data from graphs of Ref. [60]; in the
(n = 10)	1200	1.84	178.9	0.950	8.91	spectral range 1200–1600 nm:
	1250	0.71	177.9	0.951	8.67	$\mu_s'=2.242 \times 10^8\lambda^{-2.736} + 31.258\lambda^{-0.192}$, [λ] in nm
	1300	0.54	177.1	0.952	8.46	
	1350	0.86	176.4	0.953	8.37	
	1400	3.43	175.0	0.952	8.40	
	1450	4.48	173.5	0.952	8.29	
	1500	3.06	172.3	0.953	8.12	
	1550	1.83	170.9	0.953	7.99	
	1600	1.29	169.8	0.953	7.91	
	1650	1.38	168.4	0.953	7.90	
	1700	5.76	167.0	0.952	8.08	
	1750	7.59	165.7	0.952	8.01	
	1800	5.21	163.8	0.952	7.86	
	1850	4.43	162.1	0.952	7.78	
	1900	11.25	157.8	0.948	7.93	
	1950	13.70	155.0	0.949	7.86	
	2000	9.75	155.8	0.949	7.92	
	2050	7.02	156.0	0.949	8.01	
	2100	5.62	156.0	0.948	8.17	
	2150	6.12	156.0	0.945	8.54	
	2200	5.19	156.0	0.943	8.86	
	2250	8.31	156.6	0.939	9.49	
Human subcutaneous	400	2.26(0.24)	–	–	13.4(2.8)	IS, IAD; in the spectral range of 600–1500 nm:
(n = 6)	500	1.49(0.06)	–	–	13.8(4.0)	$\mu_s' = 1.05 \times 10^3\lambda^{-0.68}$, [λ] in nm; [89]
	600	1.18(0.02)	–	–	13.4(4.7)	
	700	1.11(0.05)	–	–	12.2(4.4)	
	800	1.07(0.11)	–	–	11.6(4.6)	
	900	1.07(0.07)	–	–	10.0 (3.4)	
	1000	1.06(0.06)	–	–	9.39(3.3)	
	1100	1.01(0.05)	–	–	8.74(3.3)	
	1200	1.06(0.07)	–	–	7.91(3.2)	
	1300	0.89(0.07)	–	–	7.81(3.2)	
	1400	1.08(0.03)	–	–	7.51(3.3)	
	1500	1.05(0.02)	–	–	7.36(3.4)	
	1600	0.89(0.04)	–	–	7.16(3.2)	
	1700	1.26(0.07)	–	–	7.53(3.3)	
	1800	1.21(0.01)	–	–	7.5(3.48)	
	1900	1.62(0.06)	–	–	8.72(4.2)	
	2000	1.43(0.09)	–	–	8.24(4.0)	
Rat subcutaneous	400	2.25(1.34)	–	–	19.8(6.3)	IS, IAD; in the spectral range of 600–1400 nm:
(n = 10)	500	0.64(0.34)	–	–	14.3(4.1)	$\mu_s' = 25.51\lambda^{-0.12}$, [λ] in nm; [90]
	600	0.64(0.33)	–	–	12.2(3.5)	
	700	0.75(0.36)	–	–	11.4(3.2)	
	800	1.05(0.47)	–	–	11.0(3.1)	
	900	1.25(0.55)	–	–	10.8(3.0)	
	1000	1.43(0.61)	–	–	10.9(3.0)	
	1100	1.43(0.61)	–	–	10.6(2.8)	
	1200	2.07(0.99)	–	–	11.2(2.9)	
	1300	1.43(0.64)	–	–	10.5(2.8)	
	1400	2.29(1.20)	–	–	10.7(3.0)	
	1500	2.03(1.07)	–	–	10.3(3.0)	
	1600	1.40(0.72)	–	–	9.33(2.7)	
	1700	3.04(1.69)	–	–	11.6(3.4)	
	1800	2.67(1.43)	–	–	10.8(3.2)	
	1900	4.55(2.65)	–	–	13.8(4.5)	
	2000	3.99(2.28)	–	–	12.7(4.2)	
	2100	2.76(1.53)	–	–	11.3(3.6)	
	2200	2.65(1.48)	–	–	12.2(3.6)	
	2300	6.92(3.67)	–	–	22.7(6.0)	
	2400	6.54(3.52)	–	–	24.0(6.1)	
	2500	5.58(3.04)	–	–	23.9(6.4)	

(Continued)

TABLE 27.1 (CONTINUED)

Optical properties of human and animal adipose tissue (AT) measured *in vitro* and *ex vivo* (rms values are given in parentheses). IS – single integrating sphere; DIS – double integrating sphere; IMC – inverse Monte Carlo method; IAD – inverse adding-doubling method.

AT	λ, nm	μ_a, cm^{-1}	μ_s, cm^{-1}	g	μ_s', cm^{-1}	Remarks
Human subcutaneous	400	15.98(3.2)	–	–	49.5(6.5)	IS, IMC, in the spectral range of 370–1300 nm:
($n = 10$)	500	5.50(0.69)	–	–	35.4(4.5)	$\mu_s'=1.08 \times 10^8\lambda^{-2.525} + 157.494\lambda^{-0.345}$,
	600	1.89(0.40)	–	–	27.0(3.2)	[λ] in nm; [91]
	700	1.27(0.24)	–	–	23.0(2.5)	
	800	1.08(0.23)	–	–	20.2(2.1)	
	900	0.95(0.22)	–	–	18.5(1.8)	
	1000	0.89(0.25)	–	–	17.4(1.7)	
	1100	0.74(0.22)	–	–	16.6(1.5)	
	1200	1.65(0.30)	–	–	16.1(1.5)	
	1300	1.05(0.27)	–	–	15.8(1.4)	
	1400	6.27(0.88)	–	–	16.8(1.6)	
	1500	8.52(1.46)	–	–	17.6(1.8)	
	1600	3.60(0.61)	–	–	15.7(1.6)	
Subdermis (primarily	633	0.12	–	–	12.58	IS, IMC; data from graphs of Ref. [92]; in the
globular fat cells)	700	0.09	–	–	12.10	spectral range of 620–1000 nm $\mu_s' =$
($n = 12$)	750	0.09	–	–	11.75	$139.24\lambda^{-0.373}$, [λ] in nm
	800	0.08	–	–	11.40	
	850	0.09	–	–	11.17	
	900	0.12	–	–	10.95	
	950	0.15	–	–	10.81	
	1000	0.12	–	–	10.71	
Human WAT (n = 7)	250	103.91	–	–	168.98	IS, IMC; data from graphs of Ref. [93]; in the
	300	17.54	–	–	37.01	spectral range of 420–800 nm $\mu_s' = 5.73 \times$
	350	12.88	–	–	25.80	$10^{10}\lambda^{-3.635} + 23.7\lambda^{-0.175}$, [$\lambda$] in nm
	400	25.93	–	–	27.64	
	450	12.19	–	–	18.97	
	500	4.19	–	–	16.03	
	550	5.08	–	–	14.92	
	600	1.86	–	–	12.49	
	650	0.68	–	–	11.20	
	700	1.02	–	–	9.38	
	750	0.68	–	–	9.45	
	800	0.34	–	–	9.08	
Human subcutaneous	400	6.09	–	–	25.61	DIS, IMC; data from graphs of Ref. [94]; in the
(n = 15)	500	2.61	–	–	19.63	spectral range of 400–1100 nm $\mu_s' =$
	600	0.94	–	–	17.51	$3939\lambda^{-0.846}$, [λ] in nm
	700	0.65	–	–	15.76	
	800	0.48	–	–	13.54	
	900	0.49	–	–	12.35	
	1000	0.43	–	–	11.58	
	1100	0.53	–	–	10.06	

to the dissociation of collagen fibers [113–115] and the physiological effect of dehydration on the vascular network of the dermis. Thus, in order to avoid or reduce the negative effect on the tissue, it is critical to find the optimal OCAs, their concentration, permissible exposure time, and number of applications on the tissue *in vivo*.

Despite the fact that significant progress has been achieved in TOC of tissues for different medical applications, unresolved issues remain on the choice of the optimal TOC technique for AT because it presents a great challenge due to the large concentration of lipids. For three-dimensional visualization of AT, some OC protocols have been suggested. For example, it has been shown that Murray's Clear (a mixture of benzyl alcohol and benzyl benzoate) with refractive index 1.55 can be used as an OCA for AT [116]. A modified iDISCO+ protocol named Adipo-Clear was recently suggested to remove lipids

from mouse AT without altering its structure in order to study beige fat biogenesis [48].

For WAT clearing, a sample incubation with lipase was perceived as a potentially valuable step to be added to TOC protocols [106, 117].

However, such approaches cannot be used *in vivo*, although the development of a suitable protocol is in demand for a number of medical applications [118]. For example, in the case of a surgical procedure, the surgical site may be covered with AT and therefore cannot be observed. In order to visualize the surgical site, the AT must be removed. At the same time, in the process of removing AT, it is important to see the position of a blood vessel that passes inside or under the AT; therefore, OC of AT is necessary so as not to damage it. In relation to such a goal, an *in vivo* OC of the AT could be useful.

TABLE 27.2

Optical properties of human and animal AT measured *in vivo* and *ex vivo* using diffuse reflectance spectroscopy (rms values are given in parentheses)

AT	λ, nm	μ_a, cm^{-1}	μ_s, cm^{-1}	g	μ_s', cm^{-1}	Remarks
Human abdominal	650	0.05	–	–	12.0	*In vivo*; data from graphs of Ref. [8]; $\mu_s' =$
(n = 10)	700	0.03	–	–	11.41	$1.243 \times 10^3 \lambda^{-0.716}$, [$\lambda$] in nm
	750	0.04	–	–	10.86	
	800	0.03	–	–	10.37	
	850	0.04	–	–	9.93	
	900	0.08	–	–	9.53	
	950	0.11	–	–	9.17	
	1000	0.13	–	–	8.85	
Swine	900	0.66	–	–	7.58	*Ex vivo*; data from graphs of Ref. [59]; $\mu_s' =$
subcutaneous	950	0.78	–	–	7.05	$5.019 \times 10^5 \lambda^{-1.629}$, [$\lambda$] in nm
	1000	0.78	–	–	6.57	
	1050	0.81	–	–	6.02	
	1100	0.58	–	–	5.57	
	1150	1.64	–	–	5.22	
	1200	3.57	–	–	4.84	
	1250	1.79	–	–	4.53	
	1300	1.44	–	–	4.24	
	1350	2.08	–	–	3.97	
	1400	5.16	–	–	3.73	
	1450	6.69	–	–	3.54	
	1500	5.36	–	–	3.41	
	1550	3.94	–	–	3.19	
	1600	3.21	–	–	3.02	
Swine visceral fat	900	0.09	–	–	9.17	*Ex vivo*; data from graphs of Ref. [59]; $\mu_s' =$
	950	0.16	–	–	8.38	$1.12 \times 10^6 \lambda^{-1.721}$, [$\lambda$] in nm
	1000	0.18	–	–	7.70	
	1050	0.13	–	–	7.07	
	1100	0.10	–	–	6.53	
	1150	0.58	–	–	5.96	
	1200	1.65	–	–	5.69	
	1250	0.59	–	–	5.29	
	1300	0.53	–	–	4.83	
	1350	1.11	–	–	4.65	
	1400	5.25	–	–	4.30	
	1450	8.87	–	–	4.05	
	1500	6.06	–	–	3.80	
	1550	3.50	–	–	3.64	
	1600	2.42	–	–	3.37	
Bovine fat	633	0.026(0.007)	–	–	12.0(0.7)	*Ex vivo*; Ref. [95]
	751	0.021(0.006)	–	–	10.0(0.5)	
Porcine	650	0.4	110.24	0.96	45.2	*Ex vivo*; Ref. [96]
Bovine	650	0.17	124.0	0.95	6.2	*Ex vivo*; Ref. [96]
Human (n = 39)	400	16.19	–	–	13.66	*Ex vivo*; IMC; data from graphs of Ref. [97]; in
	450	21.8	–	–	12.89	the spectral range of 400–600 nm $\mu_s' =$
	500	5.56	–	–	11.6	$5.085 \times 10^9 \lambda^{-3.496} + 14.872 \lambda^{-0.066}$, [$\lambda$] in nm
	550	1.21	–	–	11.12	
	600	0.15	–	–	10.97	
Porcine	450	0.55	–	–	27.14	*Ex vivo*; Diffusion approximation; data from
	500	0.28	–	–	23.69	graphs of Ref. [98]; in the spectral range of
	550	0.45	–	–	20.19	450–1000 nm $\mu_s' = 4.2284 \times 10^3 \lambda^{-0.8369}$, [$\lambda$]
	600	0.07	–	–	21.0	in nm
	650	0.02	–	–	19.58	
	700	0.01	–	–	18.21	
	750	0.02	–	–	17.18	
	800	0.02	–	–	15.85	
	850	0.02	–	–	15.43	
	900	0.07	–	–	14.68	
	950	0.08	–	–	14.16	
	1000	0.07	–	–	13.63	

TABLE 27.3

Optical properties of human and animal AT measured *in vivo* and *ex vivo* using time-domain technique (rms values are given in parentheses)

AT	λ, nm	μ_a, cm⁻¹	μ_s, cm⁻¹	g	μ_s', cm⁻¹	Remarks
Human	600	0.04	–	–	13.52	*In vivo*; Time-domain; data from
subcutaneous	650	0.01	–	–	12.87	graphs of Ref. [40]; in the spectral
(n = 10)	700	0.01	–	–	12.26	range of 600–1100 nm $\mu_s' = 3.161 \times$
	750	0.01	–	–	11.65	$10^3\lambda^{-0.848}$, [λ] in nm
	800	0.01	–	–	11.09	
	850	0.01	–	–	10.55	
	900	0.05	–	–	10.03	
	950	0.06	–	–	9.48	
	1000	0.04	–	–	8.99	
	1050	0.06	–	–	8.57	
	1100	0.03	–	–	8.16	
Human	610	0.17(0.07)	–	–	8.44(1.76)	*In vivo*; Time-domain; data from
abdominal	650	0.08(0.08)	–	–	8.03(1.74)	graphs of Ref. [58]; in the spectral
(n = 3)	700	0.06(0.05)	–	–	7.57(1.72)	range of 610–1010 nm $\mu_s' = 1.307 \times$
	750	0.06(0.04)	–	–	7.17(1.7)	$10^3\lambda^{-0.786}$, [λ] in nm
	800	0.06(0.05)	–	–	6.81(1.68)	
	850	0.08(0.05)	–	–	6.49(1.67)	
	900	0.13(0.04)	–	–	6.21(1.66)	
	950	0.17(0.08)	–	–	5.96(1.65)	
	1010	0.17(0.09)	–	–	5.68(1.63)	
Porcine (n = 9)	650	0.013(0.002)	–	–	16.7(1.13)	*Ex vivo*; Time-domain; data from Ref.
	700	0.006(0.005)	–	–	15.44(1.36)	[99]; in the spectral range of
	750	0.013(0.005)	–	–	14.39(1.34)	650–1100 nm $\mu_s' = 1.027 \times$
	800	0.006(0.006)	–	–	13.4(1.38)	$10^4\lambda^{-0.993}$, [λ] in nm
	850	0.008(0.008)	–	–	12.7(1.5)	
	900	0.048(0.006)	–	–	11.88(1.42)	
	950	0.042(0.012)	–	–	11.42(1.53)	
	1000	0.042(0.013)	–	–	10.74(1.48)	
	1050	0.06(0.009)	–	–	10.33(1.46)	
	1100	0.028(0.007)	–	–	9.98(1.37)	

Earlier, we have shown that the method based on the combined photochemical and photothermal effect induces OC of the AT cells [119–121]. In such a technique, laser radiation is used to control the optical scattering properties of the AT. Depending on the intensity of laser radiation, the biological response of the cell may lead to irreversible or reversible injury of the cell membrane, which results in the creation of new pores or enlargement of the already existing ones, through which efficient exchange between the cell content and the environment takes place. For adipocytes, the presence of pores promotes lipolysis, as a result of which the intercellular space is filled with the content of the cells and their decay products (free fatty acids, water, and glycerol) [120, 122, 123]. The appearance in the intercellular space of such an immersion fluid contributes to the process of OC of AT [120]. However, this technique is accompanied by necrosis of irradiated fat cells. Thus, it is necessary to decrease the side effects of such an OC procedure.

The propagation of light in a tissue differs with change in its morphologic as well as physiological properties, which can be controlled by tissue mechanical compression or stretching, leading to reduction of light scattering [55]. This is the basic idea of the compression OC methods [111, 124] (see Chapter 10). Various studies have proved that the induced mechanical force can almost produce the similar OC efficiency as the immersion OC [125] including a change in optical properties and an increase in the penetration depth [126]. The OC efficiency of the novel mechanical tissue OC device that combined a pin array and vacuum pressure source applied directly to the skin surface was tested for *ex vivo* porcine skin and AT during laser irradiation (980 and 1210 nm) [126].

Selective thermolysis of AT was reported by Anderson et al. [127], who noted many difficulties, such as low optical contrast of AT and the fact that its thermal conductivity is substantially lower, so it retains heat longer [128]. Fat can be melted or burned by lasers as an irreversible effect; however, it is hard to reach fat damage *in vivo* without usage of external dyes for better optical contrast. During heating of AT, an OC effect can also be obtained once the fat melting temperature has been achieved [129–131]. Further analysis showed that such an increase of the transmittance is provided by reduction of the scattering [132].

In addition to chemicals to enhance tissue permeability, a large number of different physical methods have been proposed, such as fractional tube [133] or laser microablation [134]. When using microablation, cell damage occurs, and their contents flow into the intercellular space. Therefore, self-optical clearing of the tissue may take place.

Recently, the possibility of temperature monitoring within AT in a wide temperature range from room to physiological temperatures and above by use of thermosensitive luminescent upconversion nanoparticles (UCNPs) [NaYF$_4$:Yb^{3+}, Er^{3+}] has been shown [135]. The increase of luminescent intensity at high temperatures can be explained by TOC associated with the phase transition of lipids. Application of OCAs led to a significant increase of detected intensity of laser-induced luminescence from the UCNPs. The most effective OCAs in this study were 50%-fructose and 100%-propylene glycol: fructose was faster (5 min for saturation of intensity) with 1.5 fold efficiency, while propylene glycol was slower (30 min for saturation) but with 2.9 fold efficiency. The obtained results confirm a high sensitivity of the luminescent UCNPs to the temperature variations within tissues and show a strong potential for the controlled tissue thermolysis. Therefore we can propose a more reliable temperature measurement technique in tissues by using luminescence from laser beam excited UCNPs at TOC. These studies were done for a rat visceral AT that is a part of the greater omentum, which is found in the lower abdominal area. It contains two layers of AT and both supports and covers the organs and intestines in this area of the body. The greater omentum is responsible for fat deposition and contains amounts of AT that vary from person to person [17, 34, 136–138].

Histological investigation of adipose tissue changes after optical clearing

The most commonly used OCAs (e.g. glucose, glycerol, and propylene glycol) are generally nontoxic agents. However, prolonged treatment with a highly concentrated OCA can have a negative effect on tissue, such as local hemostasis, shrinkage, and even tissue necrosis. For example, glycerol can cause alteration in the skin morphology due to a dissociation of the collagen fibers [114, 139]. Glycerol has also been found to have a dehydrating effect on the cutaneous vasculature [140]. This study revealed that glycerol's effect on vessels is reversible with hydration. In addition, the transition from oxygenated form of Hb to deoxygenated form in rat skin related to local hemostasis during 84.4% glycerol treatment *in vivo* has been investigated [141]. It was found that the topical application of a 75% glycerol solution on the mesenteric microvessels of a rat *in vivo* within 1–3 s led to reduction of blood flow velocity in all microvessels and to stasis appearing after 20 s of application [142]. At the same time, it was shown that a 75% glycerol solution does not cause a loss of collagen organization [143]. Nevertheless, irreversible changes in the collagen structure under the action of anhydrous glycerol are possible [139].

In a number of studies, some side effects of OCAs in terms of irritation and edema were noted [110, 143–146]. Omnipaque™ has an acceptable OC effect without noticeable sample shrinkage [100, 147] because of less dehydration.

It was found that tissue dehydration and packing density increased under action of glycerol [148–151], glucose [151–153], ethylene glycol [154], polyethylene glycols [155], and other hyperosmotic OCAs [156–158].

However, according to the literature, there is no information of structural changes in AT *ex vivo* and *in vivo* during OC.

Methods and materials

OCAs

The list of OCAs used is presented in Table 27.4. All components of the OCAs are biocompatible as substances themselves. All solutions were in weight concentrations. The fructose

TABLE 27.4

Properties of agents and mode of application, refractive index at 589 nm

Agent	Refractive index	OCA composition (%-weight concentration)	*Ex vivo*	*In vivo*
Saline (control)	1.3330	0.09% NaCl in water		+
DMSO	1.4790	100%	+	
Diatrizoic acid	1.5110	21% diatrizoic acid, 66% DMSO, 13% N-methyl-glucamine	+	+
Metrizoic acid	1.5290	30% metrizoic acid, 58% DMSO, 12% N-methyl-glucamine	+	+
Sucrose	1.5090	60% sucrose, 40% DMSO	+	+
Fructose	1.4080	50% fructose*, 30% ethanol, 20% water	+	+
PEG-200	1.4294	73.9% PEG200, 26.1% water		+
Trehalose	1.4183	52.6% trehalose, 47.4% water		+
Maltose	1.4290	49.7% maltose, 50.3% water		+
Mannose	1.4340	57.7% mannose, 42.3% water		+
Ribitol	1.4170	52.6% ribitol, 47.4% water		+
Glucose	1.4215	50.8% glucose, 49.2% water		+
Altrose	1.4170	48.9% altrose, 51.1% water		+
Erythrose	1.3970	53.7% erythrose, 46.3% Water		+
Fucose	1.4038	43.0% fucose, 57.0% water		+
Tagatose	1.4170	49.6% tagatose, 50.4% water		+
Omnipaque™-300	1.4363	100%		+

*All solutions are in weight/weight concentrations. The fructose solution is in weight/volume concentration.

solution was an exception and had a weight–volume concentration. Omnipaque™ and diatrizoic and metrizoic acids are radiopaque X-ray contrast agents used in diagnostic radiography [159–161]. Omnipaque™ and diatrizoic acid are used for OC [162, 163]. DMSO and ethanol are used as enhancers of tissue permeability [164, 165]. N-methyl-glucamine is an amino sugar (derived from sorbitol). It is an U.S. Food and Drug Administration (FDA)–approved excipient in pharmaceuticals, and it is used for increasing the aqueous solubility and stability of pharmaceutical compounds [166]. Such saccharides as sucrose, maltose, and fructose were used as OCAs for skin and brain OC [158, 167–169]. Polyethylene glycol 200 (PEG-200) has been reported as an OCA for skin *in vitro* [157]. The saccharides trehalose, maltose, mannose, altrose, erythrose, fucose, and tagatose and alcohol ribitol were tested as potential OCAs for AT *in vivo*. Refractive indices of the OCAs were measured at the wavelength 589 nm using the Abbe refractometer (IRF-454B2M, Russia).

Experimental setup

Figure 27.5(a) shows the experimental setup for *ex vivo* and *in vivo* measurements. Spatially resolved backreflectance measurements were carried out using a specially designed Olympus device [see Figure 27.5(b, c)].

The cuvette was also specially designed for the experiment. It was a round black cell with matt bottom to exclude a contribution of specular reflection to the backreflected signal. It was placed on the heating element, which was fixed on the experimental table to provide a constant temperature for the visceral adipose tissue (VAT) under study (about 38°C). IR imager IRISYS 4010 (InfraRed Integrated System Ltd, UK) located at a distance of 30 cm from the table was used for noncontact measurement of the surface temperature of the sample.

The configuration of the Olympus device for measurement of tissue optical properties is shown in Figure 27.5 (c). This device consists of a laser light source (wavelength 635 nm), an optical probe, photo detectors, a signal amplifier, a DAQ (data acquisition), and a PC (personal computer). The optical probe has a multimode fiber array. The outermost fiber was connected to the laser light source and the other fibers were connected to the photodetectors. The signal detected by photodetectors was amplified and recorded by DAQ and PC. The backscattering intensity was measured when the optical probe and tissue surface were contacted. The recorded data

was processed using inverse Monte Carlo simulation to get the values of a reduced scattering coefficient (RSC).

OC protocols

Ex vivo OC protocol

All OCAs were tested in *ex vivo* measurements. Each experiment was performed on two samples. The mean area of the samples was 1.0×1.5 cm^2, and the thickness of the tissue samples was 1.5 ± 0.5 mm. VAT samples from white outbreed rat were used. The samples were placed into the cuvette. The RSC was measured in five different points of the samples immersed in saline and then every 5 min after the OCA application during 90 min. Saline and the OCAs were pre-warmed up to 38°C before application. The volume of the OCA applied was 1.25 mL. Before each RSC measurement, the OCA was removed from cuvette, and the refractive index of the solution was measured. Then a new portion of OCA was poured into the cuvette with the VAT sample. The temperature of the VAT samples was measured before the OCA application and then every 5 min during OC to keep it constant at about 38°C throughout the experiment. The thicknesses of the tissue samples were measured before and after OC.

In vivo OC protocol

All procedures with animals were done in accordance with the European Convention for the Protection of Animals used for experimental and other scientific purposes [170]. The experimental study was conducted in Centre of Collective Use of Saratov State Medical University (Russia) and according to the guidelines of the University's Animal Ethics Committee (Russia) and the relevant national agency regulating experiments with animals.

Rats were anaesthetized intramuscularly by "Zoletil" (Virbac, France) and placed on the experimental table 5–7 min after the anesthetic injection. The laparotomy allowed for a part of VAT *in vivo* to be placed into cuvette filled up with saline or OCA heated up to 38°C.

Procedure of *in vivo* measurement of RSC was similar to *ex vivo* one. Each experiment was performed on two animals. The control measurements of RSC were performed with saline immersion. Total duration of OC was 120 min. The treated part of VAT with thickness of 1.5–2.5 mm was cut off and

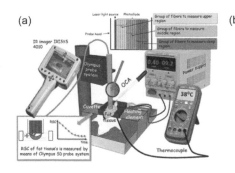
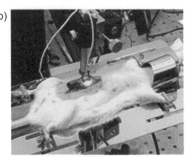
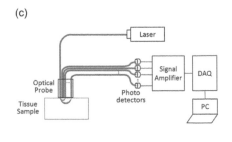

FIGURE 27.5 Experimental setup (a); animal under study (b); schematics of Olympus optical probe system (c).

fixed by 10% solution of formaldehyde just after experiment for following histological analysis.

The OC efficiency OC_{eff} of tissue sample was determined as the relative change of RSC of the tissue under the action of the OCA:

$$OC_{eff} = \frac{\mu'_{s0} - \mu'_{smin}}{\mu'_{s0}} \times 100\%, \qquad (27.1)$$

where μ'_{s0} and μ'_{smin} are the initial and the minimal value of the RSC.

In the case of using saline for sample immersion μ'_{smin} was replaced by maximal value of the RSC due to tissue sample swelling.

Histological analysis of samples

Histological studies were performed to find agents that cause minimal morphological changes in the VAT *in vivo*. The tissue samples were subject to a standard procedure of histological preparation. After fixation, the 3–5-μm slices across all layers of a tissue sample were made and stained by haemotoxylin and eosin (H&E) [171] and analyzed using microvisor mVizo-103 (LOMO, Russia). A morphological study was conducted independently by two experts in the field of pathological anatomy. Thus, it was suited to a double-blind manner.

Treatments of adipose tissue *ex vivo* and *in vivo*

Fractional laser microablation (FLMA)

To overcome the protective tissue barrier, the technique of fractional laser microablation (FLMA) of the surface layers of tissue was developed. The multi-beam Palomar Lux2940 erbium laser (Palomar Medical Products Ltd., USA) with a wavelength 2940 nm, a pulse energy 1.12 J, and a pulse duration of 5 ms was used as a light source. The FLMA in two modes was applied to two parts of the sample:

Mode 1: 64 round-shaped channels were made in the VAT sample within the area of 6×6 mm^2. The separation between the centers of the channels was of 1.3 mm, and the diameter of their openings at the sample surface was 0.2 mm.

Mode 2: 169 round-shaped channels were made with the similar area 6×6 mm^2 with a separation between the centers of the channels of 0.5 mm; the diameter of their openings at the sample surface was 0.1 mm, and the depth was evaluated from histological analysis as 1.2 mm.

US exposure in combination with application of OCA

The treatment of two VAT samples was implemented by means of the Dynatron 125 US transducer (Dynatronics, USA). The US probe was displaced on the sample surface and then turned to 1 MHz frequency and 0.5 W power. Four modes of sonophoresis were applied for *ex vivo* and three for *in vivo* studies. After the cuvette with the sample was filled up by an OCA, it was exposed to US (Tables 27.5 and 27.6).

Combination of US and FLMA with OCA application

The FLMA was applied to two parts of the sample after measurement of the optical properties of a sample placed in saline. After that the cuvette with the sample was filled up by the OCA, and US with 1 MHz frequency and 0.5 W power was applied during 1 min. The different combinations of US and FLMA modes that were used are presented in Table 27.7.

Results

Optical clearing of VAT

The kinetics of RSC during OC produced by different OCAs were measured. Prior to *in vivo* studies, *ex vivo* measurements were done for all OCAs. After OC, the histological analysis

TABLE 27.5

Modes of US waveform (power, 0.5 W; frequency, 1 MHz) in combination with OCA exposure parameters

	Mode 1	Mode 2	Mode 3	Mode 4
Waveform	CW	Pulse mode (Duty cycle 50%)	CW	Pulse mode (Duty cycle 50%)
Exposure, min	2	2	1	1
Number of US exposures [time after OCA application, min]	1 [0]	1 [0]	2 [0 and 15]	2 [0 and 15]

TABLE 27.6

Three combinations of US waveform (pulse mode, duty cycle 50%; power, 0.5 W; frequency, 1 MHz) and OCA exposure parameters

	Combination 1	Combination 2	Combination 3
US exposure, min	1	1	2
Number of US exposures [time after OCA application, min]	1 [0]	3 [0,15, and 30]	3 [0,15, and 30]

TABLE 27.7

The combinations of US and FLMA modes

FLMA	US	Channel diameter	Number of samples
Mode 1	CW	0.2 mm	2
Mode 1	Pulse	0.2 mm	2
Mode 2	CW	0.1 mm	2
Mode 2	Pulse	0.1 mm	2

was done for each sample to reveal possible tissue injury. The trial measurements were performed with fructose solution with US or FLMA *ex vivo*.

Table 27.8 presents results for OC of *ex vivo* VAT samples using different OCAs.

Table 27.9 presents results for OC of *in vivo* VAT samples using different OCAs.

Typical digital images of rat VAT *in vivo* before and after application of fructose are presented in Figure 27.6 (a) and (b), respectively. The part of the tissue sample was put on the paper with a printed figure. It is seen that VAT became more transparent after immersion in fructose. The thickness of the sample decreased during the OC due to water diffusion from tissue to the surrounding hyperosmotic solution. Figure 27.6 (c) shows the typical kinetics of RSC of the VAT samples during the OC. It shows the decrease of RSC of the VAT sample during its immersion in the OCA.

The obtained results on the OC efficiency for *ex vivo* and *in vivo* investigations are presented in Figures 27.7(a) and 27.7 (b), respectively. In *ex vivo* investigations, the OC efficacy of three agents – diatrizoic acid, metrizoic acid, and sucrose – was approximately equal (about 50%). However, the maximal

OC effect (~77%) was observed for the samples subjected to the fructose action.

Firstly, the molecular weight of the fructose (180.16 g/mol) is smaller than that of other used agents [diatrizoic acid (613.91 g/mol), metrizoic acid (627.94 g/mol) and sucrose (342.2965 g/mol)]. Since smaller molecules diffuse faster in tissues, the highest OC efficiency during examination time can be due to fast fructose diffusion in VAT. Also the use of ethanol as a solvent for the preparation of the fructose solution may help to provide cell membrane permeability. Tissue dehydration and corresponding tissue shrinkage caused by hyperosmotic pressure of fructose solution led to smaller thickness compared to the initial one, with more light transmitted through.

In vivo investigations demonstrate 50% and more OC efficiency for three agents: diatrizoic acid (65%), fructose (50%), and sucrose (58%). The maximal OC effect in *in vivo* investigation (65%) was observed for the samples subjected to the diatrizoic acid action during 40 min.

The cohesive nature of the bilayer decreased due to the binding of DMSO molecules at the water–bilayer interface; consequently, the entrance of water molecules into the hydrophobic core was facilitated [164]. Moreover, bound DMSO molecules contributed with an electric dipole to the local electric field and altered the electrostatic profile across the membrane. With the amphiphilic structure, bound DMSO increased the amount of hydrophilic molecules at the interface. These properties enabled a facilitated water penetration into the bilayer, resulting in a significantly increased tendency for pore formation in the presence of DMSO molecules. Apart from DMSO, alkanols also act as membrane permeation enhancers. However, their mechanism differs in that ethanol thins phospholipid bilayers and may lead at high concentrations to micellar-like defects.

TABLE 27.8

The OC effects measured *ex vivo*: initial RSC (μ'_{s0}), minimum RSC (μ'_{smin}), and OC efficiency OC_{eff} (%) with rms values given in parentheses. Averaged data of several rat AT samples. Total duration of OC was 90 min. Thickness of AT samples was 1.5 ± 0.5 mm

OCA	Physical enhancement	μ'_{s0}, cm^{-1}	μ'_{smin}, cm^{-1}	OC_{eff}, %
Fructose (50%)*	-	16.8	8.5	49.5(0.7)
Fructose (50%)*	FLMA (Mode 2)	13.0	6.0	68.0
Fructose (50%)*	FLMA (Mode 1)	12.4	5.4	70.0(4.9)
Fructose (50%)*	US (Mode 1)	10.3	3.4	68.5(6.4)
Fructose (50%)*	US (Mode 2)	11.3	2.5	87.5(3.5)
Fructose (50%)*	US (Mode 3)	12.2	6.0	61.5(3.5)
Fructose (50%)*	US (Mode 4)	10.2	2.5	77.0(5.7)
Fructose (50%)*	Combined US (CW) and FLMA (Mode 1)	11.0	2.7	83.5(2.1)
Fructose (50%)*	Combined US (pulse) and FLMA (Mode 1)	13.5	4.6	89.5(3.5)
Fructose (50%)*	Combined US (CW) and FLMA (Mode 2)	10.2	1.5	87.0
Fructose (50%)*	Combined US (pulse) and FLMA (Mode 2)	9.9	3.9	65.0(4.2)
Sucrose (60%)	-	9.0	2.9	62.0
Sucrose (60%)	FLMA (Mode 2)	9.6	2.3	73.0
Metrizoic acid (30%)	-	8.1	2.0	61.0
Metrozoic acid (30%)	FLMA (Mode 2)	12.9	4.2	86.5(4.9)
DMSO (100%)	-	10.3	3.4	68.5(2.1)
Diatrizoic acid (21%)	-	11.5	3.3	82.0(2.8)
Diatrizoic acid (21%)	FLMA (Mode 2)	10.0	2.7	73.0

* All solutions are in weight/weight concentrations. The fructose solution is in weight/volume concentration.

TABLE 27.9

The OC effects measured *in vivo*: initial RSC (μ'_{s0}), minimum RSC (μ'_{smin}), and OC efficiency OC_{eff} (%) with rms values given in parentheses. Averaged data of several rat AT samples. Total duration of OC was 120 min. Thickness of AT samples was 1.5 ± 0.5 mm

OCA	μ'_{s0}, cm^{-1}	μ'_{smin}, cm^{-1}	OC_{eff}, %
Saline (100%)	10.3	8.4	18.0 (13.6))
PEG-200 (73.9%)	10.1	5.1	47.9
Trehalose (52.6%)	8.9	4.3	51.4
Maltose (49.7%)	10.7	6.4	39.9
Mannose (57.7%)	9.2	3.8	57.7(39.1)
Ribitol (52.6%)	9.8	3.0	69.3(11.8)
Glucose (50.8%)	10.2	2.7	74.0(11.1)
Altrose (48.9%)	9.5	2.5	74.0
Tagatose (49.6%)	9.3	4.3	53.9
Fucose (43.0%)	10.8	8.4	22.4
Erythrose (53.7%)	11.2	1.8	84.3
Fructose (50%)*	11.5	1.3	89.0(1.6)
Omnipaque™ (100%)	10.3	6.5	37.8(9.0)
Sucrose (60%)	16.5	9.5	58.0
Metrizoic acid (30%)	14.0	11.0	23.0
Diatrizoic acid (21%)	10.5	3.0	65.0

*All solutions are in weight/weight concentrations. The fructose solution is in weight/volume concentration.

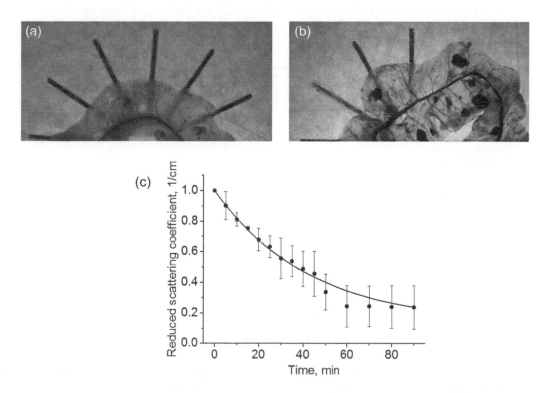

FIGURE 27.6 Typical digital images of rat VAT before (thickness is 1.3 mm) (a) and after application of fructose solution (50%) (thickness is 1.0 mm) (b) in *in vivo* measurements. Kinetics of reduced scattering coefficient of the *ex vivo* fat samples during the optical clearing by fructose solution (50%) (c).

In turn, long-chain alkanols are expected to thin membranes by an increased cohesion between the lipid leaflets if only applied on one side of the membrane.

The difference in OC efficiency in *ex vivo* and *in vivo* studies is due to the absence of blood and lymph flow, and innervations in *ex vivo* tissue. Thus in *ex vivo* conditions, there is no systemic reaction of the organism on molecular diffusion.

As for application of OC enhancers, the least destructive effect on tissue was found for combined use of US and FLMA.

TABLE 27.10

OCA exposure to *ex vivo* AT. Total duration of OC was 90 min. Thickness of tissue samples was 1.5 ± 0.5 mm

OCA	Physical enhancement	Necrosis			
		No	Light	Moderate	Total
Sucrose (60%)	-				+
DMSO (100%)	-			+	
Metrizoic acid (30%)	-				+
Diatrizoic acid (21%)	-			+	
Fructose (50%)*	-	+			
Fructose (50%)*	FLMA (Mode 2)		+		
Fructose (50%)*	FLMA (Mode 1)		+		
Sucrose (60%)	FLMA (Mode 2)				+
Metrozoic acid (30%)	FLMA (Mode 2)				+
Fructose (50 %)*	US (2 min CW or pulse mode)		+		
Fructose (50%)*	US (twice for 1 min CW mode)		+		
Fructose (50%)*	US (twice for 1 min pulse mode)				+
Fructose (50%)*	Combined US (2 min CW mode) and FLMA (Mode 1)	+			
Fructose (50%)*	Combined US (2 min pulse mode) and FLMA (Mode 1)		+		
Fructose (50%)*	Combined US (2 min CW mode) and FLMA (Mode 2)		+		
Fructose (50%)*	Combined US exposure (2 min pulse mode) and FLMA Mode 2				+

*All solutions are in weight/weight concentrations. The fructose solution is in weight/volume concentration.

TABLE 27.11

OCA exposure to *in vivo* AT. Total duration of OC was 120 min. Thickness of tissue samples was 1.5 ± 0.5 mm

OCA	Physical enhancement	Necrosis		
		Light	Moderate	Total
Saline (100%)	-	+		
PEG-200 (73.9%)	-			+
Trehalose (52.6%)	-			+
Maltose (49.7%)	-		+	
Mannose (57.7%)	-		+	
Ribitol (52.6%)	-		+	
Glucose (50.8%)	-	+		
Fructose (50%)*	-	+		
Altrose (48.9%)	-			+
Erythrose (53.7%)	-			+
Fucose (43.0%)	-		+	
Tagatose (49.6%)	-			+
Omnipaque™ (100%)	-	+		
Glucose (50.8%)	US exposure (once for 1 min pulse mode)			+
Glucose (50.8%)	US exposure (thrice for 1 min pulse mode)			+
Glucose (50.8%)	US exposure (thrice for 2 min pulse mode)		+	

*All solutions are in weight/weight concentrations. The fructose solution is in weight/volume concentration.

The signal from samples immersed in saline is reduced compared to the initial value (negative OC efficiency).

Results of histological analysis of VAT samples *ex vivo* and *in vivo* are presented (see Table 27.10 and 27.11).

Ex vivo optical clearing of adipose tissue (results of histology)

Ex vivo, control (without any exposure): VAT sample is represented by cells of approximately the similar size. Adipocytes

are with preserved nuclei and clear contours of the membranes (see arrows in Figure 27.8a).

After topical application of sucrose (60%), VAT is 90% damaged. Membranes of adipocytes are thickened and the membranes with a violation of the fibrous structure are observed. Nuclei in cells are absent, which is a sign of necrosis (see arrow in Figure 27.8b).

After topical application of DMSO (100%), VAT is mostly damaged and the membranes with a violation of the fibrous structure are observed. Membranes are thinned (see arrow in Figure 27.9a) and basophilic.

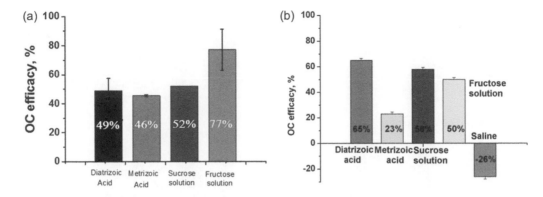

FIGURE 27.7 OC efficacy of rat VAT for topical application different OCAs in *ex vivo* (a) and *in vivo* (b) measurements. Mean temperature of measurements was around 37°C.

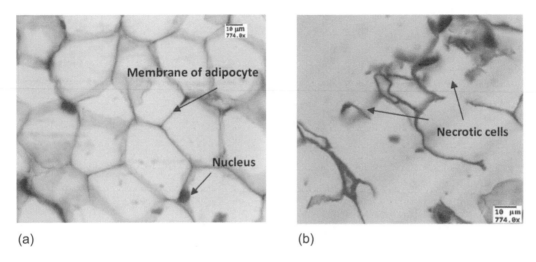

FIGURE 27.8 Histology of rat VAT: control sample (a); after topical application of sucrose (60%) (b). H&E staining. Magnification 774×.

After topical application of fructose (50%), VAT is preserved. Nuclei in cells are presented. The thickness of the membrane is uniform (see arrow in Figure 27.9, b).

After topical application of metrizoic acid (30%): In VAT sample center, tissue is completely damaged; on the periphery, crenulated membranes are noted (see arrows in Figure 27.10a), being partially thickened, eosinophilic, and thinned. The nuclei are absent, and some adipocytes are in a state of necrosis.

Basophilicity and eosinophilicity of the membrane are characterized by a change in the properties of its staining, due to an increase in the sorption properties of damaged cells and an increase in the permeability of their membrane.

There are signs of damage in VAT; the border between cells is thinned and hyaline-crenulated. This term was proposed by Segura and Pujol [172] to describe lipomembranous fat necrosis of the subcutaneous tissue. It is characterized by thinning, folding, and undulating membranes and can be found under certain conditions of damage, including degenerate changes in membranes of adipocytes and macrophages [172–175]. Similar changes with adipocytes were observed with obesity and were associated with thickening of the basement membrane (a thin noncellular layer located between adipocytes and consisting of collagen and other proteins) [173, 176].

After topical application of diatrizoic acid (21%), the membranes of adipocytes are thickened and eosinophilic. It is possible that the plasma permeation takes place which is caused by chemical exposure. A part of membranes is thinned and crenulated (see arrow in Figure 27.10b).

After FLMA (Mode 2) with preliminary topical application of fructose (50%), necrosis of the single adipocytes is observed (see arrows in Figure 27.11a). The VAT structure is kept intact for the most part. Membranes are thickened a little.

After FLMA (Mode 1) with preliminary topical application of fructose (50%), the structure of VAT is preserved, but some adipocytes have crenulated and disrupted membranes and their nuclei are absent (see arrows in Figure 27.11b).

The FLMA (Mode 2) with topical application of sucrose (60%) caused severe damage of VAT, and the membranes with a violation of the fibrous structure are observed. In the center part of the sample, membranes are thickened and membrane eosinophilia is increased. However, several membranes of adipocytes are thinned, crenulated, and broken, and 80%–90% of adipocytes are in a state of necrosis (see arrow in Figure 27.11c).

After FLMA (Mode 2) with preliminary topical application of diatrizoic acid (21%), a part of adipocyte membranes are thickened and eosinophilic, and others are thinned and crenulated. In some cases, necrosis of VAT is detected (see arrow in Figure 27.12a).

After FLMA (Mode 2) with preliminary topical application of metrozoic acid (30%), the thickened membranes of

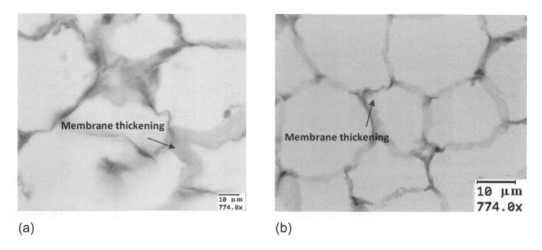

(a)   (b)

FIGURE 27.9 Histology of rat VAT: after topical application of DMSO (100%) (a); after topical application of fructose (50%) (b). H&E staining. Magnification 774×.

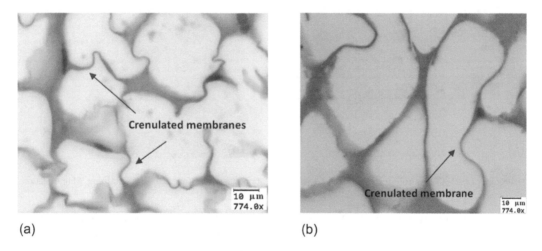

(a)   (b)

FIGURE 27.10 Histology of rat VAT: after topical application of metrizoic acid (30%) (a); after topical application of diatrizoic acid (21%) (b). H&E staining. Magnification 774×.

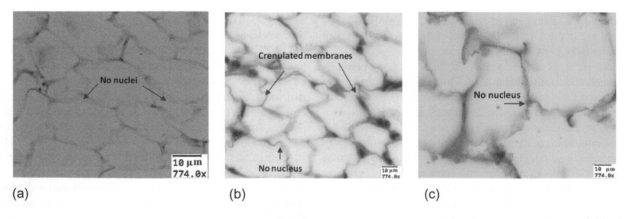

(a)   (b)   (c)

FIGURE 27.11 Histology of rat VAT: after FLMA (Mode 2) with preliminary topical application of fructose (50%) (a); after FLMA (Mode 1) with preliminary topical application of fructose (50%) (b); after FLMA (Mode 2) with topical application of sucrose (60%) (c). H&E staining. Magnification 774×.

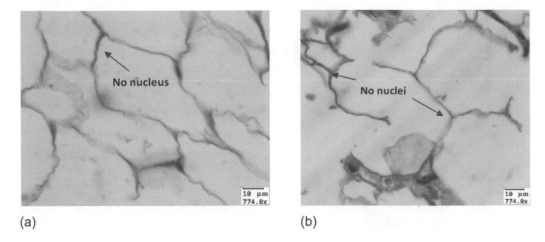

(a) (b)

FIGURE 27.12   Histology of rat VAT: after FLMA (Mode 2) with preliminary topical application of diatrizoic acid (21%) (a); after FLMA (Mode 2) with preliminary topical application of metrozoic acid (30%) (b). H&E staining. Magnification 774×.

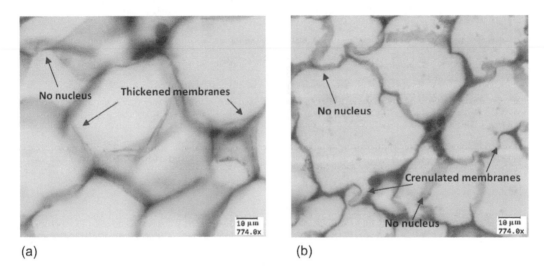

(a) (b)

FIGURE 27.13   Histology of rat VAT after US exposure with preliminary topical application of fructose (50%): mode 1 (a); mode 2 (b). H&E staining. Magnification 774×.

adipocytes are noted with high eosinophilia on the periphery areas. Cell membranes have been saved in the sample center, but no nuclei in cells are found (see arrows in Figure 27.12b). The VAT is completely damaged for 90% of the sample section.

It has been shown that at fractional ablation, membranes of cells around the created channel can be damaged [177–182].

After US exposure in Mode 1 (once for 2 min in CW mode) and Mode 2 (once for 2 min in pulsed mode) with preliminary topical application of fructose (50%), the VAT have the same morphology. The adipocytes have crenulated membranes, a part of which is thinned and another part is thickened. The adipocytes in the state of necrosis are found (see arrows in Figure 27.13), but the bulk of the cells have undergone a moderate damage.

After US exposure in Mode 3 (twice for 1 min in CW mode) with preliminary topical application of fructose (50%), adipocytes have smooth, regular shape of the membrane, sometimes thickened. Plasma permeation of several adipocytes is slightly expressed. The membranes are highly crenulated (see arrow in Figure 27.14a) and thinned. Cells have folded form on the sample periphery.

After US exposure in Mode 4 (twice for 1 min in pulsed mode) with preliminary topical application of fructose (50%), a significant necrosis of VAT is observed (see arrow in Figure 27.14b). Membranes of adipocytes are crenulated and disrupted (see arrows in Figure 27.14b). Cells lost their round shape, and the compression of cells are noted. It has been shown that ultrasonic cavitation allows for direct damage of the body fat via cell necrosis [183–185].

At combined US with topical application of fructose (50%) and FLMA (CW US, FLMA Mode 1), adipocytes have a roughly round shape. The cell membranes are thickened, sometimes crenulated (see arrow in Figure 27.15a). In some adipocytes, nuclei are absent.

At combined US with topical application of fructose (50%) and FLMA (pulsed US, FLMA Mode 1), adipocyte membranes are crenulated (see arrows in Figure 27.15 b) and cells have nuclei.

After combined US with topical application of fructose (50%) and FLMA (CW US, FLMA Mode 2), most of the adipocytes are preserved. A small area of tissue has cells with disrupted membranes (see arrow in Figure 27.16a).

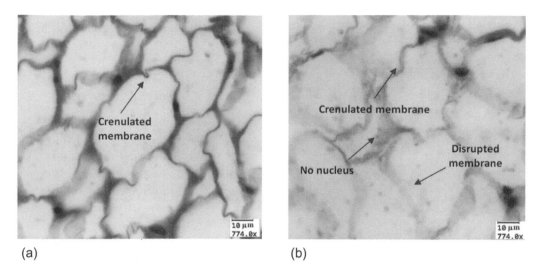

FIGURE 27.14   Histology of rat VAT after US exposure with preliminary topical application of fructose (50%): mode 3 (a); mode 4 (b). H&E staining. Magnification 774×.

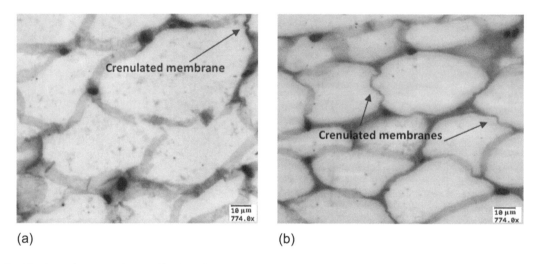

FIGURE 27.15   Histology of rat VAT after combined application of (CW/pulse) US with fructose (50%) and FLMA (mode 1/2): mode 1 (a); mode 2 (b). H&E staining. Magnification 774×.

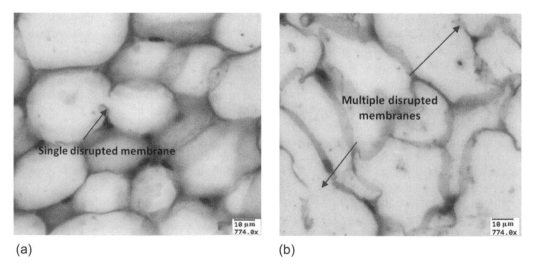

FIGURE 27.16   Histology of rat VAT after combined application of US (CW/pulse) with fructose (50%) and FLMA (mode 2): CW (a); pulse (b). H&E staining. Magnification 774×.

After combined pulsed US with topical application of fructose (50%) and FLMA (Mode 2), the VAT is completely damaged. Cells formed cavities (Figure 27.16b). More severe damage with smaller holes (0.1 mm) can be explained by the higher power density than for FLMA mode 1 (the difference is 4 times).

In vivo measurements

Without any exposure, VAT is represented by cells of approximately the same size. Each adipocyte appears as a thin ring of cytoplasm surrounding the vacuole left by the dissolved lipid droplet. Cells have preserved eccentric and flattened nuclei and clear contours of the membranes (Figure 27.17).

After topical application of sucrose (60%), interstitial edema in VAT and expressed diapedesis of erythrocytes (see arrows in Figure 27.18a) are observed. Cells similar to mast cells have appeared. The appearance of mast cells indicates a possible allergic reaction to the introduced agent as the fat cells contain a large amount of biologically active substances.

After topical application of erythrose (53.7%), the structure of VAT is mostly preserved, but some areas of hemorrhage, plasma permeation, and edema are presented (see arrows in Figure 27.18b). Some areas of necrosis are observed due to the influence of erythrose.

The development of diapedesis hemorrhages, plasma permeation, and edema in tissues adjacent to VAT may be due to disturbance of microcirculation, an increase in the permeability of the vascular wall, and a change in the blood vessel tonus at agent application.

Topical application of fructose (50%) caused complete destruction of VAT, which is observed at the depth of 0.572 mm at the periphery. The *in vivo* action of fructose is more aggressive than the same action *ex vivo*. VAT is damaged and areas of necrotic cells are observed (see arrow in Figure 27.19a). Plethora of large vessels is noted due to plasma mainly. Membranes of cells are folded, and parts of the membranes are thinned. The massive output of the cellular elements, mainly lymphocytes from blood vessels, is noted in the sample periphery (see arrows in Figure 27.19b). Expressed membrane permeation is observed with up to 70% of adipocytes with thinned membranes.

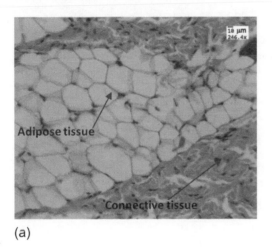

(a)

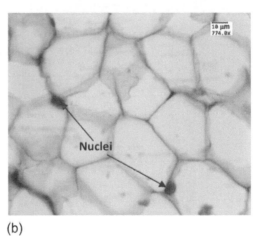

(b)

FIGURE 27.17 Histology of rat VAT of control sample for *in vivo* study: magnification 246.4× (a); magnification 774× (b). H&E staining.

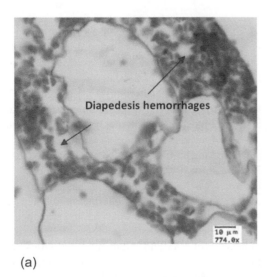

(a)

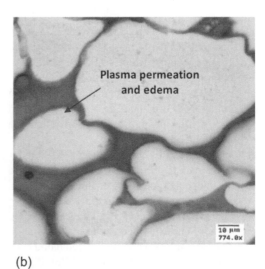

(b)

FIGURE 27.18 Histology of rat VAT after *in vivo* topical application of: sucrose (60%) (a); erythrose (53.7%) (b). H&E staining. Magnification 774×.

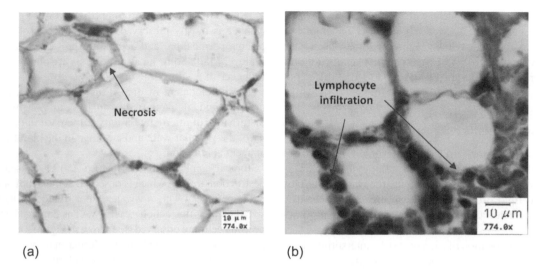

(a)   (b)

FIGURE 27.19   Histology of rat VAT after *in vivo* topical application of fructose (50%): membranes of adipocytes are folded (a); the massive output of the cellular elements, mainly lymphocytes from blood vessels (b). H&E staining. Magnification 774×.

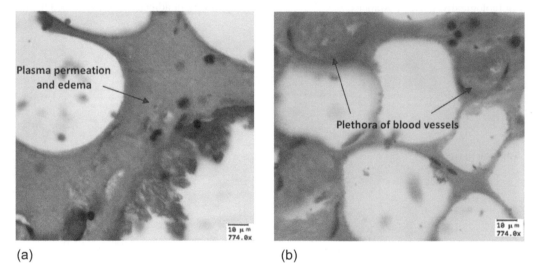

(a)   (b)

FIGURE 27.20   Histology of rat VAT after *in vivo* topical application of diatrizoic acid (21%), expressed interstitial edema is seen (a); metrozoic acid (30%) (b). H&E staining. Magnification 774×.

After topical application of diatrizoic acid (21%), a moderately expressed interstitial edema in VAT is observed (see arrow in Figure 27.20a). Around the vessels, the buildup of fluid is noted, which moved away the adjacent VAT.

After topical application of metrozoic acid (30%), plethora of blood vessels and expressed diapedesis of erythrocytes in VAT is observed (see arrows in Figure 27.20, b). Adipocyte membranes are broken and crenulated.

After topical application of PEG-200 (73.9%), VAT is damaged and extensive necrosis is observed. Cell membranes are semitransparent and thickened and parts of them are broken. This indicates dystrophic changes and membrane damage. For example, when a cell dies, all the nucleus becomes pale-colored and then it cannot be seen. The expressed plethora of vessels and phenomenon of separation of blood and mainly red blood cells in the vessels are noted (see arrows in Figure 27.21a). Plasma permeation with hemorrhagic component is seen. Expressed infiltration of lymphocytes is seen on the sample periphery.

After topical application of D-(+)-maltose (49.7%), three large plethoric vessels are observed at the center of the VAT sample. Adipocyte membranes are thickened and crenulated (see arrow in Figure 27.21b).

After topical application of D-(+)-mannose (57.7%), almost total necrosis of VAT (see arrow in Figure 27.22a) is observed and membranes of preserved adipocytes are crenulated.

After topical application of ribitol (52.6%), plethora of small and medium-sized vessels is noted (see arrow in Figure 27.22b). Basically, the VAT structure is preserved, but there are areas where it seems as if fat has been dissolved. Of the adipocytes, 50% are intact, but their membranes are modified, mostly folded with a small part being crenulated.

After topical application of D-(+)-trehalose (52.6%), areas of necrosis of VAT and cell membranes with a violation of the fibrous structure are observed (see arrows in Figure 27.23a).

After topical application of altrose (48.9%), VAT structure is significantly damaged. The cells are flattened and membranes

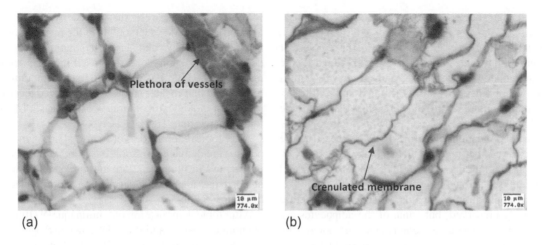

(a)

(b)

FIGURE 27.21 Histology of rat VAT after *in vivo* topical application of: PEG-200 (73.9%) (a); D-(+)-maltose (49.7%) (b). H&E staining. Magnification 774×.

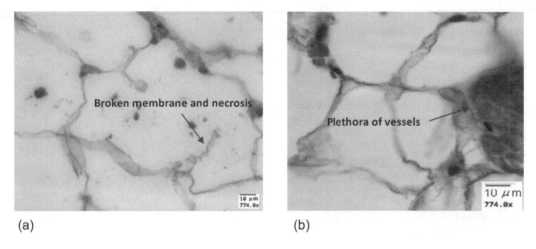

(a)

(b)

FIGURE 27.22 Histology of rat VAT after *in vivo* topical application of: D-(+)-mannose (57.7%) (a); ribitol (52.6%) (b). H&E staining. Magnification 774×.

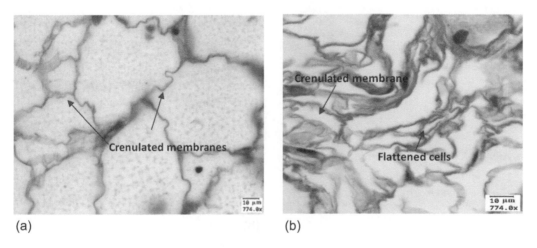

(a)

(b)

FIGURE 27.23 Histology of rat VAT after *in vivo* topical application of: D-(+)-trehalose (52.6%) (a); altrose (48.9%) (b). H&E staining. Magnification 774×.

are crenulated, but nuclei are preserved (see arrows in Figure 27.23b).

After topical application of glucose (50.8%), the structure of VAT is preserved, but some areas of necrosis are found. There are areas impregnated with glucose, and adipocytes with a round shape. Membranes are sharply thinned and semitransparent, and some membranes are thickened and disrupted. We are observed plethora of vessels (see arrow in Figure 27.24a).

After topical application of saline (100%), the structure of VAT is preserved, but areas of hemorrhage are observed. Semitransparent membranes are folded and twisted, and necrosis is found (see arrow in Figure 27.24b).

After topical application of tagatose (49.6%), the structure of VAT is mainly preserved, but some of the adipocytes are in the necrotic state. Diapedetic hemorrhages are noted. The adipocytes are elongated, and parts of them have thickened or broken membranes (see arrow in Figure 27.25a). The stretching of adipocytes in various directions can be hypothetically associated with lipid droplet loss, and therefore cells do not retain their shape [127, 186].

Similar features were discovered by the groups led by Mordon [185] and Anderson [187] for lipolysis of AT caused by radiofrequency and laser irradiation, respectively. To describe the phenomenon, they used the term "reduplicated" membranes, which was often used in early literature [188]. It is remarkable that in later studies it was proved that the modified regions of biological membranes can consist of long stretches folded many times [188].

After topical application of fucose (43.0%), most of the adipocytes are damaged. In some cases membranes are thickened (see arrows in Figure 27.25b) and broken. The areas of small hemorrhages and plasma permeation are presented (see arrows in Figure 27.25b).

After topical application of Omnipaque™ (100%), moderate plethora of vessels and diapedetic hemorrhage is observed in VAT (see arrow in Figure 27.26a). Cell membranes are thickened, a little crenulated, and folded. Infiltration of lymphocytes is noted (see arrow in Figure 27.26b).

After topical application of glucose (50.8%) and US exposure (Mode 1), total damage of VAT is observed (see arrow in

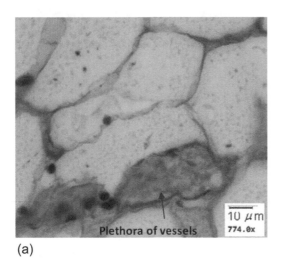
(a)

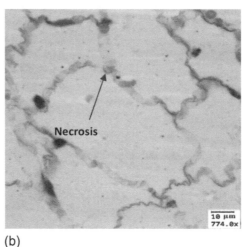
(b)

FIGURE 27.24 Histology of rat VAT after *in vivo* topical application of: glucose (50.8%) (a); saline (b). H&E staining. Magnification 774×.

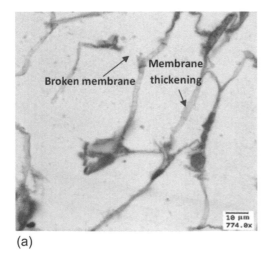
(a)

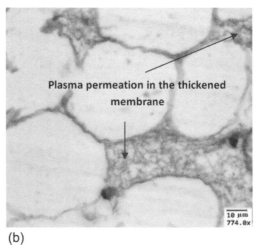
(b)

FIGURE 27.25 Histology of rat VAT after *in vivo* topical application of: tagatose (49.6%) (a); fucose (43.0%) (b). H&E staining. Magnification 774×.

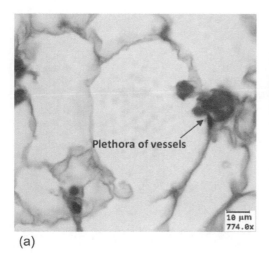

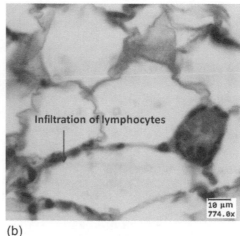

FIGURE 27.26 Histology of rat VAT after *in vivo* topical application of Omnipaque™-300: plethora of vessels is seen (a); infiltration of lymphocytes is seen (b). H&E staining. Magnification 774×.

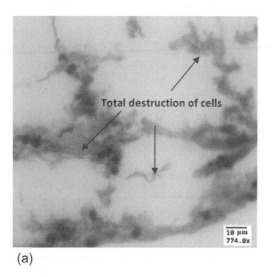

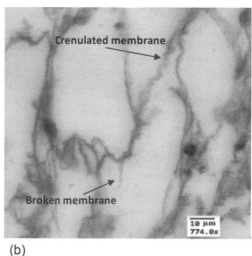

FIGURE 27.27 Histology of rat VAT after *in vivo* topical application of glucose (50.8%) and US exposure (mode 1): rat 1(a); rat 2(b). H&E staining. Magnification 774×.

Figure 27.27a). Cells with broken and crenulated membranes are seen (see arrows in Figure 27.27b).

After topical application of glucose (50.8%) and US exposure (Mode 2), total damage of VAT is observed, and parts of membranes are soaked in plasma. Thickened, broken, and crenulated membranes are presented in some cases due to the immersion solution (see arrows in Figure 27.28a).

After US exposure (Mode 3) with topical application of glucose (50.8%), areas of hemorrhage are presented in VAT. Infiltration of lymphocytes is noted. Some membranes are thickened due to the glucose solution and impregnation of plasma (see arrow in Figure 27.28b).

For *in vivo* OCA/enhancer exposure, a more complex response of VAT is found, which is caused by the physiological reaction of the animal, i.e. in response of the microcirculatory bed, there was more plethora and plasma of blood permeation; and in response to damage, there was tissue infiltration by white blood cells. That was a systemic reaction, in contrast to *ex vivo* exposure, where mainly signs of damage and changes only in the cell membranes were observed.

Conclusion

To conclude, it should be noted that TOC is still a novel, rapidly expanding set of methodologies that require standardization of the described protocols. Comparative studies, in which many TOC methods are applied side-by-side to the organ of interest, will also serve as a valuable reference for choosing the most optimal approaches from the plethora of already existing techniques.

The maximal OC efficiency of adipose tissue *ex vivo* (77%) was observed for the samples subjected to the fructose action during 40 min.

In *in vivo* experiments, at least a low level of necrosis was observed for all samples and OCAs except those subjected to the fructose (no necrosis).

The selected mode has shown no necrosis for samples of adipose tissue after fructose solution impact and a low level of necrosis for samples after sucrose or fructose solution action.

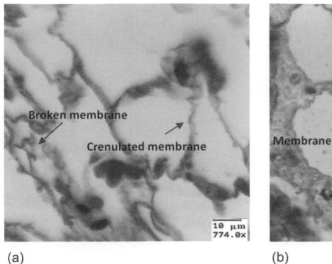

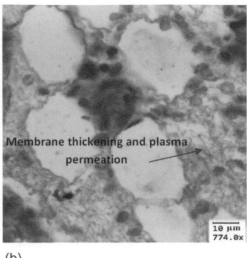

(a)   (b)

FIGURE 27.28 Histology of rat VAT after *in vivo* topical application of glucose (50.8%) and US exposure (mode 2/3): mode 2(a); mode 3(b). H&E staining. Magnification 774×.

The maximal OC efficiency in *in vivo* investigations (65%) was observed for the samples subjected to the diatrizoic acid action during 120 min.

Acknowledgments

This research was supported by joint research agreement (1.10.2013) between Olympus Corporation and Saratov State University.

REFERENCES

1. W. Sun, J. Fang, Q. Yong, S. Li, Q. Xie, J. Yin, L. Cui, "Subcutaneous construction of engineered adipose tissue with fat lobule-like structure using injectable poly-benzyl-l-glutamate microspheres loaded with adipose-derived stem cells," *PLOS ONE* **10**(8), e0135611 (2015).
2. C.W. Patrick, "Breast tissue engineering,"*Annu. Rev. Biomed. Eng.* **6**, 109–130 (2004).
3. A. Wronska, Z. Kmiec, "Structural and biochemical characteristics of various white adipose tissue depots," *Acta Physiol.* **205**(2), 194–208 (2012).
4. A. Cryer, R.L.R. Van, *New Perspectives in Adipose Tissue: Structure, Function and Development*, Butterworth-Heinemann, Oxford (2014), 490 p.
5. B.S. Wiseman, Z. Werb, "Stromal effects on mammary gland development and breast cancer," *Science* **296**(5570), 1046–1049 (2002).
6. D.T. Janigan, R. Durning, B. Perey, A.S. MacDonald, G. Klassen, "Structural changes in the subcutaneous compartment in morbid obesity," *Obes. Res.* **1**(5), 384–389 (1993).
7. J.S. Petrofsky, M. Prowse, E. Lohman, "The influence of ageing and diabetes on skin and subcutaneous fat thickness in different regions of the body," *Res. J. Appl. Sci.* **8**(1), 55–61 (2008).
8. G. Ganesan, R.V. Warren, A. Leproux, et al., "Diffuse optical spectroscopic imaging of subcutaneous adipose tissue metabolic changes during weight loss," *Int. J. Obes.* **40**(8), 1292–1300 (2016).
9. M.J. Legato (ed.), *Principles of Gender-Specific Medicine. Gender in the Genomic Era*, 3rd ed., Academic Press, Cambridge, MA (2017), 792 p.
10. N.H. Anaizi, "Fat facts: An overview of adipose tissue and lipids," *Ibnosina J. Med. Biomed. Sci.* **11**(1), 5–15 (2019).
11. M.M. Ibrahim, "Subcutaneous and visceral adipose tissue: Structural and functional differences," *Obes. Rev.* **11**(1), 11–18 (2010).
12. B. Querleux, C. Cornillon, O. Jolivet, J. Bittoun, "Anatomy and physiology of subcutaneous adipose tissue by in vivo magnetic resonance imaging and spectroscopy: Relationships with sex and presence of cellulite," *Skin Res. Technol.* **8**(2), 118–124 (2002).
13. D.-S. Ho, E.-H. Kim, I.D. Hwang, K. Shin, J.-T. Oh, B.-M. Kim, "Optical skin-fat thickness measurement using miniaturized chip LEDs: A preliminary human study," *J. Opt. Soc. Korea* **13**(3), 304–309 (2009).
14. M. DiGirolamo, J.L. Owens, "Water content of rat adipose tissue and isolated adipocytes in relation to cell size," *Am. J. Physiol. Cell Physiol.* **231**(5), 1568–1572 (1976).
15. G. Vargas, M. Chandalia, Y. Jiang, H. Davila, M. Motamedi, N. Abate, "Heterogeneity in subcutaneous adipose tissue morphology and metabolic complications in overweight and obese women," *Metab. Syndr. Relat. Disord.* **11**(4), 276–282 (2013).
16. G. Geri, B.M. Poli, A. Zappa, G. Campodoni, O. Franci, "Relationships between adipose tissue characteristics of newborn pigs and subsequent performance: III. Histological and chemical characteristics of backfat," *J. Anim. Sci.* **68**(7), 1936–1943 (1990).
17. L.B. Salans, S.W. Cushman, R.E. Weismann, "Studies of human adipose tissue. Adipose cell size and number in non-obese and obese patients," *J. Clin. Invest.* **52**(4), 929–941 (1973).
18. J. Han, J.-E. Lee, J. Jin et al., "The spatiotemporal development of adipose tissue," *Development* **138**(22), 5027–5037 (2011).
19. M. Novak, E. Monkus, V. Pardo, "Human neonatal subcutaneous adipose tissue. Function and ultrastructure," *Biol. Neonat.* **19**(4), 306–321 (1971).

20. M.J. Dauncey, D. Gairdner, "Size of adipose cells in infancy," *Arch. Dis. Childh.* **50**(4), 286–290 (1975).

21. G. Enzi, E.M. Inelmen, F. Caretta, F. Villani, V. Zanardo, F. DeBiasi, "Development of adipose tissue in newborns of gestational-diabetic and insulin-dependent diabetic mothers," *Diabetes* **29**(2), 100–104 (1980).

22. F.P. Bonnet, D. Rocour-Brumioul, A. Heuskin, "Regional variations of adipose cell size and local cellularity in human subcutaneous fat during normal growth," *Acta Paediatr. Belg.* **32**(1), 17–27 (1979).

23. J. Hirsch, J.L. Knittle, "Cellularity of obese and non-obese human adipose tissue," *Fed. Proc.* **29**(4), 1516–1521 (1970).

24. P. Tandon, R. Wafer, J.E.N. Minchin, "Review. Adipose morphology and metabolic disease," *J. Exp. Biol.* **221**(Suppl 1), jeb164970 (2018).

25. S.W. Coppack, "Adipose tissue changes in obesity," *Biochem. Soc. Trans.* **33**(5), 1049–1052 (2005).

26. S.D. Parlee, S.I. Lentz, H. Mori, O.A. MacDougald, "Quantifying size and number of adipocytes in adipose tissue," *Methods Enzymol.* **537**, 93–122 (2014).

27. L.B. Sardinha, A.M. Silva, C.S. Minderico, P.J. Teixeira, "Effect of body surface area calculations on body fat estimated in non-obese and obese subjects," *Physiol. Meas.* **27**(11), 1197–1209 (2006).

28. M.I. Gurr, R.T. Jung, M.P. Robinson, W.P.T. James, "Adipose tissue cellularity in man: The relationship between fat cell size and number, the mass and distribution of body fat and the history of weight gain and loss," *Int. J. Obes.* **6**(5), 419–436 (1982).

29. Y. Taton', *Obesity. Pathphysiology, Diagnostics, Therapy*, Medical Press, Warsaw (1981).

30. D.Y. Shurigin, P.O. Vyazitskiy, K.A. Sidorov, *Obesity*, Medicine, Leningrad (1975).

31. L. Fang, F. Guo, L. Zhou, R. Stahl, J. Grams, "The cell size and distribution of adipocytes from subcutaneous and visceral fat is associated with type 2 diabetes mellitus in humans," *Adipocyte* **4**(4), 273–279 (2015).

32. T. McLaughlin, C. Lamendola, N. Coghlan, T.C. Liu, K. Lerner, A. Sherman, S.W. Cushman, "Subcutaneous adipose cell size and distribution: Relationship to insulin resistance and body fat," *Obesity* **22**(3), 673–680 (2014).

33. P. Bjorntorp, L. Sjostrom, "Number and size of adipose tissue fat cells in relations to metabolism in human obesity," *Metabolism* **20**(7), 703–713 (1971).

34. F.P. de Heredia, E. Larque, M. del Puy Portillo, M. Canteras, S. Zamora, M. Garaulet, "Age-related changes in fatty acids from different adipose depots in rat and their association with adiposity and insulin," *Nutrition* **24**(10), 1013–1022 (2008).

35. A.E. Renold, G.F. Cahill (Eds.), *Handbook of Physiology, Adipose Tissue/Comprehensive Physiology*, Wiley Online Library, Hoboken, NJ (2011).

36. T.J. Ryan, "Cutaneous circulation," *Physiology, Biochemistry, and Molecular Biology of the Skin*, ed. Goldsmith, L.A., Vol II, Oxford University Press, Oxford, pp. 1019–1084 (1991).

37. I.M. Braverman, A. Keh-Yen, "Ultrastructure of the human dermal microcirculation IV. Valve-containing collecting veins at the dermal-subcutaneous junction," *J. Invest. Dermatol.* **81**(5), 438–442 (1983).

38. A. Caduff, M.S. Talary, P. Zakharov, "Cutaneous blood perfusion as a perturbing factor for noninvasive glucose monitoring," *Diabetes Technol. Ther.* **12**(1), 1–9 (2010).

39. F.H. Mustafa, P.W. Jones, A.L. McEwan, "Near infrared spectroscopy for body fat sensing in neonates: Quantitative analysis by GAMOS simulations," *Biomed. Eng. OnLine* **16**(14) (2017).

40. P. Lanka, A. Segala, A. Farina et al., "Non-invasive investigation of adipose tissue by time domain diffuse optical spectroscopy," *Biomed. Opt. Express* **11**(5), 2779–2793 (2020).

41. R. Nachabe, B.H.W. Hendriks, M. van der Voort, A.E. Desjardins, H.J.C.M. Sterenborg, "Estimation of biological chromophores using diffuse optical spectroscopy: Benefit of extending the UV-vis wavelength range to include 1000 to 1600 nm," *Biomed. Opt. Express* **18**(24), 1432–1442 (2010).

42. C. Sun, S. Mao, S. Chen, W. Zhang, C. Liu, "PPARs-Orchestrated Metabolic Homeostasis in the Adipose Tissue," *Int. J. Mol. Sci.* **22** (16), 8974 (2021).

43. S. Cinti, A. Giordano, "The adipose organ," in *The First Outstanding 50 Years of "Università Politecnicadelle Marche*, eds. S. Longhi et al., Springer, Cham (2020).

44. A.T. Ali, W.E. Hochfeld, R. Myburgh, M.S. Pepper, "Adipocyte and adipogenesis," *Eur. J. Cell Biol.* **92**(6–7), 229–236 (2013).

45. M. Esteve Ràfols, "Adipose tissue: Cell heterogeneity and functional diversity," *Endocrinol. Nutr.* **61**(2), 100–112 (2014).

46. C. Barreau, E. Labit, C. Guissard et al., "Regionalization of browning revealed by whole subcutaneous adipose tissue imaging," *Obesity* **24**(5), 1081–1089 (2016).

47. S. Altshuler-Keylin, K. Shinoda, Y. Hasegawa et al., "Beige adipocyte maintenance is regulated by autophagy-induced mitochondrial clearance," *Cell Metab.* **24**(3), 402 (2016).

48. J. Chi, Z. Wu, C.H.J. Choi et al., "Three-dimensional adipose tissue imaging reveals regional variation in beige fat biogenesis and PRDM16-dependent sympathetic neurite density," *Cell Metab.* **27**(1), 226 (2018).

49. W. Zeng, R.M. Pirzgalska, M.M.A. Pereira et al., "Sympathetic neuro-adipose connections mediate leptin-driven lipolysis," *Cell* **163**(1), 84–94 (2015).

50. J. Chi, A. Crane, Z. Wu, P. Cohen, "Adipo-Clear: A tissue clearing method for three-dimensional imaging of adipose tissue," *J. Vis. Exp.* **137**(137), e58271 (2018).

51. Y. Cao, H. Wang, Q. Wang, X. Han, W. Zeng, "Three-dimensional volume fluorescence-imaging of vascular plasticity in adipose tissues," *Mol. Metab.* **14**, 71 (2018).

52. K. Dev, U.S. Dinish, S. Chakraborty, R. Bi, S. Andersson-Engels, S. Sugii, M. Olivo, "Quantitative in vivo detection of adipose tissue browning using diffuse reflectance spectroscopy in near-infrared II window," *J. Biophoton.* **11**(12), e201800135 (2018).

53. S.L. Jacques, "Optical properties of biological tissues: A review," *Phys. Med. Biol.* **58**(11), R37–R61 (2013).

54. T.N. Bydlon, R. Nachabe, N. Ramanujam, H.J.C.M. Sterenborg, B.H.W. Hendriks, "Chromophore based analyses of steady-state diffuse reflectance spectroscopy: Current status and perspectives for clinical adoption," *J. Biophoton.* **8**(1–2), 9–24 (2015).

55. V.V. Tuchin, *Tissue Optics: Light Scattering Methods and Instruments for Medical Diagnostics*, 3rd ed., SPIE Press, Bellingham (2015).

56. N. Bosschaart, G.J. Edelman, M.C.G. Aalders, T.G. van Leeuwen, D.J. Faber, "A literature review and novel theoretical approach on the optical properties of whole blood," *Lasers Med. Sci.* **29**(2), 453–479 (2014).

57. H.Q. Woodard, D.R. White, "The composition of body tissues," *Br. J. Radiol.* **59**(708), 1209–1219 (1986).

58. A. Torricelli, A. Pifferi, P. Taroni, E. Giambattistelli, R. Cubeddu, "*In vivo* optical characterization of human tissues from 610 to 1010 nm by time-resolved reflectance spectroscopy," *Phys. Med. Biol.* **46**(8), 2227–2237 (2001).

59. R. Nachabe, B.H.W. Hendriks, A.E. Desjardins, M. van der Voort, M.B. van der Mark, H.J.C.M. Sterenborg, "Estimation of lipid and water concentrations in scattering media with diffuse optical spectroscopy from 900 to 1600 nm," *J. Biomed. Opt.* **15**(3), 037015 (2010).

60. E. Zamora-Rojas, B. Aernouts, A. Garrido-Varo, D. Pérez-Marín, J.E. Guerrero-Ginel, W. Saeys, "Double integrating sphere measurements for estimating optical properties of pig subcutaneous adipose tissue," *Innov. Food Sci. Emerg. Technol.* **19**, 218–226 (2013).

61. E.D. Aberle, T.D. Etherton, C.E. Allen, "Predicting of pork carcass compositions using subcutaneous adipose tissue moisture or lipid concentration," *J. Anim. Sci.* **45**(3), 449–456 (1977).

62. B. Lebret, J. Mourot, "Caractéristiques et qualité des tissue adipeux chez le porc. Facteurs de variation non génétiques," *INRA Prod. Anim.* **11**(2), 131–143 (1998).

63. C.-L. Tsai, J.-C. Chen, W.-J. Wang, "Near-infrared absorption properties of biological soft tissue constituents," *J. Med. Biol. Eng.* **21**(1), 7–14 (2001).

64. R. Nachabe, J.W.A. van der Hoorn, R. van de Molengraafetal et al., "Validation of interventional fiber optic spectroscopy with MR spectroscopy, MAS-NMR spectroscopy, high-performance thin-layer chromatography, and histopathology for accurate hepatic fat quantification," *Invest. Radiol.* **47**(4), 209–216 (2012).

65. P. Ballard, D.E. Leahy, M. Rowland, "Prediction of in vivo tissue distribution from in vitro data. 1. Experiments with markers of aqueous spaces," *Pharm. Res.* **17**(6), 660–663 (2000).

66. R.F. Reinoso, B.A. Telfer, M. Rowland, "Tissue water content in rats measured by desiccation," *J. Pharmacol. Toxicol. Meth.* **38**(2), 87–92 (1997).

67. G. Nishimura, I. Kida, M. Tamura, "Characterization of optical parameters with a human forearm at the region from 1.15 to 1.52 μm using diffuse reflectance measurements," *Phys. Med. Biol.* **51**(11), 2997–3011 (2006).

68. J.A. Curcio, C.C. Petty, "The near infrared absorption spectrum of liquid water," *J. Opt. Soc. Am.* **41**(5), 302–304 (1951).

69. M. Daimon, A. Masumura, "Measurement of the refractive index of distilled water from the near-infrared region to the ultraviolet region," *Appl. Opt.* **46**(18), 3811–3820 (2007).

70. A. Djurisic, B.V. Stanic, "Modeling the wavelength dependence of the index of refraction of water in the range 200 nm to 200 μm," *Appl. Opt.* **37**(13), 2696–2698 (1998).

71. E.S. Fry, "Studies on some of the inherent optical properties of natural waters [invited]," *Appl. Opt.* **52**(5), 930–939 (2013).

72. G. Hale, M.R. Querry, "Optical constants of water in the 200-nm to 200-μm wavelength region," *Appl. Opt.* **12**(3), 555–563 (1973).

73. A.H. Harvey, J.S. Gallagher, J.M.H.L. Sengers, "Revised formulation for the refractive index of water and steam as a function of wavelength, temperature and density," *J. Phys. Chem. Ref. Data* **27**(4), 761–774 (1998).

74. L. Kou, D. Labrie, P. Chylek, "Refractive indices of water and ice in the 0.65- to 2.5-μm spectral range," *Appl. Opt.* **32**(19), 3531–3540 (1993).

75. R.A.J. Litjens, T.I. Quickenden, C.G. Freeman, "Visible and near-ultraviolet absorption spectrum of liquid water," *Appl. Opt.* **38**(7), 1216–1223 (1999).

76. K.F. Palmer, D. Williams, "Optical properties of water in the near infrared," *J. Opt. Soc. Am.* **64**(8), 1107–1110 (1974).

77. W.S. Pegau, D. Gray, J.R.V. Zaneveld, "Absorption and attenuation of visible and near-infrared light in water: Dependence on temperature and salinity," *Appl. Opt.* **36**(24), 6035–6046 (1997).

78. R.M. Pope, E.S. Fry, "Absorption spectrum (380–700 nm) of pure water. II. Integrating cavity measurements," *Appl. Opt.* **36**(33), 8710–8723 (1997).

79. P.S. Ray, "Broadband complex refractive indices of ice and water," *Appl. Opt.* **11**(8), 1836–1844 (1972).

80. P. Schiebener, J. Straub, J.M.H.L. Sengers, J.S. Gallagher, "Refractive index of water and steam as function of wavelength, temperature and density," *J. Phys. Chem. Ref. Data* **19**(3), 677–717 (1990).

81. D.J. Segelstein, "The complex refractive index of water," MS thesis, University of Missouri-Kansas City (1981).

82. R.C. Smith, K.S. Baker, "Optical properties of the clearest natural water (200–800 nm)," *Appl. Opt.* **20**(2), 177–184 (1981).

83. F.M. Sogandares, E.S. Fry, "Absorption spectrum (340–640 nm) of pure water. I. Photothermal measurements," *Appl. Opt.* **36**(33), 8699–8709 (1997).

84. S.A. Sullivan, "Experimental study of the absorption in distilled water, artificial sea water, and heavy water in the visible region of the spectrum," *J. Opt. Soc. Am.* **53**(8), 962–968 (1963).

85. D.M. Wieliczka, S. Weng, M.R. Querry, "Wedge shaped cell for highly absorbent liquids: Infrared optical constants of water," *Appl. Opt.* **28**(9), 1714–1719 (1989).

86. D.W. Leonard, K.M. Meek, "Refractive indices of the collagen fibrils and extrafibrillar material of the corneal stroma," *Biophys. J.* **72**(3), 1382–1387 (1997).

87. C. Morin, C. Hellmich, P. Henits, "Fibrillar structure and elasticity of hydrating collagen: A quantitative multiscale approach," *J. Theor. Biol.* **317**, 384–393 (2013).

88. S.K.V. Sekar, I. Bargigia, A.D. Mora et al., "Diffuse optical characterization of collagen absorption from 500 to 1700 nm," *J. Biomed. Opt.* **22**(1), 015006 (2017).

89. A.N. Bashkatov, E.A. Genina, V.I. Kochubey, V.V. Tuchin, "Optical properties of human skin, subcutaneous and mucous tissues in the wavelength range from 400 to 2000 nm," *J. Phys. D* **38**(15), 2543–2555 (2005).

90. A.N. Bashkatov, E.A. Genina, V.I. Kochubey, V.V. Tuchin, "Optical properties of the subcutaneous adipose tissue in the spectral range 400–2500 nm," *Opt. Spectrosc.* **99**(5), 836–842 (2005).

91. E. Salomatina, B. Jiang, J. Novak, A.N. Yaroslavsky, "Optical properties of normal and cancerous human skin in the visible and near-infrared spectral range," *J. Biomed. Opt.* **11**(6), 064026 (2006).

92. C.R. Simpson, M. Kohl, M. Essenpreis, M. Cope, "Near-infrared optical properties of *ex vivo* human skin and subcutaneous tissues measured using the Monte Carlo inversion technique," *Phys. Med. Biol.* **43**(9), 2465–2478 (1998).

93. E.L. Wisotzky, F.C. Uecker, S. Dommerich, A. Hilsmann, P. Eisert, P. Arens, "Determination of optical properties of human tissues obtained from parotidectomy in the spectral range of 250 to 800 nm," *J. Biomed. Opt.* **24**(12), 125001 (2019).

94. Y. Shimojo, T. Nishimura, H. Hazama, T. Ozawa, K. Awazu, "Measurement of absorption and reduced scattering coefficients in Asian human epidermis, dermis, and subcutaneous fat tissues in the 400- to 1100-nm wavelength range for optical penetration depth and energy deposition analysis," *J. Biomed. Opt.* **25**(4), 045002 (2020).

95. A. Kienle, L. Lilge, M.S. Patterson, R. Hibst, R. Steiner, B.C. Wilson, "Spatially resolved absolute diffuse reflectance measurements for noninvasive determination of the optical scattering and absorption coefficients of biological tissue," *Appl. Opt.* **35**(13), 2304–2314 (1996).

96. P. Sun, Y. Wang, "Measurements of optical parameters of phantom solution and bulk animal tissues *in vitro* at 650 nm," *Opt. Laser Technol.* **42**(1), 1–7 (2010).

97. C. Zhu, G.M. Palmer, T.M. Breslin, J. Harter, N. Ramanujam, "Diagnosis of breast cancer using diffuse reflectance spectroscopy: Comparison of a Monte Carlo versus partial least squares analysis based feature extraction technique," *Lasers Surg. Med.* **38**(7), 714–724 (2006).

98. F. Foschum, A. Kienle, "Broadband absorption spectroscopy of turbid media using a dual step steady-state method," *J. Biomed. Opt.* **17**(3), 037009 (2012).

99. S. Mosca, P. Lanka, N. Stoneetal, S. Konugolu Venkata Sekar, P. Matousek, G. Valentini, A. Pifferi, "Optical characterization of porcine tissues from various organs in the 650–1100 nm range using time-domain diffuse spectroscopy," *Biomed. Opt. Express* **11**(3), 1697–1706 (2020).

100. A.Y. Sdobnov, M.E. Darvin, E.A. Genina, A.N. Bashkatov, J. Lademann, V.V. Tuchin, "Recent progress in tissue optical clearing for spectroscopic application," *Spectrochim. Acta A Mol. Biomol. Spectrosc.* **197**, 216–229 (2018).

101. I. Costantini, R. Cicchi, L. Silvestri, F. Vanzi, F.S. Pavone, "In-vivo and ex-vivo optical clearing methods for biological tissues: Review," *Biomed. Opt. Express* **10**(10), 5251–5267 (2019).

102. M. Inyushin, D. Meshalkina, L. Zueva, A. Zayas-Santiago, "Tissue transparency *in vivo*," *Molecules* **24**(13), E2388–E2413 (2019).

103. T. Son, B. Jung, "Cross-evaluation of optimal glycerol concentration to enhance optical tissue clearing efficacy," *Skin Res. Technol.* **21**(3), 327–332 (2015).

104. K. Cu, R. Bansal, S. Mitragotri, R.D. Fernandez, "Delivery strategies for skin: Comparison of nanoliter jets, needles and topical solutions," *Ann. Biomed. Eng.* 1573 (2019).

105. D.S. Richardson, J.W. Lichtman, "Clarifying tissue clearing," *Cell* **162**(2), 246–257 (2015).

106. P. Matryba, L. Kaczmarek, J. Gołąb, "Advances in ex situ tissue optical clearing," *Laser Photon. Rev.* **13**(8), 1800292 (2019).

107. I.Yu. Yanina, J. Schleusener, J. Lademann, V.V. Tuchin, M.E. Darvin, "Confocal Raman microspectroscopy for evaluation of optical clearing efficiency of the skin ex vivo," *Proc. SPIE* **11239**, 112390W (2020).

108. I.Yu. Yanina, J. Schleusener, J. Lademann, V.V. Tuchin, M.E. Darvin, "The effectiveness of glycerol solutions for optical clearing of the intact skin as measured by confocal Raman spectroscopy," *Opt. Spectrosc.* **128**(6), 759–765 (2020).

109. A.N. Bashkatov, E.A. Genina, Y.P. Sinichkin, V.V. Tuchin, "Influence of glycerol on the transport of light in the skin," *Proc. SPIE* **4623**, 144–153 (2002).

110. E.A. Genina, A.N. Bashkatov, Y.P. Sinichkin, V.V. Tuchin, "Optical clearing of the eye sclera *in vivo* caused by glucose," *Quantum Electron.* **36**(12), 1119–1124 (2006).

111. E.A. Genina, A.N. Bashkatov, Yu.P. Sinichkin, I.Yu. Yanina, V.V. Tuchin, "Optical clearing of biological tissues: Prospects of application in medical diagnostics and phototherapy," *J. Biomed. Photon. Eng.* **1**(1), 22–58 (2015).

112. K. Tainaka, A. Kuno, S.I. Kubota, T. Murakami, H.R. Ueda, "Chemical principles in tissue clearing and staining protocols for whole-body cell profiling," *Annu. Rev. Cell Dev. Biol.* **32**, 713–741 (2016).

113. K.V. Berezin, K.N. Dvoretski, M.L. Chernavina et al., "*In vivo* optical clearing of human skin under the effect of some monosaccharides," *Opt. Spectr.* **127**(2), 329–336 (2019).

114. J.M. Hirshburg, K.M. Ravikumar, W. Hwang, A.T. Yeh, "Molecular basis for optical clearing of collagenous tissues," *J. Biomed. Opt.* **15**(5), 055002 (2010).

115. N.J. Yang, M.J. Hinner, "Getting across the cell membrane: An overview for small molecules, peptides, and proteins," *Meth. Mol. Biol.* **1266**, 29–53 (2015).

116. R. Dickie, R.M. Bachoo, M.A. Rupnick et al., "Three-dimensional visualization of microvessel architecture of whole-mount tissue by confocal microscopy," *Microvasc. Res.* **72**(1–2), 20–26 (2006).

117. M. Lai, X. Li, J. Li, Y. Hu, D.M. Czajkowsky, Z. Shao, "Improved clearing of lipid droplet-rich tissues for three-dimensional structural elucidation," *Acta Biochim. Biophys. Sin.* **49**(5), 465–467 (2017).

118. O. Hamdy, R.M. Abdelazeem, "Toward better medical diagnosis: Tissue optical clearing," *Int. J. Public Health* **2**(1), 13–21 (2020).

119. V.V. Tuchin, I.Y. Yanina, G.V. Simonenko, "Destructive fat tissue engineering using photodynamic and selective photothermal effects," *Proc. SPIE*, **7179**, 71790C (2009).

120. V.A. Doubrovskii, I.Yu. Yanina, V.V. Tuchin, "Kinetics of changes in the coefficient of transmission of the adipose tissue in vitro as a result of photodynamic action," *Biophys.* **57**(1), 94–97 (2012).

121. I.Yu. Yanina, N.A. Trunina, V.V. Tuchin, "Photoinduced cell morphology alterations quantified within adipose tissues by spectral optical coherence tomography," *J. Biomed. Opt.* **18**(11), 111407 (2013).

122. V.A. Dubrovskii, B.A. Dvorkin, I.Yu. Yanina, V.V. Tuchin, "Photoaction upon adipose tissue cells in vitro," *Cell Tissue Biol.* **5**(5), 520–529 (2011).

123. B. Alberts, A. Johnson, J. Lewis, D. Morgan, M. Raff, K. Roberts, P. Walter, *Molecular Biology of the Cell*, 6th ed., Garland Science, Taylor & Francis Group, LLC, New York (2015), 1465 p.

124. E.K. Chan, B. Sorg, D. Protsenko, M.O. Neil, M. Motamedi, A.J. Welch, "Effects of compression on soft tissue optical properties," *IEEE J. Sel. Top. Quantum Electron.* **2**(4), 943–950 (1996).

125. C.W. Drew, C.G. Rylander, "Mechanical Compression for Dehydration and Optical Clearing of Skin." *Proceedings of the ASME 2008 Summer Bioengineering Conference. ASME 2008 Summer Bioengineering Conference, Parts A and B. Marco Island*, Florida, USA. June 25–29, 2008, 803–804 (2008).

126. C.G. Rylander, T.E. Milner, S.A. Baranov, J.S. Nelson, "Mechanical tissue optical clearing devices: Enhancement of light penetration in ex vivo porcine skin and adipose tissue," *Lasers Surg. Med.* **40**(10), 688–694 (2008).

127. R.R. Anderson, W. Farinelli, H. Laubach et al., "Selective photothermolysis of lipid-rich tissues: A free electron laser study," *Lasers Surg. Med.* **38**(10), 913–919 (2006).

128. H.F. Bowman, E.G. Cravalho, M. Woods, "Theory, measurement, and application of thermal properties of biomaterials,"*Annu. Rev. Biophys. Bioeng.* **4**(1), 43–80 (1975).

129. A.V. Belikov, K.V. Prikhod'ko, O.A. Smolyanskaya, V.A. Protasov, "Temperature dynamics of the optical properties of lipids in vitro," *J. Opt. Technol.* **70**(11), 811–814 (2003).

130. B.C. Wilson, V.V. Tuchin, S. Tanev, *Advances in Biophotonics. Series I: Life and Behavioural Sciences*, vol. 369, IOS Press, Amsterdam (2005), 296 p.

131. A.V. Belikov, O.A. Smolyanskaya, "Optical model of thermo-sensitive heterophase medium (adipose tissue)," *Proc. SPIE* **6535**, 65351F (2007).

132. H. Jelínková (Ed.), *Lasers for Medical Applications: Diagnostics, Therapy and Surgery*, Woodhead Publishing, Oxford et al. (2013), 832 p.

133. R. Seckel, S.T. Doherty, J.J. Childs, M.Z. Smirnov, R.H. Cohen, G.B. Altshuler, "The role of laser tunnels in laser-assisted lipolysis," *Lasers Surg. Med.* **41**(10), 728–737 (2009).

134. E.V. Ross, Y. Domankevitz, R.R. Anderson, "Effects of heterogeneous absorption of laser radiation in biotissue ablation: Characterization of ablation of fat with a pulsed CO_2 laser," *Lasers Surg. Med.* **21**(1), 59–64 (1997).

135. I.Yu. Yanina, E.K. Volkova, D.K. Tuchina, Ju. G. Konyukhova, V.I. Kochubey, V.V. Tuchin, "Controlling of upconversion nanoparticle luminescence at heating and optical clearing of adipose tissue," *Proc. SPIE*, 10417–10415, 1–7 (2017).

136. A. Häger, "Adipose cell size and number in relation to obesity," *Postgrad. Med. J.* **53**(Supplement 2), 101–110 (1977).

137. M. Garaulet, J. Hernandez-Morante, J. Lujan, F.J. Tebar, S. Zamora, "Relationship between fat cell size and number and fatty acid composition in adipose tissue from different fat depots in overweight/obese humans," *Int. J. Obes. Lond* **30**(6), 899–905 (2006).

138. S.M. Rabson, "The fatty tissues: Normal anatomy and changes in obesity (a review)," *Am. J. Clin. Pathol.* **5**(6), 240–245 (1945).

139. A.T. Yeh, B. Choi, J.S. Nelson, B.J. Tromberg, "Reversible dissociation of collagen in tissues," *J. Invest. Dermatol.* **121**(6), 1332–1335 (2003).

140. G. Vargas, A. Readinger, S.S. Dozier, A.J. Welch, "Morphological changes in blood vessels produced by hyperosmotic agents and measured by optical coherence tomography," *Photochem. Photobiol.* **77**(5), 541–549 (2003).

141. K.V. Larin, M.G. Ghosn, A.N. Bashkatov, E.A. Genina, N.A. Trunina, V.V. Tuchin, "Optical clearing for OCT image enhancement and in-depth monitoring of molecular diffusion," *IEEE J. Sel. Top. Quantum Electr.* **18**(3), 1244–1259 (2012).

142. E.I. Galanzha, V.V. Tuchin, A.V. Solovieva, T.V. Stepanova, Q. Luo, H. Cheng, "Skin backreflectance and microvascular system functioning at the action of osmotic agents," *J. Phys. D: Appl. Phys.* **36**(14), 1739–1746 (2003).

143. X. Wen, Z. Mao, Z. Han, V.V. Tuchin, D. Zhu, "*In vivo* skin optical clearing by glycerol solutions: Mechanism," *J. Biophoton.* **3**(1–2), 44–52 (2010).

144. G. Vargas, E.K. Chan, J.K. Barton, H.G. Rylander III, A.J. Welch, "Use of an agent to reduce scattering in skin," *Lasers Surg. Med.* **24**(2), 133–141 (1999).

145. E.A. Genina, A.N. Bashkatov, V.V. Tuchin, "Glucose-induced optical clearing effects in tissues and blood," in *Handbook of Optical Sensing of Glucose in Biological Fluids and Tissues*, ed. V.V. Tuchin, Taylor & Francis Group LLC, CRC Press, Boca Raton, FL (2009), pp. 657–692.

146. A.N. Bashkatov, A.N. Korolevich, V.V. Tuchin et al., "*In vivo* investigation of human skin optical clearing and blood microcirculation under the action of glucose solution," *Asian J. Phys.* **15**, 1–14 (2006).

147. A. Bykov, T. Hautala, M. Kinnunen et al., "Imaging of subchondral bone by optical coherence tomography upon optical clearing of articular cartilage," *J. Biophoton.* **9**(3), 270–275 (2016).

148. C.G. Rylander, O.F. Stumpp, T.E. Milner, N.J. Kemp, J.M. Mendenhall, K.R. Diller, A.J. Welch, "Dehydration mechanism of optical clearing in tissue," *J. Biomed. Opt.*, **11**, 041117 (2006).

149. T. Yu, X. Wen, V.V. Tuchin, Q. Luo, D. Zhu, "Quantitative analysis of dehydration in porcine skin for assessing mechanism of optical clearing," *J. Biomed. Opt.* **16**(9), 095002 (2011).

150. D.K. Tuchina, A.N. Bashkatov, A.B. Bucharskaya, E.A.Genina, V.V. Tuchin, "Study of glycerol diffusion in skin and myocardium ex vivo under the conditions of developing alloxan-induced diabetes," *J. Biomed. Photon. Eng.* **3**(2), 020302 (2017).

151. D.K. Tuchina, A.N. Bashkatov, E.A. Genina, V.V. Tuchin, "Quantification of glucose and glycerol diffusion in myocardium," *J. Innov. Opt. Health Sci.* **8**(3), 1541006 (2015).

152. D.K. Tuchina, A.N. Bashkatov, E.A. Genina, V.V. Tuchin, "Investigation of the impact of immersion agents on weight and geometric parameters of myocardial tissue *in vitro*," *Biophysics* **63**(5), 791–797 (2018).

153. L. Oliveira, M.I. Carvalho, E.M. Nogueira, V.V. Tuchin, "The characteristic time of glucose diffusion measured for muscle tissue at optical clearing," *Laser Phys.* **23**(7), 075606 (2013).

154. L. Oliveira, M.I. Carvalho, E. Nogueira, V.V. Tuchin, "Optical clearing mechanisms characterization in muscle," *J. Innov. Opt. Health Sci.* **9**(5), 1650035 (2016).

155. D.K. Tuchina, V.D. Genin, A.N. Bashkatov, E.A. Genina, V.V. Tuchin, "Optical clearing of skin tissue ex vivo with polyethylene glycol," *Opt. Spectrosc.* **120**(1), 28–37 (2016).

156. R. Shi, W. Feng, C. Zhang, Z. Zhang, D. Zhu, "FSOCA-induced switchable footpad skin optical clearing window for blood flow and cell imaging in vivo," *J. Biophoton.* **10**(12), 1647 (2017).

157. Z. Mao, D. Zhu, Y. Hu, X. Wen, Z. Han, "Influence of alcohols on the optical clearing effect of skin *in vitro*," *J. Biomed. Opt.* **13**(2), 021104 (2008).

158. W. Feng, R. Shi, N. Ma, D.K. Tuchina, V.V. Tuchin, D. Zhu, "Skin optical clearing potential of disaccharides," *J. Biomed. Opt.* **21**(8), 081207 (2016).

159. PubChem. Open Chemistry Database (https://pubchem.ncbi.nlm.nih.gov/#query=Omnipaque) (August 19 2020).

160. PubChem. Open Chemistry Database (https://pubchem.ncbi.nlm.nih.gov/compound/Diatrizoic_acid#section=Top) (August 7 2019).

161. PubChem. Open Chemistry Database (https://pubchem.ncbi.nlm.nih.gov/compound/2528#section=Top) (August 7 2019).

162. A.Yu. Sdobnov, M.E. Darvin, J. Lademann, V.V. Tuchin, "A comparative study of *ex vivo* skin optical clearing using two-photon microscopy," *J. Biophoton.* **10**(9), 1115–1123 (2017).

163. A.S. Chiang, W.Y. Lin, H.P. Liu, M.A. Pszczolkowski, T.F. Fu, S.L. Chiu, G.L. Holbrook, "Insect NMDA receptors mediate juvenile hormone biosynthesis," *Proc. Natl. Acad. Sci. U.S.A.* **99**(1), 37–42 (2002).

164. A.K. Bui, R.A. McClure, J. Chang, C. Stoianovici, J. Hirshburg, A.T. Yeh, B. Choi, "Revisiting optical clearing with dimethyl sulfoxide (DMSO)," *Lasers Surg. Med.* **41**(2), 142–148 (2009).

165. M. Sznitowska, "The influence of ethanol on permeation behavior of the porous pathway in the stratum corneum," *Int. J. Pharmac.* **137**(1), 137–140 (1996).

166. Meglumine active pharmaceutical ingredient (http://www.merckmillipore.com/RU/) (25 April 2018).

167. J. Hirshburg, B. Choi, J.S. Nelson, A.T. Yeh, "Correlation between collagen solubility and skin optical clearing using sugars," *Lasers Surg. Med.* **39**(2), 140–144 (2007).

168. J. Wang, J. Wang, N. Ma et al., "Sugar-induced skin optical clearing: From molecular dynamics simulation to experimental demonstration," *IEEE J. Sel. Top. Quantum Electron.* **20**(2), 7101007 (2014).

169. T. Yu, Y. Qi, J. Wang et al., "Rapid and Prodium iodide-compatible optical clearing method for brain tissue based on sugar/sugar-alcohol," *J. Biomed. Opt.* **21**(8), 081203 (2016).

170. D.W. Straughan, "First European Commission report on statistic of animal use," *ATLA* **22**(4), 289–292 (1994).

171. J.H. Thompson, W.R. Richter, "Hematoxylin-eosin staining adapted to automatic tissue processing," *Stain Technol.* **35**(3), 145–148 (1960).

172. S. Segura, R.M. Pujol, "Lipomembranous fat necrosis of the subcutaneous tissue," *Dermatol. Clin.* **26**(4), 509 (2008).

173. Academic Dictionaries and Encyclopedias, Medical dictionary (2011) [Online] (accessed www.enacademic.com).

174. L. Orci, W.S. Cook, M. Ravazzola, M.Y. Wang, B.H. Park, R. Montesano, R.H. Unger, "Rapid transformation of white adipocytes into fat-oxidizing machines," *Proc. Natl. Acad. Sci. U.S.A.* **101**(7), 2058 (2004).

175. I.Y. Yanina, N.A. Navolokin, A.B. Bucharskaya, G.N. Maslyakova, V.V. Tuchin, "Skin and subcutaneous fat morphology alterations under the LED or laser treatment in rats *in vivo*," *J. Biophoton.* **12**(12), e201900117 (2019).

176. S. Reggio, C. Rouault, C. Poitou, et al., "Increased basement membrane components in adipose tissue during obesity: Links with TGFβ and metabolic phenotypes," *J. Clin. Endocrinol. Metab.* **101**(6), 2578 (2016).

177. E. van Sonnenberg, W. McMullen, L. Solbiati (Eds.), *Tumor Ablation: Principles and Practice*, Springer, Berlin/Heidelberg (2008), 542 p.

178. E.A. Genina, N.S. Ksenofontova, A.N. Bashkatov, G.S. Terentyuk, V.V. Tuchin, "Study of the epidermis ablation effect on the efficiency of optical clearing of skin *in vivo*," *Quantum Electron.* **47**(6), 561 (2017).

179. I.A. Aljuffali, C.-H. Lin, J.-Y. Fang, "Skin ablation by physical techniques for enhancing dermal/transdermal drug delivery," *J. Drug Deliv. Sci. Technol.* **24**(3), 277–287 (2014).

180. E.A. Genina, L.E. Dolotov, A.N. Bashkatov et al., "Fractional laser microablation of skin aimed at enhancing its permeability for nanoparticles," *Quantum Electron.* **41**(5), 396–401 (2011).

181. G.S. Terentyuk, E.A. Genina, A.N. Bashkatov et al., "Use of fractional laser microablation and ultrasound to facilitate the delivery of gold nanoparticles into skin *in vivo*," *Quantum Electron.* **42**(6), 471–477 (2012).

182. E.A. Genina, L.E. Dolotov, A.N. Bashkatov, V.V. Tuchin, "Fractional laser microablation of skin: Increasing the efficiency of transcutaneous delivery of particles," *Quantum Electron.* **46**(6), 502–509 (2016).

183. M.D. Liang, K. Narayanan, "Minimally invasive fat cavitation method," *Patent US US7374551B2* (2003).

184. K.M. Coleman, W.P. Coleman 3rd, A. Benchetrit, "Non-invasive, external ultrasonic lipolysis," *Semin. Cutan. Med. Surg.* **28**(4), 263–267 (2009).

185. M.A. Trelles, S.R. Mordon, "Adipocyte membrane lysis observed after cellulite treatment is performed with radiofrequency," *Aesthet. Plast. Surg.* **33**(1), 125 (2009).

186. J.T. Alander, O.M. Villet, T. Pätilä et al., "Review of indocyanine green imaging in surgery," in *Fluorescence Imaging for Surgeons: Concepts and Applications*, ed. F.D. Dip, T. Ishizawa, N. Kokudo, and R. Rosenthal, Springer, Switzerland, Cham, Heidelberg, New York, Dordrecht, London, pp. 35–53 (2015).

187. M. Wanner, M. Avram, D. Gagnon et al., "Effects of non-invasive, 1,210 nm laser exposure on adipose tissue: Results of a human pilot study," *Lasers Surg. Med.* **41**(6), 401 (2009).

188. F. Becker, *Cancer a Comprehensive Treatise 2: Etiology: Viral Carcinogenesis*, Springer, New York (1975), 439 p.

28

Diabetes mellitus-*induced alterations of tissue optical properties, optical clearing efficiency, and molecular diffusivity*

Daria K. Tuchina and Valery V. Tuchin

CONTENTS

Introduction

Diabetes mellitus (DM) is a chronic disease that presents a serious socioeconomic problem in modern society due to its widespread prevalence and the risks of developing a number of complications. Therefore, it is necessary to improve existing methods and search for new ones for diagnosing and treating DM, which requires a detailed study of the mechanisms of the development of the disease, as well as the development of reliable and simple methods and criteria for detecting precursors of complications. The solution to these problems is made possible by combining numerous results and knowledges.

DM is a chronic endocrine disease caused by an increased content of free glucose in the body, resulting in a metabolic disorder in the organism. According to the World Health Organization, the move of DM from eighth to seventh place in the list of the most common causes of death worldwide is predicted by 2030 due to an increase in the death rate of the world's population from this disease by approximately 54% compared to 2010 [1]. At the same time, the number of patients with DM has increased since 1980 (108 million people) by 2014 by about 4 times (422 million) [2, 3].

Glucose is a monosaccharide that plays an important role in the functioning of the living body. Glucose is a source of energy for metabolic processes in the organism [4, 5]. The normal level of free glucose in blood (glycaemia) is 3.3–5.5 mmol/l (60–100 mg/dl) [4]. Beta cells in the pancreas produce the peptide hormone insulin to regulate glucose metabolism. The insufficient production of insulin by the pancreas or an abnormal reaction of the organism cells to the produced insulin leads to

an excess content of glucose in the blood. Long-term elevated glucose content in blood (hyperglycemia) (above 11 mmol/l (200 mg/dl)) leads to excess glucose content in the interstitial fluid and then violates the metabolism in the organism, causing DM development [2, 6–13].

It was supposed that hyperglycemia could cause the increased oxidation of glucose and formation of superoxide radicals in mitochondria [14–16]. Anions of superoxide radicals are benign and nonreactive; however they turn into highly reactive free radicals under some conditions, mainly in the presence of reductive-oxidative transition metals, such as copper or iron [14, 17]. Such transformation occurs via metal-mediated Fenton or Haber-Weiss reactions (Equations (28.1) and (28.2) respectively) [14]:

$$Fe^{2+} / Cu^+ + H_2O_2 \rightarrow Fe^{3+} / Cu^{2+} + HO^- + HO^\bullet \quad (28.1)$$

$$O_2^{\bullet-} + H_2O_2 \xrightarrow{Fe/Cu} O_2 + HO^- + HO^\bullet \quad (28.2)$$

Hyperglycemia-induced overproduction of superoxide activates hyperglycemic damage in endothelium cells of aorta. The damaging action of superoxide results in the redirection of glucose metabolism from the glycolytic path to alternative metabolic paths including the formation of the advanced glycation end products (AGEs).

Since the glucose level in blood is related to its level in the interstitial fluid, hyperglycemia leads to the metabolic disbalance and disorder of the organs functioning [13, 18–21]. Thus many papers are aimed at determining the concentration of free glucose in blood, interstitial fluid, or other fluids of the body [22–36] and in tissues [37]; detecting morphological changes in vital organs [11, 13, 19, 38]; designing biosensors for continuous monitoring of free glucose in the body [22, 34, 39, 40]; studying the impact of glucose on proteins in tissues and blood and developing methods for DM monitoring based on the difference of optical properties between diabetic and healthy tissues [18, 22, 41–57]; and assessing the glucose diffusion rate in healthy [22, 56, 58–63] and pathological tissues (the walls of atherosclerotic blood vessels [56, 64] and malignant tumours [65]).

Optical methods are developed for the determination of glucose levels in blood; for example, optical coherence tomography (OCT) [36, 66–68], Raman [69], and terahertz (THz) [70] spectroscopy are used for this purpose, while near-infrared spectroscopy (NIR) is used for measuring the glucose content in skin [37].

One third of all DM patients have skin injury as a complication, the most severe of them being chronic ulcers that recover slowly, increasing the risk of infection [20, 21, 71–73]. DM development increases the risk of cardiovascular diseases, neuropathy, nephropathy, bone fragility, and retinopathy leading to blindness [5, 9–14, 20, 21, 38, 74–79]. According to data from the World Health Organization, DM is the main cause of heart attacks, brain stroke, myocardial infarction, renal failure, blindness, and amputation of lower extremities, and it is a risk factor for many diseases including dementia and diabetic foot [5, 10, 14, 20, 21, 74, 77, 78, 80].

Thus, DM is a cause of serious complications that should be prevented by inventing new methods of DM treatment and diagnostics and improving the existing ones, which requires careful study of the development mechanisms of this disease, as well as the development of reliable and simple methods and criteria for detecting the complication precursors.

DM development and its complications are widely studied using different methods, including optical ones, which often allow a noninvasive monitoring of the key parameters of tissues in real time, which explains the promise and constant expansion of the scope of application of optical methods in medicine.

Classification of DM

DM has two main types: type I is insulin-dependent diabetes ("immune-mediated diabetes," "diabetes of the young"), while type II is insulin-independent diabetes ("diabetes of the aged") [9, 10, 20, 71]. Gestational DM is separately considered, being diagnosed during pregnancy and not observed before it. There are also specific types of DM induced by other causes, for example, monogenic diabetes syndrome, when the mutant genes are inherited (diabetes of newborns); diabetes induced by the diseases of exocrine pancreas (such as cystic fibrosis); and diabetes caused by chemical agents (for example, when medicines are used in HIV/AIDS treatment or after transplantation of organs) [9, 21].

DM of types I and II is the most widespread. Insulin-dependent DM of type I develops due to destruction of the pancreas beta cells, which disturbs insulin production. A reduced amount of produced insulin means that it is insufficient to process the amount of glucose received by the organism. The glucose content increases in the organism, chronical hyperglycemia arises, and the metabolism is disturbed in many organs [2, 6–12, 20, 71, 78, 81–83]. Type I DM is more widespread among the young people, which is why it is also called the "diabetes of the young." Type I of DM accounts for about 5%–10% of the total number of diabetes cases [2, 9]. Type I DM is characterized by such symptoms as polydipsia (thirst), polyuria (excess excretion of urine), loss of weight, permanent hunger, tiredness, and visual impairment [2, 9, 19, 20, 83].

Type II insulin-independent DM develops due to cell resistance to the produced insulin, and insulin deficiency also takes place in the organism. Type II is more widespread and accounts for about 90%–95% of all diabetes. In this case, the disease develops in middle and old age, and that is why this type of disease is called "the diabetes of aged"; however, this type of diabetes has been appearing more often children in recent times. DM is largely caused by physical inactivity and overweight [2, 9, 10, 20, 78]. The symptoms of types I and II of DM are similar, but they are often less expressed in the second case, which is why the disease can be diagnosed a few years after it has started, when complications have already arisen [2]. An additional symptom of DM of type II is obesity [10]. It is important to determine the diabetes type in time for prescription of the most efficient therapy [9].

A diabetes-related hypoglycaemic coma (crisis) can happen under the excess injection of insulin or its excess production caused by medical preparations, when the glucose content in blood falls below 3.9 mmol/l (70 mg/dl). In this case, the body has not enough energy to support normal vital activity, leading to trembling, weakness, tachycardia, irritability, hunger, and loss of consciousness, and a fatal outcome is possible [9]. Genetic predisposition is the main cause of DM development [9, 20]; in other cases, diabetes is mainly caused by different diseases disturbing the metabolism in the body, as well as low physical activity [9].

The development of DM accelerates the processes occurring with organism aging [21, 77]. Water content in skin decreases with age, and, correspondingly, the ratio of fibrous structures and the base substance changes. The ratio decreases due to an increase in collagen content and decrease of glycosaminoglycans concentration. The content of hyaluronic acid greatly reduces. The chemical and physical properties of collagen change: the number and strength of intra and inter-molecular crosslinks increases, the swelling ability and elasticity reduces, the resistance to collagenase develops, and the structural stability of collagen fibers rises, i.e., the maturation of fibrillar structures of the connective tissue progresses. Collagen aging is caused by metabolic processes that occur in the organism and affect the molecular structure of collagen. When connective tissue lesions take place, all structural parts of the tissue are damaged: fibrils, cells, and intercellular substance [84].

The structure of collagen fibers in studies of the human sclera aging was investigated using wide-angle X-ray scattering. The increase of rigidity of the scleral matrix, the decrease in mechanical anisotropy and the degree of fiber straightening in the peripapillary sclera with age were observed [85]. The organism aging is accompanied by functional and structural modifications of macromolecules [86]. The changes occurring in organism tissues under the development of DM are investigated in laboratory animals using experimental models of diabetes [10–12, 19, 47, 48, 50, 53–55, 87–93] and directly in patients in the course of clinical examination [13, 43, 94–103].

Experimental models of DM

The chemically induced models of diabetes are widely used due to their relatively cheapness and simple induction of diabetes development in rodents, and similar models can also be implemented in highly developed animals [10–12, 19, 47, 48, 50, 53, 87–93, 104]. The most widely used and well-known methods of chemical diabetes induction are implemented by two compounds: alloxan [19, 47, 48, 50, 53, 54, 87, 105] and streptozotocin [55, 89–93, 104, 106–108]. They accumulate mainly in the pancreas and provoke the production of radicals that lead to the destruction of the pancreas beta cells and, therefore, disorder of the insulin production [10, 11, 12, 91, 106]. Since streptozotocin and alloxan have a similar structure to glucose, their accumulation occurs by means of glucose transporters in the pancreas, which is why animals are more susceptible to them after fasting [10, 12, 109]. Thus, streptozotocin or alloxan injection in animals induces insulin-dependent (type I) diabetes [12, 106]. Since alloxan and streptozotocin solutions are relatively unstable, they should be prepared immediately before the injection. It should be noted that these substances can be toxic for other organs [10, 109, 110]. Besides destruction of the beta cells, alloxan and streptozotocin are able to modify biological macromolecules and to fragment DNA [10, 11, 12, 106].

Dunn et al. were first to describe the alloxan model of diabetes in 1943 [12, 106, 111]. A single dose of alloxan varies from 40 to 200 mg/kg of body mass for rats and from 50 to 200 mg/kg for mice, depending on the breed [10, 11, 50]. Alloxan is injected subcutaneously, intraperitoneally, or intravenously [11, 50, 92].

Rakieten published a paper in 1963 reporting that streptozotocin can induce diabetes [12, 91, 112]. The maximal single dose of streptozotocin varies from 35 to 65 mg/kg of body mass for rats [10, 113] and from 100 to 200 mg/kg for mice [10, 113, 114] depending on the breed [10, 115]. There is also a method for injecting streptozotocin in animals in small doses (from 20 to 40 mg/kg) over 5 days [10].

In contrast to animals, human beta cells of the pancreas are resistant to the toxicity of alloxan and streptozotocin [116]. It is interesting that the variation of glucose content in blood during the first week after alloxan or streptozotocin injection is nonlinear (Figure 28.1); the glucose content in blood increases during about 3–4 hours after the injection. It is accompanied by morphological changes in beta cells: dilatation of rough endoplasmic reticulum, intercellular vacuolization, diminution of insulin content and secretor granules, reduction of Golgi apparatus area, and mitochondria swelling. Then the hypoglycemic stage starts, when the glucose content in blood falls in a few hours, and hypoglycemia can lead to a fatal outcome.

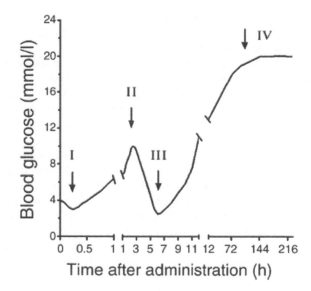

FIGURE 28.1 Phases of free glucose variation in blood after the injection of alloxan (I-IV) and streptozotocin (II-IV) [12].

Such serious transient hypoglycemia appears during insulin accumulation as a result of the cell membrane rupture and the secretory granule intoxication. In addition to the morphological changes, the nuclei of beta cells become irreversibly pyknotic [12]. The permanent diabetic hyperglycemic phase comes about 6 days after the injection, characterized by loss of beta cells integrity during 12–48 hours and morphologically full degranulation [10, 12]. Thus, it is important to choose the time for measuring the glucose content in blood in diabetes modeling.

Animals with the model diabetes have such symptoms as polydipsia, polyuria [83], and loss of body weight [10, 83], which is typical for type I DM.

Besides the listed models of type I diabetes, there are other ones which are used less widely, for example, the autoimmune modes. The lines of biobreeding (BB) rats and mice with "spontaneous" DM ("nonobese diabetes"), in which the prediabetic state of the organism develops from birth or in some time after birth, are created [10, 82, 83, 117–119]. In Akita mouse lines, in which diabetes genetically develops a few weeks after birth, diabetes is also induced using viruses [10].

Mono- and polygenic models of obesity, the rats of the Zucker line, etc. are used to model type II diabetes [10, 120]. The most widely used monogenic models of obesity are characterized by defects in the transmission of signals of the hormone leptin. Since leptin is responsible for the feeling of satiety, the absence of functional leptin in animals leads to hyperphagia (overeating) and subsequent obesity. The Lep[ob/ob] and Lepr[DM/DM] line of mice, the Zucker line rats have a lack of leptin and have no leptin receptor. These models are often used for testing new methods for treatment of type II diabetes [10, 120–123].

Leptin-deficient (Lep[ob/ob]) mice begin increasing in mass from 2 weeks of age; hyperinsulinemia develops, hyperglycemia appears within 4 weeks, and the concentration of free glucose in the blood continues to grow, achieving its maximum in 3–5 months [10, 124]. The volume of the pancreas significantly increases [10, 125], temperature regulation is disturbed, hyperlipidemia progresses, physical activity reduces [10, 124], and infertility develops [10, 126].

Lepr$^{DM/DM}$ mice have the autosomal recessive mutation in the leptin receptor [10, 127], and they have obesity, hyperphagia, hyperglycemia, and hyperinsulinemia. The hyperinsulinemia of such mice manifests itself at the age of about 2 weeks, obesity from the age of 2–4 weeks, and hyperglycemia at 4–8 weeks [10].

The Zucker rats have an expressed resistance to insulin that cannot be compensated for due to the increased level of apoptosis in beta cells [10, 128], which leads hyperinsulinemia to develop at the age of about 8 weeks with a subsequent decrease in insulin level [10, 129]. The diabetes in Zucker rats usually develops at about 8–10 weeks in males, while in females, explicit diabetes does not develop [10, 113]; the signs of diabetic complications are observed in rats of this line [10, 129].

It is necessary to choose a suitable model of diabetes for experimental studies taking in account the specific features of the models depending on the set task. It is also important to note that a particular model of diabetes could develop in different ways in animals of different lines. For this reason, the authors of Reference [89] obtained different results for line and outbred mice.

Glycation of proteins

The main complications of DM are connected with glycation of proteins. Glycation is a result of interaction between proteins and glucose molecules, which leads to a change in protein structure and the restriction of tissue functioning [20, 41–44, 46, 131]. Protein glycation is initiated by nonenzymatic reaction between the carbonyl group of sugars and the amino group of proteins, which correspondingly, leads to the formation of crosslinks between protein molecules [18, 42, 45, 132]. The result of glycation is the Mallard reaction [133, 134], which includes two stages. The interaction of protein with glucose via the adducts of the Schiff's base leads to the formation of a stable Amadori product. Further incubation with glucose leads to transformation of the Amadori product into the advanced glycation end product (AGE) [18, 134–136]. The glycation can be accompanied by both attachment of the AGE to the protein receptor and protein modification [86]. The mechanisms of collagen glycation involve complex processes accompanying the occurring reactions [42, 45, 135, 137, 138]. Such AGEs as N(ε)-carboximetillysine (CML) and pentosidine accumulate in skin collagen with age and DM development. Pentosidine is a fluorescent molecule, which can be detected by the methods of fluorescence analysis [45, 86, 135, 137, 139–142] or confocal Raman spectroscopy [143]. Glucosepane is related to retinopathic, neuropathic, and nephropathic complications [98, 141, 142]. Pyrroline [86], glyoxal [139], and glucosepane [143] are also defined as AGEs; other AGEs and the substances facilitating their accumulation are also known [86, 142].

Since proteins constitute a considerable part of many tissues, glycation of proteins leads to changes in tissue structure, restriction of tissue functioning [7, 42–46, 86, 131, 139], metabolic disbalance and, as a result, malfunction of organs [18].

Since the structure of tissues determines their optical properties, it is possible to monitor their changes by optical methods, such as fluorescence spectroscopy [42, 44, 50, 68, 70, 95, 100–103, 137, 139, 140, 142, 144–150], spectrophotometry in a wide spectral range [47–49, 55, 70, 100, 102, 151, 152], refractometry [43, 154], electron microscopy [19, 89, 94], confocal microscopy [96], laser speckle contrast imaging (LSCI) [53, 54, 70, 106, 155], multiphoton microscopy [131], Raman spectroscopy [51, 52, 68, 143, 150, 156], laser Doppler flowmetry [46, 68, 100–102, 157], optoacoustic spectroscopy [65, 158], OCT [79], and THz spectroscopy [70].

To study protein glycation, tissue and cell samples taken from objects with natural or model DM (*in vivo* glycation), as well as samples glycated in *in vitro* conditions, are used. For example, studies of *in vitro* glycation are conducted for the human placental IV type collagen, which is performed using fluorescence analysis, densitometry, and electrophoresis [159]; for the collagen of bovine skin, using multiphoton microscopy [42]; for hemoglobin, using OCT [161–163], IR spectroscopy, refractometry [154], and biochemical analysis [160]; for albumin, using fluorescence spectroscopy [86], THz spectroscopy [164] and refractometry [154]; for collagen of tendon, using biomechanical and biochemical analyzes [165]; and for collagen hydrogels, using multispectral fluorescence life time imaging (FLIM) [166]. Glycation was provided by incubation tissue and liquid samples in solutions of ribose [45, 159, 165, 166], glucose [86, 154, 159, 160, 161, 164], fructose [86, 164], glycolaldehyde [132], or glyoxal [139]. All solutions show a sufficiently effective protein glycation during 10–11 days of incubation and change in tissues' mechanical properties due to the formation of collagen cross-links. Increase of glucosapane concentration in the skin with age and with the development of hyperglycemia in patients with type I DM was described [98].

Studies of the fluorescent properties of the glycated hemoglobin and tissues have been performed to determine the degree of protein glycation [42, 44, 50, 137, 139, 140, 144–148]. A refractometry-based biosensor was offered for studying the glycated hemoglobin in human blood [167]. The modification of vascular walls at protein glycation [168], the refractive properties of erythrocytes of patients with DM and healthy volunteers [43], and the optical properties of skin in diabetes [46, 147, 148] were also investigated. Many papers are focused on the development of methods of monitoring of DM and its different complications, which are based on the detection of changes of the tissue optical properties with diabetes development [7, 18, 42–46, 49, 56, 57, 68, 93].

The increased deformation, maximum load, tension, Young modulus of elasticity, and viscosity indicating that glycation increases the stiffness of the tendon matrix were observed after 8 months of glycation of the rabbit Achilles tendon in ribose solution [165]. The tendon glycation led to significant reduction in the soluble collagen content and a significant increase in insoluble collagen and pentosidine content. Thus, the obtained results showed that the collagen crosslinks formed by glycation directly increase the stiffness of the matrix and change other mechanical properties of the tendon.

The literature analysis shows that the study of glycated tissues is demanded and is a promising field of research in terms of developing of new technologies for noninvasive or least-invasive monitoring of severe complications, their prevention, and treatment maintenance.

Optical and structural properties of tissues at glycation and DM development

Blood and cardiovascular system

The DM development is related to increased risk of macro- and microvascular complications (angiopathy) [9, 10, 14, 20, 21, 76–78, 98], such as nephropathy and neuropathy [9, 10, 14, 21, 74] and retinopathy that leads to blindness [9–11, 14, 21, 38, 74, 79]. DM significantly enhances the risk of vascular diseases that finally lead to brain stroke [5, 14, 20, 21, 74, 77]. Hyperglycemia accelerates the atherosclerotic processes [14, 75, 77]; atherosclerosis and the corresponding loss of elasticity in coronary artery walls induce stenosis and, therefore, the reduction of blood supply to the cardiac muscle, which finally leads to cardiomyopathy, angina pectoris, and increased risk of myocardial infarction [13, 14, 77, 78]. The cause of serious complications is often the glycation of proteins in blood and in vascular walls permanently washed by blood in the presence of high level of free glucose in it [49, 156, 159, 161].

For determination of glycated albumin or hemoglobin, different methods are used such as ion exchange, liquid or affine chromatography, electrophoresis, immunochemical, and colorimetric methods [86, 169]. Firstly, these methods are invasive, because blood collection from the patient's vein is needed. Since the mean lifetime of erythrocytes is 3 months, only information about glycation accumulated during those 3 months can be collected. Moreover, the measurement accuracy is affected by the increased content of glucose, lipids, bilirubin, and other substances [169]. The level of free glucose in the blood is also influenced by external factors, for example, physical activity and food intake, which limits the applicability of many diagnostic methods where blood measurements are required [8, 9]; therefore, it is necessary to develop new algorithms for processing the obtained data to reduce the influence of external factors.

Optical methods for diagnosing the state of tissues and organs are widely used in biology and medicine, for example, to determine the degree of their oxygenation and blood perfusion [170, 171]. The safety of these methods and the ability to obtain information in real time explain their widespread use and the constant expansion of the field of application in medicine [22, 46, 56, 172–175]. It is possible to get information about the composition, structure, and properties of tissues to investigate metabolic processes, avoiding the negative effect on them using optical radiation [170]. An essential advantage of visible and NIR optical radiation is its capability to penetrate sufficiently deep into tissues. The wavelength range from 600 to 2500 nm includes four "transparency windows," within which the light attenuation by tissues is minimal. Since many tissues contain a lot of water, the spectral "transparency windows" correspond to the wavelengths at which the light absorption by water is minimal. The first "transparency window" includes the wavelengths from 650 to 950 nm;

the absorption of light by water is relatively weak in it, but the light absorption by hemoglobin and myoglobin is high. The second "transparency window" is placed between two bands water absorption and corresponds to the wavelengths from 1100 to 1350 nm. The third "window of transparency" is in the wavelength range from 1600 to 1870 nm and the fourth "window" is from 2100 to 2300 nm, which is convenient for the investigation of collagen-containing tissues [171, 176–178].

The possibility of Raman spectroscopy to determine the degree of hemoglobin glycation was shown [156]. The significant difference in the refractive properties of erythrocytes in diabetic and healthy patients was observed in microscopic studies using the Nomarsky interference microscope combining a two-beam interferometer and a polarization microscope for enhancing the contrast of phase images [43]. The quantitative analysis of the glycated hemoglobin concentration in human blood using the spectral analysis in the infrared wavelength range from 780 to 2498 nm was proposed in the Reference [49]. The authors of Reference [50] measured the fluorescence of blood plasma 12 days after the alloxan injection in rats. It was shown that the shape of the fluorescence band at an excitation wavelength of 320 nm is most indicative of hyperglycemia in blood plasma samples due to the formation of a protein fluorescent bond due to nonenzymatic glycation. In Reference [154], the optical properties of albumin and hemoglobin in aqueous solutions of glucose were investigated. The increase in the refractive index and the reduction of the absorption of the solutions with the increase in glucose concentration in the solution were obtained. The probable cause is protein–glucose binding. A decrease in the albumin absorbance in the THz range of wavelengths with increase in the incubation time of albumin in glucose and fructose solutions (glycation *in vitro*) was observed [164]. Also the slow glycation in the glucose solution in comparison with the fructose one and the dependence of albumin glycation rate on the pH of the sugar solution were obtained in this work.

The applicability of Raman spectroscopy for the assessment of glycated hemoglobin *in vivo*, carried out for ear, hand, and forehead skin sites, was shown by the authors of Reference [52]. The porphyrin conformations of erythrocyte hemoglobin were observed in diabetes using Raman spectroscopy [51]. A reduction in erythrocyte membrane viscosity was obtained in diabetes by spectroscopy of electronic paramagnetic resonance. A change in permeability of the erythrocyte plasmatic membrane was found, namely, a higher rate of Na-H exchange, the activity of Ca^{2+}-dependent K^+-channels, and a decrease of the Ca-ATPase activity in patients with II type of DM [51]. It was supposed that the change of permeability and viscosity of the erythrocyte plasmatic membrane can cause the conformation change of the hemoglobin porphyrin, the reduction of oxygen transport by hemoglobin, and binding activity under DM [51].

A noninvasive method and a device for assessing the glycated hemoglobin concentration based on light reflection spectra measured from blood samples were proposed in the patent [151]. The device consists of two measuring units, one of which contains a Raman spectrometer, and the other a spectrophotometer recording the absorption spectra. The possibility of using absorption spectroscopy in the spectral range 200–850 nm for the assessment of glycated hemoglobin amount was demonstrated in Reference [152]. A correlation between results obtained by the spectroscopic method and the standard method of high-efficiency liquid chromatography was found.

The method of quantitative analysis of the glycated hemoglobin concentration in hemolysate samples of human blood using NIR spectroscopy and subsequent processing the spectra by the method of moving window partial least squares (MWPLS) was proposed in Reference [49]. This method allows one to find the optimal spectral ranges for determining the content of nonglycated and glycated hemoglobin. The optimal wavelengths for analysis were in the NIR from 958 to 1036 nm for native hemoglobin and from 1492 to 1858 nm for glycated hemoglobin.

The optical characteristics of erythrocytes from diabetic patients were noninvasively studied by three-dimensional quantitative phase imaging based on common-path diffraction optical tomography (cDOT) [179]. The morphological (surface area, volume, and sphericity), mechanical (membrane fluctuations), and biochemical (concentration of hemoglobins) parameters of individual cells were quantitatively determined from the measured three-dimensional tomograms of the refractive index and two-dimensional time-dependent phase images. Statistically significant changes in biochemical and morphological parameters of erythrocytes of diabetic and nondiabetic patients were not obtained; however, the lower deformability of the membrane of diabetic erythrocytes was demonstrated.

Since the AGEs are dissolved in the blood plasma, they interact with the endothelium, and, hence, affect the endothelium functioning with tissue hypoxia and hypoperfusion [21, 77, 132, 135]. Thus, the AGEs accumulate in vascular walls with formation of the crosslinks [135]. The corresponding damage to the endothelium leads to the development of atherosclerosis [14, 21, 135, 168]. The risk of ischemia increases in the presence of DM [21, 78, 97].

In comparison to hemoglobin, which is subject to glycation over the course of 3 months, the glycation of other proteins lasts longer; therefore, the diagnostics of such proteins should provide more information about organism condition when DM develops.

The exhaustion of lectin content in the mice endothelium with 4-week streptozotocin-induced diabetes was found using the histological sections stained by the lectin antibody (Figure 28.2) [89]. Poorer perfusion recovery after ischemia was obtained in diabetic outbred mice compared to linear mice. Thus, the result can depend on the animals that are used in the study. The study [97] of human atrium samples *in vitro*, carried out after reoxygenation and perfusion modelled ischemia, showed stronger cell necrosis and apoptosis expressed in patients with type I and II DM. The apoptosis of myocardial cells was observed in rats with streptozotocin-induced diabetes [90].

DM disturbs reperfusion mechanisms, i.e., the activation of earlier existing arterial collaterals and the generation of new vessels (angiogenesis, arteriogenesis), which hampers recovery after an ischemic stroke [77, 78]. A smaller number of vessels and the accumulation of the AGE CML in muscles and

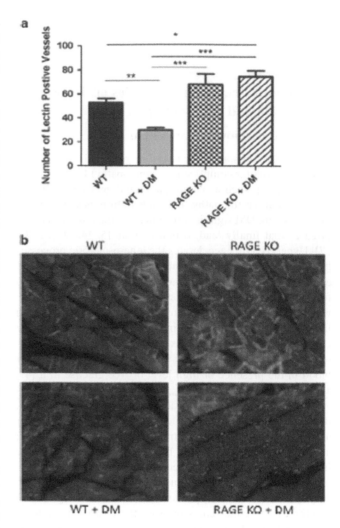

FIGURE 28.2 Histogram (a) and images (b) of histological sections reflecting the content of lectin (red) in outbred mice of the control (WT) and diabetic (WT+DM) groups, and in linear mice of the control (RAGE KO) and diabetic (RAGE KO+DM) groups, respectively [89].

blood of all outbred mice with diabetes was found [89]. A topical application of an iohexol solution (X-ray contrast solution "Omnipaque®") on blood flow in the pancreas of rats with alloxan-induced diabetes studied using laser speckle contrast imaging (LSCI) showed an increase in blood flow rate compared to the control group of rats, indicating the increase of endothelial permeability at the development of disease [53]. The endothelium-dependent skin microvascular vasodilator response was found to be significantly impaired in patients with type I diabetes compared to healthy patients using LSCI coupled with physiological postocclusive reactive hyperemia and pharmacological iontophoresis of acetylcholine as a local vasodilator stimulus [155]. An increase in cutaneous microvascular permeability was obtained in mice with alloxan diabetes by spectral imaging and the optical clearing skin window [131].

A decrease in the elasticity of arterial walls under the development of type II DM was demonstrated by the rate of ultrasound propagation and elastic modulus of arteries [77, 180]. A decreased blood flow rate during diabetes development was observed by laser Doppler flowmetry [46, 157]. An anomalous thickening of vascular walls in skin was also found [46].

Studies of assessment of noradrenalin impact on the blood flow of mice with alloxan diabetes were carried out [106]. The decrease of arterial and venous blood flow without subsequent recovery caused by noradrenalin injection was observed using the LSCI in mice with 2- and 4-week alloxan diabetes (see Chapter 29, Figure 29.9).

Since the heart is the main organ of the cardiovascular system, through which the process of blood circulation is implemented in the organism, it is directly affected by the negative effect of the disease with numerous complications [13, 14, 77, 78]. Changes in adipocyte size and an excess of lipid accumulation on the heart epicardium in patients with DM were found [13]. Such changes lead to the development of heart disease. The optical measurements demonstrate the reduction of the glycerol diffusion rate in *ex vivo* myocardium of rats with 2 weeks [48] and 1 month [105] of alloxan DM, which indicates the change of cardiac muscle tissue structure in the first weeks of the development of alloxan diabetes in rats.

The possibility of assessment of the functional condition of the microcirculatory system in patients with DM using such noninvasive optical methods as laser Doppler flowmetry, diffuse reflection spectroscopy, and fluorescence spectroscopy was shown [100, 102]. The obtained data have demonstrated that the combined application of these three diagnostic technologies allows one to reveal and predict the development of trophic disorders and the syndrome of diabetic foot at early stages. The use of wavelet analysis for evaluating the regulatory mechanisms of peripheral blood flow during the heat tests makes it possible to study the change of vascular tonus autoregulation and the regulation of bypass blood flow by sympathic fibers. This allows for indirect consideration of blood flow innervation and can indicate the presence of neuropathies [100]. The variation of hemoglobin concentration under the impairments in the microcirculatory layer of the foot skin due to diabetic microangiopathy can be detected in the skin optical reflection spectra due to changes in the scattering and absorption properties of skin [102].

Diabetic abnormalities in the structure of various tissues

Type I DM develops due to a disorder of the pancreas functioning that leads to development of chronical hyperglycemia and to glucose excess in the interstitial fluid, and then to disturbance of the metabolism and functioning of many organs [2, 6–12, 18, 21, 69, 81–83, 106, 111]. Since all vital organs, such as cerebral tissues, myocardium, and eye retina, are strongly supplied with blood and therefore with glucose, so they are glycated first in patients with DM. Multiple studies of the impact of DM on different properties and functioning of tissues and organs have been conducted for finding new effective methods for DM diagnostics and treatment and preventing its serious complications.

Since the pancreas is the primary organ to suffer from diabetes, many pancreatic diseases develop in diabetes [21]. For example, the swelling of the pancreas and liver mitochondria

was fixed 12 days after intra-abdominal injection of streptozotocin to rats, leading to damage of mitochondria and ATP deficiency [92]. The influence of DM on rat internal organs was studied [19]. Hyperglycemia, polyuria, and polydipsia were fixed in the rats after the intra-abdominal injection of alloxan. The histological sections of the spleen, liver, pancreas, and kidney tissues were taken on the 15th day after alloxan injection. Morphological changes of different degrees of severity in comparison with the control group of rats were observed in tissues after the histological analysis of tissues. The authors obtained a significant decrease of glycogen accumulation in the liver, the presence of perivascular fibrous indurations, and a decrease in the size and the number of pancreatic islands in the pancreas [19]. The necrosis of individual cells in kidneys was revealed with the development of diabetes. These results confirm that alloxan diabetes results in the structural changes of tissues and organs caused by an accumulation of glycogen 2 weeks after the intra-abdominal injection of alloxan in rats.

Since the high concentration of free blood glucose leads to the growth of its level in interstitial fluid, glycation affects not only the blood proteins, such as hemoglobin and albumin [86, 156, 160, 161, 164, 181], but also other proteins of the tissues. For example, collagen glycation leads to fibrosis development [72, 182]. In its turn, the structural changes of tissue proteins lead to changes in their optical properties.

Studies of the fluorescence properties of the AGEs of skin, hemoglobin, cornea, articular tendon, and aorta [42, 44, 137, 139, 140, 145, 183] and corresponding nonlinear susceptibilities of tissue compounds [42, 144, 184] showed that glycation facilitates an increase in tissue fluorescence intensity [42, 137, 139, 140, 145, 146, 183] and a decrease in second harmonic generation (SHG) intensity [42, 144, 184] (see Figure 28.3).

The authors of Reference [95] have studied skin *in vivo* autofluorescence and brain magnetic resonance (MRI) images in patients with type II DM. An increase in autofluorescence was found both with age and with DM development. The corresponding reduction of the gray-matter volume in the brain was observed in DM patients, which is the cause of cognitive defects of various degrees.

It was found that the length and density of nerve fibers are significantly smaller in patients with recently diagnosed II type DM than in the control group [96]. Such studies have been performed by skin biopsy and confocal microscopy of the eye cornea, for which the loss of nerve fibers and the accompanying injure of nerve conductivity were found, indicating early pathological changes in both large and small nerve fibers.

In turn, the damage of nerve fibers (neuropathy) can lead to a neuropathic foot ulcer, the most severe skin injuries arising when the reduced recovery capability of diabetic skin gives rise to infection, gangrene formation, and, finally, leads to the amputation of lower extremities [72, 73]. The authors of Reference [95] suggested that skin autofluorescence can indicate the formation of AGEs of other cell proteins, e.g., in neurons.

Since proteins are the main components of many tissues, protein glycation leads to significant changes in tissue structure [42, 45]. Since tissue permeability for chemicals is largely determined by tissue structure and its changes caused

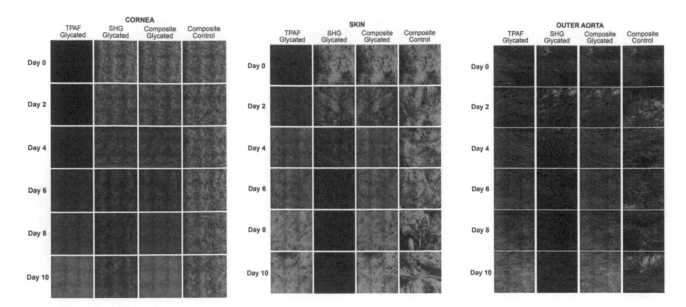

FIGURE 28.3 Images of glycated and control samples of bovine cornea, skin, and outer surface of aorta obtained by multiphoton autofluorescence (green) and the SHG (red); "Composite" means two superimposed images: the autofluorescence one (green) and the SHG one (red). The laser wavelength is 700 nm and the power is 180 mW, the image size is $620 \times 620 \ \mu m^2$ [42].

by pathological processes, such as glycation, the change of the molecule diffusion rate in tissue during a certain time interval can therefore indicate the change of the tissue structure and, thus, can be used as a biomarker of the degree of tissue glycation [48, 114, 185]. The study of the permeability of tissues for different molecules is aimed at getting information about the mechanisms of tissue interaction with different chemicals, drug transport in tissues, and agent impact on the morphological, optical, functional, and diffusion properties of tissues [56, 57, 61, 105, 107, 172, 173, 186–188]. These data are necessary for effective application of different pharmacological preparations to treat DM and for developing noninvasive optical methods of disease diagnostics and monitoring [56, 189], since efficiency of treatment and diagnostics is determined by drug (agent) diffusion rate, i.e., the time needed for drug (agent) molecules to reach the target part of the organism.

The slower glucose diffusion in a kidney sample of a diabetic mouse as compared to a nondiabetic one was obtained [55]. It was supposed that the diabetic kidney has a denser structure due to tissue glycation. The optical clearing efficiency of kidney, cornea, and skin samples in THz wavelength range was higher for nondiabetic samples than for diabetic ones at application of glycerol solutions of different concentration [55], which can be associated with lesser water flux in glycated tissues, as tissue dehydration is the major mechanism of optical clearing for THz waves.

About 80% of glucose in the organism is transported to muscles [20]. The reduction of insulin production under DM leads to the malfunction of glucose absorption by muscle cells, which leads to muscle dysfunction and reduction in muscle mass [190]. A 35% reduction of mitochondria size in skeletal muscles of patients with obesity and type II DM was observed using electron microscopy [94]. Vacuoles growth in muscle fibers and lowering of $NADH:O_2$-oxydoreductase was also found in patients with type II DM [94]. The authors concluded

that the bioenergetic capability of the mitochondria of skeletal muscles is disturbed in type II DM.

Larger fat cells were found in patients with DM as compared to healthy people [99]. It was demonstrated that the main phenotype of white fat tissue in humans with type II DM without obesity is adipocytes hypertrophy, which can lead to the inflammation of fat tissue, the release of fatty acids, the deposition of ectopic fat, and stronger sensitivity to insulin.

Using a fluorescence microscope, it was found that the mean cross-section of adipocytes of the fat tissue extracted from patients during surgical operation is higher in diabetes than in the control group (see Figure 28.4) [103]. A reduction of fibrous tissue formation and an increase of adipocyte hypertrophy was also obtained in patients with DM and obesity. It is noted that hypertrophy is a cause of fat tissue dysfunction [103].

Observed changes in ocular tissue autofluorescence in patients with type II DM having no evident signs of diabetic retinopathy indicates the accumulation of AGEs in the eye tissues [191].

The studies of air expired by patients with type II DM, prediabetic patients, and a control group of patients observed a correlation between the index of insulin sensitivity $(1/ISI_{0,120})$ and the concentration of the isotope ^{13}C $(\delta_{DOB}{}^{13}C(‰))$ in the expired air (Figure 28.5), which could serve as a marker for noninvasive assessment of both the development of type II DM and prediabetic condition in humans [192]. To obtain such a marker, the patient was orally administered ^{13}C-labelled D-glucose, which is metabolized and produces ^{13}C-labelled carbon dioxide $(^{13}CO_2)$, which gets to the lungs via the blood flow and then is expired [192].

Skin pathologies induced by DM

Skin is subjected to various infections during DM development; skin fibrosis is observed, the skin becomes more dry and vulnerable due to functioning disorder of leucocytes,

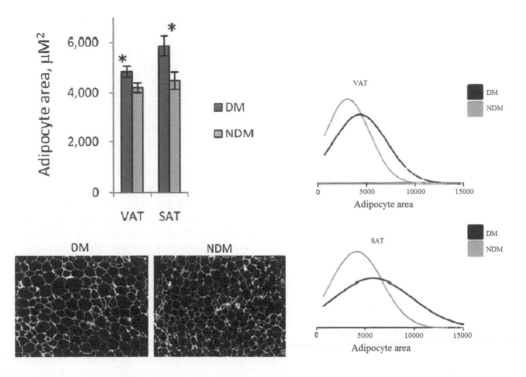

FIGURE 28.4 Histogram, microscopic images, and plots describing the increased size of adipocytes of the visceral (VAT) and subcutaneous (SAT) adipose tissue of the patients suffering from DM and control group (NDM) [103].

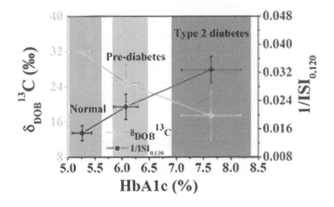

FIGURE 28.5 Distribution of ^{13}C isotope concentration in the expired air ($\delta_{DOB}{}^{13}C(\permil)$) and the insulin sensitivity index ($1/ISI_{0,120}$) depending on the level of glycated hemoglobin HbA1c (%) in the blood of the control group (Normal), prediabetic patients (Pre-diabetes), and of the patients with type II DM (Type 2 diabetes) [192].

and patients develop skin itching. These symptoms mean the presence of DM complications caused by hyperglycemia and protein glycation. There are many skin diseases indicated by DM [20, 72, 73]. Such structural changes as pore enlargement, dermis thinning, volume loss of collagen, flattening of the dermis–epidermis junction, and atrophy of blood vessels occur in the skin with age [193]. These age-related changes are enhanced by the development of DM [77]. Increased skin autofluorescence can indicate age-related changes [147], which are enhanced at DM with nephropathy and retinopathy complications [148]. The excitation light wavelength varied from 300 to 420 nm with the excitation peak at 370 nm with autofluorescence in the range of 420–600 nm [148].

As seen from the Refs. [95, 147], the increase of fluorescence intensity is associated with DM development and aging, and thus it is important to distinguish the contribution of DM and age to the signal. An increase in the glycated collagen amount of only 33% was obtained in patients not suffering from DM at the age of 20–85 years [194], while the fluorescence of the AGEs in DM increased five-fold, strongly correlating with age. Collagen glycation increased by three-fold in patients with DM compared to nondiabetic subjects, strongly correlating with the concentration of glycated hemoglobin in blood, but not with age. Thus, appropriate glycation levels in tissues can be identified for a certain age, but these indicators are significantly increased with DM development. Therefore, it is possible to have criteria differentiating changes associated with DM or age.

Elevated skin autofluorescence in patients with insulin resistance and/or DM, as well as in healthy elderly and middle-aged subjects, was observed [150]. Changes in the skin hydration state, degradation of type I collagen, and greater glycation related to DM and aging were evaluated by Raman spectroscopy. A weak positive correlation between the Raman peaks ratio (855/876) related to glycated proteins and the skin autofluorescence was found [150]. Using confocal Raman spectroscopy, the increase in the amount of such AGEs as pentosidine and glucosepane in the skin dermis of elderly women with type II DM and healthy elderly women in comparison with healthy young women was obtained [143].

Many papers are devoted to the study of structural and optical parameters of skin in DM development and its modeling. Unfortunately, the conclusions based on different studies are often contradictory [69]. The reduction of stratum corneum hydration was shown *in vivo* in humans and mice during DM development [195, 196], which was not related to disorder of the barrier function of the epidermis. THz imaging applied for early screening of diabetic foot syndrome showed less water concentration in diabetic subjects [80]. However, it was noted that the transepidermal water loss (TEWL) does not increase in the case of DM [195, 197]. However, long-term hyperglycemia disturbs the barrier function of the skin and its permeability [196].

A reduction in keratinocyte proliferation in the epidermis of mice with modeled DM was observed [69, 197, 198], while no such changes were obtained in the experiments in rats and mice of other ages [196, 198]. Contradictory results were obtained in the studies of epidermal thickness at DM development [69]. The authors of [196, 198, 199] conclude that epidermis thickness does not change under the conditions of long-term hyperglycemia in rats and DM in humans and mice; however, for another DM model and ages, the thickness decreases [197, 198, 200] or increases [201]. The increase of skin thickness in patients with DM was measured ultrasonically [202]. The different conclusions could be caused by different ages of mice under study and DM animal model. Using microscopy and histology, changes in collagen distribution in the dermis of DM patients with complications were observed in contrast to those having no complications [203]. The inconsistent results require one to specify the experimental conditions.

AGEs formation reduces the solubility and elasticity of collagen, thereby enhancing its stiffness. [72, 204]. The results presented in Reference [205] confirm that the accumulation of fragmented skin collagen and the presence of molecular cross-links during DM development impairs the structural integrity and mechanical properties of skin collagen. The pronounced crosslinks of collagen fibers were demonstrated in the studies of diabetic skin aging. Atomic force microscopy (AFM) has shown that the collagen fibrils of skin are disordered and fragmented (see Figure 28.6(a, b)), and their key mechanical properties are essentially changed in the case of diabetes [205]. The quantitative analysis of AFM data has demonstrated that the mean roughness of collagen fibrils as a measure of their arrangement increases in diabetic dermis by 176% (Figure 28.6(c)) from 16 nm to 29 nm; this can be related to the increased concentration of matrix metalloproteinases, which provoke fragmentation of collagen fibrils. Fibril fragmentation impairs the structural integrity of collagen and, thus, changes the mechanical properties of skin dermis. The basic mechanical properties of the dermis, such as tensile strength (Figure 28.6(d)) and traction force (Figure 28.6(e)), were increased by 197% and 182% respectively, while the deformation of collagen fibrils was reduced by 58% (Figure 28.6(f)) in the diabetic dermis in comparison with the nondiabetic one.

A significant reduction in the number of collagen fibers and disorder of skin collagen fibers caused by alloxan diabetes was observed using two-photon imaging [131]. Classical histological analysis also showed that DM led to a change in skin filamentous structure.

The mechanical properties of collagen hydrogel were studied using FLIM after its incubation in ribose and glutaraldehyde solutions, which contribute to the formation of collagen cross-links [166]. Correlations between the mechanical properties of collagen and the fluorescence lifetime were observed

in the case of collagen incubation in glutaraldehyde in contrast to ribose. It has been found that the degree and nature of collagen cross-linking significantly affects tissue elasticity also under the action of ribose.

In vitro glycation of murine skin in the glyoxal solution showed an increase of TEWL, and significant increase of saturated fatty acids concentration in the epidermis; moreover, the barrier function of the epidermis was impaired [139]. The skin immersed in the glyoxal solution became a yellow color. The presence of AGEs in the skin was detected by increased autofluorescence [139]. The yellow color could be caused by nonenzymatic glycation of dermal collagen. Yellowish nails and skin are observed in diabetic patients [21].

The results of a study of glucose diffusion in *ex vivo* skin of mice with alloxan diabetes by measuring the collimated transmittance of visible light through the skin samples are presented in Reference [47]. The glucose diffusion rate in the skin of mice with alloxan diabetes was up to 2.5 times slower as compared to the control group of mice (Figure 28.7). Similar results were received for the aqueous glucose solutions of three different concentrations: 30%, 43%, and 56%.

Reduction of glycerol molecule diffusion rate in *ex vivo* skin samples taken from rats with 2 [48] and 4 [105] weeks of alloxan diabetes was observed. Two weeks of the streptozotocin diabetes caused the reduction of glycerol diffusion rate in skin *ex vivo* [107]. *In vivo* OCT studies also showed slower OC of skin under action of glycerol solution in DM animals [105]. These results are illustrated by data presented in Figure 28.8 and summarized in Table 28.1. The typical kinetic curves of collimated transmittance spectra of rat skin and myocardium *ex vivo* samples of nondiabetic and diabetic animals measured during tissue sections immersion in a 70%-glycerol solution are shown in Figure 28.8. The collimated transmittance of both tissues increases with time and reaches saturation.

Transmittance is significantly increased for all wavelengths of both tissues and much faster for myocardium. It is seen that collimated transmittance of diabetic tissues grows much slowly than for nondiabetic tissues. In Table 28.1, mean values of thickness and weight of *ex vivo* rat tissue samples before and after immersion in a 70%-glycerol solution and tissue permeability (P) and diffusion (D) coefficients for glycerol in the skin and myocardium of nondiabetic and diabetic groups, evaluated from collimated transmittance kinetic curves as shown in Figure 28.8, are presented. Both skin and myocardium permeability for glycerol is decreased in diabetic groups. Transverse shrinkage (less sample thickness) and dehydration (less sample weight) in all tissue samples caused by osmotic pressure induced by glycerol are seen.

The characteristic time τ and OC efficiency of *in vivo* rat skin under the action of a 70%-glycerol solution was estimated [105]. The degree of the OC was defined as the ratio of tissue optical attenuation coefficient μ_t before and after glycerol action, for nondiabetic animals μ_t was reduced up to $(46 \pm 19)\%$ and for diabetic (alloxan type), was up $(37 \pm 16)\%$. The OC characteristic time was found as $\tau = (2.7 \pm 0.4)$ min for nondiabetic and (8.9 ± 7.7) min for diabetic animals.

The increase of light penetration into tissue under the action of glycerol solution is observed for both *ex vivo* and *in vivo* studies. The slowdown of skin OC in the diabetic group was

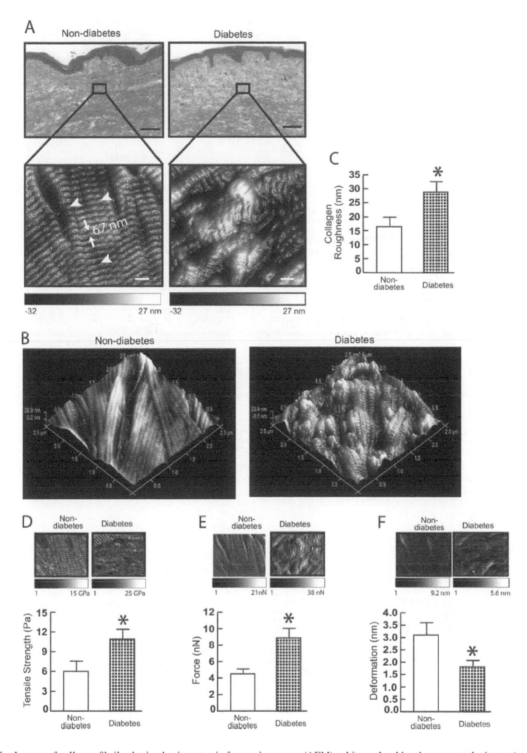

FIGURE 28.6 Images of collagen fibrils obtained using atomic force microscopy (AFM): white and red head-arrows at the intact (nondiabetes) and fragmented/disordered (diabetes) fibrils, respectively (A); 3D images of collagen fibrils obtained using the AFM (B); roughness of collagen fibrils was estimated using the Nanoscope Analysis software (Nanoscope_Analysis_v120R1sr3, Bruker-AXS, Santa Barbara, CA) (C); the tensile strength (D); the traction force (E); the deformation of collagen fibrils determined using the AFM PeakForce™ Quantitative NanoMechanics mode and Nanoscope Analysis software (F) [205].

established in two independent sets of measurements *ex vivo* (spectral collimated transmittance) and *in vivo* (OCT).

The reduction of tissue permeability to glycerol molecules during development of diabetes must be related to alterations in tissue morphology [19] and modification of structure,

including the degree of fibril packing, loss of axial packing of the collagen fibrils due to the twisting and distortion of the matrix by the glycation adducts [144, 170, 171], the cross-linking of proteins, changes in free and bound water content in tissues, and increase in sarcoplasm viscosity [19, 41, 170].

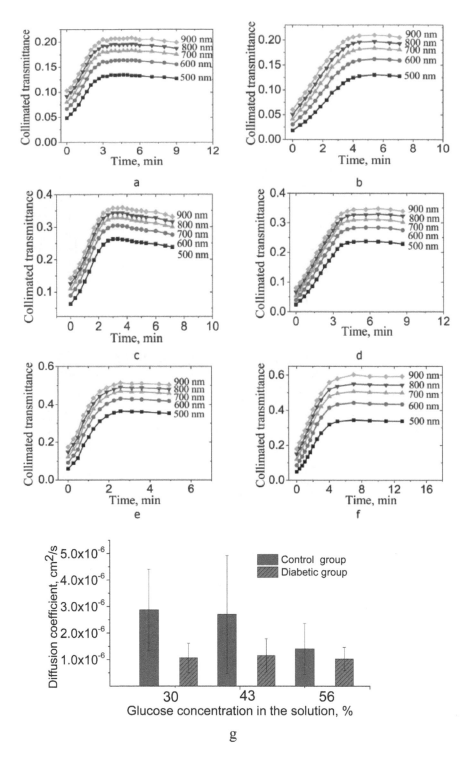

FIGURE 28.7 Kinetic curves of optical clearing (a–f) and diffusion coefficients of glucose (g) in the skin of mice of the control group (a, c, e) and alloxan diabetic group (b, d, f) obtained with the 30%- (a, b), 43%- (c, d), and 56%- (e, f) aqueous solutions of glucose [47].

These structural modifications also lead to changes of tissue optical properties, specifically to the increase of scattered light intensity [144]. A decrease in tissue permeability for glycerol molecules during the development of diabetes should be associated with changes in tissue morphology [19], including the degree of fibril packing, loss of axial packing of collagen fibrils due to twisting, and distortion of the matrix due to the appearance of glycation adducts [144, 170, 171], protein crosslinking, changes in the content of free and bound water in tissues, and an increase in the viscosity of sarcoplasma [19, 41, 170]. These structural modifications also lead to changes in the optical properties of tissues, and in particular to an increase in the intensity of scattered light [144].

Optoacoustic spectroscopy was used to study AGEs in skin at *in vitro* glycation [183]. The principle of optoacoustic spectroscopy is based on the transformation of energy of

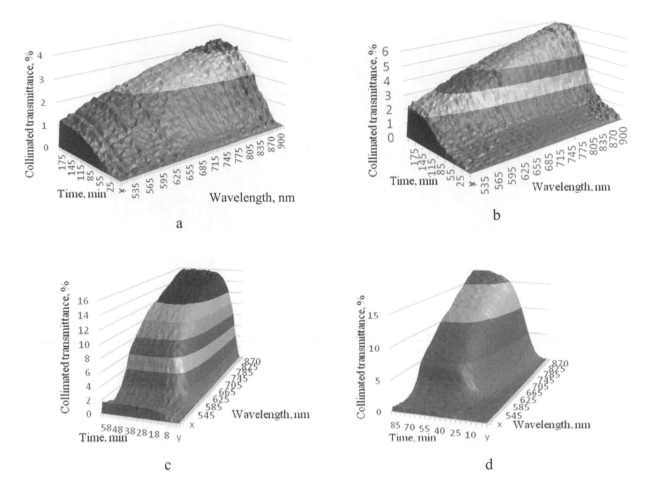

FIGURE 28.8 Typical time dependences of collimated transmittance spectra of *ex vivo* Wistar rat tissue samples for nondiabetic (a, c) and diabetic (alloxan model) (b, d) animal groups during OC using a 70%-glycerol solution; sections with the initial thicknesses (0.79 ± 0.04) mm of nondiabetic (a) and (0.81 ± 0.01) mm of diabetic (b) skin; and sections with the initial thicknesses (0.58 ± 0.02) mm of nondiabetic (c) and (0.76 ± 0.03) mm of diabetic (d) myocardium [105].

TABLE 28.1

Mean values of the thickness and weight of *ex vivo* Wistar rat tissue sections before (l_0 and W_0) and after (l and W) immersion in 70%-glycerol solution; tissue permeability (P) and diffusion (D) coefficients of glycerol measured in skin and myocardium samples of nondiabetic and alloxan/streptozotocin diabetic animal groups

Group	l_0/l, mm	W_0/W, mg	P, cm/sec	D, cm²/sec
Skin				
Nondiabetic [48]	$(0.53\pm0.11)/(0.55\pm0.13)$	$(271\pm68)/(183\pm21)$	$(1.68\pm0.88) \times 10^{-5}$	$(8.33\pm2.60) \times 10^{-7}$
Diabetic/alloxan [107]	$(0.56\pm0.04)/(0.57\pm0.07)$	$(270\pm32)/(203\pm73)$	$(1.20\pm0.33) \times 10^{-5}$	$(6.77\pm2.11) \times 10^{-7}$
Diabetic/ streptozotocin [107]	$(0.47\pm0.05)/(0.57\pm0.06)$	$(198\pm17)/(161\pm1)$	$(1.36\pm0.82) \times 10^{-5}$	$(6.97\pm4.37) \times 10^{-7}$
Myocardium				
Non-diabetic [48]	$(0.68\pm0.11)/(0.51\pm0.09)$	$(210\pm37)/(146\pm35)$	$(1.18\pm0.61) \times 10^{-5}$	$(7.90\pm3.61) \times 10^{-7}$
Diabetic/alloxan [48]	$(0.58\pm0.06)/(0.47\pm0.07)$	$(214\pm41)/(153\pm36)$	$(0.86\pm0.32) \times 10^{-5}$	$(5.14\pm2.10) \times 10^{-7}$
Diabetic/alloxan [105]	$(0.96\pm0.30)/(0.80\pm0.26)$	$(377\pm139)/(315\pm138)$	$(0.88\pm0.50) \times 10^{-5}$	$(7.74\pm4.40) \times 10^{-7}$

time-modulated or pulsed optical radiation interacting with object into thermal and further acoustic waves [206]. A spectrally nonselective detector of absorbed energy at different wavelengths is a microphone or ultrasonic transducer [206]. The conditions of physiological hyperglycemia were provided by immersion of porcine skin samples in ribose solution over

the course of 17 days [183]. An increase in optoacoustic signal of glycated skin at light absorption in the range 540–620 nm with time of incubation was observed [183]. The increase in the optoacoustic signal can be related not only to high light absorption by glycated skin, but also to the increased efficiency of light-to-sound conversion due to greater tissue elasticity caused

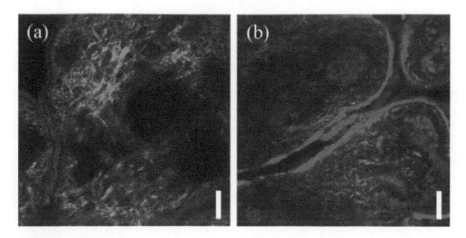

FIGURE 28.9 Microscopic images of porcine skin incubated during 17 days in phosphate buffer (a) and ribose (b) solutions. Superimposed images of autofluorescence (green) and second harmonic (red) intensity are shown. The images were obtained at the depth 35 μm, the scale bars correspond to 50 μm [183].

by the crosslinks that appear between the collagen molecules. It was shown [158] that photoacoustic tomography can be used to detect vascular dysfunction in diabetes due to their unique peripheral hemodynamic response to occlusion and a lower level of SO_2, compared to that for healthy patients. In Reference [183], optoacoustic measurements were compared with nonlinear microscopic imaging of glycated skin (see Figure 28.9). An increase of intensity of autofluorescence and decrease of SHG-signal were demonstrated for the glycated skin and other collagenous tissues (see Figure 28.3 and Figure 28.9) [42, 183], which demonstrate modification of tissue structural properties and corresponding increase of optoacoustic signal.

A patent [149] proposes a device with an operation principle based on autofluorescence detection, by which it is possible to determine the collagen AGEs and to measure their concentration in sclera, oral mucosa, and skin for diagnosing DM or prediabetes. The degree of tissue glycation is quantified from the autofluorescence spectra.

The possibility of combined measurements of blood perfusion and skin fluorescence was evaluated using laser Doppler flowmetry for diagnostics of complications of type II DM in the lower extremities of patients [101]. The blood microcirculation was topically stimulated by heating. The studies showed that the patients with DM had an enhanced fluorescence and lower perfusion response to local heating, which can be used as markers for evaluation of diabetic complications.

A method of noninvasive *in vivo* monitoring of hyperglycemic state in mice with streptozotocin diabetes based on mm-wave spectroscopy and experimentally confirmed using live animal models as objects is described in Reference [104]. The transmittance coefficient of the skin fold at the nape of mice of various lines was measured at 25 uniformly located frequencies in the range of 0.075–0.110 THz. Skin transmittance was several times higher for animals in a hyperglycemic state and depended a little on the presence of white or black hair or its absence in mice.

Conclusion

The influence of diabetes mellitus on the optical properties of biological fluids and tissues has been studied quite widely, and

the development of unique optical methods is further stimulating interest in a deeper study of this formidable disease. From the point of view of diagnosis, it is important to observe the earliest changes in tissue properties caused by the initial stage of the disease. The optical methods "see" the presence of free glucose in the blood, in the interstitial fluid, and in cells through optical clearing, which is associated with the matching of the refractive indices of the scatterers and the environment, as well as with the dehydration of tissue and cells. At the same time, tissue glycation leads to a change in tissue structure and also to a change in the optical and diffusion properties of tissues, which makes it possible to assess the degree of tissue glycation.

As one can see from this chapter, many optical methods are applicable for diagnostic and monitoring of the degree of tissue and fluid glycation during the development of diabetes. Such optical techniques include fluorescence spectroscopy, refractometry, Raman spectroscopy, broad-wavelength spectrophotometry, OCT, confocal microscopy, multiphoton fluorescence microscopy, and harmonic generation, optoacoustics, laser speckle contrast imaging, and laser Doppler flowmetry. Among the many optical methods, one of the most promising is fluorescence spectroscopy, which makes it possible to register the presence of the end products of glycation in tissues, is simple to implement, and makes possible noninvasive diagnostics. Refractometric and phase methods are promising, including the use of three-dimensional diffraction tomography and OCT. Terahertz spectroscopy and imaging are also promising methods for studying glycated tissues and cells. The problem of noninvasive monitoring of early structural changes in tissues associated with diabetes is gradually losing its "unsolved" status.

Acknowledgments

The authors are grateful to A.N. Bashkatov, A.B. Bucharskaya, P.A. Dyachenko, E.A. Genina, and D. Zhu for collaboration. The work was supported by grant No 13.2251.21.0009 of the Ministry of Science and Higher Education of the Russian Federation. D.K.T. was supported by the RF Presidential grant No SP-5422.2021.4 for Scholarships of young scientists and postgraduate students (2021–2023).

REFERENCES

1. J.E. Shaw, R.A. Sicree, P.Z. Zimmet, "Global estimates of the prevalence of diabetes for 2010 and 2030," *Diabet. Res. Clin.. Pract.* **87**(1), 4–14 (2010).
2. Diabetes Fact Sheet, World Health Organization, http://www.who.int/mediacentre/factsheets/fs312/en/ (2017).
3. C.D. Mathers, D. Loncar, "Projections of global mortality and burden of disease from 2002 to 2030," *PLOS Med.* **3**(11), 2011–2030 (2006).
4. D. LeRoith, J.M. Olefsky, S.I. Taylor (Eds.), *Diabetes Mellitus: A Fundamental and Clinical Text*, 3rd ed., Lippincott Williams & Wilkins, Philadelphia, PA, 2003.
5. Y. Nakabeppu, T. Ninomiya (Eds.), *Diabetes Mellitus: A Risk Factor for Alzheimer's Disease*, 1128, Springer, New York, 2019.
6. K.T. Patton, G.A. Thibodeau, *The Human Body in Health & Disease*, 6th ed., Elsevier Inc., Amsterdam, 2014.
7. D.G. Gardner, D.M. Shoback, *Greenspan's Basic & Clinical Endocrinology*, 9th ed., McGraw-Hill Medical, New York, 2011.
8. B.B. Tripathy, *RSSDI Textbook of Diabetes Mellitus*, 2nd ed., Jaypee Brothers Medical Publishers, New Delhi, 2012.
9. T. Danne, R. Nimri, T. Battelino, R. M. Bergenstal, K. L. Close, J. H. DeVries, at al., "International consensus on use of continuous glucose monitoring," *Diabetes care* **40**(12), 1631–1640 (2017).
10. A.J. King, "The use of animal models in diabetes research," *Br. J. Pharmacol.* **166**(3), 877–894 (2012).
11. T. Szkudelski, "The mechanism of alloxan and streptozotocin action in B cells of the rat pancreas," *Physiol. Res.* **50**(6), 537–546 (2001).
12. S. Lenzen, "The mechanisms of alloxan- and streptozotocin-induced diabetes," *Diabetologia* **51**(2), 216–226 (2008).
13. A. Burgeiro, A. Fuhrmann, S. Cherian, et al., "Glucose uptake and lipid metabolism are impaired in epicardial adipose tissue from heart failure patients with or without diabetes," *Am. J. Physiol. Endocrinol. Metab.* **310**(7), E550–E564 (2016).
14. V. Vinokur, G. Leibowitz, L. Grinberg, R. Eliashar, E. Berenshtein, M. Chevion, "Diabetes and the heart: Could the diabetic myocardium be protected by preconditioning?" *Redox Rep.* **12**(6), 246–256 (2007).
15. T. Nishikawa, D. Edelstein, X.L. Du, et al., "Normalizing mitochondrial superoxide production blocks three pathways of hyperglycemic damage," *Nature* **404**(6779), 787–790 (2000).
16. X. Du, T. Matsumura, D. Edelstein, L. Rossetti, Z. Zsengellér, C. Szabó, M. Brownlee, "Inhibition of GAPDH activity by poly(ADP-ribose) polymerase activates three major pathways of hyperglycemic damage in endothelial cells," *J. Clin. Invest.* **112**(7), 1049–1057 (2003).
17. M. Chevion, "A site-specific mechanism for free radical induced biological damage: The essential role of redox-active transition metals," *Free Radic. Biol. Med.* **5**(1), 27–37 (1988).
18. N. Vigneshwaran, G. Bijukumar, N. Karmakar, S. Anand, A. Misra, "Autofluorescence characterization of advanced glycation end products of hemoglobin," *Spectrochim. Acta A Mol. Biomol. Spectrosc.* **61**(1), 163–170 (2005).
19. N.I. Dikht, A.B. Bucharskaya, G.S. Terentyuk, et al., "Morphological study of the internal organs in rats with alloxan diabetes and transplanted liver tumour after intravenous injection of gold nanorods," *Russ. Open Med. J.* **3**(3), 0301 (2014).
20. D. McGuire, N. Marx, *Diabetes in Cardiovascular Disease: A Companion to Braunwald's Heart Disease*, Elsevier Health Sciences, Amsterdam, 2014.
21. E. Bonora, R.A. DeFronzo (Eds.), *Diabetes Complications, Comorbidities and Related Disorders*, Springer, New York, 2018.
22. V.V. Tuchin (Ed.), *Handbook of Optical Sensing of Glucose in Biological Fluids and Tissues*, Taylor & Francis Group LLC, CRC Press, Boca Raton, FL, 2009.
23. M., Shokrekhodaei, S. Quinones, "Review of non-invasive glucose sensing techniques: Optical, electrical and breath acetone," *Sensors* **20**(5), 1251 (2020).
24. H. Ullah, A. Mariampillai, M. Ikram, I.A. Vitkin, "Can temporal analysis of optical coherence tomography statistics report on dextrorotatory-glucose levels in blood? " *Laser Phys.* **21**(11), 1962–1971 (2011).
25. G. Purvinis, B.D. Cameron, D.M. Altrogge, "Noninvasive polarimetric-based glucose monitoring: An in vivo study," *J. Diabet. Sci. Technol.* **5**(2), 380–387 (2011).
26. N.C. Dingari, I. Barman, G.P. Singh, J.W. Kang, R.R. Dasari, M.S. Feld, "Investigation of the specificity of Raman spectroscopy in non-invasive blood glucose measurements," *Analyt. Bioanalyt. Chem.* **400**(9), 2871–2880 (2011).
27. Y. Zhang, G. Wu, H. Wei, et al., "Continuous noninvasive monitoring of changes in human skin optical properties during oral intake of different sugars with optical coherence tomography," *Biomed. Opt. Express* **5**(4), 990–999 (2014).
28. R.Y. He, H.J. Wei, H.M. Gu, Z. Zhu, Y. Zhang, X. Guo, T. Cai, "Effects of optical clearing agents on noninvasive blood glucose monitoring with optical coherence tomography: A pilot study," *J. Biomed. Opt.* **17**(10), 101513 (2012).
29. M.A. Pleitez, T. Lieblein, A. Bauer, O. Hertzberg, H. von Lilienfeld-Toal, W. Mäntele, "Windowless ultrasound photoacoustic cell for in vivo mid-IR spectroscopy of human epidermis: Low interference by changes of air pressure, temperature, and humidity caused by skin contact opens the possibility for a non-invasive monitoring of glucose in the interstitial fluid," *Rev. Sci. Instrum.* **84**(8) 084901 (2013).
30. N.C. Dingari, I. Barman, J.W. Kang, C.R. Kong, R.R. Dasari, M.S. Feld, "Wavelength selection-based nonlinear calibration for transcutaneous blood glucose sensing using Raman spectroscopy," *J. Biomed. Opt.* **16**(8), 087009 (2011).
31. J.M. Yuen, N.C. Shah, J.T. Walsh, M.R. Glucksberg, R.P. Van Duyne, "Transcutaneous glucose sensing by surface-enhanced spatially offset Raman spectroscopy in a rat model," *Anal. Chem.* **82**(20), 8382–8385 (2010).
32. S. Firdous, M. Nawaz, M. Ahmed, S. Anwar, A. Rehman, R. Rashid, A. Mahmood, "Measurement of diabetic sugar concentration in human blood using Raman spectroscopy," *Laser Phys.* **22**(6), 1090–1094 (2012).
33. X.X. Guo, A. Mandelis, B. Zinman, "Noninvasive glucose detection in human skin using wavelength modulated differential laser photothermal radiometry," *Biomed. Opt. Express* **3**(11), 3012–3021 (2012).
34. M.K. Chowdhury, A. Srivastava, N. Sharma, S. Sharma, "Challenges & countermeasures in optical noninvasive blood glucose detection," *Int. J. Innov. Res. Sci. Eng. Techn.* **2**(1), 324–329 (2013).

35. N. Li, H. Zang, H. Sun, X. Jiao, K. Wang, T.C. Liu, Y. Meng, "A noninvasive accurate measurement of blood glucose levels with Raman spectroscopy of blood in microvessels," *Molecules* **24**(8), 1500 (2019).

36. S. Zhu, S. Huang, J. Feng, "Non-invasive blood glucose detection with free-space SD-OCT system," *Proc. SPIE*, **11020**, 110200W (2019).

37. M. Katsuhiko, Y. Yukio, "Near-infrared noninvasive blood glucose prediction without using multivariate analyses: Introduction of imaginary spectra due to scattering change in the skin, " *J. Biomed. Opt.* **20**(4), 047003 (2015).

38. S.A.A. Shah, A. Laude, I. Faye, T.B. Tang, "Automated microaneurysm detection in diabetic retinopathy using curvelet transform," *J. Biomed. Opt.* **21**(10), 101404 (2016).

39. Y.J. Heo, S. Takeuchi, "Towards smart tattoos: Implantable biosensors for continuous glucose monitoring," *Adv. Healthc. Mater.* **2**(1), 43–56 (2013).

40. C. Stark, C.A.C. Arrieta, R. Behroozian, B. Redmer, F. Fiedler, S. Müller, "BroaDMand polarimetric glucose determination in protein containing media using characteristic optical rotatory dispersion," *Biomed. Opt. Express* **10**(12), 6340–6350 (2019).

41. E. Selvin, M.W. Steffes, H. Zhu, et al., "Glycated hemoglobin, diabetes, and cardiovascular risk in nondiabetic adults," *N. Engl. J. Med.* **362**(9), 800–811 (2010).

42. J.-Y. Tseng, A.A. Ghazaryan, W. Lo, et al., "Multiphoton spectral microscopy for imaging and quantification of tissue glycation," *Biomed. Opt. Express* **2**(2), 218–230 (2011).

43. G. Mazarevica, T. Freivalds, A. Jurka, "Properties of erythrocyte light refraction in diabetic patients," *J. Biomed. Opt.* **7**(2), 244–247 (2002).

44. J. Blackwell, K.M. Katika, L. Pilon, K.M. Dipple, S.R. Levin, A. Nouvong, *In vivo* time-resolved autofluorescence measurements to test for glycation of human skin," *J. Biomed. Opt.* **13**(1), 014004 (2008).

45. U. Kanska, J. Boratynski, "Thermal glycation of proteins by D-glucose and D-fructose," *Arch. Immunol. Ther. Exp.* **50**(1), 61–66 (2002).

46. O.S. Khalil, "Non-invasive glucose measurement technologies: An update from 1999 to the dawn of the new millennium," *Diabet. Technol. Ther.* **6**(5), 660–697 (2004).

47. D.K. Tuchina, R. Shi, A.N. Bashkatov, E.A. Genina, D. Zhu, Q. Luo, V.V. Tuchin, "Ex vivo optical measurements of glucose diffusion kinetics in native and diabetic mouse skin," *J. Biophoton.* **8**(4), 332–346 (2015).

48. D.K. Tuchina, A.N. Bashkatov, A.B. Bucharskaya, E.A. Genina, V.V. Tuchin, "Study of glycerol diffusion in skin and myocardium ex vivo under the conditions of developing alloxan-induced diabetes," *J. Biomed. Photon. Eng.* **3**(2), 020302 (2017).

49. T. Pan, M. Li, J. Chen, H. Xue, "Quantification of glycated hemoglobin indicator HbA1c through near-infrared spectroscopy," *J. Innov. Opt. Health Sci.* **7**(4), 1350060 (2014).

50. E. Shirshin, O. Cherkasova, T. Tikhonova, E. Berlovskaya, A. Priezzhev, V. Fadeev, "Native fluorescence spectroscopy of blood plasma of rats with experimental diabetes: Identifying fingerprints of glucose-related metabolic pathways," *J. Biomed. Opt.* **20**(5), 051033 (2015).

51. G.V. Maksimov, O.G. Luneva, N.V. Maksimova, E. Matettuchi, E.A. Medvedev, V.Z. Pashchenko, A.B. Rubin, "Role of viscosity and permeability of the erythrocyte plasma membrane in changes in oxygen-binding properties of hemoglobin during *Diabetes Mellitus*," *Bull. Exp. Biol. Med.* **140**(5), 510–513 (2005).

52. J.F. Villa-Manríquez, J. Castro-Ramos, F. Gutiérrez-Delgado, M.A. Lopéz-Pacheco, A.E. Villanueva-Luna, "Raman spectroscopy and PCA-SVM as a non-invasive diagnostic tool to identify and classify qualitatively glycated hemoglobin levels *in vivo*," *J. Biophoton.* **10**(8), 1074–1079 (2016).

53. P.A. Timoshina, A.B. Bucharskaya, D.A. Alexandrov, V.V. Tuchin, "Study of blood microcirculation of pancreas in rats with alloxan diabetes by Laser Speckle Contrast Imaging," *J. Biomed. Photon. Eng.* **3**(2), 020301–020307 (2017).

54. F. Wei, S. Rui, D. Zhu, "Monitoring skin microvascular dysfunction of type 1 diabetic mice using *in vivo* skin optical clearing," *Proc. SPIE* **10493**, 104931O (2018).

55. O.A. Smolyanskaya, I.J. Schelkanova, M.S. Kulya, et al., "Glycerol dehydration of native and diabetic animal tissues studied by THz-TDS and NMR methods," *Biomed. Opt. Express* **9**(3), 6467–6476 (2018).

56. V.V. Tuchin, *Optical Clearing of Tissues and Blood*, **PM 154**, SPIE Press, Bellingham, WA, 2006.

57. D.K. Tuchina, V.V. Tuchin, "Optical and structural properties of biological tissues under *Diabetes Mellitus*," *J. Biomed. Photon. Eng.* **4**(2), 020201 (2018).

58. D.Zhu, K.V., Q. Larin Luo, V.V. Tuchin, "Recent progress in tissue optical clearing," *Laser Photonics Rev.* **7**(5), 732–757 (2013).

59. L.M. Oliveira, M.I. Carvalho, E. Nogueira, V.V. Tuchin, "The characteristic time of glucose diffusion measured for muscle tissue at optical clearing," *Laser Phys.* **23**(7), 075606 (2013).

60. L. Oliveira, V.V. Tuchin, *The Optical Clearing Method: A New Tool for Clinical Practice and Biomedical Engineering*, Springer Nature Switzerland AG, Basel, 2019, 177 p.

61. M.G. Ghosn, N. Sudheendran, M. Wendt, A. Glasser, V.V. Tuchin, K.V. Larin, "Monitoring of glucose permeability in monkey skin in vivo using Optical Coherence Tomography," *J. Biophoton.* **3**(1–2), 25–33 (2010).

62. A.N. Bashkatov, E.A. Genina, Yu.P. Sinichkin, V.I. Kochubey, N.A. Lakodina, V.V. Tuchin, "Glucose and mannitol diffusion in human dura mater," *Biophys. J.* **85**(5), 3310–3318 (2003).

63. M. Kreft, M. Luksic, T.M. Zorec, M. Prebil, R. Zorec, "Diffusion of D-glucose measured in the cytosol of a single astrocyte," *Cell. Mol. Life Sci.* **70**(8), 1483–1492 (2012).

64. M.G. Ghosn, E.F. Carbajal, N.A. Befrui, K.V. Larin, J.F. Granada, K.V. Larin, "Permeability of hyperosmotic agent in normal and atherosclerotic vascular tissues," *J. Biomed. Opt.* **13**(1), 010505 (2008).

65. X. Guo, G. Wu, H. Wei, et al., "Quantification of glucose diffusion in human lung tissues by using Fourier domain optical coherence tomography," *Photochem. Photobiol.* **88**(2), 311–316 (2012).

66. R.O. Esenaliev, K.V. Larin, I.V. Larina, M. Motamedi, "Noninvasive monitoring of glucose concentration with optical coherence tomography," *Opt. Lett.* **26**(13), 992–994 (2001).

67. K.V. Larin, M.S. Eledrisi, M. Motamedi, R.O. Esenaliev, "Noninvasive blood glucose monitoring with optical coherence tomography: A pilot study in human subjects," *Diabetes Care* **25**(12), 2263–2267 (2002).

68. E. Zharkikh, V. Dremin, E. Zherebtsov, A. Dunaev, I. Meglinski, Biophotonics methods for functional monitoring of complications of diabetes mellitus, *J. Biophoton.* **13**(10), e202000203 (2020).

69. B.M., Chege, Z. Birech, P.W. Mwangi, F.O. Bukachi, "Utility of Raman spectroscopy in diabetes detection based on biomarker Raman bands and in antidiabetic efficacy studies of herbal extract Rotheca myricoides Hochst," *J. Raman Spectrosc.* **50**(10), 1358–1366 (2019).

70. O.A. Smolyanskaya, E.N. Lazareva, S.S. Nalegaev, et al., "Multimodal optical diagnostics of glycated biological tissues," *Biochemistry* **84**(Suppl 1), S124–S143 (2019).

71. F. Quondamatteo, "Skin and diabetes mellitus: What do we know?" *Cell Tissue Res.* **355**(1), 1–21 (2014).

72. G.M. Campos de Macedo, S.N., T. Barreto, "Skin disorders in diabetes mellitus: An epidemiology and physiopathology review," *Diabetol. Metab. Syndr.* **8**(1), 63 (2016).

73. J.A. Suaya, D.F. Eisenberg, C. Fang, L.G. Miller, "Skin and soft tissue infections and associated complications among commercially insured patients aged 0–64 years with and without diabetes in the U.S.," *PLOS ONE* **8**(4), e60057 (2013).

74. M. Marre, "Genetics and the prediction of complications in type 1 diabetes," *Diabetes Care* **22**(Supplement 2), B53–B58 (1999).

75. I.J. Goldberg, "Why does diabetes increase atherosclerosis? I don't know!" *J. Clin. Invest.* **114**, 613–615 (2004).

76. W.W. Song, A. Ergul, "Type-2 diabetes-induced changes in vascular extracellular matrix gene expression: Relation to vessel size," *Cardiovasc. Diabetol.* **5**, 3-1-7 (2006).

77. G. Spinetti, N. Kraenkel, C. Emanueli, P. Madeddu, "Diabetes and vessel wall remodelling: From mechanistic insights to regenerative therapies," *Cardiovasc. Res.* **78**(2), 265–273 (2008).

78. D. Pedicino, A.F. Giglio, V.A. Galiffa, F. Trotta, G. Liuzzo, "Type 2 diabetes, immunity and cardiovascular risk: A complex relationship," Chap. 3 in *The Pathophysiology and Complications of Diabetes Mellitus*, O.O. Oguntibeju (Ed.), InTech, London, 2012.

79. J. Lee, R. Rosen, "Optical coherence tomography angiography in diabetes," *Curr. Diabetes Rep.* **16**(12), 123 (2016).

80. G.G. Hernandez-Cardoso, S.C. Rojas-Landeros, M. Alfaro-Gomez, et al., "Terahertz imaging for early screening of diabetic foot syndrome: A proof of concept," *Sci. Rep.* **7**, 42124 (2017).

81. M.S. Anderson, J.A. Bluestone, "The NOD mouse: A model of immune dysregulation," *Annu. Rev. Immunol.* **23**, 447–485 (2005).

82. J.A. Bluestone, K. Herold, G. Eisenbarth, "Genetics, pathogenesis and clinical interventions in type 1 diabetes," *Nature* **464**(7293), 1293–1300 (2010).

83. S. Makino, K. Kunimoto, Y. Muraoka, Y. Mizushima, K. Katagiri, Y. Tochino, "Breeding of a non-obese, diabetic strain of mice," *Exp. Anim.* **29**(1), 1–13 (1980).

84. T.T. Berezov, B.F. Korovkin, *Biological Chemistry*, Meditsina, Moscow, 1998 (in Russian).

85. B. Coudrillier, J. Pijanka, J. Jefferys, T. Sorensen, H.A. Quigley, C. Boote, T.D. Nguyen, "Collagen structure and mechanical properties of the human sclera: Analysis for the effects of age," *J. Biomech. Eng.* **137**(4), 041006 (2015).

86. M. Maciążek-Jurczyk, A. Szkudlarek, M. Chudzik, J. Pożycka, A. Sułkowska, "Alteration of human serum albumin binding properties induced by modifications: A review," *Spectrochim. Acta A Mol. Biomol. Spectrosc.* **188**, 675–683 (2018).

87. S.F. Diniz, F.P.L.G. Amorim, F.F. Cavalcante-Neto, A.L. Bocca, A.C. Batista, G.E. Simm, T.A. Silva, "Alloxan-induced diabetes delays repair in a rat model of closed tibial fracture," *Brazil. J. Med. Biol. Res.* **41**(5), 373–379 (2008).

88. D. Dufrane, M. van Steenberghe, Y. Guiot, R.M. Goebbels, A. Saliez, P. Gianello, "Streptozotocin-induced diabetes in large animals (pigs/primates): Role of GLUT2 transporter and beta-cell plasticity," *Transplantation* **81**(1), 36–45 (2006).

89. L.M. Hansen, D. Gupta, G. Joseph, D. Weiss, W.R. Taylor, "The receptor for advanced glycation end products impairs collateral formation in both diabetic and non-diabetic mice," *Lab. Inv.* **97**(1), 34–42 (2017).

90. H. Yu, J. Zhen, B. Pang, J. Gu, S. Wu, Ginsenoside, "Rg1 ameliorates oxidative stress and myocardial apoptosis in streptozotocin-induced diabetic rats," *Univ-Sci B, (Biomed. Biotechnol.)* **16**(5), 344–354 (2015).

91. J. Wu, L. Yan, "Streptozotocin-induced type 1 diabetes in rodents as a model for studying mitochondrial mechanisms of diabetic β cell glucotoxicity," *Diabet. Metab. Syndr. Obes. Targ Ther.* **8**, 181–188 (2015).

92. M.I. Asrarov, M.K. Pozilov, N.A. Ergashev, M.M. Rakhmatullaeva, "The influence of the hypoglycemic agent glycorazmulin on the functional state of mitochondria in the rats with streptozotocin-induced diabetes," *Prob. Endocrinol.* **3**, 38–42 (2014).

93. L. Dancakova, T. Vasilenko, I. Kova, et al., "Low-level laser therapy with 810nm wavelength improves skin wound healing in rats with streptozotocin-induced diabetes," *Photomed. Laser Surg.* **32**(4), 198–204 (2014).

94. D.E. Kelley, J. He, E.V. Menshikova, V.B. Ritov, "Dysfunction of mitochondria in human skeletal muscle in Type 2 diabetes," *Diabetes* **51**(10), 2944–2950 (2002).

95. C. Moran, G. Münch, J.M. Forbes, et al., "Type 2 diabetes, skin autofluorescence, and brain atrophy," *Diabetes* **64**(1), 279–283 (2015).

96. D. Ziegler, N. Papanas, A. Zhivov, et al., "Early detection of nerve fiber loss by corneal confocal microscopy and skin biopsy in recently diagnosed Type 2 diabetes," *Diabetes* **63**(7), 2454–2463 (2014).

97. M.F. Chowdhry, H.A. Vohra, M. Galiñanes, "Diabetes increases apoptosis and necrosis in both ischemic and non-ischemic human myocardium: Role of caspases and poly–adenosine diphosphate–ribose polymerase," *J. Thorac. Cardiovasc. Surg.* **134**(1), 124–131 (2007).

98. V.M. Monnier, D.R. Sell, C. Strauch, et al., "The association between skin collagen glucosepane and past progression of microvascular and neuropathic complications in type 1 diabetes," *Diabet. Compls* **27**(2), 141–149 (2013).

99. J.R. Acosta, I. Douag, D.P. Andersson, J. Bäckdahl, M. Rydén, P. Arner, J. Laurencikiene, "Increased fat cell size: A major phenotype of subcutaneous white adipose tissue in non-obese individuals with type 2 diabetes," *Diabetol.* **59**(3), 560–570 (2016).

100. E.V. Zharkikh, V.V. Dremin, M.A. Filina, et al., "Application of optical non-invasive methods to diagnose the state of the lower limb tissues in patients with diabetes mellitus," *J. Phys. Conf. Ser.* **929**, 012069 (2017).

101. V. Dremin, E. Zherebtsov, V. Sidorov, et al., "Multimodal optical measurement for study of lower limb tissue viability in patients with diabetes mellitus," *J. Biomed. Opt.* **22**(8), 085003 (2017).

102. E.V. Potapova, V.V. Dremin, E.A. Zherebtsov, et al., "A complex approach to noninvasive estimation of microcirculatory tissue impairments in feet of patients with diabetes mellitus using spectroscopy," *Opt. Spectrosc.* **123**(6), 955–964 (2017).

103. L.A. Muir, C.K. Neeley, K.A. Meyer, et al., "Adipose tissue fibrosis, hypertrophy, and hyperplasia: Correlations with diabetes in human obesity," *Obesity* **24**(3), 597–605 (2016).

104. P. Martín-Mateos, F. Dornuf, B. Duarte, et al., "In-vivo, non-invasive detection of hyperglycemic states in animal models using mm-wave spectroscopy," *Sci. Reps* **6**, 34035 (2016).

105. D.K. Tuchina, A.N. Bashkatov, A.B. Bucharskaya, V.V. Tuchin, "Exogenous agent diffusivity in tissues as a biomarker of diabetes mellitus pathology," *Proc. SPIE* **10877**, 108770X (2019).

106. W. Feng, R. Shi, C. Zhang, S. Liu, T. Yu, D. Zhu, "Visualization of skin microvascular dysfunction of type 1 diabetic mice using *in vivo* skin optical clearing method," *J. Biomed. Opt.* **24**(3), 031003 (2018).

107. D.K. Tuchina, A.B. Bucharskaya, V.V. Tuchin, "Pilot study of glycerol diffusion in ex vivo skin: A comparison of alloxan and streptozotocin diabetes models," *Proc. SPIE* **11457**, 114570J (2020).

108. A. Rohilla, S. Ali, "Alloxan induced diabetes: Mechanisms and effects," *Int. J. Res. Pharmac. Biomed. Sci.* **3**(2), 819–823 (2012).

109. R. Bansal, N. Ahmad, J.R. Kidwai, "Alloxan-glucose interaction: Effect on incorporation of 14C-leucine into pancreatic islets of rat," *Acta Diabetol. Lat.* **17**(2), 135–143 (1980).

110. J.H. Lee, S.H. Yang, J.M. Oh, M.G. Lee, "Pharmacokinetics of drugs in rats with diabetes mellitus induced by alloxan or streptozocin: Comparison with those in patients with type I diabetes mellitus," *J. Pharm. Pharmacol.* **62**(1), 1–23 (2010).

111. J.R. Garrett, J. Ekström, L.C. Anderson (Eds.), *Glandular Mechanisms of Salivary Secretion (Frontiers of Oral Biology,* vol. 10), Karger, 1998.

112. N. Rakieten, M.L. Rakieten, M.V. Nadkarni, "Studies on the diabetogenic action of streptozotocin," *Cancer Chemother. Rep.* **29**, 91–98 (1963).

113. K. Srinivasan, P. Ramarao, "Animal models in type 2 diabetes research: An overview," *Indian J. Med. Res.* **125**(3), 451–472 (2007).

114. Y. Dekel, Y. Glucksam, I. Elron-Gross, R. Margalit, "Insights into modeling streptozotocin-induced diabetes in ICR mice," *Lab Anim.* **38**(2), 55–60 (2009).

115. K. Hayashi, R. Kojima, M. Ito, "Strain differences in the diabetogenic activity of streptozotocin in mice," *Biol. Pharm. Bull.* **29**(6), 1110–1119 (2006).

116. M. Elsner, M. Tiedge, S. Lenzen, "Mechanism underlying resistance of human pancreatic beta cells against toxicity of streptozotocin and alloxan," *Diabetologia* **46**(12), 1713–1714 (2003).

117. Y. Yang, P. Santamaria, "Lessons on autoimmune diabetes from animal models," *Clin. Sci. (Lond)* **110**(6), 627–639 (2006).

118. T. Hanafusa, J. Miyagawa, H. Nakajima, K. Tomita, M. Kuwajima, Y. Matsuzawa, S. Tarui, "The NOD mouse," *Diabetes Res. Clin. Pract.* **24**(Supplement), S307–S311 (1994).

119. J.P. Mordes, R. Bortell, E.P. Blankenhorn, A.A. Rossini, D.L. Greiner, "Rat models of type 1 diabetes: Genetics, environment, and autoimmunity," *ILAR J.* **45**(3), 278–291 (2004).

120. X. Wang, D.C. DuBois, S. Sukumaran, V. Ayyar, W.J. Jusko, R.R. Almon, "Variability in Zucker diabetic fatty rats: Differences in disease progression in hyperglycemic and normoglycemic animals," *Diabet. Metab. Syndr. Obes. Targ Ther.* **7**, 531–541 (2014).

121. S. Yoshida, H. Tanaka, H. Oshima et al., "AS1907417, a novel GPR119 agonist, as an insulinotropic and beta-cell preservative agent for the treatment of type 2 diabetes," *Biochem. Biophys. Res. Commun.* **400**(4), 745–751 (2010).

122. V.A. Gault, B.D. Kerr, P. Harriott, P.R. Flatt, "Administration of an acylated GLP-1 and GIP preparation provides added beneficial glucose-lowering and insulinotropic actions over single incretins in mice with Type 2 diabetes and obesity," *Clin. Sci.* **121**(3), 107–117 (2011).

123. J.S. Park, S.D. Rhee, N.S. Kang et al., "Anti-diabetic and anti-adipogenic effects of a novel selective 11beta-hydroxysteroid dehydrogenase type 1 inhibitor, 2-(3-benzo yl)-4-hydroxy-1,1-dioxo-2H-1,2-benzothiazine-2-yl-1-phe nylethanone (KR-66344)," *Biochem. Pharmacol.* **81**(8), 1028–1035 (2011).

124. P. Lindstrom, "The physiology of obese-hyperglycemic mice [ob/ob mice]," *Sci. World J.* **7**, 666–685 (2007).

125. T. Bock, B. Pakkenberg, K. Buschard, "Increased islet volume but unchanged islet number in ob/ob mice," *Diabetes* **52**(7), 1716–1722 (2003).

126. F.F. Chehab, M.E. Lim, R. Lu, "Correction of the sterility defect in homozygous obese female mice by treatment with the human recombinant leptin," *Nat. Genet.* **12**(3), 318–320 (1996).

127. H. Chen, O. Charlat, L.A. Tartaglia et al., "Evidence that the diabetes gene encodes the leptin receptor: Identification of a mutation in the leptin receptor gene in DM/DM mice," *Cell* **84**(3), 491–495 (1996).

128. A. Pick, J. Clark, C. Kubstrup, M. Levisetti, W. Pugh, S. Bonner-Weir, K.S. Polonsky, "Role of apoptosis in failure of beta-cell mass compensation for insulin resistance and beta-cell defects in the male Zucker diabetic fatty rat," *Diabetes* **47**(3), 358–364 (1998).

129. T. Shibata, S. Takeuchi, S. Yokota, K. Kakimoto, F. Yonemori, K. Wakitani, "Effects of peroxisome proliferator-activated receptor-alpha and -gamma agonist, JTT-501, on diabetic complications in Zucker diabetic fatty rats," *Br. J. Pharmacol.* **130**(3), 495–504 (2000).

131. W. Feng, C. Zhang, T. Yu, D. Zhu, "Quantitative evaluation of skin disorders in type 1 diabetic mice by *in vivo* optical imaging," *Biomed. Opt. Express* **10**(6), 2996–3008 (2019).

132. M. Kuzuya, S. Satake, S. Ai et al., "Inhibition of angiogenesis on glycated collagen lattices," *Diabetologia* **41**(5), 491–499 (1998).

133. Q.A. Kleter, J.J.M. Damen, M.J. Buijs, J.M. Ten Cate, "The Maillard reaction in demineralized dentin *in vitro*," *Eur. J. Oral Sci.* **105**(3), 278–284 (1997).

134. A. Ioannou, C. Varotsis, "Modifications of hemoglobin and myoglobin by Maillard reaction products (MRPs)," *PLOS ONE* **12**(11), e0188095 (2017).

135. R.D.G. Leslie, D.C. Robbins (eds.), *Diabetes: Clinical Science in Practice*, Cambridge University Press, 1995.

136. A.M. Schmidt, O. Hori, J.X. Chen et al., "Advanced glycation endproducts interacting with their endothelial receptor induce expression of vascular cell adhesion Molecule-1 (VCAM-1) in cultured human endothelial cells and in mice. A potential mechanism for the accelerated vasculopathy of diabetes," *J. Clin. Invest.* **96**(3), 1395–1403 (1995).

137. J. Kinnunen, H.T. Kokkonen, V. Kovanen et al., "Nondestructive fluorescence-based quantification of threose-induced collagen cross-linking in bovine articular cartilage," *J. Biomed. Opt.* **17**(9), 097003 (2012).

138. G.T. Wondrak, M.J. Roberts, D. Cervantes-Laurean, M.K. Jacobson, E.L. Jacobson, "Proteins of the extracellular matrix are sensitizers of photo-oxidative stress in human skin cells," *J. Invest. Dermatol.* **121**(3), 578–586 (2003).

139. M. Yokota, Y. Tokudome, "The effect of glycation on epidermal lipid content, its metabolism and change in barrier function," *Skin Pharmacol. Physiol.* **29**(5), 231–242 (2016).

140. E.L. Hull, M.N. Ediger, A.N.T. Unione, E.K. Deemer, M.L. Stroman, J.W. Baynes, "Noninvasive, optical detection of diabetes: Model studies with porcine skin," *Opt. Express* **12**(19), 4496–4510 (2004).

141. V.M. Monnier, W. Sun, X. Gao et al., "Skin collagen advanced glycation endproducts (AGEs) and the long-term progression of sub-clinical cardiovascular disease in type 1 diabetes," *Cardiovasc. Diabetol.* **14**, 118-1-9 (2015).

142. S. Genuth, W. Sun, P. Cleary et al., "Skin advanced glycation end products glucosepane and methylglyoxal hydroimidazolone are independently associated with long-term microvascular complication progression of type 1 diabetes," *Diabetes* **64**(1), 266–278 (2015).

143. L. Pereira, C.A. Soto, L.D. Santos, P.P. Favero, A.A. Martin, "Confocal Raman spectroscopy as an optical sensor to detect advanced glycation end products of the skin dermis," *Sens. Lett.* **13**(9), 791–801 (2015).

144. B.-M. Kim, J. Eichler, K.M. Reiser, A.M. Rubenchik, L.B. Da Silva, "Collagen structure and nonlinear susceptibility: Effects of heat, glycation, and enzymatic cleavage on second harmonic signal intensity," *Lasers Surg. Med.* **27**(4), 329–335 (2000).

145. B. Gopalkrishnapillai, V. Nadanathangam, N. Karmakar, S. Anand, A. Misra, "Evaluation of autofluorescent property of hemoglobin-advanced glycation end product as a long-term glycemic index of diabetes," *Diabetes* **52**(4), 1041–1046 (2003).

146. Y.-J. Hwang, J. Granelli, J. Lyubovitsky, "Multiphoton optical image guided spectroscopy method for characterization of collagen-based materials modified by glycation," *Anal. Chem.* **83**(1), 200–206 (2011).

147. P.A. Cleary, B.H. Braffett, T. Orchard et al., "Clinical and technical factors associated with skin intrinsic fluorescence in subjects with type 1 diabetes from the diabetes control and complications trial/epidemiology of diabetes interventions and complications study," *Diabetes Technol. Ther.* **15**(6), 466–474 (2013).

148. E. Sugisawa, J. Miura, Y. Iwamoto, Y. Uchigata, "Skin autofluorescence reflects integration of past long-term glycemic control in patients with type 1 diabetes," *Diabetes Care* **36**(8), 2339–2345 (2013).

149. J.D. Maynard, M.N. Ediger, R.D. Johnson, M.R. Robinson, "Determination of a measure of a glycation end-product or disease state using a flexible probe to determine tissue fluorescence of various sites," *Patent Application Publication*, Pub. No.: US 2012/0179010 A1, Pub. Date: Jul. 12, 2012.

150. F.R. Paolillo, V.S. Mattos, A.O. de Oliveira, F.E. Guimarães, V.S. Bagnato, J.C. de Castro Neto, "Noninvasive assessments of skin glycated proteins by fluorescence and Raman techniques in diabetics and nondiabetics," *J. Biophoton.* **12**(1), e201800162 (2019).

151. K. Sangkyu, L. Joonhyung, "Noninvasive apparatus and method for testing glycated hemoglobin, Assignee," *Patent* 9841415, Samsung Electronics Co., ltd., Suwon-si, KR, 2017.

152. M. Mallya, R. Shenoy, G. Kodyalamoole, M. Biswas, J. Karumathil, S. Kamath, "Absorption spectroscopy for the estimation of glycated hemoglobin (HbA1c) for the diagnosis and management of DM: A pilot study," *Photomed. Laser Surg.* **31**(5), 219–224 (2013).

154. O.S. Zhernovaya, V.V. Tuchin, I.V. Meglinski, "Monitoring of blood proteins glycation by refractive index and spectral measurements," *Laser Phys. Lett.* **5**(6), 460–464 (2008).

155. A.S.D.M. Matheus, E.L.S. Clemente, M.D.L.G. Rodrigues, D.C.T. Valença, M.B. Gomes, "Assessment of microvascular endothelial function in type 1 diabetes using laser speckle contrast imaging," *J. Diabet. Compl.* **31**(4), 753–757 (2017).

156. J. Lin, J. Lin, Z. Huang, P. Lu, J. Wang, X. Wang, R. Chen, "Raman spectroscopy of human hemoglobin for diabetes detection," *J. Innov. Opt. Health Sci.* **7**(1), 1350051 (2014).

157. M. Rendell, T. Bergman, G. O'Donnell, E. Drobny, J. Borgos, R. Bonnor, "Microvascular blood flow, volume, and velocity, measured by laser Doppler techniques in IDDM," *Diabetes* **38**(7), 819–824 (1989).

158. J. Yang, G. Zhang, Q. Shang, M. Wu, L. Huang, H. Jiang, "Detecting hemodynamic changes in the foot vessels of diabetic patients by photoacoustic tomography," *J. Biophotonics*, e202000011 (2020).

159. H.M. Raabe, H. Molsen, S.-M. Mlinaric et al., "Biochemical alterations in collagen IV induced by *in vitro* glycation," *Biochem. J.* **319**(3), 699–704 (1996).

160. P.J. Higgins, H.F. Bunn, "Kinetic analysis of the nonenzymatic glycosylation of hemoglobin," *J. Biol. Chem.* **256**(10), 05204–05208 (1981).

161. O.S. Zhernovaya, A.N. Bashkatov, E.A. Genina et al., "Investigation of glucose–hemoglobin interaction by optical coherence tomography," *Proc. SPIE* **6535** (2007).

162. E.I. Galanzha, A.V. Solovieva, V.V. Tuchin, R.K. Wang, S.G. Proskurin, "Application of optical coherence tomography for diagnosis and measurements of glycated hemoglobin," *Proc. SPIE* **5140** (2003).

163. V.V. Tuchin, R.K. Wang, E.I. Galanzha, J.B. Elder, D.M. Zhestkov, "Monitoring of glycated hemoglobin by OCT measurement of refractive index," *Proc. SPIE* **5316** (2004).

164. M. Mernea, A. Ionescu, I. Vasile, C. Nica, G. Stoian, T. Dascalu, D.F. Mihailescu, "*In vitro* human serum albumin glycation monitored by Terahertz spectroscopy," *Opt. Quant. Electron.* **47**(4), 961–973 (2015).

165. G.K. Reddy, "Cross-linking in collagen by nonenzymatic glycation increases the matrix stiffness in rabbit Achilles tendon," *Exp. Diab. Res.* **5**(2), 143–153 (2004).

166. B.E. Sherlock, J.N. Harvestine, D. Mitra et al., "Nondestructive assessment of collagen hydrogel cross-linking using time-resolved autofluorescence imaging," *J. Biomed. Opt.* **23**(3), 036004 (2018).

167. A. Tavousi, M.R. Rakhshani, M.A. Mansouri-Birjandi, "High sensitivity label-free refractometer based biosensor applicable to glycated hemoglobin detection in human blood using all-circular photonic crystal ring resonators," *Opt. Commun.* **429**, 166–174 (2018).

168. A. Goldin, J.A. Beckman, A.M. Schmidt, M.A. Creager, "Advanced glycation end products sparking the development of diabetic vascular injury," *Circulation* **114**(6), 597–605 (2006).

169. V.L. Emanuel, I.Yu. Karyagina, V. Emanuel Yu, "Comparison of method for determining glycosylated hemoglobin, *Laboratornaya Meditsina*" **5**, 98–104 (2002) (in Russian).

170. A.N. Bashkatov, K.V. Berezin, K.N. Dvoretskiy, et al., "Measurement of tissue optical properties in the context of tissue optical clearing," *J. Biomed. Opt.* **23**(9), 091416 (2018).

171. V.V. Tuchin, *Tissue Optics: Light Scattering Methods and Instruments for Medical Diagnostics*, 3rd ed., **PM 254**, SPIE Press, Bellingham, WA, 2015, 988 p.

172. J. Wang, N. Ma, R. Shi, Y. Zhang, T. Yu, D. Zhu, "Sugar-induced skin optical clearing: From molecular dynamics simulation to experimental demonstration," *IEEE J. Sel. Top. Quantum Electron.* **20**(2) (2014).

173. E.A. Genina, A.N. Bashkatov, V.V. Tuchin, "Tissue optical immersion clearing," *Expert Rev. Med. Devices* **7**(6), 825–842 (2010).

174. F.S. Pavone, P.J. Campagnola (eds.), *Second Harmonic Generation Imaging*, CRC Press, Taylor & Francis Group, Boca Raton, FL, 169–189, 2014.

175. F.S. Pavone (ed.), *Laser Imaging and Manipulation in Cell Biology*, Wiley-VCH Verlag GmbH & Co. KGaA, Weinheim, 2010.

176. L. Shi, L.A. Sordillo, A. Rodriguez-Contreras, R. Alfano, "Transmission in near-infrared optical windows for deep brain imaging," *J. Biophoton.* **9**(1–2), 38–43 (2016).

177. D.C. Sordillo, L.A. Sordillo, P.P. Sordillo, R.R. Alfano, "Fourth near-infrared optical window for assessment of bone and other tissues," *Proc. SPIE* **9689**, 96894J (2016).

178. L.A. Sordillo, Y. Pu, S. Pratavieira, Y. Budansky, R.R. Alfano, "Deep optical imaging of tissue using the second and third near-infrared spectral windows," *J. Biomed. Opt.* **19**(5), 056004 (2014).

179. S.Y. Lee, H.J. Park, K. Kim, Y.H. Sohn, S. Jang, Y.K. Park, "Refractive index tomograms and dynamic membrane fluctuations of red blood cells from patients with diabetes mellitus," *Sci. Rep.* **7**(1), 1039 (2017).

180. R.M.A. Henry, P.J. Kostense, A.M.W. Spijkerman et al., "Arterial stiffness increases with deteriorating glucose tolerance status: The Hoorn Study," *Circulation* **107**(16), 2089–2095 (2003).

181. E. Danese, M. Montagnana, A. Nouvenne, G. Lippi, "Advantages and pitfalls of fructosamine and glycated albumin in the diagnosis and treatment of diabetes," *J. Diabetes Sci. Technol.* **9**(2), 169–176 (2015).

182. A.Yuen, C. Laschinger, I. Talior et al., "Methylglyoxalmodified collagen promotes myofibroblast differentiation," *Matrix Biol.* **29**(6), 537–548 (2010).

183. A. Ghazaryan, M. Omar, G.J. Tserevelakis, V. Ntziachristos, "Optoacoustic detection of tissue glycation," *Biomed. Opt. Express* **6**(9), 3149–3156 (2015).

184. M. Gniadecka, O.F. Nielsen, S. Wessel, M. Heidenheim, D.H. Christensen, H.C. Wulf, "Water and protein structure in photoaged and chronically aged skin," *J. Investig. Dermatol.* **111**(6), 1129–1133 (1998).

185. D.K. Tuchina, A.N. Bashkatov, E.A. Genina, V.V. Tuchin, "Biosensor for noninvasive optical monitoring of the pathology of biological tissues," *Patent* RF No. 2633494 MPK A61B 5/05, G01N(21/01), Patent holder: N.G. Chernyshevsky Saratov State University, Application No. 2016102046, 22.01.2016, Bul. No. 29. (2017).

186. L.M. Oliveira, M.I. Carvalho, E.M. Nogueira, V.V. Tuchin, "Diffusion characteristics of ethylene glycol in skeletal muscle," *J. Biomed. Opt.* **20**(5), 051019 (2014).

187. J.M. Andanson, K.L.A. Chan, S.G. Kazarian, "High-throughput spectroscopic imaging applied to permeation through the skin," *Appl. Spectrosc.* **63**(5), 512–517 (2009).

188. M.J. Choi, H.I. Maibach, "Elastic vesicles as topical/trans-dermal drug delivery systems," *Int. J. Cosmet. Sci.* **27**(4), 211–221 (2005).

189. N. Akhtar, "Vesicles: A recently developed novel carrier for enhanced topical drug delivery," *Curr. Drug Deliv.* **11**(1), 87–97 (2014).

190. L.C. Freitas Lima, V. Andrade Braga, M.S. França Silva, J.C. Cruz, S.H. Sousa Santos, M.M. de Oliveira Monteiro, C.M. Balarini, "Adipokines, diabetes and atherosclerosis: An inflammatory association," *Front. Physiol.* **6**, 304 (2015).

191. D. Schweitzer, L. Deutsch, M. Klemm et al., "Fluorescence lifetime imaging ophthalmoscopy in type 2 diabetic patients who have no signs of diabetic retinopathy," *J. Biomed. Opt.* **20**(6), 061106 (2015).

192. C. Ghosh, P. Mukhopadhyay, S. Ghosh, M. Pradhan, "Insulin sensitivity index ($ISI_{0,120}$) potentially linked to carbon isotopes of breath CO_2 for prediabetes and type 2 diabetes," *Sci. Rep.* **5**, 11959 (2015).

193. C.-M. Cheng, Y.-F. Chang, H.-C. Chiang, C.-W. Chang, "Optical coherence tomography for the structural changes detection in aging skin," *Proc. SPIE* **10456** 104565B (2017).

194. D.G. Dyer, J.A. Dunn, S.R. Thorpe, K.E. Bailie, T.J. Lyons, D.R. McCance, J.W. Baynes, "Accumulation of Maillard reaction products in skin collagen in diabetes and aging," *J. Clin. Invest.* **91**(6), 2463–2469 (1993).

195. S. Sakai, K. Kikuchi, J. Satoh, H. Tagami, S. Inoue, "Functional properties of stratum corneum in patients with diabetes mellitus: Similarities to senile xerosis," *Br. J. Dermatol.* **153**(2), 319–323 (2005).

196. H.Y. Park, H.J. Kim, M. Jung, C.H. Chung, R. Hasham, C.S. Park, E.H. Choi, "A long-standing hyperglycaemic condition impairs skin barrier by accelerating skin ageing process," *Clin. Exp. Dermatol.* **20**(12), 969–974 (2011).

197. S. Sakai, Y. Endo, N. Ozawa et al., "Characteristics of the epidermis and stratum corneum of hairless mice with experimentally induced diabetes mellitus," *J. Invest. Dermatol.* **120**(1), 79–85 (2003).

198. K.R. Taylor, A.E. Costanzo, J.M. Jameson, "Dysfunctional γδ T cells contribute to impaired keratinocyte homeostasis in mouse models of obesity," *J. Invest. Dermatol.* **131**(12), 2409–2418 (2011).

199. P. Zakharov, M.S. Talary, I. Kolm, A. Caduff, "Full-field optical coherence tomography for the rapid estimation of epidermal thickness: Study of patients with diabetes mellitus type 1," *Physiol. Meas.* **31**(2), 193–205 (2010).

200. X. Chen, W. Lin, S. Lu et al., "Mechanistic study of endogenous skin lesions in diabetic rats," *Exp. Dermatol.* **19**(12), 1088–1095 (2010).

201. U. Bertheim, A. Engstorm-Laurent, P. Hofer, P. Hallgren, J. Asplund, S. Hellstrom, "Loss of hyaluronan in the basement membrane zone of the skin correlates to the degree of stiff hands in diabetic patients," *Acta Derm.-Vener.* **82**(5), 329–334 (2002).

202. J.G.B. Derraik, M. Rademaker, W.S. Cutfield et al., "Effects of age, gender, BMI, and anatomical site on skin thickness in children and adults with diabetes," *PLOS ONE* **9**(1), e86637 (2014).

203. A.A. Tahrani, W. Zeng, J. Sakher, M.K. Piya, S. Hughes, K. Dubb, M.J. Stevens, "Cutaneous structural and biochemical correlates of foot complications in high-risk diabetes," *Diabetes Care* **35**(9), 1913–1918 (2012).

204. N.C. Avery, A.J. Bailey, "The effects of the Maillard reaction on the physical properties and cell interactions of collagen," *Pathol. Biol.* **54**(7), 387–395 (2006).

205. A.J. Argyropoulos, P. Robichaud, R.M. Balimunkwe, G.J. Fisher, C. Hammerberg, Y. Yan, T. Quan, "Alterations of dermal connective tissue collagen in diabetes: Molecular basis of aged-appearing skin," *PLOS ONE* **11**(4), e0153806 (2016).

206. L.V. Wang (ed.), *Photoacoustic Imaging and Spectroscopy*, CRC Press, Boca Raton, FL, 2009.

29

Tissue optical clearing for in vivo detection and imaging diabetes induced changes in cells, vascular structure, and function

Dongyu Li, Wei Feng, Rui Shi, Valery V. Tuchin, and Dan Zhu

CONTENTS

Introduction

Diabetes mellitus is a systemic disease characterized by hyperglycemia. The high blood glucose caused by abnormal insulin secretion or disordered insulin biological function can lead to an increase in the glucose level in the tissue interstitial fluid of the living body, which contributes to metabolic imbalance, vascular dysfunction, and immune disorders [1–3], and even severe complications to various organs, including brain, eyes, and skin [4–6]. The main complications in diabetics at the advanced stages are related to protein glycation, which is the result of the interaction between glucose molecules and proteins, leading to changes in protein structure and limitations in tissue function [7–9].

All vital organs and tissues, such as the myocardium, eye retina, and brain tissues, are well supplied with blood, i.e. with glucose in diabetes, and thus are primarily exposed to glycation. Since diabetes is a chronic metabolic disease [10], it is necessary to monitor changes in biological structure and dysfunction of organs *in vivo* during the different developmental stages of diabetes. However, it is a big challenge to monitor early structural and functional changes in these tissues

noninvasively. Fortunately, as a superficial organ that is also well supplied by blood and contains a big amount of proteins, skin can be easily investigated optically and used for monitoring early diabetic changes to predict the course of the disease development for the whole body, including vital organs.

It has been known that diabetes can cause various skin vascular changes, including microvascular structure, vascular density and diameter, blood flow and blood oxygen, and vascular permeability [11–14]. In addition, the immune response of monocytes/macrophages (MMs) can be impaired by diabetes [15–17], and the footpad skin of rodents provides a classical physiological site for investigating cellular recruitment and motility during inflammations [18–20]. Various noninvasive *in vivo* optical imaging techniques have been proposed to monitor the microenvironment of skin, but the high scattering of skin greatly limits optical imaging performance, leading to poor imaging resolution and imaging depth. Therefore, some methods for making surgery-based skin imaging windows were established to improve the performance of optically imaging cutaneous blood vessels or cells. However, the surgery leads to not only proinflammatory stimulus but also impact on vascular structure and function.

DOI: 10.1201/9781003025252-33

Apart from skin, monitoring changes in cerebrovascular dysfunction with high resolution *in vivo* during different developmental stages of diabetes is also crucial because diabetes frequently induces cerebral ischemic injury, which mainly manifests as small vessel diseases and increases the risk of stroke, cognitive decline, and dementia [21–23]. However, in order to overcome the scattering of skull above the cortex, traditional skull windows such as open skull glass windows and thinned skull windows are usually performed before observation, which may change intracranial pressure and even cause bleeding or inflammation.

Fortunately, the novel *in vivo* tissue optical clearing technique could reduce the scattering of skin and skull by topical application of chemical agents, providing a skin window and a skull window without causing vascular changes or immune response, which has been proved to improve the imaging performance of various optical systems, including optical coherence tomography (OCT) [24–26], photoacoustic microscopy [27–29], laser speckle contrast imaging (LSCI) [30–36], hyperspectral imaging (HSI) [37, 38], confocal microscopy [39, 40], and two-photon microscopy [41]. Thus, optical clearing windows are suitable for monitoring diabetes-induced changes in the cutaneous and cerebrovascular microenvironment.

In this chapter, we firstly introduce how skin structure and microenvironment are impacted by diabetes and how tissue optical clearing techniques can be used for such investigations. Then, footpad skin optical clearing window–assisted visualization of immune response dysfunction is introduced. Finally, we state that the skull optical clearing technique could be utilized to *in vivo* monitor cortical vascular response dysfunction with the development of diabetes.

Maladjustment of skin structure induced by diabetes

Since high glucose level in blood results in the elevation of its level in interstitial fluid, glycation can occur not only with blood proteins, but also with proteins of tissues. Indeed, proteins are the main components of many tissues, and their glycation causes the change of tissue structure significantly [8, 42]. Since skin is well supplied by blood and contains a large amount of proteins, diabetes can affect the skin structure and components [43, 44]. In order to quantitatively evaluate skin disorders, various optical methods have been utilized, where skin optical clearing technique provides an effective window and good enhancement.

Feng et al. [45] used two-photon imaging to realize label-free *in vivo* 3D visualization of skin, and quantitatively evaluated the changes in skin structure of T1D mice. In their study, two-photon automatic fluorescence (TPAF) and SHG signals of skin from epidermis layer to dermis layer were imaged in normal and T1D mice. As shown in Figure 29.1, compared to the normal mice, the integrated TPAF intensity of the T1D mice was significantly increased, which could be related to the oxidative stress-induced increase of NAD(P)H [46–48], while the integrated SHG intensity was significantly decreased, which indicated that the T1D may cause the decrease of skin collagen.

From the SHG signal, it can be found that the dermal collagen structure of the normal mice was relatively ordered and fiber-rich, while it obviously changed in T1D mice. Feng et al. [45] analyzed the proportion of collagen fibers. The result demonstrated that the collagen fibers are almost all distributed in a certain direction in non-T1D mice, while T1D could lead to changes in the degree of disorder in fibril orientation (Figure 29.2).

In addition, two statistics, energy, and homogeneity, were calculated to further describe the directional consistency of collagens at different dermal depths in mice skin [49, 50]. Energy reflects the uniformity of grayscale distribution and the regular pattern of texture. Homogeneity can be used for evaluating the degree of local changes in image texture. The larger the value of homogeneity, the less change in image texture in different regions, and the more homogeneous the local image is. As shown in Figure 29.3, the energy values of the T1D group were lower than that of the normal group, suggesting there were some uniformities of gray distribution in SHG images caused by the T1D. The green color represents the autofluorescence of epidermis, and the blue color represents the SHG signals of the collagenous fibers in the dermis.

In summary, diabetes can lead to changes in skin structure, including water formation and arrangement of collagens. Such influence is certainly related to the development of the disease, which is worth tracking. With the tissue optical clearing technique, it is possible to evaluate the content of different forms of water in skin, providing evidence to predict the severity of diabetes, which is of importance to biological and clinical research.

Skin optical clearing for imaging diabetes-induced immune response dysfunction

The footpad of rodents provides a classical physiological site for monitoring cellular behavior during inflammation that can be impaired by diabetes, but is limited to the superficial dermis due to the strong scattering of footpad. Shi et al. [51] combined an *in vivo* footpad skin optical clearing technique with confocal microscopy to visualize the immune dysfunction of MMs in deeper tissue around inflammatory foci with the development of type 1 diabetic (T1D) mice. In their study, fructose, ethanol, dimethylsulfoxide (DMSO), polyethylene-glycol-400 (PEG-400), and thiazone were used together on footpad skin OCA (FSOCA), and a delayed-type hypersensitivity (DTH) model was elicited on the footpad of T1D mice with green fluorescent protein expression in MMs [51].

DTH mice model

Mice were sensitized by subcutaneously injecting 50 μg of ovalbumin (OVA) emulsified in 20 μL of complete Freund's adjuvant (CFA) into each side of the tail base. About 7 days later, DTH reaction was elicited by challenging the mice with 30 μL of 2% AOVA (heat-aggregated ovalbumin) via subcutaneous injection in the center of the right hind footpad, as

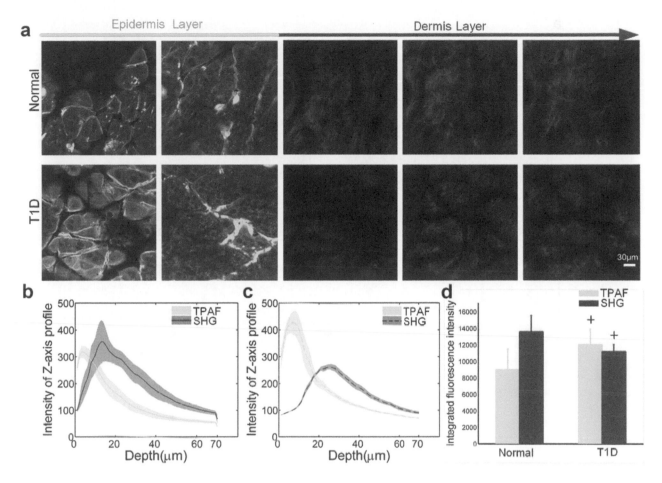

FIGURE 29.1 Two-photon imaging of skin in normal and T1D mice. (a) The image of each layer for the dorsal skin in normal and T1D mice from the skin surface to deep layer of the dermis. (b) and (c) were the fluorescence intensity profiles of normal and T1D mice, respectively (The green and blue curves are mean values and represent the autofluoresence of the epidermis (green) and the SHG signals of the collagenous fibers in the dermis (blue), and the shadows represent standard error, n = 6). (d) The integrated fluorescence intensity profiles of normal and T1D mice (error bars represent the standard error, n = 6). + p < 0.05 compared with the normal mice. Reprinted from Ref [45].

shown in Figure 29.4a. The footpad incurred 23.6% incrassation at AOVA-4 hours, which gradually increased to 42.5% at AOVA-24 hours; reached its maximum at AOVA-48 hours, with 73.9% incrassation; and then decreased to 52.1% incrassation at AOVA-72 hours (Figure 29.4b–c). In addition, the histological results demonstrated that AOVA-elicited DTH reaction is characterized by a gradually increasing cellular infiltration, with about 2.4-fold and 4.5-fold relative increases at AOVA-48 hours and AOVA-72 hours, respectively (Figure 29.4d–e).

Cellular recruitment during DTH with the development of T1D

Confocal microscopy was then used to monitor the MM recruiting through the switchable footpad skin optical clearing window at various time points after AOVA injection in T1D mice and non-T1D mice, and it was found the distributional density of MMs became increasingly larger with time after AOVA injection, as well as with the development of T1D for the same time point after AOVA injection (Figure 29.5a–d). however, compared to non-T1D mice, the number of MMs recruiting to the imaging area gradually increases in connection with progressive T1D develops (Figure 29.5e–f). Furthermore, Figure 29.5g

shows that, with a large imaging area, it was observed that progressive T1D leads to the much stronger MM recruitment, and even the formation of some dense cellular clusters, seen as the accumulated clouds marked with white arrows. Such clusters appeared earlier with the development of diabetes.

In addition, it was observed in the non-T1D mice that a few monocytes patrolled in blood vessels disregard the direction of blood flow at AOVA-4 hours, and the majority of GFP cells are outside blood vessels (Figure 29.6a). However, monocyte patrolling behaviors decreased at AOVA-24 hours, demonstrated by the sessile cells in vasculature (Figure 29.6b).

Motility trajectory and displacement during DTH with the development of T1D

Further, MM motility at different depths in skin during DTH as progressive T1D develops was monitored by dynamic time-lapse confocal microscopy with the help of a switchable footpad skin optical clearing window. To clarify whether there are differences between the motility of MMs at different depths in tissue around the inflammatory foci, time-lapse dynamic imaging at depth of the superficial dermis and the deeper dermis was also performed at each indicated time-point for 16 min

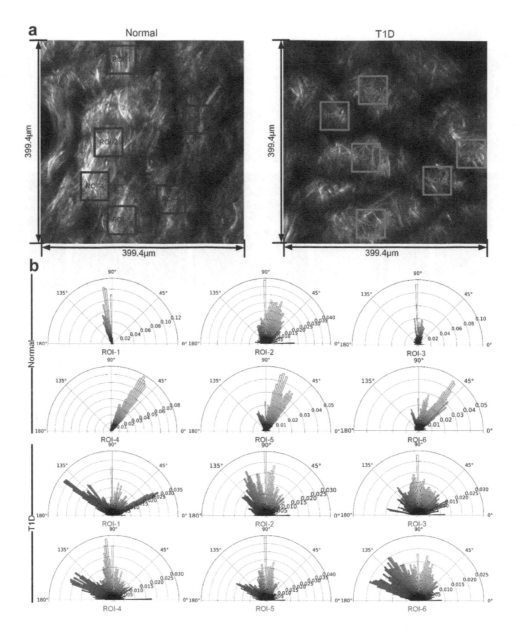

FIGURE 29.2 The disorder degree of longitudinal orientated collagen of skin. (a) Typical maximum intensity projection maps (from 20 μm to 60 μm) of dermal SHG collagen in normal and T1D mice. (b) Quantitative evaluation of the skin collagen orientation from 0° to 180°corresponding to the blue and red regions of interest as indicated in (a) for normal mice and T1D mice. (The polar axis represents the proportion of collagen in each direction.). Reprinted from reference [45].

with 1-min intervals. As shown in Figure 29.7, the superficial dermis is the epidermis/dermis bounder in the depth range of 35–45 μm, and the deeper dermis is in the depth range of 85–95 μm.

MM motion trajectories at depths of superficial dermis and the deeper dermis in tissue during DTH on mice with development of T1D were tracked based on the images. The results indicate that the number of MMs that could move during DTH decreases gradually as progressive T1D develops. In addition, MMs at depth of the deeper dermis clearly exhibit a much larger migration displacement than those at superficial dermis at early stages (AOVA-4 hours, AOVA-24 hours) of DTH on non-T1D mice, while this phenomenon becomes obscure as progressive T1D develops, which attenuates the migration displacement of MMs at both depths.

The motility parameters of cells during DTH as progressive T1D develops

Finally, several motility parameters of cells were measured from the images during DTH as progressive T1D develops. The results show that the mean velocity and mean motility coefficient of MMs both in superficial dermis and in deeper dermis decrease with time after injection of AOVA. These two parameters of deeper MMs are both higher than those of superficial MMs at early stages (AOVA-4 hours, AOVA-24 hours) of DTH, while the parameters of deeper MMs decrease much faster with time and the difference is inconspicuous at an advanced stage of DTH. However, this phenomenon becomes obscure as progressive T1D develops. As for arrest coefficient of MMs, it exhibits a gradual increasing tendency

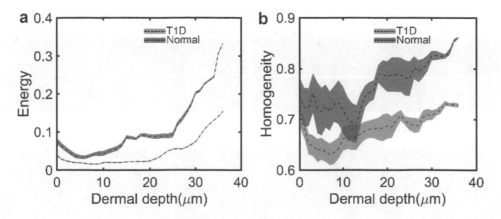

FIGURE 29.3 The texture analysis of skin SHG collagen images. The energy (a) and homogeneity (b) of SHG images at different dermal depths in normal and T1D mice. (The dotted lines represent the average values of gray-level co-occurrence matrix of SHG images in four directions. The red and blue shadows represent standard deviations.) Reprinted from reference [45].

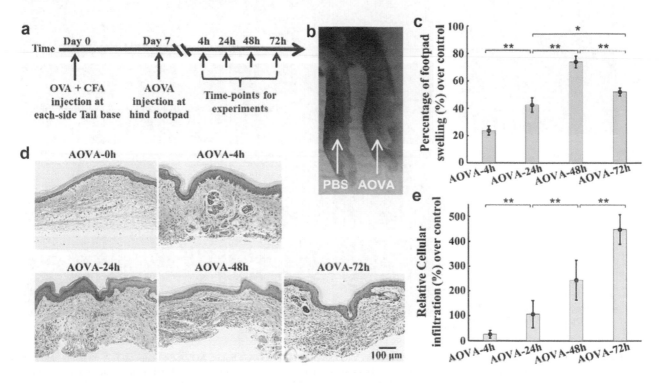

FIGURE 29.4 Footpad swelling and histological examination. (a) Time flow schematic for model establishment of DTH and time points selected for experiments. (b) Photograph of footpad at 24-hour post-injection of AOVA (indicated with AOVA) compared to that of PBS (indicated with PBS).(c) Percentage of footpad swelling (d) Histological examination at AOVA-0 hour, AOVA-4 hours, AOVA-24 hours, AOVA-48 hours, and AOVA-72 hours. (e) The relative changes of cellular infiltration per unit area of tissue section. Mean values ± SEM (data were pooled from 6 mice per group). * and ** refer to P < .05 and P < .01, respectively. Reprinted from reference [51].

at both depths as progressive T1D develops. Further comparisons demonstrate that the arrest coefficient is about 0.2–0.3 for non-T1D. As for T1D-2 weeks, the parameter is about 0.5, and this changing tendency is further exacerbated as progressive T1D develops, which indicates that most MMs are kept in an arresting status even at early stages of DTH.

In conclusion, with the switchable footpad skin optical clearing window, it was visualized that there is indeed different intensity in immune response at different depths in tissue around the inflammatory foci, and this difference becomes gradually indistinct as progressive T1D develops.

Visualization of T1D-induced changes in skin microvascular function with the assistance of skin optical clearing technique

Diabetes not only leads to structural changes in skin, but also causes cutaneous vascular dysfunction, including changes in permeability and blood flow/blood oxygen response. The skin optical clearing technique provides a window to dynamically monitor such changes with the development of diabetes.

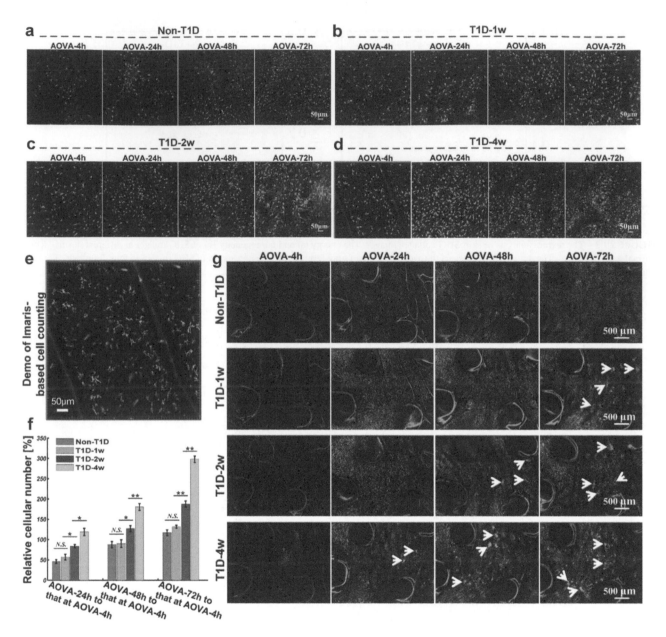

FIGURE 29.5 Cell recruitment single-imaging field demonstrates MM recruitment at AOVA-4 hours, AOVA-24 hours, AOVA-48 hours, and AOVA-72 hours on non-T1D (a), T1D-1 week (b), T1D-2 weeks (c) and T1D-4 weeks (d) mice. (e) Demo of Imaris-based cell counting, and (f) Relative cellular number in single-imaging field at AOVA-24 hours, AOVA-48 hours and AOVA-72 hours to that at AOVA-4 hours. (g) Large-scale imaging at each indicated time points demonstrates the shifting ahead of time for forming the dense cellular accumulated clusters as progressive T1D develops. Mean values ± SEM (data were pooled from six mice per group). N.S., *, and ** refer to no significant difference, P < .05 and P < .01, respectively. Reprinted from reference [51].

T1D-induced enhancement of cutaneous vascular permeability

Feng et al. [45] monitored skin microvascular permeability *in vivo* in T1D mice. In their study, spectral imaging was used to quantitatively evaluate the Evans Blue dye (EBd) concentration outside blood vessels in skin. In order to improve the imaging quality, skin OCA [39] was topically applied before imaging. After establishing the optical clearing skin window, spectral images before and after intravenous injection of EBd were acquired, and 620-nm reflectance imaging was used to measure the concentration of EBd. With the help of the optical clearing skin window, the cutaneous microvessels

can be monitored clearly by spectral imaging. As shown in Figure 29.8, for the normal mice, distribution of the EBd concentration changed little over time, while for the T1D mice, the vascular and extravascular EBd concentration started to significantly increase at 10 min, indicating that T1D caused an increase in cutaneous microvascular permeability.

T1D-mediated dysfunction of cutaneous vascular response to stimulations

In their further study, Feng et al. [52] visualized T1D-induced cutaneous vascular response dysfunction through *in vivo* skin

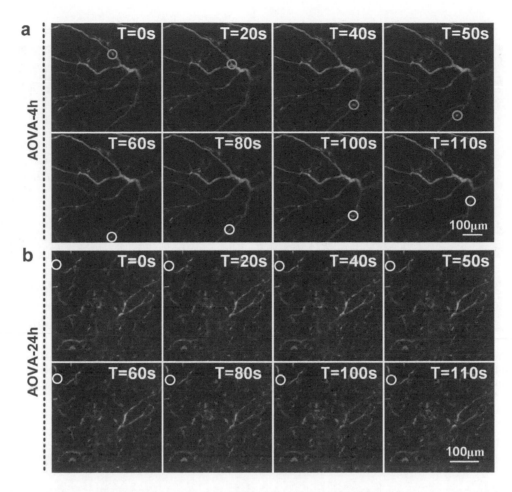

FIGURE 29.6 Time-lapse dynamic imaging demonstrating the location relationship between MMs and vasculature. The time-lapse distribution of MMs (green) and vasculature (red) at AOVA-4 hours (a) and that at AOVA-24 hours (b) within 2 min. By means of labeling, the vasculature of non-T1D mice with 2% tetramethylrhodamine-conjugated dextran (70 kDa, Invitrogen), which was excited by a 561-nm laser and the fluorescence emission was recorded using a 570–620 bandpass filter, which can be references for reflecting the location relationship of MMs and vasculature. Circles indicate the changes of the location of monocytes in blood vessels with time. It not only indicates a few monocytes patrol in blood vessels disregard for the direction of blood flow at AOVA-4 hours, but also demonstrates that the majority of GFP cells are out of blood vessels (a). However, monocyte patrolling behaviors decreased at AOVA-24 hours, as demonstrated by the sessile cells in vasculature. Reprinted from reference [51].

optical clearing window. After topical treatment of skin OCA (a mixture of PEG, thiazone, and sucrose), a dual mode optical imaging system with a combination of LSCI and HSI was used to monitor blood flow and blood oxygen responses in dorsal dermis vasculature to intravenous noradrenaline (NE) injection.

In their study, dynamic responses of blood flow to NE were monitored with the development of T1D. As shown in 29.9a, the skin vascular blood flow distribution is nearly invisible for intact skin, while after establishment of the optical clearing skin window, the vascular distribution can be clearly observed. For non-T1D and T1D-1 week (T1D-1w), the blood flow evidently decreases in the first 5 or 10 min after injecting NE (as indicated by the white arrows), and then recovers to its initial level. However, for T1D-2w and T1D-4w, NE induces an evident decrease in blood flow, evening leads to little blood flow perfusion in some vessels. and the blood flow does not recover to the initial level after injection of NE. In addition, quantitative analyzation demonstrates that the venous blood flow velocity of T1D-1w and T1D-2w decreases more than non-T1D mice, while for

T1D-4w, venous blood flow perfusion almost disappears and cannot be analyzed after NE injection. Furthermore, interruption to blood flow response is also observed in arteries as T1D develops (Figure 29.9b–e).

Similarly, dysfunction of dynamic response of blood oxygen with the development of T1D was also investigated through skin optical clearing windows. As shown in Figure 29.10a, the optical clearing skin window also allows much clearer distribution maps of skin vascular blood oxygen to be obtained, compared to images from intact skin. For the non-T1D and T1D-1w, the blood oxygen decreases after injection of NE at the first several minutes and then almost recovers. For T1D-2w and T1D-4w, blood oxygen saturation of some veins decreases and keeps at a very low level after injection of NE. Figure 29.10b–e shows that After injection of NE, the arterious blood oxygen saturation of non-T1D, T1D-1w, T1D-2w, and T1D-4w decreases by 27.6% ± 7.3%, 30.6% ± 2.3%, 10.9% ± 1%, and 3.9% ± 0.4%, respectively, indicating that NE-induced the response of arterious blood oxygen becomes weaker for T1D-2w and T1D-4w than for non-T1D and T1D-1w. Moreover, for T1D-2w and T1D-4w,

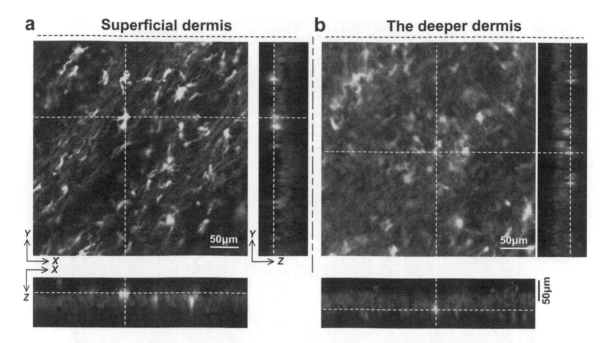

FIGURE 29.7 Three-dimensional intravital imaging indicating the imaging volume we focused on. The distribution of MMs (green) and dermis-derived the second-harmonic generation (SHG) (blue) at depth of the superficial dermis (35–45 μm) (A) and that of the deeper dermis (85–95 μm) (B). SHG signals were excited by an 840-nm laser, and the fluorescence emission was recorded using a 410–430 bandpass filter, which can be referenced to locate the imaging depth we focused on. Reprinted from reference [51].

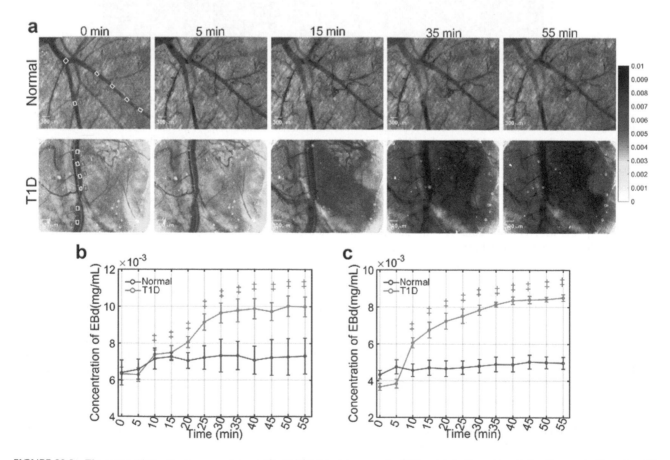

FIGURE 29.8 The permeability of cutaneous microvessels. (a) The skin microvascular EBd concentration maps obtained by spectral imaging with the help of *in vivo* skin optical clearing method (The colorbar represents the EBd concentration). The time-dependent changes of EBd concentration at vascular positions (b) and extravascular positions (c) (The vascular positions and extravascular positions were white and red rectangular areas as indicated in (a), respectively. Mean ± standard deviations). + + p < 0.01 compared with EBd concentration at 5 min. Reprinted from reference [45].

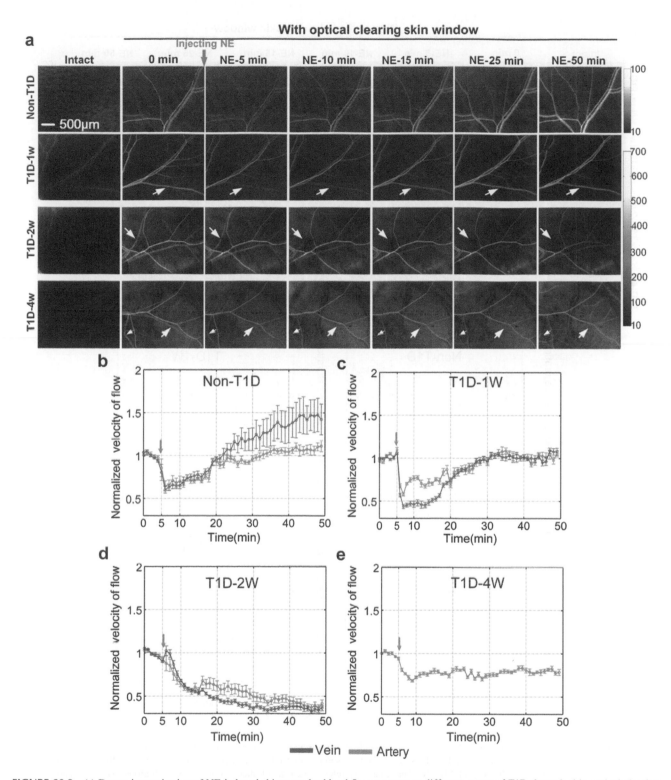

FIGURE 29.9 (a) Dynamic monitoring of NE-induced skin vascular blood flow response at different stages of T1D through skin optical clearing window. (b–e) The time-lapse blood flow dynamic responses in artery (red line) and vein (blue line) at different stages of T1D: (b) non-T1D, (c) T1D-1w, (d) T1D-2w, and (e) T1D-4w. The red arrows refer to the time of injection (n = 8 for each group, mean ± standard error). Reprinted from reference [52].

the blood oxygen saturation of the vein decreases without recovery after injection of NE.

Such results indicate that diabetes can lead to abnormal blood flow and blood oxygen response in cutaneous vascular structures, and the development of diabetes will exacerbate the disfunction.

In summary, diabetes affects the function of blood vessels in the skin. It can not only cause vascular permeability increase, but can also lead to abnormal responses of blood flow and blood oxygen to drugs. With the *in vivo* skin optical clearing window, it is possible to monitor such changes in the long term as diabetes develops.

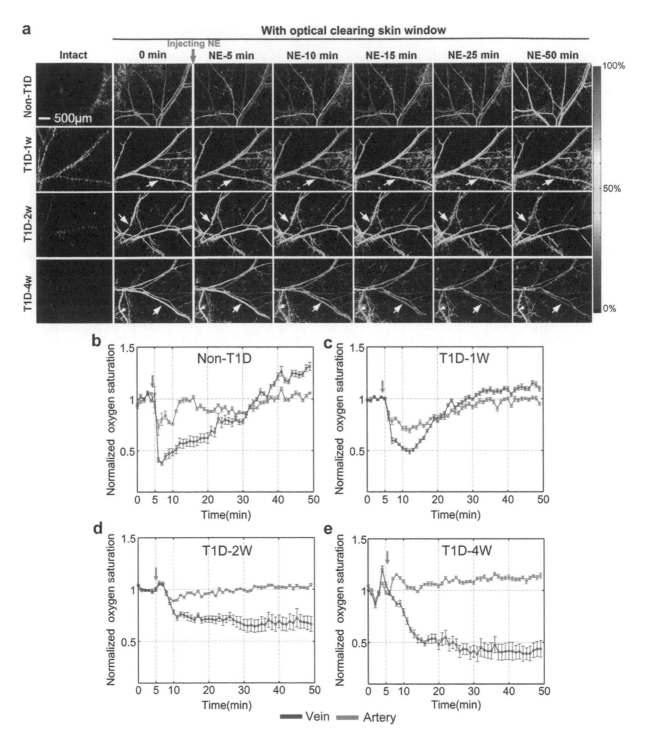

FIGURE 29.10   (a) Dynamic monitoring of NE-induced corresponding skin vascular blood oxygen response at different stages of T1D through skin optical clearing window. (b–e) The time-lapse blood oxygen dynamic responses in artery (red line) and vein (blue line) at different stages of T1D: (b) non-T1D, (c) T1D-1w, (d) T1D-2w, and (e) T1D-4w. The red arrows refer to the time of injection (n = 8 for each group, mean ± standard error). Reprinted from reference [52].

Comparison of diabetes-mediated cortical and cutaneous vascular response dysfunctions through in vivo optical clearing windows

Diabetes can lead not only to cutaneous vascular dysfunction, but also to cerebral vascular dysfunction. Feng et al. [53] used an optical clearing skin window and skull window to

investigate vascular function changes in both skin and cortex with the development of T1D, and further analyzed their distinction and connection.

In their study, a mixture of PEG-400, thiazone, and sucrose was used as skin OCA, and a urea-based skull OCA [38] was used to establish an optical clearing skull window. Using a lab-built dual mode system with a combination of LSCI and HSI, blood flow and blood oxygen responses in cutaneous and cortical vasculature were monitored after injection of nitro-prusside (SNP, a common vascular stimulant) or acetyl choline (Ach, a common dilator in most vascular beds) as progressive T1D develops.

In vivo optical clearing window for dynamic monitoring of blood flow and blood oxygen

As shown in Figure 29.11, it is hard to distinguish blood vessels due to the strong scattering of turbid skull and skin. However, with the help of optical clearing skin and skull windows, clear vascular blood flow and oxygen maps can be obtained.

Cerebral microvascular dysfunction as T1D develops

Then, SNP- and Ach-induced cerebral arteriovenous functional responses were dynamically monitored at different stages of diabetes development, and the cerebral arteriovenous △Flow and △SO₂, recorded before and after the injection

of SNP and Ach, were quantitatively analyzed. As shown in Figure 29.12, in the non-T1D mice, cortical microvascular △Flow quickly increased after an initial SNP-induced decrease, and the shapes of the time-lapse curves of △SO₂ were slightly similar to those observed for △Flow. However, in the T1D mice, the cerebral arteriovenous △Flow increased directly, and the initial decrease almost disappeared after the injection of SNP in the T1D mice at 1, 2, 3, and 4 weeks. In the 3- and 4-week T1D mice, the range of blood flow response induced by SNP became dramatically weak. In addition, in the T1D 1-, 2-, 3-, and 4-week mice, the increasing SNP-induced cerebral arteriovenous range of △SO2 was larger than that observed in the non-T1D mice

Figure 29.13 shows the Ach-induced cerebral arteriovenous functional responses. Ach caused △Flow and △SO₂ to decrease in nondiabetic mice. In the T1D mice, the diminished blood flow response after injection of Ach was observed in the T1D mice, and this change was similar to that in the SNP response. Furthermore, the Ach-induced decrease in △SO₂ occurred only in veins, while the changes in △SO2 in artery

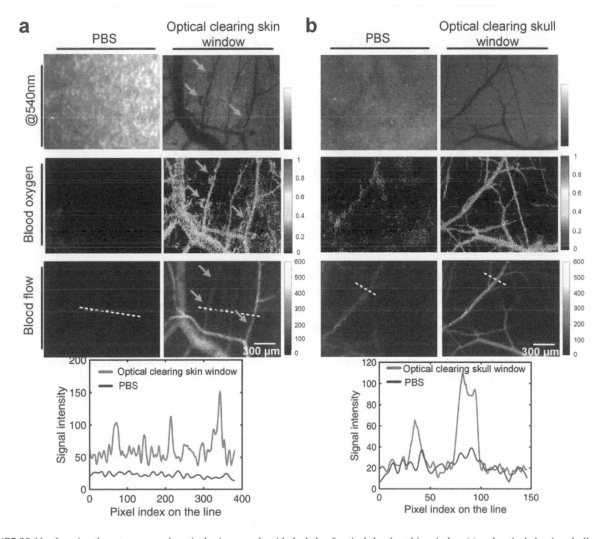

FIGURE 29.11 Imaging the cutaneous and cortical microvessels with the help of optical clearing skin window (a) and optical clearing skull window (b). First row: the images at 540 nm; second row: blood oxygen maps obtained by HSI; third row: blood flow maps obtained by LSCI; fourth row: the signal intensity along the white dashed lines in the blood flow images. First column: cutaneous imaging through skin treated with PBS; second column: imaging cutaneous vessels through optical clearing skin window; third column: cortical imaging through skull treated with PBS; fourth column: imaging cortical vessels through optical clearing skull window. (The green arrows in (a) indicate the capillaries of the skin). Reprinted from reference [53].

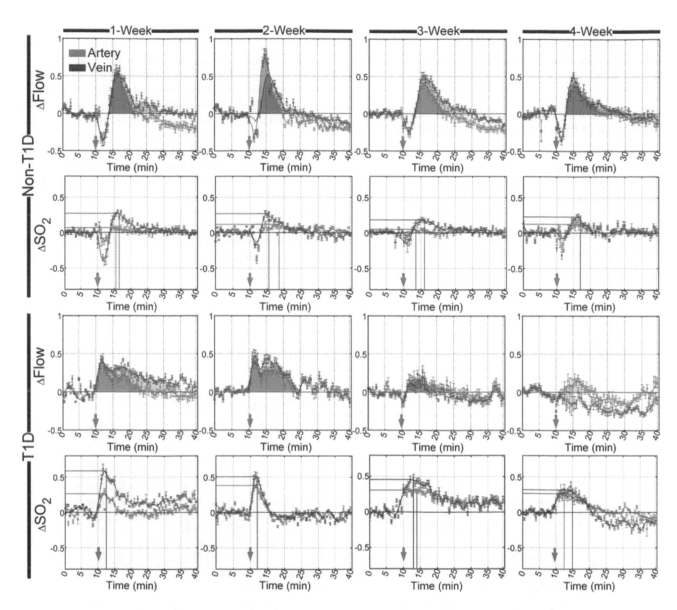

FIGURE 29.12 Time-lapse data showing the relative changes in cerebral vascular blood flow and corresponding blood oxygen saturation that occurred in arteries (red) and veins (blue) after the injection of SNP in different stages of T1D. The red arrows indicate the time of the injection. The shadowed areas indicate the areas under curves of relative changes in blood flow (red and blue represent arteries and veins, respectively). The lines perpendicular to the *x*- and *y*- axes represent the position of the maximum value of relative changes in blood oxygen saturation (red and blue lines indicate arteries and veins, respectively) (n=8, mean ± standard error) Reprinted from reference [53].

almost disappeared in the T1D 1-, 2-, 3-, and 4-week mice. Such results indicate that T1D can cause abnormal changes in cerebrovascular blood flow and blood oxygen responses at the early stage of T1D. In addition, the blood flow and the blood oxygen responses to SNP and Ach in cerebral microvasculature are observed to be similar, indicating there is a close correlation between cerebrovascular blood flow and blood oxygen, confirming previous reports [54].

Quantitative comparison of cerebral and cutaneous microvascular functional response during the development of T1D

Similarly, quantitative analysis of cutaneous arteriovenous △Flow and the corresponding △SO₂ before and after the

injection of SNP and Ach was performed based on the dual mode images at different stages of diabetic development. Furthermore, the sums of cerebral and cutaneous arteriovenous areas under/over the curves of △Flow and the sums of cerebral and cutaneous arteriovenous peak values of △SO₂ were used for statistical analysis. As shown in Figure 29.14a, following SNP injection, there was no significant change in cerebral or cutaneous vascular responses between non-T1D and T1D mice after 1 week and 2 weeks, while the sums of cerebral and cutaneous arteriovenous areas of △Flow in the 3-week and 4-week T1D groups were significantly lower than those in the non-T1D groups. As for vascular blood oxygen response to SNP, the arteriovenous peak values of cutaneous microvascular △SO₂ for T1D groups significantly decreased compared to the non-T1D groups from 1 to 4 weeks, and the

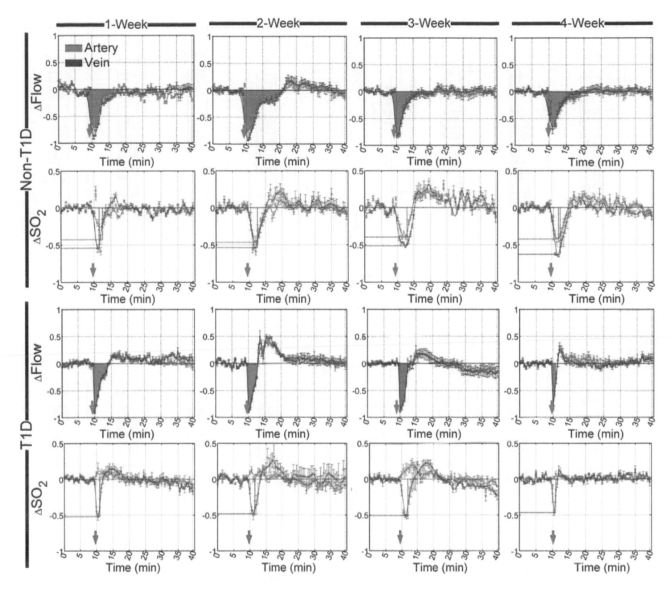

FIGURE 29.13 Time-lapse data showing the relative changes in cerebral vascular blood flow and corresponding blood oxygen saturation that occurred in arteries (red) and veins (blue) after the injection of Ach in different stages of T1D. The red arrows refer to the time of injection. The shadowed areas indicate the areas under curves of relative changes in blood flow (red and blue represent arteries and veins, respectively). The lines perpendicular to the *x*- and *y*- axes represent the position of the maximum value of relative changes in blood oxygen saturation (red and blue lines indicate arteries and veins, respectively) (n=6, mean ± standard error). Reprinted from reference [53].

sums of the arteriovenous peak values of $\triangle SO_2$ for cerebral blood oxygen responses in the T1D groups significantly increased.

Figure 29.14b shows that, after injection of Ach, there was a significant difference in the cerebral blood flow response between non-T1D and T1D mice from 1 week to 4 weeks, whereas in the 3-week and 4-week groups, the sums of cutaneous arteriovenous areas of $\triangle$Flow were significantly lower in the T1D groups than in the non-T1D groups. The arteriovenous peak values of cutaneous microvascular $\triangle SO_2$ for T1D groups significantly decreased compared to the non-T1D groups from 1 to 4 weeks, while the sums of the arteriovenous peak values of $\triangle SO2$ for cerebral blood oxygen responses in the T1D groups were significantly lower than those found in the non-T1D groups from 1 to 4 weeks.

Insulin-mediated cerebral and cutaneous function improvement in T1D mice

Furthermore, in their study, the 2-week T1D mice were managed with insulin to regulate blood sugar level for one week, and SNP/ACh-induced changes in cerebral and cutaneous vascular blood flow and blood oxygen responses were monitored to investigate the relationship between the abnormal vascular function and hyperglycemia. The result shows that after insulin treatment for 1 week, the SNP- and Ach-induced cerebral and cutaneous vascular responses of 2-week T1D mice did not change in a manner similar to that observed in the 3-week T1D groups, indicating that that blood sugar control with insulin plays an important role in restraining the vascular dysfunction caused by T1D.

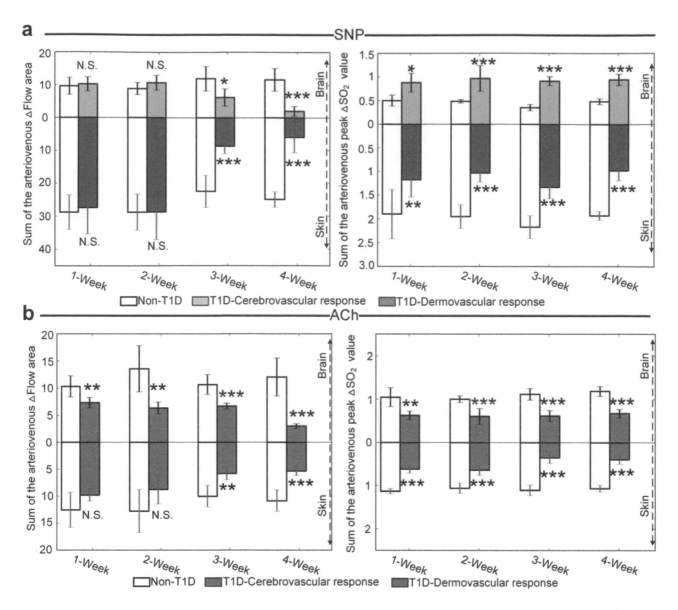

FIGURE 29.14 Quantitative evaluation of cerebral and cutaneous microvascular function responses. (a) Sum of the arteriovenous area of $\triangle$Flow and sum of the arteriovenous peak values of $\triangle SO_2$ for SNP-induced microvascular response at different T1D stages (n = 8). (b) Sum of the arteriovenous area of $\triangle$Flow and sum of the arteriovenous peak values at decline of $\triangle SO_2$ for Ach-induced microvascular responses at different T1D stages (n=6). N.S., *, **, and *** indicate not significant, $p < 0.05$, $p < 0.01$, and $p < 0.001$, respectively, versus non-T1D (mean $\pm$ standard deviation). Reprinted from reference [53].

In summary, *in vivo* monitoring of SNP- and Ach-induced cerebral and cutaneous microvascular responses during the development of T1D can be performed with the assistance of *in vivo* skull and skin optical clearing techniques. It was found that in the early stage of diabetes (1 week), abnormal changes occurred in cerebral vascular blood flow and blood oxygen responses as well as in cutaneous vascular blood oxygen. With the development of T1D, the cutaneous vascular blood flow response also became abnormal. Such results indicate that cutaneous vascular blood oxygen response was more sensitive than blood flow to the reactivity test, and therefore has the potential to serve as a good assessment indicator for revealing cerebrovascular dysfunction in the early stage of diabetes.

Summary

The glycosylation of tissue proteins brought about by chronic hyperglycemia can change the normal structure of organs and thus affect their function. The tissue optical clearing technique has been used to study diabetes-mediated biological structural and functional changes.

Skin, as the largest and most superficial organ that can be affected by diabetes, has the advantage of being easy to monitor to predict the course of the disease development for the whole body, including vital organs. The skin structure, including especially the content and distribution of collagen, can be significantly influenced by T1D. With the assistance of the skin optical clearing technique, the abnormal immune response of MMs in footpad caused by T1D was also observed. In addition, through a skin optical clearing window, vascular permeability, blood flow response, blood oxygen response, and inflammatory response can also be visualized *in vivo*. It was found that T1D caused not only an increase in cutaneous microvascular permeability, but also abnormal changes in cutaneous microvascular blood flow and blood oxygen response to several chemicals, including NE, SNP, and Ach.

Furthermore, novel *in vivo* skull optical clearing technique provides a craniotomy-free skull window for diabetes-mediated cortical vascular dysfunction visualization *in vivo*. It was observed that the vascular response to stimulations in the cortex became abnormal as diabetes developed. Therefore, tissue optical clearing can serve to investigate the connection and difference between cortical and cutaneous microvascular changes with disease development.

It is worth mentioning that, other than skin and cortex, the tissue optical clearing technique has been used for high-resolution three-dimensional (3D) kidney imaging, which provides every detail of diabetes-induced glomeruli structural change. Zhu et al. [55] developed an m-xylylenediamine-based aqueous clearing system (MACS), which was a rapid and effective optical clearing method and was capable of visualizing large tissues. After optical clearing, the glomerulus trees and branches in healthy kidneys were clearly visible with light-sheet fluorescence microscopy and the fine structure could be visualized with confocal microscopy. It was found that kidneys of diabetic models at 4 weeks presented a glomerulus distribution different from that of normal kidneys, and the capillary tufts of the glomerulus in the 4-week diabetic glomerulus were clustered and seriously damaged.

In conclusion, the combination of modern optical imaging techniques and the tissue optical clearing technique allows the visualization of physiological changes with a high resolution, and monitoring of biological dysfunctions with minimal invasiveness in the pathological state, providing a powerful tool for various biomedical investigations.

Acknowledgments

This chapter was supported by the National Science Foundation of China (Grant Nos. 61860206009, 81870934) and China Postdoctoral Science Foundation funded project (Grant Nos. BX20190131, 2019M662633).

REFERENCES

1. N.E. Cameron, and M.A. Cotter, "The relationship of vascular changes to metabolic factors in diabetes-mellitus and their role in the development of peripheral-nerve complications," *Diabetes/Metabolism Reviews* **10**(3), 189–224 (1994).

2. D. Roosterman, T. Goerge, S.W. Schneider, N.W. Bunnett, and M. Steinhoff, "Neuronal control of skin function: The skin as a neuroimmunoendocrine organ," *Physiological Reviews* **86**(4), 1309–1379 (2006).

3. N. Vigneshwaran, G. Bijukumar, N. Karmakar, S. Anand, and A. Misra, "Autofluorescence characterization of advanced glycation end products of hemoglobin," *Spectrochimica Acta. Part A: Molecular and Biomolecular Spectroscopy* **61**(1–2), 163–170 (2005).

4. M.M. Sanches, A. Roda, R. Pimenta, P.L. Filipe, and J.P. Freitas, "Cutaneous manifestations of diabetes mellitus and prediabetes," *Acta Medica Portuguesa* **32**(6), 459–465 (2019).

5. F. Quondamatteo, "Skin and diabetes mellitus: What do we know?" *Cell and Tissue Research* **355**(1), 1–21 (2014).

6. N.P. Gonçalves, C.B. Vægter, H. Andersen et al., "Schwann cell interactions with axons and microvessels in diabetic neuropathy," *Nature Reviews Neurology* **13**(3), 135–147 (2017).

7. E. Selvin, M.W. Steffes, H. Zhu et al., "Glycated hemoglobin, diabetes, and cardiovascular risk in nondiabetic adults," *New England Journal of Medicine* **362**(9), 800–811 (2010).

8. J.Y. Tseng, A.A. Ghazaryan, W. Lo et al., "Multiphoton spectral microscopy for imaging and quantification of tissue glycation," *Biomedical Optics Express* **2**(2), 218–230 (2011).

9. G. Mazarevica, T. Freivalds, and A. Jurka, "Properties of erythrocyte light refraction in diabetic patients," *Journal of Biomedical Optics* **7**(2), 244–247 (2002).

10. L.P. Reagan, "Diabetes as a chronic metabolic stressor: Causes, consequences and clinical complications," *Experimental Neurology* **233**(1), 68–78 (2012).

11. S.P.M. Hosking, R. Bhatia, P.A. Crock et al., "Non-invasive detection of microvascular changes in a paediatric and adolescent population with type 1 diabetes: A pilot cross-sectional study," *BMC Endocrine Disorders* **13**(1), 41 (2013).

12. C. Schaefer, T. Biermann, M. Schroeder et al., "Early microvascular complications of prediabetes in mice with impaired glucose tolerance and dyslipidemia," *Acta Diabetologica* **47**(Supplement 1), S19–S27 (2010).

13. A. Krumholz, L.D. Wang, J.J. Yao, and L.H.V. Wang, "Functional photoacoustic microscopy of diabetic vasculature," *Journal of Biomedical Optics* **17**(6), 060502 (2012).

14. L. Khaodhiar, T. Dinh, K.T. Schomacker et al., "The use of medical hyperspectral technology to evaluate microcirculatory changes in diabetic foot ulcers and to predict clinical outcomes," *Diabetes Care* **30**(4), 903–910 (2007).

15. L. Toma, G.M. Sanda, M. Deleanu, C.S. Stancu, and A.V. Sima, "Glycated LDL increase VCAM-1 expression and secretion in endothelial cells and promote monocyte adhesion through mechanisms involving endoplasmic reticulum stress," *Molecular and Cellular Biochemistry* **417**(1–2), 169–179 (2016).

16. P. Fang, D.Q. Zhang, Z.J. Cheng et al., "Hyperhomocysteinemia potentiates hyperglycemia-induced inflammatory monocyte differentiation and atherosclerosis," *Diabetes* **63**(12), 4275–4290 (2014).

17. T. Komura, Y. Sakai, M. Honda et al., "CD14(+) monocytes are vulnerable and functionally impaired under endoplasmic reticulum stress in patients with type 2 diabetes," *Diabetes* **59**(3), 634–643 (2010).

18. M.J. Luo, Z.H. Zhang, H. Li et al., "Multi-scale optical imaging of the delayed type hypersensitivity reaction attenuated by rapamycin," *Theranostics* **4**(2), 201–214 (2014).

19. T.R. Mempel, M.L. Scimone, J.R. Mora, and U.H. von Andrian, "In vivo imaging of leukocyte trafficking in blood vessels and tissues," *Current Opinion in Immunology* **16**(4), 406–417 (2004).

20. S. McArdle, Z. Mikulski, and K. Ley, "Live cell imaging to understand monocyte, macrophage, and dendritic cell function in atherosclerosis," *Journal of Experimental Medicine* **213**(7), 1117–1131 (2016).

21. H.G. Zhou, X.M. Zhang, and J.F. Lu, "Progress on diabetic cerebrovascular diseases," *Bosnian Journal of Basic Medical Sciences* **14**(4), 185–190 (2014).

22. J.C. Khoury, D. Kleindorfer, K. Alwell et al., "Diabetes mellitus: A risk factor for ischemic stroke in a large biracial population," *Stroke* **44**(6), 1500–1504 (2013).

23. L.A. Profenno, A.P. Porsteinsson, and S.V. Faraone, "Meta-analysis of Alzheimer's disease risk with obesity, diabetes, and related disorders," *Biological Psychiatry* **67**(6), 505–512 (2010).

24. L. Guo, R. Shi, C. Zhang et al., "Optical coherence tomography angiography offers comprehensive evaluation of skin optical clearing in vivo by quantifying optical properties and blood flow imaging simultaneously," *Journal of Biomedical Optics* **21**(8), 081202 (2016).

25. K.V. Larin, M.G. Ghosn, A.N. Bashkatov et al., "Optical clearing for OCT image enhancement and in-depth monitoring of molecular diffusion," *IEEE Journal of Selected Topics in Quantum Electronics* **18**(3), 1244–1259 (2012).

26. X. Wen, S.L. Jacques, V.V. Tuchin, and D. Zhu, "Enhanced optical clearing of skin in vivo and optical coherence tomography in-depth imaging," *Journal of Biomedical Optics* **17**(6), 066022 (2012).

27. Y. Zhou, J.J. Yao, and L.H.V. Wang, "Optical clearing-aided photoacoustic microscopy with enhanced resolution and imaging depth," *Optics Letters* **38**(14), 2592–2595 (2013).

28. Y.Y. Liu, X.Q. Yang, D. Zhu, R. Shi, and Q.M. Luo, "Optical clearing agents improve photoacoustic imaging in the optical diffusive regime," *Optics Letters* **38**(20), 4236–4239 (2013).

29. X.Q. Yang, Y. Zhang, K. Zhao et al., "Skull optical clearing solution for enhancing ultrasonic and photoacoustic imaging," *IEEE Transactions on Medical Imaging* **35**(8), 1903–1906 (2016).

30. D. Zhu, J. Wang, Z.W. Zhi, X. Wen, and Q.M. Luo, "Imaging dermal blood flow through the intact rat skin with an optical clearing method," *Journal of Biomedical Optics* **15**(2), 026008 (2010).

31. J. Wang, R. Shi, and D. Zhu, "Switchable skin window induced by optical clearing method for dermal blood flow imaging," *Journal of Biomedical Optics* **18**(6), 061209 (2013).

32. R. Shi, M. Chen, V.V. Tuchin, and D. Zhu, "Accessing to arteriovenous blood flow dynamics response using combined laser speckle contrast imaging and skin optical clearing," *Biomedical Optics Express* **6**(6), 1977–1989 (2015).

33. J. Wang, N. Ma, R. Shi et al., "Sugar-induced skin optical clearing: From molecular dynamics simulation to experimental demonstration," *IEEE Journal of Selected Topics in Quantum Electronics* **20**(2), 256–262 (2014).

34. J. Wang, Y. Zhang, P.C. Li, Q.M. Luo, and D. Zhu, "Review: Tissue optical clearing window for blood flow monitoring," *IEEE Journal of Selected Topics in Quantum Electronics* **20**(2), 6801112 (2014).

35. J. Wang, D. Zhu, M. Chen, and X.J. Liu, "Assessment of optical clearing induced improvement of laser speckle contrast imaging," *Journal of Innovative Optical Health Sciences* **3**(3), 159–167 (2010).

36. J. Wang, Y. Zhang, T.H. Xu, Q.M. Luo, and D. Zhu, "An innovative transparent cranial window based on skull optical clearing," *Laser Physics Letters* **9**(6), 469–473 (2012).

37. W. Feng, R. Shi, C. Zhang, T.T. Yu, and D. Zhu, "Lookup-table-based inverse model for mapping oxygen concentration of cutaneous microvessels using hyperspectral imaging," *Optics Express* **25**(4), 3481–3495 (2017).

38. C. Zhang, W. Feng, Y.J. Zhao et al., "A large, switchable optical clearing skull window for cerebrovascular imaging," *Theranostics* **8**(10), 2696–2708 (2018).

39. R. Shi, L. Guo, C. Zhang et al., "A useful way to develop effective in vivo skin optical clearing agents," *Journal of Biophotonics* **10**(6–7), 887–895 (2017).

40. R. Shi, W. Feng, C. Zhang, Z. Zhang, and D. Zhu, "FSOCA-induced switchable footpad skin optical clearing window for blood flow and cell imaging in vivo," *Journal of Biophotonics* **10**(12), 1647–1656 (2017).

41. Y.J. Zhao, T.T. Yu, C. Zhang et al., "Skull optical clearing window for in vivo imaging of the mouse cortex at synaptic resolution," *Light: Science and Applications* **7**(2), 17153 (2018).

42. U. Kańska, and J. Boratyński, "Thermal glycation of proteinS by D-glucose and D-fructose," *Archivum Immunologiae et Therapiae Experimentalis* **50**(1), 61–66 (2002).

43. Z. Sun, X. Li, S. Massena et al., "VEGFR2 induces c-Src signaling and vascular permeability in vivo via the adaptor protein TSAd," *Journal of Experimental Medicine* **209**(7), 1363–1377 (2012).

44. A. Goldin, J.A. Beckman, A.M. Schmidt, and M.A. Creager, "Advanced glycation end products: Sparking the development of diabetic vascular injury," *Circulation* **114**(6), 597–605 (2006).

45. W. Feng, C. Zhang, T. Yu, and D. Zhu, "Quantitative evaluation of skin disorders in type 1 diabetic mice by in vivo optical imaging," *Biomedical Optics Express* **10**(6), 2996–3008 (2019).

46. L.H. Laiho, S. Pelet, T.M. Hancewicz, P.D. Kaplan, and P.T. So, "Two-photon 3-D mapping of ex vivo human skin endogenous fluorescence species based on fluorescence emission spectra," *Journal of Biomedical Optics* **10**(2), 024016 (2005).

47. P. So, H. Kim, and I. Kochevar, "Two-photon deep tissue ex vivo imaging of mouse dermal and subcutaneous structures," *Optics Express* **3**(9), 339–350 (1998).

48. M.P. Wautier, O. Chappey, S. Corda et al., "Activation of NADPH oxidase by AGE links oxidant stress to altered gene expression via RAGE," *American Journal of Physiology: Endocrinology and Metabolism* **280**(5), E685–E694 (2001).

49. L. Soh, and C. Tsatsoulis, "Texture analysis of SAR sea ice imagery using gray level co-occurrence matrices," *IEEE Transactions on Geoscience and Remote Sensing* **37**(2), 780–795 (1999).

50. R.M. Haralick, K. Shanmugam, and I. Dinstein, "Textural features for image classification" *IEEE Transactions on Systems, Man, and Cybernetics* **SMC-3**(6), 610–621 (1973).

51. R. Shi, W. Feng, C. Zhang et al., "In vivo imaging the motility of monocyte/macrophage during inflammation in diabetic mice," *Journal of Biophotonics* **11**(5), e201700205 (2018).

52. W. Feng, R. Shi, and C. Zhang, "Visualization of skin microvascular dysfunction of type 1 diabetic mice using in vivo skin optical clearing method," *Journal of Biomedical Optics* **24**(3), 1–9 (2018).

53. W. Feng, S. Liu, C. Zhang et al., "Comparison of cerebral and cutaneous microvascular dysfunction with the development of type 1 diabetes, " *Theranostics* **9**(20), 5854–5868 (2019).

54. L. Ostergaard, T.S. Engedal, R. Aamand et al., "Capillary transit time heterogeneity and flow-metabolism coupling after traumatic brain injury," *Journal of Cerebral Blood Flow and Metabolism* **34**(10), 1585–1598 (2014).

55. J. Zhu, T. Yu, Y. Li et al., "MACS: Rapid aqueous clearing system for 3D mapping of intact organs," *Advanced Science* **7**(8), 1903185 (2020).

30

Light operation on cortex through optical clearing skull window

Dongyu Li, Chao Zhang, Oxana Semyachkina-Glushkovskaya, Yanjie Zhao, and Dan Zhu

CONTENTS

Introduction

In addition to cortical imaging, modern optical techniques can be used to manipulate cortical activity and environment. Models such as cerebral arterial thrombosis and brain tissue injury can be established by light irradiation [1–3]. In addition, the optical process can take control of blood–brain barrier (BBB) opening and other neurological functions [4–9]. However, the turbid skull above the cortex strongly attenuates light, causing a limitation of the efficiency and targeting of optical manipulation. To overcome the scattering of skull, targeted optical manipulation is usually performed after craniotomy or by embedding optical fibers in the cortex, which changes the natural environment of the brain and may even further lead to bleeding or inflammation.

The *in vivo* skull optical clearing technique not only allows optical imaging systems to achieve cortex through the intact skull, but also provides a craniotomy-free optical window for cortical operation [10, 11]. Up to now, an optical clearing skull window–based BBB opening and precisely positioned cortical ablation has been already realized [11–13], suggesting that the *in vivo* skull optical clearing technique has great potential in noninvasive or minimally invasive cortical modeling.

In this chapter, the applications of optical clearing skull window–based light operation on cortex were introduced, including their methods and progress. In addition, the possibility of using *in vivo* skull optical clearing technique for a wide variety of optical manipulations were discussed.

Combination of skull optical clearing and PDT for blood–brain barrier opening

The blood-brain barrier (BBB) plays an important role in the central nervous system (CNS), which is formed by endothelial cells that line cerebral microvessels [14, 15]. BBB acts as a selective "physical barrier," a "transport barrier," and a "metabolic barrier" [16], controlling penetration of blood-borne agents into the brain and protecting the CNS from toxins and pathogens [17]. However, while BBB blocks harmful substances, it also dramatically reduces the chance that drugs will enter the brain, creating a challenge for effective therapy for the majority of CNS diseases [18], which account for 30% of all diseases [19]. Thus, temporarily opening the BBB for drug delivery has important medical implications, and various methods have been developed in the last decades [20–27]. Among them, photodynamic effect–induced BBB opening has attracted much attention and has been widely used [28, 29].

Photodynamic process is induced by combining light irradiation with photosensitizers. The excited photosensitizer directly oxidizes biomolecules and/or interacts with molecular triplet oxygen ($^3O_2$), producing singlet oxygen ($^1O_2$) that causes cells apoptosis and/or necrosis through plasma and mitochondria membrane rupture. Therefore, the photodynamic effect will induce edema, which happens at the region surrounding the site of light treatment, suggesting a local degradation of the BBB [29]. It means, theoretically, that targeted BBB opening can be realized using the photodynamic effect, as long as the light reaches a local location precisely. However, the severe scattering of skull tissue causes nonnegligible attenuation when light passes through, and makes it impossible to focus it on a small area of cortex. Thus, targeted BBB opening is hard to perform through the turbid skull. Even for nontargeted BBB opening, it is necessary to increase light dose as well as photosensitizer concentration, which may lead to severe vasogenic edema [29, 30]. To overcome the circumstance, a craniotomy needs to be implemented before applying the photodynamic effect [4, 28]. But this will unavoidably induce changes in intracranial pressure and cortical inflammation, which may cause some misinterpretations of mechanisms underlying photodynamic effect–related BBB opening [31].

Zhang et al. used their newly developed urea-based skull clearing agent (USOCA) to open a switchable optical skull window without craniotomy on mice [10], with which they realized photodynamic effect–induced BBB opening through the optical clearing skull window without causing obvious side effects [12]. In addition, they further investigated age differences in photodynamic effect–induced BBB opening [32]. It is strong evidence that optical clearing skull window will be a promising tool for noninvasive photodynamic effect–related BBB opening.

Methods

In their study [12], USOCA was used to open the optical clearing skull window. Briefly, USOCA consists of two solutions, named solution 1 (S1) and solution 2 (S2), respectively. After the skull was exposed, S1 (a saturated supernatant solution of 75% (vol/vol) ethanol) was topically applied to the skull for 10 min, and then removed. Then S2 (a high-concentration sodium dodecylbenzenesulfonate (SDBS) solution) was topically

applied to the skull for 5 min, and the mice could be used for further experiment.

For BBB opening through the established optical clearing skull window, 635-nm laser irradiation was performed with a 30 min intravenous injection of photosensitizer 5-ALA (20 mg/kg). The treated light dose was adjusted by adjusting the irradiation duration (10, 20, 30, 40 J/cm²).

To evaluate the BBB permeability for molecules and liposomes, Evans Blue, rhodamine-dextran (70 kDa), and fluorescently labelled GM1-liposomes were used as tracer agents, respectively. The tracer was intravenously injected 20 min before the injection of 5-ALA. The brain was then irradiated with laser, and followed by *in vivo* or *ex vivo* assessment with two-photon/confocal microscopy, respectively. Figure 30.1 shows the schematic diagram of BBB opening through the optical clearing skull window.

Ex vivo assessment of BBB opening induced by photodynamic effect through optical clearing skull window

They firstly analyzed laser dose–dependent photodynamic effects on BBB permeability using spectrofluorimetric assay of Evans Blue dye (EBd) content in the mouse brain [33, 34]. EBd is a 961 Da dye that binds to serum albumin, becoming a high molecular weight complex (68.5 kDa) in the blood, which cannot pass through intact BBB. The laser doses were 10, 20, 30, and 40 J/cm², respectively. As shown in Table 30.1, it was found that there was no EBd leakage with BBB with different doses of irradiation through the intact skull, while strong EBd leakage through the opened BBB was observed with irradiation through the USOCA treated skull.

In addition, histological images suggested that the low laser doses (10 J/cm² and 20 J/cm²) caused mild accumulation of solutes around microvessels, while higher laser doses (30 J/cm² and 40 J/cm²) induced stronger vasogenic edema (Figure 30.2). Considering the fact that the EBd leakage was more pronounced with laser dose of 20 J/cm² versus 10 J/cm² but caused comparable morphological changes in the brain

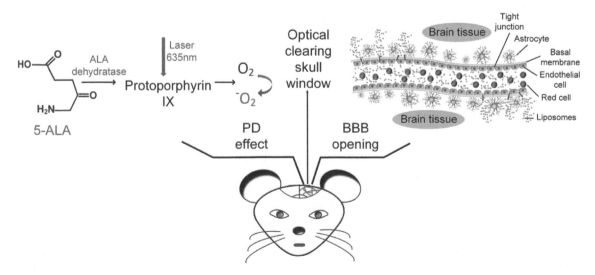

FIGURE 30.1 Photodynamic effect induced opening of BBB through optical clearing skull window. Reprinted from Reference [12].

TABLE 30.1

Photodynamic effect induced BBB opening for the EBd albumin complex. Reprinted with permission from Reference [12]

Groups	Content of Evans Blue in the brain (µg/g tissue)			
Untreated mice	0.685 ± 0.012			
5-ALA	0.650 ± 0.011			
Laser 10, 20, 30, 40 J/cm²	0.708 ± 0.007	0.673 ± 0.003	0.700 ± 0.006	0.801 ± 0.011
PD (5-ALA + laser) without optical clearing skull window				
10, 20, 30, 40 J/cm²	0.741 ± 0.017	0.687 ± 0.013	0.807 ± 0.015	0.796 ± 0.010
PD (5-ALA + laser) with optical clearing skull window				
10, 20, 30, 40 J/cm²	7.618 ± 0.496 ***	13.863 ± 0.712 ***	18.017 ± 1.171 ***	18.295 ± 1.466 ***

***-$p < 0.001$ the comparison between untreated mice and mice underwent PD through optical clearing skull window, n = 6 in each group and sub-group.

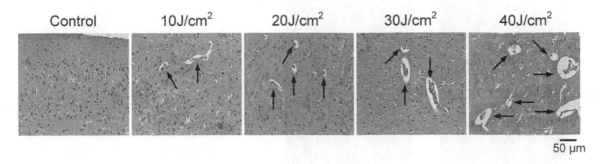

FIGURE 30.2 The histological analysis of photodynamic effect–related changes in the brain tissues and cerebral vessels. The black arrows show vasogenic edema. Reprinted from Reference [12].

tissues, 20 J/cm² was chosen in other experiments as having an optimal photodynamic effect on the BBB.

In vivo observation of BBB opening through optical clearing skull window

For *in vivo* imaging of BBB permeability, two-photon laser scanning microscopy (2PLSM) was used to image the extravasation of rhodamine-dextran through the USOCA treated skull. Figure 30.3a shows the timeline of the experiment. Firstly, 5-ALA (20 mg/kg) was intravenously injected into the anaesthetized mice and circulated for 30 min; during this time, the optical clearing skull window was established by topical application of USOCA. Then the cerebral vessels were imaged through the optical clearing skull window using 2PLSM. After that, a 635 nm laser was applied to irradiate the mouse brain through the window, with the constant power fluence at 165 mW/cm² and 2 min irradiation durations. 1 h after laser irradiation (BBB was opened), the cerebral vessels at the same region were imaged through the optical clearing skull window again.

Figure 30.3b–e shows a strong leakage of rhodaminedextran (70 kDa) from the cerebral vessels into the perivascular space resulting in a high fluorescent signal around the cerebral microvessels after 1 hour of irradiation. In addition, it was calculated that the fluorescent signal in the perivascular space increased 2.0-fold, suggesting a high leakage of rhodaminedextran from the cerebral vessels into the brain parenchyma via the opened BBB, and the fluorescent signal in the cerebral vessels decreased 2.3-fold due to rhodamine extravasation into the brain tissues, metabolism by the kidney, and even probably some photobleaching (Figure 30.3f).

BBB opening for GM1-liposomes

Liposomes have a phospholipid bilayer structure that makes them more compatible than other nanoparticles with the lipoidal layer of the BBB, which makes them promising for drug delivery to brain tissue [35–39]. In the study, fluorescently labeled GM1-liposomes were constructed on the basis of a matrix of egg yolk phosphatidylcholine, which contained mol.% BODIPY-phosphatidylcholine in the bilayer (λ_{ex} = 497 nm, λ_{em} = 504 nm). In addition, three different markers were used to reveal the BBB integrity: 1) the endothelial barrier antigen conjugated with antibodies SMI-71 as a marker of cerebrovascular endothelium; 2) the anti-glial fibrillary acidic protein (GFAP) labeling astrocytes; and 3) the laminin, labeling the basal membranes.

Figure 33.4 demonstrates effective extravasation of liposomes from the cerebral vessels into the brain parenchyma via opened BBB. The distribution of liposomes was observed among the astrocytes (Figure 30.4a) and outside of the cerebrovascular endothelium and the basal membrane (Figure 30.4b and c). Such results show that the BBB can be open for GM1-liposomes (100 nm) so they could go through all elements of BBB including the vascular endothelium, the basal membrane, and the astrocyte feet.

Age differences in photodynamic effect–induced BBB opening through optical clearing skull window

Zhang et al. [32] further investigated whether there were age differences in photodynamic effects through optical clearing skull window on the BBB, and found more pronounced

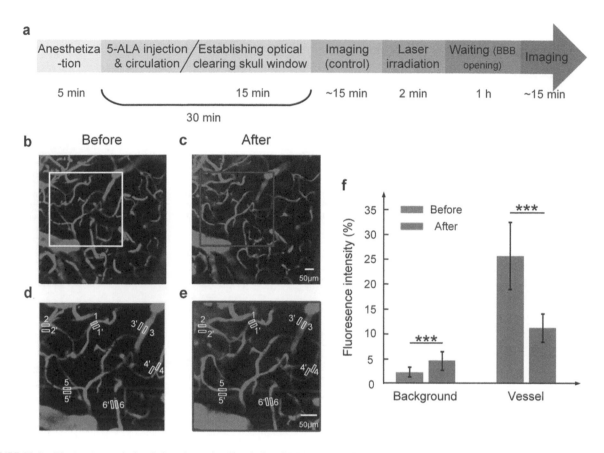

FIGURE 30.3 The *in vivo* analysis of photodynamic effect induced BBB opening for rhodamine-dextran. (a) The timeline of the experiment. (b) and (c) is the same area imaged by 2PLSM before and after PD. (d) and (e) are the magnified images corresponding to the areas boxed in (b) and (c), respectively. (f) The bar graph of the average signal intensity inside (1–6) and outside (1'–6') vessels. Reprinted from Reference [12].

photodynamic effect–induced BBB disruption in juvenile mice compared with adult mice. In their work, they introduced photodynamic effect on cortex through the USOCA-based optical clearing window, and studied photodynamic effect–mediated opening of the BBB in a radiant exposure manner (635 nm, 10/20/30/40 J/cm^2, and 5-ALA, 20 mg/kg) in healthy 4- and 8-week-old mice by using quantitative and qualitative tests for BBB permeability.

The BBB permeability to EBd was analyzed every 30 min after 5-ALA-mediated PDT over a 4-hour duration. As shown in Table 30.2, there were no changes in BBB permeability to the EBd albumin complex in mice of both ages in all groups, including laser irradiation itself or 5-ALA injection without laser, as well as in the untreated mice, while significant age differences in photodynamic effect–related opening of the BBB to EBd were observed. For 10 and 20 J/cm^2 laser irradiation to cause photodynamic effect, the increase of BBB permeability was 1.7 and 1.6 times higher in 4-week-old mice compared with 8-week-old ones. However, application of higher radiant exposures (30 and 40 J/cm^2) was accompanied by a significant EBd leakage that was similar for both ages. In addition, the content of EBd in the brain for both 4- and 8-week-old mice with different radiant exposures returned to the normal state 4 hours after photodynamic therapy, indicating that the BBB opening was reversible.

The FITC-dextran (70 kDa) leakages in different conditions were also quantitatively evaluated by confocal imaging brain slices. As shown in Figure 30.5 and Table 30.3, for low and medium radiant exposures (10–20 J/cm^2), 4-week-old mice demonstrated more significant changes in BBB permeability to FITC-dextran compared with 8-week-old mice, while high radiant exposures (30–40 J/cm^2) caused significant BBB disruption in both age groups. In addition, further experiments demonstrated that FITC-dextran effectively crossed all elements of the BBB, including the cerebral endothelium, the basal membrane, and astrocytes after photodynamic effect performance.

In addition, confocal imaging of photodynamic effect induced BBB opening to FITC-dextran was performed using specific markers of neurovascular unit (NVU) at radiant exposures of 10 and 20 J/cm^2 for 4- and 8-week-old mice, respectively. The result demonstrated that in both conditions, the selected tracer of FITC-dextran effectively crossed all elements of the BBB, including the cerebral endothelium, the basal membrane, and astrocytes after PDT (Figure 30.6).

To analyze the morphological changes in the brain tissues and cerebral vessels after photodynamic effect performance using different radiant exposures, Zhang et al. performed histological studies in mice of both ages. The results demonstrated that the photodynamic effect–induced BBB opening was accompanied by the formation of vasogenic edema, and it was positively related with the exposure dose. Moreover, changes were more pronounced in juvenile mice compared with adult ones (Figure 30.7).

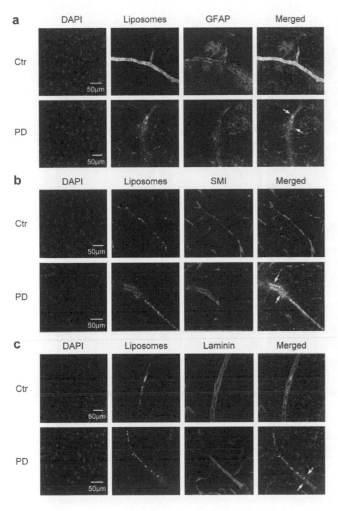

FIGURE 30.4 The confocal imaging of noninvasive photodynamic effect induced BBB opening for GM1-liposomes with usage of markers of neurovascular unit: (a) liposome distributions between the astrocytes labeled by antibodies of GFAP; (b) liposome leakage outside the vascular endothelial cells labeled by antibodies of SMI; (c) liposome distribution outside the basal membrane labeled by antibodies of laminin. Approximately 10–12 brain slices per animal (n = 6) were imaged. The white arrows show the sites of liposome leakage. Reprinted from Reference [12].

In summary, topical application of skull optical clearing agent allows photodynamic effect-induced BBB opening with low doses of photosensitizer and laser irradiation to be performed. Thanks to its noninvasiveness, the mouse brain does not have to suffer secondary injuries and cortical inflammations caused by craniotomy. With the optical clearing skull window, which permits cortical optical operation and *in vivo* observation with high resolution, age differences in photodynamic effect–induced BBB opening can be well studied. It was found that young mice demonstrated a more pronounced photodynamic effect–induced increase in BBB permeability to high weight molecules.

Since the optical clearing skull window has been proved to allow BBB opening with photodynamic effect, it has high potential for other cortical optical operations related to the photodynamic effect, including photothrombosis and photodynamic therapy.

TABLE 30.2

The photodynamic effect–related changes in the BBB permeability to EBd. Reprinted with permission from Reference [32]

Groups	Content of Evans Blue in the brain (μg/g tissue)	
Age	4 weeks	8 weeks
Control (with optical clearing skull window)		
Untreated mice	0.60 ± 0.01 (n = 6)	0.69 ± 0.01 (n = 6)
5-ALA	0.64 ± 0.03 (n = 6)	0.66 ± 0.01 (n = 6)
Laser 10 J/cm^2	0.59 ± 0.01 (n = 6)	0.62 ± 0.02 (n=6)
Laser 20 J/cm^2	0.62 ± 0.02 (n = 6)	0.61 ± 0.01 (n = 6)
Laser 30 J/cm^2	0.60 ± 0.01 (n = 6)	0.65 ± 0.03 (n = 6)
Laser 40 J/cm^2	0.64 ± 0.02 (n=6)	0.67 ± 0.03 (n = 6)

Ischemic stroke is a group of common cerebrovascular diseases with high disability and mortality rate, so it has always been a research hotspot in the field of biomedicine [40–43]. Photothrombosis using photodynamic effects can establish an ischemic stroke model, which has been widely used in the research of repair mechanism and long-term functional recovery after stroke [44, 45]. Photothrombosis involves first injecting a photosensitizer, and then irradiating the brain with light of a specific wavelength to stimulate the photodynamic effect of the photosensitizer to form singlet oxygen, causing endothelial damage, platelet activation, and aggregation, thus forming a vascular embolism [46–48]. By controlling the position and dose of light, a controlled light plug model with a controlled degree can be easily established [49]. Tang et al. [45] applied photothrombosis to rats and performed long-term observation of cortical blood perfusion and tissue damage with optical coherent tomographic angiography (OCTA). In their study, the focus of laser irradiation was adjusted to 500 *μm* beneath the cortex surface in the photothrombosis experiment, and the ischemic region was observed to spread from the deep irradiation focus up to the surface area. Clark et al. [50] created a single vessel–targeted photothrombosis model by using digital micromirror device (DMD). In their study, targeted vessels were irradiated with the patterned laser. It was found that confining laser illumination to individual arteries on the cortical surface with a digital micromirror device expanded the ischemic penumbra, supporting its usefulness for examining the impact of remodeling events within the penumbra on mechanisms of recovery from ischemia. However, due to the severe light scattering of the skull, it is impossible to perform targeted photothrombosis through the skull in its original state, thus it was only achievable with the skull removed in the previous studies. Fortunately, the *in vivo* skull optical clearing technique provides a new idea. Since the scattering of skull can be reduced by topically applying optical clearing agents, it is worth trying to combine optical clearing skull window with focused photodynamic effect for targeted photothrombosis modeling.

Besides photothrombosis, the photodynamic effect can also be used for therapy. Photodynamic therapy (PDT) is a method to use photodynamic effect–induced $^1O_2$ to perform tumor damage with several accesses: (1) to kill tumor cells by direct injury, leading to tumor cell necrosis or inducing apoptosis;

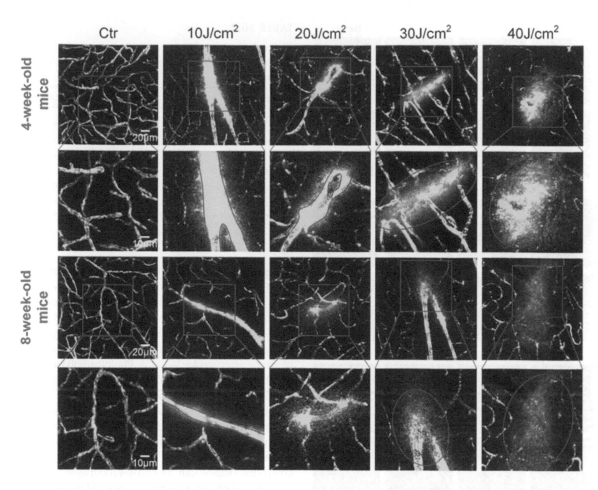

FIGURE 30.5 The PDT-related opening of the BBB evaluated by confocal imaging of FITC-dextran (70kDa) extravasation from cerebral vessels into the brain parenchyma. In each group, images in the second row are magnified maps of the area enclosed in the first row, and the profiles of vessels are outlined. Reprinted from Reference [32].

TABLE 30.3

Photodynamic effect related opening of BBB evaluated by FITC-Dextran extravasation. Reprinted with permission from Reference [32]

Light doses Age	Classification of FITC-dextran extravasation	
	4 weeks	8 weeks
No irradiation (control, n = 6)	/	/
10 J/cm² (n = 6)	+	+
20 J/cm² (n = 6)	+++	++
30 J/cm² (n = 6)	++++	+++
40 J/cm² (n = 6)	++++	++++

(2) to damage tumor blood vessels, leading to tumor blood stasis, collapse, contracture, occlusion, etc., as well as tumor ischemia and hypoxia, thereby indirectly killing tumor cells; (3) to promote the release of cytokines, inflammatory mediators, and immune antigens by target cells, induce inflammatory and immune responses, and damage tumor cells [51–53].

PDT has been widely used for neuroglioma therapy [54–56]. Since the higher the grade of glioma, the greater the damage to the blood–brain barrier and normal brain parenchyma, the grade of glioma is directly related to the level of photosensitizer in the tumor. As a consequence, PDT could damage the tumor with less injury to the normal tissue. It has been reported that

PDT can be used in conjunction with other therapies, including radiotherapy, chemotherapy, thermal therapy, immunotherapy, and boron neutron capture therapy [57–60]. In addition, some fluorescent photosensitizers such as 5-ALA could also perform optical-guided therapy. Since the *in vivo* skull optical clearing technique can reduce the scattering of the skull and the attenuation of light, it holds potential for minimally invasive, precise, and optical guided PDT for neuroglioma.

Laser ablation of neuronal dendrites

Laser-induced injury is a widely used injury model because the extent and site of the injury are easily controlled [2, 3]. However, the skull is a barrier to avoid cortical laser irradiation. Other than a traditional surgery-based cranial window, the novel optical clearing skull window provides a minimally invasive tool to perform targeted laser ablation in cortex, as well as dynamic optical observation of brain response with synaptic resolution [11].

Methods

Zhao et al. [11] used transgenic mice whose dendritic spines and microglia were labeled with yellow fluorescent protein (YFP) and enhanced green fluorescent protein (EGFP), respectively.

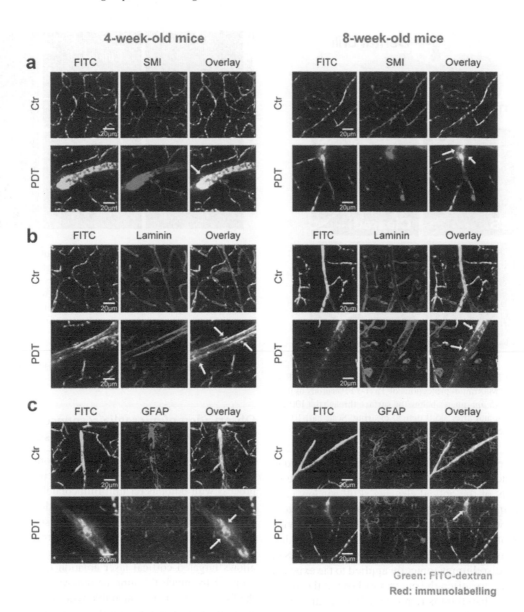

FIGURE 30.6 The confocal imaging of PDT-induced opening of the BBB to FITC-dextran using specific markers of NVU: (a) FITC-dextran leakage outside of the endothelial cells of cerebral vessels labeled by SMI; (b) FITC-dextran leakage outside the basal membrane labeled by the antibodies for the laminin; (c) FITC-dextran extravasation from the cerebral vessels into the brain tissues among astrocytes. The white arrows points to the sites of FITC-dextran leakage. Reprinted from Reference [32].

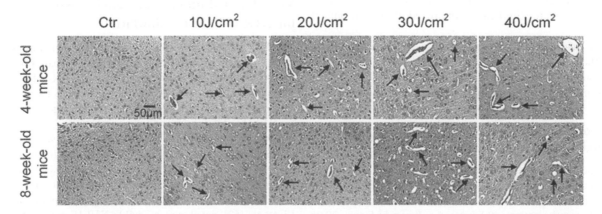

FIGURE 30.7 Photodynamic effect–induced opening of the BBB evaluated by histological analysis of solute extravasation in 4- and 8-week-old mice. Ctr = the control group, where there was no solute leakage; 10, 20, 30, and 40 J/cm2 caused the perivascular edema (black arrows), which appears as empty spaces around cerebral vessels. Reprinted from Reference [32].

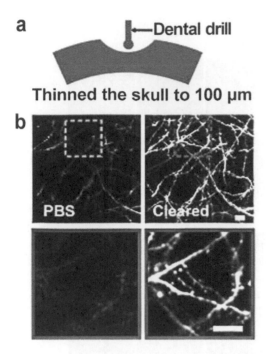

FIGURE 30.8 (a) The dental drill is used to thin the mouse skull (>P30) to about 100 *μm*. (b) Maximum projections of Thy1-YFP neurons of mice aged P60 across 10–15 *μm* images below the surface through the 100-*μm* thinned skull, before and after skull optical clearing. Scale bar, 10 *μm*. Reprinted from Reference [11].

They used the optical clearing skull window technique to perform cortical laser ablation. Briefly, the procedure of the establishment of the optical clearing skull window is as follows. After anesthesia, hair removal, and scalp incision, mice are fixed with a custom-built immobilization device that consisted of a custom-built plate and a skull holder. Then, the skull is thinned to around 100 *μm* (Figure 30.8a), and 10% EDTA disodium is topically applied to the exposed skull for 5–10 min to soften the outermost layer of the skull. Lastly, the EDTA disodium is removed using a clean cotton ball, and 80% glycerol is topically applied to the skull for matching the refractive index. In their study, 2PLSM was used for laser positioning and dynamic observation. Before imaging, a layer of plastic wrap was placed over the cleared skull to separate the water-immersion objective from the glycerol. As shown in Figure 30.8b, the signal intensity and contrast were significantly improved after skull optical clearing.

Once ensuring the position under 2PLSM to perform ablation, the 780-nm femtosecond laser beam was focused on the position of interest for approximately 60 s with power of 60–80 mW to create a tiny injury site.

Monitoring dendrites after laser ablation through optical clearing skull window

With the optical clearing skull window, Zhao et al. [11] kept the laser beam focused on a dendrite to cause ablation. After laser irradiation, continuous two-photon imaging was performed over the course of 1 hour. As shown in Figure 30.9a and b, the dendrites on the laser injury side formed bead-like

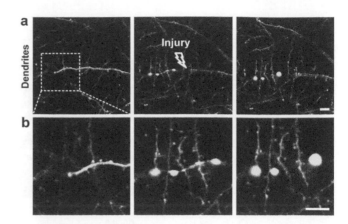

FIGURE 30.9 Changes induced by laser ablation in dendrites. (a) Morphology of dendrites after laser injury, obtained using the optical clearing skull window technique (P30, n = 4 mice). After creating a localized ablation inside the cortex with a two-photon laser, nearby dendrites form bead-like structures. (b) Magnified images corresponding to the rectangle areas shown in a. Scale bar = 25 *μm*. Reprinted from Reference [11].

structures, while the sites that did not suffer damage remained in a normal state.

Observation of microglia response after laser ablation through optical clearing skull window

During a 1-hour observation, microglia soma did not show any significant movement after focused laser irradiation, while the process with bulbous termini immediately moved toward the site of injury (Figure 30.10). The results demonstrate that optical clearing skull windows enable scientists to characterize the effects of laser injury on cortical structures.

In conclusion, the *in vivo* skull optical clearing technique allows targeted cortical laser ablation to be performed with no need for much thinning or removal of the skull. Despite the fact that it is not comparable with traditional cranial windows in terms of imaging depth, it can still serve as an alternative technique for cortical imaging and manipulation with the brain in its normal state.

Vis-NIR-II skull optical clearing window for NIR-II light manipulation

The NIR-II light (> 900 nm) has higher penetration depth in the cortex due to its lower scattering compared to visible and NIR-I light (700–900 nm). Therefore, NIR-II based optical manipulation holds great potential for deep-tissue cortical operation, which requires an optical clearing agent compatible for NIR-II light.

Actually, water, as the main component of various optical clearing reagents, shows strong absorption in the wavelength range of longer than 1300 nm [61]. Thus, in the NIR-II region, the previous optical clearing window might induce extra attenuation of light due to the strong absorption of water. Li et al. [62] developed a visible-NIR-II compatible skull optical clearing agent (VNSOCA) and used it for 1560-nm

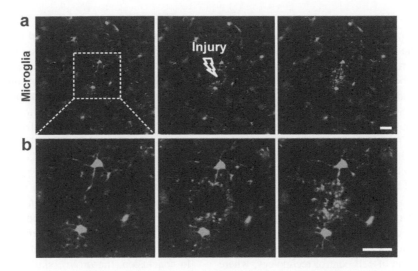

FIGURE 30.10 Changes induced by laser ablation in microglia. (a) Morphology of microglia after laser injury, obtained using the optical clearing skull window technique (P30, n = 4 mice). After creating a localized injury, nearby microglial processes respond immediately with bulbous termini. (b) Magnified images corresponding to the rectangular areas shown in c. The arrows show the brain injury locations. Scale bar = 25 μm. Reprinted from Reference [11].

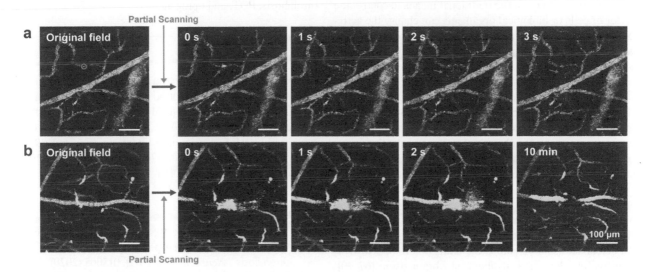

FIGURE 30.11 Dynamically observing cerebral hemorrhage using THG scanning imaging. The cerebral hemorrhage was made by partially scanning the region showed by the red circles for (a) 10 s and (b) 15 s. Reprinted from Reference [62]

fs laser-induced single cortical blood vessel injury through intact skull.

Methods

VNSOCA was developed based on USOCA, but the water was replaced by D_2O. Briefly, VNSOCA consists of two mixtures, named solution 1 (S1) and solution 2 (S2). S1 is a mixture of urea and ethanol in D_2O, and S2 is a high-concentration sodium dodecyl benzenesulfonate (SDBS) in D_2O solution.

For skull optical clearing, S1 was dropped onto the exposed skull so that the skull was immersed for 10 min. After that, S1 was removed with medical cotton and S2 was applied to the same area for 5 min, at which point skull optical clearing was established.

For vasculature injury, firstly, Li et al. performed cortical vascular THG microscopy through the VNSOCA-induced

optical clearing window with the assistance of a THG probe, and then chose a small capillary in the imaging vision. Secondly, the 1560-nm fs laser (80 mW after objective) was focused onto the vessel for 10 seconds to make ablation, followed by a 3-second dynamic THG observation of the cortical hemorrhage. In addition, another larger blood vessel was chosen and laser was then focused onto the vessel for 15 s, followed by observation for 10 min.

NIR-II Laser-induced single blood vessel injury in cortex

As shown in Figure 30.11a, a small vessel was chosen to be irradiated by the 1560-nm fs laser for 10 s, after which the THG probe quickly diffused into the surrounding tissue. As shown in Figure 30.11b, a larger blood vessel was also partially scanned by the laser, but for 15 s. The fracture of the vessel wall was clearly observed. 10 min later, the broken blood vessel was blocked, and the THG probes still remained in the surrounding tissue. The results indicated that precise NIR-II light manipulation could be performed through the established skull optical clearing window.

In conclusion, the VNSOCA, due to its high transparency, allows NIR-II light (1560 nm) to realize targeted optical manipulation in cortex. The previous skull optical clearing windows were demonstrated to be suitable for visible/NIR-I optical manipulating, such as photodynamic opening of the blood–brain barrier (635 nm) [12, 32], and laser ablation of neurons (720 nm) [11].

Apart from photodynamic effect–related applications and laser-induced targeted injury, the skull optical clearing technique can also provide a tool for optogenetic research. Optogenetics is a new type of cell control technology, where light-sensitive ion channel proteins are expressed on excitable target cells or organs, and light of a corresponding wavelength activates the light-sensitive channels to achieve fine regulation of physiological functions of cells, tissues, organs, and animals [6, 63–66]. Although the traditional means to control the behavior of cells or biological organisms can change the activity of cells, they cannot be precisely located in a certain cell, so it has a wide range of effects and a large number of toxic and side effects. Optogenetics technology can achieve higher spatial control accuracy by using advanced light-feeding technology. For example, by changing the power intensity of the light, the light stimulation can reach a certain depth on the tissue, so as to achieve precise spatial positioning [7]. Another advantage of optogenetics technology is that it can make the occurrence of light stimulation and specific behavioral effects almost synchronous, without waiting for the blood or other media to transport the stimulus to a specific site, so that the stimulation occurs when light occurs, and the stimulation immediately stops when light ceases; thus the time-control accuracy is very high. However, due to obstruction of the skull, in the previous study, scientists usually remove the skull before optical irradiation, or introduce optical fibers into brain tissue [67–70]. Serving as a noninvasive light-transparent skull window, the optical clearing skull window holds great potential to be combined with optogenetics to perform optical manipulations.

Summary and prospect

Recent reports have clearly shown that some light operations can be performed well through optical clearing skull windows, including BBB opening and laser-induced targeted cortical injury.

Photodynamic effect–related BBB opening with low doses of photosensitizer and laser irradiation can be realized through the intact skull with topical treatment of USOCA [12], which permits minimizing of brain tissue injuries after modeling. Using an optimal laser dose, the BBB can be opened not only for high weight molecules, but also for 100 nm GM1-liposomes that passed through the vascular endothelium and the basal membrane and were distributed among astrocytes. In addition, the *in vivo* optical clearing skull window was used

to investigate age differences in photodynamic effect–mediated BBB opening, and presented a novel understanding that young mice demonstrated a more pronounced photodynamic effect induced increase in BBB permeability to high weight molecules as well as increase in vasogenic edema [32],.

The *in vivo* skull optical clearing technique can also be used for laser ablation of dendrites and vasculature, as well as dynamic visualization of dendrites, microglia, and micro-blood vessels after injury [11, 71]. Furthermore, compared to an open skull window, the optical clearing skull window technique was safe, with less risk of producing inadvertent damage or inflammation, providing a window for cortical optical operation extremely similar to environments within the normal state of the brain, which would be more suitable for the study of immune cells (such as microglia), which are highly sensitive to the microenvironment.

In conclusion, the *in vivo* skull optical clearing technique has been used as an alternative to craniotomy but with lower invasiveness, providing a skull window for various optical cortical operations, including BBB opening and laser ablation. In the future, it might attract more attention in the fields of photo-thrombosis, PDT, and optogenetics.

Acknowledgments

D.Z. was supported by the National Science Foundation of China (Grant Nos. 61860206009, 81870934); D.L. was supported by a China Postdoctoral Science Foundation funded project (Grant Nos. BX20190131, 2019M662633); O. S.-G. was supported by RF Governmental Grant No. 075-15-2019-1885, Grant from RSF No. 20-15-00090; Grant from RFBR 19-515-55016 China a, 20-015-00308-a.

REFERENCES

1. S. Sunil, S.E. Erdener, B.S. Lee et al., "Awake chronic mouse model of targeted pial vessel occlusion via photothrombosis," *Neurophotonics* **7**(1), 015005 (2020).
2. D. Davalos, J. Grutzendler, G. Yang *et al.*, "ATP mediates rapid microglial response to local brain injury in vivo," *Nat. Neurosci.* **8**(6), 752–758 (2005).
3. A. Nimmerjahn, F. Kirchhoff, and F. Helmchen, "Resting microglial cells are highly dynamic surveillants of brain parenchyma in vivo," *Science* **308**(5726), 1314–1318 (2005).
4. O. Semyachkina-Glushkovskaya, J. Kurths, E. Borisova et al., "Photodynamic opening of blood–brain barrier," *Biomed. Opt. Express* **8**(11), 5040–5048 (2017).
5. O. Semyachkina-Glushkovskaya, V. Chehonin, E. Borisova et al., "Photodynamic opening of the blood–brain barrier and pathways of brain clearing," *J. Biophotonics* **11**(8), e201700287 (2018).
6. K. Deisseroth, "Optogenetics," *Nat. Methods* **8**(1), 26–29 (2011).
7. O. Yizhar, L.E. Fenno, T.J. Davidson, M. Mogri, and K. Deisseroth, "Optogenetics in neural systems," *Neuron* **71**(1), 9–34 (2011).
8. D. Oron, E. Papagiakoumou, F. Anselmi, and V. Emiliani, "Two-photon optogenetics," *Optogenetics Tool. Controlling Monit. Neuronal Act.* **196**, 119–143 (2012).

9. A.M. Packer, D.S. Peterka, J.J. Hirtz et al., "Two-photon optogenetics of dendritic spines and neural circuits," *Nat. Methods* **9**(12), 1202 (2012).

10. C. Zhang, W. Feng, Y.J. Zhao et al., "A large, switchable optical clearing skull window for cerebrovascular imaging," *Theranostics* **8**(10), 2696–2708 (2018).

11. Y.J. Zhao, T.T. Yu, C. Zhang et al., "Skull optical clearing window for in vivo imaging of the mouse cortex at synaptic resolution," *Light Sci. Appl.* **7**(2), 17153 (2018).

12. C. Zhang, W. Feng, E. Vodovozova et al., "Photodynamic opening of the blood–brain barrier to high weight molecules and liposomes through an optical clearing skull window," *Biomed. Opt. Express* **9**(10), 4850–4862 (2018).

13. W. Feng, C. Zhang, T.T. Yu, O. Semyachkina-Glushkovskaya, and D. Zhu, "In vivo monitoring blood–brain barrier permeability using spectral imaging through optical clearing skull window," *J. Biophoton.* **12**(4), e201800330 (2019).

14. W. Risau, and H. Wolburg, "Development of the blood–brain barrier," *Trends Neurosci.* **13**(5), 174–178 (1990).

15. N.J. Abbott, and I.A. Romero, "Transporting therapeutics across the blood–brain barrier," *Mol. Med. Today* **2**(3), 106–113 (1996).

16. N.J. Abbott, L. Ronnback, and E. Hansson, "Astrocyte-endothelial interactions at the blood–brain barrier," *Nat. Rev. Neurosci.* **7**(1), 41–53 (2006).

17. R.S. el-Bacha, and A. Minn, "Drug metabolizing enzymes in cerebrovascular endothelial cells afford a metabolic protection to the brain," *Cell. Mol. Biol.* **45**(1), 15–23 (1999).

18. W.M. Pardridge, "Molecular Trojan horses for blood–brain barrier drug delivery," *Curr. Opin. Pharmacol.* **6**(5), 494–500 (2006).

19. D. Silberberg, N.P. Anand, K. Michels, and R.N. Kalaria, "Brain and other nervous system disorders across the lifespan: Global challenges and opportunities," *Nature* **527**(7578), S151–S154 (2015).

20. M.M. Patel, and B.M. Patel, "Crossing the blood–brain barrier: Recent advances in drug delivery to the brain," *CNS Drugs* **31**(2), 109–133 (2017).

21. S. Mitragotri, "Devices for overcoming biological barriers: The use of physical forces to disrupt the barriers," *Adv. Drug Deliv. Rev.* **65**(1), 100–103 (2013).

22. D.S. Hersh, A.S. Wadajkar, N.B. Roberts et al., "Evolving drug delivery strategies to overcome the blood brain barrier," *Curr. Pharm. Des.* **22**(9), 1177–1193 (2016).

23. V. Kiviniemi, V. Korhonen, J. Kortelainen et al., "Real-time monitoring of human blood–brain barrier disruption," *PLOS ONE* **12**(3), e0174072 (2017).

24. C. Poon, D. McMahon, and K. Hynynen, "Noninvasive and targeted delivery of therapeutics to the brain using focused ultrasound," *Neuropharmacology* **120**, 20–37 (2017).

25. P.S. Fishman, and V. Frenkel, "Focused ultrasound: An emerging therapeutic modality for neurologic disease," *Neurotherapeutics* **14**(2), 393–404 (2017).

26. S.V. Dhuria, L.R. Hanson, and W.H. Frey, "Intranasal delivery to the central nervous system: Mechanisms and experimental considerations," *J. Pharm. Sci.* **99**(4), 1654–1673 (2010).

27. I. Karaiskos, L. Galani, F. Baziaka, and H. Giamarellou, "Intraventricular and intrathecal colistin as the last therapeutic resort for the treatment of multidrug-resistant and extensively drug-resistant Acinetobacter baumannii ventriculitis and meningitis: A literature review," *Int. J. Antimicrob. Agents* **41**(6), 499–508 (2013).

28. H. Hirschberg, F.A. Uzal, D. Chighvinadze et al., "Disruption of the blood–brain barrier following ALA-mediated photodynamic therapy," *Lasers Surg. Med.* **40**(8), 535–542 (2008).

29. S.J. Madsen, and H. Hirschberg, "Site-specific opening of the blood–brain barrier," *J. Biophotonics* **3**(5–6), 356–367 (2010).

30. S.J. Madsen, H.M. Gach, S.J. Hong et al., "Increased nanoparticle-loaded exogenous macrophage migration into the brain following PDT-induced blood–brain barrier disruption," *Lasers Surg. Med.* **45**(8), 524–532 (2013).

31. H.T. Xu, F. Pan, G. Yang, and W.B. Gan, "Choice of cranial window type for in vivo imaging affects dendritic spine turnover in the cortex," *Nat. Neurosci.* **10**(5), 549–551 (2007).

32. C. Zhang, W. Feng, Y. Li et al., "Age differences in photodynamic therapy-mediated opening of the blood–brain barrier through the optical clearing skull window in mice," *Lasers Surg. Med.* **51**(7), 625–633 (2019).

33. J.C.M. Stewart, "Colorimetric determination of phospholipids with ammonium Ferrothiocyanate," *Anal. Biochem.* **104**(1), 10–14 (1980).

34. N.R. Kuznetsova, C. Sevrin, D. Lespineux et al., "Hemocompatibility of liposomes loaded with lipophilic prodrugs of methotrexate and melphalan in the lipid bilayer," *J. Control. Release* **160**(2), 394–400 (2012).

35. F. Olson, C. Hunt, F. Szoka, W. Vail, and D. Papahadjopoulos, "Preparation of liposomes of defined size distribution by extrusion through polycarbonate membranes," *Biochim. Biophys. Acta* **557**(1), 9–23 (1979).

36. M. Agrawal, D.K. Tripathi, S. Saraf et al., "Recent advancements in liposomes targeting strategies to cross blood-brain barrier (BBB) for the treatment of Alzheimer's disease," *J. Control. Release* **260**, 61–77 (2017).

37. Z. Belhadj, C. Zhan, M. Ying et al., "Multifunctional targeted liposomal drug delivery for efficient glioblastoma treatment," *Oncotarget* **8**(40), 66889 (2017).

38. L. Zhang, A.A. Habib, and D. Zhao, "Phosphatidylserine-targeted liposome for enhanced glioma-selective imaging," *Oncotarget* **7**(25), 38693–38706 (2016).

39. D.B. Vieira, and L.F. Gamarra, "Getting into the brain: Liposome-based strategies for effective drug delivery across the blood–brain barrier," *Int. J. Nanomedicine* **11**, 5381–5414 (2016).

40. J.C. Baron, "Protecting the ischaemic penumbra as an adjunct to thrombectomy for acute stroke," *Nat. Rev. Neurol.* **14**(6), 325–337 (2018).

41. B.C. Campbell, C.B. Majoie, G.W. Albers et al., "Penumbral imaging and functional outcome in patients with anterior circulation ischaemic stroke treated with endovascular thrombectomy versus medical therapy: A meta-analysis of individual patient-level data," *Lancet Neurol.* **18**(1), 46–55 (2019).

42. A. Vupputuri, S. Ashwal, B. Tsao, and N. Ghosh, "Ischemic stroke segmentation in multi-sequence MRI by symmetry determined superpixel based hierarchical clustering," *Comput. Biol. Med.* **116**, 103536 (2020).

43. B.K. Menon, C.D. d'Esterre, E.M. Qazi et al., "Multiphase CT angiography: A new tool for the imaging triage of patients with acute ischemic stroke," *Radiology* **275**(2), 510–520 (2015).

44. K. Frauenknecht, K. Diederich, P. Leukel et al., "Functional improvement after photothrombotic stroke in rats is associated with different patterns of dendritic plasticity after G-CSF treatment and G-CSF treatment combined with concomitant or sequential constraint-induced movement therapy," *PLOS ONE* **11**(1), e0146679 (2016).

45. S. Yang, K. Liu, H. Ding et al., "Longitudinal in vivo intrinsic optical imaging of cortical blood perfusion and tissue damage in focal photothrombosis stroke model," *J. Cereb. Blood Flow Metab.* **39**(7), 1381–1393 (2019).

46. H. Yuan, W. Jiang, Y. Chen, and B.Y.S. Kim, "Study of osteocyte behavior by high-resolution intravital imaging following photo-induced ischemia," *Molecules* **23**(11), 2874 (2018).

47. B.D. Watson, W.D. Dietrich, R. Busto, M.S. Wachtel, and M.D. Ginsberg, "Induction of reproducible brain infarction by photochemically initiated thrombosis," *Ann. Neurol.* **17**(5), 497–504 (1985).

48. G.W. Kim, T. Sugawara, and P.H. Chan, "Involvement of oxidative stress and caspase-3 in cortical infarction after photothrombotic ischemia in mice," *J. Cereb. Blood Flow Metab.* **20**(12), 1690–1701 (2000).

49. Z.M. Lv, R.J. Zhao, X.S. Zhi et al., "Expression of DCX and transcription factor profiling in photothrombosis-induced focal ischemia in mice," *Front. Cell. Neurosci.* **12**, 455 (2018).

50. T.A. Clark, C. Sullender, S.M. Kazmi et al., "Artery targeted photothrombosis widens the vascular penumbra, instigates peri-infarct neovascularization and models forelimb impairments," *Sci. Rep.* **9**(1), 2323 (2019).

51. J. Akimoto, "Photodynamic therapy for malignant brain tumors," *Neurol. Med. Chir.* **56**(4), 151–157 (2016).

52. S.L. Hu, P. Du, R. Hu, F. Li, and H. Feng, "Imbalance of Ca2+ and K+ fluxes in C6 glioma cells after PDT measured with scanning ion-selective electrode technique," *Lasers Med. Sci.* **29**(3), 1261–1267 (2014).

53. C. Zavadskaya capital Te, "Photodynamic therapy in the treatment of glioma," *Exp. Oncol.* **37**(4), 234–241 (2015).

54. H. Stepp, T. Beck, T. Pongratz et al., "ALA and malignant glioma: Fluorescence-guided resection and photodynamic treatment," *J. Environ. Pathol. Toxicol. Oncol.* **26**(2), 157–164 (2007).

55. K. Mahmoudi, K.L. Garvey, A. Bouras et al., "5-aminolevulinic acid photodynamic therapy for the treatment of high-grade gliomas," *J. Neurooncol.* **141**(3), 595–607 (2019).

56. D. Ma, X. Chen, Y. Wang et al., "Benzyl ester dendrimer silicon phthalocyanine based polymeric nanoparticle for in vitro photodynamic therapy of glioma," *J. Lumin.* **207**, 597–601 (2018).

57. K.T. Chen, D.M. Hau, J.S. You, H.C. Pan, and R.W. Wong, "Therapeutic effects of photosensitizers in combination with laser and ACNU on an in vivo or in vitro model of cerebral glioma," *Chin. Med. J.* **108**(2), 98–104 (1995).

58. O.A. Gederaas, A. Hauge, P.G. Ellingsen et al., "Photochemical internalization of bleomycin and temozolomide: In vitro studies on the glioma cell line F98," *Photochem. Photobiol. Sci.* **14**(7), 1357–1366 (2015).

59. W. Wang, K. Tabu, Y. Hagiya et al., "Enhancement of 5-aminolevulinic acid-based fluorescence detection of side population-defined glioma stem cells by iron chelation," *Sci. Rep.* **7**, 42070 (2017).

60. H.A. Leroy, M. Vermandel, B. Leroux et al., "MRI assessment of treatment delivery for interstitial photodynamic therapy of high-grade glioma in a preclinical model," *Lasers Surg. Med.* **50**(5), 460–468 (2018).

61. C.C. Petty, and J.A. Curcio, "The near infrared absorption spectrum of liquid water," *J. Opt. Soc. Am.* **41**(5), 302–304 (1951).

62. D. Li, Z. Zheng, T. Yu. et al., "Visible-near infrared-II skull optical clearing window for in vivo cortical vasculature imaging and targeted manipulation," *J. Biophotonics*, e202000142 (2020).

63. E.S. Boyden, F. Zhang, E. Bamberg, G. Nagel, and K. Deisseroth, "Millisecond-timescale, genetically targeted optical control of neural activity," *Nat. Neurosci.* **8**(9), 1263–1268 (2005).

64. J.H. Lee, R. Durand, V. Gradinaru et al., "Global and local fMRI signals driven by neurons defined optogenetically by type and wiring," *Nature* **465**(7299), 788–792 (2010).

65. A. Arrenberg, B., and H. Baier, "Optical control of zebrafish behavior with halorhodopsin," *Proc. Natl Acad. Sci. U. S. A.* **106**(42), 17968–17973 (2009).

66. X. Borue, S. Cooper, J. Hirsh, B. Condron, and B.J. Venton, "Quantitative evaluation of serotonin release and clearance in Drosophila," *J. Neurosci. Methods* **179**(2), 300–308 (2009).

67. A.R. Adamantidis, F. Zhang, A.M. Aravanis, K. Deisseroth, and L. de Lecea, "Neural substrates of awakening probed with optogenetic control of hypocretin neurons," *Nature* **450**(7168), 420–424 (2007).

68. F. Zhang, V. Gradinaru, A.R. Adamantidis et al., "Optogenetic interrogation of neural circuits: Technology for probing mammalian brain structures," *Nat. Protoc.* **5**(3), 439–456 (2010).

69. J.A. Cardin, M. Carlén, K. Meletis et al., "Targeted optogenetic stimulation and recording of neurons in vivo using cell-type-specific expression of channelrhodopsin-2," *Nat. Protoc.* **5**(2), 247–254 (2010).

70. X. Liu, S. Ramirez, P.T. Pang et al., "Optogenetic stimulation of a hippocampal engram activates fear memory recall," *Nature* **484**(7394), 381–385 (2012).

71. D.Y. Li, Z. Zheng, T.T. Yu et al., "Vis-NIR-II skull optical clearing window for in vivo cortical vasculature imaging and targeted manipulation," *J. Biomed. Opt.*, 202000142 (2020).

31

The role of optical clearing to enhance the applications of in vivo *OCT and photodynamic therapy: Towards PDT of pigmented melanomas and beyond*

Layla Pires, Michelle Barreto Requena, Valentin Demidov, Ana Gabriela Salvio, I. Alex Vitkin, Brian C. Wilson, and Cristina Kurachi

CONTENTS

Introduction

This chapter presents an overview of the use of optical clearing of tumors, considering the effects *in vivo* on both optical coherence tomography (OCT) imaging and photodynamic therapy (PDT), particularly in melanoma.

Tumor tissue and light scattering. Given the limited penetration of visible/near-infrared light in tissues [1], high-resolution studies of the tumor microenvironment, including the microvasculature, have required the use of several *ex vivo* analytic optical techniques, such as frozen-section histopathology or optical sectioning through confocal laser-scanning microscopy. However, the need for real-time *in vivo* diagnostics and the use of light-based therapies such as PDT have motivated the development of techniques to overcome the limited light penetration in tissue.

Light attenuation in tissue is caused by absorption and (elastic) scattering, with the latter contributing particularly in non-pigmented lesions. Optical clearing agents (OCA) have been successfully used to decrease the light scattering in tissue and to improve imaging quality in-depth, mostly in normal skin or benign cutaneous lesions [2]. Its use to enhance the efficacy of both photodiagnostics and phototherapeutics in pigmented or thicker malignant lesions is of interest, given the potential clinical applications.

Optical clearing for *in vivo* tumor spectroscopy/imaging

Photonics-based approaches to assessing tumors and their microenvironment are actively being explored for cancer staging and prognosis, both in preclinical research and in the clinic [3–5]. An non-exhaustive list of noninvasive optical modalities includes diffuse reflectance spectroscopy, bioluminescence imaging, photoacoustic imaging, steady-state or time-resolved fluorescence spectroscopy/imaging [6], confocal and/or multi-photon approaches in white-light reflectance or fluorescence mode, and optical coherence tomography (OCT).

OCT is arguably one of the most explored techniques for tissue structural analysis and microvasculature visualization. Using the principles of electromagnetic interference and coherence gating to enable depth-resolved imaging of tissue morphology and architecture on the micron scale, OCT has emerged as viable and robust contrast-agent-free "in-vivo microscope." In addition to its impressive volumetric visualization of fine tissue microstructures (often down to the cellular level), a variety of additional OCT contrast mechanisms have been developed to measure other important functional characteristics of tissues, for example blood and lymphatic microvasculature (e.g., speckle-based OCT methods) and tissue biomechanics (OCT elastography). Its various scientific underpinnings, technological advances, preclinical studies, and clinical uses have been summarized in recent reviews [7–10]. However, while generating cross-sectional or volumetric images of high spatial resolution (typically ~10 μm), the imaging depth of OCT is only ~1–2 mm, due to rapid attenuation of the coherently backscattered light. For tumor imaging, this has prevented imaging of the whole lesion, often necessitating biopsy or resection, followed by *ex vivo* analysis.

Using both OCT and diffuse reflectance spectroscopy (DRS), we have recently demonstrated that optical clearing agents can approximately double the effective depth of light penetration in a mouse model of pigmented cutaneous melanoma [11]. Tape stripping to remove the stratum corneum was used prior to topical application of the OCA (PEG 400 and 1,2 propanediol mix, 19:1) to enhance its penetration. White-light DRS demonstrated diffusion of the OCA through the skin layers over time: for short wavelengths (<~500 nm) the signal decreased within the first few minutes, while for longer wavelengths (>~600 nm) up to 45 min was required to achieve the DRS-measured maximum clearing effect. Further, OCT imaging was performed using a swept-source system centered at 1300 nm; here, the maximum clearing effect was observed considerably later than DRS (at 4 h after the OCA application), and enabled the visualization of tumor microvasculature up to a depth of 750 μm, or about double that without clearing, in this challenging pigmented tumor model [11]. These initial results show the potential of optical clearing agents to improve the utility of optical spectroscopy and imaging techniques to study tumor microenvironment, even in the highly absorbing

milieu of pigmented melanoma. If translated to the clinic, such OCA approaches could enhance the utility of optical diagnostics, particularly in the context of tumor staging and treatment response assessment (Figure 31.1).

Photodynamic therapy of tumors

Photodynamic therapy (PDT) induces (tumor) cell death due to cytotoxic photoproducts such as singlet-state oxygen generated by light-activated photosensitizers [12]. In most non-dermatological applications, the PDT procedure is initiated by systemic administration of the photosensitizer. After a suitable time interval (minutes to days), depending on the photosensitizer uptake and clearance kinetics [13], this is followed by local irradiation using laser or LED light. In order to achieve the desired photodynamic effect, a minimum light fluence

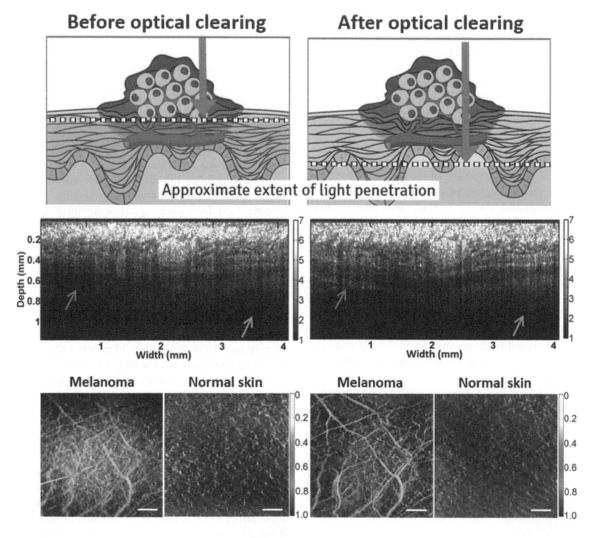

FIGURE 31.1 The effect of optical clearing on OCT imaging of melanoma. Top row: a schematic of light penetration through a tumor and its enhancement following OCA administration. Middle row: Color-coded OCT B-scan cross-sectional images of a pigmented melanoma (murine B16-F10 tumors grown in nude athymic mice), before and after optical clearing (t = 4 h). As seen on these structural images, depth visualization is considerably improved after OCA application (blue and green arrows indicate the deeper tissue structures not seen before optical clearing). Bottom row: Depth encoded tissue microvasculature in the same tumors and in the adjacent normal skin obtained in speckle-variance OCT mode (depth scale in mm: green = top tissue layers, black = deepest tissues). Note the improved visualization of deeper-lying disorganized and thick vessels of melanoma after optical clearing. Scale bars are 1 mm.

at the target tumor depth is required and typically shows a threshold dose response. The spatial distribution of both the photosensitizer and the light throughout the full depth and volume of the tumor target is critical to a successful outcome [12].

The effective treatment depth (or radius in the case of interstitial fiberoptic light delivery) depends on the photosensitizer concentration, optical extinction coefficient, and quantum yield of the cytotoxic photoproduct(s), as well as on the intrinsic photodynamic sensitivity of the treated tissue to the specific photosensitizer. Considering the light distribution, the treatment depth depends exponentially on the effective light penetration depth, d_{eff}, which is the reciprocal of the effective attenuation coefficient μ_{eff}, $d_{eff} = 1 / \mu_{eff}$,

$$\mu_{eff} = \sqrt{3\mu_a\mu_s'} \qquad (31.1)$$

where μ_a and μ_s' are the optical absorption and transport scattering coefficients of the tissue at the treatment wavelength. This approximation works when the scattering is significantly greater than absorption, which is true in most tissues at longer visible/near-infrared wavelengths; this may or may not be valid in the context of pigmented melanomas, as we discuss below. However, regardless of the light propagation regime, the penetration depth and effective treatment depth depend on both absorption and scattering properties of tissue.

Topical PDT is approved in several countries for basal cell carcinoma and non-oncological conditions [14, 15]. It is recommended for actinic keratosis (pre-malignant lesions), Bowen's disease (squamous cell *carcinoma in situ*), superficial basal cell carcinoma (BCC), and nodular BCC [16]. Currently, three commercial prodrugs are approved for topical use, all based on 5-aminolevulinic acid (ALA) formulated as an emulsion (Levulan®, DUSA Pharmaceutics, Inc., USA), a nanoemulsion (nc-ALA: Ameluz® Biofrontera Inc., USA) [17], or a more lipid-soluble metyle derivative (Metvix®, Galderma, USA). When thus supplied in excess, ALA leads to increased production of the fluorescent photosensitizer protoporphyrin IX (PpIX) through the heme biosynthesis pathway. This treatment has low systemic toxicity, can be repeated without resistance or hypersensitivity, and preserves normal tissue structure and nerve function [17]. Additionally, PDT (whether ALA-based or with other photosensitizers) can have significant antitumor effects through immune stimulation, reducing or preventing tumor metastasis, and potentially altering tumor progression and reducing tumor recurrence [18–20]. These immune effects are particularly relevant in tumors where surgery is the primary treatment, since there is increasing evidence that surgery-related stress can reduce immune system function, leading to increased metastatic spread. This has been demonstrated preclinically in melanoma and breast and lung tumors [21–23], while immune dysfunction has been implicated in postsurgical metastatic spread of prostate [18] and pancreatic [23] cancers in patients. For highly metastatic cancers such as melanoma, these positive PDT-induced immune responses make it, in principle, an attractive alternative or adjuvant to surgery.

Topical application of ALA-based photosensitizers is recommended for superficial BCC, particularly in sites that show poor healing after surgery, or are important for good cosmetsis,

or involve treatment of multiple or large superficial lesions. It is also considered as an option for nodular BCC lesions <~ 2 mm thick [17]. Cure rates of >90% can be achieved in superficial BCC [24]. However, for nodular lesions, the response rate is only around 50% [25]. PDT is also not currently recommended for pigmented tumors such as pigmented BCC or melanoma. Preclinical studies of melanoma have explored the use of infrared light and more potent photosensitizers to improve response rates, but in most cases have not achieved full tumor eradication [26]. While there may also be differences in the intrinsic photodynamic sensitivity of melanoma cells, these poor results are likely the result of limited penetration of the photosensitizer into the deeper tumor layers and/or the high light attenuation due to melanin absorption.

To address the former, several chemical and physical pretreatment procedures have been explored, including different drug formulations, the use of agents to enhance PpIX biosynthesis, and physical manipulations such as the use of microneedles, iontophoresis and temperature modulation [27]. Deep curettage to debulk lesions before topically applying the photosensitizer or the ALA prodrug has improved outcomes [28, 29]. Bay *et al.* compared physical methods to improve methyl aminolevulinate (MAL) delivery, including the use of laser ablation of the tissue surface [30]. To overcome the photosensitizer depth penetrance limitation, systemic PDT can be an option for nonmelanoma skin cancer, but general skin photosensitivity must be minimized [31, 32]. Further, light penetration remains an intrinsic limitation for pigmented lesions.

Optical clearing in PDT treatment of melanoma

Melanoma, a highly pigmented skin cancer, continues to present significant challenges for light-based techniques and other treatments. The strong light absorption of melanin and scattering by the melanosome granules present in high concentration within the tumor both limit light penetration, preventing full lesion imaging assessment and treatment using optical techniques. Globally, more than 55,000 patients die from melanoma every year [33]. The prognosis for primary cutaneous melanoma depends on tumor thickness [34]. Importantly, melanoma metastasizes in 30% of patients following primary tumor excision, with consequent high fatality rates [35] even in thin (<1 mm) lesions (5%–15%) [36, 37]. Local recurrence rates after resection range from 3%–5% (Stage 0) to >13% (Stage 4). Hence, there are significant unmet clinical needs, both to treat the primary tumor more effectively and to reduce the risk of life-threatening tumor progression and metastatic spread.

Two main factors may be advantageous in the use of optical clearing to improve the effective treatment depth in cutaneous melanoma, namely improved optical coupling at the lesion surface and reduced attenuation through refractive index matching within the tissue. The first mechanism is well known, for example through its use in high-resolution microscopy utilizing oil-immersion lenses. In topical application at the tumor surface, OCA gel fills micro-irregularities to enhance light coupling into the tissue. For the second mechanism, light scattering within the tissue is reduced through the reduction of the refractive index contrast, based on OCAs with a refractive index of~1.4. The overall result is deeper light penetration with depth and more homogeneous light distribution, both of which can improve the PDT efficacy.

We have recently reported [38] the use of optical clearing combined with PDT in intradermal pigmented melanoma in a mouse model using the clinical photosensitizers Visudyne (vascular-targeted) and Photodithazine (tumor cell-targeted), either alone or in combination. The photosensitizers were administered systemically to circumvent the limited penetrance of topical applications (see below). Tumors grown from amelanotic (nonpigmented) cells that were otherwise essentially identical to the pigmented cells were used as controls to isolate the role of melanin in the treatment response *in vivo*.

Optical clearing had no significant effect on the PDT response of the amelanotic tumors in any treatment group. This was expected, since the tumor thickness (<1 mm) and lack of significant absorption allowed adequate (i.e., above threshold) light dose throughout the full lesion depth. Encouragingly, in the pigmented tumors, optical clearing significantly improved the PDT response in all treatment groups. Most importantly, complete tumor eradication was achieved when optical clearing was combined with dual-photosensitizer PDT (Figure 31.2). This demonstrated the critical role that melanin plays in attenuating the light and affecting the treatment outcome [38]. It should be emphasized that optical clearing did not reduce the high light absorption of melanin *per se* but did likely reduce the contribution of the melanin granules to light scattering, which impacts the overall light attenuation. It is possible that further improvements could be achieved by combining optical clearing and optical whitening agents, such as kojic acid, that destroy the melanin granules and potentially would reduce the absorption of the PDT treatment light [40].

We have also carried out additional studies with sub-therapeutic PDT doses combined with optical clearing to determine the maximum response depth in the melanoma tumor [39]. Immediately after PDT with and without optical clearing of pigmented tumors, *ex vivo* Raman spectra were taken using a confocal Raman microscope (Alpha 300 RAS microscope, WITec, Ulm, Germany), with 785 nm excitation and 100–3200 cm^{-1} wavenumber range detection. The images and spectra were collected at 20× and 50× magnifications, and spectra were taken at different depths in the PDT-treated tumor (from 25 μm to 2 mm in 100 μm increments). With the addition of optical clearing, PDT-induced spectral changes were observed up to a depth of 725 μm (Figure 31.2, bottom panel) and the spectra were similar to those at superficial depths of 25–125 μm where the light attenuation was low [39]. In contrast, the PDT group without optical clearing showed marked heterogeneity in the Raman spectra even at intermediate depths of 225–325 μm. That is, it appears that the use of optical clearing promoted a more homogenous and deeper PDT response, in agreement with the direct histopathological evaluation (Figure 31.2).

Complementing these ongoing preclinical studies, we are currently carrying out the first clinical trial at Amaral Carvalho Hospital, Brazil, applying optical clearing in patients receiving MAL-PDT treatment of thin (<2 mm) BCC lesions [41].

The preliminary results in 12 patients demonstrate that the technique is safe, with no side effects observed and no negative impact on PDT treatment. The next step will be to use optical clearing in thicker lesions that have poor PDT response [41].

Despite the initial indications of improvement in the PDT outcomes, optical clearing still requires topical application in most cases, so that methods to improve OCA diffusion into the tumor tissue (e.g., increase the amount and reduce the time) are needed to translate the concept into clinical practice. Hence, we are currently investigating the use of polymer microneedle patches that perforate the superficial layers of the tissue to improve the diffusion of both OCA and photosensitizer into the tumor tissue [42]. Emerging OCT-based tumor delineation methods [43–46] may substantially improve tumor boundary detection and its longitudinal response monitoring posttreatment. Our recently proposed fully automatic volumetric tumor delineation method based on quantitative speckle analysis of OCT images has been demonstrated *in vivo* and validated in two different biological models of human cancers grown in experimental mice, including pigmented melanoma (Figure 31.3). The same microstructural OCT datasets were also used to simultaneously obtain volumetric images of tissue microvasculature, furnishing a more complete functional tumor picture. The method was shown to (1) robustly delineate tumor from normal tissue, enabling a variety of biomedical applications (e.g., early disease detection/diagnosis or therapy response quantification), and (2) reflect some of the tumor changes following ionizing irradiation. We are currently investigating the capability of these novel techniques for detection of early signs of melanoma response to PDT and possible effects of optical clearing, which will be reported in separate publications.

Other potential optical clearing applications for PDT beyond the skin include endoscopic and interstitial scenarios [49, 50]. In the former, optical fibers placed through an endoscope instrument channel allow either surface or interstitial light delivery in hollow organs such as the esophagus and bronchus [51]. Nontoxic optical clearing agents could also be delivered directly to the tumor surface via the instrument channel and reapplied if necessary during the light irradiation; the smooth mucosal surface and absence of a significant physical barrier (e.g., stratum corneum as in skin) may enable easier penetration of OCAs.

In interstitial PDT, including endoscopic approaches, optical fibers are inserted using needles or catheters directly into tissues, typically for treating thicker and larger-volume lesions; often the use of cylindrical diffusing-tip fibers improves the homogeneity of light distribution throughout the entire lesion volume. This approach involves more complex dosimetry with image-based treatment planning [52], but has shown promising results for the treatment of cancers in prostate [52, 53], pancreas [54], head and neck [55, 56], and brain [57, 58]. In this scenario, the optical clearing agent could be injected intratumorally prior to the irradiation to increase the effective treatment volume per fiber and improve the light homogeneity.

Optical clearing agents could be also used in other phototherapeutic modalities, such as photothermal therapy, in which the absorbed light leads to tissue heating above ~55°C to destroy tumors by coagulative necrosis [59]. Again, the use

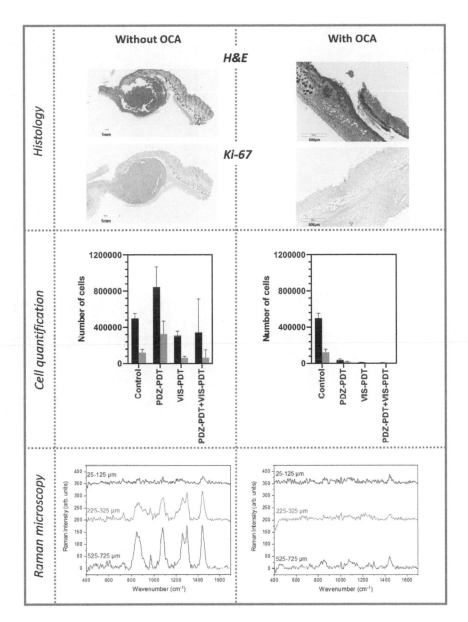

FIGURE 31.2 The effect of optical clearing on the PDT outcome in melanotic melanoma. Top row: Histology images stained with Hematoxylin and Eosin (H&E) and Ki-67 marker, the latter staining viable melanoma cells. Without optical clearing, the tumor is still visible in the H&E section and also viable as indicated by the red Ki-67 staining. Middle row: Tumor cell counts at t = 10 days post-PDT show the marked reduction in viable tumor using optical clearing for PDT treatments using either the single vascular-targeted photosensitizer Visudyne (VIZ) or tumor cell-targeted photosensitizer Photodithazine (PDZ) or both (means ± standard deviation: n = 12 tumors). In each panel, black bars indicate the total number of cells, and gray bars the number of positively stained cells for each of the treatment protocols. The dual-agent treatment completely eradicated the tumor. Bottom row: Raman microspectroscopy of pigmented melanoma that received a sub-therapeutic PDT dose, showing that photodithazine-mediated PDT without optical clearing damaged the tumor up to ~125 μm depth. With the addition of optical clearing, an encouraging increase in PDT effectiveness of up to ~700 μm depth is observed. Adapted from [38, 39].

of OCAs could improve the effective treatment volume and homogeneity, especially in larger lesions.

Conclusions

In summary, optical clearing is a safe and inexpensive method to modify tissue optical properties, not only increasing the light penetration depth but also improving the light distribution throughout the target lesion, with potential to enable improved PDT response. OCA may expand the use of light-based diagnostic and treatment techniques to different tumors that are commonly considered unsuitable for biophotonic approaches. The case in point demonstrated here is pigmented melanoma, a dangerous and difficult-to-treat malignancy. This chapter highlighted initial *in vivo* preclinical and clinical studies using OCAs for optical imaging and PDT treatments and has discussed the overall prospects of OCA-enabled PDT in refractory lesions and beyond.

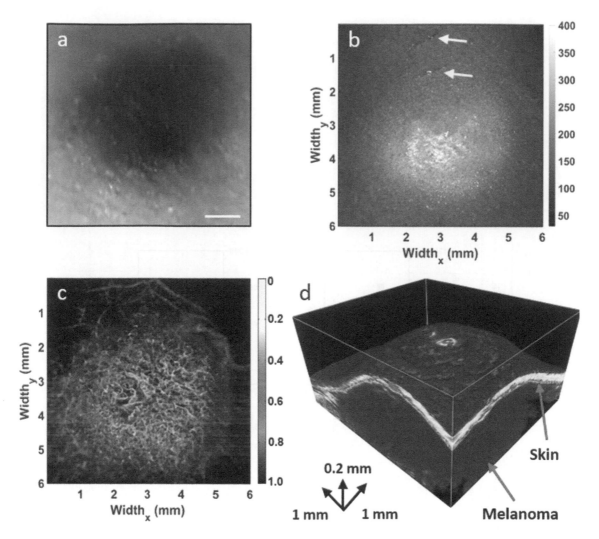

FIGURE 31.3 Recently developed OCT volumetric tumor delineation method[47] applied *in vivo* for detection of pigmented melanoma grown in a nude mouse skin. (a) Microphotograph of melanoma. Scale bar is 1mm; (b) OCT-scanned skin surface of the tumor shown in (a) 90 min after OCA application. Yellow arrows indicate the OCA residuals on the skin surface; (c) Depth-encoded tumor microvasculature obtained with conventional speckle variance method [48] from the same microstructural OCT dataset (depth scale in mm: green = top tissue layers, black = deepest tissues); (d) OCT-derived 3D parametric image of the tumor and surrounding nonmalignant tissues. The sharp tumor-skin "interface" at specific delineation parameter range allows for accurate tumor delineation through parametric thresholding for further analysis.

Acknowledgments

The authors wish to thank the Princes Margaret Foundation (Invest in Research program) and the Cancer Research Society, Canada, and FAPESP and CNPq, Brazil, for financial support of this work.

REFERENCES

1. K. Plaetzer, B. Krammer, J. Berlanda, F. Berr, T. Kiesslich, Photophysics and photochemistry of photodynamic therapy: Fundamental aspects, *Lasers in Medical Science* **24**(2), 259–268 (2009).
2. M. Mogensen, T.M. Joergensen, B.M. Nürnberg, H.A. Morsy, J.B. Thomsen, L. Thrane, G.B. Jemec, Assessment of optical coherence tomography imaging in the diagnosis of non-melanoma skin cancer and benign lesions Versus normal skin: Observer-blinded evaluation by dermatologists and pathologists, *Dermatologic Surgery* **35**(6), 965–972 (2009).
3. A. Sieroń, K. Sieroń-Stołtny, A. Kawczyk-Krupka, W. Latos, S. Kwiatek, D. Straszak, A.M. Bugaj, The role of fluorescence diagnosis in clinical practice, *OncoTargets and Therapy* **6**, 977–982 (2013).
4. S.H. Yun, S.J.J. Kwok, Light in diagnosis, therapy and surgery, *Nature Biomedical Engineering* **1**, 1–6 doi:10.1038/s41551-016-0008 (2017).
5. E.V. Gubarkova, F.I. Feldchtein, E.V. Zagaynova, *et al.*, Optical coherence angiography for pre-treatment assessment and treatment monitoring following photodynamic therapy: A basal cell carcinoma patient study, *Scientific Reports* **9**(1), 1–13 (2019).
6. J. Unger, C. Hebisch, J.E. Phipps, *et al.*, Real-time diagnosis and visualization of tumor margins in excised breast specimens using fluorescence lifetime imaging and machine learning, *Biomedical Optics Express* **11**(3), 1216–1230 (2020).
7. D. Huang, E.A. Swanson, C.P. Lin, *et al.*, Optical coherence tomography, *Science* **254**(5035), 1178–1181 (1991).

8. J.F. de Boer, R. Leitgeb, M. Wojtkowski, Twenty-five years of optical coherence tomography: The paradigm shift in sensitivity and speed provided by Fourier domain OCT [Invited], *Biomedical Optics Express* **8**(7), 3248–3280 (2017).

9. H. Li, B.A. Standish, A. Mariampillai, *et al.*, Feasibility of interstitial Doppler optical coherence tomography for in vivo detection of microvascular changes during photodynamic therapy, *Lasers in Surgery and Medicine* **38**(8), 754–761 (2006).

10. R.F. Spaide, J.G. Fujimoto, N.K. Waheed, S.R. Sadda, G. Staurenghi, Optical coherence tomography angiography, *Progress in Retinal and Eye Research* **64**, 1–55 (2018).

11. L. Pires, V. Demidov, I.A. Vitkin, V. Bagnato, C. Kurachi, B.C. Wilson, Optical clearing of melanoma *in vivo*: Characterization by diffuse reflectance spectroscopy and optical coherence tomography, *Journal of Biomedical Optics* **21**(8), 081210 (2016).

12. B.C. Wilson, M.S. Patterson, The physics, biophysics and technology of photodynamic therapy, *Physics in Medicine and Biology* **53**(9), R61–R109 (2008).

13. J.C. Kennedy, R.H. Pottier, D.C. Pross, Photodynamic therapy with endogenous protoporphyrin IX: Basic principles and present clinical experience, *Journal of Photochemistry and Photobiology, Part B: Biology* **6**(1–2), 143–148 (1990).

14. D.E.J.G.J. Dolmans, D. Fukumura, R.K. Jain, Photodynamic therapy for cancer, *Nature Reviews Cancer* **3**(5), 380–387 (2003).

15. R.-M. Szeimies, M. Landthaler, S. Karrer, Non-oncologic indications for ALA-PDT, *Journal of Dermatological Treatment* **13** Suppl 1, S13–S18 (2002).

16. C.A. Morton, K.E. McKenna, L.E. Rhodes, British Association of Dermatologists Therapy Guidelines and Audit Subcommittee and the British Photodermatology Group, Guidelines for topical photodynamic therapy: Update, *British Journal of Dermatology* **159**(6), 1245–1266 (2008).

17. T.H. Wong, C.A. Morton, N. Collier, *et al.*, British Association of Dermatologists and British Photodermatology Group guidelines for topical photodynamic therapy 2018, *British Journal of Dermatology* **180**(4), 730–739 (2019).

18. I.J. Fidler, Biological behavior of malignant melanoma cells correlated to their survival in vivo, *Cancer Research* **35**(1), 218–224 (1975).

19. G. Tosti, A.D. Iacobone, E.P. Preti, S. Vaccari, A. Barisani, E. Pennacchioli, C. Cantisani, The role of photodynamic therapy in the treatment of vulvar intraepithelial neoplasia, *Biomedicines* **6**(1), 1–11 doi:10.3390/biomedicines6010013 (2018).

20. G. Dragieva, J. Hafner, R. Dummer, *et al.*, Topical photodynamic therapy in the treatment of actinic keratoses and Bowen's disease in transplant recipients, *Transplantation* **77**(1), 115–121 (2004).

21. E. Reginato, P. Wolf, M.R. Hamblin, Immune response after photodynamic therapy increases anti-cancer and anti-bacterial effects, *World Journal of Immunology* **4**(1), 1–11 (2014).

22. R.A. Valdez, T.C. McGuire, W.C. Brown, W.C. Davis, D.P. Knowles, Long-term in vivo depletion of functional CD4+ T lymphocytes from calves requires both thymectomy and anti-CD4 monoclonal antibody treatment, *Immunology* **102**(4), 426–433 (2001).

23. D.L. Morton, D.G. Davtyan, L.A. Wanek, L.J. Foshag, A.J. Cochran, Multivariate analysis of the relationship between survival and the microstage of primary melanoma by Clark level and Breslow thickness, *Cancer* **71**(11), 3737–3743 (1993).

24. R.M. Szeimies, S. Ibbotson, D.F. Murrell, *et al.*, A clinical study comparing methyl aminolevulinate photodynamic therapy and surgery in small superficial basal cell carcinoma (8–20 mm), with a 12-month follow-up, *Journal of the European Academy of Dermatology and Venereology* **22**(11), 1302–1311 (2008).

25. M. Fernández-Guarino, A. Harto, B. Pérez-García, A. Royuela, P. Jaén, Six years of experience in photodynamic therapy for basal cell carcinoma: Results and fluorescence diagnosis from 191 lesions, *Journal of Skin Cancer*, **2014**, 1–7 doi:10.1155/2014/849248 (2014).

26. K.W. Woodburn, Q. Fan, D. Kessel, Y. Luo, S.W. Young, Photodynamic therapy of B16F10 murine melanoma with lutetium texaphyrin, *Journal of Investigative Dermatology* **110**(5), 746–751 (1998).

27. M. Champeau, S. Vignoud, L. Mortier, S. Mordon, Photodynamic therapy for skin cancer: How to enhance drug penetration? *Journal of Photochemistry and Photobiology, Part B: Biology* **197**, 111544 (2019).

28. E. Christensen, C. Mørk, O.A. Foss, Pre-treatment deep curettage can significantly reduce tumour thickness in thick basal cell carcinoma while maintaining a favourable cosmetic outcome when used in combination with topical photodynamic therapy, *Journal of Skin Cancer*, **2001**, 1–6 doi:10.1155/2011/240340 (2011).

29. M.R.T.M. Thissen, C.A. Schroeter, H.A.M. Neumann, Photodynamic therapy with delta-aminolaevulinic acid for nodular basal cell carcinomas using a prior debulking technique, *British Journal of Dermatology* **142**(2), 338–339 (2000).

30. C. Bay, C.M. Lerche, B. Ferrick, P.A. Philipsen, K. Togsverd-Bo, M. Haedersdal, Comparison of physical pre-treatment regimens to enhance protoporphyrin IX uptake in photodynamic therapy: A randomized clinical trial, *JAMA Dermatology* **153**(4), 270–278 doi:10.1001/jamadermatol.2016.5268 (2017).

31. H. Lui, L. Hobbs, W.D. Tope, *et al.*, Photodynamic therapy of multiple nonmelanoma skin cancers with verteporfin and red light-emitting diodes: Two-year results evaluating tumor response and cosmetic outcomes, *Archives of Dermatology* **140**(1), 26–32 doi:10.1001/archderm.140.1.26 (2004).

32. P. Calzavara-Pinton, R.M. Szeimies, B. Ortel (eds.) *Photodynamic Therapy and Fluorescence Diagnosis in Dermatology*, Elsevier, (2001).

33. J. Ferlay, M. Colombet, I. Soerjomataram, *et al.*, Estimating the global cancer incidence and mortality in 2018: GLOBOCAN sources and methods, *International Journal of Cancer* **144**(8), 1941–1953 (2019).

34. J.E. Gershenwald, R.A. Scolyer, K.R. Hess, *et al.*, Melanoma staging: Evidence-based changes in the American Joint Committee on Cancer eighth edition cancer staging manual, *CA: A Cancer Journal for Clinicians* **67**(6), 472–492 (2017).

35. R. Essner, J.H. Lee, L.A. Wanek, H. Itakura, D.L. Morton, Contemporary surgical treatment of advanced-stage melanoma, *Archives of Surgery* **139**(9), 961–966; discussion 966–967 (2004).

36. J.M. Ranieri, J.D. Wagner, S. Wenck, C.S. Johnson, J.J. Coleman 3rd, The prognostic importance of sentinel lymph node biopsy in thin melanoma., *Annals of Surgical Oncology* **13**(7), 927–932 (2006).

37. A. Sandru, S. Voinea, E. Panaitescu, A. Blidaru, Survival rates of patients with metastatic malignant melanoma, *Journal of Medicine and Life* **7**(4), 572–576 (2014).

38. L. Pires, V. Demidov, B.C. Wilson, *et al.*, Dual-agent photodynamic therapy with optical clearing eradicates pigmented melanoma in preclinical tumor models, *Cancers* **12**(7), 1956 (2020).

39. L.P. Martinelli, I. Iermak, L.T. Moriyama, M.B. Requena, L. Pires, C. Kurachi, Optical clearing agent increases effectiveness of photodynamic therapy in a mouse model of cutaneous melanoma: An analysis by Raman microspectroscopy, *Biomedical Optics Express* **11**(11), 6516–6527 (2020).

40. A.F.B. Lajis, M. Hamid, A.B. Ariff, A.F.B. Lajis, M. Hamid, A.B. Ariff, Depigmenting effect of kojic acid esters in hyperpigmented B16F1 melanoma cells, *BioMed Research International*, **2012**, 1–9 doi:10.1155/2012/952452 (2012).

41. L. Pires, M.B. Requena, A.Z. Freitas, *et al.*, Clinical application of optical clearing agent combined with PDT for BCC treatment, *Abstracts*, (2019) (available at https://repositorio.usp.br/item/002945796).

42. R.F. Donnelly, D.I.J. Morrow, P.A. McCarron, *et al.*, Microneedle arrays permit enhanced intradermal delivery of a preformed photosensitizer, *Photochemistry and Photobiology* **85**(1), 195–204 (2009).

43. A. Moiseev, L. Snopova, S. Kuznetsov, *et al.*, Pixel classification method in optical coherence tomography for tumor segmentation and its complementary usage with OCT microangiography, *Journal of Biophotonics* **11**(4), e201700072 (2018).

44. A.A. Lindenmaier, L. Conroy, G. Farhat, R.S. DaCosta, C. Flueraru, I.A. Vitkin, Texture analysis of optical coherence tomography speckle for characterizing biological tissues in vivo, *Optics Letters* **38**(8), 1280–1282 (2013).

45. P.B. Garcia-Allende, I. Amygdalos, H. Dhanapala, R.D. Goldin, G.B. Hanna, D.S. Elson, Morphological analysis of optical coherence tomography images for automated classification of gastrointestinal tissues, *Biomedical Optics Express* **2**(10), 2821–2836 (2011).

46. M. Sugita, R.A. Brown, I. Popov, A. Vitkin, K-distribution three-dimensional mapping of biological tissues in optical coherence tomography, *Journal of Biophotonics* **11**(3), 1–18 doi:10.1002/jbio.201700055 (2018).

47. V. Demivod, N. Demidova, L. Pires, *et al.*, Volumetric tumor delineation and assessment of its early response to radiotherapy with optical coherence tomography, *Biomed. Opt. Express* **12**, 2952–2967 (2021).

48. A. Mariampillai, M.K.K. Leung, M. Jarvi, *et al.*, Optimized speckle variance OCT imaging of microvasculature, *Optics Letters* **35**(8), 1257 (2010).

49. G. Shafirstein, D. Bellnier, E. Oakley, S. Hamilton, M. Potasek, K. Beeson, E. Parilov, Interstitial photodynamic therapy: A focused review, *Cancers* **9**(2), doi:10.3390/cancers9020012 (2017).

50. N. Davies, B.C. Wilson, Interstitial in vivo ALA-PpIX mediated metronomic photodynamic therapy (mPDT) using the CNS-1 astrocytoma with bioluminescence monitoring, *Photodiagnosis and Photodynamic Therapy* **4**(3), 202–212 (2007).

51. H. Wu, T. Minamide, T. Yano, Role of photodynamic therapy in the treatment of esophageal cancer, *Digestive Endoscopy* **31**(5), 508–516 (2019).

52. S.R.H. Davidson, R.A. Weersink, M.A. Haider, *et al.*, Treatment planning and dose analysis for interstitial photodynamic therapy of prostate cancer, *Physics in Medicine and Biology* **54**(8), 2293–2313 (2009).

53. J. Trachtenberg, R.A. Weersink, S.R.H. Davidson, *et al.*, Vascular-targeted photodynamic therapy (padoporfin, WST09) for recurrent prostate cancer after failure of external beam radiotherapy: A study of escalating light doses, *BJU International* **102**(5), 556–562 (2008).

54. M.T. Huggett, M. Jermyn, A. Gillams, *et al.*, Phase I/II study of verteporfin photodynamic therapy in locally advanced pancreatic cancer, *British Journal of Cancer* **110**(7), 1698–1704 (2014).

55. P.J. Lou, L. Jones, C. Hopper, Clinical outcomes of photodynamic therapy for head-and-neck cancer, *Technology in Cancer Research and Treatment* **2**(4), 311–317 (2003).

56. P.J. Lou, H.R. Jäger, L. Jones, T. Theodossy, S.G. Bown, C. Hopper, Interstitial photodynamic therapy as salvage treatment for recurrent head and neck cancer, *British Journal of Cancer* **91**(3), 441–446 (2004).

57. A. Johansson, F. Kreth, W. Stummer, H. Stepp, Interstitial photodynamic therapy of brain tumors, *IEEE Journal of Selected Topics in Quantum Electronics* **16**(4), 841–853 (2010).

58. S. Krishnamurthy, S.K. Powers, P. Witmer, T. Brown, Optimal light dose for interstitial photodynamic therapy in treatment for malignant brain tumors, *Lasers in Surgery and Medicine* **27**(3), 224–234 (2000).

59. C.M. MacLaughlin, L. Ding, C. Jin, *et al.*, Porphysome nanoparticles for enhanced photothermal therapy in a patient-derived orthotopic pancreas xenograft cancer model: A pilot study, *Journal of Biomedical Optics* **21**(8), 84002 (2016).

32

Combination of tissue optical clearing and OCT for tumor diagnosis via permeability coefficient measurements

Qingliang Zhao

CONTENTS

Quantifying optical clearing agent (OCAs) permeability and clearing on human tissue with optical coherence tomography (OCT)

Permeability coefficient measurements (PCM) and OCT human tissue imaging *in vivo* and *in vitro*

Development of noninvasive imaging methods for functional monitoring and quantification of molecular transport in epithelial tissues as well as controlling of tissues' optical properties are extremely important for tumor, arteriosclerosis, diabetic retinopathy, and glaucoma research [1]. Many diseases can alter the physiological structure of the tissue and, thus, could affect the permeability rate of the molecules. Knowing this, the change in the permeability of chemicals and analytes could be used to differentiate abnormal from healthy tissues and could potentially be utilized for the development of novel early diagnostic methods [2]. Therefore, exploring the value of permeability coefficients of hyperosmotic agents could be helpful for the diagnosis of disease.

Recently some studies have monitored and quantified the diffusion of an aqueous solution of glucose in normal esophageal epithelium and esophageal squamous cell carcinoma (ESCC) human tissues [3]. Figure 32.1 demonstrated the typical OCTSS graph acquired from normal esophageal epithelium and ESCC tissues of human during glucose diffusion experiments, respectively. From Figure 32.1, we can see that 40% glucose took less time to reach the monitored region of the ESCC tissues than the normal esophageal tissue. In addition, the permeability coefficient of glucose 40% between the normal esophageal tissue and ESCC tissues

was $(1.74 \pm 0.04) \times 10^{-5}$ cm/s and $(2.45 \pm 0.06) \times 10^{-5}$ cm/s respectively. The optical backscattering from the tumor tissue appears more heterogeneous compared with that of normal tissue [4]. It means that there is stronger scattering in the ESCC tissues than in normal esophageal tissue.

Figure 32.2 shows the relative 1/e light penetration depth value change graph for human normal esophageal tissues and ESCC tissues after application of glucose at 120 min, respectively. It can be seen that light penetration depth for the normal esophageal treated with 40% glucose was gradually enhanced after 40 min treatment and, after 80 min, penetration depth reached its highest. But for the ESCC tissues, the penetration depth shows a significant increase approximately 20 min after the application of 40% glucose and reached the highest (around 70 min) value with the increasing of time. Such an effect is believed due to the agent migrating into the extracellular and intracellular space; a refractive index matching environment is created by matching the chemical agents with the main scattering components within the tissue, leading to the enhanced light penetration together with a dehydration effect [5, 6]. The light penetration depth for the ESCC tissues is relatively lower than that of normal esophagus tissues in the same time range. It is likely relate to the tumor tissue having stronger scattering due to larger nuclei, the higher nuclear-to-cytoplasmic ratio in tumor cells, and the higher regional tumor cell density of the tumor tissues [7].

At the same time, noninvasive optical methods of early detection of diseases in human organ tissues have been a hot topic for the research of biomedical photonics. In contrast, an optical diagnostic probe could be moved from site to site in succession, with each measurement being recorded in a fraction

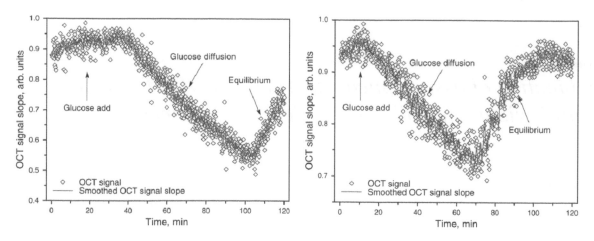

FIGURE 32.1   OCTSS graphs as a function of time from normal and ESCC esophageal tissues during glucose diffusion. Reprinted with permission from Reference [3].

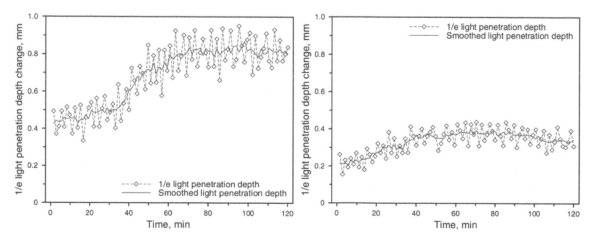

FIGURE 32.2   1/e light penetration depth change as a function of time. Reprinted with permission from Reference [3].

of a second, by simply moving the location of the probe tip. In summary, optical spectrometry offers the potential to improve disease management, with reduced risks for the patient and the potential for earlier diagnosis and immediate treatment. Fluorescence and reflectance spectroscopy are promising techniques for early diagnosis of neoplasia, as numerous diseases are associated with alterations in tissue and cell structure and/or mitochondrial energy metabolic state, reflected in changes in their optical properties [8, 9]. Some studies explore the potential of using the diffuse reflectance (DR) spectral ratio R540/R575 of HbO_2 absorption bands at 540 and 575 nm for in vitro detection of esophageal cancer [10]. Figure 32.3 shows changes in the average of the DR spectra for the normal epithelial tissues and epithelial tissues of ESCC at different heat treatment temperatures of 37, 42, 50, and 60°C in the range of 400–650 nm, and a control group at 20°C was also plotted for comparison respectively. It can be seen from Figure 32.3a that there were three dips in average of DR spectra for the epithelial tissues of normal esophagus around 417, 540, and 575 nm, and for the epithelial tissues of ESCC around 423, 540, and 575 nm. Furthermore, it can be seen that the mean R540/R575 ratios for the epithelial tissues of ESCC were always smaller

than that for the normal esophagus at the same heat treatment temperature. In addition, the mean R540/R575 ratios for the epithelial tissues of normal esophagus and ESCC decreased with the increase of heat treatment temperature. This phenomenon may be related to the cancers and precancerous tissues are characterized by increased microvascular volume, and hence increased blood content [11–14].

In addition, the combination of OCT and OCAs can be used for local quantitative measurement of attenuation coefficient, which can provide additional information for the identification of different tissues [3, 15–21]. OCT has been used to discriminate between different structural features of the normal and atherosclerotic vascular tissues [22], apoptosis, and necrosis in human fibroblasts. This shows that OCT is sensitive to the changes of the attenuation coefficient (AC) caused by analyte diffusion, which is induced by morphological changes in biological tissue [23]. Therefore, OCT techniques were used to assess the difference in permeability coefficient and AC between normal and cancerous tissues caused by hyperosmotic agents. This may help distinguish between cancerous and nonmalignant tissues and hold the promise of early diagnosis of colon cancer.

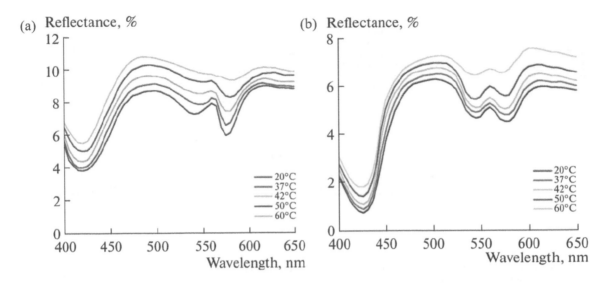

FIGURE 32.3 The average DR spectroscopy with temperature for the normal and ESCC esophageal tissue. Reprinted with permission from Reference [10].

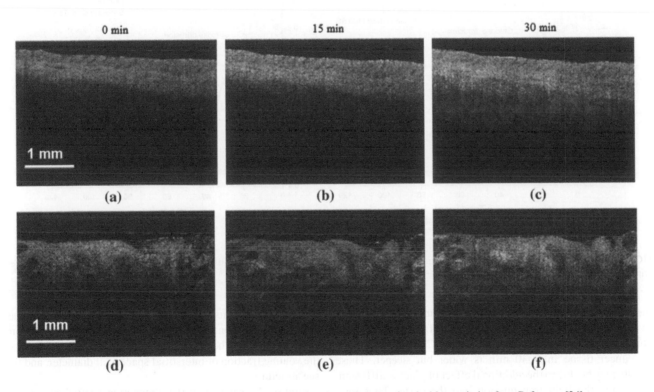

FIGURE 32.4 OCT images of normal (a–c) and the adenomatous colon tissue (d–f). Reprinted with permission from Reference [24].

OCT has been also used to conduct real-time monitoring, identification, and quantification of the diffusion of glucose solutions in human normal and adenoma colon tissues [24]. Figure 32.4a–c showed OCT images of normal colon tissue 0, 15, and 30 min after the topical application of 30% glucose, and Figure 32.4 d–f showed images of the cancerous colon tissue. Normal colon tissue had a regular and compact appearance and layers were clearly visible, but the cancerous tissue structure appeared disorganized and nonuniform and had many dark crypts. Figure 32.4 also showed that the visibility, contrast, and imaging depth in both tissues were significantly

improved after the application of glucose. This change is due to the diffusion of locally applied analytes into the extracellular and intracellular spaces. By matching chemical reagents with the main scattering components in the tissue, a refractive index matching environment is created, and the light penetration ability is enhanced. Glucose dehydration reduces light scattering and further enhances light penetration [25, 26].

The typical dynamic changes of the 1-D OCT normalized signal intensity curve and the corresponding exponential best-fit curve in the experiment are shown in Figure 32.5. The degree of change in the OCT signal intensity is consistent with

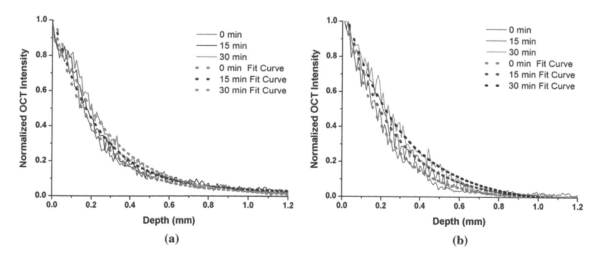

FIGURE 32.5 Normalized OCT intensity profiles with their corresponding exponential best fit curves (a) normal colon tissue and (b) adenomatous colon tissue. Reprinted with permission from Reference [24].

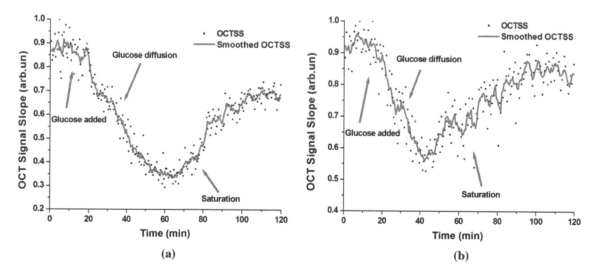

FIGURE 32.6 The typical OCTSS graphs for human normal (a) and adenocarcinoma (b) colon tissues during a 30% glucose diffusion. Reprinted with permission from Reference [24].

the increase in glucose penetration and glucose concentration in the internal structure. As can be seen from Figure 32.5b, compared with normal colon tissue, the OCT signal produced by cancer tissue in 0–30 min is wider but steeper. These results may be due to the additional effect of glucose diffusion, enhancing the light transmission into the tissue. Therefore, more photons propagate to the deep reflective surface below the tissue, producing a stronger antireflection signal.

As shown in Figure 32.6, a similar trend was observed in the normal and malignant colon tissue. In Figure 32.6a, after the glucose reaches the monitoring area in approximately 20 min, it takes another 47 min to completely finish the diffusion process. In contrast, for the cancer tissue, it only took about 13 min to reach the monitored region, and then another 41 min to completely diffuse the whole region, as seen in Figure 32.6b. The permeability coefficient of 30% glucose for the normal colon tissue in Figure 32.7 was significantly slower at $(3.37 \pm 0.17) \times 10^{-6}$ cm/s compared with the permeability coefficient in the adenomatous colon tissue $(5.65 \pm 0.24) \times 10^{-6}$ cm/s ($p < 0.05$).

This may be due to the diffusion of glucose into the tissue, where the tissue contrast is mainly caused by the tissue attenuation coefficient. The tissue attenuation coefficient depends on the volume fraction of interstitial space, cell diameter, and tissue structure.

Figure 32.8 shows that the attenuation coefficients of normal tissue were found to differ significantly from other cancer components with the continuous diffusion of glucose into the tissues. The comparison of Figure 32.8a and Figure 32.8b shows that the decrease in light attenuation was much more prominent in the cancer tissue than that of normal tissue in the same region where we determined the OCTSS. This may be related to some exchange processes in the glucose and differently sized and hydrated (collagen, elastin) structures of tissue because of glucose osmotic impact inducing water flux from these structures (dehydration) and back to them (rehydration) inside tissue [27].

Therefore, these results indicate that OCT technology can assist in the early detection and identification of tumors by

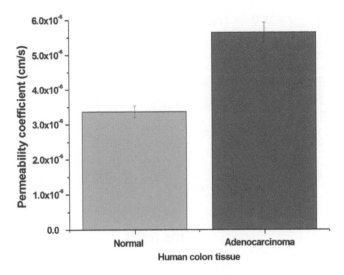

FIGURE 32.7 Comparison of the mean permeability coefficient of 30% glucose diffusion in normal human and adenocarcinoma colon tissues. Reprinted with permission from Reference [24].

monitoring and quantifying the diffusion of hyperosmotic agents and the difference in optical communication between normal and cancer tissues by OCT imaging. During the OCT imaging process, due to the complex morphological characteristics of the tissue, it provides a high-scattering medium for the visible and near-infrared bands, reducing the penetration of light into the tissue and limiting the imaging depth [28–32]. It applies OCAs to biological tissues, mainly to reduce random scattering in tissues through better refractive index matching and dehydration [33]. Currently, glycerol [34–36], glucose [37–39], and dimethyl sulfoxide [40, 41] as OCAs can effectively improve the light transmission depth of various biological tissues.

In order to understand how to distinguish normal tissue from diseased tissue by the permeability of 30% glucose solution during OCT imaging, a study was used to monitor and quantify the differences in permeability coefficients of 30% glucose diffusion by the OCTSS method in four kinds of human lung tissue *in vitro*: normal lung tissue, benign granulomatosis lung tissue, squamous cell carcinoma, and adenocarcinoma tumor [42]. As can be seen from Figure 32.9a, the average permeability coefficient of 30% glucose in normal lung tissue is $(1.35 \pm 0.13) \times 10^{-5}$ cm/s. In addition, about 85 min into the application on normal human lung tissue, the 30% glucose solution had entirely penetrated through the monitored region. Compared with Figure 32.9a, Figure 32.9b, Figure 32.9c, and Figure 32.9d showed the OCT signal slope as a function of time recorded from benign granulomatosis, adenocarcinoma, and squamous cell carcinoma lung tissue during the 30% glucose diffusion experiment, respectively. The permeability coefficients of 30% glucose solution in Figures 32.9b, 32.9c, and 32.9d are $(1.78 \pm 0.21) \times 10^{-5}$ cm/s, $(2.88 \pm 0.19) \times 10^{-5}$ cm/s, and $(3.53 \pm 0.25) \times 10^{-5}$ cm/s respectively. The obvious difference in permeability coefficient of the same hypertonic agent in normal, benign, and malignant lung tissue may be due to several pathological and disease conditions that can change the mechanical properties and microstructure of the tissue. This method illustrated that OCT can distinguish normal tissue from diseased tissue by quantifying the diffusion coefficient of glucose in human lung tissue.

Glucose solutions can also be used as OCA of the skin. Figure 32.10 shows a typical OCTSS graph for skin tissue after application of 40% glucose. 40% glucose took 0.26 h to reach the monitored region, and then another 3.02 h to completely diffuse through the whole region. The permeability coefficient of 40% glucose in skin tissues was found to be $(1.94 \pm 0.05) \times 10^{-5}$ cm/s. The permeability coefficient of hyperosmotic agent diffusion in human skin might be related to the microstructures, refractive indexes, and molecular sizes of the OCAs.

In summary, applying glucose solutions as hyperosmotic agents to biotissue has been inferred to reduce random scattering within tissue primarily by better refractive index matching and a dehydration action. In addition, by combining the penetration of several therapeutic or diagnostic reagents in normal and abnormal tissues, OCT can become a noninvasive imaging method to distinguish normal tissues from diseased tissues.

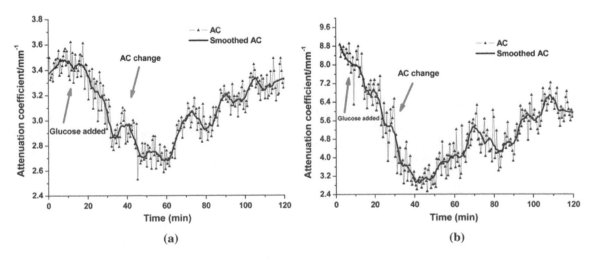

FIGURE 32.8 Mean attenuation coefficients of normal (a) and adenomatous colon tissues (b) after topical application of 30% glucose. Reprinted with permission from Reference [24].

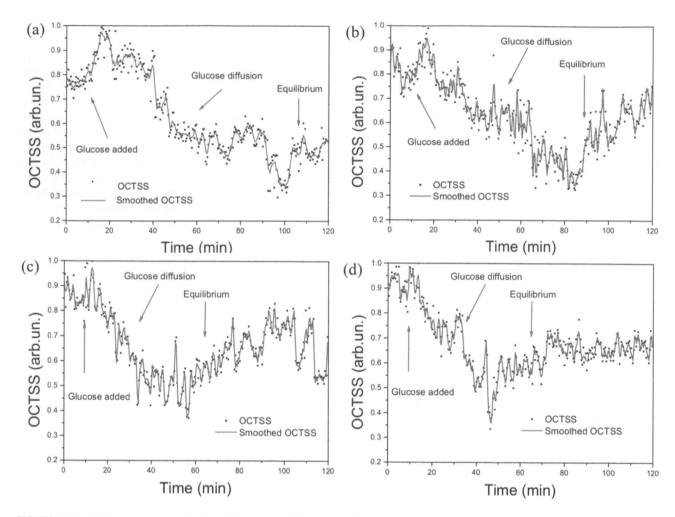

FIGURE 32.9 OCT signal slope as a function of time recorded from normal, benign granulomatosis, adenocarcinoma, and squamous cell carcinoma of human lung tissue (a–d) during 30% glucose diffusion. Reprinted with permission from Reference [42].

Combination of physical and nanoparticles strategy for enhancing optical imaging performance

Evaluation of synergy efficacy with OCAs on tissue using OCT

In the process of optical imaging, accurately acquiring the internal structural characteristics of biological tissues and controlling the optical properties of various biological tissues *in vivo* and *in vivo* are of great significance for many medical applications. However, due to the low absorption and high scattering of most biological tissues and biolipids in the visible and near infrared (NIR) wavelength regions, the spatial resolution and light penetration depth of optical diagnostic and therapeutic methods are limited [27, 31, 43, 44]. Therefore, in order to increase the light penetration depth and obtain more imaging information, it is necessary to reduce multiple scattering in biological tissues. A series of recent studies have shown that OCA as an enhancer with different physical and chemical properties can effectively improve the optical clarity of tissues *in vivo* and *in vitro*, improve imaging depth, enhance contrast, and increase turbid tissue [45–47]. However, due to the barrier

function of OCAs in tissues, many physical methods have been proposed, such as electroporation [48], iontophoresis [49], and microneedling [50]. Recently, it has been reported that the optical cleaning of ultrasound tissue and the synergistic effect of OCAs can be used in the spectral domain optical coherence tomography of human normal tissue and malignant tissue.

A study has reported that 30% glucose solution (G) alone and 30% G combined with 15 min ultrasound (sonophoretic delivery, SP) were used to observe normal and malignant colon tissues [51]. Figure 32.11a and b show OCT images of 30% G and 30% normal colon tissue alone and G/SP from left to right for 0, 10, 30, and 45 min, respectively. Figure 32.12a and b show the OCT images of adenoma colon specimens with 30% G and 30% G/SP applied from 0–10, 30, and 45 min intervals from left to right, respectively. Figure 32.11 and 32.12 showed the 2D OCT images investigations, after application of 30% G and 30% G/SP; the structural features and optical clearing effect are both more obvious for normal and malignant colon tissue samples as time progresses. Figure 32.11a and b also showed that after 30% G combined with ultrasound treatment, the structure of normal colon tissue and malignant colon tissue are more loose and irregular than that after 30% G treatment alone, especially for malignant colon organization. This phenomenon may be caused by the cavitation effect of the

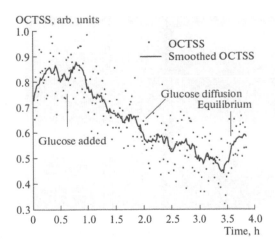

OCTSS, arb. units

FIGURE 32.10 OCTSS as a function of time recorded from the human skin *in vivo* during glucose diffusion. Reprinted with permission from Reference [18].

ultrasound, which has also been reported in transdermal drug delivery and cancer treatment experiments [52, 53].

Figure 32.13 and Figure 32.14 are the 1D OCT signal normalized to the maximum value of the normal and malignant colon tissue intensity curves, respectively, and the corresponding exponential best-fit curve, showing the dynamics at 30% G and 30% G/SP, respectively. From Figure 32.13a and Figure 32.14a, the intensity of the OCT signal gradually increases with time with the 30% G alone in the upper layers of normal and malignant colon tissues. However, Figures 32.13b and 32.14b indicate that the intensity of the OCT signal and light-penetration depth is greatly enhanced after treatment with 30% G/SP in normal and malignant colon tissues.

Figure 32.15 quantifies the penetration-depth enhancement of light into the tissue with 30% G and ultrasound combination. As can be seen from Figure 32.15a, the corresponding relative 1/e light-penetration depth is limited for the normal colon tissue with 30% G alone, compared with 30% G and the ultrasound. Figure 32.15b shows the same trend in malignant colon tissue. These results also indicate that the depth of light penetration increases when ultrasound is applied. This may be

due to the high concentration of chemical enhancer inside the tissue under the action of ultrasonic disturbance [54–56].

Figure 32.16 indicated the dynamic changes of the normalized intensity of the OCT signal as a function of time for the 30% G and 30% G and ultrasound with about 100 min for the normal and malignant colon tissues were summarized. The results showed that in the 30% G and ultrasound groups, the signal intensity of OCT was significantly enhanced compared to 30% G alone in both benign and malignant colon tissues.

Comparing the permeability coefficients of 30% G and 30% G/SP for the normal and cancerous colon tissues presented in Figure 32.17, the permeability coefficients with the 30% G and ultrasound in the normal and cancerous colon tissues are 1.87-fold and 2.18-fold of that without the ultrasound, respectively. These results indicate that the permeability coefficient of hypertonic agent in cancerous colon tissue after ultrasound treatment is greater than that without ultrasound. The reason may be that the intervention of low-intensity ultrasound destroys the characteristics of the tissue structure, resulting in enhanced cell membrane porosity, 30% G permeability, and diffusion in deep tissue.

These findings illustrated that the synergistic effect of ultrasound with glucose can be utilized as an aid to increase glucose permeability and tissue optical clearing and light penetration into deeper biological tissue. Furthermore, more detailed *in vivo* investigations are required to fully assess the microscopic mechanisms by simultaneous application of ultrasound and OCAs, which can enhance the tissue optical clearing effect and light penetration in tissues. As an important part of the human body, skin can be divided into the epidermal layer, dermis layer, and subcutaneous tissue layer. Its main role is to act as a powerful shield to protect the body from extreme temperatures, sunlight, and chemical hazards in the environment. The stratum corneum (SC) is the outermost layer of the skin, which is an important barrier to prevent skin moisture loss and drug percutaneous absorption [57–59]. Failure in protection of the skin will cause wrinkles, pore bulk, acne spots, yellow calluses, or even more serious diseased states such as seborrheic dermatitis and psoriasis [60, 61]. Thus, the development of chemical peeling (CP) technology has been widely used for facial rejuvenation in the past decade [62–66].

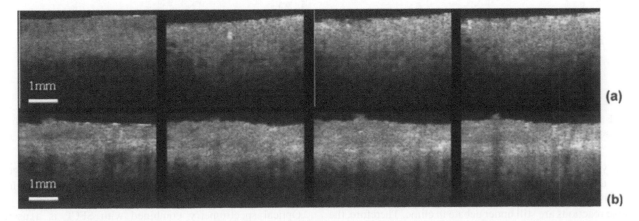

FIGURE 32.11 2D OCT images of human normal colon tissue at different times after topical application of (a) 30% G alone and (b) 30% G/SP, respectively. Reprinted with permission from Reference [51].

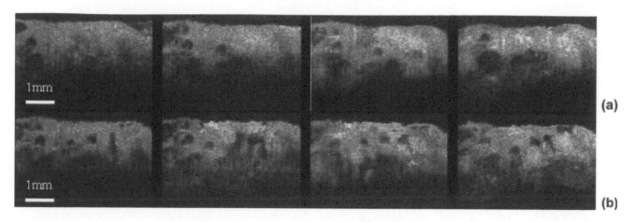

FIGURE 32.12 2D OCT images of human adenomatous colon tissue at different times after topical application of (a) 30% G alone and (b) 30% G/SP, respectively. Reprinted with permission from Reference [51].

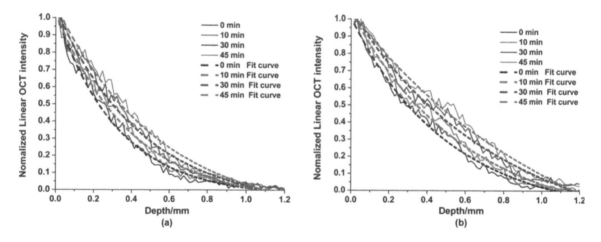

FIGURE 32.13 Normalized OCT intensity profiles (a) 30% G alone; (b) 30% G/SP in normal colon tissue. Reprinted with permission from Reference [51].

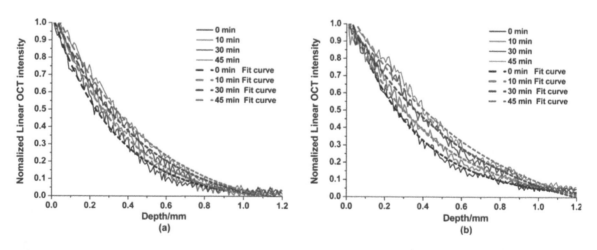

FIGURE 32.14 Normalized OCT intensity profiles (a) 30% G alone and (b) 30% G/SP in adenomatous colon tissue. Reprinted with permission from Reference [51].

Salicylic acid (SA) as a superficial facial CP (SFCP) has been widely used in clinical treatment of acne, freckles, and other facial diseases [67–69]. However, when SA is used as SFCP, its adverse reactions are still under debate in clinic. Therefore, the use of penetration enhancers can optimize the efficacy of SA at the minimum dose, which is of great significance in clinical facial applications. There are currently reports of azone as a chemical permeation enhancer for effective SC peeling-agent delivery into skin. FCPAs can effectively cause morphological structure and optical property variations in skin tissue.

Optical spectrometry combined with SFCP is a useful technology for skin disease diagnosis and monitoring. This technology is mainly based on the differences of various endogenous substances with characteristic optical bands [70].

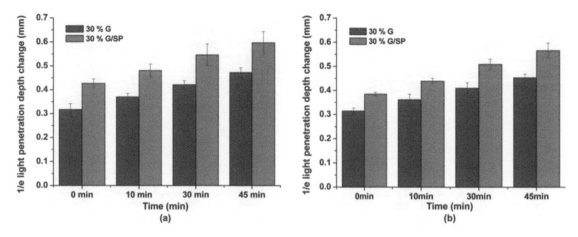

FIGURE 32.15 The dynamic changes of 1/e light-penetration depth (a) normal colon tissue (b) malignant colon tissue after 30% G and 30% G/SP at different times, respectively. Reprinted with permission from Reference [51].

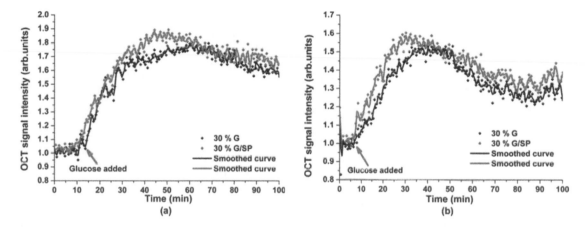

FIGURE 32.16 Normalized intensity of OCT signal as a function of time during addition of 30% G and 30% G/SP (a) normal (b) malignant colon tissue, respectively. Reprinted with permission from Reference [51].

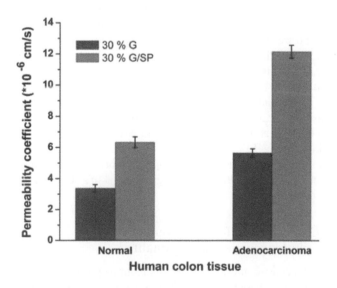

FIGURE 32.17 Comparison of the mean permeability coefficients of 30% G alone and 30% G/SP in normal and malignant colon tissues. Reprinted with permission from Reference [51].

Salicylic acid (SA) has been frequently used as a facial chemical peeling agent (FCPA) in various cosmetics for facial rejuvenation and dermatological treatments in clinic. However,

there is a tradeoff between therapeutic effectiveness and possible adverse effects caused by this agent for cosmetologists. To optimize cosmetic efficacy with minimal concentration, we proposed the use of the chemical permeation enhancer (CPE) azone to synergistically work with SA on human skin *in vivo*. The optical properties of human skin after being treated with SA alone and SA combined with azone (SA@ azone) were successively investigated by diffuse reflectance spectroscopy (DRS) and OCT [71]. To illustrate the dynamics of skin optical properties after the application of chemical permeation enhancer, a set of dynamic spectral changes (from 400 nm to 800 nm) of human skin sample was recorded at regular time intervals over a period of 80 min before and after application of the SA group (S_1, S_2, S_3, and S_4 represent 0.5%, 1%, 1.5%, and 2% SA solutions respectively) and SA@ azone group (0.5%, 1%, 1.5%, and 2% SA solution combined 1% azone mixture solution marked as S_1@A, S_2@A, S_3@A, and S_4@A, respectively) in Figure 32.18 and Figure 32.19 respectively. To quantify the optical characteristic variations of skin resulted by S_1-S_4 and S_1@A-S_4@A, the reduction of DR was calculated at three characteristic wavelengths, 420, 540, and 580 nm, at the different time points in Figure 32.20, Figure 32.21, and Figure 32.22, respectively. The results not only revealed that the DR decreased linearly with SA concentration, but also implied that light transmission significantly

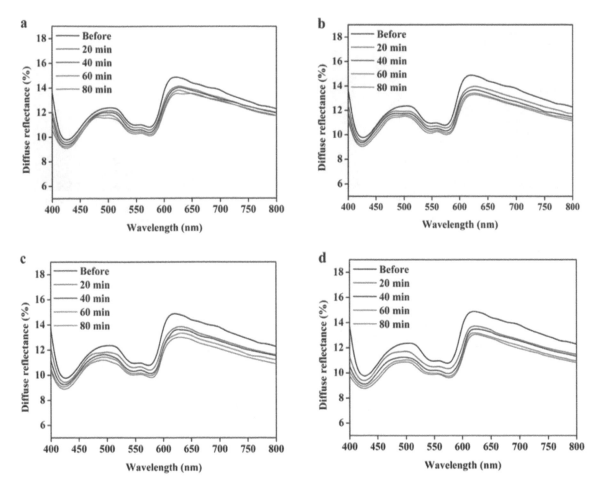

FIGURE 32.18 Dynamic spectral changes of the human skin treated with SA. (a–d) the spectral changes of skin before and after application of the S_1, S_2, S_3, and S_4 at the different time intervals, respectively. Reprinted with permission from Reference [71].

increased in skin with SA concentration increment, which suggested SA at higher concentration had greater desmolytic ability on the SC [72, 73]. In addition, with azone as a penetration enhancer, the greatest decrease in DR was not in the S_4@A but the S_2@A group, since the linear concentration dependent effect occurred in SA group alone was not applicable for synergy group. Therefore, the greatest DR decrease in S_2@A may be attributable to the formulation of SA and azone in equal proportions, which indicated that the two components are transported through the same microenvironment of the skin.

As shown in Figure 32.23, the epidermal structure of skin appears compact and hierarchical before treatment. Structural clarity of tissue and imaging depth of OCT can be improved by S_1–S_4 treatment with time. Furthermore, it also can be seen from Figure 32.23 that there is a remarkable difference in skin treated with SA alone and SA@azone at different concentrations, and the structure of skin becomes looser and the imaging depth of OCT has significant enhancement in S_1@A–S_4@A at the same time intervals, respectively. Figure 32.24 quantified the OCT in-depth reflectance profiles for the human skin topically applied with S1–S4 alone and S1@A–S4@A at time intervals of 0, 20, 40, 60, and 80 min. As seen in Figure 32.24 a–d, the intensity of the OCT signal gradually increased from the upper layers caused by increase in SA contents over time. In addition, Figure 32.24 e–h indicate

that the intensity of the OCT signal is greatly enhanced after being treated with S_1@A-S_4@A. This phenomenon caused by azone was through interaction with the lipid domains of the SC, which increased skin absorption by reversibly damaging or altering the physicochemical nature of the SC to reduce its diffusional resistance [74, 75].

These findings implied that SA mixed with azone may also be considered by commercial cosmetics to improve the exfoliation ability of SA, retard skin wrinkling, and maintain skin vitality. In addition, chemical permeation enhancers of skin tissue may have some important biomedical applications connected with the investigation of skin's structure and functioning.

TOC technique for improvement of photoacoustic imaging quality

Photoacoustic microscopy (PAM) technology has played an irreplaceable role in biomedical imaging. Nevertheless, in strong scattering tissue such as skin, breast, bioliquids, etc., the optical focusing capability degrades due to optical scattering. In order to overcome these challenges, various physical and chemical methods have been deployed to enhance optical imaging depth, such as the utilization of OCAs [76–79]. In recent years, the optical tissue clearing (OTC) technique

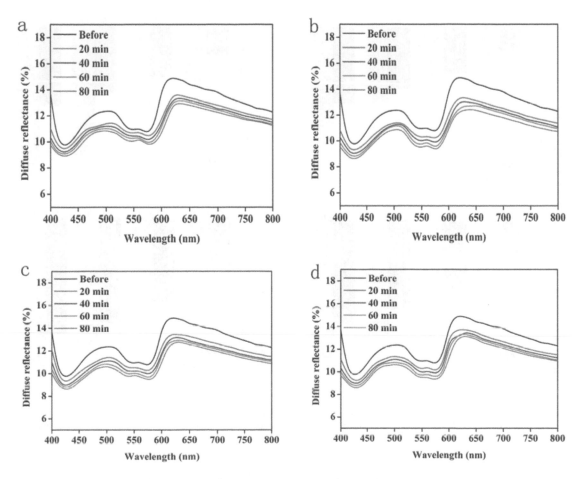

FIGURE 32.19 Dynamic spectral changes of the human skin treated with SA@azone. (a–d) The spectral changes of skin before and after application of the S_1@A, S_2@A, S_3@A, and S_4@A at the different time intervals, respectively. Reprinted with permission from Reference [71].

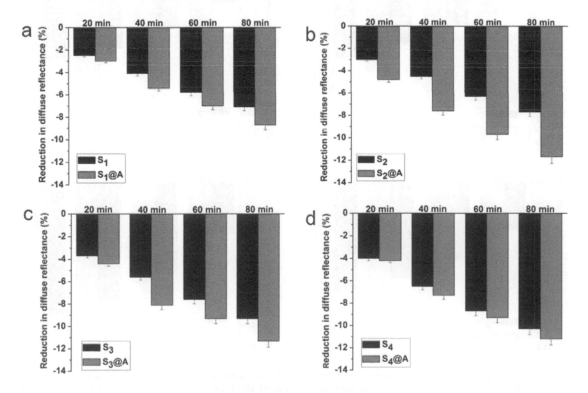

FIGURE 32.20 Representative diffuse reflectance reduction of human skin treated with (a) S_1 and S_1@A, (b) S_2 and S_2@A, (c) S_3 and S_3@A, and (d) S_4 and S4@A at 420 nm, respectively. Reprinted with permission from Reference [71].

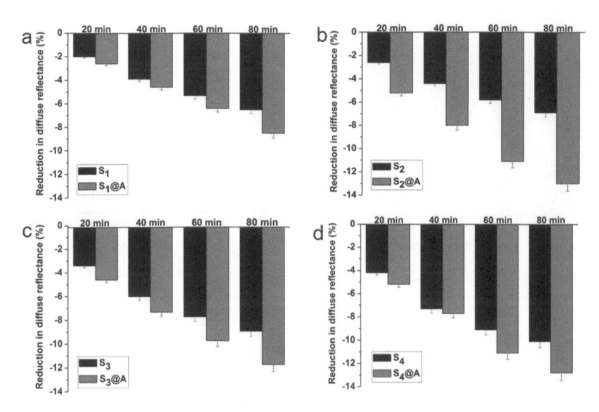

FIGURE 32.21 Diffuse reflectance reduction of human skin at 540 nm. Diffuse reflectance reduction at 540 nm for human skin after treatment with (a) S_1 and $S_1@A$, (b) S_2 and $S_2@A$, (c) S_3 and $S_3@A$, and (d) S_4 and $S_4@A$ at 20, 40, 60, and 80 min, respectively. Reprinted with permission from Reference [71].

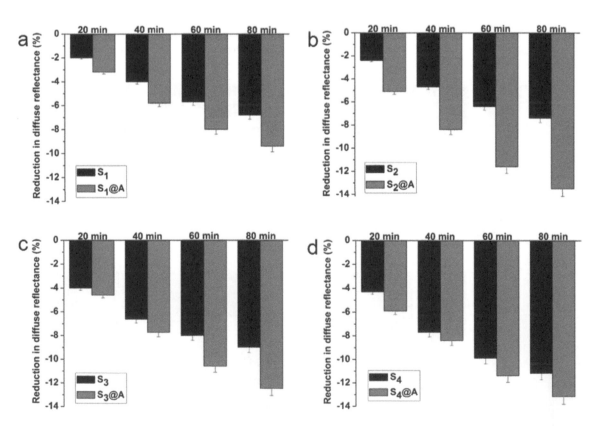

FIGURE 32.22 Diffuse reflectance reduction of human skin at 580 nm. Reduction in diffuse reflectance at 580 nm for human skin after treatment with (a) S_1 and $S_1@A$, (b) S_2 and $S_2@A$, (c) S_3 and $S_3@A$, and (d) S_4 and $S_4@A$ at 20, 40, 60, and 80 min, respectively. Reprinted with permission from Reference [71].

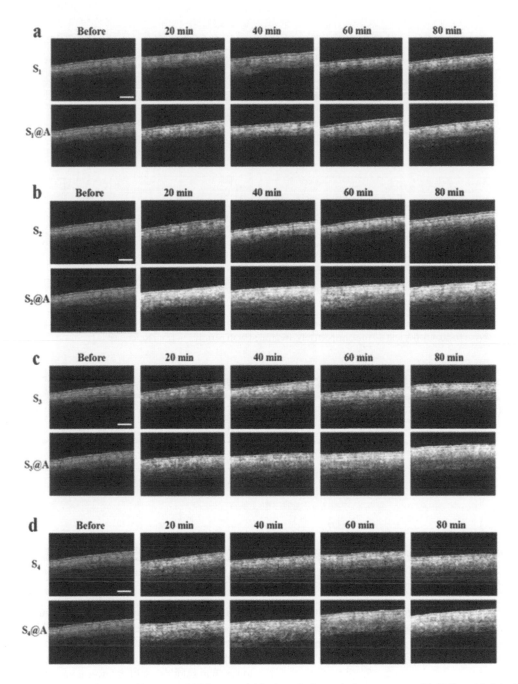

FIGURE 32.23 2D OCT images of human skin tissue. OCT images of skin tissue before and after treatment with (a) S_1 and $S_1@A$, (b) S_2 and $S_2@A$, (c) S_3 and $S_3@A$, and (d) S_4 and $S_4@A$ at 20, 40, 60, and 80 min, respectively. Scale bar: 1 mm. Reprinted with permission from Reference [71].

has shown great potential in inducing optical clearing effects (OCEs) to reduce scattering in tissues using hyperosmotic and biocompatible chemicals agents [25, 41, 80, 81]. In addition, more and more OCAs are being used to investigate whether they can improve the resolution of PAM imaging, such as glucose solution, dimethyl sulfoxide solution, glycerin, and propylene glycol solution, which have refractive index close to that of collagen and are applicable for altering the scattering properties of tissues [41, 80, 82–85]. Among these OCAs, glycerol is one of the most common and efficient OCAs in skin *in vitro* and *in vivo* and it has been proven in medical applications such as tooth therapy and cosmetics study [31, 86, 87].

To increase the efficiency of the topical application of OCAs, different concentrations (0%, 20%, 40%, and 60% respectively) of glycerol solution were applied to a piece of fresh pigskin with a thickness of 0.5 mm [88]. The solutions have a mean refractive index of 1.36, 1.39, and 1.41 respectively [45]. Figure 32.25 shows a photo of the registered trademark logo after covering porcine skin tissue with different concentrations of glycerin solution for 15 min. This finding demonstrated that with the increasing of concentration of glycerol, the OCE of skin is dramatically improved. The greatest increase of light transmittance and sign integrality was found with 60% glycerol, indicating that more light may be allowed through the skin tissue and delivered into the sign by applying

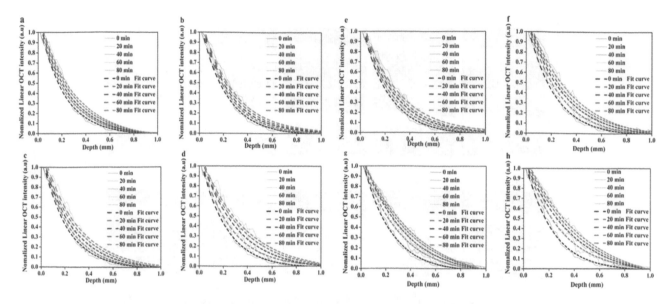

FIGURE 32.24 Normalized OCT intensity profiles of human skin. (a–d) OCT in-depth reflectance profiles for the human skin topically applied with S_1–S_4. (e–h) S_1@A-S_4@A at 0, 20, 40, 60, and 80 min, respectively. Reprinted with permission from Reference [71].

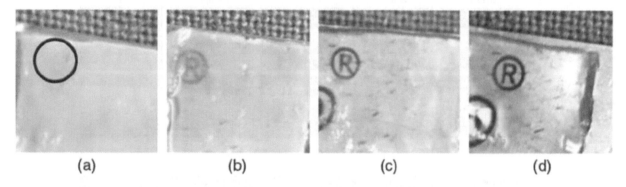

FIGURE 32.25 Photographs of the registered trademark sign covered with a porcine skin tissue under 0% (a), 20% (b), 40% (c), and 60% (d) glycerol at 15 min, respectively. The black circle is the location of the treatment site of the imaging sign. Reprinted with permission from Reference [88].

60% glycerol than other concentrations. Visibility differences are mainly due to the hyperosmotic glycerol resulting in refractive index matching between cellular tissue components: cell membrane, cell nucleus, cell organelles, melanin granules, and extracellular fluid.

Glycerol is applied to fresh pigskin to decrease skin reflectance and increase tissue transmittance. Figure 32.26 shows the change of PAM images of the sign covered with a porcine skin tissue at different times (0, 15, 30, 45, and 60 min) immersed in 20%, 40%, and 60% glycerol solution. There were significant differences between different concentration groups. The clarity of PAM imaging is enhanced as the concentration of glycerol solution and the soaking time increase. The photoacoustic (PA) signal amplitude can be caused by light penetration differences or ultrasound transmitting change. From Figure 32.27, it can be seen that PA signal amplitudes of the sign show significant enhancement after applying glycerol solutions with different concentrations. Since the high-concentration glycerin solution has higher sound attenuation, the PA signal amplitude is higher in the group with higher glycerin concentration. This is because the group with higher glycerin

concentration improves optical clearing effects, allowing more light to penetrate the skin tissue. The difference in the tendency of PA signal amplitude change possibly reflects the fact that the diffusion of high concentrated glycerol solution is lower than solutions with a lower concentration. This phenomenon is consistent with previous studies on the relationship of the concentration of aqueous sucrose solution and its diffusion through semipermeable membranes [89, 90].

This evidence indicates that glycerol as an optical transparency agent can significantly improve the quality of PA imaging. The PA signal amplitudes and the visibility of a phantom light absorber covered by skin tissues were found to increase with the concentration of the glycerol solution. Therefore, this method may become an effective tool for enhancing PA imaging of biological tissues.

Nanoparticles for OCT contrast enhancement imaging

Gold nanorods (GNRs) as an optical contrast agent and photothermal agent have been widely used in the field of

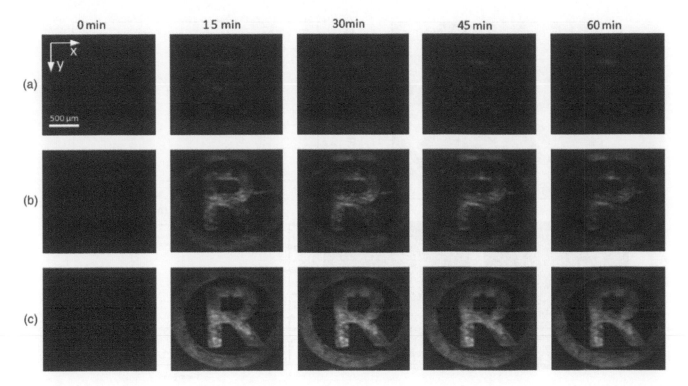

FIGURE 32.26 The OR-PAM MAP images of the sign covered with the skin immersed in glycerol solutions imaged at different time intervals. (a) MAP images with the 20% glycerol; (b) MAP images with 40% glycerol; (c) MAP images with 60% glycerol. Reprinted with permission from Reference [88].

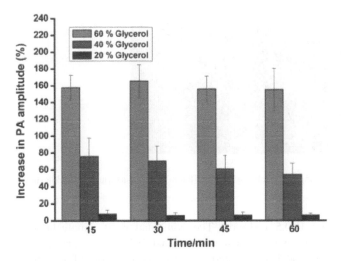

FIGURE 32.27 Dynamic changes of the PA amplitude enhancement after application of 20%, 40%, and 60 % glycerol at 0, 15, 30, 45, and 60 min respectively. Reprinted with permission from Reference [88].

biomedical photonics due to their excellent biocompatibility and biocompatibility, which has aroused widespread concern [91]. Figure 32.28 illustrated the highly efficient absorption of GNRs in the near-infrared (NIR) region, a spectral window which permits photons to penetrate biological tissues, allowing deeper imaging depth inside specimens. Therefore, the combination of pure OCT system and GNRs can effectively enhance the structural information of living embryonic tissue. As we all know, embryonic development is a very important stage in life. Early embryonic development will have an important

impact on the survival, foraging, movement, and reproduction of individuals. Therefore, the development of a noninvasive, high-resolution, and strong penetration depth imaging technology is essential for the study of embryonic development. A large number of studies have reported that OCT is used as a noninvasive, high-resolution, and convenient imaging technique in the three-dimensional (3D) imaging research on mammalian embryos [92–97]. However, due to the inherent multiple optical scattering characteristics of embryonic tissue, it is still a challenge for pure OCT to obtain satisfactory spatial resolution to display embryonic structure [91, 98]. Contrast agents which enhance the OCT images allow for a wider usage of OCT systems [99–101].

In *in vitro* experiments of living embryonic tissues, GNRs are added to the embryo culture medium and transported into the embryo as the nutrients are transported through the placenta. In order to quantify the sensitivity of contrast agents in embryonic tissue, the corresponding OCT imaging was applied (see Figure 32.28). Two images of living embryonic tissue sample without and with topical application of GNRs solution were captured. It is clear that OCT signal intensity reached the highest when living embryonic tissue was treated with GNRs. By comparison of images of Figures 32.29a and 32.29b, OCT signal intensity reached the highest after the topical application of GNR solution with 20 μg/mL in both 3 and 6 h culture in E 9.5 embryos, whereas without application of GNR solution, the light is almost blocked by the tissue. This is due to the fact that the embryonic tissue highly scatters the incoming light, which degrades the imaging performance.

In addition to enhancing OCT signal intensity, GNRs may have an impact on three-dimensional imaging of the organs of

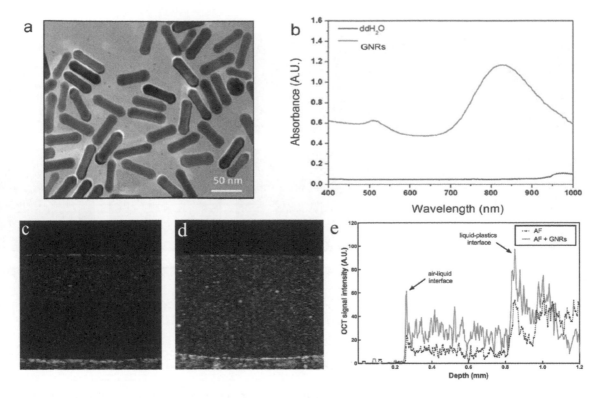

FIGURE 32.28 (a) TEM image of the GNRs. (b) Absorbance spectrum of the GNRs. The peak extinction is around 835 nm. (c) OCT B-scan images of 1 cm glass dish filled with amniotic fluid (AF). (d) OCT B-scan images of 1 cm glass dish filled with 10 μg/mL^{-1} GNRs in AF. (e) Average A-lines profile of these two B-scan images. Reprinted with permission from Reference [97].

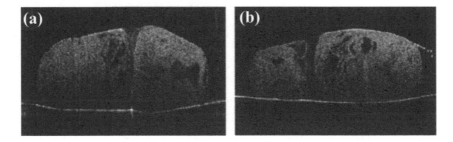

FIGURE 32.29 OCT signal analysis for E 9.5 embryos. 2D OCT images of living E9.5 embryos in the different GNR treatments for (a) 3 h, (b) 6 h with 20 μg/mL GNRs at E 9.5. Reprinted with permission from Reference [97].

living embryos. At day E10.5, when the embryo was not treated with GNRs, only a few faint anatomical structures appeared in the three-dimensional reconstruction (see Figure 32.30a). However, after treatment with 10 μg/mL GNRs for 6 h, the 3D reconstruction imaging had more accurate spatial localization with structural features, and better contrasting of the borders among organs can be observed on the GNR-treated embryo (see Figure 32.30b). For GNRs as a contrast agent, these findings confirmed that with an exogenous OCT contrast agent, greater 3D structural detail was obtained as compared to the control embryos. Moreover, the outlines of internal organs are clearer in the GNR-treated embryos.

In summary, GNRs could be delivered to the embryos with a short period of *ex vivo* culture. The utility of GNRs as contrast agent in OCT system decreased the high scattering of tissues and improved the contrast and penetration depth of images. GNRs also allowed reconstructing clear 3D images of E9.5

and E10.5 embryos. The combination of nanotechnology allows us to improve conventional OCT and assess the morphological changes and structural abnormalities of developing organs in real-time and in situ images.

Challenges and perspectives

In summary, optical clearing techniques have enriched our understanding of life sciences and innovated the way we study biological tissues, which are quite crucial for various optical imaging techniques. Evaluating the diffusion and permeability of OCAs in different tissue by OCT is an effective method for differentiating the normal and disease situation. However, further research still has to be done to find the optimal and safest type, concentration, and method of OCAs for the imaging of living biotissue. As mentioned above, optical

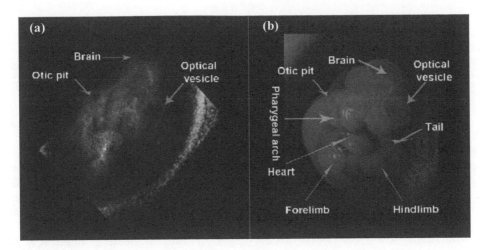

FIGURE 32.30 3D reconstruction images of E10.5 embryos (a) 3D reconstruction image of an E10.5 embryo without GNR treatment. (b) 3D reconstruction image of an E10.5 embryo treated with 10 μg mL^{-1} GNRs for 6 h. Reprinted with permission from Reference [97].

clearing techniques enable us to differentiate normal from pathological tissues based on the molecular permeability rates method, but the specific mechanisms of tissue optical clearing *in vivo* still remain a challenge. Thus, in practical clinical applications, specific type, concentration, and clearing method of the OCAs should be chosen according to the research requirements.

For future optical imaging applications, if we want to make them more effective, it is very important to understand the optical properties of the research object and screening optimal OCAs. Understanding the optical properties of the sample is not only the basis for screening or designing OCAs, but is also helpful for exploring the mechanisms of TOC. In addition, we believe that increasing numbers of new and safe enhancement permeability methods and other delivery techniques for OCAs diffusion will be developed and applied in biomedicine imaging.

Acknowledgments and funding

This work was supported by the National Natural Science Foundation of China (81701743, 81601530, 60778047, 81171377, 61273368, 81571744, 81301257, and 61307015), National Basic Research Program of China (2011CB707504, 2015AA020502, and 2010CB933903), Xiamen Science and Technology Plan Project (3502Z20183018), the Scientific Research Foundation of State Key Laboratory of Molecular Vaccinology and Molecular Diagnostics (2017KF05, 2017KF06), the Natural Science Foundation of Guangdong Province (06025080), the Key Science and Technology Project of Guangdong Province of China (2005B50101015, 2008B090500125), Open Research Fund of State Key Laboratory of Transient Optics and Photonics, Chinese Academy of Sciences (SKLST201501).

REFERENCES

1. M.G. Ghosn, E.F. Carbajal, N.A. Befrui, et al., "Concentration effect on the diffusion of glucose in ocular tissues," *Opt. Lasers Eng.* 46(12), 911–914 (2008).

2. M.G. Ghosn, E.F. Carbajal, N.A. Befrui, et al., "Permeability of hyperosmotic agent in normal and atherosclerotic vascular tissues," *J. Biomed. Opt.* 13(1), 010505 (2008).

3. Q.L. Zhao, J.L. Si, Z.Y. Guo, et al., "Quantifying glucose permeability and enhanced light penetration in ex vivo human normal and cancerous esophagus tissues with OCT," *Laser Phys. Lett.* 8(1), 71–77 (2010).

4. H.Q. Zhong, Z.Y. Guo, H.J. Wei, et al., "Quantification of glycerol diffusion in human normal and cancer breast tissues in vitro with optical coherence tomography," *Laser Phys. Lett.* 7(4), 315–320 (2010).

5. C. Buranachai, P. Thavarungkul, P. Kanatharana, I.V. Meglinski, "Application of wavelet analysis in optical coherence tomography for obscured pattern recognition," *Laser Phys. Lett.* 6(12), 892–895 (2009).

6. R.K. Wang, "Role of mass diffusion and water desorption on optical clearing of biological tissue immersed with the hyperosmotic agents," *Proc. SPIE* 5330, 160–166 (2004).

7. S.A. Boppart, W. Luo, D.L. Marks, K.W. Singletary, "Optical coherence tomography: Feasibility for basic research and image-guided surgery of breast cancer," *Breast Cancer Res. Treat.* 84(2), 85–97 (2004).

8. D. Chorvat, A. Chorvatova, "Multi-wavelength fluorescence lifetime spectroscopy: A new approach to the study of endogenous fluorescence in living cells and tissues," *Laser Phys. Lett.* 6(3), 175–193 (2009).

9. N.N. Bulgakova, N.I. Kazachkina, V.V. Sokolov, V.V. Smirnov, "Local fluorescence spectroscopy and detection of malignancies using laser excitation at various wavelengths," *Laser Phys.* 16(5), 889–895 (2006).

10. Q.L. Zhao, Z.Y. Guo, J.L. Si, et al., "Heat treatment of human esophageal tissues: Effect on esophageal cancer detection using oxygenated hemoglobin diffuse reflectance ratio," *Laser Phys.* 21(3), 559–565 (2011).

11. W.G. Zijlstra, A. Buursma, W.P.M. Roest, "Absorption spectra of human fetal and adult oxyhemoglobin, de-oxyhemoglobin, carboxyhemoglobin, and methemoglobin," *Clin. Chem.* 37(9), 1633–1638 (1991).

12. R.K. Jain, "Determinants of tumor blood flow: A review," *Cancer Res.* 48(10), 2641–2658 (1988).

13. G.P. Petrova, A.V. Boiko, K.V. Fedorova, et al., "Optical properties of solutions consisting of albumin and γ-globulin molecules in different ratio modeling blood serum," *Laser Phys.* 19(6), 1303–1307 (2009).

14. D.C.G.D. Veld, M. Skurichina, M.J.H. Witjes, et al., "Autofluorescence characteristics of healthy oral mucosa at different anatomical sites," *Lasers Surg. Med.* 32(5), 367–376 (2003).

15. K.V. Larin, M.G. Ghosn, S.N. Ivers, et al., "Quantification of glucose diffusion in arterial tissues by using optical coherence tomography," *Laser Phys. Lett.* 4(4), 312–317 (2007).

16. H.L. Xiong, Z.Y. Guo, C.C. Zeng, et al., "Application of hyperosmotic agent to determine gastric cancer with optical coherence tomography ex vivo in mice," *J. Biomed. Opt.* 14(2), 024029 (2009).

17. M.G. Ghosn, V.V. Tuchin, K.V. Larin, "Nondestructive quantification of analyte diffusion in cornea and sclera using optical coherence tomography," *Invest. Ophthalmol. Vis. Sci.* 48(6), 2726–2733 (2007).

18. X. Guo, Z.Y. Guo, H.J. Wei, et al., "In vivo quantification of propylene glycol, glucose and glycerol diffusion in human skin with optical coherence tomography," *Laser Phys.* 20(9), 1849–1855 (2010).

19. D.J. Faber, F.J. van der Meer, M.C.G. Aalders, T. van Leeuwen, "Quantitative measurement of attenuation coefficients of weakly scattering media using optical coherence tomography," *Opt. Express* 12(19), 4353–4365 (2004).

20. C.Y. Xu, J.M. Schmitt, S.G. Carlier, R. Virmani, "Characterization of atherosclerosis plaques by measuring both backscattering and attenuation coefficients in optical coherence tomography," *J. Biomed. Opt.* 13(3), 034003 (2008).

21. V.M. Kodach, D.J. Faber, J.V. Marle, et al., "Determination of the scattering anisotropy with optical coherence tomography," *Opt. Express* 19(7), 6131–6140 (2011).

22. F.J. van der Meer, D.J. Faber , M.C.G. Aalders, et al., "Apoptosis and necrosis-induced changes in light attenuation measured by optical coherence tomography," *Lasers Med. Sci.* 25(2), 259–267 (2010).

23. K.V. Larin, M.S. Eledrisi, M. Motamedi, R.O. Esenaliev, "Noninvasive blood glucose monitoring with optical coherence tomography: A pilot study in human subjects," *Diabetes Care* 25(12), 2263–2267 (2002).

24. Q.L. Zhao, C.Q. Zhou, H.J. Wei, et al., "Ex vivo determination of glucose permeability and optical attenuation coefficient in normal and adenomatous human colon tissues using spectral domain optical coherence tomography," *J. Biomed. Opt.* 17(10), 105004 (2012).

25. R.K. Wang, V.V. Tuchin, "Enhance light penetration in tissue for high resolution optical imaging techniques by use of biocompatible chemical agents," *J. X-Ray Sci. Technol.* 10(3), 167–176 (2002).

26. V.V. Tuchin, I.L. Maksimova, D.A. Zimnyakov, et al., "Light propagation in tissues with controlled optical properties," *J. Biomed. Opt.* 2(4), 401–417 (1997).

27. V.V. Tuchin, "Coherent optical techniques for the analysis of tissue structure and dynamics," *J. Biomed. Opt.* 4(1), 100–125 (1999).

28. M.H. Khan, B. Choi, S. Chess, et al., "Optical clearing of in vivo human skin: Implications for light-based diagnostic imaging and therapeutics," *Lasers Surg. Med.* 34(2), 83–85 (2004).

29. V.V. Tuchin, "Optical immersion as a new tool for controlling the optical properties of tissues and blood," *Laser Phys.* 15(8), 1109–1136 (2005).

30. V.V. Tuchin, "Optical clearing of tissues and blood using the immersion method," *J. Phys. D: Appl. Phys.* 38(15), 2497–2518 (2005).

31. A.N. Bashkatov, E.A. Genina, Y.P. Sinichkin, et al., "Glucose and mannitol diffusion in human dura mater," *Biophys. J.* 85(5), 3310–3318 (2003).

32. L. Tsu, E. Chen, T.S. Ishak, et al., "Determination of chemical agent optical clearing potential using in vitro human skin," *Lasers Surg. Med.* 36(2), 72–75 (2005).

33. X.Q. Xu, Q.H. Zhu, "Feasibility of sonophoretic delivery for effective skin optical clearing," *IEEE Biomed. Eng.* 55(4), 1432–1437 (2008).

34. J.Y. Jiang, M. Boese, P. Turner, R.K. Wang, "Penetration kinetics of dimethyl sulphoxide and glycerol in dynamic optical clearing of porcine skin tissue in vitro studied by Fourier transform infrared spectroscopic imaging," *J. Biomed. Opt.* 13(2), 021105 (2008).

35. D. Zhu, J. Zhang, H. Cui, et al., "Short-term and long-term effects of optical clearing agents on blood vessels in chick chorioallantoic membrane," *J. Biomed. Opt.* 13(2), 40–47 (2008).

36. R.J. McNichols, M.A. Fox, A. Gowda, et al., "Temporary dermal scatter reduction: Quantitative assessment and implications for improved laser tattoo removal," *Lasers Surg. Med.* 36(4), 289–296 (2005).

37. M. Kinnunen, R. Myllylae, S. Vainio, "Detecting glucose-induced changes in in vitro and in vivo experiments with optical coherence tomography," *J. Biomed. Opt.* 13(2), 78–84 (2008).

38. A.N. Bashkatov, A.N. Korolevich, V.V. Tuchin, et al., "In vivo investigation of human skin optical clearing and blood microcirculation under the action of glucose solution," *Asian J. Phys.* 15, 1 (2006).

39. R.K. Wang, X.Q. Xu, J.B. Elder, et al., "Possible mechanisms for optical clearing of whole blood by dextrans," *Proc. SPIE* 4965(1), 84–94 (2003).

40. Y.H. He, R.K. Wang, "Dynamic optical clearing effect of tissue impregnated with hyperosmotic agents and studied with optical coherence tomography," *J. Biomed. Opt.* 9(1), 200 (2004).

41. W.K. den Otter, R. Notman, J. Anwar, et al., "Modulating the skin barrier function by DMSO: Molecular dynamics simulations of hydrophilic and hydrophobic transmembrane pores," *Chem. Phys. Lipids* 154, S2–S3 (2008).

42. X. Guo, G.Y. Wu, H.J. Wei, et al., "Quantification of glucose diffusion in human lung tissues by using Fourier domain optical coherence tomography," *Photochem. Photobiol.* 88(2), 311–316 (2012).

43. S. Mccoy, A. Evans, C. James, "Histological study of hair follicles treated with a 3-msec pulsed ruby laser," *Lasers Surg. Med.* 24(2), 133–141 (1999).

44. G. Vargas, K.F. Chan, S.L. Thomsen, A.J. Welch, "Use of osmotically active agents to alter optical properties of tissue: Effects on the detected fluorescence signal measured through skin," *Lasers Surg. Med.* 29(3), 213–220 (2001).

45. R.K. Wang, X.Q. Xu, V.V. Tuchin, J.B. Elder, "Concurrent enhancement of imaging depth and contrast for optical coherence tomography by hyperosmotic agents," *J. Opt. Soc. Am. B* 18(7), 948–953 (2001).

46. V.V. Tuchin. *Tissue Optics: Light Scattering Methods and Instruments for Medical Diagnosis* 2nd edn., SPIE Press, Bellingham (2007).

47. I.L. Maksimova, D.A. Zimnyakov, V.V. Tuchin, "Control of optical properties of biotissues: I. spectral properties of the eye sclera," *Opt. Spectrosc.* 89(1), 78–86 (2000).

48. M.R. Prausnitz, V.G. Bose, R. Langer, J.C. Weaver, "Electroporation of mammalian skin: A mechanism to enhance transdermal drug delivery," *Proc. Natl Acad. Sci. U.S.A.* 90(22), 10504–10508 (1993).

49. E.H. Choi, S.H. Lee , S.K. Ahn, S.M. Hwang, "The pre-treatment effect of chemical skin penetration enhancers in transdermal drug delivery using iontophoresis," *Skin Pharmacol. Appl. Skin Physiol.* 12(6), 326–335 (1999).

50. S. Henry, D.V. Mcallister, M.G. Allen, et al., "Microfabricated microneedles: A novel approach to transdermal drug delivery," *J. Pharm. Sci.* 87(8), 922–925 (1998).

51. Q.L. Zhao, H.J. Wei, Y.H. He, et al., "Evaluation of ultrasound and glucose synergy effect on the optical clearing and light penetration for human colon tissue using SD-OCT, " *J. Biophoton.* 7(11–12), 938–947 (2014).

52. I. Lavon, J. Kost, "Ultrasound and transdermal drug delivery," *Drug Discov. Today* 9(15), 670–676 (2004).

53. G.A. Husseini, W.G. Pitt, "Ultrasonic-activated micellar drug delivery for cancer treatment," *J. Pharm. Sci.* 98(3), 795–811 (2009).

54. N. Kim, A.F. El-Kattan, C.S. Asbill, ct al., "Evaluation of derivatives of 3-(2-oxo-1-pyrrolidine) hexahydro-1H-az-epine-2-one as dermal penetration enhancers: Side chain length variation and molecular modeling," *J. Controlled Release* 73(2–3), 183–196 (2001).

55. R.H. Guy, Y.N. Kalia, M.B. Delgado-Charro, et al., "Iontophoresis: Electrorepulsion and electroosmosis," *J. Controlled Release* 64(1–3), 129–132 (2000).

56. N. Dujardin, P.V. Der, V. Préat Smissen, V. Preat. "Topical gene transfer into rat skin using electroporation," *Pharm. Res.* 18(1), 61–66 (2001).

57. A.S. Michaels, S.K. Chandrasekaran, J.E. Shaw, "Drug permeation through human skin: Theory and *in vitro* experimental measurement," *J. Am. Inst. Chen. Eng.* 21(5), 985–996 (1975).

58. P.M. Elias, "Epidermal lipids, barrier function and desquamation," *J. Invest. Dermatol.* 80(1), 44–49 (1983).

59. L. Landmann, "The epidermal permeability barrier," *Anat. Embryol.* 178(1), 1–13 (1998).

60. M.E. Hartstein, G.G. Massry, J.B. Holds, et al., *Pearls and Pitfalls in Cosmetic Oculoplastic Surgery Books* 2nd edn., Springer, Kaufman (2015).

61. A.D. Katsambas, M.L. Torello. *European Handbook of Dermatological Treatments Books*, Springer, Berlin Heidelberg (2015).

62. B.K. Kang, J.H. Choi, K.H. Jeong, et al., "A study of the effects of physical dermabrasion combined with chemical peeling in porcine skin," *J. Cosmet. Laser Ther.* 17(1), 24–30 (2015).

63. N. Zakopoulou, G. Kontochristopoulos, "Superficial chemical peels," *J. Cosmet. Dermatol.* 5(3), 246–253 (2006).

64. M.R.A. Hussein, E.A.D. Eman, A.A.M. Amira, et al., "Chemical peeling and microdermabrasion of the skin: Comparative immunohistological and ultrastructural studies," *J. Dermatol. Sci.* 52(3), 205–209 (2008).

65. B. Marczyk, P. Mucha, E. Budzisz, H. Rotsztejn, "Comparative study of the effect of 50% pyruvic and 30% salicylic peels on the skin lipid film in patients with acne vulgaris," *J. Cosmet. Dermatol.* 13(1), 15–21 (2014).

66. J.A. Bouwstra, M.Ponec, "The skin barrier in healthy and diseased state," *Biochim. Biophys. Acta Biomembr.* 1758(12), 2080–2095 (2006).

67. H.S. Lee, I.H. Kim, "Salicylic acid peels for the treatment of acne vulgaris in Asian patients," *Dermatol. Surg.* 29(12), 1196–1199 (2003).

68. C.M. Burgess, *Cosmetic. Dermatology Books*, Springer, Berlin Heidelberg (2005).

69. A. Tosti, P.E. Grimes, P.E. De Padova. *Color Atlas of Chemical Peels*, Springer, Berlin Heidelberg (2006).

70. W.P. Bowe, A.R. Shalita, "Effective over-the-counter acne treatments. Seminars in cutaneous medicine and surgery," *WB Saunders* **27**(3), 170–176 (2008).

71. Q.L. Zhao, C.X.Dai, S.H. Fan, et al., "Synergistic efficacy of salicylic acid with a penetration enhancer on human skin monitored by OCT and diffuse reflectance spectroscopy," *Sci. Rep.* 6, 34954 (2016).

72. C. Thomas, "Cosmeceutical agents: A comprehensive review of the literature clinical medicine," *Dermatology* 1, 1–20 (2008).

73. C. Huber, E. Christophers, "'Keratolytic' effect of salicylic acid," *Arch. Dermatol. Res.* 257(3), 293–297 (1977).

74. A. Hussain, P.H. Andrzej, et al., "Potential enhancers for transdermal drug delivery: A review," *Int. J. Bas. Med. Sci. Pharm.* 4(181), 19–22 (2014).

75. K. Swain, S. Pattnaik, S.C. Sahu, et al., "Drug in adhesive type transdermal matrix systems of ondansetron hydrochloride: Optimization of permeation pattern using response surface methodology," *J. Drug Target.* 18(2), 106–114 (2010).

76. Y. Zhou, J. Yao, L.V. Wang, "Optical clearing-aided photoacoustic microscopy with enhanced resolution and imaging depth," *Opt. Lett.* 38(14), 2592–2595 (2013).

77. G. Ku, L.V. Wang, "Deeply penetrating photoacoustic tomography in biological tissues enhanced with an optical contrast agent," *Opt. Lett.* 30(5), 507–509 (2005).

78. X.D. Wang, G. Ku, M.A. Wegiel, et al., "Noninvasive photoacoustic angiography of animal brains in vivo with near-infrared light and an optical contrast agent," *Opt. Lett.* 29(7), 730–732 (2004).

79. R.K. Wang, "Signal degradation by coherence tomography multiple scattering in optical of dense tissue: A Monte Carlo study towards optical clearing of biotissues," *Phys. Med. Biol.* 47(13), 2281–2299 (2002).

80. V.V. Tuchin, "Tissue optics: Light scattering methods and instruments for medical diagnosis," in *SPIE Tutorial Texts in Optical Engineering*, SPIE Press, Bellingham, Washington (2000).

81. V.V. Tuchin, D.A. Zimnyakov, I.L. Maksimova, et al., "The coherent, low-coherent and polarized light interaction with tissues undergo the refractive indices matching control," *Proc. SPIE* 3251, 12–21 (1998).

82. A.N. Bashkatov, E.A. Genina, V.I. Kochubey, et al., "In vivo and in vitro study of control of rat skin optical properties by acting of 40%-glucose solution," *Proc. SPIE* 4241, 223–230 (2001).

83. A.N. Bashkatov, E.A. Genina, V.I. Kochubey, et al., "In vivo and in vitro study of control of rat skin optical properties by acting of osmotical liquid," *Proc. SPIE* 4224, 300–311 (2000).

84. A.N. Bashkatov, E.A. Genina, V.I. Kochubey, et al., "Study of osmotical liquids diffusion within sclera," *Proc. SPIE* 3908, 266–276 (2000).

85. V.V. Tuchin, "Controlling of tissue optical properties," *Proc. SPIE* 4001, 30–53 (2000).

86. X. Xu, Q. Zhu, C. Sun, "Combined effect of ultrasound-SLS on skin optical clearing," *IEEE Photon. Technol. Lett.* 20(24), 2117–2119 (2008).

87. N.A. Trunina, V.V. Lychagov, V.V. Tuchin, "OCT monitoring of diffusion of water and glycerol through tooth dentine in different geometry of wetting," *Proc. SPIE* 7563, 75630U (2010).

88. Q.L. Zhao, L. Lin, L. Qian, et al., "Concentration dependence of optical clearing on the enhancement of laser scanning optical resolution photoacoustic microscopy (LSOR-PAM) imaging," *J. Biomed. Opt.* 19(3), 36019 (2014).

89. P.N. Henrion, "Diffusion in the sucrose + water system," *Trans. Faraday Soc.* 60, 72–74 (1964).

90. L.G. Longsworth, "Diffusion measurements, at 25, of aqueous solutions of amino acids, peptides and sugars," *J. Am. Chem. Soc.* 75(22), 5705–5709 (1953).

91. K.V. Larin, I.V. Larina, L. Michael, et al., "Live imaging of early developmental processes in mammalian embryos with optical coherence tomography," *J. Innov. Opt. Health Sci.* 02(03), 253–259 (2009).

92. M. Singh, C. Wu, D. Mayerich, et al., "Multimodal embryonic imaging using optical coherence tomography, selective plane illumination microscopy, and optical projection tomography," *IEEE Eng. Biol. Soc.* 2016, 3922–3925 (2016).

93. C. Wu, N. Sudheendran, M. Singh, et al., "Rotational imaging optical coherence tomography for full-body mouse embryonic imaging," *J. Biomed. Opt.* 21(2), 026002 (2016).

94. C. Wu, L. Henry, S.H. Ran, et al., "Comparison and combination of rotational imaging optical coherence tomography and selective plane illumination microscopy for embryonic study," *Biomed. Opt. Express* 8(10), 4629–4639 (2017).

95. M. Singh, R. Raghunathan, V. Piazza, et al., "Applicability, usability, and limitations of murine embryonic imaging with optical coherence tomography and optical projection tomography," *Biomed. Opt. Express* 7(6), 2295 (2016).

96. J. Zhu, K.T. Yong, R. Indrajit, et al., "Additive controlled synthesis of gold nanorods (GNRs) for two-photon luminescence imaging of cancer cells," *Nanotechnology* 21(28), 285106 (2010).

97. Y.L. Huang, M.H. Li, D.D. Huang, et al., "Depth-resolved enhanced spectral-domain OCT imaging of live mammalian embryos using gold nanoparticles as contrast agent," *Small* 15(35), 1902346 (2019).

98. S.A. Boppart, M.E. Brezinski, B.E. Bouma, et al., "Investigation of developing embryonic morphology using optical coherence tomography," *Dev. Biol.* 177(1), 54–63 (1996).

99. C.H. Yang, "Molecular contrast optical coherence tomography: A review," *Photochem. Photobiol.* 81(2), 215–237 (2005).

100. Q.L. Zhao, Z.Y. Guo, H.J. Wei, et al., "Depth-resolved monitoring of diffusion of hyperosmotic agents in normal and malignant human esophagus tissues using OCT *in-vitro*," *Quantum Electron.* 41(10), 950–955 (2011).

101. S.A. Boppart, A.L. Oldenburg, C.Y. Xu, et al., "Optical probes and techniques for molecular contrast enhancement in coherence imaging," *J. Biomed. Opt.* 10(4), 041208 (2005).

33

Optical clearing for cancer diagnostics and monitoring

Luís M. Oliveira and Valery V. Tuchin

CONTENTS

Introduction

Amongst the many diseases that can affect the human body, cancer-type pathologies are a particular branch, where the disease development is associated with multiple changes in the health level of cells and tissues. Such changes are space- and time-dependent, leading ultimately to the development of malignant tumors [1]. The abnormal cell growth (neoplasia) that occurs in tumor development may lead to the invasion of the surrounding tissues, or the occurrence of metastasis (spread to distant organs), which is the main cause of morbidity and mortality for most patients [1–6]. Without a clear and precise explanation of how cancer arises, it becomes difficult to establish effective prevention and long-term management of the disease [1].

The concern about cancer pathologies is high since it can develop in different anatomical areas of the human body without any apparent reason, or spread to distant organs, which makes its explanation, detection, and control difficult. A large number of cancers are known, and classifications related to the anatomical area where they occur or to the degree of fatality have been reported [7–9]. Due to such high cancer diversity, different oncologic specialties have been developed throughout the history of medicine to study the various types of cancers. Many cancer-related studies have been developed to characterize the various stages of cancer development and to propose potential diagnostic and treatment methods. In recent years, the number of studies that use optical methods to diagnose cancer has grown significantly, as demonstrated by Figure 33.1, where the number of publications that contain the words "optical cancer diagnostics" are presented according to Pubmed, Web of Science, and ScienceDirect.

Noninvasive spectroscopic or imaging methods can be used directly or through endoscopy to diagnose cancer when the cancer polyps or tumors are located in the outermost tissue layers. If tumors are located deep in the tissue, their detection through noninvasive optical methods is not easy due to the high light scattering that occurs in biological tissues [10].

The combination of optical clearing treatments with imaging or spectroscopic methods provides new approaches for cancer detection at deep-tissue layers and may allow its detection at early stages of development and monitoring of its progress [10, 11]. As described in Chapter 1, the study of the diffusion properties of optical clearing agents (OCAs) can be made from *ex vivo* tissues using collimated transmittance, T_c, and thickness measurements performed during treatments with different osmolarities of an agent. Figure 1.10 of Chapter 1 shows that those studies allowed researchers to discriminate between human normal and pathological colorectal mucosa by identifying a shift in the peak of the diffusion time dependence on the OCA concentration in the treating solution [12, 13]. Such a shift indicates that the mobile water content in cancer tissue is about 5% higher than in normal tissue. As indicated in Chapter 1, those studies and the discriminated evaluation of mobile water content in normal and cancer tissues can be reproduced for *in vivo* tissues if, instead of T_c measurements, diffuse reflectance, R_d, measurements are performed. The calculation of the diffusion coefficients for water and OCA, which is also discriminated between normal and cancer tissues [12, 13], can be made for *in vivo* tissues based on thickness measurements performed via optical coherence tomography (OCT) or confocal microscopy measurements [14, 15]. Similar studies were performed with skin and myocardium tissues from rats, allowing the diagnosis of diabetes [16], which shows that such a method is reliable for diagnosing other pathologies as well.

Considering the combination of the OCT method with optical clearing (OC) treatments, it is also possible to perform cancer diagnosis. A study has reported a change in the slope of the OCT signal during the diffusion of glycerol into breast tissues, which is different between normal and cancer tissues [17].

DOI: 10.1201/9781003025252-37

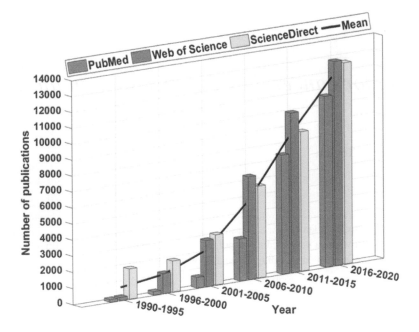

FIGURE 33.1   Number of publications about optical cancer diagnostics from 1990 and 2020.

Recent progress in optical methods, such as optical imaging and spectroscopy, will be presented and discussed in the following sections.

Enhanced tumor imaging through optical clearing

The combination of OC treatments with imaging methods also allows improved tumor detection. By OC-assisted enhancement of image contrast, resolution, and visualization depth, the detection and monitoring of hidden tumors have become possible. Some examples were reported regarding the laser speckle contrast imaging of prostate cancer [18] and microvessels [19]; Raman spectroscopy of subcutaneous tumors [20]; and 3D multimodal high-resolution imaging of tumor microvasculature structure through optical computed tomography and optical emission computed tomography methods [21, 22].

Wei et al. [23] used OC-assisted label-free stimulated Raman scattering microscopy to reveal significant differences in lipid-to-protein ratios between normal and glioblastoma brain tissues and also between glioblastoma and tumor cells. Quantification of the 3D microvascular characteristics in normal mouse brains and in tissues with diffuse, infiltrative growing glioblastoma brain tumors has been possible using fluorescent confocal imaging, two-photon imaging, photoacoustic imaging, and image reconstruction of optically cleared tissues [24]. White-light diffuse reflectance spectroscopy and near infrared OCT were used by Pires et al. to evaluate the clearing effect of topically applied OCA (19:1 mixture of polyethylene glycol – PEG-400 and 1,2-propanediol) in induced cutaneous melanoma of *in vivo* mice skin and to image the microvascular network [25]. The enhancement in spatial contrast and resolution obtained in this study at depths of 300 and 450 μm is presented in Figure 33.2.

As represented in Figure 33.2, and by comparing images from left to right at the same tissue depth, we see an increase in contrast and resolution as a result from OCA application to a pigmented melanoma tissue. Reference [25] presents complimentary images and data for tissue depths between 0 and 750 μm.

Other studies involving OC-assisted microscopy have opened new possibilities for visualizing the entire central nervous system, the blood microvessel system, whole organs, and the whole body with subcellular resolution in healthy and tumor tissues, allowing for single-cell studies and new biomedical engineering improvements [26]. One such study was reported by Cui et al. [27], where OCA-assisted hyper spectral dark-filed microscopy was used to quantitatively map the dimerization-activated receptor kinase HER2 (human epidermal growth fact receptor 2) in a single cancer cell using a nonfluorescence approach. As a result of OC, high spatiotemporal resolution at the characterization of intracellularly grown gold particulates has also been demonstrated in this study [27].

A pioneering lung clearing study was reported by Scott et al. [28]. Using benzyl alcohol/benzyl benzoate (BABB) to clear rat lung tissues, a confocal microscope with automated scanning was used to acquire images that presented a detailed view of the relationship between nerves, vessels, and airway architecture. Novel patterns of pleural innervations and connectivity patterns of pleural nerves were also detected in this study [28].

Lung clearing was proved useful for light-sheet microscopy detection and analysis of tumor metastases in rats [29], or for confocal microscopy to analyze tumor-immune microenvironment, such as a 3D pattern of tumor-associated macrophages (TAM) distribution [30, 31]. The authors of Reference 30 have used fluorescence imaging to detect both large and small tumors of pulmonary carcinoma. Considering Figure 33.3(d), both large and small tumors are visible after optical clearing,

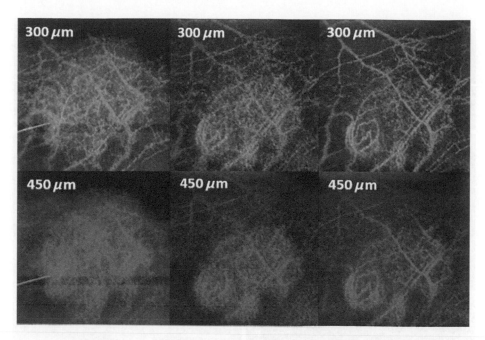

FIGURE 33.2 OCT images of melanoma vasculature obtained at 300 *μm* (top) and 450 *μm* (bottom) for mice skin at 0 min (left), 120 min (middle), and 250 min (right). Reproduced from Reference 25 under creative commons license.

while in Figure 33.3(a), only the large tumor can be seen in the conventional micro-CT image. The position of the large tumor is identified with a big arrow in Figures 33.3(a), 33.3(b), and 33.3(d) and the positions of small tumors are identified with small arrows in Figure 33.3(d). Applying the confocal microscopy technique to the optically cleared lung tissues, it was possible to perform a complete observation of the tumor distribution in the lung and obtain a cellular resolution throughout the entire tumor mass (see Figure 33.3(e)–(g)). Figure 33.3 (f) shows a computational 3D-rendering image of a whole-lung tumor nodule.

The combination of OC treatments with light-sheet microscopy was also found beneficial for detecting and monitoring prostate and bladder cancer. Treatment of human prostate biopsy specimens [32] and bladder tumors [33] with either CUBIC (Clear, Unobstructed Brain Imaging cocktails and Computational analysis) or PACT (PAssive Clarity Technique) resulted in GFP (Green Fluorescence Protein) fluorescence preservation, minimal tissue deformation, and feasibility of whole-mount two-photon imaging. The combination of Murray's Clear (same as BABB – see above) with immunochemistry allowed van Royen et al. to shed light on architectural differences between grades of prostate cancer [34]. Later, the same group used iDISCO (immunolabeling-enabled three-Dimensional Imaging of Solvent-Cleared Organs) protocols to characterize 3D architecture of both benign and precancerous prostate lesions [35] and described two distinct growth patterns of prostate cancer in patients [36].

Regarding human breast tumor biopsy, Hume et al. used multiphoton microscopy to prove the effectiveness of CUBIC as a clearing agent, and the combination of this method with immunostaining allowed monitoring structural integrity of an engineered 3D *in vitro* adipose-tissue model [37]. Migratory effects of adipocytes on breast cancer cells were seen in this

study, indicating that such tissue models can be used to plan cancer therapy strategies.

Spectroscopy methods are also useful for detecting and monitoring cancer development, provided that OC treatments are used to uncover the hidden oncologic information from tissues. The following two sections describe the recent developments obtained in this field at our lab in Porto as a result of our cooperation with the Portuguese Oncology Institute of Porto.

OC-induced ultraviolet tissue windows as a new means for potential cancer diagnosis and treatment

Multiple light scattering in tissues is the major obstacle for the application of optical methods in clinical practice, since due to the refractive index (RI) mismatch between tissue components and fluids, small light penetration depth is obtained [10]. Due to the decreasing behavior of the RI with increasing wavelength for tissue components and fluids, such a mismatch is higher for the ultraviolet (UV) range, which suggests that when clearing treatments are applied to tissues, a higher magnitude for the RI matching mechanism should be obtained in that range [38]. In addition to such high scattering in the UV range, some tissue components present significant absorption bands in that range: 200 to 230 nm (proteins), 260 nm (DNA and RNA), 275 and 345 (HbO_2), and 275 and 360 nm (Hb) [39–41]. Such combination of absorption bands and multiple scattering makes light transmission significantly low between 200 and 400 nm, as recently reported by Carneiro et al. [41].

As discussed in Chapter 1, the third OC mechanism is designated as protein dissociation, and since proteins present an absorption band in the deep-UV, the combination of the three

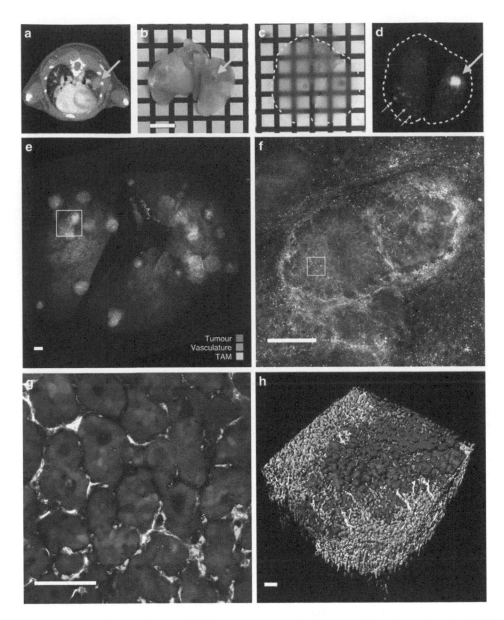

FIGURE 33.3 Micro-CT scan of mouse lungs (a), intact lungs (b), cleared lungs (c), fluorescence of cleared lungs (d), wide-field fluorescence of whole lung with color identification of tumors, vasculature and TAM (e), ×4 magnification of box in (e) showing a tumor (f), ×20 magnification of box in (f) showing the tumor cells (g), and computational 3D-rendering of a whole-lung tumor nodule (h). Scale bars of 5 *mm* in (b) of 1 mm in (e) and (f) and of 0.1 mm in (g) and (h). Reproduced from Reference 30 under creative commons license.

OC mechanisms should provide a better clearing in the UV range. To study the efficiency of the OC treatments in this range, Carneiro et al. have performed T_c measurements from human colorectal muscle tissues during treatments with glycerol solutions with different osmolarities [38]. In this study, muscle samples having approximated circular slab-form ($\phi = 1$ cm, $d = 0.5$ mm) were treated by immersion in aqueous solutions containing glycerol in the concentrations of 20%, 40%, or 60% [10, 38]. Figure 33.4 presents the 3D kinetics obtained for the T_c measurements from each treatment.

A comparison between all the graphs in Figure 33.4 shows that the magnitude of the increase in the T_c spectra grows with the glycerol concentration in the treating solution. Such behavior is seen for the visible to near infrared (NIR) range, but no significant variations are seen in the UV range. To verify if any variations occur at the deep-UV range, a magnification of

the graphs in Figure 33.4 are presented in Figure 33.5 for the wavelength range between 200 and 340 nm.

The graphs presented in Figure 33.5 show that tissue transparency also increases in the UV range, since in all treatments we see an increase in T_c on both sides of the absorption band of DNA (260 nm). Once again, such variations increase in magnitude as the concentration of glycerol increases in the treating solution. However, these variations are small in magnitude when compared to the variations seen for the visible-NIR range (see Figure 33.4).

To effectively evaluate what happens in the UV range, the absolute variations are not useful. Instead, one needs to calculate the relative variations (or the OC efficiency) in the entire spectral range to perform an accurate comparison between the UV and the visible-NIR ranges. To perform such a calculation, Equation (33.1) [10, 38] was used:

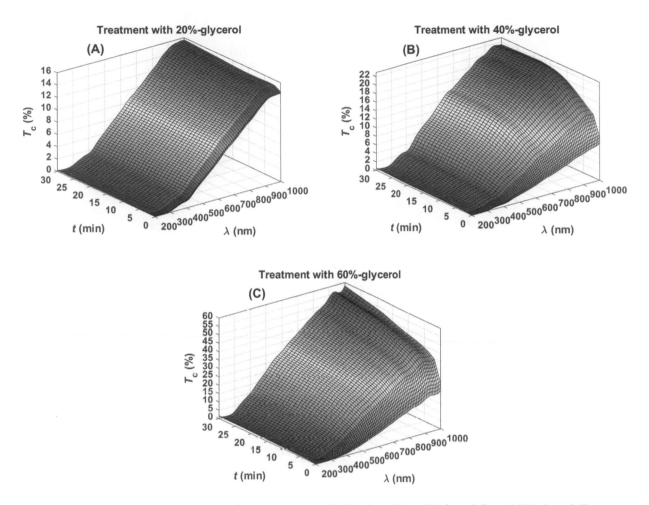

FIGURE 33.4 T_c kinetics for human colorectal muscle under treatment with 20% glycerol (A), 40% glycerol (B), and 60% glycerol (C).

$$OC_{eff}(\lambda) = \frac{T_c(\lambda,t) - T_c(\lambda,t=0)}{T_c(\lambda,t=0)} \times 100\%, \qquad (33.1)$$

where $T_c(\lambda, t)$ represents a spectrum measured at a time of treatment t and $T_c(\lambda, t = 0)$ is the spectrum for the untreated tissue ($t = 0$).

As a result of performing this calculation for all treatments, we see very important information in graphs of Figure 33.6, where only the range between 200 and 450 nm was considered for a better-detailed representation.

The optical clearing efficiency presented in graphs of Figure 33.6 shows that the RI matching mechanism is more efficient in the UV range. In fact, such matching in the visible (also in the NIR) range only seems to have approximate magnitude to the one observed in the deep-UV, as the glycerol concentration in the treating solution increases above 40%. These graphs show that the increase in magnitude of the OC efficiency also grows with glycerol concentration in the treating solution. For the treatment with 20% glycerol, we see the occurrence of a transparency window with a peak at 230 nm that has much higher magnitude than the transparency created for wavelengths above 260 nm. Considering the treatment with 40% glycerol, the same happens, but with higher magnitude. In this treatment, a second transparency window seems to occur with smaller magnitude between the absorption band of DNA (260 nm) and the Soret band (418 nm). As we move to the

treatment with 60% glycerol, this second transparency window seems to follow the behavior and magnitude of the window in the deep-UV. In fact, for this treatment, if the absorption band of DNA did not exist, the created transparency window would be one, with a peak at 230 nm and decreasing in magnitude for higher wavelengths. This means that due to the occurrence of the absorption band of DNA, the OC treatment creates two transparency windows in the muscle – one between 200 and 260 nm, with peak at 230 nm; and the second between 260 and NIR wavelengths, with a peak at 300 nm. This last window was already known for visible and NIR wavelengths [42–44], but the occurrence of its peak at 300 nm was previously not known.

This discovery was recently patented by the authors of Reference [38] since the creation of OC-induced windows opens new possibilities for diagnosis and treatment of diseases by using light at their central wavelengths.

The following section shows one diagnostic application that makes use of one of these clearing-induced windows.

Evaluation of the third OC mechanism as a cancer diagnostic and monitoring procedure

As discussed in the previous section, the OC-induced transparency windows in the UV range can open new diagnostic and/or treatment possibilities. To use UV light in clinical procedures,

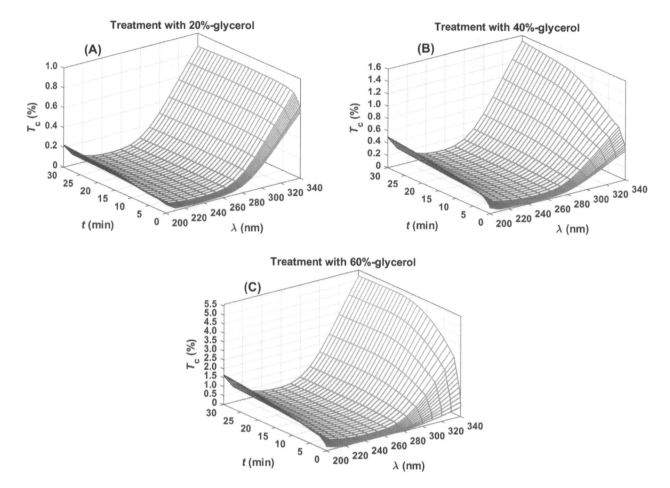

FIGURE 33.5   Magnification of the T_c kinetics in Figure 33.4 for the treatments with 20% glycerol (A), 40% glycerol (B), and 60% glycerol (C).

some attention should be paid to the time of exposure and light dosage, since it is known that UV light is cytotoxic [45]. Light toxicity in tissue is higher at deep-UV, but by applying low dose and at short time durations, such effect should be minimized and clinical procedures that use UV light may be applied for clinical purposes.

One of the procedures of interest is the diagnosis of cancer and other pathologies. Cancer tissues are known to have higher protein content than healthy tissues [46], meaning that when OC treatments are applied, the detection and evaluation of protein dissociation (third OC mechanism – see Chapter 1) should be differentiated between normal and pathological tissues. To develop such a procedure, UV light is necessary, since as discussed above, proteins present a significant absorption band at 200 nm [39–41].

To verify if differentiated data concerning protein dissociation could be obtained from spectral measurements during OC treatments, our research group in Porto has performed some studies with human colorectal mucosa tissues, under treatment with glycerol. The glycerol used in these studies was 93% pure, and such high degree of purity for glycerol was selected to maximize the protein dissociation mechanism. Five normal and five pathological (colorectal carcinoma) mucosa tissues were prepared with slab-form, as indicated in the previous section, to be used in this study. Each of these tissue samples was submitted to T_c measurements between 200 and 1000 nm

during immersion in glycerol. The data acquisition and calculation procedures were the same as described in the previous section for the colorectal muscle, but in this study, mean T_c spectra were calculated for each time of treatment, both for normal and pathological mucosa.

Since the mean measured data did not show significant variations in the deep-UV, the calculation of the OC efficiency, as described by Equation (33.1) was adopted for both tissues. Figure 33.7 presents the result of this calculation for both tissues, between 200 and 450 nm.

Both kinetics presented in graphs of Figure 33.7 show the occurrence of two transparency windows, but with central peaks located at different wavelengths than the ones observed for the colorectal muscle (see Figure 33.6). Due to different tissue components, those central peaks for mucosa tissues are seen at 200 and 370 nm (normal mucosa) and 200 and 380 nm (pathological mucosa). As in the case of the colorectal muscle, the second OC-induced UV window is located between two absorption bands: the DNA band (260 nm) and the Soret band (418 nm). Considering that the colorectal mucosa and muscle tissues have different components, this second window is not as high as the one observed for the colorectal muscle (Figure 33.6(c)).

For the first OC-induced window, and comparing the normal and pathological mucosa in Figure 33.7, we see that higher values are reached in the case of pathological mucosa.

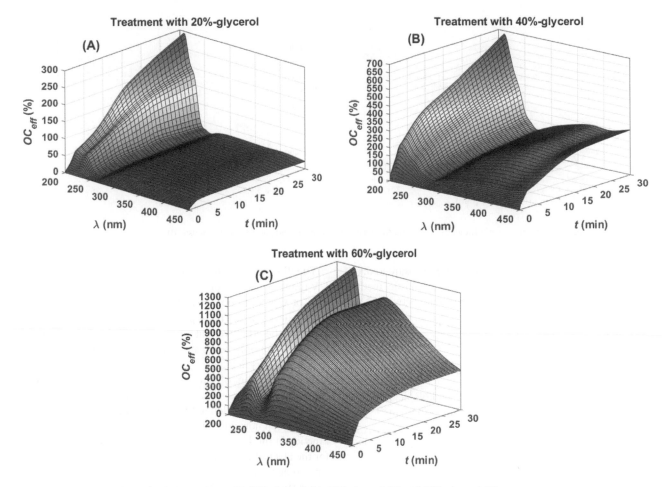

FIGURE 33.6 OC efficiency for the treatments with 20% glycerol (A), 40% glycerol (B), and 60% glycerol (C).

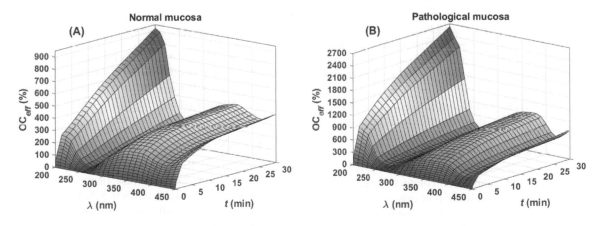

FIGURE 33.7 Mean OC efficiency for the treatments with 93% glycerol for normal mucosa (A) and pathological mucosa (B).

Such difference in the OC efficiency shows that cancer tissues have higher protein content, as described in the literature [46]. Since proteins have a significant absorption band at 200–230 nm [39–41], and due to the fact that the highest value observed in both graphs of Figure 33.7 occurs at 200 nm, we have represented the OC efficiency kinetics for this wavelength in Figure 33.8.

In graphs of Figure 33.8, we have represented the OC efficiency values at each time of treatment with the corresponding

standard-deviation bars. The red line represents the best fitting of the kinetics data in the entire time-span, which is described by an equation such as:

$$OC_{eff}\left(\lambda = 200, t\right) = \sqrt{\frac{t}{A}} + B \times t, \qquad (33.2)$$

where a combination of square root and linear dependencies occurs. Looking into the graphs of Figure 33.8, this fitting is not accurate for the time interval between 1 and 5 min.

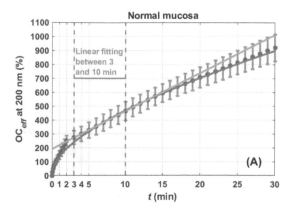
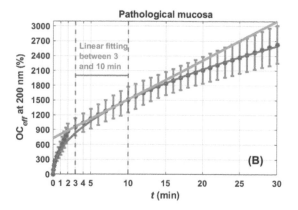

FIGURE 33.8   Linear fitting of the OC efficiency at 200 nm for normal mucosa (A) and pathological mucosa (B).

Due to the objective of minimizing tissue exposure to deep-UV cytotoxic light, and also because a diagnostic method should be fast, we found a linear fitting of data between 3 and 10 min for both tissues. For the normal mucosa, such fitting is described as:

$$OC_{\text{eff-nm}}\left(\lambda = 200, t\right) = 27.4 \times t + 189.6, \qquad (33.3)$$

where t is time to be used in minutes. For pathological mucosa, the linear fitting obtained for the same time range was:

$$OC_{\text{eff-pm}}\left(\lambda = 200, t\right) = 79.1 \times t + 721.5, \qquad (33.4)$$

using also t in minutes.

Since the kinetics presented in the graphs shown in Figure 33.8 are mainly due to protein dissociation in the mucosa tissues, one can define the slope of such kinetics as a dissociation parameter, which can be used for discrimination between normal and pathological tissues. By comparison between these values in Equations (33.3) and (33.4), we see about a three-fold increase from the normal to the pathological mucosa.

Reproduction of these studies can be made *in vivo* using R_d measurements, and two or three measurements made between 3 and 10 min can be used to estimate such a slope and allow for cancer diagnosis and monitoring.

This discovery of different slopes in the OC efficiency at 200 nm for normal and pathological colorectal tissues is subject to a patent pendent by the authors of this chapter, and new studies with other types of cancer are now being conducted for cancer-calibration purposes and disease progress monitoring.

Future perspectives

According to the various studies presented in the previous sections, the application of OC treatments has provided new tools for cancer diagnosis and monitoring, both with imaging and spectral techniques. Since the research in this field is very active, with an increasing number of publications per year, new OCAs and clearing protocols for cancer detection are to be expected in the near future.

The discovery of OC-induced transparency windows in the UV range opens new possibilities for oncology. Due to the toxicity of UV light, not only diagnostic/monitoring procedures can be developed. The application of UV light to kill cancer cells may be effective, but adequate procedures and treatment protocols can only be developed after conducting appropriate research in this field.

The evaluation of protein dissociation presented in the previous section can be used as a diagnostic method for other pathologies, depending on the different protein content between normal and pathological tissues. As the clearing kinetics in the UV range is observed in different proportions for the absorption bands of proteins and DNA, it may be possible to quantify the ratio of OC efficiency at 200 and 260 nm to discriminate other pathologies.

Acknowledgments

The work of L.O. was supported by the Portuguese Science Foundation, grant # FCT-UIDB/04730/2020.

The work of V.V.T. was supported by the Government of the Russian Federation grant # 075-15-2021-615.

REFERENCES

1. T.N. Seyfried, R.E. Flores, A.M. Poff, and D.P. D'Agostino, "Cancer as a metabolic disease: Implications for novel therapeutics," *Carcinogenesis* 35(3), 515–527 (2014).
2. M.B. Sporn, "The war on cancer," *Lancet* 347(9012), 1377–1381 (1996).
3. T.N. Seyfried, *Cancer as a Metabolic Disease: On the Origin, Management, and Prevention of Cancer* (Wiley, Hoboken, NJ, 2012).
4. I.J. Fidler, "The pathogenesis of cancer metastasis: The 'seed and soil' hypothesis revisited," *Nat. Rev. Cancer* 3(6), 453–458 (2003).
5. Y. Lazebnik, "What are the hallmarks of cancer?" *Nat. Rev. Cancer* 10(4), 232–233 (2010).
6. D. Tarin, "Cell and tissue interactions in carcinogenesis and metastasis and their clinical significance," *Semin. Cancer Biol.* 21(2), 72–82 (2011).

7. D.M. Parkin, J. Stjernswärd, and C.S. Muir, "Estimates of the worldwide frequency of twelve major cancers," *Bull. World Health Organ.* 62(2), 163–182 (1984).

8. J. Ferlay, I. Soerjomataram, R. Dikshit, et al., "Cancer incidence and mortality worldwide: Sources, methods and major patterns in GLOBOCAN 2012," *Int. J. Cancer* 136(5), E359–E386 (2015).

9. F. Bray, J. Ferlay, I. Soerjomataram, R. Siegel, L.A. Torre, and A. Jemal, "Global cancer Statistics 2018: GLOBOCAN estimates of incidence and mortality worldwide for 36 cancers in 185 countries," *CA: Cancer J. Clin.* 68(8), 394–424 (2018).

10. L. Oliveira, and V.V. Tuchin, *The Optical Clearing Method: A New Tool for Clinical Practice and Biomedical Engineering* (Springer, Cham-Switzerland, 2019).

11. E.A. Genina, L.M. Oliveira, A.N. Bashkatov, and V.V. Tuchin, "Optical clearing of biological tissues: Prospects of application for multimodal malignancy diagnostics," in *Multimodal Optical Diagnostics of Cancer*, eds. V.V. Tuchin, J. Popp, and V. Zakharov (Springer, Cham-Switzerland, 2020).

12. S. Carvalho, N. Gueiral, E. Nogueira, R. Henrique, L. Oliveira, and V.V. Tuchin, "Glucose diffusion in colorectal mucosa: A comparative study between normal and cancer tissues," *J. Biomed. Opt.* 22(9), 091506 (2017).

13. I. Carneiro, S. Carvalho, R. Henrique, L.M. Oliveira, and V.V. Tuchin, "A robust ex vivo method to evaluate the diffusion properties of agents in biological tissues," *J. Biophoton.* 12(4), e201800333 (2019).

14. D. Zhu, K.V. Larin, Q. Luo, and V.V. Tuchin, "Recent progress in tissue optical clearing," *Laser Photonics Rev.* 7(5), 732–757 (2013).

15. K. Schilling, V. Janve, Y. Gao, I. Stepniewska, B.A. Landman, and A.W. Anderson, "Comparison of 3D orientation distribution functions measured with confocal microscopy and diffusion MRI," *NeuroImage* 129, 185–197 (2016).

16. D.K. Tuchina, A.N. Bashkatov, A.B. Bucharskaya, E.A. Genina, and V.V. Tuchin, "Study of glycerol diffusion in skin and myocardium ex vivo under the conditions of developing alloxan-induced diabetes," *J. Biomed. Photon. Eng.* 3(2), 020302-1-9 (2017).

17. H.Q. Zhong, Z.Y. Guo, H.J. Wei, et al., "Quantification of glycerol diffusion in human normal and cancer breast tissues in vitro with optical coherence tomography," *Laser Phys. Lett.* 7(4), 315–320 (2010).

18. D. Abookasis, and T. Moshe, "Reconstruction enhancement of hidden object using speckle contrast projections and optical clearing agents," *Opt. Commun.* 300, 58–64 (2013).

19. D. Zhu, W. Lu, Y. Weng, H. Cui, and Q. Luo, "Monitoring thermal-induced changes in tumor blood flow and microvessels with laser speckle contrast imaging," *Appl. Opt.* 46(10), 1911–1917 (2007).

20. Y. Zhang, H. Liu, J. Tang, et al., "Non-invasively imaging subcutaneous tumor xenograft by handheld Raman detector, with assistance of optical clearing agent," *ACS Appl. Mater. Interfaces* 9(21), 17769–17776 (2017).

21. M. Oldham, H. Sakhalkar, T. Oliver, et al., "Three-dimensional imaging of xenograft tumors using optical computed and emission tomography," *Med. Phys.* 33(9), 3193–3202 (2006).

22. M. Oldham, H. Sakhalkar, T. Oliver, G. Allan Johnson, and M. Dewhirst, "Optical clearing of unsectioned specimens for three-dimensional imaging via optical transmission and emission tomography," *J. Biomed. Opt.* 13(2), 021113 (2008).

23. M. Wei, L. Shi, Y. Shen, et al., "Volumetric chemical imaging by clearing-enhanced stimulated Raman scattering microscopy," *Proc. Natl. Acad. Sci. U.S.A.* 116(14), 6608–6617 (2019).

24. T. Lagerweijt, S.A. Dusoswa, A. Negrean, et al., "Optical clearing and fluorescence deep-tissue imaging for 3D quantitative analysis of the brain tumor microenvironment," *Angiogenesis* 20(4), 533–546 (2017).

25. L. Pires, V. Demidov, I.A. Vitkin, V.S. Bagnato, C. Kurachi, and B.C. Wilson, "Optical clearing of melanoma in vivo: Characterization by diffuse reflectance spectroscopy and optical coherence tomography," *J. Biomed. Opt.* 21(8), 081210 (2016).

26. P. Matryba, L. Kaczmarek, and J. Golab, "Advances in ex situ tissue optical clearing," *Laser Photonics Rev.* 13(8), 1800292 (2019).

27. Y. Cui, X. Wang, W. Ren, J. Liu, and J. Irudayaraj, "Optical clearing delivers ultrasensitive hyperspectral dark-field imaging for single-cell evaluation," *ACS Nano* 10(3), 3132–3143 (2016).

28. G.D. Scott, E.D. Blum, A.D. Fryer, and D.B. Jacoby, "Tissue optical clearing, three-dimensional imaging, and computer morphometry in whole mouse lungs and human airways," *Am. J. Respir. Cell Mol. Biol.* 51(1), 43–55 (2014).

29. B. von Neubeck, G. Gondi, C. Riganti, et al., "A inhibitory antibody targeting carbonic anhydrase XIII abrogates chemoresistance and significantly reduces lung metastases in an orthotopic breast cancer model *in vivo*," *Int. J. Cancer* 143(8), 2065–2075 (2018).

30. M.F. Cuccarese, J.M. Dubach, C. Pfirschke, et al., "Heterogeneity of macrophage infiltration and therapeutic response in lung carcinoma revealed by 3D organ imaging," *Nat. Commun.* 8, 14293 (2017).

31. N.S. Joshi, E.H. Akama-Garren, Y. Lu, et al., "Regulatory T cells in tumor-associated tertiary lymphoid structures suppress anti-tumor T cell responses," *Immunity* 43(3), 579–590 (2015).

32. A.K. Glaser, N.P. Reder, Y. Chen, et al., "Light-sheet microscopy for slide-free non-destructive pathology of large clinical specimens," *Nat. Biomed. Eng.* 1(7), 0084 (2017).

33. N. Tanaka, D. Kaczynska, S. Kanatani, et al., "Mapping of the three-dimensional lymphatic microvasculature in bladder tumours using light-sheet microscopy," *Br. J. Cancer* 118(7), 995–999 (2018).

34. M.E. van Royen, E.I. Verhoef, C.F. Kweldam, et al., "Three-dimensional microscopic analysis of clinical prostate specimens," *Histopathology* 69(6), 985–992 (2016).

35. E.I. Verhoef, W.A. van Cappellen, J.A. Slotman, et al., "Three-dimensional architecture of common benign and precancerous prostate epithelial lesions," *Histopathology* 74(7), 1036–1044 (2019).

36. E.I. Verhoef, W.A. van Cappellen, J.A. Slotman, et al., "Three-dimensional analysis reveals two major architectural subgroups of prostate cancer growth patterns," *Mod. Pathol.* 32(7), 1032–1041 (2019).

37. R.D. Hume, L. Berry, S. Reichelt, et al., "An engineered human adipose/collagen model for in vitro breast cancer cell migration studies," *Tissue Eng. A* 24(17–18), 1309–1319 (2018).

38. I. Carneiro, S. Carvalho, R. Henrique, L. Oliveira, and V.V. Tuchin, "Moving spectral window to the deep-ultraviolet via optical clearing," *J. Biophoton.* 12(2), e201900181 (2019).

39. Y. Zhou, J. Yao, and L.V. Wang, "Tutorial on photoacoustic tomography," *J. Biomed. Opt.* 21(6), 061007 (2016).

40. P. Brescia, "Micro-volume purity assessment of nuclei acids using A260/A280 ratio and spectral scanning protein and nucleic acid quantification," http://www.biotek.com/res ources/application-notes/micro-volume-purity-assessment-of-nucleic-acids-using-asub260/sub/asub280/sub-ratio-and-spectral-scanning/ (accessed: May 2020).

41. I. Carneiro, S. Carvalho, R. Henrique, L. Oliveira, and V.V. Tuchin, "Measurement of optical properties of normal and pathological human liver tissue from deep-UV to NIR," *Proc. SPIE* **11363**, 113630G (2020).

42. L. Oliveira, A. Lage, M. Pais Clemente, and V.V. Tuchin, "Rat muscle opacity decrease due to the osmosis of a simple mixture," *J. Biomed. Opt.* 15(5), 055004 (2010).

43. L. Oliveira, M.I. Carvalho, E.N. Nogueira, and V.V. Tuchin, "The characteristic time of glucose diffusion measured for muscle tissue at optical clearing," *Laser Phys.* 23(7), 075606 (2013).

44. L.M. Oliveira, M.I. Carvalho, E.M. Nogueira, and V.V. Tuchin, "Diffusion characteristics of ethylene glycol in skeletal muscle," *J. Biomed. Opt.* 20(5), 051019 (2015).

45. K. Kim, H. Park, and K.-M. Lim, "Phototoxicity: Its mechanisms and animal alternative test methods," *Toxicol. Res.* 31(2), 97–104 (2015).

46. S. Peña-Llopis, and J. Brugarolas, "Simultaneous isolation of high-quality DNA, RNA, miRNA and proteins from tissues for genomic applications," *Nat. Protoc.* 8(11), 2240–2255 (2013).

34

Contrast enhancement and tissue differentiation in optical coherence tomography with mechanical compression

Pavel D. Agrba and Mikhail Yu. Kirillin

CONTENTS

Introduction: Mechanisms of controlling tissue optical properties by mechanical compression

Due to strong coherence between the physical state of biotissue and its optical properties, mechanical compression can serve as an efficient tool for controlling tissue optical properties and, in particular, for optical clearing. Optical clearing techniques traditionally aim to decrease scattering and/or the absorption coefficient of biotissue to suppress attenuation of probing radiation in tissues, thus enhancing probing/imaging depth. Pressure applied to tissue surface induces redistribution of biological liquids within the tissue, thus causing changes in optical properties. Blood is known to be the essential tissue chromophore in the visible and NIR ranges. Tissue compression is usually accompanied by blood-vessel shrinkage, which causes the ejection of blood from the vessels under pressure, thus decreasing local absorbance due to a dramatic drop in the local blood content.

In reference [1], the authors study a label-free methodology for estimating and visualizing tissue perfusion rates. They used a technique based on tissue compression and imaging with a high-frequency photoacoustic-ultrasound system. Full blood ejection occurred when compressive pressure of 50 kPa was applied. It was shown that when a soft tissue layer is compressed, blood is removed from the compression volume, thus reducing the PA signal in the compressed area. In tissue, these changes correspond to the reduction of blood content in the capillary bed and small arteries and veins. A collapse of a vessel tracked by photoacoustic–ultrasound dual imaging approach is also reported (Figure 34.1).

These results are in agreement with results presented in references [2–4], in which the decrease in blood flow during external pressure is demonstrated using different optical techniques. Reference [2] reports on the disappearance of blood in capillaries of a nailfold in angiographic images when external pressure is applied. In reference [3], the observed effect of the increased spectral reflectance is explained by removal of blood from dermis layers and confirmed by numerical Monte Carlo simulations. Reference [4] employed laser Doppler flowmetry to demonstrate the decrease in the microcirculatory activity upon pressure application and its further restoration and temporal activation after compression release, which is in line with the results of reference [2].

When considering classical theory of light transport in tissues, scattering and absorption properties of tissue are determined by scattering and absorption cross-sections, σ_s and σ_a, of microscopic optical inhomogeneities (also attributed as scattering centers) within the tissue, respectively. Macroscopic optical properties, namely scattering and absorption coefficients, μ_s and μ_a, are determined as products of corresponding cross-sections and scattering centers concentration C:

$$\mu_s = C\sigma_s, \tag{34.1}$$

$$\mu_a = C\sigma_a. \tag{34.2}$$

Under mechanical compression, extracellular liquid is removed from the area under pressure, resulting in local increase of scatterers concentration C and, hence, scattering and absorption coefficients. This effect is observed when imaging tissues under compression with optical coherence tomography (OCT) [5, 6]. Since the OCT-signal is usually proportional to transport coefficient $\mu_s' = \mu_s(1 - g)$ (given that absorption at the probing wavelength is much smaller than scattering), the increase in OCT signal under compression indicates the increase in local backscattering (Figure 34.2).

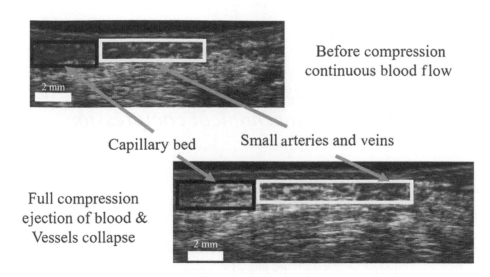

FIGURE 34.1 Photoacoustic signals of the blood before and during full compression [1].

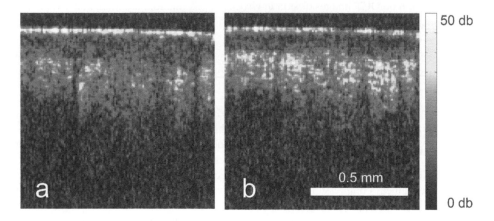

FIGURE 34.2 OCT images of human thin skin layer immediately after (a) and in 5 min after (b) pressure ($p = 210$ kPa) application. [7].

Note that Equation (34.1) is valid for the case of independent scattering event, when the distance between scatterers exceeds mean photon free pathlength. Under further compression, the concentration C continues to increase resulting in dense package situation, when two scattering events cannot be considered as independent, and two smaller scatterers are to be considered as a larger single scatterer. Dependence of blood optical properties on hematocrit (volume concentration of red blood cells in whole blood) [8] can serve as an illustration of this effect (Figure 34.3). Absorption coefficient μ_a linearly depends on concentration, since it is determined by the total amount of chromophores within a particular volume. Particle concentration increase causes a corresponding increase in the total amount of chromophores and, hence, in μ_a. Scattering coefficient μ_s demonstrates liner dependence of concentration only for relatively small concentrations, and reaches a plateau with the concentration increase. Reduced scattering coefficient μ_s' demonstrates linear dependence on concentration only for hematocrits below 45%, which is around the upper limit for normal physiological value, while for higher values the dependence is quite weak.

This description is valid for media where the scatterers are suspended in a surrounding bulk matter. Biotissues have a more complex structure; although cellular nuclei can be treated as primary scatterers, refractive index mismatches at the extracellular liquid–membrane–cytoplasm interfaces also contribute to scattering. In this connection, the mechanism of optical clearing by mechanical compression consists in the removal of extracellular liquid from the area under pressure inducing the decrease in scattering at the liquid–membrane interfaces, since after liquid ejection, the scattering interfaces disappear. A similar effect on a larger scale is demonstrated in reference [9], which reports on dual modality photoacoustic-ultrasound monitoring of blood capillary shrinkage under pressure (Figure 34.4).

A further mechanism consists of the decrease of the tissue thickness in the area under compression, which leads to a decrease in sample optical thickness, thus increasing probing depth in the compressed sample.

A simple model of compression effect on sample scattering properties

In order to demonstrate the effect of compression on the scattering properties of a tissue sample consisting in a decrease of scattering interfaces within the tissue, a simple model considering a collimated beam incident orthogonally on an ensemble

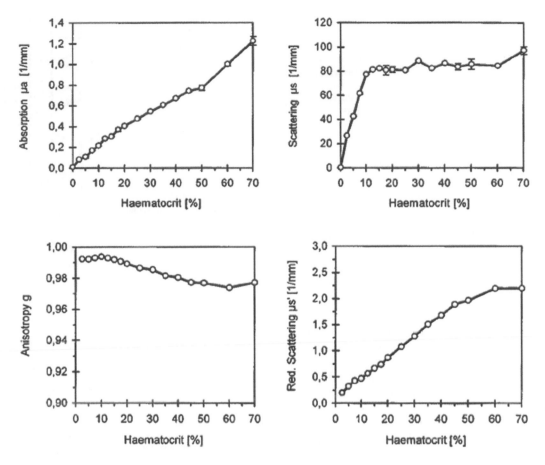

FIGURE 34.3 Mean values of blood absorption, scattering, and reduced scattering coefficients, and anisotropy factor vs hematocrit ($\pi = 300$ mosmol/L, $\gamma = 500s^{-1}$, SatO$_2$>98%, $\lambda = 633$ nm) [8].

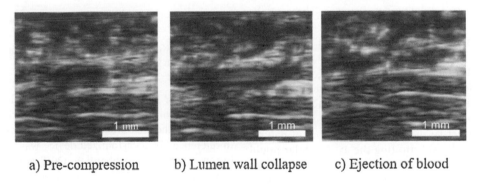

a) Pre-compression   b) Lumen wall collapse   c) Ejection of blood

FIGURE 34.4 Collapse of a vessel tracked by photoacoustic–ultrasound dual imaging [9].

of refractive index jump interfaces can be employed. The model can be implemented using the Monte Carlo technique by considering the probabilities of reflection and transmission of the incident light beam at each interface, thus allowing sample transmittance and reflectance to be calculated. The reflection probability is governed by the difference in refractive indices n_1 and n_2 in accordance with the Snell law for normal beam incidence:

$$R = \left(\frac{n_1 - n_2}{n_1 + n_2}\right)^2$$

The model accounts for a particular number of scattering interfaces, and further, it is assumed that their number decreases as a result of mechanical compression, which mimics extracellular liquid removal under pressure. Since absorption plays an important role in light propagation in tissues, accounting for absorption within the sample is also added to the model by decreasing the weight P of the light beam in accordance with Beer's law:

$$P = P_0 \exp(-\mu_a l),$$

where P_0 is the light beam weight before the step, μ_a is absorption coefficient, and l is the distance between interfaces. Moreover, since the concentration of chromophores may increase as a result of physical compression of a sample, it is assumed that the optical density $\mu_a L$, where L is the sample thickness, is preserved.

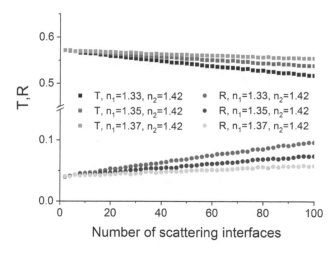

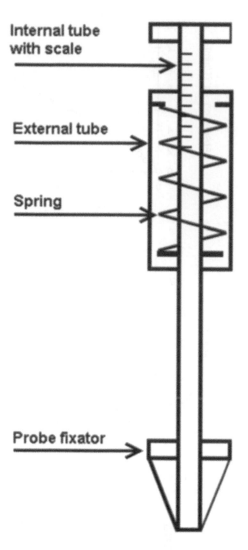

FIGURE 34.5 The effect of mechanical compression and optical clearing on transmittance T and reflectance R of a biotissue sample: simulation results.

Below, the results of the simulations for parameters close to those of biotissues are given. The refractive index values determining reflection probability at each interface are equal to $n_1 = 1.33$ and $n_2 = 1.42$, which correspond to values for extracellular liquid and cell nuclei, respectively. The refractive index of the outer medium $n_{out} = 1$, which corresponds to air. Total sample optical density $\mu_a L = 0.5$, which corresponds to a sample with thickness $L=5$ mm and $\mu_a = 0.1$ mm^{-1}. The total number of interfaces in an uncompressed sample is 100, which approximately corresponds to $\mu_s = 20$ mm^{-1}.

Results of the simulations are shown in Figure 34.5. One can see that transmittance T demonstrates a monotonous increase with the decrease in the number of interfaces, while reflectance R demonstrates a monotonous decrease. These results confirm the proposed mechanism of tissue optical clearing as a result of compression and give the estimate of a 10% transmittance increase and 50% reflectance decrease as a result of sample mechanical compression.

It is worth noting that the described simple model can be employed for simulation of the effect of traditional optical clearing performed by application of an optical clearing agent with refractive index higher than that of natural extracellular liquid. By substituting the extracellular liquid within the tissue, the clearing agent decreases the corresponding refractive index difference, thus reducing scattering within sample. Figure 34.6 shows the results for higher values of $n_1 = 1.35$ and 1.37 corresponding to those for typical clearing agents. Simulation results evidently demonstrate an optical clearing effect manifested by an increase in transmittance and a decrease in reflectance of the sample, and the role of this effect becomes smaller with the decrease in the number of scattering interfaces within the tissue sample.

Compression as a tool for controlling biotissue optical properties: A review

As demonstrated above, the application of mechanical compression may serve as a tool for controlling optical properties

FIGURE 34.6 Customized device for compression control in course of OCT inspection [7].

of biotissues. It is important to mention that high pressures may cause pain or even injuries to living tissues and, in this connection, the pressure level is to be controlled. For example, in reference [7], a customized device was developed combining the channel for a fiberoptic OCT probe with a compression control dynamometer (Figure 34.6). The dynamometer has a channel similar to that of an endoscope with a fixator for the endoscopic OCT probe. The flexible fiberoptic OCT probe could be inserted into the channel and fixed together with the inner tube of the dynamometer. The outer tube of the dynamometer is connected to the inner one with a graduated spring. By pressing the outer tube towards the inspected tissue, one can control the pressure of the probe.

In general, the studies of the effect of compression on biotissue optical properties consider the pressure range of 0.1–2 MPa [10–16]. It is worth noting that to ensure a patient's comfort, the pressure applied is to be below the pain threshold. References [17–20] report on the studied of pain thresholds and pain tolerance under applied pressure for different localizations. Generalizing reported data, the pain threshold is about 1.1 MPa [17, 18].

From a retrospective point of view, the study of the effect of mechanical compression on biotissue optical properties progressed in line with the development of optical diagnostics techniques themselves.

A pioneering reference [10] reported on revealing the compression abilities for enhancing probing depth in biotissues for optical techniques. Radiative transport of light in turbid media, both tissue mimicking phantoms and real biotissues, with thickness significantly exceeding photon free pathlength was experimentally studied. Compression of a soft scattering tissue sample was demonstrated to significantly increase its diffuse transmittance (probing photons fraction that passed through the sample). In particular, the total thickness of a human palm was demonstrated to decrease from 2.7 to 2.0 cm, which was accompanied by a weak pain manifestation. An effect of relaxation of optical clearing induced by mechanical compression was also shown; the tissue remained optically cleared for 1–3 s after compression release, which was associated with temporal blood removal from the area under compression.

In reference [11], the results of the study of mechanical compression on *ex vivo* biotissue optical properties are described. The study was performed on samples of human skin, human aorta, and cattle and porcine sclera. The studies with human tissues were performed with post-mortem samples excised 24 h after death; applied pressure varied between 10 and 100 kPa. Contrary to reference [10], diffuse reflectance (probing photons fraction that was backscattered by the sample) was measured in addition to diffuse transmittance. In the majority of measurements, a decrease in diffuse reflectance was observed as a result of compression, while diffuse transmission increased, which is in line with results of [10] and modeling described in paragraph 2 of this chapter. Measured transmittance and reflectance values allowed the reconstruction of the absorption coefficient, which was demonstrated to increase under compression. Since the blood redistribution role may be lower in post-mortem samples, this effect is primarily explained by an increase in the concentration of absorbing elements as a result of compression.

Reference [12] analyzes the mechanisms of changes in diffuse reflectance spectra in the range of 400–2000 nm induced by long time compression with subsequent compression release. The studies were performed for human forearm *in vivo* for the pressures range from 0 to 1 MPa. Diffuse reflectance in the range 700–2000 nm is shown to monotonously decrease under compression reaching a minimum after 6 min after compression start. The authors suppose these changes to be associated with redistribution of free and bound water within skin. Two stages of tissue relaxation were demonstrated after compression release: the fast stage lasts for 5–12 s after release, while the slow one last for up to 160 s.

A study of the dependence of skin reflectance spectrum on compression level is reported in reference [13]. The study is performed on human skin of different localizations *in vivo* in five volunteers aged between 18 and 30 years old. An increase in scattering coefficient accompanied by a decrease in absorption coefficient with pressure increase from 9 to 152 kPa are demonstrated for the considered localizations. Reduced scattering coefficient is shown to decrease for those localizations where the backscattering from the underlying bones does not affect the registered reflectance spectra.

These studies confirm the above described mechanisms of the effect of mechanical compression on biotissue optical properties consisting in blood removal, extracellular liquid redistribution, and dense scatterers package, which finally lead to decrease in scattering and absorption coefficient value, thus providing optical clearing.

Compression as a tool in optical coherence tomography ex vivo

Optical coherence tomography [21] is one of the major applications of mechanical compression to enhance images among optical imaging modalities. OCT is a commonly employed technique providing imaging of the inner structure of biotissue with spatial resolution of units of microns at depth up to 2 mm. It is based on principles of low-coherent detection of photons backscattered from the inspected tissue, and hence, its imaging depth is limited primarily by multiple scattering of probing radiation in tissues disturbing its coherence. In this connection, optical clearing techniques that aim to decrease scattering within tissues have high potential for OCT imaging.

In reference [5], mechanical compression was employed to improve differential diagnostics of mucosa pathologies using traditional OCT. The traditional approach consists in detecting mucosa layered structure in an OCT image indicating the absence of morphological alterations, since for tumors, unstructured OCT images are typical. However, the presence of manifested edema may cause disturbance of the layered structure of mucosa, which, in turn, leads to the absence of evident layer boundaries in an OCT image. This may lead to a misinterpretation of the image by a specialist and false positive tumor diagnosis. Tissue compression leads to removal of extracellular liquid from edema region resulting in revealing layered structure in an OCT image.

Reference [5] demonstrates this effect for differential diagnostics of edema and carcinoma in human rectum on *ex vivo* samples. In addition, an increase of the layer boundary contrast in OCT image from 1 to 10 dB with an increase in pressure from 0 to 450 kPa was shown (Figure 34.7a–c). This result is in line with the results of reference [15], which demonstrated the increase of layer boundary contrast under pressure. Moreover, it is demonstrated that mechanical compression allows OCT imaging depth to be increased due to a decrease in the total sample thickness after compression. Conclusions on the mechanisms of the experimental study are confirmed by Monte Carlo simulations of OCT images (Figure 34.7d–f) that allowed changes in mucosa optical properties change to be evaluated as 1.5–2 or 4–10 fold increases in absorption and scattering coefficients for pressures of 90 or 450 kPa, respectively.

Effect of mechanical compression on OCT images of skin *in vivo*

The effect similar to that observed in OCT imaging of biotissues *ex vivo* under compression can also be observed *in vivo*.

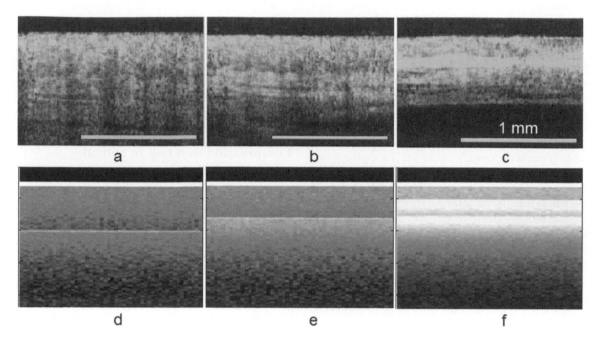

FIGURE 34.7 OCT images of human rectum with inflammation ex vivo (a–c) and corresponding Monte Caro simulated images (d–f) without pressure (a, d), under pressure of 170 (b, e), and 440 (c, f) kPa [5].

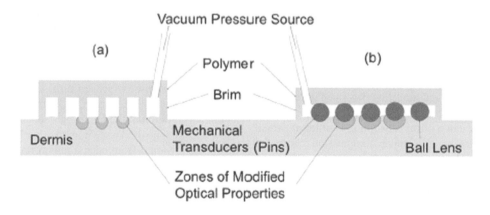

FIGURE 34.8 Simplified cross-sectional schematic diagram of (a) first and (b) second TOCDs applied to skin [15].

References [14, 15] studied the effect of mechanical compression on OCT imaging depth, which served as a measure of optical clearing level. The study employed an OCT system that does not require contact between the OCT probe and a sample to acquire an OCT image. A custom-designed lattice of optically transparent pins or ball lenses was employed for tissue compression for systems with probing wavelengths of 850 and 1310 nm, respectively (Figure 34.8). Control of the compression force was performed using a vacuum pump. Comparison of OCT imaging depth under the pins and/or ball lenses with that under free space demonstrated the effect of image depth enhancement. Imaging depth under the pins and ball lenses was enhanced three- and two-fold, respectively, as compared to peripheral tissue regions.

Figure 34.9 clearly shows that OCT signal level increases in the areas of pressure application together with an increase in imaging depth. The difference in imaging depth change

is explained by the difference in tissue optical properties at 850 and 1310 nm. Increased OCT image contrast of the epidermal–dermal junction was also observed in regions under the pins as a result of compression. The study also analyzes the dynamics of refractive index value and tissue water content after compression start; an analytical model demonstrates an increase in tissue refractive index from 1.38 to 1.46 due to water content decrease from 70% to 30% as a result of its removal, which is in agreement with the results of experiments on *ex vivo* samples. The same conclusions are further extended to *in vivo* experiments.

Reference [7] reports on study of the effect of mechanical compression on OCT images of human skin *in vivo*. The study included eight volunteers and was performed with cross-polarization OCT modality; the contrasts of stratum corneum–epidermis and epidermis–dermis junctions in parallel and orthogonal channels of CP-OCT system were assessed.

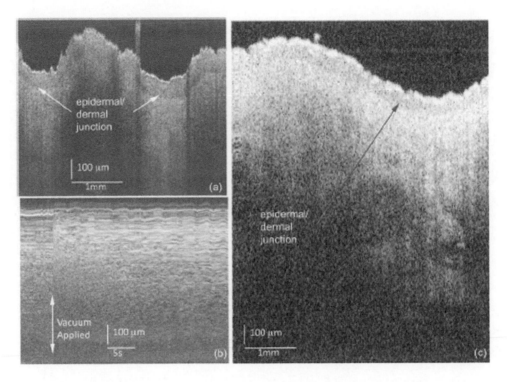

FIGURE 34.9 In vivo human skin: (a) OCT B-scan image (850nm) following application of the first tissue optical clearing device (TOCD), (b) OCT M-scan (horizontal axis represents time) image (1310nm) before and during application of the first TOCD. (c) OCT B-scan image (1310nm) following application and removal of the second TOCD [15].

The contrast was calculated as the difference between averaged OCT signal levels for corresponding layers expressed in dB [6]. Figure 34.10a–c demonstrates the evolution of OCT image of human skin under compression, revealing the increase in the contrast of interlayer junctions. Results of corresponding Monte Carlo simulations confirming the increase in scattering properties of particular skin layers as a result of pressure (see Table 34.1) are shown in Figure 34.10d–f. More evidently, the effect mechanical compression on skin-layer backscattering properties can be observed from comparison of OCT A-scans immediately and 5 min after compression start (Figure 34.11).

Figure 34.12 demonstrates the evolution of corresponding contrasts as a result of compression. This figure shows that the contrast of stratum corneum–epidermis junction monotonously decreases, which can be associated with slow compression and ordering of dry cells or inflow of extracellular liquid, providing a clearing effect. The decrease is manifested both for parallel and orthogonal polarizations. On the contrary, epidermis–dermis junction contrast demonstrates a monotonous increase, which is associated both with increase in scattering coefficient of dermis and ordering of collagen fibers, resulting in stronger birefringence. It is confirmed by a faster increase in the contrast in orthogonal polarization compared to that in the parallel one.

An earlier pilot study [6] compared the dynamics of OCT images under weak ($p = 70$ kPa) and strong ($p = 350$ kPa) compression for volunteers of different ages (23, 29, and 49 years). Qualitatively, the results are in agreement with the studies described above; however, typical rates of contrast change differ for different ages, presumably due to the different elasticity of skin of various age groups.

Combined effect of mechanical compression and temperature on OCT images of human skin

An additional optical clearing effect can be achieved when applying alternative actions affecting local blood contact in the image biotissue. Several authors studied the effect of temperature on biotissue properties *in vivo*. The results of an *in vivo* study on rat skin are presented by Quyang et al. [23]. The authors demonstrated that temperature rise induced a decrease in the attenuation coefficient and an increase in the penetrability of light.

However, results for human skin demonstrate the opposite effect. Khalil et al. [24] investigated the temperature effect on optical properties of human forearm skin *in vivo*. The cutaneous reduced scattering coefficient increases linearly with temperature, and the absorption coefficient also increases with temperature rise; however, these changes are complex for temperatures above normal human body temperature. Changes in optical properties of skin *in vivo* can be induced by changes in blood saturation of biotissue, which leads to a local variation of the absorption at particular wavelengths. Variations in temperature regime may also stimulate sweating, thus decreasing the scattering of upper layers of biotissue and changing absorption within them.

Reference [25] reports on the combined effect of compression and temperature variation on optical properties of skin *in*

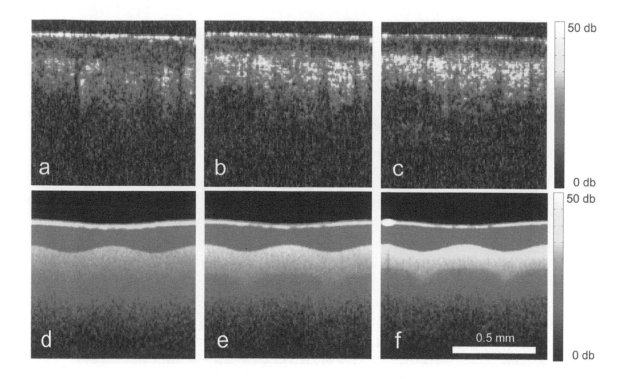

FIGURE 34.10 Experimental (a–c) and Monte Carlo simulated (d–f) OCT images of thin skin (forearm) of a 26-year-old volunteer in parallel polarization immediately after (A, D), 3 min after (B, E), and 6 min after compression (p = 210 kPa) start (C, F) [7]. Each figure is 1 × 1 mm; in-depth axis presents optical distance; transversal axis presents physical distance.

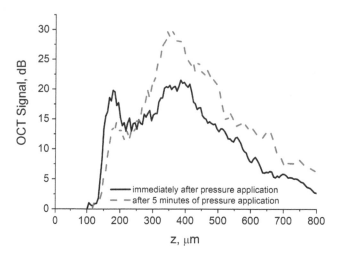

FIGURE 34.11 OCT A-scans of human skin *in vivo* immediately and 5 min after compression (p = 350 kPa) start [22].

vivo assessed with OCT. The contrast of stratum corneum–epidermis and epidermis–dermis junctions in OCT images of skin were studied. Proposed mechanisms of the skin optical properties control by compression and cooling/heating are shown in Figure 34.13. Compression alone is assumed to lead to adhesion of horny dry cells and redistribution of extracellular liquid (Figure 34.5a). The first factor induces a decrease in scattering from the stratum corneum and a decrease in contrast of the stratum corneum–epidermis junction. The second factor leads to an increase in scattering from the dermis layer and an increase in contrast of the epidermis–dermis junction, respectively.

The mechanisms of the effect of compression with preliminary cooling or heating is presented in Figures 34.13b and c. In addition to changes observed in case of compression alone, a spasm of blood vessels in dermis occurs in the case of cooling. This factor causes a decrease in relative blood content and a decrease in absorption in the region under action. A decrease in absorption in its turn causes an increase in the OCT signal from the dermis layer. Heating induces two mechanisms affecting both considered dependencies. The former is vasodilatation, causing an increase in relative blood content and a corresponding increase in local absorption. The latter is sweating of skin in the area of action, which leads to optical clearing due to sweat filling spaces between horny dry cells. This mechanism causes a decrease in contrast of the stratum corneum–epidermis junction.

For measurements, a thermostat at room temperature (reference measurement), at a temperature of –18°C (preliminary cooling), or at a temperature of 50°C (heating) was placed on the forearm for 2 min. Then the thermostat was removed and the probe was pressed to the human skin with a pressure of about 210 kPa and was held in this position for 5–7 min. OCT images of skin were obtained every 5 s after compression start. Since the probe surface area is about 4 mm², a skin area of 20 cm² was a subject to thermal action.

Figure 34.14 shows typical reference OCT images of compressed human thin skin without preliminary action in comparison with precooling and preheating images at three different time intervals: immediately after, 3 min after, and 6 min after ending the thermostat application. This figure reveals the changes in junction contrast as a result of the combined action.

TABLE 34.1

Optical and physical properties of model skin layers under pressure of 210 kPa at λ = 910 nm immediately after compression start/in 3 min/in 6 min employed for Monte Carlo simulations [7]. For those values which are not supposed to be affected by compression, only one value is shown

Skin layer	Stratum corneum	Epidermis	Upper dermis	Lower dermis
μ_s (mm^{-1})	12/8/6	8	12/16/20.5	12
μ_a (mm^{-1})	0.02	0.015	0.02	0.1
g	0.9	0.9	0.9	0.9
n	1.54	1.38	1.38	1.39
Thickness (μm)	15/12/8	100	180/135/105	700

μ_s, scattering coefficient; μ_a, absorption coefficient, g, anisotropy factor; n, refractive index

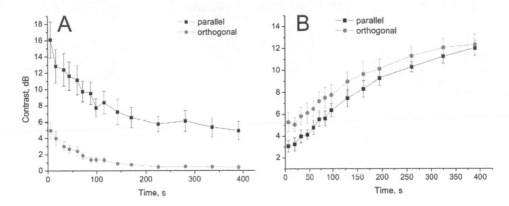

FIGURE 34.12 Temporal evolution of stratum corneum–epidermis (A) and epidermis dermis (B) junction contrast in OCT images (mean and standard deviation of the mean for eight volunteers) obtained in parallel and orthogonal polarizations under pressure of 210 kPa [7].

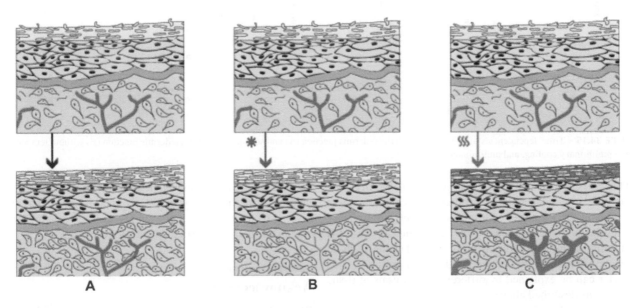

FIGURE 34.13 Skin response to mechanical compression (a), compression with cooling (b), and compression with heating (c) [25].

Figure 34.15 presents temporal dependencies of epidermis–dermis and stratum corneum–epidermis junction contrasts in OCT images. A decrease in absorption in case of preliminary cooling caused by spasm of blood vessels leads to an increase in OCT signal from the dermis layer and, consequently, an increase in contrast of the epidermis–dermis junction. As a result, during the whole observation time, the contrast in the case of preliminary cooling is higher than that without cooling. The vasodilatation is induced by a preliminary heating increase in local absorption, which leads to a decrease in epidermis–dermis junction contrast. The sweating of skin in the area of action that leads to sweat filling spaces between horny dry cells also causes a decrease in contrast of the stratum corneum–epidermis junction. Accordingly, during the whole observation time, the contrast in the case of preliminary heating is lower compared to that without heating.

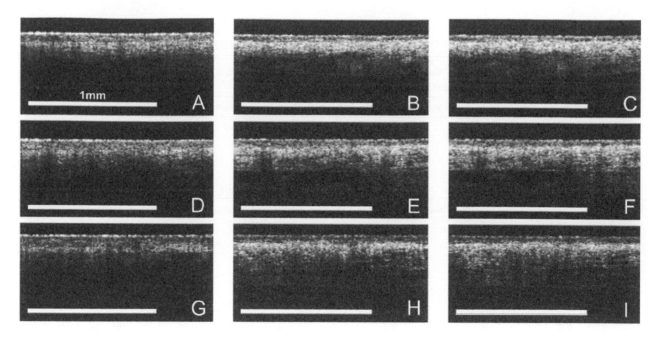

FIGURE 34.14 OCT images of compressed thin skin (forearm) without temperature variation (A–C), after preliminary cooling (D–F), and heating (G–I) immediately after (A, D, G), 3 min after (B, E, H), and 7 min (C, F, I) after compression ($p = 210$ kPa) start/thermostat removal [25].

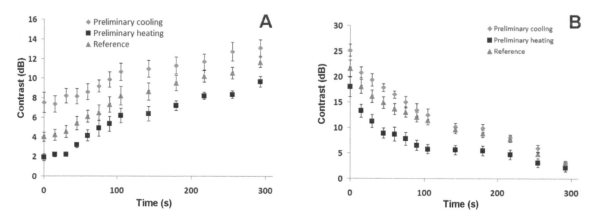

FIGURE 34.15 Time dependencies of contrast of epidermis–dermis junction (A) and stratum corneum–epidermis junction (B) without temperature effect, preliminary cooling, and preliminary heating [25].

Thus, it is demonstrated that preliminary change in biotissue temperature induces additional changes in contrast to these junctions in OCT images with opposite effects. Preliminary cooling leads to an additional increase (about 2–5 dB at the beginning of observation) and preliminary heating leads to a decrease (also 2–5 dB) of contrast for both junctions. This behavior can be explained by different mechanisms of tissue response, as described above.

Conclusion

This chapter reviews an approach to enhancement of biotissue OCT images by means of mechanical compression. A short narrative review of the studies on the effect of tissue compression outlines it as a tool for controlling tissue optical properties. The main mechanisms behind this effect are blood removal, which causes a decrease in absorption coefficient; and a close package of scatterers within biotissue, which affects scattering

coefficient. Pressure application allows for an increase in the contrast of particular layers in OCT images when imaging a layered biotissue, and it can even reveal layered tissue structure in particular important diagnostic cases.

Acknowledgments

The authors are grateful to Dr Anna Orlova for useful discussions. The chapter preparation is supported by the Ministry of Science and Higher Education of Russian Federation (IAP RAS governmental project #0030-2021-0014)

REFERENCES

1. M. Choi, A.M. James Shapiro, R. Zemp "Tissue perfusion rate estimation with compression-based photoacoustic-ultrasound imaging", *Journal of Biomedical Optics*, 23(1), 016010, doi: 10.1117/1.JBO.23.1.016010 (2018).

2. W.J. Choi, H. Wang, R.K. Wang "Optical coherence tomography microangiography for monitoring the response of vascular perfusion to external pressure on human skin tissue", *Journal of Biomedical Optics*, 19(5), 056003 (2014).

3. A.P. Popov, A.V. Bykov, I.V. Meglinski "Influence of probe pressure on diffuse reflectance spectra of human skin measured in vivo", *Journal of Biomedical Optics*, 22(11), 110504, doi: 10.1117/1.JBO.22.11.110504 (2017).

4. E.C. Herrman, C.F. Knapp, J.C. Donofrio, R. Salcido "Skin perfusion responses to surface pressure-induced ischemia: Implication for the developing pressure ulcer", *Journal of Rehabilitation Research and Development*, 36(2), 109–120 (1999).

5. P.D. Agrba, M.Yu. Kirillin, A.I. Abelevich, E.V.Zagaynova, V.A.Kamensky "Compression as a method for increasing the informativity of optical coherence tomography of biotissues", *Optics and Spectroscopy*, 107(6), 853–858 (2009).

6. M.Yu. Kirillin, P.D. Agrba, V.A. Kamensky "In vivo study of the effect of mechanical compression on formation of OCT images of human skin", *Journal of Biophotonics*, 3(12), 752–758 (2010).

7. M. Kirillin, P. Agrba, V. Kamensky "Mechanical compression in cross-polarization OCT imaging of skin: In vivo study and Monte Carlo simulation", *Photonics and Lasers in Medicine*, 3(4), 363–372 (2014).

8. A. Roggan, M. Friebel, K. Doerschel, A. Hahn, G.J. Mueller "Optical properties of circulating human blood in the wavelength range 400–2500 nm", *Journal of Biomedical Optics*, 4(1), 36–46 (1999).

9. M. Choi, R. Zemp "Towards microvascular pressure estimation using ultrasound and photoacoustic imaging", *Photoacoustics*, 14, 99–104 (2019).

10. G.A. Askaryan "Increase of laser and other radiation transmissions through soft opaque physical and biological media", *Kvantovaya Elektronika*, 9(7), 1379–1383 (1982).

11. E.K. Chan, B. Sorg, D. Protsenko, M. O'Neil, M. Motamedi, A.J. Welch "Effects of compression on soft tissue optical properties", *IEEE Journal of Selected Topics in Quantum Electronics*, 2(4), 943–950 (1996).

12. O.A. Zyuryukina, Yu.P. Sinichkin "The dynamics of optical and physiological characteristics of human skin in vivo during its compression", *Optics and Spectroscopy*, 127(3), 555–563 (2019).

13. L. Lim, B. Nichols, N. Rajaram, J.W. Tunnell "Probe pressure effects on human skin diffuse reflectance and fluorescence spectroscopy measurements", *Journal of Biomedical Optics*, 16(1), 011012 (2011).

14. C.G. Rylander, T.E. Milner, J.S. Nelson "Mechanical tissue optical clearing devices: Enhancement of light penetration and heating of ex vivo porcine skin and adipose tissue", *Lasers in Surgery and Medicine*, 40(10), 688–694 (2008).

15. C. Drew, T.E. Milner, C.G. Rylander "Mechanical tissue optical clearing devices: Evaluation of enhanced light penetration in skin using optical coherence tomography", *Journal of Biomedical Optics*, 14(6), 064019 (2009).

16. A. Izquierdo-Roma, W.C. Vogt, L. Hyacinth, C.G. Rylander "Mechanical tissue optical clearing technique increases imaging resolution and contrast through ex vivo porcine skin", *Lasers in Surgery and Medicine*, 43(8), 814–823 (2011).

17. W.C. Lee, M. Zhang, A.F. Mak "Regional differences in pain threshold and tolerance of the transtibial residual limb: Including the effects of age and interface material", *Archives of Physical Medicine and Rehabilitation*, 86(4), 641–649 (2005).

18. S. Xiong, R.S. Goonetilleke, C.P. Witana, W.D.A.S. Rodrigo "An indentation apparatus for evaluating discomfort and pain thresholds in conjunction with mechanical properties of foot tissue in vivo", *Journal of Rehabilitation Research and Development*, 47(7), 629–641 (2010).

19. A.A. Fischer "Pressure tolerance over muscles and bones in normal subjects", *Archives of Physical Medicine and Rehabilitation*, 67(6), 406–409 (1986).

20. G. Pickering, D. Jourdan, A. Eschalier, C. Dubray "Impact of age, gender and cognitive functioning on pain perception", *Gerontology*, 48(2), 112–118 (2002).

21. B.E. Bouma, G.J. Tearney (eds.) *Handbook of Optical Coherence Tomography*. New York: Marcel Dekker, 2002.

22. P.D. Agrba, E.A. Bakshaeva, D.O. Ellinsky, I.L. Shlivko, M.Yu. Kirillin "The role of mechanical compression in human skin imaging using cross-polarization optical coherence tomography", *CTM*, 6(1), 75–82 (2014).

23. Q. Ouyang, D. Zhu, Q. Luo, H. Gong, Q. Luo "Modulation of temperature on optical properties of rat skin in vivo", *Proceedings of the SPIE*, 6534, 65343I, doi: 10.1117/12. 741499 (2007).

24. O.S. Khalil, S.J. Yeh, M.G. Lowery, X. Wu, C.F. Hanna, S. Kantor, T.W. Jeng, J.S. Kanger, R.A. Bolt, F.F. de Mul "Temperature modulation of the visible and near infrared absorption and scattering coefficients of human skin", *Journal of Biomedical Optics*, 8(2), 191–205 (2003).

25. P.D. Agrba, M.Yu. Kirillin "Effect of temperature regime and compression in OCT-imaging of skin", *Photonics and Lasers in Medicine*, 5(2), 161–168 (2016).

35

Measurement of the dermal beta-carotene in the context of multimodal optical clearing

Mohammad Ali Ansari and Valery V. Tuchin

CONTENTS

Introduction

Carotenoids are a large water-repelling pigment group of isoprenoid metabolites covering more than 700 compounds [1, 2]. They have essential roles in all photosynthetic organisms due to their important photoprotective and antioxidant properties [3]. Studies have shown that plants, algae, cyanobacteria, some fungi, and nonphotosynthetic bacteria can synthesize carotenoids as well, and many animals rely on food-borne carotenoids as visual pigments, antioxidants, or colorants [3, 4]. The carotenoid molecule is typically a C_{40} hydrocarbon and forms a chromophore of double bonds in conjugation, which absorbs light in the blue light spectrum. This causes these compounds to be colored ranging from yellow to orange and red [1, 5–8].

In 2018, Widjaja-Adhi et al. investigated the role of carotenoids in the visual procedure. They applied green (532 nm) and blue (405 nm) low-energy lasers to damage the mouse retina and observed that blue laser light treatment triggered the formation of aberrant retinaldehyde isomers in the retina. In addition, they found that zeaxanthin supplementation of mice shielded retinoids from these photochemical modifications. These pigments also reduced the extent of the damage to the retina after the blue laser light insult [6].

Several studies have shown that carotenoids also play a key role in the prevention of many oxidative stress–mediated degenerative diseases such as Alzheimer's disease (AD). For example, some results have shown that low plasma β-carotene concentrations have been estimated in AD subjects compared with cognitively healthy subjects as presented by Boccardi et al. [9]. They investigated the relation between beta-carotene and accelerated cellular aging such as leucocyte telomere length (LTL) and peripheral mononuclear cell (PBMC) telomerase activity in a cohort of old-age subjects. Their results confirmed that lower plasma beta-carotene levels are associated with AD diagnosis independent of multiple covariates [9]. Based on some animal studies (2004–2009), Obulesu et al. emphasized the dietary supplementation of carotenoids to help fight Alzheimer's. and they recommended further animal studies to understand their mechanism of neuroprotection [10]. Hira et al. applied 50 male mice (3 months old, weighing 30–40 g) categorized into five groups. Groups I, II, and III served as the control group, disease group (streptozotocin 3 mg/kg i.c.v. – intracerebroventricular injection), and standard (piracetam 200 mg/kg. i.p.—intraperitoneally) groups, respectively. Groups IV and V served as treatment groups (beta-carotene at a dose of 1.02 and 2.05 mg/kg, respectively) [11]. Their results showed the administration of beta-carotene reduced streptozotocin-induced cognitive deficit via its antioxidative effects, the inhibition of acetylcholinesterase, and the reduction of amyloid β-protein fragments [11]. Li et al. categorized five studies (2002–2010) to evaluate the effects of dietary intake of vitamin E, vitamin C, and beta-carotene to decrease the relative risk of AD [12], while Dover and her colleagues, during a mean fellow-up period of 9.5 years on 5395 participants (55 years and older), statistically studied the consumption of major dietary antioxidants relative to long-term risk of AD and they compared the effects of dietary intake of vitamin E, vitamin C, and beta-carotene on risk of AD [13]. In summary, these results suggest that dietary intakes of vitamin E, vitamin C, and beta-carotene can help lower the risk of AD [11]. Moreover, Muscogiuri et al. reported that beta-carotene can enhance the immune system, since it can increase the

DOI: 10.1201/9781003025252-39

number of T-cell subsets and enhance lymphocyte response to mitogen, increase interleukin-2 production, and potentiate natural killer cell activity, which is vital parameter during sessional influenza or the pandemic of COVID-19 [14, 15].

Carotenoids can be classified into subgroups: carotenes (having the formula $C_{40}H_x$, such as alpha-, beta-, delta-, and gamma-carotene and lycopene) and xanthophylls (oxygenated derivatives of carotenes with formula $C_{40}H_xO_y$, such as lutein, zeaxanthin, and neoxanthin). Some of the carotenoids are called *provitamin A*, which yield vitamin A and retinoids such as alpha- and beta-carotene and beta-cryptoxanthin, while lycopene, lutein, zeaxanthin, and meso-zeaxanthin are called *non-provitamin A* carotenoids [16], and provide significant protection against the potential damage caused by light striking this portion of the retina. In the eye, lutein and zeaxanthin have been shown to filter high-energy wavelengths of visible light and act as antioxidants to protect against the formation of reactive oxygen species and subsequent free radicals [16–19].

The epidemiological observations very consistently show that people who consume higher dietary levels of fruits and vegetables have a lower risk of certain types of cancer; e.g., in 2003, Chao et al. showed that there is a relationship between lower risk of breast cancer and increased dietary intake of beta-carotene [15, 17]. Based on several studies, it seems that beta-carotene is useful for maintaining normal redox, immunological, and probably cell-to-cell communication activity in the general population [8–25]. So, the dermal beta-carotene concentration can be a good indicator of physical health.

Three methods have been introduced for detection and measurement of beta-carotene and other carotenoids: (a) mass spectroscopy and nuclear magnetic resonance (NMR), (b) high-performance liquid chromatography (HPLC), and (c) optical methods including Raman spectroscopy and diffuse reflectance spectroscopy (DRS) or absorption spectroscopy. Measuring beta-carotene using DRS provides three advantages: (I) low-cost instrumentation, (II) a portable device, and (III) high discrimination power to distinguish different carotenoids. Here, we overview how one can apply low-cost DRS to measure dermal beta-carotene. The presence of blood, melanin, and water and strong scattering of light in the skin hide the absorption spectra of beta-carotene. Hence, one can apply OC methods to detect beta-carotene spectra.

Optical properties of beta-carotene

Beta-carotene is made up of π-electron conjugated carbon chain molecules (its molar mass is 536.87 g/mol, and its density is 1 g/cm³). The chemical structures of beta-carotene and some carotenoids are shown in Figure 35.1. The beta-carotene molecule has nine conjugated carbon double bonds along its backbone and four methyl groups attached as side groups [27]. Furthermore, the molecule is terminated on each end by an ionone ring, each having an additional internal carbon bond, and each adding to the effective conjugation length of the molecule [26–28]. As shown in Figure 35.2, beta-carotene in ethanol has two peaks at wavelengths of 460 and 488 nm.

Beta-carotene and related polyenes are also known to exhibit extremely strong enhancement in resonance Raman scattering. A large number of studies have reported on different aspects of the resonance Raman spectrum of the ground state of beta-carotene [30]. In 2005, Darvin et al. developed a portable resonance Raman microscope to *in vivo* detect beta-carotene and lycopene within human skin [31]. Using multiline Ar+ laser excitation (wavelengths of 488 and 514.5 nm), the Raman signals are characterized by two prominent Stokes lines at 1160 and 1525 cm⁻¹, which have nearly identical relative intensities. Both substances were detected simultaneously [31, 32]. According to Ref. [31], the Raman shift of beta-carotene may be derived from the Raman spectrum of human skin, as shown in Figure 35.3. Raman shifts on the forehead of three groups of healthy volunteers – vegetarians, nonsmokers, and heavy smokers – were measured to estimate concentration of beta-carotene, and the beta-carotene contents are measured to be 0.63 ± 0.05, 0.39 ± 0.07, and 0.22 ± 0.08 nmol/g, respectively [31]. Based on several studies using HPLC and resonance Raman spectroscopy, the concentrations of different carotenoids in human skin are depicted in Table 35.1. It was shown that beta-carotene in human skin strongly depends on the anatomic site and can drastically change inter-individually [31, 33].

Results presented in Refs. [28, 34] show that Raman spectroscopy is sensitive and reproducible and strongly correlates with the HPLC analysis of carotenoids, where a single carotenoid is presented. But for skin containing a complex mixture of similar compounds, Raman spectroscopy is not able to clearly discriminate between the individual carotenoids [28], or, due to high background autofluorescence in resonance

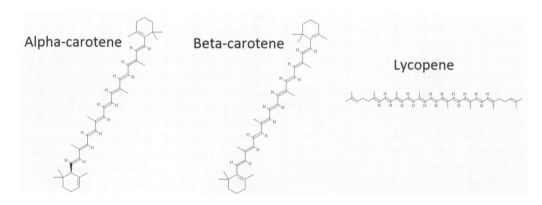

FIGURE 35.1 Chemical structures of alpha- and beta-carotene ($C_{40}H_{56}$) and lycopene ($C_{40}H_{56}$) based on data depicted in Reference [26].

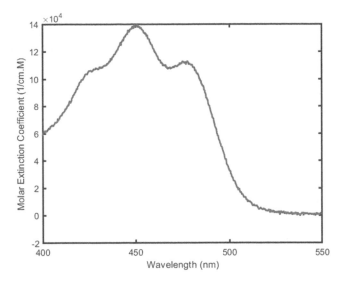

FIGURE 35.2 Molar extinction coefficient of beta-carotene in ethanol based on data presented in Reference [29].

Raman microscopy, this method can underestimate the beta-carotene concentration in skin.

Detection of dermal beta-carotene using multimodal OC

As shown in Figure 35.4(a), the presence of blood, melanin, water, and other chromophores inside skin masks the beta-carotene spectrum (see Figure 35.4(b)). Three OC methods have been applied to detect absorption spectra of beta-carotene: (a) compression, (b) chemical agent, and computational OC methods, which are presented in the following sections.

Compression OC

The optical properties of skin can be changed due to the application of mechanical forces, such as compression and

TABLE 35.1

Carotenoids composition in human skin (ng per g tissue and based on data presented in references [31, 33])

	Skin source			Mean value (ng/g)
	#1	#2	#3	
Lycopene + Z-isomers	105	9	93	69 ± 52
Beta-carotene	38	2	37	26 ± 20
Alpha + gamma + delta + zeta carotene	58	6	18	41 ± 17
Lutein + Zeaxanthin	26	ND.	ND.	26*
Phytoene	92	45	57	65 ± 24
Phytofluene	21	6	17	15 ± 8
TOTAL	**340**	**68**	**222**	**210 ± 36**

ND: nondetectable data

stretching. Compression removes the interstitial water and blood from the area of interest, leading to an increase in optical tissue homogeneity. Due to the interstitial water loss, closer packing of tissue components causes less light scattering (refractive index matching) due to cooperative interference effects and thinner tissue [36–38]. In 2011, Gurjarpadhye et al. showed that changes in the thickness of *ex vivo* porcine skin during air dehydration results in an increase of refractive index from 1.4 to 1.5 (at a wavelength of 1310 nm) [39]. The optical depth of diffuse reflectance spectroscopy is increased by lowering the scattering and absorption coefficient. In addition, transport of water and blood away from the compressed region helps us to see the absorption spectrum of beta carotene (in the spectrum of 450–500 nm) as shown in Figure 35.5.

In 2012, Ermakov and Gellermann reported a new reflection-spectroscopy method for the quantitative detection of carotenoid in the living human skin [40]. They used topical pressure in the reflectivity approach presented in Figure 35.6. In this way, blood is squeezed out of the measured tissue volume and the resupply is temporarily blocked. As a consequence, overlapping HbO_2 absorptions are effectively removed and any residual HbO_2 is converted into Hb, which has about a factor of 2.5 lower absorption strength in the spectral window

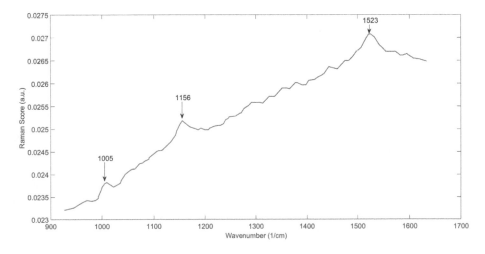

FIGURE 35.3 A typical Raman shift spectrum obtained from the human skin with an excitation wavelength at 514.5 nm based on data presented in Reference [32]. The Raman lines at 1005 cm⁻¹, 1156 cm⁻¹, and 1523 cm⁻¹ originate, respectively, from the rocking motions of the methyl groups, and from carbon–carbon single bond and carbon–carbon double-bond stretch vibrations of the conjugated backbone.

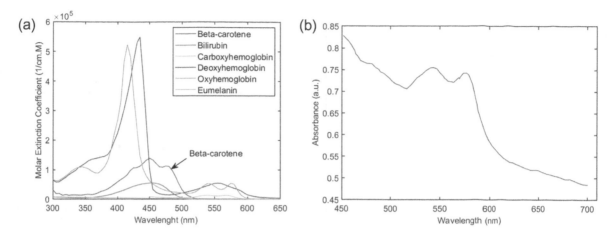

FIGURE 35.4 Absorbance of some skin components. Molar extinction coefficients of beta-carotene (blue), melanin (cyan), bilirubin (red), and oxy-(green), deoxy-, (purple) and carboxyhemoglobin (orange) based on data are shown in Refs. [29, 35]. (a) and normal absorbance of skin of palm of hand (b).

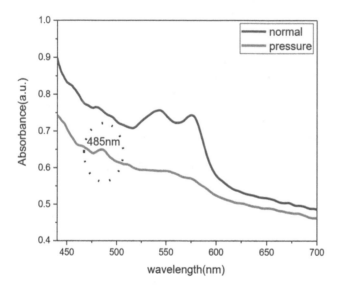

FIGURE 35.5 Effect of mechanical compression on the absorbance of skin. Black line shows normal spectrum and red line shows the variation of absorbance during applying compression. The dotted oval depicts the absorbance spectra of beta-carotene around 485 nm (based on data shown in Reference [8]).

relative to HbO$_2$ [40, 41]. They applied a piezo-resistive load cell that is mounted onto the lens assembly to monitor and control the topical pressure. They also compared the results obtained by compression OC method with results obtained via resonance Raman spectroscopy. A high correlation with a squared correlation coefficient of R^2 = 0.85 is obtained for the two methods (see Figure 35.6).

It seems that application of reflectance spectroscopy to detect dermal beta-carotene has two advantages compared to resonance Raman microscopy: (a) low-cost instrumentation and (b) providing accurate results, because Raman spectroscopy may underestimate dermal beta-carotene due to the high background autofluorescence. Recently, dermal beta-carotene has been measured by compression mediated reflectance spectroscopy using low-cost LEDs (490 and 505 nm) [41]. They were able to measure mean dermal beta-carotene to be approximately 2.5 ± 0.6 nmol/g.

They applied topical pressure to temporarily squeeze blood out of the illuminated tissue volume.

Immersion OC

As mentioned, the optical immersion of tissues in exogenous chemical agents is the most recent technique for controlling the optical properties of biological tissue [36], and the one with the most potential. In 2018, a solution of DMSO 50% with ethanol as the solvent was applied to detect beta-carotene content in a two-layer phantom [8]. The underside layer consisted of Intralipid® 20% (3.3 ml), distilled water (13 ml), agar (0.7 g), and beta-carotene [≥ 97.0% (UV) Sigma-Aldrich] in different concentrations of 0.2, 0.3, 0.4, and 0.6 μM. The upper layer is a mouse ear (2 months old, 20 g) with an average thickness of 250 μm (one-half of its full thickness is composed of two layers – the epidermis and the dermis – and below the skin layers, cartilage forms the structural support for the mouse ear). Ear specimens were immersed in DMSO solution for 15 min, immersion OC [8, 22]. Figure 35.7 shows the schematic of setup and phantom in this study.

In this study, the authors first applied compression OC and studied the variation of diffuse reflectance and absorbance. The results, including error bars, show that the removal of blood dip between 500 and 570 nm depends on the amount of mechanical compression. This is because the transport of blood is a function of the amount of compression and Young's module of tissue [23]. One can see in Figure 35.8 that, after the application of pressure and when blood is removed, the absorbance decreases due to less blood absorption. Indeed, scattering is also changed a little in this range (scattering by red blood cells and lipids), but, in this wavelength range, blood absorption prevails [8]. Then, the authors simultaneously studied the influence of both compression and a chemical clearing agent on tissue phantoms. Immersion of ears in DMSO leads to a decrease in their optical scattering, so one can see an increase in optical depth (diffusion of DMSO in phantom causes a reduction in the scattering coefficient by refractive index matching) [41]. In addition to DMSO, a combination of 70% glycerol + 30% liquid paraffin as a chemical OC agent (OCA) was also applied to measure dermal beta-carotene as stated in Reference [41].

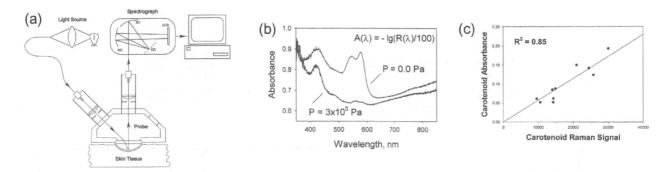

FIGURE 35.6 Schematics of skin reflectivity apparatus and compression mediated DRS. The device consists of a "white" light source, light delivery and collection module that is placed in contact with the tissue site of interest, a spectrograph or two-wavelength detection scheme, and data acquisition, processing and display electronics. The module's lens is pressed against the tissue site of interest for about 10–15 seconds for skin carotenoid measurements (a). Absorbance spectra of human skin tissues is obtained with the instrument's contact lens touching the skin tissue site with zero pressure and applied pressure of 3×10^5 Pa, respectively. A desired optical "blood clearing" effect is achieved in the tissue site for carotenoid measurements with minimal interference from HbO_2 (b), and correlation between reflectivity- and Raman-derived skin carotenoid levels, is measured with both methods for ten volunteer subjects (C). Reprinted with permission from Reference [40].

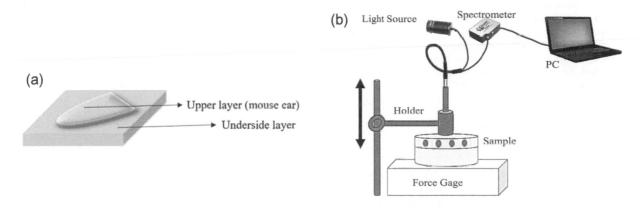

FIGURE 35.7 A schematic of phantom and compression mediated diffuse reflectance spectroscopy. Underside layer consists of intralipid, agar- and beta-carotene, and ear of mouse applied as an upper layer [8](a), and the scheme of experiment setup for measuring beta-carotene inside tissue phantom. It consists of a tungsten halogen lamp with 20-W power (HL2000; Ocean Optics Inc., Dunedin, Florida) and spectrometer (USB200, Ocean Optics) with a spectral range of 350 to 1100 nm and an optical fiber probe (R200-REF, Ocean Optics, Dunedin). This probe consists of a bundle of seven optical fibers – six illumination fibers around one read fiber – each of which is 200 μm in diameter (NA =0.22, length = 2 m) and a distance of source–detector separation of 200 μm [8, 20] (b). Reprinted with permission from Reference [8].

Computational OC

In addition to immersion OC, a more sophisticated method based on optical diffusion equation can be applied to detect dermal beta-carotene. Recent studies show that computational OC can be used as a noninvasive method to estimate dermal chromophores [8, 24, 29, 41–45]. In this method, the diffuse reflectance (or absorbance) is used to reconstruct distribution of a particular chromophore of interest that is normally hidden in the background of absorption of hemoglobin, water, and melanin [25, 43]. This method has been demonstrated in application to estimate the concentration of beta-carotene inside a tissue phantom, as depicted in Reference [8].

In 2006, Zonios and Dimou presented a simple way to calculate diffuse reflectance, $R(\lambda)$, from a homogeneous semi-infinite turbid medium. The forward and backscattered light are exponentially attenuated according to the sum of the absorption and reduced scattering coefficients, $(\mu_s'(\lambda) + \mu_a(\lambda))$ [43], where $\mu_s(\lambda)$ is the wavelength-dependent reduced scattering

coefficient, and $\mu_a(\lambda)$ is the wavelength-dependent absorption coefficient of tissue. It is well established that exponential solutions of this type constitute an acceptable approximation, especially in one-dimensional geometries [8, 29, 43]. Figure 35.9 depicts the geometry of this problem, e.g., a light with an intensity I_0 at the top surface of skin (at $z = 0$) is exponentially attenuated as it propagates inside skin, $I = I_0 e^{-(\mu_s'(\lambda) + \mu_a(\lambda))z}$. At a given depth z, light is scattered back from a thin layer of thickness dz. The backscattered light is assumed to be proportional to the reduced scattering coefficient $\mu_s(\lambda)$ and travels back to the surface while being exponentially attenuated by a factor $e^{-(\mu_s'(\lambda) + \mu_a(\lambda))z}$. The total diffuse reflectance is then given by [43]:

$$R(\lambda) = 2\mu_s'(\lambda) \int_0^\infty e^{-(\mu_s'(\lambda) + \mu_a(\lambda))z} dz = \frac{\mu_s'(\lambda)}{\mu_s'(\lambda) + \mu_a(\lambda)} \quad (35.1)$$

By taking into account of refractive index mismatch, the diffuse reflectance $R(\lambda)$ from skin can be stated as the following [44]:

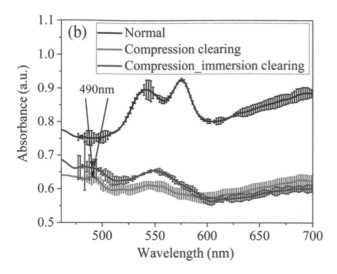

FIGURE 35.8 The effect of applying experimental clearing on absorbance of tissue phantom depicted in Figure 35.7. The normal spectrum is depicted via the black line. The red line depicts the variations of absorption due to compression OC (300 N/m²). The blue line presents the effects of the combination of immersion and compression OC. Eight ears are used while doing the experiment in order to investigate the repeatability of this method. Arrows point out the absorption peak of beta-carotene. Reprinted with permission from Reference [8].

$$R\left(\lambda\right)=\frac{\mu_s'\left(\lambda\right)}{k_1+k_2\mu_a\left(\lambda\right)} \quad (35.2)$$

Where k_1 and k_2 are the parameters depended on the probe geometry and refractive indices of studied sample; for example, for the setup shown in Figure 38.7(b), they were determined to be 0.025 and 0.075, respectively [8]. The absorption coefficient and reduced scattering of skin can be given as [45]:

$$\mu_a\left(\lambda\right)=C_{mel}\,\varepsilon_{mel}\left(\lambda\right)+C_{beta}\,\varepsilon_{beta}\left(\lambda\right)+C_{bl}\,\varepsilon_{blood}\left(\lambda\right)$$
$$+C_{bili}\,\varepsilon_{bili}\left(\lambda\right)+C_{water}\,\varepsilon_{water}\left(\lambda\right) \quad (35.3a)$$

$$\mu_s'\left(\lambda\right)=\left(1-c_1\frac{\lambda-\lambda_{max}}{\lambda_{max}-\lambda_{min}}\right)\mu_s'\left(\lambda_{min}\right) \quad (35.3b)$$

where C_{mel}, C_{beta}, C_{blood}, C_{bili}, and C_{water} are concentrations of melanin, beta-carotene, blood, bilirubin, and water in the tissue, respectively; and $\varepsilon_{mel}\left(\lambda\right)$, $\varepsilon_{beta}\left(\lambda\right)$, $\varepsilon_{blood}\left(\lambda\right)$, $\varepsilon_{bili}\left(\lambda\right)$, and $\varepsilon_{water}\left(\lambda\right)$ are molar extinctions of melanin, beta-carotene, blood, bilirubin, and water, respectively. The parameter c_1 is a parameter related to effective scatter size, $\lambda_{min}=400$ nm and $\lambda_{max}=800$ nm [29]. Based on the data presented in Reference [35], the molar extinction coefficients mentioned in the previous relationship can be modeled via the following relations:

$$\varepsilon_{mel}\left(\lambda\right)\cong 1.7\times 10^{12}\times\lambda^{-3.4} \quad (35.4a)$$

$$\varepsilon_{beta}\left(\lambda\right)\cong 1.06\times 10^{6}\times e^{-\left(\frac{222.8-\lambda}{5.23}\right)^2}+1.26\times 10^{5}\times e^{-\left(\frac{\lambda-443}{49}\right)^2}$$
$$\quad (35.4b)$$
$$+3.84\times 10^{4}\times e^{-\left(\frac{\lambda-481}{11.29}\right)^2}$$

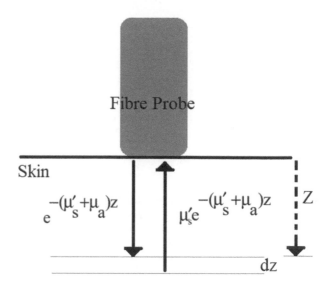

FIGURE 35.9 A schematic one-dimensional diffuse reflectance model based on depicted scheme in Reference [43].

$$\varepsilon_{blood}\left(\lambda\right)\cong 0.4\times\left(4.75\times 10^{4}\times e^{-\left(\frac{\lambda-552}{48}\right)^2}\right)$$
$$+0.6\times\left(\begin{array}{l}3.8\times 10^{18}\times e^{-\left(\frac{\lambda+2069}{447}\right)^2}+5.5\times 10^{4}\times e^{-\left(\frac{\lambda-576}{11.29}\right)^2}\\ +5.08\times 10^{4}\times e^{-\left(\frac{\lambda-542}{19}\right)^2}\end{array}\right)$$
$$\quad (35.4c)$$

$$\varepsilon_{bili}\left(\lambda\right)\cong 2.86\times 10^{4}\times e^{-\left(\frac{\lambda-463}{23.8}\right)^2}+7.65\times 10^{5}\times e^{-\left(\frac{\lambda+502}{370.5}\right)^2}$$
$$\quad (35.4d)$$
$$+4.06\times 10^{4}\times e^{-\left(\frac{\lambda-432}{40.8}\right)^2}$$

In 2009, Tseng et al. applied a more sophisticated methodology to measure the absorption coefficient of a blood sample. In this study, the probe has been adjusted into multiple source-detector pairs. The normalized reflectance versus source-detector separation is then fit to a diffusion model by a least-square minimization algorithm to determine the absorption and reduced scattering spectra. The reconstructed absorption spectra are fit linearly with known chromophore absorption spectra to extract chromophore concentrations, and the reduced scattering spectra are fit to a scattering power law to obtain the scattering power [46, 47]. The authors applied the diffusion equation to calculate diffuse reflectance from a two-layer tissue phantom:

$$R\left(\lambda,\rho\right)=\frac{1}{4\pi}\int_{2\pi}\left(1-R_{fres}\left(\theta\right)\right)\cos\theta\,L\,d\Omega \quad ,(35.5)$$

where $\rho=\sqrt{x^2+y^2}$, and $R_{fres}\left(\theta\right)$ is the Fresnel reflection coefficient for a photon with an incident angle θ relative to the normal to the boundary. The parameter L indicates the fluence at the top surface, $L=\varphi+3D\left(\partial\varphi/\partial z\right)\cos\theta$, where

TABLE 35.2

Chromophore concentrations of dorsal forearm of 18 subjects reconstructed with two regional fitting (600–1000 nm). Data from Reference [47]

Skin Type	Deoxyhemoglobin [μM]	Oxyhemoglobin [μM]	Melanin [%]	Water [%]	Lipid [%]
I-II	0.12 ± 0.51	7.74 ± 3.15	0.87 ± 0.18	21.42 ± 2.56	27.74 ± 5.01
III-IV	0.08 ± 0.26	9.47 ± 3.41	1.15 ± 0.09	22.46 ± 2.46	26.12 ± 6.04
V-VI	0.02 ± 0.10	2.72 ± 2.40	1.65 ± 0.23	18.66 ± 3.17	16.56 ± 8.23

$D = 1/3\left(\mu'_s + \mu_a\right)$ and φ are diffusion coefficient and fluence rate, respectively. In this study, the reflectance curve obtained from the volunteer skin (in a spectrum range of 500–1000 nm) was then fit to Equation 35.5. The "lsqcurvefit" function in MATLAB (MathWorks, MA, USA) was used to perform the least-squares fittings to recover absorption and reduced scattering coefficients. Table 35.2 shows the mean value of dorsal chromophore concentration within the forearms of 18 volunteers.

On the other hand, in Equation 35.2 and assuming R_0 indicates the diffuse reflectance for $\mu_a(\lambda) = 0$, so $\mu'_s(\lambda) = k_1 R_0$. This diffuse reflectance R_0 can be estimated from the wavelength region 600–900 nm using the scattering power law as mentioned in Reference [8, 45]. Finally, the absorption coefficient is obtained from

$$\mu_a(\lambda) = \frac{k_1}{k_2} \cdot \left(\frac{R_0}{R(\lambda)} - 1\right). \qquad (35.6)$$

Since, in the visible spectrum, blood and melanin are the main chromophores that affect absorption coefficient of our phantom, while variations in water and lipids do not have significant effects in this spectrum interval. Figure 35.10 shows that analytical OC could extract and quantify the absorption coefficient of each chromophore within a tissue-like optical phantom (as depicted in Figure 35.6) without immersion or mechanical OC methods for detection of beta-carotene.

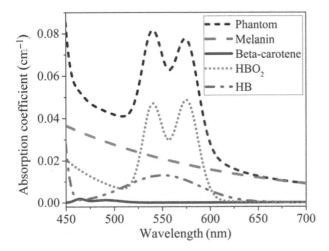

FIGURE 35.10 Extracted absorption coefficient of a tissue phantom. The absorption coefficient of skin components inside the tissue phantom are shown based on computational OC. Reprinted with permission from Reference [8].

As presented in Refs. [8, 35, 44], the simulated absorbance can be applied to measured absorbance for estimating the concentration of beta-carotene (see Figure 35.11). Recently, this method has been applied to measure dermal beta-carotene after a 1-week beta-carotene diet (at least 250 cc/day carrot juice). To do this, one requires a calibrating relation between diffuse reflectance and concentration of beta-carotene inside the tissue phantom. For example, in Reference [41] and similar to Reference [23], a tissue phantom based on the mouse ear alone and together with immersion (70% glycerol + 30% liquid paraffin as a chemical OCA) and compression OC was used. The authors defined delta parameter, $\Delta = A_{490} - B_{505}$, to clarify the effect of OC on absorption spectrum (A_{490} is the absorbance at the wavelength of 490 nm and B_{505} is the absorbance at the wavelength of 505 nm); the parameter Δ depends on the beta-carotene content of the tissue and the efficacy of OC. Because the first factor (concentration of beta-carotene) increases the amount of Δ parameter and the second one (efficiency of OC) decreases this parameter. Since, the compression OC can displace the blood (and water) and decrease the absorption spectrum around wavelength of 490 nm (absorption peak of beta-carotene) and hence it reduces the amount of parameter of Δ. The water transport (dehydration) and displacement of blood help us to see the spectra of beta-carotene [35]. Moreover, applying an OC agent such as DMSO or glycerol causes an intrinsic refractive index matching, which may also enhance the efficacy of OC (to decrease absorbance spectra via reduction of light scattering inclusion) [24, 41, 46] and consequently reduce the parameter Δ. The relation between parameter Δ and the concentration of beta-carotene in the tissue phantom was obtained as follows:

$$\Delta = 17 \times 10^{-4} C + 0.009. \qquad (35.7)$$

Based on the above relation, the value of dermal beta-carotene increases from 0.9 nmol/g to 1.7 nmol/g (after a 1-week beta-carotene diet).

In addition, artificial neural networks (ANNs) were also used to obtain absorption and scattering coefficients, estimated using diffuse reflectance [48, 49]. Monte Carlo (MC) simulations can be also performed for the sake of possibly obtaining simulated diffuse reflectance to contribute to the ANN as input parameters [48]. They taught ANN using different MC simulation to estimate the absorption coefficient of liquid phantoms (including clinoleic intravenous lipid emulsion (CILE) and Evans Blue (EB) dye into a distilled water). They compared the real value of absorption coefficients with the ANN method as depicted in Figure 35.12.

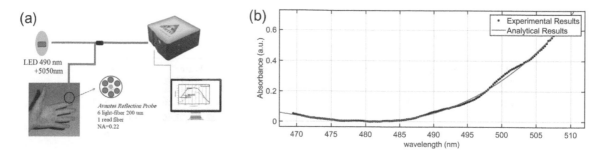

FIGURE 35.11 A schematic setup to measure dermal beta-carotene using multimodal OC including OCA, compression, and computational OC(a). The skin absorbance is modeled based on computational OC and compared with measured absorbance ($R^2 = 0.98$) as shown in (b).

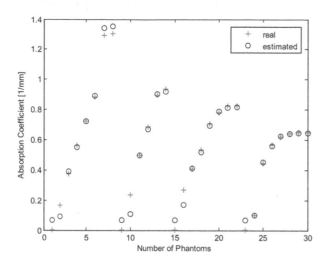

FIGURE 35.12 Estimation of absorption coefficient μ_a using MC simulation-based diffuse reflectance. All MC results were used to estimate absorption coefficient values by ANN. During learning mode, the Levenberg–Marquardt algorithm was used such that 60% of diffuse reflectance values was for testing and 35% was for training; finally, the authors validated with the remaining 5% part of all values using a cross-validation technique. Reprinted with permission from Reference [48].

Some researchers combined spectroscopy and HPLC with ANN methods to reconstruct beta-carotene and lycopene content within fruits and agricultural products [49–51]. For example, Torrecilla et al. applied a neural method to estimate the carotene content in tomato products in standard samples [50]. This method could estimate the beta-carotene content with an error rate of less than 10%. ANN and deep learning have been applied to recover beta-carotene content in food or agricultural products, and it seems that this approach can be used for dermal screening as a new computational OC.

Summary

Recent studies show that carotenoids have important role in preventing or decreasing the risk of some diseases such as immunity diseases, neurological disorders, Alzheimer's disease, photosensitive disorders, and prostate, colon, and oral cancers. Therefore, carotenoid monitoring in skin can be a good indicator of physical health. This chapter aimed to apply low-cost DRS to estimate dermal beta-carotene content. We introduced combinations of compression and immersion OC methods to decrease both light scattering and the effect of blood spectra. The recent presented results depict that the absorption spectrum of beta-carotene can easily be seen after applying experimental OC methods. Then, one can apply computational OC to determine beta-carotene content. We noted that the computational OC can be applied independently or in combination with experimental OC methods. Hence, a combination of experimental OC and computational OC provides a low-cost method to measure dermal beta-carotene *in vivo*. Finally, it is proposed that a powerful approach to estimate beta-carotene content based on artificial neural networks could open new avenues in this area.

Acknowledgments

V.V.T. is grateful for support from the Government of the Russian Federation (grant # 075-15-2019-1885). MAA gratefully acknowledges the support of the Iranian National Science Foundation (grant # 98029460).

REFERENCES

1. M. Rodríguez-Concepción and R. Welsc, *Plant and Food Carotenoid*, Human Press, New York (2020).
2. H. Croft and J.M. Chen, "Leaf pigment content," *Reference Module in Earth Systems and Environmental Sciences*, pp. 1–22, Elsevier Inc, Oxford (2017).
3. M. Lohr, "Carotenoids," in: *The Chlamydomonas Sourcebook, Organellar and Metabolic Processes*, Eds. D.B. Stern, E.H. Harris, Vol. 2, 2nd ed., pp. 799–817, Academic Press, San Diego (2009), doi: 10.1016/B978-0-12-370873-1.00029-0.
4. N.I. Krinsky and E.J. Johnson, "Carotenoid actions and their relation to health and disease," *Molec. Asp. Med.* **26**(6), 459–516 (2005).
5. A.V. Popova, "Spectral characteristics and solubility of beta-carotene and zeaxanthin in different solvents," *Compt. Rend. Acad. Bulg. Sci.* **70**(1), 55–62 (2017).
6. M.A.K. Widjaja-Adhi, S. Ramkumar and J. von Lintig, "Protective role of carotenoids in the visual cycle," *FASEB J.* **32**(11), 6305–6315 (2018).
7. V. Shete and Q. Loredana, "Mammalian metabolism of β-carotene: Gaps in knowledge," *Nutrients* **5**(12), 4849–4868 (2013).
8. S. Masoumi, M.A. Ansari, E. Mohajerani, E.A. Genina and V.V. Tuchin, "Combination of analytical and experimental optical clearing of rodent specimen for detecting beta-carotene: Phantom study," *J. Biomed. Opt.* **23**(9), 095002 (2018).

9. V. Boccardi, B. Arosio, L. Cari et al., "Beta-carotene, telomerase activity and Alzhimer's disease in old age subjects," *Eur. J. Nutr.* **59**(1), 119–126 (2020).

10. M. Obulesu, M.R. Dowlathabad and P.V. Bramhachari, "Carotenoids and Alzheimer's disease: An insight into therapeutic role of retinoids in animal models," *Neurochem. Int.* **59**(5), 535–541 (2011).

11. S. Hira, U. Saleem, F. Anwar, M.F. Sohail, Z. Raza and B. Ahmad, "β-carotene: A natural compound improves cognitive impairment and oxidative stress in a mouse model of streptozotocin-induced Alzheimer's disease," *Biomolecules* **9**(9), 441 (2019).

12. F.J. Li, L. Shen and H.F. Ji, "Dietary intakes of vitamin E, vitamin C, and β-carotene and risk of Alzheimer's disease: A meta-analysis," *J. Alzheimers Dis.* **31**(2), 253–258 (2012).

13. E.E. Devore, F. Grodstein, F.J. van Rooij et al., "Dietary antioxidants and long-term risk of dementia," *Arch. Neurol.* **67**(7), 819–825 (2010).

14. G. Muscogiuri, L. Barrea, S. Savastano and A. Colao, "Nutritional recommendations for COVID-19 quarantine," *Eur. J. Clin. Nutr.* **74**(6) 1–2. doi: 10.1038/s41430-020-0635-2 (2020).

15. S. Fakhri, Z. Nouri, S.Z. Moradi and M.H. Farzaei, "Astaxanthin, COVID-19 and immune response: Focus on oxidative stress, apoptosis and autophagy," *Phytother. Res.* **34**(11) 1–3. doi: 10.1002/ptr.6797 (2020).

16. E. Cho, D. Spiegelman, D.J. Hunter et al., "Premenopausal intakes of vitamins A, C, and E, folate, and carotenoids, and risk of breast cancer," *Cancer Epidemiol. Biomarkers Prev.* **12**(8), 713–720 (2003).

17. S.B. Nimse and D. Pal, "Free radicals, natural antioxidants, and their reaction mechanisms," *RSC Adv.* **5**(35), 27986–28006 (2015).

18. B.J. Burri, "Beta-carotene and human health: A review of current research," *Nutr. Res.* **17**(3), 547–580 (1997).

19. R.L. Roberts, J. Green and B. Lewis, "Lutein and zeaxanthin in eye and skin health," *Clin. Dermatol.* **27**(2), 195–201 (2009).

20. M.M.V. Naves and F.S. Moreno, "β-carotene and cancer chemoprevention: From epidemiological associations to cellular mechanisms of action," *Nutr. Res.* **18**(10), 1807–1824 (1998).

21. T. Strandin, S.A. Babayan and K.M. Forbes, "Reviewing the effects of food provisioning on wildlife immunity," *Philos. Trans. R. Soc. B: Boil. Sci.* **373**(1745), 20170088 (2018).

22. W. Stahl, H. Sies and H., "Nutritional protection against photooxidative stress in human skin and eye," in: *Oxidative Stress*, pp. 389–402, Academic Press, Cambridge, MA (2020).

23. S. Masoumi, M.A. Ansari, E. Mohajerani, E.A. Genina and V.V. Tuchin, "Estimation of beta-carotene using calibrated reflection spectroscopy method: Phantom study," in: International Conference Laser Optics (ICLO), St. Petersburg, Russia, p. 496, IEEE (2018).

24. A.N. Bashkatov, K.V. Berezin, K.N. Dvoretskiy et al., "Measurement of tissue optical properties in the context of tissue optical clearing," *J. Biomed. Opt.* **23**(9), 091416 (2018).

25. M.E. Darvin, J.W. Fluhr, M.C. Meinke, L. Zastrow, W. Sterry and J. Lademann, "Topical beta-carotene protects against infra-red-light–induced free radicals," *Exp. Dermatol.* **20**(2), 125–129 (2011).

26. National Center for Biotechnology Information, "PubChem database. All-trans-alpha-carotene," CID=4369188, https://pubchem.ncbi.nlm.nih.gov/compound/All-trans-alpha-Carotene (accessed on Apr. 15, 2020).

27. I.V. Ermakov, M.R. Ermakova, W. Gellermann and J. Lademann, "Noninvasive selective detection of lycopene and β-carotene in human skin using Raman spectroscopy," *J. Biomed. Opt.* **9**(2), 332–339 (2004).

28. J. Jehlička, H.G. Edwards, K. Osterrothová et al., "Potential and limits of Raman spectroscopy for carotenoid detection in microorganisms: Implications for astrobiology," *Philos. Trans. R. Soc. A* **372**(2030), 20140199 (2014).

29. G. Zonios, A. Dimou, I. Bassukas, D. Galaris, A. Tsolakidis and E. Kaxiras, "Melanin absorption spectroscopy: New method for noninvasive skin investigation and melanoma detection," *J. Biomed. Opt.* **13**(1), 014017 (2008).

30. N.H. Jensen, R. Wilbrandt, P.B. Pagsberg, A.H. Sillesen and K.B. Hansen, "Time-resolved resonance Raman spectroscopy: The excited triplet state of all-trans-beta-carotene," *J. Americ Chem. Soc.* **102**(25), 7441–7444 (1980).

31. M.E. Darvin, I. Gersonde, M. Meinke, W. Sterry and J. Lademann, "Non-invasive in vivo determination of the carotenoids beta-carotene and lycopene concentrations in the human skin using the Raman spectroscopic method," *J. Phys. D* **38**(15), 2696 (2005).

32. M.E. Darvin, A. Patzelt, F. Knorr, U. Blume-Peytavi, W. Sterry and J. Lademann, "One-year study on the variation of carotenoid antioxidant substances in living human skin: Influence of dietary supplementation and stress factors," *J. Biomed. Opt.* **13**(4), 044028 (2008).

33. T.R. Hata, T.A. Scholz, I.V. Ermakove et al., "Non-invasive Raman spectroscopic detection of carotenoids in human skin," *J. Invest. Dermatol.* **115**(3), 441–448 (2000).

34. M.E. Darvin, C. Sandhagen, W. Koecher, W. Sterry, J. Lademann and M.C. Meinke, "Comparison of two methods for noninvasive determination of carotenoids in human and animal skin: Raman spectroscopy versus reflection spectroscopy," *J. Biophoton.* **5**(7), 550–558 (2012).

35. http://www.npsg.uwaterloo.ca/index.php.

36. V.V. Tuchin, *Tissue Optics: Light Scattering Methods and Instruments for Medical Diagnostics*, 3rd ed., SPIE Press, Bellingham (2015).

37. L.M.C. Oliveira and V.V. Tuchin, *The Optical Clearing Method: A New Tool for Clinical Practice and Biomedical Engineering*, Springer, Berlin (2019).

38. C.G. Rylander, T.E. Milner, S.A. Baranov and J.S. Nelson, "Mechanical tissue optical clearing devices: Enhancement of light penetration in ex vivo porcine skin and adipose tissue," *Lasers Surg. Med.* **40**(10), 688–694 (2008).

39. A.A. Gurjarpadhye, W.C. Vogt, Y. Liu and C.G. Rylander, "Effect of localized mechanical indentation on skin water evaluated using OCT," *Int. J. Biomed. Imaging*, 817250 (2011).

40. I.V. Ermakov and W. Gellermann, "Dermal carotenoid measurements via pressure mediated reflection spectroscopy," *J. Biophoton.* **5**(7), 559–570 (2012).

41. M.A. Ansari, A. Morovati and V.V. Tuchin, "Low-cost measurement of the dermal beta-carotene in the context of optical clearing," *Proc. SPIE* **11457**, 1145704 (2020).

42. A.K. Bui, R.A. McClure, J. Chang et al., "Revisiting optical clearing with dimethyl sulfoxide (DMSO)," *Lasers Surg. Med.* **41**(2), 142–148 (2009).

43. G. Zonios and A. Dimou, "Modeling diffuse reflectance from semi-infinite turbid media: Application to the study of skin optical properties," *Opt. Express* **14**(19), 8661–8674 (2006).

44. S. Andree, C. Reble and J. Helfmann, "Spectral in vivo signature of carotenoids in visible light diffuse reflectance from skin in comparison to ex vivo absorption spectra," *Photon. Lasers Med.* **2**(4), 323–335. doi: 10.1515/plm-2013-0032 (2013).

45. W.F. Cheong, S.A. Prahl and A.J. Welch, "A review of the optical properties of biological tissues," *IEEE J. Quantum Electron.* **26**(12), 2166–2185. doi: 10.1109/3.64354 (1990).

46. A. Cerussi, N. Shah, D. Hsiang, A. Durkin, J. Butler and B.J. Tromberg, "In vivo absorption, scattering, and physiologic properties of 58 malignant breast tumors determined by broadband diffuse optical spectroscopy," *J. Biomed. Opt.* **11**(4), 044005 (2006).

47. S.H. Tseng, P. Bargo, A. Durkin and N. Kollias, "Chromophore concentrations, absorption and scattering properties of human skin in-vivo," *Opt. Express* **17**(17), 14599–14617 (2009).

48. M.O. Gökkan and M. Engin, "Artificial neural networks based estimation of optical parameters by diffuse reflectance imaging under in vitro conditions," *J. Innov. Opt. Heal. Sci.* **10**(1), 1650027 (2017).

49. L. Zhang, Z. Wang and M. Zhou, "Determination of the optical coefficients of biological tissue by neural network," *J. Mod. Opt.* **57**(13), 1163–1170 (2010).

50. J.C. Torrecilla, M. Cámara, V. Fernández-Ruiz, G. Piera and J.O. Caceres, "Solving the spectroscopy interference effects of β-carotene and lycopene by neural networks," *J. Agric. Food Chem.* **56**(15), 6261–6266 (2008).

51. M. Soltani, M. Omid and R. Alimardani, "Egg volume prediction using machine vision technique based on pappus theorem and artificial neural network," *J. Food Sci. Technol.* **52**(5), 3065–3071 (2015).

36

Optical clearing and molecular diffusivity of hard and soft oral tissues

Alexey A. Selifonov and Valery V. Tuchin

CONTENTS

Introduction

The scientific and technological progress of recent decades has taken medicine to a new and higher level. The improvement of optical systems, the discovery of laser and LED sources, the widespread use of computers and microprocessors, and the development of new technologies for obtaining three-dimensional images have made it possible to make huge changes in imaging technology. Optical methods for tissue imaging and disease diagnosis in humans are increasingly used in various fields of medicine [1, 2]. An overview of emerging novel optical imaging techniques, light scattering, nonlinear optics, and optical tomography of tissues and cells is presented elsewhere [3].

For example, optical imaging of the deep layers of tissues is used to differentiate malignant neoplasms in their early stages, including optical coherence tomography, fluorescence and Raman spectroscopies, and fiberoptic technologies for delivering radiation to pathological sites and back in portable imaging systems [4]. Within the light scattering and autofluorescence (AF) methods, it becomes possible to quantify noninvasively the content of endogenous chromophores of tissues *in vivo* (oxy- and deoxyhemoglobin, melanin, bilirubin, etc.) [5]. Since therapeutic effects are achieved due to absorption of the optical radiation by tissue chromophores, a more detailed study of the optical properties of tissues and cellular structures at specific wavelengths is needed. One of the frequently used methods for calculating tissue scattering and absorption coefficients is the inverse adding-doubling (IAD) technique. It was developed by Dr Scott Prahl et al. [6]. It is one of the most used algorithms for the solution of inverse optical problem for reconstruction reduced (transport) scattering and absorption coefficients, and it is quite fast and accurate. The IAD

method is widely used for processing spectrophotometry data obtained with integrating spheres, including *in vitro* studies of pathologically altered mucous membrane of the human maxillary sinus in the spectral range of 350–2000 nm [7], sclera of the eye in the spectral range of 370–2500 nm [8], human eye lenses with various stages of cataract [9], peritoneal tissues in the spectral range of 350–2500 nm [10], human stomach mucosa in the spectral range of 400–2000 nm [11], human colon tissues in the spectral range 350–2500 [12], human colon mucosa and colon precancerous polyps in the spectral range of 400–1000 nm [13], and also a rather large number of other tissues, such as skin, muscles, skull, etc. [1, 14–16]. Optical properties (absorption and scattering coefficients, anisotropy factor, and refractive index) of various biological tissues measured for selected wavelengths or in a wide spectral range are summarized in Ref [1].

For semi-quantitative assessment of light permeability through photosensitized dentin *in vitro*, the method of photoacoustic spectroscopy was used [17]. An analysis of the AF spectra recorded *in vivo* at excitation of a 325 nm CW He-Cd laser made it possible to distinguish qualitatively normal, potentially malignant, and malignant sites of the oral mucosa. When collecting the optimal amount of data and compiling an appropriate database, it can be used in clinical diagnostics [18]. Authors [19] have proposed a survey algorithm for more accurate identification of the pathological process in the oral mucosa using the direct AF imaging of tissue and microscopy of biopsy material. Using the AF imaging method of the oral mucosa, realized as an AF stomatoscopy with an AFS-400 LED illuminator (400 ± 10 nm) and special glasses, the authors were able to detect verrucous leukoplakia, lichen planus, and squamous cell carcinomas [20]. The use of color spaces, known as CIELAB, allowed the color parameters of oral tissues in

DOI: 10.1201/9781003025252-40

normal and pathological conditions to be evaluated [21–26]. The authors determined the ratio of color coordinates of natural teeth *in vivo* and tooth tabs after treatment [27, 28].

Intravital noninvasive high-resolution visualization of the structure of the tissues of the oral cavity using optical coherence tomography (OCT) in its various modes demonstrates excellent ability to detect and diagnose precancer, early cancer, dysplasia, and malignancy of the mucous membrane epithelium [29–32]. The authors used optical coherence microscopy (OCM) to determine the scattering coefficient of normal oral epithelium *ex vivo* as $\mu_s = 27 \pm 11$ cm^{-1} at 855 nm [33].

The optical thickness of the human gingival layers measured using OCT (1310 nm) with a probe for *in vivo* measurements, was ~237 μm for epithelium propria (EP) and 830 μm for lamina propria (LP) (attached gingiva) for a 30-year-old Asian female volunteer [34]. The geometric thicknesses of tissue layers can be evaluated using the mean refractive index $n \cong 1.4$ at 1310 nm [1] as 169 μm for EP and 593 μm for LP, a total of 762 μm.

The hybrid Raman spectroscopy and OCT technique are capable of simultaneously acquiring both morphological and biochemical information about the oral tissue *in vivo*, facilitating real-time diagnosis and characterization within the oral cavity [35]. OCT increases the depth of optical penetration when using optical clearing agents (OCAs) and improves the visibility of subsurface occlusal lesions and dentin-enamel junction in the foci of demineralization, which makes it possible to diagnose caries in the white spot stage [36–38].

An automated mobile microscope and simplified protocols for staining the oral mucosa were used as a tool for screening of the oral cavity for cancer based on a collection of digital images and remote assessment of images by doctors, which showed compliance with existing histology and cytology methods [39]. Oral tissue optical transmittance was investigated in [40–44]. The epithelial scattering coefficient of a healthy oral mucosa in *in vivo* studies using diffuse reflection spectroscopy was defined as $\mu_s = 42$ cm^{-1} at 810 nm and $\mu_s = 39$ cm^{-1} at 855 nm [45].

The optical properties of the new organotypic substitute for the oral mucosa based on fibrin–agarose scaffolds, which is proposed to be used to replace excised tissues of the oral cavity and the native mucous membrane of the oral cavity, were determined using the integrating sphere measurements and IAD software [46]. The optical properties of pig gums used for the manufacture of material for the human gum phantom for subsequent use in the prototype of a robotic system for laser maxillofacial surgery were studied [47]. Significant interest in studying the optical properties of tissues of the oral cavity of a subject is due to the trend of modern medicine, and dentistry in particular, toward the use of less invasive diagnostic and therapeutic methods that can be achieved by application of optical methods. Laser diagnostic technologies, photodynamic therapy, light physiotherapy, and aesthetic dentistry require accurate knowledge of the optical properties of oral tissues for successful implementation in clinical practice. While a significant number of studies are devoted to the solution of dental optical diagnostic and therapeutic problems, they do not always demonstrate quantitative approaches characterizing the optical properties of oral tissues and how to control them using optical clearing (OC) technology. In this regard,

the determination of absorption and scattering coefficients of such turbid media as gums and dentin and the design of robust technologies to control these coefficients are urgent tasks.

Materials and methods

The samples for the *in vitro* study were tooth dentin cuts (sections) obtained from orthodontically extracted human teeth (molars) as well as sections (slices) of the human gingival mucosa obtained from gingival tissue after surgery. The gingiva (gums) is the portion of oral mucosa covering the alveolar bone ridge surrounding the tooth. Before cutting, the teeth were stored in saline in a dark place at a temperature of 4–6°C, and the gum tissue was stored in a frozen state. The thickness of the samples was measured using a micrometer, placed between two glass slides, and the measurement accuracy was ± 10 μm. The measurements were carried out at five points and the values were then averaged. The average thickness of the cuts of human dentin was (0.50 ± 0.07) mm and of the gums was (0.40 ± 0.08) mm. In total, the measurements were provided for ten dentinal samples and ten gum samples obtained from different subjects. The area of dentin cuts was on average 150–210 mm^2, and that of the gingival sections was 150–220 mm^2.

To measure the total transmittance and diffuse reflectance of tissue samples in the spectral range of 200–800 nm, a Shimadzu UV-2550 dual-beam spectrophotometer (Japan) with an integrating sphere was used. A halogen lamp with radiation filtering in the studied spectral range served as a radiation source. The spectrometer resolution was 0.1 nm. The spectra were calibrated before the measurements using a BaSO$_4$ reference reflector, which has the best properties in UV [48]. Figure 36.1 shows the location of the sample of tissue during recording spectra. All measurements were carried out at room temperature (~25°C) and normal atmospheric pressure. Each sample of the studied tissue was fixed in a special frame with a window of 0.5 × 0.5 cm, and fixed in a quartz cuvette so that the tissue sample was pressed against the wall of the cuvette and turned to an optical measurement. To measure the total transmission spectra, a quartz cuvette with a tissue sample was mounted directly in front of the integrating sphere collecting all the radiation transmitted through the tissue sample (Figure 36.1). When measuring the diffuse reflection spectra, a cuvette with a sample was placed behind an integrating sphere, which collected all the radiation backscattered by the sample. The light beam diameter falling onto a sample was 3 mm. Prior to measurements, a quartz cuvette with a fixed sample was filled with saline to wet the sample and to bring the measurements closer to *ex vivo*.

To process the experimental results and determine the optical properties of the human gums and dentin, the combined method was used, at the first stage of which the measurement data were processed using the IAD [6]. The IAD method allows one to determine the absorption coefficient μ_a and transport scattering coefficient μ'_s of tissue using experimental data for diffuse reflectance and total transmittance. The transport (or reduced) scattering coefficient is defined as [1]

$$\mu'_s = \mu_s \cdot (1 - g), \tag{36.1}$$

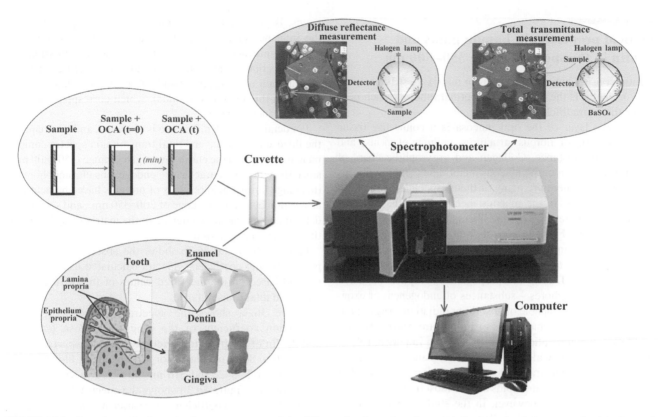

FIGURE 36.1 Experimental schematics for measurements of the diffuse reflection and total transmission spectra of tissue samples (gingiva and dentin).

where μ_s is the scattering coefficient and g is the scattering anisotropy factor. During calculations, the anisotropy factor is fixed. For the tissues (gums and dentin of the human tooth) studied in this work, g was assumed to be 0.9. However, this value of the anisotropy factor (g) is characteristic for only the visible and NIR spectral ranges, and in UV it strongly differs; therefore, we took a restriction when calculating the optical properties of the studied tissues over the range from 350 to 800 nm. The dentin refractive index of a human tooth was taken as 1.49 [49]. The refractive index of the human oral mucosa was taken as 1.45 [50]. The main limitation of the IAD method is related to the possible loss of scattered radiation through the sides of the tissue sample, which is possible in the case of the dimensions of the sample being relatively small compared to the dimensions of the beam incident on the tissue sample, or when the tissue is characterized by relatively low absorption and scattering coefficients. Not accounting the lateral losses of the probe radiation leads to an overestimation of the determined absorption coefficient [1, 5]. For the correct application of the IAD method, it is necessary that the distance from the edge of the probe beam incident on the tissue sample to the nearest sample boundary is greater than the transport mean free path of the photons, which is defined as $1/(\mu_a + \mu'_s)$ [1]. The calculation of the optical parameters was performed separately for each spectral point. The algorithm used includes the following steps:

1) Setting the initial values and using the following expressions [5]:

$$\frac{\mu'_s}{\mu_a + \mu'_s} = \begin{cases} 1 - \left(\dfrac{1-4R_d-T_t}{1-T_t}\right)^2, & \text{if } \dfrac{R_d}{1-T_t} < 0.1 \\[2ex] 1 - \dfrac{4}{9}\left(\dfrac{1-R_d-T_t}{1-T_t}\right)^2, & \text{if } \dfrac{R_d}{1-T_t} \ge 0.1 \end{cases}$$

$$(\mu_a + \mu'_s) \times l = \begin{cases} -\dfrac{\ln T_t \ln(0.05)}{\ln R_d}, & \text{if } R_d < 0.1 \\[2ex] 2^{1+5(R_d+T_t)}, & \text{if } R_d \ge 0.1. \end{cases}$$

Here R_d and T_t are the measured diffuse reflection and total transmission coefficients, and l is the thickness of tissue sample.

2) Calculation of diffuse reflection and total transmission coefficients basing on the initial values of μ_a and μ'_s and the "adding-doubling" method [4].

3) Comparison of the calculated R_d and T_t values with experimentally measured ones.

4) As a criterion for completing the iterative procedure, the following condition was used [5]:

$$\frac{\left| R_d^{exp} - R_d^{calc} \right|}{R_d^{exp}} + \frac{\left| T_t^{exp} - T_t^{calc} \right|}{T_t^{exp}} < 0.001,$$

where R_d^{exp}, R_d^{calc}, R_t^{exp}, R_t^{calc} are the experimental (*exp*) and calculated (*calc*) values of the diffuse reflection and total transmission coefficients, respectively.

Optical properties of human gums and dentin in the spectral range 350–800 nm

Dentin consists mainly of a collagen matrix. A structural feature of dentin is the presence of dentinal tubules, which penetrate the entire thickness of dentin.

The own layer of the oral mucosa is a connective tissue which consists of fibrous structures, cellular elements, and intercellular substance. Collagen and argyrophilic fibers of the lamina propria of the mucous membrane comprise fibrous structures, and there are many of them especially in the hard palate and gums. In the mucous membrane of the oral cavity, there are more argyrophilic fibers and less collagen fibers than in the skin. From an optical point of view, the gums and dentin of a human tooth can be attributed to optically turbid media in which, along with absorption, strong light scattering is observed. During the propagation of optical radiation in tissue, chromophores – substances of endogenous or exogenous origin – are capable of absorbing radiation energy (photons). The main chromophores of tissues are water, proteins, and lipids, the absorption of which plays an important role in determining the optical properties of tissue and, especially, in determining the penetration depth of radiation into tissue. For the soft tissues, water and lipids are the most important endogenous chromophores; however, in the studied optical range, their absorption is negligible. Water transmission in the measured range of 350–800 nm is negligible and begins to have an effect in the range of 1200–2500 nm [1]. The gingival mucosa and dentin of a human tooth can be attributed to fibrous tissues, which are based on collagen and argyrophilic fibers and hemoglobins in the gum tissue. Figure 36.2a shows the diffuse reflectance spectra (DRS) of the gingival mucosa (curve 1) and dentin of a human tooth (curve 2).

In the region from 350 to 650 nm, the shape of the diffuse reflectance spectra correlates quite well with the shape of the transmission spectrum of the gingiva, since in this wavelength range, the shape of the spectra is determined by strong absorption bands of oxyhemoglobin and the effect of light scattering by the main scatterers of the gingival mucosa – collagen and elastin fibers. The DRS and total transmission spectra clearly show the dips corresponding to the absorption bands of oxyhemoglobin at wavelengths 415, 542, and 576 nm. The presence of strong absorption bands reduces the number of both transmitted and backscattered photons within the absorption

bands. Beginning from 650 nm and further, up to 800 nm, the influence of the absorption bands of hemoglobin is no longer significant, and the spectra of total transmission and back reflection are formed mainly due to scattering, since the contribution of absorption bands of all chromophores of soft tissues in this region is minimal, which corresponds to their "transparency window."

For dentin, no such difference is observed, and the shapes of the diffuse reflectance and total transmission spectra complement each other in the entire investigated range (350–800 nm), since there are no characteristic endogenous chromophores in this range (amino acid residues of proteins have characteristic absorption bands in the range of 200–350 nm), and spectra are determined mainly by scattering in the matrix of hydroxyapatite and dentinal collagen.

Figures 36.3a and 36.3b shows the absorption and transport scattering coefficients spectra calculated using the IAD method based on the measured values of the diffuse reflectance and total transmittance.

Figure 36.3a shows the spectral dependences of the absorption coefficient of the gingival mucosa (curve 1) and dentin of a human tooth (curve 2) in the spectral range from 350 to 800 nm. The absorption bands of blood oxyhemoglobin (415, 542, and 576 nm) are clearly visible in the spectrum [1]. Figure 36.3b shows the spectral dependences of the transport coefficient of scattering of the gingival mucosa (curve 1) and dentin of a human tooth (curve 2). These dependences were obtained by averaging the spectra of absorption and transport scattering coefficient for ten samples of each tissue, the mucous membrane and dentin. It is clearly seen that the transport scattering coefficient decreases rather smoothly towards large wavelengths, which corresponds to the general nature of the spectral behavior of the scattering properties of tissues [1]. However, in the region of strong absorption bands (i.e., 415, 542, and 576 nm), the shape of the scattering spectrum is distorted, i.e. it is deviated from monotonous dependence, which could be a manifestation of IAD algorithm drawback that allows for crosstalk between absorption and scattering when absorption is strong. However, some influence of physical factors, related to light diffraction on particles with a high absorption, is also possible. Our estimations using a Mie calculator for an ensemble of spherical particles [52] showed that the drop in reduced scattering coefficient at 415 nm can be of 12%–20% in comparison with a case of no absorption.

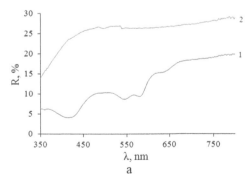

a

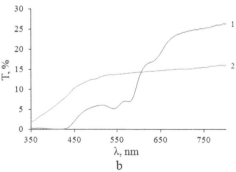

b

FIGURE 36.2 Spectra of diffuse reflection (a) and total transmission (b) of human gums (curve 1) and dentin (curve 2) samples [51].

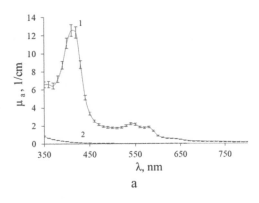
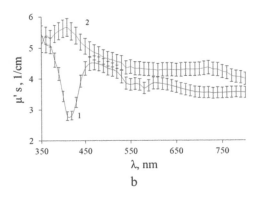

a

b

FIGURE 36.3 The absorption coefficient (a) and transport scattering coefficient (b) of the human gums (1) and dentin (2), calculated by IAD method using experimental data [51]. Bars present standard deviation.

The scattering coefficient of a healthy oral mucosa in *in vivo* studies using diffuse reflection spectroscopy was determined as μ_s= 42 cm^{-1} at 810 nm and μ_s= 39 cm^{-1} at 855 nm [45]. Using OCM in *ex vivo* measurements of normal oral epithelium, scattering coefficient was found to be μ_s = (27 ± 11) cm^{-1} at 855 nm [33]. The scattering coefficient of the gingival mucosa at 800 nm (Figure 36.3b) μ_s= 36 cm^{-1} for g=0.9 (Equation36.1) correlates well with these results [33, 45].

The penetration depth of light is one of the most important characteristics for the correct determination of the radiation dose during photochemical and photodynamic therapy of various diseases [1]. The penetration depth of tissue (δ) was estimated using the formula obtained in the diffusion approximation [6], $1/\sqrt{3\mu_a(\mu_a+\mu'_s)}$ (see Figure 36.4 and Table 36.1) [51].

From Figure 39.4 and Table 39.1, it is clearly seen that depending on the wavelength of the probe radiation, the depth of its penetration into the studied tissues varies significantly. The maximum effect is observed in the spectral range from 600 to 800 nm, where the radiation penetrates to a depth of 3–7 mm in the human gums and 19–55 mm in dentin. We determined the penetration depth of light in dentin of (18.7 ± 0.6) mm at 650 nm and (55 ± 3) mm at 800 nm. As for cemented tooth root in healthy molars, the penetration depth for a laser source at 780 nm was in the range from 4.3 mm to 13.3 mm [53]. The difference in light transmission can be attributed to significant differences in the structural organization of different parts of the tooth. Cement is a highly mineralized tissue and transmits less radiation, and dentin is a porous light-conducting tissue, where the dentinal tubular structures are waveguides [54].

OC efficiency and kinetics of human gingival tissues

The mucous membrane of the gum consists of two layers: the epithelium propria (EP) and the lamina propria (LP) (Figure 36.5). There is no submucous layer in the gingiva. The LP on which EP is located consists of connective tissue, represented by fibrous structures – collagen and reticular fibers and cellular elements [34, 55].

In the study [56, 57], the mean thickness of ten samples of the gingival mucosa was equal to 590 ± 60 μm, thus samples consisted of an epithelial layer and the major part of LP layer. After biopsy, samples were placed between two glasses and stored frozen until studied. Before measurement, the samples were thawed. The thickness of the samples was measured with an accuracy of ± 10 μm using a micrometer.

The determination of the diffusion coefficient of an agent in biological tissue is based on measuring the kinetics of changes in the DRS (Figure 36.1). To carry out these measurements, each sample of tissue was fixed in a special clamp in the form of a frame with a window of 0.5–0.5 cm placed in a cuvette with an immersion agent (OCA). Measurements of the diffuse reflectance of the samples were carried out *in vitro* over time until the saturation of reflection temporal dependence occurred. All experiments were carried out at room temperature.

The process of molecular transport in a section of biological tissue can be described in terms of the model of free diffusion [58, 59]. Geometrically, a sample of tissue can be represented by a plane-parallel plate of a finite thickness, taking into account some limitations inherent in the free diffusion model: 1 – only concentration diffusion takes place; 2 – the diffusion coefficient is constant at all points inside the tissue sample,

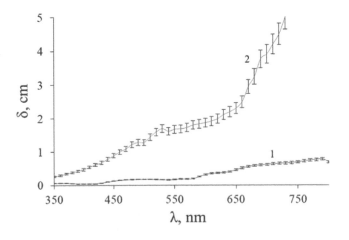

FIGURE 36.4 The penetration depth of the human gums (1) and dentin (2), calculated using data obtained from measurements by the IAD method (Figure36.3). Bars present standard deviation.

TABLE 36.1

The absorption coefficient μ_a, transport scattering coefficient μ'_s and the penetration depth δ of the human gums and dentin. Mean $\pm$ standard deviation are presented

λ, nm	350	400	450	500	550	600	650	700	750	800
Gums										
μ_a,	6.6	11.2	3.4	1.8	2.1	0.9	0.5	0.2	0.2	0.2
cm^{-1}	± 0.5	± 0.2	± 0.3	± 0.4	± 0.4	± 0.2	± 0.1	± 0.1	± 0.1	± 0.1
μ'_s,	5.4	3.4	4.3	4.3	3.8	3.9	3.7	3.6	3.5	3.6
cm^{-1}	± 0.8	± 0.5	± 0.4	± 0.4	± 0.3	± 0.5	± 0.4	± 0.3	± 0.4	± 0.3
μ_s,	54	34	43	43	38	39	37	36	35	36
cm^{-1}	± 8	± 5	± 4	± 4	± 3	± 5	± 4	± 3	± 4	± 3
δ,	0.10	0.04	0.10	0.18	0.17	0.32	0.46	0.62	0.69	0.69
cm	± 0.01	±0.02	±0.01	± 0.02	± 0.02	± 0.02	± 0.2	± 0.03	± 0.03	± 0.03
Dentin										
μ_a,	0.8	0.3	0.08	0.04	0.03	0.02	0.02	0.01	0	0
cm^{-1}	± 0.1	± 0.1	± 0.1	± 0.1	± 0.1	± 0.1	± 0.1	± 0.1		
μ'_s,	5.1	5.6	5.1	4.7	4.4	4.3	4.3	4.3	4.2	4.1
cm^{-1}	± 0.9	± 0.9	± 0.7	± 0.5	± 0.5	± 0.6	± 0.5	± 0.5	± 0.4	± 0.4
μ_s,	51	56	51	47	44	43	43	43	42	41
cm^{-1}	± 9	± 9	± 7	± 5	± 5	± 6	± 5	± 5	± 4	± 4
δ,	0.26	0.47	0.89	1.28	1.68	1.87	1.87	3.9	5.5	5.5
cm	± 0.02	± 0.02	± 0.02	± 0.02	± 0.04	± 0.05	± 0.06	± 0.1	± 0.3	± 0.3

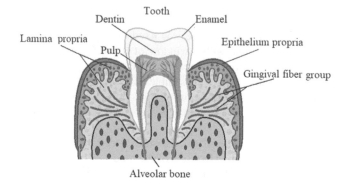

FIGURE 36.5 Structure of human gingival tissue.

3 – the volume of the OCA significantly exceeds the volume of the sample, and 4 – the absence of the OCA at all internal points of the sample is assumed before the measurements.

Using the second Fick law and carrying out transformations based on the use of the modified Beer–Bouguer–Lambert law described in detail [60], we obtain the final expression:

$$A(t,\lambda) = \log R(t,\lambda);$$

$$\Delta A(t,\lambda) = A(t,\lambda) - A(t = 0,\lambda)$$
$$= \Delta\mu_{\mathrm{eff}}(t,\lambda)L \sim C_0\left\{1 - \exp(-\pi^2 D / 4l^2)\right\}L \quad (36.2)$$

$$\mu_{\mathrm{eff}}(t,\lambda) = \sqrt{3\mu_a\left(\mu_a + \mu'_s\right)} \rightarrow \Delta\mu_{\mathrm{eff}}(t,\lambda)$$

where $\Delta A(t, \lambda)$ is the difference between the effective absorbance at the current time $A(t, \lambda)$ and at the initial time $A(t=0, \lambda)$; t is the time during which the diffusion process occurs, s; λ is the wavelength, nm; $\Delta\mu_{\mathrm{eff}}(t, \lambda)$ is the difference between the effective attenuation coefficient in tissue at the current time

and at the initial time, 1/cm; L is the average path length of photons, which in the backscattering mode is $L \approx 2l$; and for transmission $L \approx l$; l is the sample thickness, cm; D is the diffusion coefficient of the OCA molecules, cm²/s; and C_0 is the initial concentration of the OCA, mol/l. From the analysis of the kinetics of changes in the effective absorbance (ΔA), the diffusion coefficient (D) was determined by fitting experimental data to Equation 36.2.

The experimental DRS of human gingival samples is shown in Figure 36.6a. They have obvious dips typical for the absorption bands of the proteins of the connective tissue in the form of collagen and reticular fibers, as well as hemoglobin (in the UV range). In the region of about 415–420 nm and 540–580 nm, dips observed on the DRS correspond to the absorption bands of oxyhemoglobin (415, 542, and 576 nm). The absorption of water in the measured range of 200–800 nm is negligible [61]. Figures 36.6a, b show the DRS of the samples upon immersion in a 99.5%-glycerol solution at different time intervals. It is seen that the interaction of glycerol with the samples leads to a gradual decrease in the reflection coefficient. When the sample is saturated with a hyperosmotic agent, smoothing of typical dips on the DRS occurs, although the shape of the spectra does not change. After immersion of gingival samples in a 99.5%-glycerol solution, the absorption bands of the endogenous chromophores (oxyhemoglobin) on DRS become less pronounced, which is associated with a lower probability of effective absorption of photons when they pass through an optically cleared (less scattering) tissue sample. In this case, the shape of the spectra remains practically unchanged. The values of the diffuse reflection coefficients are reduced in relation to the initial state of the sample, which indicates a decrease in light scattering in the samples.

Using *in vitro* experimental kinetic curves for all ten samples like those shown in Figure 36.6b, the effective diffusion coefficient of the tissue fluid under action of 99.5%

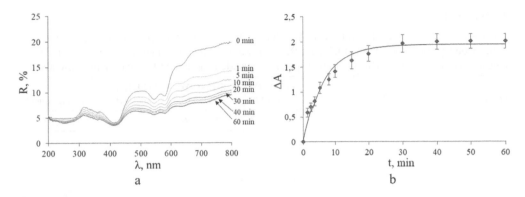

FIGURE 36.6 The healthy human gingival tissue sample at immersion in 99.5% glycerol: DRS measured for different time of immersion, from 0 to 60 min (a); kinetics of differential effective absorbance $\Delta A(t)$ (b); symbols correspond to experimental data, the solid curves represents an approximation of the experimental data using free diffusion model. Bars present standard deviation [56].

glycerol in healthy human gingival tissue was found as $(1.78 \pm 0.22) \times 10^{-6}$ cm²/s, which is consistent with the literature data for other soft tissues [57] and mostly related to diffusion of water molecules through two semipermeable biological membranes with thicknesses $l_1 = 169\ \mu m$ (EP) and $l_2 = 593\ \mu m$ (LP) and diffusion coefficients D_1 and D_2, respectively. The total diffusion coefficient can be calculated as

$$D = \left(l_1 + l_2\right)^2 / \left[\left(l_1\right)^2 / D_1 + \left(l_2\right)^2 / D_2\right]. \quad (36.3)$$

For D_1 as an estimation, the diffusion coefficient of water in a lipid membrane $D_1 = D_{\text{lipid}} \cong 3.0 \cdot 10^{-7}$ cm²/s could be appropriate [62]. To evaluated D_2 for less dense and more permeable LP layer, the concept of hindered molecule mobility in tissues relative to diffusion water in water D_w, which is quantified by such parameter as tortuosity of tissue λ, can be used [63]:

$$S = l_d / L = \sqrt{D_w / D_2}, \quad (36.4)$$

which is the ratio of the pathlength of the molecular flow between two points l_d to the direct distance between these points L. Here D_2 is the effective diffusion coefficient accounting for elongation of diffusion path of water molecules. Tortuosity of tissues was found in the range from 1.2 for brain to 3–3.5 for skin dermis [63, 64]. For gingival LP layer $S \cong 3.9$ should be good, thus for $D_w = 3 \cdot 10^{-5}$ cm²/s [36] from Equation 36.4 $D_2 = 0.2 \cdot 10^{-5}$ cm²/s. Finally, from Equation 36.3 the total diffusion coefficient is calculated as $D \cong 2.1 \cdot 10^{-6}$ cm²/s, corresponding well to experimental value $D = (1.78 \pm 0.22) \times 10^{-6}$ cm²/s (Figure 36.6b).

Similarly, the effective diffusion coefficient of OCA (binary diffusion coefficients of tissue water and OCA) in the human gingival tissue sample *in vitro* were determined, which are presented in Table 36.2 [56, 65].

The spectra of total transmission (T) of the gingival sample are shown in Figure 36.7. In the UV range, the transmission spectra of the gingiva have pronounced dips characteristic to the absorption bands of proteins and blood hemoglobin and correlate with diffuse reflection spectra (Figure 36.7b).

The total transmittance increases with time over the entire wavelength range (Figure 36.7a) with respect to the initial state of the sample (before immersion in glycerol), which indicates

a decrease in light scattering of the sample as a result of its immersion in glycerol. Since glycerol does not have absorption bands in the entire studied range from 200 to 800 nm, the increase in transmission is not accompanied by changes in the shape of the spectra. Although the absolute transmittance in deep UV is small due to the strong absorption of DNA and proteins and is only 0.4%, the increase in the number of transmitted photons is significant, which contributes to the successful detection of UV signals from the larger depths of tissue [58, 66].

Glycerol is a highly effective agent for OC of tissues, and its mechanism of action is mainly determined by the osmotic dehydration of tissues. Being a hygroscopic agent, glycerol first draws water from the tissue due to osmosis, and then penetrates into the tissue and binds interstitial and intercellular water, thereby increasing the concentration of soluble components in the remaining water and increasing its refractive index. The main source of light scattering in tissues is the spatial inhomogeneity of the refractive index caused by different tissue and cell components, i.e., mitochondria, nuclei, other organelles, cytoplasm, collagen, and elastin fibers, and interstitial fluid. When an immersion OCA with a higher refractive index than that of the cytoplasm and interstitial fluid is introduced into tissue and water leaves the tissue, refractive index matching takes place, and as a result light scattering significantly reduces. To quantifying the achieved tissue OC, optical clearing efficiency can be introduced as

$$OC_{\text{eff}}(\%) = \left\{T(t) - T(t=0)\right\} / T(t=0), \quad (36.5)$$

where $T(t=0)$ is the initial transmittance and $T(t)$ is the transmittance at time of interest t.

Using Equation 36.5, the OC_{eff} of human gingival tissue samples was calculated (Figure 36.8, Table 36.3). It follows that the greatest efficiency is achieved at a wavelength of 210–230 nm for 60 min of glycerol action and amounts to 3500%. Thus, the efficiency of OC is higher, the lower the initial intensity before OC. On average, the efficiency changes in accordance with the behavior of the transmission spectrum (Figure 36.7). In the visible and NIR, within the so-called "transparent window" [1], the OC_{eff} is significantly lower, around 100–200%; however, due to the absence of strong absorption bands of

TABLE 36.2

The effective diffusion coefficients of OCAs (glucose (G), glycerol (GL), propylene glycol (PG), and Omnipaque™-300 (iohexol) and their mixtures with water (W)) in human gingival tissue samples measured *in vitro* using diffuse reflection spectroscopy. Mean ± standard deviation are presented

OCA	99.5% GL	40% G	PG/GL/W) (30/70/0)	PG/GL/W) (50/50/0)	PG/GL/W) (55/35/10)	Omnipaque™
$D \times 10^6$, cm²/s	1.78 ±0.22	4.08 ± 0.82	2.3 ± 0.4	2.6 ± 0.6	3.2 ± 0.8	4.7 ± 0.9

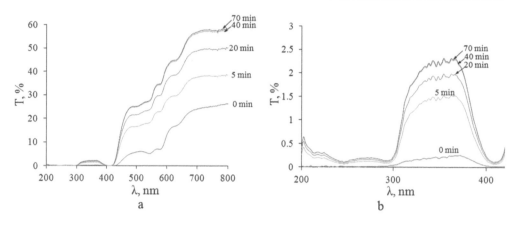

FIGURE 36.7 The total transmittance spectra of the sample of healthy human gingiva measured for different time of immersion in 99.5% glycerol, from 0 to 70 min: spectra from 200 to 800 nm (a); magnified spectra from 200 to 400 nm (b).

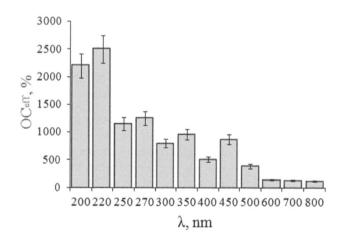

FIGURE 36.8 The histogram of the mean optical clearing efficiency OC_{eff} (λ) (n=5) at different wavelengths of healthy human gingiva after immersion in a 99.5 % glycerol solution. Bars present standard deviation.

endogenous chromophores in this spectral region, the absolute transmission is quite large and reaches 60%.

The total transmission spectra of healthy human gums after immersion in various OCAs are shown in Figure 36.9. By combining the immersion OC technique with UV-spectroscopy, it was possible to verify and study one of the major OC mechanisms – tissue dehydration – and to show that OC efficiency is higher in the deep-UV than in the visible-NIR range. These results prove that the refractive index mismatch in tissues is strong in this wavelength range. In Chapter 33, it was shown that by fitting the efficiency kinetics at 200 nm for colorectal tissues, it was possible to distinguish between healthy and malignant mucosa. Such pathology discrimination needs to be further studied with other types of cancer with

the view that noninvasive optical methods can be developed for cancer diagnosis. The applied OC treatments with highly concentrated glycerol have induced two tissue windows in the UV range in colorectal tissues – from 200 to 260 nm and from 260 to 418 nm [58, 66] (Chapter 33); and three tissue windows in gingival tissues – from 200 to 250 nm, from 250 to 300 nm, and from 300 to 400 nm. By treating gingival tissues with a solution containing 40% glucose, only one tissue window was observed between 300 and 400 nm with a lower efficiency than those observed for tissues treated with glycerol. The efficiency of creating these windows is higher in the deep-UV range.

Determination of the diffusion coefficient of dyes and OCAs in dentin of a human tooth *in vitro*

Quantitative parameters of the transport and interaction of photosensitizers in tissues, which are necessary for implementation in clinical practice, are not always given. In this section, the effective diffusion coefficients (binary diffusion coefficients of tissue water and the agents) of rivanol, methylene blue, glucose, and glycerol in the dentin of a human tooth determined *in vitro* from DRS measurements will be presented. Samples of dentinal sections were installed in the integrating sphere port, as shown in Figure 36.1, to collect all reflected radiation from the sample while recording the diffuse reflectance spectra of sample immersed in an agent solution.

For *in vitro* study, human teeth (molars) removed from patients aged 15–25 years in a dental clinic according to orthodontic indications were used. The extracted teeth were stored in saline at 4°C in a dark place. Wet teeth were cut with a diamond disk into sections about 1 mm thick along the growth

TABLE 36.3

The OC_{eff} (%) of human gingival tissue during immersion in OCAs (glucose (G), glycerol (GL), mixtures glycerol (GL), propylene glycol (PG) and water (W), Omnipaque™-300 (iohexol). Mean ± standard deviation are presented

λ, nm/OCA	200	250	262	300	330	400	500	600	700	800
99.5 % GL	1847 ± 60	855 ± 45	1920 ± 70	450 ± 25	1166 ± 70	457 ± 26	420 ± 28	230 ± 23	180 ± 15	180 ± 15
40% G	80 ± 10	320 ± 15	42 ± 7	26 ± 4	146 ± 11	241 ± 14	15 ± 7	10 ± 4	10 ± 4	10 ± 4
PG/GL/W (30/70/0)	864 ± 42	1740 ± 54	7750 ± 218	1725 ± 59	780 ± 43	1691 ± 48	113 ± 26	63 ± 11	42 ± 7	28 ± 3
PG/GL/W (55/35/10)	1309 ± 54	300 ± 28	525 ± 37	170 ± 43	835 ± 51	57 ± 6	1761 ± 67	575 ± 27	305 ± 29	279 ± 24
PG/GL/W (50/50/0)	100 ± 11	17 ± 3	1200 ± 78	324 ± 23	1352 ± 79	842 ± 43	117 ± 23	70 ± 12	34 ± 6	22 ± 7
Omnipaque™	81 ± 9	500 ± 36	25 ± 6	81 ± 10	136 ± 12	162 ± 13	247 ± 18	104 ± 11	81 ± 11	69 ± 6

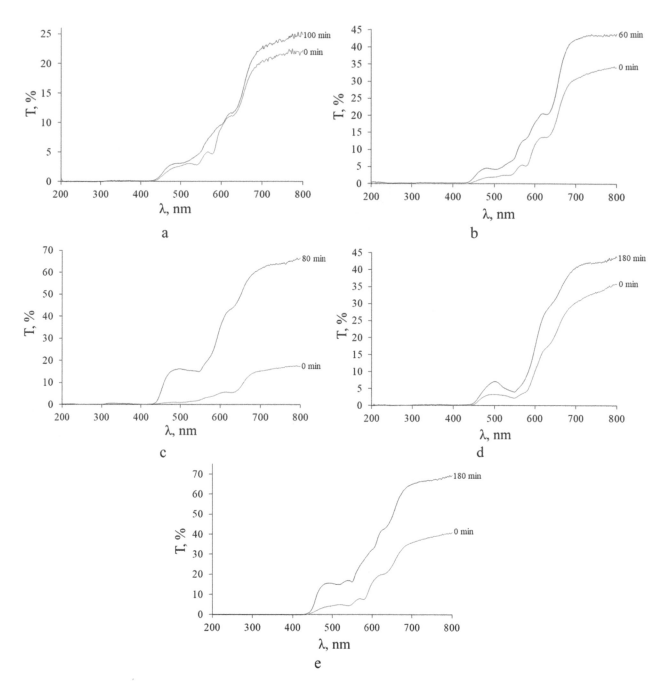

FIGURE 36.9 The total transmittance spectra 200 to 800 nm of the sample of healthy human gingiva measured for different time of immersion in: 40% -glucose (a); 30 % PG and 70% GL (b); 55% PG, 35% GL and 10% W (c); 50% G and 50% PG (d); and Omnipaque™-300 (iohexol) (e).

axis and etched with 35% phosphoric acid for 15 s, after which the acid was removed using a stream of water. Then, using a brush and 95% ethanol, the section surface was cleaned from sawn products and other external contaminants by placing in a Techsonic UD100 SH-45 L ultrasonic bath for 10 min. Samples were wiped with a lint-free cloth moistened with alcohol and dried in air for 24 hours. The thickness of the sections was measured with a micrometer. On average, the thickness of the samples was (650 ± 90) μm. The accuracy of each measurement is $\pm$ 10 μm. SEM images were used to evaluate the degree of purification of samples from sawing products and evaluation of density of dentinal channels, which can clearly be seen on the images (Figure 36.10).

The solid part of the tooth consists of enamel and dentin, which have an uneven surface and are firmly adhered to each other due to the dentinal–enamel junction (Figure 36.10a). Optical properties of the tooth are mainly determined by dentin, which is composed of the main substance penetrated by the tubules (diameter of 2–5 μm) (Figure 36.10b), which begins in the pulp, near the inner surface of the dentin, and is very much different to its external surface [41, 44, 67]. Protein-like molecules are major absorbers in dentin, and hydroxyapatite crystals are scatterers. Reflection of light on the periodic structures of the dental tissue gives information about its structure. It can be considered that the tooth presents clusters of coordinated natural waveguides. Regardless of how the light hits

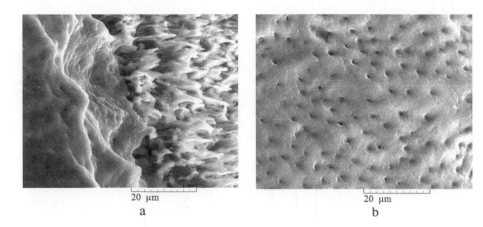

a   b

FIGURE 36.10 Electron micrographs (SEM) of a longitudinal dentinal section of a human tooth (magnification 5000 times): the region of dentinal–enamel junction (a) and tubular structure of dentin (b).

TABLE 36.4

The parameters of the human tooth sections and dentinal tubules (DTs). Mean ± standard deviation are presented

Parameter	Section							
	#1	#2	#3	#4	#5	#6	#7	#8
Section thickness, mm	0.92	0.97	0.93	0.57	0.62	0.58	0.76	0.82
Diameter of DTs, μm	2.94 ± 0.81	2.32 ± .07	3.08 ± 0.94	1.98 ± 0.34	1.50 ± 0.41	3.08 ± 0.94	3.86 ± 1.02	2.56 ± 0.73
Density of DTs, $1/mm^2$	20661	9120	21125	10023	6790	27000	35076	19962

the external surface of the tooth, pulp illumination is always effective due to the crimped shape of the waveguides. The waveguide effect is significantly more pronounced in dentin than in enamel [44, 67, 68]. The number and diameter of the dentinal tubules vary significantly depending on the part of the tooth. Their number decreases per unit volume of dentin in the direction from the horn of the pulp to the apex of the root and from the cavity of the tooth to the dentinal enamel. In coronal dentin, the number of tubes on average is $5.5 \cdot 10^4$ mm^{-2} with a diameter of 2–3 μm, and the distance between them is about 6 μm. In the area of the dentinal–enamel junction, the number of tubules is about $1.7 \cdot 10^4$ mm^{-2}, the diameter is 0.7–1.0 μm, and the distance between them is about 15 μm. In the middle third (between the pulp and the dentinal–enamel junction), the number of tubules on average is about $3.2 \cdot 10^4$ mm^{-2}, and their diameter is 1–2 μm [44, 69]. The average number of dentinal tubules per unit volume (density) in the middle part of the dentin of the root of the tooth is much lower than in the middle part of the dentin of the tooth crown. It is noted that at a low density of the dentinal tubules, they have a more branched structure. The size of the branches of the dentinal tubules varies depending on the location: the branches of the dentinal tubules located on the periphery have a diameter of 0.5–1 μm; small branches with a diameter of 300–700 nm are located at an angle of 45° and are abundant in areas such as the root, where the density of tubules is relatively low; the smallest branches with a diameter of 25–200 nm are elongated at right angles from the tubules in all parts of the dentin [44, 69]. The density of the dentinal tubules of the studied human teeth is presented in Table 36.4.

The main mineral components of the tooth are hydroxyapatites, which are presented in dental tissue in the nanocrystalline state and contain a large number of external elements of impurities in the structure [70]. The elemental compositions of the intact human tooth samples demineralized by application of 40% HCl for 20 min are presented in Table 36.5.

The DRS of dentin samples of a human tooth during the diffusion of the agents (methylene blue (MB) aqueous solution, MB solution in 40% glucose, MB in a 0.05% aqueous solution of cationic surfactant cetylpyridinium chloride (CPC), rivanol, 99.5% glycerol, and 40% glucose) are presented in Figure 36.11. They show pronounced decreases characteristic of the absorption bands of the proteins of the connective tissue in the form of collagen and reticular fibers in the region of 200–320 (Figures 36.11a, 36.11d, 36.11g, and 36.11j).

In Figure 36.11c, the absorption spectrum of rivanol (pH = 5.45 ± 0.02) is shown, which has an intense peak in the ultraviolet (UV) range of 200–300 nm, as well as absorption peaks at 363 and 405 nm. In the process of diffusion of ethacridine lactate molecules (the active component of rivanol) in dentin, the diffusion reflection coefficient R of dentin decreases in the entire spectral region from 200 to 800 nm. However, with the passage of time of the diffusion process, the form of DRS changes in shape. In the region of rivanol inherent to the absorption peaks at 363 and 405 nm (Figure 36.11c), an increase in characteristic dips is observed in the recorded DRS, which indicates achieved staining of the dentin sample.

A solution of MB in water (pH = 7.37 ± 0.02), in CPC (pH = 7.03 ± 0.02), and in a 40% glucose solution (pH = 7.11 ± 0.02) has several absorption peaks in the spectral region 200–800 nm. The main dye absorption peak (MB concentration of 4.5×10^{-5} mol/L) has two maxima: the first at 668 nm, which corresponds to the monomeric form of the dye; and the second, much less pronounced, at a wavelength

TABLE 36.5

The elemental composition of the intact human tooth samples and after demineralization. Mean ± standard deviation are presented

Intact tooth samples

Method	Element	S	Cl	K	Sc	Fe	Ni	Cu	Zn	As	Rb	Sr	W
X-ray fluorescent analysis	ppm	33170 ± 925	3332 ± 125	401 ± 96	66067±2500	5583 ± 148	664 ± 58	220 ± 8	4432 ± 136	113 ± 10	8817 ± 250	4491 ± 210	771 ± 25
SEM	Element	C			N		O		P			Ca	
SEM	%	51 ± 12			10 ± 5		24 ± 7		5 ± 2			10 ± 2	

Demineralized samples

Method	Element	S	Cl	K	Sc	Fe	Ni	Cu	Zn	As	Rb	Sr	W
X-ray fluorescent analysis	ppm	22950 ± 1500	5570 ± 98	4485 ± 126	55103 ± 2500	4426 ± 58	771 ± 20	335 ± 19	2230± 85	110 ± 6	5598 ± 147	4412 ± 256	885 ± 31
SEM	Element	C			N		O		P			Ca	
SEM	%	50 ± 9			11 ± 4		35 ± 9		2 ± 1			2 ± 1	

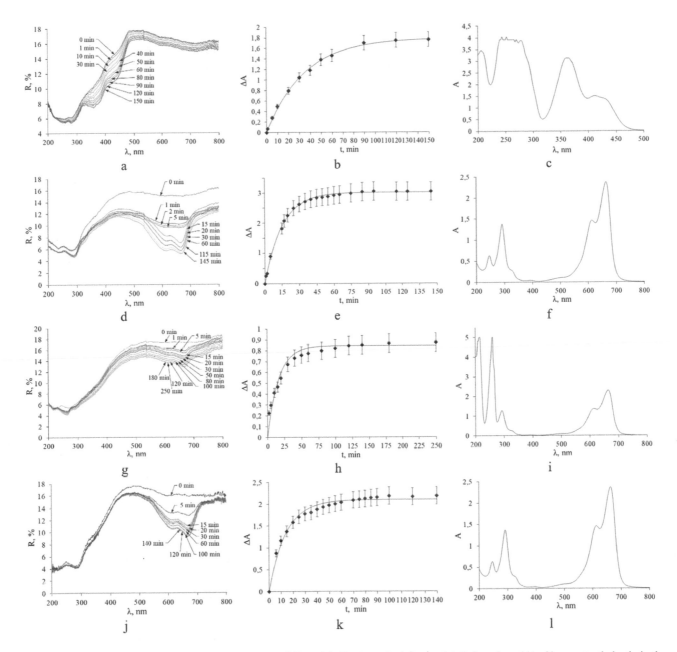

FIGURE 36.11 DRS (*R*) (a, d, g, and j) and kinetics of the differential effective optical density Δ*A* (*t*) (b, e, h, and k) of human tooth dentin in the course of diffusion of rivanol (a, b); MB aqueous solution (d, e); MB in CPC (g, h); and MB solution in 40% glucose (j, k). Absorption spectra of the initial solutions: rivanol (c); MB aqueous solution (f); MB in CPC (i); and MB solution in 40% glucose (l). MB concentration is 4.5 × 10⁻⁵mol/L. For (b, e, h, and k), symbols correspond to experimental data, the solid curve represents an approximation of the experimental data in the framework of the free diffusion model; bars present standard deviation [71–74].

of 612 nm, corresponding to the dimeric form (Figures 36.11d and 36.11f). In the UV, the MB has two significantly smaller absorption peaks at 246 and 295 nm (aqueous solution of MB) (Figure 36.11f). MB in CPC has additional pronounced absorption peaks at 215 and 250 nm, which are characteristic for CPC (Figure 36.11i). By penetrating into dentin, the MB dye changes the shape of the reflection spectra of the samples and manifests itself in the form of the absorption bands of the dimeric and monomeric molecular forms localized in the spectral region of 580–700 nm (Figures 36.11d, 36.11g and 36.11j).

The kinetics of the differential effective optical density Δ*A* (*t*) (b, e, h, and k) of human tooth dentin in the course

of diffusion of rivanol (b), MB aqueous solution (e), MB in CPC (h), and MB solution in 40% glucose (k) are different. For rivanol, the diffusion process is saturated to 90 min, for MB aqueous solution to 60 min, for MB in CPC to 50 min, and for MB solution in 40% glucose to 60 min.

Glycerol and glucose are highly effective agents for OC of tissues. The mechanisms of action are mainly associated with refractive index matching and osmotic dehydration of tissues. Being a hygroscopic agent, glycerol first draws water from the tissue due to osmosis, and then, penetrating into the tissue, binds interstitial and intercellular water, thereby increasing the concentration of soluble components in the remaining

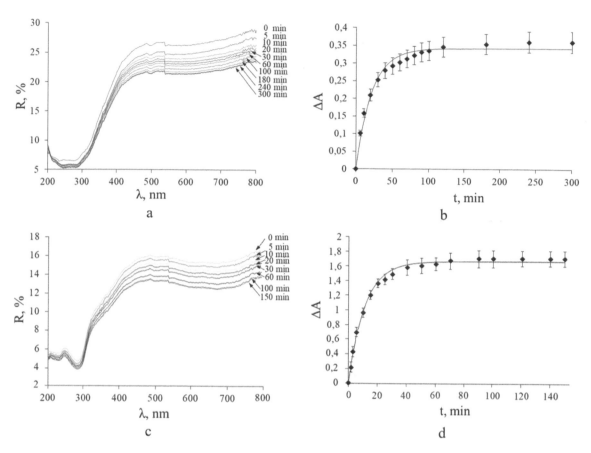

FIGURE 36.12   DRS (*R*) (a, c) and kinetics of the differential effective optical density ΔA (*t*) (b, d) of human tooth dentin in the course of diffusion of 99.5% glycerol (a, b); 40% glucose (c, d); symbols correspond to experimental data, the solid curve represents an approximation of the experimental data in the framework of the free diffusion model; bars present standard deviation [60, 75].

TABLE 36.6

The effective (binary) diffusion coefficient $D \times 10^6$ (cm²/s) of tissue water and agent molecules in the human dentin measured *in vitro* [56, 71–75]. Mean ± standard deviation are presented

OCAs			Dyes		
99.5% GL	40% G	Rivanol	MB in W	MB in CPC	MB in 40% G
0.59 ± 0.04	5.4 ± 0.8	2.27 ± 0.32	(5.29 ± 1.33)*	(1.5 ± 1.33)*	(3.1 ± 0.85)*

*Data from References [72–74] for diffusion of MB were corrected accordingly to Equation (36.2) by multiplying on π/4, which was originally missed.

water and increasing its refractive index. A similar action is characteristic for glucose. After dentin immersion in 99.5% glycerol or 40% glucose, a decrease in the scattering of the sample occurs, which is expressed as a reduction of diffuse reflectance (Figures 36.12a and 36.12c). Glycerol and glucose do not have noticeable absorption peaks in the range of 200–800 nm and therefore do not change the shape of the DRS. In Figures 36.12b and 36.12d, typical kinetic curves for the differential effective optical density ΔA (*t*) of human tooth dentin samples in the course of glycerol and glucose diffusion are shown, respectively. For glycerol, the diffusion process is saturated totally to approximately 180 min, and for an aqueous solution of 40% glucose to approximately 100 min.

The diffusion coefficients of all agents presented in Figures 36.11 and 36.12 were determined from the analysis of the experimental kinetic curves ΔA (*t*) by fitting to Equations

36.2. The obtained values of mean and standard deviation of the effective diffusion coefficients are summarized in Table 36.6.

The obtained values of the diffusion coefficient are in the range of 10^{-6}–10^{-7}cm²/s [59], which is typical for the diffusion of such molecules in tissues. The diffusion coefficients of low concentrated glycerol and glucose in water are $7.2 \cdot 10^{-6}$ cm²/s and $5.2 \cdot 10^{-6}$ cm²/s, respectively [76]. Due to the high viscosity of glycerol, its diffusion coefficient should be much lower for high concentrations. For 40% aqueous glycerol diffusion in human colorectal muscle, the diffusion coefficient is $0.33 \cdot 10^{-6}$ cm²/s [2, 58] and in human liver tissue $0.82 \cdot 10^{-6}$ cm²/s [58, 77]. In aqueous saturated solutions, the diffusion coefficient of glucose according to various studies is $0.68 \cdot 10^{-6}$ см²/c [78] or $0.85 \cdot 10^{-6}$ см²/c [79]. The diffusion coefficient of MB in water is $3.8 \cdot 10^{-6}$ cm²/s [80]. The mean diffusion coefficient for water in dentin is $0.27 \cdot 10^{-6}$ cm²/s, and in the case of diffusion of a 44% aqueous glycerol solution, it was $1.2 \cdot 10^{-6}$ cm²/s [81]. All these data show that, depending on the experimental conditions and the structure of the studied samples, the diffusion coefficient of water, aqueous solutions of dyes, and glycerol and glucose molecules in dentin can vary significantly, but lies in the range of $(0.27-5.4) \cdot 10^{-6}$ cm²/s.

Conclusions and perspectives

This chapter presents the optical properties of human gums and dentin experimentally studied in the spectral range 200–800 nm. Based on *in vitro* measured spectra of diffuse reflectance and total transmittance, absorption, and transport scattering coefficients of the gums and dentin and their optical penetration depths were estimated using the IAD method. Knowledge of tissue optical properties is important for the development of optical methods for early *in vivo* diagnostics of pathological changes in tissues of the gingival mucous and dentin, as well as for the preparation of the correct clinical protocols for the photodynamic and photothermal therapy of various dental diseases.

By combining the immersion OC technique in human tissues with UV-spectroscopy, it was possible to verify and study one of the major OC mechanisms – tissue dehydration – and to show that the OC efficiency is higher in the deep-UV than in the visible-NIR range. The applied OC treatments with highly concentrated glycerol solutions have induced three tissue transparency windows in gingiva – 200–250 nm, 250–300 nm, and 300–400 nm. By treating gingival tissues with a solution containing 40% glucose, only one transparency window was observed between 300 and 400 nm with a lower OC efficiency than the ones observed for tissues treated with glycerol.

The use of UV light has also allowed the evaluation of the diffusion coefficient of water and OCAs in gingival tissues, showing that it is possible to use UV light to study molecule diffusivity in tissues. The chapter presents data on the effective diffusion coefficients (binary diffusion coefficients of tissue water and agent) of such agents as glycerol, glucose, and mixtures of glycerol and propylene glycol in the human gingival mucosa, as well as glycerol, glucose, mixtures with methylene blue dye, and medicine rivanol in human dentin. The possibility of evaluating diffusion coefficients of water molecules in human gingival tissue layers separately was demonstrated with the usage of independent OCT measurements of the thickness of these tissue layers. Also, the tortuosity coefficient for water diffusion in gingival lamina propria layer $S \approx 3.9$ was evaluated for the first time [82].

The fundamentally new result in this paper is the development of OC technology for a new short-wavelength range, up to 200 nm, where the OC of tissue is the only way to go into depth of the tissue to study molecular structures using UV spectroscopy. UV technologies to study protein structure and dynamics such as deep-UV Raman spectroscopy [83], UV resonance spectroscopy [84, 85], and deep-UV fluorescence microscopy for detection gliomas [86, 87] or tissue optical histology [88] can benefit from the OC technology we have developed to improve tissue transparency and light penetration depth in the UV range.

This technology is adaptable for use with any current UV technologies or others to be developed in the future [2, 58].

In addition, we should note that glycerol is the main component of therapeutic drugs for the mucous membranes of the oral cavity, such as Lugol's iodine solution (94% glycerol) [89]. Thus, the received results can be used to improve existing and develop new optical methods for the diagnosis and treatment of dental diseases basing on glycerol interaction with oral tissues by making them transparent.

Acknowledgments

This work was carried out with the financial support of the project No. 13.2251.21.0009 of the Ministry of Science and Higher Education of the Russian Federation and RFBR grant # 18-52-16025.

REFERENCES

1. V.V. Tuchin, *Tissue Optics: Light Scattering Methods and Instruments for Medical Diagnostics*, 3rd ed., SPIE Press, Bellingham, 2015.

2. I. Carneiro, S. Carvalho, R. Henrique, L. Oliveira, V.V. Tuchin, "A robust ex vivo method to evaluate the diffusion properties of agents in biological tissues," *J. Biophoton.* 12(4), e201800333 (2019).

3. L. Shi, R.R. Alfano (Eds.), *Deep Imaging in Tissue and Biomedical Materials: Using Linear and Nonlinear Optical Methods*, Taylor & Francis Group, Pan Stanford Publishing Pte. Ltd., Singapore, 2017.

4. F.J. Bolton, A.S. Bernat, K. Bar-Am, D. Levitz, S. Jacques, "Portable, low-cost multispectral imaging system: Design, development, validation, and utilization," *J. Biomed. Opt.* 23(12). 121612 (2018).

5. P. Valdes, V. Jacobs, B. Wilson, F. Leblond, D.W. Roberts, K.D. Paulsen, "System and methods for wide-field quantitative fluorescence imaging during neurosurgery," *Opt. Lett.* 38(15), 2786–2788 (2013).

6. S.A. Prahl, M.J.C. Gemert, A.J. Welch, "Determining the optical properties of turbid media by using the adding-doubling method," *Appl. Opt.* 32(4), 559–568 (1993).

7. A.N. Bashkatov, E.A. Genina, V.I. Kochubey, et al., "Optical properties of mucous membrane in the spectral range 350–2000 nm," *Opt. Spectrosc.* 97(6), 978–983 (2004).

8. A.N. Bashkatov, E.A. Genina, V.I. Kochubey, V.V. Tuchin, "Optical properties of human sclera in spectral range 370–2500 nm," *Opt. Spectrosc.* 109(2), 197–204 (2010).

9. A.V. Belikov, A.M. Zagorul'ko, S.N. Smirnov, A.N. Sergeev, A.A. Mikhailova, A.A. Shimko, "Optical properties of human eye cataractous lens in vitro in the visible and NIR ranges of the spectrum," *Opt. Spectrosc.* 126(5), 574–579 (2019).

10. A.N. Bashkatov, E.A. Genina, M.D. Kozintseva, V.I. Kochubey, S.Y. Gorofkov, V.V. Tuchin, "Optical properties of peritoneal biological tissues in the spectral range of 350–2500 nm," *Opt. Spectrosc.* 120(1), 1–8 (2016).

11. A.N. Bashkatov, E.A. Genina, V.I. Kochubey et al., "Optical properties of human stomach mucosa in the spectral range from 400 to 2000 nm: Prognosis for gastroenterology," *Med. Laser Appl.* 22(2), 95–104 (2007).

12. A.N. Bashkatov, E.A. Genina, V.I. Kochubey, V.S. Rubtsov, E.A. Kolesnikova, V.V. Tuchin, "Optical properties of human colon tissues in the 350–2500 spectral range," *Quantum Electron.* 44(8), 779–784 (2014).

13. S. Carvalho, N. Gueiral, E. Nogueira, R. Henrique, L. Oliveira, V. V. Tuchin, "Glucose diffusion in colorectal mucosa—a comparative study between normal and cancer tissues," *J. Biomed. Opt.* 22(9), 091506 (2017). DOI: 10.1117/1.JBO.22.12.125002.

14. S. Carvalho, N. Gueiral, E. Nogueira, R. Henrique, L. Oliveira, V.V. Tuchin, "Comparative study of the optical properties of colon mucosa and colon precancerous polyps between 400 and 1000 nm," *Proc. SPIE* 10063, L-1–16, 100631 (2017).

15. S.L. Jacques, "Optical properties of biological tissues: A review," *Phys. Med. Biol.* 58(11), 37–61 (2013).

16. W.F. Cheong, S.A. Prahl, A.J. Welch, "A review of the optical properties of biological tissues," *IEEE J. Quantum Electron.* 26(12), 2166–2185 (1990).

17. C. Nogueira, A.X. Graciano, J.Y. Nagata, et al., "Photosensitizer and light diffusion through dentin in photodynamic therapy," *J. Biomed. Opt.* 18(5), 1125–1132. DOI: 10.1117 / 1.JBO.18.5.055004 (2013).

18. A. Patil, V.K. Unnikrishnan, R. Ongole, K.M. Pai, V.B. Kartha, S. Chidangil, "Non-invasive *in vivo* screening of oral malignancy using laser-induced fluorescence based system," *CTM* 10(1), 15–26. DOI: 10.17691/stm2018.10.1.02 (2018).

19. O.V. Shkarednaya, T.P. Goryacheva, A.A. Chunikhin, E.A. Bazikyan, S.I. Gazhva, "Optimizing the early diagnosis of oral mucosal pathologies," *CTM* 9(3), 119–124 (2017).

20. N.N. Bulgakova, E.A. Volkov, T.I. Pozdnyakova, "Autofluorescent somatoscope as a method of oncoscience diseases of the oral mucosa," *Ross. Stomatol. Zh.* 19(1), 27–30 (2015).

21. M.M. Perez, R. Ghinea, L.J. Herrera, F. Carrillo, A.M. Ionescu, R.D. Paravina, "Color difference thresholds for computer-simulated human gingiva," *J. Esthet. Restor. Dent.* 30(2), E24–30 (2018).

22. I. Sailer, "Threshold values for the perception of color changes in human teeth," *Int. J. Periodon Restor. Dent.* 36, 777–783 (2016).

23. N.D. Sarmast, N. Angelov, R. Ghinea, J.M. Powers, R.D. Paravina, "Color compatibility of gingival shade guides and gingiva-colored dental materials with healthy human gingiva," *Int. J. Periodon Restor. Dent.* 38(3), 397–403 (2018).

24. D.K. Ho, R. Ghinea, L.J. Herrera, N. Angelov, R.D. Paravina, "Color range and color distribution of healthy human gingiva: A prospective clinical study," *Sci. Reps* 5, 18498. DOI: 10.1038/srep18498 (2015).

25. C.G. Polo, J. Montero, A.M.M. Casado, "Proposal for a gingival shade guide based on in vivo spectrophotometric measurements," *J. Adv. Prosthodont.* 11(5), 239–246. DOI: 10.4047/jap.2019.11.5.239 (2019).

26. R. Ghinea, L.J. Herrera, M.M. Perez, A.M. Ionescu, R.D. Paravina, "Gingival shade guides: Colorimetric and spectral modeling," *J. Esthet. Restor. Dent.* 30(2), E31–38 (2018).

27. C. Gomez-Polo, M. Gomez-Polo, J.A. Martinez Vazquez de Parga, A. Celemin-Vinuela, "Clinical study of the 3D-master color system among the Spanish population," *J. Prosthodont.* 27(8), 708–715 (2018).

28. C. Gomez-Polo, J. Montero, M. Gomez-Polo, J.A. de Parga, A. Celemin-Vinuela, "Natural Tooth color estimation based on age and gender," *J. Prosthodont.* 26(2), 107–114 (2017).

29. B. Baumann, "Polarization sensitive optical coherence tomography: A review of technology and applications," *Appl. Sci.* 7(5), 474–481. DOI: 10.3390/app7050474 (2017).

30. Z. Hamdoon, W. Jerjes, G. McKenzie, A. Jay, C. Hopper, "Optical coherence tomography in the assessment of oral squamous cell carcinoma resection margins," *Photodiagn. Photodyn. Ther.* 13, 211–217 (2016).

31. N.M. Le, Sh. Song, H. Zhou, et al., "A noninvasive imaging and measurement using optical coherence tomography angiography for the assessment of gingiva: An in vivo study," *J. Biophoton.* 11(12), e201800242 (2018). DOI: 10.1002/jbio.201800242.

32. M.-T. Tsai, Y. Chen, Ch.-Yu. Lee, et al., "Noninvasive structural and microvascular anatomy of oral mucosae using handheld optical coherence tomography," *Biomed. Opt. Express* 8(11), 5001–5012 (2017).

33. A.L. Clark, A. Gillenwater, R. Alizadeh-Naderi, A. Elnaggar, R.R. Kortum, "Detection and diagnosis of oral neoplasia with an optical coherence microscope," *J. Biomed. Opt.* 9(6), 1271–1280. DOI: 10.1117/1.1805558 (2004).

34. K. Li, Z. Yang, W. Liang, J. Shang, Y. Liang, S. Wan, "Low-cost, ultracompact handheld optical coherence tomography probe for in vivo oral maxillofacial tissue imaging," *J. Biomed. Opt.* 25(4), 046003. DOI: 10.1117/1.JBO.25.4.046003 (2020).

35. J. Wang, W. Zheng, K. Lin, Zh. Huang, "Development of a hybrid Raman spectroscopy and optical coherence tomography technique for real-time in vivo tissue measurements," *Opt. Lett.* 41(13), 3045–3048. DOI: 10.1364/OL.41.003045 (2016).

36. K.J. Park, H. Schneider, R. Haak, "Assessment of defects at tooth/self-adhering flowable composite interface using swept-source optical coherence tomography (SS-OCT)," *Dent. Mater.* 31(5), 534–541 (2015).

37. K. Horie, Y. Shimada, K. Matin et al., "Monitoring of cariogenic demineralization at the enamel-composite interface using swept-source optical coherence tomography," *Dent. Mater.* 32(9), 1103–1112 (2016).

38. H. Kang, C.L. Darling, D. Fried, "Use of an optical clearing agent to enhance the visibility of subsurface structures and lesions from tooth occlusal surfaces," *J. Biomed. Opt.* 21(8), 081206. DOI: 10.1117/1.JBO.21.8.081206 (2016).

39. A. Skandarajah, S.P. Sunny, P. Gurpur, et al., "Mobile microscopy as a screening tool for oral cancer in India: A pilot study," *PLOS ONE* 27(12), e0188440 (2017).

40. F. Jiang, L. Luo, S.S. Alauddin, J. Glande, J. Chen, "Light transmittance of the periodontium," *Laser. Dent. Sci.* 1(2–4), 107–115. DOI: 10.1007/s41547-017-0015-y (2017).

41. V.N. Grisimov, "Assessment of the enamel mineralization dynamics by the manifestation of Fraunhofer diffraction," *The Dental Institute* 4(85), 111–113 (2019).

42. M. Villarroel, N. Fahl, A.M. De Sousa, O.B. De Oliveira Jr., "Direct esthetic restorations based on translucency and opacity of composite resins," *J. Esthet. Restor. Dent.* 23(2), 73–87 (2011).

43. G.B. Altshuler, A.V. Belikov, F.I. Feldchtein, V.V. Tuchin, A.G. Vybornov, "Method and apparatus for diagnostic and treatment using hard tissue or material microperforation," US Patent US20100015576A1, Published 21 January 2010.

44. V.V. Tuchin, G.B. Altshuler Chapter 9, "Dental and oral tissue optics," In: *Fundamentals and Applications of Biophotonics in Dentistry, Series on Biomaterials and Bioengineering*, 4, A. Kishen and A. Asundi (Eds.), Imperial College Press, London, 2007.

45. F. Ko, G. Tien, M. Chuang, T. Huang, M. Hung, K. Sung, *In-Vivo Diffuse Reflectance Spectroscopy (DRS) of Oral Mucosa of Normal Volunteers*, Optical Tomography and Spectroscopy, Fort Lauderdale, FL, ISBN: 978-1-943580-10-1, JTu3A.45, 25–28 April, 2016.

46. A.M. Ionescu, J.C. Cardona, I. Garzón, et al., "Integrating-sphere measurements for determining optical properties of tissue-engineered oral mucosa," *J. Eur. Opt. Soc.* 10, 15012. DOI: 10.2971/jeos.2015.15012 (2015).

47. D. Gekelman, J.M. White, "Optical properties and color of porcine gingiva," *Proc. SPIE* 4610, 31–38 (2002). DOI: 10.1117/12.469330.

48. J.B. Schutt, J.F. Arens, C.M. Shai, E. Stromberg, "Highly reflecting stable white paint for the detection of ultraviolet and visible radiations," *Appl. Opt.* 13(10), 2218–2221 (1974).

49. J.R. Zijp, J.J.T. Bosch, "Theoretical model for the scattering of light by dentin and comparison with measurements," *Appl. Opt.* 32(4), 411–415 (1993).

50. A. Kienle, L. Lilge, M.S. Patterson, R. Hibst, R. Steiner, B.C. Wilson, "Spatially resolved absolute diffuse reflectance measurements for noninvasive determination of the optical scattering and absorption coefficients of biological tissue," *Appl. Opt.* 32(13), 2304–2314 (1996).

51. A.A. Selifonov, O.A. Zyuryukina, E.N. Lazareva et al., "Measurement of optical properties of human gums and dentin in the spectral range from 350 to 800 nm," *Izv. Saratov Univ. (N. S.) Ser. Phys.* 34(4), 258–267 (2020).

52. https://omlc.org/calc/mie_calc.html.

53. P. Serkan, Er. Kürşat, T.P. Nilüfer, "Penetration depth of laser Doppler flowmetry beam in teeth," *Oral Surg. Oral Med. Oral Pathol. Oral Radiol. Endodontol.* 100(1), 125–129. DOI: 10.1016/j.tripleo.2004.11.018 (2005).

54. A.V. Belikov, G.B. Altshuler, K.V. Shatilova et al., "Peroxide dental bleaching via laser microchannels and tooth color measurements," *J. Biomed. Opt.* 21(12), 125001 (2016).

55. S.J. Nelson, *Wheeler's Dental Anatomy, Physiology and Occlusion*, Elsevier Inc., Canada, 2020.

56. A.A. Selifonov, V.V. Tuchin, "Determination of the kinetic parameters of glycerol diffusion in the gingival and dentinal tissue of a human tooth using optical method: *In vitro* studies," *Opt. Quantum Electron.* 52(2), 1–10 (2020).

57. A.A. Selifonov, V.V. Tuchin, "Control of the optical properties of gum and dentin tissue of a human tooth at laser spectral lines in the range of 200–800 nm," *Quantum Electron.* 50(1), 47–54 (2020).

58. L. Oliveira, V.V. Tuchin, *The Optical Clearing Method: A New Tool for Clinical Practice and Biomedical Engineering*, Springer Nature Switzerland AG, Basel, 2019, 177.

59. A. Kotyk, K. Janacek, *Membrane Transport: An Interdisciplinary Approach*, Plenum Press, London, 1977; Mir Moscow, 1980, 344 p.

60. E.A. Genina, A.N. Bashkatov, E.E. Chikina, V.V. Tuchin, "Diffusion of methylene blue in human maxillary sinus mucosa," *Biophys.* 52(6), 1104–1111 (2007).

61. A. Vogel, V. Venugopalan, "Mechanisms of pulsed laser ablation of biological tissues," *Chem. Rev.* 103(2), 577–644 (2003).

62. D.A. Schwindt, K.P. Wilhelm, H.I. Maibach, "Water diffusion characteristics of human stratum corneum at different anatomical sites in vivo," *J. Invest. Dermatol.* 111(3), 385–389 (2016).

63. S. Mériaux, A. Conti, B. Larrat, "Assessing diffusion in the extracellular space of brain tissue by dynamic MRI mapping of contrast agent concentrations," *Front. Phys.* 6, 38 (2018).

64. D.K. Tuchina, I.G. Meerovich, O.A. Sindeeva et al., "Magnetic resonance contrast agents in optical clearing: Prospects for multimodal tissue imaging," *J. Biophoton.* 13(11), e201960249. DOI: 10. 1002/jbio.201960249 (2020).

65. A.A. Selifonov, V.V. Tuchin, "Determination of the diffusion coefficient of 40%-glucose in human gum tissue by optical method," *Opt. Spectrosc.* 128(6), 766–770 (2020). DOI: 10.1134/S0030400X20060211.

66. I. Carneiro, S. Carvalho, R. Henrique, L. Oliveira, V.V. Tuchin, "Moving tissue spectral window to the deep-ultraviolet via optical clearing," *J. Biophoton.* 12(12), e201900181 (2019).

67. V. Grisimov, S. Radlinsky, "Biomechanics of teeth and restorations," *Dent Art: J. Sci. Art Dent.* 2, 42-482006 (2018).

68. A. Kienle, R. Michels, R. Hibst, "Magnification: A new look at a long-known optical property of dentin," *J. Dent. Res.* 85(10), 955 (2006).

69. A. Kishen, S. Vedantam, "Hydromechanics in dentine: Role of dentinal tubules and hydrostatic pressure on mechanical stress-strain distribution," *Dent. Mater.* 23(10), 1296 (2007).

70. W. Xia, C. Lindahl, C. Persson, P. Thomsen, J. Lausmaa, H. Engqvist, "Changes of surface composition and morphology after incorporation of ions into biomimetic apatite coatings," *J. Biomater. Nanobiotechnol.* 1(1), 7–16 (2010).

71. A.A. Selifonov, V.V. Tuchin, "Determination of the diffusion coefficient of rivanol in dentin of a human tooth in vitro," *Proc. SPIE* 11359, 113591S. DOI: 10.1117/12.2556085 (2020).

72. A.A. Selifonov, V.V. Tuchin, "Investigation of the diffusion of methylene blue through dentin from a human tooth," *Biophysics* 63(6), 981–988. DOI: 10.1134/S0006350918060222 (2018).

73. A.A. Selifonov, O.G. Shapoval, A.N. Mikerov, V.V. Tuchin, "Determination of the diffusion coefficient of methylene blue solutions in dentin of a human tooth using reflectance spectroscopy and their antibacterial activity during laser exposure," *Opt. Spectrosc.* 126(6), 758–768. DOI: 10.1134/S0030400X19060213 (2019).

74. A.A. Selifonov, V.V. Tuchin, "Diffusion of methylene blue in human dentin in the presence of glucose: In vitro study," *Proc. SPIE* 11065, 110651Y. DOI: 10.1117/12.2527992 (2019).

75. A.A. Selifonov, V.V. Tuchin, "Optical properties of human dentin when it is immersed in glucose in vitro and the kinetics of this process," *J. Opt. Technol.* 87(3), 168–174. DOI: 10.1364/JOT.87.000168 (2020).

76. I.S. Grigoriev, E.Z. Meilikhov, *Physical Quantities. Handbook*, Energoatomizdat, Moscow, 1232 p, 1991.

77. I. Carneiro, S. Carvalho, R. Henrique, L. Oliveira, V.V. Tuchin, "Simple multimodal optical technique for evaluation of free/bound water and dispersion of human liver tissue," *J. Biomed. Opt.* 22(12), 125002. DOI: 10.1117/1. JBO.22.12.125002 (2017).

78. F.J. Gutter, G. Kegeles, "Diffusion in supersaturated solutions II. Glucose solutions," *J. Am. Chem. Soc.* 75(15), 3900–3904 (1953).

79. E. McLaughlin, "Diffusion in a mixed dense fluid," *J. Chem. Phys.* 50(3), 1254–1262 (1969).

80. G.L. Derek, "The diffusion coefficient of MB in water," *J. Chem.* 66, 2452 (1988).

81. N.A. Trunina, V.V. Lychagov, V.V. Tuchin, "OCT monitoring of diffusion of water and glycerol through tooth dentine in different geometry of wetting," *Proc. SPIE* 7563, 7563OU (2010).

82. I. Carneiro, S. Carvalho, R. Henrique, A. Selifonov, L. Oliveira, V.V. Tuchin, Enhanced ultraviolet spectroscopy by optical clearing for biomedical applications, *IEEE J. Sel. Top. Quantum Electron.* 27(4), 1–8. DOI: 10.1109/ JSTQE.2020.3012350 (2021).

83. B.R. Arnold, C.E. Cooper, M.R. Matrona, D.K. Emge, J.B. Oleske, "Stand-off deep-UV Raman spectroscopy," *Can. J. Chem.* 96(7), 614–620 (2018).

84. R.S. Jakubek, J. Handen, S.E. White, S.A. Asher, I.K. Lednev, "Ultraviolet resonance Raman spectroscopic markers for protein structure and dynamics," *Trends Anal. Chem.* 103, 223–229 (2018).

85. J.D. Handen, I.K. Lednev, "Deep UV resonance Raman spectroscopy for characterizing amyloid aggregation," *Meth. Mol. Biol.* 1345, 89–100. DOI: 10.1007/978-1-4939-2978-8_6 (2016).

86. H. Mehidine, A. Chalumeau, F. Poulon, et al., "Optical signatures derived from deep UV to NIR excitation discriminates healthy samples from low and high grades glioma," *Sci. Rep.*, 9, 8786 (2019).

87. S. Demos, R. Levenson, *System and Method for Controlling Depth of Imaging in Tissues Using Fluorescence Microscopy under Ultraviolet Excitation Following Staining with Fluorescing Agents* U.S. Patent 9 625 387, 16, 2014.

88. F. Jamme, S. Kascakova, S. Villette, et al., "Deep UV autofluorescence microscopy for cell biology and tissue histology," *Biol. Cell* 105(7), 277–288 (2013).

89. M. Petruzzi, A. Lucchese, E. Baldoni, F.R. Grassi, R. Serpico, "Use of Lugol's iodine in oral cancer diagnosis: An overview," *Oral Oncol.* 46(11), 811–813 (2010).

37

Optical clearing and Raman spectroscopy: In vivo *applications*

Qingyu Lin, Ekaterina N. Lazareva, Vyacheslav I. Kochubey, Yixiang Duan, and Valery V. Tuchin

CONTENTS

Introduction

Raman spectroscopy has major benefits in various research areas. The Raman scattering technique requires virtually no sample preparation and is nondestructive, and thus Raman is a popular tool for *in vivo* biological samples studies [1, 2]. Raman spectroscopy is based on the inelastic scattering of a sample. Long described the detailed theory of Raman scattering [3]. In practice, two types of light scattering (elastic and inelastic) exist when incident monochromatic light interacts with the sample. The elastic scattering (Rayleigh scattering) causes no change in the frequency of the photons. The inelastic scattering (Raman scattering) is accompanied by a shift in wavelength. The frequency difference between the scattered photons and incident photons is the Raman shift. The Raman shift represents the modes of molecular vibration or rotation which are related to the corresponding molecular structure [4]. Compared with Rayleigh scattering, Raman scattering is extremely weak. However, the weak vibrational information is specific to chemical bonds, the atomic mass in the bond, the electronic environment, and the symmetry of molecules, so the Raman signal represents the molecular fingerprint of a sample in terms of its chemical composition [5].

As a potential nondestructive measurement technique, Raman spectroscopy has many attractive advantages, including high sensitivity, ease of use, and absence of photobleaching [6]. Therefore, Raman spectroscopy is a promising technique for *in situ* and *in vivo* analysis in biomedical applications. However, Raman signals from biological samples are inherently weak due to the low portion of incident photons being inelastically scattered. Therefore, the growing demand for noninvasive diagnostic and nondestructive structure analysis tools to identify and investigate tissues has led to great progress in the development of Raman spectroscopic methods [7]. Several modified technologies have been developed to increase the intensity of Raman signals. Resonance Raman scattering [8], surface-enhanced Raman scattering [9], and tip-enhanced

Raman scattering [10] are the most representative methods. However, strong elastic scattering by tissue markedly decreases both the resolution and intensity of light propagating into deep tissues [11]. Thus, current applications of such Raman-enhanced methods in clinical settings are restricted to lesions located in superficial tissues. The scattering effect further limits the light penetration depth in tissues such that it remains difficult to obtain complete structural information for tissues.

Optical clearing (OC) provides an improvement in image quality and precision of spectroscopic and microscopic information from tissue depth [12]. The OC of tissues has been of great interest owing to its great potential in enhancing the capabilities of noninvasive light-based diagnostics and laser therapy [13]. OC enables light to penetrate more deeply and effectively reduces light scattering via the application of special optical clearing agents (OCAs) into tissues [14]. Therefore, a detailed discussion on the Raman technique combined with OC will be reported in this section. We review the advantages and disadvantages of Raman spectroscopy for *in vivo* studies, especially the *in vivo* application of a portable Raman system.

Raman spectroscopy for *in vivo* studies: Advantages and disadvantages

Currently, histopathology is the "gold standard" investigation in cancer diagnostics. However, this method is invasive, requiring removal of tissue, followed by tissue fixation, sectioning, and staining which can be costly and slow [15]. Raman spectroscopy is beginning to gain recognition as a potential adjunct to histopathology due to the fact the method can overcome the limitations of current diagnostic techniques. Generally, diseases always lead to chemical or structural changes on the molecular level. The changes can be used as potential markers of the disease. Raman can provide a rapid, reproducible,

DOI: 10.1201/9781003025252-41

and nondestructive measurement of the specific fingerprint changes without extrinsic contrast-enhancing agents [16]. Thus, detailed chemical information about a biological sample can be used to distinguish the changes that accompany different diseases [17]. Complementary to MIR and NIR spectroscopy, Raman spectroscopy holds considerable potential as both a qualitative and quantitative biochemical analysis tool for tissue engineering because of its nondestructive nature and minimal sample preparation [18]. Body fluids and tissue, as well as single cells, can be easily investigated via Raman spectroscopy [19].

At present, Raman spectroscopy is gaining more popularity in the research field of cancer diagnostics [20]. Several Raman diagnostic methods have been developed for use in cancer [21, 22]. It offers the possibility of measuring tumor tissue nondestructively without labeling *in vivo*. It offers direct spectroscopic information about the biomolecular composition of the tissue without or with minimal preparation. Haka et al. investigated Raman spectroscopy as a clinical tool for the diagnosis of a variety of breast pathologies [17]. In contrast to fluorescence, there are a large number of Raman active molecules in tumor tissue, and the spectral signatures are sharp and well delineated. When combined with powerful multivariate algorithms, the technique can potentially provide an automated and reproducible classification of pathology. Bratchenko et al. established that specific features within the pathologies impose certain requirements on the choice of statistical analysis methods [23]. Stone et al. used NIR Raman spectroscopy to interrogate epithelial tissue biochemistry and investigated the distinction between normal and abnormal tissues [24]. Six different epithelial tissues from the larynx, tonsil, esophagus, stomach, bladder, and prostate were measured. Spectral diagnostic models were constructed using multivariate statistical analysis of the spectra to classify samples of epithelial cancers and precancers. A support vector machine with a linear kernel and a sequential minimal optimization solver was applied to generate a Raman classification of normal and tumoral renal tissue [25]. The results demonstrated that Raman can accurately differentiate normal and tumoral renal tissue, low-grade and high-grade renal tumors, and histologic subtypes of renal cell carcinoma. With developing technology and complex analytical methods, there has been a move towards the diagnosis of neoplastic change at much earlier stages [15]. New techniques in cancer diagnostics play an important role in improving the chances of early detection of cancer.

However, Raman spectroscopy has poor sensitivity and a small volume of tissue is sampled, as only 1 in 10^8 photons is Raman scattered [26]. The weak signals usually need rather long measurement times. However, the potential of damaging the sample due to laser exposure, which depends on the excitation wavelength, has to be taken into account during the planning of the measurements [20]. By using plasmonic nanoplatforms as Raman signal amplifiers, surface-enhanced Raman scattering (SERS) provides an ultrasensitive spectroscopic tool [27]. The electromagnetic and chemical enhancement mechanism both make contributions to field enhancement in SERS. When the laser excited on a rough surface, the rough surface structures could lead to stronger localized electric fields, and larger

enhancement of Raman effects might appear. The rough surface was more appropriate to provide a larger enhancement compared with flat surfaces [26]. The two kinds of enhancement mechanisms contributed to the enhancement of about 10^9 and 10^3, respectively [28]. With the improvement and deep understanding of SERS, this technology has been extensively applied in different areas. However, the extensive application of SERS has been limited by some obstacles. Au, Ag, and Cu provided large enhancement for its surface plasmon properties. Other transition metals merely exhibit weak SERS enhancement [29]. Its practical application needs extremely stable and repeatable instruments [30]. To overcome the bottlenecks in SERS, TERS [31], SHINERS [32], ultraviolet SERS [33], and NIR SERS [34] were extensively proposed and applied.

At present, SERS imaging is attracting interest in the biomedical community because of its quite high sensitivity, signal specificity, and unequivocal detection without issues of autofluorescence [27]. Improving the time resolution of Raman imaging is essential for the observation of dynamic processes in biological systems. Wang et al. proposed a signal processing algorithm for fast Raman imaging by reconstructing the Raman spectrum with the aid of weak signal processing, extracting the reliable Raman signal of the target under the low SNR, and then determining the suitable scanning time to obtain the Raman image with trustworthy image quality [35].

Another problem encountered in the Raman of biological tissues is the significant fluorescence signal produced under visible excitation. Fluorescence is a competing emission process often seen in Raman spectra, and especially in Raman spectra of biological samples. The Raman effect is inherently weak, and fluorescence is often of orders of magnitude stronger than the Raman scattering. Thus, the Raman scattered light is obscured by the broad bands of fluorescence. Fluorescence may be omitted using lasers in the NIR or ultraviolet (UV) region. However, by employing NIR lasers, it is necessary to increase the experimental exposure times. By employing UV lasers, the risk of heating and destruction of the sample is increased [36]. Therefore, various baseline correction methods have been used for removing the effects of fluorescence baselines, such as fitting polynomials, standard normal variate [37] transformation, and multiplicative signal correction [38] methods. The Fourier transform and wavelets [39], extended inverted signal correction [40], and orthogonal signal correction [41] combinations of the preprocessing methods are also applied. Human breast tissues do not show very strong fluorescence of other tissues when their Raman spectra are excited with radiation of 780 nm. However, when excited with radiation of 1064 nm, no background is visible [42].

Shim et al. declared that there are three main obstacles to successful *in vivo* Raman spectroscopy : (1) tissue fluorescence is several orders of magnitude greater than the Raman signal; (2) silica Raman and fiber fluorescence, due to crystalline impurities in the glass, contaminate tissue spectra when the excitation light travels through the delivery fiber and Rayleigh scattered light is passed through collection fibers; (3) the Raman signal from tissue may require prohibitive excitation powers or collection times to obtain spectra with acceptable SNR [43].

The portable Raman system

In recent years, an increasing number of portable Raman spectrometers have become available on the market. These spectrometers are generally designed for a broad range of applications [44]. Briefly, *in vivo*, a portable Raman system is a good candidate for the real-time analysis carried out using medical diagnostic tools [45]. Commercial off-the-shelf components such as a semiconductor laser, a fiberoptic probe, and a mini spectrometer can be used to construct a portable, low-cost dispersive Raman system [46]. However, the detection sensitivity of portable Raman systems is inferior to that of large Raman spectroscopic systems, because all the spectroscopic components, including the laser source, monochromator, and detector, must be miniaturized [47]. Raman-based microspectroscopy holds great promise to advance skull and tissue OC methods [48]. Kang et al. directly observed the glucose fingerprint using *in vivo* Raman spectroscopy for the first time. Their goal was to detect changes in blood glucose levels by glucose fingerprint peaks. In their measurement, a noncontact off-axis system was proposed to increase stability. Moreover, tissue distortion and the specular Rayleigh reflection from the skin surface were largely eliminated in this system. The results indicate the possibility of Raman spectroscopy for monitoring blood glucose concentration *in vivo*. By increasing sensitivity, reducing the integration time, and miniaturizing the system, this method will show excellent application prospects [49]. Mo et al. developed a rapid-acquisition NIR Raman spectroscopy system associated with a ball-lens fiberoptic Raman probe for *in vivo* spectroscopic measurements at 785 nm excitation. This study demonstrates for the first time that high-wavenumber (2800–3700 cm^{-1}) Raman spectroscopy has the potential for the noninvasive, *in vivo* diagnosis and detection of precancer of the cervix [50]. The probable Raman can be used as a noninvasive preterm birth risk assessment tool to reduce the incidence of and morbidity and mortality caused by preterm birth. O'Brien et al. measured *in vivo* Raman spectra of the cervix throughout pregnancy for 68 healthy pregnant women to detect changes in biochemical markers [51]. They recorded six components that comprise key biochemical changes in the human cervix throughout pregnancy.

To extend the applications of a handheld Raman detector to deep tissues, a gold nanostar–based SERS nanoprobe with robust colloidal stability, a fingerprint-like spectrum, and extremely high sensitivity was developed [11]. A multicomponent optical clearing agent (OCA) efficiently suppressed light scattering from the turbid dermal tissues, and the handheld Raman detector noninvasively visualized the subcutaneous tumor xenograft with a high target-to-background ratio after intravenous injection of the gold nanostar–based SERS nanoprobe. The combination of OC technology and SERS is a promising strategy for the extension of the clinical applications of the handheld Raman detector [11].

We investigated the OC effect of glycerol, Omnipaque™ (iohexol), and polyethylene glycol in human skin *in vivo* using a portable fiberoptic Raman spectroscopy system (Figure 37.1) [52]. Raman spectra in the fingerprint regions of human skin were obtained and analyzed.

We used an innovative statistics-sensitive nonlinear iterative peak-clipping fitting algorithm to preprocess the skin Raman spectra. Kinetics curves were constructed for different concentrations of OCAs when applied topically for different times. To study the OC process of different OCAs topically applied to human skin, the intensity of the six analytical lines of the *in vivo* Raman spectra was recorded at 0, 5, 10, 20, 30, and 60 min. Figure 37.2(a–d) shows the kinetic curves that reveal the effect of contact time on the intensity of the target Raman peaks following the application of different concentrations of glycerol. The intensity of all of the targeted Raman peaks increased following treatment OC with glycerol. Application of an OCA monotonically increased Raman signal intensities with increasing OCA application time at 50% and 70% concentration, with the saturation of the dependence occurring after 20–60 min. This time interval can be considered as the optically transparent equilibrium time and was 30 min

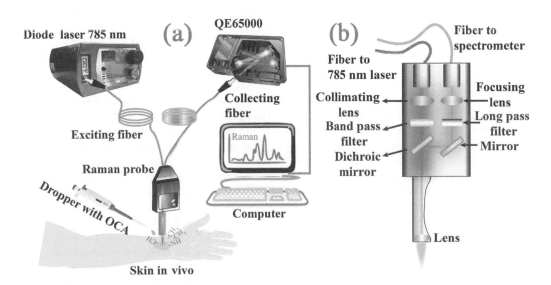

FIGURE 37.1 Diagram of the Raman experimental device (a) and detail of Raman probe (b).

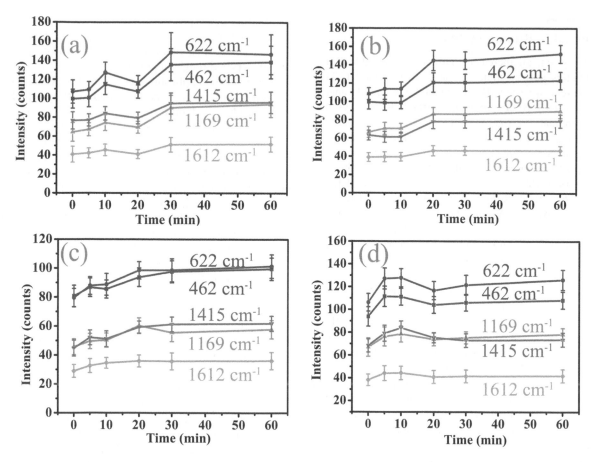

FIGURE 37.2   The time dependencies of skin-based Raman peaks after application of glycerol in different concentrations: 30% (a), 50% (b), 70% (c), and 99.9% (d). Reprinted with permission from Reference [52].

when processed with 30% glycerol, but only 20 min when the concentration was 50%. At 70%, the optimum contact time was 30 min for the peaks at 462 cm⁻¹ (δ(CCC), proteins), and 1415 cm⁻¹ (δ(CH₃), lipids). The spectral maximum values appeared after 10 min with 99.9% glycerol. Our results demonstrate the enhanced effect of glycerol on the *in vivo* Raman spectra. Meanwhile, the optically transparent equilibrium time decreased with increasing concentrations of glycerol.

SERS-based diagnosis mediated by AuNS provides an accurate measure of multiple biomarkers *in vivo*. Ou et al. designed a composite probe of AuNS labeled with Raman tags and antibodies to detect epidermal growth factor receptor in breast cancer tumors [53]. Ju et al. reported a novel SERS sensor based on a low-cost polymethyl methacrylate microneedle array for the *in vivo* intradermal detection of glucose (Figure 37.3). The functional microneedle array was able to achieve *in vivo* intradermal measurements of glucose from interstitial fluid in the mouse model of type I diabetes [54]. This polymeric array–based SERS biosensor has the potential to be used in painless glucose monitoring of diabetic patients in the future. The array had enough strength to penetrate the skin while minimizing damage. In addition, this technique was also highly appropriate for navigation of the margin of solid tumors during surgery [55].

Silica core–silica clad fibers, with an acrylate coating and a black nylon jacket, proved to be one of the best candidates

for Raman spectroscopy measurements [56]. A single hollow optical fiber was also developed for performing a high axial resolution and high sensitivity remote Raman analysis of biomedical tissues [57]. An *in vivo* Raman system was further described which can obtain spectra *in vivo* from tissue in less than 30 s [43]. The system was compact and mobile, making it suitable for clinical applications. Spectra were collected *in vivo* from human skin, buccal cheek epithelium, fingernail, and tooth. Ding et al. utilized an endoscopic Raman spectroscopy as a noninvasive technique to determine ulcerative colitis status [58]. A special fiberoptic probe in the portable endoscopic Raman system consisted of a 200 μm core fiber to deliver the laser light and seven 300 μm core fibers to collect the signals. A miniature converging lens was built into the tip of the probe, which allows for a maximum effective depth of tissue up to 1000 μm. Multivariate analysis with a support vector machine classified the spectra of the controls or the inactive colitis from those of inflamed tissue with a sensitivity of 83.5% and 97.1%, respectively. This work suggested that the endoscopic Raman spectroscopy has the potential for noninvasive disease assessment in situ (Figure 37.4). Similarly, the endoscopic Raman system was utilized to evaluate potential physiological factors affecting tissue biochemistry in the other work [59]. Fifty-six healthy patients were scheduled for colonoscopy screening to acquire spectra from the colon. With analyses and comparisons, the variability of spectra resulted from different

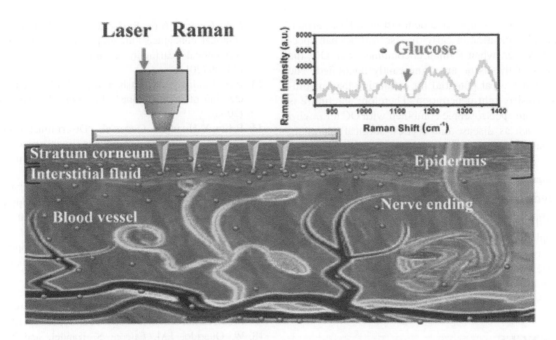

FIGURE 37.3 Schematic illustration of the glucose measurement using the F-PMMA MN array for in vivo transdermal detection based on surface-enhanced Raman spectroscopy. Reprinted with permission from Ref [54].

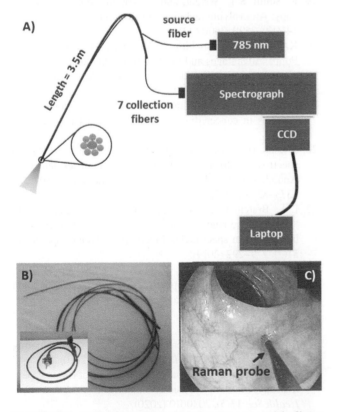

FIGURE 37.4 (A) Scheme of the portable Raman system and the fiberoptic Raman probe for endoscopic analysis of tissue composition. (B) Photo of the Raman probe and the standard endoscope (inset). (C) Raman probe in contact with the surface of the colon after traveling through the accessory channel of the endoscope. Reprinted with permission from Ref [59].

abundance in lipids and proteins. These changes mainly came from the difference in physiological factors, including gender, ethnicity, age, and anatomical locations. Hence, the results

suggested the significances of considering physiological features as an important factor when applying *in vivo* Raman spectroscopy for optical diagnosis. The Raman endoscopy technique also has promising clinical potential for real-time, *in vivo* diagnosis and detection of malignancies in the upper gastrointestinal at the molecular level [60].

Pence et al. used colonoscopy-coupled fiberoptic probe-based Raman spectroscopy to diagnose inflammatory bowel disease (IBD) of the colon [61]. *In vivo* Raman spectra for IBD could be acquired nondestructively and in real time in human subjects from the colon during routine surveillance endoscopy with high signal-to-noise ratio. The resultant spectra had a wealth of information content that can discriminate between IBD and normal colon, as well as indicating the level of disease activity present. Raman spectroscopy has the potential for providing previously unobtainable biochemical information that may be useful as a diagnostic adjunct for IBD. Brinkmann et al. presented an all-fiber light source for coherent Raman scattering microscopy, tunable across more than 2700 cm^{-1} in less than 5 ms [62]. The system ran at a high repetition rate around 40 MHz for rapid imaging. Compared with a commercial laboratory-bound reference system, the equivalent imaging of mouse and human skin tissue can be acquired from the system. These results verified that the portable light source was well suited for coherent Raman microscopy. Moreover, it can be applied for *in vivo* investigations in pharmaceutical research and for bedside virtual optical biopsies.

Summary

There is a high demand for clinical tools to nondestructively diagnose disease *in vivo*. Raman spectroscopy has been attractive for various biological applications. The portable Raman system is a good candidate for the *in vivo* analysis of medical diagnostics. Various groups have demonstrated this, particularly for the identification of early disease with Raman spectroscopy. Moreover, when Raman is combined with OC, the Raman peaks of the tissue are significantly intensified. The OC of tissues has great potential for enhancing the capabilities of noninvasive light penetration. Therefore, the future is bright for Raman spectroscopic combined with OC for *in vivo* application, such as disease diagnostics, surgical operation, and even treatment monitoring.

Acknowledgments

Q.L. and Y.D. are grateful for financial support from the Sichuan Applied Basic Research Project (2019YJ0078); E.N.L. and V.I.K., are grateful for financial support from RFBR grant # 18-52-16025 and from the project No. 13.2251.21.0009 of the Ministry of Science and Higher Education of the Russian Federation (V.V.T.).

REFERENCES

1. I. Pence, and A. Mahadevan-Jansen, "Clinical instrumentation and applications of Raman spectroscopy," *Chem. Soc. Rev.* **45**(7), 1958–1979 (2016).
2. S. Latka, C. Dochow, B. Krafft, B. Dietzek, and J. Popp, "Fiber optic probes for linear and nonlinear Raman applications: Current trends and future development," *Laser Photonics Rev.* **7**(5), 698–731 (2013).
3. D.A. Long, *The Raman Effect: A Unified Treatment of the Theory of Raman Scattering by Molecules*, Wiley (2002).
4. C. Krafft, "Bioanalytical applications of Raman spectroscopy," *Anal. Bioanal. Chem.* **378**(1), 60–62 (2004).
5. X. Zhu, T. Xu, Q. Lin, and Y. Duan, "Technical development of Raman spectroscopy: From instrumental to advanced combined technologies," *Appl. Spectrosc. Rev.* **49**(1), 64–82 (2014).
6. W.L. Liao, Q.Y. Lin, S.C. Xie et al., "A novel strategy for rapid detection of bacteria in water by the combination of three-dimensional surface-enhanced Raman scattering (3D SERS) and laser induced breakdown spectroscopy (LIBS)," *Anal. Chim. Acta* **1043**, 64–71 (2018).
7. C. Kendall, M. Isabelle, F. Bazant-Hegemark et al., "Vibrational spectroscopy: A clinical tool for cancer diagnostics," *Analyst* **134**(6), 1029–1045 (2009).
8. W. Smith, "Practical understanding and use of surface enhanced Raman scattering/surface enhanced resonance Raman scattering in chemical and biological analysis," *Chem. Soc. Rev.* **37**(5), 955–964 (2008).
9. L.E. Jamieson, S.M. Asiala, K. Gracie, K. Faulds, and D. Graham, "Bioanalytical measurements enabled by surface-enhanced Raman scattering (SERS) probes," *Annu. Rev. Anal. Chem.* **10**(1), 415–437 (2017).
10. R. Zhang, X. Zhang, H. Wang et al., "Distinguishing individual DNA bases in a network by non-resonant tip-enhanced Raman scattering," *Angew. Chem.* **129**(20), 5653–5656 (2017a).
11. Y. Zhang, H. Liu, J. Tang et al., "Noninvasively imaging subcutaneous tumor xenograft by a handheld Raman detector, with the assistance of an optical clearing agent," *ACS Appl. Mater. Interfaces* **9**(21), 17769–17776 (2017b).
12. E.A. Genina, A.N. Bashkatov, E.A. Kolesnikova et al., "Optical coherence tomography monitoring of enhanced skin optical clearing in rats in vivo," *J. Biomed. Opt.* **19**(2) 021109 (2014).
13. E.A. Genina, A.N. Bashkatov, and V.V. Tuchin, "Optical clearing of cranial bone," *Adv. Opt. Technol.* **2008**, 1–8 (2008).
14. P. Liu, Y. Huang, Z. Guo et al., "Discrimination of dimethyl sulphoxide diffusion coefficient in the process of optical clearing by confocal micro-Raman spectroscopy," *J. Biomed. Opt.* **18**(2), 020507 (2013).
15. C. Kallaway, L.M. Almond, H. Barr et al., "Advances in the clinical application of Raman spectroscopy for cancer diagnostics," *Photodiagn. Photodyn. Ther.* **10**(3), 207–219 (2013).
16. C. Krafft, and V. Sergo, "Biomedical applications of Raman and infrared spectroscopy to diagnose tissues," *Spectroscopy* **20**(5–6), 195–218 (2006).
17. S. Haka, K.E. Shafer-Peltier, M. Fitzmaurice et al., "Diagnosing breast cancer by using Raman spectroscopy," *Proc. Natl. Acad. Sci. U.S.A.* **102**(35), 12371–12376 (2005).
18. W. Querido, J.M. Falcon, S. Kandel, and N. Pleshko, "Vibrational spectroscopy and imaging: Applications for tissue engineering," *Analyst* **142**(21), 4005–4017 (2017).
19. R. Smith, K.L. Wright, and L. Ashton, "Raman spectroscopy: An evolving technique for live cell studies," *Analyst* **141**(12), 3590–3600 (2016).
20. K. Eberhardt, C. Stiebing, C. Matthaus, M. Schmitt, and J. Popp, "Advantages and limitations of Raman spectroscopy for molecular diagnostics: An update," *Expert Rev. Mol. Diagn.* **15**(6), 773–787 (2015).
21. L.A. Austin, S. Osseiran, and C.L. Evans, "Raman technologies in cancer diagnostics," *Analyst* **141**(2), 476–503 (2016).
22. J. Zhao, H. Lui, S. Kalia, and H. Zeng, "Real-time Raman spectroscopy for automatic in vivo skin cancer detection: An independent validation," *Anal. Bioanal. Chem.* **407**(27), 8373–8379 (2015).
23. L.A. Bratchenko, I.A. Bratchenko, A.A. Lykina et al., "Comparative study of multivariate analysis methods of blood Raman spectra classification," *J. Raman Spectrosc.* **51**(2), 279–292 (2019).
24. N. Stone, C. Kendall, N. Shepherd, P. Crow, and H. Barr, "Near-infrared Raman spectroscopy for the classification of epithelial pre-cancers and cancers," *J. Raman Spectrosc.* **33**(7), 564–573 (2002).
25. K. Bensalah, J. Fleureau, D. Rolland et al., "Raman spectroscopy: A novel experimental approach to evaluating renal tumours," *Eur. Urol.* **58**(4), 602–608 (2010).
26. P. Yin, Q. Lin, and Y. Duan, "Applications of Raman spectroscopy in two-dimensional materials," *J. Innov. Opt. Health Sci.* **13**(5), 2030010 (2020).
27. I. Henry, B. Sharma, M.F. Cardinal, D. Kurouski, and R.P. Van Duyne, "Surface-enhanced Raman spectroscopy biosensing: In vivo diagnostics and multimodal imaging," *Anal. Chem.* **88**(13), 6638–6647 (2016).
28. M. Osawa, N. Matsuda, K. Yoshii, and I. Uchida, "Charge transfer resonance Raman process in surface-enhanced Raman scattering from p-aminothiophenol adsorbed on silver: Herzberg-Teller contribution," *J. Phys. Chem.* **98**(48), 12702–12707 (1994).

29. Z.Q. Tian, B. Ren, and D.Y. Wu, "Surface-enhanced Raman scattering: From noble to transition metals and from rough surfaces to ordered nanostructures," *J. Phys. Chem. B* **106**(37), 9463–9483 (2002).

30. R. Panneerselvam, G.-K. Liu, Y.-H. Wang et al., "Surface-enhanced Raman spectroscopy: Bottlenecks and future directions," *Chem. Commun.* **54**(1), 10–25 (2018).

31. R. Treffer, R. Böhme, T. Deckert-Gaudig et al., "Advances in TERS (tip-enhanced Raman scattering) for biochemical applications," *Biochem. Soc. Trans.* **40**(4), 609–614 (2012).

32. H. Zhang, C. Wang, H.L. Sun et al., "In situ dynamic tracking of heterogeneous nanocatalytic processes by shell-isolated nanoparticle-enhanced Raman spectroscopy," *Nat. Commun.* **8**(1), 1–8 (2017c).

33. B. Ren, X.F. Lin, Z.L. Yang et al., "Surface-enhanced Raman scattering in the ultraviolet spectral region: UV-SERS on rhodium and ruthenium electrodes," *J. Am. Chem. Soc.* **125**(32), 9598–9599 (2003).

34. B. Moll, T. Tichelkamp, S. Wegner et al., "Near-infrared (NIR) surface-enhanced Raman spectroscopy (SERS) study of novel functional phenothiazines for potential use in dye sensitized solar cells (DSSC)," *RSC Adv.* **9**(64), 37365–37375 (2019).

35. X. Wang, G. Liu, M. Xu, B. Ren, and Z. Tian, "Development of weak signal recognition and an extraction algorithm for Raman imaging," *Anal. Chem.* **91**(20), 12909–12916 (2019).

36. N.K. Afseth, V.H. Segtnan, and J.P. Wold, "Raman spectra of biological samples: A study of preprocessing methods," *Appl. Spectrosc.* **60**(12), 1358–1367 (2006).

37. D. Archibald, S. Kays, D. Himmelsbach, and F. Barton, "Raman and NIR spectroscopic methods for determination of total dietary fiber in cereal foods: A comparative study," *Appl. Spectrosc.* **52**(1), 22–31 (1998).

38. T.M. Hancewicz, and C. Petty, "Quantitative analysis of vitamin A using Fourier transform Raman spectroscopy," *Spectrochim. Acta A* **51**(12), 2193–2198 (1995).

39. M. Pelletier, "Quantitative analysis using Raman spectrometry," *Appl. Spectrosc.* **57**(1), 20A–42A (2003).

40. M. Dyrby, R. Petersen, and J. Larsen, "Towards on-line monitoring of the composition of commercial carrageenan powders," *Carbohydr. Polym.* **57**(3), 337–348 (2004).

41. M. Sohn, D.S. Himmelsbach, and F.E. Barton, "A comparative study of Fourier transform Raman and NIR spectroscopic methods for assessment of protein and apparent amylose in rice," *Cereal Chem.* **81**(4), 429–433 (2004).

42. B. Schrader, B. Dippel, S. Fendel et al., "Nir FT Raman spectroscopy-a new tool in medical diagnostics," *J. Mol. Struct.* **408**, 23–31 (1997).

43. M.G. Shim, and B.C. Wilson, "Development of an in vivo spectroscopic Raman system for diagnostic applications," *J. Raman Spectrosc.* **28**(2–3), 131–142 (1997).

44. D. Lauwers, A.G. Hutado, V. Tanevska et al., "Characterisation of a portable Raman spectrometer for in situ analysis of art objects," *Spectrochim. Acta A Mol. Biomol. Spectrosc.* **118**, 294–301 (2014).

45. H. Zeng, J. Zhao, Michael Short et al., "Raman spectroscopy for in vivo tissue analysis and diagnosis, from instrument development to clinical applications," *J. Innov. Opt. Health Sci.* **1**(1), 95–107 (2008).

46. B.S. Luo, and M. Lin, "A portable Raman system for the identification of foodborne pathogenic bacteria," *J. Rapid Methods Autom. Microbiol.* **16**(3), 238–255 (2008).

47. L.X. Quang, C. Lim, G.H. Seong et al., "A portable surface-enhanced Raman scattering sensor integrated with a lab-on-a-chip for field analysis," *Lab Chip* **8**(12), 2214–2219 (2008).

48. Y. Chen, S. Liu, H. Liu et al., "Coherent Raman scattering unravelling mechanisms underlying skull optical clearing for through-skull brain imaging," *Anal. Chem.* **91**(15), 9371–9375 (2019).

49. J.W. Kang, Y.S. Park, H. Chang et al., "Direct observation of glucose fingerprint using in vivo Raman spectroscopy," *Sci. Adv.* **6**(4), 5206 (2020).

50. J. Mo, W. Zheng, J.J.H. Low et al., "High wavenumber Raman spectroscopy for in vivo detection of cervical dysplasia," *Anal. Chem.* **81**(21), 8908–8915 (2009).

51. C.M. O'Brien, E. Vargis, A. Rudin et al., "In vivo Raman spectroscopy for biochemical monitoring of the human cervix throughout pregnancy," *Am. J. Obstet. Gynecol.* **218**(5), e1–e18 (2018).

52. Q.Y. Lin, E.N. Lazareva, V.I. Kochubey, Y.X. Duan, and V.V. Tuchin, "Kinetics of optical clearing of human skin studied in vivo using portable Raman spectroscopy," *Laser Phys. Lett.* **17**,105601 (2020).

53. Y.C. Ou, J.A. Webb, C.M. O'Brien et al., "Diagnosis of immunomarkers in vivo via multiplexed surface enhanced Raman spectroscopy with gold nanostars," *Nanoscale* **10**(27), 13092–13105 (2018).

54. J. Ju, C.M. Hsieh, Y. Tian et al., "Surface enhanced Raman spectroscopy based biosensor with a microneedle array for minimally invasive in vivo glucose measurements," *ACS Sens.* **5**(6), 1777–1785 (2020).

55. Z. Bao, Y. Zhang, Z. Tan et al., "Gap-enhanced Raman tags for high-contrast sentinel lymph node imaging," *Biomaterials* **163**, 105–115 (2018).

56. L.S.F. Santos, R. Wolthuis, S. Koljenovic, R.M. Almeida, and G.J. Puppels, "Fiber-optic probes for in vivo Raman spectroscopy in the high-wavenumber region," *Anal. Chem.* **77**(20), 6747–6752 (2005).

57. T. Katagiri, Y.S. Yamamoto, Y. Ozaki, Y. Matsuura, and H. Sato, "High axial resolution Raman probe made of a single hollow optical fiber," *Appl. Spectrosc.* **63**(1), 103–107 (2009).

58. H. Ding, A.W. Dupont, S. Singhal et al., "In vivo analysis of mucosal lipids reveals histological disease activity in ulcerative colitis using endoscope-coupled Raman spectroscopy," *Biomed. Opt. Express* **8**(7), 3426–3439 (2017a).

59. H. Ding, A.W. Dupont, S. Singhal et al., "Effect of physiological factors on the biochemical properties of colon tissue – an in vivo Raman spectroscopy study," *J. Raman Spectrosc.* **48**(7), 902–909 (2017b).

60. M.S. Bergholt, W. Zheng, K. Lin et al., "Characterizing variability in in vivo Raman spectra of different anatomical locations in the upper gastrointestinal tract toward cancer detection," *J. Biomed. Opt.* **16**(3) (2011).

61. J. Pence, D.B. Beaulieu, S.N. Horst et al., "Clinical characterization of in vivo inflammatory bowel disease with Raman spectroscopy," *Biomed. Opt. Express* **8**(2), 524–535 (2017).

62. M. Brinkmann, A. Fast, T. Hellwig et al., "Portable all-fiber dual-output widely tunable light source for coherent Raman imaging," *Biomed. Opt. Express* **10**(9), 4437–4449 (2019).

Index